Forum for Interdisciplinary Mathematics

The *Forum for Interdisciplinary Mathematics* is a Scopus-indexed book series. It publishes high-quality textbooks, monographs, contributed volumes and lecture notes in mathematics and interdisciplinary areas where mathematics plays a fundamental role, such as statistics, operations research, computer science, financial mathematics, industrial mathematics, and bio-mathematics. It reflects the increasing demand of researchers working at the interface between mathematics and other scientific disciplines.

Lam Quoc Anh • Phan Quoc Khanh •
Nguyen Hong Quan

Equilibrium Problems

Existence, Stability, and Approximation

Springer

Lam Quoc Anh
Department of Mathematics
Can Tho University
Can Tho, Vietnam

Phan Quoc Khanh
Faculty of Mathematics and Statistics
Ton Duc Thang University
Ho Chi Minh City, Vietnam

Nguyen Hong Quan
Department of Mathematics
Posts and Telecommunications Institute of
Technology Ho Chi Minh City Campus
Ho Chi Minh City, Vietnam

ISSN 2364-6748 ISSN 2364-6756 (electronic)
Forum for Interdisciplinary Mathematics
ISBN 978-3-031-89002-4 ISBN 978-3-031-89003-1 (eBook)
https://doi.org/10.1007/978-3-031-89003-1

Mathematics Subject Classification: 90C30, 90C33, 90C31, 90C29, 49K40, 49K35, 65K10, 65K05

This Springer imprint is published by the registered company Springer Nature Switzerland AG
The registered company address is: Gewerbestrasse 11, 6330 Cham, Switzerland

Preface

This book aims to present some major topics in the theory of equilibrium problems. These problems are the object of increasingly intensive studies in applied mathematics in connection with various other areas of science, technology, management, and economics. This fact is easily understood since equilibrium phenomena appear so frequently in our life and so it is natural that the concept of equilibrium plays a central role in many scientific areas. Any secondary school student can understand that an object is in a mechanical equilibrium when the impact of all the forces on it is equal to zero. It is also common to consider even the much more abstract concept of "ecological balance," also a notion of equilibrium. In this book, we attempt to discuss in detail only equilibrium models suitable for socioeconomic sciences. However, very roughly speaking, one can imagine "equilibrium" in various areas as the impact of all the "forces" being zero, i.e., like in classical mechanics, the simplest natural science. The socioeconomic equilibrium models were traditionally developed in economics, with contributions from Nobel Prize winners and other big names such as K. J. Arrow, G. Debreu, L.V. Kantorovich, T. Koopmans, J. F. Nash, and Leontief. The formulations of equilibrium problems in the main parts of this book originate from the works of H. Nikaido and K. Isoda (1955), K. Fan (1972), and F. Blum and W. Oettli (1994) (sometimes referred to as the Ky Fan inequality in the literature). We restrict ourselves to the most developed themes in both theory and application, namely, the existence of solutions, stability, well-posedness, and approximation. Numerical algorithms are not discussed. We intend to give a comprehensive and coherent exploration of the addressed topics, but make no attempt to cover all the developments available in the literature; this would be impossible, since thousands of relevant papers have been published.

Here we would like to mention some practical examples to argue that the socioeconomic meanings of equilibrium concepts are natural counterparts of those in mechanics. Let us look at the simplest market where consumers and producers buy and sell, respectively, the only commodity. Assume that their reaction depends on its (current) price p, i.e., the consumers determine their total demand $D(p)$, and the producers their total supply $S(p)$. If we consider a certain amount of transactions between the two groups, then there exists an equality between the supply and demand at each price. The problem is to find the *equilibrium price* p^* which implies the zero excess demand $D(p^*)-S(p^*) = 0$. So, "equilibrium" here can be imagined as the impact of buying and producing, like that of two forces on the market, equals zero. Now look at the Nash equilibrium game/problem, probably the most popular equilibrium model studied in applied mathematics. In this case, we have a game (more general than the above commodity market) with m agents and each agent i has an objective function $c_i(x_i, x_{-i})$ which expresses the loss of agent i. He, let's view all the agents as males for convenience, chooses a strategy $x_i \in A_i$, a subset of a metric space X_i of available strategies, where x_{-i} denotes the vector of strategies chosen by the remaining agents, except i. Then, $c_i : A_i \times \prod_{j \in I \setminus \{i\}} A_j \to \mathbb{R}$ for $i \in I$. The vector strategy $\bar{x} := (\bar{x}_i)_{i \in I}$ determines a NASH EQUILIBRIUM of our problem if, for all $i \in I$,

$$\bar{x}_i \in \operatorname{argmin}_{x_i \in A_i} c_i(x_i, \bar{x}_{-i}).$$

This equilibrium concept was introduced by John Forbes Nash Jr. (although this concept can be traced back to the early insights of Émile Borel and John von Neumann). His name is really appropriately attached to the above worldwide recognized definition. It states that all possible unilateral actions of an arbitrary agent in the Nash equilibrium state lead to states that are no better than the original one. Roughly speaking, this concept shows the state that the impact of all the agents, like their forces in the game, is zero. For this socioeconomic problem, this concept also reflects an equilibrium state between the losses of the agents, such a meaning obviously does not exist in a mechanical problem.

To express this equilibrium state as a generalization of the above equation $D(p^*) - S(p^*) = 0$ for the simplest market, we leverage the following bifunction $N : A \times A \to \mathbb{R}$, with $A := \prod_{i \in I} A_i$, which is called the NIKAIDO-ISODA BIFUNCTION

$$N(x, y) := \Sigma_{i \in I}[c_i(x_i, x_{-i}) - c_i(y_i, x_{-i})].$$

It is proved that (see, e.g., Proposition 7.110 in Chap. 7) the considered Nash equilibrium problem can be formulated in terms of the bifunction N as follows. Find $\bar{x} \in A$ such that $N(\bar{x}, y) \leq 0$ for all $y \in A$. So, the inequality $N(\bar{x}, y) \leq 0$ is the generalization for the Nash equilibrium problem of the aforementioned equality $D(p^*) - S(p^*) = 0$. The same formulation as this Nash equilibrium problem but for a general bifunction Φ on general product $A \times A$:

$$\text{find } \bar{\mathrm{x}} \in \mathrm{A} \text{ such that } \mathrm{N}(\bar{\mathrm{x}}, \mathrm{y}) \leq 0 \text{ for all } y \in A$$

is just the basic general equilibrium problem, denoted by (EP) in Chaps. 1, 4, and 7. Furthermore, the reader will also be acquainted with the following more realistic extension of the above Nash equilibrium problem. Assume now that the set of available strategies for agent i is $\mathscr{D}_i(x_{-i})$, which depends on the strategies chosen by the other players $x_{-i} := (x_1, \ldots, x_{i-1}, x_{i+1}, \ldots, x_m)$. Then, $\mathscr{D}_i : X_{-i} \rightrightarrows X_i$.

We set $\mathscr{D} := \prod_{i\in I} \mathscr{D}_{-i} : \prod_{i\in I} X_i \rightrightarrows \prod_{i\in I} X_i =: X$. However, $\mathscr{D}$ can be restricted to the following set, which is the domain of $\mathscr{D}$,

$$A := \{x \in X \mid x_i \in \mathscr{D}_i(x_{-i}),\ \forall i \in I\},$$

since its outside points are irrelevant, i.e., $\mathscr{D} : A \rightrightarrows X$. Obviously, for any $x \in A$, $x \in \mathscr{D}(x)$ and so $(x, x) \in \text{gph}\mathscr{D}$. Hence, $\mathscr{D}$ is defined on the whole A and maps onto $\mathscr{D}(A)$. Then, $\bar{x} \in A$ is a NASH QUASIEQUILIBRIUM if for all $i \in I$,

$$\bar{x} \in \text{argmin}_{x_i \in \mathscr{D}_i(\bar{x}_{-i})} c_i(x_i, \bar{x}_{-i})$$

and the problem becomes that of finding the so-called Nash quasiequilibria. Leveraging the Nikaido-Isoda bifunction: $N : \text{gph}\mathscr{D} \to \mathbb{R}$ defined by

$$N(x, y) := \sum_{i=1}^{m} [c_i(x_i, x_{-i}) - c_i(y_i, x_{-i})],$$

we convert the problem to

$$\text{finding } \bar{x} \in A \text{ such that } N(\bar{x}, y) \leq 0 \text{ for all } y \in \mathscr{D}(\bar{x}),$$

see, e.g., Sect. 8.5. Nevertheless, this problem is an important practical special case of our generalized quasiequilibrium model

$$\text{(QEP)} \qquad \text{find } \bar{x} \in A \text{ such that } \Phi(\bar{x}, y) \leq 0 \text{ for all } y \in \mathscr{D}(\bar{x}),$$

which is explored in detail in Chaps. 2, 5, and 8.

Let us briefly describe the main contents of each of the ten chapters.

Chapter 1 deals with the scalar equilibrium problem (EP) and the existence of its solutions systematically. In Sect. 1.1, we state problem (EP) and clarify the relationships between (EP) and some other problems. This section also provides some basic concepts and results such as semicontinuity of scalar functions, Sperner's lemma, Brouwer's fixed-point theorem, etc. Section 1.2 presents existence conditions for solutions of (EP) under convexity assumptions. Here, a number of notions closely related to the convexity structure such as hemicontinuity and sign continuity of bifunctions, generalized convexity of scalar functions, and monotonicity properties of bifunctions are also discussed systematically. Section 1.3 provides existence results in nonconvex cases under pure topological assumptions. The results are established for classes of bifunctions, including the one of bifunctions satisfying the triangle inequality, the class of generalized cyclically monotone bifunctions, that of KKM-bifunctions, and the one of connectedness-bifunctions.

Many important problems cannot be described by (EP) since they involve an inequality of a bifunction involving constraints that depend on the decision variable, such as quasioptimization problems, quasivariational inequalities, generalized Nash equilibria, traffic networks with elastic demands, etc. To study these problems within a common framework, we merge them into a general problem, called quasiequilibrium problem (QEP).

Chapter 2 deals with this formalization of many equilibrium and optimization-related models and existence results for its solutions. Section 2.1 introduces some basic notions and properties of semicontinuity for set-valued mappings, which are necessary for stating problems and establishing existence results. Section 2.2 presents the definition of quasiequilibrium problem and several related problems. Section 2.3 provides some fixed-point theorems for set-valued mappings, which are used to establish existence results for (QEP). Section 2.4 presents conditions for the existence of solutions of (QEP). While Sect. 2.6 applies the results of the preceding section to several particular cases of (QEP).

Chapter 3 extends the scalar equilibrium problem to the vector case with existence results. Section 3.1 presents some basic concepts such as the algebraic interior and the vectorial closure of a set, cones, nonconvex separation functionals, binary relations and Zorn's lemma, and continuity of vector functions. Section 3.2 introduces a vector equilibrium problem (VEP)(f, C), concepts of its solutions, relationships between the kinds of solutions, and particular cases of (VEP). Existence conditions for the considered kinds of solutions of (VEP)(f, C) are examined in Sect. 3.3. The results are divided into two groups. The first one includes the results that are established based on assumptions related to convexity. The second group consists of results without convexity assumptions. The last part of this section provides some forms of vectorial Ekeland' s variational principle.

Chapter 4 considers stability of parametric scalar equilibrium problems. Sufficient conditions for the semicontinuity and continuity of solution maps under relaxed continuity and generalized convexity assumptions are provided. The chapter is not only an attempt to provide a complete and comprehensive presentation of the results in the literature so far, but also some new facts or refinements of known results are given. Hölder (or Lipschitz) continuity of solution maps of these problems has been investigated from the beginning, along with types of (semi)continuity. Combining assumptions on Hölder continuity with conditions related to (strong) monotonicity and (strong) convexity, we introduce Hölder/Lipschitz properties of approximate and exact solution maps of (EP). The results presented here could be regarded as an endeavor to unify the corresponding contributions in the literature. For well-posedness, one of the most important topics in both theory and numerical methods in optimization, we discuss a concept, which can cover many important ones, including Tikhonov and Hadamard well-posedness, under suitable adjustments. In this section, for the results found in the literature, we seek a unified exploration, not repeating the presentation of any article.

Stability and well-posedness conditions for quasiequilibrium models constitute the content of Chap. 5. It can be seen that in most known contributions to this topic, topological properties of constraint sets are imposed directly on the fixed points of the set-valued maps. Such assumptions are difficult to check. To overcome this, in the chapter we first study separately many important properties of solution maps of fixed-point problems of set-valued maps, and then apply them to the above stability and well-posedness conditions. In this way, the assumptions are verifiable being imposed directly on the data of our problem.

Furthermore, an overview of sufficient conditions for Hölder and Lipschitz properties of solution maps of quasiequilibrium problems via assumptions related to monotonicity and convexity is provided. The related calmness property of such maps is also derived under relaxed conditions in this chapter. Note that for solution maps of quasiequilibrium problems, the difference between stability and well-posedness in the sense of semicontinuity-continuity and in the sense of Hölder and Lipschitz properties is so big that many methods and techniques for dealing with equilibrium problems cannot be employed. So, many counterparts of conclusions for equilibrium problems have not yet been obtained for quasiequilibrium problems, and hence Hölder and Lispchitz conditions of the solution maps of such models are still open problems.

Chapter 6 deals with stability in various senses for vector equilibrium problems (VEPs), including their weak and strong variants. Two approaches are presented: the first is direct on the vector formulation and the second is via scalarization techniques. Regarding the first approach, for stability in the sense of semicontinuity, we apply cone continuity/semicontinuity properties to get upper semicontinuity results for solution maps, while to receive lower semicontinuity ones, generalized convexity is additionally needed. Finally, for continuity of these solution maps, we also aim at some results independent from the above ones about semicontinuity. For stability in the Hölder/Lipschitz continuity or calmness sense, we resort to relaxed strong monotonicity or strong convexity properties of vector maps. When studying approximate solution maps, these properties are omitted or mitigated. Besides stability, a unified form of the Hadamard and Tikhonov well-posedness concepts in terms of semicontinuity, which is popular in the literature, is investigated as well in the first approach. Moreover, we also establish relationships between well-posedness and stability of approximate solutions.

Regarding the second approach (using scalarization), for weak VEPs we can use linear scalarizations by adding quasiconvexity hypotheses. But, for strong VEPs, we define many extensions of the Gerstewitz/Tammer nonlinear scalarization functions, and show that as a byproduct they can be used also for weak VEPs. Furthermore, through these nonlinear scalarization techniques, we find interesting relationships between VEPs and set optimization problems, and then discuss stability for set optimization problems.

Chapter 7 and the next chapter are devoted to approximations of equilibrium and quasiequilibrium problems. The topic is rather different from what we discuss in the previous part of the book. Here approximations are in terms of variational convergence. In Sects. 7.1 and 7.2, the four types of each of the traditional variational convergences: epi-convergence for functions and epi/hypo- and lopsided convergences for bifunctions are examined in detail, with a focus on characterizations and variational properties together with relationships between the convergence types, especially with those related to their suitability for the study of variational properties. Applications of the obtained results to approximations of the equilibrium and strong equilibrium problems are considered in Sect. 7.3, including discussions on two kinds of approximate solutions and also their convergence for pairs of dual problems. The subsequent five sections are devoted to important special cases: multiobjective optimization, approximations of optimization problems in terms of optimality functions, Nash equilibria, Walras economic equilibria, estimators of solutions of stochastic optimization models. In each sections, convergence results for optimal values and solutions/approximate solutions in terms of appropriate types of variational convergence are discussed.

Chapter 8 contains approximation results for quasiequilibrium problems. Since the bifunctions of these models are defined on general (nonrectangular) domains, Sect. 8.1 highlights variational objects of such bifunctions and solutions of quasiequilibrium problems since they are remarkably different from the special rectangular case in Chap. 7. Section 8.2 considers characterizations and variational properties of the types of variational convergence following the same line as in Sects. 7.1 and 7.2 in order to shorten the presentation, including clear comparisons. Section 8.3 deals with the recent topic of quantification of variational convergence of bifunctions. In Sect. 8.4, the convergence results for approximate solutions of quasiequilibrium problems are obtained in a short presentation, including a highlight of the differences with the case in Sect. 7.3. Then, the new topic of quantitative approximations of quasiequilibrium problems is addressed by combining the above convergence results and the quantification of variational convergence of bifunctions. Approximations of the generalized noncooperative game both in terms of variational convergence of the Nikaido-Isoda bifunction of the game and directly in terms of its data are considered in Sect. 8.5. Section 8.6 provides very recent results on approximations of set-valued quasivariational inequalities, which constitute an important particular case of quasiequilibrium problems, and applications to a modern model of traffic networks. Section 8.7 deals with optimization problems under stochastic ambiguity with a focus on their distribution-based formulations. Examples of such formulations and applications of lop-convergence in practical problems are provided.

Chapter 9 considers approximations of vector quasiequilibrium problems and of their special case of vector Nash quasiequilibrium problems, which are among the most important models in applied mathematics. Since there are no studies of variational convergence of vector bifunctions in the literature, our approach is first scalarizing the problems by the Tammer and Hiriart-Urruty methods and then applying variational convergence of scalar bifunctions. Sufficient conditions for the convergence of three types of approximate solutions of our problems under suitable types of variational convergence of the scalarized ones are established.

Final Chap. 10 is devoted to extending equilibrium/quasiequilibrium problems to two general models: a variational inclusion problem and the variational relation problem. We discuss some basic existence and stability issues for these general models. Concerning the existence of solutions, first we prove several existence results for the variational relation problem. Then, applying these results, we obtain some existence conditions for variational inclusion problems. We employ semicontinuity properties of set-valued maps to establish sufficient conditions for the (semi)continuity of solution maps and for well-posedness of the variational inclusion problem. As for the variational relation problem, we also provide stability results related to (semi)continuity, convergence, and Hölder calmness of solution maps. Although the results presented in this chapter encompass many contributions to these models, there remain many topics worthy of further study in the future.

On the one hand, this book has the character of a monograph as we seek to provide a mathematically rigorous treatment of the subjects, encompassing major results in the literature and including also a number of our own contributions. On the other hand, it may also be considered a textbook as we have endeavored to make the presentation easy to follow and suitable for teaching as some courses for advanced or graduate students.

Acknowledgments Our first gratitude goes to the coauthors of our joint papers whose results constitute significant parts of this book: Huynh Thi Hong Diem, Tran Quoc Duy, Nguyen Xuan Hai, Vo Si Trong Long, Le Minh Luu, and Tran Ngoc Tam. Of course, the warm thanks must be addressed also to the other members of our team of Southern Vietnam working on optimization, variational analysis, and related areas during about the last 30 years: Nguyen Le Hoang Anh, Nguyen Xuan Duy Bao, Truong Quang Bao, Nguyen Huu Danh, Pham Thanh Duoc, Le Phuoc Hai, Dinh Vinh Hien, Tran Hue Nuong, Dinh Ngoc Quy, Ho Thuc Quyen, Tran Trinh Minh Son, Nguyen Hieu Thao, Nguyen Dinh Tuan, Nguyen Minh Tung, Le Thanh Tung, and Dang Thi My Van. Our joint publications with them are not involved directly in the scope of this book, but create an important knowledge base of the authors. Furthermore, the joint research activities of our team, including very lively seminars, have encouraged us to do research for more than 30 years now.

We are very grateful to generations of Vietnamese colleagues, who have greatly influenced our research results: Pham Ky Anh, Pham Ngoc Anh, Nguyen Dinh, Truong Xuan Duc Ha, Dinh The Luc, Do Van Luu, Le Dung Muu, Huynh Van Ngai, Van Hien Nguyen, Hoang Xuan Phu, Pham Huu Sach, Pham Tien Son, Nguyen Xuan Tan, Duong Viet Thong, Hoang Tuy, Nguyen The Vinh, and Nguyen Dong Yen (some of them also have joint works with us). Among them, the pioneering role of Hoang Tuy in modern Vietnamese mathematics must be mentioned.

International mathematicians have also greatly influenced our mathematical development and this monograph in particular. Our first special thanks are devoted to Stefan Rolewicz, the PhD thesis supervisor of the second author and had a long influence also on the two coauthors of this book, who were PhD students of the second author. Our thanks go to the coauthors of our joint works: Ali P. Farajzadeh, Lidia Huerga, Somaye Zafari, Alexander Ya. Kruger, Daishi Kuroiwa, Lai-Jiu Lin, Sirous Moradi, Vicente Novo, Narin Petrot, Somyot Plubtieng, Antoine Soubeyran, Kamonrat Sombut, Michel Théra, Rabian Wangkeeree, and Jen-Chih Yao.

We also do not have enough words to thank very well-known mathematicians, including world "big names," whose suggestions, discussions, and encouragements have been so important for our mathematical research: Adil Bagirov, Ewa Bednarczuk, Szymon Dolecki, Marius Durea, Alexander D. Ioffe, Alejandro Jofré, Abderrahim Jourani, Akhtar A. Khan, Do Sang Kim, Gue Myung Lee, Nelson F. Maculan, Juan-Enrique Martinez-Legaz, Boris S. Mordukhovich, Jean-Paul

Penot, Tyrrel R. Rockafellar, Wen Song, Jean-Jacques Strodiot, Martin Studniarski, Christiane Tammer, Roger J-B Wets, and Constantin Zalinescu.

The authors wish to thank Alexander Kruger for his assistance with improving the presentation when preparing the final version of the book.

We are so grateful to the reviewers for their invaluable remarks and suggestions helping us to remarkably improve the original version of the monograph. We are also indebted to the nice people at Springer, especially Donna Chernyk, Editor Mathematics Editorial, Springer Nature in New York, who encouraged the second author in 2015 to write the monograph and continuously discussed, commented, and patiently waited for the completion of our book, which is late due to the influence of the Covid-19 pandemic.

With all questions/suggestions concerning the content of the Analytical and Algebraic Methods in Optimization Group, the readers are welcome to contact the corresponding author Phan Quoc Khanh, Analytical and Algebraic Methods in Optimization Group, Faculty of Mathematics and Statistics, Ton Duc Thang University, Ho Chi Minh City, Vietnam, email: phanquockhanh@tdtu.edu.vn, pqkhanhus@gmail.com.

Can Tho, Vietnam — Lam Quoc Anh
Ho Chi Minh City, Vietnam — Phan Quoc Khanh
Ho Chi Minh City, Vietnam — Nguyen Hong Quan
November 2024

Contents

Chapter 1
Solution Existence for Scalar Equilibrium Problems

Abstract This chapter deals with the scalar equilibrium problem (EP) and the existence of solutions to this problem systematically. The main points of this chapter are as follows. In Sect. 1.1 we state problem (EP) and clarify the relationships between (EP) and some other problems. This section also provides some basic concepts and results such as semicontinuity of scalar functions, Sperner's lemma, and Brouwer's fixed-point theorem. Section 1.2 presents existence conditions for solutions of (EP) under convexity assumptions. Here, we also systematize several notions closely related to the convexity structure such as hemicontinuity and sign continuity of bifunctions, generalized convexity of scalar functions, and monotonicity properties of bifunctions. Section 1.3 provides existence results in nonconvex settings under pure topological assumptions. The results are established for certain classes of bifunctions, including the class of bifunctions satisfying the triangle inequality, the class of generalized cyclically monotone bifunctions, the class of KKM-bifunctions, and the one of connectedness-bifunctions.

1.1 Problem Formulation, Notation, and Preliminaries

Throughout this chapter we use the following notation. Let $\mathbb{N}$ be the set of natural numbers, $\mathbb{R} = (-\infty, +\infty)$ be the set of real numbers, and $\overline{\mathbb{R}} = [-\infty, +\infty] = \mathbb{R} \cup \{-\infty, +\infty\}$.

- Let A be any nonempty set.
  - By $\mathscr{F}(A)$ we denote the class of the finite subsets of A.
  - For a bifunction $\Phi : A \times A \to \mathbb{R}$ and $B \subset A$, $\Phi|_{B\times B}$ is the restriction of Φ to the set $B \times B$.
  - $\mathscr{A}_A^=$, $\mathscr{A}_A^\geq$, and $\mathscr{A}_A$ denote the classes of bifunctions Φ defined on $A \times A$ satisfying $\Phi(x, x) = 0$ for all $x \in A$, $\Phi(x, x) \geq 0$ for all $x \in A$, and for each $y \in A$, $\Phi(x, y) \geq 0$ for some $x \in A$, resp. It is clear that $\mathscr{A}_A^= \subset \mathscr{A}_A^\geq \subset \mathscr{A}_A$.
- Let $\varphi : A \to \overline{\mathbb{R}}$.

L. Q. Anh et al., *Equilibrium Problems*, Forum for Interdisciplinary Mathematics, https://doi.org/10.1007/978-3-031-89003-1_1

- The effective domain of φ is $\text{dom}\varphi := \{x \in A \mid \varphi(x) < +\infty\}$.
- The graph, epigraph, and hypograph of φ are, resp.

$$\text{gph}\varphi := \{(x, \beta) \in A \times \mathbb{R} \mid \varphi(x) = \beta\},$$

$$\text{epi}\varphi := \{(x, \beta) \in A \times \mathbb{R} \mid \varphi(x) \leq \beta\},$$

$$\text{hyp}\varphi := \{(x, \beta) \in A \times \mathbb{R} \mid \varphi(x) \geq \beta\}.$$

- The β-level sets of φ are

$$\text{lev}_{\leq\beta}\varphi := \{x \in A \mid \varphi(x) \leq \beta\},$$

$$\text{lev}_{<\beta}\varphi := \{x \in A \mid \varphi(x) < \beta\},$$

$$\text{lev}_{\geq\beta}\varphi := \{x \in A \mid \varphi(x) \geq \beta\},$$

$$\text{lev}_{>\beta}\varphi := \{x \in A \mid \varphi(x) > \beta\}.$$

If $\beta = 0$, we write simply $\text{lev}_{\leq}\varphi$ ($\text{lev}_{<}\varphi$, $\text{lev}_{\geq}\varphi$, $\text{lev}_{>}\varphi$, resp.).

1.1.1 Equilibrium Problem and Applications

In this chapter, we study the existence of solutions of the following SCALAR EQUILIBRIUM PROBLEM:

$$\text{(EP)} \qquad \text{find } \bar{x} \in A \text{ such that } \Phi(\bar{x}, y) \geq 0 \text{ for all } y \in A,$$

where A is a nonempty set and $\Phi : A \times A \to \mathbb{R}$ is a bifunction. We denote the solution set of (EP) by Sol(EP).

In the literature, bifunction Φ satisfying $\Phi(x, x) = 0$ for all $x \in A$ is called an equilibrium bifunction, and problem (EP) is often considered for equilibrium bifunctions. Here, we investigate problem (EP) with arbitrary bifunctions. For Sol(EP) to be nonempty, we need conditions concerning A and Φ. The following sections of this chapter present certain conditions on A and Φ to ensure the solvability of the problem.

The equilibrium problem (EP) includes many problems in optimization as particular cases. Here are some important special cases of (EP).

(a) **Minimization problem**. Let $\varphi : A \longrightarrow \mathbb{R} \cup \{+\infty\}$. The minimization problem, denoted MP(A, φ), is the problem

$$\text{find } \bar{x} \in A \text{ such that } \varphi(\bar{x}) \leq \varphi(y) \text{ for all } y \in A.$$

This problem is also written as

$$\min \varphi(x) \quad \text{subject to} \quad x \in A.$$

The solution set of MP(A, φ) is usually denoted by $\arg\min_A \varphi$. Problem MP(A, φ) is a particular case of problem (EP) with $\Phi(x, y) = \varphi(y) - \varphi(x)$ for all $x, y \in A$.

(b) **Minimax and saddle points**. Given sets C, D and a bifunction $\phi : C \times D \longrightarrow \overline{\mathbb{R}}$, we say that the MINIMAX EQUALITY holds for ϕ if

$$\phi^* := \inf_{x \in C} \sup_{y \in D} \phi(x, y) = \sup_{y \in D} \inf_{x \in C} \phi(x, y) =: \phi_*. \tag{1.1}$$

A point $(\bar{x}, \bar{y}) \in C \times D$ is called a SADDLE POINT of ϕ if

$$\phi(\bar{x}, y) \leq \phi(\bar{x}, \bar{y}) \leq \phi(x, \bar{y}) \text{ for all } (x, y) \in C \times D. \tag{1.2}$$

The problem of finding a point $(\bar{x}, \bar{y}) \in C \times D$ satisfying (1.2) is called a SADDLE POINT PROBLEM. The equality (1.1) says that the supremum and infimum can be interchanged. The inequalities (1.2) show that the point $\bar{x}$ is a minimizer of the function $\phi(\cdot, \bar{y})$ on C and the point $\bar{y}$ is a maximizer of the function $\phi(\bar{x}, \cdot)$ on D. Note that we always have the following inequality: $\phi^* \geq \phi_*$.

As $C \subset \mathbb{R}$ and $D \subset \mathbb{R}$, in a neighborhood $U \subset C \times D$ of the point $(\bar{x}, \bar{y})$ the shape of the graph $\text{gph}_U \phi := \{(x, y, \phi(x, y)) \mid (x, y) \in U\}$ evokes a saddle or a pass connecting two mountains (see Fig. 1.1).

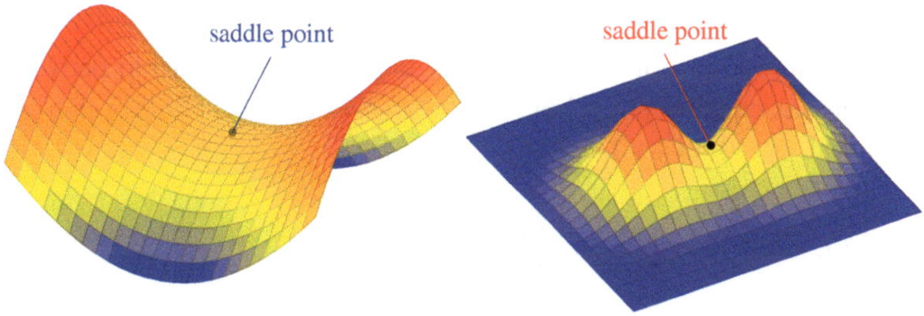

Fig. 1.1 An illustration of saddle points

Proposition 1.1 *A point* $(\bar{x}, \bar{y})$ *is a saddle point of* ϕ *if and only if the minimax equality (1.1) holds and*

$$\bar{x} \in \operatorname{argmin}_{x\in C} \sup_{y\in D} \phi(x, y), \;\; \bar{y} \in \operatorname{argmax}_{y\in D} \inf_{x\in C} \phi(x, y). \tag{1.3}$$

Hence, if ϕ *has a saddle point, then* $\min_{x\in C} \sup_{y\in D} \phi(x, y) = \max_{y\in D} \inf_{x\in C} \phi(x, y)$.

Proof If $(\bar{x}, \bar{y})$ is a saddle point, then $\phi^* = \inf_{x\in C} \sup_{y\in D} \phi(x, y) \leq \sup_{y\in D} \phi(\bar{x}, y) = \phi(\bar{x}, \bar{y}) = \inf_{x\in C} \phi(x, \bar{y}) \leq \sup_{y\in D} \inf_{x\in C} \phi(x, y) = \phi_*$. Hence, $\phi^* = \phi_*$ and (1.3) holds. Conversely, when (1.3) holds, we have $\sup_{y\in D} \inf_{x\in C} \phi(x, y) = \inf_{x\in C} \phi(x, \bar{y}) \leq \phi(\bar{x}, \bar{y}) \leq \sup_{y\in D} \phi(\bar{x}, y) = \inf_{x\in C} \sup_{y\in D} \phi(x, y)$. Using the minimax equality (1.1) we deduce that $\inf_{x\in C} \phi(x, \bar{y}) = \phi(\bar{x}, \bar{y}) = \sup_{y\in D} \phi(\bar{x}, y)$, and hence, $(\bar{x}, \bar{y})$ is a saddle point. □

Let $A = C \times D$ and $\Phi : A \times A \to \mathbb{R}$ be defined, for all $(x, y), (x', y') \in A$, by

$$\Phi((x, y), (x', y')) = \phi(x', y) - \phi(x, y'). \tag{1.4}$$

Then, we have the relationship between the saddle points of ϕ and the solutions of (EP) as follows.

Proposition 1.2 *A point* $(\bar{x}, \bar{y})$ *is a saddle point of* ϕ *if and only if it is a solution of equilibrium problem (EP) with* Φ *defined by (1.4).*

Proof If $(\bar{x}, \bar{y})$ is a saddle point of ϕ, then it is obviously a solution of (EP) by the definition of saddle points. Conversely, suppose that $(\bar{x}, \bar{y})$ solves (EP), i.e.,

$$\Phi((\bar{x}, \bar{y}), (x', y')) = \phi(x', \bar{y}) - \phi(\bar{x}, y') \geq 0, \;\; \text{for all } (x', y') \in A = C \times D.$$

Taking in turn $x' = \bar{x}$ and $y' = \bar{y}$, we have $\phi(\bar{x}, y') \leq \phi(\bar{x}, \bar{y}) \leq \phi(x', \bar{y})$ for all $(x', y') \in C \times D$, which means that $(\bar{x}, \bar{y})$ is a saddle point of ϕ. □

The following proposition gives another look at the relationship between (EP) and the minimax equality (1.1). For $\lambda \in \mathbb{R}$, let $A = C \times D$ and $\Phi_\lambda : A \times A \to \mathbb{R}$ defined, for all $(x, y), (x', y') \in A$, by

$$\Phi_\lambda((x, y), (x', y')) = \begin{cases} 0 \text{ if } (x, y) \in A \setminus (\mathrm{lev}_{>\lambda}\phi(\cdot, y') \times \mathrm{lev}_{<\lambda}\phi(x', \cdot)), \\ -1 \text{ otherwise.} \end{cases} \tag{1.5}$$

Proposition 1.3 *If for all* $\lambda > \phi_*$ *sufficiently close to* ϕ_**, problem* (EP) *with* $\Phi \equiv \Phi_\lambda$ *has solutions, then the minimax equality (1.1) holds.*

Proof Suppose that $\phi^* > \phi_*$. Then, there exists $\lambda > \phi_*$ sufficiently close to ϕ_* such that $\phi^* > \lambda > \phi_*$. These inequalities give that

$$\text{lev}_{<\lambda}\phi(\cdot, y) \times \text{lev}_{>\lambda}\phi(x, \cdot) \neq \emptyset \text{ for all } (x, y) \in A. \tag{1.6}$$

On the other hand, by assumption, there exists $(\bar{x}, \bar{y}) \in A$ such that $\Phi_\lambda((\bar{x}, \bar{y}), (x, y)) \geq 0$ for all $(x, y) \in A$. By the definition of Φ_λ,

$$(\bar{x}, \bar{y}) \notin \text{lev}_{>\lambda}\phi(\cdot, y) \times \text{lev}_{<\lambda}\phi(x, \cdot) \text{ for all } (x, y) \in A,$$

or equivalently,

$$(x, y) \notin \text{lev}_{<\lambda}\phi(\cdot, \bar{y}) \times \text{lev}_{>\lambda}\phi(\bar{x}, \cdot) \text{ for all } (x, y) \in A,$$

i.e., $\text{lev}_{<\lambda}\phi(\cdot, \bar{y}) \times \text{lev}_{>\lambda}\phi(\bar{x}, \cdot) = \emptyset$. This contradicts (1.6). Thus, $\phi^* = \phi_*$. □

Now, consider the case where $C = D := A$ and $\phi : A \times A \longrightarrow \mathbb{R}$. The following problem is called a MINIMAX PROBLEM:

$$\text{minimize } \sup_{y \in A} \phi(x, y) \text{ subject to } x \in A.$$

We set $\varphi_\phi(x) := \sup_{y \in A} \phi(x, y)$ and $\text{mis}\phi := \text{argmin}_{x \in A}\varphi_\phi(x) = \{x \in A \mid \varphi_\phi(x) = \phi^*\}$. Any $\bar{x} \in \text{mis}\phi$ is called a MINSUP POINT of ϕ.

For $\alpha \in \mathbb{R}$, let $\Phi_{(\phi,\alpha)}, \Phi'_{(\phi,\alpha)} : A \times A \to \mathbb{R}$ be defined, for all $x, y \in A$, by

$$\Phi_{(\phi,\alpha)}(x, y) = -\phi(x, y) + \alpha, \tag{1.7}$$

$$\Phi'_{(\phi,\alpha)}(x, y) = \phi(y, x) - \alpha. \tag{1.8}$$

Let $\text{Sol}(\text{EP}_{\Phi_{(\phi,\alpha)}})$ ($\text{Sol}(\text{EP}_{\Phi'_{(\phi,\alpha)}})$, resp.) be the solution set of (EP) with $\Phi \equiv \Phi_{(\phi,\alpha)}$ ($\Phi \equiv \Phi'_{(\phi,\alpha)}$, resp.).

Proposition 1.4

(a) *ϕ possesses a saddle point if and only if there exists $\alpha \in \mathbb{R}$ such that both* $\text{Sol}(\text{EP}_{\Phi_{(\phi,\alpha)}})$ *and* $\text{Sol}(\text{EP}_{\Phi'_{(\phi,\alpha)}})$ *are nonempty.*
(b) *If $\phi(x, x) = \alpha$ for all $x \in A$, then $\bar{x} \in \text{Sol}(\text{EP}_{\Phi_{(\phi,\alpha)}})$ if and only if $\bar{x} \in \text{mis}\phi$ and $\phi^* = \alpha$.*

Proof

(a) Assume that $(\bar{x}, \bar{y})$ is a saddle point of ϕ. Then, by taking $\alpha = \phi(\bar{x}, \bar{y})$, we have

$$\phi(\bar{x}, y) \leq \alpha \text{ for all } y \in A \Leftrightarrow \Phi_{(\phi,\alpha)}(\bar{x}, y) = -\phi(\bar{x}, y) + \alpha \geq 0 \text{ for all } y \in A,$$

and

$$\alpha \leq \phi(x, \bar{y}) \text{ for all } x \in A \quad \Leftrightarrow \Phi'_{(\phi,\alpha)}(\bar{y}, x) = \phi(x, \bar{y}) - \alpha \geq 0 \text{ for all } x \in A.$$

Thus, $\bar{x} \in \mathrm{Sol}(\mathrm{EP}_{\Phi_{(\phi,\alpha)}})$ and $\bar{y} \in \mathrm{Sol}(\mathrm{EP}_{\Phi'_{(\phi,\alpha)}})$. Conversely, assume that there exist $\alpha \in \mathbb{R}$, $\bar{x} \in \mathrm{Sol}(\mathrm{EP}_{\Phi_{(\phi,\alpha)}})$, and $\tilde{x} \in \mathrm{Sol}(\mathrm{EP}_{\Phi'_{(\phi,\alpha)}})$. Then,

$$\phi(\bar{x}, y) \leq \alpha \text{ for all } y \in A \text{ and } \alpha \leq \phi(y, \tilde{x}) \text{ for all } y \in A.$$

These inequalities imply that

$$\phi(\bar{x}, y) \leq \alpha \leq \phi(x, \tilde{x}) \text{ for all } (x, y) \in A \times A.$$

In particular, for $x = \bar{x}$ and $y = \tilde{x}$, one has $\phi(\bar{x}, \tilde{x}) \leq \alpha \leq \phi(\bar{x}, \tilde{x})$, i.e., $\alpha = \phi(\bar{x}, \tilde{x})$. Thus, $(\bar{x}, \tilde{x})$ is a saddle point of ϕ.

(b) Let $\bar{x} \in \mathrm{Sol}(\mathrm{EP}_{\Phi_{(\phi,\alpha)}})$, i.e., $\Phi_{(\phi,\alpha)}(\bar{x}, y) \geq 0$ for all $y \in A$. Then,

$$\phi^* = \inf_{x\in A} \big(\sup_{y\in A} \phi(x, y)\big) \leq \sup_{y\in A} \phi(\bar{x}, y) = -\inf_{y\in A} \Phi_{(\phi,\alpha)}(\bar{x}, y) + \alpha \leq \alpha.$$

On the other hand, $\sup_{y\in A} \phi(x, y) \geq \phi(x, x) = \alpha$ for all $x \in A$. Hence, $\sup_{y\in A} \phi(\bar{x}, y) \geq \alpha$ and $\phi^* = \inf_{x\in A} \sup_{y\in A} \phi(x, y) \geq \alpha$. Thus, $\sup_{y\in A} \phi(\bar{x}, y) = \alpha$ and $\phi^* = \alpha$, and hence, $\bar{x} \in \mathrm{mis}\phi$. Conversely, taking $\bar{x} \in \mathrm{mis}\phi$, we have $\sup_{y\in A} \phi(\bar{x}, y) = \varphi_\phi(\bar{x}) = \phi^* = \alpha$. It follows that $-\Phi_{(\phi,\alpha)}(\bar{x}, y)+\alpha = \phi(\bar{x}, y) \leq \alpha$ for all $y \in A$, or equivalently, $\Phi_{(\phi,\alpha)}(\bar{x}, y) \geq 0$ for all $y \in A$, and so $\bar{x} \in \mathrm{Sol}(EP_{\Phi_{(\phi,\alpha)}})$.

□

(c) **Variational inequalities**. Given a map $T : A \subset \mathbb{R}^n \longrightarrow \mathbb{R}^n$, the classical variational inequality is the problem of finding a point $\bar{x} \in A$ such that

$$\langle T(\bar{x}), y - \bar{x}\rangle \geq 0 \text{ for all } y \in A,$$

where $\langle\cdot, \cdot\rangle$ is the symbol of the inner product in $\mathbb{R}^n$.

Because $\langle T(\bar{x}), y-\bar{x}\rangle = \|T(\bar{x})\|\|y-\bar{x}\| \cos \angle(T(\bar{x}), y-\bar{x})$, where $\angle(T(\bar{x}), y-\bar{x})$ is the angle between vectors $T(\bar{x})$ and $y - \bar{x}$, $\langle T(\bar{x}), y - \bar{x}\rangle \geq 0$ if only if $\cos \angle(T(\bar{x}), y - \bar{x}) \geq 0$. Therefore, geometrically, the variational inequality problem is finding a vector $\bar{x} \in A$ such that for every vector $y \in A$, $T(\bar{x})$ forms a non-obtuse angle with $y-\bar{x}$ (see Fig. 1.2). If further requiring that A is a cone and $T(\bar{x})$ is orthogonal to $\bar{x}$, then the variational inequality becomes a complementarity problem: find a vector $\bar{x} \in \mathbb{R}^n$ such that

$$\bar{x} \in A, \quad \bar{x}\perp T(\bar{x}), \quad \langle T(\bar{x}), y\rangle \geq 0 \text{ for all } y \in A,$$

where the notation $\perp$ means "perpendicular."

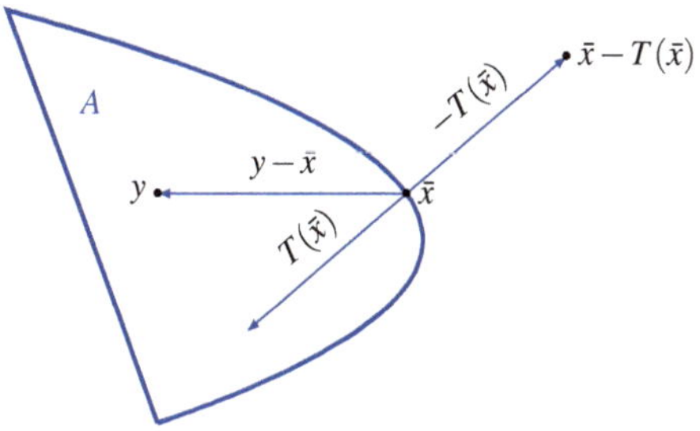

Fig. 1.2 Geometric illustration of the variational inequality

When A is a convex set, we can prove that "$\langle T(\bar{x}), y - \bar{x}\rangle \geq 0$ for all $y \in A$" if and only if "$\|T(\bar{x}))\| = \|(\bar{x} - T(\bar{x})) - \bar{x}\| \leq \|(\bar{x} - T(\bar{x})) - y\|$ for all $y \in A$" (see Exercise 1.1). Hence, $\bar{x}$ is a solution of the variational inequality problem if and only if $\bar{x}$ is the projection of $\bar{x} - T(\bar{x}))$ on the feasible set A (see Fig. 1.2).

In general, let A be a subset of a topological vector space X, $T : A \longrightarrow X^*$ be a map; the VARIATIONAL INEQUALITY, denoted VI(A, T), is to find $\bar{x} \in A$ such that

$$\langle T(\bar{x}), y - \bar{x}\rangle \geq 0 \text{ for all } y \in A,$$

where $\langle x^*, x\rangle = x^*(x)$ for $x^* \in X^*$ and $x \in X$. In the special case where A is a cone, with

$$A^* = \{x^* \in X^* : \langle x^*, y\rangle \geq 0 \text{ for all } y \in A\}$$

denoting its polar cone, VI(A, T) is equivalent to the COMPLEMENTARITY PROBLEM CP(A, T), which is stated as follows: find $\bar{x} \in A$ such that

$$T(\bar{x}) \in A^*, \quad \langle T(\bar{x}), \bar{x}\rangle = 0.$$

Indeed, if $\bar{x}$ is a solution of CP(A, T), then obviously it is also a solution of VI(A, T). As $\bar{x}$ solves VI(A, T), taking $y := 2\bar{x}$ and $y := 0$, we obtain $\langle T(\bar{x}), \bar{x}\rangle \geq 0$ and $\langle T(\bar{x}), -\bar{x}\rangle \geq 0$, which imply that $\langle T(\bar{x}), \bar{x}\rangle = 0$. Therefore, $\langle T(\bar{x}), y\rangle = \langle T(\bar{x}), y - \bar{x}\rangle \geq 0$ for all $y \in A$, i.e., $T(\bar{x}) \in A^*$. Hence, $\bar{x}$ is a solution of CP(A, T).

Now, we set $\Phi(x, y) = \langle T(x), y - x\rangle$ for all $x, y \in A$. Then, VI(A, T) (and hence CP(A, T)) becomes a special case of (EP).

(d) **Nash equilibria**. A game with m players where the players are not able to make binding commitments to each other is called an m-PLAYER NONCOOPERATIVE GAME, denoted by $\mathscr{G}$. For the game $\mathscr{G}$, let $I := \{1, 2, \ldots, m\}$ be the set of the players. An ith player's strategy x_i is a complete plan of action. The strategy x_i specifies an action player i takes at all actions in the game in which he/she is called on to act. Let A_i be the set of all strategies available to ith player, called the strategy set of ith player. Set $A = A_1 \times \cdots \times A_m$. Each $x = (x_1, \ldots, x_m) \in A$ is a combination of strategies, each of which is a strategy for one player in the game. For $(x_1, \dots, x_m) \in A$, let $c_i(x_1, \dots, x_m) \in \mathbb{R}$ be the payoff of player i when he/she chooses the strategy x_i and each of the other players k chooses the strategy x_k $(k \neq i)$. The ith player's payoff is usually the profit he/she receives when the game is over. The function $c_i : A \to \mathbb{R}$, $x = (x_1, \dots, x_m) \mapsto c_i(x_1, \dots, x_m)$ is called the payoff function of the ith player. A point $(\bar{x}_1, \dots, \bar{x}_m) \in A$ is called a NASH EQUILIBRIUM of $\mathcal{G}$ if, for all $y_i \in A_i$ and $i = 1, 2, \dots, m$, we have

$$c_i(\bar{x}_1, \dots, \bar{x}_{i-1}, \bar{x}_i, \bar{x}_{i+1}, \dots, \bar{x}_m) \geq c_i(\bar{x}_1, \dots, \bar{x}_{i-1}, y_i, \bar{x}_{i+1}, \dots, \bar{x}_m).$$

At the Nash equilibrium, for each $i \in I$, ith player's strategy is the best response to all other players' strategies. If the ith player gives up his/her equilibrium strategy while the other players still retain their equilibrium strategies, then he/she will be disadvantaged.

A NASH EQUILIBRIUM PROBLEM is understood as the problem of finding a Nash equilibrium of the game $\mathcal{G}$.

We use the following notation: $A_{-i} := \prod_{k \neq i} A_k$, x_{-i} is the projection of $x = (x_1, \dots, x_m)$ onto A_{-i} and $(x_{-i}, y_i) := (x_1, \dots, x_{i-1}, y_i, x_{i+1}, \dots, x_m)$. Let $N : A \times A \to \mathbb{R}$ be defined by

$$N(x, y) = \sum_{i=1}^{n} [c_i(x) - c_i(x_{-i}, y_i)]. \tag{1.9}$$

This bifunction is often called the NIKAIDO-ISODA BIFUNCTION.

Proposition 1.5 *A point $\bar{x} = (\bar{x}_1, \dots, \bar{x}_m) \in A$ is a Nash equilibrium of $\mathcal{G}$ if and only if it is a solution of the problem* (EP) *with $\Phi = N$ defined by (1.9).*

Proof It is obvious that if $\bar{x} = (\bar{x}_1, \dots, \bar{x}_m) \in A$ is a Nash equilibrium of the game $\mathcal{G}$, then $N(\bar{x}, y) \geq 0$ for all $y \in A$. Conversely, let $\bar{x} \in A$ be a solution of (EP), i.e., $N(\bar{x}, y) \geq 0$ for all $y \in A$. Suppose that $\bar{x} = (\bar{x}_1, \dots, \bar{x}_m)$ is not a Nash equilibrium. Then, there exist $i \in I$ and $y_i \in A_i$ such that $c_i(\bar{x}) < c_i(\bar{x}_{-i}, y_i)$. Taking $y = (\bar{x}_{-i}, y_i)$, by the definition of bifunction N, we have $N(\bar{x}, y) = c_i(\bar{x}) - c_i(\bar{x}_{-i}, y_i) < 0$. This contradicts the fact that $\bar{x}$ is a solution of (EP). □

(e) **Price equilibria**. We consider an economy $\mathcal{E}$ with a finite number of agents dealing with n commodities. Let $\mathbf{A}$ be the set of agents with $\mathbf{A} = \mathbf{B} \cup \mathbf{C}$, where $\mathbf{B}$ is the set of the producers and $\mathbf{C}$ is the set of consumers. For a given price vector $x \in \mathbb{R}^n_+$, the supply of producer b is $s_b(x) \in \mathbb{R}^n_+$ and the demand of consumer c is $d_c(x) \in \mathbb{R}^n_+$. Let $f : \mathbb{R}^n_+ \to \mathbb{R}^n$ be defined by

$$f(x) = \sum_{b \in \mathbf{B}} s_b(x) - \sum_{c \in \mathbf{C}} d_c(x).$$

This mapping f is called the EXCESS DEMAND MAPPING. A price vector $\bar{x} \in \mathbb{R}^n_+$ is called an EQUILIBRIUM PRICE if $f(\bar{x}) \in \mathbb{R}^n_+$ and $\langle f(\bar{x}), \bar{x} \rangle = 0$. At an equilibrium price, each producer can sell all the units he/she wants to produce, and each consumer can buy all the units he/she wants.

Let $A := \Delta := \{x = (x_1, \ldots, x_n) \in \mathbb{R}^n_+ \mid \sum_{j=1}^n x_j = 1\}$, the price simplex, and let $\Phi : A \times A \to \mathbb{R}$ be defined by

$$\Phi(x, y) = \langle f(x), y - x \rangle \quad \text{for all} \quad x, y \in A.$$

Suppose that the WALRAS LAW holds on A, i.e., $\langle f(x), x \rangle = 0$ for all $x \in A$. Then, *any* $\bar{x} \in A$ *satisfying* $\bar{x} \in \text{Sol(EP)}$ *is also an equilibrium price of* $\mathscr{E}$. Indeed, if $\bar{x} \in \text{Sol(EP)}$ then $\Phi(\bar{x}, y) \geq 0$ for all $y \in A$, or equivalently,

$$\langle f(\bar{x}), y \rangle \geq 0 \quad \text{for all} \quad y \in A.$$

Suppose $f(\bar{x}) = (f_i(\bar{x}))$ and let $e^i = (e^i_j) \in A$, where

$$e^i_j = \begin{cases} 1 \text{ if } j = i, \\ 0 \text{ if } j \neq i. \end{cases}$$

Then, $f_i(\bar{x}) = \langle f(\bar{x}), e^i \rangle \geq 0$ for all $i = 1, 2, \ldots, n$, which means that $f(\bar{x}) \in \mathbb{R}^n_+$. Thus, $\bar{x}$ is an equilibrium price of $\mathscr{E}$.

(f) **Traffic network problem**. Consider a TRAFFIC NETWORK $G = (\mathbf{N}, \mathbf{A}, \mathbf{W}, c, d)$ with capacity constraints on arcs, where $\mathbf{N}$ denotes the set of nodes, $\mathbf{A}$ is the set of directed arcs, $\mathbf{W}$ is the set of the ORIGIN-DESTINATION O/D PAIRS of nodes that represent the origin and the destination of paths, $c = (c_a)_{a \in \mathbf{A}}$ is the capacity vector with $c_a > 0$ being the CAPACITY OF ARC a, and $d = (d_w)_{w \in \mathbf{W}}$ is the demand vector with $d_w > 0$ being the DEMAND OF THE TRAFFIC FLOW on O/D pair w (d_w representing the rate of traffic entering and exiting the network at the origin and the destination of w, resp.). Let P_w be the set of the paths connecting the pair $w \in \mathbf{W}$, $P = \bigcup_{w \in \mathbf{W}} P_w$, and $m = |P|$. Figure 1.3 illustrates a network with eight nodes, eleven arcs (c_{ij} is the capacity of the arc (ij)), three origin-destination O/D pairs ($\mathbf{W} = \{$O1/D7, O2/D6, O4/D8$\}$), and seven paths (for $w = O2/D6$, P_w includes two paths that are colored blue).

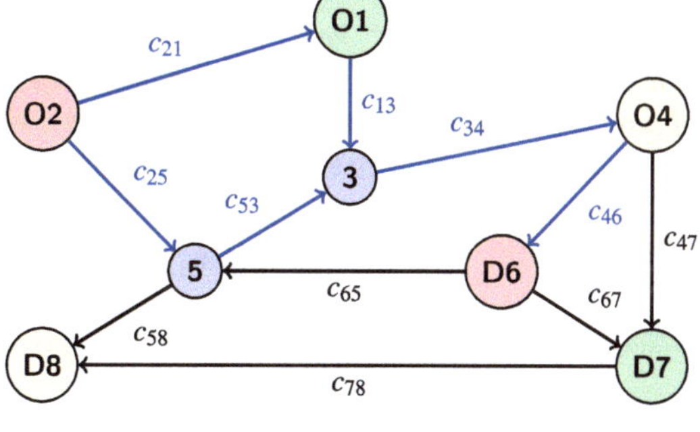

Fig. 1.3 An illustration of the traffic network problem

Then, a point $\bar{x} \in A$ solves IP($\mathscr{K}$) *if and only if it is a solution of* (EP) *with Φ defined by (1.11).*

The following principle is among the most fundamental results of continuous mathematics.

Finite Intersection Principle *Let K be a compact subset of a topological space A and $\mathscr{K} = \{K_y \mid y \in \Lambda\}$ $(\Lambda \subset A)$ be a family of nonempty closed subsets of K. Then, $\bigcap_{y\in\Lambda} K_y$ is nonempty if and only if the family $\mathscr{K}$ has the finite intersection property, i.e., $\bigcap_{y\in N} K_y$ is nonempty for each finite subset N of Λ.*

Definition 1.9 Problem (EP) is said to be FINITELY SOLVABLE on $B \subset A$ if for each finite subset N of A, there is some point $x_N \in B$ such that $\Phi(x_N, y) \geq 0$ for every $y \in N$. If (EP) is finitely solvable on A, then we say that it is finitely solvable.

The following finite solvability principle for (EP) is equivalent to the above finite intersection principle.

Finite Solvability Principle *Suppose that A is a topological space. Then, equilibrium problem* (EP) *has solutions if and only if it is finitely solvable on a compact subset K of A and $K \cap \operatorname{lev}_{\geq}\Phi(\cdot, y)$ is closed in K for every $y \in A$.*

Proof *The finite intersection principle implies the finite solvability principle.* If (EP) is solvable, then obviously it is finitely solvable on the compact subset $\{\bar{x}\}$ and $\{\bar{x}\} = \{\bar{x}\} \cap \operatorname{lev}_{\geq}\Phi(\cdot, y)$ is closed in $\{\bar{x}\}$ for every $y \in A$, where $\bar{x} \in \operatorname{Sol(EP)}$. Conversely, if (EP) is finitely solvable on a compact subset K of A, then, for all $N \in \mathscr{F}(A)$, $\Phi(\bar{x}, y) \geq 0$ for some $\bar{x} \in K$ and for all $y \in N$, i.e., $\bigcap_{y\in N}(K \cap \operatorname{lev}_{\geq}\Phi(\cdot, y)) \neq \emptyset$. Thus, the family $\{K \cap \operatorname{lev}_{\geq}\Phi(\cdot, y) | \, y \in A\}$ has the finite intersection property. Moreover, this family is a family of closed subsets in K. Hence, by the finite intersection principle, $\bigcap_{y\in A}(K \cap \operatorname{lev}_{\geq}\Phi(\cdot, y)) \neq \emptyset$ which means that (EP) has solutions.

The finite solvability principle implies the finite intersection principle. If the family $\mathscr{K} = \{K_y \mid y \in \Lambda\}$ $(\Lambda \subset A)$ has a nonempty intersection, then obviously this family has the finite intersection property. Conversely, assume that the family $\mathscr{K}$ has the finite intersection property. Let $\Phi : A \times A \to \mathbb{R}$ be defined, for all $(x, y) \in A \times A$, by

$$\Phi(x, y) = \begin{cases} 0 & \text{if } x \in K_y \text{ and } y \in \Lambda, \\ 1 & \text{if } x \in K \text{ and } y \notin \Lambda, \\ -1 & \text{otherwise.} \end{cases}$$

Then, (EP) with this bifunction Φ is finitely solvable on the compact set K. Moreover, for all $y \in A$,

$$\operatorname{lev}_{\geq}\Phi(\cdot, y) = \begin{cases} K_y & \text{if } y \in \Lambda, \\ K & \text{otherwise} \end{cases}$$

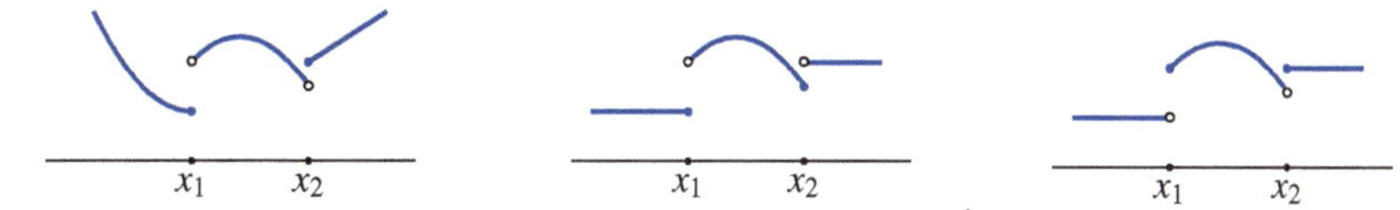

Fig. 1.4 Semicontinuity

is closed in K. Hence, by the finite solvability principle, there exists $\bar{x} \in A$ such that $\Phi(\bar{x}, y) \geq 0$ for all $y \in A$, i.e., $\emptyset \neq \bigcap_{y \in A} \operatorname{lev}_{\geq} \Phi(\cdot, y) \subset \bigcap_{y \in A} K_y$. □

1.1.3 Semicontinuity of Scalar Functions

Definition 1.10 Let X be a topological space. A function $\varphi : X \to \overline{\mathbb{R}}$ is called LOWER SEMICONTINUOUS (UPPER SEMICONTINUOUS FUNCTIONS, resp.) at $\bar{x} \in X$ if for each $\epsilon > 0$, there exists a neighborhood U of $\bar{x}$ such that $\varphi(x) > \varphi(\bar{x}) - \epsilon$ ($\varphi(x) < \varphi(\bar{x}) + \epsilon$, resp.) for all $x \in U$. If φ is both lower semicontinuous and upper semicontinuous at $\bar{x}$, then ϕ is CONTINUOUS at $\bar{x}$. φ is called lower semicontinuous (upper semicontinuous, continuous, resp.) if the aforementioned property is satisfied at all $x \in X$.

Figure 1.4 provides some examples of semicontinuity.

Recall that lower and upper limits of φ at $\bar{x}$ are defined by

$$\liminf_{x \to \bar{x}} \varphi(x) := \sup_{U \in \mathscr{U}(\bar{x})} \inf_{x \in U} \varphi(x), \quad \limsup_{x \to \bar{x}} \varphi(x) := \inf_{U \in \mathscr{U}(\bar{x})} \sup_{x \in U} \varphi(x),$$

where $\mathscr{U}(\bar{x})$ is the collection of all neighborhoods of $\bar{x}$. If (X, d) is a metric space, then

$$\liminf_{x \to \bar{x}} \varphi(x) = \lim_{\delta \to 0} \inf_{x \in \mathbb{B}(\bar{x}, \delta)} \varphi(x) \text{ and } \limsup_{x \to \bar{x}} \varphi(x) = \lim_{\delta \to 0} \sup_{x \in \mathbb{B}(\bar{x}, \delta)} \varphi(x).$$

When $\liminf_{x \to \bar{x}} \varphi(x) = \limsup_{x \to \bar{x}} \varphi(x) = \ell$, we write $\lim_{x \to \bar{x}} \varphi(x) = \ell$ and call this value the limit of φ at $\bar{x}$. In terms of the lower/upper limit and limit, we have:

- φ is lower semicontinuous at $\bar{x}$ if and only if $\liminf_{x \to \bar{x}} \varphi(x) \geq \varphi(\bar{x})$.
- φ is upper semicontinuous at $\bar{x}$ if and only if $\limsup_{x \to \bar{x}} \varphi(x) \leq \varphi(\bar{x})$.
- φ is continuous at $\bar{x}$ if and only if $\lim_{x \to \bar{x}} \varphi(x) = \varphi(\bar{x})$.

Semicontinuity of a function φ corresponds geometrically to the closedness of its epigraph, hypograph and level sets as follows.

Proposition 1.11 *Let A be a subset of a topological space X and $\varphi : A \to \overline{\mathbb{R}}$. The following statements are equivalent:*

(a) *φ is lower semicontinuous (upper semicontinuous, resp.).*
(b) $\mathrm{epi}\varphi$ *(*$\mathrm{hypo}\phi$*, resp.) is closed.*
(c) *Any lower-level set* $\mathrm{lev}_{\leq\beta}\varphi$ *(any upper level set* $\mathrm{lev}_{\geq\beta}\varphi$*, resp.) of φ is closed.*

Proof We prove for the lower semicontinuous case. The upper semicontinuous case is treated similarly.

(a) $\Rightarrow$ (b). Let $(\bar{x}, \bar{\beta}) \in (A \times \mathbb{R}) \setminus \mathrm{epi}\varphi$. Then, $\varphi(\bar{x}) > \bar{\beta}$, and hence, $\epsilon = \frac{\varphi(\bar{x})-\bar{\beta}}{4} > 0$. By the lower semicontinuity of φ, there exists a neighborhood U of $\bar{x}$ such that $\varphi(x) > \varphi(\bar{x}) - \epsilon = \frac{3\varphi(\bar{x})+\bar{\beta}}{4} > \frac{4\bar{\beta}+\varphi(\bar{x})-\bar{\beta}}{4} = \bar{\beta} + \epsilon > \beta$ for all $x \in U$ and $\beta \in \Lambda := (\bar{\beta} - \epsilon, \bar{\beta} + \epsilon)$. It follows that $(\bar{x}, \bar{\beta}) \in U \times \Lambda \subset (A \times \mathbb{R}) \setminus \mathrm{epi}\varphi$. Thus, $(A \times \mathbb{R}) \setminus \mathrm{epi}\varphi$ is an open set.

(b) $\Rightarrow$ (c). Take any $\bar{\beta} \in \mathbb{R}$ and $\bar{x} \in A \setminus \mathrm{lev}_{\leq\bar{\beta}}\varphi$. Then, $(\bar{x}, \bar{\beta}) \in (A \times \mathbb{R}) \setminus \mathrm{epi}\varphi$. Since $(A \times \mathbb{R}) \setminus \mathrm{epi}\varphi$ is open, there exist a neighborhood U of $\bar{x}$ and a neighborhood Λ of $\bar{\beta}$ such that $(\bar{x}, \bar{\beta}) \in U \times \Lambda \subset (A \times \mathbb{R}) \setminus \mathrm{epi}\varphi$. This means that $\varphi(x) > \beta$ for all $x \in U$ and $\beta \in \Lambda$. In particular, $\varphi(x) > \bar{\beta}$ for all $x \in U$, i.e., $U \subset A \setminus \mathrm{lev}_{\leq\bar{\beta}}\varphi$. Thus, $A \setminus \mathrm{lev}_{\leq\bar{\beta}}\varphi$ is an open set.

(c) $\Rightarrow$ (a). Let $\bar{x} \in A$ and $\epsilon > 0$ be arbitrary. Since $\bar{x} \in A \setminus \mathrm{lev}_{\leq\varphi(\bar{x})-\epsilon}\varphi$ which is an open set, there exists a neighborhood U of $\bar{x}$ such that $\bar{x} \in U \subset A \setminus \mathrm{lev}_{\leq\varphi(\bar{x})-\epsilon}\varphi$. It follows that $\varphi(x) > \varphi(\bar{x}) - \epsilon$ for all $x \in U$. Thus, f is lower semicontinuous at $\bar{x}$. □

1.1.4 Simplices, Sperner's Lemma, and Brouwer's Fixed-Point Theorem

Simplices are the simplest convex sets and n-dimensional generalizations of a triangle. A point is a 0-simplex, a line segment is a 1-simplex, a triangle is a 2-simplex, a tetrahedron is a 3-simplex, etc. The formal definition is as follows.

Definition 1.12 (Simplices) An n-SIMPLEX $\mathscr{T}$ in $\mathbb{R}^m$ is the convex hull of an affinely independent set of $n + 1$ vectors $x^0, x^1, \dots, x^n$, where the set $\{x^0, x^1, \dots, x^n\}$ is called affinely independent if from $\sum_{i=0}^{n} \lambda_i x^i = 0$ and $\sum_{i=0}^{n} \lambda_i = 0$ it follows that $\lambda_0 = \lambda_1 = \dots = \lambda_n = 0$.

Each x^i, $i = 0, 1, \dots, n$, is a vertex of the n-simplex $\mathscr{T}$. For $\{i_0, i_1, \dots, i_k\} \subset \{0, 1, \dots, n\}$, the k-simplex $\mathrm{conv}\{x^{i_0}, x^{i_1}, \dots, x^{i_k}\}$ is a k-face of $\mathscr{T}$. Each vertex also is a face and $\mathscr{T}$ is a face of itself. For each $x \in \mathscr{T}$, there exist unique numbers $\lambda_0, \lambda_1, \dots, \lambda_n$ such that $x = \sum_{i=0}^{n} \lambda_i x^i$ and $\sum_{i=0}^{n} \lambda_i = 1$. Hence, x belongs to exactly one face of $\mathscr{T}$. The numbers $\lambda_0, \lambda_1, \dots, \lambda_n$ are called the barycentric coordinates of x.

The STANDARD n-SIMPLEX, denoted by Δ_n, is the n-simplex in $\mathbb{R}^{n+1}$ with vertices being the unit vectors $e^0 = (1, 0, \dots, 0)$, $e^1 = (0, 1, 0, \dots, 0)$,...,

$e^n = (0, 0, \ldots, 1)$ of $\mathbb{R}^{n+1}$:

$$\Delta_n := \text{conv}\{e^i,\ i = 1, \ldots, n\} = \left\{ e \in \mathbb{R}^{n+1} \mid e = \sum_{i=0}^{n} \lambda_i e^i,\ \lambda_i \geq 0,\ \sum_{i=0}^{n} \lambda_i = 1 \right\}.$$

Definition 1.13 A SIMPLICIAL SUBDIVISION of an n-simplex $\mathcal{T}$ is a finite collection of subsimplices $\{\mathcal{T}_i \mid i \in I\}$ satisfying $\bigcup_{i \in I} \mathcal{T}_i = \mathcal{T}$ such that for any $i, j \in I$, $\mathcal{T}_i \cap \mathcal{T}_j$ is either empty or a common face of $\mathcal{T}_i$ and $\mathcal{T}_j$. The diameter of the largest subsimplex is called the MESH of a subdivision.

For example, with a given $\epsilon > 0$, we can subdivide the standard n-simplex Δ_n into subsimplices of diameter $< \epsilon$. Let m be a positive integer such that $\frac{\sqrt{2}}{m} < \epsilon$ and

$$V = \{e = (e_0, e_1, \ldots, e_n) \in \mathbb{R}_+^{n+1} \mid e_i = \frac{k_i}{m},$$
$$\sum_{i=0}^{n} k_i = m,\ k_i \text{ integers},\ i = 0, 1, \ldots, n\}$$

be the collection of all vertices of all the subsimplices of a simplicial subdivision (see Fig. 1.5). For this subdivision, Δ_n is subdivided into m^n n-subsimplices of diameter $\frac{\sqrt{2}}{m}$ and assorted lower-dimensional subsimplices.

Definition 1.14 Suppose that an n-simplex $\mathcal{T} = [x^0, x^1, \ldots, x^n]$ is simplicially subdivided. For $x = \sum_{i=0}^{n} \lambda_i x^i \in \mathcal{T}$, we set $\chi(x) = \{i \mid \lambda_i > 0\}$. Let V be the collection of all the vertices of all the subsimplices. A PROPER LABELING of the subdivision is a function $\lambda : V \to \{0, 1, \ldots, n\}$ satisfying $\lambda(x) \in \chi(x)$. A subsimplex of the subdivision is called COMPLETELY LABELED if λ takes all the values $0, 1, \ldots, n$ on its set of vertices.

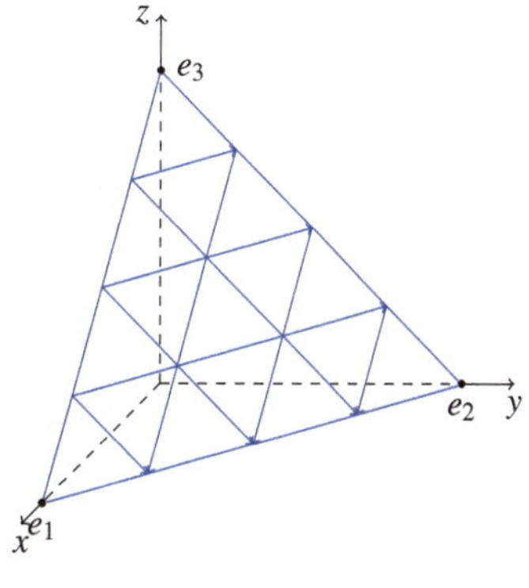

Fig. 1.5 A simplicial subdivision

Theorem 1.15 (Sperner's lemma) *Let an n-simplex be simplicially subdivided and properly labeled. Then, there are an odd number of completely labeled subsimplices in the subdivision. In particular, there exists at least one completely labeled subsimplex.*

Proof Suppose that $\mathcal{T} = [x^0, x^1, \dots, x^n]$ is simplicially subdivided and properly labeled by a function λ. If $n = 0$, $\mathcal{T} = [x^0]$ only contains the point x^0, and hence, it must bear the label 0. In this case, x^0 itself is the unique completely labeled subsimplex. Supposing that the statement is true for $n-1$, we prove it for n. Let:

C be the set of the completely labeled n-subsimplices in the subdivision
A be the set of the n-subsimplices in the subdivision such that for each $\mathcal{T}_A \in A$, the range of its set of vertices under λ is exactly $\{0, 1, \dots, n-1\}$
B be the set of the $(n-1)$-subsimplices in the subdivision such that for each $\mathcal{T}_B \in B$, $\mathcal{T}_B$ lies on the boundary and bear all the labels $\{0, 1, \dots, n-1\}$
E be the set of the $(n-1)$-simplices in the subdivision which bear all the labels $\{0, 1, \dots, n-1\}$

In the simplicial subdivision, an $(n-1)$-subsimplex either is a common face of two n-subsimplices or lies on the boundary and is the face of a single n-subsimplex. We view each subsimplex belonging to $D = C \cup A \cup B$ as a NODE and each subsimplex belonging to E as an EDGE, and so we obtain a collection of nodes and edges joining them, i.e., we have a graph. We call an edge $\mathcal{T}_E \in E$ INCIDENT at node $\mathcal{T}_D \in D$ if either $\mathcal{T}_D \in A \cup C$ and $\mathcal{T}_E$ is a face of $\mathcal{T}_D$ or $\mathcal{T}_E = \mathcal{T}_D \in B$. The number of edges incident at the node $\mathcal{T}_D$, denoted by $\delta(\mathcal{T}_D)$, is called the DEGREE of $\mathcal{T}_D$. If $\mathcal{T}_D \in B \cup C$, $\delta(\mathcal{T}_D) = 1$. If $\mathcal{T}_D \in A$, there are two vertices of $\mathcal{T}_D$ bearing one label. Hence, there are exactly two faces of $\mathcal{T}_D$ belonging to E, so $\delta(\mathcal{T}_D) = 2$. Moreover, each edge $\mathcal{T}_E \in E$ is incident at exactly two nodes. If $\mathcal{T}_E \in B$, then it is incident at itself (as a node in B) and at an n-subsimplex $\mathcal{T}_D$ (which must belong to $A \cup C$). If $\mathcal{T}_E$ is a common face of two n-subsimplices, then these n-subsimplices belong to $A \cup C$. Thus,

$$\delta(\mathcal{T}_D) = \begin{cases} 1 \text{ if } \mathcal{T}_D \in B \cup C, \\ 2 \text{ if } \mathcal{T}_D \in A. \end{cases}$$

This equality implies that $\sum_{\mathcal{T}_D \in D} \delta(\mathcal{T}_D) = 1|B \cup C| + 2|A| = |B| + |C| + 2|A|$. Since each edge joins two nodes, we obtain $\sum_{\mathcal{T}_D \in D} \delta(\mathcal{T}_D) = 2|E|$. Thus, $|B| + |C| + 2|A| = 2|E|$. Hence, $|B| + |C|$ is even. It follows that $|C|$ is odd because $|B|$ is odd by the induction hypothesis. In particular, because the zero is not an odd number, there exists at least one completely labeled subsimplex. □

Theorem 1.16 (Brouwer's fixed-point theorem) *Let $T : \Delta_n \to \Delta_n$ be continuous. Then, T has a fixed point $\bar{e} \in \Delta_n$, i.e., $T(\bar{e}) = \bar{e}$.*

Proof We simplicially subdivide Δ_n into subsimplices of diameter $\le \frac{1}{\nu}$, $\nu \in \mathbb{N}$, and let V be the set of vertices of the subdivision. For $e = (e_0, e_1, \dots, e_n) = \sum_{j=0}^{k} \lambda_{i_j} e^{i_j} \in V$ $(\lambda_{i_j} > 0)$ and $T(e) = (T_0(e), T_1(e), \dots, T_n(e)) = \sum_{i=0}^{n} T_i(e) e^i$, the set

$$M(e) := \chi(e) \cap \{i \mid T_i(e) \le e_i\} = \{i_0, i_1, \ldots, i_k\} \cap \{i \mid T_i(e) \le e_i\}$$

is nonempty. Indeed, otherwise, $T_{i_j}(e) > e_{i_j}$ for all $j \in \{0, 1, \ldots, k\}$, and one has

$$1 = \sum_{i=0}^{n} T_i(e) \ge \sum_{j=0}^{k} T_{i_j}(e) > \sum_{j=0}^{k} e_{i_j} = \sum_{j=0}^{k} \lambda_{i_j} = 1,$$

which is a contradiction. Now we define a labeling function λ satisfying $\lambda(e) \in M(e)$. Then, λ satisfies the assumption of Sperner's lemma. Hence, there exists a completely labeled subsimplex $[e^{0\nu}, e^{1\nu}, \ldots, e^{n\nu}]$ with $e^{j\nu} = (e_0^{j\nu}, e_1^{j\nu}, \ldots, e_n^{j\nu})$. This subsimplex satisfies that $T_j(e^{j\nu}) \le e_j^{j\nu}$ for all $j = 0, 1, \ldots, n$. Fixing an $\bar{j} \in \{0, 1, \ldots, n\}$, since Δ_n is compact, we can suppose that the sequence $s = \{e^{\bar{j}\nu}\}_\nu$ converges to a point $\bar{e} = (\bar{e}_0, \bar{e}_1, \ldots, \bar{e}_n) \in \Delta_n$ as $\nu \to \infty$. Then, for each $j \in \{0, 1, \ldots, n\}$ we have $\| e^{j\nu} - \bar{e} \| \le \| e^{j\nu} - e^{\bar{j}\nu} \| + \| e^{\bar{j}\nu} - \bar{e} \| \le \frac{1}{\nu} + \| e^{\bar{j}\nu} - \bar{e} \| \to 0$ as $\nu \to \infty$. Thus, $\{e^{j\nu}\}_\nu$ converges to $\bar{e}$ as $\nu \to \infty$ for all $j = 0, 1, \ldots, n$. Since T is continuous, we must have $T_j(\bar{e}) \le \bar{e}_j$ for all $j = 0, 1, \ldots, n$. Therefore, $T(\bar{e}) = \bar{e}$ (for if $T(\bar{e}) \ne \bar{e}$, then $T_j(\bar{e}) < \bar{e}_j$ for some j, and hence, $1 = \sum_{j=0}^{n} T_j(\bar{e}) < \sum_{j=0}^{n} \bar{e}_j = 1$, a contradiction). □

1.2 Equilibria over Convex Regions of Topological Vector Spaces

In this section, the standard assumption reads as follows:

$$\begin{cases} A \text{ is a nonempty convex subset of a real topological vector space } X \\ \text{and } \Phi : A \times A \to \mathbb{R} \text{ is a given bifunction.} \end{cases} \tag{1.12}$$

Under assumption (1.12), we look for conditions concerning Φ which ensure the existence of solutions to the equilibrium problem (EP).

1.2.1 *Hemicontinuity and Sign Continuity of Bifunctions*

Definition 1.17 Bifunction Φ is said to be UPPER HEMICONTINUOUS in the first component at x if for all $y \in A$, the function $t \to \Phi(x + t(y - x), y)$ from $[0, 1]$ to $\mathbb{R}$ is upper semicontinuous at 0^+, i.e., $\Phi(x, y) \ge \limsup_{t \to 0^+} \Phi(x + t(y - x), y)$. If $-\Phi$ is upper hemicontinuous in the first component at x, then Φ is called LOWER HEMICONTINUOUS in the first component at x. If Φ is both upper hemicontinuous and lower hemicontinuous in the first component at x, then Φ is called HEMICONTINUOUS in the first component at x. φ is called upper hemicontinuous (lower hemicontinuous, hemicontinuous) in the first component if the aforementioned property is satisfied at all $x \in A$. A similar definition can be stated for the second component.

If Φ is upper semicontinuous (lower semicontinuous, continuous) in the ith component at x $(i = 1, 2)$, then Φ is upper hemicontinuous (lower hemicontinuous, continuous) in the ith component at x. The converse is not true. For example, the bifunction $\Phi : \mathbb{R} \times \mathbb{R} \longrightarrow \mathbb{R}$ defined by

$$\Phi(x, y) = \begin{cases} 1 \text{ if } x, y \in [-1, 1], \\ 0 \text{ otherwise} \end{cases}$$

is lower hemicontinuous in each component, but Φ is not lower semicontinuous in the ith component at two points -1 and $1 (i = 1, 2)$.

Definition 1.18 Bifunction Φ is called UPPER-SIGN CONTINUOUS in the first component at x if for all $y \in A$, $\Phi(x + t(y - x), y) \geq 0$ for all $t \in (0, 1)$ implies that $\Phi(x, y) \geq 0$. Φ is called LOWER-SIGN CONTINUOUS in the first component at x if $-\Phi$ is UPPER-SIGN CONTINUOUS in the first component at x. A similar definition can be stated for the second component.

Sign continuity of bifunctions is a very weak type of continuity. Any bifunction Φ which is upper/lower hemicontinuous in the first (second) component is also upper-/lower-sign continuous in the first (second) component. The converse is not true in general. For instance, the bifunction $\Phi : \mathbb{R} \times \mathbb{R} \longrightarrow \mathbb{R}$ defined by

$$\Phi(x, y) = \begin{cases} 1 \text{ if } x, y \in (-1, 1], \\ 0 \text{ otherwise} \end{cases}$$

is upper sign continuous in each component, but Φ is not upper hemicontinuous in the first component at $x = -1$ (because, for $y = 1$, the function

$$t \to \Phi(x + t(y - x), y) = \Phi(-1 + 2t), 1) = \begin{cases} 1 \text{ if } t > 0, \\ 0 \text{ if } t = 0 \end{cases}$$

is not upper semicontinuous at 0^+).

1.2.2 *Generalized Convexity*

Definition 1.19 Let $\varphi : A \to \mathbb{R}$ be a real-valued function.

(a) φ is called CONVEX if for all $x, y \in A$ and $t \in (0, 1)$, $\varphi(tx + (1-t)y) \le t\varphi(x) + (1-t)\varphi(y)$, and φ is called STRICTLY CONVEX if this inequality is strict for $x \neq y$.
(b) φ is called EXPLICITLY QUASICONVEX if, for all $x, y \in A$ and $t \in (0, 1)$, inequality $\varphi(x) \le \varphi(tx + (1-t)y)$ implies that $\varphi(tx + (1-t)y) \le \varphi(y)$.
(c) φ is called QUASICONVEX if, for all $x, y \in A$ and $t \in (0, 1)$, $\varphi(tx+(1-t)y) \le \max\{\varphi(x), \varphi(y)\}$. When this inequality is strict for $x \neq y$, φ is called STRICTLY QUASICONVEX.
(d) φ is called SEMISTRICTLY QUASICONVEX if, for all $x, y \in A$ with $\varphi(x) \neq \varphi(y)$ and $t \in (0, 1)$, we have $\varphi(tx + (1-t)y) < \max\{\varphi(x), \varphi(y)\}$.

By definition, φ is explicitly quasiconvex if and only if it is both quasiconvex and semistrictly quasiconvex. Indeed, the "only if"" part is obvious. For the "if"" part, assume that φ is not explicitly quasiconvex, i.e., there exist $x, y \in A$ and $t \in (0, 1)$ such that $\varphi(x) \le \varphi(tx + (1-t)y)$ and $\varphi(y) < \varphi(tx + (1-t)y)$. Then, $\max\{\varphi(x), \varphi(y)\} \le \varphi(tx + (1-t)y) \le \max\{\varphi(x), \varphi(y)\}$ (by the quasiconvexity). Hence, $\varphi(tx + (1-t)y) = \max\{\varphi(x), \varphi(y)\}$. If $\varphi(x) = \varphi(y)$, we have $\varphi(y) < \varphi(tx + (1-t)y) = \max\{\varphi(x), \varphi(y)\} = \varphi(y)$, a contradiction. If $\varphi(x) \neq \varphi(y)$, we have $\max\{\varphi(x), \varphi(y)\} = \varphi(tx + (1-t)y) < \max\{\varphi(x), \varphi(y)\}$ (by the semistrict quasiconvexity), another contradiction.

Definition 1.19 can be explained geometrically as follows. If the line segment connecting any two points in gphφ lies on or above gphφ, then φ is convex, and if that line segment lies above gphφ, then φ is strictly convex. Thus, φ is convex if and only if its epigraph is convex, φ is strictly convex if and only if its epigraph is convex, and its graph does not contain any line segment.

If, for any two points in the domain, the graph of the restriction of φ to the line joining them either lies under both endpoints or lies below at least one of the endpoints, then φ is explicitly quasiconvex; if that restricted part lies below at least one of the endpoints, then φ is strictly quasiconvex; and if that restricted part lies under at least one of the endpoints, then φ is quasiconvex. Moreover, we can see that the quasiconvexity coincides with the convexity either of the lower-level set $\text{lev}_{\le\beta}\varphi = \{x \in A \mid \varphi(x) \le \beta\}$ or of the strict lower-level set $\text{lev}_{<\beta}\varphi = \{x \in A \mid \varphi(x) < \beta\}$ ($\beta \in \mathbb{R}$). Figure 1.6 provides some examples of generalized convexity.

Each of the above convexity properties can be characterized as follows.

- φ is convex if and only if for all $x^1, \ldots, x^n \in A$ and $t_1, \ldots, t_n \in (0, 1)$ with $\sum_{i=1}^n t_i = 1$, $\varphi(\sum_{i=1}^n t_i x^i) \le \sum_{i=1}^n t_i \varphi(x^i)$, and φ is strictly convex if and only if this inequality is strict when $x^1, \ldots, x^n$ are not all equal.
- φ is explicitly quasiconvex if and only if for all $x^1, \ldots, x^n \in A$ and $t_1, \ldots, t_n \in (0, 1)$ with $\sum_{i=1}^n t_i = 1$, from $\varphi(x^j) \le \varphi(\sum_{i=1}^n t_i x^i)$ for some $x^j \in \{x^1, \ldots, x^n\}$, it follows that there exists $x^i \in \{x^1, \ldots, x^n\}$ such that $\varphi(x^i) \ge \varphi(\sum_{i=1}^n t_i x^i)$.
- φ is quasiconvex if and only if for all $x^1, \ldots, x^n \in A$ and all $t_1, \ldots, t_n \in (0, 1)$ with $\sum_{i=1}^n t_i = 1$, $\varphi(\sum_{i=1}^n t_i x^i) \le \max\{\varphi(x^1), \varphi(x^2), \ldots, \varphi(x^n)\}$. φ is strictly

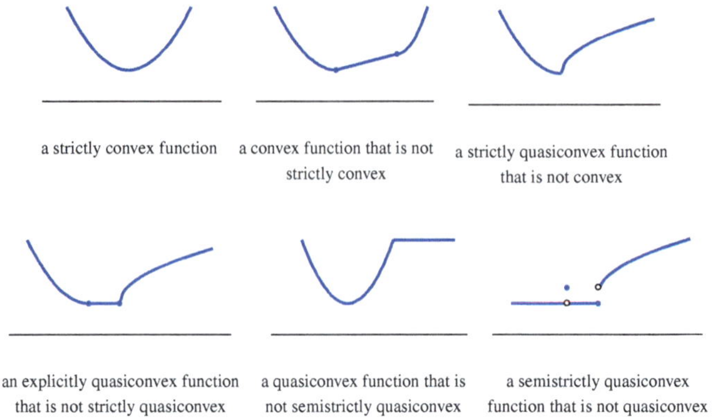

Fig. 1.6 Some generalized convex functions

quasiconvex if and only if the inequality is strict whenever $x^1, \ldots, x^n$ are not all equal.

If $-\varphi$ is convex (strictly convex, explicitly quasiconvex, quasiconvex, strictly quasiconvex, semistrictly quasiconvex, resp.), then we say that φ is concave (strictly concave, explicitly quasiconcave, quasiconcave, strictly quasiconcave, semistrictly quasiconcave, resp.).

Example 1.20

(a) Let A and B be nonempty convex subsets of a normed space. The norm function $||\cdot||$, the distance function: $\text{dist}(\cdot, B) : A \to \mathbb{R}$, defined by $\text{dist}(x, B) = \inf_{y \in B} ||x - y||$, and the function $d_B : A \to \mathbb{R}$, defined by $d_B(x) = \sup_{y \in B} ||x - y||$, are all convex. If A and B are nonempty convex subsets of a Hilbert space, then the support function $S_B : A \to \mathbb{R}$ defined by $S_B(x) = \sup_{y \in B} \langle x, y \rangle$ is convex.

(b) If $\varphi : A \to \mathbb{R}$ is strictly concave and $h : \mathbb{R} \to \mathbb{R}$ is strictly convex and nondecreasing, then $h \circ \varphi : A \to \mathbb{R}$ is explicitly quasiconcave.

(c) The COBB-DOUGLAS FUNCTION is $\varphi : A := \mathbb{R}^n_+ \to \mathbb{R}$, $\varphi(x) = r x_1^{\alpha_1} x_2^{\alpha_2} \cdots x_n^{\alpha_n}$ for all $x = (x_1, \ldots, x_n) \in \mathbb{R}^n_+$, where $\mathbb{R}^n_+ = \{(x_1, x_2, \ldots, x_n) \in \mathbb{R} \mid x_i \geq 0, i = 1, 2, \ldots, n\}$. The Cobb-Douglas function is one of the most important utility functions with applications in consumer theory. It is easy to check that if φ is strictly convex (strictly concave), then $e^{\varphi(\cdot)}$ is explicitly quasiconvex (explicitly quasiconcave). For the Cobb-Douglas function φ, we have the following.

+ If $r > 0$ and $\alpha_i > 0, i = 1, \ldots, n$, then $\log \phi(x) = \log r + \sum_{i=1}^n \alpha_i \log x_i$ is a strictly concave function. Hence, $\varphi(x) = e^{\log \varphi(x)}$ is explicitly quasiconcave. It follows that if $r < 0$ and $\alpha_i > 0, i = 1, \ldots, n$, then $\varphi(x) = -(-r x_1^{\alpha_1} x_2^{\alpha_2} \cdot \cdot \cdot x_n^{\alpha_n})$ is explicitly quasiconvex. Moreover, in this case if $\sum_{i=1}^n \alpha_i > 1$, then φ is not convex and φ is convex if and only if $\sum_{i=1}^n \alpha_i \leq 1$.

+ If $r > 0$ and $\alpha_i < 0, i = 1, \ldots, n$, then $\log \varphi(x) = \log r + \sum_{i=1}^{n} \alpha_i \log x_i$ is a strictly convex function. Hence, $\varphi(x) = e^{\log \varphi(x)}$ is explicitly quasiconvex.

(d) In economics, another important utility function is the CONSTANT ELASTICITY OF THE SUBSTITUTION (C.E.S) FUNCTION $\varphi : \mathbb{R}^n_+ \to \mathbb{R}$, which is defined as follows: for $x = (x_1, \ldots, x_n) \in \mathbb{R}^n_+$, $\varphi(x) = (a_1 x_1^{\beta} + a_2 x_2^{\beta} + \cdots + a_n x_n^{\beta})^{\frac{1}{\beta}}$, $a_i > 0$, $i = 1, 2, \ldots, n$, and $\beta \neq 0$. It follows directly from Definition 1.19 that if $\varphi(x)$ is strictly convex, then $\frac{1}{\varphi(x)}$ is explicitly quasiconcave. Using this fact, we easily show that the C.E.S. function is explicitly quasiconcave if $\beta < 0$ and convex if and only if $\beta \geq 1$. The C.E.S. function is quasiconcave if $\beta \leq 1$, but, in general, this function is not explicitly quasiconcave when $0 \leq \beta \leq 1$.

(e) THE LENGTH OF VECTOR FUNCTIONS. The length of a vector $x \in \mathbb{R}^n$ is defined as the largest index of a nonzero component: $\ell : \mathbb{R}^n \to \mathbb{R}$ defined by $\ell(x) = \max\{i \mid x_i \neq 0\}$ for all $x = (x_1, \ldots, x_n) \in \mathbb{R}^n$. This function is quasiconvex but not explicitly quasiconvex.

(f) Let x^* and y^* be two fixed elements in the dual space of a topological vector space X and $a, b \in \mathbb{R}$. Let $A = \{x \in X \mid \langle y^*, x\rangle + b > 0\}$. Then, the linear-fractional function $\varphi : A \to \mathbb{R}$, defined by $\varphi(x) = (\langle x^*, x\rangle + a)(\langle y^*, x\rangle + b)^{-1}$, is a quasiconvex function.

Definition 1.21 Let $\varphi : A \to \mathbb{R}$ be a real-valued function.

(a) φ is called QUASICONVEX-LIKE (qcl) if it has the property: for all $x, y \in A$ and $t \in (0, 1)$, $\max\{\varphi(x), \varphi(y)\} \leq 0$ implies $\varphi(tx + (1-t)y) \leq 0$, or equivalently, the lower-level set $\mathrm{lev}_{\leq}\varphi = \{x \in A \mid \varphi(x) \leq 0\}$ is convex. If the latter inequality is strict for $x \neq y$, then we call φ STRICTLY QUASICONVEX-LIKE (s.qcl).

(b) φ is called STRONGLY QUASICONVEX-LIKE (st.qcl) if for all $x, y \in A$ and $t \in (0, 1)$, $\max\{\varphi(x), \varphi(y)\} < 0$ implies $\varphi(tx + (1-t)y) < 0$, or equivalently, the strict lower-level set $\mathrm{lev}_{<}\varphi = \{x \in A \mid \varphi(x) < 0\}$ is convex.

Example 1.22 The function $\varphi : [-\frac{5\pi}{4}, \frac{5\pi}{4}] \to \mathbb{R}$, defined by

$$\varphi(x) = \begin{cases} 0 & \text{if } x \in (-\frac{\pi}{4}, \frac{\pi}{4}), \\ -\cos x & \text{if } \notin (-\frac{\pi}{4}, \frac{\pi}{4}), \end{cases}$$

is quasiconvex-like, but neither quasiconvex nor strongly quasiconvex-like (see Fig. 1.7). The function $\psi : [-\frac{5\pi}{4}, \frac{5\pi}{4}] \to \mathbb{R}$, defined by

$$\psi(x) = \begin{cases} -2\cos 2x & \text{if } x \in [-\frac{3\pi}{4}, \frac{3\pi}{4}], \\ 0 & \text{if } x \notin [-\frac{3\pi}{4}, \frac{3\pi}{4}], \end{cases}$$

is strongly quasiconvex-like, but neither quasiconvex nor quasiconvex-like (see Fig. 1.7).

Figure 1.8 illustrates the relationship between the concepts in Definitions 1.19 and 1.21, without any continuity assumption.

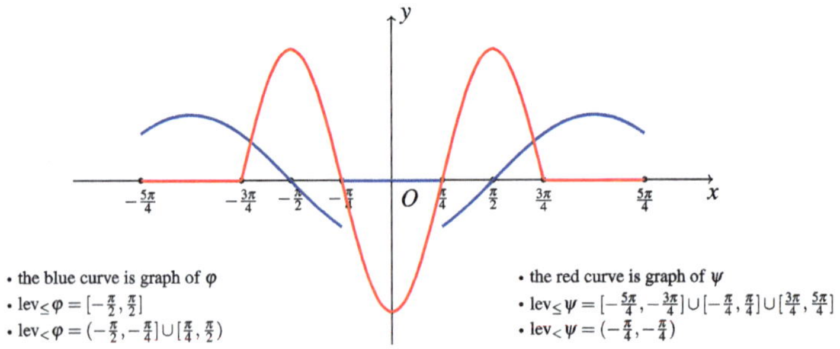

Fig. 1.7 Generalized convex functions in Example 1.22

Fig. 1.8 The relationship between the concepts of convexity

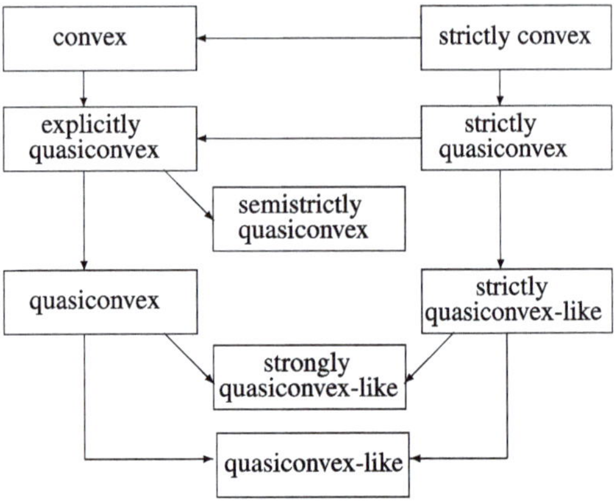

The following proposition can be easily checked.

Proposition 1.23 *Let ℓ be one of the properties of convexity in Definitions 1.19 and 1.21. Then, φ has the property ℓ if and only if $\varphi|_B$ has the property ℓ for every convex subset B of A, if and only if $\varphi|_{[x,y]}$ has the property ℓ for every $x, y \in A$, where $[x, y] := \{tx + (1-t)y \mid t \in [0, 1]\}$ is the line segment joining x and y.*

To show complete relationships, we prove the following proposition.

Proposition 1.24 *If function φ is lower semicontinuous and semistrictly quasiconvex, then it is explicitly quasiconvex and hence quasiconvex.*

Proof For $x, y \in A$ assume without loss of generality that $\varphi(x) \leq \varphi(y)$. If $\varphi(x) < \varphi(y)$, then the conclusion follows from the semistrict quasiconvexity of φ. Consider the case $\varphi(x) = \varphi(y)$. Suppose to the contrary that a $\bar{t} \in (0, 1)$ exists such that $\varphi(\bar{x}) > \varphi(x)$, where $\bar{x} = (1-\bar{t})x + \bar{t}y$. Then, the lower semicontinuity of φ implies that for $0 < \epsilon < \varphi(\bar{x}) - \phi(x)$, there exists $0 < \delta < 1$ such that $\varphi(x_s) \geq \varphi(\bar{x}) - \epsilon > \varphi(x)$ for all $s \in (0, \delta)$, where $x_s = sx + (1-s)\bar{x} \in (x, \bar{x}) \subset [x, y]$. Fixing $s \in (0, \delta)$, we have $\varphi(x_s) > \varphi(x)$. Since $x_s \in (x, \bar{x})$, $\bar{x} \in [x_s, y]$. Hence, we obtain a contradiction from the semistrict quasiconvexity of $\varphi|_{[x_s,y]}$ if $\varphi(x_s) < \varphi(\bar{x})$ or of $\varphi|_{[x,\bar{x}]}$ if $\varphi(x_s) \geq \varphi(\bar{x})$.

□

1.2.3 Monotonicity of Bifunctions

Definition 1.25

(a) Bifunction Φ is called MONOTONE if, for each $x, y \in A$, we have $\Phi(x, y) + \Phi(y, x) \leq 0$. If the inequality is strict for $x \neq y$, then Φ is called STRICTLY MONOTONE.
(b) Bifunction Φ is called PSEUDOMONOTONE if for each $x, y \in A$, $\Phi(x, y) \geq 0$ implies $\Phi(y, x) \leq 0$, or equivalently, $\Phi(x, y) > 0$ implies $\Phi(y, x) < 0$. If $\Phi(x, y) \geq 0$ implies $\Phi(y, x) < 0$ for each $x, y \in X$ with $x \neq y$, then Φ is called STRICTLY PSEUDOMONOTONE.
(c) Bifunction Φ is called QUASIMONOTONE if for each $x, y \in A$, $\Phi(x, y) > 0$ implies $\Phi(y, x) \leq 0$.

The concepts in Definition 1.25 imply the corresponding properties for an operator $T : A \to X^*$ from A into the dual space X^* as follows.

- T is called monotone if $\langle T(y) - T(x),\ y - x\rangle \geq 0$ for all $x, y \in A$. If this inequality is strict for $x \neq y$, then T is said to be strictly monotone.
- T is called pseudomonotone if for each $x, y \in A$, $\langle T(x),\ y - x\rangle \geq 0$ implies $\langle T(y),\ y-x\rangle \geq 0$, or equivalently, $\langle T(x),\ y-x\rangle > 0$ implies $\langle T(y),\ y-x\rangle > 0$. If $\langle T(x),\ y - x\rangle \geq 0$ implies $\langle T(y),\ y - x\rangle > 0$ for each $x, y \in A$ with $x \neq y$, then T is called strictly pseudomonotone.
- T is called quasimonotone if for each $x, y \in A$, $\langle T(x),\ y - x\rangle > 0$ implies $\langle T(y),\ y - x\rangle \geq 0$.

Define $\Phi : A \times A \to \mathbb{R}$ by $\Phi(x, y) = \langle T(x),\ y - x\rangle$. Then, Φ is monotone (strictly monotone, pseudomonotone, strictly pseudomonotone, quasimonotone, resp.) if and only if T is monotone (strictly monotone, pseudomonotone, strictly pseudomonotone, quasimonotone, resp.).

Example 1.26

(a) Let $\Phi : \mathbb{R} \times \mathbb{R} \to \mathbb{R}$. If $\Phi(x, y) = x - y - 1$ for all $x, y \in \mathbb{R}$, Φ is strictly monotone. If $\Phi(x, y) = x - y$ for all $x, y \in \mathbb{R}$, then it is monotone but not strictly monotone. While if Φ is defined by

$$\Phi(x, y) = \begin{cases} x - y \text{ if } x > y, \\ -1 \quad \text{if } x \leq y, \end{cases}$$

then Φ is strictly pseudomonotone but not monotone.

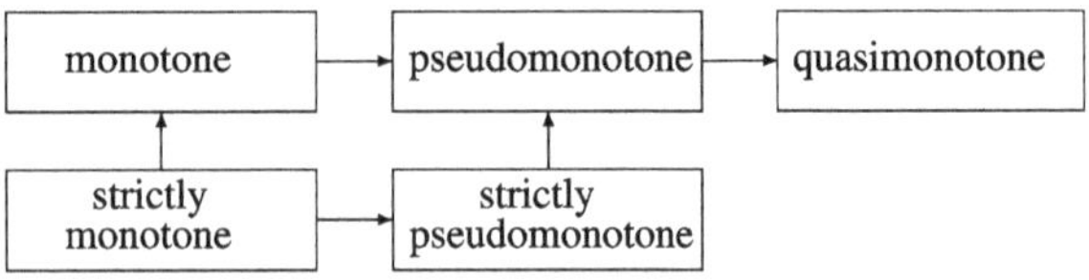

Fig. 1.9 The relationship between the concepts of monotonicity

(b) Let $\Phi : [0, 1] \times [0, 1] \to \mathbb{R}$ be defined by

$$\Phi(x, y) = \begin{cases} x - y & \text{if } x > y, \\ \frac{1}{2}(x - y) & \text{if } x \leq y. \end{cases}$$

Then, Φ is pseudomonotone, but neither monotone nor strictly pseudomonotone. If

$$\Phi(x, y) = \begin{cases} 1 \text{ if } x + y < 1, \;\; y > x, \;\; \text{or } x + y > 1, \;\; y < x, \\ 0 \text{ otherwise}, \end{cases}$$

then Φ is quasimonotone, but not pseudomonotone.

The relationships between these concepts are illustrated in Fig. 1.9.

1.2.4 *Proper Monotonicity of Bifunctions*

Definition 1.27 Bifunction Φ is said to be PROPERLY QUASIMONOTONE (pr.qm) in the first component (in the second component, resp.) on A if, for all $N \in \mathscr{F}(A)$ and $x \in \operatorname{conv} N$,

$$\max_{x^j \in N} \Phi(x^j, x) \geq 0 (\max_{x^j \in N} \Phi(x, x^j) \geq 0, \text{resp.}).$$

Example 1.28

(a) Let $T : A \to X^*$. Then, bifunction Φ, defined by $\Phi(x, y) = \langle T(x), y - x \rangle$ for all $x, y \in A$, is properly quasimonotone in the second argument.
(b) Let $\Phi : A \times A \to \mathbb{R}$ be defined by $\Phi(x, y) = \varphi(y) - \varphi(x)$ for all $x, y \in A$, where $\varphi : A \to \mathbb{R}$. Then, if φ is a quasiconcave (quasiconvex, resp.) function, then Φ is properly quasimonotone in the first component (properly quasimonotone in the second component, resp.). Indeed, for all $N \in \mathscr{F}(A)$ and $x \in \operatorname{conv} N$, taking $x^{i_0} \in N$ such that $\varphi(x^{i_0}) = \min\{\varphi(x^i) \mid x^i \in N\}$ ($\varphi(x^{i_0}) = \max\{\varphi(x^i) \mid x^i \in N\}$, resp.), we have $\Phi(x^{i_0}, x) = \varphi(x) - \varphi(x^{i_0}) \geq \varphi(x) - \varphi(x) = 0$ ($\Phi(x, x^{i_0}) = \varphi(x^{i_0}) - \varphi(x) \geq \varphi(x^{i_0}) - \varphi(x^{i_0}) = 0$, resp.).

(c) Let $\Phi : A \times A \to \mathbb{R}$ be defined for all $x, y \in A$ by

$$\Phi(x, y) = \begin{cases} \Omega(x, y) & \text{if } \varphi(x) < \varphi(y), \\ |\Omega(x, y)| & \text{if } \varphi(x) \geq \varphi(y), \end{cases}$$

where $\varphi : A \to \mathbb{R}$ and $\Omega : A \times A \to \mathbb{R}$. Then, if φ is quasiconvex (quasiconcave, resp.), then Φ is properly quasimonotone in the first component (properly quasimonotone in the second component, resp.). Indeed, for all $N \in \mathscr{F}(A)$ and $x \in \operatorname{conv} N$, taking $x^{i_0} \in N$ such that $\varphi(x^{i_0}) = \max\{\varphi(x^i) \mid x^i \in N\}$ ($\varphi(x^{i_0}) = \min\{\varphi(x^i) \mid x^i \in N\}$, resp.), we have $\varphi(x^{i_0}) \geq \varphi(x)$ ($\varphi(x) \geq \varphi(x^{i_0})$, resp.), and hence, $\Phi(x^{i_0}, x) = |\Omega(x^{i_0}, x)| \geq 0$ ($\Phi(x, x^{i_0}) = |\Omega(x, x^{i_0})| \geq 0$, resp.).

(d) The bifunction $\Phi : [0, 1] \times [0, 1] \to \mathbb{R}$, defined by

$$\Phi(x, y) = \begin{cases} -1 & \text{if } x = 0,\ y > 0, \text{ or } y = 0,\ x > 0, \\ 0 & \text{otherwise}, \end{cases}$$

is properly quasimonotone in both the first component and the second component.

(e) The bifunction $\Phi : [-1, 1] \times [-1, 1] \to \mathbb{R}$, defined by

$$\Phi(x, y) = \begin{cases} -1 & \text{if } x^2 < y^2, \\ 0 & \text{otherwise}, \end{cases}$$

is properly quasimonotone in the first component, but it is not properly quasimonotone in the second component (since, for $N = \{-\frac{1}{2}, \frac{1}{2}\}$ and $x = 0 \in \operatorname{conv} N$, we have $\Phi(0, -\frac{1}{2}) = -1 < 0$ and $\Phi(0, \frac{1}{2}) = -1 < 0$). While the bifunction $\Phi : [-1, 1] \times [-1, 1] \to \mathbb{R}$, defined by

$$\Phi(x, y) = \begin{cases} -1 & \text{if } x^2 > y^2, \\ 0 & \text{otherwise}, \end{cases}$$

is properly quasimonotone in the second component but not properly quasimonotone in the first component.

From now on, we say that a bifunction Φ has some property (such as convexity, quasiconvexity, etc.) in the first (second, resp.) component if for all $y \in A$ (for all $x \in A$, resp.), $\Phi(\cdot, y)$ ($\Phi(x, \cdot)$, resp.) has this property. We use $\mathscr{A}_{\mathscr{R}}$ ($\mathscr{A}^i_{\mathscr{R}}$, resp.) to denote the class of all bifunctions defined on $A \times A$ which have a property $\mathscr{R}$ (a property $\mathscr{R}$ in the first or second component, resp.). For a class of bifunctions $\mathscr{A}$, we define $-\mathscr{A} := \{\Phi \mid -\Phi \in \mathscr{A}\}$. The abbreviations of the above classes of bifunctions are listed in Table 1.1.

Figure 1.10 illustrates the relationships between the classes of bifunctions in Table 1.1.

Table 1.1 Table of the abbreviations of some classes of bifunctions

$\mathscr{A}^i_\ell / \mathscr{A}_\ell$	ℓ	$\mathscr{A}^i_\ell / \mathscr{A}_\ell$	ℓ
$\mathscr{A}^i_c$	c (convex)	$\mathscr{A}^i_{s.c}$	s.c (strictly convex)
$\mathscr{A}^i_{eqc}$	eqc (explicitly quasiconvex)	$\mathscr{A}^i_{s.qc}$	s.qc (strictly quasiconvex)
$\mathscr{A}^i_{qc}$	qc (quasiconvex)	$\mathscr{A}^i_{s.qcl}$	s.qcl (strictly quasiconvex-like)
$\mathscr{A}^i_{qcl}$	qcl (quasiconvex-like)	$\mathscr{A}^i_{st.qcl}$	st.qcl (strongly quasiconvex-like)
$\mathscr{A}^i_{ss.qc}$	ss.qc (semistrictly quasiconvex)		
$\mathscr{A}_m$	m (monotone)	$\mathscr{A}_{s.m}$	s.m (strictly monotone)
$\mathscr{A}_{pm}$	pm (pseudomonotone)	$\mathscr{A}_{s.pm}$	s.pm (strictly pseudomonotone)
$\mathscr{A}_{qm}$	qm (quasimonotone)	$\mathscr{A}^i_{pr.qm}$	pr.qm (properly quasimonotone)

$$
\begin{array}{ccccccc}
 & & \mathscr{A}^i_c & \supset & \mathscr{A}^i_{s.c} & & \\
 & & \cap & & \cap & & \\
\mathscr{A}^i_{qcl} & \supset & \mathscr{A}^i_{eqc} & \supset & \mathscr{A}^i_{s.qc} & & \\
 & & \cap & & \cap & & \\
\mathscr{A}^i_{st.qcl} & \supset & \mathscr{A}^i_{qc} & & \mathscr{A}^i_{s.qcl} & \subset & \mathscr{A}^i_{ss.qc} \\
 & & \cap & & \cap & & \\
 & & \mathscr{A}^i_{qcl} & & \mathscr{A}^i_{st.qcl} & &
\end{array}
$$

(a) The relationships between the classes of bifunctions which have the given convexities in the ith component.

$$
\begin{array}{ccccc}
-\mathscr{A}_{qm} & \subset & \mathscr{A}^{\geq}_A & & \\
 & & & & \\
\mathscr{A}_{s.m} & \subset & \mathscr{A}_{s.pm} & & \\
\cap & & \cap & & \\
\mathscr{A}_m & \subset & \mathscr{A}_{pm} & \subset & \mathscr{A}_{qm}
\end{array}
$$

(b) The relationships between the classes of bifunctions which have the given monotonicities.

Fig. 1.10 The relationships between the classes of bifunctions

Proposition 1.29 *The following inclusions always hold:*

(a) $\mathscr{A}^{\geq}_A \cap \mathscr{A}^i_{st.qcl} \subset \mathscr{A}^i_{pr.qm} \subset \mathscr{A}^{\geq}_A$.
(b) $-\mathscr{A}_{qm} \cap \mathscr{A}^i_{st.qcl} \subset \mathscr{A}^i_{pr.qm}$.
(c) $\mathscr{A}^{=}_A \cap (-\mathscr{A}_{pm}) \cap (-\mathscr{A}^1_{st.qcl}) \subset \mathscr{A}^2_{pr.qm}$.
(d) $\mathscr{A}^{=}_A \cap \mathscr{A}_{pm} \cap \mathscr{A}^2_{st.qcl} \subset -\mathscr{A}^1_{pr.qm}$.
(e) $\mathscr{A}^{=}_A \cap (-\mathscr{A}_{qm}) \cap (-\mathscr{A}^1_{s.qcl}) \subset \mathscr{A}^2_{pr.qm}$.
(f) $\mathscr{A}^{=}_A \cap \mathscr{A}_{qm} \cap \mathscr{A}^2_{s.qcl} \subset -\mathscr{A}^1_{pr.qm}$.

Proof

(a) We prove this for $i = 2$; the case $i = 1$ can be proved similarly. Let $\Phi \in \mathscr{A}^{\geq}_A \cap \mathscr{A}^2_{st.qcl}$. Reasoning ad absurdum, suppose that $\Phi \notin \mathscr{A}^2_{pr.qm}$, i.e., there exist $N \in \mathscr{F}(A)$ and $x \in \text{conv}N$ such that $\max_{x^j \in N} \Phi(x, x^j) < 0$. Since $\Phi \in \mathscr{A}^2_{st.qcl}$, i.e., $\Phi(x, \cdot)$ is strongly quasiconvex-like, we have $\Phi(x, x) < 0$, which contradicts the assumption that $\Phi \in \mathscr{A}^{\geq}_A$. The inclusion $\mathscr{A}^i_{pr.qm} \subset \mathscr{A}^{\geq}_A$ is obvious.
(b) It is easy to see that $-\mathscr{A}_{qm} \subset \mathscr{A}^{\geq}_A$. Hence, (b) is a special case of (a).
(c) Let $\Phi \in \mathscr{A}^{=}_A \cap (-\mathscr{A}_{pm}) \cap (-\mathscr{A}^1_{st.qcl})$ and suppose that $\Phi \notin \mathscr{A}^2_{pr.qm}$. Then, there exist $N \in \mathscr{F}(A)$ and $x \in \text{conv}N$ such that $\max_{x^j \in N} \Phi(x, x^j) < 0$, i.e., for all

$x^j \in N$, $\Phi(x, x^j) < 0$, and so $\Phi(x^j, x) > 0$ by the pseudomonotonicity of $-\Phi$. Hence, $\min_{x^j \in N} \Phi(x^j, x) > 0$. By the strong quasiconvex-like property of $-\Phi(\cdot, x)$, $\Phi(x, x) > 0$. This contradicts the fact that $\Phi \in \mathscr{A}_A^=$.
The proofs of (d), (e), and (f) are similar to that of (c).

□

Moreover, it is easy to check the following.

Proposition 1.30 *Let B be a subset of A. If $\Phi \in \mathscr{A}_\ell$ $(-\mathscr{A}_\ell, resp.)$, $\ell \in$* {m, sm, pm, qm, pr.qm} *and $i = 1, 2$, then $\Phi|_{B\times B} \in \mathscr{B}_\ell$ $(-\mathscr{B}_\ell, resp.)$.*

Proof This is obvious. (The abbreviation $\mathscr{B}_\ell$ $[-\mathscr{B}_\ell$, resp.] is defined for B in the same way as $\mathscr{A}_\ell$ $[-\mathscr{A}_\ell$, resp.] for A.) □

1.2.5 Existence of Solutions

First, we prove some lemmas for the finite solvability of (EP). The following statement is equivalent to the Brouwer fixed-point theorem.

Lemma 1.31 *If $\Phi \in \mathscr{A}^2_{pr.qm}$ and $\mathrm{lev}_{\geq}\Phi(\cdot, y)$ is closed for every $y \in A$, then* (EP) *is finitely solvable.*

Proof Suppose that (EP) is not finitely solvable. Then, there exists a finite subset $N := \{y^0, y^1, \ldots, y^n\}$ of A such that for each $x \in A$, there is $y^i \in N$ satisfying $\Phi(x, y^i) < 0$. Because conv$N \subset A$, the finite family $\mathscr{P} := \{U_{y^i} \mid y^i \in N\}$, where $U_{y^i} = \mathrm{conv}N \cap \mathrm{lev}_{<}\Phi(\cdot, y^i) = \{x \in \mathrm{conv}N \mid \Phi(x, y^i) < 0\}$, is a cover of conv$N$. Since the set $\mathrm{lev}_{<}\Phi(\cdot, y^i)$ is open, U_{y^i} is open in conv$N$. Hence, $\mathscr{P}$ is an open cover of conv$N$. Let Δ_n be the standard n-simplex of $\mathbb{R}^{n+1}$. We define a map $\sigma : \Delta \to A$ by $\sigma(e) = \sum_{i=0}^n t_i y^i$ for all $e = \sum_{i=0}^n t_i e^i \in \Delta_n$. It is clear that σ is continuous, and so conv$N = \sigma(\Delta_n)$ is compact. Then, there exists a continuous partition of unity associated with the cover $\mathscr{P}$, i.e., there exist functions ϕ_i : conv$N \to [0, 1]$ $(i = 0, 1, \ldots, n)$ satisfying the conditions

$$\begin{cases} \phi_i \text{ is continuous for each } i = 0, 1, \ldots, n; \\ \phi_i(x) > 0 \text{ implies } x \in U_{y_i}; \\ \sum\limits_{i=0}^{n} \phi_i(x) = 1 \text{ for each } x \in \mathrm{conv}N. \end{cases}$$

Let ϕ : conv$N \to \Delta_n$ be defined by $\phi(x) = \sum_{i=0}^n \phi_i(x)e^i$ for all $x \in$ conv$N$. Then, ϕ is continuous and so is $\phi \circ \sigma : \Delta_n \to \Delta_n$. Brouwer's fixed-point theorem gives us a point $\bar{e} \in \Delta_n$ such that $\bar{e} = \phi(\sigma(\bar{e})) = \phi(\bar{x})$, where $\bar{x} = \sigma(\bar{e})$. Let $J(\bar{x}) := \{j \in \{0, 1, \ldots, n\} | \phi_j(\bar{x}) > 0\}$ and $M(\bar{x}) = \{y^j \in N \mid j \in J(\bar{x})\}$. We have $\bar{e} = \phi(\bar{x}) = \sum_{j\in J(\bar{x})} \phi_j(\bar{x})e^j$, and hence, $\bar{x} = \sigma(\bar{e}) = \sum_{j\in J(\bar{x})} \phi_j(\bar{x})y^j \in$ conv$M(\bar{x})$. Because $\Phi \in \mathscr{A}^2_{pr.qm}$, there exists $\bar{j} \in J(\bar{x})$ such that $\Phi(\bar{x}, y^{\bar{j}}) \geq 0$.

On the other hand, as $\phi_{\bar{j}}(\bar{x}) > 0$, $\bar{x} \in U_{\bar{j}} = \{x \in \text{conv}N : \Phi(x, y^{\bar{j}}) < 0\}$, i.e., $\Phi(\bar{x}, y^{\bar{j}}) < 0$, a contradiction. □

Lemma 1.31 is proved based on Brouwer's fixed-point theorem and the existence of a partition of unity for a finite covering of a compact space. To show the equivalence between Lemma 1.31 and Brouwer's fixed-point theorem, we deduce from Lemma 1.31 the following assertion which implies Brouwer's fixed-point theorem.

Corollary 1.32 *Assume that A is compact. Then:*

(a) *if $T : A \to X^*$ is a continuous mapping, then the variational inequality* VI(A, T) *has solutions, i.e., there exists $\bar{x} \in A$ such that $\langle T(\bar{x}), y - \bar{x}\rangle \geq 0$ for all $y \in A$.*
(b) *if X is a Hilbert space and $T : A \to A$ is a continuous mapping, then T has a fixed point.*

Proof

(a) By assumption, the bifunction $\Phi : A \times A \to \mathbb{R}$, defined by $\Phi(x, y) = \langle T(x), y - x\rangle$ for all $x, y \in A$, satisfies the conditions of Lemma 1.31. Thus, by Lemma 1.31 and the finite solvability principle, an $\bar{x} \in A$ exists satisfying $\Phi(\bar{x}, y) = \langle T(\bar{x}), y - \bar{x}\rangle \geq 0$ for all $y \in A$.
(b) Applying the conclusion of (a) with $x - T(x)$ in place of $T(x)$, we have an $\bar{x} \in A$ such that $\langle \bar{x} - T(\bar{x}), y - \bar{x}\rangle \geq 0$ for all $y \in A$. If $\bar{x} \neq T(\bar{x})$, then there exists a $t \in (0, 1)$ such that $y_t = t\bar{x} + (1 - t)T(\bar{x})$ is different from $\bar{x}$ and $-(1 - t)\langle \bar{x} - T(\bar{x}), \bar{x} - T(\bar{x})\rangle = \langle \bar{x} - T(\bar{x}), y_t - \bar{x}\rangle \geq 0$. Since $t < 1$, we must have $\bar{x} = T(\bar{x})$.

□

The proof of the following lemma uses a basic proof technique.

Lemma 1.33 *If $\Phi \in (-\mathscr{A}^1_{qcl}) \cap \mathscr{A}^2_{st.qcl} \cap \mathscr{A}_A$ and the sets* lev$_{\geq}\Phi(x, \cdot)$ *and* lev$_{\geq}\Phi(\cdot, y)$ *are closed for every $x, y \in A$, then* (EP) *is finitely solvable.*

Proof We prove the claim "for any finite subset N of A, there exists $x_N \in A$ such that $\Phi(x_N, y) \geq 0$ for all $y \in N$" by induction on the number of elements of N. The claim is evident for N being a singleton because $\Phi \in \mathscr{A}_A$, i.e., for each $y \in A$, $\Phi(x, y) \geq 0$ for some $x \in A$. Assume that it is true for any N containing n elements and now $N := \{y^0, y^1, \ldots, y^n\}$, and let $\Phi' : \text{conv}N \times \text{conv}N \to \mathbb{R}$ be defined by $\Phi'(x, y) = \min\{\Phi(x, y), \Phi(x, y^n)\}$ for all $x, y \in \text{conv}N$. Then, clearly the assertion holds for Φ and N if the following assertion is true:

(∗) For any N' consisting of n elements of conv$N$, there exists $x_N \in \text{conv}N$ such that $\Phi'(x_N, y) \geq 0$ for all $y \in N'$.

We will show that the assumptions of Lemma 1.33 with Φ' and $B = \text{conv}N$ in place of Φ and A are also satisfied, and thus that (∗) is true by the induction assumption. By Proposition 1.23, $\Phi' \in -\mathscr{B}^1_{qcl} \cap \mathscr{B}^2_{st.qcl}$. Moreover, lev$_{\geq}\Phi'(x, \cdot)$

and $\mathrm{lev}_{\geq}\Phi'(\cdot, y)$ are closed in B for all $x, y \in B$. Thus, it remains to show that $\Phi' \in \mathscr{B}_B$. For this, it is sufficient to check that for any $y, y' \in B$, there exists $x \in B$ such that $\min\{\Phi(x, y), \Phi(x, y')\} \geq 0$. Suppose to the contrary that there exist $y, y' \in B$ such that for all $x \in B$, either $\Phi(x, y) < 0$ or $\Phi(x, y') < 0$. This implies that

$$\mathrm{lev}_{B\geq}\Phi(\cdot, y) \cap \mathrm{lev}_{B\geq}\Phi(\cdot, y') = \emptyset, \tag{1.13}$$

where $\mathrm{lev}_{B\geq}\Phi(\cdot, z) = \{x \in B \mid \Phi(x, z) \geq 0\} = \mathrm{lev}_{\geq}\Phi(\cdot, z) \cap B$ for each $z \in A$. Note that for each $z \in B$, the set $\mathrm{lev}_{B\geq}\Phi(\cdot, z)$ is convex and closed in B because $\Phi(\cdot, z)$ is quasiconcave-like and $\mathrm{lev}_{\geq}\Phi(\cdot, z)$ is closed in A.

We set

$$M_y = \{t \in [0, 1] \mid \mathrm{lev}_{B\geq}\Phi(\cdot, z_t) \subset \mathrm{lev}_{B\geq}\Phi(\cdot, y)\},$$

$$M_{y'} = \{t \in [0, 1] \mid \mathrm{lev}_{B\geq}\Phi(\cdot, z_t) \subset \mathrm{lev}_{B\geq}\Phi(\cdot, y')\},$$

where $z_t = (1 - t)y + ty' \in A$, $t \in [0, 1]$. Obviously, $0 \in M_y$ and $1 \in M_{y'}$. By (1.13), $M_y \cap M_{y'} = \emptyset$. When $\Phi \in \mathscr{A}^2_{st.qcl}$, i.e., $\Phi(x, \cdot)$ is strongly quasiconvex-like for each $x \in A$, we have for all $t \in [\bar{t}, \bar{t}'] \subset [0, 1]$,

$$\mathrm{lev}_{B\geq}\Phi(\cdot, z_t) \subset \mathrm{lev}_{B\geq}\Phi(\cdot, z_{\bar{t}}) \cup \mathrm{lev}_{B\geq}\Phi(\cdot, z_{\bar{t}'}). \tag{1.14}$$

In particular,

$$\mathrm{lev}_{B\geq}\Phi(\cdot, z_t) \subset \mathrm{lev}_{B\geq}\Phi(\cdot, y) \cup \mathrm{lev}_{B\geq}\Phi(\cdot, y') \quad \text{for all } t \in [0, 1]. \tag{1.15}$$

Indeed, for any $t \in [\bar{t}, \bar{t}']$ and $x \in \mathrm{lev}_{B\geq}\Phi(\cdot, z_t)$, $0 \leq \Phi(x, z_t)$. The strong quasiconvex-like property of $\Phi(x, \cdot)$ implies that $0 \leq \max\{\Phi(x, z_{\bar{t}}), \Phi(x, z_{\bar{t}'})\}$. Hence, either $\Phi(x, z_{\bar{t}}) \geq 0$ or $\Phi(x, z_{\bar{t}'}) \geq 0$, i.e., either $x \in \mathrm{lev}_{B\geq}\Phi(\cdot, z_{\bar{t}})$ or $x \in \mathrm{lev}_{B\geq}\Phi(\cdot, z_{\bar{t}'})$. Thus, (1.14) is true. Because the sets $\mathrm{lev}_{B\geq}\Phi(\cdot, y)$ and $\mathrm{lev}_{B\geq}\Phi(\cdot, y')$ are closed and the sets $\mathrm{lev}_{B\geq}\Phi(\cdot, z_t)$ ($t \in [0, 1]$) are convex, (1.13) and (1.15) together imply that for each $t \in [0, 1]$, either $\mathrm{lev}_{B\geq}\Phi(\cdot, z_t) \subset \mathrm{lev}_{B\geq}\Phi(\cdot, y)$ or $\mathrm{lev}_{B\geq}\Phi(\cdot, z_t) \subset \mathrm{lev}_{B\geq}\Phi(\cdot, y')$, which means that $M_y \cup M_{y'} = [0, 1]$.

We prove that the sets M_y and $M_{y'}$ are open in $[0, 1]$. For any $t_0 \in M_y$, we have $\mathrm{lev}_{B\geq}\Phi(\cdot, z_{t_0}) \subset \mathrm{lev}_{B\geq}\Phi(\cdot, y) \subset B \setminus \mathrm{lev}_{B\geq}\Phi(\cdot, y')$. Then, for each $x \in \mathrm{lev}_{B\geq}\Phi(\cdot, y')$, $\Phi(x, z_{t_0}) < 0$, i.e., $z_{t_0} \in \mathrm{lev}_{B<}\Phi(x, \cdot) := \mathrm{lev}_{<}\Phi(x, \cdot) \cap B$ which is an open set in B by the openness of $\mathrm{lev}_{<}\Phi(x, \cdot)$. Since the map $\phi(t) = z_t$ is continuous, $\phi^{-1}(\mathrm{lev}_{B<}\Phi(x, \cdot))$ is an open neighborhood of t_0 in $[0, 1]$. Hence, there exists an open set $V_{t_0 x} = (t_x, t'_x) \subset [0, 1]$ with $t_0 \in V_{t_0 x}$ and $[t_x, t'_x] \subset \phi^{-1}(\mathrm{lev}_{B<}\Phi(x, \cdot))$. This and (1.14) give that $x \in B \setminus \bigcup_{t \in [t_x, t'_x]} \mathrm{lev}_{B\geq}\Phi(\cdot, z_t) = B \setminus (\mathrm{lev}_{B\geq}\Phi(\cdot, z_{t_x}) \cup \mathrm{lev}_{B\geq}\Phi(\cdot, z_{t'_x})) =: U_x$, which is an open set in B. Then, the family $\{U_x \mid x \in \mathrm{lev}_{B\geq}\Phi(\cdot, y')\}$ is an open cover (in B) of $\mathrm{lev}_{B\geq}\Phi(\cdot, y')$.

Since $\mathrm{lev}_{B\geq}\Phi(\cdot, y')$ is compact in B, there exist $x^1, \ldots, x^k \in \mathrm{lev}_{B\geq}\Phi(\cdot, y')$ such that $\mathrm{lev}_{B\geq}\Phi(\cdot, y') \subset \bigcup_{i=1}^k U_{x^i}$. Let $V_{t_0} = \bigcap_{i=1}^k V_{t_0 x^i}$. Then, V_{t_0} is an open neighborhood of t_0 in $[0, 1]$. Moreover, $\mathrm{lev}_{B\geq}\Phi(\cdot, z_t) \subset \mathrm{lev}_{B\geq}\Phi(\cdot, y) \subset B \setminus \mathrm{lev}_{B\geq}\Phi(\cdot, y')$ for all $t \in V_{t_0}$, i.e., $V_{t_0} \subset M_y$. Thus, M_y is open. The openness of the set $M_{y'}$ can be proved similarly. Thus, $[0, 1]$ is the union of two nonempty disjoint open subsets M_y and $M_{y'}$, which is impossible. □

In Lemma 1.33, if we replace $\mathscr{A}_A$ by $\mathscr{A}_A^{\geq}$, then Lemma 1.33 becomes a particular case of Lemma 1.31 because $\mathscr{A}^2_{st.qcl} \cap \mathscr{A}_A^{\geq} \subset \mathscr{A}^2_{pr.qm}$ (see Proposition 1.29 (a)). Moreover, in this case, the condition "$\Phi \in -\mathscr{A}^1_{qcl}$ and the set $\mathrm{lev}_{\geq}\Phi(x, \cdot)$ is closed for every $x \in A$" is not needed. However, in general, Lemmas 1.31 and 1.33 are different because $\mathscr{A}^2_{pr.qm} \neq -\mathscr{A}^1_{qcl} \cap \mathscr{A}^2_{st.qcl} \cap \mathscr{A}_A$. For instance, let $A = [0, 1]$ and $\Phi : [0, 1] \times [0, 1] \to \mathbb{R}$ be defined by $\Phi(x, y) = -2x + y$ for $x, y \in [0, 1]$. Then, $\Phi \notin \mathscr{A}^2_{pr.qm}$ (since $\Phi(\frac{1}{2}, \frac{1}{2}) < 0$). Hence, A and Φ do not satisfy the assumptions of Lemma 1.31. While it is not hard to see that $-\Phi \in \mathscr{A}^1_{qcl}$, $\Phi \in \mathscr{A}^2_{st.qcl}$, $\Phi \in \mathscr{A}_A$ (for each $y \in A$, taking $x = 0$ we have $\Phi(0, y) \geq 0$), and the sets $\mathrm{lev}_{\geq}\Phi(x, \cdot)$ and $\mathrm{lev}_{\geq}\Phi(\cdot, y)$ are closed for all $x, y \in A$. Thus, A and Φ fulfill the assumptions of Lemma 1.33. It is clear that for each $N \in \mathscr{F}(A)$, there exists $\bar{x} = 0$ satisfying $\Phi(0, y) \geq 0$ for all $y \in N$.

If the feasible set A in the assumption (1.12) is compact, then by the finite solvability principle, Lemmas 1.31 and 1.33 yield sufficient conditions for the existence of solutions to (EP). In noncompact settings, we use coercivity conditions to replace the compactness of A.

Theorem 1.34 *Assume that* $\Phi \in \mathscr{A}^2_{pr.qm}$, $\mathrm{lev}_{\geq}\Phi(\cdot, y)$ *is closed for every* $y \in A$, *and the following coercivity condition is satisfied:*

(C_1) *There exists a nonempty compact subset* K *of* X *such that, for each finite subset* N *of* A, *there exists a compact convex subset* L_N *of* A *containing* N *such that for each* $x \in L_N \setminus K$, $\Phi(x, y) < 0$ *for some* $y \in L_N$.

Then, (EP) *has solutions.*

Proof For the compact set K in condition (C_1), $A \cap K$ is compact in A. By the finite solvability principle, it is sufficient to show that (EP) is finitely solvable on $A \cap K$. Suppose to the contrary that there exists a finite subset N of A such that for each $x \in A \cap K$, there is some $y \in N$ with $\Phi(x, y) < 0$. Let L_N be the compact convex subset of A containing N given in (C_1). Then, by the finite solvability principle and Lemma 1.31 for L_N and $\Phi|_{L_N \times L_N}$, there exists $x_N \in L_N$ such that $\Phi(x_N, y) \geq 0$ for all $y \in L_N$. Because for every $x \in L_N \setminus K$, $\Phi(x, y) < 0$ for some $y \in L_N$, we derive that $x_N \in L_N \cap K \subset K$. Hence, $\Phi(x_N, y) < 0$ for every $y \in N \subset L_N$, a contradiction. □

Condition (C_1) is automatically satisfied when A is compact. Moreover, (C_1) can be replaced by any of the following conditions.

(C_{11}) There exist a nonempty compact subset K of X and $\bar{y} \in K \cap A$ such that, for every $x \in A \backslash K$, $\Phi(x, \bar{y}) < 0$.
(C_{12}) There exist a nonempty compact subset K of X and a finite subset $\overline{N}$ of A such that, for each $x \in A \backslash K$, $\Phi(x, y) < 0$ for some $y \in \overline{N}$.
(C_{13}) There exists a nonempty compact convex subset K of A such that, for each $x \in A \backslash K$, $\Phi(x, y) < 0$ for some $y \in K$.

Indeed, we will show that (C_{11}) $\Rightarrow$ (C_{12}) $\Rightarrow$ (C_1) and (C_{13}) $\Rightarrow$ (C_1). That (C_{11}) implies (C_{12}) is obvious (we take $\overline{N}$ in (C_{12}) being $\{\bar{y}\}$ in (C_{11})). To see that (C_{12}) implies (C_1), in (C_1) we take $L_N = \text{conv}(N \cup \overline{N})$ for each finite subset N of A, where $\overline{N}$ is as in (C_{12}). To show that (C_{13}) implies (C_1), when K is given in (C_{13}), for each finite subset N of A, we choose L_N in (C_1) being $L_N = \text{conv}(N \cup K)$.

Corollary 1.35 *Assume that A is a closed convex subset of a reflexive Banach space, $\Phi \in \mathscr{A}^2_{pr.qm}$ and $\text{lev}_{\geq}\Phi(\cdot, y)$ is weakly closed for every $y \in A$. Assume further that either of the following coercivity conditions is satisfied:*

(C_{14}) *There exists $r > 0$ such that for all $x \in A$ with $\|x\| > r$, there is $y \in A$ with $\|y\| \leq r$ such that $\Phi(x, y) < 0$.*
(C_{15}) *For some $\hat{y} \in A$, $\Phi(x, \hat{y}) \to -\infty$ as $\|x - \hat{y}\| \to +\infty$ for any $x \in X$.*

Then, (EP) *has solutions.*

Proof To prove the case of (C_{14}), let $K = \{x \in A \mid \|x\| \leq r\}$. Then, K is weakly compact and convex, and the assumptions of Theorem 1.34 with the coercivity condition (C_{13}) are satisfied in the weak topology. Thus, (EP) has solutions.

For the case of (C_{15}) choose $R > 0$ big enough such that

$$\|x - \hat{y}\| > R \quad \Rightarrow \quad \Phi(x, \hat{y}) < 0.$$

Taking $r = R + \|\hat{y}\|$, we easily see that condition (C_{14}) is satisfied. □

Corollary 1.36 *Assume that A is a closed convex subset of a reflexive Banach space, $\Phi \in \mathscr{A}^{=}_{A} \cap \big([(-\mathscr{A}_{pm}) \cap (-\mathscr{A}^1_{qc})] \cup [(-\mathscr{A}_{qm}) \cap (-\mathscr{A}^1_{s.qcl})]\big)$, $\text{lev}_{\geq}\Phi(\cdot, y)$ is closed for every $y \in A$, and either of the coercivity conditions* (C_{14}) *and* (C_{15}) *is satisfied. Then,* (EP) *has solutions.*

Proof Note that every nonempty closed convex subset of a normed space is weakly closed. When $\Phi \in -\mathscr{A}^1_{qc}$ or $\Phi \in -\mathscr{A}^1_{s.qcl}$, $\text{lev}_{\geq}\Phi(\cdot, y)$ is convex. Hence, the assumption "$\Phi \in \mathscr{A}^{=}_{A} \cap \big([(-\mathscr{A}_{pm}) \cap (-\mathscr{A}^1_{qc})] \cup [(-\mathscr{A}_{qm}) \cap (-\mathscr{A}^1_{s.qcl})]\big)$, $\text{lev}_{\geq}\Phi(\cdot, y)$ *is closed for every $y \in A$*" is a particular case of the assumption "$\Phi \in \mathscr{A}^2_{pr.qm}$ *and* $\text{lev}_{\geq}\Phi(\cdot, y)$ *is weakly closed for every $y \in A$*" (see also Proposition 1.29 (c), (e) and Fig. 1.10). Thus, Corollary 1.36 is a special case of Corollary 1.35. □

Example 1.37

(a) Let $A = [0, +\infty)]$ and

$$\Phi(x, y) = \begin{cases} y - x & \text{if } x > y + 1, \\ 0 & \text{if } y \le x \le y + 1, \\ 1 & \text{if } x < y. \end{cases}$$

Then, $\Phi \in \mathscr{A}_A^{=} \cap (-\mathscr{A}_{pm}) \cap (-\mathscr{A}_{qc}^1)$ and $\text{lev}_{\geq}\Phi(\cdot, y) = [0, y+1]$ is closed for each $y \in A$. Moreover, for $\hat{y} = 0$, (C_{15}) is satisfied. Thus, by Corollary 1.36, (EP) has solutions. Observe that in this example $\Phi \notin \mathscr{A}_A^{=} \cap (-\mathscr{A}_{qm}) \cap (-\mathscr{A}_{s.qcl}^1)$ (since $\Phi \notin -\mathscr{A}_{s.qcl}^1$).

(b) Let $A = [0, +\infty)]$ and

$$\Phi(x, y) = \begin{cases} -\frac{1}{2} & \text{if } x > y + 1, \\ y - x & \text{if } y \le x \le y + 1, \\ 1 & \text{if } x < y. \end{cases}$$

Then, $\Phi \in \mathscr{A}_A^{=} \cap (-\mathscr{A}_{qm}) \cap (-\mathscr{A}_{s.qcl}^1)$ and $\text{lev}_{\geq}\Phi(\cdot, y) = [0, y]$ is closed for each $y \in A$. Choosing $r > 3$ we see that (C_{14}) holds. Thus, by Corollary 1.36, (EP) has solutions. In this example $\Phi \notin \mathscr{A}_A^{=} \cap (-\mathscr{A}_{pm}) \cap (-\mathscr{A}_{qc}^1)$ (since $\Phi \notin -\mathscr{A}_{qc}^1$).

Definition 1.38 A function $\varphi : A \to \mathbb{R}$ is said to HAVE PROPERTY $\mathbb{P}$ if for all $x, y \in A$, $\varphi(x) = 0$ and $\varphi(y) < 0$ imply $\varphi((1-t)x + ty) < 0$ for all $t \in (0, 1)$. Bifunction Φ is said to HAVE PROPERTY $\mathbb{P}^1$ ($\mathbb{P}^2$, resp.) if $\Phi(\cdot, y)$ ($\Phi(x, \cdot)$, resp.) has property $\mathbb{P}$ for all $y \in A$ ($x \in A$, resp.).

Example 1.39

(a) For $i = 1, 2$, every bifunction $\Phi \in \mathscr{A}_{s.qcl}^i \cup \mathscr{A}_{ss.qc}^i$ has property $\mathbb{P}^i$. Hence, every $\Phi \in \mathscr{A}_{eqc}^i$ and $\Phi \in \mathscr{A}_c^i$ also have property $\mathbb{P}^i$.

(b) Let $A = \mathbb{R}$ and $\Phi : A \times A \to \mathbb{R}$ be defined for all $x, y \in A$ by

$$\Phi(x, y) = \begin{cases} x - y & \text{if } x \neq 0, \\ y^2 - y & \text{if } x = 0, \ y \le 1, \\ 1 & \text{if } x = 0, \ y > 1. \end{cases}$$

Then, Φ has property $\mathbb{P}^2$ and $\Phi \in \mathscr{A}_{st.qcl}^2 \cap \mathscr{A}_A^{=}$, but $\Phi \notin \mathscr{A}_{s.qcl}^2 \cup \mathscr{A}_{ss.qc}^2$ (because $\Phi(0, \cdot)$ is not semistrictly quasiconvex, and $\Phi(x, \cdot)$ is not strictly quasiconvex-like for all $x \neq 0$).

(c) Let $A = [0, +\infty)$ and $\Phi : A \times A \to \mathbb{R}$ be defined by

$$\Phi(x, y) = \begin{cases} (x + y)^2 & \text{if } x \leq y, \\ -(y - \frac{x}{2})^2 & \text{if } y < x. \end{cases}$$

Then, Φ has property $\mathbb{P}^2$ and $\Phi \in \mathscr{A}^2_{pr.qm}$, but $\Phi \notin \mathscr{A}^2_{st.qcl}$.

Definition 1.40 Bifunction Φ is said to HAVE DUAL PROPERTY if for all $x, y \in A$, from $\Phi((1-t)x + ty, x) \leq 0$ for all $t \in (0, 1]$, it follows that $\Phi(x, y) \geq 0$.

Example 1.41 Let $A = \mathbb{R}$ and $\Phi : A \times A \to \mathbb{R}$ be defined by

$$\Phi(x, y) = \begin{cases} (y - x)(x - 1)^2 & \text{if } y > x, \\ -(x - y)^2 & \text{if } y \leq x. \end{cases}$$

Then, Φ has dual property and $\Phi \in -\mathscr{A}^1_{pr.qm}$, but Φ does not have property $\mathbb{P}^2$.

Proposition 1.42 *Let $\Phi \in \mathscr{A}^2_{st.qcl} \cap \mathscr{A}^{\geq}_A$ have property $\mathbb{P}^2$ and be upper-sign continuous in the first component. Then, Φ has dual property.*

Proof Take any $x, y \in A$. Assume that $\Phi(y_t, x) \leq 0$ for all $t \in (0, 1]$, where $y_t = x + t(y - x)$. Then, there are two cases: either $\Phi(y_t, x) < 0 \leq \Phi(y_t, y_t)$ or $\Phi(y_t, x) = 0 \leq \Phi(y_t, y_t)$. Suppose that $\Phi(y_t, y) < 0$. If $\Phi(y_t, x) < 0$, then $\Phi(y_t, y_t) < 0$ because $\Phi \in \mathscr{A}^2_{st.qcl}$, a contradiction. If $\Phi(y_t, x) = 0$, then $\Phi(y_t, y_t) < 0$ by property $\mathbb{P}^2$, another contradiction. Thus, we have $\Phi(y_t, y) \geq 0$. Finally, by the upper-sign continuity in the first component of Φ, $\Phi(x, y) \geq 0$. □

Lemma 1.43 *Assume that Φ has dual property. Then, if for some $\bar{x} \in A$, $\Phi(y, \bar{x}) \leq 0$ for all $y \in A$, then $\Phi(\bar{x}, y) \geq 0$ for all $y \in A$.*

Proof Let $y \in A$. Then, $\Phi(y_t, \bar{x}) \leq 0$ for all $t \in (0, 1]$, where $y_t = \bar{x} + t(y - \bar{x})$. By dual property of Φ, $\Phi(\bar{x}, y) \geq 0$. □

In the literature, the problem

(DEP) $\quad$ find $\bar{x} \in A$ such that $\Phi(y, \bar{x}) \leq 0$ for all $y \in A$

is called the DUAL EQUILIBRIUM PROBLEM of the problem (EP). ((EP) is also called the STAMPACCHIA EQUILIBRIUM PROBLEM, and then (DEP) is called the MINTY EQUILIBRIUM PROBLEM.) Lemma 1.43 gives a sufficient condition for the solution set of (DEP) to be contained in the solution set of (EP). If in Lemma 1.43 we add the condition "$\Phi \in \mathscr{A}_{pm}$," and then the solution sets of (DEP) and (EP) coincide.

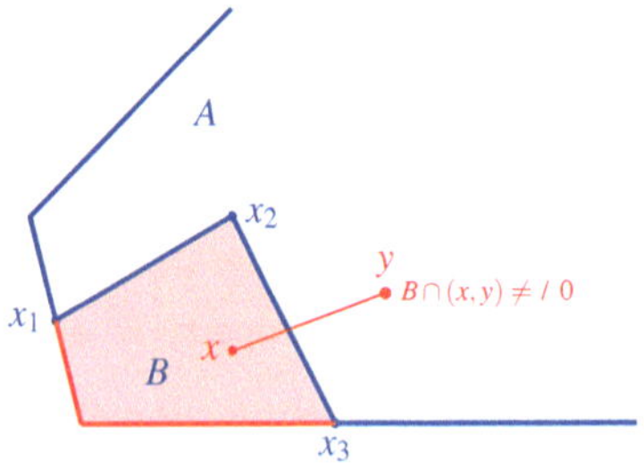

Fig. 1.11 Core of a subset $B \subset A$ relative to A

Definition 1.44 Let B be a subset of A. $\text{core}_A B$, the CORE OF B RELATIVE TO A, is defined as the set of all $x \in B$ satisfying: for each $y \in A \setminus B$, there exists $t \in (0, 1)$ with $tx + (1 - t)y \in B$. In other words,

$$x \in \text{core}_A B \quad \Leftrightarrow \quad x \in B \text{ and } B \cap (x, y) \neq \emptyset \text{ for all } y \in A \setminus B.$$

In Fig. 1.11, $\text{core}_A B$ includes points of B that do not belong to the line segments $[x_1, x_2]$ and $[x_2, x_3]$.

Lemma 1.45 *Suppose that Φ has the property $\mathbb{P}^2$ and $B \subset A$. If for some $\bar{x} \in A$, $\Phi(\bar{x}, y) \geq 0$ for all $y \in B$ and there exists $\bar{y} \in \text{core}_A B$ such that $\Phi(\bar{x}, \bar{y}) = 0$, then $\Phi(\bar{x}, y) \geq 0$ for all $y \in A$.*

Proof Take any $y \in A \setminus B$. By the definition of $\text{core}_A B$, there exists $\bar{t} \in (0, 1)$ such that $y_{\bar{t}} = \bar{t}\bar{y} + (1 - \bar{t})y \in B$, and hence, $\Phi(\bar{x}, y_{\bar{t}}) \geq 0$. Assume that $\Phi(\bar{x}, y) < 0$. Then, by the property $\mathbb{P}^2$, $\Phi(\bar{x}, t\bar{y} + (1 - t)y) < 0$ for all $t \in (0, 1)$, including $\Phi(\bar{x}, y_{\bar{t}}) < 0$ for $t = \bar{t}$, a contradiction. □

Theorem 1.46 *Assume that $\Phi \in \mathscr{A}^2_{pr.qm}$ such that Φ has property $\mathbb{P}^2$, $\text{lev}_{\geq}\Phi(\cdot, y)$ is closed for every $y \in A$, and the following coercivity condition is satisfied:*

(C_2) *There exists a nonempty compact convex subset K of A such that for every $x \in A$, there is $y \in \text{core}_A K$ satisfying $\Phi(x, y) \leq 0$.*

Then, (EP) *has solutions.*

Proof For the nonempty compact convex subset K of A in condition (C_2), applying Lemma 1.31 and the finite solvability principle for K and $\Phi|_{K \times K}$, we obtain an $\bar{x} \in K$ such that $\Phi(\bar{x}, y) \geq 0$ for all $y \in K$. By (C_2), there exists $\bar{y} \in \text{core}_A K$ such that $\Phi(\bar{x}, \bar{y}) \leq 0$, and hence, $\Phi(\bar{x}, \bar{y}) = 0$. Finally, Lemma 1.45 shows that $\bar{x}$ is a solution of (EP). □

If in the statement of Theorem 1.46, we add the condition $\Phi \in \mathscr{A}^{=}_A$, then ($C_2$) can be replaced by:

(C'_2) There exists a nonempty compact convex subset K of A such that for every $x \in K \setminus \text{core}_A K$, there is $y \in \text{core}_A K$ satisfying $\Phi(x, y) \leq 0$.

Indeed, looking at the proof of Theorem 1.46, it is sufficient to prove that there exists $\bar{y} \in \text{core}_A K$ such that $\Phi(\bar{x}, \bar{y}) \leq 0$. It is clear that such a $\bar{y}$ exists if $\bar{x} \in K \setminus \text{core}_A K$ (by (C_2')). Since $\Phi \in \mathscr{A}_A$, $\Phi(\bar{x}, \bar{x}) = 0$. Hence, if $\bar{x} \in \text{core}_A K$, then taking $\bar{y} = \bar{x}$ we have $\Phi(\bar{x}, \bar{y}) \leq 0$.

Including the additional condition "Φ has property $\mathbb{P}^2$," Theorem 1.46 still differs from Theorem 1.34, because the coercivity conditions (C_1) and (C_2) are different. For a simple example, let $A = [0, +\infty)$ and Φ be defined by $\Phi(x, y) = 0$ for all $x, y \in A$. Then, (C_1) does not hold because for any compact subset K of A, $N \in \mathscr{F}(K)$, compact convex subset L_N of A containing N, and all $x \in L_N \setminus K$, we have $\Phi(x, y) \geq 0$ for all $y \in L_N$. Condition (C_2) holds with $K = [0, a]$ and $\text{core}_A K = [0, a)$ $(a > 0)$.

Theorem 1.46 is not true if we remove the condition "Φ has property $\mathbb{P}^2$." For example, let $A = [0, +\infty)$ and Φ be defined by

$$\Phi(x, y) = \begin{cases} -1 & \text{if } x < y, \\ 0 & \text{if } x \geq y. \end{cases}$$

Then, $\Phi \in \mathscr{A}^2_{pr.qm}$ (since for all $N \in \mathscr{F}(A)$ and $x \in \text{conv}N = [\min N, \max N]$, one has $\Phi(x, \min N) \geq 0$) and (C_2) holds with $K = [0, a]$ $(a > 0)$. Moreover, $\text{lev}_{\geq}\Phi(\cdot, y) = [y, +\infty)$ is closed in A for each $y \in A$. But, (EP) in this case does not have solutions because for all $x \in A$, there exists $y > x$ satisfying $\Phi(x, y) < 0$. In this example, Φ does not have property $\mathbb{P}^2$.

Corollary 1.47 *Assume that A is a closed and convex subset of a reflexive Banach space and $\Phi \in \mathscr{A}^2_{pr.qm}$ such that Φ has property $\mathbb{P}^2$ and $\text{lev}_{\geq}\Phi(\cdot, y)$ is weakly closed for every $y \in A$. Then,* (EP) *has solutions if the following coercivity condition is satisfied:*

(C_{21}) *There exists $r > 0$ such that for all $x \in A$ with $\|x\| \leq r$, there is $y \in A$ with $\|y\| < r$ such that $\Phi(x, y) \leq 0$.*

Moreover, if we add the condition $\Phi \in \mathscr{A}^{=}_A$, then the coercivity condition (C_{21}) *can be replaced by either of the following coercivity conditions:*

(C_{22}) *For any sequence $\{x^k\}_k \subset A$ satisfying $\|x^k\| \to +\infty$, there exists $y \in A$ and k_0 such that $\Phi(x^k, y) \leq 0$ for all $k \geq k_0$.*

(C_{23}) *There exists $r > 0$ such that for all $x \in A$ with $\|x\| = r$, there is $y \in A$ with $\|y\| < r$ such that $\Phi(x, y) \leq 0$.*

Proof For the case of (C_{21}), let $K = \{x \in A \mid \|x\| \leq r\}$. Then, K is weakly compact and convex, and $K^0 = \{x \in A \mid \|x\| < r\} \subset \text{core}_A K$. Then, the coercivity condition (C_{21}) implies the coercivity condition (C_2) with the weak topology. Thus, Theorem 1.46 yields that (EP) has solutions.

In the case of (C_{22}), for each k, let $A^k = \{x \in A \mid \|x\| \leq k\}$. Because A^k is nonempty when k is sufficiently large, we suppose without loss of generality that A^k is nonempty for all k. The sets A^k, $k \in \mathbb{N}$, are weakly compact and convex. Therefore, it is not hard to check that A^k and $\Phi|_{A^k \times A^k}$ satisfy the conditions of

Theorem 1.46 with the weak topology. Thus, for each $k \in \mathbb{N}$, there exists $\bar{x}^k$ such that $\Phi(\bar{x}^k, y) \geq 0$ for all $y \in A^k$. If there exists $\hat{k}$ such that $\| \bar{x}^{\hat{k}} \| < \hat{k}$, i.e., $\bar{x}^{\hat{k}} \in A_0^{\hat{k}} = \{x \in A \mid \|x\| < \hat{k}\} \subset \mathrm{core}_A A^{\hat{k}}$, then taking $\bar{y} = \bar{x}^{\hat{k}} \in A_0^{\hat{k}}$ we have $\Phi(\bar{x}^{\hat{k}}, \bar{y}) = \Phi(\bar{x}^{\hat{k}}, \bar{x}^{\hat{k}}) = 0$. If $\|\bar{x}^k\| = k$ for all $k \in \mathbb{N}$, then $\|\bar{x}^k\| \to +\infty$. Then, by ($C_{22}$), there exist $\bar{y} \in A$ and k_0 such that $\Phi(\bar{x}^k, \bar{y}) \leq 0$ for all $k \geq k_0$. Choosing $\hat{k} > k_0$ such that $\|\bar{y}\| < \hat{k}$, we have $\bar{y} \in A_0^{\hat{k}}$ and $\Phi(\bar{x}^{\hat{k}}, \bar{y}) \leq 0$ (hence $\Phi(\bar{x}^{\hat{k}}, \bar{y}) = 0$). Thus, there always exist $\hat{k}$ and $\bar{y}$ such that

$$\bar{y} \in A_0^{\hat{k}} \subset \mathrm{core}_A A^{\hat{k}}, \quad \Phi(\bar{x}^{\hat{k}}, \bar{y}) = 0, \quad \text{and} \quad \Phi(\bar{x}^{\hat{k}}, y) \geq 0 \ \text{ for all } y \in A^{\hat{k}}.$$

This and Lemma 1.45 show that $\bar{x}_{\hat{k}}$ is a solution of (EP).

For the case of (C_{23}), this condition is a particular case of the condition (C_2') with $K = \{x \in A \mid \|x\| \leq r\}$ considered in the weak topology. □

Corollary 1.48 *Assume that A is a closed convex subset of a reflexive Banach space, $\Phi \in \mathscr{A}_A^{=} \cap \big([(-\mathscr{A}_{pm}) \cap (-\mathscr{A}_{qc}^1)] \cup [(-\mathscr{A}_{qm}) \cap (-\mathscr{A}_{s.qcl}^1)]\big)$, Φ has property $\mathbb{P}^2$, $\mathrm{lev}_{\geq}\Phi(\cdot, y)$ is closed for every $y \in A$, and at least one of the coercivity conditions (C_{21}), (C_{22}), and (C_{23}) is satisfied. Then, (EP) has solutions.*

Proof Because $\Phi \in \mathscr{A}_A^{=} \cap \big([(-\mathscr{A}_{pm}) \cap (-\mathscr{A}_{qc}^1)] \cup [(-\mathscr{A}_{qm}) \cap (-\mathscr{A}_{s.qcl}^1)]\big) \subset \mathscr{A}_{pr.qm}^2$ and $\mathrm{lev}_{\geq}\Phi(\cdot, y)$ is convex when $\Phi \in (-\mathscr{A}_{qc}^1) \cup (-\mathscr{A}_{s.qcl}^1)$, Corollary 1.48 is a special case of Corollary 1.47. □

Theorem 1.49 *Assume that $\Phi \in -\mathscr{A}_{pr.qm}^1$, Φ has dual property, $\mathrm{lev}_{\leq}\Phi(x, \cdot)$ is closed for every $x \in A$, and the following coercivity condition is satisfied.*

(C_3) *There exists a nonempty compact subset K of X such that, for each finite subset N of A, there exists a compact convex subset L_N of A containing N such that for each $x \in L_N \backslash K$, $\Phi(y, x) > 0$ for some $y \in L_N$.*

Then, (EP) has solutions.

Proof Let $\Phi' : A \times A \to \mathbb{R}$ be defined by $\Phi'(x, y) = -\Phi(y, x)$ for $x, y \in A$. Since $\Phi \in -\mathscr{A}_{pr.qm}^1$ and $\mathrm{lev}_{\leq}\Phi(x, \cdot)$ is closed for every $x \in A$, we deduce that $\Phi' \in \mathscr{A}_{pr.qm}^2$ and $\mathrm{lev}_{\geq}\Phi'(\cdot, y)$ is closed for every $y \in A$. The condition (C_3) implies that there exists a nonempty compact subset K of X such that, for each finite subset N of A, there exists a compact convex subset L_N of A containing N such that for every $x \in L_N \backslash K$, $\Phi'(x, y) < 0$ for some $y \in L_N$. This is just condition (C_1) in Theorem 1.34 for A and Φ'. Hence, we have an $\bar{x} \in A$ such that $\Phi'(\bar{x}, y) \geq 0$ for all $y \in A$, or equivalently, $\Phi(y, \bar{x}) \leq 0$ for all $y \in A$. Because Φ has dual property, Lemma 1.43 shows that $\bar{x}$ is a solution of (EP). □

Theorem 1.50 *Assume that $\Phi \in (-\mathscr{A}_{pr.qm}^1) \cap \mathscr{A}_{st.qcl}^2 \cap \mathscr{A}_A^{=}$, Φ has the property $\mathbb{P}^2$ and is upper-sign continuous in the first argument, $\mathrm{lev}_{\leq}\Phi(x, \cdot)$ is closed for every $x \in A$, and either of the following coercivity conditions is satisfied.*

(C3) *There exists a nonempty compact subset K of X such that, for each finite subset N of A, there exists a compact convex subset L_N of A containing N such that for each $x \in L_N \backslash K$, $\Phi(y, x) > 0$ for some $y \in L_N$.*

(C4) *There exists a nonempty compact convex subset K of A such that for every $x \in K \backslash \mathrm{core}_A K$, there is $y \in \mathrm{core}_A K$ satisfying $\Phi(x, y) \leq 0$.*

Then, (EP) has solutions.

Proof Theorem 1.50 in the case of (C3) is a direct consequence of Theorem 1.49 and Proposition 1.42.

In the case of (C4), it is not hard to see that $\Phi|_{K\times K} \in (-\mathscr{K}^1_{pr.qm}) \cap \mathscr{K}^{=}_K$ ($\mathscr{K}^1_{pr.qm}$ ($\mathscr{K}^{=}_K$, resp.) is defined similarly as $\mathscr{A}^1_{pr.qm}$ ($\mathscr{A}^{=}_A$, resp.)), $\Phi|_{K\times K}$ has property $\mathbb{P}$ and is upper-sign continuous in the first component, and $\mathrm{lev}_{\leq}\Phi|_{K\times K}(x, \cdot)$ is closed in K for every $x \in K$. Applying the case of (C3) (which is obviously satisfied because K is compact), we have $\bar{x} \in K$ such that $\Phi(\bar{x}, y) \geq 0$ for all $y \in K$. If $\bar{x} \in K \backslash \mathrm{core}_A K$, by (C4), an $\bar{y} \in \mathrm{core}_A K$ exists such that $\Phi(\bar{x}, \bar{y}) \leq 0$, and hence, $\Phi(\bar{x}, \bar{y}) = 0$. If $\bar{x} \in \mathrm{core}_A K$, then taking $\bar{y} = \bar{x}$ we have $\Phi(\bar{x}, \bar{y}) = 0$. Thus, there always exists $\bar{y} \in \mathrm{core}_A K$ such that $\Phi(\bar{x}, \bar{y}) = 0$. Using Lemma 1.45, we conclude that $\bar{x}$ is a solution of (EP). □

It is not hard to check that each of the following conditions is sufficient for condition (C3) to be fulfilled.

(C31) There exist a nonempty compact subset K of X and an $\bar{y} \in K \cap A$ such that $\Phi(\bar{y}, x) > 0$ for every $x \in A \backslash K$.

(C32) There exist a nonempty compact subset K of X and a finite subset $\bar{N}$ of A such that for each $x \in A \backslash K$, $\Phi(y, x) > 0$ for some $y \in \bar{N}$.

(C33) There exists a nonempty compact convex subset K of A such that for each $x \in A \backslash K$, $\Phi(y, x) > 0$ for some $y \in K$.

Corollary 1.51 *Assume that A is a closed and convex subset of a reflexive Banach space, $\Phi \in (-\mathscr{A}^1_{pr.qm}) \cap \mathscr{A}^2_{st.qcl} \cap \mathscr{A}^{=}_A$, Φ has property $\mathbb{P}^2$ and is upper-sign continuous in the first component, $\mathrm{lev}_{\leq}\Phi(x, \cdot)$ is weakly closed for every $x \in A$, and one of the following coercivity conditions is satisfied.*

(C34) *There exists $r > 0$ such that for all $x \in A$ with $\|x\| > r$, there is $y \in A$ with $\|y\| \leq r$ such that $\Phi(y, x) > 0$.*

(C35) *For some $\hat{y} \in A$, $\Phi(\hat{y}, x) \to +\infty$ as $\|x - \hat{y}\| \to +\infty$ for all $x \in A$.*

(C36) *For any sequence $\{x^k\}_k \subset A$ satisfying $\|x^k\| \to +\infty$, there exists $y \in A$ and k_0 such that $\Phi(x^k, y) \leq 0$ for all $k \geq k_0$.*

Then, (EP) *has solutions.*

Proof In the case of (C34), the coercivity condition (C34) is just (C33) considered in the weak topology with $K = \{x \in A \mid \|x\| \leq r\}$.

The case of (C35) is a particular case of (C34) with $r = R + \|\hat{y}\|$, where $R > 0$ is big enough such that $\|x - \hat{y}\| > R$ implies $\Phi(\hat{y}, x) > 0$.

The proof of the case of (C_{36}) is similar to that of the case of (C_{22}) in Corollary 1.47, using Theorem 1.50 instead of Theorem 1.46. □

Corollary 1.52 *Assume that X is a closed convex subset of a reflexive Banach space, $\Phi \in (-\mathscr{A}^1_{pr.qm}) \cap \left(\mathscr{A}^2_{eqc} \cup \mathscr{A}^2_{s.qcl}\right) \cap \mathscr{A}^{=}_{A}$, Φ is upper-sign continuous in the first component, $\mathrm{lev}_{\leq}\Phi(x,\cdot)$ is closed for every $x \in A$, and at least one of the coercivity conditions* (C_{34}), (C_{35}), *and* (C_{36}) *is satisfied. Then,* (EP) *has solutions.*

Proof We observe that for every $x \in A$, $\mathrm{lev}_{\leq}\Phi(x,\cdot)$ is weakly closed because $\Phi \in \mathscr{A}^2_{eqc} \cup \mathscr{A}^2_{s.qcl}$ (hence $\mathrm{lev}_{\leq}\Phi(x,\cdot)$ is convex) and $\mathrm{lev}_{\leq}\Phi(x,\cdot)$ is closed. Moreover, when $\Phi \in \mathscr{A}^2_{eqc} \cup \mathscr{A}^2_{s.qcl}$, Φ has the property $\mathbb{P}^2$ (see Example 1.39). Hence, Corollary 1.52 is a special case of Corollary 1.51. □

Some Applications

As examples of applications, we deduce some results in minimax theory and Nash equilibrium problems.

(a) **Minimax and Saddle Points**. When considering minimax and saddle point problems for certain cases, we can use results in this section for (EP) with Φ defined by (1.4) or (1.5). Here is an example.

Example 1.53 Let $\phi : \mathbb{R}^m \times \mathbb{R}^n \to \mathbb{R}$ be defined for all $(x, y) \in \mathbb{R}^m \times \mathbb{R}^n$ by $\phi(x, y) = \langle p, x\rangle + \frac{1}{2}\langle x, Px\rangle + \langle q, y\rangle - \frac{1}{2}\langle y, Qy\rangle - \langle y, Sx\rangle$, where $p \in \mathbb{R}^m$, $q \in \mathbb{R}^n$, $P \in \mathbb{R}^{m\times m}$, $Q \in \mathbb{R}^{n\times n}$, $S \in \mathbb{R}^{n\times m}$, P and Q are symmetric positive definite matrices. We consider the problem of finding a saddle point of ϕ.

For $A = \mathbb{R}^m \times \mathbb{R}^n$ and Φ defined by (1.4), by setting

$$a = (-p, q),\quad M = \begin{pmatrix} P & 0 \\ 0 & Q \end{pmatrix},\quad N = \begin{pmatrix} 0 & S^t \\ -S & 0 \end{pmatrix},$$

we calculate: for $u = (x, y)$, $v = (x', y') \in A$, $\Phi(u, v) = \phi(x', y) - \phi(x, y') = \langle a, u - v\rangle - \frac{1}{2}\langle u, Mu\rangle + \frac{1}{2}\langle v, Mv\rangle + \langle u, Nv\rangle$. It is easy to check that $\Phi \in (-\mathscr{A}^1_{eqc}) \cap \mathscr{A}^2_{eqc} \cap \mathscr{A}^{=}_{A} \subset (-\mathscr{A}^1_{pr.qm}) \cap \left(\mathscr{A}^2_{eqc} \cup \mathscr{A}^2_{s.qcl}\right) \cap \mathscr{A}^{=}_{A}$ and Φ is continuous in each component (hence is upper-sign continuous in the first component and $\mathrm{lev}_{\leq}\Phi(u,\cdot)$ is closed for each $u \in A$). Obviously matrix M is symmetric and positive definite. Hence, for $\hat{v} = 0$, $0 \neq u_0 \in A$, and $t \in \mathbb{R}$, one has $\Phi(0, tu_0) = -t\langle a, u_0\rangle + \frac{1}{2}t^2\langle u_0, Mu_0\rangle \to +\infty$ as $t \to \pm\infty$. This implies that $\Phi(0, u)\rangle \to +\infty$ as $\|u\| \to +\infty$. Thus, Corollary 1.52 with the coercivity condition (C_{35}) shows that an $\bar{u} = (\bar{x}, \bar{y})$ exists such that $\Phi(\bar{u}, v) \geq 0$ for all $v \in A$, i.e., ϕ has a saddle point.

The following corollaries are typical consequences of Theorem 1.34 on solvability of (EP), which give conditions on ϕ such that the minimax equality or saddle points of ϕ.

Corollary 1.54 *Let C, D be compact convex subsets of topological vector spaces, and $\phi : C \times D \to \mathbb{R}$. Assume that for all $\lambda > \phi_*$ sufficiently close to ϕ_* such that for each $(x, y) \in C \times D$, $\mathrm{lev}_{<\lambda}\phi(\cdot, y)$ and $\mathrm{lev}_{>\lambda}\phi(x,\cdot)$ are convex, and $\mathrm{lev}_{\leq\lambda}\phi(\cdot, y)$ and $\mathrm{lev}_{\geq\lambda}\phi(x,\cdot)$ are closed in C and D. Then, the minimax equality (1.1) holds for ϕ.*

Proof By assumptions, for $A = C \times D$, and the bifunction Φ_λ defined by (1.5):

$$\Phi_\lambda((x, y), (x', y')) = \begin{cases} 0 \text{ if } (x, y) \in A \setminus (\text{lev}_{>\lambda}\phi(\cdot, y') \times \text{lev}_{<\lambda}\phi(x', \cdot)), \\ -1 \text{ otherwise} \end{cases}$$

for all $(x, y), (x', y') \in A$, we have the following observations.

- For all $(x, y) \in A$, $\text{lev}_{<}\Phi_\lambda((x, y), \cdot) = \text{lev}_{<\lambda}\phi(\cdot, y) \times \text{lev}_{>\lambda}\phi(x, \cdot)$ is convex, i.e., Φ_λ is strongly quasiconvex-like in the second component. Moreover, $(x, y) \notin \text{lev}_{>\lambda}\phi(\cdot, y) \times \text{lev}_{<\lambda}\phi(x, \cdot)$, i.e., $\Phi_\lambda((x, y), (x, y)) = 0$. Thus, by Proposition 1.29 (a), Φ_λ is properly quasimonotone in the second argument.
- For all $(x', y') \in A$,

$$\text{lev}_{>\lambda}\phi(\cdot, y') \times \text{lev}_{<\lambda}\phi(x', \cdot) = (C \setminus \text{lev}_{\leq\lambda}\phi(\cdot, y')) \times (D \setminus \text{lev}_{\geq\lambda}\phi(x', \cdot))$$

  is open in A. Hence, $\text{lev}_{\geq}\Phi_\lambda(\cdot, (x', y')) = A \setminus \text{lev}_{>\lambda}\phi(\cdot, y') \times \text{lev}_{<\lambda}\phi(x', \cdot)$ is closed in A.

Thus, applying Theorem 1.34 we see that the problem (EP) with $\Phi \equiv \Phi_\lambda$ has solutions. Hence, by Proposition 1.3, the minimax equality (1.1) holds for ϕ. □

Corollary 1.55 (Sion's minimax theorem) *Let C, D be compact convex subsets of two topological vector spaces, and let $\phi : C \times D \to \mathbb{R}$ be a bifunction that is upper semicontinuous and quasiconcave in the second variable and lower semicontinuous and quasiconvex in the first variable. Then, ϕ has a saddle point, and* $\min_{x \in C} \max_{y \in D} \phi(x, y) = \max_{y \in D} \min_{x \in C} \phi(x, y)$.

Proof Under assumptions of Corollary 1.55, all the assumptions of Corollary 1.54 are satisfied. Hence, the minimax equality (1.1) holds for ϕ. Under the topological assumptions of Corollary 1.55, $\phi(x, \cdot)$ has a maximum for each $x \in A$ and $\phi(\cdot, y)$ has a minimum for each $x \in A$. Moreover, the function $x \mapsto \max_{y \in D} \phi(x, y)$ ($y \mapsto \min_{x \in C} \phi(x, y)$, resp.) is lower semicontinuous (upper semicontinuous, resp.) on the compact set C (D, resp.), and hence, it has a minimum $\bar{x}$ (a maximum $\bar{y}$, resp.). By Proposition 1.1, ϕ has a saddle point and $\min_{x \in C} \max_{y \in D} \phi(x, y) = \max_{y \in D} \min_{x \in C} \phi(x, y)$. □

Corollary 1.56 *Let C, D be compact convex subsets of topological vector spaces, and let $\phi : C \times D \to \mathbb{R}$ be a bifunction such that $\{(x', y') \in C \times D \mid \phi(x', y) - \phi(x, y') < 0\}$ is convex for each $(x, y) \in C \times D$ and $\{(x, y) \in C \times D \mid \phi(x', y) - \phi(x, y') \geq 0\}$ is closed for each $(x', y') \in C \times D$. Then, ϕ has a saddle point.*

Proof Corollary 1.56 is a consequence of Theorem 1.34 and Proposition 1.2 for Φ defined by (1.4): $\Phi((x, y), (x', y')) = \phi(x', y) - \phi(x, y')$ for all $(x, y), (x', y') \in A =: C \times D$. Here we observe that $\Phi((x, y), (x, y)) = 0$ and $\Phi((x, y), \cdot)$ is strongly quasiconvex-like, and hence, by Proposition 1.29 (a), Φ is properly quasimonotone in the second argument. □

If $\phi(x', \cdot)$ is upper semicontinuous and $\phi(\cdot, y')$ is lower semicontinuous for each $(x', y') \in C \times D$, then the function $(x, y) \to \phi(x', y) - \phi(x, y')$ is upper semicontinuous and hence the set $\{(x, y) \in C \times D \mid \phi(x', y) - \phi(x, y') \geq 0\}$ is closed. Moreover, if $\phi(x, \cdot)$ is concave for each $x \in C$ and $\phi(\cdot, y)$ is convex for each $y \in D$, then $\{(x', y') \in C \times D \mid \phi(x', y) - \phi(x, y') < 0\}$ is convex for each $(x, y) \in C \times D$ (see Exercise 1.2). Hence, both Corollaries 1.55 and 1.56 include the following corollary as a particular case.

Corollary 1.57 (von Neumann's minimax theorem) *Let C, D be compact convex subsets of topological vector spaces, and let $\phi : C \times D \to \mathbb{R}$ be a bifunction that is upper semicontinuous and concave in the second variable and lower semicontinuous and convex in the first variable. Then, ϕ has a saddle point, and* $\min_{x \in C} \max_{y \in D} \phi(x, y) = \max_{y \in D} \min_{x \in C} \phi(x, y)$.

(b) **Nash Equilibria**.

Example 1.58 Let $A = \mathbb{R}^m_+$, and for each $i \in I := \{1, 2, \ldots, m\}$, let $c_i : A \to \mathbb{R}$ be defined by $c_i(x) = x_i p(\theta_x) - h_i(x_i)$, where $x = (x_1, \ldots, x_m) \in A$ and $\theta_x = \sum_{i=1}^m x_i$, $p, h_i : \mathbb{R}_+ \to \mathbb{R}$. We consider the problem of finding a point $\bar{x} = (\bar{x}_1, \ldots, \bar{x}_m) \in A$ such that

$$c_i(\bar{x}) = \max_{y_i \in \mathbb{R}_+} c_i(\bar{x}_1, \ldots, \bar{x}_{i-1}, y_i, \bar{x}_{i+1}, \ldots, \bar{x}_m) \quad \text{for all } i = 1, 2, \ldots, m.$$

This problem is an imperfectly competitive equilibrium model in a market. A homogeneous product is supplied by m firms and $p = p(\theta)$ is the price at which consumers will purchase a quantity θ and is called the INVERSE DEMAND FUNCTION. The value $h_i(x_i)$ is the ith firm's total cost of supplying x_i units of the product and $\theta_x = \sum_{i=1}^m x_i$ is the total supply in the market where each ith firm supplies x_i units of the product. The ith firm's profit is $c_i(x)$ and depends on the outputs of all the firms, and the ith firm seeks to maximize the value $c_i(x)$ by choosing the corresponding production level. The point $\bar{x}$ is called a NASH EQUILIBRIUM SOLUTION for the OLIGOPOLISTIC MARKET.

We suppose that h_i is nondecreasing, lower semicontinuous, and pseudoconvex, p is nonincreasing and twice continuously differentiable such that $\mu(\theta) = -\theta p(\theta)$ is convex on $\mathbb{R}_+$, and there exists $r_0 > 1$ such that for each $x = (x_1, \ldots, x_m) \in A$, there is $i \in \{1, 2, \ldots, m\}$ satisfying $p(\theta_x) \leq \frac{1}{r_0} p(\theta_x - \frac{1}{r_0}(r_0 - 1)x_i)$. (This means that if the ith firm's quantity of the product increases [decreases] r_0 times, then the price decreases [increases] more than r_0 times.)

Let the Nikaido-Isoda bifunction $N : A \times A \to \overline{\mathbb{R}}$ be defined for $x, y \in A$ by

$$N(x, y) = \sum_{i=1}^m [c_i(x) - \varphi_i(x_1, \ldots, x_{i-1}, y_i, x_{i+1}, \ldots, x_m)]$$

$$= \theta_x p(\theta_x) - \sum_{i=1}^m h_i(x_i) - \sum_{i=1}^m y_i p(\theta_x - x_i + y_i) + \sum_{i=1}^m h_i(y_i).$$

When p is nonincreasing and twice continuously differentiable and μ is convex, the function $t \mapsto -tp(t+\beta)$, $\beta \geq 0$, is convex. This and the pseudoconvexity of h_i imply that $N \in \mathscr{A}^2_{pc}$. Obviously, $N(x,x) = 0$ and N is continuous in the first component. Now we check the condition (C_{23}). Choose $r = r_0$. Then, for any $x = (x_1, \ldots, x_m) \in A$ with $\|x\| = r$, by assumption, there exists i such that $p(\theta_x) \leq \frac{1}{r} p(\theta_x - \frac{r-1}{r} x_i)$. Taking $y = (x_1, \ldots x_{i-1}, \frac{x_i}{r}, x_{i+1}, \ldots, x_m)$, we have $\|y\| < r$ and $N(x,y) = x_i p(\theta_x) - \frac{x_i}{r} p(\theta_x - \frac{r-1}{r} x_i) + h_i(\frac{x_i}{2}) - h_i(x_i) \leq 0$. Thus, by Corollary 1.47, there exists $\bar{x}$ such that $N(\bar{x}, y) \geq 0$ for all $y \in A$, i.e., a solution of the Nash equilibrium problem exists (by Proposition 1.5).

Corollary 1.59 *For the m-player noncooperative game $\mathscr{G}$, assume that for each $i \in I$, A_i is a compact convex subset of a topological vector space, c_i is continuous, and $c_i(x_{-i}, \cdot)$ is concave for each $x \in A$. Then, $\mathscr{G}$ has Nash equilibrium.*

Proof Under the assumptions, the Nikaido-Isoda bifunction, $N(x,y) = \sum_{i=1}^{n} [c_i(x) - c_i(x_{-i}, y_i))]$, satisfies the assumptions of Theorem 1.34 (see Exercise 1.3), and therefore, the conclusion of Corollary 1.59 follows from Theorem 1.34 and Proposition 1.5. □

Example 1.60 Consider the game $\mathscr{G} = (A_i, c_i)_{i \in I}$ with $A_i = [a_i, b_i] \subset \mathbb{R}$ and $c_i(x) = -x_i + \sum_{j \neq i} x_j$ for $x = (x_i)_{i \in I} \in A$. For $i \in I$, obviously c_i is continuous. For $i \in I$ and $x \in A$, $y_i \mapsto c_i(x_{-i}, y_i) = -y_i + \sum_{j \neq i} x_j$ is concave. Thus, by Corollary 1.59, $\mathscr{G}$ has Nash equilibrium.

1.3 Equilibria over Nonconvex Regions

When the feasible region has no linear structure, bifunctions which have certain convexity properties are of no use. Therefore, in this section we use some other classes of bifunctions to replace those considered in Sect. 1.2.

1.3.1 Existence of Equilibria for Bifunctions Satisfying the Triangle Inequality

In this subsection, we study the existence of solutions to (EP) with a bifunction Φ satisfying the so-called triangle inequality.

Definition 1.61 Let A be a nonempty set and $\Phi : A \times A \to \mathbb{R}$ be a bifunction. We say that Φ satisfies the TRIANGLE INEQUALITY if

$$\Phi(x,z) \leq \Phi(x,y) + \Phi(y,z) \text{ for all } x, y, z \in A. \tag{1.16}$$

We denote by $\mathscr{A}_\Delta$ the class of the bifunctions defined on $A \times A$ which satisfy (1.16).

For any nonempty set A, the class $\mathscr{A}_\Delta$ is nonempty because $\Phi = c \in \mathscr{A}_\Delta$ for any $c \geq 0$, where $\Phi = c$ means that $\Phi(x, y) = c$ for all $x, y \in A$. Furthermore, the class $\mathscr{A}_\Delta$ is wide. Below are some subclasses of $\mathscr{A}_\Delta$:

- $\mathscr{A}_{1\Delta} := \big\{\Phi(x, y) = \psi(y) - \psi(x) \mid \psi : A \to \mathbb{R}\big\}$.
- $\mathscr{A}_{2\Delta} := \big\{\Phi(x, y) = \psi(x) + \xi(y) \mid \psi, \xi : A \to \mathbb{R}$ with $\psi(x) + \xi(x) \geq 0\ \forall x \in A\big\}$.
- $\mathscr{A}_{3\Delta} := \big\{\Phi(x, y) = \psi(x, y) + r \mid \psi : A \times A \to [m, M] \subset \mathbb{R},\ r \geq M - 2m\big\}$.

Let $\mathbb{C}(A)$ be the set of all bifunctions defined on $A \times A$. For $\mathscr{A} \subset \mathbb{C}(A)$, $\mathscr{B} \subset \mathbb{C}(A)$, and $\lambda \in \mathbb{R}$, we define

$$\lambda\mathscr{A} := \{\lambda\Phi \mid \Phi \in \mathscr{A}\} \text{ and } \mathscr{A} + \mathscr{B} := \{\Phi + \Phi' \mid \Phi \in \mathscr{A}, \Phi' \in \mathscr{B}\}.$$

Then, $\mathbb{C}(A)$ is a linear space. The following proposition states some properties of the class $\mathscr{A}_\Delta$.

Proposition 1.62 *For any nonempty set A, the following statements are true:*

(a) $\mathscr{A}_\Delta + \mathscr{A}_\Delta \subset \mathscr{A}_\Delta$ *and* $\lambda\mathscr{A}_\Delta \subset \mathscr{A}_\Delta$ *for all* $\lambda \geq 0$, *i.e.,* $\mathscr{A}_\Delta$ *is a cone in the space* $\mathbb{C}(A)$.
(b) *For any* $\Phi \in \mathscr{A}_\Delta$, $\Phi(x, x) \geq 0$ *and* $\Phi(x, y) + \Phi(y, x) \geq 0$ *for all* $x, y \in A$. *This means that* $\mathscr{A}_\Delta = -\mathscr{A}_m \cap \mathscr{A}_A^{\geq}$.
(c) *If* $\Phi \in \mathscr{A}_\Delta$ *and* $\emptyset \neq B \subset A$, *then* $\Phi|_{B\times B} \in \mathscr{B}_\Delta$.
(d) *For any* $\Phi \in \mathscr{A}_\Delta$, *we have*

$$\bigcup_{x\in A} \arg\min_A \Phi(x, \cdot) \subset \mathrm{Sol(EP)} \ \text{ and } \ \bigcup_{y\in A} \arg\min_A -\Phi(\cdot, y) \subset \mathrm{Sol(EP)}.$$

If, in addition, $\Phi \in \mathscr{A}_A^{=}$, *then the first inclusion becomes equality.*
(e) *Assume that A is a subset of a topological space and* $\Phi \in \mathscr{A}_\Delta \cap \mathscr{A}_A^{=}$. *Then, if* Φ *is upper semicontinuous in the first component, then* Φ *is lower semicontinuous in the second component.*

Proof (a) and (c) are obvious.

(b) Let $\Phi \in \mathscr{A}_\Delta$. The triangle inequality (1.16) implies that $\Phi(x, y) \leq \Phi(x, x)+ \Phi(x, y)$ for all $x, y \in A$. Hence, $\Phi(x, x) \geq 0$. This and (1.16) imply that $0 \leq \Phi(x, x) \leq \Phi(x, y) + \Phi(y, x)$.

(d) Let $x \in A$ and $\bar{x} \in \arg\min_A \Phi(x, \cdot)$. Then, for all $y \in A$, $\Phi(x, y) \leq \Phi(x, \bar{x}) + \Phi(\bar{x}, y)$, or equivalently $\Phi(\bar{x}, y) \geq \Phi(x, y) - \Phi(x, \bar{x})$. Since $\Phi(x, y) - \Phi(x, \bar{x}) \geq 0$, it follows that $\Phi(\bar{x}, y) \geq 0$. Thus, $\bar{x} \in \mathrm{Sol(EP)}$. Note that if $\Phi \in \mathscr{A}_A^{=}$, then $x \in \arg\min\limits_A \Phi(x, \cdot)$, and the reverse of the first inclusion follows. The second inclusion is proved in the same way as follows. Let $x \in A$ and $\bar{x} \in \arg\min_A(-\Phi(\cdot, x))$. Then, for all $y \in A$, $\Phi(\bar{x}, y) \geq -\Phi(y, x) - (-\Phi(\bar{x}, x))$. As $-\Phi(y, x) - (-\Phi(\bar{x}, x)) \geq 0$, $\Phi(\bar{x}, y) \geq 0$.

(e) For any $\epsilon > 0$, by the upper semicontinuity of $\Phi(\cdot, y)$, there exists a neighborhood U of y such that $\Phi(z, y) \leq \Phi(y, y) + \epsilon = \epsilon$ for all $z \in U \cap A$.

This and the triangle inequality give that $\Phi(x, y) - \epsilon \leq \Phi(x, y) - \Phi(z, y) \leq \Phi(x, z) + \Phi(z, y) - \Phi(z, y) = \Phi(x, z)$ for all $z \in U \cap A$. □

Theorem 1.63 (Weierstrass' theorem for (EP)) *Assume that A is a compact subset of a topological space and $\Phi \in \mathscr{A}_\Delta$ such that for some $x \in A$, either $\Phi(x, \cdot)$ is lower semicontinuous or $\Phi(\cdot, x)$ is upper semicontinuous. Then,* (EP) *has solutions.*

Proof Suppose that for some $x \in A$, $\Phi(x, \cdot)$ is lower semicontinuous. Set $\beta_y = \Phi(x, y)$ for each $y \in A$. Then, $\{\text{lev}_{\leq \beta_y} \Phi(x, \cdot) \mid y \in A\}$ is a family of nonempty closed subsets and has the finite intersection property. By the compactness of A, $\bigcap_{y \in A} \text{lev}_{\leq \beta_y} \Phi(x, \cdot) \neq \emptyset$, i.e., there exists $\bar{x} \in A$ such that $\Phi(x, \bar{x}) \leq \beta_y = \Phi(x, y)$ for all $y \in A$. Because $\Phi \in \mathscr{A}_\Delta$, $\Phi(\bar{x}, y) \geq \Phi(x, y) - \Phi(x, \bar{x}) \geq 0$ for all $y \in A$.

If for some $x \in A$, $\Phi(\cdot, x)$ is upper semicontinuous, then $\widehat{\Phi}(x, \cdot) := -\Phi(\cdot, x)$ is lower semicontinuous. Similarly as above, we find an $\bar{x} \in A$ such that $\widehat{\Phi}(x, \bar{x}) \leq \widehat{\Phi}(x, y)$, or equivalently, $\Phi(\bar{x}, x) \geq \Phi(y, x)$ for all $y \in A$. When $\Phi \in \mathscr{A}_\Delta$, $\Phi(\bar{x}, y) \geq \Phi(\bar{x}, x) - \Phi(y, x) \geq 0$ for all $y \in A$. □

If we take $\Phi(x, y) = \varphi(y) - \varphi(x)$ in Theorem 1.63, then we obtain the following classical result.

Corollary 1.64 (Weierstrass' minimization theorem) *If A is a compact subset of a topological space and a function $\varphi : A \to \mathbb{R}$ is lower semicontinuous, then φ has a minimum point, i.e.,* $\arg\min_A \varphi \neq \emptyset$.

Conversely, Weierstrass' minimization theorem and Proposition 1.62(d) imply Theorem 1.63. This theorem is quite general because it holds under purely topological assumptions, without additional structures. In order to obtain existence results in a noncompact setting, we need coercivity conditions.

Theorem 1.65 *Assume that A is a closed subset of a topological space X, $\Phi \in \mathscr{A}_\Delta$ is lower semicontinuous in the second component, and the following coercivity condition is satisfied:*

(C_Φ) *There exists a nonempty compact subset K of X such that, for each finite subset N of A, there exists a compact subset L_N of A containing N such that for every $x \in L_N \backslash K$, $\Phi(x, y) < 0$ for some $y \in L_N$.*

Then, (EP) *has solutions.*

Proof For K in the condition (C_Φ), $A \cap K$ is compact in K. We will show that (EP) is finitely solvable on $A \cap K$. Then, (EP) has solutions by the finite solvability principle. Suppose to the contrary that there exists a finite subset N of A such that

$$\text{for each } x \in A \cap K, \text{ there is some } y \in N \text{ satisfying } \Phi(x, y) < 0. \tag{1.17}$$

For this N, by condition (C_Φ), there exists a compact subset L_N of A containing N such that

$$\text{for each } x \in L_N \setminus K, \text{ there is some } y \in L_N \text{ satisfying } \Phi(x, y) < 0. \tag{1.18}$$

Moreover, using Theorem 1.63 for L_N and $\Phi|_{L_N \times L_N}$ we have

$$\text{there exists } \bar{x} \in L_N \text{ such that } \Phi(\bar{x}, y) \geq 0 \text{ for all } y \in L_N. \tag{1.19}$$

If $L_N \cap K = \emptyset$, then (1.18) and (1.19) give a contradiction. If $L_N \cap K \neq \emptyset$, then (1.18) and (1.19) imply that $\bar{x} \in L_N \cap K \subset A \cap K$. By (1.17), there exists $y \in N \subset L_N$ such that $\Phi(\bar{x}, y) < 0$, which contradicts (1.19). The proof is complete. □

To obtain other coercivity conditions, we need specific structures of the feasible region. One class of spaces which helps us to obtain a coercivity condition is as follows. Let $\mathscr{M}$ be the class of the metric spaces which have the following property:

(P) Each $(X, d) \in \mathscr{M}$ admits a topology τ_X (possibly different from the original topology induced by d) such that the closed balls are τ_X-compact and the distance d is τ_X-lower semicontinuous in the first component.

For each space $(X, d) \in \mathscr{M}$, we write (X, d, τ_X) instead of (X, d). The class $\mathscr{M}$ is wide enough to contain important spaces such as finite-dimensional spaces and infinite-dimensional normed spaces. Note that for each $(X, d, \tau_X) \in \mathscr{M}$, the symmetry of the distance function d implies that d is τ_X-lower semicontinuous in the second component.

Theorem 1.66 *Assume that A is a τ_X-closed subset of a metric space $(X, d, \tau_X) \in \mathscr{M}$, $\Phi \in \mathscr{A}_\Delta$ is τ_X-lower semicontinuous in the second component, and there exist $x_0 \in X$ and a nonempty τ_X-compact subset K of A such that for each $x \in A \setminus K$, there exists $y \in A$ with $d(x_0, y) < d(x_0, x)$ and $\Phi(x, y) \leq 0$. Then,* (EP) *has solutions.*

Proof Applying Theorem 1.63 to K and $\Phi|_{K \times K}$, we obtain $\bar{x} \in K$ such that

$$\Phi(\bar{x}, y) \geq 0 \text{ for all } y \in K. \tag{1.20}$$

We show that $\bar{x}$ is a solution of (EP). Suppose that there exists $x \in A \setminus K$ such that $\Phi(\bar{x}, x) < 0$. Let $B = \mathrm{cl}B(x_0, d(x_0, x)) \cap \mathrm{lev}_{\leq}\Phi(x, \cdot)$. Since $x \in A \setminus K$, by assumption, an $y \in A$ exists with $d(x_0, y) < d(x_0, x)$ and $\Phi(x, y) \leq 0$, and so this y belongs to B, i.e., $B \neq \emptyset$. Moreover, B is a τ_X-closed subset of the τ_X-compact set $\mathrm{cl}B(x_0, d(x_0, x))$. Hence, B is τ_X-compact. The τ_X-compactness of B and the τ_X-lower semicontinuity in the first component of the distance function d give an $\bar{y} \in B$ with

$$d(\bar{y}, x_0) \leq d(y, x_0) \text{ for all } y \in B. \tag{1.21}$$

If $\bar{y} \in A \setminus K$, then by assumption, an $\hat{y} \in A$ exists with $d(\hat{y}, x_0) = d(x_0, \hat{y}) < d(x_0, \bar{y}) = d(\bar{y}, x_0)$ and $\Phi(\bar{y}, \hat{y}) \leq 0$. Since $d(x_0, \bar{y}) \leq d(x_0, x)$ and $\Phi(x, \hat{y}) \leq \Phi(x, \bar{y}) + \Phi(\bar{y}, \hat{y}) \leq 0$, we have $\hat{y} \in B$ and $d(\hat{y}, x_0) < d(\bar{y}, x_0)$. This contradicts (1.21). If $\bar{y} \in K$, then $\Phi(\bar{x}, \bar{y}) \leq \Phi(\bar{x}, x) + \Phi(x, \bar{y}) \leq 0$ which contradicts (1.20).

□

Corollary 1.67 *Assume that A is a weakly closed subset of a Banach space X and $\Phi \in \mathscr{A}_\Delta$ such that Φ is weakly lower semicontinuous in the second component, there exists a nonempty weakly compact subset K of A such that for each $x \in A \setminus K$, there exists $y \in A$ with $\|y\| < \|x\|$ and $\Phi(x, y) \leq 0$. Then,* (EP) *has solutions.*

Proof Corollary 1.67 is a special case of Theorem 1.66 with τ_X being the weak topology in X. (Note that the norm function $x \rightarrow \|x\|$ is weakly lower semicontinuous.) □

Example 1.68 Let B be a nonempty subset of a real normed space $(X, \|\cdot\|)$. The set B is called PROXIMINAL if for every $\hat{x} \in X$, there exists $\bar{x} \in B$ such that $\|\hat{x} - \bar{x}\| \leq \|\hat{x} - y\|$ for all $y \in B$. This $\bar{x}$ is called a best approximation to $\hat{x}$ in B. We have the following assertion: *every nonempty closed convex subset B of a reflexive Banach space* $(X, \|\cdot\|)$ *is proximinal.* Indeed, let $\hat{x} \in X$ be arbitrary and $\Phi : B \times B \rightarrow \mathbb{R}$ be defined by $\Phi(x, y) = \|\hat{x} - y\| - \|\hat{x} - x\|$ for all $x, y \in B$. Then, $\bar{x}$ is a best approximation to $\hat{x}$ in B if and only if $\Phi(\bar{x}, y) \geq 0$ for all $y \in B$. Fix an $\hat{y} \in B$ and set $A = \{y \in B \mid \|\hat{x} - y)\| \leq \|\hat{x} - \hat{y}\|\}$. Since B is closed and convex, it is easy to check that A is a nonempty closed convex subset of X. Moreover, for all $y \in A$ we have $\|y\| = \|(\hat{x} - y) + \hat{x}\| \leq \|\hat{x} - y\| + \|\hat{x}\| \leq \|\hat{x} - \hat{y}\| + \|\hat{x}\|$. Thus, A is bounded and hence weakly compact. Now, we easily see that $\Phi|_{A\times A} \in \mathscr{A}_\Delta$ and $\Phi|_{A\times A}$ is weakly lower semicontinuous in the second component (this follows from the weak lower semicontinuity of the norm function $\|\cdot\|$). By Corollary 1.67, there exists $\bar{x} \in A$ such that $\Phi(\bar{x}, y) \geq 0$ for all $y \in A$. For all $y \in B \setminus A$, we have $\Phi(\bar{x}, y) = \|\hat{x} - y\| - \|\hat{x} - \bar{x}\| \geq \|\hat{x} - \hat{y}\| - \|\hat{x} - \bar{x}\| \geq 0$. Thus, $\Phi(\bar{x}, y) \geq 0$ for all $y \in B$.

Recall that the DIAMETER of a subset S of a metric space (X, d), denoted by $\operatorname{diam}(S)$, is defined as the largest distance between any two points of S:

$$\operatorname{diam}(S) := \sup\{d(x, y) \mid x, y \in S\}.$$

Theorem 1.69 (Cantor's principle for (EP)) *Assume that A is a closed subset of a complete metric space (X, d) and $\Phi \in \mathscr{A}_\Delta \cap \mathscr{A}_A^=$ is such that $\operatorname{lev}_\leq \Phi(x, \cdot)$ is closed for every $x \in A$. Then,* (EP) *has a solution if and only if there exists a sequence $\{x^k\}_k$ such that for each k, $\Phi(x^k, x^{k+1}) \leq 0$ and $\lim_{k\to\infty} \operatorname{diam}(\operatorname{lev}_\leq \Phi(x^k, \cdot)) = 0$. Furthermore, $\bar{x} = \lim_{k\to\infty} x^k$ is a solution of* (EP) *and* $\operatorname{lev}_\leq \Phi(\bar{x}, \cdot) = \{\bar{x}\}$.

Proof ($\Rightarrow$) If $\bar{x}$ is a solution of (EP), then taking $x^k = \bar{x}$ for all k, we have $\Phi(x^k, x^{k+1}) = 0$ and $\operatorname{lev}_\leq \Phi(\bar{x}, \cdot) = \{\bar{x}\}$ for all k.

($\Leftarrow$) When $\Phi \in \mathscr{A}_\Delta \cap \mathscr{A}_A^=$, we have $\operatorname{lev}_\leq \Phi(x, \cdot) \neq \emptyset$ for all $x \in A$ and $\operatorname{lev}_\leq \Phi(y, \cdot) \subset \operatorname{lev}_\leq \Phi(x, \cdot)$ for all $y \in \operatorname{lev}_\leq \Phi(x, \cdot)$ and $x \in A$. For each k, since $\Phi(x^k, x^{k+1}) \le 0$, we have $x^{k+1} \in \mathrm{lev}_{\le}\Phi(x^k, \cdot)$, and so $\mathrm{lev}_{\le}\Phi(x^{k+1}, \cdot) \subset \mathrm{lev}_{\le}\Phi(x^k, \cdot)$. It follows that $x^k, x^{k+1} \in \mathrm{lev}_{\le}\Phi(x^{k-1}, \cdot)$ for all $k > 1$. Hence, $d(x^{k+1}, x^k) \le \mathrm{diam}(\mathrm{lev}_{\le}\Phi(x^{k-1}, \cdot)) \to 0$, i.e., $\{x^k\}$ is a Cauchy sequence in the complete metric space (X, d). Therefore, $x^k \to \bar{x} \in X$. For each k, since $\mathrm{lev}_{\le}\Phi(x^k, \cdot)$ is closed and $x^{k'} \in \mathrm{lev}_{\le}\Phi(x^k, \cdot)$ for all $k' > k$, we have $\bar{x} \in \mathrm{lev}_{\le}\Phi(x^k, \cdot)$. Thus, $\bar{x} \in \bigcap_{k=1}^{\infty} \mathrm{lev}_{\le}\Phi(x^k, \cdot)$. If $\bar{y} \in \bigcap_{k=1}^{\infty} \mathrm{lev}_{\le}\Phi(x^k, \cdot)$, then $\bar{x}$ and $\bar{y}$ belong to $\mathrm{lev}_{\le}\Phi(x^k, \cdot)$ for all k. Hence, $d(\bar{x}, \bar{y}) \le \mathrm{diam}(\mathrm{lev}_{\le}\Phi(x^k, \cdot)) \to 0$, i.e., $\bar{x} = \bar{y}$. Finally, $\mathrm{lev}_{\le}\Phi(\bar{x}, \cdot) \subset \bigcap_{k=1}^{\infty} \mathrm{lev}_{\le}\Phi(x^k, \cdot) = \{\bar{x}\}$, which implies that $\bar{x}$ is a solution of (EP). □

Example 1.70 Consider (EP) with $A = [0, 1]$ and $\Phi : [0, 1] \times [0, 1] \to \mathbb{R}$ defined by $\Phi(x, y) = y - x$ for $x, y \in A$. We have $\Phi \in \mathscr{A}_\Delta \cap \mathscr{A}_A^{=}$, and $\mathrm{lev}_{\le}\Phi(x, \cdot) = [0, x]$ is closed for all $x \in A$. For $x^k = \frac{1}{k} \in A$, $k = 1, 2, \ldots$, we have $\Phi(x^k, x^{k+1}) = -\frac{1}{k(k+1)} \le 0$ and $\lim_{k\to\infty} \mathrm{diam}(\mathrm{lev}_{\le}\Phi(x^k, \cdot)) \le \lim_{k\to\infty} \frac{1}{k} = 0$. Thus, $\bar{x} = \lim_{k\to\infty} x^k = 0$ is a solution.

In the particular case where the equilibrium bifunction has the form $\Phi(x, y) + \epsilon d(x, y)$ with $\epsilon > 0$, a sequence as in Theorem 1.69 always exists, as shown in the proof of the following version of Ekeland variational principle (EVP, in brief).

Theorem 1.71 (Ekeland's variational principle for (EP)) *Assume that (A, d) is a complete metric space, $\epsilon > 0$, and $\Phi \in \mathscr{A}_\Delta \cap \mathscr{A}_A^{=}$ is such that $\{y \in A \mid \Phi(x, y) + \epsilon d(x, y) \le 0\}$ is closed and $\inf_{y\in A} \Phi(x, y) > -\infty$ for each $x \in A$. Then, for all $x^0 \in A$, there exists $\bar{x} \in A$ such that:*

(a) $\Phi(x^0, \bar{x}) + \epsilon d(x^0, \bar{x}) \le 0$.
(b) $\Phi(\bar{x}, y) + \epsilon d(\bar{x}, y) > 0$ *for all $y \in A$ different from $\bar{x}$.*

Proof Define $\Phi' : A \times A \to \mathbb{R}$ by $\Phi'(x, y) = \Phi(x, y) + \epsilon d(x, y)$ for all $x, y \in A$. It is easy to check that $\Phi' \in \mathscr{A}_\Delta \cap \mathscr{A}_A^{=}$ and $\mathrm{lev}_{\le}\Phi'(x, \cdot)$ is closed for every $x \in A$.

For each $x \in A$, define

$$\theta(x) = \inf_{y\in \mathrm{lev}_{\le}\Phi'(x,\cdot)} \Phi(x, y). \tag{1.22}$$

Then, for all $x \in A$ and $y, y' \in \mathrm{lev}_{\le}\Phi'(x, \cdot)$, we have

$$d(y, y') \le d(x, y) + d(x, y') \le -\frac{1}{\epsilon}\Phi(x, y) - \frac{1}{\epsilon}\Phi(x, y')$$

$$\le \frac{1}{\epsilon}\Big(\sup_{y\in \mathrm{lev}_{\le}\Phi'(x,\cdot)} -\Phi(x, y) + \sup_{y'\in \mathrm{lev}_{\le}\Phi'(x,\cdot)} -\Phi(x, y')\Big) = -\frac{2}{\epsilon}\theta(x).$$

Hence, $\operatorname{diam}(\operatorname{lev}_{\le}\Phi'(x,\cdot)) \le -\frac{2}{\epsilon}\theta(x)$ for all $x \in A$. Now, fix an arbitrary point $x^0 \in A$. We construct inductively a sequence $\{x^k\}_k$, starting with x^0 and with x^{k+1} defined by

$$\begin{cases} x^{k+1} \in \operatorname{lev}_{\le}\Phi'(x^k,\cdot), \\ \Phi(x^k,x^{k+1}) \le \theta(x^k) + 2^{-k}. \end{cases}$$

Such a sequence is completely defined by (1.22) and the nonemptiness of $\operatorname{lev}_{\le}\Phi'(x,\cdot)$ (for $x \in A$). The sequence $\{x^k\}_k$ satisfies $\operatorname{lev}_{\le}\Phi'(x^{k+1},\cdot) \subset \operatorname{lev}_{\le}\Phi'(x^k,\cdot)$ (easily checked) and $\Phi'(x^k,x^{k+1}) \le 0$. Furthermore,

$$\operatorname{diam}(\operatorname{lev}_{\le}\Phi'(x^{k+1},\cdot)) \le -\frac{2}{\epsilon}\theta(x^{k+1}) = -\frac{2}{\epsilon}\inf_{y \in \operatorname{lev}_{\le}\Phi'(x^{k+1},\cdot)} \Phi(x^{k+1},y)$$

$$\le -\frac{2}{\epsilon}\inf_{y \in \operatorname{lev}_{\le}\Phi'(x^k,\cdot)} \Phi(x^{k+1},y) \le -\frac{2}{\epsilon}\inf_{y \in \operatorname{lev}_{\le}\Phi'(x^k,\cdot)} \Phi(x^k,y) + \frac{2}{\epsilon}\Phi(x^k,x^{k+1})$$

$$= \frac{2}{\epsilon}\Big(-\theta(x^k) + \Phi(x^k,x^{k+1})\Big) \le \frac{2}{\epsilon}\Big(-\theta(x^k) + \theta(x^k) + 2^{-k}\Big) = \frac{2\cdot 2^{-k}}{\epsilon}.$$

Hence, $\lim_{k\to\infty}\operatorname{diam}(\operatorname{lev}_{\le}\Phi'(x^k,\cdot)) = 0$. Thus, applying Theorem 1.69 for A and Φ', one has $\bar{x} = \lim_{k\to\infty} x^k$ with $\operatorname{lev}_{\le}\Phi'(\bar{x},\cdot) = \{\bar{x}\}$ such that $\Phi'(\bar{x},y) \ge 0$ for all $y \in A$, i.e., $\Phi(\bar{x},y) + \epsilon d(\bar{x},y) \ge 0$ for all $y \in A$. Now, if for some $y \in A$, $\Phi'(\bar{x},y) = \Phi(\bar{x},y) + \epsilon d(\bar{x},y) = 0$, then $y \in \operatorname{lev}_{\le}\Phi'(\bar{x},\cdot)$ and hence $y = \{\bar{x}\}$. Thus, $\Phi(\bar{x},y) + \epsilon d(\bar{x},y) > 0$ for all $y \in A$, $y \ne \bar{x}$.

Because $x^k \in \operatorname{lev}_{\le}\Phi'(x^0,\cdot)$ for all k and $\operatorname{lev}_{\le}\Phi'(x^0,\cdot)$ is closed, we obtain $\bar{x} \in \operatorname{lev}_{\le}\Phi'(x^0,\cdot)$ which means that $\Phi(x^0,\bar{x}) + \epsilon d(x^0,\bar{x}) \le 0$. □

Corollary 1.72 *Assume that* (A,d) *is a complete metric space and* $\Phi \in \mathscr{A}_\Delta \cap \mathscr{A}_A^=$ *is such that* Φ *is lower semicontinuous in the second component, and* $\inf_{y\in A}\Phi(x,y) > -\infty$ *for each* $x \in A$. *Then, for all* $\epsilon > 0$ *and* $x^0 \in A$, *there exists* $\bar{x} \in A$ *such that:*

(a) $\Phi(x^0,\bar{x}) + \epsilon d(x^0,\bar{x}) \le 0$.
(b) $\Phi(\bar{x},y) + \epsilon d(\bar{x},y) > 0$ *for all* $y \in A$ *different from* $\bar{x}$.

Proof By the lower semicontinuity of Φ and d in the second variable, the set $\{y \in A \mid \Phi(x,y) + \epsilon d(x,y) \le 0\}$ is closed for all $\epsilon > 0$ and $x \in A$. Thus, all the conditions of Theorem 1.71 are satisfied for all $\epsilon > 0$, and by this theorem we obtain the conclusion of Corollary 1.72. □

Corollary 1.72 is equivalent to the following well-known Ekeland variational principle.

Corollary 1.73 (Classical Ekeland's variational principle) *Let (A, d) be a complete metric space and $\varphi : A \to \mathbb{R}$ be lower semicontinuous and bounded from below on A. Then, for all $\epsilon > 0$ and $x^0 \in A$, there exists $\bar{x} \in A$ such that:*

(a) $\varphi(\bar{x}) \leq \varphi(x^0) - \epsilon d(\bar{x}, x^0)$.
(b) $\varphi(y) > \varphi(\bar{x}) - \epsilon d(\bar{x}, y) > 0$ *for all* $y \in A$ *different from* $\bar{x}$.

Proof of the equivalence. Corollary 1.72 implies the Ekeland variational principle with $\Phi(x, y) = \varphi(y) - \varphi(x)$ for all $x, y \in A$. Conversely, we assume that Φ satisfies the conditions of Corollary 1.72. Then, for all $x^0 \in A$, $\Phi(x^0, \cdot)$ satisfies the assumptions of Corollary 1.73. Hence, for all $\epsilon > 0$ and $x^0 \in A$, there exists $\bar{x} \in A$ such that:

(a′) $\Phi(x^0, \bar{x}) \leq \Phi(x^0, x^0) - \epsilon d(\bar{x}, x^0)$.
(b′) $\Phi(x^0, y) > \Phi(x^0, \bar{x}) - \epsilon d(\bar{x}, y) > 0$ for all $y \in A$ different from $\bar{x}$.

Because $\Phi \in \mathscr{A}_\Delta \cap \mathscr{A}_A^=$, we have $\Phi(x^0, x^0) = 0$ and for all $y \in A$, $\Phi(x^0, y) \leq \Phi(x^0, \bar{x}) + \Phi(\bar{x}, y)$, or equivalently, $\Phi(\bar{x}, y) \geq \Phi(x^0, y) - \Phi(x^0, \bar{x})$. Then, (a′) and (b′) imply that $\Phi(x^0, \bar{x}) + \epsilon d(x^0, \bar{x}) \leq 0$ and $\Phi(\bar{x}, y) + \epsilon d(\bar{x}, y) > 0$, respectively, for all $y \in A$ different from $\bar{x}$. □

Theorem 1.74 *Let (A, d) be a complete metric space, $\Phi \in \mathscr{A}_\Delta \cap \mathscr{A}_A^=$ be lower semicontinuous in the second component, and $\inf_{y \in A} \Phi(x, y) > -\infty$ for each $x \in A$. Assume that there exists $\epsilon > 0$ such that for every $x \in A$ with $\inf_{y \in X} \Phi(x, y) < 0$, there is $y \in A$ different from x with $\Phi(x, y) + \epsilon d(x, y) \leq 0$. Then,* (EP) *has solutions.*

Proof By Corollary 1.72, an $\bar{x} \in A$ exists such that

$$\Phi(\bar{x}, y) + \epsilon d(\bar{x}, y) > 0 \text{ for all } y \in A, \ y \neq \bar{x}. \tag{1.23}$$

Suppose that $\bar{x}$ is not a solution of (EP). Then, there exists $y' \in A$ such that $\Phi(\bar{x}, y') < 0$. This implies that $\inf_{y \in A} \Phi(\bar{x}, y) < 0$. By assumption, there is $y'' \in A$, $y'' \neq \bar{x}$, such that

$$\Phi(\bar{x}, y'') + \epsilon d(\bar{x}, y'') \leq 0. \tag{1.24}$$

The relations (1.23) and (1.24) give a contradiction. □

Theorem 1.74 is equivalent to the following result.

Corollary 1.75 (Takahashi's minimization theorem) *Let (A, d) be a complete metric space and $\varphi : A \to \mathbb{R}$ be lower semicontinuous with $\inf_{x \in A} \varphi(x) > -\infty$. Assume that there exists $\epsilon > 0$ such that that for every $x \in A$ with $\inf_{y \in A} \varphi(y) < \varphi(x)$, there is $y \in A$ different from x with $\varphi(y) + \epsilon d(x, y) \leq \varphi(x)$. Then, there exists $\bar{x} \in A$ such that $\varphi(\bar{x}) \leq \varphi(y)$ for all $y \in A$.*

Proof of the equivalence. Corollary 1.75 follows from Theorem 1.74 with $\Phi(x, y) = \varphi(y) - \varphi(x)$. Conversely, Corollary 1.75 implies Theorem 1.74 with $\varphi(\cdot) = \Phi(\hat{x}, \cdot)$ for some fixed $\hat{x} \in A$. □

As examples, we provide some applications in fixed-point theory.

Example 1.76

(a) ***Contractive maps***. Let (A, d) be a complete metric space. A map $f : A \to A$ is called CONTRACTIVE if for some $\gamma \in (0, 1)$, $d(f(x), f^2(x)) \leq (1 - \gamma)d(x, f(x))$. Here $f^2(x) = f(f(x))$. Applying Theorem 1.74, we deduce the following result: *every contractive and continuous mapping $f : A \to A$ has a fixed point.* Indeed, let $\Phi : A \times A \to \mathbb{R}$ be defined by $\Phi(x, y) = \frac{1}{\gamma}(d(y, f(y)) - d(x, f(x)))$ for all $x, y \in A$. Then, for $\bar{x} \in A$, if $\Phi(\bar{x}, y) \geq 0$ for all $y \in A$, then $\bar{x} = f(\bar{x})$, since otherwise

$$\begin{aligned}\Phi(\bar{x}, f(\bar{x})) &= \frac{1}{\gamma}(d(f(\bar{x}), f^2(\bar{x})) - d(\bar{x}, f(\bar{x})))\\ &< \frac{1}{\gamma}(d(f(\bar{x}), f^2(\bar{x})) - (1-\gamma)d(\bar{x}, f(\bar{x}))) \leq 0.\end{aligned}$$

Clearly, $\Phi \in \mathscr{A}_\Delta \cap \mathscr{A}_A^=$, Φ is lower semicontinuous in the second component, and $\inf_{y \in A} \Phi(x, y) > -\infty$ for each $x \in A$. If for some $x \in A$ with $\inf_{y \in A} \Phi(x, y) < 0$ and $x = f(x)$, then x is a fixed point of f. Thus, we can assume that for every $x \in A$ with $\inf_{y \in A} \Phi(x, y) < 0$, $x \neq f(x)$. Then, taking $\epsilon = 1$, we have for every $x \in A$ with $\inf_{y \in A} \Phi(x, y) < 0$, there exists $y = f(x)$ satisfying $\Phi(x, y) + d(x, y) = \frac{1}{\gamma}(d(f(x), f^2(x)) - (1-\gamma)d(x, f(x))) \leq 0$. Thus, by Theorem 1.74, there exists $\bar{x} \in A$ such that $\Phi(\bar{x}, y) \geq 0$ for all $y \in A$. Note that if f satisfies $d(f(x), f(y)) \leq \gamma d(x, y)$ for all $x, y \in A$ and for some $\gamma \in (0, 1)$, then f is continuous and contractive. In this case we call f a CONTRACTION MAPPING and we obtain the Banach contraction principle for complete metric spaces.

(b) ***Caristi's fixed-point theorem***. *Let (A, d) be a complete metric space, $\varphi : A \to \mathbb{R}$ be a lower semicontinuous function bounded from below on A, and $f : A \to A$ be a mapping. Assume that $\varphi(f(x)) - \varphi(x) + d(x, f(x)) \leq 0$ for all $x \in A$. Then, f has a fixed point.*

We deduce this theorem from Theorem 1.74. Assume that for all $x \in A$, $x \neq f(x)$, i.e., $d(x, f(x)) > 0$. Let $\Phi(x, y) = \varphi(y) - \varphi(x)$ for all $x, y \in A$. Then, by assumption, $\Phi \in \mathscr{A}_\Delta \cap \mathscr{A}_A^=$, Φ is lower semicontinuous in the second component, and $\inf_{y \in A} \Phi(x, y) > -\infty$ for each $x \in A$. Moreover, for some $x \in A$ satisfying $\inf_{y \in X} \Phi(x, y) < 0$, taking $y = f(x)$ we have $\Phi(x, y) + d(x, y) = \varphi(f(x)) - \varphi(x) + d(x, f(x)) \leq 0$. Applying Theorem 1.74 with $\epsilon = 1$, one has $\bar{x} \in A$ such that $\Phi(\bar{x}, y) = \varphi(y) - \varphi(\bar{x}) \geq 0$ for all $y \in A$. In particular, for $y = f(\bar{x})$, $\varphi(f(\bar{x})) - \varphi(\bar{x}) \geq 0$, which implies that $\varphi(f(\bar{x})) - \varphi(\bar{x}) + d(\bar{x}, f(\bar{x})) > 0$, contradicting the assumption of the theorem.

(c) ***Clarke's fixed-point theorem***. *Let A be a closed subset of a Banach space and $f : A \to A$ be a continuous mapping. Assume that there exists $\gamma \in (0, 1)$ such that for any $x \in A$ with $x \neq f(x)$, there exists $y \in \{z \in A \mid \|x - z\| + \|z - f(x)\| = \|x - f(x)\|\}$ different from x, such that $\|f(x) - f(y)\| \leq \gamma\|x - y\|$. Then, f has a fixed point.*

The set $\{z \in A \mid \|x - z\| + \|z - f(x)\| = \|x - f(x)\|\}$ represents the line segment $[x, f(x)]$ connecting x and $f(x)$. Therefore, the condition in the Clarke theorem is called the DIRECTIONAL CONTRACTION CONDITION. Clarke's theorem is an extension of Banach's contraction principle because every contraction mapping is a directional contraction mapping. Clarke's theorem follows from Corollary 1.72 as follows. Let $\Phi(x, y) = \|y - f(y)\| - \|x - f(x)\|$ for all $x, y \in A$. Then, all the assumptions of Corollary 1.72 are satisfied. Taking any $\varepsilon > 0$ with $\varepsilon < 1 - \gamma$, by Corollary 1.72, there exists $\bar{x} \in A$ such that $\Phi(\bar{x}, y) + \varepsilon d(\bar{x}, y) = \|y - f(y)\| - \|\bar{x} - f(\bar{x})\| + \varepsilon\|\bar{x} - y\| > 0$ for all $y \neq \bar{x}$. If $\bar{x} \neq f(\bar{x})$, by assumption, there exists $\bar{y} \in \{z \in A \mid \|x - z\| + \|z - f(x)\| = \|x - f(x)\|\}$ with $\bar{x} \neq \bar{y}$ such that $\|\bar{y} - f(\bar{y})\| - \|\bar{y} - f(\bar{x})\| \leq \|f(\bar{x}) - f(\bar{y})\| \leq \gamma\|\bar{x} - \bar{y}\|$. Then,

$$
\begin{aligned}
-(1 - \gamma - \varepsilon)\|\bar{x} - \bar{y}\| &\geq \|\bar{y} - f(\bar{y})\| - \|\bar{y} - f(\bar{x})\| - \|\bar{x} - \bar{y}\| + \varepsilon\|\bar{x} - \bar{y}\| \\
&= \|\bar{y} - f(\bar{y})\| - \|\bar{x} - f(\bar{x})\| + \varepsilon\|\bar{x} - \bar{y}\| = \Phi(\bar{x}, \bar{y}) + \varepsilon d(\bar{x}, \bar{y}) > 0.
\end{aligned}
$$

We obtain a contradiction, because $-(1 - \gamma - \epsilon) < 0$ and $\bar{x} \neq \bar{y}$.

1.3.2 Existence of Equilibria for Generalized Cyclically Monotone Bifunctions

In this subsection, we introduce some classes of bifunctions which are broader than the class $\mathscr{A}_\Delta$ and establish initial results for the existence of solutions of (EP) for these bifunctions.

Definition 1.77 Let A be a nonempty set and $\Phi : A \times A \to \mathbb{R}$ be a bifunction.

(a) Φ is said to be CYCLICALLY ANTI-MONOTONE if, for any $x^1, x^2, \ldots, x^m \in A$ ($m \geq 2$), we have $\Phi(x^1, x^2) + \Phi(x^2, x^3) + \ldots + \Phi(x^{m-1}, x^m) + \Phi(x^m, x^1) \geq 0$.
(b) Φ is called CYCLICALLY ANTI-QUASIMONOTONE if, for any $x^1, x^2, \ldots, x^m \in A$, there exists an $i \in \{1, 2, \ldots, m\}$ such that $\Phi(x^i, x^{i+1}) \geq 0$, where $x^{m+1} := x^1$.

By $\mathscr{A}_{cam}$ ($\mathscr{A}_{caqm}$, resp.), we denote the class of all cyclically anti-monotone bifunctions (the class of all cyclically anti-quasimonotone bifunctions, resp.). If $-\Phi \in \mathscr{A}_{cam}$ ($-\Phi \in \mathscr{A}_{caqm}$, resp.), then Φ is called CYCLICALLY MONOTONE (CYCLICALLY QUASIMONOTONE, resp.).

Example 1.78

(a) The bifunction $\Phi(x, y) = \varphi(y) - \varphi(x)$, where $\varphi : A \to \mathbb{R}$, is always cyclically anti-monotone.
(b) Let $\Phi : A \times A \to \mathbb{R}$ be defined for all $x, y \in A$ by

$$\Phi(x, y) = \begin{cases} \Psi(x, y) \text{ if } \varphi(x) < \varphi(y), \\ \Omega(x, y) \text{ if } \varphi(x) \geq \varphi(y), \end{cases}$$

where $\varphi : A \to \mathbb{R}$, $\Psi : A \times A \to \mathbb{R}$ and $\Omega : A \times A \to \mathbb{R}_+$. Then, Φ is always cyclically anti-quasimonotone. Indeed, for any $x^1, x^2, \ldots, x^m \in A$ with $x^{m+1} := x^1$, by choosing $i \in \{1, 2, \ldots, m\}$ satisfying $\varphi(x^i) = \max\{\varphi(x^j) \mid j \in \{1, 2, \ldots, m\}\}$ we have $\Phi(x^i, x^{i+1}) = \Omega(x^i, x^{i+1}) \geq 0$.

The following properties follow directly from Definition 1.77 (b).

(a) If $\Phi \in \mathscr{A}_{caqm}$, then:
   - $\Phi(x, x) \geq 0$ for all $x \in X$, i.e., $\Phi \in \mathscr{A}_A^{\geq}$.
   - For any $x, y \in A$, $\Phi(x, y) < 0$ implies $\Phi(y, x) \geq 0$, i.e., $-\Phi \in \mathscr{A}_{qm}$.
   - For any $B \subseteq A$, $\Phi|_{B\times B} : B \times B \to \mathbb{R}$ is cyclically anti-quasimonotone.
   - Let $\Phi' : A \times A \to \mathbb{R}$ be defined by $\Phi'(x, y) = \Phi(y, x)$ for all $x, y \in A$, then $\Phi \in \mathscr{A}_{caqm}$.
   - If the bifunction $\Omega : A \times A \to \mathbb{R}$ satisfies $\Omega(x, y) \geq \Phi(x, y)$ for all $x, y \in A$, then $\Omega \in \mathscr{A}_{caqm}$.
   - If $\gamma : \mathbb{R} \to \mathbb{R}$ is a nondecreasing function and $\gamma(0) \geq 0$, then $\gamma \circ \Phi : X \times X \to \mathbb{R}$, defined by $(\gamma \circ \Phi)(x, y) = \gamma(\Phi(x, y))$, belongs to $\mathscr{A}_{caqm}$.

(b) If $\{\Phi_\alpha : A \times A \to \mathbb{R} \mid \alpha \in I\}$ is a family of bifunctions such that $\Phi_{\alpha_0} \in \mathscr{A}_{caqm}$ for some $\alpha_0 \in I$, then the bifunction Φ, defined by $\Phi(x, y) = \sup_{\alpha\in I} \Phi_\alpha(x, y)$, also belongs to $\mathscr{A}_{caqm}$.

The following characterization can be used to replace Definition 1.77(b) and to check cyclic anti-quasimonotonicity in many cases.

Proposition 1.79 *Let A be a nonempty set and $\Phi : A \times A \to \mathbb{R}$ be a bifunction. Then, Φ is cyclically anti-quasimonotone if and only if whenever x^1, x^2,..., $x^m \in A$ $(m \geq 2)$ satisfy $\Phi(x^i, x^{i+1}) < 0$ for all $i = 1, 2, \ldots, m - 1$, we have $\Phi(x^m, x^1) \geq 0$.*

Proof Let (C) be the condition "whenever x^1, x^2,..., $x^m \in A$ $(m \geq 2)$ satisfy $\Phi(x^i, x^{i+1}) < 0$ for all $i = 1, 2, \ldots, m - 1$, we have $\Phi(x^m, x^1) \geq 0$." The part "if" is obvious. For the part "only if," we take any $x_1, x_2, \ldots, x_m \in X$. If $m \geq 2$, the condition (C) gives an $i \in \{1, 2, \ldots, m\}$ such that $\Phi(x^i, x^{i+1}) \geq 0$. For the case $m = 1$, suppose that $\Phi(x^1, x^{1+1}) = \Phi(x^1, x^1) < 0$ (where $x^{1+1} := x^1$). We set $x^2 = x^1$. Then, the condition (C) for the cycle of length 2 $\{x^1, x^2, x^{2+1} := x^1\}$ gives $\Phi(x^1, x^1) = \Phi(x^2, x^1) \geq 0$, a contradiction. Thus, $\Phi(x^1, x^{1+1}) \geq 0$. □

Let $\mathscr{A}_{tr}$ be the class of the bifunctions which have the TRANSITIVE PROPERTY: for $x, y, z \in A$, if $\Phi(x, y) < 0$ and $\Phi(y, z) < 0$, then $\Phi(x, z) < 0$.

The following inclusions are valid:

- $\mathscr{A}_\Delta \subset \mathscr{A}_{cam} \subset \mathscr{A}_{caqm}$.
- $\mathscr{A}_\Delta \subset \mathscr{A}_{tr} \cap \mathscr{A}_A^{\geq} \subset \mathscr{A}_{caqm}$.

Indeed, the inclusions $\mathscr{A}_\Delta \subset \mathscr{A}_{cam} \subset \mathscr{A}_{caqm}$ and $\mathscr{A}_\Delta \subset \mathscr{A}_{tr} \cap \mathscr{A}_A^{\geq}$ are obvious. To show $\mathscr{A}_{tr} \cap \mathscr{A}_A^{\geq} \subset \mathscr{A}_{caqm}$, we observe that $\Phi \in \mathscr{A}_{tr} \cap \mathscr{A}_A^{\geq}$ if and only if "$\Phi(x, x) \geq 0$ for each $x \in A$, and whenever $x^1, x^2, \ldots, x^m \in A$ satisfy $\Phi(x^i, x^{i+1}) < 0$ for all $i = 1, 2, \ldots, m-1$, we have $\Phi(x^1, x^m) < 0$." Now let $\Phi \in \mathscr{A}_{tr} \cap \mathscr{A}_A^{\geq}$ and suppose that $\Phi \notin \mathscr{A}_{caqm}$, i.e., there exist $x^1, x^2, \ldots, x^m \in A$ satisfying $\Phi(x^i, x^{i+1}) < 0$ for all $i = 1, 2, \ldots, m-1$, but $\Phi(x^m, x^1) < 0$. Then, $\Phi(x^1, x^m) < 0$ and $\Phi(x^m, x^1) < 0$, which implies that $\Phi(x^1, x^1) < 0$ because $\Phi \in \mathscr{A}_{tr}$, a contradiction.

The reverse inclusions are not true. Moreover, $\mathscr{A}_{cam} \neq \mathscr{A}_{tr} \cap \mathscr{A}_A^{\geq}$. The following example verifies these facts.

Example 1.80

(a) Let $\Phi : [a, b] \times [a, b] \to \mathbb{R}$ (with $a \leq -3$, $b \geq 3$) be defined by $\Phi(x, y) = -x^3 + y^3 + x^2y^2$ for $x, y \in [a, b]$. Then, $\Phi \in \mathscr{A}_{cam}$. Indeed, for any $x^1, x^2, \ldots, x^m \in [a, b]$ with $x^{m+1} := x^1$, we have $\sum_{i=1}^m \Phi(x^i, x^{i+1}) = \sum_{i=1}^m (x^i)^2(x^{i+1})^2 \geq 0$. $\Phi \notin \mathscr{A}_{tr} \cap \mathscr{A}_A^{\geq}$ because for $x = 3$, $y = -1$, $z = -2$, $\Phi(x, y) = -19$ and $\Phi(y, z) = -3$, but $\Phi(x, z) = 1$.

(b) The bifunction $\Phi : [0, 1] \times [0, 1] \to \mathbb{R}$, defined by $\Phi(x, y) = xy - x^3$, belongs to $\mathscr{A}_{tr} \cap \mathscr{A}_A^{\geq}$(and hence to $\mathscr{A}_{caqm}$), but $\Phi \notin \mathscr{A}_{cam}$. Indeed, $\Phi(x, x) = x^2(1 - x) \geq 0$. For any $x, y, z \in [0, 1]$ satisfying $\Phi(x, y) = xy - x^3 < 0$ and $\Phi(y, z) = yz - y^3 < 0$, we have $y < x^2$ and $z < y^2$. Hence, $\Phi(x, z) = xz - x^3 < xy^2 - x^3 < xx^4 - x^3 = x^3(x^2 - 1) \leq 0$. Thus, $\Phi \in \mathscr{A}_{tr} \cap \mathscr{A}_A^{\geq}$. Because $\Phi(0, 1) + \Phi(1, 0) = -1$, $\Phi \notin \mathscr{A}_{cam}$.

(c) The bifunction $\Phi : [a, b] \times [a, b] \to \mathbb{R}$ (with $a \leq -1$, $b \geq 2$), defined by

$$\Phi(x, y) = \begin{cases} y^2 & \text{if } x \geq y, \\ y - 3x^2 & \text{if } x < y, \end{cases}$$

belongs to $\mathscr{A}_{caqm}$, but $\Phi \notin \mathscr{A}_{tr} \cap \mathscr{A}_A^{\geq}$ and $\Phi \notin \mathscr{A}_{cam}$. Indeed, take any $x^1, x^2, \ldots, x^m \in [a, b]$ such that $\Phi(x^i, x^{i+1}) < 0$, $i = 1, 2, \ldots, m-1$. By the definition of Φ, we have $x^1 < x^2 < \cdots < x^m$ and $\Phi(x^m, x^1) = (x^1)^2 \geq 0$. Thus, $\Phi \in \mathscr{A}_{caqm}$. For $x = -2^{-1/2}$, $y = 1$ and $z = 2$, we have $\Phi(x, y) = -1/2$, $\Phi(y, z) = -1$, $\Phi(x, z) = \Phi(z, x) = 1/2$, and $\Phi(x, y) + \Phi(y, z) + \Phi(z, x) = -1$. Thus, $\Phi \notin \mathscr{A}_{tr} \cap \mathscr{A}_A^{\geq}$ and $\Phi \notin \mathscr{A}_{cam}$.

Lemma 1.81 *If A is a nonempty set and $\Phi \in \mathscr{A}_{caqm}$, then* (EP) *is finitely solvable.*

Proof Reasoning ad absurdum, suppose that there exists a finite subset $M = \{y^1, y^2, \ldots, y^m\}$ of A such that for each $x \in A$, there is $y^i \in M$ satisfying $\Phi(x, y^i) < 0$. Let $V_{y^i} = \{x \in X \mid \Phi(x, y^i) < 0\}$. Then, $A = \bigcup_{i=1}^m V_{y^i}$.

Because $\Phi \in \mathscr{A}_{caqm}$, $\Phi(y^i, y^i) \geq 0$, i.e., $y^i \notin V_{y^i}$ for all $i = 1, 2, \ldots, m$. When $y^1 \notin V_{y^1}$, we can assume that $y^1 \in V_{y^2}$ without loss of generality. Then, $\Phi(y^1, y^2) < 0$, and by the cyclic anti-quasimonotonicity of Φ, $\Phi(y^2, y^1) \geq 0$, i.e., $y^2 \notin V_{y^1}$. Thus, $y^2 \notin V_{y^1} \cup V_{y^2}$. We can assume further that $y^2 \in V_{y^3}$. Then, $\Phi(y^1, y^2) < 0$, $\Phi(y^2, y^3) < 0$, and the cyclic anti-quasimonotonicity of Φ imply that $\Phi(y^3, y^1) \geq 0$ and $\Phi(y^3, y^2) \geq 0$, i.e., $y^3 \notin V_{y^1}$ and $y^3 \notin V_{y^2}$. Hence, $y^3 \notin V_{y^1} \cup V_{y^2} \cup V_{y^3}$. Continuing this process yields that $y^i \notin \bigcup_{j=1}^{i} V_{y^j}$ for $i = 1, 2, \ldots, m$. In particular, $y^m \notin \bigcup_{j=1}^{m} V_{y^j} = A$, a contradiction. □

Theorem 1.82 (Generalized Weierstrass' theorem for (EP)) *Let A be a nonempty closed subset of a topological space X and $\Phi \in \mathscr{A}_{caqm}$ such that* $\text{lev}_{\geq}\Phi(\cdot, y)$ *is closed for each $y \in A$. Assume that the following coercivity condition is satisfied:*

(C_Φ) *There exists a nonempty compact subset K of X such that, for each finite subset N of A, there exists a compact subset L_N of A containing N such that for every $x \in L_N \backslash K$, $\Phi(x, y) < 0$ for some $y \in L_N$.*

Then, (EP) *is solvable.*

Proof If A is compact, then (EP) is solvable by Lemma 1.81 and the finite solvability principle.

If A is noncompact and the condition (C_Φ) holds, then using the above case and arguing similarly to the proof of Theorem 1.65, we obtain the conclusion. □

Example 1.83 Let $A = \{(u, v) \in \mathbb{R}^2 \mid (u+v)^2 + 4(u-v)^2 \leq 32 \leq 4(u+v)^2 + 16(u-v)^2\}$ and $\Phi : A \times A \to \mathbb{R}$ be defined for all $x = (u_1, v_1)$ and $y = (u_2, v_2) \in A$ by

$$\Phi(x, y) = \min\left\{\left(u_1 - \frac{1}{2}u_2\right)(u_1 - u_2), |v_1 - v_2|\right\}.$$

The set A is not convex because $x = (2, 0) \in A$ and $y = (0, 2) \in A$, but $\frac{1}{2}x + \frac{1}{2}y = (1, 1) \notin A$. Thus, all results based on the convexity cannot be used. A is closed (obviously) and bounded (since $\|x\|^2 = u^2 + v^2 = \frac{1}{2}(5u^2 + 5v^2 - 3(u^2 + v^2)) \leq \frac{1}{2}(5u^2 + 5v^2 - 6uv) = \frac{1}{2}((u+v)^2 + 4(u-v)^2) \leq 16$ for all $x = (u, v) \in A$). Hence, A is compact, and therefore, (C_Φ) holds. For any $x_1, x_2, \ldots, x_m \in A$ ($m \geq 2$), where $x_i = (u_i, v_i)$, satisfying $\Phi(x_i, x_{i+1}) < 0$, $i = 1, 2, \ldots, m-1$, we have $(u_i - \frac{1}{2}u_{i+1})(u_i - u_{i+1}) < 0$, which is equivalent to $\min\{u_{i+1}, \frac{1}{2}u_{i+1}\} < u_i < \max\{u_{i+1}, \frac{1}{2}u_{i+1}\}$, $i = 1, 2, \ldots, m-1$. It follows that $0 < u_1 < u_2 < \ldots < u_m$ if $u_m \geq 0$, and $u_m < u_{m-1} < \ldots < u_1 < 0$ if $u_m < 0$. This implies that $\Phi(x_m, x_1) = \min\{(u_m - \frac{1}{2}u_1)(u_m - u_1), |v_m - v_1|\} \geq 0$. Thus, $\Phi \in \mathscr{A}_{caqm}$. For all $y = (u_y, v_y) \in A$, an elementary calculation shows that $\text{lev}_{\geq}\Phi(\cdot, y) = \{(u, v) \in A \mid u \leq \min\{\frac{1}{2}u_y, u_y\}\} \cup \{(u, v) \in A \mid u \geq \max\{\frac{1}{2}u_y, u_y\}\}$ which is a closed subset of A. Thus, by Theorem 1.82, (EP) has solutions. Note that $\Phi \notin \mathscr{A}_{cam}$(to see this take $x = (\frac{3}{2}, 0)$ and $y = (2, \frac{1}{8})$ being two points in A, $\Phi(x, y) + \Phi(y, x) = -\frac{1}{8} < 0$).

Let $\mathscr{N}$ be the class of the metric spaces which have the following property:

(Q) Each $(X, d) \in \mathscr{N}$ admits a topology τ_X (possibly different from the original topology induced by d) such that the closed balls are τ_X-compact.

For each space $X \in \mathscr{N}$, we write (X, d, τ_X) instead of (X, d). Class $\mathscr{N}$ contains the class $\mathscr{M}$ introduced in Sect. 1.3.1 because it does not require any condition on the distance function.

Corollary 1.84 *Let A be a closed subset of a metric space $(X, d, \tau_X) \in \mathscr{N}$ and $\Phi \in \mathscr{A}_{caqm}$ be such that* $\mathrm{lev}_{\geq}\Phi(\cdot, y)$ *is τ_A-closed for each $y \in A$. Assume that the following coercivity condition is satisfied:*

(C'$_\Phi$) *There exist $x_0 \in X$ and a nonempty τ_X-compact subset K of A such that for each $x \in A \setminus K$, there exists $y \in A$ with $d(x_0, y) \leq d(x_0, x)$ and $\Phi(x, y) < 0$.*

Then, (EP) *has solutions.*

Proof It is sufficient to show that condition (C$_\Phi$) in Theorem 1.82 holds for the topology τ_X. For each finite subset N of A, let $r_N = \max\{d(x_0, x) \mid x \in N\}$ and $L_N = \{x \in A \mid d(x_0, x) \leq r_N\} = \mathrm{cl}B(x_0, r_N) \cap A$. Then, L_N is a compact subset of A containing N. For each $x \in L_N \setminus K \subset A \setminus K$, by (C$'_\Phi$), there exists $y \in A$ with $d(x_0, y) \leq d(x_0, x) \leq r_N$ and $\Phi(x, y) < 0$, i.e., $y \in L_N$ and $\Phi(x, y) < 0$. □

Corollary 1.85 *Assume that A is a weakly closed subset of a reflexive Banach space X, $\Phi \in \mathscr{A}_{caqm}$,* $\mathrm{lev}_{\geq}\Phi(\cdot, y)$ *is weakly closed for each $y \in A$, and there exists a nonempty weakly compact subset K of X such that for each $x \in A \setminus K$, there exists $y \in A$ with $\|y\| \leq \|x\|$ and $\Phi(x, y) < 0$. Then,* (EP) *has solutions.*

Proof This is a special case of Corollary 1.84 with τ_X being the weak topology in X. We note that the norm function $x \to ||x||$ is weakly lower semicontinuous. □

Definition 1.86 Bifunction Φ is said to HAVE PROPERTY $\mathbb{Q}$ if from $\Phi(y, x) \leq 0$ and U being an open neighborhood of x, it follows that there exist $z \in U$ and $\alpha, \beta \in (0, +\infty)$ such that $\alpha\Phi(z, y) + \beta\Phi(z, x) \geq 0$.

Example 1.87 Assume that A is a convex subset of a topological vector space and Φ has the following properties:

- For each $x \in A$, $\Phi(x, x) = 0$.
- $\Phi(x, \cdot)$ is explicitly quasiconvex, i.e., for all $y, z \in X$ and $t \in (0, 1)$, it follows from $\Phi(x, y) \leq \Phi(x, ty + (1 - t)z)$ that $\Phi(x, ty + (1 - t)z) \leq \Phi(x, z)$.

Then, Φ has property $\mathbb{Q}$. Indeed, take $x, y \in A$ and an open neighborhood U of x. If $x = y$, then for $z = x$ and any $\alpha, \beta \in (0, +\infty)$, one has $\alpha\Phi(z, y) + \beta\Phi(z, x) = 0$. If $x \neq y$, then there exists $t \in (0, 1)$ such that $z = tx + (1 - t)y \in U$. If $\min\{\Phi(z, y), \Phi(z, x)\} < \max\{\Phi(z, y), \Phi(z, x)\} \leq 0$, then by the explicitly quasiconvexity of $\Phi(z, \cdot)$ we have $0 = \Phi(z, z) \leq \min\{\Phi(z, y), \Phi(z, x)\}$ which is impossible. Thus, either $0 \leq \min\{\Phi(z, y), \Phi(z, x)\} \leq \max\{\Phi(z, y), \Phi(z, x)\}$ or $\min\{\Phi(z, y), \Phi(z, x)\} \leq 0 < \max\{\Phi(z, y), \Phi(z, x)\}$. In both cases, we can find $\alpha, \beta \in (0, +\infty)$ such that $\alpha\Phi(z, y) + \beta\Phi(z, x) \geq 0$.

Lemma 1.88 *Assume that Φ has property $\mathbb{Q}$ and $\text{lev}_{\geq}\Phi(\cdot, y)$ is closed for each $y \in A$. Then, if for some $\bar{x} \in A$, $\Phi(y, \bar{x}) \leq 0$ for all $y \in A$, then $\Phi(\bar{x}, y) \geq 0$ for all $y \in A$.*

Proof Suppose that there exists $\bar{y} \in A$ such that $\Phi(\bar{x}, \bar{y}) < 0$, i.e., $\bar{x} \in \text{lev}_{<}\Phi(\cdot, \bar{y}) = A \setminus \text{lev}_{\geq}\Phi(\cdot, \bar{y})$. Since $\Phi(\bar{y}, \bar{x}) \leq 0$, by property $\mathbb{Q}$, there exist $z \in \text{lev}_{<}\Phi(\cdot, \bar{y})$ and $\alpha, \beta \in (0, +\infty)$ such that $\alpha\Phi(z, y) + \beta\Phi(z, x) \geq 0$. This is impossible because $\Phi(z, \bar{y}) < 0$ and $\Phi(z, \bar{x}) \leq 0$. Thus, $\Phi(\bar{x}, y) \geq 0$ for all $y \in A$. □

Lemma 1.88 (which differs from Lemma 1.43) gives a sufficient condition for the solution set of the dual equilibrium problem (DEP): "*find $\bar{x} \in A$ such that $\Phi(y, \bar{x}) \leq 0$ for all $y \in A$*" to be contained in the solution set of (EP).

Theorem 1.89 *Assume that $\Phi \in -\mathscr{A}_{caqm}$, Φ has property $\mathbb{Q}$, $\text{lev}_{\leq}\Phi(x, \cdot)$, and $\text{lev}_{\geq}\Phi(\cdot, x)$ are closed for each $x \in A$, and the following condition is satisfied:*

(C_Φ) *There exists a nonempty compact subset K of X such that, for each finite subset N of A, there exists a compact subset L_N of A containing N such that for every $x \in L_N \setminus K$, $\Phi(y, x) > 0$ for some $y \in L_N$.*

Then, (EP) *has solutions.*

Proof Let $\Phi' : A \times A \to \mathbb{R}$ be defined by $\Phi'(x, y) = -\Phi(y, x)$ for all $x, y \in A$. Then, $\Phi' \in \mathscr{A}_{caqm}$ and $\text{lev}_{\geq}\Phi'(\cdot, y) = \text{lev}_{\leq}\Phi(y, \cdot)$ is closed for each $y \in A$. By Theorem 1.82, an $\bar{x} \in A$ exists such that $\Phi'(\bar{x}, y) \geq 0$ for all $y \in A$, i.e., $\Phi(y, \bar{x}) \leq 0$ for all $y \in A$. Now, by Lemma 1.88, $\Phi(\bar{x}, y) \geq 0$ for all $y \in A$. □

Example 1.90 Let $A = \{(u, v) \in \mathbb{R}^2 \mid 0 \leq uv,\ u^2+v^2 \leq 25\}$ and $\Phi : A\times A \to \mathbb{R}$ be defined for all $x = (u_1, v_1),\ y = (u_2, v_2) \in A$ by

$$\Phi(x, y) = \Omega(x, y)(u_2 - u_1) + \Psi(x, y)(v_2 - v_1),$$

where $\Omega(x, y) = 5u_2^2+5v_2^2+u_1-u_2+10$ and $\Psi(x, y) = 5u_2^2+5v_2^2+v_1-v_2+10$. The set A is compact but not convex. We have

$$\Phi(x, y) = (5u_2^2 + 5v_2^2 + 10)((u_2 + v_2) - (u_1 + v_1)) - (u_2 - u_1)^2 - (v_2 - v_1)^2.$$

Then, we easily check that $\Phi \in -\mathscr{A}_{caqm}$ and $\text{lev}_{\leq}\Phi(x, \cdot)$ and $\text{lev}_{\geq}\Phi(\cdot, x)$ are closed for each $x \in A$. Consider the property $\mathbb{Q}$. If U is an open neighborhood of x, then there exists $0 < \delta$ such that $B(x, \delta) \cap A \subset U$. For $x, y \in A$, $-5 \leq u_1, v_1, u_2, v_2 \leq 5$. Hence, $\Omega(x, y) = 4u_2^2+5v_2^2+(u_2-\frac{1}{2})^2+u_1+\frac{39}{4} > 0$ and, similarly, $\Omega(y, x) > 0$, $\Psi(x, y) > 0$, $\Psi(y, x) > 0$ for any $x, y \in A$. So, we have three possibilities for $\Phi(y, x) = \Omega(y, x)(u_1 - u_2) + \Psi(y, x)(v_1 - v_2)$ to be nonpositive.

Case 1: $u_1 \leq u_2$ and $v_1 \leq v_2$. Choose $z = x$. Then, $\Phi(z, x) = 0$ and $\Phi(z, y) \geq 0$.

Case 2: $u_1 < u_2$ and $v_1 > v_2$. For $\varepsilon = \min\{\frac{\delta}{2}, \frac{v_1 - v_2}{2}\}$ and $z = (u_1, v_1 - \varepsilon)$, we have $z \in B(x, \delta) \cap A$ and $\Phi(z, x) > 0$.

Case 3: $u_1 > u_2$ and $v_1 < v_2$. Take $\epsilon = \min\{\frac{\delta}{2}, \frac{u_1 - u_2}{2}\}$ and $z = (u_1 - \varepsilon, v_1)$. Then, $z \in B(x, \delta) \cap A$ and $\Phi(z, x) > 0$.

In all the three cases, we can find $\alpha, \beta \in]0, +\infty[$ such that $\alpha\Phi(z, y) + \beta\Phi(z, x) \geq 0$, and hence, Φ has the property $\mathbb{Q}$. Thus, by Theorem 1.89, (EP) has solutions.

Corollary 1.91 *Let A be a closed subset of a metric space $(X, d, \tau_X) \in \mathcal{N}$ and $\Phi \in -\mathscr{A}_{caqm}$ be such that Φ has the property $\mathbb{Q}$, and* $\mathrm{lev}_{\leq}\Phi(x, \cdot)$ *and* $\mathrm{lev}_{\geq}\Phi(\cdot, x)$ *are τ_A-closed for each $x \in A$. Assume that the following coercivity condition is satisfied:*

(C'_Φ) *There exist $x_0 \in X$ and a nonempty τ_X-compact subset K of A such that for each $x \in A \backslash K$, there exists $y \in A$ with $d(x_0, y) \leq d(x_0, x)$ and $\Phi(y, x) > 0$.*

Then, (EP) *has solutions.*

Proof The argument is similar to the one in the proof of Corollary 1.84. □

Corollary 1.92 *Assume that A is a weakly closed subset of a reflexive Banach space X and $\Phi \in -\mathscr{A}_{caqm}$ has the property $\mathbb{Q}$,* $\mathrm{lev}_{\leq}\Phi(x, \cdot)$ *and* $\mathrm{lev}_{\geq}\Phi(\cdot, x)$ *is weakly closed for each $y \in A$, and there exists a nonempty weakly compact subset K of X such that for each $x \in A \setminus K$, there exists $y \in A$ with $\|y\| \leq \|x\|$ and $\Phi(y, x) > 0$. Then,* (EP) *has solutions.*

Proof Corollary 1.92 is a special case of Corollary 1.91 with τ_X being the weak topology in X. □

Some Applications

As examples of applications of results in this subsection, we deduce conditions for the existence of solutions to minimax problems, saddle point problems, noncooperative games, price equilibria, and variational inequalities.

Corollary 1.93 (Saddle points) *Let C, D be nonempty compact subsets of two topological spaces, and let $\phi : C \times D \to \mathbb{R}$ be a bifunction such that $\{(x, y) \in C \times D \mid \phi(x', y) \geq \phi(x, y')\}$ is closed for each $(x', y') \in C \times D$, and for any $(x_1, y_1), (x_2, y_2), \ldots, (x_m, y_m) \in C \times D$ $(m \geq 2)$, $\phi(x_i, y_{i+1}) \leq \phi(x_{i+1}, y_i)$ for some $i \in \{1, 2, \ldots, m\}$, where $(x_{m+1}, y_{m+1}) := (x_1, y_1)$). Then, ϕ has a saddle point.*

Proof Let $A = D \times C$ and Φ be defined by (1.4): for all $(x, y), (x', y') \in A \times A$,

$$\Phi((x, y), (x', y')) = \phi(x', y) - \phi(x, y').$$

By assumption, it is easy to see that all the assumptions of Theorem 1.82 are satisfied for Φ. Hence, there exists $(\bar{x}, \bar{y}) \in A = C \times D$ such that $\Phi((\bar{x}, \bar{y}), (x', y')) \geq 0$ for all $(x', y') \in A$. By the relationship between (EP) and the saddle point problem (see Sect. 1.1.1, Proposition 1.2), we obtain the conclusion of Corollary 1.93. □

Observe that if for each $(x', y') \in C \times D$, $\phi(x', \cdot)$ is upper semicontinuous and $\Phi : \phi(\cdot, y')$ is lower semicontinuous, then the function $(x, y) \to \phi(x, y') - \phi(x', y)$ is lower semicontinuous and hence the set $\{(x, y) \in C \times D \mid \phi(x', y) \geq \phi(x, y')\}$ is closed. Moreover, the condition of cyclic anti-quasimonotonicity in Corollary 1.93 is fulfilled if either "*for all* $(x_i, y_i) \in C \times D$, $i = 1, 2, \ldots, m$, $\sum_{i=1}^{m} \phi(x_i, y_{i-1}) \geq \sum_{i=1}^{m} \phi(x_i, y_{i+1})$, *where* $y_0 := y_m$ *and* $y_{m+1} := y_1$" or "*whenever* $(x, y), (x', y'), (x", y") \in C \times D$ *satisfy* $\phi(x, y') > \phi(x', y)$ *and* $\phi(x', y") > \phi(x", x')$, *it holds that* $\phi(x, y") > \phi(x", y)$."

Example 1.94 Let C, D be compact topological spaces and $\varphi : C \to \mathbb{R}$ and $\psi : D \to \mathbb{R}$ be semicontinuous functions. Let $\phi : C \times D \to \mathbb{R}$ be defined by $\phi(x, y) = \varphi(x) - \psi(y)$ for all $(x, y) \in C \times D$. Then, ϕ is upper semicontinuous in the second component and lower semicontinuous in the first component. Moreover, for any $(x_1, y_1), (x_2, y_2), \ldots, (x_m, y_m) \in C \times D$ $(m \geq 2)$ such that $\phi(x_i, y_{i+1}) > \phi(x_{i+1}, y_i)$ $(\Leftrightarrow \varphi(x_i) + \psi(y_i) > \varphi(x_{i+1}) + \psi(y_{i+1}))$ for $i = 1, 2, \ldots, m-1$, one has $\varphi(x_1) + \psi(y_1) > \varphi(x_m) + \psi(y_m)$, i.e., $\phi(x_m, y_1) < \phi(x_1, y_m)$. Thus, all conditions of Corollary 1.93 hold. Hence, ϕ has a saddle point.

Corollary 1.95 (Minimax problems) *Let A be a nonempty compact subset of a topological space and $\phi : A \times A \to \mathbb{R}$ be a bifunction. Set* $\phi^* := \inf_{x \in A} \left(\sup_{y \in A} \phi(x, y) \right)$ *and* $\text{mis}\phi := \text{argmin}_{x \in A} \left(\sup_{y \in A} \phi(x, y) \right)$. *Assume that $\phi(x, x) = \alpha$ for all $x \in A$, $\phi - \alpha$ is cyclically quasimonotone and* $\text{lev}_{\leq \alpha} \phi(\cdot, y)$ *is closed for each $y \in A$. Then,* $\text{mis}\phi \neq \emptyset$ *and* $\phi^* = \alpha$.

Proof Let $\Phi_{(\phi, \alpha)} : A \times A \to \mathbb{R}$ be defined by $\Phi_{(\phi, \alpha)}(x, y) = -\phi(x, y) + \alpha$ for all $x, y \in A$. By assumption, all the assumptions of Theorem 1.82 are satisfied. Hence, there exists $\bar{x} \in A$ such that $\Phi(\bar{x}, y) \geq 0$ for all $y \in A$. From the relationship between (EP) and the minimax problem (see Sect. 1.1.1, Proposition 1.4), we obtain the conclusion of Corollary 1.95. □

Example 1.96 Let $A = \{x = (u, v) \in \mathbb{R}^2 \mid uv \geq 0, 1 \leq \|x\| \leq 10\}$, $k, \ell \in \mathbb{N}$, $k < \ell$, and for all $x, y \in A$, $\phi(x, y) = \|x\|^k \|y\|^{\ell - k} - \|y\|^\ell + \|x\|^k \|y\|^{-k}$. Then, $\phi(x, x) = 1$ for all $x \in A$. For $\alpha = 1$, we have $\phi(x, y) - 1 = (\|y\|^\ell + 1)\|y\|^{-k}\left(\|x\|^k - \|y\|^k\right)$ and $\text{lev}_{\leq 1} \phi(\cdot, y) = \{x \in A \mid \|x\|^k \leq \|y\|^k\}$. We easily check the assumptions of Corollary 1.95. Thus, $\text{mis}\phi \neq \emptyset$ and $m_\phi = 1$.

Consider the m-player noncooperative game $\mathscr{G} = (A_i, c_i)_{i \in I}$ in Sect. 1.1.1 (d). Applying Theorem 1.82 for the Nikaido-Isoda bifunction, we obtain the following result.

Corollary 1.97 (Nash equilibria) *For the m-player noncooperative game $\mathscr{G} = (A_i, c_i)_{i \in I}$, suppose that A_i is a compact topological space for each $i \in I$, the*

Definition 1.101 (KKM-structure) Let A be a nonempty set. A pair (Σ, τ) is called a KKM-STRUCTURE of A if τ is a topology on A and $\Sigma := \{\sigma_N : \Delta_N \to A \mid N \in \mathscr{F}(A)\}$ is a family of maps such that each $\sigma_N \in \Sigma$ is τ-continuous. If τ is compact, i.e., A is τ-compact, then (Σ, τ) is called a COMPACT KKM-STRUCTURE.

Example 1.102

(a) Any family of maps $\Sigma = \{\sigma_N : \Delta_N \to A \mid N \in \mathscr{F}(A)\}$ together with the coarse topology $\{A, \emptyset\}$ of A forms a KKM-structure of A. Setting

$$\tau_\Sigma = \bigcap_{N \in \mathscr{F}(A)} \{U \subset A \mid \sigma_N^{-1}(U) \text{ is open in } \Delta_N\},$$

then (Σ, τ_Σ) is also a KKM-structure of A. (This τ_Σ is the smoothest topology on A such that the maps of Σ are continuous.) If Σ consists of constant maps, then (Σ, τ) is a KKM-structure of A for any topology τ of A.

(b) Let B be a nonempty subset of A, $\tau = \{U \subset A \mid U \subset A \setminus B\} \cup \{A\}$, and $\Sigma = \{\sigma_N : \Delta_N \to B \mid N \in \mathscr{F}(A)\}$ be any family of maps. Then, (Σ, τ) is a compact KKM-structure of A.

(c) Let A be a subspace of a topological vector space X. For each $N \in \mathscr{F}(A)$, fix an $\bar{x}_N \in A$ and a $\gamma_N \in \mathbb{R}$. Let τ be the topology on A induced by that of X and Σ consist of maps $\sigma_N : \Delta_N \to A$ $(N \in \mathscr{F}(A))$, where each σ_N is defined by $\sigma_N(e) = \bar{x}^N + \gamma_N \sum_{x^i \in N} \lambda_i x^i$ for all $e = \sum \lambda_i e^i \in \Delta_N$. Then, (Σ, τ) is a KKM-structure of A. If A is a convex subset of X, $\bar{x}^N = 0$, and $\gamma_N = 1$ for all $N \in \mathscr{F}(A)$, then this KKM-structure is the usual convex structure of X and is called the NATURAL KKM-STRUCTURE of A.

The notions of convex subsets and convex hulls are generalized to KKM-structures as follows.

Definition 1.103 (Σ-convex subsets)

(a) Let (Σ, τ) be a KKM-structure of A. We say that a subset B of A is Σ-CONVEX if, for all $\sigma_N \in \Sigma$ and $M \subset N \cap B$, $\sigma_N(\Delta_M^{\mathscr{F}}) \subset B$.

(b) For $B \subset A$, the smallest Σ-convex set containing B, denoted by $\operatorname{conv}_\Sigma B$, is called the Σ-CONVEX HULL of B.

It is not hard to check that $\operatorname{conv}_\Sigma B$ is equal to the intersection of all Σ-convex sets containing B. In the case where A is a topological vector space and (Σ, τ) is the natural KKM-structure, the set B is Σ-convex if and only if it is convex, and the Σ-convex hull of a set becomes the usual convex hull.

Observe that a set A can be equipped with many KKM-structures, and a subset of B may be Σ-convex in a certain KKM-structure (Σ, τ) and not in another KKM-structure. For example, let B be the regular pentagram in $A = \mathbb{R}^2$ with the center at the origin. Then, B is not a convex set in the natural KKM-structure of $\mathbb{R}^2$, but B is Σ-convex in (Σ, τ), where τ is the usual topology of $\mathbb{R}^2$ and Σ consists of mappings $\sigma_N : \Delta_N \to \mathbb{R}^2$, $N \in \mathscr{F}(\mathbb{R}^2)$, defined by $\sigma_N(e) = \frac{1}{2\mu\sqrt{1+\mu^2}} \sum_{x_i \in N} \lambda_i x_i$ for all $e := \sum \lambda_i e^i \in \Delta_N$, where $\mu = \frac{1+\sqrt{5}}{2}$ is called the golden ratio.

Proposition 1.104 $\operatorname{conv}_\Sigma B = \bigcup_{N \in \mathscr{F}(B)} \operatorname{conv}_\Sigma N$ *for any KKM-structure* (Σ, τ) *of* A *and* $B \subset A$.

Proof Clearly, $\operatorname{conv}_\Sigma N \subset \operatorname{conv}_\Sigma B$ for any $N \in \mathscr{F}(B)$. Hence, $B \subset \bigcup_{N \in \mathscr{F}(B)} \operatorname{conv}_\Sigma N \subset \operatorname{conv}_\Sigma B$. Therefore, it suffices to show that the union in these inclusions, which is now denoted by B', is a Σ-convex subset of A. Assume that $N \in \mathscr{F}(A)$ and $M \subset N \cap B'$. By the definition of B', there are $N_0, N_1, \ldots, N_l \in \mathscr{F}(B)$ such that $M \subset \bigcup_{j=0}^{l} \operatorname{conv}_\Sigma N_j \subset B'$. Since $\bigcup_{j=0}^{l} N_j \in \mathscr{F}(B)$, one has $\operatorname{conv}_\Sigma(\bigcup_{j=0}^{l} N_j) \subset B'$. As $\operatorname{conv}_\Sigma N_j \subset \operatorname{conv}_\Sigma(\bigcup_{j=0}^{l} N_j)$, one has further $M \subset \operatorname{conv}_\Sigma(\bigcup_{j=0}^{l} N_j)$. Since the Σ-convex hull of a set is a Σ-convex subset, by Definition 1.103, $\sigma_N(\Delta_M^{\mathscr{F}}) \subset \operatorname{conv}_\Sigma(\bigcup_{j=0}^{l} N_j) \subset B'$. Again by this definition, B' is a Σ-convex subset of A. □

2. **Existence of Solutions**. First, we extend several notions of convexity in Sect. 1.2.2 to KKM-structures.

Definition 1.105 Let (Σ, τ) be a KKM-structure of A and $\varphi : A \to \mathbb{R}$ be a function.

(a) φ is said to be Σ-QUASICONVEX if for all $N \in \mathscr{F}(A)$, $M \subset N$, and $x \in \sigma_N(\Delta_M^{\mathscr{F}})$, we have $\varphi(x) \le \max_{x^j \in M} \varphi(x^j)$. φ is called STRICTLY Σ-QUASICONVEX if the inequality is strict.
(b) φ is said to be Σ-QUASICONVEX-LIKE if for all $N \in \mathscr{F}(A)$, $M \subset N$, and $x \in \sigma_N(\Delta_M^{\mathscr{F}})$, we have the following implication: $\max_{x^j \in M} \varphi(x^j) \le 0$ implies $\varphi(x) \le 0$. φ is called STRICTLY Σ-QUASICONVEX-LIKE if the latter inequality is strict.
(c) φ is termed STRONGLY Σ-QUASICONVEX-LIKE if for all $N \in \mathscr{F}(A)$, $M \subset N$, and $x \in \sigma_N(\Delta_M^{\mathscr{F}})$, $\max_{x^j \in M} \varphi(x^j) < 0$ implies $\varphi(x) < 0$.

When (Σ, τ) is the usual convexity structure, the properties in Definition 1.105 collapse to the corresponding usual properties in Definitions 1.19 and 1.21. We give an example where φ satisfies some properties in Definition 1.105 for a KKM-structure, but not the corresponding ones in Definitions 1.19 and 1.21. Let $A = \{x = (u, v) \in \mathbb{R}^2 \mid \|x - (0,2)\| \ge 1, 0 \le v \le 2\}$ and $\varphi : A \to \mathbb{R}$ be defined by $\varphi(x) = \|x\|$ for all $x \in A$. Because A is not convex, φ does not satisfy any property in Definitions 1.19 and 1.21. We equip A with the KKM-structure (Σ, τ) with τ being the topology on A induced by that of $\mathbb{R}^2$ and Σ consisting of the mappings $\sigma_N : \Delta_N \to \mathbb{R}^2$, $N \in \mathscr{F}(\mathbb{R}^2)$, defined by $\sigma_N(e) = \frac{1}{2} \sum_{x^i \in N} \lambda_i x^i$ for all $e = \sum \lambda_i e^i \in \Delta_N$. Then, φ is Σ-quasiconvex.

We use $\Sigma\mathscr{A}^i_{qc}$ ($\Sigma\mathscr{A}^i_{s.qc}$, $\Sigma\mathscr{A}^i_{qcl}$, $\Sigma\mathscr{A}^i_{s.qcl}$, $\Sigma\mathscr{A}^i_{st.qcl}$, resp.) to denote the class of the bifunctions which are Σ-quasiconvex (strictly Σ-quasiconvex, Σ-quasiconvex-like, strictly Σ-quasiconvex-like, strongly Σ-quasiconvex-like, resp.) in the ith component, $i = 1, 2$. Then,

$$\begin{array}{ccccc} \Sigma\mathscr{A}^i_{qcl} \supset & \Sigma\mathscr{A}^i_{qc} & \subset & \Sigma\mathscr{A}^i_{st.qcl} & \\ & \cup & & \cup & \\ & \Sigma\mathscr{A}^i_{s.qc} & \subset & \Sigma\mathscr{A}^i_{s.qcl} & \subset \Sigma\mathscr{A}^i_{qcl}. \end{array}$$

Definition 1.106 (Proper Σ-quasimonotonicity) Let (Σ, τ) be a KKM-structure of A. The bifunction $\Phi : A \times A \to \mathbb{R}$ is said to be PROPERLY Σ-QUASIMONOTONE in the first component (in the second component, resp.) on A if for all $N \in \mathscr{F}(A)$, $M \subset N$, and $x \in \sigma_N(\Delta^{\mathscr{F}}_M)$, we have $\max_{x^j \in M} \Phi(x^j, x) \geq 0$ ($\max_{x^j \in M} \Phi(x, x^j) \geq 0$, resp.).

For $i = 1, 2$, we denote by $\Sigma\mathscr{A}^i_{pr.qm}$ the class of bifunctions defined on $A \times A$ which are properly Σ-quasimonotone in the ith component, and $-\Sigma\mathscr{A}^i_{pr.qm}$ the class of bifunctions Φ which satisfy $-\Phi \in \Sigma\mathscr{A}^i_{pr.qm}$. Note that $\mathscr{A}^i_{pr.qm} \subset \mathscr{A}^{\geq}_A$ (see Proposition 1.29(a)), but $\Sigma\mathscr{A}^i_{pr.qm} \not\subset \mathscr{A}^{\geq}_A$. If (Σ, τ) is the usual convexity structure, then $\Sigma\mathscr{A}^i_{pr.qm} = \mathscr{A}^i_{pr.qm}$. The following proposition is a natural extension of Proposition 1.29 to KKM-structures. Its proof is similar to that of Proposition 1.29 and hence is omitted.

Proposition 1.107 *Let (Σ, τ) be a KKM-structure of A. Then, the following inclusions hold.*

(a) $\mathscr{A}^{\geq}_A \cap \Sigma\mathscr{A}^i_{st.qcl} \subset \Sigma\mathscr{A}^i_{pr.qm}$.
(b) $\mathscr{A}^{=}_A \cap -\mathscr{A}_{pm} \cap -\Sigma\mathscr{A}^1_{st.qcl} \subset \Sigma\mathscr{A}^2_{pr.qm}$.
(c) $\mathscr{A}^{=}_A \cap \mathscr{A}_{pm} \cap \Sigma\mathscr{A}^2_{st.qcl} \subset -\Sigma\mathscr{A}^1_{pr.qm}$.
(d) $\mathscr{A}^{=}_A \cap -\mathscr{A}_{qm} \cap -\Sigma\mathscr{A}^1_{s.qcl} \subset \Sigma\mathscr{A}^2_{pr.qm}$.
(e) $\mathscr{A}^{=}_A \cap \mathscr{A}_{qm} \cap \Sigma\mathscr{A}^2_{s.qcl} \subset -\Sigma\mathscr{A}^1_{pr.qm}$.

By suitable modifications, we can extend the existence results in Sect. 1.2.4 for the usual convexity structure to the case where the feasible region is equipped with a KKM-structure. Next, using KKM-structures, we establish a full characterization (necessary and sufficient condition) for the existence of solutions of (EP).

Definition 1.108 (KKM-bifunctions) A bifunction $\Phi : A \times A \to \mathbb{R}$ is said to be a KKM-BIFUNCTION if there exists a KKM-structure (Σ, τ) of A such that $\Phi \in \Sigma\mathscr{A}^2_{pr.qm}$ and $\mathrm{lev}_{\geq}\Phi(\cdot, y)$ is closed (in τ) for each $y \in A$.

We denote by $\mathscr{A}_{KKM}$ the class of KKM-bifunctions defined on $A \times A$. For each $\Phi \in \mathscr{A}_{KKM}$, we denote by $(\Sigma, \tau)_\Phi$ a KKM-structure such that $\Phi \in \Sigma\mathscr{A}^2_{pr.qm}$ and $\mathrm{lev}_{\geq}\Phi(\cdot, y)$ is closed (in τ) for each $y \in A$. Obviously, each $\Phi \in \mathscr{A}_{KKM}$ can have various $(\Sigma, \tau)_\Phi$.

Lemma 1.109 *Equilibrium problem* (EP) *is finitely solvable if and only if* Φ *is a KKM-bifunction.*

Proof *Necessity*. Assume that (EP) is finitely solvable. Then, for each $N \in \mathscr{F}(A)$, an $x^N \in A$ exists such that $\Phi(x^N, y) \geq 0$ for all $y \in N$. We set

$$\Sigma = \{\sigma_N : \Delta_N \to A \mid \sigma_N(e) = x^N \text{ for all } e \in \Delta_N\}.$$

Let τ be any topology such that $\text{lev}\Phi_{\geq}(\cdot, y)$ is τ-closed for each $y \in A$. Such a topology always exists (e.g., the discrete topology). Then, (Σ, τ) is a KKM-structure of A and it is not hard to check that $\Phi \in \Sigma\mathscr{A}^2_{pr.qm}$.

Sufficiency. Suppose that Φ is a KKM-bifunction, i.e., there exists a KKM-structure $(\Sigma, \tau)_\Phi$ of A such that $\Phi \in \Sigma\mathscr{A}^2_{pr.qm}$ and $\text{lev}_{\geq}\Phi(\cdot, y)$ is closed (in τ) for each $y \in A$, but problem (EP) is not finitely solvable. Then, there exists $N = \{y^0, y^1, \dots, y^n\} \in \mathscr{F}(A)$ such that for each $x \in A$, there is $y^i \in N$ with $\Phi(x, y^i) < 0$.

Let $\sigma_N \in \Sigma$ correspond to N. We set

$$E_i = \{e \in \Delta_N \mid \Phi(\sigma_N(e), y^i) \geq 0\} = \sigma_N^{-1}(\text{lev}_{\geq}\Phi(\cdot, y^i)).$$

For each $\nu \in \mathbb{N}$, we subdivide Δ_N into subsimplices of diameter $\leq \frac{1}{\nu}$ and let V be the set of vertices of the subdivision. For $e = \sum_{j=0}^{k} \lambda_{i_j} e^{i_j} \in V$ $(\lambda_{i_j} > 0)$, we have $e \in \Delta^{\mathscr{F}}_{M(e)}$, where $M(e) = \{y^{i_j} \mid j = 0, 1, \dots, k\} \subset N$. This implies that the set $J(e) := \chi(e) \cap \{i \mid e \in E_i\} = \{i_0, i_1, \dots, i_k\} \cap \{i \mid e \in E_i\}$ is nonempty because $\Phi \in \Sigma\mathscr{A}^2_{pr.qm}$. Now we label V by the function λ satisfying $\lambda(e) \in J(e)$. Then, the condition of Sperner's lemma is satisfied, and so a completely labeled subsimplex $[e^{0\nu}, e^{1\nu}, \dots, e^{n\nu}]$ exists such that $e^{i\nu} \in E_i$ for all $i = 0, 1, \dots, n$. Let $\{z^k\}$ be the sequence $e^{01}, \dots, e^{n1}, e^{02}, \dots, e^{n2}, e^{03}, \dots$. Since Δ_N is compact, without loss of generality we can assume that the sequence $\{z^k\}_k$ converges to $\bar{e} \in \Delta_N$. Then, for each $i \in \{0, 1, \dots, n\}$, $\|e^{i\nu} - \bar{e}\| \leq \|z^k - \bar{e}\| + \|e^{i\nu} - z^k\| \leq \|z^k - \bar{e}\| + \text{diam}([e^{0\nu}, e^{1\nu}, \dots, e^{n\nu}]) \to 0$ as $\nu, k \to +\infty$. Hence, $e^{i\nu} \to \bar{e}$ as $\nu \to +\infty$. For all $i \in \{0, 1, \dots, n\}$, E_i is closed because $\text{lev}_{\geq}\Phi(\cdot, y^i)$ is closed. Hence, $\bar{e} \in E_i$ for all $i \in \{0, 1, \dots, n\}$, i.e., $\Phi(\sigma_N(\bar{e}), y^i) \geq 0$ for all $y^i \in N$. Denoting $\bar{x} = \sigma_N(\bar{e})$, we have $\Phi(\bar{x}, y^i) \geq 0$ for all $y^i \in N$, a contradiction. □

The following theorem gives a characterization of solvability of (EP) in terms of KKM-bifunctions.

Theorem 1.110 *Equilibrium problem* (EP) *has solutions if and only if* Φ *is a KKM-bifunction and the following coercivity condition is satisfied:*

(C_Σ) *For some* $(\Sigma, \tau)_\Phi$, *there exist a finite subset* N *of* A *and a nonempty and* τ*-compact subset* K *of* A *such that for each* $x \in A \backslash K$, *there is* $y \in N$ *such that* $\Phi(x, y) < 0$.

the inclusion $\Im_A \subset \Im_\Gamma$, in many situations, connectedness structures are more useful than classical path-connected spaces.

(b) Let A be an arbitrary topological vector space. For each $x, y \in A$, let $s_{xy} : [0, 1] \to A$ be defined by $s_{xy}(t) = (1 - t)x + ty$ for all $t \in [0, 1]$, and let $\delta_{xy} : [0, 1] \to (0, +\infty)$ be a function defined on $[0, 1]$ such that $\delta_{xy}(0) = \delta_{xy}(1) = 1$. We define Γ as the collection of all paths of the form $\gamma_{xy} = [x, y]$, defined by the function $\gamma_{xy}(\cdot) = \delta_{xy}(\cdot)s_{xy}(\cdot)$. Then, Γ is a connectedness structure on X. In this connectedness structure, we say that δ_{xy} breaks the segment line $[x, y]$. If δ_{xy} is continuous, then we say that it curves the segment line $[x, y]$. In the case that A is a convex subset of a vector space and, for all $x, y \in X$, $\delta_{xy}(t) = 1$ for all $t \in [0, 1]$, then Γ is the usual convexity structure.

(c) Let (A, d) be a metric space. A subset S of A is called a METRIC SEGMENT between two points x and y in A if there exists an isometry $h_{xy} : [0, 1] \to A$ such that $h_{xy}([0, 1]) = S$, $h_{xy}(0) = x$ and $h_{xy}(1) = y$. If (A, d) admits metric segments between any two points in the space, then it is called a segment-convex metric space. In a segment-convex metric space (A, d), for each $x, y \in A$, take the path $[x, y]$ defined by h_{xy}. Then, the family $\Gamma := \{[x, y] \mid x, y \in A\}$ forms a connectedness structure on A.

(d) A TAKAHASHI-CONVEX METRIC SPACE is a metric space (A, d) which is endowed additionally with a continuous function $h : A \times A \times [0, 1] \to A$ satisfying $d(s, h(u, v, t)) \leq (1 - t)d(s, u) + td(s, v)$ for all $(s, u, v, t) \in A \times A \times A \times [0, 1]$. For each $x, y \in A$, take the path $\gamma_{xy} = [x, y]$ with $\gamma_{xy}(\cdot) = h(x, y, \cdot)$. Then, the family $\Gamma := \{[x, y] \mid x, y \in A\}$ is a connectedness structure on A.

2. **Existence of Solutions**. The following definition extends some notions of the usual convexity structure to connectedness structures.

Definition 1.115 Let Γ be a connectedness structure on X and $\varphi : A \to \mathbb{R}$.

(a) φ is called Γ-QUASICONNECTED if for every $x, y \in A$, $\varphi(z) \leq \max\{\varphi(x), \varphi(y)\}$ for all $z \in [x, y]$. When this inequality is strict for $x \neq y$, we say that φ is STRICTLY Γ-QUASICONNECTED.

(b) φ is called Γ-QUASICONNECTED-LIKE if for all $x, y \in A$ and $z \in [x, y]$, $\max\{\varphi(x), \varphi(y)\} \leq 0$ implies $\varphi(z) \leq 0$. If the latter inequality is strict for $x \neq y$, then φ is called STRICTLY Γ-QUASICONNECTED-LIKE.

(c) φ is called STRONGLY Γ-QUASICONNECTED-LIKE if for all $x, y \in A$ and $z \in [x, y]$, $\max\{\varphi(x),$
$\varphi(y)\} < 0$ implies $\varphi(z) < 0$.

If A is equipped with the usual convexity structure Γ, then the class of Γ-quasiconnected functions (strictly Γ-quasiconnected functions, Γ-quasiconnected-like functions, strictly Γ-quasiconnected-like functions, strongly Γ-quasiconnected-like functions, resp.) becomes the class of the quasiconvex functions (strictly quasiconvex functions, quasiconvex-like functions, strictly quasiconvex-like functions, strongly quasiconvex-like functions, resp.). The following example yields a function which has some properties in a connectedness

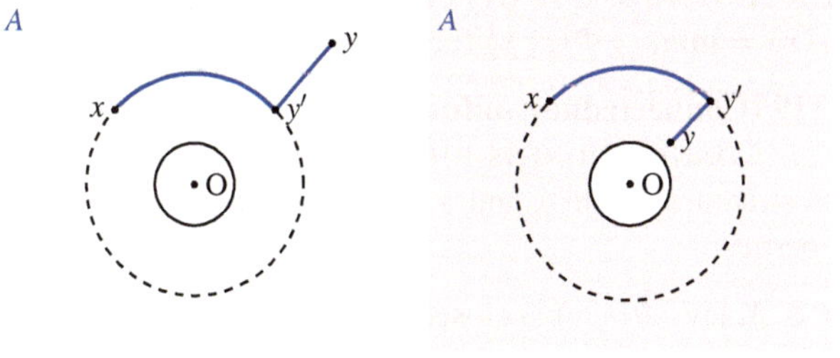

Fig. 1.12 The connectedness structure of the set A in Example 1.116

structure but does not have the corresponding properties in the usual convexity structure.

Example 1.116 Let $A = \{x = (x_1, x_2) \in \mathbb{R}^2 \mid \|x\| \geq 1\}$ and $\varphi : A \to \mathbb{R}$ be defined by $\varphi(x) = \|x\|$ for all $x \in A$. Since A is not convex, φ is not quasiconvex. Let $\Gamma = \{[x, y] \mid x, y \in A\}$ where each $[x, y] \in \Gamma$ is defined as follows: y' is the orthogonal projection of y to the circular disc C_x with the center at the origin and radius $\|x\|$, $[x, y]$ is the union of the line segment joining y and y' and the shortest arc belonging to the boundary of C_x joining y' and x (see Fig. 1.12). Then, for every $x, y \in A$ and $z \in [x, y]$, we have $\|z\| \leq \max\{\|x\|, \|y\|\}$, i.e., φ is Γ-quasiconnected.

The following statements give characterizations of the above notions.

Proposition 1.117 *Let Γ be a connectedness structure of A and $\varphi : A \to \mathbb{R}$.*

(a) *φ is Γ-quasiconnected if and only if the lower-level sets $\text{lev}_{\leq\beta}\varphi$ and $\text{lev}_{<\beta}\varphi$ are path Γ-connected for every $\beta \in \mathbb{R}$.*
(b) *φ is Γ-quasiconnected-like if and only if the lower-level set $\text{lev}_{\leq}\varphi$ is path Γ-connected.*
(c) *φ is strongly Γ-quasiconnected-like if and only if the lower-level set $\text{lev}_{<}\varphi$ is path Γ-connected.*

Proof This is an elementary exercise. □

Proposition 1.118 *Let Γ be a connectedness structure of A and $\Phi : A \times A \to \mathbb{R}$. Then, $\Phi(x, \cdot)$ is strongly Γ-quasiconnected-like for any $x \in A$ if and only if, for any path $[y, y'] \in \Gamma$ and $z \in [y, y']$, one has $\text{lev}_{\geq}\Phi(\cdot, z) \subset \text{lev}_{\geq}\Phi(\cdot, y) \cup \text{lev}_{\geq}\Phi(\cdot, y')$.*

Proof Suppose $\Phi(x, \cdot)$ is strongly Γ-quasiconnected-like for all $x \in A$. Taking any path $[y, y'] \in \Gamma$, $z \in [y, y']$, and $x \in \text{lev}_{\geq}\Phi(\cdot, z)$, we have $\Phi(x, z) \geq 0$. The strong Γ-quasiconnectedness-likeness of $\Phi(x, \cdot)$ implies that either $\Phi(x, y) \geq 0$ or $\Phi(x, y') \geq 0$, i.e., either $x \in \text{lev}_{\geq}\Phi(\cdot, y)$ or $x \in \text{lev}_{\geq}\Phi(\cdot, y')$. Conversely, for all $x, y, y' \in A$ and $z \in [y, y']$, if $\max\{\Phi(x, y), \Phi(x, y')\} < 0$ then $\Phi(x, z) < 0$. Indeed, suppose that $\Phi(x, z) \geq 0$. Then, $x \in \text{lev}_{\geq}\Phi(\cdot, z) \subset \text{lev}_{\geq}\Phi(\cdot, y) \cup \text{lev}_{\geq}\Phi(\cdot, y')$, i.e., either $\Phi(x, y) \geq 0$ or $\Phi(x, y') \geq 0$. Hence, $\max\{\Phi(x, y), \Phi(x, y')\} \geq 0$, a contradiction. □

Let $\Phi : A \times A \to \mathbb{R}$, $J \in \mathscr{F}(A)$ be a finite subset of A, and $\varphi_J : A \to \mathbb{R}$ be defined by $\varphi_J(x) = \min_{y \in J} \Phi(x, y)$ for all $x \in A$.

Definition 1.119 (Connectedness-bifunctions) Bifunction $\Phi : A \times A \to \mathbb{R}$ is said to be a CONNECTEDNESS-BIFUNCTION if $\Phi \in \mathscr{A}_A$ and there exist a connectedness structure Γ on A and a topology τ on A such that Φ satisfies the following properties:

(a) For any $x \in A$, $\mathrm{lev}_{\geq}\Phi(\cdot, y)$ is closed in τ.
(b) For any $x \in A$, $\Phi(x, \cdot)$ is strongly Γ-quasiconnected-like.
(c) For any $J \in \mathscr{F}(A)$, $\mathrm{lev}_{\geq}\varphi_J$ is τ-connected.
(d) For any $y \in A$ and τ-open set U such that $\mathrm{lev}_{\geq}\Phi(\cdot, y) \subset U$, there exists a τ_Γ-open neighborhood V of y such that $\mathrm{lev}_{\geq}\Phi(\cdot, y') \subset U$ for all $y' \in V$.

We denote by $\mathscr{A}_\Gamma$ the class of connectedness-bifunctions defined on $A \times A$. Each $\Phi \in \mathscr{A}_\Gamma$ can have many pairs (Γ, τ) such that (a)–(d) hold. We denote each such pair by $(\Gamma, \tau)_\Phi$. We provide some sufficient conditions for the properties in Definition 1.119 to be fulfilled.

Proposition 1.120 *Let Γ and τ be defined as at the beginning of Definition 1.119. Then, the property (d) in Definition 1.119 is equivalent to the following property:*

(d$'$) *For any path $\gamma_{xy}(\cdot) = [x, y] \in \Gamma$, $t \in [0, 1]$, and τ-open set U with $\mathrm{lev}_{\geq}\Phi(\cdot, \gamma_{xy}(t)) \subset U$, there exists an open neighborhood V_t of t in $[0, 1]$ such that $\mathrm{lev}_{\geq}\Phi(\cdot, \gamma_{xy}(t')) \subset U$ for all $t' \in V_t$.*

Proof (d) $\Rightarrow$ (d$'$). This is obvious.

(d$'$) $\Rightarrow$ (d). This follows from the equality $\gamma_{xy}^{-1}\big(\{z \in A \mid \mathrm{lev}_{\geq}\Phi(\cdot, z) \subset U\}\big) = \{t \in [0, 1] \mid \mathrm{lev}_{\geq}\Phi(\cdot, \gamma_{xy}(t)) \subset U\}$. □

Proposition 1.121 *Let Γ and τ be as in Proposition 1.120 with τ being a compact topology. Assume that:*

(d$_1$) *For each $x, y \in A$, $\mathrm{lev}_{\geq}\Phi(\cdot, y)$ is closed in τ and $\mathrm{lev}_{\geq}\Phi(x, \cdot)$ is closed in τ_Γ.*
(d$_2$) *For all $x, y^1, y^2 \in A$, $y'^1, y'^2 \in [y^1, y^2]$, $z \in [y'^1, y'^2]_{y^1y^2}$, if $\max\{\Phi(x, y'^1), \Phi(x, y'^2)\} < 0$, then $\Phi(x, z) < 0$.*

Then, properties (a), (b), and (d) in Definition 1.119 hold.

Proof Obviously, Conditions (d$_1$) and (d$_2$) imply properties (a) and (b), resp., of Definition 1.119. For property (d), using Proposition 1.120, we prove that for any path $\gamma_{y^1y^2} = [y^1, y^2] \in \Gamma$, $t \in [0, 1]$, and τ-open set U such that $\mathrm{lev}_{\geq}\Phi(\cdot, \gamma_{y^1y^2}(t)) \subset U$, there exists an open neighborhood V_t of t in $[0, 1]$ such that $\mathrm{lev}_{\geq}\Phi(\cdot, \gamma_{y^1y^2}(t')) \subset U$ for all $t' \in V_t$. Suppose to the contrary that there exist a path $\gamma_{y^1y^2} = [y^1, y^2] \in \Gamma$, $\bar{t} \in [0, 1]$, and a τ-open set U such that $\mathrm{lev}_{\geq}\Phi(\cdot, \gamma_{y^1y^2}(\bar{t})) \subset U$, but for any $V_{\bar{t}}$ of $\bar{t}$ in $[0, 1]$, there exists $t \in V_{\bar{t}}$ such that $\mathrm{lev}_{\geq}\Phi(\cdot, \gamma_{y^1y^2}(t)) \not\subset U$. Then, there exist nets $\{t_\alpha\}_{\alpha \in (I, \preceq)} \subset [0, 1]$ and $\{x^\alpha\}_{\alpha \in (I, \preceq)} \subset A$ such that $t_\alpha \to \bar{t}$, $x^\alpha \notin U$ for all α, and

$$\Phi(x^\alpha, \gamma_{y^1y^2}(t_\alpha)) \geq 0 \text{ for all } \alpha. \tag{1.25}$$

As $t_\alpha \to \bar{t}$, there exists a subnet $\{t_{\alpha_\beta}\}$ converging to $\bar{t}$ such that either $0 \leq t_{\alpha_\beta} \leq \bar{t}$ for all β and $t_{\alpha_\beta} \leq t_{\alpha_\sigma}$ for all $\beta \preceq \sigma$, or $\bar{t} \leq t_{\alpha_\beta} \leq 1$ for all β and $t_{\alpha_\sigma} \leq t_{\alpha_\beta}$ for all $\beta \preceq \sigma$. We discuss only the former case. (The latter one is similar.) Since τ is compact, we can suppose $x^{\alpha_\beta} \to \bar{x} \in A$. From $\mathrm{lev}_{\geq}\Phi(\cdot, \gamma_{y^1y^2}(\bar{t})) \subset U$ and $x^{\alpha_\beta} \notin U$ for all β, it follows that $\bar{x} \notin \mathrm{lev}_{\geq}\Phi(\cdot, \gamma_{y^1y^2}(\bar{t}))$, i.e.,

$$\bar{x} \in \mathrm{lev}_{<}\Phi(\cdot, \gamma_{y^1y^2}(\bar{t})), \tag{1.26}$$

or equivalently, $\gamma_{y^1y^2}(\bar{t}) \in \mathrm{lev}_{<}\Phi(\bar{x}, \cdot)$. Because $\mathrm{lev}_{<}\Phi(\bar{x}, \cdot)$ is an open set in τ_Γ (by (d_1)) and $\gamma_{y^1y^2}(t_{\alpha_\beta}) \to \gamma_{y^1y^2}(\bar{t})$ (in τ_Γ), there exists $\bar{\beta}$ such that $\gamma_{y^1y^2}(t_{\alpha_\beta}) \in \mathrm{lev}_{<}\Phi(\bar{x}, \cdot)$ for all $\beta \succeq \bar{\beta}$. In particular, $\gamma_{y^1y^2}(t_{\alpha_{\bar{\beta}}}) \in \mathrm{lev}_{<}\Phi(\bar{x}, \cdot)$, i.e., $\bar{x} \in \mathrm{lev}_{<}\Phi(\cdot, \gamma_{y^1y^2}(t_{\alpha_{\bar{\beta}}}))$. This and (1.26) imply that $\bar{x} \in U_1 = \mathrm{lev}_{<}\Phi(\cdot, \gamma_{y^1y^2}(\bar{t})) \cap \mathrm{lev}_{<}\Phi(\cdot, \gamma_{y^1y^2}(t_{\alpha_{\bar{\beta}}}))$ which is an open set in τ by condition (d_1). Moreover, $t_{\alpha_\beta} \in [t_{\alpha_{\bar{\beta}}}, \bar{t}]$ for all $\beta \succeq \bar{\beta}$. Condition ($d_2$) and (1.25) imply that $x^{\alpha_\beta} \notin U_1$ for all $\beta \succeq \bar{\beta}$. This contradicts the fact that $x^{\alpha_\beta} \to \bar{x}$. □

Lemma 1.122 *If Φ is a connectedness-bifunction, then equilibrium problem* (EP) *is finitely solvable.*

Proof Suppose that $\Phi \in \mathscr{A}_\Gamma$, but (EP) is not finitely solvable. Then, there exists $N \in \mathscr{F}(A)$ such that for each $x \in A$, there is $y \in N$ such that $\Phi(x, y) < 0$ or, equivalently,

$$\mathrm{lev}_{\geq}\varphi_N = \bigcap_{y \in N} \mathrm{lev}_{\geq}\Phi(\cdot, y) = \emptyset. \tag{1.27}$$

Since $\Phi \in \mathscr{A}_A$, the cardinality $|N| \geq 2$. With no loss of generality, we can assume that among the finite subsets of A satisfying (1.27), the cardinality of N is the smallest. Since $\Phi \in \mathscr{A}_\Gamma$, a pair $(\Gamma, \tau)_\Phi$ exists such that (a)–(d) of Definition 1.119 hold. Fix $y^1, y^2 \in N$ with $y^1 \neq y^2$, and set $\widehat{N} = N \setminus \{y^1, y^2\}$ and $J_y = \{y\} \cup \widehat{N}$ for all $y \in A$. Properties (a) and (c) in Definition 1.119 imply that

$$\mathrm{lev}_{\geq}\varphi_{J_y} \text{ is nonempty, closed in } \tau, \text{ and connected in } \tau. \tag{1.28}$$

By property (b) in Definition 1.119, the set $D = \{y \in A \mid \mathrm{lev}_{\geq}\varphi_{J_y} \subset \mathrm{lev}_{\geq}\varphi_{J_{y^1}} \cup \mathrm{lev}_{\geq}\varphi_{J_{y^2}}\} = \bigcap_{x \in B} \mathrm{lev}_{<}\Phi(x, \cdot)$, where $B = \mathrm{lev}_{\geq}\varphi_{\widehat{N}} \setminus (\mathrm{lev}_{\geq}\varphi_{J_{y^1}} \cup \mathrm{lev}_{\geq}\varphi_{J_{y^2}})$ is path Γ-connected and hence connected in τ_Γ. Let

$$M_i = \{y \in D \mid \mathrm{lev}_{\geq}\varphi_{J_y} \subset \mathrm{lev}_{\geq}\varphi_{J_{y^i}}\}, \quad i \in \{1, 2\}.$$

Then, $y^1 \in M_1$ and $y^2 \in M_2$. The definition of D and M_i, (1.27), and (1.28) give that $M_1 \cap M_2 = \emptyset$ and $M_1 \cup M_2 = D$. To have a contradiction, it is sufficient to prove that M_i is open in D with respect to topology τ_D in D being induced by τ_Γ. We prove this for M_1. (The case of M_2 is similar.) Take any $\bar{y} \in M_1$. By (1.27) we have $\text{lev}_{\geq}\varphi_{J_{\bar{y}}} \subset \text{lev}_{\geq}\varphi_{J_{y1}} \subset A \setminus \text{lev}_{\geq}\varphi_{J_{y2}}$. Hence, $\text{lev}_{\geq}\Phi(\cdot, \bar{y}) \subset (A \setminus \text{lev}_{\geq}\varphi_{J_{y2}}) \cup (A \setminus \text{lev}_{\geq}\varphi_{J_{\bar{y}}}) =: U$ which is an open set in τ. By (d) of Definition 1.119, there exists a τ_Γ-open set V containing $\bar{y}$ such that $\text{lev}_{\geq}\Phi(\cdot, y) \subset U$ for all $y \in V$. Setting $V_{\bar{y}} = V \cap M_1$, we have $\bar{y} \in V_{\bar{y}} \subset M_1$. Thus, M_1 is a τ_D-open set in D. □

The following theorem gives a characterization of solvability of (EP) in terms of connectedness-bifunctions.

Theorem 1.123 *Equilibrium problem* (EP) *has solutions if and only if* Φ *is a connectedness-bifunction and the following coercivity condition is satisfied:*

(C_Γ) *For some* $(\Gamma, \tau)_\Phi$*, there exist a finite subset* N *of* A *and a nonempty and* τ*-compact subset* K *of* A *such that, for each* $x \in A \setminus K$*, there is* $y \in N$ *such that* $\Phi(x, y) < 0$.

Proof *Necessity*. If (EP) has a solution $\bar{x} \in A$, then we take $\tau := \{U \subset A \mid \bar{x} \notin U\} \cup \{A\}$ for a topology on A and $\Gamma := \{\gamma_{xx'} = [x, y] \mid x, y \in A\}$ for a connectedness structure on A, where each $\gamma_{xy} \in \Gamma$ is defined by $\gamma_{xy}(0) = x$ and $\gamma_{xy}(t) = y$ for all $t \in (0, 1]$. Then, for these τ and Γ, properties (a)–(d) of Definition 1.119 and the coercivity condition (C_Γ) are checked easily.

Sufficiency. If Φ is a connectedness-bifunction, then (EP) is finitely solvable by Lemma 1.122. For any $N' \in \mathscr{F}(A)$, $(\Gamma, \tau)_\Phi$, N and K provided in (C_Γ), we have $\bar{x} \in A$ such that $\Phi(\bar{x}, y) \geq 0$ for all $y \in N \cup N'$. If $\bar{x} \in A \setminus K$, then $\Phi(\bar{x}, y) < 0$ for some $y \in N$, which is impossible. Hence, $\bar{x} \in K$. In particular, $\Phi(\bar{x}, y) \geq 0$ for all $y \in N'$. Thus, (EP) is finitely solvable on K. By the finite solvability principle, (EP) has solutions. □

Example 1.124 Let $A = [0, 1] \times [0, 1]$ and $\Phi : A \times A \to \mathbb{R}$ be defined by

$$\Phi(x, y) = \min\{x_1 - y_1, x_2 + y_2\} \text{ for all } x = (x_1, x_2) \in A \text{ and } y = (y_1, y_2) \in A.$$

Let τ be the usual topology of A and the connectedness structure Γ consist of paths $\gamma_{xy}(\cdot) = [x, y]$ defined as follows: for all $x = (x_1, x_2)$, $y = (y_1, y_2) \in A$ (assume that $x_1 \leq y_1$), $\gamma_{xy}(t) = x$ if $t = 0$ and $\gamma_{xy}(t) = y$ if $t \neq 0$. For any $x, y, y' \in X$ and any $z \in [x, y]$, we have

$$\Phi(x, z) = \begin{cases} \Phi(x, y) & \text{if } t = 0, \\ \Phi(x, y') & \text{if } t \neq 0. \end{cases}$$

Then, $\max\{\Phi(x, y), \Phi(x, y')\} < 0$ and so $\Phi(x, z) < 0$. Hence, $\Phi(x, \cdot)$ is strongly Γ-quasiconnected-like. For all $y = (y_1, y_2) \in A$, we have $\text{lev}_{\geq}\Phi(\cdot, y) = \{x = (x_1, x_2) \in A \mid \min\{x_1 - y_1, x_2 + y_2\} \geq 0\} = [y_1, 1] \times [0, 1]$. Moreover, for any $J = \{(y_1^j, y_2^j) \mid j = 1, 1 \ldots, n\} \in \mathscr{F}(A)$, we have $\text{lev}_{\geq}\varphi_J = \bigcap_{j=1}^n \text{lev}_{\geq}\Phi(\cdot, y^j) = [\max_{1 \leq j \leq n} y_1^j, 1] \times [0, 1]$. These equalities show that (a) and (c) of Definition 1.119

hold. For any path $\gamma_{xy}(\cdot) = [x, y] \in \Gamma$ and $t \in [0, 1]$, we have

$$\mathrm{lev}_{\geq} \Phi(\cdot, \gamma_{xy}(t)) = \begin{cases} [x_1, 1] \times [0, 1] \text{ if } t = 0, \\ [y_1, 1] \times [0, 1] \text{ if } t \neq 0. \end{cases}$$

Since we assume that $x_1 \leq y_1$, this equality implies that (d′) in Proposition 1.120 holds. Thus, $\Phi \in \mathscr{A}_\Gamma$. By Theorem 1.123, problem (EP) has solutions.

1.4 Exercises

Exercise 1.1 Let A be a convex subset of $\mathbb{R}^n$ and $T : A \to \mathbb{R}^n$. Prove that "$\langle T(\bar{x}), y - \bar{x}\rangle \geq 0$ for all $y \in A$" if and only if "$\|T(\bar{x}))\| = \|(\bar{x} - T(\bar{x})) - \bar{x}\| \leq \|(\bar{x} - T(\bar{x})) - y\|$ for all $y \in A$."

Exercise 1.2 Let C, D be convex subsets of topological vector spaces and $\phi : C \times D \to \mathbb{R}$. Prove that if $\phi(x, \cdot)$ is concave for each $x \in C$ and $\phi(\cdot, y)$ is convex for each $y \in D$, then for each $(x, y) \in C \times D$, the set $A_\phi(x, y) := \{(x', y') \in C \times D \mid \phi(x', y) - \phi(x, y') < 0\}$ is convex.

Exercise 1.3 Let I be an index set. For each $i \in I$, A_i is a convex subset of a topological vector space, $A_{-i} := \prod_{k \neq i} A_k$, x_{-i} is the projection of $x = (x_1, \ldots, x_m)$ onto A_{-i}, $(x_{-i}, y_i) := (x_1, \cdots, x_{i-1}, y_i, x_{i+1}, \cdots x_m)$, and $c_i : A \to \mathbb{R}$. Prove that if for each $i \in I$ and $x \in A$, $c_i(x_{-i}, \cdot)$ is concave, then $N(x, \cdot)$ is convex for each $x \in A$, where $N : A \times A \to \mathbb{R}$ is the Nikaido-Isoda bifunction: $N(x, y) = \sum_{i=1}^n [c_i(x) - c_i(x_{-i}, y_i))]$.

Exercise 1.4 Let A be a convex subset of a topological vector space, $\alpha \in \mathbb{R}$, and $\Phi : A \times A \to \mathbb{R}$. We say that:

(a) Φ is α-properly quasimonotone in the first component (in the second component, resp.) on A if, for all $N \in \mathscr{F}(A)$ and $x \in \mathrm{conv}N$, $\max_{x^j \in N} \Phi(x^j, x) \geq \alpha$ ($\max_{x^j \in N} \Phi(x, x^j) \geq \alpha$, resp.).
(b) Bifunction Φ is diagonally quasiconvex in the first component (in the second component, resp.) on A if, for all $N \in \mathscr{F}(A)$ and $x \in \mathrm{conv}N$, $\max_{x^j \in N} \Phi(x^j, x) \geq \Phi(x, x)$ ($\max_{x^j \in N} \Phi(x, x^j) \geq \Phi(x, x)$, resp.).

By $\mathscr{A}^i_{\alpha-pr.qm}$ ($\mathscr{A}^i_{dqc}$, resp.), we denote the class of all bifunctions defined on $A \times A$ having the α-proper quasimonotonicity (the diagonal quasiconvexity, resp.) in ith component. Prove that:

(a) If A is compact, $\Phi \in \mathscr{A}^2_{\alpha-pr.qm}$, and $\mathrm{lev}_{\geq\alpha}\Phi(\cdot, y)$ is closed for each $y \in A$, then there exists $\bar{x} \in A$ such that $\Phi(\bar{x}, y) \geq \alpha$ for all $y \in A$.
(b) If A is compact, $\Phi \in \mathscr{A}^2_{dqc}$, and $\Phi(\cdot, y)$ is upper semicontinuous for each $y \in A$, then there exists $\bar{x} \in A$ such that $\Phi(\bar{x}, y) \geq \inf_{x \in A} \Phi(x, x)$ for all $y \in A$.

Exercise 1.5 Show that $\mathscr{A}^i_{qc} \subset \mathscr{A}^i_{dqc}$.

Exercise 1.6 Let A be a compact convex subset of a topological vector space and $\Phi : A \times A \to \mathbb{R}$. Let $\Phi_q : A \times A \to \mathbb{R}$ be defined by

$$\Phi_q(x, y) = \inf\{\lambda \in \mathbb{R} \mid y \in \text{conv}(\text{lev}_{\le\lambda}\Phi(x, \cdot))\} \text{ for all } x, y \in A.$$

Prove that if $\Phi_q \in \mathscr{A}_A^{\geq}$ and $\Phi_q(\cdot, y)$ is upper semicontinuous for each $y \in A$, then *(EP)* has solutions.

Exercise 1.7 Let A be a compact convex subset of a topological vector space and $\Phi : A \times A \to \mathbb{R}$. Let $\Phi_{\bar{q}} : A \times A \to \mathbb{R}$ be defined by

$$\Phi_{\bar{q}}(x, y) = \inf\{\lambda \in \mathbb{R} \mid y \in \text{cl}[\text{conv}(\text{lev}_{\le\lambda}\Phi(x, \cdot))]\} \text{ for all } x, y \in A.$$

Prove that if $\Phi_{\bar{q}} \in -\mathscr{A}^1_{pr.qm} \cap \mathscr{A}_A^{=}$, $\Phi_{\bar{q}}$ has the property $\mathbb{P}^2$ and is upper-sign continuous in the first argument, then *(EP)* has solutions.

Exercise 1.8 Let A be a nonempty convex subset of a topological vector space X and $T : A \to X^*$. T is said to be upper hemicontinuous (uhc) (lower hemicontinuous [lhc], resp.) if for any fixed $x, y \in A$, the function $t \to \langle T(x+t(y-x)), y-x\rangle$ from $[0, 1]$ to $\mathbb{R}$ is upper semicontinuous (lower semicontinuous, resp.). If T is both uhc and lhc, then T is called hemicontinuous. T is called pseudomonotone if the bifunction $\Phi(x, y) = \langle T(x), y - x\rangle$ is pseudomonotone.

Prove that if A is compact and T is pseudomonotone and hemicontinuous, then there exists a solution $\bar{x} \in A$ of the variational inequality VI(A, T). Moreover, the compactness of A can be replaced by "A be a nonempty bounded closed convex subset of a reflexive Banach space."

Exercise 1.9 Let A be a subset of a topological space, $\alpha \in \mathbb{R}$, and $\Phi : A \times A \to \mathbb{R}$. We say that:

(1) Φ is α-transfer lower semicontinuous (α-transfer upper semicontinuous, resp.) in the first argument on A if, for all $x \in A$ and $y \in A$ with $\Phi(x, y) > \alpha$ ($\Phi(x, y) < \alpha$, resp.), there exist a point $z \in A$ and a neighborhood $U(x)$ of x such that $\Phi(x', z) > \alpha$ ($\Phi(x', z) < \alpha$, resp.) for all $x' \in U(x)$.
(2) Φ is transfer lower semicontinuous (transfer upper semicontinuous, resp.) in the first argument on A if it is α-transfer lower semicontinuous (α-transfer upper semicontinuous, resp.) in the first argument on A for all $\alpha \in \mathbb{R}$.

Prove the following assertions:

(a) If $\text{lev}_{\ge\alpha}\Phi(\cdot, y)$ (resp, $\text{lev}_{\le\alpha}\Phi(\cdot, y)$) is closed for each $y \in A$, then Φ is α-transfer upper semicontinuous (resp, α-transfer lower semicontinuous) in the first argument on A. Hence, if $\Phi(\cdot, y)$ is upper semicontinuous (resp, lower semicontinuous) for each $y \in A$, then Φ is transfer upper semicontinuous (resp, transfer lower semicontinuous) in the first argument on A.

(b) If Φ is α-transfer upper semicontinuous in the first argument on A, then

$$\mathrm{cl}(\bigcap_{y\in A} \mathrm{lev}_{\geq\alpha}\Phi(\cdot, y)) = \bigcap_{y\in A} \mathrm{cl}(\mathrm{lev}_{\geq\alpha}\Phi(\cdot, y)).$$

Exercise 1.10 Let A be a compact convex subset of a topological vector space, $\alpha \in \mathbb{R}$, and $\Phi : A \times A \to \mathbb{R}$. Prove the following statements:

(a) If $\Phi \in \mathscr{A}^2_{pr.qm}$ and Φ is 0-transfer upper semicontinuous in the first argument on A, then *(EP)* has solutions.
(b) If $\Phi \in \mathscr{A}^2_{\alpha-pr.qm}$ and Φ is α-transfer upper semicontinuous in the first argument on A, then there exists $\bar{x} \in A$ such that $\Phi(\bar{x}, y) \geq \alpha$ for all $y \in A$.
(c) If $\Phi \in \mathscr{A}^2_{dqc}$ and Φ is transfer upper semicontinuous in the first argument on A, then there exists $\bar{x} \in A$ such that $\Phi(\bar{x}, y) \geq \inf_{x\in A} \Phi(x, x)$ for all $y \in A$.

Exercise 1.11 Let A be a compact convex subset of a topological vector space, $\gamma \in \mathbb{R}$, and $\phi : A \times A \to \mathbb{R}$. Assume that:

(a) $\phi \in -\mathscr{A}^2_{(-\alpha)-pr.qm}$ and ϕ is α-transfer lower semicontinuous in the first argument on A.
(b) $\phi \in \mathscr{A}^1_{\alpha-pr.qm}$ and ϕ is α-transfer upper semicontinuous in the second argument on A.

Prove that ϕ has a saddle point.

Exercise 1.12 Let A be a nonempty set, $\alpha \in \mathbb{R}$, and $\Phi : A \times A \to \mathbb{R}$ be a bifunction.

(a) Φ is said to be *α-cyclically anti-monotone* if, for any $x^1, x^2, \ldots, x^m \in A$, we have $\Phi(x^1, x^2) + \Phi(x^2, x^3) + \ldots + \Phi(x^{m-1}, x^m) + \Phi(x^m, x^1) \geq \alpha$.
(b) Φ is called *α-cyclically anti-quasimonotone* if, for any $x^1, x^2, \ldots, x^m \in A$, there exists an $i \in \{1, 2, \ldots, m\}$ such that $\Phi(x^i, x^{i+1}) \geq \alpha$, where $x^{m+1} := x^1$.

By $\mathscr{A}_{\alpha-cam}$ ($\mathscr{A}_{\alpha-caqm}$, resp.), we denote the class of the α-cyclically anti-monotone bifunctions (the class of the α-cyclically anti-quasimonotone bifunctions, resp.). If $-\Phi \in \mathscr{A}_{\alpha-cam}$ ($-\Phi \in \mathscr{A}_{\alpha-caqm}$, resp.), then Φ is called *α-cyclically monotone* (*α-cyclically quasimonotone*, resp.).

Assume that A is a compact subset of a topological space. Prove the following:

(a) If $\Phi \in \mathscr{A}_{caqm}$ and Φ is 0-transfer upper semicontinuous in the first argument on A, then *(EP)* has solutions.
(b) If $\Phi \in \mathscr{A}_{\alpha-caqm}$ and Φ is α-transfer upper semicontinuous in the first argument on A, then there exists $\bar{x} \in A$ such that $\Phi(\bar{x}, y) \geq \alpha$ for all $y \in A$.

Exercise 1.13 Let A be a convex subset of a topological vector space. Prove that $\mathscr{A}^i_{st.qlc} \cap \mathscr{A}_{caqm} \subset \mathscr{A}^i_{pr.qm}$.

Exercise 1.14 Let C, D be compact subsets of topological spaces and $\phi : C \times D \to \mathbb{R}$. Assume that for all $\lambda > \phi_*$ sufficiently close to ϕ_*, $\mathrm{lev}_{\leq\lambda}\phi(\cdot, y)$ and $\mathrm{lev}_{\geq\lambda}\phi(x, \cdot)$

are closed in C and D for each $(x, y) \in C \times D$, and the following condition is satisfied:

(∗) For any $(x_1, y_1), (x_2, y_2), \ldots, (x_m, y_m) \in C \times D$ $(m \geq 2)$, there exists $i \in \{1, 2, \ldots, m\}$ such that either $\phi(x_i, y_{i+1}) \leq \lambda$ or $\phi(x_{i+1}, y_i) \geq \lambda$, where $(x_{m+1}, y_{m+1}) := (x_1, y_1)$).

Show that the minimax equality (1.1) holds for ϕ.

Exercise 1.15 Let A be a compact subset of a topological space, $\alpha \in \mathbb{R}$, and $\phi : A \times A \to \mathbb{R}$. Assume that ϕ is both α-cyclically quasimonotone and α-cyclically anti-quasimonotone, and ϕ is both α-transfer lower semicontinuous in the first argument and α-transfer upper semicontinuous in the second argument. Prove that ϕ has a saddle point.

Exercise 1.16 Let C, D be compact subsets of topological spaces, and $\phi : C \times D \to \mathbb{R}$. Consider the existence of saddle points of ϕ in the following cases:

(a) $\phi(x, y) = \varphi(x)\xi(y)$ for all $x, y \in C \times D$, where $\varphi : C \to \mathbb{R}$, $\xi : D \to (0, +\infty)$ are continuous functions.
(b) $\phi(x, y) = \frac{\varphi(x)}{\varphi(x)+\xi(y)}$, where $\varphi : C \to \mathbb{R}$ and $\xi : D \to \mathbb{R}$ are continuous functions such that either $\min_{x \in X} \xi(x) + \min_{y \in Y} \varphi(y) > 0$ or $\max_{x \in X} \xi(x) + \max_{y \in Y} \varphi(y) < 0$.
(c) $\phi(x, y) = \sum_{i=0}^{n} (\varphi(x)\xi(y))^{2i+1}$ for all $x, y \in X \times Y$, where $\varphi : C \to \mathbb{R}$, $\xi : D \to (0, +\infty)$ are continuous functions.
(d) $C \equiv D$, $\phi(x, y) = \varphi(\Omega(x, y) + \alpha)$, where $\varphi : \mathbb{R} \to \mathbb{R}$ is strictly increasing and continuous such that $\varphi(0) \geq 0$ (for instance, $\varphi(t) = t^{2k+1}$ or $\varphi(t) = e^{kt} - 1$), $\Omega : C \times C \to \mathbb{R}$ is both cyclically quasimonotone and cyclically anti-quasimonotone such that $\Omega(x, \cdot)$ is upper semicontinuous and $\Omega(\cdot, y)$ is lower semicontinuous for each $(x, y) \in C \times C$ (for instance, $\Omega(x, y) = \xi(x) - \xi(y)$, $\xi : C \to \mathbb{R}$ is lower semicontinuous), and $\alpha \in \mathbb{R}$.

1.5 Commentary

Because the solution set of (EP) is the intersection of a family of level sets of Φ, the nonempty intersection theorems and equivalent results such as many fixed-point theorems, coincidence-point theorems, etc. are crucial tools in studying the existence of solutions of (EP). The results in this chapter are presented in two sections based on whether or not the employed tools require the convexity structure.

In Sect. 1.2, existence results are established under two groups of assumptions, the first one being topological in nature (the compactness of A and the closedness of level sets of Φ) and the second group being algebraic (the convexity of A and the generalized convexity of Φ). Brouwer's fixed-point theorem and the finite intersection principle are the main tools in this subsection. The first basic result to use both these groups of assumptions was obtained by Fan [196]: *problem* (EP) *has solutions under the following assumptions:*

(a) *A is compact and convex.*
(b) $\Phi(x,x) \geq 0$ *for all* $x \in A$.
(c) $\Phi(\cdot, y)$ *is upper semicontinuous for each* $y \in A$ *and* $\Phi(x,\cdot)$ *is convex for each* $x \in A$.

Fan proved the above result using his/her KKM lemma [195] which is equivalent to Brouwer's fixed-point theorem (this KKM lemma is an infinite-dimensional extension of the classical KKM theorem [301] by Knaster, Kuratowshi, and Mazurkiewicz, which is the starting point of KKM theory, including various theorems of the KKM type; see, for example, [132–135, 289, 292, 316, 332, 335, 386]). Many researchers have weakened assumptions (a) and (b) above to obtain stronger results, by replacing the compactness of A by coercivity conditions and the convexity of $\Phi(x,\cdot)$ by generalized convexity conditions (e.g., [98, 100, 102, 200, 202, 203, 234, 261, 262, 309, 473]). Moreover, some coercivity and generalized convexity conditions are satisfied under monotonicity properties of Φ. Hence, some assumptions on monotonicity have also been considered (e.g., [94, 102, 108, 130, 131, 193, 194, 203, 233]). These concepts and their relationships are systematized and refined in Definitions 1.19, 1.21 1.25, and 1.27 and Proposition 1.29.

In studies of existence of solutions, one main technique (which originated from Minty [366]) is as follows. First, one searches for conditions ensuring that the solutions of the dual problem (DEP) are also solutions of (EP). Next, one restates the obtained existence conditions for (EP) with $\Phi'(x,y) = -\Phi(y,x)$ to get existence conditions for (DEP). By combining the conditions in the two steps above, we obtain new results for (EP). This technique is demonstrated in Theorems 1.49–1.50.

In Sect. 1.3, the imposed assumptions are purely topological, with no connection to linear or (generalized) convexity structures. Regarding the approaches in this section, we make the following remarks. First, because the model (EP) generalizes the minimization problem, it is natural to extend existence results for the minimization problem to (EP). In [108], Blum and Oettli extended the result of Takahashi [423] (Corollary 1.75 in this section) to a bifunction Φ satisfying the triangle inequality property (Theorem 1.74 here) on complete metric spaces. Some other authors (e.g., [251], [95], and [337]) generalized the famous Ekeland variational principle to (EP) and existence results for (EP) were obtained as a consequence. However, those results required some strict conditions such as completeness of spaces or lower boundedness of $\Phi(x,\cdot)$. Following an approach in Weierstrass' minimization theorem in [125], an existence result has been obtained without these strict conditions (Theorem 1.63 here). In all the abovementioned results, the triangle inequality property of Φ, a relatively restrictive property, was needed. Some attempts to develop existence results for (EP) for broader classes of bifunctions can be found in [123] and [291]. Section 1.3.2 presents the results obtained in [291]. The approach in Sect. 1.3.3 comes from Horvarth [135], replacing a convex hull by an image of a simplex through a continuous map. The concept of KKM-structure was introduced in [287] and [290] to replace the convexity structures. The approach to avoid linear structures in Sect. 1.3.4 is originated in Wu [456] with connectedness conditions replacing convexity conditions.

Chapter 2
Existence of Solutions to Scalar Quasiequilibrium Problems

Abstract There are many problems described by an inequality of a bifunction involving constraints that depend on the decision variable, such as quasioptimization problems, quasivariational inequalities, generalized Nash equilibria, traffic networks with elastic demands, etc. To deeply study these problems within a common framework, they are merged into a general problem, called quasiequilibrium problem. This chapter deals with the quasiequilibrium problem (QEP) and results on existence of its solutions. Section 2.1 introduces some basic notions and properties of semicontinuity for set-valued mappings which are necessary to state problems and to establish the existence results. Section 2.2 presents the definition of the quasiequilibrium problem and several related problems. Section 2.3 provides some fixed-point theorems for set-valued mappings which are used to establish existence results for (QEP). Section 2.4 presents conditions for the existence of solutions of (QEP). Section 2.5 discusses some set-valued versions of equilibrium/quasiequilibrium problems, where objective bifunctions are set-valued bifunctions. Section 2.6 contains applications of the results of Sects. 2.4–2.5 to several particular cases of (QEP) including Browder variational inclusions.

2.1 Continuity of Set-Valued Mappings

Let X, Y be nonempty sets. A SET-VALUED MAPPING S from X to Y is a correspondence that maps a point in $x \in X$ to a subset $S(x)$ of Y. We denote a set-valued mapping by $S : X \rightrightarrows Y$. For each $x \in X$, the set $S(x)$ is called the IMAGE of x under S. The DOMAIN, RANGE, and GRAPH of S are defined, resp., as

$$\mathrm{dom}S = \{x \in X \mid S(x) \neq \emptyset\},$$

$$\mathrm{rang}S = S(X) = \bigcup_{x \in \mathrm{dom}S} S(x),$$

$$\mathrm{gph}S = \{(x, y) \in X \times Y \mid y \in S(x)\}.$$

L. Q. Anh et al., *Equilibrium Problems*, Forum for Interdisciplinary Mathematics, https://doi.org/10.1007/978-3-031-89003-1_2

For a nonempty subset A of X, the image of A under S is the set

$$S(A) = \bigcup_{x \in A} S(x).$$

When $A = \emptyset$, we write $S(\emptyset) = \emptyset$. For each $y \in Y$, the set $S^{-1}(y) := \{x \in X \mid y \in S(x)\}$ is called an inverse image of y under S. The set-valued mapping $S^{-1} : Y \rightrightarrows X$ that associates with any $y \in Y$ the subset $S^{-1}(y)$ of X is called the INVERSE MAPPING of S. For a set $B \subset Y$, the STRONG INVERSE IMAGE $S^{+}(B)$ and the WEAK INVERSE IMAGE $S^{-}(B)$ of B under S are defined by

$$S^{+}(B) := \{x \in X \mid S(x) \subset B\}$$

and

$$S^{-}(B) := \{x \in X \mid S(x) \cap B \neq \emptyset\}.$$

Definition 2.1 (Closed and open mappings) Let X, Y be topological spaces. The set-valued mapping S: $X \rightrightarrows Y$ is said to be CLOSED at $x \in X$ if for any net $\{(x^{\alpha}, y^{\alpha})\} \subset \mathrm{gph}S$ converging to $(x, y) \in X \times Y$, we have $y \in S(x)$. S is called CLOSED if it is closed at every $x \in X$. It is clear that S is closed if and only if its graph, gph$S$, is a closed subset of $X \times Y$. S is called OPEN if its graph is an open subset of $X \times Y$.

An immediate consequence of Definition 2.1 is that if S is closed (open, resp.), then the images of S are closed (open, resp.) in Y and the inverse images of S are closed (open, resp.) in X.

Definition 2.2 (Types of continuous mappings) Let X, Y be topological spaces and S: $X \rightrightarrows Y$ be a set-valued mapping.

(a) S is called UPPER SEMICONTINUOUS at x if for every open subset V in Y satisfying $V \supset S(x)$, there is a neighborhood U of x such that $V \supset S(x')$ for all $x' \in U$.
(b) S is called LOWER SEMICONTINUOUS at x if for every open subset V in Y satisfying $V \cap S(x) \neq \emptyset$), there is a neighborhood U of x such that $V \cap S(x') \neq \emptyset$ for all $x' \in U$.
(c) S is called CONTINUOUS at x if S is both upper semicontinuous and lower semicontinuous at x.
(d) S is called upper semicontinuous on $A \subset X$ if S is upper semicontinuous at all $x \in A$. If $A = \mathrm{dom}S$ we omit "on A." We adopt this convention for each property of a set-valued mapping.

The following propositions provide characterizations of upper semicontinuity and lower semicontinuity, resp. They immediately follow from Definition 2.2.

Proposition 2.3 *Let X, Y be topological spaces and $S\colon X \rightrightarrows Y$ be a set-valued mapping. The following are equivalent:*

(a) *S is upper semicontinuous.*
(b) *For every open set $V \subset Y$, $S^+(V)$ is open.*
(c) *For every closed set $C \subset Y$, $S^-(C)$ is closed.*
(d) *For any net $\{x^\alpha\}_{\alpha \in (I,\succeq)} \subset X$ converging to $x \in X$ and any open set $V \subset Y$ such that $S(x) \subset V$, there exists $\alpha^0 \in I$ such that $S(x^\alpha) \subset V$ for all $\alpha \succeq \alpha^0$.*

Proposition 2.4 *Let X, Y be topological spaces and $S\colon X \rightrightarrows Y$ be a set-valued mapping. The following are equivalent:*

(a) *S is lower semicontinuous.*
(b) *For every open set $V \subset Y$, $S^-(V)$ is open.*
(c) *For every closed set $C \subset Y$, $S^+(C)$ is closed.*
(d) *For any net $\{x^\alpha\}_{\alpha \in (I,\succeq)} \subset X$ converging to $x \in X$ and any open set $V \subset Y$ such that $S(x) \cap V \neq \emptyset$, there exists $\alpha^0 \in I$ such that $S(x^\alpha) \cap V \neq \emptyset$ for all $\alpha \succeq \alpha^0$.*
(d) *For any net $\{x^\alpha\}_{\alpha \in (I,\succeq)} \subset X$ converging to $x \in X$ and any $y \in S(x)$, there exists a net $\{y^\alpha\}_{\alpha \in (I,\succeq)} \subset Y$ such that $y^\alpha \in S(x^\alpha)$ for all α and $\{y^\alpha\}$ converges to y.*

The following result gives another characterization of upper semicontinuity.

Proposition 2.5 *Let X, Y be topological spaces and $S\colon X \rightrightarrows Y$ be a set-valued mapping. Then, S is upper semicontinuous and has compact images if and only if for any net $\{x^\alpha\}_{\alpha \in (I,\succeq)} \subset X$ converging to $x \in X$ and any net $\{y^\alpha\}_{\alpha \in (I,\succeq)} \subset Y$ with $y^\alpha \in S(x^\alpha)$ for all α, there exists a subnet $\{y^\beta\}_{\alpha \in (J,\succeq)}$ of $\{y^\alpha\}_{\alpha \in (I,\succeq)}$ such that $y^\beta \to y \in S(x)$. Consequently, if A is a compact subset of X and S is upper semicontinuous with compact images, then $S(A)$ is compact.*

Proof *Necessity*. Suppose to the contrary that there exists a net $\{x^\alpha\}_{\alpha \in (I,\succeq)} \subset X$ converging to $x \in X$ and a net $\{y^\alpha\}_{\alpha \in (I,\succeq)} \subset Y$ with $y^\alpha \in S(x^\alpha)$ for all α such that for all $y \in S(x)$, $\{y^\alpha\}_{\alpha \in (I,\succeq)}$ does not have any subnet converging to y. Then, for each $y \in S(x)$ we can find an open neighborhood $V(y)$ of y and $\alpha_0(y)$ such that $y^\alpha \notin V(y)$ for all $\alpha \succeq \alpha^0(y)$. The family $\{V(y)\}_{y \in S(x)}$ forms an open cover of the compact set $S(x)$. So there exists a finite subcover $\{V(y_k)\}_{k=1}^n$. Set $V = \bigcup_{k=1}^n V(y_k) \supset S(x)$. Clearly, there exists $\alpha^1 \in I$ such that $y^\alpha \notin V$ for all $\alpha \succeq \alpha^1$. On the other hand, as S is upper semicontinuous, we can find an open neighborhood U of x such that $S(U) \subset V$. Because $x^\alpha \to x$, we have $x^\alpha \in U$ for all $\alpha \succeq \alpha^2 \succeq \alpha^1$ and so $y^\alpha \in V$ for all $\alpha \succeq \alpha^2 \succeq \alpha^1$, a contradiction.

Sufficiency. It is clear that the hypothesis implies that $S(x)$ is compact for each $x \in X$. Let $\{x^\alpha\}_{\alpha \in (I,\succeq)}$ be a net in X such that $x^\alpha \to x \in X$ and $V \subset Y$ be an open set such that $S(x) \subset V$. Assume that for each $\alpha \in I$, there is a $\beta \in I$ with $\beta \succeq \alpha$ such that $S(x^\beta) \not\subset V$. Then, there exist a subnet $\{x^\beta\}_{\beta \in (J,\succeq)}$ of $\{x^\alpha\}_{\alpha \in (I,\succeq)}$ and a net $\{y^\beta\}_{\beta \in (J,\succeq)}$ with $y^\beta \in S(x^\beta)$ and $y^\beta \in Y \setminus V$ for all $\beta \in J$. By the hypothesis, there exists a subnet $\{y^\gamma\}_{\gamma \in (L,\succeq)}$ of $\{y^\beta\}_{\beta \in (J,\succeq)}$ with $y^\gamma \in S(x^\gamma)$ such that $y^\gamma \to y \in S(x) \subset V$. Moreover, $\{y^\gamma\}_{\gamma\in(L,\succeq)} \subset Y \setminus V$ and $Y \setminus V$ is closed; we have that $y \in Y \setminus V$, a contradiction. Thus, there must exist $\alpha^0 \in I$ such that $S(x^\beta) \subset V$ for all $\alpha \succeq \alpha^0$. This and Proposition 2.3 show that S is upper semicontinuous.

Finally, assume that A is a compact subset of X and S is upper semicontinuous with compact images. Let $\{y^\alpha\}_{\alpha\in(I,\succeq)}$ be a net in $S(A)$. Then, there exists a net $\{x^\alpha\}_{\alpha\in(I,\succeq)} \subset X$ such that $y^\alpha \in S(x^\alpha)$ for all α. Since A is compact, there exists a subnet $\{x^\beta\}_{\beta\in(J,\succeq)}$ of $\{x^\alpha\}_{\alpha\in(I,\succeq)}$ such that $x^\beta \to x \in A$. The above part tells us that the net $\{y^\beta\}_{\beta\in(J,\succeq)}$ has a subnet $\{y^\gamma\}_{\gamma\in(L,\succeq)}$ converging to $y \in S(x) \subset S(A)$. Thus, $S(A)$ is compact. □

The next proposition states a relationship between the upper semicontinuity and the closedness of a set-valued mapping. Let X, Y be topological spaces. We say that a set-valued mapping $S : X \rightrightarrows Y$ is LOCALLY COMPACT if for every $x \in X$, there is an open neighborhood U of x such that $\mathrm{cl}(S(U))$ is a compact subset of Y. In particular, S is locally compact if either $S(X)$ or Y is compact.

Proposition 2.6 *Let X, Y be topological spaces and S: $X \rightrightarrows Y$ be a set-valued mapping.*

(a) *If Y is a regular topological space and S is upper semicontinuous and has closed images, then S is closed.*
(b) *If Y is a Hausdorff topological space and S is upper semicontinuous and has compact images, then S is closed.*
(c) *If S is closed and locally compact, then S is upper semicontinuous.*

Proof To prove the closedness of S in (a) and (b), we take any net $\{(x^\alpha, y^\alpha)\}_{\alpha\in(I,\succeq)} \subset \mathrm{gph}S$ converging to $(x, y) \in X \times Y$ and show that $(x, y) \in \mathrm{gph}S$, i.e., $y \in S(x)$.

(a) Suppose that $y \notin S(x)$. Since $S(x)$ is closed and Y is regular, there exist two open sets V, $V' \subset Y$ such that $y \in V$, $S(x) \subset V'$ and $V \cap V' = \emptyset$. By the upper semicontinuity of S, there exists $\alpha^0 \in I$ such that $y^\alpha \in S(x^\alpha) \subset V' \subset Y \setminus V$ for all $\alpha \succeq \alpha^0$ (see Proposition 2.3). This contradicts the fact that $y^\alpha \to y \in V$. Thus, $(x, y) \in \mathrm{gph}S$.
(b) Suppose that $y \notin S(x)$. By Proposition 2.5, there exists a subnet $\{y^\beta\}_{\alpha\in(J,\succeq)}$ of $\{y^\alpha\}_{\alpha\in(I,\succeq)}$ such that $y^\beta \to y' \in S(x)$. Since Y is Hausdorff, we can find open sets V, $V' \subset Y$ such that $y \in V$, $y' \subset V'$ and $V \cap V' = \emptyset$. As $y^\alpha \to y$, $y^\beta \to y$. Hence, there exists $\beta^1 \in J$ such that $y^\beta \in V$ for all $\beta \succeq \beta^1$. Moreover, since $y^\beta \to y'$, there exists $\beta^2 \in J$ such that $y^\beta \in V'$ for all $\beta \succeq \beta^2$. Letting $\beta^3 \in J$ be such that $\beta^3 \succeq \beta^1$ and $\beta^3 \succeq \beta^2$, we have that $y^\beta \in V \cap V' = \emptyset$ for all $\beta \succeq \beta^3$, a contradiction.
(c) To prove the upper semicontinuity of S, by Proposition 2.3, we need to show that if a net $\{x^\alpha\}_{\alpha\in(I,\succeq)}$ in X converges to $x \in X$ and a set $V \subset Y$ is open such that $S(x) \subset V$, then there exists $\alpha^0 \in I$ such that $S(x^\alpha) \subset V$ for all $\alpha \succeq \alpha^0$. Suppose the opposite. Then, there exist a subnet $\{x^\beta\}_{\beta\in(J,\succeq)}$ of $\{x^\alpha\}_{\alpha\in(I,\succeq)}$ and a net $\{y^\beta\}_{\beta\in(J,\succeq)}$ with $y^\beta \in S(x^\beta)$ and $y^\beta \in Y \setminus V$ for all $\beta \in J$. When S is

locally compact, there exists an open neighborhood U of x such that $\mathrm{cl}(S(U))$ is compact. Since $x^\beta \to x$, $x^\beta \in U$ for all $\beta \succeq \beta^0$ with some $\beta^0 \in J$. Then, $y^\beta \in \mathrm{cl}(S(U))$ for all $\beta \succeq \beta^0$, and therefore, we can assume that $y^\beta \to y \in Y$. Because $(x^\beta, y^\beta) \to (x, y)$, S is closed, and $Y \setminus V$ is closed; we have that $y \in S(x) \subset V$ and $y \in Y \setminus V$, a contradiction.

□

Proposition 2.7 (Lower semicontinuity of intersections of maps) *Let X, Y be topological spaces and S, H: $X \rightrightarrows Y$. If S is lower semicontinuous and H has an open graph, then, F: $X \rightrightarrows Y$, defined by $F(x) = (S \cap H)(x) := S(x) \cap H(x)$ is lower semicontinuous.*

Proof Let $x \in X$ and V be an open set in Y such that $F(x) \cap V \neq \emptyset$ and let $y \in F(x) \cap V$. Then, $(x, y) \in \mathrm{gph}H \cap X \times V$. Since $\mathrm{gph}H$ is open, $\mathrm{gph}H \cap X \times V$ is open in $X \times Y$. Hence, there exists open sets $U(x) \subset X$ and $V(y) \subset Y$ such that $x \in U(x)$, $y \in V(y)$ and $U(x) \times V(y) \subset \mathrm{gph}H \cap X \times V$. Because $y \in S(x) \cap V(y)$ and S is lower semicontinuous, there exists an open set $U'(x) \subset X$ such that $x \in U'(x)$ and $S(x') \cap V(y) \neq \emptyset$ for all $x' \in U'(x)$. Set $U^*(x) = U(x) \cap U'(x)$. Then, $U^*(x) \times V(y) \subset \mathrm{gph}H \cap X \times V$ and $S(x') \cap V(y) \neq \emptyset$ for all $x' \in U^*(x)$, implying $\emptyset \neq S(x') \cap H(x') \cap V = F(x') \cap V$ for all $x' \in U^*(x)$. This proves the lower semicontinuity of F. □

Proposition 2.8 (Lower semicontinuity of convex hulls) *Let X be a topological space and Y be a normed space. If $S : X \rightrightarrows Y$ is lower semicontinuous, then the set-valued mapping* $\mathrm{conv}S : X \rightrightarrows Y$*, defined by* $\mathrm{conv}S(x) = \mathrm{conv}(S(x))$ *for all $x \in X$, is lower semicontinuous*

Proof Let $V \subset Y$ be open. We show that the set $(\mathrm{conv}S)^-(V) = \{x \in X \mid \mathrm{conv}S(x) \cap V \neq \emptyset\}$ is open. Let $x \in (\mathrm{conv}S)^-(V)$ and $y \in \mathrm{conv}S(x) \cap V$. Then, there exist $\{y^1, y^2, \dots, y^n\} \subset S(x)$ and $\{t_1, t_2, \dots, t_n\} \subset [0, 1]$ such that $y = \sum_{i=1}^n t_i y^i$. Since $y \in V$, there exists $\epsilon > 0$ such that $\mathbb{B}(y, \epsilon) := \{y \in Y \mid \|y' - y\| < \epsilon\} \subset V$. For each i, by the lower semicontinuity of S and the fact that $S(x) \cap \mathbb{B}(y^i, \epsilon) \neq \emptyset$, there exists an open neighborhood $U_i(x) \subset X$ of x such that $S(x') \cap \mathbb{B}(y^i, \epsilon) \neq \emptyset$ for all $x' \in U_i(x)$. Set $U(x) = \bigcap_{i=1}^n U_i(x)$. For $x' \in U(x)$, to show that $S(x') \cap V \neq \emptyset$, let $y'^i \in S(x') \cap \mathbb{B}(y^i, \epsilon)$ and $y' = \sum_{i=1}^n t_i y'^i$. We have $\|y' - y\| = \|\sum_{i=1}^n t_i y'^i - \sum_{i=1}^n t_i y^i\| \le \sum_{i=1}^n t_i \|y'^i - y^i\| < \epsilon$. Hence, $y' \in \mathrm{conv}S(x) \cap V$. Thus, $U(x) \subset (\mathrm{conv}S)^-(V)$ and so $(\mathrm{conv}S)^-(V)$ is open. □

Proposition 2.9 *Let X, Y be topological spaces and A a closed subset of X. Let $S : X \rightrightarrows Y$, $H : A \rightrightarrows Y$ be two lower semicontinuous set-valued mappings such that, for every $x \in A$, $H(x) \subset S(x)$. Let F be the set-valued mapping from X into Y defined for all $x \in X$ by*

$$F(x) = \begin{cases} H(x) \text{ if } x \in A, \\ S(x) \;\text{ if } x \in X \setminus A. \end{cases}$$

Then, F is lower semicontinuous.

Proof Let $x \in X$ and $V \subset Y$ be open such that $F(x) \cap V \neq \emptyset$. If $x \in A$, then $H(x) \cap V \neq \emptyset$ and $S(x) \cap V \neq \emptyset$ (as $H(x) \subset S(x)$). By the lower semicontinuity of H and S, there exist open neighborhoods $U'(x) \subset X$ and $U''(x)$ of x such that $H(x') \cap V \neq \emptyset$ for all $x' \in U'(x) \cap A$ and $S(x') \cap V \neq \emptyset$ for all $x' \in U''(x)$. Setting $U(x) = U'(x) \cap U''(x)$, we have $F(x') \cap V \neq \emptyset$ for all $x' \in U(x)$. If $x \notin A$, then $S(x) \cap V \neq \emptyset$. By the lower semicontinuity of S, there exists an open neighborhood $U'(x) \subset X$ of x such that $S(x') \cap V \neq \emptyset$ for all $x' \in U'(x)$. Taking $U(x) = U'(x) \cap (X \setminus A)$, we have $F(x') \cap V = S(x') \cap V \neq \emptyset$ for all $x' \in U(x)$. Thus, we can always find an open neighborhood $U(x)$ of x such that $F(x') \cap V \neq \emptyset$ for all $x' \in U(x)$, and therefore, F is lower semicontinuous. □

2.2 Scalar Quasiequilibrium Problems

A QUASIEQUILIBRIUM PROBLEM is a constrained equilibrium problem in which the constraint set depends on the point considered as a candidate solution. The problem reads as

$$\text{(QEP)} \qquad \text{find } \bar{x} \in A \text{ such that } \bar{x} \in \mathscr{D}(\bar{x}) \text{ and } \Phi(\bar{x}, y) \geq 0 \text{ for all } y \in \mathscr{D}(\bar{x}),$$

where A is a nonempty set, $\mathscr{D} : A \rightrightarrows A$ is a set-valued mapping with nonempty images, and $\Phi : A \times A \to \mathbb{R}$ is a bifunction. In this problem, Φ is the objective bifunction, and set-valued mapping $\mathscr{D}$ describes the dependence of the constraint set on the decision variable and is called the constraint mapping. By Sol(QEP), we denote the solution set of (QEP).

Remark 2.10

(a) It is clear that when $\mathscr{D}(x) = A$ for all $x \in A$, (QEP) becomes the equilibrium problem (EP).
(b) For each $y \in A$, we set

$$T(y) = [A \setminus \mathscr{D}^{-1}(y)] \cup [\{x \in A \mid x \in \mathscr{D}(x)\} \cap \operatorname{lev}_{\geq} \Phi(\cdot, y)]$$

and define $\Phi_T : A \times A \to \mathbb{R}$ by

$$\Phi_T(x, y) = \begin{cases} 0 & \text{if } x \in T(y), \\ -1 & \text{otherwise.} \end{cases}$$

Then, $\bar{x}$ is a solution of (QEP) if and only if $\bar{x}$ is a solution of (EP_{Φ_T}), the equilibrium problem defined by Φ_T. Indeed, if $\bar{x} \in$ Sol(QEP), then for each $y \in A$, either $y \in \mathscr{D}(\bar{x})$ or $y \notin \mathscr{D}(\bar{x})$. If $y \in \mathscr{D}(\bar{x})$, then $\bar{x} \in \{x \in A \mid x \in \mathscr{D}(x)\} \cap \operatorname{lev}_{\geq} \Phi(\cdot, y) \subset T(y)$, implying $\Phi_T(\bar{x}, y) \geq 0$. If $y \notin \mathscr{D}(\bar{x})$, then $\bar{x} \in A \setminus \mathscr{D}^{-1}(y) \subset T(y)$, and hence, $\Phi_T(\bar{x}, y) \geq 0$. Thus, $\bar{x} \in (\text{EP}_{\Phi_T})$. Conversely, assume that $\bar{x} \in \text{Sol}(\text{EP}_{\Phi_T})$. Then, $\bar{x} \in T(y)$ for all $y \in A$. First, we show that $\bar{x} \in \mathscr{D}(\bar{x})$. Indeed, if not, one would have $\bar{x} \in A \setminus \mathscr{D}^{-1}(y)$ for all $y \in A$, which implies that $\mathscr{D}(\bar{x}) = \emptyset$, a contradiction. Now, let $y \in \mathscr{D}(\bar{x})$. Then, $\bar{x} \notin A \setminus \mathscr{D}^{-1}(y)$. By the definition of $T(y)$, one has that $\bar{x} \in \text{lev}_{\geq}\Phi(\cdot, y)$, i.e., $\Phi(\bar{x}, y) \geq 0$. Thus, $\bar{x} \in \text{Sol(QEP)}$.

From the above remark, we can say that the class of quasiequilibrium problems can be considered as a subclass of the class of equilibrium problems. Therefore, we can use results in Chap. 1 to investigate the existence of solutions of (QEP). However, in most cases, the calculation of $T(y)$ and Φ_T is not easy because it involves the calculation of the set of fixed points of $\mathscr{D}$. Hence, we need to find more specific conditions on A, $\mathscr{D}$, and Φ to ensure the solvability of (QEP). Moreover, the main reason for studying (QEP) is that this problem contains some important problems. Here are several special cases of (QEP).

(a) **Quasioptimization**. Given a real-valued function $\varphi : A \to \mathbb{R} \cup \{+\infty\}$ and a set-valued map $\mathscr{D} : A \rightrightarrows A$, the QUASIOPTIMIZATION PROBLEM (QOP) is described by

$$\text{find } \bar{x} \in \mathscr{D}(\bar{x}) \text{ such that } \min_{x \in \mathscr{D}(\bar{x})} \varphi(x) = \varphi(\bar{x}).$$

Problem (QOP) is not a standard minimization problem since the constraint set depends on the solution; it is a particular case of quasiequilibrium problem with $\Phi(x, y) = \varphi(y) - \varphi(x)$.

(b) **Quasivariational Inequalities**. Let A, T be as in Sect. 1.1.1 (c) and $\mathscr{D} : A \rightrightarrows A$. The problem of

$$\text{finding } \bar{x} \in \mathscr{D}(\bar{x}) \text{ such that } \langle T(\bar{x}), y - \bar{x} \rangle \geq 0 \text{ for } \text{ all } y \in \mathscr{D}(\bar{x})$$

is called a QUASIVARIATIONAL INEQUALITY (QVI). This problem is a special case of (QEP) with $\Phi(x, y) = \langle T(x), y - x \rangle$ for all $x, y \in A$.

Now, let $T : A \rightrightarrows X^*$ be a set-valued mapping. Consider the following set-valued quasivariational inequalities:

(SwQVI) Find $\bar{x} \in \mathscr{D}(\bar{x})$ such that there exists $x^* \in T(\bar{x})$ satisfying

$$\langle x^*, y - \bar{x} \rangle \geq 0 \text{ for } \text{ all } y \in \mathscr{D}(\bar{x}).$$

(SQVI) Find $\bar{x} \in \mathscr{D}(\bar{x})$ such that for all $x^* \in T(\bar{x})$,

$$\langle x^*, y - \bar{x} \rangle \geq 0 \text{ for } \text{ all } y \in \mathscr{D}(\bar{x}).$$

We define bifunctions $\Phi_T^s, \Phi_T^i : A \times A \to \mathbb{R}$ by, resp.,

$$\Phi_T^s(x, y) = \sup_{x^* \in T(x)} \langle x^*, y - x \rangle,$$

$$\Phi^i_T(x, y) = \inf_{x^* \in T(x)} \langle x^*, y - x \rangle.$$

Φ^s_T and Φ^i_T are called *the representative bifunctions* of T. It is not hard to see that every solution of (QEP) associated with Φ^i_T is a solution of the generalized quasivariational inequality (SQVI).

Proposition 2.11 *Let A be a subset of a topological vector space X, $\mathscr{D} : A \rightrightarrows A$ and $T : A \rightrightarrows X^*$. Assume that $\bar{x}$ is a solution of (QEP) associated with Φ^s_T. Then, if $\mathscr{D}(\bar{x})$ is convex and compact and $T(\bar{x})$ is convex and weak* compact, then $\bar{x}$ is a solution of the set-valued quasivariational inequality (SwQVI).*

Proof Suppose that $\bar{x}$ is a solution of (QEP) associated with Φ^s_T, i.e., $\bar{x} \in \mathscr{D}(\bar{x})$ and $\Phi^s_T(\bar{x}, y) = \sup_{x^* \in T(\bar{x})} \langle x^*, y - \bar{x} \rangle \geq 0$ for all $y \in \mathscr{D}(\bar{x})$. As $\mathscr{D}(\bar{x})$ is convex and compact and $T(\bar{x})$ is convex and weak* compact, by Sion's minimax theorem (Corollary 1.55), $\max_{x^* \in T(\bar{x})} \min_{y \in \mathscr{D}(\bar{x})} \langle x^*, y - \bar{x} \rangle = \min_{y \in \mathscr{D}(\bar{x})} \Phi^s_T(\bar{x}, y) \geq 0$. Hence, there exists $x^* \in T(\bar{x})$ satisfying $\langle x^*, y - \bar{x} \rangle \geq 0$ for all $y \in \mathscr{D}(\bar{x})$, i.e., $\bar{x}$ is a solution of (SwQVI). □

(c) **Generalized Nash Equilibria**. With the same notations as in Sect. 1.1.1(d), consider the game $\mathscr{G}$ where each player i chooses strategy $x_i \in A_i$ in order to maximize his/her payoff $c_i(x_1, \ldots, x_i, \ldots, x_m)$. The possible strategies of the player i belong to a set $\mathscr{D}_i(x_{-i})$ (with $x_{-i} := (x_1, x_2, \ldots, x_{i-1}, x_{i+1}, \ldots, x_n) \in A_{-i} = \prod_{k \neq i} A_k$) which depends on the other players' strategies. A point $\bar{x} = (\bar{x}_1, \ldots, \bar{x}_m) \in A$ is called a GENERALIZED NASH EQUILIBRIUM of $\mathscr{G}$ if, for all $y_i \in \mathscr{D}_i(\bar{x}_{-i})$ and $i = 1, 2, \ldots, m$, we have

$$c_i(\bar{x}) \geq c_i(\bar{x}_{-i}, y_i).$$

Similarly to Sect. 1.1.1(d), it can be proved that a point $\bar{x} \in A$ is a generalized Nash equilibrium of $\mathscr{G}$ if and only if it is a solution of the problem (QEP) with Φ being the Nikaido-Isoda bifunction and $\mathscr{D}$ defined by $\mathscr{D}(x) := \mathscr{D}_1(x_{-1}) \times \mathscr{D}_2(x_{-2}) \times \ldots \times \mathscr{D}_m(x_{-m})$.

(d) **Traffic Network Problem with Elastic Demands**. With $\mathbf{N}$, $\mathbf{A}$, $\mathbf{W}$, c, d, $P = \bigcup_{w \in W} P_w$, $m = |P|$, x_p $(p \in P)$, $x = (x_1, \ldots, x_m) \in \mathbb{R}^m_+$, v_a, and δ_{ap}, T_p, T defined as in Sect. 1.1.4(f), we consider the traffic network $G = (\mathbf{N}, \mathbf{A}, \mathbf{W}, c, d)$ with the following phenomena of elasticity of the demand which may occur in dynamic settings when information on the state of the network is available. For each O/D pair w, the demand d_w is not fixed but depends upon the flow on the network. Then, the demand constraint on O/D pair w is $\sum_{p \in P_w} x_p = d_w(x)$. An equilibrium flow of $\mathscr{G}$ is defined as in Definition 1.6.

Let

$$A = \{x \in \mathbb{R}^m_+ \mid 0 \leq v_a \leq c_a \ \text{ for } \ a \in \mathbf{A}\}$$

be the set of path flow vectors satisfying capacity constraints of arcs and, for $x \in A$,

$$\mathscr{D}(x) = \{y \in A \mid \sum_{p \in P_w} y_p = d_w(x)\}.$$

Let $\Phi_{net} : A \times A \to \mathbb{R}$ be defined by

$$\Phi_{net}(x, y) := \langle T(x), y - x \rangle = \sum_{w \in \mathbf{W}} \sum_{l \in P_w} (y_l - x_l) T_l(x) \quad \text{for all } x, y \in A$$

and (QEP$_{\Phi_{net}}$) denote the quasiequilibrium problem defined by Φ_{net}.

Proposition 2.12 *If x is a solution of the problem* (QEP) *with $\Phi \equiv \Phi_{net}$, then x is an equilibrium flow of G.*

Proof The proof is the same as that of Proposition 1.7 with a slight modification as follows. Let $x \in A$ be a solution of (QEP). For ε, $y_{p,p'}$ and the condition (∗) as in the proof of Proposition 1.7, if (∗) holds, then $y_{p,p'}$ belongs to $\mathscr{D}(x)$. □

2.3 Fixed-Point Theorems

The existence of solutions of problem (QEP) is related to the existence of fixed points of the constraint mapping $\mathscr{D}$. Moreover, the proof of existence results is often based on transforming the problem into a fixed-point problem. Therefore, fixed-point theorems play an important role in the study of quasiequilibrium problem (QEP). This section presents some fixed-point theorems which will be used in this chapter. For a set-valued mapping $\mathscr{D} : A \rightrightarrows A$, a point $\bar{x} \in A$ is called a FIXED POINT of $\mathscr{D}$ if $\bar{x} \in \mathscr{D}(\bar{x})$.

Theorem 2.13 (Browder's fixed-point theorem) *Let A be a nonempty compact convex subset of a topological vector space X and $\mathscr{D} : A \rightrightarrows A$ be a set-valued mapping with convex values. If $A = \bigcup_{x \in A} \operatorname{int}\mathscr{D}^{-1}(x)$, then $\mathscr{D}$ has a fixed point.*

Proof Let $\Phi : A \times A \to \mathbb{R}$ be defined for $x, y \in A$ by

$$\Phi(x, y) = \begin{cases} 0 \text{ if } x \in A \setminus \operatorname{int}\mathscr{D}^{-1}(y), \\ -1 \text{ otherwise.} \end{cases}$$

Then, for each $x \in A$, by $A = \bigcup_{x \in A} \operatorname{int}\mathscr{D}^{-1}(x)$, there exists $y \in A$ such that $x \in \operatorname{int}\mathscr{D}^{-1}(y)$, i.e., $x \notin A \setminus \operatorname{int}\mathscr{D}^{-1}(y)$, implying $\Phi(x, y) = -1 < 0$. Hence, for this Φ, Sol(EP) $= \emptyset$. Moreover, $\operatorname{lev}_{\geq}\Phi(\cdot, y) = A \setminus \operatorname{int}\mathscr{D}^{-1}(y)$ is closed for each $y \in A$. It follows from Theorem 1.34 that $\Phi \notin \mathscr{A}^2_{pr.qm}$. Then, there exist $N = \{x^1, x^2, \ldots, x^m\} \subset A$ and $x \in \operatorname{conv}N$ such that $\Phi(x, x^i) < 0$ for all $i = 1, 2, \ldots, m$. By the definition of Φ, $x \in \operatorname{int}\mathscr{D}^{-1}(x^i) \subset \mathscr{D}^{-1}(x^i)$ for all

$i = 1, 2, \ldots, m$. Hence, $x^i \in \mathscr{D}(x)$ for all $i = 1, 2, \ldots, m$. Since $\mathscr{D}(x)$ is convex, $x \in \mathrm{conv}N \subset \mathscr{D}(x)$. Thus, x is a fixed point of $\mathscr{D}$. □

If $\mathscr{D}$ has nonempty images and $\mathscr{D}^{-1}(y)$ is open for each $y \in A$, then condition $A = \bigcup_{x\in A} \mathrm{int}\mathscr{D}^{-1}(x)$ in Theorem 2.13 is satisfied. As a consequence of Theorem 2.13, we obtain the Kakutani theorem.

Theorem 2.14 (Kakutani theorem) *Let A be a nonempty compact convex subset of a Hausdorff locally convex space X and $\mathscr{D} : A \rightrightarrows A$ be a set-valued mapping. If $\mathscr{D}$ is upper semicontinuous and nonempty-closed-convex valued, then $\mathscr{D}$ has a fixed point.*

Proof Let $\mathscr{U}$ be a basis of neighborhoods of the origin in X consisting of the open balanced convex neighborhoods. Let $W \in \mathscr{U}$. As $\mathscr{D}$ is upper semicontinuous, for each $x \in A$, there exists an open neighborhood U_x of the origin 0 in X such that $U_x \subset W$ and $\mathscr{D}((x + U_x) \cap A) \subset (\mathscr{D}(x) + W) \cap A$. Then, the family $\{(x + U_x) \cap A\}_{x\in A}$ is an open cover (in A) of A. Because A is compact, there exist $x^1, x^2, \ldots, x^m \in A$ such that $A = \bigcup_{i=1}^m ((x^i + U_{x^i}) \cap A)$. For each $x \in A$, set $I(x) = \{i \in \{1, 2, \ldots, m\} \mid x \in (x^i + U_{x^i}) \cap A\}$ and define $\mathscr{D}_W : A \rightrightarrows A$ by

$$\mathscr{D}_W(x) = \bigcap_{i\in I(x)} ((\mathscr{D}(x^i) + W) \cap A) \quad \text{for all } x \in A. \tag{2.1}$$

Clearly, $\emptyset \neq \mathscr{D}(x) \subset \mathscr{D}_W(x)$ and $\mathscr{D}_W(x)$ is convex for all $x \in A$. Moreover, for any $x \in A$, letting $O_x := \bigcap_{i\in I(x)}((x^i + U_{x^i}) \cap A)$ and $x' \in O_x$, one has $I(x) \subset I(x')$, and therefore, $\mathscr{D}_W(x) \subset \mathscr{D}_W(O_x) \subset \mathscr{D}_W(x)$. This implies that $\{x\} \times \mathscr{D}_W(x) \subset O_x \times \mathscr{D}_W(x) \subset \mathrm{gph}\mathscr{D}_W$. Thus, $\mathrm{gph}\mathscr{D}_W$ is open in $A \times A$, and hence, $\mathscr{D}_W^{-1}(y)$ is open in A for each $y \in A$. By Theorem 2.13, there exists $x_W \in A$ such that

$$\begin{aligned} x_W \in \mathscr{D}_W(x_W) &\subset \bigcap_{i\in I(x_W)} [(\mathscr{D}((x_W - U_{x^i}) \cap A) + W) \cap A] \\ &\subset (\mathscr{D}((x_W + W) \cap A) + W) \cap A. \end{aligned}$$

It follows that $\emptyset \neq \mathscr{D}((x_W + W) \cap A) \cap (x_W + W)$. Hence, there exist $x'_W \in A$ and $x''_W \in A$ such that $x''_W \in x_W + W$, $x'_W \in \mathscr{D}(x''_W)$ (i.e., $(x''_W, x'_W) \in \mathrm{gph}\mathscr{D}$), and $x'_W \in x_W + W$. Now, by the compactness of A, we may assume that the net $\{x_W\}_{W\in\mathscr{U}}$ converges to some $\bar{x} \in A$. Since X is Hausdorff, $\{x'_W\}_{W\in\mathscr{U}}$ and $\{x''_W\}_{W\in\mathscr{U}}$ also converge to $\bar{x}$. When $\mathscr{D}$ is upper semicontinuous and has compact images, $\mathrm{gph}\mathscr{D}$ is closed. Hence, $(\bar{x}, \bar{x}) \in \mathrm{gph}\mathscr{D}$, which means that $\bar{x}$ is a fixed point of $\mathscr{D}$. □

Corollary 2.15 (Schauder-Tychonoff theorem) *Every continuous function ϕ from a convex compact subset A of a Hausdorff locally convex space to A itself has a fixed point.*

Definition 2.16 (Selections of set-valued maps) Given a set-valued mapping $\mathscr{D}: X \rightrightarrows Y$, a single-valued mapping $s : X \to Y$ is called a SELECTION of $\mathscr{D}$ if $s(x) \in \mathscr{D}(x)$ for all $x \in X$.

The SELECTION PROBLEM is formulated as follows: what conditions on X, Y and $\mathscr{D}$ ensure that S admits a continuous selection? This problem was handled by Michael. A possible solution is indicated in the following result ([365], Theorem 3.1).

Theorem 2.17 (Michael selection theorem) *Every lower semicontinuous set-valued map $\mathscr{D}$ from a perfectly normal space to a separable Banach space with nonempty convex images admits a continuous selection.*

Here a PERFECTLY NORMAL SPACE is a topological space in which every two disjoint nonempty closed sets can be precisely separated by a continuous function from X to $\mathbb{R}$. Metric spaces are perfectly normal Hausdorff. As a consequence of Corollary 2.15 and Theorem 2.17, we have the following result.

Theorem 2.18 *Let A be a nonempty compact convex subset of a separable Banach space X and $\mathscr{D} : A \rightrightarrows A$ be a set-valued mapping. If $\mathscr{D}$ is lower semicontinuous and has nonempty and convex images, then $\mathscr{D}$ has a fixed point.*

Theorem 2.19 *Let A be a nonempty compact subset of a topological space X and $\mathscr{D} : A \rightrightarrows X$ be a set-valued map. Assume that $\mathscr{D}(y) \subset \mathscr{D}(x)$ for any $x \in A$ and $y \in \mathscr{D}(x) \cap A$, and $A = \bigcup_{x\in A} \text{int}\mathscr{D}^{-1}(x)$. Then, $\mathscr{D}$ has a fixed point.*

Proof Define $\Phi : A \times A \to \mathbb{R}$ for $x, y \in A$ by

$$\Phi(x, y) = \begin{cases} 0 \text{ if } x \in A \setminus \text{int}\mathscr{D}^{-1}(y), \\ -1 \text{ otherwise.} \end{cases}$$

Similar to the proof of Theorem 2.13, for this Φ, Sol(EP) $= \emptyset$ and $\text{lev}_{\geq}\Phi(\cdot, y) = A \setminus \text{int}\mathscr{D}^{-1}(y)$ is closed for each $y \in A$. It follows from Theorem 1.82 that $\Phi \notin \mathscr{A}_{caqm}$. Then, there exists $N = \{x^1, x^2, \ldots, x^m\} \subset A$ such that $\Phi(x^i, x^{i+1}) < 0$ for all $i = 1, 2, \ldots, m$ with $x^{m+1} := x^1$. By the definition of Φ, it follows that $x^1 \in \text{int}\mathscr{D}^{-1}(x^2) \subset \mathscr{D}^{-1}(x^2)$, $x^2 \in \text{int}\mathscr{D}^{-1}(x^3) \subset \mathscr{D}^{-1}(x^3), \ldots,$ $x^{m-1} \in \text{int}\mathscr{D}^{-1}(x^m) \subset \mathscr{D}^{-1}(x^m)$, and $x^m \in \text{int}\mathscr{D}^{-1}(x^1) \subset \mathscr{D}^{-1}(x^1)$. Hence,

$$x^2 \in \mathscr{D}(x^1), \ x^3 \in \mathscr{D}(x^2), \ \ldots, \ x^m \in \mathscr{D}(x^{m-1}), \text{ and } x^1 \in \mathscr{D}(x^m). \tag{2.2}$$

Since $\mathscr{D}(y) \subset \mathscr{D}(x)$ for any $x \in A$ and $y \in \mathscr{D}(x) \cap A$, (2.2) implies that $\mathscr{D}(x^1) \subset \mathscr{D}(x^m) \subset \mathscr{D}(x^{m-1}) \subset \ldots \subseteq \mathscr{D}(x^2) \subset \mathscr{D}(x^1)$. Hence, $\mathscr{D}(x^1) = \mathscr{D}(x^2) = \ldots = \mathscr{D}(x^m)$ and so each x^i, $i = 1, 2, \ldots, m$, is a fixed point of $\mathscr{D}$. □

2.4 Existence of Solutions

For quasiequilibrium problem (QEP), Fix$\mathscr{D}$ denotes the set of fixed points of $\mathscr{D}$. For each $y \in A$, we set

$$T(y) := [A \setminus \mathscr{D}^{-1}(y)] \cup [\mathrm{Fix}\mathscr{D} \cap \mathrm{lev}_{\geq}\Phi(\cdot, y)]$$

and define the bifunction $\Phi_T : A \times A \to \mathbb{R}$ by

$$\Phi_T(x, y) := \begin{cases} 0 \text{ if } x \in T(y), \\ -1 \text{ otherwise.} \end{cases}$$

When A is a subset of a topological space, let $\overline{\Phi}_T : A \times A \to \mathbb{R}$ be defined by

$$\overline{\Phi}_T(x, y) := \begin{cases} 0 \text{ if } x \in \mathrm{cl}(T(y)), \\ -1 \text{ otherwise.} \end{cases}$$

We consider the following equilibrium problems:

(EP_{Φ_T}) find $\bar{x} \in A$ such that $\Phi_T(\bar{x}, y) \geq 0$ for all $y \in A$;

$(\mathrm{EP}_{\overline{\Phi}_T})$ find $\bar{x} \in A$ such that $\overline{\Phi}_T(\bar{x}, y) \geq 0$ for all $y \in A$.

By $\mathrm{Sol}(\mathrm{EP}_{\Phi_T})$ and $\mathrm{Sol}(\mathrm{EP}_{\overline{\Phi}_T})$, we denote the solution sets of (EP_{Φ_T}) and $(\mathrm{EP}_{\overline{\Phi}_T})$, resp. By Remark 2.10, $\mathrm{Sol(QEP)} = \mathrm{Sol}(\mathrm{EP}_{\Phi_T})$. A relationship between $\mathrm{Sol}(\mathrm{EP}_{\Phi_T})$ and $\mathrm{Sol}(\mathrm{EP}_{\overline{\Phi}_T})$ is given by the following proposition.

Proposition 2.20 *Let A be a subset of a topological space. Assume that $\mathscr{D}^{-1}(y)$ and $\mathrm{lev}_{<}\Phi(\cdot, y)$ are open (in A) for each $y \in A$ and the following condition is satisfied:*

$$\delta(\mathrm{Fix}\mathscr{D}) \setminus \mathrm{Fix}\mathscr{D} \subset \bigcup_{y \in A} [\mathscr{D}^{-1}(y) \cap \mathrm{lev}_{<}\Phi(\cdot, y)], \tag{2.3}$$

where $\delta(\mathrm{Fix}\mathscr{D})$ is the boundary of Fix$\mathscr{D}$. *Then,* $\mathrm{Sol}(\mathrm{EP}_{\overline{\Phi}_T}) = \mathrm{Sol}(\mathrm{EP}_{\Phi_T})$.

Proof By the definition of Φ_T and $\overline{\Phi}_T$, it is clear that $\mathrm{Sol}(\mathrm{EP}_{\overline{\Phi}_T}) \supset \mathrm{Sol}(\mathrm{EP}_{\Phi_T})$. Suppose that $x \in \mathrm{Sol}(\mathrm{EP}_{\overline{\Phi}_T})$ but $x \notin \mathrm{Sol}(\mathrm{EP}_{\Phi_T})$. Then, there is an $y_x \in A$ such that $x \in T^c(y_x) := A \setminus T(y_x) = \mathscr{D}^{-1}(y_x) \cap [(A \setminus \mathrm{Fix}\mathscr{D}) \cup \mathrm{lev}_{<}\Phi(\cdot, y_x)]$. If $x \in A \setminus \mathrm{cl}(\mathrm{Fix}\mathscr{D})$, we have $x \in \mathscr{D}^{-1}(y_x) \cap (A \setminus \mathrm{cl}(\mathrm{Fix}\mathscr{D})) \subset \mathrm{int}T^c(y_x)$. If $x \in \mathrm{Fix}\mathscr{D}$, we have $x \in \mathscr{D}^{-1}(y_x) \cap \mathrm{lev}_{<}\Phi(\cdot, y_x) \subset \mathrm{int}T^c(y_x)$. If $x \in \delta(\mathrm{Fix}\mathscr{D}) \setminus \mathrm{Fix}\mathscr{D}$, then, by (2.3), there exists $y'_x \in A$ such that $x \in \mathscr{D}^{-1}(y'_x) \cap \mathrm{lev}_{<}\Phi(\cdot, y'_x) \subset \mathrm{int}T^c(y'_x)$. Thus, we can always find $y \in A$ such that $x \in \mathrm{int}T^c(y)$, i.e., $x \notin \mathrm{cl}(T(y))$. This contradicts the fact that $x \in \mathrm{Sol}(\mathrm{EP}_{\overline{\Phi}_T})$. □

Theorem 2.21 *Assume that A is a nonempty compact convex subset of a topological vector space X, $\Phi \in \mathscr{A}^2_{pr.qm}$, $\mathscr{D}$ has convex images, $\mathscr{D}^{-1}(y)$, and $\mathrm{lev}_{<}\Phi(\cdot, y)$ are open for each $y \in A$. Then, at least one of the following two assertions is true:*

(a) (QEP) *has solutions.*
(b) *There exists $\bar{x} \in \delta(\mathrm{Fix}\mathscr{D}) \setminus \mathrm{Fix}\mathscr{D}$ such that $\Phi(\bar{x}, y) \geq 0$ for all $y \in \mathscr{D}(\bar{x})$.*

Consequently, if in addition, the following condition is satisfied:

(B) *For each $x \in \delta(\mathrm{Fix}\mathscr{D}) \setminus \mathrm{Fix}\mathscr{D}$, $\Phi(x, y) < 0$ for some $y \in \mathscr{D}(x)$,*

then (QEP) *has solutions.*

Proof Theorem 2.13 ensures the nonemptiness of the set $\mathrm{Fix}\mathscr{D}$. First, we show that $\mathrm{Sol}(\mathrm{EP}_{\overline{\Phi}_T})$ is nonempty. It is clear that $\mathrm{lev}_{\geq}\overline{\Phi}_T(\cdot, y) = \mathrm{cl}(T(y))$ is closed. We show next that $\overline{\Phi}_T \in \mathscr{A}^2_{pr.qm}$. To this purpose, let $N = \{x^1, x^2, \ldots, x^n\} \in \mathscr{F}(A)$ and $x \in \mathrm{conv}N$. If $x \in A \setminus \mathscr{D}^{-1}(x^i)$ for some $x^i \in N$, then $x \in T(x^i) \subset \mathrm{cl}(T(x^i))$ and hence $\max_{x^i \in N} \overline{\Phi}_T(x, x^i) \geq 0$. If $x \notin A \setminus \mathscr{D}^{-1}(x^i)$ for all $x^i \in N$, then $x^i \in \mathscr{D}(x)$ for all $x^i \in N$. Since $\mathscr{D}(x)$ is convex, $x \in \mathrm{conv}N \subset \mathscr{D}(x)$, i.e., $x \in \mathrm{Fix}\mathscr{D}$. Moreover, as $\Phi \in \mathscr{A}^2_{pr.qm}$, there exists $x^i \in N$ such that $\Phi(x, x^i) \geq 0$, i.e., $x \in \mathrm{lev}_{\geq}\Phi(\cdot, x^i)$. This implies that x belongs to $\mathrm{Fix}\mathscr{D} \cap \mathrm{lev}_{\geq}\Phi(\cdot, x^i) \subset T(x^i) \subset \mathrm{cl}(T(x^i))$, and hence, $\max_{x^i \in N} \overline{\Phi}_T(x, x^i) \geq 0$. Thus, by Theorem 1.34, $\mathrm{Sol}(\mathrm{EP}_{\overline{\Phi}_T}) \neq \emptyset$.

Now, assume that assertion (b) in Theorem 2.21 is not true. Then, for each $x \in \delta(\mathrm{Fix}\mathscr{D}) \setminus \mathrm{Fix}\mathscr{D}$, $\Phi(x, y_x) < 0$ for some $y_x \in \mathscr{D}(x)$, i.e.,

$$x \in \mathscr{D}^{-1}(y_x) \cap \mathrm{lev}_{<}\Phi(\cdot, y_x) \subset \bigcup_{y \in A} [\mathscr{D}^{-1}(y) \cap \mathrm{lev}_{<}\Phi(\cdot, y)],$$

which implies that Condition (2.3) is satisfied. Hence, by Proposition 2.20, $\emptyset \neq \mathrm{Sol}(\mathrm{EP}_{\overline{\Phi}_T}) = \mathrm{Sol}(\mathrm{EP}_{\Phi_T}) = \mathrm{Sol}(\mathrm{QEP})$. □

Remark 2.22

(a) From Theorem 2.21, we deduce that if the conditions of Theorem 2.21 are satisfied, then there exists $\bar{x} \in \mathrm{cl}(\mathrm{Fix}\mathscr{D})$ such that $\Phi(\bar{x}, y) \geq 0$ for all $y \in \mathscr{D}(\bar{x})$.
(b) Theorem 2.21 does not guarantee that problem (QEP) has solutions. However, by checking condition (B), we can check the existence of solutions in many situations. If $\mathrm{Fix}\mathscr{D} = A$ or $\mathrm{Fix}\mathscr{D}$ is closed (in particular, $\mathscr{D}$ is closed), then condition (B) is obviously satisfied. Furthermore, if

$$A = \bigcup_{y \in A} [\mathscr{D}^{-1}(y) \cap \mathrm{lev}_{<}\Phi(\cdot, y)],$$

then condition (B) is also satisfied.
(c) If $\mathscr{D}(x) = A$ for all $x \in A$, then $\mathrm{Fix}\mathscr{D} = A$ and Theorem 2.21 becomes Theorem 1.34 (in the compact case).

Example 2.23 Let $A = [-1, 1]$, $\mathscr{D}(x) = (x^2 - \frac{3}{4}, 1 - \frac{1}{2}x^2)$, and $\Phi(x, y) = xy - x^2$. Then, $\Phi \in \mathscr{A}_A^{=} \cap \mathscr{A}^2_{st.qcl} \subset \mathscr{A}^2_{pr.qm}$ (easily to check). For each $y \in A$,

$$\mathrm{lev}_{<}\Phi(\cdot, y) = \begin{cases} [-1, 0) \cup (y, 1] \text{ if } y \geq 0, \\ [-1, y) \cup (0, 1] \text{ if } y < 0, \end{cases}$$

and

$$\mathscr{D}^{-1}(y) = \begin{cases} \emptyset & \text{if } y \leq -\frac{3}{4}, \\ (-\sqrt{y + \frac{3}{4}}, \sqrt{y + \frac{3}{4}}) & \text{if } -\frac{3}{4} < y \leq \frac{1}{4}, \\ [-1, 1] & \text{if } \frac{1}{4} < y \leq \frac{5}{12}, \\ (-\sqrt{2(1-y)}, \sqrt{2(1-y)}) & \text{if } \frac{5}{12} < y \leq \frac{1}{2}, \\ [-1, 1] & \text{if } \frac{1}{2} < y \end{cases}$$

are open. We have $\mathrm{Fix}\mathscr{D} = \{x \in A \mid x^2 - \frac{3}{4} < x < 1 - \frac{1}{2}x^2\} = (-\frac{1}{2}, \sqrt{3} - 1)$, $\delta(\mathrm{Fix}\mathscr{D}) \setminus \mathrm{Fix}\mathscr{D} = \{-\frac{1}{2}, \sqrt{3} - 1\}$. Checking the condition (B), we see that for $y \in (0, \frac{7}{8}) \subset (-\frac{1}{2}, \frac{7}{8}) = \mathscr{D}(-\frac{1}{2})$, $\Phi(-\frac{1}{2}, y) = -\frac{1}{2}(y + \frac{1}{2}) < 0$, and for $y \in (\frac{13-8\sqrt{3}}{4}, \sqrt{3} - 1) = \mathscr{D}(\sqrt{3} - 1)$, $\Phi(\sqrt{3} - 1, y) = (\sqrt{3} - 1)(y - \sqrt{3} + 1) < 0$. Thus, by Theorem 2.21, (QEP) has solutions.

Corollary 2.24 *Assume that A is a nonempty compact convex subset of a Hausdorff locally convex space X, $\Phi \in \mathscr{A}^2_{pr.qm}$, $\mathscr{D}$ is closed with convex images, $\mathrm{lev}_{\geq}\Phi(\cdot, y)$ is closed for all $y \in A$, and the set $\{x \in A \mid \mathscr{D}(x) \cap \mathrm{lev}_{<}\Phi(x, \cdot) = \emptyset\}$ is closed. Then,* (QEP) *has solutions.*

Proof Since A is compact and $\mathscr{D}$ is closed, $\mathscr{D}$ is upper semicontinuous by Proposition 2.6(c). Let $\mathscr{U}$ be a basis of neighborhoods of the origin in X consisting of open balanced convex neighborhoods of the origin. For each $W \in \mathscr{U}$, let $\mathscr{D}_W : A \rightrightarrows A$ be set-valued mapping defined by (2.1) as in the proof of Theorem 2.14. From the proof of Theorem 2.14, we have that $\mathscr{D}_W^{-1}(y)$ is open in A for all $y \in A$. For each $W \in \mathscr{U}$, consider the quasiequilibrium problem with the map $\mathscr{D}_W$ instead of $\mathscr{D}$. Then, all the conditions of Theorem 2.21 are satisfied. By this theorem, there exists $x_W \in \mathrm{cl}(\mathrm{Fix}\mathscr{D}_W)$ such that $\Phi(x_W, y) \geq 0$ for all $y \in \mathscr{D}_W(x_W)$, which means that $\mathscr{D}_W(x_W) \cap \mathrm{lev}_{<}\Phi(x_W, \cdot) = \emptyset$. Now, for each $W \in \mathscr{U}$, let $\mathscr{D}^*_W : A \rightrightarrows A$ be defined by $\mathscr{D}^*_W(x) := (\mathscr{D}((x + W) \cap A) + W) \cap A$. Then, $\mathscr{D}(x) \subset \mathscr{D}_W(x) \subset \mathscr{D}^*_W(x)$ for all $x \in X$. Set $A_W = \{x \in \mathrm{cl}(\mathrm{Fix}\mathscr{D}^*_W) \mid \mathscr{D}(x) \cap \mathrm{lev}_{<}\Phi(x, \cdot) = \emptyset\}$. Since $\mathrm{cl}(\mathrm{Fix}\mathscr{D}_W) \subset \mathrm{cl}(\mathrm{Fix}\mathscr{D}^*_W)$ and $\mathscr{D}(x) \cap \mathrm{lev}_{<}\Phi(x, \cdot) \subset \mathscr{D}_W(x) \cap \mathrm{lev}_{<}\Phi(x, \cdot)$ for all $x \in A$, we have that $A_W \neq \emptyset$. Moreover, as the set $\{x \in A \mid \mathscr{D}(x) \cap \mathrm{lev}_{<}\Phi(x, \cdot) = \emptyset\}$ is closed, A_W is closed. It is clear that A_W is decreasing as W decreases. So, the family of compact sets A_W, $W \in \mathscr{U}$, has a common point, say $\bar{x}$. Then, $\mathscr{D}(\bar{x}) \cap \mathrm{lev}_{<}\Phi(\bar{x}, \cdot) = \emptyset$, i.e., $\Phi(\bar{x}, y) \geq 0$ for all $y \in \mathscr{D}(\bar{x})$. We will show that $\bar{x} \in \mathrm{Fix}\mathscr{D}$, and hence, $\bar{x}$ is a solution of (QEP). For all $W \in \mathscr{U}$, since $\bar{x} \in \mathrm{cl}(\mathrm{Fix}\mathscr{D}^*_W)$, $(\bar{x} + W) \cap \mathrm{Fix}\mathscr{D}^*_W \neq \emptyset$ implying that $\bar{x} \in \mathrm{Fix}\mathscr{D}^*_W + W$. Then, there exists $x_W, x'_W, x''_W \in A$ such that $\bar{x} - x_W \in W$, $x_W - x'_W \in W$, $x'_W \in \mathscr{D}(x''_W)$ and $x''_W - x_W \in W$. By the compactness of A, we may assume that the net $\{x_W\}_{W\in\mathscr{U}}$ converges to some $\hat{x} \in A$. Since X is Hausdorff, $\hat{x} = \bar{x}$, and the nets $\{x'_W\}_{W\in\mathscr{U}}$ and $\{x''_W\}_{W\in\mathscr{U}}$ also converge to $\hat{x} = \bar{x}$. Because $\mathscr{D}$ is closed, $\bar{x} \in \mathscr{D}(\bar{x})$. □

Example 2.25 Let $A = [0, 1]$, $\Phi(x, y) = -x^2 + \frac{1}{2}y + \frac{1}{2}$, and

$$\mathscr{D}(x) = \begin{cases} [\frac{1-x}{3}, \frac{1}{3}] & \text{if } 0 \le x < \frac{1}{2}, \\ [0, 1] & \text{if } x = \frac{1}{2}, \\ [1 - \frac{x}{3}, 1] & \text{if } \frac{1}{2} < x \le 1. \end{cases}$$

Then, $\mathscr{D}$ is closed. For all $x, y \in A$, $\text{lev}_{\ge}\Phi(\cdot, y) = \{x \in A \mid x^2 \le \frac{1}{2}(y + 1)$ and $\{x \in A \mid \mathscr{D}(x) \cap \text{lev}_{<}\Phi(x, \cdot) = \emptyset\} = [0, \frac{\sqrt{145}-1}{12}]$ are closed. To see $\Phi \in \mathscr{A}^2_{pr.qm}$, for all $N \in \mathscr{F}(A)$ and $x \in \text{conv}N$, by taking $x^{j_0} = \max N$, we have $\Phi(x, x^{j_0}) = -x^2 + \frac{1}{2}x^{j_0} + \frac{1}{2} \ge -(x^{j_0})^2 + \frac{1}{2}x^{j_0} + \frac{1}{2} \ge 0$. Thus, by Corollary 2.24, (QEP) has solutions.

If the set-valued mapping $S_A : A \rightrightarrows A$, defined by $S_A(x) := \mathscr{D}(x) \cap \text{lev}_{<}\Phi(x, \cdot)$ for all $x \in A$, is lower semicontinuous, then the set $\{x \in A \mid \mathscr{D}(x) \cap \text{lev}_{<}\Phi(x, \cdot) = \emptyset\}$ is closed. Moreover, S_A is lower semicontinuous if $\mathscr{D}$ is lower semicontinuous and the set-valued mapping $S_\Phi : x \longmapsto \text{lev}_{<}\Phi(x, \cdot)$ has an open graph (by Proposition 2.7). In particular, when the set $\text{lev}_{\ge}\Phi := \{(x, y) \in A \times A \mid \Phi(x, y) \ge 0\}$ is closed, S_Φ is open and $\text{lev}_{\ge}\Phi(\cdot, y)$ is closed for all $y \in A$. Thus, we can state the following corollary.

Corollary 2.26 *Assume that A is a nonempty compact convex subset of a Hausdorff locally convex space X, $\Phi \in \mathscr{A}^2_{pr.qm}$, $\mathscr{D}$ is closed with convex images,* $\text{lev}_{\ge}\Phi(\cdot, y)$ *is closed for all $y \in A$, and the set-valued mapping $S_A : A \rightrightarrows A$, defined by $S_A(x) := \mathscr{D}(x) \cap \text{lev}_{<}\Phi(x, \cdot)$ for all $x \in A$, is lower semicontinuous (in particular, $\mathscr{D}$ is lower semicontinuous and the set* $\text{lev}_{\ge}\Phi := \{(x, y) \in A \times A \mid \Phi(x, y) \ge 0\}$ *is closed). Then* (QEP) *has solutions.*

Note that when A is compact and $\mathscr{D}$ is closed and lower semicontinuous, $\mathscr{D}$ is continuous actually (since $\mathscr{D} : A \rightrightarrows A$ is upper semicontinuous if A is compact and $\mathscr{D}$ is closed, see Proposition 2.6(c)).

Example 2.27 Let $A = [0, 2]$, $\Phi(x, y) = x - y^2 + 2$, and

$$\mathscr{D}(x) = \begin{cases} [-\frac{3}{2}x + \frac{3}{2}, 2] & \text{if } 0 \le x \le 1, \\ [0, -\frac{3}{2}x + \frac{7}{2}] & \text{if } 1 < x \le 2. \end{cases}$$

Then, $\mathscr{D}$ is closed and lower semicontinuous and the images of $\mathscr{D}$ are convex. To see $\Phi \in \mathscr{A}^2_{pr.qm}$, for all $N \in \mathscr{F}(A)$ and $x \in \text{conv}N$, by taking $x^{j_0} = \min N$, we have $\Phi(x, x^{j_0}) = x - (x^{j_0})^2 + 2 \ge x^{j_0} - (x^{j_0})^2 + 2 \ge 0$. Thus, by Corollary 2.26, (QEP) has solutions.

When X is a separable Banach space, the closedness of $\mathscr{D}$ can be relaxed to the closedness of the set of all fixed points of $\mathscr{D}$.

Theorem 2.28 *Assume that A is a nonempty compact convex subset of a separable Banach space X, $\Phi \in \mathscr{A}^2_{pr.qm}$, $\mathscr{D}$ is lower semicontinuous with nonempty convex images,* $\mathrm{Fix}\mathscr{D}$ *is closed, and the set-valued mapping* $S_{\mathrm{Fix}\mathscr{D}} : \mathrm{Fix}\mathscr{D} \rightrightarrows A$*, defined by* $S_{\mathrm{Fix}\mathscr{D}}(x) = \mathscr{D}(x) \cap \mathrm{lev}_{<}\Phi(x, \cdot)$ *for all* $x \in \mathrm{Fix}\mathscr{D}$*, is lower semicontinuous. Then,* (QEP) *has solutions.*

Proof By Theorem 2.18, $\mathrm{Fix}\mathscr{D}$ is nonempty. Let set-valued map $H : A \rightrightarrows A$ be defined for all $x \in A$ by

$$H(x) := \begin{cases} \mathrm{conv}(\mathscr{D}(x) \cap \mathrm{lev}_{<}\Phi(x, \cdot)) & \text{if } x \in \mathrm{Fix}\mathscr{D}, \\ \mathscr{D}(x) & \text{if } x \notin \mathrm{Fix}\mathscr{D}. \end{cases}$$

Because $S_{\mathrm{Fix}\mathscr{D}}$ is lower semicontinuous, the mapping $\mathrm{conv}S_{\mathrm{Fix}\mathscr{D}} : \mathrm{Fix}\mathscr{D} \rightrightarrows A$, defined by

$$\mathrm{conv}S_{\mathrm{Fix}\mathscr{D}}(x) := \mathrm{conv}(\mathscr{D}(x) \cap \mathrm{lev}_{<}\Phi(x, \cdot)) \text{ for all } x \in \mathrm{Fix}\mathscr{D},$$

is lower semicontinuous by Proposition 2.8. Moreover, $\mathrm{conv}S_{\mathrm{Fix}\mathscr{D}}(x) \subset \mathscr{D}(x)$ for all $x \in \mathrm{Fix}\mathscr{D}$. Hence, H is lower semicontinuous by Proposition 2.9. As the images of $\mathscr{D}$ are convex, $H(x)$ is clearly convex for each $x \in A$. If $H(x)$ is nonempty for all $x \in A$, then, by Theorem 2.18, there exists $\hat{x} \in A$ such that $\hat{x} \in H(\hat{x})$. By the definition of H, $\hat{x} \in \mathrm{Fix}\mathscr{D}$ and $\hat{x} \in \mathrm{conv}(\mathscr{D}(\hat{x}) \cap \mathrm{lev}_{<}\Phi(\hat{x}, \cdot))$. It follows that there exists a finite subset N of $\mathscr{D}(\hat{x}) \cap \mathrm{lev}_{<}\Phi(\hat{x}, \cdot)$ such that $\hat{x} \in convN$. Then, $\hat{x} \in convN$ and $\Phi(\hat{x}, x^i) < 0$ for all $x^i \in N$. This contradicts the fact that $\Phi \in \mathscr{A}^2_{pr.qm}$. Thus, there must exist $\bar{x} \in A$ such that $H(\bar{x}) = \emptyset$. Since $\mathscr{D}(\bar{x}) \neq \emptyset$, again by the definition of H we have $\bar{x} \in \mathrm{Fix}\mathscr{D}$ and $\mathscr{D}(\bar{x}) \cap \mathrm{lev}_{<}\Phi(\bar{x}, \cdot) \subset \mathrm{conv}(\mathscr{D}(\hat{x}) \cap \mathrm{lev}_{<}\Phi(\hat{x}, \cdot)) = \emptyset$, which implies that $\bar{x}$ is a solution of (QEP). □

Corollary 2.29 *Assume that A is a nonempty compact convex subset of a separable Banach space X, $\Phi \in \mathscr{A}^2_{pr.qm}$, $\mathscr{D}$ is lower semicontinuous with convex images,* $\mathrm{Fix}\mathscr{D}$ *is closed, and* $\mathrm{lev}^{fix}_{\geq}\Phi := \{(x, y) \in \mathrm{Fix}\mathscr{D} \times A \mid \Phi(x, y) \geq 0\}$ *is closed. Then,* (QEP) *has solutions.*

Proof By Theorem 2.18, $\mathrm{Fix}\mathscr{D}$ is nonempty. Let set-valued map $S : \mathrm{Fix}\mathscr{D} \rightrightarrows A$ be defined by

$$S(x) := \mathrm{lev}_{<}\Phi(x, \cdot) \text{ for all } x \in \mathrm{Fix}\mathscr{D}.$$

Because $\mathscr{D}|_{\mathrm{Fix}\mathscr{D}}$ is lower semicontinuous and $\mathrm{gph}S = \mathrm{Fix}\mathscr{D} \times A \setminus \mathrm{lev}_{\geq}\Phi$ is open in $\mathrm{Fix}\mathscr{D} \times A$, $S_{\mathrm{Fix}\mathscr{D}} = \mathscr{D}|_{\mathrm{Fix}\mathscr{D}} \cap S : \mathrm{Fix}\mathscr{D} \rightrightarrows A$ is lower semicontinuous by Proposition 2.7. Thus, by Theorem 2.28, (QEP) has solutions. □

In the following, for the notions of a bifunction having property $\mathbb{P}^2$ and dual property, see Definitions 1.38–1.40.

Lemma 2.30 (An extension of Lemma 1.43) *Assume that Φ has dual property. Then, if for some $\bar{x} \in A$, $\bar{x} \in \mathscr{D}(\bar{x})$, $\mathscr{D}(\bar{x})$ is convex, and $\Phi(y, \bar{x}) \leq 0$ for all $y \in \mathscr{D}(\bar{x})$, then $\Phi(\bar{x}, y) \geq 0$ for all $y \in \mathscr{D}(\bar{x})$.*

Proof The proof is similar to that of Lemma 1.43. Assume that $\bar{x} \in \mathscr{D}(\bar{x})$ and $\Phi(y, \bar{x}) \leq 0$ for all $y \in \mathscr{D}(\bar{x})$. Take any $y \in \mathscr{D}(\bar{x})$, $t \in [0, 1]$, and $y_t = \bar{x}+t(y-\bar{x})$. Then, $y_t \in \mathscr{D}(\bar{x})$ (since $\mathscr{D}(\bar{x})$ is convex) and hence $\Phi(y_t, \bar{x}) \leq 0$. This and the dual property of Φ imply that $\Phi(\bar{x}, y) \geq 0$. □

Theorem 2.31 *Assume that A is a compact convex subset of a topological vector space, $\Phi \in -\mathscr{A}^1_{pr.qm}$, and has dual property, $\mathscr{D}$ has convex images, and $\mathscr{D}^{-1}(y)$ and $\mathrm{lev}_{>}\Phi(y, \cdot)$ are open for each $y \in A$. Then, at least one of the following two assertions is true:*

(a) (QEP) *has solutions.*
(b) *There exists $\bar{x} \in \delta(\mathrm{Fix}\mathscr{D}) \setminus \mathrm{Fix}\mathscr{D}$ such that $\Phi(y, \bar{x}) \leq 0$ for all $y \in \mathscr{D}(\bar{x})$.*

Consequently, if additionally, the following condition is satisfied:

(B′) *For each $x \in \delta(\mathrm{Fix}\mathscr{D}) \setminus \mathrm{Fix}\mathscr{D}$, $\Phi(y, x) > 0$ for some $y \in \mathscr{D}(x)$,*

then (QEP) *has solutions.*

Proof Consider $\Phi' : A \times A \to \mathbb{R}$ defined by $\Phi'(x, y) := -\Phi(y, x)$ for $x, y \in A$. As $\Phi \in (-\mathscr{A}^1_{pr.qm}) \cap \mathscr{A}^{=}_{A}$, $\Phi' \in \mathscr{A}^2_{pr.qm}) \cap \mathscr{A}^{=}_{A} \subset \mathscr{A}^2_{pr.qm}$. Moreover, $\mathrm{lev}_{<}\Phi'(\cdot, y) = \mathrm{lev}_{>}\Phi(y, \cdot)$. Thus, it is not hard to see that when the conditions of Theorem 2.31 are satisfied for $\mathscr{D}$ and Φ, the conditions of Theorem 2.21 are fulfilled for $\mathscr{D}$ and Φ'. By Theorem 2.21, at least one of the following two assertions is true:

(a’) There exists $\bar{x} \in A$ such that $\bar{x} \in \mathscr{D}(\bar{x})$ and $\Phi'(\bar{x}, y) \geq 0$ for all $y \in \mathscr{D}(\bar{x})$, i.e., $\Phi(y, \bar{x}) \leq 0$ for all $y \in \mathscr{D}(\bar{x})$.
(b’) There exists $\bar{x} \in \delta(\mathrm{Fix}\mathscr{D}) \setminus \mathrm{Fix}\mathscr{D}$ such that $\Phi'(\bar{x}, y) \geq 0$ for all $y \in \mathscr{D}(\bar{x})$, i.e., $\Phi(y, \bar{x}) \leq 0$ for all $y \in \mathscr{D}(\bar{x})$.

By Lemma 2.30, (a’) implies that (QEP) has solutions. □

Remark 2.32 Condition (B′) is satisfied if at least one of the following conditions is satisfied:

(i) $\mathrm{Fix}\mathscr{D} = A$.
(ii) $\mathrm{Fix}\mathscr{D}$ is closed (in particular, $\mathscr{D}$ is closed).
(iii) $A = \bigcup_{y\in A}[\mathscr{D}^{-1}(y) \cap \mathrm{lev}_{>}\Phi(y, \cdot)]$.

Corollary 2.33 *Assume that A is a nonempty compact convex subset of a Hausdorff locally convex space X, $\Phi \in -\mathscr{A}^1_{pr.qm}$, and has dual property, $\mathscr{D}$ is closed with convex images, and for all $x \in A$, the sets $\mathrm{lev}_{\leq}\Phi(x, \cdot)$ and $\{x \in A \mid \mathscr{D}(x) \cap \mathrm{lev}_{>}\Phi(\cdot, x) = \emptyset\}$ are closed. Then,* (QEP) *has solutions.*

Proof Argue as in the proof of Theorem 2.31, using Corollary 2.24 instead of Theorem 2.21. □

Corollary 2.34 *Assume that A is a compact convex subset of Hausdorff locally convex space X, $\Phi \in -\mathscr{A}^1_{pr.qm}$, and has dual property, $\mathscr{D}$ is closed with convex images, and for all $x \in A$, $\text{lev}_{\leq}\Phi(x, \cdot)$ is closed and the set-valued mapping $S^\bullet_A : A \rightrightarrows A$, defined by $S^\bullet_A(x) := \mathscr{D}(x) \cap \text{lev}_{>}\Phi(\cdot, x)$, is lower semicontinuous (in particular, $\mathscr{D}$ is lower semicontinuous and the set $\text{lev}^{\leq}\Phi := \{(x, y) \in A \times A \mid \Phi(y, x) \leq 0\}$ is closed). Then,* (QEP) *has solutions.*

Proof Argue as in the proof of Theorem 2.31, using Corollary 2.26 instead of Theorem 2.21. □

Theorem 2.35 *Assume that A is a compact convex subset of a topological vector space, $\Phi \in (-\mathscr{A}^1_{pr.qm}) \cap \mathscr{A}^2_{st.qcl} \cap \mathscr{A}^{=}_A$ has property $\mathbb{P}^2$ and is upper-sign continuous in the first argument, $\mathscr{D}$ has convex images, and $\mathscr{D}^{-1}(y)$ and $\text{lev}_{>}\Phi(y, \cdot)$ are open for each $y \in A$. Then, at least one of the following two assertions is true:*

(a) (QEP) *has solutions.*
(b) *There exists $\bar{x} \in \delta(\text{Fix}\mathscr{D}) \setminus \text{Fix}\mathscr{D}$ such that $\Phi(y, \bar{x}) \leq 0$ for all $y \in \mathscr{D}(\bar{x})$.*

Consequently, if additionally, the following condition is satisfied:

(B′) *For each $x \in \delta(\text{Fix}\mathscr{D}) \setminus \text{Fix}\mathscr{D}$, $\Phi(y, x) > 0$ for some $y \in \mathscr{D}(x)$,*

then (QEP) *has solutions.*

Proof When $\Phi \in \mathscr{A}^2_{st.qcl} \cap \mathscr{A}^{\geq}_A$ has property $\mathbb{P}^2$ and is upper-sign continuous in the first component, Φ has dual property by Proposition 1.42. Thus, Theorem 2.35 is a direct consequence of Theorem 2.31 and Proposition 1.42. □

Corollary 2.36 *Assume that A is a nonempty compact convex subset of a Hausdorff locally convex space X, $\Phi \in (-\mathscr{A}^1_{pr.qm}) \cap \mathscr{A}^2_{st.qcl} \cap \mathscr{A}^{=}_A$ has property $\mathbb{P}^2$, Φ is upper-sign continuous in the first argument, $\mathscr{D}$ is closed with convex images, and for all $x \in A$, the sets $\text{lev}_{\leq}\Phi(x, \cdot)$ and $\{x \in A \mid \mathscr{D}(x) \cap \text{lev}_{>}\Phi(\cdot, x) = \emptyset\}$ are closed. Then,* (QEP) *has solutions.*

Proof Corollary 2.36 is a direct consequence of Corollary 2.33 and Proposition 1.42. □

Corollary 2.37 *Assume that A is a compact convex subset of Hausdorff locally convex space X, $\Phi \in (-\mathscr{A}^1_{pr.qm}) \cap \mathscr{A}^2_{st.qcl} \cap \mathscr{A}^{=}_A$ has property $\mathbb{P}^2$, Φ is upper-sign continuous in the first argument, $\mathscr{D}$ is closed with convex images, and for all $x \in A$, $\text{lev}_{\leq}\Phi(x, \cdot)$ is closed and the set-valued mapping $S^\bullet_A : A \rightrightarrows A$, defined by $S^\bullet_A(x) := \mathscr{D}(x) \cap \text{lev}_{>}\Phi(\cdot, x)$, is lower semicontinuous (in particular, $\mathscr{D}$ is lower semicontinuous and the set $\text{lev}^{\leq}\Phi := \{(x, y) \in A \times A \mid \Phi(y, x) \leq 0\}$ is closed). Then,* (QEP) *has solutions.*

Proof Corollary 2.37 is a direct consequence of Corollary 2.34 and Proposition 1.42. □ □

Note that if $\Phi \in (-\mathscr{A}^1_{pr.qm}) \cap \mathscr{A}^2_{s.qcl} \cap \mathscr{A}^{=}_A$, then the condition "$\Phi \in (-\mathscr{A}^1_{pr.qm}) \cap \mathscr{A}^2_{st.qcl} \cap \mathscr{A}^{=}_A$ has property $\mathbb{P}^2$" in Theorem 2.35 and Corollaries 2.36–2.37 is satisfied because $\mathscr{A}^2_{s.qcl} \subset \mathscr{A}^2_{st.qcl}$ (see, Fig. 1.10). Moreover, when $\Phi \in \mathscr{A}^2_{s.qcl}$, Φ has property $\mathbb{P}^2$ (see Example 1.39).

Example 2.38

(a) Let A, $\mathscr{D}$ be as in Example 2.23 and $\Phi(x, y) = x(x - y)$. Then, $\mathscr{D}^{-1}(y)$ is open for each $y \in A$, $\delta(\text{Fix}\mathscr{D}) \setminus \text{Fix}\mathscr{D} = \{-\frac{1}{2}, \sqrt{3} - 1\}$, $\mathscr{D}(-\frac{1}{2}) = (-\frac{1}{2}, \frac{7}{8})$, and $\mathscr{D}(\sqrt{3} - 1) = (\frac{13-8\sqrt{3}}{4}, \sqrt{3} - 1)$. It is not hard to see that $\Phi \in (-\mathscr{A}^1_{pr.qm}) \cap \mathscr{A}^2_{s.qcl} \cap \mathscr{A}^=_A$ and for each $y \in A$,

$$\text{lev}_{>}\Phi(y, \cdot) = \begin{cases} [-1, y) & \text{if } y \geq 0, \\ (y, 1] & \text{if } y < 0 \end{cases}$$

is open. Checking condition (B′), we see that $\Phi(y, -\frac{1}{2}) = y(y + \frac{1}{2}) > 0$ for $y \in (0, \frac{7}{8}) \subset \mathscr{D}(-\frac{1}{2})$ and $\Phi(y, \sqrt{3} - 1) = y(y - \sqrt{3} + 1) > 0$ for $y \in (\frac{13-8\sqrt{3}}{4}, 0) \subset \mathscr{D}(\sqrt{3} - 1)$. Thus, by Theorem 2.35, (QEP) has solutions. Note that in this example, the assumptions of Theorem 2.21 are also satisfied, but condition (B) does not hold.

(b) Let A, $\mathscr{D}$ be as in Example 2.27 and $\Phi : [0, 2] \times [0, 2] \to \mathbb{R}$ be defined by

$$\Phi(x, y) := \begin{cases} -1 & \text{if } x = 0, \ y > 0, \ \text{ or } 0 \leq y < x, \\ 0 & \text{if } x = y, \\ 1 & \text{if } 0 < x < y. \end{cases}$$

It is easy to check that $\Phi \in (-\mathscr{A}^1_{pr.qm}) \cap \mathscr{A}^2_{s.qcl} \cap \mathscr{A}^=_A$ and $\text{lev}^{\leq}\Phi := \{(x, y) \in A \times A \mid \Phi(y, x) \leq 0\} = \{(x, 0) \mid x \in [0, 2]\} \cup \{(x, y) \mid 0 \leq x \leq y \leq 2\}$ is closed. Thus, by Corollary 2.37, (QEP) has solutions. In this example, $\text{lev}_{\geq}\Phi := \{(x, y) \in A \times A \mid \Phi(x, y) \geq 0\} = \{(x, x) \mid x \in [0, 2]\} \cup \{(x, y) \mid 0 < x < y \leq 2\}$ is not closed, and hence, Corollary 2.26 is not used.

We state some results from Theorem 2.28 and its corollaries with proof technique as in the proof of Theorem 2.31.

Theorem 2.39 *Assume that A is a compact convex subset of a separable Banach space X, $\Phi \in -\mathscr{A}^1_{pr.qm}$, and has dual property, $\mathscr{D}$ is lower semicontinuous with nonempty convex images,* Fix$\mathscr{D}$ *is closed, and the set-valued mapping $S^{\text{Fix}\mathscr{D}} : \text{Fix}\mathscr{D} \rightrightarrows A$, defined by $S^{\text{Fix}\mathscr{D}}(x) := \mathscr{D}(x) \cap \text{lev}_{>}\Phi(\cdot, x)$ for all $x \in \text{Fix}\mathscr{D}$, is lower semicontinuous and has convex images. Then,* (QEP) *has solutions.*

Proof Argue as in the proof of Theorem 2.31, using Theorem 2.28 instead of Theorem 2.21. □

Corollary 2.40 *Assume that A is a compact convex subset of a separable Banach space X, $\Phi \in -\mathscr{A}^1_{pr.qm}$, and has dual property, $\mathscr{D}$ is lower semicontinuous with convex images,* Fix$\mathscr{D}$ *is closed, and $\text{lev}^{\leq}_{fix}\Phi := \{(x, y) \in A \times \text{Fix}\mathscr{D} \mid \Phi(x, y) \leq 0\}$ is closed. Then,* (QEP) *has solutions.*

Proof Argue as in the proof of Theorem 2.31, using Corollary 2.29 instead of Theorem 2.21. □

Theorem 2.41 *Assume that A is a compact convex subset of a separable Banach space X, $\Phi \in (-\mathscr{A}^1_{pr.qm}) \cap \mathscr{A}^2_{st.qcl} \cap \mathscr{A}^{=}_{\mathrm{A}}$ has property $\mathbb{P}^2$ and is upper-sign continuous in the first argument, $\mathscr{D}$ is lower semicontinuous with nonempty convex images, $\mathrm{Fix}\mathscr{D}$ is closed, and the set-valued mapping $S^{\mathrm{Fix}\mathscr{D}} : \mathrm{Fix}\mathscr{D} \rightrightarrows A$, defined by $S^{\mathrm{Fix}\mathscr{D}}(x) = \mathscr{D}(x) \cap \mathrm{lev}_{>}\Phi(\cdot, x)$ for all $x \in \mathrm{Fix}\mathscr{D}$, is lower semicontinuous and has convex images. Then,* (QEP) *has solutions.*

Proof Theorem 2.41 is a consequence of Theorem 2.39 and Proposition 1.42. □

Corollary 2.42 *Assume that A is a compact convex subset of a separable Banach space X, $\Phi \in (-\mathscr{A}^1_{pr.qm}) \cap \mathscr{A}^2_{st.qcl} \cap \mathscr{A}^{=}_{\mathrm{A}}$ has property $\mathbb{P}^2$ and is upper-sign continuous in the first argument, $\mathscr{D}$ is lower semicontinuous with convex images, $\mathrm{Fix}\mathscr{D}$ is closed, and $\mathrm{lev}^{\leq}_{fix}\Phi := \{(x, y) \in A \times \mathrm{Fix}\mathscr{D} \mid \Phi(x, y) \leq 0\}$ is closed. Then,* (QEP) *has solutions.*

Proof Corollary 2.42 is deduced from Corollary 2.40 and Proposition 1.42. □

Next, we establish some results in nonconvex cases.

Proposition 2.43 *Suppose that A is a closed subset of a complete metric space (X, d), $\mathscr{D} : A \rightrightarrows A$, and $\Phi \in \mathscr{A}_{\Delta} \cap \mathscr{A}^{=}_{\mathrm{A}}$ is lower semicontinuous in the second component and $\inf_{y \in A} \Phi(x, y) > -\infty$ for each $x \in A$. Assume further that for every $\epsilon > 0$ and $x^0 \in A$, the following implication holds: for all $x \in A$,*

$$\Phi(x^0, x) + \epsilon d(x^0, x) \leq 0 \Rightarrow \exists y \in \mathscr{D}(x),\;\; \Phi(x, y) + \epsilon d(x, y) \leq 0. \tag{2.4}$$

Then, there exists $\bar{x} \in \mathrm{Fix}\mathscr{D}$ such that:

(a) $\Phi(x^0, \bar{x}) + \epsilon d(x^0, \bar{x}) \leq 0.$
(b) $\Phi(\bar{x}, y) + \epsilon d(\bar{x}, y) > 0$ *for all $y \in A$ different from $\bar{x}$.*

Proof By Theorem 1.71, for each $\epsilon > 0$ and $x^0 \in A$, there exists $\bar{x} \in A$ such that conclusions (a) and (b) of Proposition 2.43 are satisfied. We only need to show that $\bar{x} \in \mathrm{Fix}\mathscr{D}$. From conclusion (a) and implication (2.4), there exists $y \in \mathscr{D}(\bar{x})$ such that

$$\Phi(\bar{x}, y) + \epsilon d(\bar{x}, y) \leq 0.$$

This inequality and conclusion (b) imply that $y = \bar{x}$, i.e, $\bar{x} \in \mathrm{Fix}\mathscr{D}$. □

Theorem 2.44 *Assume that A is a closed subset of a complete metric space (X, d), $\text{Fix}\mathscr{D}$ is compact and nonempty, $\Phi \in \mathscr{A}_\Delta \cap \mathscr{A}_A^{=}$ is lower semicontinuous in the second component and upper semicontinuous in the first component, $\inf_{y \in A} \Phi(x, y) > -\infty$ for each $x \in A$, and the following implication holds for all $\varepsilon > 0$, $x^0 \in A$, and $x \in A$:*

$$\Phi(x^0, x) + \varepsilon d(x^0, x) \leq 0 \Rightarrow \exists y \in \mathscr{D}(x), \quad \Phi(x, y) + \varepsilon d(x, y) \leq 0. \tag{2.5}$$

Then, (QEP) *has solutions.*

Proof By Proposition 2.43 and the fact that $\Phi \in \mathscr{A}_A^{=}$, for each $n \in \mathbb{N}$, there exists $x^n \in \text{Fix}\mathscr{D}$ such that

$$\Phi(x^n, y) + \frac{1}{n} d(x^n, y) \geq 0 \text{ for all } y \in A.$$

Since $\text{Fix}\mathscr{D}$ is compact, without loss of generality, we can assume that $\{x^n\}_n$ converges to $\bar{x} \in \text{Fix}\mathscr{D}$. Because the distance function is continuous and $\Phi(\cdot, y)$ is lower semicontinuous, we have

$$\Phi(\bar{x}, y) \geq \limsup(\Phi(x^n, y) + \frac{1}{n} d(x^n, y)) \geq 0 \text{ for all } y \in A.$$

This implies that $\bar{x} \in \text{Sol(QEP)}$. □

Theorem 2.45 *Assume that A is a closed subset of a complete metric space (X, d), $\text{Fix}\mathscr{D}$ is compact and nonempty, there exists a lower semicontinuous and bounded from below function $\varphi : A \to \mathbb{R}$ such that $\Phi(x, y) \geq \varphi(y) - \varphi(x)$ for all $x, y \in A$, and for every $\epsilon > 0$ and for any $x^0 \in A$, the following implication holds: for all $x \in A$,*

$$\varphi(x) + \epsilon d(x^0, x) \leq \varphi(x^0) \Rightarrow \exists y \in \mathscr{D}(x), \quad \varphi(y) + \epsilon d(x, y) \leq \varphi(x). \tag{2.6}$$

Then, (QEP) *has solutions.*

Proof Let $\Phi' : A \times A \to \mathbb{R}$ be defined by $\Phi'(x, y) := \varphi(y) - \varphi(x)$ for $x, y \in A$. Then, from the conditions of Theorem 2.45, it is easy to see that all the assumptions of Theorem 2.44 hold for $\mathscr{D}$ and Φ'. Hence, by Theorem 2.44, an $\bar{x} \in \text{Fix}\mathscr{D}$ exists such that

$$\Phi(\bar{x}, y) \geq \Phi'(\bar{x}, y) = \varphi(y) - \varphi(\bar{x}) \geq 0 \quad \text{for all } x, y \in \mathscr{D}(\bar{x}).$$

This means that $\bar{x} \in \text{Sol(QEP)}$. □

Definition 2.46 Let A be a nonempty set and $\mathscr{D} : A \rightrightarrows A$ be a set-valued mapping. $\mathscr{D}$ is called QUASICYCLIC if for any $x^1, x^2, \ldots, x^m \in A$ with $x^i \neq x^{i+1}$, there exists an $i \in \{1, 2, \ldots, m\}$ such that $x^{i+1} \notin \mathscr{D}(x^i)$, where $x^{m+1} := x^1$.

Example 2.47 (Quasicyclic maps)

(a) Let $A = \mathbb{X}$ be a normed space. The maps $\mathscr{D}, \mathscr{D}' : A \rightrightarrows A$ defined for all $x \in A$ by

$$\mathscr{D}(x) = \{y \in A \mid \| y \| < \| x \|\}, \qquad \mathscr{D}'(x) = \{y \in A \mid \| y \| > \| x \|\},$$

are quasicyclic.

(b) Let (X, d) be a metric space, $A \subseteq \mathbb{X}$, and $\theta : A \to X$ be a single-valued map. Then, the map $\mathscr{D} : A \rightrightarrows A$, defined as below, is quasicyclic:

$$\mathscr{D}(x) = \{y \in A \mid d(x, \theta(x) > d(y, \theta(y)\} \text{ for } x \in A.$$

For the class of bifunctions $\mathscr{A}_{caqm}$ (see Definition 1.77), using Theorem 1.82 we establish the following result.

Theorem 2.48 *Assume that A is a nonempty compact subset of a topological space X, $\mathscr{D}^{-1}(y)$ and $\mathrm{lev}_{<}\Phi(\cdot, y)$ are open for each $y \in A$, and either "$\Phi \in \mathscr{A}_A^{\geq}$ and $\mathscr{D}$ is quasicyclic" or "$\Phi \in \mathscr{A}_{caqm}$ and $\mathscr{D}(y) \subset \mathscr{D}(x)$ for any $x \in A$ and $y \in \mathscr{D}(x)$." Then, we obtain the conclusion of Theorem 2.21.*

Proof The set $\mathrm{Fix}\mathscr{D}$ is nonempty by Theorem 2.19. From the proof of Theorem 2.21, it is sufficient to show that $\mathrm{Sol}(\mathrm{EP}_{\overline{\Phi}_T})$ is nonempty. Obviously $\mathrm{lev}_{\geq}\overline{\Phi}_T(\cdot, y) = \mathrm{cl}(T(y))$ is closed for each $y \in A$. We show that $\overline{\Phi}_T \in \mathscr{A}^2_{caqm}$. Take any $x^1, x^2, \ldots, x^m \in A$ such that $\overline{\Phi}_T(x^i, x^{i+1}) < 0$ for all $i = 1, 2, \ldots, m-1$. Assume that $\overline{\Phi}_T(x^m, x^1) < 0$. Then, by the definition of T we have:

$(*)$ $x^{i+1} \in \mathscr{D}(x^i)$ for all $i = 1, 2, \ldots, m$ $(x^{m+1} := x^1)$.
$(**)$ $x^i \notin \mathrm{Fix}\mathscr{D} \cap \mathrm{lev}_{\geq}\Phi(\cdot, x^{i+1})$ for all $i = 1, 2, \ldots, m$ $(x^{m+1} := x^1)$.

If (a) holds, then $(*)$ and the anti-cyclicity of $\mathscr{D}$ imply that there must exist an i such that $x^{i+1} = x^i$. Then, $x^i \in \mathrm{Fix}\mathscr{D}$. Moreover, $\Phi(x^i, x^i) \geq 0$, i.e., $x^i \in \mathrm{lev}_{\geq}\Phi(\cdot, x^i)$. Hence, $x^i \in \mathrm{Fix}\mathscr{D} \cap \mathrm{lev}_{\geq}\Phi(\cdot, x^i) \subset T(x^i) \subset \mathrm{cl}T(x^i)$. Then, $\overline{\Phi}_T(x^i, x^{i+1}) = \overline{\Phi}_T(x^i, x^i) \geq 0$, a contradiction. If (b) holds, then by the assumption "$\mathscr{D}(y) \subset \mathscr{D}(x)$ for any $x \in A$ and $y \in \mathscr{D}(x)$," $(*)$ implies that $\mathscr{D}(x^1) = \mathscr{D}(x^2) = \cdots = \mathscr{D}(x^m)$, and hence, $x^i \in \mathrm{Fix}\mathscr{D}$ for all $i = 1, 2, \ldots, m$. This and $(**)$ show that $x^i \notin \mathrm{lev}_{\geq}\Phi(\cdot, x^{i+1})$ for all $i = 1, 2, \ldots, m$, i.e., $\Phi(x^i, x^{i+1}) < 0$ for all $i = 1, 2, \ldots, m$. This contradicts the cyclic anti-quasimonotonicity of Φ. Thus, $\overline{\Phi}_T(x^m, x^1) \geq 0$. Hence, $\overline{\Phi}_T \in \mathscr{A}_{caqm}$. Applying Theorem 1.82, we have that $\mathrm{Sol}(\mathrm{EP}_{\overline{\Phi}_T}) \neq \emptyset$. □

Example 2.49 Let $A = \{x \in \mathbb{R}^n \mid r \leq \|x\| \leq 1\}$ $(0 \leq r < 1)$, $k \in \mathbb{N}$, $\mathscr{D} : A \rightrightarrows A$, and $\Phi : A \times A \to \mathbb{R}$ be defined by

$$\mathscr{D}(x) := \begin{cases} \{y \in A \mid \|y\| < \frac{\|x\|+r}{2}\} & \text{if } \|x\| > r, \\ \{y \in A \mid \|y\| = r\} & \text{if } \|x\| = r \end{cases} \quad \text{for all } x \in A;$$

$$\Phi(x, y) := \|x\|\|y\| - \|x\|^k \quad \text{for all } x, y \in A.$$

Then, for each $y \in A$,

$$\mathscr{D}^{-1}(y) = \begin{cases} A \cap \{x \in A \mid 2\|y\| - r < \|x\|\} & \text{if } \|y\| > r, \\ A & \text{if } \|y\| = r, \end{cases}$$

and

$$\text{lev}_{<}\Phi(\cdot, y) = \{x \in A \mid \sqrt[k-1]{\|y\|} < \|x\|\}$$

are open in A. Let any $x \in A$, $y \in \mathscr{D}(x)$ and $z \in \mathscr{D}(y)$. If $\|x\| > r$, then $\|z\| < \frac{\|y\|+r}{2} \leq \frac{\|x\|+r}{2}$. If $\|x\| = r$, then $\|z\| = \|y\| = \|x\|$. Hence, $\mathscr{D}(y) \subset \mathscr{D}(x)$. To verify the cyclic anti-quasimonotonicity of Φ, take arbitrarily $x^1, x^2, \ldots, x^m \in A$ such that

$$\Phi(x^i, x^{i+1}) = \|x\|^i \|x\|^{i+1} - (\|x\|^i)^k < 0 \quad \text{for } i = 1, 2, \ldots, m-1. \tag{2.7}$$

Suppose to the contrary that $\Phi(x^m, x^1) = \|x^m\| \|x^1\| - (\|x^m\|)^k < 0$. This together with (2.7) implies that

$$(\|x^1\| \|x^2\|)(\|x^2\| \|x^3\|) \ldots (\|x^m\| \|x^1\|) < (\|x^1\|)^k (\|x^2\|)^k \ldots (\|x^m\|)^k,$$

i.e.,

$$1 < (\|x^1\|)^{k-2} (\|x^2\|)^{k-2} \ldots (\|x^m\|)^{k-2}.$$

Since $(\|x^1\|)^{k-2}(\|x^2\|)^{k-2} \ldots (\|x^m\|)^{k-2} \leq 1$, we have a contradiction. Thus, Φ is cyclically anti-quasimonotone. We have that $\text{Fix}\mathscr{D} = \{x \in A \mid \|x\| = r$ is closed in A. Thus, by Theorem 2.48, the problem (QEP) has a solution.

2.5 Set-Valued Versions of Equilibrium and Quasiequilibrium Problems

In this section we discuss several versions of equilibrium/quasiequilibrium problems, where objective bifunctions are set-valued bifunctions. These problems are called set-valued equilibrium/quasiequilibrium problems. Let A be a nonempty set, $\mathscr{D} : A \rightrightarrows A$ be a set-valued mapping with nonempty images, and $F : A \times A \rightrightarrows \mathbb{R}$ be a set-valued bifunction. We consider the following problems.

- *Strong set-valued equilibrium problem*:

$$(\text{SSEP}_F) \qquad \text{find } \bar{x} \in A \text{ such that } F(\bar{x}, y) \subset \mathbb{R}_+ \text{ for all } y \in A.$$

- *Weak set-valued equilibrium problem*:

$$(\mathrm{WSEP}_F) \qquad \text{find } \bar{x} \in A \text{ such that } F(\bar{x}, y) \cap \mathbb{R}_+ \neq \emptyset \text{ for all } y \in A.$$

- *Strong set-valued quasiequilibrium problem*:

$$(\mathrm{SSQEP}_{(F,\mathscr{D})}) \qquad \text{find } \bar{x} \in A \text{ such that } \bar{x} \in \mathscr{D}(\bar{x}) \text{ and}$$
$$F(\bar{x}, y) \subset \mathbb{R}_+ \text{ for all } y \in \mathscr{D}(\bar{x}).$$

- *Weak set-valued quasiequilibrium problem*:

$$(\mathrm{WSQEP}_{(F,\mathscr{D})}) \qquad \text{find } \bar{x} \in A \text{ such that } \bar{x} \in \mathscr{D}(\bar{x}) \text{ and}$$
$$F(\bar{x}, y) \cap \mathbb{R}_+ \neq \emptyset \text{ for all } y \in \mathscr{D}(\bar{x}).$$

By Sol(SSEP_F) (Sol(WSEP_F), Sol($\mathrm{SSQEP}_{(F,\mathscr{D})}$), and Sol($\mathrm{WSQEP}_{(F,\mathscr{D})}$), resp.), we denote the solution set of (SSEP_F) ((WSEP_F), ($\mathrm{SSQEP}_{(F,\mathscr{D})}$), ($\mathrm{WSQEP}_{(F,\mathscr{D})}$), resp.). It is obvious that Sol(SSEP_F) $\subset$ Sol(WSEP_F) and Sol($\mathrm{SSQEP}_{(F,\mathscr{D})}$) $\subset$ Sol($\mathrm{WSQEP}_{(F,\mathscr{D})}$).

When $F \equiv \Phi : A \times A \to \mathbb{R}$, both ($\mathrm{SSEP}_F$) and ($\mathrm{WSEP}_F$) (both ($\mathrm{SSQEP}_{(F,\mathscr{D})}$ and ($\mathrm{WSQEP}_{(F,\mathscr{D})}$, resp.) collapse into (EP) ((QEP), resp.).

We transform the above problems to equilibrium and quasiequilibrium problems. Define bifunctions $\Phi^F_{\subset}, \Phi^F_{\cap} : A \times A \to \mathbb{R}$ by, resp.,

$$\Phi^F_{\subset}(x, y) = \begin{cases} 1 & \text{if } F(x, y) \subset \mathbb{R}_{++}, \\ 0 & \text{if } F(x, y) \cap (-\mathbb{R}_+) = \{0\}, \\ -1 & \text{if otherwise;} \end{cases}$$

$$\Phi^F_{\cap}(x, y) = \begin{cases} 1 & \text{if } F(x, y) \cap \mathbb{R}_{++} \neq \emptyset, \\ 0 & \text{if } F(x, y) \cap \mathbb{R}_+ = \{0\}, \\ -1 & \text{if otherwise.} \end{cases}$$

For $\mathscr{T} \in \{\subset, \cap\}$, we consider the following equilibrium and quasiequilibrium problems:

$$(\mathrm{EP}_{\Phi^F_{\mathscr{T}}}) \qquad \text{find } \bar{x} \in A \text{ such that } \Phi^F_{\mathscr{T}}(x, y) \geq 0 \text{ for all } y \in A;$$

$$(\mathrm{QEP}_{(\Phi^F_{\mathscr{T}},\mathscr{D})}) \quad \text{find } \bar{x} \in A \text{ such that } \bar{x} \in \mathscr{D}(\bar{x}) \text{ and } \Phi^F_{\mathscr{T}}(x, y) \geq 0 \text{ for all } y \in \mathscr{D}(\bar{x}).$$

The solution set of ($\mathrm{EP}_{(\Phi^F_{\mathscr{T}},\mathscr{D})}$) (($\mathrm{QEP}_{\Phi^F_{\mathscr{T}}}$), resp.) ($\mathscr{T} \in \{\subset, \cap\}$) is denoted by Sol($\mathrm{EP}_{\Phi^F_{\mathscr{T}}}$) (Sol($\mathrm{QEP}_{(\Phi^F_{\mathscr{T}},\mathscr{D})}$), resp.). The following proposition easily follows from the definitions of problems. It states the equivalence between equilibrium/quasiequilibrium problems and set-valued equilibrium/quasiequilibrium problems.

Proposition 2.50 *For the above problems, we have that:*

(a) $\mathrm{Sol}(\mathrm{SSEP}_F) = \mathrm{Sol}(\mathrm{EP}_{\Phi^F_{\subset}})$ *and* $\mathrm{Sol}(\mathrm{SSQEP}_{(F,\mathscr{D})}) = \mathrm{Sol}(\mathrm{QEP}_{(\Phi^F_{\subset},\mathscr{D})})$.
(b) $\mathrm{Sol}(\mathrm{WSEP}_F) = \mathrm{Sol}(\mathrm{EP}_{\Phi^F_{\cap}})$ *and* $\mathrm{Sol}(\mathrm{WSQEP}_{(F,\mathscr{D})}) = \mathrm{Sol}(\mathrm{QEP}_{(\Phi^F_{\cap},\mathscr{D})})$.
(c) *for* $F^{cl} : A \times A \rightrightarrows \mathbb{R}$ *defined by* $F^{cl}(x, y) := \mathrm{cl}F(x, y)$ *for all* $x, y \in A$, $\mathrm{Sol}(\mathrm{SSEP}_F) = \mathrm{Sol}(\mathrm{SSEP}_{F^{cl}})$ *and* $\mathrm{Sol}(\mathrm{SSQEP}_{(F,\mathscr{D})}) = \mathrm{Sol}(\mathrm{SSQEP}_{(F^{cl},\mathscr{D})})$.

Through problems $(\mathrm{EP}_{\Phi^F_{\mathscr{T}}})$ and $(\mathrm{QEP}_{(\Phi^F_{\mathscr{T}},\mathscr{D})})$ we can convert results in Chap. 1 and Sect. 2.2 of this chapter into their set-valued versions. In the next part we will convert some results. For the remaining results, the readers can make a conversion in a similar way.

First, we have the following remark.

Remark 2.51

(a) Looking at the definitions of the bifunctions $\Phi^F_{\mathscr{T}}$, $\mathscr{T} \in \{\subset, \cap\}$, we see that, for all $x, y \in A$,

(a1) $\mathrm{lev}_{\geq}\Phi^F_{\subset}(\cdot, y) = \{x \in A \mid F(x, y) \subset \mathbb{R}_+\} := \mathrm{Lev}^{\geq}_{\subset}F(\cdot, y)$.
(a2) $\mathrm{lev}_{<}\Phi^F_{\subset}(\cdot, y) = \{x \in A \mid F(x, y) \cap (-\mathbb{R}_{++}) \neq \emptyset\} := \mathrm{Lev}^{<}_{\subset}F(\cdot, y)$.
(a3) $\mathrm{lev}_{\leq}\Phi^F_{\subset}(x, \cdot) = \{y \in A \mid F(x, y) \cap (-\mathbb{R}_+) \neq \emptyset\} := \mathrm{Lev}^{\leq}_{\subset}F(x, \cdot)$.
(a4) $\mathrm{lev}_{>}\Phi^F_{\subset}(\cdot, x) = \{y \in A \mid F(y, x) \cap (-\mathbb{R}_+) = \emptyset\} := \mathrm{Lev}^{>}_{\subset}F(\cdot, x)$.
(a5) $\mathrm{lev}_{<}\Phi^F_{\subset}(x, \cdot) = \{y \in A \mid F(x, y) \cap (-\mathbb{R}_{++}) \neq \emptyset\} := \mathrm{Lev}^{<}_{\subset}F(x, \cdot)$.
(a6) $\mathrm{lev}_{\geq}\Phi^F_{\cap}(\cdot, y) = \{x \in A \mid F(x, y) \cap \mathbb{R}_+ \neq \emptyset\} := \mathrm{Lev}^{\geq}_{\cap}F(\cdot, y)$.
(a7) $\mathrm{lev}_{<}\Phi^F_{\cap}(\cdot, y) = \{x \in A \mid F(x, y) \subset -\mathbb{R}_{++}\} := \mathrm{Lev}^{<}_{\cap}F(\cdot, y)$.
(a8) $\mathrm{lev}_{<}\Phi^F_{\cap}(x, \cdot) = \{y \in A \mid F(x, y) \cap \mathbb{R}_+ = \emptyset\} := \mathrm{Lev}^{<}_{\cap}F(x, \cdot)$.
(a9) $\mathrm{lev}_{\leq}\Phi^F_{\cap}(x, \cdot) = \{y \in A \mid F(x, y) \subset -\mathbb{R}_+\} := \mathrm{Lev}^{\leq}_{\cap}F(x, \cdot)$.
(a10) $\mathrm{lev}_{>}\Phi^F_{\cap}(\cdot, x) = \{y \in A \mid F(y, x) \cap \mathbb{R}_{++} \neq \emptyset\} := \mathrm{Lev}^{>}_{\cap}F(\cdot, x)$.

(b) When A is a subset of a topological space, it follows from Propositions 2.3–2.4 that:

(b1) If $F(\cdot, y)$ is lower semicontinuous (upper semicontinuous, resp.), then $\mathrm{Lev}^{\geq}_{\subset}F(\cdot, y)$ is closed and $\mathrm{Lev}^{<}_{\subset}F(\cdot, y)$ is open ($\mathrm{Lev}^{\geq}_{\cap}F(\cdot, y)$ is closed and $\mathrm{Lev}^{<}_{\cap}F(\cdot, y)$ is open, resp.).
(b2) If $F(x, \cdot)$ is lower semicontinuous (upper semicontinuous, resp.), then $\mathrm{Lev}^{\leq}_{\cap}F(x, \cdot)$ is closed and $\mathrm{Lev}^{<}_{\subset}F(x, \cdot)$ is open ($\mathrm{Lev}^{\leq}_{\subset}F(x, \cdot)$ is closed and $\mathrm{Lev}^{<}_{\cap}F(x, \cdot)$ is open, resp.).
(b3) If $F(\cdot, x)$ is lower semicontinuous (upper semicontinuous, resp.), then $\mathrm{Lev}^{>}_{\cap}F(\cdot, x)$ ($\mathrm{Lev}^{>}_{\subset}F(\cdot, x)$, resp.,) is open.

The following definition extends some concepts on convexity in Definitions 1.19 and 1.21 to set-valued maps.

Definition 2.52 Let A be a nonempty convex subset of vector space and $S : A \rightrightarrows \mathbb{R}$ be a set-valued map.

(a) S is called $\mathbb{R}_+$-CONVEX OF TYPE 1 ($\mathbb{R}_+$-CONVEX OF TYPE 2, resp.) if for all $x, y \in A$ and $t \in (0, 1)$,

$$tS(x) + (1-t)S(y) \subset S(tx + (1-t)y) + \mathbb{R}_+$$

$$(S(tx + (1-t)y) \subset tS(x) + (1-t)S(y) - \mathbb{R}_+, \text{ resp.}).$$

(b) S is called $\mathbb{R}_+$-QUASICONVEX OF TYPE 1 ($\mathbb{R}_+$-QUASICONVEX OF TYPE 2, resp.) if, for all $x, y \in A$ and $t \in (0, 1)$, we have that

$$\text{either } S(x) \subset S(tx + (1-t)y) + \mathbb{R}_+, \text{ or } S(y) \subset S(tx + (1-t)y) + \mathbb{R}_+$$

$$(\text{ either } S(tx+(1-t)y) \subset S(x)-\mathbb{R}_+, \text{ or } S(tx+(1-t)y) \subset S(y)-\mathbb{R}_+, \text{ resp.}).$$

(c) S is called STRONGLY $\mathbb{R}_+$-QUASICONVEX-LIKE OF TYPE 1 (STRONGLY $\mathbb{R}_+$-QUASICONVEX-LIKE OF TYPE 2, resp.) if for all $x, y \in A$ and $t \in (0, 1)$,

$$\text{from } S(x) \not\subset \mathbb{R}_+ \text{ and } S(y) \not\subset \mathbb{R}_+, \text{ it implies that } S(tx + (1-t)y) \not\subset \mathbb{R}_+$$

$$(\text{from } S(x) \subset -\mathbb{R}_{++} \text{ and } S(y) \subset -\mathbb{R}_{++}, \text{ it implies that } S(tx + (1-t)y) \subset -\mathbb{R}_{++}, \text{ resp.}).$$

Thus, S is strongly $\mathbb{R}_+$-quasiconvex-like of type 1 (of type 2, resp.) if and only if the level set $\{x \in A \mid S(x) \cap (-\mathbb{R}_{++}) \neq \emptyset\}$ ($\{x \in A \mid S(x) \subset -\mathbb{R}_{++}\}$, resp.) is convex.

If $F \equiv \Phi$, the $\mathbb{R}_+$-convexity (the $\mathbb{R}_+$-quasiconvexity, the strong $\mathbb{R}_+$-quasiconvexity-like, resp.) of type ℓ ($\ell \in \{1, 2\}$) recovers to the convexity (quasiconvexity, strong quasiconvexity-like, resp.) in Definitions 1.19 and 1.21.

We say that a set-valued bifunction $F : A \times A \rightrightarrows \mathbb{R}$ has some property (such as $\mathbb{R}_+$-convexity of type 1, $\mathbb{R}_+$-quasiconvexity of type 1, etc.) in the first (second, resp.) component if for all $y \in A$ (for all $x \in A$, resp.), $F(\cdot, y)$ ($F(x, \cdot)$, resp.) has this property.

For $\ell \in \{1, 2\}$ and $i \in \{1, 2\}$, we set

$$\mathscr{S}^\ell \mathscr{A}_c^i = \{F : A \times A \rightrightarrows \mathbb{R} \mid F \text{ is } \mathbb{R}_+\text{-convex of type } \ell \text{ in ith component}\};$$

$$\mathscr{S}^\ell \mathscr{A}_{qc}^i = \{F : A \times A \rightrightarrows \mathbb{R} \mid F \text{ is } \mathbb{R}_+\text{-quasiconvex of type } \ell \text{ in ith component}\};$$

$$\mathscr{S}^\ell \mathscr{A}^i_{st.qcl} = \{F : A \times A \rightrightarrows \mathbb{R} \mid F \text{ is strongly } \mathbb{R}_+\text{-quasiconvex-like of type } \ell \text{ in ith component}\}.$$

Remark 2.53

(a) It is not hard to check that $\mathscr{S}^\ell \mathscr{A}^i_c \subset \mathscr{S}^\ell \mathscr{A}^i_{qc} \subset \mathscr{S}^\ell \mathscr{A}^i_{st.qcl}$ (see Exercise 2.14).
(b) $\Phi^F_\subset \in \mathscr{A}^2_{st.qlc} \Leftrightarrow \forall x \in A, \{y \in A \mid \Phi^F_\subset(x, y) < 0\} = \{y \in A \mid F(x, y) \cap (-\mathbb{R}_{++}) \neq \emptyset\}$ is convex $\Leftrightarrow F \in \mathscr{S}^1 \mathscr{A}^2_{st.qlc}$.
(c) $\Phi^F_\cap \in \mathscr{A}^2_{st.qlc} \Leftrightarrow \forall x \in A, \{y \in A \mid \Phi^F_\cap(x, y) < 0\} = \{y \in A \mid F(x, y) \subset -\mathbb{R}_{++}\}$ is convex $\Leftrightarrow F \in \mathscr{S}^2 \mathscr{A}^2_{st.qlc}$.

Definition 2.54 Let A be a nonempty convex subset of vector space and $F : A \times A \rightrightarrows \mathbb{R}$ be a set-valued bifunction.

(a) F is said to be PROPERLY $\mathbb{R}_+$-QUASIMONOTONE OF TYPE 1 in the first component (in the second component, resp.) on A if, for all $N \in \mathscr{F}(A)$ and $x \in \text{conv}N$, there exists $x^j \in N$ such that $F(x^j, x) \subset \mathbb{R}_+$ ($F(x, x^j) \subset \mathbb{R}_+$, resp.).
(b) F is said to be PROPERLY $\mathbb{R}_+$-QUASIMONOTONE OF TYPE 2 in the first component (in the second component, resp.) on A if, for all $N \in \mathscr{F}(A)$ and $x \in \text{conv}N$, there exists $x^j \in N$ such that $F(x^j, x) \cap \mathbb{R}_+ \neq \emptyset$ ($F(x, x^j) \cap \mathbb{R}_+ \neq \emptyset$, resp.).

When $F \equiv \Phi$, the properly $\mathbb{R}_+$-quasimonotonicity of type ℓ ($\ell \in \{1, 2\}$) recovers to the properly quasimonotonicity in Definitions 1.27.

For $\ell \in \{1, 2\}$ and $i \in \{1, 2\}$, by $\mathscr{S}^\ell \mathscr{A}^i_{pr.qm}$ we denote the class of set-valued bifunctions defined on $A \times A$ which are properly $\mathbb{R}_+$-quasimonotone of type ℓ:

$$\mathscr{S}^\ell \mathscr{A}^i_{pr.qm} = \{F : A \times A \rightrightarrows \mathbb{R} \mid F \text{ is properly } \mathbb{R}_+\text{-quasimonotone of type } \ell \text{ in ith component}\}.$$

For $F : A \times A \rightrightarrows \mathbb{R}$, we define $-F : A \times A \rightrightarrows \mathbb{R}$ for all $x, y \in A$ by, $(-F)(x, y) = -F(x, y) = \{-t \mid t \in F(x, y)\}$.

We set

$$\mathscr{S}^1 \mathscr{A}^{\geq}_A = \{F : A \times A \rightrightarrows \mathbb{R} \mid F(x, x) \subset \mathbb{R}_+ \text{ for all } x \in A\},$$

$$\mathscr{S}^2 \mathscr{A}^{\geq}_A = \{F : A \times A \rightrightarrows \mathbb{R} \mid F(x, x) \cap \mathbb{R}_+ \neq \emptyset \text{ for all } x \in A\},$$

$$\mathscr{S}^1 \mathscr{A}^{=}_A = \{F : A \times A \rightrightarrows \mathbb{R} \mid F(x, x) \cap (-\mathbb{R}_+) = \{0\} \text{ for all } x \in A\},$$

$$\mathscr{S}^2 \mathscr{A}^{=}_A = \{F : A \times A \rightrightarrows \mathbb{R} \mid F(x, x) \cap \mathbb{R}_+ = \{0\} \text{ for all } x \in A\}.$$

For a class of bifunctions $\mathscr{S}\mathscr{A}$, we define $-\mathscr{S}\mathscr{A} := \{F : A \times A \rightrightarrows \mathbb{R} \mid -F \in \mathscr{S}\mathscr{A}\}$.

Remark 2.55

(a) For $\ell \in \{1, 2\}$ and $i \in \{1, 2\}$, $\mathscr{S}^{\ell}\mathscr{A}^{i}_{st.qcl} \cap \mathscr{S}^{\ell}\mathscr{A}^{\geq}_{A} \subset \mathscr{S}^{\ell}\mathscr{A}^{i}_{pr.qm}$ (see Exercise 2.14). Moreover, it is clear that $\mathscr{S}^{1}\mathscr{A}^{i}_{pr.qm} \subset \mathscr{S}^{2}\mathscr{A}^{i}_{pr.qm}$.

(b) By definitions of $\Phi^F_{\subset}$ and $\Phi^F_{\cap}$, $\Phi^F_{\subset}(x, y) \geq 0$ if and only if $F(x, y) \subset \mathbb{R}_+$, and $\Phi^F_{\cap}(x, y) \geq 0$ if and only if $F(x, y) \cap \mathbb{R}_+ \neq \emptyset$. Therefore, it follows that $\Phi^F_{\subset} \in \mathscr{A}^{i}_{pr.qm}$ if and only if $F \in \mathscr{S}^{1}\mathscr{A}^{i}_{pr.qm}$, and $\Phi^F_{\cap} \in \mathscr{A}^{i}_{pr.qm}$ if and only if $F \in \mathscr{S}^{2}\mathscr{A}^{i}_{pr.qm}$.

(c) By definitions of $\Phi^F_{\subset}$ and $\Phi^F_{\cap}$, $-\Phi^F_{\subset}(x, y) \geq 0$ if and only if $(-F(x, y)) \cap \mathbb{R}_+ \neq \emptyset$, and $-\Phi^F_{\cap}(x, y) \geq 0$ if and only if $(-F(x, y)) \subset \mathbb{R}_+$. It follows that $\Phi^F_{\subset} \in -\mathscr{A}^{i}_{pr.qm}$ if and only if $F \in -\mathscr{S}^{2}\mathscr{A}^{i}_{pr.qm}$, and $\Phi^F_{\cap} \in -\mathscr{A}^{i}_{pr.qm}$ if and only if $F \in -\mathscr{S}^{1}\mathscr{A}^{i}_{pr.qm}$.

(d) Similarly as above, $\Phi^F_{\subset} \in \mathscr{A}^{\geq}_{A}$ if and only if $F \in \mathscr{S}^{1}\mathscr{A}^{\geq}_{A}$, $\Phi^F_{\subset} \in \mathscr{A}^{=}_{A}$ if and only if $F \in \mathscr{S}^{1}\mathscr{A}^{=}_{A}$, $\Phi^F_{\cap} \in \mathscr{A}^{\geq}_{A}$ if and only if $F \in \mathscr{S}^{2}\mathscr{A}^{\geq}_{A}$, and $\Phi^F_{\cap} \in \mathscr{A}^{=}_{A}$ if and only if $F \in \mathscr{S}^{2}\mathscr{A}^{=}_{A}$.

Definition 2.56 Let A be a nonempty convex subset of a vector space and $S : A \rightrightarrows \mathbb{R}$ be a set-valued map.

(a) S is said to HAVE PROPERTY $\mathbb{SP}$ OF TYPE 1 if for all $x, y \in A$, $S(x) \cap (-\mathbb{R}_+) = \{0\}$ and $S(y) \cap (-\mathbb{R}_{++}) \neq \emptyset$ imply $S((1-t)x + ty) \cap (-\mathbb{R}_{++}) \neq \emptyset$ for all $t \in (0, 1)$.

(b) S is said to HAVE PROPERTY $\mathbb{SP}$ OF TYPE 2 if for all $x, y \in A$, $S(x) \cap \mathbb{R}_+ = \{0\}$ and $S(y) \subset -\mathbb{R}_{++}$ imply $S((1-t)x + ty) \subset -\mathbb{R}_{++}$ for all $t \in (0, 1)$.

Set-valued bifunction $F : A \times A \rightrightarrows \mathbb{R}$ is said to HAVE PROPERTY $\mathbb{SP}^1$ of type ℓ ($\mathbb{SP}^2$ of type ℓ, resp.) ($\ell \in \{1, 2\}$) if $F(\cdot, y)$ ($F(x, \cdot)$, resp.) has property $\mathbb{SP}$ of type ℓ for all $y \in A$ ($x \in A$, resp.).

Remark 2.57

(a) If $S \equiv \varphi : A \to \mathbb{R}$, then properties $\mathbb{SP}$ of type ℓ ($\ell \in \{1, 2\}$) collapse to property $\mathbb{P}$ in Definition 1.38.

(b) It is not hard to check that:

- F has property $\mathbb{SP}^i$ of type 1 if and only if $\Phi^F_{\subset}$ has property $\mathbb{P}^i$ (see Definition 1.38).
- F has property $\mathbb{SP}^i$ of type 2 if and only if $\Phi^F_{\cap}$ has property $\mathbb{P}^i$.

Definition 2.58 Let A be a nonempty convex subset of a vector space and $F : A \times A \rightrightarrows \mathbb{R}$ be a set-valued bifunction.

(a) F is called $\mathbb{R}_+$-UPPER SIGN-CONTINUOUS OF TYPE 1 in the first component if for all $x, y \in A$, $F(x + t(y - x), y) \subset \mathbb{R}_+$ for all $t \in (0, 1)$ implies that $F(x, y) \subset \mathbb{R}_+$.
(b) F is called $\mathbb{R}_+$-UPPER SIGN-CONTINUOUS OF TYPE 2 in the first component if for all $x, y \in A$, $F(x + t(y - x), y) \cap \mathbb{R}_+ \neq \emptyset$ for all $t \in (0, 1)$ implies that $F(x, y) \cap \mathbb{R}_+ \neq \emptyset$.

Similar definitions can be stated for the second component.

Remark 2.59

(a) If $F \equiv \Phi : A \times A \to \mathbb{R}$ then, for $\ell \in \{1, 2\}$ and $i \in \{1, 2\}$, the $\mathbb{R}_+$-upper sign-continuity of type ℓ in ith component collapses to the upper sign-continuity in ith component in Definition 1.18.
(b) It is not hard to check that:
   - F is $\mathbb{R}_+$-upper sign-continuous of type 1 in ith component if and only if $\Phi^F_{\subset}$ is upper sign-continuity in ith component.
   - F is $\mathbb{R}_+$-upper sign-continuous of type 2 in ith component if and only if $\Phi^F_{\cap}$ is upper sign-continuity in ith component.

Let A be a nonempty convex subset of a topological vector space and $F : A \times A \rightrightarrows \mathbb{R}$ be a set-valued bifunction. We summarize some relations between F, $\Phi^F_{\subset}$, and $\Phi^F_{\cap}$ which have been stated in Proposition 2.50 and Remarks 2.51, 2.53–2.55, and 2.57–2.59.

• The relationships between $\Phi^F_{\subset}$ and F:

$$\begin{cases} \circ\ \Phi^F_{\subset} \in \mathscr{A}^2_{pr.qm} \ \Leftrightarrow\ F \in \mathscr{S}^1\mathscr{A}^2_{pr.qm}, \\ \circ\ \Phi^F_{\subset}(x, y) < 0 \ \Leftrightarrow\ F(x, y) \cap (-\mathbb{R}_{++}) \neq \emptyset, \\ \circ\ \mathrm{lev}_{\geq}\Phi^F_{\subset}(\cdot, y) = \mathrm{Lev}^{\geq}_{\subset}F(\cdot, y) := \{x \in A \mid F(x, y) \subset \mathbb{R}_+\}, \\ \circ\ \mathrm{lev}_{<}\Phi^F_{\subset}(\cdot, y) = \mathrm{Lev}^{<}_{\subset}F(\cdot, y) := \{x \in A \mid F(x, y) \cap (-\mathbb{R}_{++}) \neq \emptyset\}, \\ \circ\ \mathrm{lev}_{<}\Phi^F_{\subset}(x, \cdot) = \mathrm{Lev}^{<}_{\subset}F(x, \cdot) := \{y \in A \mid F(x, y) \cap (-\mathbb{R}_{++}) \neq \emptyset\}, \\ \circ\ \mathrm{Sol}(\mathrm{EP}_{\Phi^F_{\subset}}) = \mathrm{Sol}(\mathrm{SSEP}_F), \\ \circ\ \mathrm{Sol}(\mathrm{QEP}_{(\Phi^F_{\subset}, \mathscr{D})}) = \mathrm{Sol}(\mathrm{SSQEP}_{(F, \mathscr{D})}), \end{cases} \tag{2.8}$$

and

$$\begin{cases} \circ\ \Phi^F_{\subset} \in (-\mathscr{A}^1_{pr.qm}) \cap \mathscr{A}^2_{st.qcl} \cap \mathscr{A}^=_A \\ \quad \Leftrightarrow\ F \in (-\mathscr{S}^2\mathscr{A}^1_{pr.qm}) \cap \mathscr{S}^1\mathscr{A}^2_{st.qcl} \cap \mathscr{S}^1\mathscr{A}^=_A, \\ \circ\ \Phi^F_{\subset}(y,x) > 0 \ \Leftrightarrow\ F(y,x) \subset \mathbb{R}_{++}, \\ \circ\ \Phi^F_{\subset} \text{ has the property } \mathbb{P}^2 \ \Leftrightarrow\ F \text{ has the property } \mathbb{SP}^2 \text{ of type 1}, \\ \circ\ \Phi^F_{\subset} \text{ is upper-sign continuous in the first argument} \\ \quad \Leftrightarrow\ F \text{ is } \mathbb{R}_+\text{-upper-sign continuous of type 1 in the first argument}, \\ \circ\ \mathrm{lev}_{\leq}\Phi^F_{\subset}(x,\cdot) = \mathrm{Lev}^{\leq}_{\subset}F(x,\cdot) := \{y \in A \mid F(x,y) \cap (-\mathbb{R}_+) \neq \emptyset\}, \\ \circ\ \mathrm{lev}_{>}\Phi^F_{\subset}(\cdot,x) = \mathrm{Lev}^{>}_{\subset}F(\cdot,x) := \{y \in A \mid F(y,x) \cap (-\mathbb{R}_+) = \emptyset\}, \\ \circ\ \mathrm{Sol}(\mathrm{EP}_{\Phi^F_{\subset}}) = \mathrm{Sol}(\mathrm{SSEP}_F), \\ \circ\ \mathrm{Sol}(\mathrm{QEP}_{(\Phi^F_{\subset},\mathscr{D})}) = \mathrm{Sol}(\mathrm{SSQEP}_{(F,\mathscr{D})}). \end{cases} \tag{2.9}$$

- The relationships between $\Phi^F_{\cap}$ and F:

$$\begin{cases} \circ\ \Phi^F_{\cap} \in \mathscr{A}^2_{pr.qm} \ \Leftrightarrow\ F \in \mathscr{S}^2\mathscr{A}^2_{pr.qm}, \\ \circ\ \Phi^F_{\cap}(x,y) < 0 \ \Leftrightarrow\ F(x,y) \subset -\mathbb{R}_{++}, \\ \circ\ \mathrm{lev}_{\geq}\Phi^F_{\cap}(\cdot,y) = \mathrm{Lev}^{\geq}_{\cap}F(\cdot,y) := \{x \in A \mid F(x,y) \cap \mathbb{R}_+ \neq \emptyset\}, \\ \circ\ \mathrm{lev}_{<}\Phi^F_{\cap}(\cdot,y) = \mathrm{Lev}^{<}_{\cap}F(\cdot,y) := \{x \in A \mid F(x,y) \cap \mathbb{R}_+ = \emptyset\}, \\ \circ\ \mathrm{lev}_{<}\Phi^F_{\cap}(x,\cdot) = \mathrm{Lev}^{<}_{\cap}F(x,\cdot) := \{y \in A \mid F(x,y) \cap \mathbb{R}_+ = \emptyset\}, \\ \circ\ \mathrm{Sol}(\mathrm{EP}_{\Phi^F_{\cap}}) = \mathrm{Sol}(\mathrm{WSEP}_F), \\ \circ\ \mathrm{Sol}(\mathrm{QEP}_{(\Phi^F_{\cap},\mathscr{D})}) = \mathrm{Sol}(\mathrm{WSQEP}_{(F,\mathscr{D})}), \end{cases} \tag{2.10}$$

and

$$\begin{cases} \circ\ \Phi^F_{\cap} \in (-\mathscr{A}^1_{pr.qm}) \cap \mathscr{A}^2_{st.qcl} \cap \mathscr{A}^=_A \\ \quad \Leftrightarrow\ F \in (-\mathscr{S}^1\mathscr{A}^1_{pr.qm}) \cap \mathscr{S}^2\mathscr{A}^2_{st.qcl} \cap \mathscr{S}^2\mathscr{A}^=_A, \\ \circ\ \Phi^F_{\cap}(y,x) > 0 \ \Leftrightarrow\ F(y,x) \cap \mathbb{R}_{++} \neq \emptyset, \\ \circ\ \Phi^F_{\cap} \text{ has the property } \mathbb{P}^2 \ \Leftrightarrow\ F \text{ has the property } \mathbb{SP}^2 \text{ of type 2}, \\ \circ\ \Phi^F_{\cap} \text{ is upper-sign continuous in the first argument} \\ \quad \Leftrightarrow\ F \text{ is } \mathbb{R}_+\text{-upper-sign continuous of type 2 in the first argument}, \\ \circ\ \mathrm{lev}_{\leq}\Phi^F_{\cap}(x,\cdot) = \mathrm{Lev}^{\leq}_{\cap}F(x,\cdot) := \{y \in A \mid F(x,y) \subset -\mathbb{R}_+\}, \\ \circ\ \mathrm{lev}_{>}\Phi^F_{\cap}(\cdot,x) = \mathrm{Lev}^{>}_{\cap}F(\cdot,x) := \{y \in A \mid F(y,x) \cap \mathbb{R}_{++} \neq \emptyset\}, \\ \circ\ \mathrm{Sol}(\mathrm{EP}_{\Phi^F_{\cap}}) = \mathrm{Sol}(\mathrm{WSEP}_F), \\ \circ\ \mathrm{Sol}(\mathrm{QEP}_{(\Phi^F_{\cap},\mathscr{D})}) = \mathrm{Sol}(\mathrm{WSQEP}_{(F,\mathscr{D})}). \end{cases} \tag{2.11}$$

By stating Theorem 1.34 and its corollaries for the problem $(\mathrm{EP}_{\Phi^F_{\subset}})$ ($(\mathrm{EP}_{\Phi^F_{\cap}})$, resp.), then converting to the problem (SSEP) ((WSEP), resp.) via (2.8) ((2.10), resp.) we obtain set-valued versions of this theorem and its corollaries.

Below are two versions of Theorem 1.34 with the coercivity condition (C_{13}).

Theorem 2.60 *Let A be a nonempty convex subset of a topological vector space and $F : A \times A \rightrightarrows \mathbb{R}$ be a set-valued bifunction. Assume that $F \in \mathscr{S}^1\mathscr{A}^2_{pr.qm}$, $\mathrm{Lev}^{\geqq}_{\subset}F(\cdot, y)$ is closed for every $y \in A$, and the following coercivity condition is satisfied:*

(ssC_{13}) *There exists a nonempty compact convex subset K of A such that, for each $x \in A\backslash K$, $F(x, y) \cap (-\mathbb{R}_{++}) \neq \emptyset$ for some $y \in K$.*

Then, (SSEP_F) *has solutions.*

Proof Under the conditions of Theorem 2.60, by (2.8), all the conditions of Theorem 1.34 (with the coercivity condition (C_{13})) hold for $\Phi^F_{\subset}$. Hence, by Theorem 1.34 and (2.8), $\mathrm{Sol}(\mathrm{SSEP_F}) = \mathrm{Sol}(\mathrm{EP}_{\Phi^{\mathrm{F}}_{\subset}}) \neq \emptyset$. □

Example 2.61 Let $\{0\} \neq A \subset \mathbb{R}^n$ be a nonempty closed convex set and $F : A \times A \rightrightarrows \mathbb{R}$ be defined for all $x, y \in A$ by

$$F(x, y) = (\|y\| - \|x\|, \|y\| + \|x\|).$$

Then, $F \in \mathscr{S}^1\mathscr{A}^2_{pr.qm}$. Indeed, for all $N \in \mathscr{F}(A)$ and $x \in \mathrm{conv}N$, by taking $x^{j_0} \in N$ with $\|x^{j_0}\| = \max\{\|x^j\|, x^j \in N\}$ we have that $\|x\| \leq \|x^{j_0}\|$, and hence, $F(x, x^{j_0}) = (\|x^{j_0}\| - \|x\|, \|x^{j_0}\| + \|x\|) \subset \mathbb{R}_+$. Moreover, for every $y \in A$, $\mathrm{Lev}^{\geqq}_{\subset}F(\cdot, y) = \{x \in A \mid (\|y\| - \|x\|, \|y\| + \|x\|) \subset \mathbb{R}_+\} = \{x \in A \mid \|x\| \leq \|y\|\}$ is closed. We check condition (ssC_{13}). Fix $0 \neq y^0 \in A$ and set $K = \mathbb{B}[0, \|y^0\|] \cap A$ ($\mathbb{B}[0, \|y^0\|]$ is the closed ball of radius $\|y^0\|$ centered at 0). Then, K is a nonempty compact convex subset of A. For each $x \in A\backslash K$, we have that $\|y^0\| < \|x\|$, and hence, $F(x, y^0) \cap (-\mathbb{R}_{++}) = (\|y^0\| - \|x\|, \|y\| + \|x\|) \cap (-\mathbb{R}_{++}) \neq \emptyset$. Thus, (ssC_{13}) holds. By Theorem 2.60, (SSEP_F) has solutions.

Theorem 2.62 *Let A be a nonempty convex subset of a topological vector space and $F : A \times A \rightrightarrows \mathbb{R}$ be a set-valued bifunction. Assume that $F \in \mathscr{S}^2\mathscr{A}^2_{pr.qm}$, $\mathrm{Lev}^{\geqq}_{\cap}F(\cdot, y)$ is closed for every $y \in A$, and the following coercivity condition is satisfied:*

(wsC_{13}) *There exists a nonempty compact convex subset K of A such that, for each $x \in A\backslash K$, $F(x, y) \subset -\mathbb{R}_{++}$ for some $y \in K$.*

Then, (WSEP_F) *has solutions.*

Proof Theorem 2.62 is a consequence of Theorem 1.34 (with the coercivity condition (C_{13})) and (2.10). □

Example 2.63 Let $\{0\} \neq A \subset \mathbb{R}^n$ be a nonempty convex set and $F : A \times A \rightrightarrows \mathbb{R}$ be defined for all $x, y \in A$ by

$$F(x, y) = (-\infty, \|y\| - \|x\|].$$

For all $N \in \mathscr{F}(A)$ and $x \in \text{conv}N$, by taking $x^{j_0} \in N$ as in part (a) we have that $F(x, x^{j_0}) \cap \mathbb{R}_+ = (-\infty, \|x^{j_0}\| - \|x\|] \cap \mathbb{R}_+ \neq \emptyset$. Therefore, $F \in \mathscr{S}^2\mathscr{A}^2_{pr.qm}$. For every $y \in A$, $\text{Lev}^{\geq}_{\cap} F(\cdot, y) = \{x \in A \mid (-\infty, \|y\| - \|x\|] \cap \mathbb{R}_+ \neq \emptyset\} = \{x \in A \mid \|x\| \leq \|y\|\}$ is closed. For K as in Example 2.61, we check easily that the condition (wsC_{13}) holds. Thus, by Theorem 2.62, (WSEP_F) has solutions.

Now, through ($\text{EP}_{\Phi^F_{\subset}}$) and (2.9) (($\text{EP}_{\Phi^F_{\cap}}$) and (2.11), resp.) we restate Theorem 1.50 (with the coercivity condition (C_{33})) in terms of (SSEP_F) ((WSEP_F), resp.) and obtain the following versions of Theorem 1.50.

Theorem 2.64 *Let A be a nonempty convex subset of a topological vector space and* $F : A \times A \rightrightarrows \mathbb{R}$ *be a set-valued bifunction. Assume that* $F \in (-\mathscr{S}^2\mathscr{A}^1_{pr.qm}) \cap \mathscr{S}^1\mathscr{A}^2_{st.qcl} \cap \mathscr{S}^1\mathscr{A}^{=}_{A}$, *F has the property* $\mathbb{SP}^2$ *of type 1 and is* $\mathbb{R}_+$*-upper-sign continuous of type 1 in the first argument,* $\text{Lev}^{\leq}_{\subset} F(x, \cdot)$ *is closed for every* $x \in A$, *and the following coercivity condition is satisfied:*

(ssC_{33}) *There exists a nonempty compact convex subset K of A such that for each* $x \in A\backslash K$, $F(y, x) \subset \mathbb{R}_{++}$ *for some* $y \in K$.

Then, (SSEP_F) *has solutions.*

Proof Theorem 2.64 is a direct consequence of Theorem 1.50 (with the coercivity condition (C_{33})) and (2.9). □

Theorem 2.65 *Let A be a nonempty convex subset of a topological vector space and* $F : A \times A \rightrightarrows \mathbb{R}$ *be a set-valued bifunction. Assume that* $F^{cl} \in (-\mathscr{S}^2\mathscr{A}^1_{pr.qm}) \cap \mathscr{S}^1\mathscr{A}^2_{st.qcl} \cap \mathscr{S}^1\mathscr{A}^{=}_{A}$, F^{cl} *has the property* $\mathbb{SP}^2$ *of type 1 and is* $\mathbb{R}_+$*-upper-sign continuous of type 1 in the first argument,* $\text{Lev}^{\leq}_{\subset} F^{cl}(x, \cdot)$ *is closed for every* $x \in A$, *and the following coercivity condition is satisfied:*

$\overline{(\text{ssC}_{33})}$ *There exists a nonempty compact convex subset K of A such that for each* $x \in A\backslash K$, $\text{cl}(F(y, x)) \subset \mathbb{R}_{++}$ *for some* $y \in K$.

Then, (SSEP_F) *has solutions.*

Proof By applying Theorem 2.64 for F^{cl} we have that $\text{Sol}(\text{SSEP}_{F^{cl}}) \neq \emptyset$. By Proposition 2.50 $\text{Sol}(\text{SSEP}_F) = \text{Sol}(\text{SSEP}_{F^{cl}}) \neq \emptyset$. □

Example 2.66 Let $A = \mathbb{R}$, $F : A \times A \rightrightarrows \mathbb{R}$ be defined for all $x, y \in A$ by

$$F(x, y) = (xy - x^2, x^2 + y^2).$$

Then, $F^{cl}(x, y) = [xy - x^2, x^2 + y^2]$ for all $x, y \in A$. For all $x \in A$,

$$\mathrm{Lev}_{\subseteq}^{\leq} F^{cl}(x, \cdot) = \{y \in A \mid F^{cl}(x, y) \cap (-\mathbb{R}_{++}) \neq \emptyset\} = \begin{cases} (-\infty, x) & \text{if } x > 0, \\ \emptyset & \text{if } x = 0, \\ (x, +\infty) & \text{if } x < 0. \end{cases}$$

is convex. Hence, $F^{cl} \in \mathscr{S}^1\mathscr{A}^2_{st.qcl}$. For all $x \in A$, $\mathrm{cl}(F(x, x)) \cap (-\mathbb{R}_+) = [0, 2x^2] \cap (-\mathbb{R}_+) = \{0\}$. Therefore, $F^{cl} \in \mathscr{S}^1\mathscr{A}^{=}_{A}$. $F^{cl} \in -\mathscr{S}^2\mathscr{A}^1_{pr.qm}$. Indeed, for all $N \in \mathscr{F}(A)$ and $x \in \mathrm{conv}N$, by taking $x^{j_0} = \min N$ if $x \leq 0$ and $x^{j_0} = \max N$ if $x > 0$ we have that $-\mathrm{cl}(F(x^{j_0}, x)) \cap \mathbb{R}_+ = [-(x^{j_0})^2 - x^2, x^{j_0}(x^{j_0} - x)] \cap \mathbb{R}_+ \neq \emptyset$. Thus, $F^{cl} \in (-\mathscr{S}^2\mathscr{A}^1_{pr.qm}) \cap \mathscr{S}^1\mathscr{A}^2_{st.qcl} \cap \mathscr{S}^1\mathscr{A}^{=}_{A}$. Because, $F^{cl}(0, \cdot) = [0, y^2] \subset \mathbb{R}_+$ for all $y \in A$, $F^{cl}(0, \cdot)$ has property $\mathbb{SP}$ of type 1. For $x, y, y' \in A$ $(x \neq 0)$, assume that $\mathrm{cl}(F(x, y)) \cap (-\mathbb{R}_+) = [xy - x^2, x^2 + y^2] \cap (-\mathbb{R}_+) = \{0\}$ and $\mathrm{cl}(F(x, y')) \cap (-\mathbb{R}_{++}) = [xy' - x^2, x^2 + y'^2] \cap (-\mathbb{R}_+) \neq \emptyset$. It follows that $y = x$, $y' > x$ if $x < 0$, and $y' < x$ if $x > 0$. Hence, $x((1-t)y + ty') - x^2 = tx(y' - x) < 0$. This implies that $\mathrm{cl}(F(x, (1-t)y + ty')) \cap (-\mathbb{R}_{++}) = [tx((y' - x), x^2 + ((1-t)x + ty')^2] \cap (-\mathbb{R}_{++}) \neq \emptyset$ for all $t \in (0, 1)$. Thus, F^{cl} has property $\mathbb{SP}^2$ of type 1. For all $x \in A$, assume that $F^{cl}(x + t(y - x), y) = [(x + t(y - x))(1 - t)(y - x), (x + t(y - x))^2 + y^2] \subset \mathbb{R}_+$ for all $t \in (0, 1)$. Then, either $0 \leq x \leq y$ or $0 \geq x \geq y$, which implies that $F^{cl}(x, y) \subset \mathbb{R}_+$. Thus, F^{cl} is $\mathbb{R}_+$-upper-sign continuous of type 1 in the first argument. Moreover, for every $x \in A$,

$$\mathrm{Lev}_{\subseteq}^{\leq} F^{cl}(x, \cdot) = \{y \in A \mid [xy - x^2, x^2 + y^2] \cap (-\mathbb{R}_+) \neq \emptyset\} = \begin{cases} (-\infty, x] & \text{if } x > 0, \\ A & \text{if } x = 0, \\ [x, +\infty) & \text{if } x < 0. \end{cases}$$

is closed. Fix $0 < y^0 \in \mathbb{R}$ and set $K = [-2y^0, 2y^0]$. Then, for each $x \in A \backslash K$, by taking $y = y^0$ if $x > 2y^0$, and $y = -y^0$ if $x < -2y^0$, we have that $\mathrm{cl}(F(y, x)) \subset \mathbb{R}_{++}$. Hence, condition $\overline{(\mathrm{ssC}_{33})}$ holds. Thus, by Theorem 2.65, (SSEP_F) has solutions.

Theorem 2.67 *Let A be a nonempty convex subset of a topological vector space and $F : A \times A \rightrightarrows \mathbb{R}$ be a set-valued bifunction. Assume that $F \in (-\mathscr{S}^1\mathscr{A}^1_{pr.qm}) \cap \mathscr{S}^2\mathscr{A}^2_{st.qcl} \cap \mathscr{S}^2\mathscr{A}^{=}_{A}$, F has property $\mathbb{SP}^2$ of type 2 and is $\mathbb{R}_+$-upper-sign continuous of type 2 in the first argument, $\mathrm{Lev}_{\cap}^{\leq} F(x, \cdot)$ is closed for every $x \in A$, and the following coercivity condition is satisfied*

(wsC_{33}) *There exists a nonempty compact convex subset K of A such that for each $x \in A \backslash K$, $F(y, x) \cap \mathbb{R}_{++} \neq \emptyset$ for some $y \in K$.*

Then, (WSEP_F) *has solutions.*

Proof Theorem 2.67 is a direct consequence of Theorem 1.50 (with the coercivity condition (C_{33})) and (2.11). □

Example 2.68 Let $A = (-\infty, 0]$, $F : A \times A \rightrightarrows \mathbb{R}$ be defined for all $x, y \in A$ by

$$F(x, y) = (-\infty,\ x^3 - x^2 y + x - y].$$

For all $x \in A$,

$$\mathrm{Lev}_{\cap}^{<} F(x, \cdot) = \{y \in A \mid F(x, y) \subset -\mathbb{R}_{++}\} = (x, 0]$$

is convex. Hence, $F \in \mathscr{S}^2\mathscr{A}^2_{st.qcl}$. For all $x \in A$, $F(x, x) \cap \mathbb{R}_+ = (-\infty, 0] \cap \mathbb{R}_+ = \{0\}$. Therefore, $F \in \mathscr{S}^2\mathscr{A}^{=}_{\mathrm{A}}$. $F \in -\mathscr{S}^1\mathscr{A}^1_{pr.qm}$. Indeed, for all $N \in \mathscr{F}(A)$ and $x \in \mathrm{conv}N$, by taking $x^{j_0} = \min N$ we have that $-F(x^{j_0}, x) = [((x - x^{j_0})((x^{j_0})^2 + 1), +\infty) \subset \mathbb{R}_+$. Thus, $F \in (-\mathscr{S}^1\mathscr{A}^1_{pr.qm}) \cap \mathscr{S}^2\mathscr{A}^2_{st.qcl} \cap \mathscr{S}^2\mathscr{A}^{=}_{\mathrm{A}}$. For $x, y, y' \in A$, if $\mathrm{cl}(F(x, y)) \cap \mathbb{R}_+) = \{0\}$ and $\mathrm{cl}(F(x, y')) \subset -\mathbb{R}_{++}$. It follows that $y = x$ and $y' > x$. Hence, $\mathrm{cl}(F(x, (1 - t)y + ty')) = (-\infty,\ t((x - y')(x^2 + 1)] \subset -\mathbb{R}_{++}$ for all $t \in (0, 1)$. Thus, F has property $\mathbb{SP}^2$ of type 2. For all $x \in A$, assume that $F(x + t(y - x), y) \cap \mathbb{R}_+ = (-\infty,\ (1 - t)(x - y)((x + t(y - x))^2 + 1)] \cap \mathbb{R}_+ \neq \emptyset$ for all $t \in (0, 1)$. Then, $y \leq x$, which implies that $F(x, y) \cap \mathbb{R}_+ \neq \emptyset$. Thus, F is $\mathbb{R}_+$-upper-sign continuous of type 2 in the first argument. For every $x \in A$,

$$\mathrm{Lev}_{\cap}^{\leq} F(x, \cdot) = \{y \in A \mid (-\infty,\ x^3 - x^2 y + x - y] \subset -\mathbb{R}_+\} = [x,\ 0]$$

is closed. Fix $y^0 < 0$ and set $K = [-2y^0, 0]$. Then, for each $x \in A \backslash K$, by taking $y = y^0$ we have that $F(y, x) \cap \mathbb{R}_{++} = (-\infty\ (y^0 - x)((y^0)^2 + 1)] \cap \mathbb{R}_{++} \neq \emptyset$. Hence, the condition (wsC$_{33}$) holds. Thus, by Theorem 2.67, (WSEP$_F$) has solutions.

Restating Theorems 2.21 and 2.28 and their corollaries for the problem (QEP$_{(\Phi^F_{\subset}, \mathscr{D})}$) ((QEP$_{(\Phi^F_{\cap}, \mathscr{D})}$), resp.) and, by using (2.8) ((2.10), resp.), transferring the obtained statements to the problem (SSQEP$_{(F, \mathscr{D})}$) ((WSQEP$_{(F, \mathscr{D})}$), resp.), we get set-valued versions of these theorems and their corollaries.

Theorem 2.21 and Corollary 2.24 read for (SSQEP$_{(F, \mathscr{D})}$) as follows.

Theorem 2.69 *Let A be a nonempty compact convex subset of a topological vector space, $\mathscr{D} : A \rightrightarrows A$ be a set-valued map with nonempty images, and $F : A \times A \rightrightarrows \mathbb{R}$ be a set-valued bifunction. Assume that $F \in \mathscr{S}^1\mathscr{A}^2_{pr.qm}$, $\mathscr{D}$ has convex images, $\mathscr{D}^{-1}(y)$ and* $\mathrm{Lev}_{\subset}^{<} F(\cdot, y)$ *are open for each $y \in A$. Then, at least one of the following two assertions is true:*

(a) (SSQEP$_{(F, \mathscr{D})}$) *has solutions.*
(b) *There exists $\bar{x} \in \delta(\mathrm{Fix}\mathscr{D}) \setminus \mathrm{Fix}\mathscr{D}$ such that $F(\bar{x}, y) \subset \mathbb{R}_+$ for all $y \in \mathscr{D}(\bar{x})$.*

Consequently, if in addition, the following condition is satisfied:

(SB) *For each $x \in \delta(\mathrm{Fix}\mathscr{D}) \setminus \mathrm{Fix}\mathscr{D}$, $F(x, y) \not\subset \mathbb{R}_+$ for some $y \in \mathscr{D}(x)$,*

then (SSQEP$_{(F, \mathscr{D})}$) *has solutions.*

Corollary 2.70 *Let A be a nonempty compact convex subset of a Hausdorff locally convex space, $\mathscr{D} : A \rightrightarrows A$ be a set-valued map with nonempty images, and $F : A \times A \rightrightarrows \mathbb{R}$ be a set-valued bifunction. Assume that $F \in \mathscr{S}^1\mathscr{A}^2_{pr.qm}$, $\mathscr{D}$ is closed with convex images,* $\mathrm{Lev}^{\geq}_{\subset} F(\cdot, y)$ *is closed for all $y \in A$, and the set $\{x \in A \mid \mathscr{D}(x) \cap \mathrm{Lev}^{\leq}_{\subset} F(x, \cdot) = \emptyset\}$ is closed. Then,* (SSQEP$_{(F,\mathscr{D})}$) *has solutions.*

Proof Theorem 2.69 (Corollary 2.70, resp.) follows directly from Theorem 2.21 (Corollary 2.24, resp.) and (2.8). □

Example 2.71 Let $A = [-7, 7] \subset \mathbb{R}$, $\mathscr{D} : A \rightrightarrows A$ and $F : A \times A \rightrightarrows \mathbb{R}$ be defined for all $x, y \in A$ by, resp.,

$$\mathscr{D}(x) = [-|x|, |x|] \text{ and } F(x, y) = (x^2 - y^2, x^2 + y^2).$$

It is clear that $\mathscr{D}$ is closed with convex images. For each $y \in A$, $\mathrm{Lev}^{\geq}_{\subset} F(\cdot, y) = \{x \in A \mid (x^2 - y^2, x^2 + y^2) \subset \mathbb{R}_+\} = [-7, -|y|] \cup [|y|, 7]$ is closed. For all $x \in A$, we have $\mathscr{D}(x) \cap \mathrm{Lev}^{\leq}_{\subset} F(x, \cdot) = [-|x|, |x|] \cap ([-7, -|x|) \cup (|x|, 7]) = \emptyset$. Hence, $\{x \in A \mid \mathscr{D}(x) \cap \mathrm{Lev}^{\leq}_{\subset} F(x, \cdot) = \emptyset\} = A$ is closed. $F \in \mathscr{S}^1\mathscr{A}^2_{pr.qm}$. Indeed, for all $N \in \mathscr{F}(A)$ and $x \in \mathrm{conv}N$, by taking $x^{j_0} \in N$ with $|x^{j_0}| = \min\{|x^j|, x^j \in N\}$ we have that $|x^{j_0}| \leq |x|$, and hence, $F(x, x^{j_0}) = (x^2 - (x^{j_0})^2, x^2 + (x^{j_0})^2) \subset \mathbb{R}_+$. Thus, by Corollary 2.70, (SSQEP$_{(F,\mathscr{D})}$) has solutions.

Theorem 2.21 and Corollary 2.24 read for (WSQEP$_{(F,\mathscr{D})}$) as follows.

Theorem 2.72 *Let A be a nonempty compact convex subset of a topological vector space, $\mathscr{D} : A \rightrightarrows A$ be a set-valued map with nonempty images, and $F : A \times A \rightrightarrows \mathbb{R}$ be a set-valued bifunction. Assume that $F \in \mathscr{S}^2\mathscr{A}^2_{pr.qm}$, $\mathscr{D}$ has convex images, $\mathscr{D}^{-1}(y)$ and* $\mathrm{Lev}^{<}_{\cap} F(\cdot, y)$ *are open for each $y \in A$. Then, at least one of the following two assertions is true:*

(a) (WSQEP$_{(F,\mathscr{D})}$) *has solutions.*
(b) *There exists $\bar{x} \in \delta(\mathrm{Fix}\mathscr{D}) \setminus \mathrm{Fix}\mathscr{D}$ such that $F(\bar{x}, y) \cap \mathbb{R}_+ \neq \emptyset$ for all $y \in \mathscr{D}(\bar{x})$.*

Consequently, if in addition, the following condition is satisfied:

(WB) *For each $x \in \delta(\mathrm{Fix}\mathscr{D}) \setminus \mathrm{Fix}\mathscr{D}$, $F(\bar{x}, y) \subset -\mathbb{R}_{++}$ for some $y \in \mathscr{D}(x)$,*

then (WSQEP$_{(F,\mathscr{D})}$) *has solutions.*

Corollary 2.73 *Let A be a nonempty compact convex subset of a Hausdorff locally convex space, $\mathscr{D} : A \rightrightarrows A$ be a set-valued map with nonempty images, and $F : A \times A \rightrightarrows \mathbb{R}$ be a set-valued bifunction. Assume that $F \in \mathscr{S}^2\mathscr{A}^2_{pr.qm}$, $\mathscr{D}$ is closed with convex images,* $\mathrm{Lev}^{\geq}_{\cap} F(\cdot, y)$ *is closed for all $y \in A$, and the set $\{x \in A \mid \mathscr{D}(x) \cap \mathrm{Lev}^{<}_{\cap} F(x, \cdot) = \emptyset\}$ is closed. Then,* (WSQEP$_{(F,\mathscr{D})}$) *has solutions.*

Proof Theorem 2.72 (Corollary 2.73, resp.) follows directly from Theorem 2.21 (Corollary 2.24, resp.) and (2.10). □

Example 2.74 Let $A = [-17, 17] \subset \mathbb{R}$, $\mathscr{D} : A \rightrightarrows A$ and $F : A \times A \rightrightarrows \mathbb{R}$ be defined for all $x, y \in A$ by, resp.,

$$\mathscr{D}(x) = [\frac{1}{7}|x|, |x|] \text{ and } F(x, y) = (-x^2 - y^2 - 1, xy].$$

Clearly, $\mathscr{D}$ is closed with convex images. For all $x, y \in A$,

$$\text{Lev}_{\cap}^{\geq} F(\cdot, y) = \begin{cases} [0, 17] & \text{if } y > 0, \\ [-17, 17] & \text{if } y = 0, \\ [-17, 0] & \text{if } y < 0 \end{cases}$$

and

$$\{x \in A \mid \mathscr{D}(x) \cap \text{Lev}_{\cap}^{<} F(x, \cdot) = \emptyset\} = \begin{cases} \emptyset & \text{if } x \geq 0, \\ [\frac{1}{7}|x|, |x|] & \text{if } x < 0 \end{cases}$$

are closed. For all $N \in \mathscr{F}(A)$ and $x \in \text{conv}N$, by taking $x^{j_0} = \min N$ if $x \leq 0$ and $x^{j_0} = \max N$ if $x > 0$, we have that $F(x, x^{j_0}) \cap \mathbb{R}_+ = (-x^2 - (x^{j_0})^2 - 1, xx^{j_0}] \cap \mathbb{R}_+ \neq \emptyset$. Therefore, $F \in \mathscr{S}^2\mathscr{A}^2_{pr.qm}$. Thus, by Corollary 2.73, (WSQEP$_{(F,\mathscr{D})}$) has solutions.

Set-valued versions of Corollary 2.36 are stated as follows.

Corollary 2.75 *Let A be a nonempty compact convex subset of a Hausdorff locally convex space X, $\mathscr{D} : A \rightrightarrows A$ be a set-valued map with nonempty images, and $F : A \times A \rightrightarrows \mathbb{R}$ be a set-valued bifunction. Assume that $F \in (-\mathscr{S}^2\mathscr{A}^1_{pr.qm}) \cap \mathscr{S}^1\mathscr{A}^2_{st.qcl} \cap \mathscr{S}^1\mathscr{A}^{=}_{A}$ has property $\mathbb{SP}^2$ of type 1, F is $\mathbb{R}_+$-upper-sign continuous of type 1 in the first argument, $\mathscr{D}$ is closed with convex images, and for all $x \in A$, the sets $\text{Lev}_{\subset}^{\leq} F(x, \cdot)$ and $\{x \in A \mid \mathscr{D}(x) \cap \text{Lev}_{\subset}^{>} F(\cdot, x) = \emptyset\}$ are closed. Then, (SSQEP$_{(F,\mathscr{D})}$) has solutions.*

Proof Corollary 2.75 follows directly from Corollary 2.36 and (2.9). □

Corollary 2.76 *Let A be a nonempty compact convex subset of a Hausdorff locally convex space X, $\mathscr{D} : A \rightrightarrows A$ be a set-valued map with nonempty images, and $F : A \times A \rightrightarrows \mathbb{R}$ be a set-valued bifunction. Assume that $F^{cl} \in (-\mathscr{S}^2\mathscr{A}^1_{pr.qm}) \cap \mathscr{S}^1\mathscr{A}^2_{st.qcl} \cap \mathscr{S}^1\mathscr{A}^{=}_{A}$ has property $\mathbb{SP}^2$ of type 1, F^{cl} is $\mathbb{R}_+$-upper-sign continuous of type 1 in the first argument, $\mathscr{D}$ is closed with convex images, $\text{Lev}_{\subset}^{\leq} F^{cl}(x, \cdot)$ is closed for all $x \in A$, and $\{x \in A \mid \mathscr{D}(x) \cap \text{Lev}_{\subset}^{>} F^{cl}(\cdot, x) = \emptyset\}$ is closed. Then, (SSQEP$_{(F,\mathscr{D})}$) has solutions.*

Proof Applying Corollary 2.75 for F^{cl} we obtain that (SSQEP$_{(F^{cl},\mathscr{D})}$) is nonempty. Hence, by Proposition 2.50, (SSQEP$_{(F,\mathscr{D})}$) is nonempty. □

Corollary 2.77 *Let A be a nonempty compact convex subset of a Hausdorff locally convex space X, $\mathscr{D} : A \rightrightarrows A$ be a set-valued map with nonempty images, and*

$F : A \times A \rightrightarrows \mathbb{R}$ be a set-valued bifunction. Assume that $F \in (-\mathscr{S}^1\mathscr{A}^1_{pr.qm}) \cap \mathscr{S}^2\mathscr{A}^2_{st.qcl} \cap \mathscr{S}^2\mathscr{A}^{=}_{A}$ has property $\mathbb{SP}^2$ of type 2, F is $\mathbb{R}_+$-upper-sign continuous of type 2 in the first argument, $\mathscr{D}$ is closed with convex images, $\mathrm{Lev}^{\leq}_{\cap}F(x,\cdot)$ is closed for all $x \in A$, and $\{x \in A \mid \mathscr{D}(x) \cap \mathrm{Lev}^{\geq}_{\cap}F(\cdot,x) = \emptyset\}$ is closed. Then, (WSQEP$_{(F,\mathscr{D})}$) has solutions.

Proof Corollary 2.77 follows directly from Corollary 2.36 and (2.11). □

2.6 Some Applications

In this section, we particularize results in Sects. 2.4–2.5 to obtain some existence conditions for special cases of (QEP).

2.6.1 *Quasioptimization Problem*

Recall that the quasioptimization problem (QOP) is to find $\bar{x} \in \mathscr{D}(\bar{x})$ such that

$$\min_{x \in \mathscr{D}(\bar{x})} \varphi(x) = \varphi(\bar{x}),$$

where $\mathscr{D} : A \rightrightarrows A$ and $\varphi : A \to \mathbb{R} \cup \{+\infty\}$. Let $\Phi_\varphi : A \times A \to \mathbb{R}$ be the bifunction associated with φ, defined by $\Phi_\varphi(x, y) := \varphi(y) - \varphi(x)$ for all $x, y \in A$. We have some remarks as follows.

- When A is a convex subset of a topological vector space, $\Phi_\varphi \in \mathscr{A}^2_{pr.qm}$ if and only if $\Phi_\varphi \in \mathscr{A}^{\geq}_{A} \cap \mathscr{A}^2_{st.qlc}$ if and only if φ is quasiconvex.
- Φ_φ always belongs to $\mathscr{A}^2_{caqm}$.
- The set $\mathrm{lev}_{\geq}\Phi := \{(x, y) \in A \times A \mid \Phi_\varphi(x, y) \geq 0\}$ is closed if A is a subset of a topological space and φ is continuous.

By concretizing some results in Sect. 2.4 for the bifunction Φ_φ, we state several existence conditions for (QOP) as follows.

Corollary 2.78 *For quasioptimization problem (QOP), we have the following assertions:*

(a) *If A is a compact convex subset of a topological vector space X, $\mathscr{D}$ has convex images and $\mathscr{D}^{-1}(y)$ is open for each $y \in A$, and φ is quasiconvex and lower semicontinuous, then at least one of the following assertions is true:*

  (a_1) (QOP) *has a solution, i.e., there exists $\bar{x} \in \mathrm{Fix}\mathscr{D}$ such that $\min_{x\in\mathscr{D}(\bar{x})} \varphi(x) = \varphi(\bar{x})$.*
  (a_2) There exists $\bar{x} \in \delta(\mathrm{Fix}\mathscr{D}) \setminus \mathrm{Fix}\mathscr{D}$ such that $\min_{x\in\mathscr{D}(\bar{x})} \varphi(x) = \varphi(\bar{x})$.

Consequently, if in addition, $\inf_{x'\in\mathscr{D}(x)}\varphi(x') < \varphi(x)$ *for each* $x \in \delta(\mathrm{Fix}\mathscr{D}) \setminus \mathrm{Fix}\mathscr{D}$, *then* (QOP) *has solutions.*

(b) *If A is a compact convex subset of a Hausdorff locally convex space X, $\mathscr{D}$ is lower semicontinuous with convex images, $\mathscr{D}$ is closed, and φ is quasiconvex and continuous, then* (QOP) *has solutions. In the case that X is a separable Banach space, condition "$\mathscr{D}$ is closed" can be replaced by condition "*Fix$\mathscr{D}$ *is closed."*

(c) *If A is a compact subset of a topological space X, $\mathscr{D}^{-1}(y)$ is open for each $y \in A$, $\mathscr{D}(y) \subset \mathscr{D}(x)$ for any $x \in A$ and $y \in \mathscr{D}(x)$, and φ is lower semicontinuous, then we have the conclusion of part (a).*

Proof Parts (a) and (c) are consequences of Theorems 2.21 and 2.48. Part (b) is a consequence of Corollary 2.26 and Theorem 2.28. □

Example 2.79 Let $A = [0, 1]$, $\mathscr{D}(x) = (0, -x^2 + x + \frac{1}{4})$, and $\varphi(x) = x^2 - \frac{1}{2}x$. Then, φ is lower semicontinuous and quasiconvex and $\mathscr{D}$ has convex images. For each $y \in A$,

$$\mathscr{D}^{-1}(y) = \begin{cases} \emptyset & \text{if } y \geq \frac{1}{2}, \\ (\frac{1-\sqrt{2(1-2y)}}{2}, \frac{1+\sqrt{2(1-2y)}}{2}) & \text{if } \frac{1}{4} \leq y < \frac{1}{2}, \\ [0, 1] & \text{if } y < \frac{1}{4} \end{cases}$$

is open. We have $\mathrm{Fix}\mathscr{D} = \{x \in A \mid 0 < x < -x^2 + x + \frac{1}{4}\} = (0, \frac{1}{2})$, $\delta(\mathrm{Fix}\mathscr{D}) \setminus \mathrm{Fix}\mathscr{D} = \{0, \frac{1}{2}\}$. For $x' = \frac{1}{8} \in \mathscr{D}(0) = (0, \frac{1}{4}) \subset (0, \frac{1}{2}) = \mathscr{D}(\frac{1}{2})$, we have $\varphi(x') = -\frac{3}{64} < 0 = \varphi(0) = \varphi(\frac{1}{2})$. Thus, (QOP) has solutions by part (a) of Corollary 2.78.

2.6.2 *Generalized Nash Equilibrium Problem*

From Sect. 2.2 we already know that this generalized Nash equilibrium problem can be converted into the following quasiequilibrium problem: find $\bar{x} \in \mathscr{D}(\bar{x})$ such that $N(\bar{x}, y) = \sum_{i=1}^{n}[c_i(\bar{x}) - c_i(\bar{x}_{-i}, y_i)))] \geq 0$ for all $y \in \mathscr{D}(\bar{x})$, where $\mathscr{D} = \prod_{i=1}^{m} \mathscr{D}_i : A = \prod_{i=1}^{m} A_i \rightrightarrows A$, $\mathscr{D}_i : A_{-i} = \prod_{k\neq i} A_k \rightrightarrows A_i$ is the strategy constraint of player i, $c_i : A \to \mathbb{R}$ is their payoff function, and $N : A \times A \to \mathbb{R}$ is the Nikaido-Isoda bifunction.

We have several remarks.

- If for each $i \in I$, A_i is a convex subset of a topological vector space and $c_i(x_{-i}, \cdot)$ is concave for each $x \in A$, then $N(x, \cdot)$ is convex for each $x \in A$ (see Exercise 1.3).
- Assume that for each $i \in I$, A_i is a subset of a topological space. Then, for each $y \in Y$, we have

$$\mathscr{D}^{-1}(y) = \{x \in A \mid y_i \in \mathscr{D}_i(x_{-i}) \;\; \forall i = 1, 2, \ldots, m\}$$

$$= \bigcap_{i=1}^{m} \{x \in A \mid y_i \in \mathscr{D}_i(x_{-i})\}$$

$$= \bigcap_{i=1}^{m} A_i \times \{x_{-i} \in A_{-i} \mid y_i \in \mathscr{D}_i(x_{-i})\} = \bigcap_{i=1}^{m} A_i \times \mathscr{D}_i^{-1}(y_i).$$

Therefore, if $\mathscr{D}_i^{-1}(y_i)$ is open for all $y_i \in A_i$ and $i \in I$, then $\mathscr{D}^{-1}(y)$ is open for each $y \in A$.

For an open set $V = \prod_{i=1}^m V_i \subset A$, where each $V_i \subset A_i$ is open, we have that

$$\mathscr{D}^{-1}(V) = \{x \in A \mid \mathscr{D}_i(x_{-i}) \cap V_i \neq \emptyset \;\; \forall i = 1, 2, \ldots, m\}$$

$$= \bigcap_{i=1}^{m} \{x \in A \mid \mathscr{D}_i(x_{-i}) \cap V_i \neq \emptyset\}$$

$$= \bigcap_{i=1}^{m} A_i \times \{x_{-i} \in A_{-i} \mid \mathscr{D}_i(x_{-i}) \cap V_i \neq \emptyset\} = \bigcap_{i=1}^{m} A_i \times \mathscr{D}_i^{-1}(V_i)$$

is open. Hence, if each $\mathscr{D}_i$ is lower semicontinuous, then $\mathscr{D}$ is lower semicontinuous.

Additionally, if $\mathscr{D}_i$ is closed for each i, then $\mathscr{D}$ is closed. Indeed, let $\{(x^\alpha, y^\alpha)\}_{\alpha \in (I, \succeq)} \subset \operatorname{gph}\mathscr{D}$ be a net converging to (x, y). Then, for all $i \in I$, $y_i^\alpha \in \mathscr{D}_i(x_{-i}^\alpha)$ for all α and $(x_{-i}^\alpha, y_i^\alpha) \to (x_{-i}, y_i)$. Since $\mathscr{D}_i$ is closed, $y_i \in \mathscr{D}_i(x_{-i})$. Hence, $y \in \prod_{i=1}^m \mathscr{D}_i(x_{-i}) = \mathscr{D}(x)$.

From Theorem 2.21, Corollaries 2.26 and 2.29, and the above remarks, we obtain the following corollary.

Corollary 2.80 *Assume that for all $i \in I$, A_i is a compact convex subset of a Hausdorff locally convex space X_i. Then, the following assertions hold:*

(a) *If the set $\{x \in A \mid x_i \in \mathscr{D}(x_{-i}),\ \forall i\}$ is closed, and for all $i \in I$, $\mathscr{D}_i$ has convex images and $\mathscr{D}_i^{-1}(y_i)$ is open for all $y_i \in A_i$, the function c_i is continuous and the function $c_i(x_{-i}, \cdot)$ is concave for each $x_{-i} \in A_{-i}$, then $\mathscr{G}$ has a generalized Nash equilibrium.*

(b) *If for all $i \in I$, $\mathscr{D}_i$ is closed and lower semicontinuous with convex images, the function c_i is continuous and the function $c_i(x_{-i}, \cdot)$ is concave for each $x_{-i} \in A_{-i}$, then $\mathscr{G}$ has a generalized Nash equilibrium. When all X_i are separable Banach spaces, the condition "each $\mathscr{D}_i$ is closed" can be replaced by " the set $\{x \in A \mid x_i \in \mathscr{D}(x_{-i}),\ \forall i\}$ is closed."*

Example 2.81 Consider the generalized Nash equilibrium problem with data as follows. $A_1 = A_2 = \cdots = A_m = [0, 1]$, $\mathscr{D}_i : A_{-i} \rightrightarrows A_i$ is defined by

$$\mathscr{D}_i(x_{-i}) := \left[0, \sqrt{\frac{1}{m-1} \sum_{j \neq i} (x_i)^2}\right] \text{ for } x_{-i} = (x_1, \ldots, x_{i-1}, x_{i+1}, \ldots, x_m) \in A_{-i},$$

and $c_i : A \to \mathbb{R}$ is defined by $c_i(x) := -(x_i)^2 + \sum_{j \neq i} x_j$ for $x \in A$. It is not hard to verify that all conditions of part (a) of Corollary 2.80 hold.

2.6.3 *Browder Variational Inclusions*

In this subsection, let A be a subset of a topological vector space X and X^* be the dual space of X. We consider X^* with the weak* topology which is Hausdorff. In the literature, problems of the form: "find $\bar{x} \in A$ such that $\langle C, \bar{x} \rangle \subset \mathbb{R}_+$" or "find $\bar{x} \in A$ such that $\langle C, \bar{x} \rangle \cap \mathbb{R}_+ \neq \emptyset$," where $C \subset X^*$ and $\langle C, x \rangle = \{\langle x^*, x \rangle \mid x^* \in C\}$, are called Browder variational inclusions. In the literature, Browder variational inclusions are also considered as set-valued variational inequalities.

Definition 2.82 Let A be a nonempty convex subset of a topological vector space X and $T : A \rightrightarrows X^*$ be a set-valued mapping with nonempty images. T is said to be UPPER/LOWER HEMICONTINUOUS if for any $x, y \in A$, the set-valued mapping

$$t \mapsto \langle T(x + t(y - x)), y - x \rangle = \bigcup_{x^* \in T(x+t(y-x))} \langle x^*, y - x \rangle$$

from $[0, 1]$ to $\mathbb{R}$ is upper/lower semicontinuous at 0^+.

Definition 2.83 Let A be a nonempty convex subset of a topological vector space X and $T : A \rightrightarrows X^*$ be a set-valued mapping.

(a) T is called MONOTONE, if for all $x, y \in A$ and all $x^* \in T(x)$, $y^* \in T(y)$ one has

$$\langle y^* - x^*, y - x \rangle \geq 0.$$

(b) T is called PSEUDOMONOTONE, if for all $x, y \in A$ and all $x^* \in T(x)$, $y^* \in T(y)$ one has

$$\langle x^*, y - x \rangle \geq 0 \Longrightarrow \langle y^*, y - x \rangle \geq 0.$$

(c) T is said to be PROPERLY QUASIMONOTONE if, for all $N \in \mathscr{F}(A)$ and $x \in \text{conv}N$, there exists $x^i \in N$ such that $\langle (x^i)^*, x - x^i \rangle \leq 0$ for all $(x^i)^* \in T(x^i)$.

Remark 2.84

(a) Obviously, monotonicity implies pseudomonotonicity. Moreover, pseudomonotonicity implies proper quasimonotonicity. Indeed, suppose to the contrary that there exist $N \in \mathscr{F}(A)$ and $x = \sum_{x^i \in N} \lambda_i x^i \in \text{conv}N$ such that for each $x^i \in N$, $\langle (x^i)^*, x - x^i \rangle > 0$ for some $(x^i)^* \in T(x^i)$. By fixing a $x^* \in T(x)$, we have that $\langle x^*, x - x^i \rangle > 0$ for all $x^i \in N$ (by the pseudomonotonicity). It follows that $0 = \langle x^*, x - x \rangle = \sum_{x^i \in N} \langle x^*, x - x^i \rangle > 0$, a contradiction.
(b) It is clear that T is properly quasimonotone if, and only if, $F : A \times A \rightrightarrows \mathbb{R}$ defined by $F(x, y) := \langle T(y), y - x \rangle$ for all $x, y \in A$, is properly $\mathbb{R}_+$-quasimonotone of type 1 in the second component.

Theorem 2.85 *Let A be a nonempty convex subset of a topological vector space X and $T : A \rightrightarrows X^*$ be a set-valued mapping. Assume that T is properly quasimonotone, and the following coercivity condition is satisfied:*

(sBC$_{13}$) *There exists a nonempty compact convex subset K of A such that, for each $x \in A \backslash K$, $\langle T(y), y - x \rangle \cap (-\mathbb{R}_{++}) \neq \emptyset$ for some $y \in K$.*

Then, there exists $\bar{x} \in A$ such that $\langle T(y), y - \bar{x} \rangle \subset \mathbb{R}_+$ for all $y \in A$. If addition, T is lower hemicontinuous, then there exists $\bar{x} \in A$ such that $\langle T(\bar{x}), y - \bar{x} \rangle \subset \mathbb{R}_+$ for all $y \in A$.

Proof Define the set-valued bifunction $F : A \times A \rightrightarrows \mathbb{R}$ by

$$F(x, y) = \langle T(y), y - x \rangle \quad \text{for all } x, y \in A. \tag{2.12}$$

Then, $F \in \mathscr{S}^1 \mathscr{A}^2_{pr.qm}$ by Remark 2.84(b). Moreover, it is not hard to check that $\text{Lev}^{\geq}_{\subset} F(\cdot, y) := \{x \in A \mid \langle T(y), y - x \rangle \subset \mathbb{R}_+\}$ is closed for each $y \in A$. Thus, F satisfies all conditions in Theorem 2.60. Hence, there exists $\bar{x} \in A$ such that $F(\bar{x}, y) = \langle T(y), y - \bar{x} \rangle \subset \mathbb{R}_+$ for all $y \in A$.

Now, suppose additionally that T is lower hemicontinuous. We show that $\langle T(\bar{x}), y - \bar{x} \rangle \subset \mathbb{R}_+$ for all $y \in A$. Let $y \in A$ be arbitrary, $t \in (0, 1]$, and $y_t = \bar{x} + t(y - \bar{x})$. Then, $t\langle T(y_t), y - \bar{x} \rangle = \langle T(y_t), t(y - \bar{x}) \rangle = \langle T(y_t), y_t - \bar{x} \rangle \subset \mathbb{R}_+$. Hence, $\langle T(y_t), y - \bar{x} \rangle \subset \mathbb{R}_+$. If $\langle T(\bar{x}), y - \bar{x} \rangle \not\subset \mathbb{R}_+$, i.e., $\langle T(\bar{x}), y - \bar{x} \rangle \cap (-\mathbb{R}_{++}) \neq \emptyset$. Then, by the lower hemicontinuity of T, there exists $\epsilon > 0$ such that $\langle T(y_t), y - \bar{x} \rangle \cap (-\mathbb{R}_{++}) \neq \emptyset$ for all $t \in (0, \epsilon)$, a contradiction. Thus, $\langle T(\bar{x}), y - \bar{x} \rangle \subset \mathbb{R}_+$. Since y is arbitrary, $\langle T(\bar{x}), y - \bar{x} \rangle \subset \mathbb{R}_+$ for all $y \in A$. □

Theorem 2.86 *Let A be a nonempty convex subset of a topological vector space X and $T : A \rightrightarrows X^*$ be an upper semicontinuous mapping with compact images. Assume that the following coercivity condition is satisfied:*

(wBC$_{13}$) *There exists a nonempty compact convex subset K of A such that, for each $x \in A \backslash K$, $\langle T(x), y - x \rangle \subset -\mathbb{R}_{++}$ for some $y \in K$.*

Then, there exists $\bar{x} \in A$ such that $\langle T(\bar{x}), y - \bar{x} \rangle \cap \mathbb{R}_+ \neq \emptyset$ for all $y \in A$.

Proof Define the set-valued bifunction $F : A \times A \rightrightarrows \mathbb{R}$ by

$$F(x, y) = \langle T(x), y - x\rangle \text{ for all } x, y \in A. \tag{2.13}$$

It is clear that $F(x, x) = \{0\} \subset \mathbb{R}_+$ for all $x \in A$, i.e., $F \in \mathscr{S}^1\mathscr{A}_A^{\geq}$. Moreover, $F \in \mathscr{S}^2\mathscr{A}^2_{st.qlc}$, i.e., F is strongly $\mathbb{R}_+$-quasiconvex-like of type 2 in the second component. Indeed, suppose to the contrary that $F \notin \mathscr{S}^2\mathscr{A}^2_{st.qlc}$. Then, there exist $x \in A$, $y, y' \in A$ and $t \in (0, 1)$ such that $F(x, y) = \langle T(x), y - x\rangle \subset -\mathbb{R}_{++}$ and $F(x, y') = \langle T(x), y' - x\rangle \subset -\mathbb{R}_{++}$, but $F(x, ty + (1-t)y') \cap \mathbb{R}_+ = \langle T(x), ty + (1-t)y' - x\rangle \cap \mathbb{R}_+ \neq \emptyset$. It follows that there exists $x^* \in T(x)$ such that $0 \leq \langle x^*, ty+(1-t)y'-x\rangle = t\langle x^*, y-x\rangle+(1-t)\langle x^*, y'-x\rangle < 0$, a contradiction. Thus, $F \in \mathscr{S}^1\mathscr{A}_A^{\geq} \cap \mathscr{S}^2\mathscr{A}^2_{st.qlc} \subset \mathscr{S}^2\mathscr{A}^2_{pr.qm}$. For $y \in A$, we show that $\mathrm{Lev}^{\geq}_{\cap} F(\cdot, y) := \{x \in A \mid \langle T(x), y - x\rangle \cap \mathbb{R}_+ \neq \emptyset\}$ is closed. Assume that $\{x^\alpha\}_{\alpha\in(I,\succeq)}$ is a net in $\mathrm{Lev}^{\geq}_{\cap} F(\cdot, y)$ converging to x. Then, for all α, there exists $x^{*\alpha} \in T(x^\alpha)$ such that

$$\langle x^{*\alpha}, y - x^\alpha\rangle \geq 0.$$

When T is upper semicontinuous with compact values, by Proposition 2.5, there exists a subnet $\{x^{*\beta}\}_{\beta\in(J,\succeq)}$ of the net $\{x^{*\alpha}\}_{\alpha\in(I,\succeq)}$ with $x^{*\beta} \in T(x^\beta)$ such that $x^{*\beta} \to x^* \in T(x)$. Then, $\langle x^{*\beta}, y - x^\beta\rangle \to \langle x^*, y - x\rangle$. Since $\langle x^{*\beta}, y - x^\beta\rangle \geq 0$ for all β, we have that $\langle x^*, y - x\rangle \geq 0$, and hence, $\langle T(x), y - x\rangle \cap \mathbb{R}_+ \neq \emptyset$, i.e., $x \in \mathrm{Lev}^{\geq}_{\cap} F(\cdot, y)$. The coercivity condition (wBC$_{13}$) is the condition (wsC$_{13}$) in Theorem 2.62 for F. Thus, all the conditions in Theorem 2.62 are satisfied for F. Hence, there exists $\bar{x} \in A$ such that $\langle T(\bar{x}), y - \bar{x}\rangle \cap \mathbb{R}_+ = F(\bar{x}, y) \cap \mathbb{R}_+ \neq \emptyset$ for all $y \in A$. □

Theorem 2.87 *Let A be a nonempty convex subset of a topological vector space X and $T : A \rightrightarrows X^*$ be a set-valued mapping. Assume that T is upper hemicontinuous and properly quasimonotone, and the following coercivity condition is satisfied:*

(sBC$_{13}$) *There exists a nonempty compact convex subset K of A such that, for each $x \in A \backslash K$, $\langle T(y), y - x\rangle \cap \mathbb{R}_{++} \neq \emptyset$ for some $y \in K$.*

Then, there exists $\bar{x} \in A$ such that $\langle T(\bar{x}), y - \bar{x}\rangle \cap \mathbb{R}_+ \neq \emptyset$ for all $y \in A$.

Proof Let $F : A \times A \rightrightarrows \mathbb{R}$ be defined by (2.13). We check the conditions of Theorem 2.67. Clearly, (sBC$_{13}$) implies (wsC$_{13}$). $F \in \mathscr{S}^2\mathscr{A}^2_{st.qcl}$ (see the proof of Theorem 2.86). When T is properly quasimonotone, we easily check that $F \in -\mathscr{S}^1\mathscr{A}^1_{pr.qm}$. Moreover, $F(x, x) \cap \mathbb{R}_+ = \{0\}$ for all $x \in A$, i.e., $F \in \mathscr{S}^2\mathscr{A}_A^{=}$. Thus, $F \in (-\mathscr{S}^1\mathscr{A}^1_{pr.qm}) \cap \mathscr{S}^2\mathscr{A}^2_{st.qcl} \cap \mathscr{S}^2\mathscr{A}_A^{=}$. F has property $\mathbb{SP}^2$ of type 2. Indeed, for all $x, y, y' \in A$ and $t \in (0, 1)$, assume that $F(x, y) \cap \mathbb{R}_+ = \{0\}$ and $F(x, y') \subset -\mathbb{R}_{++}$. Then, $\langle T(x), y-x\rangle \subset -\mathbb{R}_+$ and $\langle T(x), y'-x\rangle \subset -\mathbb{R}_{++}$, which imply that $F((x, 1-t)y+ty') = \langle T(x), (1-t)y+ty'-x\rangle = (1-t)\langle T(x), y-x\rangle + t\langle T(x), y'-x\rangle \subset -\mathbb{R}_+ - \mathbb{R}_{++} \subset -\mathbb{R}_{++}$. F is $\mathbb{R}_+$-upper-sign continuous of type 2 in the first argument. Indeed, let $x, y \in A$ such that $F(x+t(y-x), y)\cap\mathbb{R}_+ \neq \emptyset$ for all $t \in (0, 1)$. We show that $F(x, y) \cap \mathbb{R}_+ \neq \emptyset$. We have $\emptyset \neq F(x + t(y - x), y) \cap$

$\mathbb{R}_+ = \langle T(x+t(y-x)), y-(x+t(y-x))\rangle \cap \mathbb{R}_+ = (1-t)\langle T(x+t(y-x)), y-x\rangle \cap \mathbb{R}_+$ for all $t \in (0, 1)$. Hence, $\langle T(x + t(y - x)), y - x\rangle \cap \mathbb{R}_+ \neq \emptyset$ for all $t \in (0, 1)$. If $\langle T(x), y - x\rangle \cap \mathbb{R}_+ = \emptyset$, i.e., $\langle T(x), y - x\rangle \subset -\mathbb{R}_{++}$. Then, by the upper hemicontinuity of T, there exists $\epsilon > 0$ such that $\langle T(x+t(y-x)), y-x\rangle \subset -\mathbb{R}_{++}$ for all $t \in (0, \epsilon)$, a contradiction. Thus, $F(x, y) \cap \mathbb{R}_+ = \langle T(x), y - x\rangle \cap \mathbb{R}_+ \neq \emptyset$. Finally, it is not hard to check that $\mathrm{Lev}^{\leq}_{\cap} F(x, \cdot) := \{y \in A \mid \langle T(x), y-x\rangle \subset -\mathbb{R}_+\}$ is closed for each $x \in A$. Thus, F satisfies all the conditions in Theorem 2.67. Hence, there exists $\bar{x} \in A$ such that $\langle T(\bar{x}), y - \bar{x}\rangle \cap \mathbb{R}_+ = F(\bar{x}, y) \cap \mathbb{R}_+ \neq \emptyset$ for all $y \in A$. □

Theorem 2.88 *Let A be a nonempty compact convex subset of a topological vector space X, $\mathscr{D} : A \rightrightarrows A$ be a set-valued map with nonempty convex images, and $T : A \rightrightarrows X^*$ be a set-valued mapping. Assume that T is properly quasimonotone and lower hemicontinuous, $\mathscr{D}^{-1}(y)$ is open for each $y \in A$, and*

(BSB) *for each $x \in \delta(\mathrm{Fix}\mathscr{D}) \setminus \mathrm{Fix}\mathscr{D}$, there exist $y \in \mathscr{D}(x)$ and $y^* \in T(y)$ such that $\langle y^*, y - x\rangle < 0$.*

Then, there exists $\bar{x} \in A$ such that $\bar{x} \in \mathscr{D}(\bar{x})$ and $\langle T(\bar{x}), y - \bar{x}\rangle \subset \mathbb{R}_+$ for all $y \in \mathscr{D}(\bar{x})$, i.e., the set-valued quasivariational inequality (SQVI) *has solutions.*

Proof Let $F : A \times A \rightrightarrows \mathbb{R}$ be defined by (2.12). Then, $F \in \mathscr{S}^1\mathscr{A}^2_{pr.qm}$ by Remark 2.84(b). It is not hard to check that $\mathrm{Lev}^{\leq}_{\subset} F(\cdot, y) := \{x \in A \mid \langle T(y), y - x\rangle \cap \mathbb{R}_{++} \neq \emptyset\}$ is open for each $y \in A$. Thus, F and $\mathscr{D}$ satisfy all the conditions in Theorem 2.69. Moreover, the condition (BSB) implies the condition (SB) in Theorem 2.69 for F. Thus, by Theorem 2.69, there exists $\bar{x} \in A$ such that $\bar{x} \in \mathscr{D}(\bar{x})$ and $F(\bar{x}, y) = \langle T(y), y - \bar{x}\rangle \subset \mathbb{R}_+$ for all $y \in \mathscr{D}(\bar{x})$. We show that $\langle T(\bar{x}), y - \bar{x}\rangle \subset \mathbb{R}_+$ for all $y \in \mathscr{D}(\bar{x})$. Let $y \in \mathscr{D}(\bar{x})$ be arbitrary, $t \in (0, 1]$, and $y_t = \bar{x} + t(y - \bar{x})$. Since $\mathscr{D}(\bar{x})$ is convex, $y_t \in \mathscr{D}(\bar{x})$ for all $t \in (0, 1]$. Then, $\langle T(y_t), y - \bar{x}\rangle = \frac{1}{t}\langle T(y_t), t(y - \bar{x})\rangle = \frac{1}{t}\langle T(y_t), y_t - \bar{x}\rangle \subset \mathbb{R}_+$ for all $t \in (0, 1]$. If $\langle T(\bar{x}), y - \bar{x}\rangle \not\subset \mathbb{R}_+$, i.e., $\langle T(\bar{x}), y - \bar{x}\rangle \cap (-\mathbb{R}_{++}) \neq \emptyset$. Then, by the lower hemicontinuity of T, there exists $\epsilon > 0$ such that $\langle T(y_t), y - \bar{x}\rangle \cap (-\mathbb{R}_{++}) \neq \emptyset$ for all $t \in (0, \epsilon)$, a contradiction. Thus, $\langle T(\bar{x}), y - \bar{x}\rangle \subset \mathbb{R}_+$. □

Theorem 2.89 *Let A be a nonempty compact convex subset of a topological vector space, $\mathscr{D} : A \rightrightarrows A$ be a set-valued map with nonempty convex images, and $T : A \rightrightarrows X^*$ be a set-valued mapping. Assume that T is upper semicontinuous mapping with compact images, $\mathscr{D}^{-1}(y)$ is open for each $y \in A$, and*

(BWB) *for each $x \in \delta(\mathrm{Fix}\mathscr{D})\setminus\mathrm{Fix}\mathscr{D}$, there exists $y \in \mathscr{D}(x)$ such that $\langle x^*, y-x\rangle < 0$ for all $x^* \in T(x)$.*

Then, there exists $\bar{x} \in A$ such that $\bar{x} \in \mathscr{D}(\bar{x})$ and $\langle T(\bar{x}), y - \bar{x}\rangle \cap \mathbb{R}_+ \neq \emptyset$ for all $y \in \mathscr{D}(\bar{x})$.

Proof Let $F : A \times A \rightrightarrows \mathbb{R}$ be defined by (2.13). From the proof of Theorem 2.86 we have that $F \in \mathscr{S}^2\mathscr{A}^2_{pr.qm}$ and $\mathrm{Lev}^{\geq}_{\cap} F(\cdot, y)$ is closed for $y \in A$ (which implies that $\mathrm{Lev}^{<}_{\cap} F(\cdot, y) = A \setminus \mathrm{Lev}^{\geq}_{\cap} F(\cdot, y)$ is open for $y \in A$). Thus, F and $\mathscr{D}$ satisfy all the conditions in Theorem 2.72. Moreover, the condition (BWB) implies the condition (WB) in Theorem 2.72 for F. Thus, by Theorem 2.72, there exists $\bar{x} \in A$ such that $\bar{x} \in \mathscr{D}(\bar{x})$ and $\langle T(\bar{x}), y - \bar{x}\rangle \cap \mathbb{R}_+ = F(\bar{x}, y) \cap \mathbb{R}_+ \neq \emptyset$ for all $y \in \mathscr{D}(\bar{x})$. □

Corollary 2.90 *Let A be a nonempty compact convex subset of a Hausdorff locally convex space, $\mathscr{D} : A \rightrightarrows A$ be a set-valued map with nonempty images, and $T : A \rightrightarrows X^*$ be a set-valued mapping. Assume that T is upper semicontinuous mapping with compact images, $\mathscr{D}$ is closed with convex images, and the set $\{x \in A \mid \mathscr{D}(x) \subset \{y \in A \mid \langle T(x), y - x\rangle \cap \mathbb{R}_+ \neq \emptyset\}\}$ is closed. Then, there exists $\bar{x} \in A$ such that $\bar{x} \in \mathscr{D}(\bar{x})$ and $\langle T(\bar{x}), y - \bar{x}\rangle \cap \mathbb{R}_+ \neq \emptyset$ for all $y \in \mathscr{D}(\bar{x})$. If addition, T has convex images, then the set-valued quasivariational inequality* (SwQVI) *has solutions.*

Proof Let $F : A \times A \rightrightarrows \mathbb{R}$ be defined by (2.13). Similarly to the proof of Theorem 2.89, we can check that F and $\mathscr{D}$ satisfy all the conditions of Corollary 2.73. Hence, there exists $\bar{x} \in A$ such that $\bar{x} \in \mathscr{D}(\bar{x})$ and $\langle T(\bar{x}), y - \bar{x}\rangle \cap \mathbb{R}_+ \neq \emptyset$ for all $y \in \mathscr{D}(\bar{x})$. Note that $\langle T(\bar{x}), y - \bar{x}\rangle \cap \mathbb{R}_+ \neq \emptyset$ implies $\Phi_T^s(\bar{x}, y) = \sup_{x^* \in T(\bar{x})}\langle x^*, y - \bar{x}\rangle \geq 0$. Therefore, as T has convex images, all the conditions of Proposition 2.11 hold, and hence, (SwQVI) has solutions. □

Corollary 2.91 *Let A be a nonempty compact convex subset of a Hausdorff locally convex space X, $\mathscr{D} : A \rightrightarrows A$ be a set-valued map with nonempty images, and $T : A \rightrightarrows X^*$ be a set-valued mapping. Assume that T is upper hemicontinuous and properly quasimonotone, $\mathscr{D}$ is closed with convex images, and $\{x \in A \mid \mathscr{D}(x) \subset \{y \in A \mid \langle T(x), y - x\rangle \subset -\mathbb{R}_+\}$ is closed. Then, there exists $\bar{x} \in A$ such that $\bar{x} \in \mathscr{D}(\bar{x})$ and $\langle T(\bar{x}), y - \bar{x}\rangle \cap \mathbb{R}_+ \neq \emptyset$ for all $y \in \mathscr{D}(\bar{x})$. If addition, T has compact convex images, then the set-valued quasivariational inequality* (SwQVI) *has solutions.*

Proof Let $F : A \times A \rightrightarrows \mathbb{R}$ be defined by (2.13). From the proof of Theorem 2.87 we have that $F \in (-\mathscr{S}^1\mathscr{A}^1_{pr.qm}) \cap \mathscr{S}^2\mathscr{A}^2_{st.qcl} \cap \mathscr{S}^2\mathscr{A}^{=}_{A}$, F has property $\mathbb{SP}^2$ of type 2 and is $\mathbb{R}_+$-upper-sign continuous of type 2 in the first argument, and $\mathrm{Lev}^{\leq}_{\cap} F(x, \cdot)$ is closed for each $x \in A$. Moreover, the condition "$\{x \in A \mid \mathscr{D}(x) \subset \{y \in A \mid \langle T(x), y - x\rangle \subset -\mathbb{R}_+\}$ is closed" implies that the set $\{x \in A \mid \mathscr{D}(x) \cap \mathrm{Lev}^{>}_{\cap} F(\cdot, x) = \emptyset\}$ is closed. Thus, all the conditions of Corollary 2.77 hold for F and $\mathscr{D}$. Hence, there exists $\bar{x} \in A$ such that $\bar{x} \in \mathscr{D}(\bar{x})$ and $\langle T(\bar{x}), y - \bar{x}\rangle \cap \mathbb{R}_+ = F(\bar{x}, y) \cap \mathbb{R}_+ \neq \emptyset$ for all $y \in \mathscr{D}(\bar{x})$. When T has compact convex images, all the conditions of Proposition 2.11 hold, and hence, (SwQVI) has solutions. □

2.7 Exercises

Exercise 2.1 Let A be a nonempty convex subset of a topological vector space X and $S : A \rightrightarrows A$ be a set-valued mapping. S is called a KKM MAPPING if for all $N \in \mathscr{F}(A)$, $\mathrm{conv}N \subset \bigcup_{x \in N} S(x)$.

(a) Prove the following result by using Theorem 1.34.

Theorem 2.92 (The KKM-Fan lemma) *Let A be a nonempty compact convex subset of a topological vector space. If $S : A \rightrightarrows A$ is a KKM mapping with closed values, then $\bigcap_{x \in A} S(x) \neq \emptyset$.*

(b) Prove Theorems 1.34 and 2.13 by using the KKM-Fan lemma.

Exercise 2.2 (a) Prove the following result by using Theorem 1.34 or the KKM-Fan lemma.

Theorem 2.93 (Maximal element theorem) *Let A be a nonempty compact convex subset of a topological vector space and $S : A \rightrightarrows A$. Assume that the following conditions are satisfied:*

(a) *For all $y \in A$, $S^{-1}(y)$ is open.*
(b) *For all $N \in \mathscr{F}(A)$, $\operatorname{conv} N \cap \bigcap_{y^i \in N} S^{-1}(y^i) = \emptyset$.*

Then, there exists $\bar{x} \in A$ such that $S(\bar{x}) = \emptyset$.

(b) Prove Theorems 1.34, 2.21, and 2.92 by using Theorem 2.93.

Exercise 2.3 Given a nonempty set A, a real function φ on A, and two families of real functions $g(x, \omega)$, $\omega \in \Omega$, and $h(x, \lambda)$, $\lambda \in \Lambda$, a constrained extremum problem (GOP) reads as

$$\min \varphi(x) \quad \text{subject to } g(x, \omega) \leq 0 \text{ for all } \omega \in \Omega, h(x, \lambda) = 0 \text{ for all } \lambda \in \Lambda.$$

Transform (GOP) to (QEP).

Exercise 2.4

(a) For A, $\mathscr{D}$, and Φ as in the formula of (QEP), let $S : A \rightrightarrows A$ be defined by

$$\begin{aligned} S(x) &:= \{x' \in \mathscr{D}(x) \mid \Phi(x', y) \geq 0 \text{ for all } y \in \mathscr{D}(x)\} \\ &= \mathscr{D}(x) \cap \Big(\bigcap_{y \in \mathscr{D}(x)} \operatorname{lev}_{\geq} \Phi(\cdot, y) \Big). \end{aligned}$$

Prove that x is a solution of (QEP) if and only if x is a fixed point of S.

(b) Prove the following result by using Kakutani's theorem: *Assume that A is a compact convex subset of a Hausdorff locally convex space X, $\Phi \in (-\mathscr{A}^1_{qcl}) \cap \mathscr{A}^2_{pr.qm}$, $\mathscr{D}$ is closed and lower semicontinuous with convex values, and $\operatorname{lev}_{\geq} \Phi := \{(x, y) \in X \times X \mid \Phi(x, y) \geq 0\}$ is closed. Then,* (QEP) *has solutions.* This result is a special case of Corollary 2.26 because $(-\mathscr{A}^1_{qcl}) \cap \mathscr{A}^2_{pr.qm} \subset \mathscr{A}^2_{pr.qm}$.

Exercise 2.5 Assume that A is a compact convex subset of a Hausdorff locally convex space X, $\Phi(\cdot,\cdot)$ is lower semicontinuous, $\mathscr{D}$ is closed with convex values, and the following conditions are satisfied:

(a) $\Phi(x,x) \geq 0$ for all $x \in A$.
(b) The function $m : A \to \mathbb{R}$, defined by $m(x) := \min_{y\in\mathscr{D}(x)} \Phi(x,y)$ for all $x \in A$, is upper semicontinuous.
(c) For all $x \in A$, the set $M(x) := \{y \in \mathscr{D}(x) \mid \Phi(x,y) = m(x)\}$ is convex.

Prove that (QEP) has solutions.

Exercise 2.6 There is a result on the lower semicontinuity of an intersection as follows.

Proposition 2.94 *Let X be a topological space and $S, H : X \rightrightarrows \mathbb{R}^n$. If S and H are lower semicontinuous and H has convex open images, then $F : X \rightrightarrows \mathbb{R}^n$ defined by $F(x) := (S \cap H)(x) := S(x) \cap H(x)$ is lower semicontinuous.*

For the proof of this proposition, see [253]. By using Proposition 2.94, prove that the assumption "the set-valued mapping $S : \mathrm{Fix}\mathscr{D} \rightrightarrows A$, defined by $S(x) := \mathscr{D}(x) \cap \mathrm{lev}_{<}\Phi(x,\cdot)$ for all $x \in \mathrm{Fix}\mathscr{D}$, is lower semicontinuous" in Theorem 2.28 can be replaced by the following conditions:

(a) $A \subset \mathbb{R}^n$.
(a) the set-valued mapping $S^* : A \rightrightarrows A$ defined by

$$S^*(x) := \mathrm{lev}_{<}\Phi(x,\cdot) = \{y \in A \mid \Phi(x,y) < 0\}$$

is lower semicontinuous.

Exercise 2.7 Assume that A is a nonempty compact convex subset of a topological vector space X, $\Phi \in \mathscr{A}^2_{pr.qm}$, $\mathscr{D}$ has convex images, $\mathscr{D}^{-1}(y)$ and $\mathrm{lev}_{<}\Phi(\cdot,y)$ are open for each $y \in A$, and the set-valued mapping $\overline{\mathscr{D}} : A \rightrightarrows A$, defined by $\overline{\mathscr{D}}(x) := \mathrm{cl}\mathscr{D}(x)$ for all $x \in A$, is closed. Prove that there exists $\bar{x} \in \mathrm{cl}\mathscr{D}(\bar{x})$ such that $\Phi(\bar{x},y) \geq 0$ for all $y \in \mathscr{D}(\bar{x})$.

Exercise 2.8 Prove that in Theorem 2.21 the compactness of A can be replaced by one of the following coercivity conditions:

(C^{cl}_{1Q}) There exists a nonempty compact subset K of X such that, for each finite subset N of A, there exists a compact convex subset L_N of A containing N such that $\mathscr{D}(x) \cap L_N \neq \emptyset$ for each $x \in L_N$, and for each $x \in \mathrm{cl}(\mathrm{Fix}\mathscr{D}) \cap (L_N \setminus K)$, $\Phi(x,y) < 0$ for some $y \in \mathscr{D}(x) \cap L_N$.
(C^{cl}_{2Q}) There exists a nonempty compact convex subset K of A such that $\mathscr{D}(x) \cap K \neq \emptyset$ for each $x \in A \setminus K$, and for each $x \in \mathrm{cl}(\mathrm{Fix}\mathscr{D}) \setminus K$, $\Phi(x,y) < 0$ for some $y_x \in \mathscr{D}(x) \cap K$.

Exercise 2.9 Prove that the compactness of A in Corollaries 2.24 and 2.26 can be replaced by one of the following coercivity conditions:

(C_{1Q}) There exists a nonempty compact subset K of X such that, for each finite subset N of A, there exists a compact convex subset L_N of A containing N such that $\mathscr{D}(x) \cap L_N \neq \emptyset$ for each $x \in L_N$, and for each $x \in \mathrm{Fix}\mathscr{D} \cap (L_N \backslash K)$, $\Phi(x, y) < 0$ for some $y \in \mathscr{D}(x) \cap L_N$.

(C_{2Q}) There exists a nonempty compact convex subset K of A such that $\mathscr{D}(x) \cap K \neq \emptyset$ for each $x \in A$, and for each $x \in \mathrm{Fix}\mathscr{D} \cap (A\backslash K)$, $\Phi(x, y) < 0$ for some $y \in \mathscr{D}(x) \cap K$.

Exercise 2.10 Give some coercivity conditions for the theorems and corollaries in Sect. 2.4.

Exercise 2.11 State some corollaries of the results of Sect. 2.4 for the case where A is a closed convex subset of a reflexive Banach space.

Exercise 2.12 Let A be a nonempty set, $\mathscr{D}_1, \mathscr{D}_2 : A \rightrightarrows A$ be set-valued mappings, and $\Phi : A \times A \to \mathbb{R}$. Consider the following problem:

(GQEP) find $\bar{x} \in A$ such that $\bar{x} \in \mathscr{D}_1(\bar{x})$ and $\Phi(\bar{x}, y) \geq 0$ for all $y \in \mathscr{D}_2(\bar{x})$.

(a) Find an equilibrium problem which is equivalent to (GQEP).
(b) Find a quasiequilibrium problem which is equivalent to (GQEP).
(c) Prove the following result by using Theorem 1.34 or Theorem 2.21: *Assume that A is a compact convex subset of a topological vector space; $\Phi \in \mathscr{A}^2_{pr.qm}$. Suppose that* $\mathrm{Fix}\mathscr{D}_1$ *is closed,* $\mathrm{conv}\mathscr{D}_2(x) \subset \mathscr{D}_1(x)$ *for all $x \in A$,* $\mathrm{lev}_{\geq}\Phi(\cdot, y)$ *is closed, and $\mathscr{D}_2^{-1}(y)$ is open. Then,* (GQEP) *has solutions.*

Exercise 2.13 Let A be a nonempty convex subset of a vector space and $S : A \rightrightarrows \mathbb{R}$ be a set-valued map. Assume that the following condition is satisfied: *whenever* $\{x^1, x^2, \ldots, x^n\} \subset A$ *and* $\{t_1, t_2, \ldots, t_n\} \subset \mathbb{R}_+$ *such that* $\sum_{i=1}^n t_i = 1$, *then for every* $\{\lambda_1, \lambda_2, \ldots, \lambda_n\} \subset \mathbb{R}$ *with* $\lambda_i \in S(x^i)$ *for every* $i = 1, 2, \ldots, n$, *there exists* $\lambda \in S(\sum_{i=1}^n t_i x^i)$ *such that*

$$\lambda \leq \max\{\lambda_i : i = 1, 2, \ldots, n\}.$$

Prove that S is $\mathbb{R}_+$-quasiconvex of type 1.

Exercise 2.14 Prove that $\mathscr{S}\ell\mathscr{A}^i_c \subset \mathscr{S}\ell\mathscr{A}^i_{qc} \subset \mathscr{S}\ell\mathscr{A}^i_{st.qcl}$ and $\mathscr{S}\ell\mathscr{A}^i_{st.qcl} \cap \mathscr{S}\ell\mathscr{A}^{\geq}_{\bar{A}} \subset \mathscr{S}\ell\mathscr{A}^i_{pr.qm}$ ($\ell \in \{1, 2\}$ and $i \in \{1, 2\}$).

Exercise 2.15 Restate Corollaries 1.35 and 1.36, Theorem 1.46, and Corollaries 1.47, 1.48, 1.51, and 1.52 for set-valued equilibrium problems (SSEP_F) and (WSEP_F). From which deduce some corollaries for Browder variational inclusions.

Exercise 2.16 Let $A = [0, 1]^n \subset \mathbb{R}^n$ and $\mathscr{D} : A \times A \rightrightarrows A$ be defined by $\mathscr{D}(x) := \prod_{i=1}^n [x_i, 1]$ for all $x = (x_1, \ldots, x_n) \in A$. Prove that for any $T : A \rightrightarrows \mathbb{R}^n (\equiv$

$(\mathbb{R}^n)^*$ which is upper semicontinuous and has compact convex images, (SwQVI) has solutions.

Exercise 2.17 Consider existence of solutions of (SQVI) and (SwQVI), with $A = [0, 1]^n \subset \mathbb{R}^n$, $T : A \rightrightarrows \mathbb{R}^n$ defined by

$$T(x) := \{(\alpha_1, \alpha_2, \dots, \alpha_n)\|x\| \mid \alpha_1, \alpha_2, \dots, \alpha_n \in [a, b]\} \quad \text{for all } x \in A,$$

and $\mathscr{D} : A \times A \rightrightarrows A$ defined by $\mathscr{D}(x) := \mathscr{D}_1(x_1) \times [0, 1]^{n-1}$ for all $x = (x_1, \dots, x_n) \in A$, where

$$\mathscr{D}_1(t) := \begin{cases} [0, \frac{t}{2}] \text{ if } t > \frac{2}{3}, \\ [0, \frac{1}{2}] \text{ if } t < \frac{2}{3}, \end{cases} \quad \text{for all } t \in [0, 1].$$

Exercise 2.18 Let A be a nonempty set and $F : A \times A \rightrightarrows \mathbb{R}$ be a set-valued bifunction. We say that F is $\mathbb{R}_+$-CYCLICALLY ANTI-QUASIMONOTONE OF TYPE 1 (OF TYPE 2, resp.) if, for any $x^1, x^2, \dots, x^m \in A$, there exists an $i \in \{1, 2, \dots, m\}$ such that $F(x^i, x^{i+1}) \subset \mathbb{R}_+$ ($F(x^i, x^{i+1}) \cap \mathbb{R}_+ \neq \emptyset$, resp.), where $x^{m+1} := x^1$.

(a) Prove that F is $\mathbb{R}_+$-cyclically anti-quasimonotone of type 1 (of type 2, resp.) if and only if whenever $x^1, x^2, \dots, x^m \in A$ ($m \geq 2$) satisfy $F(x^i, x^{i+1}) \cap (-\mathbb{R}_{++}) \neq \emptyset$ ($F(x^i, x^{i+1}) \subset -\mathbb{R}_{++}$, resp.) for all $i = 1, 2, \dots, m-1$, we have $F(x^m, x^1) \subset \mathbb{R}_+$ ($F(x^m, x^1) \cap \mathbb{R}_+ \neq \emptyset$, resp.).

(b) Prove the following results.

Theorem 2.95 *Let A be a nonempty compact subset of a topological space and $F : A \times A \rightrightarrows \mathbb{R}$ be a set-valued bifunction. If $F \in \mathscr{S}^1\mathscr{A}_{caqm}$ and $\mathrm{Lev}_{\subset}^{\geq} F(\cdot, y)$ is closed for every $y \in A$, then* (SSEP$_F$) *has solutions.*

Theorem 2.96 *Let A be a nonempty compact subset of a topological space and $F : A \times A \rightrightarrows \mathbb{R}$ be a set-valued bifunction. If $F \in \mathscr{S}^2\mathscr{A}_{caqm}$ and $\mathrm{Lev}_{\cap}^{\geq} F(\cdot, y)$ is closed for every $y \in A$, then* (WSEP$_F$) *has solutions.*

2.8 Commentary

We have seen that problem (QEP) can be transformed into a peculiarity equilibrium problem and we cannot simply apply existence results for (EP) to (QEP). To see the difficulty in studying of (QEP), we look at (QEP) from other angles. By defining $S : A \rightrightarrows A$ by $S(x) = \mathscr{D}(x) \cap \{x \in A \mid \inf_{y \in \mathscr{D}(x)} \Phi(z, y) \geq 0\}$, we show that the set of solutions of (QEP) coincides with the set of fixed points of S which is tricky to handle. Moreover, if we define $H : A \rightrightarrows A$ by

$$H(x) = \begin{cases} \mathscr{D}(x) \cap \mathrm{lev}_{<} \Phi(x, \cdot) & \text{if } x \in \mathrm{Fix}\mathscr{D}, \\ \mathscr{D}(x) & \text{if } x \notin \mathrm{Fix}\mathscr{D}, \end{cases}$$

then x is a solution of (QEP) if and only if $H(x) = \emptyset$, i.e., x is a maximal element of H. But, to utilize the mapping H, we have to calculate the set of fixed points of $\mathscr{D}$, and therefore, it is hard to check the existence of maximal elements of H. So, finding certain conditions for (QEP) to have solutions is really difficult. That is why so far there are not many articles related to the existence of solutions of (QEP) although some particular cases have been mentioned for a long time [168, 376].

In the literature, there are some approaches to studying of the existence of solutions of quasiequilibrium problems such as the fixed point theory approach and selection theory approach [276]. In the articles [151, 152, 385], based on fixed point theorems one searches for conditions on $\mathscr{D}$ and Φ ensuring that either the above mapping S has fixed points and hence the problem has solutions, or a contraction as the above mapping H has fixed points and its images are nonempty and hence the problem has solutions. Observing results in this direction, we see that the assumptions on the constraint mapping $\mathscr{D}$ are usually strict. For example, $\mathscr{D}$ is often assumed to be closed and lower semicontinuous; but in fact this assumption implies that $\mathscr{D}$ must be continuous as A is compact. The articles [121, 122, 124, 158–160] in the second approach, based on conditions for a set-valued mapping to have selections, give conditions on $\mathscr{D}$ and Φ such that mapping S or mapping H has continuous selections and hence they have fixed points by the Schauder-Tychonoff theorem. For this approach, the closedness of $\mathscr{D}$ can be relaxed to the closedness of the set of its fixed points, but the lower semicontinuity of $\mathscr{D}$ is required. Moreover, the underlying space is a separable Banach space or a finite dimensional space. In this direction, Michael's selection theorems are important tools.

In this chapter, based on the relationships between (EP) and (QEP), we use some results of Chap. 1 to obtain several specific existence results for (QEP). The results of this chapter contain some results in the literature as particular cases. For example, in Corollary 2.24, $\mathscr{D}$ only needs to be closed but not necessarily lower semicontinuous. Theorem 2.28 and its corollaries are stated based on the results in [121, 124, 158]. Note that from the proof of Theorem 2.28, we see that the assumption "$\Phi \in \mathscr{A}^2_{pr.qm}$" can be relaxed to "for each $x \in \mathrm{Fix}\mathscr{D}$, Φ is properly quasimonotone in the second component on $\mathscr{D}(x)$." Moreover, the results are stated with the assumption that A is a compact set. However, from the mathematical point of view, this in fact does not affect much the essence of the results because when A is not necessarily compact, we can add suitable coercivity conditions to reduce the problem to a compact case (see, e.g., Exercises 2.8, 2.9).

Chapter 3
Solution Existence for Vector Equilibrium Problems

Abstract This chapter extends the scalar equilibrium problem and develops existence results in Chap. 1 to the vector case. Section 3.1 presents some basic notions such as the algebraic interior and the vectorial closure of a set, cones, nonconvex separation functionals, and continuity of vector functions. Section 3.2 introduces a vector equilibrium problem ($\text{VEP}_{(f,C)}$), concepts of solutions, relationships between kinds of solutions, and particular cases of ($\text{VEP}_{(f,C)}$). Section 3.3 provides existence conditions for the kinds of solutions of ($\text{VEP}_{(f,C)}$). The results are divided into two groups. The first one includes the results which are established based on the vector-proper quasimonotonicity. The second group consists of results formed based on the vector-cyclic quasimonotonicity. The last part of this section provides some forms of vectorial Ekeland variational principle. Section 3.4 introduces the vector quasiequilibrium problem, which is an extension of quasiequilibrium problem (QEP) to vector bifunctions.

3.1 Some Basic Notions

Throughout this chapter, Y is a real vector space, and 0 (or 0_Y) is the zero vector in Y. For nonempty sets $C \subset Y$ and $E \subset Y$, we define

$$C + E := \{c + e \mid c \in C, \ e \in E\},$$

and for $\Lambda \subset \mathbb{R}$,

$$\Lambda E := \{\lambda e \mid \lambda \in \Lambda, \ e \in E\}.$$

- $C_0 = C \backslash \{0\}$, $E^c = Y \setminus E$ is the complement of E.
- When Y is a topological vector space, we use $\text{int}E$ and $\text{cl}E$ to denote the topological interior and the closure of E.

L. Q. Anh et al., *Equilibrium Problems*, Forum for Interdisciplinary Mathematics, https://doi.org/10.1007/978-3-031-89003-1_3

- Let A be any nonempty set, $E \subset Y$ and $f : A \times A \to Y$. For $x, y \in A$, we consider the following E-level sets of f:

∘ $\text{lev}_E^1 f(x, \cdot) := \{y \in A \mid f(x, y) \notin -E\}$.
∘ $\text{lev}_E^2 f(x, \cdot) := \{y \in A \mid f(x, y) \in E\}$.
∘ $\overline{\text{lev}}_E^1 f(x, \cdot) := \{y \in A \mid f(x, y) \in -E\}$.
∘ $\overline{\text{lev}}_E^2 f(x, \cdot) := \{y \in A \mid f(x, y) \notin E\}$.
∘ $\text{lev}_E^1 f(\cdot, y) := \{x \in A \mid f(x, y) \notin -E\}$.
∘ $\text{lev}_E^2 f(\cdot, y) := \{x \in A \mid f(x, y) \in E\}$.
∘ $\overline{\text{lev}}_E^1 f(\cdot, y) := \{x \in A \mid f(x, y) \in -E\}$.
∘ $\overline{\text{lev}}_E^2 f(\cdot, y) := \{x \in A \mid f(x, y) \notin E\}$.

For $\ell \in \{1, 2\}$, we set $\ell^c = \{1, 2\} \setminus \{\ell\}$. Thus, $\text{lev}_E^{1^c} f(x, \cdot) = \text{lev}_E^2 f(x, \cdot)$, $\text{lev}_E^{2^c} f(x, \cdot) = \text{lev}_E^1 f(x, \cdot)$, $\overline{\text{lev}}_E^{1^c} f(x, \cdot) = \overline{\text{lev}}_E^2 f(x, \cdot)$, etc.

3.1.1 Algebraic Interiors, Vectorial Closures, Cones, and Nonconvex Separation Functionals in Vector Spaces

Definition 3.1 Let $E \subset Y$ be nonempty.

(a) The set

$$\text{core}E := \{u \in Y \mid \forall v \in Y, \exists \lambda > 0 \text{ such that } u + [0, \lambda]v \subset E\}$$

is called the ALGEBRAIC INTERIOR of E. We say that E is an algebraically solid set if $\text{core}E \neq \emptyset$. When $E = \text{core}E$, E is called ALGEBRAICALLY OPEN.

(b) The set

$$\text{vcl}E := \{u \in Y \mid \exists v \in Y,\ \forall \lambda > 0, \exists \lambda' \in [0, \lambda] \text{ such that } u + \lambda' v \in E\}$$

is called the VECTORIAL CLOSURE of E. The set E with $E = \text{vcl}E$ is called VECTORIALLY CLOSED.

(c) For $q \in Y$, the set

$$\text{vcl}_q E := \{u \in Y \mid \forall \lambda > 0, \exists \lambda' \in [0, \lambda] \text{ such that } u + \lambda' q \in E\}$$

is called the VECTORIAL CLOSURE OF E IN DIRECTION q. E is called q-VECTORIALLY CLOSED if $E = \text{vcl}_q E$.

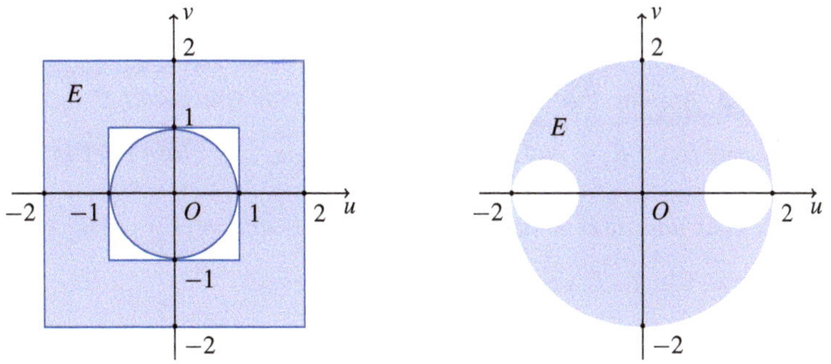

(a): The points $(-1,0)$, $(1,0)$, $(0,-1)$, $(0,1)$ belong to core$E$ but not belong to int$E$

(b): The points $(-2,0)$ and $(2,0)$ belong to cl$E$ but not belong to vcl$E$

Fig. 3.1 Illustration of Example 3.2

It is clear that

$$\mathrm{core}E \subset E \subset \mathrm{vcl}_q E \subset \mathrm{vcl}E = \bigcup_{q \in Y} \mathrm{vcl}_q E.$$

Therefore, if E is vectorially closed, then it is q-vectorially closed for every $q \in Y$.

Example 3.2

(a) Let $Y = \mathbb{R}^2$, $E_1 = \{(u, v) \mid -2 \leq u, v \leq 2\}$, $E_2 = \{(u, v) \mid -1 < u, v < 1\}$, $E_3 = \{(u, v) \mid u^2 + v^2 \leq 1\}$, and $E = (E_1 \setminus E_2) \cup E_3$ (see Fig. 3.1a). Then,

$$\mathrm{int}E = (\{(u, v) \mid -2<u, v<2\} \setminus \{(u, v) \mid -1 \leq u, v \leq 1\}) \cup \{(u, v) \mid u^2+v^2 < 1\}$$

and $\mathrm{core}E = \mathrm{int}E \cup \{(-1, 0), (1, 0), (0, -1), (0, 1)\}$ (see Fig. 3.1a). Thus, the points $(-1, 0)$, $(1, 0)$, $(0, -1)$, $(0, 1)$ belong to core$E$ but not to int$E$.

(b) Let $Y = \mathbb{R}^2$, $E_1 = \{(u, v) \mid u^2 + v^2 < 4\}$, $E_2 = \{(u, v) \mid (u + \frac{3}{2})^2 + v^2 \leq \frac{1}{4}\}$, $E_3 = \{(u, v) \mid (u - \frac{3}{2})^2 + v^2 \leq \frac{1}{4}\}$, and $E = E_1 \setminus (E_2 \cup E_3)$ (see Fig. 3.1b). Then,

$$\mathrm{cl}E = \left\{(u, v) \mid u^2 + v^2 \leq 4\right\} \setminus \left(\left\{(u, v) \mid \left(u + \frac{3}{2}\right)^2 + v^2 < \frac{1}{4}\right\} \cup \left\{(u, v) \mid \left(u - \frac{3}{2}\right)^2 + v^2 < \frac{1}{4}\right\}\right)$$

and $\mathrm{vcl}E = \mathrm{cl}E \setminus \{(-2, 0), (2, 0)\}$ (see Fig. 3.1b). Thus, the points $(-2, 0)$ and $(2, 0)$ belong to cl$E$ but not to vcl$E$.

The following proposition shows relationships between topological interior, algebraic interior, topological closure, and vectorial closure.

Proposition 3.3 *Assume that Y is a topological vector space and $E \subset Y$. Then:*

(a) $\operatorname{int}E \subset \operatorname{core}E \subset E \subset \operatorname{vcl}E \subset \operatorname{cl}E$ *and hence, any closed set is vectorially closed.*
(b) *If E is convex and* $\operatorname{int}E \neq \emptyset$, *then* $\operatorname{int}E = \operatorname{core}E$ *and* $\operatorname{vcl}E = \operatorname{cl}E$.

Proof (a) For $u, v \in Y$, $T_{uv} : \mathbb{R} \to Y$ defined by

$$T_{uv}(t) := u + tv \text{ for all } t \in \mathbb{R}$$

is continuous. Assume that $u \in \operatorname{int}E$ and $v \in Y$. Since T_{uv} is continuous and $T_{uv}(0) = u \in \operatorname{int}E$, there exists $\varepsilon > 0$ such that $T_{uv}(t) = u + tv \in \operatorname{int}E$ for all $t \in (-\varepsilon, \varepsilon)$. Taking $\lambda = \varepsilon/2$, we have $u + [0, \lambda]v = T_{uv}([0, \lambda]) \subset T_{uv}((-\varepsilon, \varepsilon)) \subset \operatorname{int}E \subset E$. This implies that $u \in \operatorname{core}E$. Thus, $\operatorname{int}E \subset \operatorname{core}E$. Let $u \in \operatorname{vcl}E$. Assume that $u \notin \operatorname{cl}E$, i.e., $u \in Y \setminus \operatorname{cl}E$. Then, for any $v \in Y$, $T_{uv}(0) = u \in Y \setminus \operatorname{cl}E$. By the continuity of T_{uv}, there exists $\epsilon > 0$ such that $u + [0, \lambda]v = T_{uv}([0, \lambda]) \subset T_{uv}((-\epsilon, \epsilon)) \subset Y \setminus \operatorname{cl}E \subset Y \setminus E$ for $\lambda = \epsilon/2$. This contradicts the fact that $u \in \operatorname{vcl}E$ (which means that there is $v \in Y$ such that for each $\lambda > 0$, $(u + [0, \lambda]v) \cap E \neq \emptyset$). Thus, $\operatorname{vcl}E \subset \operatorname{cl}E$.

(b) By part (a), we need to prove only that $\operatorname{core}E \subset \operatorname{int}E$ and $\operatorname{cl}E \subset \operatorname{vcl}E$. For $u \in \operatorname{core}E$ and $v^0 \in \operatorname{int}E$, there exists $\epsilon > 0$ such that

$$v := u + \epsilon(u - v^0) \in E.$$

Set $V = \frac{1}{1+\epsilon}v + \frac{\epsilon}{1+\epsilon}\operatorname{int}E$. Then, V is open and $V \subset \frac{1}{1+\epsilon}E + \frac{\epsilon}{1+\epsilon}E = E$. Moreover,

$$u = \frac{\epsilon}{1+\epsilon}v + \frac{1}{1+\epsilon}v \in \frac{\epsilon}{1+\epsilon}v^0 + \frac{1}{1+\epsilon}(v + \epsilon(\operatorname{int}E - v^0)) = V.$$

Thus, $u \in \operatorname{int}E$, i.e., $\operatorname{core}E \subset \operatorname{int}E$.

Assume that $u \in \operatorname{cl}E$. Fix $v^0 \in \operatorname{int}E$ and set $v = v^0 - u$. Then, for all $\lambda > 0$, there exists $y \in E \cap (u + \lambda(\operatorname{int}E - v^0))$, i.e., $y \in E$ and $v^0 \in \frac{1}{\lambda}y - \frac{1}{\lambda}u + \operatorname{int}E$. Taking $\lambda' = \frac{\lambda}{1+\lambda} \in (0, \lambda]$, we have

$$u + \lambda' v = \frac{1}{1+\lambda}u + \frac{\lambda}{1+\lambda}v^0 \in \frac{1}{1+\lambda}u + \frac{\lambda}{1+\lambda}\left(\frac{1}{\lambda}y - \frac{1}{\lambda}u + \operatorname{int}E\right)$$

$$= \frac{1}{1+\lambda}y + \frac{\lambda}{1+\lambda}\operatorname{int}E \subset \frac{1}{1+\lambda}E + \frac{\lambda}{1+\lambda}E = E.$$

Thus, $u \in \operatorname{vcl}E$, and hence $\operatorname{cl}E \subset \operatorname{vcl}E$. □

Proposition 3.4 *Let Y be a real vector space, and $q \in Y$ and $C, E \subset Y$ be nonempty sets. Then:*

(a) $E + \text{core}C \subset \text{core}(E + C)$, $\text{vcl}_q E + C \subset \text{vcl}_q(E + C)$, *and* $\text{vcl}E + C \subset \text{vcl}(E + C)$.
(b) $\bigcap_{\delta>0}(-\delta q + \text{vcl}_q E) \subset \text{vcl}_q E$.
(c) *For all* $\alpha > 0$, $\text{core}(\alpha E) = \alpha\text{core}E$ *and* $\text{vcl}(\alpha E) = \alpha\text{vcl}E$.
(d) *If* $\text{core}C = \text{core}(\text{core}C)$ *and* $E \cap \text{core}C = \emptyset$, *then* $\text{vcl}E \cap \text{core}C = \emptyset$;
(e) *if E is convex, then* $\text{core}(\text{core}E) = \text{core}E$ *and* $\text{core}E$ *is convex.*

Proof

(a) Take any $u = e + c$ with $e \in E$ and $c \in \text{core}C$. Then, for all $v \in Y$, there exists $\lambda > 0$ such that $c + [0, \lambda]v \subset C$. It follows that $u + [0, \lambda]v = e + c + [0, \lambda]v \subset E + C$. Hence, $u \in \text{core}(E + C)$. Thus, $E + \text{core}C \subset \text{core}(E + C)$. Let $u = e + c$ with $e \in \text{vcl}_q E$ and $c \in C$. Since $e \in \text{vcl}_q E$, we have that for each $\lambda > 0$, there is $0 < \lambda' < \lambda$ satisfying $e + \lambda' q \in E$. Then, for each $\lambda > 0$, there is $0 < \lambda' < \lambda$ satisfying $u + \lambda' q = e + \lambda' q + c \in E + C$. Hence, $u \in \text{vcl}_q(E + C)$ and so $\text{vcl}_q E + C \subset \text{vcl}_q(E + C)$. Now,

$$\text{vcl}E+C = \bigcup_{q\in Y} \text{vcl}_q E+C = \bigcup_{q\in Y}[\text{vcl}_q E+C] \subset \bigcup_{q\in Y} \text{vcl}_q(E+C) = \text{vcl}(E+C).$$

(b) Take any $u \in \bigcap_{\delta>0}(-\delta q + \text{vcl}_q E)$. Then, for all $\lambda > 0$, $u \in -\frac{\lambda}{2}q + \text{vcl}_q E$, i.e., $u + \frac{\lambda}{2}q \in \text{vcl}_q E$. Hence, there exists $\mu \in [0, \frac{\lambda}{2}]$ such that $(u + \frac{\lambda}{2}q) + \mu q \in E$. By choosing $\lambda' = \frac{\lambda}{2} + \mu$, we have $\lambda' \in [0, \lambda]$ and $u + \lambda' q \in E$. Thus, $u \in \text{vcl}_q E$.
(c) For $\alpha > 0$, we have

$$\begin{aligned} u \in \text{core}(\alpha E) &\Leftrightarrow \forall v \in Y, \exists \lambda > 0 \text{ such that } u + [0, \lambda]v \subset \alpha E \\ &\Leftrightarrow \forall v \in Y, \exists \lambda > 0 \text{ such that } \left(\frac{1}{\alpha}u\right) + \left[0, \frac{\lambda}{\alpha}\right]v \subset E \\ &\Leftrightarrow \left(\frac{1}{\alpha}u\right) \in \text{core}E \\ &\Leftrightarrow u \in \alpha\text{core}E. \end{aligned}$$

Thus, $\text{core}(\alpha E) = \alpha\text{core}E$. The equality $\text{vcl}(\alpha E) = \alpha\text{vcl}E$ can be proved similarly.
(d) Assume that there exists $y \in \text{vcl}E \cap \text{core}C$. Since $y \in \text{core}C = \text{core}(\text{core}C)$, for all $v \in Y$, there exists $\lambda_v > 0$ such that $y + [0, \lambda_v]v \subset \text{core}C$. As $y \in \text{vcl}E$, there exist $v' \in Y$ and $\lambda'_{v'} \in [0, \lambda_{v'}]$ such that $y + \lambda'_{v'}v' \in E$. Then, $y + \lambda'_{v'}v' \in E \cap \text{core}C$, a contradiction.

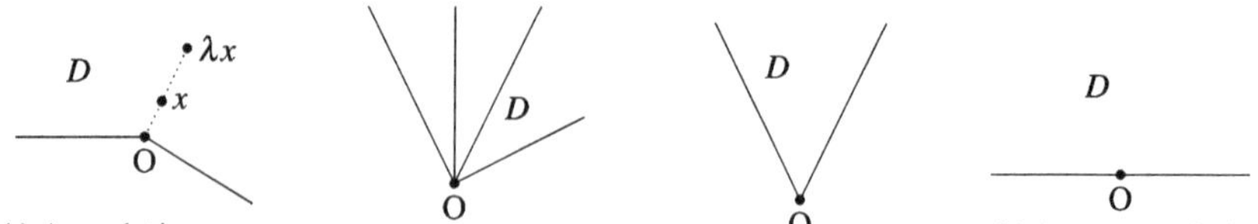

(a): A cone that is nonconvex and nonpointed (b): A pointed cone that is nonconvex (c): A pointed convex cone (b): A convex cone that is nonpointed

Fig. 3.2 Cones

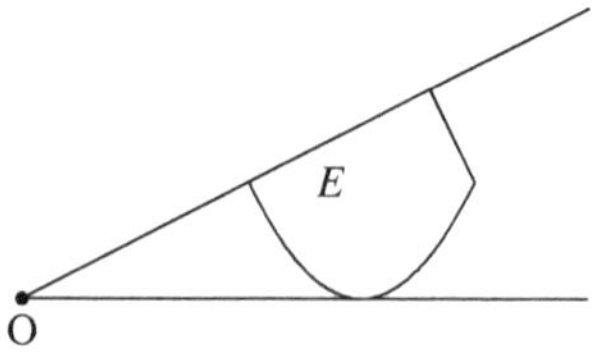

Fig. 3.3 Cone generated by E

(e) Assume that E is convex. For any $t \in (0, 1)$, by (a) and (c) we have $\mathrm{core}E \subset t\mathrm{core}E + (1-t)\mathrm{core}E = \mathrm{core}(tE) + \mathrm{core}((1-t)E) \subset \mathrm{core}(\mathrm{core}(tE) + (1-t)E) \subset \mathrm{core}(\mathrm{core}(tE + (1-t)E)) \subset \mathrm{core}(\mathrm{core}E) \subset \mathrm{core}E$. Hence, $\mathrm{core}E = t\mathrm{core}E + (1-t)\mathrm{core}E = \mathrm{core}(\mathrm{core}E)$ for all $t \in (0, 1)$, and so $\mathrm{core}E$ is convex.

□

Definition 3.5 A nonempty subset D of Y is called a CONE if $\lambda u \in D$ for all $u \in D$ and $\lambda \geq 0$ (see Fig. 3.2).

By definition, Y and $\{0\}$ are cones and are called trivial cones. A cone D is PROPER if $\{0\} \neq D \neq Y$, is POINTED if $D \cap (-D) = \{0\}$, and is a CONVEX CONE if D is a convex set (see Fig. 3.2).

By $\mathscr{K}$ we denote the set of proper cones in Y.

Definition 3.6 Let $E \subset Y$ be nonempty. The set

$$\mathrm{cone}E := \bigcup_{\lambda \geq 0} \lambda E.$$

is called the CONE GENERATED BY E (see Fig. 3.3). It is clear that $\mathrm{cone}E$ is a cone, and E is a cone if and only if $E = \mathrm{cone}E$.

Proposition 3.7 *Let $D \subset Y$ be a proper cone. Then:*

(a) $0 \notin \mathrm{core}D$ *and* $\mathrm{core}D \cup \{0\}$ *is a cone.*
(b) *If* $\mathrm{core}D \neq \emptyset$, *then* $Y = D - \mathrm{core}D = \mathrm{core}D - D = D - D$.
(c) *D is a convex cone if and only if $D + D = D$.*
(d) *if D is a convex cone, then* $\mathrm{core}D \cup \{0\}$ *is a convex cone; if addition,* $\mathrm{core}D \neq \emptyset$, *then* $D + \mathrm{core}D = \mathrm{core}D$.

Proof

(a) If $0 \in \text{core}D$, then, for all $v \in Y$, there exists $\lambda > 0$ such that $\lambda v \in 0+[0, \lambda]v \subset D$, i.e., $v \in \frac{1}{\lambda}D \subset D$. Hence, $Y = D$, which is impossible because D is a proper cone. We show that $\text{core}D \cup \{0\}$ is a cone. Take arbitrary $e \in \text{core}D \cup \{0\}$ and $\beta \geq 0$. If $e = 0$ or $\beta = 0$, then, obviously $\beta e \in \text{core}D \cup \{0\}$. Considering $e \in \text{core}D$ and $\beta > 0$, we have for all $v \in Y$, there exists $\lambda > 0$ such that $e + [0, \lambda](\frac{1}{\beta}v) \subset D$. Hence, $\beta e + [0, \lambda]v = \beta(e + [0, \lambda](\frac{1}{\beta}v)) \subset \beta D \subset D$. Thus, $\beta e \in \text{core}D$.

(b) Assume that $\text{core}D \neq \emptyset$. For any $u \in \text{core}D$ and $v \in Y$, there exists $\lambda > 0$ such that $u + \lambda v \in D$. This and Proposition 3.4 (c) imply that $v \in \frac{1}{\lambda}D - \frac{1}{\lambda}u \subset \frac{1}{\lambda}D - \frac{1}{\lambda}\text{core}D = \frac{1}{\lambda}D - \text{core}(\frac{1}{\lambda}D) \subset D - \text{core}D \subset D - D$ and so $Y = D - \text{core}D = D - D$. Moreover, $\text{core}D - D = -Y = Y$.

(c) Assume that D is convex. For any $x, y \in D$, $\frac{1}{\lambda}x \in D$ and $\frac{1}{1-\lambda}x \in D$ for $\lambda \in (0, 1)$. Then,

$$x + y = \lambda\left(\frac{1}{\lambda}x\right) + (1 - \lambda)\left(\frac{1}{1 - \lambda}y\right) \in D.$$

Hence, $D + D \subset D$. Conversely, suppose that $D + D \subset D$. For any $x, y \in D$ and $\lambda \in [0, 1]$, $\lambda x \in D$ and $(1 - \lambda)x \in D$. Hence,

$$\lambda x + (1 - \lambda)y \in D + D \subset D.$$

Thus, D is convex.

(d) The convexity of $\text{core}D \cup \{0\}$ is a consequence of Proposition 3.4 (b) and (d). The equality $D + \text{core}D = \text{core}D$ follows from the above assertion (c) and Proposition 3.4 (a).

□

Proposition 3.8 *Let $D \subset Y$ be a vectorially closed and solid convex cone and $u, v \in Y$ with $u \in -\text{core}D$ and $v \notin D$. Then, there exists $w \in D^c$ such that $w - u \in \text{core}D$ and $w - v \in D$.*

Proof Since $\text{core}D \neq \emptyset$, by Proposition 3.7 (b), there exists $d \in \text{core}D$ such that $d-v \in D$. For $t \in [0, 1]$, set $v_t = (1-t)v+td$. When D is vectorially closed convex and $d \in \text{core}D \subset D$, there exists $t_0 \in (0, 1)$ such that $v_t \notin D$ for all $t \in [0, t_0)$ and $v_t \in D$ for all $t \in [t_0, 1]$. (If this is false, then there are two possibilities: either "there exists $t_0 \in (0, 1)$ such that $v_t \notin D$ for all $t \in [0, t_0]$ and $v_t \in D$ for all $t \in (t_0, 1]$" or " for all $t \in (0, 1)$, there exists $t' \in (0, t)$ such that $v_{t'} \in D$." In the first case, $v_{t_0} \notin D = \text{vcl}D$. Hence, for $v' = d - v$, there exists $\lambda > 0$ such that $v_{t_0} + \lambda'(d - v) \notin D$ for all $\lambda' \in (0, \lambda]$. Choosing λ' sufficiently close to 0 such that $t_0+\lambda' \in (t_0, 1]$ we have $v_{t_0}+\lambda'(d-v) = v_{(t_0+\lambda')} \in D$, a contradiction. In the second case, for $v' = d - v$ and for all $\lambda > 0$ we can find $\lambda' \in (0, 1) \cap (0, \lambda]$ such that $v_{\lambda'} = v + \lambda' v' \in D$, implying that $v \in \text{vcl}D = D$, a contradiction.) In particular, $v_{t_0} \in D$, and therefore $v_{t_0} - u \in D + \text{core}D \subset \text{core}D = \text{core}(\text{core}D)$ (see Proposition 3.4

(e)). This implies that there exists $\lambda > 0$ such that $(v_{t_0} - u) + [0, \lambda](v - d) \subset \text{core}D$. We choose $\lambda_0 \in (0, \lambda]$ sufficiently close to 0 such that $0 < t_1 := t_0 - \lambda_0 < t_0$ and set $w = v_{t_1}$. Then, $w \notin D$, $w - u \in \text{core}D$ and $w - v = t_1(d - v) \in D$. □

Definition 3.9 Let $q \in Y \setminus \{0\}$ and $E \subset Y$. The NONCONVEX SEPARATION FUNCTIONAL $r_E^q : Y \to \mathbb{R} \cup \{\pm\infty\}$ is defined by

$$r_E^q(v) := \begin{cases} \inf\{t \in \mathbb{R} \mid v \in tq - E\} & \text{if } v \in \mathbb{R}q - E, \\ +\infty & \text{otherwise.} \end{cases} \tag{3.1}$$

This functional is usually called the GERSTEWITZ/TAMMER SCALARIZATION FUNCTION or the SMALLEST STRICT MONOTONIC FUNCTION or the SHORTAGE FUNCTION.

Below are some general properties of functional $r_E^q(\cdot)$. For $\gamma : Y \to \mathbb{R}$, we say that γ is E-NONDECREASING (E-INCREASING, resp.) if $v, v' \in Y$, $v \neq v'$, and $v - v' \in -E$, imply $\gamma(v) \leq \gamma(v')$ ($\gamma(v) < \gamma(v')$, resp.).

Proposition 3.10 ([226]) *Let $q \in Y \setminus \{0\}$ and $E \subset Y$.*

(a) $\text{dom}r_E^q = \mathbb{R}q - E$ *and* $\{v \in Y \mid r_E^q(v) = -\infty\} = \text{ovcl}_q^{+\infty}(-E)$, *where*

$$\text{ovcl}_q^{+\infty}(-E) := \{v \in Y \mid \forall\lambda > 0, \exists\lambda' \in [\lambda, +\infty) \text{ such that } v + \lambda' q \in -E\}.$$

(b) r_E^q *is finite if and only if* $Y = \mathbb{R}q - E$ *and* $\text{ovcl}_q^{+\infty}(-E) = \emptyset$.
(c) $r_E^q(v + rq) = r_E^q(v) + r$ *for all* $v \in Y$ *and* $r \in \mathbb{R}$.
(d) $\{v \in Y \mid r_E^q(v) < 0\} = -\mathbb{R}_{++}q - E = -\mathbb{R}_{++}q - \text{vcl}_q E$;
(e) $\{v \in Y \mid r_E^q(v) \leq 0\} = -\mathbb{R}_{+}q - \text{vcl}_q E$.
(f) r_E^q *is* E'*-nondecreasing if and only if* $E + E' \subset \mathbb{R}_{+}q + \text{vcl}_q E$.
(g) *If* r_E^q *is finite, then* r_E^q *is* E'*-increasing if and only if* $\text{vcl}_q E + E' \setminus \{0\} \subset \mathbb{R}_{++}q + \text{vcl}_q E$.

Proof

(a) It is obvious that $\text{dom}r_E^q = \mathbb{R}q - E$. For all $v \in Y$, $r_E^q(v) = -\infty$ if and only if $\forall t < 0, \exists t' \in (-\infty, t)$ such that $v \in t'q - E$ or, equivalently, $\forall\lambda > 0, \exists\lambda' \in (\lambda, +\infty)$ such that $v + \lambda' q \in -E$, i.e., $v \in \text{ovcl}_q^{+\infty}(-E)$.
(b) This is a direct consequence of part (a).
(c) For all $v \in Y$ and $r \in \mathbb{R}$, it is clear that $v \in \mathbb{R}q - E$ if and only if $v + rq \in \mathbb{R}q - E$, and so $r_E^q(v) + r = +\infty$ if and only if $r_E^q(v + rq) = +\infty$. Otherwise,

$$\begin{aligned} r_E^q(v + rq) &= \inf\{t \in \mathbb{R} \mid v + rq \in tq - E\} \\ &= \inf\{t \in \mathbb{R} \mid v \in (t - r)q - E\} \\ &= \inf\{t' + r \in \mathbb{R} \mid v \in t'q - E\} \end{aligned}$$

$$= \inf\{t' \in \mathbb{R} \mid v \in t'q - E\} + r$$

$$= r_E^q(v) + r.$$

(d) For all $v \in Y$, it is clear that $r_E^q(v) < 0 \Leftrightarrow \exists t < 0, v \in tq - E \Leftrightarrow v \in -\mathbb{R}_{++}q - E$. Therefore, $\{v \in V \mid r_E^q(v) < 0\} = -\mathbb{R}_{++}q - E$. The equality $-\mathbb{R}_{++}q - E = -\mathbb{R}_{++}q - \mathrm{vcl}_q E$ is clear.

(e) For all $v \in Y$ we have

$$v \in -\mathrm{vcl}_q E \Rightarrow \forall \lambda > 0, \exists \lambda' \in \left[0, \frac{\lambda}{2}\right], -v + \lambda' q \in E$$

$$\Rightarrow \forall \lambda > 0, \exists \lambda' \in \left[0, \frac{\lambda}{2}\right], v \in \lambda' q - E$$

$$\Rightarrow \forall \lambda > 0, r_E^q(v) \leq \frac{\lambda}{2} < \lambda$$

$$\Rightarrow r_E^q(v) \leq 0.$$

This and part (d) imply that

$$-\mathbb{R}_{+}q - \mathrm{vcl}_q E = -\mathrm{vcl}_q E \cup (-\mathbb{R}_{++}q - \mathrm{vcl}_q E)$$

$$\subset \{v \in Y \mid r_E^q(v) \leq 0\} \cup \{v \in Y \mid r_E^q(v) < 0\}$$

$$= \{v \in Y \mid r_E^q(v) \leq 0\}.$$

On the other hand, for all $v \in Y$, by definition of r_E^q, we have

$$r_E^q(v) = 0 \Rightarrow \forall \lambda > 0, \exists \lambda' \in [0, \lambda], v \in \lambda' q - E$$

$$\Rightarrow \forall \lambda > 0, \exists \lambda' \in [0, \lambda], -v + \lambda' q \in E$$

$$\Rightarrow -v \in \mathrm{vcl}_q E \Leftrightarrow v \in -\mathrm{vcl}_q E.$$

This and part (d) give that

$$\{v \in Y \mid r_E^q(v) \leq 0\} = \{v \in Y \mid r_E^q(v) = 0\} \cup \{v \in Y \mid r_E^q(v) < 0\}$$

$$\subset -\mathrm{vcl}_q E \cup (-\mathbb{R}_{++}q - \mathrm{vcl}_q E)$$

$$= -\mathbb{R}_{+}q - \mathrm{vcl}_q E.$$

Thus, $\{v \in Y \mid r_E^q(v) \leq 0\} = -\mathbb{R}_{+}q - \mathrm{vcl}_q E$.

(f) ($\Leftarrow$) Let $v, v' \in Y$ be such that $v - v' \in -E'$. Assume that $r_E^q(v) > r_E^q(v')$. Then, by the definition of r_E^q, there exists $t \in \mathbb{R}$ such that $r_E^q(v') \le t < r_E^q(v)$ and $v' \in tq - E$. It follows that $v \in v' - E' \subset tq - E - E' \subset tq - \text{vcl}_q E$. Hence, $v - tq \in -\text{vcl}_q E \subset -\mathbb{R}_+ q - \text{vcl}_q E$. This and parts (c) and (e) imply that $r_E^q(v) - t = r_E^q(v - tq) \le 0$, i.e., $r_E^q(v) \le t < r_E^q(v)$, a contradiction.
($\Rightarrow$) Taking $e \in E$ and $e' \in E'$, we have $-(e + e') - (-e) = -e' \in -E'$ and $-e \in -E \subset -\mathbb{R}_+ q - \text{vcl}_q E$. Since r_E^q is E'-nondecreasing, by part (e), we have $r_E^q(-(e+e')) \le r_E^q(-e) \le 0$. Again part (e), $-(e+e') \in -\mathbb{R}_+ q - \text{vcl}_q E$, i.e., $e + e' \in \mathbb{R}_+ q + \text{vcl}_q E$. Thus, $E + E' \subset \mathbb{R}_+ q + \text{vcl}_q E$.

(g) We show first that $v \in r_E^q(v)q - \text{vcl}_q E$ for all $v \in Y$. Suppose that $v - r_E^q(v)q \notin -\text{vcl}_q E$. Then, either $v - r_E^q(v)q \notin -\mathbb{R}_+ q - \text{vcl}_q E$ or $v - r_E^q(v)q \in -\mathbb{R}_{++} q - \text{vcl}_q E$. If $v - r_E^q(v)q \notin -\mathbb{R}_+ q - \text{vcl}_q E$, then, by part (e), $0 = r_E^q(v) - r_E^q(v) = r_E^q(v - r_E^q(v)q) > 0$, a contradiction. If $v - r_E^q(v)q \in -\mathbb{R}_{++} q - \text{vcl}_q E = \mathbb{R}_{++} q - E$, then there exists $t < 0$ such that $v - r_E^q(v)q \in -tq - \text{vcl}_q E$, which implies that $0 = r_E^q(v) - r_E^q(v) = r_E^q(v - r_E^q(v)q) \le t < 0$, another contradiction.
($\Leftarrow$) Let $v, v' \in Y$, $v \ne v'$, be such that $v - v' \in -E'$. Then, $v \in v' - E' \setminus \{0\} \subset r_E^q(v')q - \text{vcl}_q E - E' \setminus \{0\} \subset r_E^q(v')q - \mathbb{R}_{++} q - \text{vcl}_q E$. It follows that $v - r_E^q(v')q \in -\mathbb{R}_{++} q - \text{vcl}_q E$. Hence, by part (d), $r_E^q(v) - r_E^q(v') = r_E^q(v - r_E^q(v')q) < 0$.
($\Rightarrow$) The proof is similar to that of ($\Rightarrow$) in part (f).

□

From now on, we will use the following notations:

$$\mathscr{K}_c := \{D \in \mathscr{K} \mid D \text{ solid and convex}\},$$

and for nonempty $C \subset Y$,

$$\mathscr{K}_C^{core} := \{D \in \mathscr{K}_c \mid \text{core} D \supset C_0\}.$$

Let us define

$$\mathscr{E}(Y) := \{E \subset Y \mid \ E \text{ solid},\ E \ne Y, \text{ and } E + E \subset E\},$$

and for $E \subset Y$,

$$\text{Q}_E := \{q \in Y \mid \mathbb{R}_{++} q + E = \text{core} E\}.$$

Clearly, $\mathscr{K}_c \subset \mathscr{E}(Y)$ and $\text{Q}_E = \text{core} E$ when $E \in \mathscr{K}_c$ (see Exercise 3.6).

For $D \subset Y$, we set

$$\mathscr{F}_D := \{E \subset Y \mid E + D \subset E\}.$$

Each $E \in \mathscr{F}_D$ is called a D-FREE DISPOSAL SET. If in addition, $0 \notin E$, then E is said to be a D-IMPROVEMENT SET.

3.1.2 Continuity of Vector Functions

Definition 3.11 Let A be a nonempty subset of a topological space, $E \subset Y$ nonempty, $q \in Y \backslash \{0\}$, and $h : A \to Y$.

(a) h is said to be (q, E)-UPPER QUASICONTINUOUS ((q, E)-LOWER QUASICONTINUOUS, resp.) if for all $r \in \mathbb{R}$, the set $\{x \in A \mid h(x) - rq \in E\}$ ($\{x \in A \mid h(x) - rq \in -E\}$, resp.) is closed.

(b) h is said to be E-UPPER QUASICONTINUOUS (E-LOWER QUASICONTINUOUS) if for all $v \in Y$, the set $\{x \in A \mid h(x) - v \in E\}$ ($\{x \in A \mid h(x) - v \in -E\}$, resp.) is closed.

Clearly, E-upper/lower quasicontinuity implies (q, E)-upper/lower semicontinuity for any $q \in Y \backslash \{0\}$. The reverse implication is not true. Moreover, note that the E-upper/lower quasicontinuity of h implies the E-upper/lower quasicontinuity of $h'(\cdot) := h(\cdot) - h(y)$ for $y \in A$, but the (q, E)-upper/lower quasicontinuity of h does not imply the (q, E)-upper/lower quasicontinuity of $h'(\cdot)$. These facts are verified in the following example.

Example 3.12 Let $A = \mathbb{R}^2$, $Y = \mathbb{R}^2$, and $E = \mathbb{R}^2_+$. Let $h : \mathbb{R}^2 \to \mathbb{R}^2$ be defined for all $x = (a, b) \in \mathbb{R}^2$ by

$$h(x) = \begin{cases} (a, b) \text{ if } b \geq 0, \\ (b, a) \text{ if } b < 0. \end{cases}$$

For $q = (1, 1)$ and for all $r \in \mathbb{R}$, we have that

$$\operatorname{lev}^{rq}_{-\mathbb{R}^2_+} h = \{x \in \mathbb{R}^2 \mid h(x) - rq \in -\mathbb{R}^2_+\} = \{(a, b) \in \mathbb{R}^2 \mid a \leq r,\ b \leq r\}$$

is closed. Hence, h is $(q, \mathbb{R}^2_+)$-lower quasicontinuous. For $y = (1, -1) \in \mathbb{R}^2$ and $r = 1$, we have that

$$h'(x) := h(x) - h((1, -1)) = \begin{cases} (a + 1, b - 1) \text{ if } b \geq 0, \\ (b + 1, a - 1) \text{ if } b < 0 \end{cases},$$

and

$$\begin{aligned}\operatorname{lev}^{rq}_{-\mathbb{R}^2_+} h' &= \{x \in \mathbb{R}^2 \mid h'(x) - rq \in -\mathbb{R}^2_+\}\\ &= \{(a,b) \in \mathbb{R}^2 \mid a \le 0,\ 0 \le b \le 2\} \cup \{(a,b) \in \mathbb{R}^2 \mid a \le 2, b < 0\}\end{aligned}$$

is not closed. Thus, h' is not $(q, \mathbb{R}^2_+)$-lower quasicontinuous. For $v = (1, -1) \in \mathbb{R}^2$, we have that

$$\operatorname{lev}^{v}_{-\mathbb{R}^2_+} h = \{x \in \mathbb{R}^2 \mid h(x) - v \in -\mathbb{R}^2_+\} = \{(a,b) \in \mathbb{R}^2 \mid a \le -1, b < 0\}$$

is not closed. Thus, h is not $\mathbb{R}^2_+$-lower quasicontinuous.

Definition 3.13 Let A be a subset of a topological space, Y a topological vector space, $E \subset Y$ nonempty, and $h : A \to Y$. h is called E-UPPER (E-LOWER) SEMICONTINUOUS, RESP. at $x \in A$ if for each open neighborhood V of $h(x)$ in Y, there exists a neighborhood U of x in A such that $h(y) \in V - E$) ($h(y) \in V + E$), resp.) for all $y \in U$. h is called E-upper (E-lower, resp.) semicontinuous if it is E-upper (E-lower, resp.) semicontinuous at every $x \in A$.

Some characterizations of E-upper/lower semicontinuity are provided in Exercises 3.7–3.8.

Proposition 3.14 *Let A be a subset of a topological space, Y a topological vector space, $E \in \mathscr{E}(Y)$ such that E is convex, and $h : A \to Y$. Then:*

(a) *If h is E-upper semicontinuous, then h is $(\operatorname{core} E)^c$-lower quasicontinuous.*
(b) *If h is E-lower semicontinuous, then h is $(\operatorname{core} E)^c$-upper quasicontinuous.*
(c) *If E is closed and h is E-upper semicontinuous, then h is E-upper quasicontinuous.*
(d) *If E is closed and h is E-lower semicontinuous, then h is E-lower quasicontinuous.*

Proof Since E is convex, $\operatorname{core} E = \operatorname{int} E$. Let $v \in Y$ be arbitrary.

(a) Assume that h is E-upper semicontinuous. We show that $\operatorname{lev}^{v}_{-(\operatorname{core} E)^c} h = \{x \in A \mid h(x) - v \in -(\operatorname{core} E)^c\} = \{x \in A \mid h(x) \notin v - \operatorname{core} E\}$ is closed. Let $x_\alpha \in \operatorname{lev}^{v}_{-(\operatorname{core} E)^c} h$ satisfy $x_\alpha \to x_0$. If $x_0 \notin \operatorname{lev}^{v}_{-(\operatorname{core} E)^c} h$, then $h(x_0) \in v - \operatorname{core} E = v - \operatorname{int} E$. By the openness of $v - \operatorname{core} E$ and the E-upper semicontinuity of h, an open neighborhood U of x_0 exists such that $h(y) \in v - \operatorname{core} E - E \subset v - \operatorname{core} E$ for all $y \in U$. As $x_\alpha \to x_0$, there exists α such that $x_\alpha \in U$. Then, $h(x_\alpha) \in v - \operatorname{core} E$, i.e., $x_\alpha \notin \operatorname{lev}^{v}_{-(\operatorname{core} E)^c} h$, which is impossible. Thus, $x_0 \in \operatorname{lev}^{v}_{-(\operatorname{core} E)^c} h$, and hence $\operatorname{lev}^{v}_{-(\operatorname{core} E)^c} h$ is closed.
(b) Argue as in the proof of part (a).
(c) Assume that h is E-upper semicontinuous. We show that $\operatorname{lev}^{v}_{E} h = \{x \in A \mid h(x) - v \in E\} = \{x \in A \mid h(x) \in v + E\}$ is closed. Let $x_\alpha \in \operatorname{lev}^{v}_{E} h$ be such that $x_\alpha \to x_0$. Suppose that $x_0 \notin \operatorname{lev}^{v}_{E} h$, i.e., $h(x_0) \in Y \setminus (v + E)$. Since $v + E$

is closed, $V \setminus (v + E)$ is open. As h is E-upper semicontinuous, there exists an open neighborhood U of x_0 such that $h(y) \in V \setminus (v + E) - E$ for all $y \in U$. As $x_\alpha \to x_0$, there exists α such that $x_\alpha \in U$. Then, $h(x_\alpha) \in V \setminus (v + E) - E$, which means that there is $e \in E$ such that $h(x_\alpha) + e \notin v + E$. This implies that $h(x_\alpha) \notin v + E$ (because if $h(x_\alpha) \in v + E$, then $h(x_\alpha) + e \in v + E + e \subset v + E$), i.e., $x_\alpha \notin \operatorname{lev}_E^v h$, a contradiction. Thus, $x_0 \in \operatorname{lev}_E^v h$, and hence $\operatorname{lev}_E^v h$ is closed.

(d) Argue as in the proof of part (c).

□

Definition 3.15 Let A be a subset of a topological space, Y a topological vector space, $E \subset Y$ nonempty, and $f : A \times A \to Y$. f is said to be UPPER E-HEMICONTINUOUS in the first component at x if for all $y, z \in A$, the function $t \to f(x + t(z - x), y)$ from $[0, 1]$ to Y is upper E-semicontinuous at 0^+. If $-f$ is upper E-hemicontinuous in the first component at x, then f is called LOWER E-HEMICONTINUOUS in the first component at x. f is called E-HEMICONTINUOUS in the first component at x if f is both upper E-hemicontinuous and lower E-hemicontinuous in the first component at x.

A similar definition can be stated for the second component.

3.2 Vector Equilibrium Problems and Applications

3.2.1 Vector Equilibrium Problems and Concepts of Solutions

In this chapter, we extend the scalar equilibrium problem (EP) to vector bifunctions in the following way:

$$(\text{VEP}_{(f,C)}) \qquad \text{find } \bar{x} \in A \text{ such that } f(\bar{x}, y) \notin -C_0 \text{ for all } y \in A,$$

where A is a nonempty set, $C \subset Y$ is a nonempty set such that $0 \notin \operatorname{core} C$, $C_0 := C \setminus \{0\}$, and $f : A \times A \to Y$ is a given vector bifunction. We call $(\text{VEP}_{(f,C)})$ a GENERAL VECTOR EQUILIBRIUM PROBLEM. When $Y = \mathbb{R}$ and $C = \mathbb{R}_+$, the problem $(\text{VEP}_{(f,C)})$ becomes the problem (EP). In the literature, $(\text{VEP}_{(f,C)})$ is often studied in the case where C is a convex cone of a real topological vector space. However, several authors have also considered $(\text{VEP}_{(f,C)})$ with C being an improvement set or a general set which are nonconic preferences. In general, $(\text{VEP}_{(f,C)})$ can describe many situations in the real world, especially in economics where preference relations are not necessarily preorders.

Definition 3.16 For problem $(\text{VEP}_{(f,C)})$, a point $\bar{x} \in A$ is said to be:

(a) A HENIG-PROPER C-SOLUTION if there exists $D \in \mathscr{K}_C^{core}$ such that

$$(f(\bar{x}, A) + C) \cap (-\operatorname{core} D) = \emptyset \tag{3.2}$$

(b) A BENSON-PROPER C-SOLUTION if

$$\text{vclcone}(f(\bar{x}, A) + C) \cap (-C_0) = \emptyset \tag{3.3}$$

(c) A C-SOLUTION if

$$f(\bar{x}, A) \cap (-C_0) = \emptyset \tag{3.4}$$

(d) A WEAK C-SOLUTION if, assuming that core$C \neq \emptyset$,

$$f(\bar{x}, A) \cap (-\text{core}C) = \emptyset \tag{3.5}$$

The case core$C = \emptyset$ is trivial since then all $x \in A$ satisfy (3.5). Hence, when considering weak C-solutions, we can regard core$C$ to be nonempty. By $\text{Sol}^{He}(\text{VEP}_{(f,C)})$ ($\text{Sol}^{Be}(\text{VEP}_{(f,C)})$, $\text{Sol}^{e}(\text{VEP}_{(f,C)})$, $\text{Sol}^{w}(\text{VEP}_{(f,C)})$, resp.), we denote the set of all Henig-proper C-solutions (Benson-proper C-solutions, C-solutions, weak C-solutions, resp.) of ($\text{VEP}_{(f,C)}$). Note that if there exists $D \in \mathcal{K}_C^{core}$ satisfying (3.2), then for $D' := \text{core}D \cup \{0\}$, we have $D' \in \mathcal{K}_C^{core}$ and

$$(f(\bar{x}, A) + C) \cap (-D' \setminus \{0\}) = \emptyset. \tag{3.6}$$

Conversely, if (3.6) holds, then (3.2) is obviously satisfied with $D = D'$. Thus, (3.2) can also be written as (3.6).

Since a major part of the related contributions in the literature studied topological settings, we note that when Y is a topological vector space, by replacing core$(\cdot)$ and vcl$(\cdot)$ in the above formulas, resp., by int$(\cdot)$ and cl$(\cdot)$, we have the concepts of solutions in topological settings. We use the subindex "top" for this case (e.g., $\text{Sol}^{He}_{top}(\text{VEP}_{(f,C)})$. Because $\text{int}(\cdot) \subset \text{core}(\cdot)$, $\text{vcl}(\cdot) \subset \text{cl}(\cdot)$, and $\text{int}(\cdot) = \text{core}(\cdot)$ if $(\cdot)$ is convex with $\text{int}(\cdot) \neq \emptyset$, $\text{Sol}^{He}_{top}(\text{VEP}_{(f,C)}) = \text{Sol}^{He}(\text{VEP}_{(f,C)})$, $\text{Sol}^{Be}_{top}(\text{VEP}_{(f,C)}) \subset \text{Sol}^{Be}(\text{VEP}_{(f,C)})$, and $\text{Sol}^{w}(\text{VEP}_{(f,C)}) \subset \text{Sol}^{w}_{top}(\text{VEP}_{(f,C)})$. When C is convex and int$C \neq \emptyset$, $\text{Sol}^{w}(\text{VEP}_{(f,C)}) = \text{Sol}^{w}_{top}(\text{VEP}_{(f,C)})$. Note that the algebraic interior (vectorial closure, resp.) and the topological interior (topological closure, resp.) have different properties. However, the results in this chapter, which are stated in non-topological settings, are still valid in the corresponding topological settings when we replace core$(\cdot)$ and vcl$(\cdot)$ by int$(\cdot)$ and cl$(\cdot)$, resp.

In Definition 3.16 we consider Henig and Benson proper solutions as representatives for other properness notions, which have practical applications to avoid anomalous from too big sets for weak C-solutions or C-solutions.

Example 3.17 Let $A = \mathbb{R}$, $Y = \mathbb{R}^2$, $C = \mathbb{R}^2_+$, and $f : A \times A \to Y$ be defined by

$$f(x, y) := \begin{cases} (|y|, y^2) & \text{if } |x| \leq 1, \\ (y, e^{|x|y}) & \text{if } |x| > 1. \end{cases} \quad \text{for all } x, y \in A.$$

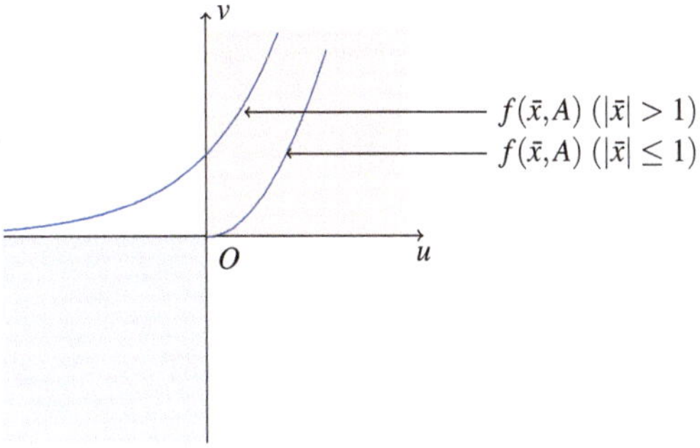

Fig. 3.4 Illustration of Example 3.18

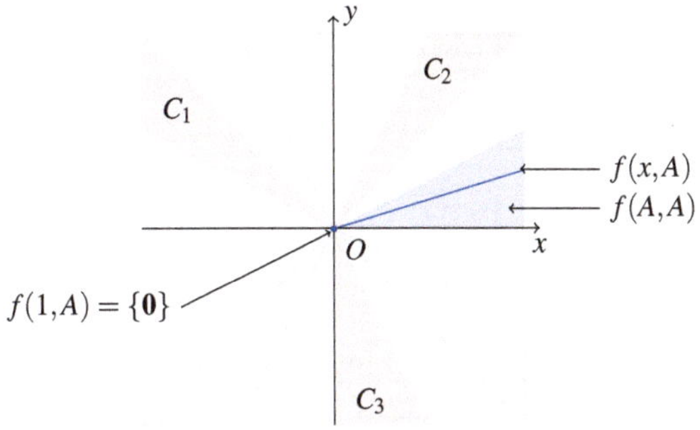

Fig. 3.5 Illustration of Example 3.18

Let $D := \{(c_1, c_2) \in \mathbb{R}^2 \mid c_1 + c_2 \leq 0\}$. Clearly, $D \in \mathscr{K}_C^{core}$. For $\bar{x} \in [-1, 1]$, we have $f(\bar{x}, A) + C \subset D$. Clearly, $D \cap (-\text{core}D) = \emptyset$. Hence, $(f(\bar{x}, A) + C) \cap (-\text{core}D) = \emptyset$. Thus, $\bar{x} \in \text{Sol}^{He}(\text{VEP}_{(f,C)})$ for all $\bar{x} \in [-1, 1]$. For all $\bar{x} \in A$, it is not hard to see that $f(\bar{x}, A) \cap (-C_0) = \emptyset$. Hence, $\text{Sol}^e(\text{VEP}_{(f,C)}) = \text{Sol}^w(\text{VEP}_{(f,C)}) = A$ (see Fig. 3.4).

Example 3.18 Let $A = \mathbb{R}$, $Y = \mathbb{R}^2$ and $C = C_1 \cup C_2 \cup C_3$, where $C_1 = \{(c_1, c_2) \in \mathbb{R}^2 \mid 0 \leq -\frac{1}{2}c_1 < c_2 < -c_1\}$, $C_2 = \{(c_1, c_2) \in \mathbb{R}^2 \mid 0 < 2c_1 < c_2\}$, $C_3 = \{(c_1, c_2) \in \mathbb{R}^2 \mid c_2 < -2c_1 < 0\}$ (see Fig. 3.5).

It is easy to see that for any proper convex cone $D \subset Y$, $C_0 \not\subset \text{core}D$. Thus, $\mathscr{K}_C^{core} = \emptyset$. So, for any $f : A \times A \to \mathbb{R}^2$, $\text{Sol}^{He}(\text{VEP}_{(f,C)})$ is empty.

Let $f : A \times A \to Y$ be defined by

$$f(x, y) := (|(x-1)y|(x^2+2), |(x-1)y|) \text{ for all } x, y \in A.$$

For $\bar{x} = 1$, $f(1, A) = \{0\}$. Hence, vcl cone$(f(1, A) + C) \cap (-C_0) = \text{vcl}C \cap (-C_0) = \emptyset$. Thus, $\bar{x} = 1 \in \text{Sol}^{Be}(\text{VEP}_{(f,C)})$. For $x \neq 1$, by choosing $y = \frac{1}{x-1}$, $c = (a, -\frac{2}{3}a + \frac{2}{3}x^2 - \frac{7}{3})$ with $0 < a < \frac{1}{2}x^2 + \frac{7}{4}$, we have $c \in C_3 \subset C$ and $f(x, y) + c = (x^2+a+2, -\frac{2}{3}(x^2+a+2)) \in -C_1 \subset -C$. Therefore, $x \notin \text{Sol}^{Be}(\text{VEP}_{(f,C)})$ for all $x \neq 1$. It is not hard to see that $f(x, y) \in \{(u, v) \mid 0 \leq u \leq v \leq \frac{1}{2}u\} \subset Y \backslash (-C_0)$ for all $x, y \in A$ (see Fig. 3.5). Thus, $\text{Sol}^e(\text{VEP}_{(f,C)}) = \text{Sol}^w(\text{VEP}_{(f,C)}) = A$.

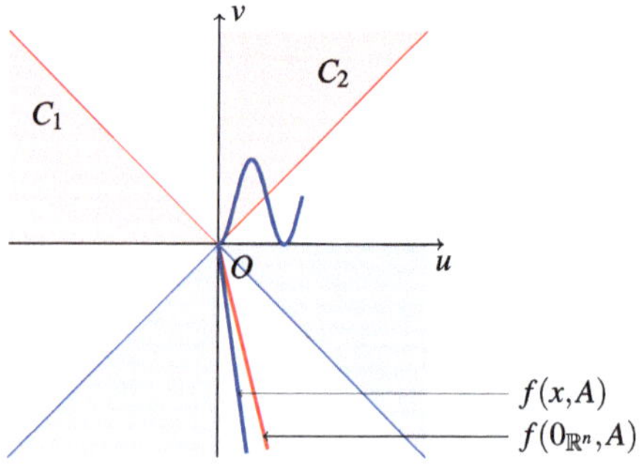

Fig. 3.6 Illustration of Example 3.19

Example 3.19 Let $A = \mathbb{R}^n$, $Y = \mathbb{R}^2$ and $C = C_1 \cup C_2$, where $C_1 = \{(c_1, c_2) \in \mathbb{R}^2 \mid 0 \le c_2 \le -c_1\}$, $C_2 = \{(c_1, c_2) \in \mathbb{R}^2 \mid 0 \le c_1 \le c_2\}$ (see Fig. 3.6). Let $f : A \times A \to Y$ be defined by

$$f(x, y) := \begin{cases} (\frac{\|y\|}{\|x\|}, 1 - (\cos(\frac{4\|y\|}{\|x\|}))^2) & \text{if } \|y\| < \|x\|, \\ (\|x + y\|, -(\|x\| + 7)\|x + y\|) & \text{if } \|y\| \ge \|x\|. \end{cases} \quad \text{for all } x, y \in A.$$

For all $x \in A$, by choosing y_x with $\|y_x\| = \|x\| + 1$, we have $c = (\|x + y_x\|, (\|x\| + 7)\|x + y_x\|) \in C_2$ and $f(x, y_x) + c = (2\|x + y_x\|, 0) \in -C_1 \setminus \{0\} \subset C_0$. Hence, $(f(x, A) + C) \cap (-C_0) \ne \emptyset$ for all $x \in A$. Thus, both $\mathrm{Sol}^{He}(\mathrm{VEP}_{(f,C)})$ and $\mathrm{Sol}^{Be}(\mathrm{VEP}_{(f,C)})$ are empty. For $\bar{x} = 0_{\mathbb{R}^n}$, $f(0_{\mathbb{R}^n}, A) = \{(\|y\|, -7\|y\|) \mid y \in A\} \subset Y \setminus (-C_0)$. Thus, $\bar{x} = 0_{\mathbb{R}^n} \in \mathrm{Sol}^e(\mathrm{VEP}_{(f,C)})$. For $x \ne 0_{\mathbb{R}^n}$, by choosing $y_x = \frac{\pi}{4}x$, we have $f(x, y_x) = (\frac{\pi}{4}, 0) \in -C_1 \subset -C$. Thus, $x \notin \mathrm{Sol}^e(\mathrm{VEP}_{(f,C)})$ for all $x \ne 0_{\mathbb{R}^n}$. It is easy to see that $f(x, y) \in \{(u, v) \mid 0 \le v \le -7u \le 0\} \cup \mathbb{R}^2_+ \subset Y \setminus (-\mathrm{core}C_0)$ for all $x, y \in A$ (see Fig. 3.6). Thus, $\mathrm{Sol}^w(\mathrm{VEP}_{(f,C)}) = A$.

Example 3.20 Let $A = \mathbb{R}^n$, $Y = \mathbb{R}^2$, $C = \mathbb{R}^2_+$ and $f : A \times A \to \mathbb{R}^2$ be defined by

$$f(x, y) := (-||x|| - 1, \; ||y|| - ||x||) \text{ for } x, y \in A.$$

Take any $x \in A$, $y_x = x \in A$, and $c = 0 \in C$. Then, $f(x, y_x) = (-||x|| - 1, 0) \in -C_0$. Hence, $\mathrm{Sol}^e(\mathrm{VEP}_{(f,C)})$ *is empty*. Moreover, $f(x, y_x) + c = (-||x|| - 1, 0) \in -C_0$. Hence, $(f(x, A) + C) \cap (-C_0) \ne \emptyset$ for any $x \in A$ (see Fig. 3.7). Thus, both $\mathrm{Sol}^{He}(\mathrm{VEP}_{(f,C)})$ and $\mathrm{Sol}^{Be}(\mathrm{VEP}_{(f,C)})$ are empty.

For $\bar{x} = 0_{\mathbb{R}^n}$, $f(0_{\mathbb{R}^n}, y) = (-1, ||y||) \notin -\mathrm{core}C$ for all $y \in A$ (see Fig. 3.7). Thus, $0_{\mathbb{R}^n} \in \mathrm{Sol}^w(\mathrm{VEP}_{(f,C)})$. For all $x \ne 0_{\mathbb{R}^n}$, there exists $y = 0_{\mathbb{R}^n}$ satisfying $f(x, 0_{\mathbb{R}^n}) = (-||x|| - 1, -||x||) \in -\mathrm{core}C$. Hence, $x \notin \mathrm{Sol}^w(\mathrm{VEP}_{(f,C)})$ for all $x \ne 0_{\mathbb{R}^n}$.

The following proposition gives relationships between the kinds of solutions in Definition 3.16.

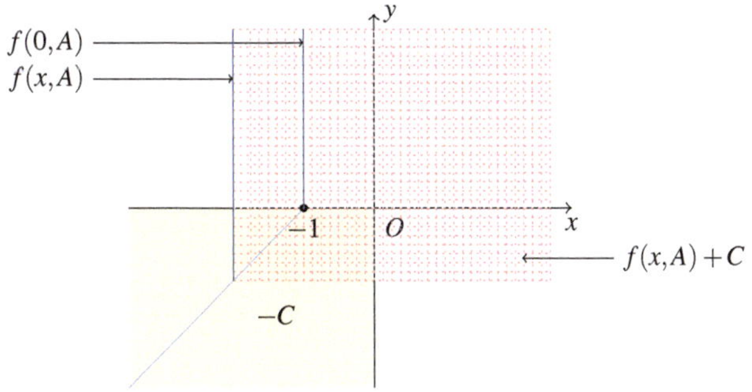

Fig. 3.7 Illustration of Example 3.20

Proposition 3.21 *For problem* $(\text{VEP}_{(f,C)})$, *we have the following relations:*

(a) $\text{Sol}^{He}(\text{VEP}_{(f,C)}) \subset \text{Sol}^{Be}(\text{VEP}_{(f,C)})$.
(b) $\text{Sol}^{e}(\text{VEP}_{(f,C)}) \subset \text{Sol}^{w}(\text{VEP}_{(f,C)})$.
(c) *If* $C_0 \subset C_0 + C$ *(in particular,* $0 \in C$*), then*

$$\text{Sol}^{He}(\text{VEP}_{(f,C)}) \subset \text{Sol}^{Be}(\text{VEP}_{(f,C)}) \subset \text{Sol}^{e}(\text{VEP}_{(f,C)}) \subset \text{Sol}^{w}(\text{VEP}_{(f,C)}).$$

Proof

(a) If $\bar{x} \in \text{Sol}^{He}(\text{VEP}_{(f,C)})$, then there exists $D \in \mathscr{K}_C^{core}$ such that $(f(\bar{x}, A) + C) \cap (-\text{core}D) = \emptyset$, which implies that $\text{cone}(f(\bar{x}, A) + C) \cap (-\text{core}D) = \emptyset$. Suppose that there exists $u \in \text{vcl}\,\text{cone}(f(\bar{x}, A) + C) \cap (-\text{core}D)$. Then, $u \in -\text{core}D = -\text{core}(\text{core}D)$ as D is convex. Hence, for all $v \in Y$, there exists $\lambda_v > 0$ such that $u + [0, \lambda_v]v \subset -\text{core}D$. Moreover, when $u \in \text{vcl}\,\text{cone}(f(\bar{x}, A) + C)$, there exists $v' \in Y$ such that $u + \lambda'_{v'}v' \in \text{cone}(f(\bar{x}, A) + C)$ for some $\lambda'_{v'} \in [0, \lambda_{v'}]$. Then, $u + \lambda'_{v'}v' \in \text{cone}(f(\bar{x}, A) + C) \cap (-\text{core}D)$, a contradiction. Thus, $\text{vcl}\,\text{cone}(f(\bar{x}, A)+C) \cap (-\text{core}D) = \emptyset$. Hence, $\text{vcl}\,\text{cone}(f(\bar{x}, A) + C) \cap (-C_0) = \emptyset$, i.e., $\bar{x} \in \text{Sol}^{Be}(\text{VEP}_{(f,C)})$.
(b) This is obvious.
(c) By parts (a) and (b), it suffices to show that $\text{Sol}^{Be}(\text{VEP}_{(f,C)}) \subset \text{Sol}^{e}(\text{VEP}_{(f,C)})$. If $\bar{x} \in \text{Sol}^{Be}(\text{VEP}_{(f,C)})$, then $\text{vclcone}(f(\bar{x}, A)+C) \cap (-C_0) = \emptyset$ which implies that $(f(\bar{x}, A) + C) \cap (-C_0) = \emptyset$. Then, for all $y \in A$, $f(\bar{x}, y) \notin -C - C_0 \supset -C_0$. Hence, $\bar{x} \in \text{Sol}^{e}(\text{VEP}_{(f,C)})$.

□

The inclusions in Proposition 3.21 may be strict (see Examples 3.17–3.20). Moreover, in general, for $t \in \{He, Be\}$ and $t' \in \{e, w\}$, $\text{Sol}^{t}(\text{VEP}_{(f,C)}) \not\subset \text{Sol}^{t'}(\text{VEP}_{(f,C)})$ and $\text{Sol}^{t'}(\text{VEP}_{(f,C)}) \not\subset \text{Sol}^{t}(\text{VEP}_{(f,C)})$.

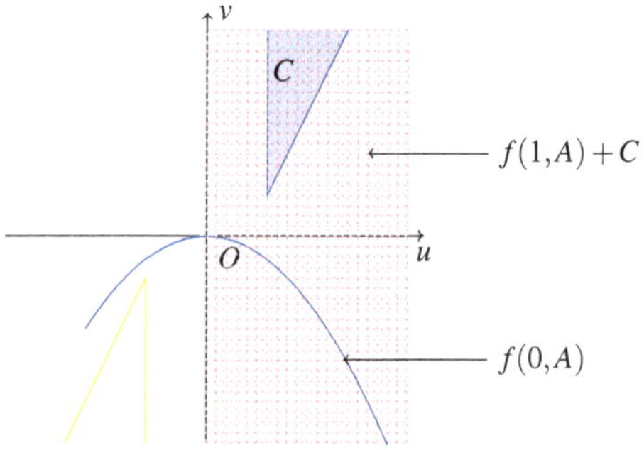

Fig. 3.8 Illustration of Example 3.22

Example 3.22 Let $A = (-\infty, 2] \subset \mathbb{R}$, $Y = \mathbb{R}^2$, $C = \{c = (c_1, c_2) \in \mathbb{R}^2 \mid 1 \leq c_1,\ 2c_1 - 1 \leq c_2\}$, and $f : A \times A \to Y$ be defined by

$$f(x, y) := \left(x - y, -\frac{1}{2}y^2\right) \text{ for all } x, y \in A.$$

Let $D = \{(c_1, c_2) \in \mathbb{R}^2 \mid 0 \leq c_1\}$. Clearly, $D \in \mathscr{K}_C^{core}$. For $\bar{x} = 1$ and any $y \in A$ and $c = (c_1, c_2) \in C$, we have $1 - y + c_1 \geq 2 - y \geq 0$. So, $f(1, A) + C \subset D$ (see Fig. 3.8). Clearly, $D \cap (-\text{core}D) = \emptyset$. Hence, $(f(1, A)+C) \cap (-\text{core}D) = \emptyset$. Thus, $1 \in \text{Sol}^{He}(\text{VEP}(f, C))$. But $1 \notin \text{Sol}^{e}(\text{VEP}(f, C))$ because $f(1, 2) = (-1, -2) \in -C_0$. For $\bar{x} = 0$, it is not hard to check that $0 \in \text{Sol}^{e}(\text{VEP}(f, C))$ (see Fig. 3.8). For $c = (1, 1) \in C$, $f(0, 2)+c = (-1, -1) \in -C_0$. Therefore, $0 \notin \text{Sol}^{He}(\text{VEP}(f, C))$.

The following proposition states relations between the solutions of the two related vector equilibrium problems.

Proposition 3.23 *If* $C \in \mathscr{E}(Y)$ *and* $f, g : A \times A \to Y$ *satisfy* $f(x, y) - g(x, y) \in C$ *for all* $x, y \in A$, *then* $\text{Sol}^{t}(\text{VEP}_{(g,C)}) \subset \text{Sol}^{t}(\text{VEP}_{(f,C)})$ *for* $t \in \{He, Be, e, w\}$.

Proof For $t = He$, let $\bar{x} \in \text{Sol}^{He}(\text{VEP}_{(g,C)})$. Then, there exists $D \in \mathscr{K}_C^{core}$ such that

$$g(\bar{x}, y) + c \notin -\text{core}D \text{ for all } y \in A \text{ and } c \in C. \tag{3.7}$$

Suppose that $\bar{x} \notin \text{Sol}^{He}(\text{VEP}_{(f,C)})$. Then, there exist $y \in A$ and $c \in C$ such that $f(\bar{x}, y) + c \in -\text{core}D$. Then, for $\hat{c} := f(\bar{x}, y) - g(\bar{x}, y) \in C$ and $c' := \hat{c} + c \in C$, we have $g(\bar{x}, y) + c' = g(\bar{x}, y) + f(\bar{x}, y) - g(\bar{x}, y) + c = f(\bar{x}, y) + c \in -\text{core}D$, contradicting (3.7).

For $t = Be$, assume that $\bar{x} \in \text{Sol}^{Be}(\text{VEP}_{(g,C)})$, i.e., $\text{vclcone}(g(\bar{x}, A) + C) \cap (-C_0) = \emptyset$. Suppose that $\bar{x} \notin \text{Sol}^{Be}(\text{VEP}_{(f,C)})$. Then, there exists $u \in \text{vcl cone}(f(\bar{x}, A) + C)$ such that $u \in -C_0$. Since $u \in \text{vcl cone}(f(\bar{x}, A) + C)$, there exists $v \in Y$ such that for all $\lambda > 0$, there exists $\lambda' \in [0, \lambda]$ such that for some

$\alpha > 0$, $y \in A$ and $c \in C$,

$$u + \lambda' v = \alpha(f(\bar{x}, y) + c) = \alpha(g(\bar{x}, y) + \bar{c} + c) \text{ (with } \bar{c} = f(\bar{x}, y) - g(\bar{x}, y) \in C)$$

$$= \alpha(g(\bar{x}, y) + c') \text{ (with } c' := \bar{c} + c \in C)$$

$$\in \text{cone}(g(\bar{x}, y) + C).$$

It follows that $u \in \text{vcl cone}(g(\bar{x}, y) + C)$, a contradiction.

For $t = e$ ($t = w$, resp.), take any $\bar{x}$ in $\text{Sol}^e(\text{VEP}_{(g,C)})$ ($\text{Sol}^w(\text{VEP}_{(g,C)})$, resp.), i.e., $g(\bar{x}, y) \notin -C_0$ ($\notin -\text{core}C$, resp.) for all $y \in A$. If $\bar{x} \notin \text{Sol}^e(\text{VEP}_{(f,C)})$ ($\text{Sol}^w(\text{VEP}_{(f,C)})$, resp.), then $f(\bar{x}, y) \in -C_0$ ($\in -\text{core}C$, resp.) for some $y \in A$. Then, $g(\bar{x}, y) = (-(f(\bar{x}, y) - g(\bar{x}, y)) + f(\bar{x}, y) \in -C - C_0$ ($\in -C - \text{core}C$, resp.) $\subset -C_0$ ($\subset -\text{core}C$, resp.), a contradiction. □

3.2.2 Applications

Let A, Y, and C be as for $(\text{VEP}_{(f,C)})$ in Sect. 3.2.1. The following are some particular cases of $(\text{VEP}_{(f,C)})$.

(a) **Vector Optimization Problems**. For $h : A \to Y$, the VECTOR OPTIMIZATION PROBLEM, denoted by $(\text{VOP}_{(h,C)})$, is to find $\bar{x} \in A$ such that

$$h(y) - h(\bar{x}) \notin -C_0 \text{ for all } y \in A.$$

This problem is also written as $\min_C\{h(x) \mid x \in A\}$.

Definition 3.24 For $(\text{VOP}_{(h,C)})$, a point $\bar{x} \in A$ is said to be:

(a) A HENIG-PROPER C-SOLUTION if there exists $D \in \mathscr{K}_C^{core}$ such that

$$(h(A) - h(\bar{x}) + C) \cap (-\text{core}D) = \emptyset \tag{3.8}$$

(b) A BENSON-PROPER C-SOLUTION if

$$\text{vclcone}(h(A) - h(\bar{x}) + C) \cap (-C_0) = \emptyset \tag{3.9}$$

(c) A C-SOLUTION if $(h(A) - h(\bar{x})) \cap (-C_0) = \emptyset$
(d) A WEAK C-SOLUTION if $(h(A) - h(\bar{x})) \cap (-\text{core}C) = \emptyset$

By $O^{He}(h, C)$ ($O^{Be}(h, C)$, $O^e(h, C)$, $O^w(h, C)$, resp.), we denote the set of the Henig-proper C-solutions (Benson-proper C-solutions, C-solutions, weak C-solutions, resp.). It is clear that if $\bar{x}$ is a Henig-proper C-solution, or a Benson-proper C-solution, or a weak C-solution of $(\text{VOP}_{(h,C)})$, then $\bar{x}$ is a C-solution of the same kind of $(\text{VEP}_{(f,C)})$ with $f(x, y) = h(y) - h(x)$.

From (3.8), we see that $C \cap (-\text{core}D) = \emptyset$ which implies $C \cap (-C_0) = \emptyset$, and from (3.9) that $\text{vcl cone}C \cap (-C_0) = \emptyset$. Hence, in Definition 3.24 (a) (Definition 3.24 (b), resp.), we only consider C satisfying $C \cap (-C_0) = \emptyset$ ($\text{vcl cone}C \cap (-C_0) = \emptyset$, resp.).

(b) **Vector Saddle-Point Problems**. Let A_1, A_2 be nonempty sets, $C \subset Y$, and $L : A_1 \times A_2 \to Y$ a vector bifunction. A point $(\bar{x}, \bar{y}) \in A_1 \times A_2$ is called a C-SADDLE POINT of L if there is no $(x, y) \in A_1 \times A_2$ such that

$$L(\bar{x}, \bar{y}) - L(x, \bar{y}) \in C_0 \tag{3.10}$$

and

$$L(\bar{x}, y) - L(\bar{x}, \bar{y}) \in C_0, \tag{3.11}$$

and it is called a WEAK C-SADDLE POINT of L if C_0 in (3.10) and (3.11) is replaced by $\text{core}C$.

It is clear that $(\bar{x}, \bar{y}) \in A_1 \times A_2$ is a C-saddle point (weak C-saddle point, resp.) of L if and only if

$$L(\bar{x}, \bar{y}) - L(x, \bar{y}) \notin C_0 \ (\notin \text{core}C, \ \text{resp.}), \ \text{for all } x \in A_1$$

and

$$L(\bar{x}, y) - L(\bar{x}, \bar{y}) \notin C_0 \ (\notin \text{core}C, \ \text{resp.}), \ \text{for all } y \in A_2.$$

A VECTOR SADDLE-POINT PROBLEM (briefly, $(\text{VSP}_{(L,C)})$) is the problem of finding weak C-saddle or C-saddle points of L.

Let $A = A_1 \times A_2$ and $f : A \times A \to Y$ be defined for all $(x, y), (x', y') \in A \times A$ by

$$f((x, y), (x', y')) = L(x', y) - L(x, y'). \tag{3.12}$$

Proposition 3.25 *If $(\bar{x}, \bar{y})$ is an C-solution (a weak C-solution, resp.) of $(\text{VEP}_{(f,C)})$ with f defined by (3.12), then it is a C-saddle point (a weak C-saddle point) of L.*

Proof Assume that $(\bar{x}, \bar{y})$ is an C-solution (a weak C-solution, resp.) of $(\text{VEP}_{(f,C)})$. Then,

$$f((\bar{x}, \bar{y}), (x', y')) = L(x', \bar{y}) - L(\bar{x}, y') \notin -C_0(-\text{core}C, \ \text{resp.}) \text{ for all } (x', y') \in A.$$

By taking $(x', y') = (x', \bar{y})$ and $(x', y') = (\bar{x}, y)$, one has

$$L(\bar{x}, \bar{y}) - L(x, \bar{y}) \notin C_0(\text{core}C, \ \text{resp.}) \text{ for all } x' \in A_1$$

and

$$L(\bar{x}, y) - L(\bar{x}, \bar{y}) \notin C_0(\text{core}C, \text{ resp.}) \text{ for all } y' \in A_2.$$

This means that $(\bar{x}, \bar{y})$ is a C-saddle point (a weak C-saddle point, resp.) of L. □

(c) **Vector Variational Inequalities and Multiobjective Traffic Networks**. Let A be a nonempty subset of a topological linear space X, $T : A \to \mathscr{L}(X, Y)$ an operator. For every $l \in \mathscr{L}(X, Y)$, by (l, x) we denote the value of l at x. The VECTOR VARIATIONAL INEQUALITY is defined as follows:

$(\text{VVI}_{(T,C)})$ Find $\bar{x} \in A$ such that $(T(\bar{x}), y - \bar{x}) \notin -C_0$ for all $y \in A$.

The concepts of solutions for $(\text{VVI}_{(T,C)})$ are considered in the sense of Definition 3.16 for $f(x, y) = (T(x), y - x)$. In this case, we use $\text{Sol}^{He}(\text{VVI}_{(T,C)})$ ($\text{Sol}^{Be}(\text{VVI}_{(T,C)})$, $\text{Sol}^{e}(\text{VVI}_{(T,C)})$, $\text{Sol}^{w}(\text{VVI}_{(T,C)})$, resp.) to denote the set of all Henig-proper E-solutions (the Benson-proper E-solutions, E-solutions, weak E-solutions, resp.) of $(\text{VVI}_{(T,C)})$.

An important particular case of (VVI) is the MULTIOBJECTIVE TRAFFIC NETWORK PROBLEM. Consider a traffic network G with capacity constraints on arcs. For $\mathbf{N}$, $\mathbf{A}$, $\mathbf{W}$, $P = \bigcup_{w \in W} P_w$, $m = |P|$, x_p $(p \in P)$, $x = (x_1, \dots, x_m) \in \mathbb{R}^m_+$, v_a, δ_{ap} are defined as in Sect. 1.1.4 (f). Here we are interested in the case of *elastic demands* and multiobjective costs. For each (O-D) pair w, the demand is not fixed but can vary in a set $\mathscr{D}_w$ in $\mathbb{R}_+$, i.e., the demand constraint is $\sum_{p \in P_w} x_p \in \mathscr{D}_w$ for $w \in \mathbf{W}$. A flow vector x is *feasible* if it satisfies the following capacity and demand constraints:

(i) $0 \le v_a \le c_a$ for $a \in \mathbf{A}$.
(j) $\sum_{p \in P_w} x_p \in \mathscr{D}_w$ for $w \in \mathbf{W}$.

Let A be the set of all feasible vector flows. For a vector flow x, let $t_a(x) \in \mathbb{R}^k$ be a VECTOR COST OF VECTOR FLOW h ON ARC a. Then, $\mathbb{T}_p(x) = \sum_{a \in \mathbf{A}} \delta_{ap} t_a(x) \in \mathbb{R}^k$ is the VECTOR COST OF x ON PATH p.

Definition 3.26 Assume that $x \in A$ and a set $C \subset \mathbb{R}^k$ satisfies $(0, \alpha]C \subset C$ for some $\alpha > 0$.

(a) x is called a HENIG-PROPER C-EQUILIBRIUM FLOW if there exists $D \in \mathscr{K}_C^{core}$ such that for all $w \in \mathbf{W}$ and $p, p' \in P_w$, we have $(\mathbb{T}_{p'}(x) - \mathbb{T}_p(x) + C) \cap (-\text{core}D) \neq \emptyset \Rightarrow x_p = 0$, or path p' is a saturated path of flow x.
(b) x is called a BENSON-PROPER C-EQUILIBRIUM FLOW if for all $w \in \mathbf{W}$ and $p, p' \in P_w$, we have $\text{vcl cone}(\mathbb{T}_{p'}(x) - \mathbb{T}_p(x) + C) \cap (-C_0) \neq \emptyset \Rightarrow x_p = 0$, or p' is a saturated path of flow x.
(c) x is called a C-EQUILIBRIUM FLOW if for all $w \in \mathbf{W}$ and all $p, p' \in P_w$, we have $\mathbb{T}_{p'}(x) - \mathbb{T}_p(x) \in -C_0 \Rightarrow x_p = 0$, or p' is a saturated path of flow x.
(d) x is called a WEAK C-EQUILIBRIUM FLOW if for all $w \in \mathbf{W}$ and all $p, p' \in P_w$, we have $\mathbb{T}_{p'}(x) - \mathbb{T}_p(x) \in -\text{core}C \Rightarrow x_p = 0$, or p' is a saturated path of flow x.

By $\mathrm{Sol}^{He}(G, C)$ ($\mathrm{Sol}^{Be}(G, C)$, $\mathrm{Sol}^{e}(G, C)$, $\mathrm{Sol}^{w}(G, C)$), resp.) we denote the set of all Henig-proper C-equilibrium flows (Benson-proper C-equilibrium flows, C-equilibrium flows, weak C-equilibrium flows, resp.) of the traffic network G.

Suppose that $\mathbb{T}_p = (\mathbb{T}_{1p}, \ldots, \mathbb{T}_{kp})$. The path cost matrix of network G is

$$\mathbb{T}(x) = \begin{pmatrix} \mathbb{T}_{11}(x) & \mathbb{T}_{12}(x) & \cdots & \mathbb{T}_{1m}(x) \\ \mathbb{T}_{21}(x) & \mathbb{T}_{22}(x) & \cdots & \mathbb{T}_{2m}(x) \\ \cdots & \cdots & \cdots & \cdots \\ \mathbb{T}_{k1}(x) & \mathbb{T}_{k2}(x) & \cdots & \mathbb{T}_{km}(x) \end{pmatrix}.$$

To convert a traffic network problem to a type of variational inequality, we define $T_j := (\mathbb{T}_{j1}, \ldots, \mathbb{T}_{jm})$ $(j = 1, 2, \ldots, k)$ and $T := (T_1, \ldots, T_k)$.

Let $x \in A$, $w \in \mathbf{W}$, and $p, p' \in P_w$ be arbitrary, and $\varepsilon = \min\{\min_{a \in A_{p'}}(c_a - v_a), x_p, \alpha\}$, where $A_{p'} = \{a \in \mathbf{A} \mid \text{path } p' \text{ goes through } a\}$. It is easy to check that if the condition $(*)$: "$x_p > 0$ *and* p' *is a nonsaturated path of flow* x" is satisfied, then $\varepsilon > 0$, $y = (y_l)$, where

$$y_l = \begin{cases} x_l & \text{if } l \neq p, p', \\ x_p - \varepsilon & \text{if } l = p, \\ x_{p'} + \varepsilon & \text{if } l = p', \end{cases}$$

belongs to A, and

$$\begin{aligned} &(T(x), y - x) \\ &= \left(\sum_{w \in \mathbf{W}} \sum_{l \in P_w} (y_l - x_l)\mathbb{T}_{1l}(x), \ldots, \sum_{w \in \mathbf{W}} \sum_{l \in P_w} (y_l - x_l)\mathbb{T}_{kl}(x) \right) \\ &= ((y_p - x_p)\mathbb{T}_{1p}(x) + (y_{p'} - x_{p'})\mathbb{T}_{1p'}(x), \ldots, (y_p - h_p)\mathbb{T}_{kp}(x) \\ &\quad + (y_{p'} - x_{p'})\mathbb{T}_{kp'}(x)) \\ &= (\varepsilon(\mathbb{T}_{1p'}(x) - \mathbb{T}_{1p}(x)), \ldots, \varepsilon(\mathbb{T}_{kp'}(x) - \mathbb{T}_{kp}(x))) = \varepsilon(\mathbb{T}_{p'}(x) - \mathbb{T}_p(x)). \end{aligned}$$

For any $D \in \mathscr{K}_C^{core}$, assume that $(\mathbb{T}_p(x) - \mathbb{T}_{p'}(x) + C) \cap (-\mathrm{core} D) \neq \emptyset$ and $(*)$ holds. Taking $d \in (\mathbb{T}_p(x) - \mathbb{T}_{p'}(x) + C) \cap (-\mathrm{core} D)$, we have $\varepsilon d \in (\varepsilon(\mathbb{T}_p(x) - \mathbb{T}_{p'}(x)) + \varepsilon C) \cap (-\mathrm{core} D) \subset ((T(x), y - x) + C) \cap (-\mathrm{core} D)$. Using this, we easily show that if $x \in \mathrm{Sol}^{He}(\mathrm{VVI}_{(T,C)})$, then x is a Henig-proper C-equilibrium flow.

Assume that there exists $e \in \mathrm{vcl\,cone}(\mathbb{T}_{p'}(x) - \mathbb{T}_p(x) + C) \cap (-C_0)$ and $(*)$ holds. Then, $\varepsilon e \in -C_0$. Moreover, there exists $v \in \mathbb{R}^k$ such that for all $\beta > 0$, there exists $\beta_v \in (0, \beta]$ satisfying $e + \beta_v v \in \mathrm{cone}(\mathbb{T}_{p'}(x) - \mathbb{T}_p(x) + C)$. Now, for any $\lambda > 0$, we choose any $\beta \in (0, \lambda]$ satisfying $\varepsilon\beta < \lambda$ and set $\lambda_v = \varepsilon\beta_v$. We see that $\lambda_v \in (0, \lambda]$ and $\varepsilon e + \lambda_v v = \varepsilon(e + \beta_v v) \in \varepsilon\mathrm{cone}(\mathbb{T}_{p'}(x) - \mathbb{T}_p(x) + C) \subset \mathrm{cone}(\varepsilon(\mathbb{T}_{p'}(x) - \mathbb{T}_p(x)) + \varepsilon C) \subset \mathrm{cone}((T(x), y - x) + C)$. Hence, $\varepsilon e \in \mathrm{vclcone}((T(x), y - x) + C)$.

Thus, $\varepsilon e \in \text{vcl cone}((T(x), y-x) + C) \cap (-C_0)$ and so $x \notin \text{Sol}^{Be}(\text{VVI}_{(T,C)})$. Therefore, if $x \in \text{Sol}^{Be}(\text{VVI}_{(T,C)})$, then x is a Benson-proper C-equilibrium flow.

Assume that $\mathbb{T}'_p(x) - \mathbb{T}_p(x) \in -C_0$ and $(*)$ holds. Then, $(T(x), y-x) = \varepsilon(\mathbb{T}_{p'}(x) - \mathbb{T}_p(x)) \in -C_0$, which implies that $x \notin \text{Sol}^e(\text{VVI}_{(T,C)})$. Thus, if $x \in \text{Sol}^e(\text{VVI}_{(T,C)})$, then x is a C-equilibrium flow. Similarly, if $x \in \text{Sol}^w(\text{VVI}_{(T,C)})$, then x is a weak C-equilibrium flow.

(d) **Noncooperative Games with Vector Payoffs**. Let $\mathscr{G} = (A_i, \mathbf{c}_i)_{i \in I}$ be a n-player noncooperative game, where $I := \{1, 2, \ldots, n\}$ is the set of players, A_i is the strategy set of the ith player, and $\mathbf{c}_i : A = \prod_{i=1}^n A_i \to Y$ is the vector payoff function of the ith player.

Definition 3.27 Let $C^i \subset Y$, $i \in I$, be solid. A point $\bar{x} = (\bar{x}_i)_{i \in I} \in A$ is said to be:

(a) A HENIG-NASH PROPER EQUILIBRIUM STRATEGY if for each $i \in I$, there exists $D^i \in \mathscr{K}_{C^i}^{core}$ such that

$$(\mathbf{c}_i(\bar{x}) - \mathbf{c}_i(\bar{x}_{-i}, A_i) + C^i) \cap (-\text{core}D^i) = \emptyset$$

(b) A BENSON-NASH PROPER EQUILIBRIUM STRATEGY if for each $i \in I$,

$$\text{vclcone}(\mathbf{c}_i(\bar{x}) - \mathbf{c}_i(\bar{x}_{-i}, A_i) + C^i) \cap (-C_0^i) = \emptyset$$

(c) A NASH EQUILIBRIUM STRATEGY if for each $i \in I$,

$$(\mathbf{c}_i(\bar{x}) - \mathbf{c}_i(\bar{x}_{-i}, A_i)) \cap (-C_0^i) = \emptyset$$

(d) A WEAK NASH EQUILIBRIUM STRATEGY if for each $i \in I$,

$$(\mathbf{c}_i(\bar{x}) - \mathbf{c}_i(\bar{x}_{-i}, A_i)) \cap (-\text{core}C^i) = \emptyset$$

By $\text{Sol}^{He}(\mathscr{G}, \{C^i\}_{i \in I})$ ($\text{Sol}^{Be}(\mathscr{G}, \{C^i\}_{i \in I})$, $\text{Sol}^e(\mathscr{G}, \{C^i\}_{i \in I})$, $\text{Sol}^w(\mathscr{G}, \{C^i\}_{i \in I})$), resp.) we denote the set of all Henig-Nash proper equilibrium strategies (Benson-Nash proper equilibrium strategies, Nash equilibrium strategies, weak Nash equilibrium strategies, resp.) of $\mathscr{G}$.

We apply the traditional method to convert a noncooperative game to a special case of equilibrium problem by using the vector bifunction $\mathscr{N} : A \times A \to Y$ defined by

$$\mathscr{N}(x, y) := \sum_{i=1}^{n} [\mathbf{c}_i(x) - \mathbf{c}_i(x_{-i}, y_i)] \tag{3.13}$$

and choosing $C = \bigcup_{i \in I} C^i$. It is not hard to see that if $\bar{x} \in \text{Sol}^{He}(\text{VEP}_{(\mathscr{N},C)})$ ($\text{Sol}^{Be}(\text{VEP}_{(\mathscr{N},C)})$, $\text{Sol}^e(\text{VEP}_{(\mathscr{N},C)})$, resp.), then $\bar{x}$ is a Henig-Nash proper equilibrium (Benson-Nash proper equilibrium, Nash equilibrium, resp.) strategy. Because $\bigcup_{i\in I} \text{core}C^i \subset \text{core}(\bigcup_{i\in I} C^i) = \text{core}C$, we easily check that if $\bar{x} \in \text{Sol}^w(\text{VEP}_{(\mathscr{N},C)})$, then $\bar{x}$ is a weak Nash equilibrium strategy. Thus, we can use existence results for the vector equilibrium problem for these $\mathscr{N}$ and C to consider the existence of the kinds of equilibrium strategies in Definition 3.27.

3.3 Existence of Solutions to Vector Equilibrium Problems

Let $E \subset Y$ be nonempty and $f : A \times A \to Y$. Associated with f, we define $\Phi^1_{Ef}, \Phi^2_{Ef} : A \times A \to \mathbb{R}$ by

$$\Phi^1_{Ef}(x, y) = \begin{cases} 0 \text{ if } f(x, y) \notin -E\}, \\ -1 \text{ otherwise.} \end{cases} \tag{3.14}$$

$$\Phi^2_{Ef}(x, y) = \begin{cases} 0 \text{ if } f(x, y) \in E\}, \\ -1 \text{ otherwise.} \end{cases} \tag{3.15}$$

Let $(\text{EP}_{\Phi^\ell_{Ef}})$ $(\ell = 1, 2)$ be the scalar equilibrium problem corresponding to Φ^ℓ_{Ef}:

$(\text{EP}_{\Phi^\ell_{Ef}})$ Find $\bar{x} \in A$ such that $\Phi^\ell_{Ef}(\bar{x}, y) \geq 0$ for all $y \in A$.

By $\text{Sol}(\text{EP}_{\Phi^\ell_{Ef}})$, we denote the set of solutions of $(\text{EP}_{\Phi^\ell_{Ef}})$.

For $C \subset Y$, we define the following classes of subsets:

- $\mathscr{E}^1_{He}(C) = \{E \subset Y \mid \text{there exist } D \in \mathscr{K}^{core}_C \text{ with } C + \text{core}D \subset E\}$.
- $\mathscr{E}^1_{Be}(C) = \{E \subset Y \mid \text{vcl cone}(C - E^c) \cap (-C_0) = \emptyset\}$.
- $\mathscr{E}^1_{e}(C) = \{E \subset Y \mid C_0 \subset E\}$.
- $\mathscr{E}^1_{w}(C) = \{E \subset Y \mid \text{core}C \subset E\}$.
- $\mathscr{E}^2_{He}(C) = \{E \subset Y \mid \text{there exist } D \in \mathscr{K}^{core}_C \text{ with } 0 \notin C + \text{core}D + E\}$.
- $\mathscr{E}^2_{Be}(C) = \{E \subset Y \mid \text{vcl cone}(C + E) \cap (-C_0) = \emptyset\}$.
- $\mathscr{E}^2_{e}(C) = \{E \subset Y \mid 0 \notin C_0 + E\}$.
- $\mathscr{E}^2_{w}(C) = \{E \subset Y \mid 0 \notin \text{core}C + E\}$.

Remark 3.28 It is easily seen that, for $t \in \{He, Be, e, w\}$, $E \in \mathscr{E}^1_t(C)$ if and only if $-E^c \in \mathscr{E}^2_t(C)$, and $E \in \mathscr{E}^2_t(C)$ if and only if $-E^c \in \mathscr{E}^1_t(C)$. Moreover, it is not hard to verify the following facts. For all $\ell \in \{1, 2\}$ and all $t \in \{He, Be, e, w\}$:

(a) $\mathscr{E}^\ell_{He}(C) \subset \mathscr{E}^\ell_{Be}(C)$ and $\mathscr{E}^\ell_{e}(C) \subset \mathscr{E}^\ell_{w}(C)$; and $\mathscr{E}^\ell_{Be}(C) \subset \mathscr{E}^\ell_{e}(C)$ if $C_0 \subset C_0 + C$.
(b) $\mathscr{K}^{core}_C \subset \mathscr{E}^\ell_t(C)$; if $D \in \mathscr{K}^{core}_C$, then $\text{core}D \in \mathscr{E}^\ell_t(C)$.
(c) $C \in \mathscr{E}^1_e(C)$; $C \in \mathscr{E}^2_e(C)$ if $C \cap (-C) \subset \{0\}$; and $C \in \mathscr{E}^2_{Be}(C)$ if vcl cone$(C) \cap (-C) \subset \{0\}$ and $C + C \subset C$.
(d) $\text{core}C \in \mathscr{E}^1_w(C)$; and $\text{core}C \in \mathscr{E}^2_w(C)$ if $C \cap (-\text{core}C) = \emptyset$.

(e) If $E \supset \text{core}D$ for some $D \in \mathcal{K}_C^{\text{core}}$, then $E \in \mathcal{E}_{He}^1(C)$; if $E \cap (-\text{core}D) = \emptyset$ for some $D \in \mathcal{K}_C^{\text{core}}$, then $E \in \mathcal{E}_{He}^2(C)$.
(f) If $C \in \mathcal{K}_c$, $C \cap (-C) = \{0\}$ and C is vectorially closed, then $C \in \mathcal{E}_e^1(C) \cap \mathcal{E}_{Be}^2(C)$ and $\text{core}C \in \mathcal{E}_{Be}^2(C)$.

Relationships between the sets of solutions of $(\text{VEP}_{(f,C)})$ and of $(\text{EP}_{\Phi_{Ef}^\ell})$ are given as follows.

Proposition 3.29 *For* $\ell \in \{1, 2\}$ *and* $t \in \{He, Be, e, w\}$, $\text{Sol}^t(\text{VEP}_{(f,C)}) \neq \emptyset$, *if and only if there exists* $E \in \mathcal{E}_t^\ell(C)$ *such that* $\text{Sol}(EP_{\Phi_{Ef}^\ell}) \neq \emptyset$.

Proof Case $\ell = 1$. If $\text{Sol}^t(\text{VEP}_{(f,C)}) \neq \emptyset$, then by taking $E = C + \text{core}D$ for $t = He$, $E = C$ for $t = e$, $E = \text{core}C$ for $t = w$ and $E = -(Y \setminus f(\bar{x}, A))$ for $t = Be$ (here $\bar{x} \in \text{Sol}^{Be}(\text{VEP}_{(f,C)})$), we easily check the necessity part.

Consider the sufficiency part. Assume that there exists $E \in \mathcal{E}_t^1(C)$ such that $\text{Sol}(EP_{\Phi_{Ef}^1}) \neq \emptyset$. Then, taking any $\bar{x} \in \text{Sol}(EP_{\Phi_{Ef}^1})$, we have $f(\bar{x}, y) \notin -E$ for all $y \in A$. Hence,

$$f(\bar{x}, A) \subset Y \setminus (-E) = -E^c. \tag{3.16}$$

For $t = He$, as $E \in \mathcal{E}_{He}^1(C)$, there exists $D \in \mathcal{K}_C^{core}$ such that $C + \text{core}D \subset E$. By (3.16), we have $f(\bar{x}, y) \notin -C - \text{core}D$ for all $y \in A$, or equivalently, $(f(\bar{x}, y) + C) \cap (-\text{core}D) = \emptyset$ for all $y \in A$, i.e., $(f(\bar{x}, A) + C) \cap (-\text{core}D) = \emptyset$. Thus, $\bar{x} \in \text{Sol}^{He}(\text{VEP}_{(f,C)})$.

For $t = Be$, by (3.16), $\text{vcl cone}(f(\bar{x}, A) + C) \cap (-C_0) \subset \text{vcl cone}(C - E^c) \cap (-C_0) = \emptyset$ (since $E \in \mathcal{E}_{Be}^1(C)$). Hence, $\bar{x} \in \text{Sol}^{Be}(\text{VEP}_{(f,C)})$.

For $t = e$ ($t = w$, resp.), $\bar{x} \in \text{Sol}^e(\text{VEP}_{(f,C)})$ ($\bar{x} \in \text{Sol}^w(\text{VEP}_{(f,C)})$, resp.) directly follows from the inclusions $C_0 \subset E$ and $f(\bar{x}, A) \subset -E^c$ ($\text{core}C \subset E$ and $f(\bar{x}, A) \subset -E^c$, resp.).

Case $\ell = 2$. By replacing E in the case $\ell = 1$ by $E' := -E^c$ we obtain the statement for the case $\ell = 2$. □

Remark 3.30

(a) By Proposition 3.29, for studying the existence of kinds of solutions of $(\text{VEP}_{(f,C)})$ we search for conditions such that $\text{Sol}(EP_{\Phi_{Ef}^\ell}) \neq \emptyset$ with $E \subset Y$ belonging to one of the above classes (depending on the considered kind of solutions).
(b) Let $E \subset Y$ be nonempty. We observe from the definition of Φ_{Ef}^ℓ, $\ell \in \{1, 2\}$, that

$\text{Sol}(EP_{\Phi_{Ef}^\ell}) \neq \emptyset$, *if and only if* $\bigcap_{y \in A} \text{lev}_E^\ell f(\cdot, y) \neq \emptyset$, *if and only if the family* $\{\overline{\text{lev}}_E^\ell f(\cdot, y) \mid y \in A\}$ *does not cover* A.

In this chapter, for simplicity and convenience, the assumptions are expressed via a set E. When applying these results to specific situations, we can take E to be a set directly related to the data of the problem such as C, core$C$, vcl$C$, cone$C$, vcl(cone$C$), etc., or a set related to C as in Remark 3.28.

Definition 3.31 Let $E \subset Y$ be a nonempty set and $f : A \times A \to Y$ be a vector bifunction. We say that bifunction f has the E-FINITE INTERSECTION PROPERTY OF TYPE 1 (the E-FINITE INTERSECTION PROPERTY OF TYPE 2, resp.) if for all $N \in \mathscr{F}(A)$, there exists $x \in A$ such that $f(x, y) \notin -E$ ($f(x, y) \in E$, resp.) for all $y \in N$.

Remark 3.32 It is clear that f has the E-finite intersection property of type ℓ ($\ell \in \{1, 2\}$) if and only if $\bigcap_{y\in N} \text{lev}^{\ell}_E f(\cdot, y) \neq \emptyset$ for all $N \in \mathscr{F}(A)$. Moreover, f has the E-finite intersection property of type ℓ if and only if the problem (EP$_{\Phi^{\ell}_{Ef}}$) is finitely solvable.

Definition 3.33 Let $E \subset Y$ be a nonempty set and $f : A \times A \to Y$ be a vector bifunction. We say that bifunction f has the E-FINITE COVER PROPERTY OF TYPE ℓ ($\ell \in \{1, 2\}$) if, when the family $\{\overline{\text{lev}}^{\ell}_E f(\cdot, y)\}_{y\in A}$ covers A, it has a finite subcover.

Proposition 3.34 *Let $E \subset Y$ be nonempty and $\ell \in \{1, 2\}$. Assume that f has both the E-finite intersection property of type ℓ and the E-finite cover property of type ℓ. For $t \in \{He, Be, e, w\}$, if $E \in \mathscr{E}^{\ell}_t(C)$, then* $\text{Sol}^t(\text{VEP}_{(f,C)}) \neq \emptyset$.

Proof We show that the family $\{\overline{\text{lev}}^{\ell}_E f(\cdot, y) \mid y \in A\}$ does not cover A, and the conclusion of Proposition 3.34 follows from Proposition 3.29 and Remark 3.30. Suppose that the family $\{\overline{\text{lev}}^{\ell}_E f(\cdot, y)\}_{y\in A}$ covers A, i.e., $A = \bigcup_{y\in A} \overline{\text{lev}}^{\ell}_E f(\cdot, y)$. Then, when f has the E-finite cover property of type ℓ, there exists a finite subset M of A such that $A = \bigcup_{y\in M} \overline{\text{lev}}^{\ell}_E f(\cdot, y)$, i.e., $\emptyset = A\setminus(\bigcup_{y\in M} \overline{\text{lev}}^{\ell}_E f(\cdot, y)) = \bigcap_{y\in N} \text{lev}^{\ell}_E f(\cdot, y)$. This contradicts the assumption that f has the E-finite intersection property of type ℓ. Thus, the family $\{\overline{\text{lev}}^{\ell}_E f(\cdot, y)\}_{y\in A}$ does not cover A. □

Following are some sufficient conditions for a bifunction having the E-finite cover property of type ℓ.

Proposition 3.35 *Let $E \subset Y$ be nonempty and $\ell \in \{1, 2\}$. Then, f has the E-finite cover property of type ℓ if at least one of the following conditions is satisfied:*

(a) *A is a compact topological space and the set $\text{lev}^{\ell}_E f(\cdot, y)$ is closed for each $y \in A$.*
(b) *There exists $M \in \mathscr{F}(A)$ such that $\bigcup_{y\in A} \overline{\text{lev}}^{\ell}_E f(\cdot, y) \subset \bigcup_{y\in M} \overline{\text{lev}}^{\ell}_E f(\cdot, y)$.*

Proof This proposition is obvious. □

To obtain more specific existence conditions we consider some certain classes of bifunctions.

3.3.1 Existence Conditions Under Vector-Proper Quasimonotonicity

In this subsection, the standard assumption is as follows:

$$\begin{cases} A \text{ is a nonempty convex subset of a real topological vector space} X; \\ C \subset Y \text{ is an (algebraically) solid set such that} 0 \notin \text{core} C \text{ and} \\ f : A \times A \to Y \text{ is a given vector bifunction.} \end{cases} \quad (3.17)$$

Definition 3.36 Let $E \subset Y$ be a nonempty set.

(a) Vector bifunction f is said to be PROPERLY E-QUASIMONOTONE OF TYPE 1 in the first component (in the second component, resp.) on A if, for all $N \in \mathcal{F}(A)$ and $x \in \text{conv}N$, there exists $x^i \in N$ such that $f(x^i, x) \notin -E$ $(f(x, x^i) \notin -E$, resp.).

(b) Vector bifunction f is said to be PROPERLY E-QUASIMONOTONE OF TYPE 2 in the first component (in the second component, resp.) on A if, for all $N \in \mathcal{F}(A)$ and $x \in \text{conv}N$, there exists $x^i \in N$ such that $f(x^i, x) \in E$ $(f(x, x^i) \in E$, resp.).

For $\ell \in \{1, 2\}$ and $i \in \{1, 2\}$, by $\mathcal{V}_E^\ell \mathcal{A}^i_{pr.qm}$, we denote the class of all vector bifunctions defined on $A \times A$ which are properly E-quasimonotone of type ℓ in the ith component.

Remark 3.37

(a) For $E \subset Y$, we set

$$\mathcal{V}_E^1 \mathcal{A}_A^{\geq} = \{f : A \times A \to Y \mid f(x, x) \notin -E \text{ for all } x \in A\},$$

$$\mathcal{V}_E^2 \mathcal{A}_A^{\geq} = \{f : A \times A \to Y \mid f(x, x) \in E \text{ for all } x \in A\}.$$

For $\ell \in \{1, 2\}$ and $i \in \{1, 2\}$, if $f \in \mathcal{V}_E^\ell \mathcal{A}^i_{pr.qm}$, then $f \in \mathcal{V}_E^\ell \mathcal{A}_A^{\geq}$.

(b) It is obvious that if $E \cap (-E) = \emptyset$ and $f \in \mathcal{V}_E^2 \mathcal{A}^i_{pr.qm}$, then $f \in \mathcal{V}_E^1 \mathcal{A}^i_{pr.qm}$.

(c) $f \in \mathcal{V}_E^\ell \mathcal{A}^i_{pr.qm} \Leftrightarrow \Phi^\ell_{Ef} \in \mathcal{A}^i_{pr.qm}$, i.e., f is properly E-quasimonotone of type ℓ in the ith component, if and only if the bifunction Φ^ℓ_{Ef}, defined by (3.14) and (3.15) for $\ell = 1$ and $\ell = 2$, respectively, is properly quasimonotone in the ith component (see Definition 1.27).

In general, for given $E \subset Y$, the notions of proper E-quasimonotonicity are different. This is verified in the following example.

Example 3.38 Let $A = [0, 1]$, $Y = \mathbb{R}^2$ and $E = \mathbb{R}^2_+$.

(a) Let $f : [0, 1] \times [0, 1] \to \mathbb{R}^2$ be defined by

$$f(x, y) := (x - y, x) \quad \text{for all } x, y \in [0, 1].$$

Then, for all $N \in \mathscr{F}([0, 1])$ and $x \in \text{conv}N$, there exists $x^i = \min N \in N$ satisfying $f(x, x^i) = (x - x^i, x) \in \mathbb{R}^2_+$. Hence, f is properly $\mathbb{R}^2_+$-quasimonotone of type 2 in the second component. f is not properly $\mathbb{R}^2_+$-quasimonotone of type 1 in the ith component because $f(0, 0) = (0, 0) \in -\mathbb{R}^2_+$.

(b) Let $f : [0, 1] \times [0, 1] \to \mathbb{R}^2$ be defined by

$$f(x, y) := (x - y + 2, -x) \quad \text{for all } x, y \in [0, 1].$$

Then, f is properly $\mathbb{R}^2_+$-quasimonotone of type 1 in the second component because for any $N \in \mathscr{F}([0, 1])$, $f(x, x^i) = (x - x^i + 2, -x) \notin -\mathbb{R}^2_+$ with $x \in \text{conv}N$ and $x^i = \min N \in N$. f is not properly $\mathbb{R}^2_+$-quasimonotone of type 2 in the ith component because $f(1, 1) = (2, -1) \notin \mathbb{R}^2_+$.

In the following definitions, we extend some concepts in Definitions 1.19, 1.21, and 1.25 to the vector case.

Definition 3.39 Let $E \subset Y$ be a nonempty set and $h : A \to Y$ be a vector function.

(a) h is called E-CONVEX if for all $x, y \in A$ and $t \in [0, 1]$, $h(tx + (1 - t)y) \in th(x) + (1 - t)h(y) - E$.
(b) h is called PROPERLY E-QUASICONVEX if, for all $x, y \in A$ and $t \in [0, 1]$, either $h(tx + (1 - t)y) \in h(x) - E$ or $h(tx + (1 - t)y) \in h(y) - E$.
(c) h is called E-QUASICONVEX if, for all $u \in Y$, and for all $x, y \in A$, from $\{h(x), h(y)\} \subset u - E$ it follows that $h(tx + (1 - t)y) \in u - E$ for all $t \in [0, 1]$. Thus, h is E-quasiconvex if and only if for all $u \in Y$, the set $\{x \in A \mid h(x) \in u - E\}$ is convex.
(d) h is called EXPLICITLY E-QUASICONVEX if it is E-quasiconvex and for all $x, y \in A$ such that $h(x) - h(y) \in -\text{core}E$, we have $h(tx + (1 - t)y) - h(y) \in -\text{core}E$ for all $t \in [0, 1]$.

If $-h$ is E-convex (properly E-quasiconvex, E-quasiconvex, explicitly E-quasiconvex), then we say that h is E-concave (properly E-quasiconcave, E-quasiconcave, explicitly E-quasiconcave).

When $Y = \mathbb{R}$ and $E = \mathbb{R}_+$, the $\mathbb{R}_+$-convexity (explicitly $\mathbb{R}_+$-quasiconvexity, resp.) is the convexity (explicitly quasiconvexity resp.) in Definition 1.19, and both the properly $\mathbb{R}_+$-quasiconvexity and the $\mathbb{R}_+$-quasiconvexity are the quasiconvexity in Definition 1.19.

Unlike the scalar case, some concepts in Definition 3.39 are not comparable in general.

Example 3.40 Let $A = [a, b]$ $(0 < a < 1, b > 4)$, $Y = \mathbb{R}^2$ and $h : [a, b] \to \mathbb{R}^2$ be defined by

$$h(x) := \left(x, \frac{1}{x}\right) \quad \text{for all } x \in [a, b].$$

(a) Let $E = \mathbb{R}^2_+$. Because $h(tx + (1 - t)y) - (th(x) + (1 - t)h(y)) = (0, -\frac{t(1-t)(x-y)^2}{xy(tx+(1-t)y)}) \in -\mathbb{R}^2_+$ for all $x, y \in [a, b]$ and $t \in [0, 1]$, h is $\mathbb{R}^2_+$-convex. For $x = 3$, $y = 1$, $t = \frac{1}{2}$ we see that $h(tx+(1-t)y)-h(x) = (-1, \frac{1}{6}) \notin -\mathbb{R}^2_+$ and $h(tx + (1 - t)y) - h(y) = (1, -\frac{1}{2}) \notin -\mathbb{R}^2_+$. Hence, h is not properly $\mathbb{R}^2_+$-quasiconvex.

(b) Let $E = \{(e_1, e_2) \in \mathbb{R}^2 \mid e_1 \geq 0, e_2 \leq 0\}$. For $x = 3$, $y = 1$, $t = \frac{1}{2}$, $h(tx + (1 - t)y) - (th(x) + (1 - t)h(y)) = (0, -\frac{1}{6}) \notin -E$. Hence, h is not E-convex. For all $x, y \in [a, b]$ and $t \in [0, 1]$, $h(tx + (1 - t)y) - h(x) = (-(1 - t)(x - y), \frac{(1-t)(x-y)}{x(tx+(1-t)y)}) \in -E$ if $x \geq y$, and $h(tx + (1 - t)y) - h(y) = (t(x-y), \frac{-t(x-y)}{y(tx+(1-t)y)}) \in -E$ if $x \leq y$. Therefore, h is properly E-quasiconvex.

(c) Let $E = \{(e_1, e_2) \in \mathbb{R}^2 \mid e_1e_2 \leq 0\}$. Then, $E = -E$. It is not hard to see that h is both E-convex and properly E-quasiconvex. For $u = (2, \frac{1}{4})$, $x = \frac{a+1}{2}$, $= \frac{b+1}{2}$ and $t \in (\frac{b-3}{b-a}, \frac{b-1}{b-a}) \subset [0, 1]$, we see that $\{h(x), h(y)\} \subset u - E$ but $h(tx + (1 - t)y) \notin u - E$. Hence, h is not E-quasiconvex.

Proposition 3.41 *Let $E \subset Y$ be a nonempty set and $h : A \to Y$ be a vector function. Then:*

(a) *If E is convex and h is E-convex, then h is E-quasiconvex.*
(b) *If $E \in \mathscr{E}(Y)$ and h is properly E-quasiconvex, then h is E-quasiconvex.*

Proof To prove (a) and (b) we take any $u \in Y$, $x, y \in A$ such that $\{h(x), h(y)\} \subset u - E$ and show that $h(tx + (1 - t)y) \in u - E$ for all $t \in 0, 1]$.

(a) For all $t \in [0, 1]$, by the convexity of E and the E-convexity of h, $h(tx + (1 - t)y) \in th(x) + (1 - t)h(y) - E \subset t(u - E) + (1 - t)(u - E) = u - E$. Thus, h is E-quasiconvex.

(b) For all $t \in [0, 1]$, since $E + E \subset E$ and h is properly E-convex, we have $h(tx+(1-t)y) \in (h(x)-E)\cup(h(y)-E) \subset (u-E-E)\cup(u-E-E) \subset u-E$. Thus, h is E-quasiconvex.

□

Definition 3.42 Let $E \subset Y$ be a nonempty set and $h : A \to Y$ be a vector function.

(a) h is called E-QUASICONVEX-LIKE OF TYPE 1 if, for all $x, y \in A$, from $\{h(x), h(y)\} \subset -E$ it follows that $h(tx + (1 - t)y) \in -E$ for all $t \in [0, 1]$. Thus, h is E-quasiconvex of type 1, if and only if the set $\{x \in A \mid h(x) \in -E\}$ is convex.

(b) h is called E-QUASICONVEX-LIKE OF TYPE 2 if, for all $x, y \in A$, from $\{h(x), h(y)\} \subset Y \setminus E$ it follows that $h(tx + (1 - t)y) \notin E$ for all $t \in [0, 1]$. Thus, h is E-quasiconvex of type 2, if and only if the set $\{x \in A \mid h(x) \notin E\}$ is convex.

When $Y = \mathbb{R}$ and $E = \mathbb{R}_+$, the $\mathbb{R}_+$-quasiconvexity-like of type 1 is the quasiconvexity-like in Definition 1.21, and the $\mathbb{R}_+$-quasiconvexity-like of type 2 is the strongly quasiconvexity-like in Definition 1.21. It is clear that *if h is E-quasiconvex, then h is E-quasiconvex-like of type 1.*

For a vector bifunction $f : A \times A \to Y$, we say that f is E-quasiconvex-like of type ℓ ($\ell \in \{1, 2\}$) in the first (second, resp.) component if for all $y \in A$ (for all $x \in A$, resp.), $f(\cdot, y)$ ($f(x, \cdot)$, resp.) is E-quasiconvex-like of type ℓ. For $\ell \in \{1, 2\}$ and $i \in \{1, 2\}$, by $\mathscr{V}_E^\ell \mathscr{A}_{qcl}^i$, we denote the class of all vector bifunctions defined on $A \times A$ which are E-quasimonotone-like of type ℓ in the ith component.

Definition 3.43 Let $E \subset Y$ be a nonempty set such that $\text{core}E \neq \emptyset$.

(a) Vector bifunction f is called E-MONOTONE if, for each $x, y \in A$, we have $f(x, y) + f(y, x) \in -E$.
(b) Vector bifunction f is called E-PSEUDOMONOTONE if for each $x, y \in A$, $f(x, y) \in E$ implies $f(y, x) \in -E$, or equivalently, $f(y, x) \notin -E$ implies $f(y, x) \notin E$.
(c) Vector bifunction Φ is called E-QUASIMONOTONE if for each $x, y \in A$, $f(x, y) \in \text{core}E$ implies $f(y, x) \in -E$.

The class of all vector bifunctions defined on $A \times A$ which are E-monotone (E-pseudomonotone, E-quasimonotone, resp.) is denoted by $\mathscr{V}_E\mathscr{A}_m$ ($\mathscr{V}_E\mathscr{A}_{pm}, \mathscr{V}_E\mathscr{A}_{qm}$, resp.). When $Y = \mathbb{R}$ and $E = \mathbb{R}_+$, the $\mathbb{R}_+$-monotonicity ($\mathbb{R}_+$-pseudomonotonicity, $\mathbb{R}_+$-quasimonotonicity, resp.) is the mononicity (pseudomonotonicity, quasimonotonicity, resp.) in Definition 1.25.

From now on, for a set $\Omega \subset Y$ (e.g., $\Omega = E, \text{core}E, \text{vcl}E, D, C, \ldots$), a property $\mathscr{R}$ (e.g., $\mathscr{R} = qcl, pr.qm, m, pm, qm, \ldots$), $\ell \in \{1, 2\}$ and $i \in \{1, 2\}$, we use $\mathscr{V}_Q\mathscr{A}_\mathscr{R}$ ($\mathscr{V}_Q^\ell\mathscr{A}_\mathscr{R}^i$, resp.) to denote the class of all vector bifunctions defined on $A \times A$ which are Ω-$\mathscr{R}$ (Ω-$\mathscr{R}$ of type ℓ in the ith, resp.). For a class of vector bifunctions $\mathscr{V}\mathscr{A}$, we define $-\mathscr{V}\mathscr{A} := \{f \mid -f \in \mathscr{V}\mathscr{A}\}$.

The following proposition extends Proposition 1.29 to the vector case.

Proposition 3.44 *For $E \subset Y$, $\ell \in \{1, 2\}$ and $i \in \{1, 2\}$, the following inclusions always hold:*

(a) $\mathscr{V}_E^\ell\mathscr{A}_A^{\geqq} \cap \mathscr{V}_E^\ell\mathscr{A}_{qcl}^i \subset \mathscr{V}_E^\ell\mathscr{A}_{pr.qm}^i \subset \mathscr{V}_E^\ell\mathscr{A}_A^{\geqq}$.
(b) $(-\mathscr{V}_E\mathscr{A}_{qm}) \cap \mathscr{V}_{\text{core}E}^1\mathscr{A}_{qcl}^i \subset \mathscr{V}_{\text{core}E}^1\mathscr{A}_{pr.qm}^i$ *if* $E \cap (-\text{core}E) = \emptyset$.
(c) $(-\mathscr{V}_E^\ell\mathscr{A}_A^{\geqq}) \cap (-\mathscr{V}_E\mathscr{A}_{pm}) \cap (-\mathscr{V}_E^\ell\mathscr{A}_{qcl}^1) \subset \mathscr{V}_E^\ell\mathscr{A}_{pr.qm}^2$.
(d) $\mathscr{V}_E^\ell\mathscr{A}_A^{\geqq} \cap \mathscr{V}_E\mathscr{A}_{pm} \cap \mathscr{V}_E^\ell\mathscr{A}_{qcl}^2 \subset -\mathscr{V}_E^\ell\mathscr{A}_{pr.qm}^1$.
(e) $(-\mathscr{V}_E^1\mathscr{A}_A^{\geqq}) \cap (-\mathscr{V}_E\mathscr{A}_{qm}) \cap (-\mathscr{V}_E^1\mathscr{A}_{qcl}^1) \subset \mathscr{V}_{\text{core}E}^1\mathscr{A}_{pr.qm}^2$.
(f) $(-\mathscr{V}_{\text{core}E}^2\mathscr{A}_A^{\geqq}) \cap (-\mathscr{V}_E\mathscr{A}_{qm}) \cap (-\mathscr{V}_{\text{core}}^2\mathscr{A}_{qcl}^1) \subset \mathscr{V}_E^2\mathscr{A}_{pr.qm}^2$.

Proof

(a) We prove this for $\ell = 1$, $i = 2$; the remaining cases can be proved similarly. Let $f \in \mathscr{V}_E^\ell\mathscr{A}_A^{\geqq} \cap \mathscr{V}_E^\ell\mathscr{A}_{qcl}^i$. Assume that $f \notin \mathscr{V}_E^1\mathscr{A}_{pr.qm}^2$. Then, there exist $N \in \mathscr{F}(A)$ and $x \in \text{conv}N$ such that $f(x, x^j) \in -E$ for all $x^j \in N$. Since $f \in$

$\mathscr{V}_E^1\mathscr{A}_{qcl}^2$, we have $f(x,x) \in -E$, which is impossible because $f \in \mathscr{V}_E^1\mathscr{A}_A^{\geq}$. The inclusion $\mathscr{V}_E^\ell\mathscr{A}_{pr.qm}^i \subset \mathscr{V}_E^\ell\mathscr{A}_A^{\geq}$ is obvious.

(b) We prove this for $i = 2$; the case $i = 2$ can be proved similarly. Let $f \in (-\mathscr{V}_E\mathscr{A}_{qm}) \cap \mathscr{V}_{\text{core}E}^1\mathscr{A}_{qcl}^2$. Suppose that $f \notin \mathscr{V}_{\text{core}E}^1\mathscr{A}_{pr.qm}^2$, i.e., there exist $N \in \mathscr{F}(A)$ and $x \in \text{conv}N$ such that $f(x, x^j) \in -\text{core}E$ for all $x^j \in N$. As $f \in \mathscr{V}_{\text{core}E}^1\mathscr{A}_{qcl}^2$, $f(x,x) \in -\text{core}E$, i.e., $-f(x,x) \in \text{core}E$. This implies that $-f(x,x) \in -E$, i.e., $f(x,x) \in E$ (Since $f \in -\mathscr{V}_E\mathscr{A}_{qm}$). Hence, $f(x,x) \in E \cap (-\text{core}E)$ which is impossible because $E \cap (-\text{core}E) = \emptyset$.

(c) We prove this for $\ell = 1$; the case $\ell = 2$ can be proved similarly. Let $f \in (-\mathscr{V}_E^1\mathscr{A}_A^{\geq}) \cap (-\mathscr{V}_E\mathscr{A}_{pm}) \cap (-\mathscr{V}_E^1\mathscr{A}_{qcl}^1)$ and suppose that $f \notin \mathscr{V}_E^1\mathscr{A}_{pr.qm}^2$. Then, there exist $N \in \mathscr{F}(A)$ and $x \in \text{conv}N$ such that $f(x, x^j) \in -E$ for all $x^j \in N$, i.e., $-f(x, x^j) \in E$ for all $x^j \in N$. Since $f \in -\mathscr{V}_E\mathscr{A}_{pm}$, it follows that $-f(x, x^j) \in -E$ for all $x^j \in N$. Because $f \in -\mathscr{V}_E^1\mathscr{A}_{qcl}^1$, we have that $-f(x,x) \in -E$. This contradicts the fact that $f \in -\mathscr{V}_E^1\mathscr{A}_A^{\geq}$ (i.e., $-f(x,x) \notin -E$).

The proofs of (d), (e), and (f) are similar to that of (c).

□

Lemma 3.45 *Let $E \subset Y$ be nonempty and $\ell \in \{1,2\}$. If $f \in \mathscr{V}_E^\ell\mathscr{A}_{pr.qm}^2$ and $\text{lev}_E^\ell f(\cdot, y)$ is closed for each $y \in A$, then f has the E-finite intersection property of type ℓ.*

Proof Consider the bifunction Φ_{Ef}^ℓ (defined by (3.14) and (3.15) for $\ell = 1$ and $\ell = 2$, respectively). Because $f \in \mathscr{V}_E^\ell\mathscr{A}_{pr.qm}^2$, $\Phi_{Ef}^\ell \in \mathscr{A}_{pr.qm}^2$ (see Remark 3.37). For each $y \in A$, $\text{lev}_{\geq}\Phi_{Ef}^\ell(\cdot, y) = \text{lev}_E^\ell f(\cdot, y)$ is closed. Thus, all the assumptions of Lemma 1.31 are satisfied for Φ_{Ef}^ℓ, and hence by this lemma, the problem $(\text{EP}_{\Phi_{Ef}^\ell})$ is finitely solvable. Finally, the conclusion follows from Remark 3.32. □

By combining Lemma 3.45 and Propositions 3.34 and 3.35, we state the following result in terms of proper E-quasimonotonicities.

Theorem 3.46 *Let $t \in \{He, Be, e, w\}$, $E \subset Y$ be nonempty, and $\ell \in \{1,2\}$. Assume that $E \in \mathscr{E}_t^\ell(C)$, $f \in \mathscr{V}_E^\ell\mathscr{A}_{pr.qm}^2$, $\text{lev}_E^\ell f(\cdot, y)$ is closed for each $y \in A$, and at least one of the following conditions is satisfied:*

(a) *A is compact.*
(b) *There exists $M \in \mathscr{F}(A)$ such that $\bigcup_{y\in A} \overline{\text{lev}}_E^\ell f(\cdot, y) \subset \bigcup_{y\in M} \overline{\text{lev}}_E^\ell f(\cdot, y)$.*

Then, $\text{Sol}^t(\text{VEP}_{(f,C)}) \neq \emptyset$.

Remark 3.47 From the proof of Lemma 3.45 we see that conditions of Theorem 3.46 also ensure that $\text{Sol}(\text{EP}_{\Phi_{Ef}^\ell}) \neq \emptyset$.

Coercivity Conditions for Theorem 3.46 We give the following coercivity conditions:

(C_1^{1E}) There exists a nonempty compact subset K of X such that, for each finite subset N of A, there exists a compact convex subset L_N of A containing N such that for each $x \in L_N \backslash K$, there exists $y \in L_N$ such that $f(x, y) \in -E$.

(C_1^{2E}) There exists a nonempty compact subset K of X such that, for each finite subset N of A, there exists a compact convex subset L_N of A containing N such that for each $x \in L_N \backslash K$, there exists $y \in L_N$ such that $f(x, y) \notin E$.

For $\ell \in \{1, 2\}$, it is not hard to see that the coercivity condition (C_1) in Theorem 1.34 is satisfied for Φ^{ℓ}_{Ef} if the above condition ($C_1^{\ell E}$) is satisfied. Therefore, as conditions of Lemma 3.45 and ($C_1^{\ell E}$) hold, all the conditions of Theorem 1.34 are satisfied for Φ^{ℓ}_{Ef} (see the proof of Lemma 3.45) and hence $\mathrm{Sol}(\mathrm{EP}_{\Phi^{\ell}_{Ef}}) \neq \emptyset$. Moreover, from Proposition 3.29 we see that when $E \in \mathscr{E}^1_t(C)$ ($t \in \{He, Be, e, w\}$), the nonemptiness of $\mathrm{Sol}(\mathrm{EP}_{\Phi^{\ell}_{Ef}})$ implies that $\mathrm{Sol}^t(\mathrm{VEP}_{(f,C)}) \neq \emptyset$. Thus, the compactness of A in the condition (a) of Theorem 3.46 can be replaced by the above coercivity condition ($C_1^{\ell E}$).

When A is compact, conditions (C_1^{1E}) and (C_1^{2E}) are automatically satisfied. Now, we give the following conditions:

(C_{11}^{1E}) There exist a nonempty compact subset K of X and a $\bar{y} \in K \cap A$ such that, for every $x \in A \backslash K$, $f(x, \bar{y}) \in -E$.

(C_{12}^{1E}) There exist a nonempty compact subset K of X and a finite subset $\bar{N}$ of A such that, for each $x \in A \backslash K$, $f(x, y) \in -E$ for some $y \in \bar{N}$.

(C_{13}^{1E}) There exists a nonempty compact convex subset K of A such that, for each $x \in A \backslash K$, $f(x, y) \in -E$ for some $y \in K$.

(C_{11}^{2E}) There exist a nonempty compact subset K of X and a $\bar{y} \in K \cap A$ such that, for every $x \in A \backslash K$, $f(x, \bar{y}) \notin E$.

(C_{12}^{2E}) There exist a nonempty compact subset K of X and a finite subset $\bar{N}$ of A such that, for each $x \in A \backslash K$, $f(x, y) \notin E$ for some $y \in \bar{N}$.

(C_{13}^{2E}) There exists a nonempty compact convex subset K of A such that, for each $x \in A \backslash K$, $f(x, y) \notin E$ for some $y \in K$.

When A is a closed convex subset of a reflexive Banach space:

(C_{14}^{1E}) There exists $r > 0$ such that for all $x \in A$ with $\|x\| > r$, there is $y \in A$ with $\|y\| \leq r$ such that $f(x, y) \in -E$.

(C_{14}^{2E}) There exists $r > 0$ such that for all $x \in A$ with $\|x\| > r$, there is $y \in A$ with $\|y\| \leq r$ such that $f(x, y) \notin E$.

For $\ell \in \{1, 2\}$, we easily check that condition $(C_{11}^{\ell E})$ ($(C_{12}^{\ell E})$, $(C_{13}^{\ell E})$, $(C_{14}^{\ell E})$, resp.) is condition (C_{11}) ((C_{12}), (C_{13}), (C_{14}), resp.) in Chap. 1 for Φ_{Ef}^{ℓ}. Therefore, from the relationships between (C_{11}), (C_{12}), (C_{13}) and (C_{14}) which were mentioned in Chap. 1, we have

$$(C_{11}^{\ell E}) \Longrightarrow (C_{12}^{\ell E}) \Longrightarrow (C_{1}^{\ell E}) \text{ and } (C_{13}^{\ell E}) \Longrightarrow (C_{1}^{\ell E}).$$

Corollary 3.48 *Let $D \subset Y$ be a vectorially closed convex cone such that $D \cap (-D) = \{0\}$.*

(a) *Assume that $f \in \mathscr{V}_{\mathrm{core}D}^{1}\mathscr{A}_{pr.qm}^{2}$, the set $\mathrm{lev}_{\mathrm{core}D}^{1} f(\cdot, y)$ is closed for each $y \in A$, and the coercivity condition $(C_1^{1\mathrm{coreD}})$ holds. If $D \in \mathscr{K}_C^{core}$, then $\mathrm{Sol}^t(\mathrm{VEP}_{(f,C)}) \neq \emptyset$ for all $t \in \{He, Be, e, w\}$, and if $D = C$, then $\mathrm{Sol}^w(\mathrm{VEP}_{(f,C)}) \neq \emptyset$.*
(b) *Assume that $f \in \mathscr{V}_{D}^{1}\mathscr{A}_{pr.qm}^{2}$, the set $\mathrm{lev}_{D}^{1} f(\cdot, y)$ is closed for each $y \in A$, and the coercivity condition (C_1^{1D}) holds. If $D \in \mathscr{K}_C^{core}$, then $\mathrm{Sol}^t(\mathrm{VEP}_{(f,C)}) \neq \emptyset$ for all $t \in \{He, Be, e, w\}$, and if $D = C$, then $\mathrm{Sol}^t(\mathrm{VEP}_{(f,C)}) \neq \emptyset$ for $t \in \{e, w\}$.*
(c) *Assume that $f \in \mathscr{V}_{D}^{2}\mathscr{A}_{pr.qm}^{2}$, the set $\mathrm{lev}_{D}^{2} f(\cdot, y)$ is closed for each $y \in A$, and the coercivity condition (C_1^{2D}) holds. If $D \in \mathscr{K}_C^{core}$, then $\mathrm{Sol}^t(\mathrm{VEP}_{(f,C)}) \neq \emptyset$ for all $t \in \{He, Be, e, w\}$, and if $D = C$, then $\mathrm{Sol}^t(\mathrm{VEP}_{(f,C)}) \neq \emptyset$ for $t \in \{Be, e, w\}$.*
(d) *Assume that $f \in \mathscr{V}_{\mathrm{core}D}^{2}\mathscr{A}_{pr.qm}^{2}$, the set $\mathrm{lev}_{\mathrm{core}D}^{2} f(\cdot, y)$ is closed for each $y \in A$, and the coercivity condition $(C_1^{2\mathrm{core}D})$ holds. If $D \in \mathscr{K}_C^{core}$, then $\mathrm{Sol}^t(\mathrm{VEP}_{(f,C)}) \neq \emptyset$ for all $t \in \{He, Be, e, w\}$, and if $D = C$, then $\mathrm{Sol}^t(\mathrm{VEP}_{(f,C)}) \neq \emptyset$ for $t \in \{Be, e, w\}$.*

Proof Corollary 3.48 directly follows from Theorem 3.46 and Remark 3.28 for the cases where $E = D$ and $E = \mathrm{core}D$. □

The sufficient conditions in Corollary 3.48 are different. The following example provides four situations, in each of which only one of the conditions of Corollary 3.48 is satisfied.

Example 3.49 Let $A = [a, b]$ with $a > 0$ and $b > 17a$, $Y = \mathbb{R}^2$, $C \subset D = \mathbb{R}^2_+$. Since A is compact, any $f : A \times A \to \mathbb{R}^2$ satisfies the above coercivity conditions.

(a) Let f be defined for all $x, y \in A$ by

$$f(x, y) := \begin{cases} (-x, -y) \text{ if } x < y, \\ (-xy, 0) \text{ if otherwise.} \end{cases}$$

Then, for all $N \in \mathscr{F}(A)$ and $x \in \mathrm{conv}N$, there exists $x^i \in N$ with $x^i = \min\{x^j, x^j \in N\}$ satisfying $f(x, x^i) = (-xx^i, 0) \notin -\mathrm{core}\mathbb{R}^2_+$. Hence, $f \in \mathscr{V}_{\mathrm{core}\mathbb{R}^2_+}^{1}\mathscr{A}_{pr.qm}^{2}$. For each $y \in A$, we have $\mathrm{lev}_{\mathrm{core}\mathbb{R}^2_+}^{1} f(\cdot, y) =$

$[y, b]$ is closed in A. Thus, all the assumptions of part (a) are fulfilled. Because $f(x, x) = (-2x^2, 0) \in -\mathbb{R}^2_+ \setminus \{(0, 0)\}$, $f \notin (\mathscr{V}^1_{\mathbb{R}^2_+}\mathscr{A}^2_{pr.qm}) \cup (\mathscr{V}^2_{\mathbb{R}^2_+}\mathscr{A}^2_{pr.qm}) \cup (\mathscr{V}^2_{\text{core}\mathbb{R}^2_+}\mathscr{A}^2_{pr.qm})$. This means that the assumption of the vector-proper quasimonotonicity in parts (b), (c), and (d) does not hold.

(b) Let f be defined for all $x, y \in A$ by

$$f(x, y) := \begin{cases} (-x, -y) & \text{if } x \leq \frac{1}{7}y, \\ (-xy, 0) & \text{if } \frac{1}{7}y < x < y, \\ (-y, x) & \text{if } y \leq x. \end{cases}$$

Then, $f \in \mathscr{V}^1_{\mathbb{R}^2_+}\mathscr{A}^2_{pr.qm}$ because for any $N \in \mathscr{F}(A)$, $f(x, x^i) = (-x^i, x) \notin -\mathbb{R}^2_+$ with $x \in \text{conv}N$ and $x^i = \min\{x^j, x^j \in N\}$. For each $y \in A$, $\text{lev}^1_{\mathbb{R}^2_+} f(\cdot, y) = [y, b]$ is closed. Thus, all the assumptions of part (b) are satisfied. The assumption of the closedness in part (a) does not hold because $\text{lev}^1_{\text{core}\mathbb{R}^2_+} f(\cdot, y) = (\frac{1}{7}y, b]$ is not closed for all $y \in (7a, b]$. Since $f(x, x) = (-x, x) \notin \mathbb{R}^2_+$, $f \notin (\mathscr{V}^2_{\mathbb{R}^2_+}\mathscr{A}^2_{pr.qm}) \cup (\mathscr{V}^2_{\text{core}\mathbb{R}^2_+}\mathscr{A}^2_{pr.qm})$. Hence, the assumption of the vector-proper quasimonotonicity in parts (c) and (d) do not hold.

(c) Let f be defined for all $x, y \in A$ by

$$f(x, y) := \begin{cases} (-x, -y) & \text{if } x \leq \frac{1}{17}y, \\ (x, -y) & \text{if } \frac{1}{17}y < x < y, \\ (0, y) & \text{if } y \leq x. \end{cases}$$

For this f, both parts (a) and (b) are not applicable because $\text{lev}^1_{\text{core}\mathbb{R}^2_+} f(\cdot, y) = \text{lev}^1_{\mathbb{R}^2_+} f(\cdot, y) = (\frac{1}{17}y, b]$ is not closed for all $y \in (17a, b]$. Part (d) is not applicable because $f \notin \mathscr{V}^2_{\text{core}\mathbb{R}^2_+}\mathscr{A}^2_{pr.qm}$ (since $f(x, x) = (0, x) \notin \text{core}\mathbb{R}^2_+$). Since for any $N \in \mathscr{F}(A)$, $f(x, x^i) = (0, x^i) \in \mathbb{R}^2_+$ with $x \in \text{conv}N$ and $x^i = \min\{x^j, x^j \in N\}$. Hence, $f \in \mathscr{V}^2_{\mathbb{R}^2_+}\mathscr{A}^2_{pr.qm}$. For each $y \in A$, $\text{lev}^2_{\mathbb{R}^2_+} f(\cdot, y) = [y, b]$ is closed. Thus, all the assumptions of part (c) are satisfied.

(d) Let f be defined for all $x, y \in A$ by

$$f(x, y) := \begin{cases} (-a, -b) & \text{if } x = a, y = b, \\ (x, y) & \text{if } y \leq x, \\ (0, y) & \text{otherwise.} \end{cases}$$

For $y = b$, $\text{lev}^1_{\text{core}\mathbb{R}^2_+} f(\cdot, b) = \text{lev}^1_{\mathbb{R}^2_+} f(\cdot, b) = \text{lev}^2_{\mathbb{R}^2_+} f(\cdot, b) = (a, b]$ is not closed. Thus, the assumptions of closedness in parts (a), (b), and (c) are not satisfied. For all $N \in \mathscr{F}(A)$ and $x \in \text{conv}N$, there exists $x^i = \min\{x^j, x^j \in N\}$

satisfying $f(x, x^i) = (x, x^i) \in \operatorname{core}\mathbb{R}^2_+$. Hence, $f \in \mathscr{V}^2_{\operatorname{core}\mathbb{R}^2_+}\mathscr{A}^2_{pr.qm}$. For each $y \in A$, $\operatorname{lev}^2_{\operatorname{core}\mathbb{R}^2_+} f(\cdot, y) = [y, b]$ is closed. Thus, all the assumptions of part (d) are fulfilled.

Corollary 3.50 *Let $t \in \{He, Be, e, w\}$, $E \subset Y$ be nonempty, and $\ell \in \{1, 2\}$. Assume that A is a closed convex subset of a reflexive Banach space, $E \in \mathscr{E}^\ell_t(C)$, $f \in \mathscr{V}^\ell_E\mathscr{A}^2_{pr.qm}$, $\operatorname{lev}^\ell_E f(\cdot, y)$ is weakly closed (in particular, $\operatorname{lev}^\ell_E f(\cdot, y)$ is convex and closed) for each $y \in A$, and the coercivity condition $(C^{\ell E}_{14})$ holds. Then,* $\operatorname{Sol}^t(\operatorname{VEP}_{(f,C)}) \neq \emptyset$.

Proof For $\ell \in \{1, 2\}$, when $(C^{\ell E}_{14})$ holds, $K = \{x \in A \mid \|x\| \leq r\}$ is weakly compact and convex, and the assumptions of Theorem 3.46 with the coercivity condition $(C^{\ell E}_{13})$ are satisfied in the weak topology. □

Definition 3.51 Let $E \subset Y$ be nonempty and $f : A \times A \to Y$ be a vector bifunction.

(a) f is said to HAVE THE E-DUAL PROPERTY OF TYPE 1 if for all $x, y \in A$, from $f((1-t)x + ty, x) \notin E$ for all $t \in (0, 1]$, it follows that $f(x, y) \notin -E$.
(b) f is said to HAVE THE E-DUAL PROPERTY OF TYPE 2 if for all $x, y \in A$, from $f((1-t)x + ty, x) \in -E$ for all $t \in (0, 1]$, it follows that $f(x, y) \in E$.

By definition, if f has the E-dual property of type 1 (of type 2, resp.), then, for all $x \in A$, $f(x, x) \notin E \Rightarrow f(x, x) \notin -E$ ($f(x, x) \in -E \Rightarrow f(x, x) \in E$, resp.).

Example 3.52

(a) Let $A = \mathbb{R}$, $Y = \mathbb{R}^2$, $E = \mathbb{R}^2_+$ and $f : A \times A \to \mathbb{R}^2$ be defined for all $x, y \in A$ by

$$f(x, y) := \begin{cases} (x - y, |y|) & \text{if } x \leq y < 0, \\ (x - y, y) & \text{otherwise.} \end{cases}$$

For all $x, y \in A$ and $t \in (0, 1]$, we have

$$f((1-t)x + ty, x) = \begin{cases} (t(y - x), |x|) & \text{if } y \leq x < 0, \\ (t(y - x), x) & \text{otherwise.} \end{cases}$$

Assume that $f((1-t)x + ty), x) \notin \mathbb{R}^2_+$ for all $t \in (0, 1]$. Then, either $y < x$, or both $x \leq y$ and $x < 0$. This and the definition of f imply that $f(x, y) \notin -\mathbb{R}^2_+$. Thus, f has the $\mathbb{R}^2_+$-dual property of type 1. For $\bar{x} = 0$ and $\bar{y} = -1$, $f((1-t)\bar{x} + t\bar{y}), \bar{x}) = (-t, 0) \in -\mathbb{R}^2_+$ for all $t \in (0, 1]$, but $f(\bar{x}, \bar{y}) = f(0, -1) = (1, -1) \notin \mathbb{R}^2_+$. Hence, f does not have the $\mathbb{R}^2_+$-dual property of type 2.

(b) Let $A = \mathbb{R}^2$, $Y = \mathbb{R}^2$, $E = \mathbb{R}^2_+$ and $f : A \times A \to \mathbb{R}^2$ be defined for all $x = (u_x, v_x), y = (u_y, v_y) \in A$ by

$$f(x, y) := \begin{cases} x - y & \text{if } x \in -\mathbb{R}^2_+, \\ (u_y - u_x, 1) & \text{otherwise.} \end{cases}$$

For all $x, y \in A$ and $t \in (0, 1]$, we have

$$f((1-t)x + ty, x) = \begin{cases} t(y - x) & \text{if } (1-t)x + ty \in -\mathbb{R}^2_+, \\ (t(u_x - u_y), 1) & \text{otherwise.} \end{cases}$$

Then, for all $x, y \in A$, from $f((1-t)x + ty, x) = t(y - x) \in -\mathbb{R}^2_+$ for all $t \in (0, 1]$, it follows that $y - x \in -\mathbb{R}^2_+$ and $(1-t)x + ty \in -\mathbb{R}^2_+$ for all $t \in (0, 1]$. This implies that $y - x \in -\mathbb{R}^2_+$ and $x \in -\mathbb{R}^2_+$, and hence $f(x, y) = x - y \in \mathbb{R}^2_+$. Thus, f has the $\mathbb{R}^2_+$-dual property of type 2. For $\bar{x} = (0, 0) \in -\mathbb{R}^2_+$ and $\bar{y} = (1, 1)$, $(1-t)\bar{x} + t\bar{y} \notin -\mathbb{R}^2_+$ for all $t \in (0, 1]$. Hence, $f((1-t)\bar{x} + t\bar{y}), \bar{x}) = (-t, 1) \notin \mathbb{R}^2_+$ for all $t \in (0, 1]$, but $f(\bar{x}, \bar{y}) = (-1, -1) \in -\mathbb{R}^2_+$. Thus, f does not have the $\mathbb{R}^2_+$-dual property of type 1.

Lemma 3.53 *Let $E \subset Y$ and $f : A \times A \to Y$. Let $f' : A \times A \to Y$ be defined by,*

$$f'(x, y) := -f(y, x) \text{ for all } x, y \in A. \tag{3.18}$$

If f has the E-dual property of type ℓ, then $\mathrm{Sol}(\mathrm{EP}_{\Phi^\ell_{Ef'}}) \subset \mathrm{Sol}(\mathrm{EP}_{\Phi^\ell_{Ef}})$.

Proof We prove this for $\ell = 1$; the case $\ell = 2$ can be proved similarly. Assume that $\bar{x} \in \mathrm{Sol}(\mathrm{EP}_{\Phi^1_{Ef'}})$. Let $y \in A$ be arbitrary. We have $f'(\bar{x}, y_t) \notin -E$ for all $y_t := (1-t)\bar{x} + ty \in (\bar{x}, y]$, or equivalently, $f(y_t, \bar{x}) \notin E$ for all $y_t \in (\bar{x}, y]$. By the E-dual property of type 1 of f, $f(\bar{x}, y) \notin -E$. Because $y \in A$ is arbitrary, $\bar{x} \in \mathrm{Sol}(\mathrm{EP}_{\Phi^1_{Ef}})$. □

Observe that for $E \subset Y, x \in A, \ell \in \{1, 2\}$ and f' defined by (3.18) we have

$$\overline{\mathrm{lev}}^{\ell c}_E f(x, \cdot) = \mathrm{lev}^\ell_E f'(\cdot, x) \text{ and } \mathrm{lev}^{\ell c}_E f(x, \cdot) = \overline{\mathrm{lev}}^\ell_E f'(\cdot, x).$$

Theorem 3.54 *Let $t \in \{He, Be, e, w\}$, $E \subset Y$ be nonempty, and $\ell \in \{1, 2\}$. Assume that $E \in \mathscr{E}^\ell_t(C)$, f has the E-dual property of type ℓ, $f \in -\mathscr{V}^\ell_E \mathscr{A}^1_{pr.qm}$, $\overline{\mathrm{lev}}^{\ell c}_E f(x, \cdot)$ is closed for each $x \in A$, and at least one of the following conditions is satisfied:*

(a) *A is compact.*
(b) *There exists $M \in \mathscr{F}(A)$ such that $\bigcup_{x \in A} \mathrm{lev}^{\ell c}_E f(x, \cdot) \subset \bigcup_{x \in M} \mathrm{lev}^{\ell c}_E f(x, \cdot)$.*

Then, $\mathrm{Sol}^t(\mathrm{VEP}_{(f,C)}) \neq \emptyset$.

Proof Let f' be defined by (3.18) (see Lemma 3.53). Then, for $\ell \in \{1, 2\}$, from the conditions in Theorem 3.54, it is not hard to see that the conditions of Theorem 3.46 are satisfied for f'. Hence, from Remark 3.47, we have that $\mathrm{Sol}(\mathrm{EP}_{\Phi^{\ell}_{Ef'}}) \neq \emptyset$. By Lemma 3.53 we conclude that $\mathrm{Sol}(\mathrm{EP}_{\Phi^{\ell}_{Ef}}) \neq \emptyset$. This and Proposition 3.29 give the conclusion. □

Coercivity Conditions for Theorem 3.54 By restating the coercivity conditions (C_1^{1E}) and (C_1^{2E}) for f' we obtain the following conditions (C_2^{1E}) and (C_2^{2E}) which can replace the compactness of A (in the condition (a) of Theorem 3.54) in the cases $\ell = 1$ and $\ell = 2$, respectively.

($\mathrm{C}_2^{1\mathrm{E}}$) There exists a nonempty compact subset K of A such that, for each finite subset N of A, there exists a compact convex subset L_N of A containing N such that for each $x \in L_N \backslash K$, $f(y, x) \in E$ for some $y \in L_N$.

($\mathrm{C}_2^{2\mathrm{E}}$) There exists a nonempty compact subset K of A such that, for each finite subset N of A, there exists a compact convex subset L_N of A containing N such that for each $x \in L_N \backslash K$, $f(y, x) \notin -E$ for some $y \in L_N$.

Consider the following coercivity conditions.

($\mathrm{C}_{21}^{1\mathrm{E}}$) There exist a nonempty compact subset K of X and $\bar{y} \in K \cap A$ such that, for every $x \in A \backslash K$, $f(\bar{y}, x) \in E$.

($\mathrm{C}_{22}^{1\mathrm{E}}$) There exist a nonempty compact subset K of X and a finite subset $\bar{N}$ of A such that, for each $x \in A \backslash K$, $f(y, x) \in E$ for some $y \in \bar{N}$.

($\mathrm{C}_{23}^{1\mathrm{E}}$) There exists a nonempty compact convex subset K of A such that, for each $x \in A \backslash K$, $f(y, x) \in E$ for some $y \in K$.

($\mathrm{C}_{21}^{2\mathrm{E}}$) There exist a nonempty compact subset K of X and a $\bar{y} \in K \cap A$ such that, for every $x \in A \backslash K$, $f(\bar{y}, x) \notin -E$.

($\mathrm{C}_{22}^{2\mathrm{E}}$) There exist a nonempty compact subset K of X and a finite subset $\bar{N}$ of A such that, for each $x \in A \backslash K$, $f(y, x) \notin -E$ for some $y \in \bar{N}$.

($\mathrm{C}_{23}^{2\mathrm{E}}$) There exists a nonempty compact convex subset K of A such that, for each $x \in A \backslash K$, $f(y, x) \notin -E$ for some $y \in K$.

When A is a closed convex subset of a reflexive Banach space:

($\mathrm{C}_{24}^{1\mathrm{E}}$) There exists $r > 0$ such that for all $x \in A$ with $\|x\| > r$, there is $y \in A$ with $\|y\| \leq r$ such that $f(y, x) \in E$.

($\mathrm{C}_{24}^{2\mathrm{E}}$) There exists $r > 0$ such that for all $x \in A$ with $\|x\| > r$, there is $y \in A$ with $\|y\| \leq r$ such that $f(y, x) \notin -E$.

It is not hard to check that

$$(\mathrm{C}_{21}^{\ell E}) \Longrightarrow (\mathrm{C}_{22}^{\ell E}) \Longrightarrow (\mathrm{C}_{2}^{\ell E}) \text{ and } (\mathrm{C}_{23}^{\ell E}) \Longrightarrow (\mathrm{C}_{2}^{\ell E}).$$

Corollary 3.55 *Let $D \subset Y$ be a vectorially closed convex cone such that $D \cap (-D) = \{0\}$.*

(a) *Assume that f has the* core$D$*-dual property of type 1, $f \in -\mathscr{V}^1_{\mathrm{core}D}\mathscr{A}^1_{pr.qm}$, the set $\overline{\mathrm{lev}}^2_{\mathrm{core}D} f(x,\cdot)$ is closed for each $x \in A$, and the coercivity condition ($C_2^{1\mathrm{core}}$D) holds. If $D \in \mathscr{K}_C^{core}$, then* $\mathrm{Sol}^t(\mathrm{VEP}_{(f,C)}) \neq \emptyset$ *for all $t \in \{He, Be, e, w\}$, and if $D = C$, then* $\mathrm{Sol}^w(\mathrm{VEP}_{(f,C)}) \neq \emptyset$.

(b) *Assume that f has the D-dual property of type 1, $f \in -\mathscr{V}^1_D\mathscr{A}^1_{pr.qm}$, the set $\overline{\mathrm{lev}}^2_D f(x,\cdot)$ is closed for each $x \in A$, and the coercivity condition (C_2^{1D}) holds. If $D \in \mathscr{K}_C^{core}$, then* $\mathrm{Sol}^t(\mathrm{VEP}_{(f,C)}) \neq \emptyset$ *for all $t \in \{He, Be, e, w\}$, and if $D = C$, then* $\mathrm{Sol}^t(\mathrm{VEP}_{(f,C)}) \neq \emptyset$ *for $t \in \{e, w\}$.*

(c) *Assume that f has the D-dual property of type 2, $f \in -\mathscr{V}^2_D\mathscr{A}^1_{pr.qm}$, the set $\overline{\mathrm{lev}}^1_D f(x,\cdot)$ is closed for each $x \in A$, and the coercivity condition (C_2^{2D}) holds. If $D \in \mathscr{K}_C^{core}$, then* $\mathrm{Sol}^t(\mathrm{VEP}_{(f,C)}) \neq \emptyset$ *for all $t \in \{He, Be, e, w\}$, and if $D = C$, then* $\mathrm{Sol}^t(\mathrm{VEP}_{(f,C)}) \neq \emptyset$ *for $t \in \{Be, e, w\}$.*

Proof Corollary 3.55 directly follows from Theorem 3.54 and Remark 3.28 for $E = D$ and $E = \mathrm{core}D$. □

Example 3.56 Let $A = [a, b]$ with $a > 0$ and $b > 7a$, $Y = \mathbb{R}^2$, $C \subset D = \mathbb{R}^2_+$. The coercivity conditions in Corollary 3.55 hold for any $f : A \times A \to \mathbb{R}^2$ because A is compact.

(a) Let f be defined for all $x, y \in A$ by

$$f(x,y) := \begin{cases} (x^2, y^2) & \text{if } x < y, \\ (-xy, 0) & \text{if otherwise.} \end{cases}$$

Then, f has the core$\mathbb{R}^2_+$-dual property of type 1 because $f(x, y) \notin -\mathrm{core}\mathbb{R}^2_+$ for all $x, y \in A$. For all $N \in \mathscr{F}(A)$ and $x \in \mathrm{conv}N$, there exists $x^i = \max\{x^j, x^j \in N\}$ satisfying $-f(x^i, x) = (x^i x, 0) \notin -\mathrm{core}\mathbb{R}^2_+$. Hence, $f \in -\mathscr{V}^1_{\mathrm{core}\mathbb{R}^2_+}\mathscr{A}^1_{pr.qm}$. For each $x \in A$, we have $\overline{\mathrm{lev}}^2_{\mathrm{core}\mathbb{R}^2_+} f(x,\cdot) = [a, x]$ is closed in A. Thus, the assumptions of part (a) are fulfilled. Those in (b) and (c) are not because f does not have both $\mathbb{R}^2_+$-dual property of type 1 and $\mathbb{R}^2_+$-dual property of type 2.

(b) Let f be defined for all $x, y \in A$ by

$$f(x,y) := \begin{cases} (-x, -y) & \text{if } x > y, \\ (x, -x) & \text{if } x = y, \\ (x, 0) & \text{if } x < y < 2x, \\ (x, y) & \text{if } 2x \le y. \end{cases}$$

Then, f has the $\mathbb{R}^2_+$-dual property of type 1. Indeed, for all $x, y \in A$, assume that $f((1-t)x + ty), x) \notin \mathbb{R}^2_+$ for all $t \in (0, 1]$. By the definition of f, $x \le (1-t)x + ty$ for all $t \in (0, 1]$, which implies that $x \le (1-t)x + ty \le y$.

Again by the definition of f,

$$f(x, y) = \begin{cases} (x, -x) & \text{if } x = y, \\ (x, 0) & \text{if } x < y < 2x, \\ (x, y) & \text{if } 2x \leq y, \end{cases}$$

$$\notin -\mathbb{R}^2_+.$$

For any $N \in \mathscr{F}(A)$ and $x \in \text{conv}N$, by taking $x^i = \max\{x^j, x^j \in N\}$ we have

$$-f(x^i, x) = \begin{cases} (x^i, x) & \text{if } x < x^i, \\ (-x, x) & \text{if } x = x^i \end{cases} \notin -\mathbb{R}^2_+.$$

Hence, $f \in -\mathscr{V}^1_{\mathbb{R}^2_+}\mathscr{A}^1_{pr.qm}$. For each $x \in A$, $\overline{\text{lev}}^2_{\mathbb{R}^2_+} f(x, \cdot) = [a, x]$ is closed. Thus, the conditions of the part (b) are satisfied. The one in (a) does not use because $\overline{\text{lev}}^2_{\text{core}\mathbb{R}^2_+} f(x, \cdot) = [a, 2x)$ can be not closed. Because $-f(x, x) = (-x, x) \notin \mathbb{R}^2_+$, $f \notin -\mathscr{V}^2_{\mathbb{R}^2_+}\mathscr{A}^1_{pr.qm}$. Thus, the assumption of the vector-proper quasimonotonicity in part (c) does not hold.

(c) Let f be defined for all $x, y \in A$ by

$$f(x, y) := \begin{cases} (-y, -x) & \text{if } x > y, \\ (0, 0) & \text{if } x = y, \\ (x, 0) & \text{if } x < y < 2x, \\ (y, x) & \text{if } 2x \leq y. \end{cases}$$

Part (a) is not applicable because $\text{lev}^1_{\text{core}\mathbb{R}^2_+} f(x, \cdot) = [a, 2x)$ can be not closed. Part (b) is not applicable because $f \notin -\mathscr{V}^1_{\mathbb{R}^2_+}\mathscr{A}^1_{pr.qm}$ (since $-f(x, x) = (0, 0) \in -\mathbb{R}^2_+$). For all $x, y \in A$, if $f((1-t)x + ty), x) \in -\mathbb{R}^2_+$ for all $t \in (0, 1]$, then, by the definition of f, $x \leq (1-t)x + ty$ for all $t \in (0, 1]$, which implies that $x \leq (1-t)x + ty \leq y$. Again by the definition of f,

$$f(x, y) = \begin{cases} (0, 0)) & \text{if } x = y, \\ (x, 0) & \text{if } x < y < 2x, \\ (y, x) & \text{if } 2x \leq y. \end{cases} \in \mathbb{R}^2_+.$$

Thus, f has the $\mathbb{R}^2_+$-dual property of type 2. For all $N \in \mathscr{F}(A)$ and $x \in \text{conv}N$, there exists $x^i = \max\{x^j, x^j \in N\}$ satisfying

$$-f(x^i, x) = \begin{cases} (x, x^i) & \text{if } x < x^i, \\ (0, 0) & \text{if } x = x^i \end{cases}$$

belonging to $\mathbb{R}^2_+$. Hence, $f \in -\mathscr{V}^2_{\mathbb{R}^2_+}\mathscr{A}^1_{pr.qm}$. For each $x \in A$, $\overline{\mathrm{lev}}^{1}_{\mathbb{R}^2_+} f(x, \cdot) = [a, x]$ is closed. Thus, all the assumptions of (c) are fulfilled.

Next, we provide some conditions for a vector bifunction to have the E-dual property.

Definition 3.57 Vector bifunction $f : A \times A \to Y$ is called SIGN-E-CONTINUOUS OF TYPE 1 IN THE FIRST COMPONENT at x if for all $y \in A$, $f(x+t(y-x), y) \notin -E$ for all $t \in (0, 1)$ implies that $f(x, y) \notin -E$. If in this implication "$\notin -E$" is replaced by "$\in E$," then f is called SIGN-E-CONTINUOUS OF TYPE 2 IN THE FIRST COMPONENT at x. Similar definitions can be stated for the second component.

By the definition we have the following: f is sign-E-continuous of type 1 (of type 2, resp.) in the first component at x if and only if f is sign-$(-E^c)$-continuous of type 2 (of type 1, resp.) in the first component at x.

Example 3.58 Let A be a convex subset of a topological vector space, $Y = \mathbb{R}^2$, and $\varphi : A \to \mathbb{R}$ be continuous.

(a) Let $E = \mathbb{R}^2_{++}$, $\gamma : A \to (-\infty, 0)$, and $f : A \times A \to \mathbb{R}^2$ be defined by

$$f(x, y) := (\varphi(x), \gamma(y)) \text{ for all } x, y \in A.$$

Fix an arbitrary $x \in A$. Take any $y \in A$ such that $f(x + t(y - x), y) = (\varphi(x + t(y - x)), \gamma(y)) \notin -E = -\mathbb{R}^2_{++}$ for all $t \in (0, 1)$. Since $\gamma(y) < 0$, it follows that $\varphi(x + t(y - x)) \geq 0$ for all $t \in (0, 1)$. As φ is continuous, $\varphi(x) = \lim_{t\to 0^+} \varphi(x+t(y-x)) \geq 0$. Hence, $f(x, y) = (\varphi(x), \gamma(y)) \notin -\mathbb{R}^2_{++}$. Thus, for all $x \in A$, f is sign-$\mathbb{R}^2_{++}$-continuous of type 1 in the first component at x.

(b) Let $E = \mathbb{R}^2_+$, $\xi : A \to \mathbb{R}$, and $f : A \times A \to \mathbb{R}^2$ be defined by

$$f(x, y) := (\varphi(x), \xi(y)) \text{ for all } x, y \in \mathbb{R}.$$

Consider any $x \in \mathbb{R}$. Assume that for $y \in \mathbb{R}$, $f(x + t(y - x), y) = (\varphi(x + t(y - x), \xi(y)) \in E = \mathbb{R}^2_+$ for all $t \in (0, 1)$. Then, $\varphi(x + t(y - x) \geq 0$ for all $t \in (0, 1)$, and $\xi(y) \geq 0$. It follows that $f(x, y) = (\varphi(x), \xi(y)) = (\lim_{t\to 0^+} \varphi(x + t(y - x)), \xi(y)) \in \mathbb{R}^2_+$. Thus, for all $x \in A$, f is sign-$\mathbb{R}^2_+$-continuous of type 2 in the first component at x.

Proposition 3.59 *Assume that Y is a topological vector space, $E \in \mathscr{E}(Y)$, and f is upper E-hemicontinuous in the first component. If E is convex and* core$E \neq \emptyset$, *then f is sign-*core$E$*-continuous of type 1 in the first component; if E is closed, then f is sign-E-continuous of type 2 in the first component.*

Proof Assume that f is upper E-hemicontinuous. If f is not sign-core$E$-continuous of type 1 in the first component at x, then there exists $y \in A$ such that

$f(x + t(y - x), y) \notin -\text{core}E$ for all $t \in (0, 1)$ and $f(x, y) \in -\text{core}E = -\text{int}E$. Since $t \mapsto f(x + t(y - x), y)$ is upper E-continuous at 0^+, there exists $\epsilon \in (0, 1)$ such that $f(x + t(y - x), y) \in -\text{core}E - E = -\text{core}E$ for all $t \in (0, \epsilon) \subset (0, 1)$, a contradiction.

If f is not sign-E-continuous of type 2 in the first component at x, then there exists $y \in A$ such that $f(x + t(y - x), y) \in E$ for all $t \in (0, 1)$ and $f(x, y) \notin E$. As E is closed and $t \longmapsto f(x + t(y - x), y)$ is upper E-continuous at 0^+, there exists $\epsilon \in (0, 1)$ such that $f(x + t(y - x), y) \in E^c - E$ for all $t \in (0, \epsilon)$. It follows that for each $t \in (0, \epsilon) \subset (0, 1)$, there exist $e_t \in E$ and $e'_t \in E^c$ such that $f(x + t(y - x), y) = e'_t - e_t$. Then, $e'_t = f(x + t(y - x), y) + e_t \in E + E \subset E$, a contradiction. □

Definition 3.60 A function $h : A \to Y$ is said to HAVE PROPERTY $\mathbb{P}_E$ OF TYPE 1 if for all $x, y \in A$, from $h(x) \notin E \cup (-E)$ and $h(y) \in -E$ it follows that $h((1 - t)x + ty) \in -E$ for all $t \in (0, 1)$. If in this implication " $\notin E \cup (-E)$" and " $\in -E$" are replaced by " $\in E \cap (-E)$" and " $\notin E$," respectively, then we say that h HAS PROPERTY $\mathbb{P}_E$ OF TYPE 2. Vector bifunction $f : A \times A \to Y$ is said to HAVE PROPERTY $\mathbb{P}^1_E$ ($\mathbb{P}^2_E$, resp.) OF TYPE ℓ ($\ell = 1, 2$) if $f(\cdot, y)$ ($f(x, \cdot)$, resp.) has property $\mathbb{P}_E$ of type ℓ for all $y \in A$ ($x \in A$, resp.).

Example 3.61

(a) Obviously, any $h : A \to Y$ such that either "$h(x) \notin -E$ for all $x \in A$" or "$h(x) \in -E$ for all $x \in A$" has property $\mathbb{P}_E$ of type 1. Any $h : A \to Y$ such that $h(x) \notin E \cap (-E)$ for all $x \in A$ has property $\mathbb{P}_E$ of type 2.
(b) Let $A = \mathbb{R}$, $Y = \mathbb{R}^2$, $E = \mathbb{R}^2_+$. Vector bifunction $f : A \times A \to \mathbb{R}^2$, defined by

$$f(x, y) := \begin{cases} (x, y - x) & \text{if } x \leq y, \\ (x, y^2) & \text{if } x > y, \end{cases} \quad \text{for all } x, y \in A,$$

has property $\mathbb{P}^2_E$ of type 1, while $f : A \times A \to \mathbb{R}^2$, defined by

$$f(x, y) := \begin{cases} (-x, y^2) & \text{if } y < 0, \\ (x, -y^2) & \text{if } y \geq 0, \end{cases} \quad \text{for all } x, y \in A,$$

has property $\mathbb{P}^2_E$ of type 2.

Proposition 3.62 *Let $E \subset Y$ and $\ell \in \{1, 2\}$. If $f \in \mathscr{V}^\ell_E \mathscr{A}^{\geq}_A \cap \mathscr{V}^\ell_E \mathscr{A}^2_{qcl}$ has property $\mathbb{P}^2_E$ of type ℓ and is sign E-continuous of type ℓ in the first argument, then f has the E-dual property of type ℓ.*

Proof Case $\ell = 1$. Take any $x, y \in A$. Assume that $f(y_t, x) \notin E$ for all $y_t := (1 - t)x + ty \in (x, y]$. Suppose that $f(y_{t_0}, y) \in -E$ for some $y_{t_0} \in (x, y]$. We consider two cases:

Case 1. If $f(y_{t_0}, x) \in -E$, then, by the E-quasiconvex-like property of $f(y_{t_0}, \cdot)$, $f(y_{t_0}, y_{t_0}) \in -E$, a contradiction.

Case 2. If $f(y_{t_0}, x) \notin -E$, then $f(y_{t_0}, x) \notin E \cup (-E)$. By the property $\mathbb{P}_E^2$ of type 1, $f(y_{t_0}, y_{t_0}) \in -E$, another contradiction.

Thus, we must have $f(y_t, y) \notin -E$ for all $y_t \in (x, y]$. It follows from the sign E-continuity of type 1 in the first component of f, $f(x, y) \notin -E$. Thus, f has the E-dual property of type 1.

Case $\ell = 2$. By replacing E in the case $\ell = 1$ by $E' := -E^c$, we obtain the statement for the case $\ell = 2$. □

Proposition 3.63 *Let $E \subset Y$ be a vectorially closed and solid convex cone. Assume that f is sign* core$E$*-continuous of type 1 in the first argument; $f(x, x) \in E$ and $f(x, \cdot)$ is explicitly E-quasiconvex for each $x \in A$. Then f has the* core$E$*-dual property of type 1.*

Proof Let $x, y \in A$ be such that $f(y_t, x) \notin \text{core}E$ for all $y_t := (1 - t)x + ty \in (x, y]$. Suppose that $f(y_{t_0}, y) \in -\text{core}E$ for some $y_{t_0} \in (x, y]$. We consider two cases:

Case 1. $f(y_{t_0}, x) \in E$. Then, $f(y_{t_0}, y) - f(y_{t_0}, x) \in -\text{core}E - E \subset -\text{core}E$. By the explicitly E-quasiconvexity of $f(y_{t_0}, \cdot)$, $f(y_{t_0}, y_{t_0}) - f(y_{t_0}, x) \in -\text{core}E$. Hence, $f(y_{t_0}, x) \in f(y_{t_0}, y_{t_0}) + \text{core}E \subset E + \text{core}E \subset \text{core}E$, which contradicts the assumption that $f(y_{t_0}, x) \notin \text{core}E$.

Case 2. $f(y_{t_0}, x) \notin E$. Then, by Proposition 3.8, there exists $e \in E^c$ such that $e - f(y_{t_0}, x) \in E$ and $e - f(y_{t_0}, y) \in E$. By the E-quasiconvexity of $f(y_{t_0}, \cdot)$, $e - f(y_{t_0}, y_{t_0}) \in E$, i.e., $e \in f(y_{t_0}, y_{t_0}) + E \subset E$. This contradicts the fact that $e \in E^c$.

Thus, $f(y_t, y) \notin -\text{core}E$ for all $y_t \in (x, y]$. By the sign core$E$-continuity of type 1 in the first component of f, we have $f(x, y) \notin -\text{core}E$. Thus, f has the core$E$-dual property of type 1. □

3.3.2 Existence Conditions Under Vector-Cyclic Quasimonotonicity

In this subsection, the data of the problem (VEP(f, C)) reads as follows:

$$\begin{cases} A \text{ is a nonempty set;} \\ C \subset Y \text{ is an (algebraically) solid set such that } 0 \notin \text{core}C \text{ and} \\ f : A \times A \to Y \text{ is a given vector bifunction.} \end{cases} \tag{3.19}$$

To provide sufficient conditions for the existence of general vector equilibria, we introduce some notions.

Definition 3.64 Let $E \subset Y$ be nonempty.

(a) f is called E-CYCLICALLY ANTI-QUASIMONOTONE OF TYPE 1 if for any $x^1, \ldots, x^m \in A$ and $x^{m+1} := x^1$, there exists an $i \in \{1, \ldots, m\}$ such that $f(x^i, x^{i+1}) \notin -E$.
(b) f is called E-CYCLICALLY ANTI-QUASIMONOTONE OF TYPE 2 if for any $x^1, \ldots, x^m \in A$ and $x^{m+1} := x^1$, there exists $i \in \{1, \ldots, m\}$ such that $f(x^i, x^{i+1}) \in E$.

For $\ell \in \{1, 2\}$, by $\mathscr{V}_E^\ell \mathscr{A}_{caqm}$, we denote the class of all E-cyclically anti-quasimonotone bifunctions of type ℓ.

Remark 3.65

(a) These concepts extend the cyclic quasimonotonicity proposed for scalar bifunctions in Chap. 1. When $Y = \mathbb{R}$, both the $\mathbb{R}_{++}$-cyclic anti-quasimonotonicity of type 1 and the $\mathbb{R}_+$-cyclic anti-quasimonotonicity of type 2 become the cyclic anti-quasimonotonicity in Definition 1.77.
(b) For the bifunctions Φ_{Ef}^1 and Φ_{Ef}^2 defined by (3.14) and (3.15), respectively, f is E-cyclically anti-quasimonotone of type i if and only if the bifunction Φ_{Ef}^i is cyclically anti-quasimonotone, i.e., $\Phi_{Ef}^i \in \mathscr{A}_{caqm}$ (see Definition 1.77).

By restating Proposition 1.79 for Φ_{Ef}^ℓ ($\ell \in \{1, 2\}$, we obtain the following proposition which provides a most basic characterization of the vector-cyclic quasimonotonicity.

Proposition 3.66 *$f \in \mathscr{V}_E^1 \mathscr{A}_{caqm}$ ($f \in \mathscr{V}_E^2 \mathscr{A}_{caqm}$, resp.), if and only if for any $x^1,\ldots, x^m \in A$, $m \geq 2$, satisfying $f(x^i, x^{i+1}) \in -E$ ($f(x^i, x^{i+1}) \notin E$, resp.) for all $i = 1, \ldots, m-1$, we have $f(x^m, x^1) \notin -E$ ($f(x^m, x^1) \in E$, resp.).*

Proposition 3.67 *Let $E \in \mathscr{E}(Y)$, $h : A \to Y$ and $f, g : A \times A \to Y$ be such that $f(x, y) - g(x, y) \in E$ for all $x, y \in A$. Then:*

(a) *Let $f_h : A \times A \to Y$ be defined by $f_h(x, y) := h(y) - h(x)$ for all $x, y \in A$. If $0 \notin E$, then $f_h \in \mathscr{V}_E^1 \mathscr{A}_{caqm}$.*
(b) *If $g \in \mathscr{V}_E^\ell \mathscr{A}_{caqm}$, then $f \in \mathscr{V}_E^\ell \mathscr{A}_{caqm}$ ($\ell = 1, 2$). In particular, if $0 \notin E$ and there exists $\theta : A \to Y$ such that $f(x, y) - (\theta(y) - \theta(x)) \in E$ for every $x, y \in A$, then $f \in \mathscr{V}_E^1 \mathscr{A}_{caqm}$.*

Proof

(a) Assume that $f_h \notin \mathscr{V}_E^1 \mathscr{A}_{caqm}$. Then, there exist $x^1,\ldots,x^m \in A$ with $x^{m+1} := x^1$ such that $h(x^i, x^{i+1}) \in -E$ for all $i = 1, 2, \ldots, m$. It follows that $0 = \sum_{i=1}^m f_h(x^i, x^{i+1}) \in -E - \cdots - E \subset -E$, a contradiction.
(b) Case 1: $\ell = 1$. Suppose that $f \notin \mathscr{V}_E^1 \mathscr{A}_{caqm}$, i.e., there exist $x^1,\ldots, x^m \in A$, $x^{m+1} := x^1$ such that $f(x^i, x^{i+1}) \in -E$ for all $i = 1, \ldots, m$. Then, $g(x^i, x^{i+1}) \in f(x^i, x^{i+1}) - E \subset -E - E \subset -E$ for all $i = 1, \ldots, m$, contradicting the fact that $g \in \mathscr{V}_E^1 \mathscr{A}_{caqm}$.

Case 2: $\ell = 2$. Let $x^1,\ldots, x^m \in A$ and $x^{m+1} := x^1$. Since $g \in \mathscr{V}_E^2\mathscr{A}_{caqm}$, there exists i such that $g(x^i, x^{i+1}) \in E$, which implies that $f(x^i, x^{i+1}) \in g(x^i, x^{i+1}) + E \subset E + E \subset E$, i.e., $f \in \mathscr{V}_E^2\mathscr{A}_{caqm}$.

□

Definition 3.68 Let $E \subset Y$ be nonempty.

(a) We say that f satisfies the E-TRIANGLE INEQUALITY if for all $x, y, z \in A$,

$$f(x, y) + f(y, z) - f(x, z) \in E.$$

(b) We say that f is E-CYCLICALLY MONOTONE if for any $x^1,\ldots,x^m \in A$ and $x^{m+1} := x^1$, we have $\sum_{i=1}^m f(x^i, x^{i+1}) \in E$.

By $\mathscr{V}_E\mathscr{A}_\Delta$ ($\mathscr{V}_E\mathscr{A}_{cm}$, resp.) we denote the class of all bifunctions which satisfy the E-triangle inequality (the class of all E-cyclically monotone bifunctions, resp.).

Proposition 3.69 *Let $E \subset Y$ and $D \subset Y$. Then:*

(a) *If $E \in \mathscr{E}(Y)$ and $f \in \mathscr{V}_E\mathscr{A}_\Delta$, then $f \in \mathscr{V}_E\mathscr{A}_{cm}$.*
(b) *If $D \in \mathscr{E}(Y)$, $E \cap (-D) = \emptyset$ and $f \in \mathscr{V}_E\mathscr{A}_{cm}$, then $f \in \mathscr{V}_E^1\mathscr{A}_{caqm}$.*

Proof

(a) When $f \in \mathscr{V}_E\mathscr{A}_\Delta$, clearly $f(x, x) \in E$ for all $x \in A$. Then, for any $x^1, \ldots, x^m \in A$ and $x^{m+1} := x^1$, we have $\sum_{i=1}^m f(x^i, x^{i+1}) = [f(x^1, x^2) + f(x^2, x^3) - f(x^1, x^3)] + [f(x^1, x^3) + f(x^3, x^4) - f(x^1, x^4)] + \cdots + [f(x^1, x^m) + f(x^m, x^1) - f(x^1, x^1)] + f(x^1, x^1) \in E + E + \cdots + E + E \subset E$. Hence, $f \in \mathscr{V}_E\mathscr{A}_{cm}$.
(b) Suppose that $f \notin \mathscr{V}_D^1\mathscr{A}_{caqm}$, i.e., there exist $x^1,\ldots, x^m \in A$, $x^{m+1} := x^1$ such that $f(x^i, x^{i+1}) \in -E$ for all $i = 1, \ldots, m$. Then, $f(x^1, x^2) + \cdots + f(x^m, x^1) \in -D - \cdots - D \subset -D$, contradicting $E \cap (-D) = \emptyset$ and the fact that $f \in \mathscr{V}_E\mathscr{A}_{cm}$.

□

Lemma 3.70 *Let $E \subset Y$ be nonempty. If $f \in \mathscr{V}_E^\ell\mathscr{A}_{caqm}$ ($\ell = 1, 2$), then f has E-finite intersection property of type ℓ.*

Proof For $\ell \in \{1, 2\}$, because $f \in \mathscr{V}_E^\ell\mathscr{A}_{caqm}$, $\Phi_{Ef}^\ell \in \mathscr{A}_{caqm}$ (see Remark 3.65 (b)). By Lemma 1.81, the problem (EP$_{\Phi_{Ef}^\ell}$) is finitely solvable. Finally, the conclusion follows from Remark 3.32. □

By associating Propositions 3.34 and 3.35 and Lemma 3.70 we state the following result.

Theorem 3.71 *Let $t \in \{He, Be, e, w\}$, $E \subset Y$ be nonempty, and $\ell \in \{1, 2\}$. Assume that $E \in \mathcal{E}_t^\ell(C)$, $f \in \mathcal{V}_E^\ell \mathcal{A}_{caqm}$ and at least one of the following conditions is satisfied:*

(a) *A is a compact topological space and $\mathrm{lev}_E^\ell f(\cdot, y)$ is closed for each $y \in A$.*
(b) *There exists $M \in \mathcal{F}(A)$ such that $\bigcup_{y \in A} \overline{\mathrm{lev}}_E^\ell f(\cdot, y) \subset \bigcup_{y \in M} \overline{\mathrm{lev}}_E^\ell f(\cdot, y)$.*

Then, $\mathrm{Sol}^t(\mathrm{VEP}_{(f,C)}) \neq \emptyset$.

Example 3.72 Let A be a nonempty set, $Y = \mathbb{R}^2$, and C be a nonempty set contained in $\mathbb{R}^2_{++} \cup \{(0, 0)\}$. Let $\Phi_1, \Phi_2 : A \times A \to \mathbb{R}_{++}$, $\varphi : A \to \mathbb{R}$ be such that $\varphi(x) \leq \varphi(x^*)$ for all $x \in A$ and some $x^* \in A$. Set $A_\varphi = \{(x, y) \in A \times A \mid \varphi(x) < \varphi(y)\}$, $\delta_{A_\varphi}(x, y) = -1$ as $(x, y) \in A_\varphi$, and $\delta_{A_\varphi}(x, y) = 1$ as $(x, y) \notin A_\varphi$. Let $f : A \times A \to \mathbb{R}^2$ be defined for all $x, y \in A$ by

$$f(x, y) = (\delta_{A_\varphi}(x, y)\Phi_1(x, y), -\Phi_2(x, y)).$$

Take any $x^1, \ldots, x^m \in A$ such that $f(x^i, x^{i+1}) \in -\mathbb{R}^2_+$, or equivalently, $(x^i, x^{i+1}) \in A_\varphi$, $i = 1, \ldots, m-1$. By the definition of A_φ, $\varphi(x^1) < \cdots < \varphi(x^m)$ and $(x^m, x^1) \notin A_\varphi$. Moreover, by the definition of f, $f(x^m, x^1) \notin -\mathbb{R}^2_+$. Thus, $f \in \mathcal{V}^1_{\mathbb{R}^2_+} \mathcal{A}_{caqm}$. For all $y \in A$, we have $\overline{\mathrm{lev}}^1_{\mathbb{R}^2_+} f(\cdot, y) = \{x \in A | f(x, y) \in -\mathbb{R}^2_+\} = \{x \in A | (x, y) \in A_\varphi\} \subset \{x \in A | (x, x^*) \in A_\varphi\} = \overline{\mathrm{lev}}^1_{\mathbb{R}^2_+} f(\cdot, x^*)$. This means that $\bigcup_{y \in A} \overline{\mathrm{lev}}^1_{\mathbb{R}^2_+} f(\cdot, y) \subset \overline{\mathrm{lev}}^1_{\mathbb{R}^2_+} f(\cdot, x^*)$. Since $\mathrm{core}\mathbb{R}^2_+ = \mathbb{R}^2_{++} \supset C_0$ and $C + \mathrm{core}\mathbb{R}^2_+ \subset \mathbb{R}^2_+$, $\mathbb{R}^2_+ \in \mathcal{E}^1_{He}(C)$ and $\mathbb{R}^2_+ \in \mathcal{E}^1_e(C)$. Thus, by Theorem 3.71, both $\mathrm{Sol}^{He}(\mathrm{VEP}_{(f,C)})$ and $\mathrm{Sol}^e(\mathrm{VEP}_{(f,C)})$ *are nonempty.*

Example 3.73 Let A be a nonempty set, $Y = \mathbb{R}^2$, and C be a nonconvex cone:

$$C := \{c = (c_1, c_2) \in \mathbb{R}^2 \mid 0 \leq c_2 \leq -c_1\} \cup \{c = (c_1, c_2) \in \mathbb{R}^2 \mid 0 \leq c_1 \leq c_2\},$$

$\varphi : A \to \mathbb{R}_+$ be such that $\varphi(x) \leq \varphi(x^*)$ for all $x \in A$ and some $x^* \in A$, and $f : A \times A \to \mathbb{R}^2$ be defined by

$$f(x, y) := (2\varphi(y) - \varphi(x), \varphi(x)) \text{ for all } x, y \in A.$$

Let $E = \{e = (e_1, e_2) \in \mathbb{R}^2 \mid e_2 \geq 0, e_2 \geq e_1\}$. We have $C + E \subset E$ and $\mathrm{vcl}E = E$. Hence, $\mathrm{vcl\,cone}(C + E) \cap (-C_0) \subset \mathrm{vcl}E \cap (-C_0) = E \cap (-C_0) = \emptyset$. Thus, $E \in \mathcal{E}^2_{Be}(C)$. For any $x^1, \ldots, x^m \in A$ such that $f(x^i, x^{i+1}) = (2\varphi(x^{i+1}) - \varphi(x^i), \varphi(x^i)) \notin E$ for $i = 1, \ldots, m-1$, we have $\varphi(x^i) < \varphi(x^{i+1})$. Hence, $\varphi(x^1) < \cdots < \varphi(x^m)$. Then, $\varphi(x^m) \geq 2\varphi(x^1) - \varphi(x^m)$ which implies that $f(x^m, x^1) = (2\varphi(x^1) - \varphi(x^m), \varphi(x^m)) \in E$. So, $f \in \mathcal{V}^2_E \mathcal{A}_{caqm}$. For all $y \in A$, $\overline{\mathrm{lev}}^2_E f(\cdot, y) = \{x \in A \mid f(x, y) \notin E\} = \{x \in A \mid \varphi(x) < \varphi(y)\} \subset \{x \in A \mid$

$\varphi(x) < \varphi(x^*)\} = \overline{\text{lev}}_E^2 f(\cdot, x^*)$, i.e., $\bigcup_{y\in A} \overline{\text{lev}}_E^2 f(\cdot, y) \subset \overline{\text{lev}}_E^2 f(\cdot, x^*)$. Thus, by Theorem 3.71, $\text{Sol}^{Be}(\text{VEP}_{(f,C)})$ *is not empty.*

Using Remark 3.28 and applying Theorem 3.71 for $E = D$ and $E = \text{core}D$, we obtain the following corollary.

Corollary 3.74 *Let A be a compact topological space and $D \subset Y$ be a vectorially closed convex cone such that $D \cap (-D) = \{0\}$.*

(a) *Assume that $f \in \mathscr{V}^1_{\text{core}D}\mathscr{A}_{caqm}$ and $\text{lev}^1_{\text{core}D} f(\cdot, y)$ is closed for each $y \in A$. If $D \in \mathscr{K}^{core}_C$, then $\text{Sol}^t(\text{VEP}_{(f,C)}) \neq \emptyset$ for all $t \in \{He, Be, e, w\}$, and if $D = C$, then $\text{Sol}^w(\text{VEP}_{(f,C)}) \neq \emptyset$.*
(b) *Assume that $f \in \mathscr{V}^1_D\mathscr{A}_{caqm}$ and $\text{lev}^1_D f(\cdot, y)$ is closed for each $y \in A$. If $D \in \mathscr{K}^{core}_C$, then $\text{Sol}^t(\text{VEP}_{(f,C)}) \neq \emptyset$ for all $t \in \{He, Be, e, w\}$, and if $D = C$, then $\text{Sol}^t(\text{VEP}_{(f,C)}) \neq \emptyset$ for $t \in \{e, w\}$.*
(c) *Assume that $f \in \mathscr{V}^2_D\mathscr{A}_{caqm}$ and $\text{lev}^2_D f(\cdot, y)$ is closed for each $y \in A$. If $D \in \mathscr{K}^{core}_C$, then $\text{Sol}^t(\text{VEP}_{(f,C)}) \neq \emptyset$ for all $t \in \{He, Be, e, w\}$, and if $D = C$, then $\text{Sol}^t(\text{VEP}_{(f,C)}) \neq \emptyset$ for $t \in \{Be, e, w\}$.*
(d) *Assume that $f \in \mathscr{V}^2_{\text{core}D}\mathscr{A}_{caqm}$ and $\text{lev}^2_{\text{core}D} f(\cdot, y)$ is closed for each $y \in A$. If $D \in \mathscr{K}^{core}_C$, then $\text{Sol}^t(\text{VEP}_{(f,C)}) \neq \emptyset$ for all $t \in \{He, Be, e, w\}$, and if $D = C$, then $\text{Sol}^t(\text{VEP}_{(f,C)}) \neq \emptyset$ for $t \in \{Be, e, w\}$.*

The following example illustrates effective applications of Corollary 3.74 in nonconvex circumstances.

Example 3.75 Let A be compact subset of a topological space X, $Y = \mathbb{R}^2$, and $D = \mathbb{R}^2_+$, $\gamma : X \to (0, +\infty)$, and $\beta : X \to \mathbb{R}$ be upper semicontinuous.

(a) Let $f : A \times A \to \mathbb{R}^2$ be defined for all $x \in A$ and $y \in A$ by

$$f(x, y) := \begin{cases} (-\gamma(y), -\gamma(x)) & \text{if } \beta(x) < \beta(y), \\ (-\gamma(x), |\gamma(x) - \gamma(y)|) & \text{if otherwise.} \end{cases}$$

Then, $f \in \mathscr{V}^1_{\text{core}\mathbb{R}^2_+}\mathscr{A}_{caqm}$, but $f \notin (\mathscr{V}^1_{\mathbb{R}^2_+}\mathscr{A}_{caqm}) \cup (\mathscr{V}^2_{\mathbb{R}^2_+}\mathscr{A}_{caqm}) \cup (\mathscr{V}^2_{\text{core}\mathbb{R}^2_+}\mathscr{A}_{caqm})$. For all $y \in A$, $\text{lev}^1_{\text{core}\mathbb{R}^2_+} f(\cdot, y) = A \cap \beta^{-1}([\beta(y), +\infty))$ is closed. Thus, all the conditions of part (a) hold, but the assumption of the vector-cyclic quasimonotonicity in parts (b), (c), and (d) do not hold.
(b) Let $f : A \times A \to \mathbb{R}^2$ be defined for all $x, y \in A$ by

$$f(x, y) := \begin{cases} (-\gamma(x), -\gamma(y)) & \text{if } \beta(x) < \beta(y), \\ (-\gamma(y), \gamma(x)) & \text{if } \beta(y) \le \beta(x). \end{cases}$$

Then, $\text{lev}^1_{\mathbb{R}^2_+} f(\cdot, y) = A \cap \beta^{-1}([\beta(y), +\infty))$ is closed for all $y \in A$. Moreover, $f \in \mathscr{V}^1_{\mathbb{R}^2_+}\mathscr{A}_{caqm}$, but $f \notin (\mathscr{V}^2_{\mathbb{R}^2_+}\mathscr{A}_{caqm}) \cup (\mathscr{V}^2_{\text{core}\mathbb{R}^2_+}\mathscr{A}_{caqm})$. Thus, all the conditions of part (b) hold, but the assumption of the vector-cyclic quasimonotonicity in parts (c) and (d) do not hold.

(c) Let $f : A \times A \to \mathbb{R}^2$ be defined for all $x, y \in A$ by

$$f(x, y) := \begin{cases} (-\gamma(y), -\gamma(x)) & \text{if } \beta(x) < \beta(y), \\ (0, 0) & \text{if } \beta(y) \leq \beta(x). \end{cases}$$

Then, $\text{lev}^2_{\mathbb{R}^2_+} f(\cdot, y) = A \cap \beta^{-1}([\beta(y), +\infty))$ is closed for all $y \in A$. We easily see that $f \in \mathscr{V}^2_{\mathbb{R}^2_+}\mathscr{A}_{caqm}$, but $f \notin (\mathscr{V}^1_{\mathbb{R}^2_+}\mathscr{A}_{caqm}) \cup (\mathscr{V}^2_{\text{core}\mathbb{R}^2_+}\mathscr{A}_{caqm})$. Thus, parts (b) and (d) are out of use, while part (c) is applicable.

(d) Let $f : A \times A \to \mathbb{R}^2$ be defined by, for all $x \in A$ and $y \in A$,

$$f(x, x') := \begin{cases} (-\gamma(x), -\gamma(y)) & \text{if } \beta(x) < \beta(y), \\ (\gamma(x), \gamma(y)) & \text{if } \beta(y) \leq \beta(x). \end{cases}$$

It is easy to check that $f \in \mathscr{V}^2_{\text{core}\mathbb{R}^2_+}\mathscr{A}_{caqm}$. For all $y \in A$, $\text{lev}^2_{\text{core}\mathbb{R}^2_+} f(\cdot, y) = A \cap \beta^{-1}([\beta(y), +\infty))$ is closed. Thus, all the conditions of (d) hold.

Applications to the Vector Optimization Problem Using Theorem 3.71, we deduce some results for the vector optimization problem ($\text{VOP}_{(h,C)}$). First, the following corollary directly follows from Theorem 3.71 and Proposition 3.67 (a) for $f_h : A \times A \to Y$ defined by $f_h(x, y) := h(y) - h(x)$. Note that when h is E^c-upper quasicontinuous, $\text{lev}^1_E f(\cdot, y) = \{x \in A \mid h(y) - h(x) \notin -E\}$ is closed for each $y \in A$.

Corollary 3.76 *(Weierstrass-Type Theorem for ($VOP_{(h,C)}$)) Let $t \in \{He, Be, e, w\}$ and $E \in \mathscr{E}(Y)$ with $0 \notin E$. Suppose that A is a nonempty compact subset of a topological space, $E \in \mathscr{E}^1_t(C)$ and h is E^c-upper quasicontinuous. Then, $O^t(h, C) \neq \emptyset$.*

Corollary 3.77 *Let $t \in \{He, Be, e, w\}$ and $E \subset Y$ with $0 \notin E$. Assume that A is a nonempty compact subset of a topological space, $E \in \mathscr{E}^1_t(C)$, $\gamma : Y \to \mathbb{R}$ is E-increasing, and at least one of the following conditions is satisfied:*

(a) $\mathscr{L}_{(\gamma,h,\geq)}(y) = \{x \in A \mid \gamma(-h(x)) - \gamma(-h(y)) \geq 0\}$ is closed for each $y \in A$.
(b) $\mathscr{L}_{(\gamma,h,\leq)}(y) = \{x \in A \mid \gamma(h(x)) - \gamma(h(y)) \leq 0\}$ is closed for each $y \in A$.

Then, $O^t(h, C) \neq \emptyset$.

Proof Let $f_\gamma, \tilde{f}_\gamma : A \times A \to Y = \mathbb{R}$ be defined by

$$f_\gamma(x, y) := \gamma(-h(x)) - \gamma(-h(y)) \text{ and } \tilde{f}_\gamma(x, y) := \gamma(h(y)) - \gamma(h(x)).$$

Then, by using Theorem 3.71, it is not hard to see that both $\text{Sol}^w(\text{VEP}_{(f_\gamma,\mathbb{R}_+)})$ and $\text{Sol}^w(\text{VEP}_{(\tilde{f}_\gamma,\mathbb{R}_+)})$ are nonempty.

For $\bar{x} \in \mathrm{Sol}^w(\mathrm{VEP}_{(f_\gamma,\mathbb{R}_+)})$, $\gamma(-h(\bar{x})) \geq \gamma(-h(y))$ for all $y \in A$. This and the E-increasing property of γ imply that $-h(y)-(-h(\bar{x})) \notin E$, i.e., $h(y)-h(\bar{x}) \notin -E$ for all $y \in A$. For $\bar{x} \in \mathrm{Sol}^w(\mathrm{VEP}_{(\tilde{f}_\gamma,\mathbb{R}_+)})$, $\gamma(h(y)) \geq \gamma(h(\bar{x}))$ for all $y \in A$. By the E-increasing property of γ, $h(y) - h(\bar{x}) \notin -E$ for all $y \in A$. Thus, in both cases $\bar{x} \in \mathrm{Sol}^w(\mathrm{VEP}_{(f_\gamma,\mathbb{R}_+)})$ or $\bar{x} \in \mathrm{Sol}^w(\mathrm{VEP}_{(\tilde{f}_\gamma,\mathbb{R}_+)})$,

$$h(A) - h(\bar{x}) \subset Y \setminus (-E). \tag{3.20}$$

When $E \in \mathscr{E}^1_{He}(C)$, there exists $D \in \mathscr{K}_C^{core}$ such that $C + \mathrm{core}D \subset E$. This and (3.20) imply that $(h(A) - h(\bar{x})) \cap (-C - \mathrm{core}D) = \emptyset$, or equivalently, $(h(A) - h(\bar{x}) + C) \cap (-\mathrm{core}D) = \emptyset$, i.e., $\bar{x} \in O^{He}(h, C)$.

When $E \in \mathscr{E}^1_{Be}(C)$, $\mathrm{vcl\,cone}(h(A) - h(\bar{x}) + C) \cap (-C_0) \subset \mathrm{vcl\,cone}(C - E^c) \cap (-C_0) = \emptyset$. Hence, $\bar{x} \in O^{Be}(h, C)$.

If $E \in \mathscr{E}^1_e(C)$ ($E \in \mathscr{E}^1_w(C)$, resp.), then (3.20) gives that $(h(A)-h(\bar{x}))\cap(-C_0) \subset (h(A)-h(\bar{x}))\cap(-E) = \emptyset$ ($(h(A)-h(\bar{x}))\cap(-\mathrm{core}C) \subset (h(A)-h(\bar{x}))\cap(-E) = \emptyset$, resp.). Thus, $\bar{x} \in O^e(h, C)$ ($\bar{x} \in O^w(h)$, resp.). □

Corollary 3.78 (Weierstrass-type theorem for (VOP$_{(h,C)}$)) *Let* $t \in \{He, Be, e, w\}$ *and* $E \subset Y$ *be such that* $0 \notin \mathrm{core}E$ *and* $E^* \setminus \{0^*\} \neq \emptyset$, *where* $E^* := \{e^* \in Y^* \mid \langle e^*, e\rangle \geq 0, \forall e \in E\}$. *Assume that A is a nonempty compact subset of a topological space, Y is a topological vector space,* $\mathrm{core}E \in \mathscr{E}^1_t(C)$, *and h is E-lower semicontinuous. Then,* $O^t(h, C) \neq \emptyset$.

Proof Let $\gamma \in E^* \setminus \{0^*\}$. Then, γ is $\mathrm{core}E$-increasing. Indeed, take any $u, u' \in Y$ satisfying $u - u' \in -\mathrm{core}E$. Then, there exists $v \in Y$ such that $\langle\gamma, v\rangle > 0$ and $u' - u - \lambda v \in E$ for some $\lambda > 0$. So, $\langle\gamma, u' - u - \lambda v\rangle \geq 0$, i.e., $\langle\gamma, u'\rangle \geq \langle\gamma, u\rangle + \lambda\langle\gamma, v\rangle$, which implies that $\langle\gamma, u'\rangle > \langle\gamma, u\rangle$. For each $y \in A$, we claim that the set

$$\begin{aligned}\mathscr{L}_{(\gamma,h,\geq)}(y) &= \{x \in A \mid \gamma(-h(x)) - \gamma(-h(y)) \geq 0\}\\ &= \{x \in A \mid \langle\gamma, -h(x) - (-h(y))\rangle \geq 0\}\end{aligned}$$

is closed. Let $\{x_\alpha\}$ be a net in $\mathscr{L}_{(\gamma,h,\geq)}(y)$ such that $x_\alpha \to \bar{x}$. Suppose that $\bar{x} \notin \mathscr{L}_{(\gamma,h,\geq)}(y)$. Then, $\langle\gamma, -h(\bar{x})-(-h(y))\rangle < 0$, i.e., $-h(\bar{x})-(-h(y)) \in \gamma^{-1}(-\infty, 0)$ which implies that $h(\bar{x}) \in h(y) - \gamma^{-1}(-\infty, 0) =: V$, an open set. Since h is E-lower semicontinuous, there exists a neighborhood U of $\bar{x}$ such that $h(z) \in V + E$ for all $z \in U$. As $x_\alpha \to \bar{x}$, there is α such that $x_\alpha \in U$. Then, $\varphi(x_\alpha) \in V + E = \varphi(x') - \gamma^{-1}(-\infty, 0) + E$, that is, $-\varphi(x_\alpha) - (-\varphi(x')) \in \gamma^{-1}(-\infty, 0) - E$. Hence, $\langle\gamma, -h(x_\alpha) - (-h(y))\rangle \in \langle\gamma, \gamma^{-1}(-\infty, 0) - E\rangle = (-\infty, 0) - \gamma(E) \subset (-\infty, 0)$, i.e., $x_\alpha \notin \mathscr{L}_{(\gamma,h,\geq)}(y)$, a contradiction. Thus, $\mathscr{L}_{(\gamma,h,\geq)}(y)$ is closed. Applying Corollary 3.77 completes the proof. □

Recall that

$$\mathrm{Q}_E := \{q \in Y \mid \mathbb{R}_{++}q + E = \mathrm{core}E\}.$$

When $E \in \mathscr{E}(Y)$ and $q \in \mathrm{Q}_E \cap \mathrm{core}E$, r_E^q is finite and core$E$-increasing (see Exercises 3.2–3.5).

Corollary 3.79 (Weierstrass-type theorem for (VOP$_{(h,C)}$)) *Let* $t \in \{He, Be, e, w\}$ *and* $E \in \mathscr{E}(Y)$ *with* $0 \notin E$. *Assume that* A *is a nonempty compact subset of a topological space,* $\mathrm{core}E \in \mathscr{E}_t^1(C)$ *and there exists* $q \in \mathrm{Q}_E \cap \mathrm{core}E$ *such that* h *is* (q, E^c)*-upper quasicontinuous. Then,* $O^t(h, C) \neq \emptyset$.

Corollary 3.80 (Weierstrass-type theorem for (VOP$_{(h,C)}$)) *Let* $t \in \{He, Be, e, w\}$ *and* $E \in \mathscr{E}(Y)$ *such that* E *is vectorially closed. Assume that* A *is a nonempty compact subset of a topological space,* $\mathrm{core}E \in \mathscr{E}_t^1(C)$ *and there exists* $q \in \mathrm{Q}_E \cap \mathrm{core}E$ *such that* h *is* (q, E)*-lower quasicontinuous. Then,* $O^t(h, C) \neq \emptyset$.

Proof *Case of Corollary 3.79.* Since h is (q, E^c)-upper quasicontinuous, for each $\alpha > 0$, the set

$$\mathscr{L}_\alpha := \{x \in A \mid -h(x) - r_E^q(-h(y))q \notin -\alpha q - E\}$$

is closed. Hence, for each $y \in A$,

$$\begin{aligned}\mathscr{L}_{(r_E^q,h,\geq)}(y) &= \{x \in A \mid r_E^q(-h(x)) - r_E^q(-h(y)) \geq 0\}\\ &= \{x \in A \mid r_E^q(-h(x) - r_E^q(-h(y))q) \geq 0\}\\ &= \{x \in A \mid -h(x) - r_E^q(-h(y))q \notin -\mathbb{R}_{++}q - E\}\\ &= \bigcap_{\alpha>0}\{x \in A \mid -h(x) - r_E^q(-h(y))q \notin -\alpha q - E\}\end{aligned}$$

is closed.

Case of Corollary 3.80. Since E is vectorially closed, $q \in \mathrm{Q}_E \subset \mathrm{P}_E := \{q \in Y \mid \mathbb{R}_+q + E = E\}$ (see Exercise 3.3) and h is (q, E)-lower quasicontinuous, we see that for each $y \in A$,

$$\begin{aligned}\mathscr{L}_{(r_E^q,h,\leq)}(y) &= \{x \in A \mid r_E^q(h(x)) - r_E^q(h(y)) \leq 0\}\\ &= \{x \in A \mid r_E^q(h(x) - r_E^q(h(y))q) \leq 0\}\\ &= \{x \in A \mid h(x) - r_E^q(h(y))q \in -\mathbb{R}_+q - \mathrm{vcl}_q E\}\\ &= \{x \in A \mid h(x) - r_E^q(h(y))q \in -E\}\end{aligned}$$

is closed. Therefore, Corollaries 3.79 and 3.80 follow from Corollary 3.77. □

3.3.3 Vectorial Ekeland Variational Principle

In Chap. 1, we stated Ekeland variational principle for scalar bifunctions (Theorem 1.71). In this subsection, we draw some vectorial forms of the Ekeland variational principle. All symbols in this subsection are the same as those in Sect. 3.3.1. Let $E \subset Y$ be such that $E \in \mathscr{E}(Y)$, i.e., E is solid, $E \neq Y$ and $E + E \subset E$. Recall that

$$\mathrm{Q}_E := \{q \in Y \mid \mathbb{R}_{++}q + E = \operatorname{core} E\}.$$

We set

$$\mathscr{E}_\bullet(Y) := \{E \in \mathscr{E}(Y) \mid E \cap (-E) \neq \emptyset\} \text{ and } \mathscr{Q}_E := \mathrm{Q}_E \cap \operatorname{core} E.$$

We observe that if $E \in \mathscr{E}_\bullet(Y)$, then $0 \in E$. If E is a convex cone, then $E \in \mathscr{E}_\bullet(Y)$ and $\mathrm{Q}_E = \operatorname{core} E$ (and hence $\mathscr{Q}_E = \operatorname{core} E$) (see Exercise 3.6). Let A be a nonempty set and $E \in \mathscr{E}_\bullet(Y)$. We set

$$\mathscr{V}_E\mathscr{A}_A^{=} := \{f : A \times A \to Y \mid f(x, x) \in E \cap (-E) \text{ for all } x \in A\}.$$

Recall that $\mathscr{V}_E\mathscr{A}_\Delta$ is the class of all bifunctions which satisfy the E-triangle inequality:

$$\mathscr{V}_E\mathscr{A}_\Delta = \{f : A \times A \to Y \mid f(x, y) + f(y, z) - f(x, z) \in E \text{ for all } x, y, z \in A\}.$$

Next, we collect some properties of the nonconvex separation functional r_E^q defined by (3.1) in the particular case where $E \in \mathscr{E}_\bullet(Y)$ and $q \in \mathscr{Q}_E$.

Proposition 3.81 *Let $E \in \mathscr{E}_\bullet(Y)$, $q \in \mathscr{Q}_E$ and $h : A \to Y$ be a vector function. Then:*

(a) *r_E^q is finite.*
(b) $\{v \in Y \mid r_E^q(v) < 0\} = -\operatorname{core} E$.
(c) $\{v \in Y \mid r_E^q(v) \leq 0\} = -\mathrm{vcl}_q E$.
(d) *r_E^q is E-nondecreasing.*
(e) *$r_E^q(v) = 0$ for all $v \in E \cap (-E)$. Specially, $r_E^q(0) = 0$.*
(f) *$r_E^q(u + v) \leq r_E^q(u) + r_E^q(v)$ for all $u, v \in Y$.*
(g) *If there exists $\bar{e} \in Y$ such that $h(A) \cap (\bar{e} - \operatorname{core} E) = \emptyset$, then $\varphi = r_E^q \circ h$ is bounded from below.*
(h) *If h is $(q, \mathrm{vcl}_q E)$-lower quasicontinuous, then $\varphi = r_E^q \circ h$ is lower semicontinuous, and hence, if E is vectorially closed and h is (q, E)-lower quasicontinuous, then $\varphi = r_E^q \circ h$ is lower semicontinuous.*

Proof

(a) Take any $v \in Y$. When $q \in \mathcal{Q}_E \subset \text{core}E$, there exists $n_v \in \mathbb{N}$ such that $\frac{1}{n_v}v + q \in D$ and $-\frac{1}{n_v}v + q \in E$, i.e., $v + n_v q \in n_v E \subset E$ and $v \in n_v q - n_v E \subset n_v q - E \subset \mathbb{R}q - E$. Hence, by the definition of r_E^q, $r_E^q(v) < +\infty$. If $r_E^q(v) = -\infty$, then there exists $\lambda > n_v$ such that $v \in -\lambda q - E$. Then, $v + n_v q \in -(\lambda - n_v)q - E \subset -\mathbb{R}_{++}q - E = -\text{core}E$. Thus, $v + n_v q \in E \cap (-\text{core}E) = \emptyset$ (see Exercise 3.1), we obtain a contradiction.

(b) A direct consequence of Proposition 3.10(d).

(c) It is not hard to check that $-\mathbb{R}_{++}q - \text{vcl}_q E = -\mathbb{R}_{++}q - E$. Hence, by Proposition 3.10(e), $\{v \in V \mid r_E^q(v) \leq 0\} = -\mathbb{R}_+ q - \text{vcl}_q E = (-\mathbb{R}_{++}q - \text{vcl}_q E) \cup (-\text{vcl}_q E) = (-\mathbb{R}_{++}q - E) \cup (-\text{vcl}_q E) = (-\text{core}E) \cup (-\text{vcl}_q E) = -\text{vcl}_q E$.

(d) We have that $E + E \subset E \subset E \cup (\mathbb{R}_{++}q + E) = \mathbb{R}_+ q + E \subset \mathbb{R}_+ q + \text{vcl}_q E$. Hence, by Proposition 3.10(f), r_E^q is E-nondecreasing.

(e) When $E \in \mathcal{E}_\bullet(Y)$, $E \cap (-E) = (E \backslash \text{core}E) \cap ((-E) \backslash (-\text{core}E))$ (see Exercise 3.1), and hence $E \cap (-E) \subset (-\text{vcl}_q E) \backslash (-\text{core}E)$. This and the parts (b) and (c) imply that $r_E^q(v) = 0$ for all $v \in E \cap (-E)$.

(f) For all $u, v \in Y$, by definition of r_E^q, for any $\epsilon > 0$, there exist $t, t' \in \mathbb{R}$ such that $t < r_E^q(u) + \frac{\epsilon}{2}$, $t' < r_E^q(v) + \frac{\epsilon}{2}$, and $u \in tq - E$, $v \in t'q - E$. Then, $u + v \in tq + t'q - E - E \subset (t + t')q - E$. This implies that $r_E^q(u + v) \leq t + t' < r_E^q(u) + r_E^q(v) + \epsilon$. Since ϵ is arbitrary, $r_E^q(u + v) \leq r_E^q(u) + r_E^q(v)$.

(g) Since $h(A) \cap (\bar{e} - \text{core}E) = \emptyset$, $h(x) - \bar{e} \notin -\text{core}E$ for all $x \in A$. This and the part (b) imply that $r_E^q(h(x) - \bar{e}) \geq 0$ for all $x \in A$. By the part (f), $r_E^q(h(x) - \bar{e}) \leq r_E^q(h(x)) + r_E^q(-\bar{e})$ for all $x \in A$. Hence, $r_E^q(h(x)) \geq -r_E^q(-\bar{e})$ for all $x \in A$. Thus, $\varphi = r_E^q \circ h$ is bounded from below.

(h) By Propositions 3.10(c) and 3.81(c), for any $\beta \in \mathbb{R}$, $\text{lev}_{\leq}(\beta r_E^q \circ h) = \{x \in A \mid (r_E^q \circ h)(x) \leq \beta\} = \{x \in A \mid r_E^q(h(x)) - \beta \leq 0\} = \{x \in A \mid r_E^q(h(x) - \beta q) \leq 0\} = \{x \in A \mid h(x) - \beta q \in -\text{vcl}_q E\}$ is closed (by the $(q, \text{vcl}_q E)$-lower quasicontinuity of h). Thus, $\varphi = r_E^q \circ h$ is lower semicontinuous.

□

Definition 3.82 Let $E \subset Y$ be nonempty. A vector function $h : A \to Y$ is said to be E-BOUNDED FROM BELOW if there exists $\bar{e} \in Y$ such that $h(x) \in \bar{e} + E$ for all $x \in A$.

Proposition 3.83 *Let (A, d) be a complete metric space, $E \in \mathcal{E}_\bullet(Y)$, $q \in \mathcal{Q}_E$, $h : A \to Y$ and $f : A \times A \to Y$. Then:*

(a) *If h is E-bounded from below, then there exists $\bar{e} \in Y$ such that $h(A) \cap (\bar{e} - \text{core}E) = \emptyset$.*

(b) *If h is $(q, \text{vcl}_q E)$-lower quasicontinuous, then $\{y \in A \mid h(y) + \epsilon d(x, y)q \in -\text{vcl}_q E\}$ is closed for any $x \in A$, and hence if E is vectorially closed and h is (q, E)-lower quasicontinuous, then $\{y \in A \mid h(y) + \epsilon d(x, y)q \in -E\}$ is closed for any $x \in A$ and $\epsilon > 0$.*

Proof

(a) Assume that h is E-bounded from below. Then, there exists $\bar{e} \in Y$ such that $h(x) \in \bar{e} + E$ for all $x \in A$, which means that $(h(A) - \bar{e}) \subset E$. Because $E \cap (-\text{core}E) = \emptyset$ as $E \in \mathscr{E}_\bullet(Y)$(see Exercise 3.1), we have that $(h(A) - \bar{e}) \cap (-\text{core}E) = \emptyset$, i.e., $h(A) \cap (\bar{e} - \text{core}E) = \emptyset$.

(b) Let sequence $\{y_n\} \subset E_\epsilon := \{y \in A \mid h(y) + \epsilon d(x, y)q \in -\text{vcl}_q E\}$ be such that $y_n \to \bar{y}$. We show that $\bar{y} \in E_\epsilon$, i.e., $h(\bar{y}) + \epsilon d(x, \bar{y})q \in -\text{vcl}_q E$. We have that

$$h(y_n) + \epsilon d(x, y_n)q \in -\text{vcl}_q E \text{ for all } n.$$

Let $\delta > 0$ be arbitrary. Since $d(x, y_n) \to d(x, \bar{y})$, there exists $n_0 \in \mathbb{N}$ such that $d(x, \bar{y}) - d(x, y_n) < \delta/\epsilon$ for all $n \geq n_0$. This together with Propositions 3.10(c) and 3.81(e) imply that

$$r_E^q((-\epsilon d(x, y_n) + \epsilon d(x, \bar{y}) - \delta)q) = r_E^q(0) - \epsilon d(x, y_n) + \epsilon d(x, \bar{y}) - \delta$$

$$= \epsilon(-d(x, y_n) + d(x, \bar{y}) - \delta/\epsilon) < 0 \text{ for all } n \geq n_0.$$

Then, by Proposition 3.81(b), $(-\epsilon d(x, y_n) + \epsilon d(x, \bar{y}) - \delta)q \in -\text{core}E \subset -E$ for all $n \geq n_0$, i.e., $-\epsilon d(x, y_n)q \in -\epsilon d(x, \bar{y})q + \delta q - E$ for all $n \geq n_0$. It follows that

$$h(y_n) = (h(y_n) + \epsilon d(x, y_n)q) - \epsilon d(x, y_n)q$$

$$\in -\epsilon d(x, \bar{y})q + \delta q - E - \text{vcl}_q E.$$

$$\subset (-\epsilon d(x, \bar{y}) + \delta)q - \text{vcl}_q E \text{ for all } n \geq n_0.$$

Hence, $y_n \in E_\delta := \{y \in A \mid h(y) \in (-\epsilon d(x, \bar{y}) + \delta)q - \text{vcl}_q E\}$ for all $n \geq n_0$. Since h is $(q, \text{vcl}_q E)$-lower quasicontinuous, E_δ is closed, and hence $\bar{y} \in E_\delta$. This implies that $h(\bar{y}) + \epsilon d(x, \bar{y})q \in \delta q - \text{vcl}_q E$. Thus, $h(\bar{y}) + \epsilon d(x, \bar{y})q \in \bigcap_{\delta>0}(\delta q - \text{vcl}_q E) \subset -\text{vcl}_q E$ (see Proposition 3.4(b)).

□

Theorem 3.84 *Let (A, d) be a complete metric space, $\epsilon > 0$, $E \in \mathscr{E}_\bullet(Y)$ be a vectorially closed set, and $q \in \mathscr{Q}_E$. Assume that $f \in \mathscr{V}_E\mathscr{A}_\Delta \cap \mathscr{V}_E\mathscr{A}_A^=$ is such that for each $x \in A$, the set $\{y \in A \mid f(x, y) + \epsilon d(x, y)q \in -E\}$ is closed and $f(x, A) \cap (e_x - \text{core}E) = \emptyset$ for some $e_x \in Y$. Then, for all $x^0 \in A$, there exists $\bar{x} \in A$ such that:*

(a) $f(x^0, \bar{x}) + \epsilon d(x^0, \bar{x})q \in -E$.
(b) $f(\bar{x}, y) + \epsilon d(\bar{x}, y)q \notin -E$ *for all* $y \in A$ *different from* $\bar{x}$.

Proof Let $\Phi_E^q : A \times A \to \mathbb{R}$ be defined for all $x, y \in A$ by

$$\Phi_E^q(x, y) = (r_E^q \circ f)(x, y) = r_E^q(f(x, y)). \tag{3.21}$$

We check conditions of Theorem 1.71 for Φ_E^q. When $f \in \mathscr{V}_E\mathscr{A}_\Delta$, we have for all $x, y, z \in A$,

$$f(x, z) - (f(x, y) + f(y, z)) \in -E.$$

Then, by (d) and (f) of Proposition 3.81,

$$\begin{aligned}\Phi_E^q(x, z) &= r_E^q(f(x, z)) \le r_E^q(f(x, y) + f(y, z)) \le r_E^q(f(x, y)) + r_E^q(f(y, z)) \\ &= \Phi_E^q(x, y) + \Phi_E^q(y, z).\end{aligned}$$

Hence, $\Phi_E^q \in \mathscr{A}_\Delta$. Moreover, by (e) of Proposition 3.81, $\Phi_E^q(x, x) = r_E^q(f(x, x)) = 0$ for all $x \in A$ whenever $f \in \mathscr{V}_E\mathscr{A}_A^=$. Thus, $\Phi_E^q \in \mathscr{A}_\Delta \cap \mathscr{A}_A^=$. For each $x \in A$, as $E = \text{vcl}E = \text{vcl}_q E$, it follows from Propositions 3.10(c) and 3.81(c) that

$$\begin{aligned}\{y \in A \mid \Phi_E^q(x, y)+\epsilon d(x, y) \le 0\} &= \{y \in A \mid r_E^q(f(x, y))+\epsilon d(x, y) \le 0\} \\ &= \{y \in A \mid r_E^q(f(x, y) + \epsilon d(x, y)q) \le 0\} \\ &= \{y \in A \mid r_E^q(f(x, y) + \epsilon d(x, y)q) \in -\text{vcl}_q E\} \\ &= \{y \in A \mid r_E^q(f(x, y) + \epsilon d(x, y)q) \in -E\}\end{aligned}$$

is closed. Furthermore, since $f(x, A) \cap (e_x - \text{core}E) = \emptyset$ for some $e_x \in Y$, $\Phi_E^q(x, \cdot) = r_E^q \circ (f(x, \cdot))$ is bounded from below by Proposition 3.81(g). Hence, $\inf_{y\in A} \Phi_E^q(x, y) > -\infty$ for each $x \in A$. Thus, all the conditions of Theorem 1.71 are satisfied and hence there exists $\bar{x} \in A$ such that:

(a1) $\Phi_E^q(x^0, \bar{x}) + \epsilon d(x^0, \bar{x}) \le 0$.
(b1) $\Phi_E^q(\bar{x}, y) + \epsilon d(\bar{x}, y) > 0$ for all $y \in A$ different from $\bar{x}$.

Again using Propositions 3.10(c) and 3.81(c) we have that

$$\begin{aligned}\text{(a1)} &\Leftrightarrow r_E^q(f(x^0, \bar{x})) + \epsilon d(x^0, \bar{x}) \le 0 \\ &\Leftrightarrow r_E^q(f(x^0, \bar{x}) + \epsilon d(x^0, \bar{x})q) \le 0 \\ &\Leftrightarrow f(x^0, \bar{x}) + \epsilon d(x^0, \bar{x})q) \in -\text{vcl}_q E = -E,\end{aligned}$$

which is the conclusion (a) of Theorem 3.84. Moreover,

$$\text{(b1)} \Leftrightarrow r_E^q(f(x^0,\bar{x})) + \epsilon d(x^0,\bar{x}) > 0 \text{ for all } y \in A \text{ different from } \bar{x}$$

$$\Leftrightarrow r_E^q(f(x^0,\bar{x}) + \epsilon d(x^0,\bar{x})q) > 0 \text{ for all } y \in A \text{ different from} \bar{x}$$

$$\Leftrightarrow f(x^0,\bar{x}) + \epsilon d(x^0,\bar{x})q) \notin -\mathrm{vcl}_q E = -E \text{ for all } y \in A \text{ different from} \bar{x},$$

which is the conclusion (b) of Theorem 3.84. □

Remark 3.85 In Theorem 3.84, if E is not vectorially closed and the condition "the set $\{y \in A \mid f(x,y) + \epsilon d(x,y)q \in -E\}$ is closed" is replaced by "the set $\{y \in A \mid f(x,y) + \epsilon d(x,y)q \in -\mathrm{vcl}_q E\}$ is closed," then we obtain the following conclusion: for all $x^0 \in A$, there exists $\bar{x} \in A$ such that

(a') $f(x^0,\bar{x}) + \epsilon d(x^0,\bar{x})q \in -\mathrm{vcl}_q E$.
(b') $f(\bar{x},y) + \epsilon d(\bar{x},y)q \notin -\mathrm{vcl}_q E$ for all $y \in A$ different from $\bar{x}$.

By taking $\epsilon = 1$ in Theorem 3.84, we obtain the following result.

Theorem 3.86 *Let (A,d), E, and q be as in Theorem 3.84. Assume that $f \in \mathscr{V}_E\mathscr{A}_\Delta \cap \mathscr{V}_E\mathscr{A}_A^=$ is such that for each $x \in A$, the set $\{y \in A \mid f(x,y) + d(x,y)q \in -E\}$ is closed and $f(x,A) \cap (e_x - \mathrm{core}E) = \emptyset$ for some $e_x \in Y$. Then, for all $x^0 \in A$, there exists $\bar{x} \in A$ such that:*

(a) $f(x^0,\bar{x}) + d(x^0,\bar{x})q \in -E$.
(b) $f(\bar{x},y) + d(\bar{x},y)q \notin -E$ *for all* $y \in A$ *different from* $\bar{x}$.

Theorem 3.87 *Let (A,d), E, and q be as in Theorem 3.84. Assume that $f \in \mathscr{V}_E\mathscr{A}_\Delta \cap \mathscr{V}_E\mathscr{A}_A^=$ is such that for each $x \in A$, $f(x,\cdot)$ is (q,E)-lower quasicontinuous and $f(x,A) \cap (e_x - \mathrm{core}E) = \emptyset$ for some $e_x \in Y$. Then, for all $\epsilon > 0$ and $x^0 \in A$, there exists $\bar{x} \in A$ such that:*

(a) $f(x^0,\bar{x}) + \epsilon d(x^0,\bar{x})q \in -E$.
(b) $f(\bar{x},y) + \epsilon d(\bar{x},y)q \notin -E$ *for all* $y \in A$ *different from* $\bar{x}$.

Proof By Proposition 3.83(b), as $h(\cdot) = f(x,\cdot)$ is (q,E)-lower quasicontinuous, the set $\{y \in A \mid f(x,y) + \epsilon d(x,y)q \in -E\}$ is closed for any $\epsilon > 0$. Thus, Theorem 3.87 is a consequence of Theorem 3.84. □

Remark 3.88 In Theorem 3.87, if E is not vectorially closed and the condition "$f(x,\cdot)$ is (q,E)-lower quasicontinuous" is replaced by "$f(x,\cdot)$ is (q,vcl_q)-lower quasicontinuous," then we obtain the conclusion: for all $\epsilon > 0$ and $x^0 \in A$, there exists $\bar{x} \in A$ such that (a') and (b') in Remark 3.85 are satisfied.

We state some corollaries from the above theorems.

Corollary 3.89 *Let (A,d), ϵ, E, q be as in Theorem 3.84, and $h : A \to Y$. Assume that $\{y \in A \mid h(y) - h(x) + \epsilon d(x,y)q \in -E\}$ is closed for each $x \in A$, and*

$h(A) \cap (\bar{e} - \text{core}E) = \emptyset$ for some $\bar{e} \in Y$. Then, for all $x^0 \in A$, there exists $\bar{x} \in A$ such that:

(a) $h(\bar{x}) - h(x^0) + \epsilon d(\bar{x}, x^0)q \in -E$.
(b) $h(y) - h(\bar{x}) + \epsilon d(\bar{x}, y)q \notin -E$ *for all* $y \in A$ *different from* $\bar{x}$.

Proof Corollary is a special case of Theorem 3.84 with $f(x, y) = h(y) - h(x)$ for all $x, y \in A$. Here we note that when $h(A) \cap (\bar{e} - \text{core}E) = \emptyset$, by taking $e_x = \bar{e} - h(x)$, we easily check that $f(x, A) \cap (e_x - \text{core}E) = \emptyset$. □

Corollary 3.90 *Let (A, d), q, and E be as in Theorem 3.84. Assume that $h : A \to Y$ is E-lower quasicontinuous and $h(A) \cap (\bar{e} - \text{core}E) = \emptyset$ for some $\bar{e} \in Y$. Then, for all $\epsilon > 0$ and $x^0 \in A$, there exists $\bar{x} \in A$ such that:*

(a) $h(\bar{x}) - h(x^0) + \epsilon d(\bar{x}, x^0)q \in -E$.
(b) $h(y) - h(\bar{x}) + \epsilon d(\bar{x}, y)q \notin -E$ *for all* $y \in A$ *different from* $\bar{x}$.

Proof When h is E-lower quasicontinuous, for each $x \in A$, $f(x, \cdot) = h(\cdot) - h(x)$ is E-lower quasicontinuous which implies that $f(x, \cdot)$ is (q, E)-lower quasicontinuous. Thus, Corollary 3.90 is a particular case of Theorem 3.87 with $f(x, y) = h(y) - h(x)$ for all $x, y \in A$. □

Remark 3.91 When h is E-lower quasicontinuous, $f(x, \cdot)$ is (q, E)-lower quasicontinuous and hence the set $\{y \in A \mid h(y) - h(x) + \epsilon d(x, y) \in -E\}$ is closed. Therefore, Corollary 3.90 is a consequence of Corollary 3.89. The conclusion of Corollary 3.90 may not be true if we replace the condition "h is E-lower quasicontinuous" with the condition "h is (q, E)-lower quasicontinuous." The reason is because the (q, E)-lower quasicontinuity of h does not imply the (q, E)-lower quasicontinuity of $f(x, \cdot) = h(\cdot) - h(x)$ (see Example 3.12).

Proposition 3.92 *Let $E \in \mathscr{E}_\bullet(Y)$, $q \in \mathscr{Q}_E$, $\epsilon > 0$, $\bar{x} \in A$, $f : A \times A \to Y$ and $h : A \to Y$ be such that $f(x, y) - (h(y) - h(x)) \in E$ for all $x, y \in A$. If $h(y) - h(\bar{x}) + \epsilon d(\bar{x}, y)q \notin -E$ for all $y \in A \setminus \{\bar{x}\}$, then $f(\bar{x}, y) + \epsilon d(\bar{x}, y)q \notin -E$ for all $y \in A \setminus \{\bar{x}\}$.*

Proof Suppose to the contrary that there exists $y \in A$ with $y \neq \bar{x}$ such that $f(\bar{x}, y) + \epsilon d(\bar{x}, y)q \in -E$. Then, $h(y) - h(\bar{x}) + \epsilon d(\bar{x}, y)q = f(\bar{x}, y) + \epsilon d(\bar{x}, y)q - (f(\bar{x}, y) - (h(y) - h(\bar{x})) \in -E - E \subset -E$, a contradiction. □

Corollary 3.93 *Let (A, d), ϵ, E, q be as in Theorem 3.84, and $f : A \times A \to Y$. Assume that there exists a vector function $h : A \to Y$ such that $f(x, y) - (h(y) - h(x)) \in E$ for all $x, y \in A$, $\{y \in A \mid h(y) - h(x) + \epsilon d(x, y) \in -E\}$ is closed for each $x \in A$, and $h(A) \cap (\bar{e} - \text{core}E) = \emptyset$ for some $\bar{e} \in Y$. Then, for all $x^0 \in A$, there exists $\bar{x} \in A$ such that:*

(a) $h(\bar{x}) - h(x^0) + \epsilon d(\bar{x}, x^0)q \in -E$.
(b) $f(\bar{x}, y) + \epsilon d(\bar{x}, y)q \notin -E$ *for all* $y \in A$ *different from* $\bar{x}$.

Proof Corollary 3.93 is a consequence of Corollary 3.89 and Proposition 3.92. □

Corollary 3.94 *Let (A, d), E, q be as in Theorem 3.84, and $f : A \times A \to Y$. Assume that there exists a vector function $h : A \to Y$ such that $f(x, y) - (h(y) - h(x)) \in E$ for all $x, y \in A$, h is E-lower quasicontinuous and $h(A) \cap (\bar{e} - \text{core} E) = \emptyset$ for some $\bar{e} \in Y$. Then, for all $\epsilon > 0$ and $x^0 \in A$, there exists $\bar{x} \in A$ such that:*

(a) $h(\bar{x}) - h(x^0) + \epsilon d(\bar{x}, x^0)q \in -E$.
(b) $f(\bar{x}, y) + \epsilon d(\bar{x}, y)q \notin -E$ *for all* $y \in A$ *different from* $\bar{x}$.

Proof Corollary 3.94 is a consequence of Corollary 3.90 and Proposition 3.92. □

Finally, we state a result on the existence of solutions of VEP$_{(f,C)}$) via vectorial Ekeland variational principles.

Theorem 3.95 *Let (A, d) be a complete metric space, $E \in \mathscr{E}_\bullet(Y)$ be a vectorially closed set, $q \in \mathscr{Q}_E$, $f \in \mathscr{V}_E\mathscr{A}_\Delta \cap \mathscr{V}_E\mathscr{A}_A^=$ be such that for each $x \in A$, $f(x, A) \cap (e_x - \text{core} E) = \emptyset$ for some $e_x \in Y$, and the set $\{y \in A \mid f(x, y) + d(x, y)q \in -E\}$ is closed (in particular, $f(x, \cdot)$ is (q, E)-lower quasicontinuous). Assume that there exists $x^0 \in A$ such that for every $x \in A$ satisfying $f(x^0, x) + d(x^0, x)q \in -E$ and $f(x, A) \cap (-E) \neq \emptyset$, there is $y \in A$ different from x with $f(x, y) + d(x, y)q \in -E$. For $t \in \{He, Be, e, w\}$, if $E \in \mathscr{E}_t^1(C)$, then* $\text{Sol}^t(\text{VEP}_{(f,C)}) \neq \emptyset$.

Proof Under the conditions of Theorem 3.95, all the conditions of Theorem 3.86 are satisfied. Hence, there exists $\bar{x} \in A$ such that (i) $f(x^0, \bar{x}) + d(x^0, \bar{x})q \in -E$ and (ii) $f(\bar{x}, y) + d(\bar{x}, y)q \notin -E$ for all $y \neq \bar{x}$. Assume that (iii) $f(\bar{x}, A) \cap (-E) \neq \emptyset$. Then, by the assumption, (i) and (iii) imply that a $y \in A$ exists such that $y \neq \bar{x}$ and $f(\bar{x}, y) + d(x, y)q \in -E$, which contradicts (ii). Thus, $f(\bar{x}, A) \cap (-E) = \emptyset$, i.e., $f(\bar{x}, y) \notin -E$ for all $y \in A$. This implies that $\bar{x} \in \text{Sol}(EP_{\Phi^1_{Ef}})$ (with Φ^1_{Ef} defined by (3.14)). Finally, the conclusion of Theorem 3.95 follows from Proposition 3.29. □

3.4 Existence of Solutions to Vector Quasiequilibrium Problems

In this section, we consider the following vector quasiequilibrium problem, which is an extension of quasiequilibrium problem (QEP) to vector bifunctions.

(VQEP$_{(f,\mathscr{D},C)}$) find $\bar{x} \in \mathscr{D}(\bar{x})$ such that $f(\bar{x}, y) \notin -C_0$for all $y \in A$,

where A is a nonempty set, and $C \subset Y$ is a nonempty set such that $0 \notin \text{core} C$, $C_0 := C \setminus \{0\}$, $\mathscr{D} : A \rightrightarrows A$ and $f : A \times A \to Y$. When $Y = \mathbb{R}$ and $C = \mathbb{R}_+$, (VQEP$_{(f,\mathscr{D},C)}$) becomes (QEP). If $\mathscr{D}(x) = A$ for all $x \in A$, (VQEP$_{(f,\mathscr{D},C)}$) becomes (VEP$_{(f,C)}$).

Concepts of solutions of (VEP$_{(f,C)}$) are extended to (VQEP$_{(f,\mathscr{D},C)}$) as follows.

Definition 3.96 For problem (VQEP$_{(f,\mathscr{D},C)}$), a point $\bar{x} \in A$ is said to be:

(a) A HENIG-PROPER C-SOLUTION if there exists $D \in \mathscr{K}_C^{core}$ such that

$$\bar{x} \in \mathscr{D}(\bar{x}) \text{ and } (f(\bar{x}, \mathscr{D}(\bar{x})) + C) \cap (-\text{core}D) = \emptyset \tag{3.22}$$

(b) A BENSON-PROPER C-SOLUTION if

$$\bar{x} \in \mathscr{D}(\bar{x}) \text{ and } \text{vclcone}(f(\bar{x}, \mathscr{D}(\bar{x})) + C) \cap (-C_0) = \emptyset \tag{3.23}$$

(c) A C-SOLUTION if

$$\bar{x} \in \mathscr{D}(\bar{x}) \text{ and } f(\bar{x}, \mathscr{D}(\bar{x})) \cap (-C_0) = \emptyset \tag{3.24}$$

(d) A WEAK C-SOLUTION if, assuming that core$C \neq \emptyset$,

$$\bar{x} \in \mathscr{D}(\bar{x}) \text{ and } f(\bar{x}, \mathscr{D}(\bar{x})) \cap (-\text{core}C) = \emptyset \tag{3.25}$$

By Sol^{He}(VQEP$_{(f,\mathscr{D},C)}$) (Sol^{Be}(VQEP$_{(f,\mathscr{D},C)}$), Sol^{e}(VQEP$_{(f,\mathscr{D},C)}$), Sol^{w} (VQEP$_{(f,\mathscr{D},C)}$), resp.), we denote the set of all Henig-proper C-solutions (Benson-proper C-solutions, C-solutions, weak C-solutions, resp.) of (VQEP$_{(f,\mathscr{D},C)}$).

For $E \subset Y$ and $f : A \times A \to Y$, let the bifunctions $\Phi^1_{Ef}, \Phi^2_{Ef} : A \times A \to \mathbb{R}$ be defined by (3.14) and (3.15), i.e.,

$$\Phi^1_{Ef}(x, y) = \begin{cases} \ \ 0 \text{ if } f(x, y) \notin -E\}, \\ -1 \text{ otherwise,} \end{cases}$$

$$\Phi^2_{Ef}(x, y) = \begin{cases} \ \ 0 \text{ if } f(x, y) \in E\}, \\ -1 \text{ otherwise.} \end{cases}$$

For $\ell \in \{1, 2\}$, we consider the following scalar quasiequilibrium problem:

(QEP$_{\Phi^\ell_{Ef}}$) find $\bar{x} \in A$ such that $\bar{x} \in \mathscr{D}(\bar{x})$ and $\Phi^\ell_{Ef}(\bar{x}, y) \geq 0$ for all $y \in A$.

By Sol(QEP$_{\Phi^\ell_{Ef}}$), we denote the set of all solutions of (QEP$_{\Phi^\ell_{Ef}}$).

The following proposition is an extension of Proposition 3.29 to (VQEP$_{(f,\mathscr{D},C)}$).

Proposition 3.97 *For $\ell \in \{1, 2\}$ and $t \in \{He, Be, e, w\}$,* Sol^t(VQEP$_{(f,\mathscr{D},C)}$) $\neq \emptyset$ *if and only if there exists $E \in \mathscr{E}^\ell_t(C)$ such that* Sol(QEP$_{\Phi^\ell_{Ef}}$) $\neq \emptyset$.

Proof The proof is similar to that of Proposition 3.29. Case $\ell = 1$. For the necessity part, when Sol^t(VQEP$_{(f,\mathscr{D},C)}$) $\neq \emptyset$, by taking $\bar{x} \in \text{Sol}^{Be}$(VQEP$_{(f,\mathscr{D},C)}$), $E = C + \text{core}D$ for $t = He$, $E = C$ for $t = e$, $E = \text{core}C$ for $t = w$ and $E = -(Y \setminus f(\bar{x}, \mathscr{D}(\bar{x})))$ for $t = Be$, we easily check that Sol(QEP$_{\Phi^\ell_{Ef}}$) $\neq \emptyset$.

For the sufficiency part, assume that $E \in \mathscr{E}_t^1(C)$ and $\mathrm{Sol}(\mathrm{QEP}_{\Phi^1_{Ef}}) \neq \emptyset$. For $\bar{x} \in \mathrm{Sol}(\mathrm{EP}_{\Phi^1_{Ef}})$, we have that $\bar{x} \in \mathscr{D}(\bar{x})$ and

$$f(\bar{x}, \mathscr{D}(\bar{x})) \subset Y \backslash (-E) = -E^c. \tag{3.26}$$

For $t = He$, as $E \in \mathscr{E}^1_{He}(C)$, there exists $D \in \mathscr{K}_C^{core}$ such that $C + \mathrm{core} D \subset E$, which is equivalent to $(C - E^c) \cap (-\mathrm{core} D) = \emptyset$. By (3.26), we have $(f(\bar{x}, \mathscr{D}(\bar{x})) + C) \cap (-\mathrm{core} D) \subset (C - E^c) \cap (-\mathrm{core} D) = \emptyset$. Thus, $\bar{x} \in \mathrm{Sol}^{He}(\mathrm{VQEP}_{(f,\mathscr{D},C)})$.

For $t = Be$, as $E \in \mathscr{E}^1_{Be}(C)$, (3.26) implies that $\mathrm{vcl\,cone}(f(\bar{x}, \mathscr{D}(\bar{x})) + C) \cap (-C_0) \subset \mathrm{vcl\,cone}(C - E^c) \cap (-C_0) = \emptyset$, i.e., $\bar{x} \in \mathrm{Sol}^{Be}(\mathrm{VQEP}_{(f,\mathscr{D},C)})$.

For $t = e$, as $C_0 \subset E$, (3.26) implies that $f(\bar{x}, \mathscr{D}(\bar{x})) \cap (-C_0) \subset (-E^c) \cap (-C_0) = \emptyset$. Thus, $\bar{x} \in \mathrm{Sol}^e(\mathrm{VQEP}_{(f,\mathscr{D},C)})$.

For $t = e$, as $\mathrm{core} C \subset E$, by (3.26), $f(\bar{x}, \mathscr{D}(\bar{x})) \cap (-\mathrm{core} C) \subset (-E^c) \cap (-\mathrm{core} C) = \emptyset$. Hence, $\bar{x} \in \mathrm{Sol}^w(\mathrm{VQEP}_{(f,\mathscr{D},C)})$.

Case $\ell = 2$. By replacing E in the case $\ell = 1$ by $E' := -E^c$, we obtain the statement for the case $\ell = 2$. □

In Chap. 2 we used $\mathrm{Fix}\mathscr{D}$ and $\delta(\mathrm{Fix}\mathscr{D})$ to denote the set of fixed points of $\mathscr{D}$ and the boundary of $\mathrm{Fix}\mathscr{D}$. We restate condition (B) in Theorem 2.21 for $(\mathrm{QEP}_{\Phi^\ell_{Ef}})$ as follows.

(VB_ℓ) *For each* $x \in \delta(\mathrm{Fix}\mathscr{D}) \setminus \mathrm{Fix}\mathscr{D}$, *there exists* $y \in \mathscr{D}(x)$ *such that* $x \in \overline{\mathrm{lev}}^\ell_E f(\cdot, y)$ (i.e., $f(x, y) \in -E$ *for* $\ell = 1$, *and* $f(x, y) \notin E$ *for* $\ell = 2$).

The condition (VB_ℓ) is obviously satisfied if $\mathrm{Fix}\mathscr{D}$ is closed (in particular, $\mathrm{Fix}\mathscr{D} = A$ or $\mathscr{D}$ is closed). (VB_ℓ) is also satisfied if

$$A = \bigcup_{y \in A} [\mathscr{D}^{-1}(y) \cap \overline{\mathrm{lev}}^\ell_E f(\cdot, y)].$$

Theorem 3.98 *Let A be a nonempty compact convex subset of a topological vector space X, $t \in \{He, Be, e, w\}$, $E \subset Y$ be nonempty, and $\ell \in \{1, 2\}$. Assume that $E \in \mathscr{E}^\ell_t(C)$, $f \in \mathscr{V}^\ell_E \mathscr{A}^2_{pr.qm}$, is closed for each $y \in A$, $\mathscr{D}$ has convex images, $\mathscr{D}^{-1}(y)$ and $\overline{\mathrm{lev}}^\ell_E f(\cdot, y)$ are open for each $y \in A$, and the condition (VB_ℓ) is satisfied. Then,* $\mathrm{Sol}^t(\mathrm{VQEP}_{(f,\mathscr{D},C)}) \neq \emptyset$.

Proof Condition (VB_ℓ) is condition (B) in Theorem 2.21 for $\mathscr{D}$ and Φ^ℓ_{Ef}. By Remark 3.37(c), $f \in \mathscr{V}^\ell_E \mathscr{A}^2_{pr.qm}$ if and only if $\Phi^\ell_{Ef} \in \mathscr{A}^2_{pr.qm}$. Moreover, $\mathrm{lev}_< \Phi^\ell_{Ef}(\cdot, y) = \overline{\mathrm{lev}}^\ell_E f(\cdot, y)$ for all $y \in A$. Thus, under the conditions of Theorem 3.98, all the conditions of Theorem 2.21 are satisfied for $\mathscr{D}$ and Φ^ℓ_{Ef}. Hence, by Theorem 2.21, $\mathrm{Sol}(\mathrm{QEP}_{\Phi^\ell_{Ef}}) \neq \emptyset$. Finally, the conclusion follows from Proposition 3.97. □

The following corollary is a vector version of Corollary 2.24.

Corollary 3.99 *Let A be a nonempty compact convex subset of a Hausdorff locally convex space X, $t \in \{He, Be, e, w\}$, $E \subset Y$ be nonempty, and $\ell \in \{1, 2\}$. Assume that $E \in \mathscr{E}_t^\ell(C)$, $f \in \mathscr{V}_E^\ell \mathscr{A}_{pr.qm}^2$, $\text{lev}_E^\ell f(\cdot, y)$ is closed for each $y \in A$, $\mathscr{D}$ is closed with convex images, and the set $\{x \in A \mid \mathscr{D}(x) \cap \overline{\text{lev}}_E^\ell f(x, \cdot) = \emptyset\}$ is closed. Then,* $\text{Sol}^t(\text{VQEP}_{(f,\mathscr{D},C)}) \neq \emptyset$.

Proof For all $x, y \in A$, $\text{lev}_{\geq}\Phi_{Ef}^\ell(\cdot, y) = \text{lev}_E^\ell f(\cdot, y)$ and $\text{lev}_{<}\Phi_{Ef}^\ell(x, \cdot) = \overline{\text{lev}}_E^\ell f(x, \cdot)$. Therefore, from the conditions of Corollary 3.99, we easily check the conditions of Corollary 2.24 for $\mathscr{D}$ and Φ_{Ef}^ℓ. Thus, by Corollary 2.24, $\text{Sol}(\text{QEP}_{\Phi_{Ef}^\ell}) \neq \emptyset$. This and Proposition 3.97 imply the conclusion of Corollary 3.99. □

A vector version of Theorem 2.28 reads as follows.

Theorem 3.100 *Let A be a nonempty compact convex subset of a separable Banach space X, $t \in \{He, Be, e, w\}$, $E \subset Y$ be nonempty, and $\ell \in \{1, 2\}$. Assume that $E \in \mathscr{E}_t^\ell(C)$, $f \in \mathscr{V}_E^\ell \mathscr{A}_{pr.qm}^2$, $\mathscr{D}$ is lower semicontinuous with nonempty convex images,* $\text{Fix}\mathscr{D}$ *is closed, and the set-valued mapping $S_{\text{Fix}\mathscr{D}} : \text{Fix}\mathscr{D} \rightrightarrows A$, defined by $S_{\text{Fix}\mathscr{D}}(x) = \mathscr{D}(x) \cap \overline{\text{lev}}_E^\ell f(x, \cdot)$ for all $x \in \text{Fix}\mathscr{D}$, is lower semicontinuous. Then,* $\text{Sol}^t(\text{VQEP}_{(f,\mathscr{D},C)}) \neq \emptyset$.

Proof Under the conditions of Theorem 3.100, all the conditions of Theorem 2.28 are satisfied for $\mathscr{D}$ and Φ_{Ef}^ℓ. Thus, Theorem 3.100 is a consequence of Theorem 2.28 and Proposition 3.97. □

The following lemma extends Lemma 3.53 to $(\text{QEP}_{\Phi_{Ef}^\ell})$. For the notion of a vector bifunction having the E-dual property of type ℓ, see Definition 3.51.

Lemma 3.101 *Let A be a nonempty compact convex subset of a topological vector space X, $E \subset Y$ be nonempty, $\mathscr{D} : A \rightrightarrows A$, and $f : A \times A \to Y$. Let $f' : A \times A \to Y$ be defined by (3.18), i.e.,*

$$f'(x, y) = -f(y, x) \text{ for all } x, y \in A.$$

If $\mathscr{D}$ has convex images and f has the E-dual property of type ℓ, then $\text{Sol}(\text{QEP}_{\Phi_{Ef'}^\ell}) \subset \text{Sol}(\text{QEP}_{\Phi_{Ef}^\ell})$.

Proof We prove this for $\ell = 1$; the case $\ell = 2$ can be proved similarly. Assume that $\bar{x} \in \text{Sol}(\text{QEP}_{\Phi_{Ef'}^1})$, i.e., $\bar{x} \in \mathscr{D}(\bar{x})$ and $f'(\bar{x}, y) \notin -E$ for all $y \in \mathscr{D}(\bar{x})$. Then, $\bar{x} \in \mathscr{D}(\bar{x})$ and $f(y, \bar{x}) \notin E$ for all $y \in \mathscr{D}(\bar{x})$. Take any $y \in \mathscr{D}(\bar{x})$, $t \in [0, 1]$, and $y_t = \bar{x} + t(y - \bar{x})$. Since $\mathscr{D}(\bar{x})$ is convex, $y_t \in \mathscr{D}(\bar{x})$ for all $y_t := (1 - t)\bar{x} + ty \in (\bar{x}, y]$. Hence, $f(y_t, \bar{x}) \notin E$ for all $y_t \in (\bar{x}, y]$. By the E-dual property of type 1 of f, $f(\bar{x}, y) \notin -E$. Because $y \in A$ is arbitrary, $\bar{x} \in \text{Sol}(\text{QEP}_{\Phi_{Ef}^1})$. □

For f' in Lemma 3.101, condition (B) in Theorem 2.21 is restated for $(\text{QEP}_{\Phi_{Ef'}^\ell})$ as follows.

(VB'$_\ell$) *for each* $x \in \delta(\mathrm{Fix}\mathscr{D}) \setminus \mathrm{Fix}\mathscr{D}$, *there exists* $y \in \mathscr{D}(x)$ *such that* $x \in \mathrm{lev}_E^{\ell c} f(y,\cdot)$ (i.e., $f(y,x) \in E$ *for* $\ell = 1$, *and* $f(x,y) \notin -E$ *for* $\ell = 2$).

(VB'$_\ell$) holds if either Fix$\mathscr{D}$ is closed (in particular, Fix$\mathscr{D} = A$ or $\mathscr{D}$ is closed) or

$$A = \bigcup_{y\in A} [\mathscr{D}^{-1}(y) \cap \mathrm{lev}_E^{\ell c} f(y,\cdot)].$$

Theorem 3.102 *Let A be a compact convex subset of a topological vector space,* $t \in \{He, Be, e, w\}$, $E \subset Y$ *be nonempty, and* $\ell \in \{1,2\}$. *Assume that* $E \in \mathscr{E}_t^\ell(C)$, $f \in -\mathscr{V}_E^\ell \mathscr{A}^1_{pr.qm}$, *f has the E-dual property of type* ℓ, $\mathscr{D}$ *has convex images, and* $\mathscr{D}^{-1}(y)$ *and* $\mathrm{lev}_E^{\ell c} f(y,\cdot)$ *are open for each* $y \in A$, *and condition (VB'$_\ell$) is satisfied. Then,* $\mathrm{Sol}^t(\mathrm{VQEP}_{(f,\mathscr{D},C)}) \neq \emptyset$.

Proof Let f' be as in Lemma 3.101. Condition (VB'$_\ell$) is condition (B) in Theorem 2.21 for $\mathscr{D}$ and $\Phi^\ell_{Ef'}$. Moreover, it is not hard to check that $f \in -\mathscr{V}_E^\ell \mathscr{A}^1_{pr.qm}$ if and only if $\Phi^\ell_{Ef'} \in \mathscr{A}^2_{pr.qm}$, and $\mathrm{lev}_E^{\ell c} f(y,\cdot) = \mathrm{lev}_{<}\Phi^\ell_{Ef'}(\cdot,y)$ for all $y \in A$. Thus, under the conditions of Theorem 3.102, all the conditions of Theorem 2.21 are satisfied for $\mathscr{D}$ and f'. Hence, by Theorem 2.21, $\mathrm{Sol}(\mathrm{QEP}_{\Phi^\ell_{Ef'}}) \neq \emptyset$, which implies that $\mathrm{Sol}(\mathrm{QEP}_{\Phi^\ell_{Ef}}) \neq \emptyset$ by Lemma 3.101. Finally, the conclusion follows from Proposition 3.97. □

Corollary 3.103 *Let A be a nonempty compact convex subset of a Hausdorff locally convex space X,* $t \in \{He, Be, e, w\}$, $E \subset Y$ *be nonempty, and* $\ell \in \{1,2\}$. *Assume that* $E \in \mathscr{E}_t^\ell(C)$, $f \in -\mathscr{V}_E^\ell \mathscr{A}^1_{pr.qm}$, *f has the E-dual property of type* ℓ, $\overline{\mathrm{lev}}_E^{\ell c} f(x,\cdot)$ *is closed for each* $x \in A$, $\mathscr{D}$ *is closed with convex images, and the set* $\{x \in A \mid \mathscr{D}(x) \cap \mathrm{lev}_E^{\ell c} f(\cdot,x) = \emptyset\}$ *is closed. Then,* $\mathrm{Sol}^t(\mathrm{VQEP}_{(f,\mathscr{D},C)}) \neq \emptyset$.

Proof Let f' be as in Lemma 3.101. For all $x \in A$, $\mathrm{lev}_{\geq}\Phi^\ell_{Ef'}(\cdot,x) = \overline{\mathrm{lev}}_E^{\ell c} f(x,\cdot)$ and $\mathrm{lev}_{<}\Phi^\ell_{Ef'}(x,\cdot) = \mathrm{lev}_E^{\ell c} f(\cdot,x)$. Therefore, from the conditions of Corollary 3.103, we can check the conditions of Corollary 2.24 for $\mathscr{D}$ and $\Phi^\ell_{Ef'}$. Thus, by Corollary 2.24, $\mathrm{Sol}(\mathrm{QEP}_{\Phi^\ell_{Ef'}}) \neq \emptyset$, which implies that $\mathrm{Sol}(\mathrm{QEP}_{\Phi^\ell_{Ef}}) \neq \emptyset$ by Lemma 3.101. This and Proposition 3.97 imply the conclusion of Corollary 3.103. □

Theorem 3.104 *Let A be a nonempty compact convex subset of a separable Banach space X,* $t \in \{He, Be, e, w\}$, $E \subset Y$ *be nonempty, and* $\ell \in \{1,2\}$. *Assume that* $E \in \mathscr{E}_t^\ell(C)$, $f \in -\mathscr{V}_E^\ell \mathscr{A}^2_{pr.qm}$, *f has the E-dual property of type* ℓ, $\mathscr{D}$ *is lower semicontinuous with nonempty convex images,* Fix$\mathscr{D}$ *is closed, and the set-valued mapping* $S_{\mathrm{Fix}\mathscr{D}} : \mathrm{Fix}\mathscr{D} \rightrightarrows A$, *defined by* $S_{\mathrm{Fix}\mathscr{D}}(x) = \mathscr{D}(x) \cap \mathrm{lev}_E^{\ell c} f(\cdot,x)$ *for all* $x \in \mathrm{Fix}\mathscr{D}$, *is lower semicontinuous. Then,* $\mathrm{Sol}^t(\mathrm{VQEP}_{(f,\mathscr{D},C)}) \neq \emptyset$.

Proof Under the conditions of Theorem 3.100, all the conditions of Theorem 2.28 are satisfied for $\mathscr{D}$ and $\Phi^\ell_{Ef'}$. Therefore, we easily obtain the conclusion of Theorem 3.104 from Theorem 2.28, Lemma 3.101 and Proposition 3.97. □

3.5 Exercises

Exercise 3.1 Prove that for any $E \in \mathscr{E}(Y)$, $E \cap (-\mathrm{core}E) = \emptyset$ (and hence $0 \notin \mathrm{core}E$) and $E \cap (-E) = (E \backslash \mathrm{core}E) \cap ((-E) \backslash (-\mathrm{core}E))$.

Exercise 3.2 Prove that if r_E^q is finite and $E + E' \subset \mathbb{R}_+ q + \mathrm{vcl}_q E$, then r_E^q is $\mathrm{core}E'$-increasing.

Exercise 3.3 For each $E \subset Y$, let $\mathrm{P}_E := \{q \in Y \mid \mathbb{R}_+ q + E = E\}$ and $\mathrm{Q}_E := \{q \in Y \mid \mathbb{R}_{++} q + E = \mathrm{core}E\}$. Prove that $\mathrm{Q}_E \subset \mathrm{P}_E$.

Exercise 3.4 For each $E \subset Y$, let

$$E_{\mathscr{E}(Y)} = \{E' \in \mathscr{E}(Y) \mid E' \subset E \text{ and } E' \cap (-\mathrm{core}E) = \emptyset\},$$
$$\mathscr{B}_E = \bigcup_{E' \in E_{\mathscr{E}(Y)}} \mathrm{core}E',$$
$$\mathrm{H}_E := \{q \in Y \mid E + E \subset \mathbb{R}_+ q + \mathrm{vcl}_q E\}.$$

Prove that:

(a) If $q \in \mathscr{B}_E \cap \mathrm{Q}_E$, then r_E^q is finite.
(b) If $q \in \mathscr{B}_E \cap \mathrm{Q}_E \cap \mathrm{H}_E$, then r_E^q is finite and $\mathrm{core}E$-increasing.
(c) If $q \in \mathrm{Q}_E$, then $\{v \in V \mid r_E^q(v) < 0\} = -\mathrm{core}E$.

Exercise 3.5 Let $E \in \mathscr{E}(Y)$. Prove that:

(a) $\mathrm{H}_E = Y$.
(b) $\mathscr{B}_E = \mathrm{core}E$.

Exercise 3.6 Prove that if $E \in \mathscr{K}_c$, then $\mathrm{Q}_E = \mathrm{core}E$.

Exercise 3.7 Let A be a subset of a topological vector space, Y be a topological vector space, $E \subset Y$ be nonempty such that $E + E \subset E$, and $h : A \to Y$. Consider the following statements:

(a) h is E-lower semicontinuous.
(b) For all $v \in Y$, $h^{-1}(v + \mathrm{int}E)$ is open.
(c) For each $x \in A$ and any $v \in \mathrm{int}E$ there exists an open neighborhood $U \subset A$ of x such that $h(x') \in h(x) - v + \mathrm{int}E)$ for all $x' \in U$.

Prove that (a) $\Rightarrow$ (b) $\Rightarrow$ (c), and if addition, $0 \in \delta(\mathrm{int}E)$, where $\delta(\mathrm{int}E)$ is the boundary of $\mathrm{int}E$, then (c) $\Rightarrow$ (a). Note that if E is convex, then $0 \in \delta(\mathrm{int}E)$.

Exercise 3.8 Let A be a subset of a topological vector space, Y be a topological vector space, $E \subset Y$ be nonempty such that $E + E \subset E$, and $h : A \to Y$. Consider the following statements:

(a) h is E-upper semicontinuous.
(b) For all $v \in Y$, $h^{-1}(v - \mathrm{int}E)$ is open.
(c) For each $x \in A$ and any $v \in \mathrm{int}E$ there exists an open neighborhood $U \subset A$ of x such that $h(x') \in h(x) + v - \mathrm{int}E)$ for all $x' \in U$.

Prove that (a) $\Rightarrow$ (b) $\Rightarrow$ (c), and if addition, $0 \in \delta(\text{int}E)$, where $\delta(\text{int}E)$ is the boundary of $\text{int}E$, then (c) $\Rightarrow$ (a).

Exercise 3.9 Prove assertions (b) and (d) in Proposition 3.14.

Exercise 3.10 Let A be a subset of a topological space, Y be a topological vector space, $E \subset Y$ be a convex cone, and $h, s : A \to Y$. Prove that if h and s are E-lower semicontinuous (E-upper semicontinuous, resp.), then the function $h + s : x \longmapsto h(x) + s(x)$ from A to Y is E-lower semicontinuous (E-upper semicontinuous, resp.).

Exercise 3.11 Let X, Z be topological spaces, Y be a topological vector space, $A \subset X$ and $B \subset Z$ be nonempty sets, $E \subset Y$ be a convex cone, and $h : A \to Y, s : B \to Y$. Prove that if h and s are E-lower semicontinuous (E-upper semicontinuous, resp.), then the function $f : (x, z) \longmapsto h(x) + s(z)$ from $A \times B$ to Y is E-lower semicontinuous (E-upper semicontinuous, resp.).

Exercise 3.12 Prove parts (d), (e), and (f) of Proposition 3.44.

Exercise 3.13 Let A be a convex subset of a topological vector space, Y be a vector space, $E \subset Y$ be nonempty, and $f : A \times A \to Y$. We say that:

(a) f is E-diagonally quasiconvex of type 1 in the first component (in the second component, resp.) on A if, for all $N = \{x^1, \ldots, x^n\} \in \mathscr{F}(A)$ and $x = \sum_{i=1}^n \lambda_i x^i \in \text{conv}N$, $\lambda_i \geq 0$ for all $i = 1, \ldots, n$ and $\sum_{i=1}^n \lambda_i = 1$, we have that $\sum_{i=1}^n \lambda_i f(x^i, x) - f(x, x) \notin -E$ ($\sum_{i=1}^n \lambda_i f(x, x^i) - f(x, x) \notin -E$, resp.). The bifunction f is called E-diagonally quasiconcave of type 1 in the first component (in the second component, resp.) if $-f$ is E-diagonally quasiconvex of type 1 in the first component (in the second component, resp.).
(b) f is E-diagonally quasiconvex of type 2 in the first component (in the second component, resp.) on A if, for all $N = \{x^1, \ldots, x^n\} \in \mathscr{F}(A)$ and $x = \sum_{i=1}^n \lambda_i x^i \in \text{conv}N$, $\lambda_i \geq 0$ for all $i = 1, \ldots, n$ and $\sum_{i=1}^n \lambda_i = 1$, we have that $\sum_{i=1}^n \lambda_i f(x^i, x) - f(x, x) \in E$ ($\sum_{i=1}^n \lambda_i f(x, x^i) - f(x, x) \in E$, resp.). The bifunction f is called E-diagonally quasiconcave of type 2 in the first component (in the second component, resp.) if $-f$ is E-diagonally quasiconvex of type 2 in the first component (in the second component, resp.).

Prove that:

(a) If $E \in \mathscr{E}(Y)$ is convex, $f(x, x) \notin -E$ for all $x \in A$ and f is E-diagonally quasiconvex of type 2 in the first component (in the second component, resp.), then f is properly E-quasimonotone of type 1 in the first component (in the second component, resp.).
(b) If $E^c \in \mathscr{E}(Y)$ is convex, $f(x, x) \in E$ for all $x \in A$ and f is E-diagonally quasiconvex of type 1 in the first component (in the second component, resp.), then f is properly E-quasimonotone of type 2 in the first component (in the second component, resp.).

Exercise 3.14 Let A be a convex subset of a vector space, Y be a vector space, $C \in \mathscr{E}(Y)$, and $l, g : A \times A \to Y$. Assume that:

(i) l is core$C$-diagonally quasiconvex of type 1 in the second component.
(ii) g is C-diagonally quasiconvex of type 2 in the second component.
(iii) For each $y \in A$, the set $\{x \in A \mid l(x, y) + g(x, y) - l(x, x) - g(x, x) \in -\text{core}C\}$ is open.
(iv) There exist a nonempty compact subset K of A and a $\bar{y} \in K \cap A$ such that, for every $x \in A \backslash K, l(x, \bar{y}) + l(x, \bar{y}) \in l(x, x) + g(x, x) - \text{core}C$.

Prove that there exists $\bar{x} \in A$ such that $l(\bar{x}, y)+g(\bar{x}, y) \notin l(\bar{x}, \bar{x})+g(\bar{x}, \bar{x})-\text{core}C$ for all $y \in A$.

Exercise 3.15 Let A be a convex subset of a vector space, Y be a vector space, $E \in \mathscr{K}_c$, and $h : A \times A \to Y$. Prove that if h is properly E-quasiconvex, then h is core$E$-quasiconvex.

Exercise 3.16 Let A be a convex subset of a vector space, Y be a vector space, $C \in \mathscr{K}_c$, and $f, g : A \times A \to Y$. Assume that:

(i) For all $x \in A$, $f(x, x) \notin \text{core}C$.
(ii) For all $x, y \in A$, $f(x, y) + g(x, y) \in -\text{core}C$ implies $f(y, x) \in \text{core}C$.
(iii) For all $y \in A$, $f(\cdot, y)$ is properly C-quasiconcave.

Prove that $f + g$ is properly core$C$-quasimonotone of type 1 in the second component.

Exercise 3.17 Let A be a convex subset of a topological vector space, Y be a vector space, $C \in \mathscr{K}_c$ and $f, g : A \times A \to Y$. Assume that conditions (i), (ii), and (iii) of Exercise 3.16 are satisfied, and:

(iv) For each $y \in A$, the set $\{x \in A \mid f(x, y) + g(x, y) \in -\text{core}C\}$ is open.
(v) There exists a nonempty compact subset K of X such that, for each finite subset N of A, there exists a compact convex subset L_N of A containing N such that for each $x \in L_N \backslash K$, there exists $y \in L_N$ such that $f(x, y) + g(x, y) \in -\text{core}C$.

Prove that there exists $\bar{x} \in A$ such that $f(\bar{x}, y) + g(\bar{x}, y) \notin -\text{core}C$ for all $y \in A$.

Exercise 3.18 Let A be a convex subset of a topological vector space, Y be a topological vector space, and $E \subset Y$ be nonempty. We say that a bifunction $f : A \times A \to Y$ is E-transfer lower semicontinuous (E-transfer upper semicontinuous, resp.) in the first component on A if, for all $x \in A$ and $y \in A$ and for any neighborhood V of $f(x, y)$, there exist a point $z \in A$ and a neighborhood $U(x)$ of x such that $f(x', z) \in V + E$ ($f(x', z) \in V - E$, resp.) for all $x' \in U(x)$.

Prove that the assumption "$\text{lev}_E^1 f(\cdot, y)$ is closed for each $y \in A$" in Theorem 3.46 can be replaced by "f is E-transfer upper semicontinuous in the first component on A."

Exercise 3.19 Consider ($\mathrm{VEP}_{(f,C)}$) with $Y = \mathbb{R}^k, C \subset \mathbb{R}^k$, and $f = (f_1, \ldots, f_k) : A \times A \to \mathbb{R}^k$, where $f_j : A \times A \to \mathbb{R}$ ($j = \overline{1,k}$. For $\lambda = (\lambda_1, \ldots, \lambda_k) \in \mathbb{R}^k$, let $g_\lambda : A \times A \to \mathbb{R}$ be defined by

$$g_\lambda(x, y) := \langle \lambda, f(x, y) \rangle = \sum_{j=1}^{k} \lambda_j f_j(x, y) \text{ for all } x, y \in A.$$

Prove that for $t \in \{He, Be, e, w\}$, if there exists $\lambda \in \mathbb{R}^k$ such that $E_\lambda := \{e \in \mathbb{R}^k \mid \langle \lambda, e \rangle > 0\} \in \mathscr{E}_t^1(C)$ and the following conditions are satisfied:

(i_λ) *g_λ is cyclically anti-quasimonotone.*
(ii_λ) *At least one of the following conditions is satisfied:*

(ii_λ^1) *A is a compact space and $\mathrm{lev}_{\geq} g_\lambda(\cdot, y)$ is closed for each $y \in A$.*
(ii_λ^2) *There exists a finite subset M of A such that*

$$\bigcup_{y \in A} \mathrm{lev}_{<} g_\lambda(\cdot, y) \subset \bigcup_{y \in M} \mathrm{lev}_{<} g_\lambda(\cdot, y).$$

Then, $\mathrm{Sol}^t(\mathrm{VEP}_{(f,C)}) \neq \emptyset$.

Exercise 3.20 Consider the variational inequality ($\mathrm{VVI}_{(T,C)}$) (Sect. 3.2.2). For the sake of simplicity assume that $A \subset X = \mathbb{R}^m$, $Y = \mathbb{R}^k$, $C \subset \mathbb{R}^k$. For each $j \in \{1, 2, \ldots, k\}$, $T_j = (T_{j1}, T_{j2}, \ldots, T_{jm})$ with $T_{j1}, T_{j2}, \ldots, T_{jm} : A \to \mathbb{R}$, $T = (T_1, \ldots, T_k)$.

($\mathrm{VVI}_{(T,C)}$) Find $\bar{x} \in A$ such that $(T(\bar{x}), y - \bar{x}) \notin -C \setminus \{0\}$ for all $y \in A$,

where $(T(x), y - x) = (\langle T_1(x), y - x \rangle, \ldots, \langle T_k(x), y - x \rangle) := f(x, y)$. Assume that A is compact.

(a) Prove that for $t \in \{He, Be, e, w\}$, if there exists $\lambda = (\lambda_1, \ldots, \lambda_k) \in \mathbb{R}^k$ such that $E_\lambda := \{e \in \mathbb{R}^k \mid \langle \lambda, e \rangle > 0\} \in \mathscr{E}_t^1(C)$ and the following condition holds:

$$\sum_{j=1}^{k} \lambda_j T_{ji}(x) = c_\lambda^i \gamma_\lambda(x), \; i = 1, \ldots, m, \tag{3.27}$$

where $c_\lambda^i \in \mathbb{R}$ ($i = \overline{1,m}$) and $\gamma_\lambda : A \to \mathbb{R}$ with $\gamma_\lambda(A) \subset \mathbb{R}_+$ or $\gamma_\lambda(A) \subset -\mathbb{R}_+$. Then, $\mathrm{Sol}^t(\mathrm{VVI}_{(T,C)}) \neq \emptyset$.
(b) Let $a_{ji} \in \mathbb{R}$ ($j = \overline{1,k}$, $i = \overline{1,m}$), $\gamma : A \to \mathbb{R}$ with $\gamma(A) \subset \mathbb{R}_+$ or $\gamma(A) \subset -\mathbb{R}_+$, and $T_{ji}(x) = a_{ji}\gamma(x)$ ($j = \overline{1,k}$, $i = \overline{1,m}$. Prove that for $t \in \{He, Be, e, w\}$ and $\lambda \in \mathbb{R}^k$ such that $E_\lambda \in \mathscr{E}_t^1(C)$, $\mathrm{Sol}^t(\mathrm{VVI}_{(T,C)}) \neq \emptyset$.
(c) Assume that

$$T_{ji}(x) = c^i \gamma_j(x),$$

where $c = (c^1, c^2, \ldots, c^m) \in \mathbb{R}^m$ and $\gamma_j : A \to \mathbb{R}$ $(j = \overline{1,k})$. Prove that for $t \in \{He, Be, e, w\}$, if there exists $\lambda = (\lambda_1, \ldots, \lambda_k) \in \mathbb{R}^k$ such that $E_\lambda \in \mathscr{E}_t^1(C)$ and $\gamma_\lambda(x) = \sum_{j=1}^k \lambda_j \gamma_j(x)$ satisfying "$\gamma_\lambda(x) \leq 0$ for all $x \in A$" or "$\gamma_\lambda(x) \geq 0$ for all $x \in A$," then $\mathrm{Sol}^t(\mathrm{VVI}_{(T,C)}) \neq \emptyset$.

Exercise 3.21 Consider the multiobjective traffic network $G = (\mathbf{N}, \mathbf{A}, c, \mathbf{W}, d)$ with capacity constraints on arcs (Sect. 3.2.2), $C \subset \mathbb{R}^k$ satisfying $(0, \alpha]C \subset C$ for some $\alpha > 0$.

(a) Assume that the vector cost of vector flow x on arc $a \in \mathbf{A}$ has the form $t_a(x) = r_a(\gamma_1(x), \ldots, \gamma_k(x))$, where $\gamma_1, \ldots, \gamma_k : \mathbb{R}^m \to \mathbb{R}$ $(m = |P = \bigcup_{w \in W} P_w|)$ are given; $\mathscr{D}_w$ is compact for each $w \in \mathbf{W}$; and for $t \in \{He, Be, e, w\}$, there exists $\lambda = (\lambda_1, \ldots, \lambda_k) \in \mathbb{R}^k$ such that $E_\lambda \in \mathscr{E}_t^1(C)$ and $\gamma_\lambda(x) = \sum_{j=1}^k \lambda_j \gamma_j(x)$ satisfying "$\gamma_\lambda(x) \leq 0$ for all $x \in A$" or "$\gamma_\lambda(x) \geq 0$ for all $x \in A$." Prove that $\mathrm{Sol}^t(G, C) \neq \emptyset$.
(b) Assume that the vector cost of vector flow x on arc a is $t_a(x) = (\alpha_a^1 \gamma(x), \ldots, \alpha_a^k \gamma(x))$ for $a \in \mathbf{A}$, where $\alpha_a^j \in \mathbb{R}$ and $\gamma : \mathbb{R}^m \to \mathbb{R}$ such that "$\gamma(x) \leq 0$ for all $x \in A$" or "$\gamma(x) \geq 0$ for all $x \in A$"; $\mathscr{D}_w$ is compact for each $w \in \mathbf{W}$; and for $t \in \{He, Be, e, w\}$, there exists $\lambda = (\lambda_1, \ldots, \lambda_k) \in \mathbb{R}^k$ such that $E_\lambda \in \mathscr{E}_t^1(C)$. Prove that $\mathrm{Sol}^t(G, C) \neq \emptyset$.

Exercise 3.22 Consider the multiobjective traffic network $G = (\mathbf{N}, \mathbf{A}, c, \mathbf{W}, d)$ with capacity constraints on arcs, where $\mathbf{N} = \{1, 2, 3, 4, 5\}$, $\mathbf{A} = \{a_{12}, a_{14}, a_{15}, a_{24}, a_{45}, a_{34}, a_{35}\}$, $\mathbf{W} = \{(1, 5), (3, 5)\}$ (then $m = 5$), for $w \in \mathbf{W}$, $\mathscr{D}_w$ is an arbitrary compact set. The vector cost of vector flow x on arc $a_{ij} \in \mathbf{A}$ has the form $t_{a_{ij}}(x) = (\gamma_{a_{ij}}(x), \gamma(x)) \in \mathbb{R}^2$ for $a_{ij} \in \mathbf{A}$, where $\gamma_{a_{ij}} : \mathbb{R}^5 \to \mathbb{R}$ and $\gamma : \mathbb{R}^5 \to (0, +\infty)$. Show that for any $C \setminus \{0\} \subset E := \{e = (e_1, e_2) \in \mathbb{R}^2 \mid e_2 > 0\}$ satisfying $(0, \alpha]C \subset C$ for some $\alpha > 0$, the four kinds of equilibrium flows in Definition 3.26 exist.

Exercise 3.23 Consider the noncooperative games with vector payoff functions $\mathscr{G} = (A_i, \mathbf{c}_i)_{i \in I}$ with $\mathbf{c}_i : A \to \mathbb{R}^k$ $(i \in I)$ defined by

$$\mathbf{c}_i(x) := (c_{i1}(x), \ldots, c_{ik}(x)) \quad \text{for all } x = (x_i)_{i \in I} \in A,$$

where $c_{ij} : A \to \mathbb{R}$. Assume that A_i is compact for each $i \in I$; for $t \in \{He, Be, e, w\}$, if there exists $\lambda = (\lambda_1, \ldots, \lambda_k) \in \mathbb{R}^k$ such that $E_\lambda \in \mathscr{E}_t^1(C)$ and $\gamma_\lambda^i(x) = \sum_{j=1}^k \lambda_j c_{ij}(x) = h_i(x_i) + s_i(x_{-i})$ for each $i \in I$, where $h_i : A_i \to \mathbb{R}$ and $s_i : A_{-i} \to \mathbb{R}$ such that h_i is upper semicontinuous $(i \in I)$. Prove that $\mathrm{Sol}^t(\mathscr{G}, C)$ $\neq \emptyset$.

3.6 Commentary

Motivated by the demands of practical situations where multi-criteria decisions are to be made, many scalar problems such as optimization problem, saddle point problem, and Nash equilibrium problem were extended for the vector case. In 1980, Giannessi [209] studied "vector variational inequality problem" in finite-dimensional spaces which is an extension of the variational inequality problem. Inspired by Giannessi's work, some researchers introduced vector equilibrium problems which are viewed as natural extensions of the scalar equilibrium problem and vector variational inequality problem. During the last decade these problems have been extensively studied due to the fact that they are unified models for several important problems for vector-valued functions, for example, vector optimization problems, vector complementarity problems, vector saddle point problems, and multiobjective games (see [64, 67, 68, 93, 103, 208, 215, 217, 275, 345] among others).

In the major part of the known works, vector equilibrium problems are in topological settings (i.e., with topological structures in the image space of bifunctions) and concepts of solutions are defined in terms of preference relations involving convex cones and topological tools (topological interior, topological closure). Therefore, to define these concepts of solutions, one requires both linear and topological structures on the image space of the objective bifunction. Namely, for Y being a topological vector space and C being a proper closed convex cone in Y such that $\text{int}C \neq \emptyset$, a point $\bar{x} \in A$ is called an efficient C-solution if $f(\bar{x}, y) \notin -C \setminus \{0\}$ for all $y \in A$ and $\bar{x} \in A$ is called a weak efficient C-solution if $f(\bar{x}, y) \notin -\text{int}C$ for all $y \in A$. Efficient and weak efficient solutions are frequently studied (e.g., [63, 67, 93, 97, 208, 223, 252, 275, 296]), while Henig- and Benson- proper efficient solutions are proposed and studied in [116, 215, 216].

Vector optimization problems and concepts of solutions in settings using binary relations defined by a general set have been investigated; see [88, 204, 205, 222, 224, 226, 367]. In [224] several kinds of solutions were considered without topological structures in the image spaces, using algebraic tools (algebraic interior, vectorial closure, etc.). In [226] a wide class of vector equilibrium problems in a non-topological setting with general preferences with a nonconical preference has been introduced, but only weak efficient solutions have been studied via scalarization techniques. Several existence results for weak efficient solutions in particular cases of $(\text{VEP})_{(f,C)}$ were obtained in [223] and [469] via scalarization or Ekeland-type variational principles.

In this chapter, the vector equilibrium problem $(\text{VEP})_{(f,C)}$ is stated in a general framework. The underlying space A is an arbitrary set without topological and linear structures; the image space Y is equipped only with a linear structure, without topology; and the preference C is not a preorder. This general model of $(\text{VEP})_{(f,C)}$ unifies many problems in both topological and non-topological settings. Observe that, for D being a convex cone, formulas (3.2), (3.3), and (3.5) can be used to express approximate solutions of $(\text{VEP})_{(f,D)}$ by setting $C = \varepsilon q + D$ with $\varepsilon > 0$ and

$q \in Y \setminus \{0\}$. So, $(\text{VEP})_{(f,C)}$ also permits considerations of approximate solutions of $(\text{VEP})_{(f,D)}$. Moreover, from a practical point of view, problems with general preferences in optimization and equilibria allow us to describe many situations in the real world, especially in economics where preference relations are not necessarily preorders (see, e.g., [88, 358, 409]) and the references therein).

In the aforementioned works and many subsequent ones, existence results have been established either by using KKM-type theorems (and equivalent results) and requiring convexity assumptions, or via Ekeland-type variational principles for vector bifunctions. The existence results in this chapter are developed from the results of Chap. 1 via the bifunctions $\Phi^1_{Ef}, \Phi^2_{Ef} : A \times A \to \mathbb{R}$:

$$\Phi^1_{Ef}(x, y) = \begin{cases} 0 \text{ if } f(x, y) \notin -E\}, \\ -1 \text{ otherwise.} \end{cases}$$

$$\Phi^2_{Ef}(x, y) = \begin{cases} 0 \text{ if } f(x, y) \in E\}, \\ -1 \text{ otherwise.} \end{cases}$$

In Sect. 3.2.1, the existence results are established by using convexity assumptions. This subsection is written based on many papers of many authors (mainly for efficient and weak efficient solutions). Here the results are stated in ways to unify many known results. Actually, via the bifunction Φ^E_f, we can see that some results of this subsection are vector versions of the results in Sect. 1.2 of Chap. 1. Section 3.2.1 presents some of our unpublished results [281].

Chapter 4
Stability of Scalar Equilibrium Problems

Abstract In this chapter, we consider parametric scalar equilibrium problems and discuss their stability conditions. Sufficient conditions for the semicontinuity and continuity of solution maps under relaxed continuity and generalized convexity assumptions are provided. The chapter not only is an effort to provide a complete and comprehensive presentation of the results obtained so far but also contains some new facts or refinements of known results. Hölder (or Lipschitz) continuity of solution maps of these problems has been investigated from the beginning, along with several types of (semi)continuity. Combining assumptions on Holder continuity with conditions related to (strong) monotonicity and (strong) convexity, we introduce Hölder/Lipschitz properties of approximate and exact solution maps of the reference problems. The results presented in this chapter could be regarded as an endeavor to unify the corresponding contributions in the literature. We discuss these properties in connection with well-posedness, one of the most important topics in both theory and numerical methods in optimization. Note that, under suitable adjustments, this concept can cover many important ones, including Tikhonov and Hadamard well-posedness. In this chapter, for the results found in the literature, we seek a unified exploration, not repeating the presentation of any article.

4.1 Continuity of the Solution Mappings of Scalar Equilibrium Problems

In this section, we study sufficient conditions for the solution maps of parametric equilibrium problems on Hausdorff topological spaces to be (upper/lower) semi-continuous or continuous. We often omit the word "parametric" for convenience. Unless otherwise specified, let X, Y, Z be Hausdorff locally convex topological spaces, and $A \subset X$, $\Lambda \subset Y$, $M \subset Z$ be nonempty subsets.

L. Q. Anh et al., *Equilibrium Problems*, Forum for Interdisciplinary Mathematics, https://doi.org/10.1007/978-3-031-89003-1_4

Let A, Λ, M be nonempty sets, $\Phi : A \times A \times M \to \mathbb{R}$, and $\mathscr{D} : \Lambda \rightrightarrows A$ be surjective. For each $(\lambda, \mu) \in \Lambda \times M$, we consider the following equilibrium problem:

$$\text{(EP)}(\lambda, \mu) \qquad \text{find } \bar{x} \in \mathscr{D}(\lambda) \text{ such that } \Phi(\bar{x}, y, \mu) \geq 0 \text{ for all } y \in \mathscr{D}(\lambda).$$

Its solution set Sol(EP)(λ, μ) depends on two parameters: λ and μ. The set-valued map $(\lambda, \mu) \rightrightarrows$ Sol(EP)(λ, μ) is called the SOLUTION MAP of the parametric equilibrium problem $\{(EP)_{\Lambda\times M}\} := \{(EP)(\lambda, \mu) \mid \lambda \in \Lambda, \mu \in M\}$, which is a family of equilibrium problems. In this section, we investigate some qualitative properties of this map. Because existence conditions have been studied in the preceding part of the book, we always assume the existence of the considered solutions.

Now, let us recall some definitions and some properties needed in what follows. For topological vector spaces X and Y, let $Q : X \rightrightarrows Y$ be a set-valued map.

Definition 4.1 (Concepts of semicontinuity for a set-valued (multivalued) map)

(a) Q is said to be UPPER SEMICONTINUOUS (usc) at $\bar{x} \in X$ if for any open subset U of Y with $Q(\bar{x}) \subset U$, there is a neighborhood N of $\bar{x}$ such that $Q(N) \subset U$.
(b) Q is said to be LOWER SEMICONTINUOUS (lsc) at $\bar{x} \in X$ if for any open subset U of Y with $Q(\bar{x}) \cap U \neq \emptyset$, there is a neighborhood N of $\bar{x}$ such that $\forall x \in N, Q(x) \cap U \neq \emptyset$.
(c) Q is said to be CONTINUOUS at $\bar{x} \in X$ if it is both usc and lsc at $\bar{x}$.
(d) Q is said to be closed at $\bar{x} \in X$ if for each net $\{(x^\nu, y^\nu)\}_\nu \in \text{gph}Q := \{(x, y) \mid y \in Q(x)\}$ converging to $(\bar{x}, \bar{y})$, then $\bar{y}$ belongs to $Q(\bar{x})$.

Note that in the study of stability of set-valued maps on topological spaces, there are two types of the notions related to convergence: in terms of nets and in terms of sequences. For the latter case, the word "sequential" is added in front of the relevant terms, for instance, sequentially semicontinuous, sequentially continuous, etc. (Observe that most contributions in the study deal with this case.) Hence, later on, we always consider the latter case if not otherwise specified, and hence we usually omit the term "sequential" without fear of confusion.

Lemma 4.2 (Equivalent forms of semicontinuity properties of a set-valued map) *Let X and Y be Hausdorff topological spaces and $Q : X \rightrightarrows Y$ be a set-valued map.*

1. *Q is (sequentially) lsc at $\bar{x}$ if, for any sequence $\{x^k\}_k$ converging to $\bar{x}$ and for any $\bar{y} \in Q(\bar{x})$, there exists some sequence $\{y^k\} \in Q(x^k)$ such that $y^k \to \bar{y}$.*
2. *If $Q(\bar{x})$ is compact, then Q is (sequentially) usc at x_0 if and only if for any sequence $\{x^k\}_k$ converging to $\bar{x} \in X$ and for any $y^k \in Q(x^k)$, there is a subsequence $\{y_j^{k_j}\}_j$ converging to some $y \in Q(\bar{x})$.*

Definition 4.3 (Concepts of Hausdorff semicontinuity for a set-valued map)

(a) Q is said to be Hausdorff upper semicontinuous (H-usc) at $\bar{x}$ if, for each neighborhood B of the origin in Y, there exists a neighborhood N of $\bar{x}$ such that $Q(x) \subset Q(\bar{x}) + B, \forall x \in N$.
(b) Q is said to be Hausdorff lower semicontinuous (H-lsc) at $\bar{x}$ if, for each neighborhood B of the origin in Y, there exists a neighborhood N of $\bar{x}$ such that $Q(\bar{x}) \subset Q(x) + B, \forall x \in N$.
(c) Q is said to be Hausdorff continuous at $\bar{x}$ if it is both H-usc and H-lsc at $\bar{x}$.

The map Q is said to satisfy a certain property on a subset A of X if it satisfies it at every point of A. If $A = X$, we omit "on X" in the statement.

Lemma 4.4 (Relations between semicontinuity, closedness, and Hausdorff semicontinuity of a set-valued map)

(a) *If Q is usc at $\bar{x}$, then Q is H-usc at $\bar{x}$. Conversely, if Q is H-usc at $\bar{x}$ and $Q(\bar{x})$ is compact, then Q is usc at $\bar{x}$.*
(b) *If Q is H-usc at $\bar{x}$ and if $Q(\bar{x})$ is closed, then Q is closed at $\bar{x}$.*
(c) *If $Q(A)$ is compact for any compact subset A of* $\mathrm{dom}Q := \{x \mid Q(x) \neq \emptyset\}$ *and if Q is closed at $\bar{x}$, then Q usc at $\bar{x}$.*
(d) *If Y is compact and if G is closed at $\bar{x}$, then G is usc at $\bar{x}$.*
(e) *If Q is H-lsc at $\bar{x}$, then Q is lsc at $\bar{x}$. Conversely, if Q is lsc at $\bar{x}$ and $Q(\bar{x})$ is compact, then Q is H-lsc at $\bar{x}$.*

Next, we discuss sufficient conditions for the semicontinuity of the solution map of an equilibrium problem.

Theorem 4.5 (Upper semicontinuity and closedness of solution maps)

(a) *If $\mathscr{D}$ is compact-valued and $\Phi(\cdot, y, \cdot)$ is usc for each $y \in A$, then* $\mathrm{Sol(EP)}(\lambda, \cdot)$ *is both usc and closed on M for each $\lambda \in \Lambda$.*
(b) *If $\mathscr{D}$ is continuous and compact-valued, and $\Phi(\cdot, \cdot, \mu)$ is usc for each $\mu \in M$, then* $\mathrm{Sol(EP)}(\cdot, \mu)$ *is both usc and closed on Λ for each $\mu \in M$.*
(c) *If $\mathscr{D}$ is continuous and compact-valued and Φ is usc, then* Sol(EP) *is both usc and closed on $\Lambda \times M$.*
(d) *In (a) ((b), (c), resp.), if additionally, $\Phi(\cdot, \cdot, \mu) \in \mathscr{A}_{s.pm}$ for each $\mu \in M$, then* $\mathrm{Sol(EP)}(\lambda, \cdot)$ $\big(\mathrm{Sol(EP)}(\cdot, \mu)$, Sol(EP), *resp.*$\big)$ *is a single-valued map and hence is continuous.*
(e) *In (a) ((b), (c), resp.), if additionally, $\mathscr{D}$ is convex-valued, and $\Phi(\cdot, \cdot, \mu) \in (-\mathscr{A}^1_{s.qc}) \cap \mathscr{A}^{=}_{A}$ for each $\mu \in M$, then* $\mathrm{Sol(EP)}(\lambda, \cdot)$ $\big(\mathrm{Sol(EP)}(\cdot, \mu)$, Sol(EP), *resp.*$\big)$ *is a single-valued map and so is continuous.*

Proof First, we make the following remarks.

($r1$) Under the conditions in (a), for $\lambda \in \Lambda$, if $\mu^k \to \bar{\mu}$ and $x^k \in \mathrm{Sol(EP)}(\lambda, \mu^k)$ for each k, then x^k converges (or a subsequence converges) to some $\bar{x} \in \mathrm{Sol(EP)}(\lambda, \bar{\mu})$.

Indeed, because $\mathscr{D}(\lambda)$ is compact, we can assume that x^k converges to some $\bar{x} \in \mathscr{D}(\lambda)$. For any $y \in \mathscr{D}(\lambda)$, by the upper semicontinuity of $\Phi(\cdot, y, \cdot)$, we have

$$\Phi(\bar{x}, y, \bar{\mu}) \geq \limsup_{k} \Phi(x^k, y, \mu^k) \geq 0,$$

and so $\bar{x} \in \mathrm{Sol(EP)}(\lambda, \bar{\mu})$.

$(r2)$ Under the conditions in (b), for $\mu \in M$, if $\lambda^k \to \bar{\lambda}$ and $x^k \in \mathrm{Sol(EP)}(\lambda^k, \mu)$ for each k, then x^k converges to some $\bar{x} \in \mathrm{Sol(EP)}(\bar{\lambda}, \mu)$.

Indeed, since $x^k \in \mathscr{D}(\lambda^k)$ and $\mathscr{D}$ is upper semicontinuous and compact-valued, we can assume that x^k converges to some $\bar{x} \in \mathscr{D}(\bar{\lambda})$. Let $\bar{y} \in \mathscr{D}(\bar{\lambda})$ be arbitrary. It follows from the lower semicontinuity of $\mathscr{D}$ (at $\bar{\lambda}$) that there exists $y^k \in \mathscr{D}(\lambda^k)$ such that $y^k \to \bar{y}$. Since $x^k \in \mathrm{Sol(EP)}(\lambda^k, \mu)$, we have $\Phi(x^k, y^k, \mu) \geq 0$ for all k. By the upper semicontinuity of $\Phi(\cdot, \cdot, \mu)$,

$$\Phi(\bar{x}, \bar{y}, \mu) \geq \limsup_{k} \Phi(x^k, y^k, \mu) \geq 0.$$

Hence, $\bar{x} \in \mathrm{Sol(EP)}(\bar{\lambda}, \mu)$.

Repeating this argument with $\mathrm{Sol(EP)}(\lambda^k, \mu^k)$, $\Phi(x^k, y^k, \mu^k)$, and $\Phi(\bar{x}, \bar{y}, \mu)$ replacing Sol(EP) (λ^k, μ), $\Phi(x^k, y^k, \mu)$, and $\Phi(\bar{x}, \bar{y}, \mu)$, resp., we obtain the following remark $(r3)$.

$(r3)$ Under the conditions in (c), if $(\lambda^k, \mu^k) \to (\bar{\lambda}, \bar{\mu})$, and $x^k \in \mathrm{Sol(EP)}(\lambda^k, \mu^k)$ for each k, then x^k converges to some $\bar{x} \in \mathrm{Sol(EP)}(\bar{\lambda}, \bar{\mu})$.

$(r4)$ For any $(\lambda, \mu) \in \Lambda \times M$,

$$\mathrm{Sol(EP)}(\lambda, \mu) = \mathscr{D}(\lambda) \cap \bigcap_{y \in \mathscr{D}(\lambda)} \mathrm{lev}_{\geq} \Phi(\cdot, y, \mu).$$

In (a), (b), and (c), we always have the following: $\mathscr{D}(\lambda)$ is a compact set and $\Phi(\cdot, y, \mu)$ is upper semicontinuous for each $y \in A$. Hence, $\mathscr{D}(\lambda)$ is a closed set (a compact subset of a Hausdorff topological space) and $\mathrm{lev}_{\geq} \Phi(\cdot, y, \mu)$ is closed, and so $\mathrm{Sol(EP)}(\lambda, \mu)$ is closed. Thus, the values of the maps $\mathrm{Sol(EP)}(\lambda, \cdot)$, $\mathrm{Sol(EP)}(\cdot, \mu)$, and Sol(EP) in (a), (b), and (c), resp., are closed.

Now, by $(r4)$ and the fact that an upper semicontinuous mapping with closed values is closed, it is sufficient to show that $\mathrm{Sol(EP)}(\lambda, \cdot)$ ($\mathrm{Sol(EP)}(\cdot, \mu)$, Sol(EP), resp.) in (a) ((b), (c), resp.) is upper semicontinuous. If $\mathrm{Sol(EP)}(\lambda, \cdot)$ $\big(\mathrm{Sol(EP)}(\cdot, \mu)$, Sol(EP), resp.$\big)$ in (a) $\big($(b), (c), resp.$\big)$ is not upper semicontinuous at some $\bar{\mu} \in M$ $\big(\bar{\lambda}$, $(\bar{\lambda}, \bar{\mu})$, resp.$\big)$, then there are an open set $U \supset \mathrm{Sol(EP)}(\lambda, \bar{\mu})$ $\big(\mathrm{Sol(EP)}(\bar{\lambda}, \mu)$, $\mathrm{Sol(EP)}(\bar{\lambda}, \bar{\mu})$, resp.$\big)$ and sequences $\{\mu^k\}_k$ $\big(\{\lambda^k\}_k$, $\{(\lambda^k, \mu^k)\}_k$, resp$\big)$ and $\{x^k\}_k$ such that $\mu^k \to \bar{\mu}$ $\big(\lambda^k \to \bar{\lambda}$, $(\lambda^k, \mu^k) \to (\bar{\lambda}, \bar{\mu})$, resp.$\big)$ and $x^k \in \mathrm{Sol(EP)}(\lambda, \mu^k) \setminus U$ $\big(\mathrm{Sol(EP)}(\lambda^k, \mu) \setminus U$, $\mathrm{Sol(EP)}(\lambda^k, \mu^k) \setminus U$, resp.$\big)$ for each k. By Remark $(r1)$ $\big((r2)$,

$(r3)$, resp.$\big)$, $x^k \to \bar{x} \in \mathrm{Sol(EP)}(\lambda, \bar{\mu})\big(\mathrm{Sol(EP)}(\bar{\lambda}, \bar{\mu}), \mathrm{Sol(EP)}(\bar{\lambda}, \bar{\mu})$, resp.$\big) \subset U$, which is impossible because $x^k \notin U$ for all k.

For (d), let $(\lambda, \mu) \in \Lambda \times M$ and $\bar{x} \in \mathrm{Sol(EP)}(\lambda, \mu)$. Then, for all $\hat{x} \in \mathscr{D}(\lambda)$ with $\hat{x} \neq \bar{x}$, $\Phi(\bar{x}, \hat{x}, \mu) \geq 0$. By the strict pseudomonotonicity of $\Phi(\cdot, \cdot, \mu)$, $\Phi(\hat{x}, \bar{x}, \mu) < 0$. Hence, $\hat{x} \notin \mathrm{Sol(EP)}\ (\lambda, \mu)$. Thus, $\mathrm{Sol(EP)}(\lambda, \mu)$ is a singleton.

For (e), let $(\lambda, \mu) \in \Lambda \times M$. Suppose to the contrary that there exist $\bar{x}, \bar{x}' \in \mathrm{Sol(EP)}(\lambda, \mu)$ with $\bar{x} \neq \bar{x}'$. Because $\mathscr{D}(\lambda)$ is convex, $\bar{y} = \frac{\bar{x}+\bar{x}'}{2} \in \mathscr{D}(\lambda)$. Hence, as $\Phi(\cdot, \cdot, \mu) \in (-\mathscr{A}^1_{s.qc}) \cap \mathscr{A}^=_A$, we have

$$
\begin{aligned}
0 &= \Phi\left(\frac{\bar{x}+\bar{x}'}{2}, \frac{\bar{x}+\bar{x}'}{2}, \mu\right) \\
&= \Phi\left(\frac{\bar{x}+\bar{x}'}{2}, \bar{y}, \mu\right) > \min\{\Phi(\bar{x}, \bar{y}, \mu), \Phi(\bar{x}', \bar{y}, \mu)\} \geq 0,
\end{aligned}
$$

a contradiction. □

Definition 4.6 (Sign-preserving function) A function $\phi : A \to \mathbb{R}$ is called SIGN-PRESERVING if for all $x, y \in A$ and $t \in (0, 1)$, from $\phi(x) < 0$ and $\phi(y) \leq 0$, it follows that $\phi((1-t)x + ty) < 0$. A bifunction $\Omega : A \times A \to \mathbb{R}$ is called sign-preserving in the first component (in the second component, resp.) if $\Omega(\cdot, y)$ ($\Omega(x, \cdot)$, resp.) is sign-preserving for all $y \in A$ ($x \in A$, resp.).

Proposition 4.7 (Characterizations of the sign-preserving property) *For each $\mu \in M$, $\Phi(\cdot, \cdot, \mu)$ is sign-preserving in the ith component ($i = 1, 2$) if and only if $\Phi(\cdot, \cdot, \mu) \in \mathscr{A}^i_{st.qcl}$ and $\Phi(\cdot, \cdot, \mu)$ has the property $\mathbb{P}^i$.*

Proof This follows from the definitions. □

For each $(\lambda, \mu) \in \Lambda \times M$, we consider the following dual equilibrium problem of (EP):

(DEP)(λ, μ) find $\bar{x} \in \mathscr{D}(\lambda)$ such that $\Phi(y, \bar{x}, \mu) \leq 0$ for all $y \in \mathscr{D}(\lambda)$.

Let $\mathrm{Sol(DEP)}(\lambda, \mu)$ be the solution set of (DEP)(λ, μ). Then, for each $(\lambda, \mu) \in \Lambda \times M$, and for $\Phi(\cdot, \cdot, \mu)|_{\mathscr{D}(\lambda)\times\mathscr{D}(\lambda)}$, we see that if $\mathscr{D}(\lambda)$ is convex, $\Phi(\cdot, \cdot, \mu) \in \mathscr{A}^{\geq}_A$ is sign-preserving in the second component and upper sign-continuous in the first component, then

$$
\mathrm{Sol(EP)}(\lambda, \mu) \subset \mathrm{Sol(DEP)}(\lambda, \mu).
$$

If in addition, $\Phi(\cdot, \cdot, \mu) \in \mathscr{A}_{pm}$, then $\mathrm{Sol(EP)}(\lambda, \mu) = \mathrm{Sol(DEP)}(\lambda, \mu)$, and hence the solution maps Sol(EP) and Sol(DEP) coincide. On the other hand, (DEP)(λ, μ) can be viewed as a particular case of (EP)(λ, μ) with Φ in (EP)(λ, μ) being $\hat{\Phi}$ defined by $\hat{\Phi}(x, y, \mu) := -\Phi(y, x, \mu)$. From Theorem 4.5 and what we have just mentioned, we obtain the following result.

Theorem 4.8 (Upper semicontinuity and closedness of dual solution maps) *Assume that $\mathscr{D}$ is compact-convex-valued and for each $\mu \in M$, $\Phi(\cdot, \cdot, \mu) \in \mathscr{A}_A^= \cap \mathscr{A}_{pm}$ and is sign-preserving in the second component and upper sign-continuous in the first component. Then:*

(a) *If $\Phi(x, \cdot, \cdot)$ is lsc for each $x \in A$, then* Sol(DEP)$(\lambda, \cdot)$ *is both usc and closed on M for each $\lambda \in \Lambda$.*
(b) *If $\mathscr{D}$ is continuous and $\Phi(\cdot, \cdot, \mu)$ is lsc for each $\mu \in M$, then* Sol(DEP)$(\cdot, \mu)$ *is both usc and closed on Λ for each $\mu \in M$.*
(c) *If $\mathscr{D}$ is continuous and Φ is lsc, then* Sol(DEP) *is both usc and closed on $\Lambda \times M$.*

Proof By the assumption, Sol(EP) $=$ Sol(DEP). Using this fact and restating the conclusions (a), (b), and (c) of Theorem 4.1 for $\hat{\Phi}$ defined by $\hat{\Phi}(x, y, \mu) := -\Phi(y, x, \mu)$, we obtain conclusions (a), (b), and (c) of Theorem 4.4, □

To further study the lower semicontinuity of Sol(EP), for each $(\lambda, \mu) \in \Lambda \times M$, we consider the following auxiliary problem of (EP)(λ, μ):

(EP$^+$)(λ, μ) find $\bar{x} \in \mathscr{D}(\lambda)$ such that $\Phi(\bar{x}, y, \mu) > 0$ for all $y \in \mathscr{D}(\lambda)$.

Let Sol(EP$^+$)(λ, μ) be the solution set of (EP$^+$)(λ, μ). Then,

$$\text{Sol}(\text{EP}^+)(\lambda, \mu) \subset \text{Sol}(\text{EP})(\lambda, \mu).$$

Proposition 4.9 (Relationship between original and auxiliary solution sets) *If* Sol(EP$^+$)(λ, μ) *is nonempty, $\mathscr{D}(\lambda)$ is convex, and $-\Phi(\cdot, \cdot, \mu)$ is sign-preserving in the first component, then* Sol(EP)$(\lambda, \mu) \subset$ cl(Sol(EP$^+$)(λ, μ)).

Proof Take any $x^1 \in$ Sol(EP)(λ, μ) and $x^2 \in$ Sol(EP$^+$)(λ, μ). For any $t \in (0, 1)$ and $y \in \mathscr{D}(\lambda)$, by the sign-preservation property in the first component of $-\Phi(\cdot, \cdot, \mu)$,

$$\Phi((1 - t)x^1 + tx^2, y, \mu) > 0,$$

or equivalently $(1 - t)x^1 + tx^2 \in$ Sol(EP$^+$)(λ, μ). By letting $t \to 0^+$ we conclude that x^1 belongs to cl(Sol(EP$^+$)(λ, μ)). □

Lemma 4.10 (Lower semicontinuity of auxiliary solution maps) *Assume that $\mathscr{D}$ is compact-convex-valued. The following assertions hold true:*

(a) *If Φ is lsc, then* Sol(EP$^+$)$(\lambda, \cdot)$ *is lsc on M for each $\lambda \in \Lambda$.*
(b) *If $\mathscr{D}$ is continuous and $\Phi(\cdot, \cdot, \mu)$ is lsc for each $\mu \in M$, then* Sol(EP$^+$)$(\cdot, \mu)$ *is lsc on Λ for each $\mu \in M$.*
(c) *If $\mathscr{D}$ is continuous and Φ is lsc, then* Sol(EP$^+$) *is lsc on $\Lambda \times M$.*

Proof Obviously, (a) is a particular case of (c).

We prove (b). Suppose to the contrary that there are $\bar{x} \in$ Sol(EP$^+$)$(\bar{\lambda}, \mu)$ and $\{\lambda^k\}_k \subset \Lambda$ with $\lambda^k \to \bar{\lambda}$ such that there is no $x^k \in$ Sol(EP$^+$)(λ^k, μ) with $x^k \to \bar{x}$. Since $\mathscr{D}$ is lower semicontinuous, there is $\bar{x}^k \in \mathscr{D}(\lambda^k)$ such that $\bar{x}^k \to \bar{x}$. We can assume that $\bar{x}^k \notin \mathrm{Sol(EP^+)}(\lambda^k, \mu)$ for all k. Then, for each k, there exists $y^k \in \mathscr{D}(\lambda^k)$ such that $\Phi(\bar{x}^k, y^k, \mu) \leq 0$. Since $\mathscr{D}$ is upper semicontinuous and compact-valued, we can assume that $y^k \to \bar{y}$ for some $\bar{y} \in \mathscr{D}(\bar{\lambda})$. Then, by the lower semicontinuity of $\Phi(\cdot, \cdot, \mu)$,

$$\Phi(\bar{x}, \bar{y}, \mu) \leq \liminf_k \Phi(\bar{x}^k, y^k, \mu) \leq 0,$$

which is impossible as $\bar{x} \in \mathrm{Sol(EP^+)}(\bar{\lambda}, \mu)$.

For (c), we repeat the argument in the proof of (b) with the terms $\mathrm{Sol(EP^+)}(\bar{\lambda}, \mu)$, $\{\lambda^k\}_k \subset \Lambda$ with $\lambda^k \to \bar{\lambda}$, $\mathrm{Sol(EP^+)}(\lambda^k, \mu)$, $\Phi(x^k, y^k, \mu)$, and $\Phi(\bar{x}, \bar{y}, \mu)$ in the proof of (b) replaced by $\mathrm{Sol(EP^+)}(\bar{\lambda}, \bar{\mu})$, $\{(\lambda^k, \mu^k)\}_k \subset \Lambda \times M$ with $(\lambda^k, \mu^k) \to (\bar{\lambda}, \bar{\mu})$, $\mathrm{Sol(EP^+)}(\lambda^k, \mu^k)$, $\Phi(\bar{x}^k, y^k, \mu^k)$, and $\Phi(\bar{x}, \bar{y}, \bar{\mu})$, resp. □

Theorem 4.11 (Lower semicontinuity of solution maps) *Assume that* Sol(EP$^+$) *is nonempty-valued,* $\mathscr{D}$ *is compact-convex-valued, and* $-\Phi(\cdot, \cdot, \mu)$ *is sign-preserving in the first component for each* $\mu \in M$*. Then, the following assertions hold:*

(a) *If* Φ *is lsc, then* $\mathrm{Sol(EP)}S(\lambda, \cdot)$ *is lsc on* M *for each* $\lambda \in \Lambda$.
(b) *If* $\mathscr{D}$ *is continuous and* $\Phi(\cdot, \cdot, \mu)$ *is lsc for each* $\mu \in M$*, then* $\mathrm{Sol(EP)}(\cdot, \mu)$ *is lsc on* Λ *for each* $\mu \in M$.
(c) *If* $\mathscr{D}$ *is continuous and* Φ *is lsc, then* Sol(EP) *is lsc on* $\Lambda \times M$.

Proof Clearly, (a) is a particular case of (c).

For (b), let $\bar{\lambda} \in \Lambda$ and $\{\lambda^k\}_k \subset \Lambda$ be such that $\lambda^k \to \bar{\lambda}$. By Proposition 4.5 and Lemma 4.6 (b),

$$\begin{aligned}\mathrm{Sol(EP)}(\bar{\lambda}, \mu) \subset \mathrm{cl}(\mathrm{Sol(EP^+)}(\bar{\lambda}, \mu)) &\subset \liminf_k \mathrm{Sol(EP^+)}(\lambda^k, \mu)\\ &\subset \liminf_k \mathrm{Sol(EP)}(\lambda^k, \mu),\end{aligned}$$

and hence $\mathrm{Sol(EP)}(\cdot, \mu)$ is lsc.

For (c), let $\{(\lambda^k, \mu^k)\}_k \subset \Lambda \times M$ tend to $(\bar{\lambda}, \bar{\mu}) \in \Lambda \times M$. Proposition 4.9 and Lemma 4.10(c) give that

$$\begin{aligned}\mathrm{Sol(EP)}(\bar{\lambda}, \bar{\mu}) \subset \mathrm{cl}(\mathrm{Sol(EP^+)}(\bar{\lambda}, \bar{\mu})) &\subset \liminf_k \mathrm{Sol(EP^+)}(\lambda^k, \mu^k)\\ &\subset \liminf_k \mathrm{Sol(EP)}(\lambda^k, \mu^k).\end{aligned}$$

Thus, Sol(EP) is lsc. □

Remark 4.12 (Upper and lower pseudocontinuous functions) Semicontinuity properties of a function $\varphi : X \to \mathbb{R}$ from a topological space X into $\mathbb{R}$ can be relaxed as follows. φ is said to be UPPER PSEUDOCONTINUOUS (LOWER PSEUDOCONTINUOUS, resp.) at $\bar{x} \in X$ if

$$[\varphi(x) > \varphi(\bar{x})] \Rightarrow [\varphi(x) > \limsup \varphi(x^k) \text{ for all } \{x^k\} \to \bar{x}]$$

$$\big([\varphi(x) < \varphi(\bar{x})] \Rightarrow [\varphi(x) < \liminf \varphi(x^k) \text{ for all } \{x^k\} \to \bar{x}], \text{resp.}\big).$$

φ is said to be PSEUDOCONTINUOUS at $\bar{x}$ if it is both lower and upper pseudocontinuous at this point. In Theorem 4.5 (Theorem 4.8, resp.), the assumption "Φ is upper semicontinuous (lower semicontinuous, resp.)" can be replaced by "Φ is upper pseudocontinuous (lower pseudocontinuous, resp.) and $\Phi(\cdot, \cdot, \mu) \in \mathscr{A}_A^{=}$ for each $\mu \in M$." The proof is left for the reader.

Let us proceed to the Hausdorff lower semicontinuity of Sol(EP).

Definition 4.13 (Hausdorff upper and lower semicontinuity of multifunctions) Let $\mathscr{Q}$ be a multifunction from X into Y.

(a) $\mathscr{Q}$ is said to be HAUSDORFF UPPER SEMICONTINUOUS (H-usc) at $\bar{x}$ if for each neighborhood U of the origin in Y , there exists a neighborhood N of $\bar{x}$ such that $\mathscr{Q}(x) \subset \mathscr{Q}(\bar{x}) + B$ for all $x \in N$.
(b) $\mathscr{Q}$ is said to be HAUSDORFF LOWER SEMICONTINUOUS (H-lsc) at $\bar{x}$ if for each neighborhood B of the origin in Y, there exists a neighborhood N of $\bar{x}$ such that $\mathscr{Q}(\bar{x}) \subset \mathscr{Q}(x) + B$ for all $x \in N$.

$\mathscr{Q}$ is HAUSDORFF CONTINUOUS at $\bar{x}$ if it is both H-usc and H-lsc at $\bar{x}$.

Theorem 4.14 (Hausdorff lower semicontinuity of solution maps) *Assume that the assumptions of Theorem 4.11 are satisfied and $\Phi(\cdot, \cdot, \mu)$ is usc for each $\mu \in M$. Then, the solution map* Sol(EP) *is H-lsc on $\Lambda \times M$.*

Proof Using the same techniques as in the proof of Theorem 4.8, we also obtain the fact that Sol(EP) is closed-valued, and hence compact-valued as $\mathscr{D}(\lambda)$ is compact for every $\lambda \in \Lambda$.

If there exist $\bar{\lambda} \in \Lambda$ and $\mu \in M$ such that Sol(EP) is not H-lsc at $(\bar{\lambda}, \bar{\mu})$, then there are a neighborhood B of the origin in X and sequences $\lambda^k \to \bar{\lambda}$, $\mu^k \to \bar{\mu}$ such that, for each k, there is

$$x^{0k} \in \text{Sol(EP)}(\bar{\lambda}, \bar{\mu}) \setminus (S(\lambda^k, \mu^k) + B).$$

Because $\mathscr{D}(\bar{\lambda})$ is compact, one can assume that $x^{0k} \to \bar{x} \in \text{Sol(EP)}(\bar{\lambda}, \bar{\mu})$. Using the same argument as in the proof of Theorem 4.11, we arrive at a contradiction. □

In the rest of this section, we consider a special case where $\mathscr{D}(\lambda) \equiv A$ for every $\lambda \in \Lambda$, i.e., the problem

(EP)(μ) find $\bar{x} \in A$ such that $\Phi(\bar{x}, y, \mu) \geq 0$ for all $y \in A$.

This special case is motivated and inspired by studies of bilevel problems, including bilevel equilibrium problems and optimization problems with equilibrium constraints. The following results are not directly derived from those in the preceding part of this section.

We provide sufficient conditions for the Hausdorff continuity of approximate solution maps. Note that in particular, the conditions are usually employed in studies of well-posedness for equilibrium problems (cf. Sect. 4.4). We denote the ε-approximate solution set of problem (EP)(μ) by

$$\widetilde{\mathrm{Sol}}(\mathrm{EP})(\varepsilon, \mu) := \{x \in A \mid \Phi(x, y, \mu) + \varepsilon \geq 0, \forall y \in A\}.$$

Theorem 4.15 (Hausdorff continuity of approximate solution maps) *Assume that:*

(a) *A is compact and convex;*
(b) *$\Phi(x, y, \cdot)$ is continuous on M for every $x, y \in A$;*
(c) *$\Phi(\cdot, y, \mu)$ is concave on A for every $y \in A$ and $\mu \in M$.*

Then, $\widetilde{\mathrm{Sol}}(\mathrm{EP})$ is Hausdorff continuous on $\mathbb{R}_+ \times M$.

Proof For a given $(\varepsilon^0, \bar{\mu}) \in \mathbb{R}_+ \times M$, we prove that $\widetilde{\mathrm{Sol}}(\mathrm{EP})$ is Hausdorff continuous at $(\varepsilon^0, \bar{\mu})$. Take δ with $0 < \delta < \varepsilon^0$, $V = [\varepsilon^0 - \delta, \varepsilon^0 + \delta]$, $\varepsilon^1, \varepsilon^2 \in V$ with $\varepsilon^1 < \varepsilon^2$, and η satisfying $0 < \eta < \varepsilon^2 - \varepsilon^1$. For every $x, y \in A$, employing (b), we obtain a neighborhood $N_\eta(\bar{\mu})$ of $\bar{\mu}$ such that, for all $\mu \in N_\eta(\bar{\mu})$,

$$|f(x, y, \mu^1) - f(x, y, \mu^2)| \leq \eta. \tag{4.1}$$

For every $\varepsilon^2, \mu^2 \in N_\eta(\bar{\mu})$, we claim that

$$\widetilde{\mathrm{Sol}}(\mathrm{EP})(\varepsilon^1, \varepsilon^2) \subset \widetilde{\mathrm{Sol}}(\mathrm{EP})(\varepsilon^2, \mu^2). \tag{4.2}$$

Let $x \in \widetilde{\mathrm{Sol}}(\mathrm{EP})(\varepsilon^1, \varepsilon^2)$ be arbitrary. Then, for all $y \in A$,

$$\Phi(x, y, \mu^2) + \Phi(x, y, \varepsilon^2) - \Phi(x, y, \mu^2) + \varepsilon^1 \geq 0.$$

Applying this inequality together with (4.1), we have

$$\Phi(x, y, \mu^2) + \eta + \varepsilon^1 \geq 0,$$

for all $y \in A$. Therefore, $\Phi(x, y, \mu^2)+\varepsilon^2 \geq 0$, or equivalently $x \in \widetilde{\mathrm{Sol}}(\mathrm{EP})(\varepsilon^2, \mu^2)$, and then (4.2) is obtained. Hence, for every $\mu \in N_\eta(\bar{\mu})$ and $\varepsilon \in V$,

$$\widetilde{\mathrm{Sol}}(\mathrm{EP})(\varepsilon^0 - \delta, \bar{\mu}) \subset \widetilde{\mathrm{Sol}}(\mathrm{EP})(\varepsilon, \mu) \subset \widetilde{\mathrm{Sol}}(\mathrm{EP})(\varepsilon^0 + \delta, \bar{\mu}). \tag{4.3}$$

Take a number θ satisfying $1 < \theta < \dfrac{\varepsilon^2}{\varepsilon^2 - \varepsilon^1}$ and set $\gamma = \varepsilon^2 + \theta(\varepsilon^1 - \varepsilon^2)$. Let us prove that

$$\frac{1}{\theta}\widetilde{\mathrm{Sol}}(\mathrm{EP})(\gamma, \bar{\mu}) + \left(1 - \frac{1}{\theta}\right)\widetilde{\mathrm{Sol}}(\mathrm{EP})(\varepsilon^2, \bar{\mu}) \subset \widetilde{\mathrm{Sol}}(\mathrm{EP})(\varepsilon^1, \bar{\mu}). \tag{4.4}$$

For any $\bar{x} \in \widetilde{\mathrm{Sol}}(\mathrm{EP})(\gamma, \bar{\mu})$ and $\bar{x}' \in \widetilde{\mathrm{Sol}}(\mathrm{EP})(\varepsilon^2, \bar{\mu})$, we have

$$\bar{x} + \left(1 - \frac{1}{\theta}\right)\bar{x}' \in A,$$

as A is convex. For each $y \in \mathscr{D}(\bar{\lambda})$, we have

$$\min\{\Phi(\bar{x}, y, \bar{\mu}) + \gamma,\ \Phi(\bar{x}', y, \bar{\mu}) + \varepsilon^2\} \geq 0.$$

This implies that

$$\frac{1}{\theta}f(\bar{x}, y, \bar{\mu}) + \frac{1}{\theta}\gamma + \left(1 - \frac{1}{\theta}\right)f(\bar{x}', y, \bar{\mu}) + \left(1 - \frac{1}{\theta}\right)\varepsilon^2 \geq 0.$$

By assumption (c), for all $y \in A$, we obtain

$$\Phi\left(\frac{1}{\theta}\bar{x} + \left(1 - \frac{1}{\theta}\right)\bar{x}', y, \bar{\mu}\right) + \varepsilon^1 \geq 0.$$

So, $\frac{1}{\theta}\bar{x} + \left(1 - \frac{1}{\theta}\right)\bar{x}' \in \widetilde{\mathrm{Sol}}(\mathrm{EP})(\varepsilon^1, \bar{\lambda}, \bar{\mu})$. Hence, (4.4) is true. Consequently,

$$\left(1 - \frac{1}{\theta}\right)\widetilde{\mathrm{Sol}}(\mathrm{EP})(\varepsilon^2, \bar{\mu}) \subset \widetilde{\mathrm{Sol}}(\mathrm{EP})(\varepsilon^1, \bar{\mu}) - \frac{1}{\theta}\widetilde{\mathrm{Sol}}(\mathrm{EP})(\gamma, \bar{\mu}),$$

and hence

$$\begin{aligned}\widetilde{\mathrm{Sol}}(\mathrm{EP})(\varepsilon^2, \bar{\mu}) &\subset \widetilde{\mathrm{Sol}}(\mathrm{EP})(\varepsilon^1, \bar{\mu}) + \frac{1}{\theta - 1}\left[\widetilde{\mathrm{Sol}}(\mathrm{EP})(\varepsilon^1, \bar{\mu}) - \widetilde{\mathrm{Sol}}(\mathrm{EP})(\gamma, \bar{\mu})\right] \\ &\subset \widetilde{\mathrm{Sol}}(\mathrm{EP})(\varepsilon^1, \bar{\mu}) + \frac{1}{\theta - 1}[A - A].\end{aligned}$$

By (a), $A - A$ is bounded. So, for each closed convex neighborhood B of the origin in X, there is a positive number ρ such that $A - A \subset \rho B$. Therefore,

$$\widetilde{\mathrm{Sol}}(\mathrm{EP})(\varepsilon^2, \bar{\mu}) \subset \widetilde{\mathrm{Sol}}(\mathrm{EP})(\varepsilon^1, \bar{\mu}) + \frac{\rho}{\theta - 1}B. \tag{4.5}$$

Take $0 < \delta_0 < \dfrac{\varepsilon^0}{\rho + 1}$ and consider the neighborhood $[\varepsilon^0 - \delta_0, \varepsilon^0 + \delta_0]$ of ε^0. From (4.5), taking $\varepsilon^2 = \varepsilon^0$, $\varepsilon^1 = \varepsilon^0 - \delta_0$, and $\theta = \rho + 1$, we have

$$1 < \theta < \frac{\varepsilon^2}{\varepsilon^2 - \varepsilon^1} \left(= \frac{\varepsilon^0}{\delta_0} \right),$$

and so

$$\widetilde{\mathrm{Sol}}(\mathrm{EP})(\varepsilon^0, \bar{\mu}) \subset \widetilde{\mathrm{Sol}}(\mathrm{EP})(\varepsilon^0 - \delta_0, \bar{\mu}) + B.$$

Applying (4.3), we get

$$\widetilde{\mathrm{Sol}}(\mathrm{EP})(\varepsilon^0, \bar{\mu}) \subset \widetilde{\mathrm{Sol}}(\mathrm{EP})(\varepsilon, \mu) + B,$$

and so $\widetilde{\mathrm{Sol}}(\mathrm{EP})$ is H-lsc at $(\varepsilon^0, \bar{\mu})$.

Again, taking $\varepsilon^2 = \varepsilon^0 + \delta_0$, $\varepsilon^1 = \varepsilon^0$, and $\theta = \rho + 1$ in (4.5), we obtain

$$\widetilde{\mathrm{Sol}}(\mathrm{EP})(\varepsilon^0 + \delta_0, \bar{\mu}) \subset \widetilde{\mathrm{Sol}}(\mathrm{EP})(\varepsilon^0, \bar{\mu}) + B.$$

Combining this inclusion with (4.3), we arrive at

$$\widetilde{\mathrm{Sol}}(\mathrm{EP})(\varepsilon, \mu) \subset \widetilde{\mathrm{Sol}}(\mathrm{EP})(\varepsilon^0, \bar{\mu}) + B,$$

and hence $\widetilde{\mathrm{Sol}}(\mathrm{EP})$ is H-usc at $(\varepsilon^0, \bar{\mu})$. The proof is complete. □

Example 4.16 (An example of Hausdorff continuity of approximate solution map) Let $X = \mathbb{R}$, $A = [0, 1]$, $M = [0, 1]$, and

$$\Phi(x, y, \mu) = \begin{cases} 0 & x = 0, \\ 1 & x \neq 0. \end{cases}$$

Then, the assumptions of Theorem 4.15 are satisfied, and hence the approximate solution map $\widetilde{\mathrm{Sol}}(\mathrm{EP})$ is Hausdorff continuous (in fact, it is easy to compute $\widetilde{\mathrm{Sol}}(\mathrm{EP})(\varepsilon, \mu) = [0, 1]$ for all $\varepsilon \in \mathbb{R}_+$ and $\mu \in M$).

4.2 Hölder Continuity and Hölder Calmness

4.2.1 Hölder Continuity and Hölder Calmness of Solution Maps

In this section, we investigate the Hölder/Lipschitz continuity and Hölder calmness of the solution map Sol(EP). Throughout this section if not stated otherwise, all spaces X, Λ, M are assumed to be metric spaces, Y to be a linear metric space, and A is a subset of X. We use $d(\cdot, \cdot)$ for all metrics for convenience, since the context makes it clear what metric is concerned. For subsets P and Q of a metric space, we will use the following notation:

- The distance from a point a to P: $d(a, P) = \inf_{p \in P} d(a, p)$.
- $\mathbb{D}(P, Q) = \sup_{p \in P, q \in Q} d(p, q)$.
- The excess of P from Q: $\mathrm{ex}(P, Q) = \sup_{p \in P} d(p, Q)$.
- The Hausdorff distance between P and Q: $H(P, Q) = \max\{\mathrm{ex}(P, Q), \mathrm{ex}(Q, P)\}$.

Definition 4.17 (Hölder continuity and Hölder calmness) Let $h : X \to Y$, $\mathcal{H} : X \rightrightarrows Y$, and $\sigma, \alpha > 0$.

(a) h is said to be (σ, α)-HÖLDER CONTINUOUS at $\bar{x}$ if there is a neighborhood U of $\bar{x}$ such that

$$d(h(x), h(y)) \leq \sigma d^{\alpha}(x, y)$$

for all $x, y \in U$.

(b) $\mathcal{H}$ is said to be (σ, α)-HÖLDER CONTINUOUS at $\bar{x}$ if there is a neighborhood U of $\bar{x}$ such that

$$H(\mathcal{H}(x), \mathcal{H}(y)) \leq \sigma d^{\alpha}(x, y) \text{ for all } x, y \in U.$$

(c) $\mathcal{H}$ is said to be (σ, α)-HÖLDER CALM at $\bar{x}$ if there is a neighborhood U of $\bar{x}$ such that

$$H(\mathcal{H}(x), \mathcal{H}(\overline{x})) \leq \sigma d^{\alpha}(x, \overline{x}) \text{ for all } x \in U.$$

If $H(\mathcal{H}(x), \mathcal{H}(\overline{x}))$ in this inequality is replaced by $\mathbb{D}(\mathcal{H}(x), \mathcal{H}(\overline{x}))$, then we say that $\mathcal{H}$ is $\sigma\alpha$-HÖLDER CALM WITH RESPECT TO $\mathbb{D}$ at $\bar{x}$.

If in Definition 4.17, $\mathcal{H}$ is a single-valued map, then (b) reduces to (a). When Y is a linear metric space, it is clear that $\mathcal{H}$ is (σ, α)- Hölder continuous at $\bar{x}$ if and only if there is a neighborhood U of $\bar{x}$ such that for all $x, y \in U$,

$$\mathcal{H}(x) \subset \mathcal{H}(y) + \sigma d^{\alpha}(x, y) \mathrm{cl} B.$$

It is also obvious that Hölder continuity $\Rightarrow$ Hölder calmness, and Hölder calmness with respect to $\mathbb{D}$ $\Rightarrow$ Hölder calmness.

If the neighborhood U in (a) ((b), (c), resp.) is given explicitly, then we say that $\mathcal{H}$ is $\sigma\alpha$-HÖLDER CONTINUOUS AT $\bar{x}$ WITH RESPECT TO U, ($\sigma\alpha$-HÖLDER CALM AT $\bar{x}$ WITH RESPECT TO U, resp.).

Definition 4.18 (Hölder continuity and Hölder calmness uniformly over a set) Let $\Phi : A \times A \times M \to \mathbb{R}$, $B \subset A$, $\rho, \gamma > 0$, and $\theta \geq 0$.

(a) Φ is said to be $\rho\gamma$-HÖLDER CONTINUOUS AT $\hat{\mu}$, θ-UNIFORMLY OVER B if there is a neighborhood V of $\bar{\mu}$ such that

$$|\Phi(x, y, \mu) - \Phi(x, y, \mu')| \leq \rho d^{\gamma}(\mu, \mu')d^{\theta}(x, y)$$

for each $\mu, \mu' \in V$ and $x, y \in B$.

(b) Φ is said to be $\rho\gamma$-HÖLDER CALM AT $\hat{\mu}$, θ-UNIFORMLY OVER B if there is a neighborhood V of $\bar{\mu}$ such that

$$|\Phi(x, y, \hat{\mu}) - \Phi(x, y, \mu)| \leq \rho d^{\gamma}(\hat{\mu}, \mu)d^{\theta}(x, y)$$

for all $\mu \in V$ and $x, y \in B$.

If the neighborhood V in (b) ((c), resp.) is given explicitly, then we say that Φ is (ρ, γ)-HÖLDER CONTINUOUS AT $\bar{\mu}$ WITH RESPECT TO V, θ-UNIFORMLY OVER B ((ρ, γ)-HÖLDER CALM AT $\bar{\mu}$ WITH RESPECT TO V, θ-UNIFORMLY OVER B, resp.).

It is clear that when $\theta = 0$, $\Phi(x, y, \cdot)$ is $\rho\gamma$-Hölder continuous at $\bar{\mu}$ for all $x, y \in A$. The following example illustrates the case where $\theta > 0$.

Example 4.19 (An example of Hölder continuity uniformly over a set) Let X, M be Banach spaces, X^* be the dual space of X, and $f : X \times M \to X^*$. Assume that $f(x, \cdot)$ is $h\alpha$-Hölder continuous at $\bar{\mu}$. Consider a function Φ defined by

$$\Phi(x, y, \mu) := \langle f(x, \mu), y - x\rangle.$$

Then, the function $\Phi(x, y, \cdot)$ is $h\alpha$-Hölder continuous, 1-uniformly at $\bar{\mu}$.

Definition 4.20 (Strong monotonicity, semistrong monotonicity, and strong pseudomonotonicity of a bifunction) Let A be a subset of a metric space X, $\Omega : A \times A \to \mathbb{R}$ and $\sigma, \beta > 0$.

(a) Ω is said to be (σ, β)-STRONGLY MONOTONE if, for each $x, y \in A$, we have

$$\Omega(x, y) + \Omega(y, x) + \sigma d^{\beta}(x, y) \leq 0.$$

(b) Ω is said to be (σ, β)-SEMISTRONGLY MONOTONE if, for each $x, y \in A$, we have

$$\min\{\Omega(x, y), \Omega(y, x)\} + \frac{1}{2}\sigma d^{\beta}(x, y) \leq 0.$$

(c) Ω is said to be (σ, β)-STRONGLY PSEUDOMONOTONE if, for each $x, y \in A$, $\Omega(x, y) \geq 0$ implies $\Omega(y, x) + \sigma d^{\beta}(x, y) \leq 0$, or equivalently, $\Omega(x, y) > 0$ implies $\Omega(y, x) + \sigma d^{\beta}(x, y) < 0$.

Clearly, (σ, β)-strong monotonicity $\Rightarrow$ (σ, β)-semistrong monotonicity $\Rightarrow$ $(\frac{\sigma}{2}, \beta)$-strong pseudomonotonicity. Moreover, if Ω is (σ, β)-strongly pseudomonotone and $-\Omega$ is quasimonotone, then Ω is (σ, β)-semistrongly monotone. Indeed, as $-\Omega$ is quasimonotone,

$$\max\{\Omega(x, y), \Omega(y, x)\} \geq 0.$$

Hence, by the (σ, β)-strong pseudomonotonicity of Ω,

$$\min\{\Omega(x, y), \Omega(y, x)\} + \sigma d^{\beta}(x, y) \leq 0,$$

which implies that

$$\min\{\Omega(x, y), \Omega(y, x)\} + \frac{1}{2}\sigma d^{\beta}(x, y) \leq 0.$$

We consider relationships between these properties and strong convexities. Let X be a linear metric space, $A \subset X$, and $B \subset A$ be a convex set.

A function $\phi : A \to \mathbb{R}$ is said to be (σ, β)-STRONGLY CONVEX on B if, for all $x, y \in B$ and all $t \in (0, 1)$,

$$\phi(tx + (1-t)y) \leq t\phi(x) + (1-t)\phi(y) - \sigma t(1-t)d^{\beta}(x, y).$$

If the right-hand side of this inequality is replaced by

$$\max\{\phi(x), \phi(y)\} - \sigma t(1-t)d^{\beta}(x, y),$$

then ϕ is said to be (σ, β)-STRONGLY QUASICONVEX on B.

Theorem 4.21 (Hölder continuity of solution maps) *Let $(\bar{\lambda}, \bar{\mu}) \in \Lambda \times M$, $U \subset \Lambda$, and $V \subset M$ be neighborhoods of $\bar{\lambda}$ and $\bar{\mu}$, resp.*

(a) *Assume that Φ is (ρ, γ)-Hölder continuous at $\bar{\mu}$ with respect to V, θ-uniformly $(\theta < \beta)$ over $\mathscr{D}(\lambda)$, and $\Phi(\cdot, \cdot, \mu)$ is $(\hbar, \beta)$-strongly pseudomonotone on $\mathscr{D}(\lambda)$ for each $\mu \in V$. Then,* Sol(EP)$(\lambda, \cdot)$ *is single-valued on V and satisfies the Hölder condition: for all $\mu^1, \mu^2 \in V$,*

$$d\big(\mathrm{Sol(EP)}(\lambda, \mu^1), \mathrm{Sol(EP)}(\lambda, \mu^2)\big) \leq k d^{\frac{\gamma}{\beta-\theta}}(\mu^1, \mu^2),$$

where k is a positive constant depending on ρ, $\hbar$, and β.

(b) *Assume that $\mathscr{D}$ is (ℓ, α)-Hölder continuous at $\bar{\lambda}$ with respect to U, $\Phi(x, \cdot, \mu)$ is (σ, δ)-Hölder continuous on $\mathscr{D}(U)$ for each $x \in \mathscr{D}(U)$, and $\Phi(\cdot, \cdot, \mu)$ is*

$(\hbar, \beta)$-semistrongly monotone on $\mathscr{D}(U)$. Then, Sol(EP)$(\cdot, \mu)$ *is single-valued on U and satisfies the Hölder condition: for all $\lambda^1, \lambda^2 \in U$,*

$$d\big(\text{Sol(EP)}(\lambda^1, \mu), \text{Sol(EP)}(\lambda^2, \mu)\big) \le k d^{\frac{\alpha\delta}{\beta}}(\lambda^1, \lambda^2),$$

where k is a positive constant depending on $\sigma, \ell, \hbar, \delta$, and β.

(c) *Assume that $\mathscr{D}$ is (ℓ, α)-Hölder continuous at $\bar{\lambda}$ with respect to U, Φ is (ρ, γ)-Hölder continuous at $\bar{\mu}$ with respect to V, θ-uniformly ($\theta < \beta$) over $\mathscr{D}(U)$, $\Phi(x, \cdot, \mu)$ is (σ, δ)-Hölder continuous on $\mathscr{D}(U)$ for each $x \in \mathscr{D}(U)$ and $\mu \in V$, and $\Phi(\cdot, \cdot, \mu)$ is $(\hbar, \beta)$-semistrongly monotone on $\mathscr{D}(U)$ for each $\mu \in V$. Then,* Sol(EP) *is single-valued on $U \times V$ and satisfies the Hölder condition: for all $(\lambda^1, \mu^1), (\lambda^2, \mu^2) \in U \times V$,*

$$d\big(\text{Sol(EP)}(\lambda^1, \mu^1), \text{Sol(EP)}(\lambda^2, \mu^2)\big) \le k_1 d^{\frac{\alpha\delta}{\beta}}(\lambda^1, \lambda^2) + k_2 d^{\frac{\gamma}{\beta-\theta}}(\mu^1, \mu^2),$$

where k_1, k_2 are positive constants depending on $\hbar, \ell, \rho, \sigma, \alpha, \beta, \gamma, \delta$.

Proof For any $(\lambda, \mu) \in U \times V$, the monotonicity properties in (a), (b), and (c) imply that $\Phi(\cdot, \cdot, \mu)$ is strictly quasimonotone on $\mathscr{D}(\lambda)$. Hence, Sol(EP)(λ, μ) is a singleton (see the proof of Theorem 4.5(d)).

(a) For any $\mu^1, \mu^2 \in V$, Sol(EP)$(\lambda, \mu^1) := \bar{x}^1 \in \mathscr{D}(\lambda)$ and Sol(EP)$(\lambda, \mu^2) := \bar{x}^2 \in \mathscr{D}(\lambda)$. Since Φ is $\rho\gamma$-Hölder continuous at $\bar{\mu}$ with respect to V and θ-uniformly over $\mathscr{D}(\lambda)$, we have

$$\Phi(\bar{x}^2, \bar{x}^1, \mu^2) - \Phi(\bar{x}^2, \bar{x}^1, \mu^1)| \le \rho d^{\alpha}(\mu^1, \mu^2) d^{\theta}(\bar{x}^1, \bar{x}^2).$$

Since $\Phi(\bar{x}^1, \bar{x}^2, \mu^1) \ge 0$, we have

$$\Phi(\bar{x}^2, \bar{x}^1, \mu^1) + d^{\beta}(\bar{x}^1, \bar{x}^2) \le 0,$$

by the $\hbar\beta$-strong pseudomonotonicity (on $\mathscr{D}(\lambda)$) of $\Phi(\cdot, \cdot, \mu^1)$. Moreover, $\Phi(\bar{x}^2, \bar{x}^1, \mu^2) \ge 0$. Thus,

$$\begin{aligned}\hbar d^{\beta}(\bar{x}^1, \bar{x}^2) &\le -\Phi(\bar{x}^2, \bar{x}^1, \mu^1) \le -\Phi(\bar{x}^2, \bar{x}^1, \mu^1) + \Phi(\bar{x}^2, \bar{x}^1, \mu^2) \\ &\le \rho d^{\theta}(\bar{x}^1, \bar{x}^2) d^{\gamma}(\mu^1, \mu^2).\end{aligned}$$

Hence,

$$d(\bar{x}^1, \bar{x}^2) \le \left(\frac{\rho}{\hbar}\right)^{\frac{1}{\beta-\theta}} d^{\frac{\gamma}{\beta-\theta}}(\mu^1, \mu^2).$$

(b) For any $\lambda^1, \lambda^2 \in U$, set $\text{Sol(EP)}(\lambda^1, \mu) =: \bar{x}^1 \in \mathscr{D}(\lambda^1)$ and $\text{Sol(EP)}(\lambda^2, \mu) =: \bar{x}^2 \in \mathscr{D}(\lambda^2)$. Let $i \in \{1, 2\}$ be such that

$$\Phi(\bar{x}^i, \bar{x}^{3-i}, \mu) = \min\{\Phi(\bar{x}^1, \bar{x}^2, \mu), \Phi(\bar{x}^2, \bar{x}^1, \mu)\}.$$

Then, by the $\hbar\beta$-semistrong monotonicity on $\mathscr{D}(U)$ of $\Phi(\cdot, \cdot, \mu)$, we have

$$\Phi(\bar{x}^i, \bar{x}^{3-i}, \mu) + \frac{\hbar}{2} d^\beta(\bar{x}^i, \bar{x}^{3-i}) \le 0,$$

or equivalently,

$$\frac{\hbar}{2} d^\beta(\bar{x}^i, \bar{x}^{3-i}) \le -\Phi(\bar{x}^i, \bar{x}^{3-i}, \mu).$$

For this index i, as $\mathscr{D}$ is $\ell\alpha$-Hölder continuous at $\bar{\lambda}$ with respect to U, there exists a $y^i \in \mathscr{D}(\lambda^i)$ such that $d(\bar{x}^{3-i}, y^i) \le \ell d^\alpha(\lambda^{3-i}, \lambda^i)$. Obviously, $\Phi(\bar{x}^i, y^i, \mu) \ge 0$. Moreover, since $\Phi(\bar{x}^i, \cdot, \mu)$ is $\sigma\delta$-Hölder continuous on $\mathscr{D}(U)$, one has

$$-\Phi(\bar{x}^i, \bar{x}^{3-i}, \mu) + \Phi(\bar{x}^i, y^i, \mu) \le \sigma d^\delta(\bar{x}^{3-i}, y^i).$$

Thus,

$$\begin{aligned}\frac{\hbar}{2} d^\beta(\bar{x}^i, \bar{x}^{3-i}) &\le -\Phi(\bar{x}^i, \bar{x}^{3-i}, \mu) \le -\Phi(\bar{x}^i, \bar{x}^{3-i}, \mu) + \Phi(\bar{x}^i, y^i, \mu) \\ &\le \sigma d^\delta(\bar{x}^{3-i}, y^i) \le \sigma \ell^\delta d^{\alpha\delta}(\lambda^{3-i}, \lambda^i).\end{aligned}$$

Hence,

$$d(\bar{x}^1, \bar{x}^2) \le \left(\frac{2\sigma\ell^\delta}{\hbar}\right)^{\frac{1}{\beta}} d^{\frac{\alpha\delta}{\beta}}(\lambda^1, \lambda^2).$$

(c) When the conditions in (c) hold, all the conditions in both (a) and (b) are also satisfied for any $(\lambda, \mu) \in U \times V$, but now the $\hbar\beta$-strong pseudomonotonicity of $\Phi(\cdot, \cdot, \mu)$ in (a) is $\frac{\hbar}{2}\beta$-strong pseudomonotonicity. Hence, from the proofs of (a) and (b) we obtain: for all $(\lambda^1, \mu^1), (\lambda^2, \mu^2) \in U \times V$,

$$\begin{aligned} d\big(\text{Sol(EP)}(\lambda^1, \mu^1), \text{Sol(EP)}(\lambda^2, \mu^2)\big) &\le d\big(\text{Sol(EP)}(\lambda^1, \mu^1), \text{Sol(EP)}(\lambda^2, \mu^1)\big) \\ &\quad + d\big(\text{Sol(EP)}(\lambda^2, \mu^1), \text{Sol(EP)Sol(EP)}(\lambda^2, \mu^2)\big) \\ &\le \left(\frac{2\sigma\ell^\delta}{\hbar}\right)^{\frac{1}{\beta}} d^{\frac{\alpha\delta}{\beta}}(\lambda^1, \lambda^2) + \left(\frac{2\rho}{\hbar}\right)^{\frac{1}{\beta-\theta}} d^{\frac{\gamma}{\beta-\theta}}(\mu^1, \mu^2). \end{aligned}$$

The proof is complete. □

Remark 4.22 (Strong pseudo-quasiconvexity) We can extend the strong quasiconvexity to metric spaces as follows. A function $\phi : A \to \mathbb{R}$, from a subset A of a metric space X into $\mathbb{R}$, is called STRONGLY PSEUDO-QUASICONVEX on $B \subset A$ if for all $x, y \in A$,

$$\phi(z) \leq \max\{\phi(x), \phi(y)\} - hd^{\beta}(x, y),$$

for some $z \in B$. Then, the conclusion of Theorem 4.21(a) is still valid, if the assumption "$\Phi(\cdot, \cdot, \mu)$ is $(\hbar, \beta)$-strongly pseudomonotone on $\mathscr{D}(\lambda)$ for each $\mu \in V$" is replaced by "$\Phi(x, \cdot, \mu)$ is $(\hbar, \beta)$-strongly pseudo-quasiconvex on $\mathscr{D}(\lambda)$ and $\Phi(\cdot, \cdot, \mu)$ is monotone on $\mathscr{D}(\lambda)$ for each $x \in \mathscr{D}(\lambda)$ and $\mu \in V$." Indeed, we first show that, for any $\mu^1, \mu^2 \in V, \bar{x}^1 \in \text{Sol(EP)}(\lambda, \mu^1)$, and $\bar{x}^2 \in \text{Sol(EP)}(\lambda, \mu^2)$,

$$d(\bar{x}^1, \bar{x}^2) \leq \left(\frac{\rho}{\hbar}\right)^{\frac{1}{\beta-\theta}} d^{\frac{\gamma}{\beta-\theta}}(\mu^1, \mu^2).$$

Similar to the proof of Theorem 4.21(a), observe that to prove this inequality, we need only to show that

$$\Phi(\bar{x}^2, \bar{x}^1, \mu^1) + \hbar d^{\beta}(\bar{x}^1, \bar{x}^2) \leq 0,$$

as $\Phi(\bar{x}^1, \cdot, \mu^1)$ is $(\hbar, \beta)$-strongly pseudo-quasiconvex on $\mathscr{D}(\lambda)$. Then, there is a $\bar{z} \in \mathscr{D}(\lambda)$ such that

$$\Phi(\bar{x}^1, \bar{z}, \mu^1) \leq \max\{\Phi(\bar{x}^1, \bar{x}^1, \mu^1), \Phi(\bar{x}^1, \bar{x}^2, \mu^1)\} - \hbar d^{\beta}(\bar{x}^1, \bar{x}^2).$$

Since $\bar{x}^1 \in \text{Sol(EP)}(\lambda, \mu^1)$ and $\bar{x}^1, \bar{x}^2 \in \mathscr{D}(\lambda)$, $\Phi(\bar{x}^1, \bar{x}^1, \mu^1) \geq 0$ and $\Phi(\bar{x}^1, \bar{x}^2, \mu^1) \geq 0$. This together with the monotonicity of $\Phi(\cdot, \cdot, \mu^1)$ gives that $\Phi(\bar{x}^1, \bar{x}^1, \mu^1) = 0$ and

$$-\Phi(\bar{x}^2, \bar{x}^1, \mu^1) \geq \Phi(\bar{x}^1, \bar{x}^2, \mu^1) \geq 0.$$

Moreover, $\Phi(\bar{x}^1, \bar{z}, \mu^1) \geq 0$, and hence

$$\begin{aligned} 0 &\leq \max\{\Phi(\bar{x}^1, \bar{x}^1, \mu^1), \Phi(\bar{x}^1, \bar{x}^2, \mu^1)\} - \hbar d^{\beta}(\bar{x}^1, \bar{x}^2) \\ &\leq -\Phi(\bar{x}^2, \bar{x}^1, \mu^1) - \hbar d^{\beta}(\bar{x}^1, \bar{x}^2). \end{aligned}$$

Next, use the above inequality to show that Sol(EP)(λ, μ) is a singleton for all $\mu \in V$ (in the above inequality, take $\mu^1 = \mu^2 = \mu$ to have that $d(\bar{x}^1, \bar{x}^2) = 0$).

Theorem 4.23 (Hölder calmness of solution maps) *Let $(\bar{\lambda}, \bar{\mu}) \in \Lambda \times M$, $U \subset \Lambda$, and $V \subset M$ be neighborhoods of $\bar{\lambda}$ and $\bar{\mu}$, resp.*

(a) *Assume that Φ is (ρ, γ)-Hölder calm at $\bar{\mu}$ with respect to V, θ-uniformly over $\mathscr{D}(\lambda)$, and $\Phi(\cdot, \cdot, \bar{\mu})$ is $(\hbar, \beta)$-strongly pseudomonotone on $\mathscr{D}(\lambda)$. Then,*

$\mathrm{Sol(EP)}(\lambda, \cdot)$ *satisfies the Hölder calmness: for all* $\mu \in V$,

$$\mathbb{D}\big(\mathrm{Sol(EP)}(\lambda, \bar{\mu}), \mathrm{Sol(EP)}(\lambda, \mu)\big) \leq k d^{\frac{\gamma}{\beta-\theta}}(\bar{\mu}, \mu),$$

where k *is a positive constant depending on* $\rho, \hbar$ *and* β.

(b) *Assume that* $\mathscr{D}$ *is* (ℓ, α)*-Hölder calm at* $\bar{\lambda}$ *with respect to* U, $\Phi(x, \cdot, \mu)$ *is* (σ, δ)*-Hölder continuous on* $\mathscr{D}(U)$ *for each* $x \in \mathscr{D}(U)$, *and* $\Phi(\cdot, \cdot, \mu)$ *is* $(\hbar, \beta)$*-semistrongly monotone on* $\mathscr{D}(U)$. *Then,* $\mathrm{Sol(EP)}(\cdot, \mu)$ *satisfies the condition of Hölder calmness: for all* $\lambda \in U$,

$$\mathbb{D}\big(\mathrm{Sol(EP)}(\bar{\lambda}, \mu), \mathrm{Sol(EP)}(\lambda, \mu)\big) \leq k d^{\frac{\alpha\delta}{\beta}}(\bar{\lambda}, \lambda),$$

where k *is a positive constant depending on* $\sigma, \ell, \hbar, \delta$, *and* β.

(c) *Assume that* $\mathscr{D}$ *is* (ℓ, α)*-Hölder calm at* $\bar{\lambda}$ *with respect to* U, Φ *is* (ρ, γ)*-Hölder calm at* $\bar{\mu}$ *with respect to* V, θ*-uniformly over* $\mathscr{D}(U)$, $\Phi(x, \cdot, \bar{\mu})$ *is* (σ, δ)*-Hölder continuous on* $\mathscr{D}(U)$ *for each* $x \in \mathscr{D}(U)$, *and* $\Phi(\cdot, \cdot, \bar{\mu})$ *is* $(\hbar, \beta)$*-semistrongly monotone on* $\mathscr{D}(U)$. *Then,* $\mathrm{Sol(EP)}$ *satisfies the Hölder calmness: for all* $(\lambda, \mu) \in U \times V$,

$$\mathbb{D}\big(\mathrm{Sol(EP)}(\bar{\lambda}, \bar{\mu}), \mathrm{Sol(EP)}(\lambda, \mu)\big) \leq k_1 d^{\frac{\alpha\delta}{\beta}}(\bar{\lambda}, \lambda) + k_2 d^{\frac{\gamma}{\beta-\theta}}(\bar{\mu}, \mu),$$

where k_1, k_2 *are positive constants depending on* $\hbar, \ell, \rho, \sigma, \alpha, \beta, \gamma, \delta$.

Proof

(a) For any $\mu \in V$, we prove that for all $\bar{x}^{\bar{\mu}} \in \mathrm{Sol(EP)}(\lambda, \overline{\mu})$ and $\bar{x}^{\mu} \in \mathrm{Sol(EP)}(\lambda, \mu)$,

$$d\big(\bar{x}^{\bar{\mu}}, \bar{x}^{\mu}\big) \leq \left(\frac{\rho}{\hbar}\right)^{\frac{1}{\beta-\theta}} d^{\frac{\gamma}{\beta-\theta}}(\overline{\mu}, \mu). \tag{4.6}$$

This implies that

$$\mathbb{D}\big(\mathrm{Sol(EP)}(\lambda, \bar{\mu}), \mathrm{Sol(EP)}(\lambda, \mu)\big) \leq \left(\frac{\rho}{\hbar}\right)^{\frac{1}{\beta-\theta}} d^{\frac{\gamma}{\beta-\theta}}(\overline{\mu}, \mu),$$

from which the proof is complete. Since $\Phi(\bar{x}^{\bar{\mu}}, \bar{x}^{\mu}, \bar{\mu}) \geq 0$ and $\Phi(\bar{x}^{\mu}, \bar{x}^{\bar{\mu}}, \mu) \geq 0$, the $(\hbar, \beta)$-strong pseudomonotonicity of $\Phi(\cdot, \cdot, \bar{\mu})$ implies that

$$\begin{aligned}\hbar d^{\beta}(\bar{x}^{\bar{\mu}}, \bar{x}^{\mu}) &\leq -\Phi(\bar{x}^{\mu}, \bar{x}^{\bar{\mu}}, \bar{\mu}) \leq \Phi(\bar{x}^{\mu}, \bar{x}^{\bar{\mu}}, \mu) - \Phi(\bar{x}^{\mu}, \bar{x}^{\bar{\mu}}, \bar{\mu}) \\ &\leq \rho d^{\gamma}(\mu, \bar{\mu}) d^{\theta}(\bar{x}^{\mu}, \bar{x}^{\bar{\mu}}).\end{aligned}$$

(The last inequality holds because Φ is (ρ, γ)-Hölder calm at $\bar{\mu}$ with respect to V, θ-uniformly over $\mathscr{D}(\lambda)$.) Hence, (4.6) follows.

(b) Taking any $\lambda \in U$, we show that

$$d\big(\bar{x}^{\bar{\lambda}}, \bar{x}^{\lambda}\big) \leq \Big(\frac{2\sigma \ell^{\delta}}{\hbar}\Big)^{\frac{1}{\beta}} d^{\frac{\alpha\delta}{\beta}}(\bar{\lambda}, \lambda) \tag{4.7}$$

for all $\bar{x}^{\bar{\lambda}} \in \mathrm{Sol(EP)}(\bar{\lambda}, \mu)$ and $\bar{x}^{\lambda} \in \mathrm{Sol(EP)}(\lambda, \mu)$, and hence

$$\mathbb{D}\big(\mathrm{Sol(EP)}(\bar{\lambda}, \mu), \mathrm{Sol(EP)}(\lambda, \mu)\big) \leq \Big(\frac{2\sigma \ell^{\delta}}{\hbar}\Big)^{\frac{1}{\beta}} d^{\frac{\alpha\delta}{\beta}}(\bar{\lambda}, \lambda).$$

Without loss of generality, we can suppose that $\Phi(\bar{x}^{\bar{\lambda}}, \bar{x}^{\lambda}, \mu) \leq \Phi(\bar{x}^{\lambda}, \bar{x}^{\bar{\lambda}}, \mu)$. Because $\mathscr{D}$ is (ℓ, α)-Hölder calm at $\bar{\lambda}$ with respect to U, there exists $x^{\bar{\lambda}} \in \mathscr{D}(\bar{\lambda})$ such that

$$d(x^{\bar{\lambda}}, x^{\lambda}) \leq \ell d^{\alpha}(\bar{\lambda}, \lambda).$$

Clearly, $\Phi(\bar{x}^{\bar{\lambda}}, x^{\bar{\lambda}}, \mu) \geq 0$. These inequalities together with the $(\hbar, \beta)$-semistrong monotonicity of $\Phi(\cdot, \cdot, \mu)$ (on $\mathscr{D}(U)$) and the (σ, δ)-Hölder continuity of $\Phi(\bar{x}^{\bar{\lambda}}, \cdot, \mu)$ (on $\mathscr{D}(U)$) imply that

$$\begin{aligned}\frac{\hbar}{2} d^{\beta}(\bar{x}^{\bar{\lambda}}, \bar{x}^{\lambda}) &\leq -\Phi(\bar{x}^{\bar{\lambda}}, \bar{x}^{\lambda}, \mu) \leq \Phi(\bar{x}^{\bar{\lambda}}, x^{\bar{\lambda}}, \mu) - \Phi(\bar{x}^{\bar{\lambda}}, \bar{x}^{\lambda}, \mu)\\ &\leq \sigma d^{\delta}(x^{\bar{\lambda}}, \bar{x}^{\lambda}) \leq \sigma \ell^{\delta} d^{\alpha\delta}(\bar{\lambda}, \lambda),\end{aligned}$$

and hence we obtain (4.7).

(c) First, we observe that the $(\hbar, \beta)$-semistrong monotonicity on $\mathscr{D}(U)$ of $\Phi(\cdot, \cdot, \bar{\mu})$ implies the $(\frac{\hbar}{2}, \beta)$-strong pseudomonotonicity on $\mathscr{D}(U)$ of $\Phi(\cdot, \cdot, \bar{\mu})$. Next, let $(\lambda, \mu) \in U \times V$ be arbitrary and $\bar{x}'' \in \mathrm{Sol(EP)}(\lambda, \bar{\mu})$ be given. For any $\bar{x} \in \mathrm{Sol(EP)}(\bar{\lambda}, \bar{\mu})$ and $\bar{x}' \in \mathrm{Sol(EP)}(\lambda, \mu)$, applying Theorem 4.23(a) for $\mathrm{Sol(EP)}(\lambda, \cdot)$ and Theorem 4.23(b) for $\mathrm{Sol(EP)}(\cdot, \bar{\mu})$, we obtain

$$\begin{aligned}d\big(\bar{x}, \bar{x}'\big) &\leq d\big(\bar{x}, \bar{x}''\big) + d\big(\bar{x}'', \bar{x}'\big)\\ &\leq \left(\frac{2\sigma \ell^{\delta}}{\hbar}\right)^{\frac{1}{\beta}} d^{\frac{\alpha\delta}{\beta}}(\bar{\lambda}, \lambda) + \left(\frac{2\rho}{\hbar}\right)^{\frac{1}{\beta-\theta}} d^{\frac{\gamma}{\beta-\theta}}(\bar{\mu}, \mu).\end{aligned}$$

Hence,

$$\begin{aligned}&\mathbb{D}\big(\mathrm{Sol(EP)}(\bar{\lambda}, \bar{\mu}), \mathrm{Sol(EP)}(\lambda, \mu)\big)\\ &\leq \left(\frac{2\sigma \ell^{\delta}}{\hbar}\right)^{\frac{1}{\beta}} d^{\frac{\alpha\delta}{\beta}}(\bar{\lambda}, \lambda) + \left(\frac{2\rho}{\hbar}\right)^{\frac{1}{\beta-\theta}} d^{\frac{\gamma}{\beta-\theta}}(\bar{\mu}, \mu).\end{aligned}$$

Therefore, the proof follows,

□

Example 4.24 (Hölder calmness of solution map) Let $X = \mathbb{R}$, $A = [0, 1]$, $\Lambda = M = [0, 1]$, $\mathscr{D}(\lambda) = [0, \lambda]$, $\Phi(x, y, \mu) = y - x + \mu$ for all $(x, y, \mu) \in A \times A \times M$, $\bar{\lambda} = \bar{\mu} = 0$, and $U = V = [0, 1]$.

We have $|\Phi(x, y, \mu) - \Phi(x, y, \bar{\mu})| = |\mu|$ for all $\mu \in [0, 1]$. Hence, Φ is (1, 1)-Hölder calm at $\bar{\lambda}$ with respect to V, 0-uniformly over [0, 1]. Since $|\Phi(x, y, \mu) - \Phi(x, z, \mu)| = |y - z|$ for all $y, z \in [0, 1]$, $\Phi(x, \cdot, \mu)$ is (1, 1)-Hölder continuous on [0, 1]. It is easy to see that $H(\mathscr{D}(\lambda), \mathscr{D}(\bar{\lambda})) = \lambda$ for all $\lambda \in [0, 1]$, and so $\mathscr{D}$ is (1, 1)-Hölder calm at $\bar{\lambda}$ with respect to U. For $\mu = \bar{\mu} = 0$, $\Phi(\cdot, \cdot, \bar{\mu})$ is $(\hbar, 1)$-semistrongly monotone on [0, 1], but $\Phi(\cdot, \cdot, \mu)$ is not $(\hbar, 1)$-semistrongly monotone on [0, 1] for any $\mu \neq \bar{\mu}$. Hence, Theorem 4.23 shows the Hölder calmness of Sol(EP). (In fact, direct computations yield $\text{Sol(EP)}(\lambda, \mu) = [0, \min\{\lambda, \mu\}]$ for all $(\lambda, \mu) \in [0, 1] \times [0, 1]$.)

4.2.2 Hölder Continuity of Approximate Solution Mappings

In this subsection, we study sufficient conditions for Hölder continuity of solution mappings of equilibrium problems when the solutions are not unique. Of course, we will omit the assumptions related to strong monotonicity or convexity used in the preceding section. We will establish such conditions only for approximate solution maps. Fortunately, we can study these approximate solutions as close as we want to the exact ones. It is worth noting that approximate solutions are important for many reasons. Since mathematical models only approximate real problems, their exact solutions can be used also as approximations of what we want. Hence, good approximate solutions of the models may be not less useful than their exact solutions. Moreover, in numerical methods, we always have to find approximate solutions that converge to exact ones. In this subsection, besides assuming Hölder properties of the data of the equilibrium problem, which are natural assumptions and can hardly be omitted, we impose concavity assumptions instead of the mentioned monotonicity conditions.

For $\varepsilon \geq 0$ and $\lambda \in \Lambda, \mu \in M$, we denote the approximate solution set of equilibrium problem $(\text{EP}_{\lambda,\mu})$ corresponding to $(\varepsilon, \lambda, \mu)$ by $\widetilde{\text{Sol}}(\text{EP})(\varepsilon, \lambda, \mu)$, i.e.,

$$\widetilde{\text{Sol}}(\text{EP})(\varepsilon, \lambda, \mu) := \{x \in \mathscr{D}(\lambda) \mid \Phi(x, y, \mu) + \varepsilon \geq 0, \forall y \in \mathscr{D}(\lambda)\},$$

and we always assume that $\widetilde{\text{Sol}}(\text{EP})(\varepsilon, \lambda, \mu)$ is nonempty in a neighborhood of the reference point $(\bar{\varepsilon}, \bar{\lambda}, \bar{\mu})$.

Theorem 4.25 (Hölder continuity of approximate solution maps) *Let* $0 < \bar{\varepsilon} < \hat{\varepsilon}$. *For problem* $(\text{EP}_{\lambda,\mu})$, *assume that* $\mathscr{D}$ *is convex-bounded-valued on A. Assume further that:*

(a) $\mathscr{D}$ *is* (ℓ, α)*-Hölder continuous and bounded at* $\bar{\lambda}$*;*
(b) *There are neighborhoods U of* $\bar{\lambda}$ *and V of* $\bar{\mu}$ *such that* $\Phi(\cdot, y, \mu)$ *is concave on* $\mathscr{D}(U)$ *for each* $(\lambda, \mu) \in U \times V$ *and* $y \in \mathscr{D}(U)$*;*

(c) $\Phi(x, y, \cdot)$ *is* (h, β)*-Hölder continuous on* V *for each* $x, y \in \mathscr{D}(U)$*;*
(d) $\Phi(\cdot, \cdot, \mu)$ *is* $((p, \delta), (q, \zeta))$*-Hölder on* $\mathscr{D}(U) \times \mathscr{D}(U)$ *for each* $\mu \in V$.

Then, $\widetilde{\mathrm{Sol}}(\mathrm{EP})$ *satisfies the following Hölder condition on* $[\bar{\varepsilon}, \hat{\varepsilon}] \times U \times V$*:*

$$H(\widetilde{\mathrm{Sol}}(\mathrm{EP})(\varepsilon^1, \lambda^1, \mu^1), \widetilde{\mathrm{Sol}}(\mathrm{EP})(\varepsilon^2, \lambda^2, \mu^2)) \le k_1|\varepsilon^1 - \varepsilon^2| + k_2\|\mu^1 - \mu^2\|^\beta + k_3\|\lambda^1 - \lambda^2\|^{\alpha\delta},$$

where $k_1, k_2, k_3 > 0$ *and depend on* ℓ, α, h, β, *etc.*

Proof Without loss of generality, we assume that $\varepsilon^1 < \varepsilon^2$.
Step 1. We find an estimate of $H(\widetilde{\mathrm{Sol}}(\mathrm{EP})(\varepsilon^1, \lambda^1, \mu^1), \widetilde{\mathrm{Sol}}(\mathrm{EP})(\varepsilon^2, \lambda^1, \mu^1))$. We have

$$\widetilde{\mathrm{Sol}}(\mathrm{EP})(0, \lambda^1, \mu^1) \subset \widetilde{\mathrm{Sol}}(\mathrm{EP})(\varepsilon^1, \lambda^1, \mu^1) \subset \widetilde{\mathrm{Sol}}(\mathrm{EP})(\varepsilon^2, \lambda^1, \mu^1) \subset \widetilde{\mathrm{Sol}}(\mathrm{EP})(\hat{\varepsilon}, \lambda^1, \mu^1).$$

Hence,

$$\mathrm{ex}(\widetilde{\mathrm{Sol}}(\mathrm{EP})(\varepsilon^1, \lambda^1, \mu^1), \widetilde{\mathrm{Sol}}(\mathrm{EP})(\varepsilon^2, \lambda^1, \mu^1)) = 0. \tag{4.8}$$

For each $x^2 \in \widetilde{\mathrm{Sol}}(\mathrm{EP})(\varepsilon^2, \lambda^1, \mu^1)$, $\bar{x} \in \widetilde{\mathrm{Sol}}(\mathrm{EP})(0, \lambda^1, \mu^1)$, and $y \in \mathscr{D}(\lambda^1)$, we have

$$\Phi(x^2, y, \mu^1) + \varepsilon^2 \ge 0, \tag{4.9}$$

$$\Phi(\bar{x}, y, \mu^1) \ge 0. \tag{4.10}$$

Since $\varepsilon^1, \varepsilon^2, \varepsilon^2 - \varepsilon^1$ are positive, it follows from (4.9) and (4.10) that

$$\frac{\varepsilon^1}{\varepsilon^2}\Phi(x^2, y, \mu^1) + \varepsilon^1 + \frac{\varepsilon^2 - \varepsilon^1}{\varepsilon^2}\Phi(\bar{x}, y, \mu^1) \ge 0, \forall y \in \mathscr{D}(\lambda^1). \tag{4.11}$$

Combining (4.11) and (b), we conclude that

$$x^1 := \frac{\varepsilon^1}{\varepsilon^2}x^2 + \frac{\varepsilon^2 - \varepsilon^1}{\varepsilon^2}\bar{x} \in \widetilde{\mathrm{Sol}}(\mathrm{EP})(\varepsilon^1, \lambda^1, \mu^1).$$

So,

$$\|x^2 - x^1\| = \left\|x^2 - \frac{\varepsilon^1}{\varepsilon^2}x^2 - \frac{\varepsilon^2 - \varepsilon^1}{\varepsilon^2}\bar{x}\right\| = \frac{|\varepsilon^1 - \varepsilon^2|}{\varepsilon^2}\|x^2 - \bar{x}\|.$$

As $\mathscr{D}$ is Hölder continuous and bounded at $\bar{\lambda}$, we have

$$\mathscr{D}(\lambda) \subset \mathscr{D}(\bar{\lambda}) + \ell B(0, \|\lambda - \bar{\lambda}\|^{\alpha})$$

for all $\lambda \in N$, which implies that

$$\mathrm{diam}\mathscr{D}(\lambda) \leq \rho := \mathrm{diam}\mathscr{D}(\bar{\lambda}) + 2\ell(\mathrm{diam}(N))^{\alpha}.$$

Therefore,

$$\|x^2 - x^1\| \leq \frac{|\varepsilon^1 - \varepsilon^2|}{\varepsilon^2}\rho \leq \frac{|\varepsilon^1 - \varepsilon^2|}{\bar{\varepsilon}}\rho.$$

For $k_1 := \frac{\rho}{\bar{\varepsilon}}$, we obtain

$$H^*(\widetilde{\mathrm{Sol}}(\mathrm{EP})(\varepsilon^2, \lambda^1, \mu^1), \widetilde{\mathrm{Sol}}(\mathrm{EP})(\varepsilon^1, \lambda^1, \mu^1)) \leq k_1|\varepsilon^1 - \varepsilon^2|. \tag{4.12}$$

Combining (4.8) and (4.12) yields

$$H(\widetilde{\mathrm{Sol}}(\mathrm{EP})(\varepsilon^1, \lambda^1, \mu^1), \widetilde{\mathrm{Sol}}(\mathrm{EP})(\varepsilon^2, \lambda^1, \mu^1)) \leq k_1|\varepsilon^1 - \varepsilon^2|. \tag{4.13}$$

Step 2. Now we look for an estimate of $H(\widetilde{\mathrm{Sol}}(\mathrm{EP})(\varepsilon^2, \lambda^1, \mu^1), \widetilde{\mathrm{Sol}}(\mathrm{EP})(\varepsilon^2, \lambda^1, \mu^2))$. Our consideration is trivial if $\mu^1 = \mu^2$. So, we can assume that $\mu^1 \neq \mu^2$. We consider two cases.
Case 1. $\|\mu^1 - \mu^2\|^{\beta} \leq \frac{\varepsilon^2}{h}$. Set $r := h\|\mu^1 - \mu^2\|^{\beta}$. Then, $0 < r \leq \varepsilon^2$. Firstly, we show that

$$\widetilde{\mathrm{Sol}}(\mathrm{EP})(\varepsilon^2 - r, \lambda^1, \mu^1) \subset \widetilde{\mathrm{Sol}}(\mathrm{EP})(\varepsilon^2, \lambda^1, \mu^2),$$

$$\widetilde{\mathrm{Sol}}(\mathrm{EP})(\varepsilon^2 - r, \lambda^1, \mu^2) \subset \widetilde{\mathrm{Sol}}(\mathrm{EP})(\varepsilon^2, \lambda^1, \mu^1).$$

Indeed, let $\bar{x} \in \widetilde{S}(\varepsilon^2 - r, \lambda^1, \mu^1)$, $f(\bar{x}, y, \mu^1) + \varepsilon^2 - r \geq 0$ for all $y \in \mathscr{D}(\lambda^1)$. Then,

$$\Phi(\bar{x}, y, \mu^2) + \Phi(\bar{x}, y, \mu^1) - \Phi(\bar{x}, y, \mu^2) + \varepsilon^2 - r \geq 0. \tag{4.14}$$

Applying assumption (c), we have

$$|\Phi(\bar{x}, y, \mu^1) - \Phi(\bar{x}, y, \mu^2)| \leq h\|\mu^1 - \mu^2\|^{\beta} = r. \tag{4.15}$$

Combining (4.14) and (4.15), we obtain

$$\Phi(\bar{x}, y, \mu^2) + \varepsilon^2 \geq 0 \quad \text{for all} \quad y \in \mathscr{D}(\lambda^1),$$

i.e, the vector $\bar{x}$ belongs to $\widetilde{\text{Sol}}(\text{EP})(\varepsilon^2, \lambda^1, \mu^2)$, which implies that

$$\widetilde{\text{Sol}}(\text{EP})(\varepsilon^2 - r, \lambda^1, \mu^1) \subset \widetilde{\text{Sol}}(\text{EP})(\varepsilon^2, \lambda^1, \mu^2).$$

By similar arguments, we see that

$$\widetilde{\text{Sol}}(\text{EP})(\varepsilon^2 - r, \lambda^1, \mu^2) \subset \widetilde{\text{Sol}}(\text{EP})(\varepsilon^2, \lambda^1, \mu^1).$$

Therefore,

$$\begin{aligned}
&\text{ex}(\widetilde{\text{Sol}}(\text{EP})(\varepsilon^2, \lambda^1, \mu^1), \widetilde{\text{Sol}}(\text{EP})(\varepsilon^2, \lambda^1, \mu^2)) = \\
&\quad = \sup_{y \in \widetilde{\text{Sol}}(\text{EP})(\varepsilon^2, \lambda^1, \mu^1)} d(y, \widetilde{\text{Sol}}(\text{EP})(\varepsilon^2, \lambda^1, \mu^2)) \\
&\quad = \sup_{y \in \widetilde{\text{Sol}}(\text{EP})(\varepsilon^2, \lambda^1, \mu^1)} \inf_{z \in \widetilde{\text{Sol}}(\text{EP})(\varepsilon^2, \lambda^1, \mu^2)} d(y, z) \\
&\quad \leq \sup_{y \in \widetilde{\text{Sol}}(\text{EP})(\varepsilon^2, \lambda^1, \mu^1)} \inf_{z \in \widetilde{\text{Sol}}(\text{EP})(\varepsilon^2 - r, \lambda^1, \mu^1)} d(y, z) \\
&\quad = \text{ex}(\widetilde{\text{Sol}}(\text{EP})(\varepsilon^2, \lambda^1, \mu^1), \widetilde{\text{Sol}}(\text{EP})(\varepsilon^2 - r, \lambda^1, \mu^1)) \\
&\quad \leq H(\widetilde{\text{Sol}}(\text{EP})(\varepsilon^2, \lambda^1, \mu^1), \widetilde{\text{Sol}}(\text{EP})(\varepsilon^2 - r, \lambda^1, \mu^1)) \\
&\quad \leq \frac{\rho r}{\varepsilon^2} \leq \frac{\rho h}{\varepsilon^2} \|\mu^1 - \mu^2\|^\beta \leq \frac{\rho h}{\bar{\varepsilon}} \|\mu^1 - \mu^2\|^\beta,
\end{aligned}$$

and

$$\begin{aligned}
&\text{ex}(\widetilde{\text{Sol}}(\text{EP})(\varepsilon^2, \lambda^1, \mu^2), \widetilde{\text{Sol}}(\text{EP})(\varepsilon^2, \lambda^1, \mu^1)) = \\
&\quad = \sup_{y \in \widetilde{\text{Sol}}(\text{EP})(\varepsilon^2, \lambda^1, \mu^2)} d(y, \widetilde{\text{Sol}}(\text{EP})(\varepsilon^2, \lambda^1, \mu^1)) \\
&\quad = \sup_{y \in \widetilde{\text{Sol}}(\text{EP})(\varepsilon^2, \lambda^1, \mu^2)} \inf_{z \in \widetilde{\text{Sol}}(\text{EP})(\varepsilon^2, \lambda^1, \mu^1)} d(y, z) \\
&\quad \leq \sup_{y \in \widetilde{\text{Sol}}(\text{EP})(\varepsilon^2, \lambda^1, \mu^2)} \inf_{z \in \widetilde{\text{Sol}}(\text{EP})(\varepsilon^2 - r, \lambda^1, \mu^2)} d(y, z) \\
&\quad = \text{ex}(\widetilde{\text{Sol}}(\text{EP})(\varepsilon^2, \lambda^1, \mu^2), \widetilde{\text{Sol}}(\text{EP})(\varepsilon^2 - r, \lambda^1, \mu^2)) \\
&\quad \leq H(\widetilde{\text{Sol}}(\text{EP})(\varepsilon^2, \lambda^1, \mu^2), \widetilde{\text{Sol}}(\text{EP})(\varepsilon^2 - r, \lambda^1, \mu^2)) \\
&\quad \leq \frac{\rho r}{\varepsilon^2} \leq \frac{\rho h}{\varepsilon^2} \|\mu^1 - \mu^2\|^\beta \leq \frac{\rho h}{\bar{\varepsilon}} \|\mu^1 - \mu^2\|^\beta.
\end{aligned}$$

Therefore,

$$H(\widetilde{\text{Sol}}(\text{EP})(\varepsilon^2, \lambda^1, \mu^1), \widetilde{\text{Sol}}(\text{EP})(\varepsilon^2, \lambda^1, \mu^2)) \leq \frac{\rho h}{\bar{\varepsilon}} \|\mu^1 - \mu^2\|^\beta. \tag{4.16}$$

Case 2. $\|\mu^1 - \mu^2\|^\beta > \dfrac{\varepsilon^2}{h}$. Let $\mathbb{P}$ be a partition of the segment $[\mu^1, \mu^2]$ with $\mathcal{N}+1$ nodes $v^1, \ldots, v^{\mathcal{N}+1}$ such that

$$v^1 = \mu^1, \qquad v^{\mathcal{N}+1} = \mu^2, \quad \|v^i - v^{i+1}\| = \frac{\|\mu^1 - \mu^2\|}{\mathcal{N}}.$$

Taking $\mathcal{N}$ such that $\mathcal{N} \geq \dfrac{h\text{diam}(U)}{\bar{\varepsilon}}$, we have

$$\|v^i - v^{i+1}\| = \frac{\|\mu^1 - \mu^2\|}{\mathcal{N}} \leq \frac{\bar{\varepsilon}\|\mu^1 - \mu^2\|}{h\text{diam}(U)} \leq \frac{\bar{\varepsilon}}{h} \leq \frac{\varepsilon^2}{h}.$$

From (4.16), we obtain

$$H(\widetilde{\text{Sol}}(\text{EP})(\varepsilon^2, \lambda^1, v^i), \widetilde{\text{Sol}}(\text{EP})(\varepsilon^2, \lambda^1, v^{i+1})) \leq \frac{\rho h}{\bar{\varepsilon}} \|v^i - v^{i+1}|^\beta, \quad i = 1, \ldots, \mathcal{N}.$$

So,

$$\begin{aligned}
&H(\widetilde{\text{Sol}}(\text{EP})(\varepsilon^2, \lambda^1, \mu^1), \widetilde{\text{Sol}}(\text{EP})(\varepsilon^2, \lambda^1, \mu^2)) = \\
&\quad = H(\widetilde{\text{Sol}}(\text{EP})(\varepsilon^2, \lambda^1, v^1), \widetilde{\text{Sol}}(\text{EP})(\varepsilon^2, \lambda^1, v^{\mathcal{N}+1})) \\
&\quad \leq \sum_{i=1}^{\mathcal{N}} H(\widetilde{\text{Sol}}(\text{EP})(\varepsilon^2, \lambda^1, v^i), \widetilde{\text{Sol}}(\text{EP})(\varepsilon^2, \lambda^1, v^{i+1})) \\
&\quad \leq \frac{\rho h}{\bar{\varepsilon}} \sum_{i=1}^{\mathcal{N}} \|v^i - v^{i+1}\|^\beta \leq \frac{\mathcal{N} \rho h}{2\bar{\varepsilon}} \frac{\varepsilon^2}{h} \leq \frac{\mathcal{N} \rho h}{2\bar{\varepsilon}} \|\mu^1 - \mu^2\|^\beta.
\end{aligned}$$

Choosing $k_2 = \max\{\frac{\mathcal{N}\rho h}{2\bar{\varepsilon}}, \frac{\rho h}{\bar{\varepsilon}}\}$, we have

$$H(\widetilde{\text{Sol}}(\text{EP})(\varepsilon^2, \lambda^1, \mu^1), \widetilde{\text{Sol}}(\text{EP})(\varepsilon^2, \lambda^1, \mu^2)) \leq k_2 \|\mu^1 - \mu^2\|^\beta. \tag{4.17}$$

Step 3. Let $\lambda^1, \lambda^2 \in U'$, where $U' \subset N$ is a neighborhood of $\bar{\lambda}$ with $\text{diam}U' \leq \dfrac{1}{2}$. We determine an estimate of $H(\widetilde{\text{Sol}}(\text{EP})(\varepsilon^2, \lambda^1, \mu^2), \widetilde{\text{Sol}}(\text{EP})(\varepsilon^2, \lambda^2, \mu^2))$. Set

$\vartheta := \varepsilon^2 - \bar{\varepsilon}$ and $\gamma := (p\ell^\delta + q\ell^\zeta)\|\lambda^1 - \lambda^2\|^\xi$ where $\xi = \min\{\alpha, \alpha\delta, \alpha\zeta\}$. As in Step 2, we consider two cases.

Case 1. $\gamma \leq \vartheta$. For all $x^1 \in \widetilde{\mathrm{Sol}}(\mathrm{EP})(\varepsilon^2 - \gamma, \lambda^1, \mu^2)$, we have $x^1 \in \mathscr{D}(\lambda^1)$ and

$$\Phi(x^1, y, \mu^2) + \varepsilon^2 - \gamma \geq 0$$

for all $y \in \mathscr{D}(\lambda^1)$. By (a), there exists $x^2 \in \mathscr{D}(\lambda^2)$ such that

$$\|x^1 - x^2\| \leq \ell\|\lambda^1 - \lambda^2\|^\alpha.$$

For each $y^2 \in \mathscr{D}(\lambda^2)$, also by (a), there is $y^1 \in \mathscr{D}(\lambda^1)$ satisfying $\|y^1 - y^2\| \leq \ell\|\lambda^1 - \lambda^2\|^\alpha$. Since $y^1 \in \mathscr{D}(\lambda^1)$, one has

$$\Phi(x^2, y^2, \mu^2) + \Phi(x^1, y^1, \mu^2) - \Phi(x^2, y^2, \mu^2) + \varepsilon^2 - \gamma \geq 0.$$

By (d), we get

$$\begin{aligned}|\Phi(x^1, y^1, \mu^2) - \Phi(x^2, y^2, \mu^2)| &\leq p\|x^1 - x^2\|^\delta + q\|y^1 - y^2\|^\zeta \\ &\leq p\ell^\delta\|\lambda^1 - \lambda^2\|^{\alpha\delta} + p\ell^\zeta\|\lambda^1 - \lambda^2\|^{\alpha\zeta} \leq \gamma.\end{aligned}$$

Therefore, $\Phi(x^2, y^2, \mu^2) + \varepsilon^2 \geq 0$, i.e., $x^2 \in \widetilde{\mathrm{Sol}}(\mathrm{EP})(\varepsilon^2, \lambda^2, \mu^2)$.

For each $x^0 \in \widetilde{\mathrm{Sol}}(\mathrm{EP})(\varepsilon^2, \lambda^1, \mu^2)$ and $x^1 \in \widetilde{\mathrm{Sol}}(\mathrm{EP})(\varepsilon^2 - \gamma, \lambda^1, \mu^2)$, we have

$$\begin{aligned}d(x^0, \widetilde{\mathrm{Sol}}(\mathrm{EP})(\varepsilon^2, \lambda^2, \mu^2)) &\leq \|x^2 - x^0\| \leq \|x^2 - x^1\| + \|x^1 - x^0\| \\ &\leq \ell\|\lambda^1 - \lambda^2\|^\alpha + \|x^1 - x^0\|.\end{aligned}$$

As x^1 is arbitrary, one gets

$$d(x^0, \widetilde{\mathrm{Sol}}(\mathrm{EP})(\varepsilon^2, \lambda^2, \mu^2)) \leq \ell\|\lambda^1 - \lambda^2\|^\alpha + d(x^0, \widetilde{\mathrm{Sol}}(\mathrm{EP})(\varepsilon^2 - \gamma, \lambda^1, \mu^2)),$$

which implies that

$$\begin{aligned}&\mathrm{ex}(\widetilde{\mathrm{Sol}}(\mathrm{EP})(\varepsilon^2, \lambda^1, \mu^2), \widetilde{\mathrm{Sol}}(\mathrm{EP})(\varepsilon^2, \lambda^2, \mu^2)) \\ &\quad\leq \ell\|\lambda^1 - \lambda^2\|^\alpha + \mathrm{ex}(\widetilde{\mathrm{Sol}}(\mathrm{EP})(\varepsilon^2, \lambda^1, \mu^2), \widetilde{\mathrm{Sol}}(\mathrm{EP})(\varepsilon^2 - \gamma, \lambda^1, \mu^2)) \\ &\quad\leq \ell\|\lambda^1 - \lambda^2\|^\alpha + H(\widetilde{\mathrm{Sol}}(\mathrm{EP})(\varepsilon^2, \lambda^1, \mu^2), \widetilde{\mathrm{Sol}}(\mathrm{EP})(\varepsilon^2 - \gamma, \lambda^1, \mu^2)) \\ &\quad\leq \ell\|\lambda^1 - \lambda^2\|^\alpha + \frac{\rho}{\bar{\varepsilon}}|\gamma| \\ &\quad\leq \ell\|\lambda^1 - \lambda^2\|^\alpha + \frac{\rho(p\ell^\delta + q\ell^\zeta)}{\bar{\varepsilon}}\|\lambda^1 - \lambda^2\|^\xi \\ &\quad\leq \left(\ell + \frac{\rho(p\ell^\delta + q\ell^\zeta)}{\bar{\varepsilon}}\right)\|\lambda^1 - \lambda^2\|^\xi.\end{aligned}$$

Case 2. $\gamma > \vartheta$. There is a natural number $\mathcal{N}'$ satisfying $\frac{\gamma}{\mathcal{N}'} \le \vartheta$. Let $\mathbb{P}'$ be a partition of segment $[\lambda^1, \lambda^2]$ with $\mathcal{N}' + 1$ nodes $u^1, u^2, \ldots, u^{\mathcal{N}'+1}$ such that

$$u^1 = \lambda^1, u_{\mathcal{N}'+1} = \lambda^2, \quad \|u_j - u_{j+1}\| = \frac{\|\lambda^1 - \lambda^2\|}{\mathcal{N}'}.$$

Then,

$$\|u^j - u^{j+1}\|^{\xi} \le \frac{\|\lambda^1 - \lambda^2\|^{\xi}}{\mathcal{N}'} \le \frac{\vartheta}{pl^{\delta} + ql^{\zeta}},$$

i.e., $(p\ell^{\delta} + q\ell^{\zeta})\|u^j - u^{j+1}\|^{\xi} \le \vartheta$. Applying Case 1, we get

$$\begin{aligned}
&H(\widetilde{\text{Sol}}(\text{EP})(\varepsilon^2, \lambda^1, \mu^2), \widetilde{\text{Sol}}(\text{EP})(\varepsilon^2, \lambda^2, \mu^2)) \\
&\quad \le \sum_{j=1}^{\mathcal{N}'} H(\widetilde{\text{Sol}}(\text{EP})(\varepsilon^2, u^j, \mu^2), \widetilde{\text{Sol}}(\text{EP})(\varepsilon^2, u^{j+1}, \mu^2)) \\
&\quad \le \mathcal{N}' \left(\ell + \frac{\rho(p\ell^{\delta} + q\ell^{\zeta})}{\bar{\varepsilon}}\right) \|u^j - u_{J+1}\|^{\xi} \\
&\quad \le \left(\ell + \frac{\rho(p\ell^{\delta} + q\ell^{\zeta})}{\bar{\varepsilon}}\right) \mathcal{N}'^{1-\xi} \|\lambda^1 - \lambda^2\|^{\xi}.
\end{aligned}$$

Setting $k_3 := \max\left\{\ell + \frac{\rho(p\ell^{\delta} + q\ell^{\zeta})}{\bar{\varepsilon}}, \left(\ell + \frac{\rho(p\ell^{\delta} + q\ell^{\zeta})}{\bar{\varepsilon}}\right) \mathcal{N}'^{1-\xi}\right\}$, we have

$$H(\widetilde{\text{Sol}}(\text{EP})(\varepsilon^2, \lambda^1, \mu^2), \widetilde{\text{Sol}}(\text{EP})(\varepsilon^2, \lambda^2, \mu^2)) \le k_3 \|\lambda^1 - \lambda^2\|^{\alpha\delta}. \tag{4.18}$$

Step 4. From (4.13), (4.17), and (4.18), we obtain

$$\begin{aligned}
&H(\widetilde{\text{Sol}}(\text{EP})(\varepsilon^1, \lambda^1, \mu^1), \widetilde{\text{Sol}}(\text{EP})(\varepsilon^2, \lambda^2, \mu^2)) \\
&\quad \le H(\widetilde{\text{Sol}}(\text{EP})(\varepsilon^1, \lambda^1, \mu^1), \widetilde{\text{Sol}}(\text{EP})(\varepsilon^2, \lambda^1, \mu^1)) \\
&\quad\quad + H(\widetilde{\text{Sol}}(\text{EP})(\varepsilon^2, \lambda^1, \mu^1), \widetilde{\text{Sol}}(\text{EP})(\varepsilon^2, \lambda^1, \mu^2)) \\
&\quad\quad + H(\widetilde{\text{Sol}}(\text{EP})(\varepsilon^2, \lambda^1, \mu^2), \widetilde{\text{Sol}}(\text{EP})(\varepsilon^2, \lambda^2, \mu^2)) \\
&\quad \le k_1|\varepsilon^1 - \varepsilon^2| + k_2\|\mu^1 - \mu^2\|^{\beta} + k_3\|\lambda^1 - \lambda^2\|^{\alpha\delta}.
\end{aligned}$$

Therefore, the proof follows. □

The following examples show that the assumptions of the above theorem are essential.

Example 4.26 (Essentialness of the concavity assumption) Let $\bar{\varepsilon} = \frac{1}{2}, \hat{\varepsilon} = 1, X = A = \mathbb{R}, \Lambda = M = [0, 1], A = [-2, 2], \mathcal{D}(\lambda) = [-2, 2]$, and

$$\Phi(x, y, \lambda) = \begin{cases} 0 & x = 0, \\ -1 & x \neq 0. \end{cases}$$

Then, the assumptions of Theorem 4.25 are satisfied except (b). Direct calculations yield an approximate solution

$$\widetilde{\mathrm{Sol}}(\mathrm{EP})(\varepsilon) = \begin{cases} [-2, 2] & \varepsilon = 1, \\ \{0\} & \frac{1}{2} \leq \varepsilon < 1. \end{cases}$$

So, $\widetilde{\mathrm{Sol}}(\mathrm{EP})(1) = [-2, 2]$ and $\widetilde{\mathrm{Sol}}(\mathrm{EP})(\varepsilon^k) = \{0\}$ for $\varepsilon^k = 1 - \frac{1}{2n} \to 1$. Hence,

$$H(\widetilde{\mathrm{Sol}}(\mathrm{EP})(1), \widetilde{\mathrm{Sol}}(\mathrm{EP})(\varepsilon^k)) = 2 > 0.$$

Therefore, $\widetilde{\mathrm{Sol}}(\mathrm{EP})(\cdot)$ is not Hölder continuous at $\varepsilon = 1$. The cause is that assumption (b) is not fulfilled. Indeed, taking $x^1 = 0, x^2 = 1$, and $t = 1/2$, we have

$$-1 = \Phi\left(\frac{1}{2}x^1 + \frac{1}{2}x^2\right) < \frac{1}{2}\Phi(x^1) + \frac{1}{2}\Phi(x^2) = -\frac{1}{2}.$$

Example 4.27 (Cruciality of the boundedness of constraint maps) Let $\bar{\varepsilon} = 10^{-1}, \hat{\varepsilon} = 1, X = \mathbb{R}, \Lambda = M = [0, 1], A = [0, +\infty), \mathcal{D}(\lambda) = [0, +\infty)$, and $\Phi(x, y, \lambda) = -\lambda x$. It is clear that $\mathcal{D}(\lambda)$ is not bounded but the other assumptions in Theorem 4.25 are satisfied. Direct computations give us an approximate solution:

$$\widetilde{\mathrm{Sol}}(\mathrm{EP})(\varepsilon, \lambda) = \begin{cases} [0, +\infty) & \lambda = 0, \\ \left[0, \sqrt{\frac{\varepsilon}{\lambda}}\right] & \lambda \neq 0. \end{cases}$$

Hence, $\widetilde{\mathrm{Sol}}(\mathrm{EP})(\cdot, \cdot)$ is not Hölder continuous at $(1, 0)$. Indeed, take $(\varepsilon^k, \lambda^k) = (1, \frac{1}{k})$ converging to $(1, 0)$. Then, for all k, we have

$$H(\widetilde{\mathrm{Sol}}(\mathrm{EP})(1, 0), \widetilde{\mathrm{Sol}}(\mathrm{EP})(\varepsilon^k, \lambda^k)) = +\infty.$$

Example 4.28 (Criticality of the convex values of constraint maps) Let $\bar{\varepsilon} = 10^{-2}, \hat{\varepsilon} = 1, X = \mathbb{R}, \Lambda = M = [0, 3], A = [1, 2] \cup \{3\}, \mathcal{D}(\lambda) = [1, 2] \cup \{3\}$, and $\Phi(x, y, \mu) = \lambda - x$. We see that all the assumptions of Theorem 4.25 are satisfied, except that $\mathcal{D}(\lambda)$ is not convex-valued. Direct computations show that the approximate solution set is

$$\widetilde{\mathrm{Sol}}(\mathrm{EP})(\varepsilon, \lambda) = (-\infty, \lambda + \varepsilon] \cap ([1, 2] \cup \{3\}).$$

Hence, $\widetilde{\text{Sol}}(\text{EP})(\cdot,\cdot)$ is not continuous at $(1,2)$. Indeed, $\widetilde{\text{Sol}}(\text{EP})(1,2) = [1,2] \cup \{3\}$. Let $\lambda^k = 2 - \frac{1}{k}$. Then, $\lambda^k \to 2$ and $\widetilde{\text{Sol}}(\text{EP})(1,\lambda^k) = [1,2]$. So, $\widetilde{\text{Sol}}(\text{EP})$ is not lsc at $(1,2)$, and hence it is not Hölder continuous at the reference point.

Corollary 4.29 (Lipschitz continuity of approximate solution maps) *For problem* (EP)*, let solutions exist in a neighborhood of the reference point* $(\lambda_0, \mu_0) \in \Lambda \times M$*. Let the following conditions hold true:*

(a) *$\mathcal{D}$ is ℓ-Lipschitz continuous at λ_0, i.e., there is a neighborhood N of λ_0, such that for all $\lambda_1, \lambda_2 \in N$,*

$$\mathcal{D}(\lambda_1) \subset \mathcal{D}(\lambda_2) + \ell B(0, ||\lambda_1 - \lambda_2||).$$

(b) *There is a neighborhood U of μ_0 such that, for each $(\lambda, \mu) \in N \times U$ and each $y \in \mathcal{D}(N)$, $\Phi(\cdot, y, \mu)$ is concave in $\mathcal{D}(N)$.*
(c) *For $x, y \in \mathcal{D}(N)$, $\Phi(x, y, \cdot)$ is h-Lipschitz continuous in U.*
(d) *For each $\mu \in U$ and $x \in \mathcal{D}(N)$, $\Phi(x, \cdot, \mu)$ is q-Lipschitz in $\mathcal{D}(N)$.*

Then, for each $\bar{\varepsilon} > 0$, $\widetilde{\text{Sol}}(\text{EP})$ *satisfies the following Lipschitz condition in $[\bar{\varepsilon}, +\infty) \times N \times U$:*

$$\begin{aligned} &H(\widetilde{\text{Sol}}(\text{EP})(\varepsilon_1, \lambda_1, \mu_1), \widetilde{\text{Sol}}(\text{EP})(\varepsilon_2, \lambda_2, \mu_2)) \\ &\leq k_1|\varepsilon_1 - \varepsilon_2| + k_2\|\mu_1 - \mu_2\| + k_3\|\lambda_1 - \lambda_2\|, \end{aligned}$$

where k_1, k_2, k_3 are positive and depend on $\bar{\varepsilon}, \ell$, h, and q.

The following example indicates that the assumed Lipschitz continuity in Corollary 4.29 cannot be relaxed to pseudo-Lipschitz continuity (known also as the Aubin property). We recall that a multifunction $\mathcal{D} : \Lambda \rightrightarrows A$ is called ℓ-pseudo-Lipschitz continuous at $(\lambda_0, x_0) \in \text{gph}K$ if there exist neighborhoods U of λ_0 and P of x_0 such that, for all $\lambda_1, \lambda_2 \in U$,

$$\mathcal{D}(\lambda_1) \cap P \subset \mathcal{D}(\lambda_2) + \ell B(0, ||\lambda_1 - \lambda_2||).$$

Example 4.30 (The Lipschitz condition cannot be relaxed to pseudo-Lipschitz continuity) Let $X = A = \mathbb{R}$, $\Lambda = M = [-1, 1]$, and

$$\mathcal{D}(\lambda) = \begin{cases} [-1, 2] & \lambda = 0, \\ [0, \frac{1}{|\lambda|}] & \lambda \neq 0, \end{cases}$$

$$\Phi(x, y, \lambda) = |\lambda|(x - y).$$

We see that assumptions (b), (c), and (d) of Corollary 4.29 are fulfilled. Moreover, $\mathcal{D}$ is pseudo-Lipschitz continuous at $\lambda_0 = 0$ and $x_0 = 1$ (e.g., with $U = (-\frac{1}{2}, \frac{1}{2})$

and $P = (0, 2)$), but not Lipschitz continuous at $\lambda_0 = 0$ and $x_0 = 1$. We can find directly the approximate solution map, for any positive ε,

$$\widetilde{\mathrm{Sol}}(\mathrm{EP})(\varepsilon, \lambda) = \begin{cases} [-1, 2] & \lambda = 0, \\ [\max\{\frac{1-\varepsilon}{|\lambda|}, 0\}, \frac{1}{|\lambda|}] & \lambda \neq 0, \end{cases}$$

which is not Lipschitz continuous at $\lambda = 0$.

In the special case of (EP), where the constrained set is a bounded convex subset of A and does not depend on parameter (i.e., for all $\lambda \in \Lambda$, $\mathscr{D}(\lambda) \equiv K$, where K is a nonempty convex subset of A), assumption (d) in Theorem 4.25 (assumption (d) in Corollary 4.29, resp.) can be omitted. This means that we obtain the following two results.

Corollary 4.31 (Hölder continuity of solution maps in the case of a fixed constraint set) *For problem* (EP), *assume that solutions exist in a neighborhood of the reference point* $\mu_0 \in M$. *Suppose that:*

(a) *There is a neighborhood* U *of* μ_0 *such that, for each* $\mu \in U$ *and* $y \in K$, $\Phi(\cdot, y, \mu)$ *is concave on* K.
(b) *For* $x, y \in K$, $\Phi(x, y, \cdot)$ *is* (h, β)*-Hölder continuous on* U.

Then, for each $\bar{\varepsilon} > 0$, $\widetilde{\mathrm{Sol}}(\mathrm{EP})$ *satisfies the following Hölder condition on* $[\bar{\varepsilon}, +\infty) \times U$:

$$H(\widetilde{\mathrm{Sol}}(\mathrm{EP})(\varepsilon_1, \mu_1), \widetilde{\mathrm{Sol}}(\mathrm{EP})(\varepsilon_2, \mu_2)) \leq k_1|\varepsilon_1 - \varepsilon_2| + k_2||\mu_1 - \mu_2||^{\beta},$$

where $k_1, k_2 > 0$ *and depend on* $\bar{\varepsilon}$, h, *and* β.

Corollary 4.32 (Lipschitz continuity of solution maps in the case of a fixed constraint set) *For problem* (EP), *assume that solutions exist in a neighborhood of the reference point* $\mu_0 \in M$. *Suppose that:*

(a) *There is a neighborhood* U *of* μ_0 *such that, for each* $\mu \in U$ *and* $y \in K$, $\Phi(\cdot, y, \mu)$ *is concave on* K.
(b) *For* $x, y \in K$, $\Phi(x, y, \cdot)$ *is* h*-Lipschitz continuous on* U.

Then, for each $\bar{\varepsilon} > 0$, $\widetilde{\mathrm{Sol}}(\mathrm{EP})$ *satisfies the following Lipschitz condition on* $[\bar{\varepsilon}, +\infty) \times U$:

$$H(\widetilde{\mathrm{Sol}}(\mathrm{EP})(\varepsilon_1, \mu_1), \widetilde{\mathrm{Sol}}(\mathrm{EP})(\varepsilon_2, \mu_2)) \leq k_1|\varepsilon_1 - \varepsilon_2| + k_2||\mu_1 - \mu_2||,$$

where $k_1, k_2 > 0$ *and depend on* $\bar{\varepsilon}$ *and* h.

4.3 Well-Posedness

In this section if not stated otherwise, all spaces X, Y, Λ, M are assumed to be normed spaces. For each $\delta \geq 0$ and $\varepsilon \geq 0$, we consider the following approximating solution set of (EP)$(\bar{\lambda}, \bar{\mu})$:

$$\Pi(\delta, \varepsilon, \bar{\lambda}, \bar{\mu}) := \bigcup_{\lambda \in B(\bar{\lambda}, \delta), \mu \in B(\bar{\mu}, \delta)} \{x \in \mathscr{D}(\lambda) \mid \Phi(x, y, \mu) + \varepsilon \geq 0, \forall y \in \mathscr{D}(\lambda)\}.$$

In this section, $B(\bar{x}, \delta)$ denotes the closed ball centered at $\bar{x}$ with radius δ.

Note that in the literature, this notion of approximate solution sets is usually employed in the studies of well-posedness of parametric equilibrium problems, and it is different from the sets $\widetilde{\text{Sol}}$(EP)$(\varepsilon, \lambda, \mu)$ considered in previous sections.

Definition 4.33 (Hemicontinuous functions) A bifunction $\Phi : A \times A \to \mathbb{R}$ is said to be HEMICONTINUOUS if for each $x, y \in A$, $\limsup_{t \to 0^+} \Phi(x + t(y - x), y) \leq \Phi(x, y)$.

Definition 4.34 (Approximating sequences for equilibrium problems) Let $(\bar{\lambda}, \bar{\mu}) \in \Lambda \times M$ and $\{(\lambda^k, \mu^k)\}_k \subset \Lambda \times M$ tend to $(\bar{\lambda}, \bar{\mu})$. For $x^k \in \mathscr{D}(\lambda^k)$, the sequence $\{x^k\}_k$ is said to be an APPROXIMATING SEQUENCE for (EP)$(\bar{\lambda}, \bar{\mu})$ corresponding to $\{(\lambda^k, \mu^k)\}_k$ if there exists a sequence $\{\varepsilon^k\}_k$ converging to 0^+ such that $\Phi(x^k, y, \mu^k) + \varepsilon^k \geq 0$ for all $y \in \mathscr{D}(\lambda^k)$ and all k.

Definition 4.35 (Generalized well-posedness) The parametric equilibrium problem $\{(\text{EP})_{\Lambda \times M}\}$ is said to be WELL-POSED IN THE GENERALIZED SENSE at $(\bar{\lambda}, \bar{\mu})$ if:

(a) The solution set Sol(EP)$(\bar{\lambda}, \bar{\mu})$ is nonempty.
(b) For any sequence $\{(\lambda^k, \mu^k)\}_k$ converging to $(\bar{\lambda}, \bar{\mu})$, every[(c)] corresponding approximating sequence for (EP)$(\bar{\lambda}, \bar{\mu})$ has a subsequence converging to some point of Sol(EP)$(\bar{\lambda}, \bar{\mu})$.

Definition 4.36 (Well-posedness) The parametric equilibrium problem $\{(\text{EP})_{\Lambda \times M}\}$ is called WELL-POSED at $(\bar{\lambda}, \bar{\mu})$ if Sol(EP)$(\bar{\lambda}, \bar{\mu}) = \{\bar{x}\}$ and every approximating sequence converges to $\bar{x}$.

Definition 4.37 (Generalized well-posedness under perturbations) The parametric equilibrium problem $\{(\text{EP})_{\Lambda \times M}\}$ is said to be WELL-POSED IN THE GENERALIZED SENSE (well-posed, resp.) UNDER PERTURBATIONS if it is well-posed in the generalized sense (well-posed, resp.) at each point $(\lambda, \mu) \in \Lambda \times M$.

Lemma 4.38 (Equivalence of an equilibrium problem and its dual problem) *Assume that A is convex and $\Phi : A \times A \to \mathbb{R}$ is monotone and hemicontinuous. Assume further that $\Phi \in \mathscr{A}^{\geq}$ and for every $x \in A$, $\Phi(x, \cdot)$ is convex. Then, for a given point $\bar{x} \in A$,*

$$[\Phi(\bar{x}, y) \geq 0, \forall y \in A] \Leftrightarrow [\Phi(y, \bar{x}) \leq 0, \forall y \in A].$$

4.3.1 Metric Characterizations of Well-Posedness

The following theorem shows that the well-posedness of $\{(\text{EP})_{\Lambda\times M}\}$ can be characterized by considering the behavior of the diameter of the approximating solution set.

Theorem 4.39 (Characterization of well-posedness via the diameter of the approximate solution set) *Assume that X is complete and the following conditions hold true:*

(a) *$\mathscr{D}$ is closed, lsc, and convex-valued on A.*
(b) *For every $\mu \in M$, $\Phi(\cdot,\cdot,\mu) \in \mathscr{A}^{\geq}$.*
(c) *For every $\mu \in M$, $\Phi(\cdot,\cdot,\mu)$ is monotone and hemicontinuous.*
(d) *For every $(x,\mu) \in A \times M$, $\Phi(x,\cdot,\mu)$ is convex.*
(e) *For every $x \in A$, $\Phi(x,\cdot,\cdot)$ is lower semicontinuous.*

Then, $\{(\text{EP})_{\Lambda\times M}\}$ is well-posed if and only if, for every $(\lambda,\mu) \in \Lambda \times M$,

$$\Pi(\delta,\varepsilon,\lambda,\mu) \neq \emptyset, \forall \delta,\varepsilon > 0, \quad \text{diam}\Pi(\delta,\varepsilon,\lambda,\mu) \to 0 \text{ as } (\delta,\varepsilon) \to (0,0), \tag{4.19}$$

where diam$(\cdot)$ *denotes the diameter of the set* $(\cdot)$.

Proof Suppose that $\{(\text{EP})_{\Lambda\times M}\}$ is well-posed. Then, problem (EP)(λ,μ) has a unique solution $x(\lambda,\mu)$ for all $(\lambda,\mu) \in \Lambda \times M$. Since $x(\lambda,\mu) \in \Pi(\delta,\varepsilon,\lambda,\mu)$ for all $\delta,\varepsilon > 0$, we conclude that $\Pi(\delta,\varepsilon,\lambda,\mu) \neq \emptyset$.

If diam$\Pi(\delta,\varepsilon,\lambda,\mu) \not\to 0$ as $(\delta,\varepsilon) \to (0,0)$, then there exist $\rho > 0$ and $\delta^k > 0, \varepsilon^k > 0$ with $(\delta^k,\varepsilon^k) \to (0,0)$ and $\hat{x}^k, \bar{x}^k \in \Pi(\delta^k,\varepsilon^k,\lambda,\mu)$ such that

$$d(\hat{x}^k,\bar{x}^k) > \rho, \forall k \in \mathbb{N}. \tag{4.20}$$

Since $\hat{x}^k,\bar{x}^k \in \Pi(\delta^k,\varepsilon^k,\lambda,\mu)$, for each $k \in \mathbb{N}$, there exist $\hat{\lambda}^k,\bar{\lambda}^k \in B(\lambda,\delta^k)$, $\hat{\mu}^k,\bar{\mu}^k \in B(\mu,\delta^k)$ such that

$$\Phi(\hat{x}^k, y, \hat{\mu}^k) + \varepsilon^k \geq 0, \forall y \in \mathscr{D}(\hat{\lambda}^k),$$

$$\Phi(\bar{x}^k, y, \bar{\mu}^k) + \varepsilon^k \geq 0, \forall y \in \mathscr{D}(\bar{\lambda}^k).$$

Hence, $\{\hat{x}^k\}_k$ and $\{\bar{x}^k\}_k$ are approximating sequences for (EP)(λ,μ) corresponding to the sequences $\{(\hat{\lambda}^k,\hat{\mu}^k)\}_k$ and $\{(\bar{\lambda}^k,\bar{\mu}^k)\}_k$, resp. It follows from the well-posedness of $\{(\text{EP})_{\Lambda\times M}\}$ that they have to converge to the unique solution $x(\lambda,\mu)$ of (EP)(λ,μ), which contradicts (4.20).

Conversely, suppose that condition (4.19) holds. Let $\{(\lambda^k,\mu^k)\}_k$ converge to some point $(\lambda,\mu) \in \Lambda \times M$ and $\{x^k\}_k \subset A$ be an approximating sequence for

(EP)(λ, μ) corresponding to $\{(\lambda^k, \mu^k)\}_k$. Then, there exist $\varepsilon^k > 0$ with $\varepsilon^k \to 0$ such that

$$\Phi(x^k, y, \mu^k) + \varepsilon^k \geq 0, \forall y \in \mathscr{D}(\lambda^k), \ \forall n \in \mathbb{N}. \tag{4.21}$$

Consequently, $x^k \in \Pi(\delta^k, \varepsilon^k, \lambda, \mu)$ with $\delta^k = \max\{d(\lambda^k, \lambda), d(\mu^k, \mu)\} \to 0$. By (4.19), $\{x^k\}_k$ is a Cauchy sequence, and so it converges to some point $\bar{x} \in A$.

Since $\mathscr{D}$ is closed, $\bar{x} \in \mathscr{D}(\lambda)$. Let us show that $\bar{x} \in$ Sol(EP)(λ, μ). For every $y \in \mathscr{D}(\lambda)$, by the lower semicontinuity of $\mathscr{D}$, there exists $y^k \in \mathscr{D}(\lambda^k)$ such that $y^k \to y$. Employing (4.21) and assumptions (c) and (e), we obtain

$$\Phi(y, \bar{x}, \mu) \leq \liminf_{n\to\infty} \Phi(y^k, x^k, \mu^k) \leq \liminf_{n\to\infty}\{-\Phi(x^k, y^k, \mu^k)\} \leq \liminf_{n\to\infty} \varepsilon^k = 0.$$

This together with Lemma 4.38 implies that $\bar{u}$ belongs to Sol(EP)(λ, μ).

To complete the proof, we need to show that the problem (EP)(λ, μ) has a unique solution. Indeed, if (EP)(λ, μ) has two distinct solutions x^1 and x^2, then clearly $x^1, x^2 \in \Pi(\delta, \varepsilon, \lambda, \mu)$ for all $\delta, \varepsilon > 0$. Hence,

$$0 < d(x^1, x^2) \leq \operatorname{diam}\Pi(\delta, \varepsilon, \lambda, \mu),$$

which contradicts (4.19). □

Unfortunately, the assumed well-posedness in Theorem 4.39 cannot be replaced by well-posedness in the generalized sense.

Example 4.40 (Generalized well-posedness cannot be characterized by the diameter of the approximate solution set) Let $X = \mathbb{R}$, $A = M = \Lambda = [0, 1]$, $\mathscr{D}(\lambda) = [0, 1]$, and $\Phi(x, y, \mu) = x^2 + y^2 + \mu^2$. Then, $\{(\text{EP})_{\Lambda\times M}\}$ is well-posed in the generalized sense. But $\Pi(\delta, \varepsilon, \lambda, \mu) = [0, 1]$ and hence its diameter does not converge to 0.

Now, we use the Kuratowski measure of noncompactness to obtain characterizations of well-posedness for $\{(\text{EP})_{\Lambda\times M}\}$.

Definition 4.41 (The Kuratowski measure of noncompactness) Let M be a nonempty subset of a metric space X. The Kuratowski measure of M is

$$\eta(M) := \inf\{\varepsilon > 0 \mid M \subset \bigcup_{k=1}^{n} M_k \text{ and } \operatorname{diam} M_k \leq \varepsilon, k = 1, \ldots, n\}.$$

The measure η has the following properties:

(a) $\eta(M) = +\infty$ if and only if the set M is unbounded.
(b) $\eta(M) = \eta(\operatorname{cl} M)$.
(c) It follows from $\eta(M) = 0$ that M is a totally bounded set, i.e., for any $\rho > 0$, M can be covered by a finite number of balls in X with radius ρ.

(d) If X is complete and $\{M^k\}_k$ is a sequence of closed subsets of X such that $M^{k+1} \subset M^k$ for each k and $\lim_{k\to+\infty} \eta(M^k) = 0$, then $M := \cap_k M^k$ is a nonempty compact set and $\lim_{k\to+\infty} H(M^k, M) = 0$, where H is the Hausdorff distance.
(e) It follows from $M \subset N$ that $\eta(M) \leq \eta(N)$.

By using the Kuratowski measure of noncompactness of approximate solution sets, we have the following characterization of the well-posedness in the generalized sense for $\{(\text{EP})_{\Lambda\times M}\}$.

Theorem 4.42 (A characterization of generalized well-posedness via the Kuratowski measure of noncompactness of the approximate solution set) *Suppose that X is complete, Λ, M are compact or finite dimensional, and $\mathcal{D}$ is closed and lower semicontinuous on Λ. Assume further that Φ is upper semicontinuous. Then, $\{(\text{EP})_{\Lambda\times M}\}$ is well-posed in the generalized sense if and only if for every $(\lambda, \mu) \in \Lambda \times M$ and $\delta, \varepsilon > 0$,*

$$\Pi(\delta, \varepsilon, \lambda, \mu) \neq \emptyset \text{ and } \eta(\Pi(\delta, \varepsilon, \lambda, \mu)) \to 0 \text{ as } (\delta, \varepsilon) \to (0, 0). \tag{4.22}$$

Proof Suppose that $\{(\text{EP})_{\Lambda\times M}\}$ is well-posed in the generalized sense. Then, for every $(\lambda, \mu) \in \Lambda \times M$, $\text{Sol(EP)}(\lambda, \mu)$ is nonempty and compact, and hence $\Pi(\delta, \varepsilon, \lambda, \mu) \neq \emptyset$ for all $\delta, \varepsilon > 0$. Let us show that $\eta(\Pi(\delta, \varepsilon, \lambda, \mu)) \to 0$ as $(\delta, \varepsilon) \to (0, 0)$. Since

$$\text{Sol(EP)}(\lambda, \mu) \subset \Pi(\delta, \varepsilon, \lambda, \mu),$$

for all $\delta, \varepsilon > 0$, we have

$$H(\Pi(\delta, \varepsilon, \lambda, \mu), \text{Sol(EP)}(\lambda, \mu)) = \text{ex}(\Pi(\delta, \varepsilon, \lambda, \mu), \text{Sol(EP)}(\lambda, \mu)).$$

Combining the definition of η and the compactness of $\text{Sol(EP)}(\lambda, \mu))$, we have

$$\begin{aligned}\eta(\Pi(\delta, \varepsilon, \lambda, \mu)) &\leq 2H(\Pi(\delta, \varepsilon, \lambda, \mu), \text{Sol(EP)}(\lambda, \mu)) + \eta(\text{Sol(EP)}(\lambda, \mu)) \\ &= 2\text{ex}(\Pi(\delta, \varepsilon, \lambda, \mu), \text{Sol(EP)}(\lambda, \mu)).\end{aligned}$$

Finally, we prove that

$$\text{ex}(\Pi(\delta, \varepsilon, \lambda, \mu), \text{Sol(EP)}(\lambda, \mu)) \to 0 \quad \text{as} \quad (\delta, \varepsilon) \to (0, 0).$$

Suppose that $\text{ex}(\Pi(\delta, \varepsilon, \lambda, \mu), \text{Sol(EP)}(\lambda, \mu)) \not\to 0$ as $(\delta, \varepsilon) \to (0, 0)$. Then, there exist $\rho > 0$ and $\delta^k > 0, \varepsilon^k > 0$ with $(\delta^k, \varepsilon^k) \to (0, 0)$, and $x^k \in \Pi(\delta, \varepsilon, \lambda^k, \mu^k)$ such that for all k,

$$x^k \notin \text{Sol(EP)}(\lambda, \mu) + B(0, \rho). \tag{4.23}$$

Since $x^k \in \Pi(\delta, \varepsilon, \lambda^k, \mu^k)$, $\{x^k\}_k$ is an approximating sequence for (EP)(λ, μ), it follows from the well-posedness in the generalized sense that there is a subsequence $\{x^{k_i}\}_i$ of $\{x^k\}_k$ converging to some element of Sol(EP)(λ, μ). This contradicts (4.23), and hence

$$\text{ex}(\Pi(\delta, \varepsilon, \lambda, \mu), \text{Sol(EP)}(\lambda, \mu)) \to 0 \quad \text{as} \quad (\delta, \varepsilon) \to (0, 0).$$

Conversely, assume that (4.22) is fulfilled. Firstly, we prove that $\Pi(\delta, \varepsilon, \lambda, \mu)$ is closed for all $\delta, \varepsilon > 0$. Take $x^k \in \Pi(\delta, \varepsilon, \lambda, \mu)$ with $x^k \to x$. Then, for each k, there are $\lambda^k \in B(\lambda, \delta)$ and $\mu^k \in B(\mu, \delta)$ such that $x^k \in \widetilde{\text{Sol}}(\text{EP})(\lambda^k, \mu^k, \varepsilon)$, and hence

$$\Phi(x^k, y, \mu^k) + \epsilon \geq 0, \forall y \in \mathscr{D}(\lambda^k). \tag{4.24}$$

Due to the imposed assumptions on Λ and M, without loss of generality, we can suppose that $\lambda^k \to \bar{\lambda} \in B(\lambda, \delta)$ and $\mu^k \to \bar{\mu} \in B(\mu, \delta)$. By the closedness of $\mathscr{D}$, $x \in \mathscr{D}(\bar{\lambda})$. For every $y \in \mathscr{D}(\bar{\lambda})$, by the lower semicontinuity of $\mathscr{D}$, there is $y^k \in \mathscr{D}(\lambda^k)$ converging to y. Combining the upper semicontinuity of Φ with (4.24), we obtain $\Phi(x, y, \bar{\mu}) + \varepsilon \geq 0$ for all $y \in \mathscr{D}(\bar{\lambda})$. This together with $\bar{\lambda} \in B(\lambda, \delta)$ and $\bar{\mu} \in B(\mu, \delta)$ yields $x \in \Pi(\delta, \varepsilon, \lambda, \mu)$, and so $\Pi(\delta, \varepsilon, \lambda, \mu)$ is nonempty and closed for all $\delta, \varepsilon > 0$. Now, we show that

$$\text{Sol(EP)}(\lambda, \mu) = \bigcap_{\delta>0, \varepsilon>0} \Pi(\delta, \varepsilon, \lambda, \mu).$$

Firstly, we check that

$$\bigcap_{\delta>0} \Pi(\delta, \varepsilon, \lambda, \mu) = \widetilde{\text{Sol}}(\text{EP})(\varepsilon, \lambda, \mu).$$

Clearly,

$$\bigcap_{\delta>0} \Pi(\delta, \varepsilon, \lambda, \mu) \supset \widetilde{\text{Sol}}(\text{EP})(\varepsilon, \lambda, \mu).$$

Conversely, let $x \in \bigcap_{\delta>0} \Pi(\delta, \varepsilon, \lambda, \mu)$. There exist $\lambda^k \in B(\lambda, \delta)$ and $\mu^k \in B(\mu, \delta)$ such that $\Phi(x, y, \mu^k) + \varepsilon \geq 0$ for all $y \in \mathscr{D}(\lambda^k)$. As $x \in \mathscr{D}(\lambda^k)$, λ^k converge to λ, and $\mathscr{D}$ is closed, we get $x \in \mathscr{D}(\lambda)$. Next, we show that $x \in \widetilde{\text{Sol}}(\text{EP})(\varepsilon, \lambda, \mu)$. For each $y \in \mathscr{D}(\lambda)$, by the lower semicontinuity of $\mathscr{D}$, there is $y^k \in \mathscr{D}(\lambda^k)$ with $\{y^k\} \to y$. Since $x \in \widetilde{\text{Sol}}(\text{EP})(\varepsilon, \lambda, \mu)$, $\Phi(x, y^k, \mu^k) + \varepsilon \geq 0$. As Φ is upper semicontinuous, $\Phi(x, y, \mu) + \varepsilon \geq 0$, and hence

$$\bigcap_{\delta>0} \Pi(\delta, \varepsilon, \lambda, \mu) \subset \widetilde{\text{Sol}}(\text{EP})(\varepsilon, \lambda, \mu),$$

and consequently

$$\bigcap_{\delta>0} \Pi(\delta, \varepsilon, \lambda, \mu) = \widetilde{\text{Sol}}(\text{EP})(\varepsilon, \lambda, \mu).$$

Hence, we have

$$\text{Sol(EP)}(\lambda, \mu) = \bigcap_{\varepsilon>0} \widetilde{\text{Sol}}(\text{EP})(\varepsilon, \lambda, \mu) = \bigcap_{\delta>0, \varepsilon>0} \Pi(\delta, \varepsilon, \lambda, \mu).$$

Since $\eta(\Pi(\delta, \varepsilon, \lambda, \mu)) \to 0$, the properties of η imply that Sol(EP)(λ, μ) is compact and

$$H(\Pi(\delta, \varepsilon, \lambda, \mu), \text{Sol(EP)}(\lambda, \mu)) \to 0^+ \text{ as } (\delta, \varepsilon) \to (0^+, 0^+).$$

Let $(\lambda^k, \mu^k) \to (\lambda, \mu) \in \Lambda \times M$ and $\{x^k\}_k \subset A$ be an approximating sequence for (EP)(λ, μ) corresponding to $\{(\lambda^k, \mu^k)\}_k$. Then, there are $\delta^k > 0, \varepsilon^k > 0$ with $(\delta^k, \varepsilon^k) \to (0, 0)$ such that for all k, $\Phi(x^k, y, \mu^k) + \varepsilon^k \geq 0$ for all $y \in \mathscr{D}(\lambda^k)$. This implies that $x^k \in \Pi(\delta^k, \varepsilon^k, \lambda, \mu)$ with $\delta^k = \max\{d(\lambda^k, \lambda), d(\mu^k, \mu)\}$. By (4.22),

$$d(x^k, \text{Sol(EP)}(\lambda, \mu)) \leq H^*(\Pi(\delta^k, \varepsilon^k, \lambda, \mu), \text{Sol(EP)}(\lambda, \mu)) \to 0.$$

By the compactness of Sol(EP)(λ, μ), there are $\bar{x}^k \in$ Sol(EP)(λ, μ) such that

$$d(x^k, \bar{x}^k) = d(x^k, \text{Sol(EP)}(\lambda, \mu)) \to 0,$$

and $\{\bar{x}^k\}_k$ has a subsequence $\{\bar{x}^{k_i}\}_i$ converging to $\bar{x} \in$ Sol(EP)(λ, μ). Therefore, the corresponding subsequence $\{x^{k_i}\}_i$ of $\{x^k\}_k$ also converges to $\bar{x}$. So, $\{(\text{EP})_{\Lambda \times M}\}$ is well-posed in the generalized sense. □

The assumptions of Theorem 4.42 are essential. The following examples demonstrate that we cannot dispense with the closedness of $\mathscr{D}$ (Example 4.43), the lower semicontinuity of $\mathscr{D}$ (Example 4.44), or the upper semicontinuity of Φ (Example 4.45).

Example 4.43 (Essentialness of the closedness condition of constraint maps) Let $X = \mathbb{R}$, $A = [-1, 1]$, $\Lambda \equiv M = [0, 1]$, $\mathscr{D}(\lambda) = (-\lambda, 1]$, and

$$\Phi(x, y, \lambda) = xe^{\lambda}(x - y).$$

It is easy to see that X is complete, Λ is compact, $\mathscr{D}$ is lsc on Λ, and Φ is continuous on $A \times A \times \Lambda$. Moreover, $\Pi(\delta, \varepsilon, 0) \subset [-1, 1]$, and so $\eta\Pi(\delta, \varepsilon, 0)) = 0$. But

$$\text{Sol(EP)}(0) = \{1\} \quad \text{and} \quad \text{Sol(EP)}(\lambda) = \{0, 1\},$$

for all $\lambda \in (0, 1]$. Hence, $\{(\mathrm{EP})_{\Lambda \times M}\}$ is not well-posed in the generalized sense at 0. The reason is that $\mathscr{D}$ is not closed at 0. Indeed, with $\lambda^k = \frac{1}{k}$, $z^k = \frac{1}{k}$ belongs to $\mathscr{D}(\lambda^k)$. We see that $z_k \to 0 \notin \mathscr{D}(0)$, and hence $\mathscr{D}$ is not closed at 0.

Example 4.44 (One cannot dispense with the lower semicontinuity of constraint maps) Let X, A, Λ, M be as in Example 4.43, $\Phi(x, y, \lambda) = (x + y)e^{\lambda}$, and

$$\mathscr{D}(\lambda) = \begin{cases} \{-1, 0, 1\} & \text{if} \quad \lambda = 0, \\ \{0, 1\} & \text{otherwise.} \end{cases}$$

Clearly, X is complete, Λ is compact, $\mathscr{D}$ is closed, and Φ is continuous on $A \times A \times \Lambda$. $\Pi(\delta, \varepsilon, 0) \subset [-1, 1]$ and hence $\eta(\Pi(\delta, \varepsilon, 0)) = 0$. But

$$\mathrm{Sol(EP)}(0) = \{1\}, \mathrm{Sol(EP)}(\lambda) = \{0, 1\},$$

for all $\lambda \in (0, 1]$. Thus, $\{(\mathrm{EP})_{\Lambda \times M}\}$ is not well-posed in the generalized sense at 0. The reason is that $\mathscr{D}$ is not lsc on $\mathscr{D} \times \Lambda$.

Example 4.45 (Cruciality of the upper semicontinuity of objective functions) Let X, A, Λ, M be as in Example 4.44, $\mathscr{D}(\lambda) = [\lambda, 1]$, and

$$\Phi(x, y, \lambda) = \begin{cases} x(x - y) & \text{if} \quad \lambda = 0, \\ x + y & \text{otherwise.} \end{cases}$$

Obviously, X is complete, Λ is compact, and $\eta(\Pi(\delta, \varepsilon, 0)) = 0$. But

$$\mathrm{Sol(EP)}(0) = \{0, 1\} \text{ and } \mathrm{Sol(EP)}(\lambda) = [\lambda, 1]$$

for all $\lambda \in (0, 1]$. Hence, $\{(\mathrm{EP})_{\Lambda \times M}\}$ is not well-posed in the generalized sense at 0. The reason is that Φ is not upper semicontinuous. Indeed, let $(x^k, y^k, \lambda^k) = (\frac{1}{k}, 1 - \frac{1}{k}, \frac{1}{k})$. We see that $\Phi(x^k, y^k, \lambda^k) = 1$. But $\{(x^k, y^k, \lambda^k)\}_k \to (0, 1, 0)$ and $\Phi(0, 1, 0) = 0 \ngeq 1$.

4.3.2 *Sufficient Conditions for Well-Posedness*

Firstly, we show that under suitable conditions, the well-posedness of $\{(\mathrm{EP})_{\Lambda \times M}\}$ is equivalent to the existence and uniqueness of its solutions.

Theorem 4.46 (Sufficient conditions for well-posedness) *Suppose that X is a finite-dimensional space and $\mathscr{D}$ is closed, lower semicontinuous, and convex-valued on Λ. Assume further that:*

(a) *For every $\mu \in M$, $\Phi(\cdot, \cdot, \mu) \in \mathscr{A}^{\geq}$*

(b) *For every* $\mu \in M$, $\Phi(\cdot, \cdot, \mu)$ *is monotone and hemicontinuous*
(c) *For every* $(x, \mu) \in A \times M$, $\Phi(x, \cdot, \mu)$ *is convex*
(d) Φ *is continuous*

Then, $\{(\mathrm{EP})_{\Lambda \times M}\}$ *is well-posed if and only if for every* $(\lambda, \mu) \in \Lambda \times M$, $\mathrm{Sol(EP)}(\lambda, \mu) = \{\bar{x}\}$, *a singleton.*

Proof The necessity is obvious. For sufficiency, suppose that (EP)(λ, μ) has a unique solution $x(\lambda, \mu)$ for each $(\lambda, \mu) \in \Lambda \times M$. For $(\lambda^k, \mu^k) \to (\lambda, \mu)$ and $\{x^k\}_k \subset \mathscr{D}$ being an approximating sequence for (EP)(λ, μ) corresponding to $\{(\lambda^k, \mu^k)\}_k$, there are $\varepsilon^k \to 0^+$ such that for all k,

$$\Phi(x^k, y, \mu^k) + \varepsilon^k \geq 0, \forall y \in \mathscr{D}(\lambda^k). \tag{4.25}$$

Now, we show that $\{x^k\}_k$ is bounded. If $\{x^k\}_k$ is unbounded, we can assume without loss of generality that $\|x^k\| \to +\infty$. Let $t^k = \|x^k - x(\lambda, \mu)\|^{-1}$ and $w^k = x(\lambda, \mu) + t^k(x^k - x(\lambda, \mu))$. We can assume that $w^k \to w \neq x(\lambda, \mu)$. By the monotonicity of $\Phi(\cdot, \cdot, \mu^k)$ and (4.25), we have

$$\Phi(y, x^k, \mu^k) \leq -\Phi(x^k, y, \mu^k) \leq \varepsilon^k, \forall y \in \mathscr{D}(\lambda^k).$$

For each $y \in \mathscr{D}(\lambda)$, the lower semicontinuity of $\mathscr{D}$ implies the existence of $y^k \in \mathscr{D}(\lambda^k)$ with $y^k \to y$. Combining conditions (c) and (d), we obtain

$$\begin{aligned}\Phi(y, w, \mu) &= \lim_{n\to\infty} h(y^k, w^k, \mu^k) \\ &\leq \lim_{n\to\infty} \{t^k \Phi(y^k, x^k, \mu^k) + (1 - t^k)\Phi(y^k, x(\lambda, \mu), \mu^k)\} \\ &\leq \lim_{n\to\infty} \{t^k \varepsilon^k + (1 - t^k)\Phi(y_n, x(\lambda, \mu), \mu^k)\} = \Phi(y, x(\lambda, \mu), \mu).\end{aligned} \tag{4.26}$$

As $x(\lambda, \mu)$ is a solution, applying Lemma 4.38, one has $\Phi(y, x(\lambda, \mu), \mu) \leq 0$ for all $y \in \mathscr{D}(\lambda)$, which together with (4.26) implies that $\Phi(y, w, \mu) \leq 0$ for all $y \in \mathscr{D}(\lambda)$. Employing again Lemma 4.38, one sees that w is a solution of (EP)(λ, μ), which contradicts the uniqueness of solutions. So, $\{x^k\}_k$ is bounded.

Consider any subsequence $\{x^{k_i}\}_i$ of $\{x^k\}_k$ satisfying $x^{k_i} \to \bar{x}$ as $i \to \infty$. As $\mathscr{D}$ is closed, $\bar{x} \in \mathscr{D}(\lambda)$. For each $y \in \mathscr{D}(\lambda)$, since $\mathscr{D}$ is lsc, there is $y^k \in \mathscr{D}(\lambda^k)$ such that $y^k \to y$. For every $y \in \mathscr{D}(\lambda)$, we have

$$\Phi(y, \bar{x}, \mu) = \lim_{k\to\infty} \Phi(y^{k_i}, x^{k_i}, \mu^{k_i}) \leq -\lim_{k\to\infty} \Phi(x^{k_i}, y^{k_i}, \mu^{k_i}) \leq \lim_{i\to\infty} \varepsilon_{k_i} = 0.$$

This together with Lemma 4.38 shows that $\bar{x}$ is a solution of (EP)(λ, μ). By the uniqueness, we conclude that $\bar{x} = x(\lambda, \mu)$. Therefore, x^k converge to $x(\lambda, \mu)$, and so $\{(\mathrm{EP})_{\Lambda \times M}\}$ is well-posed. □

Moving on to well-posedness in the generalized sense, we obtain the following result.

Theorem 4.47 (Sufficient conditions for generalized well-posedness) *Suppose that X is finite dimensional and $\mathscr{D}$ is closed, lower semicontinuous, and convex-valued on Λ. Assume further that:*

(a) *For every $\mu \in M$, $\Phi(\cdot, \cdot, \mu) \in \mathscr{A}^{\geq}$.*
(b) *For every $\mu \in M$, $\Phi(\cdot, \cdot, \mu)$ is monotone and hemicontinuous.*
(c) *For every $(x, \mu) \in A \times M$, $\Phi(x, \cdot, \mu)$ is convex.*
(d) *Φ is continuous.*

If for each $(\lambda, \mu) \in \Lambda \times M$, there is some $\varepsilon > 0$ such that $\Pi(\varepsilon, \varepsilon, \lambda, \mu)$ is bounded. Then, $\{(\mathrm{EP})_{\Lambda \times M}\}$ is well-posed in the generalized sense.

Proof Let $(\lambda, \mu) \in \Lambda \times M$, $(\lambda^k, \mu^k) \to (\lambda, \mu)$, and $\{x^k\}_k$ be an approximating sequence for $(\mathrm{EP}_{\lambda,\mu})$ corresponding to $\{(\lambda^k, \mu^k)\}_k$. Then, there are $\varepsilon^k \to 0^+$ such that for all k,

$$\Phi(x^k, y, \mu^k) + \varepsilon^k \geq 0, \forall y \in \mathscr{D}(\lambda^k).$$

For $\varepsilon > 0$ such that $\Pi(\varepsilon, \varepsilon, \lambda, \mu)$ is bounded, there is k_0 such that $x^k \in \Pi(\varepsilon, \varepsilon, \lambda, \mu)$ for all $k \geq k_0$, and hence $\{x^k\}_k$ is bounded. Therefore, there is a subsequence $\{x^{k_i}\}_i$ of $\{x^k\}_k$ such that $x^{k_i} \to \bar{x}$ as $i \to \infty$. Using the same arguments as in the proof of Theorem 4.46, we conclude that $\bar{x}$ solves $(\mathrm{EP})(\lambda, \mu)$. So, $\{(\mathrm{EP})_{\Lambda \times M}\}$ is well-posed in the generalized sense. □

Applying Theorems 4.46 and 4.47, we obtain the following statement for well-posedness.

Theorem 4.48 (Sufficient conditions for well-posedness under perturbations) *Suppose that X is finite dimensional, and $\mathscr{D}$ is closed, lower semicontinuous, and convex-valued on Λ. Assume further that:*

(a) *For every $\mu \in M$, $\Phi(\cdot, \cdot, \mu) \in \mathscr{A}^{=}$.*
(b) *For every $\mu \in M$, $\Phi(\cdot, \cdot, \mu)$ is monotone and hemicontinuous.*
(c) *For every $(x, \mu) \in A \times M$, $\Phi(x, \cdot, \mu)$ is convex.*
(d) *Φ is continuous.*

Then, $\{(\mathrm{EP})_{\Lambda \times M}\}$ is well-posed in the generalized sense. In addition, if $\Phi(\cdot, \cdot, \mu)$ is strictly monotone for all $\mu \in M$, then $\{(\mathrm{EP})_{\Lambda \times M}\}$ is well-posed.

Remark 4.49 (Relationships of well-posedness under perturbations)

(a) If $\mathscr{D}(\lambda) \equiv A$ and $\Phi(x, y, \mu) = \Psi(x, y)$, then the concepts of well-posedness proposed in Definitions 4.34 and 4.35 reduce to the Tikhonov well-posedness for equilibrium problems.
(b) If $\Lambda = \mathbb{R}_+$, $\mathscr{D}(\lambda) = \{x \in X \mid d(x, A) \leq \lambda\}$, $\Phi(x, y, \mu) = \Psi(x, y)$, and $\bar{\lambda} = 0$, then the concepts of well-posedness proposed in Definitions 4.34 and 4.35 reduce to the Levitin-Polyak well-posedness for equilibrium problems.

4.4 Special Cases

As mentioned in the previous sections, equilibrium problem (EP) contains many optimization-related problems as special cases. Therefore, we can derive consequences of the above results for such special cases. In this section, we take optimization problems, variational inequalities, and fixed-point problems as examples.

4.4.1 Optimization Problems

Let X be a locally convex Hausdorff topological vector space, Y, W be vector topological spaces, and $A \subset X, \Lambda \subset Y, M \subset W$ be nonempty subsets. Let $\mathscr{D} : \Lambda \rightrightarrows A$ be a set-valued map with nonempty convex values. Let $\varphi : A \times M \to \mathbb{R}$. For $(\lambda, \mu) \in \Lambda \times M$, we consider the following optimization problem:

(OP) find $\bar{x} \in \mathscr{D}(\lambda)$ such that

$$\varphi(\bar{x}, \mu) = \min_{y \in \mathscr{D}(\lambda)} \varphi(y, \mu).$$

By setting $\Phi(x, y, \mu) = \varphi(y, \mu) - \varphi(x, \mu)$, problem (OP) becomes a special case of problem (EP) discussed in the previous sections.

For each $\varepsilon \in \mathbb{R}_+, \lambda \in \Lambda$, and $\mu \in M$, we denote the solution set and the ε-approximate solution set of (OP) by Sol(OP) and $\widetilde{\text{Sol}}(\text{OP})(\varepsilon, \lambda, \mu)$, resp. That is,

$$\text{Sol(OP)}(\lambda, \mu) = \{\bar{x} \in \mathscr{D}(\lambda) \mid \varphi(\bar{x}, \mu) \leq \varphi(y, \mu), \forall y \in \mathscr{D}(\lambda)\},$$

$$\widetilde{\text{Sol}}(\text{OP})(\varepsilon, \lambda, \mu) = \{\bar{x} \in \mathscr{D}(\lambda) \mid \varphi(\bar{x}, \mu) \leq \varphi(y, \mu) + \varepsilon, \forall y \in \mathscr{D}(\lambda)\}.$$

Corollary 4.50 (Hausdorff continuity of the approximate solution maps of optimization problems) *Consider* (OP). *Assume that the following conditions hold:*

(a) *$\mathscr{D}$ is continuous at λ_0 and $\mathscr{D}(\lambda_0)$ is compact.*
(b) *There exists a neighborhood N of λ_0 such that g is continuous on $\mathscr{D}(N) \times \{\mu_0\}$.*
(c) *$\varphi(\cdot, \mu_0)$ is convex on $\mathscr{D}(N)$.*

Then, $\widetilde{\text{Sol}}(\text{OP})$ *is Hausdorff continuous at* $(\varepsilon_0, \lambda_0, \mu_0)$.

Proof We prove this corollary by checking the validity of the assumptions of Theorem 4.15. It is obvious that (a) and (b) of Theorem 4.15 are satisfied. We only show that its assumption (c) is also fulfilled. Indeed, take arbitrary $x_1, x_2 \in \mathscr{D}(N)$, $t \in [0, 1]$, and $y \in \mathscr{D}(N)$. We have

$$\Phi(tx_1 + (1-t)x_2, y, \mu_0) - tf(x_1, y, \mu_0) - (1-t)f(x_2, y, \mu_0)$$
$$= \varphi(y, \mu_0) - \varphi(tx_1 + (1-t)x_2, \mu_0) - t[\varphi(y, \mu_0) - \varphi(x_1, \mu_0)]$$
$$-(1-t)[\varphi(y, \mu_0) - \varphi(x_2, \mu_0)]$$
$$= t\varphi(x_1, \mu_0) + (1-t)\varphi(x_2, \mu_0) - \varphi(tx_1 + (1-t)x_2, \mu_0) \geq 0,$$

since $\varphi(\cdot, \mu_0)$ is convex. Hence, (c) holds. □

In the case with $\mathscr{D}(\lambda) \equiv K$, a nonempty and convex set, we have the following result.

Corollary 4.51 (Hausdorff continuity of the approximate solution maps of optimization problems with fixed constraint sets) *For* (OP)*, assume the following conditions:*

(a) *For all $x \in K$, $\varphi(x, \cdot)$ is continuous at μ_0.*
(b) *$\varphi(\cdot, \mu_0)$ is convex on K.*

Then, $\widetilde{\text{Sol}}(\text{OP})$ *is Hausdorff continuous at (ε_0, μ_0).*

Passing to the Hölder properties of the solution map Sol(OP), for convenience, we assume that X, Y, and W are normed spaces. We provide here some typical results. The first one is obtained directly from Theorem 4.21.

Corollary 4.52 (Hölder continuity of solution maps of optimization problems via strong monotonicity conditions) *Assume for* (OP) *that:*

(a) *$\mathscr{D}$ is (ℓ, α)-Hölder in a neighborhood N of $\bar{\lambda}$.*
(b) *There is a neighborhood U of $\bar{\mu}$ such that for each $x \in \mathscr{D}(N)$, $\varphi(x, \cdot)$ is (n_1, δ_1)-Hölder at $\bar{\mu}$, and for each $\forall \mu \in U$, $\varphi(\cdot, \mu)$ is (n_2, δ_2)-Hölder in $\mathscr{D}(N)$.*
(c) *For each $\mu \in U$, $\varphi(\cdot, \mu)$ is (h, β)-Hölder strongly pseudomonotone in $\mathscr{D}(N)$, i.e., for every $x, y \in \mathscr{D}(N)$ and $\mu \in U$,*

$$[\varphi(y, \mu) - \varphi(x, \mu) \geq 0] \Rightarrow [\varphi(x, \mu) - \varphi(y, \mu) + h\|x - y\|^\beta \leq 0].$$

Then, in the neighborhood $N \times U$ of $(\bar{\lambda}, \bar{\mu})$, the solution map of problem (OP) *satisfies the Hölder condition*

$$\mathbb{D}\big(\text{Sol(OP)}(\lambda_1, \mu_1), \text{Sol(OP)}(\lambda_2, \mu_2)\big) \leq k_1 d^{\alpha_2 \delta_2/\beta}(\lambda_1, \lambda_2) + k_2 d^{\delta_1/\beta}(\mu_1, \mu_2),$$

where k_1 and k_2 are positive constants depending on $\ell, \alpha, n_1, n_2, \delta_1, \delta_2$, etc.

Proof By setting $\Phi(x, y, \mu) = \varphi(y, \mu) - \varphi(x, \mu)$, we have

$$|\Phi(x, y, \mu_1) - \Phi(x, y, \mu_2)| \le |\varphi(y, \mu_1) - \varphi(y, \mu_2)| + |\varphi(x, \mu_1) - \varphi(x, \mu_2)|$$
$$\le n_1|\mu_1 - \mu_2|^{\delta_1} + n_1|\mu_1 - \mu_2|^{\delta_1} = 2n_1|\mu_1 - \mu_2|^{\delta_1},$$

$$|\Phi(x, y_1, \mu) - \Phi(x, y_2, \mu)| = |\varphi(y_1, \mu) - \varphi(x, \mu) - \varphi(y_2, \mu) + \varphi(x, \mu)|$$
$$= |\varphi(y_1, \mu) - \varphi(y_2, \mu)| \le n_2|y_1 - y_2|^{\delta_2}.$$

Moreover,

$$[\Phi(x, y, \mu) = \varphi(y, \mu) - \varphi(x, \mu) < 0] \Rightarrow [\Phi(x, y, \mu) = \varphi(x, \mu) - \varphi(y, \mu) > 0],$$

and hence $-\Phi$ is quasimonotone in $\mathscr{E}(U(\lambda_0))$.

$$[\Phi(x, y, \mu) = \varphi(y, \mu) - \varphi(x, \mu) \ge 0] \Rightarrow [\varphi(x, \mu) - \varphi(y, \mu) + h\|x - y\|^{\beta} \le 0],$$

or equivalently

$$[\Phi(y, x, \mu) + h\|x - y\|^{\beta} \le 0],$$

and hence Φ is (h, β)-Hölder strongly pseudomonotone in $\mathscr{D}(N)$. The other conditions of Theorem 4.21 are easy to check, and so the proof is complete. □

Corollary 4.53 (Hölder continuity of the solution maps of optimization problems via strong convexity conditions) *For* (OP), *let the following conditions be satisfied:*

(a) *$\mathscr{D}$ is (ℓ_1, α_1)-Hölder continuous on a neighborhood N of $\bar{\lambda}$.*
(b) *There is a neighborhood U of $\bar{\mu}$ such that, for each $\mu \in U$, $\varphi(\cdot, \mu)$ is (h_1, β_1)-strongly convex as well as $(m_1, 1)$-Hölder continuous on conv($\mathscr{D}(N)$).*
(c) *For every $x \in \mathscr{D}(N)$, $\varphi(x, \cdot)$ is (n_1, γ_1)-Hölder continuous on U.*

Then, on $N \times U$, the solution map of (OP) *is single-valued and satisfies the Hölder property: for $(\lambda_1, \mu_1), (\lambda_2, \mu_2) \in N \times U$,*

$$\mathbb{D}(\mathrm{Sol(OP)}(\lambda_1, \mu_1), \mathrm{Sol(OP)}(\lambda_2, \mu_2))$$
$$\le \left(\frac{4m_1\ell_1}{h_1}\right)^{\frac{1}{\beta_1}} d^{\frac{\alpha_1}{\beta_1}}(\lambda_1, \lambda_2) + \left(\frac{2n_1}{h_1}\right)^{\frac{1}{\beta_1}} d^{\frac{\gamma_1}{\beta_1}}(\mu_1, \mu_2).$$

Proof In order to apply Theorem 4.21, we check its assumptions. Clearly, assumption (a) confirms the corresponding one of the mentioned result and the monotonicity of Φ holds. We next show that the function Φ is strongly convex in the second variable. Indeed, for arbitrary $x \in \mathscr{D}(N)$, $\mu \in U$, $y_1, y_2 \in \text{conv}(\mathscr{D}(N))$, and $t \in (0, 1)$, we have

$$\Phi(x, (1-t)y_1+ty_2, \mu)-(1-t)\Phi(x, y_1, \mu)-t\Phi(x, y_2, \mu)+h_1t(1-t)d^{\beta_1}(y_1, y_2)$$

$$= \varphi((1-t)y_1+ty_2, \mu)-(1-t)\varphi(y_1, \mu)-t\varphi(y_2, \mu)+h_1t(1-t)d^{\beta_1}(y_1, y_2) \leq 0,$$

as $\varphi(\cdot, \mu)$ is (h_1, β_1)-strongly convex. Furthermore, it follows from assumption (c) that Φ is Hölder continuous in the second variable.

Finally, taking arbitrarily $\mu_1, \mu_2 \in U$ and $x, y \in \mathscr{D}(N)$, one sees that

$$\begin{aligned}|\Phi(x, y, \mu_1) - \Phi(x, y, \mu_2)| &= |(\varphi(y, \mu_1) - \varphi(y, \mu_2)) + (\varphi(x, \mu_2) - \varphi(x, \mu_1))| \\ &\leq 2n_1||\mu_1 - \mu_2||^{\gamma_1}.\end{aligned}$$

Therefore, Φ is semistrongly monotone. □

For the case with $\mathscr{D}(\lambda) \equiv K$, where K is a nonempty and convex subset of A, we have the following result.

Corollary 4.54 (Hölder continuity of the solution maps of optimization problems via strong convex-likeness conditions) *For* (OP) *with* $\mathscr{D}(\lambda) \equiv K$*, assume that:*

(a) *There is a neighborhood* U *of* $\bar{\mu}$ *such that, for each* $\mu \in U$, $\varphi(\cdot, \mu)$ *is* (h_1, β_1)*-strongly convex-like in* K.
(b) *For every* $x \in K$, $\varphi(x, \cdot)$ *is* (n_1, γ_1)*-Hölder continuous in* U.

Then, on U*, the solution map of* (OP) *is single-valued and satisfies the Hölder condition: for* $\mu_1, \mu_2 \in U$,

$$\mathbb{D}(\text{Sol(OP)}(\mu_1), \text{Sol(OP)}(\mu_2)) \leq \left(\frac{2n_1}{h_1}\right)^{\frac{1}{\beta_1}} d^{\frac{\gamma_1}{\beta_1}}(\mu_1, \mu_2).$$

Proof Obviously, the function Φ is monotone, and by assumption (a), this function is also semistrongly convex. It follow from (b) that for any $\mu_1, \mu_2 \in U$ and $x, y \in K$, we have

$$\begin{aligned}|\Phi(x, y, \mu_1) - \Phi(x, y, \mu_2)| &= |(\varphi(y, \mu_1) - \varphi(y, \mu_2)) + (\varphi(x, \mu_2) - \varphi(x, \mu_1))| \\ &\leq 2n_1||\mu_1 - \mu_2||^{\gamma_1}.\end{aligned}$$

Passing to the strong convexlikeness of Φ, for arbitrary $y_1, y_2 \in K$, $t \in (0, 1)$, and $\mu \in U$, one has $z \in K$ such that

$$\varphi(z, \mu) - (1-t)\varphi(y_1, \mu) - t\varphi(y_2, \mu) - h_1t(1-t)d^{\beta_1}(y_1, y_2) \leq 0.$$

Then, for all $x \in K$,

$$[\varphi(z,\mu)-\varphi(x,\mu)]-(1-t)[\varphi(y_1,\mu)-\varphi(x,\mu)]-t[\varphi(y_2,\mu)-\varphi(x,\mu)] \\ - h_1 t(1-t)d^{\beta_1}(y_1,y_2)\le 0.$$

Hence,

$$\Phi(x,z,\mu)-(1-t)\Phi(x,y_1,\mu)-t\Phi(x,y_2,\mu)-h_1 t(1-t)d^{\beta_1}(y_1,y_2)\le 0$$

which completes the proof. □

Now, we move on to consider Hölder conditions of approximate solution maps for the reference problem. The following result is obtained from Theorem 4.25.

Corollary 4.55 (Hölder continuity of the approximate solution maps of optimization problems) *For* (OP), *let solutions exist in a neighborhood of the reference point* $(\lambda_0,\mu_0)\in\Lambda\times M$. *Suppose that the following conditions are satisfied:*

(a) $\mathscr{D}$ *is* (ℓ_1,α_1)*-Hölder continuous at* λ_0, *i.e., there is a neighborhood* N *of* λ_0 *such that, for all* $\lambda_1,\lambda_2\in N$,

$$\mathscr{D}(\lambda_1)\subset\mathscr{D}(\lambda_2)+\ell_1 B(0,||\lambda_1-\lambda_2||^{\alpha_1}).$$

(b) *There is a neighborhood* U *of* μ_0 *such that, for each* $\mu\in U$, $\varphi(\cdot,\mu)$ *is convex and* (q_1,δ_1)*-Hölder continuous on* $\mathscr{D}(N)$.
(c) *For every* $x\in\mathscr{D}(N)$, $\varphi(x,.)$ *is* (h_1,β_1)*-Hölder continuous on* U.

Then, for each $\bar{\varepsilon}>0$, $\widetilde{\mathrm{Sol}}(\mathrm{OP})$ *is Hölder continuous on* $[\bar{\varepsilon},+\infty)\times N\times U$ *as follows:*

$$H(\widetilde{\mathrm{Sol}}(\mathrm{OP})(\varepsilon_1,\lambda_1,\mu_1),\widetilde{\mathrm{Sol}}(\mathrm{OP})(\varepsilon_2,\lambda_2,\mu_2)) \\ \le k_1|\varepsilon_1-\varepsilon_2|+k_2\|\mu_1-\mu_2\|^{\beta_1}+k_3\|\lambda_1-\lambda_2\|^{\alpha_1\delta_1},$$

where $k_1,k_2,k_3>0$ *and depend on* $\bar{\varepsilon},\ell_1,\alpha_1,h_1,\beta_1$, *etc.*

Proof We need to check assumptions (b), (c), and (d) of Theorem 4.25. For (b), take arbitrary $x_1,x_2\in\mathscr{D}(N)$, $t\in[0,1]$, $y\in\mathscr{D}(N)$, and $\mu\in U$. We have

$$\Phi(tx_1+(1-t)x_2,y,\mu)-t\Phi(x_1,y,\mu)-(1-t)\Phi(x_2,y,\mu)$$

$$=\varphi(y,\mu)-\varphi(tx_1+(1-t)x_2,\mu)-t[\varphi(y,\mu)-\varphi(x_1,\mu)] \\ -(1-t)[\varphi(y,\mu)-\varphi(x_2,\mu)]$$

$$=t\varphi(x_1,\mu)+(1-t)\varphi(x_2,\mu)-\varphi(tx_1+(1-t)x_2,\mu)\ge 0,$$

since $\varphi(\cdot, \mu)$ is convex. Therefore, (b) holds. For (c), taking arbitrary $\mu_1, \mu_2 \in U$ and $x, y \in \mathscr{D}(N)$, one sees that

$$\begin{aligned}|\Phi(x, y, \mu_1) - \Phi(x, y, \mu_2)| &= |(\varphi(y, \mu_1) - \varphi(y, \mu_2)) + (\varphi(x, \mu_2) - \varphi(x, \mu_1))| \\ &\leq h_1\|\mu_1 - \mu_2\|^{\beta_1} + h_1\|\mu_1 - \mu_2\|^{\beta_1} \\ &\leq 2h_1||\mu_1 - \mu_2||^{\beta_1}.\end{aligned}$$

Consequently, condition (c) is satisfied with $h = 2h_1$ and $\beta = \beta_1$.

Assumption (d) is satisfied with $q = q_1$ and $\delta = \delta_1$, since for any $y_1, y_2 \in \mathscr{D}(N)$, $x \in \mathscr{D}(N)$, and $\mu \in U$, we have

$$\begin{aligned}|\Phi(x, y_1, \mu) - \Phi(x, y_2, \mu)| &= |\varphi(y_1, \mu) - \varphi(y_2, \mu)| \\ &\leq q_1||y_1 - y_2||^{\delta_1}.\end{aligned}$$

The proof is complete. □

4.4.2 *Variational Inequalities*

Let X be a locally convex Hausdorff topological vector space, Y, W be vector topological spaces, and $A \subset X, \Lambda \subset Y, M \subset W$ be nonempty subsets. Let $\mathscr{D} : \Lambda \rightrightarrows A$ be a set-valued map with nonempty convex values. Let $T : A \times M \to X^*$ where X^* is the dual space of X. For $(\lambda, \mu) \in \Lambda \times M$, the Stampacchia variational inequality problem under our consideration is

(SVI) find $\bar{x} \in \mathscr{D}(\lambda)$ such that, for all $y \in \mathscr{D}(\lambda)$, $\langle T(\bar{x}, \mu), y - \bar{x}\rangle \geq 0$.

Putting $\Phi(x, y, \mu) = \langle T(x, \mu), y - x\rangle$, (SVI) becomes a special case of problem (EP) considered in the previous sections. The dual of (SVI) is the Minty variational inequality:

(MVI) find $\bar{x} \in \mathscr{D}(\lambda)$ such that, for all $y \in \mathscr{D}(\lambda)$, $\langle T(y, \mu), \bar{x} - y\rangle \leq 0$.

For each $\varepsilon \in \mathbb{R}_+, \lambda \in \Lambda$, and $\mu \in M$, the solution maps and the ε-approximate solution maps of (SVI) and (MVI) are denoted by Sol(SVI), Sol(MVI), $\widetilde{\text{Sol}}$(SVI), and $\widetilde{\text{Sol}}$(MVI), resp., that is

$$\text{Sol(SVI)}(\lambda, \mu) := \{\bar{x} \in \mathscr{D}(\lambda) \mid \langle T(\bar{x}, \mu), y - \bar{x}\rangle \geq 0, \forall y \in \mathscr{D}(\lambda)\},$$

$$\text{Sol(MVI)}(\lambda, \mu) := \{\bar{x} \in \mathscr{D}(\lambda) \mid \langle T(y, \mu), \bar{x} - y\rangle \leq 0, \forall y \in \mathscr{D}(\lambda)\},$$

$$\widetilde{\text{Sol}}\text{(SVI)}(\varepsilon, \lambda, \mu) := \{\bar{x} \in \mathscr{D}(\lambda) \mid \langle T(\bar{x}, \mu), y - \bar{x}\rangle + \varepsilon \geq 0, \forall y \in \mathscr{D}(\lambda)\},$$

$$\widetilde{\text{Sol}}\text{(MVI)}(\varepsilon, \lambda, \mu) := \{\bar{x} \in \mathscr{D}(\lambda) \mid \langle T(y, \mu), \bar{x} - y\rangle - \varepsilon \leq 0, \forall y \in \mathscr{D}(\lambda)\}.$$

We recall now some concepts used in studying stability of (SVI) and (MVI).

Definition 4.56 (Affine functions) A function $\tau : X \to \mathbb{R}$ is said to be AFFINE on a convex subset $A \subset X$ if for all $x_1, x_2 \in A$ and $t \in [0, 1]$,

$$\tau(tx_1 + (1-t)x_2) = t\tau(x_1) + (1-t)\tau(x_2).$$

Definition 4.57 (Monotone and antimonotone maps) A map $\tau : X \to X^*$ is said to be MONOTONE (ANTIMONOTONE, resp.) on $A \subset X$ if for all $x_1, x_2 \in A$,

$$\langle \tau(x_1) - \tau(x_2), x_1 - x_2 \rangle \geq 0,$$

$$(\langle \tau(x_1) - \tau(x_2), x_1 - x_2 \rangle \leq 0, \text{resp.}).$$

We first discuss stability in the sense of Hausdorff semicontinuity.

Corollary 4.58 (Hausdorff continuity of the solution maps of Stampacchia and Minty variational inequalities) *For problem* (SVI) ((MVI), resp.), *assume that:*

(a) $\mathcal{D}$ *is continuous at* λ_0 *and* $\mathcal{D}(\lambda_0)$ *is compact.*
(b) *There exists a neighborhood* N *of* λ_0 *such that* T *is continuous in* $\mathcal{D}(N) \times \{\mu_0\}$.
(c) $T(\cdot, \mu_0)$ *is monotone (antimonotone, resp.) and affine in* $\mathcal{D}(N)$.

Then, $\widetilde{\text{Sol}}$(SVI) ($\widetilde{\text{Sol}}$(MVI), *resp.*) *is Hausdorff continuous at* $(\varepsilon_0, \lambda_0, \mu_0)$.

Proof We only prove for (SVI), because the conclusion for (MVI) can be proved similarly. It suffices to show that assumption (c) of Theorem 4.15 is satisfied. Taking any $x_1, x_2 \in \mathcal{D}(N)$, $t \in [0, 1]$, and $y \in \mathcal{D}(N)$, we have

$$\Phi(tx_1 + (1-t)x_2, y, \mu_0) - t\Phi(x_1, y, \mu_0) - (1-t)\Phi(x_2, y, \mu_0)$$

$$= \langle tT(x_1, \mu_0) + (1-t)T(x_2, \mu_0), y - tx_1 - (1-t)x_2 \rangle$$

$$-t\langle T(x_1, \mu_0), y - x_1 \rangle - (1-t)\langle T(x_2, \mu_0), y - x_2 \rangle$$

$$= t\langle T(x_1, \mu_0), (1-t)x_1 - (1-t)x_2 \rangle + (1-t)\langle T(x_2, \mu_0), -tx_1 + tx_2 \rangle$$

$$= t(1-t)\langle T(x_1, \mu_0) - T(x_2, \mu_0), x_1 - x_2 \rangle \geq 0.$$

So, (c) of Theorem 4.15 is satisfied. □

When $\mathcal{D}(\lambda) \equiv K$ is a nonempty subset of A, we have the following corollary of Theorem 4.15.

Corollary 4.59 (Hausdorff continuity of the approximate solution maps of Stampacchia and Minty variational inequalities) *Consider* (SVI) ((MVI), resp.). *Assume that the following conditions hold:*

(a) *For all* $x \in K$, $T(x, \cdot)$ *is continuous at* μ_0.
(b) $T(\cdot, \mu_0)$ *is monotone (antimonotone, resp.) and affine on* K.

Then, $\widetilde{\text{Sol}}$(SVI) ($\widetilde{\text{Sol}}$(MVI), *resp.) is Hausdorff continuous at* (ε_0, μ_0).

Now, we move on to stability conditions of the Hölder types for the reference problems. We first consider the special case $\mathscr{D}(\lambda) \equiv K \subset A$ with K being nonempty and convex. Then, we have the following.

Corollary 4.60 (Hölder continuity of the solution maps of Stampacchia and Minty variational inequalities with fixed constraint sets) *For the problem* (SVI) ((MVI), *resp.) with* $\mathscr{D}(\lambda) \equiv K$ *and* K *being nonempty and convex, assume that:*

(a) *There is a neighborhood* U *of* $\bar{\mu}$ *such that, for all* $x \in K$ *and* $\mu \in U$, $\langle T(x, \mu), \cdot \rangle$ *is* (h_2, β_2)*-strongly convex-like* ((h_2, β_2)*-strongly concave-like, resp.) on* K *with* $\beta_2 > 1$.
(b) *For each* $\mu \in U$, $T(\cdot, \mu)$ *is monotone (antimonotone, resp.) on* K.
(c) *For* $x \in K$, $T(x, \cdot)$ *is* (n_2, γ_2)*-Hölder continuous on* U.

Then, on U*, the solution map of* (SVI) ((MVI), resp.) *is single-valued and satisfies the following Hölder property: for* $\mu_1, \mu_2 \in U$,

$$\mathbb{D}(\text{Sol(SVI)}(\mu_1), \text{Sol(SVI)}(\mu_2)) \leq \left(\frac{n_2}{h_2}\right)^{\frac{1}{\beta_2 - 1}} d^{\frac{\gamma_2}{\beta_2 - 1}}(\mu_1, \mu_2),$$

$$(\mathbb{D}(\text{Sol(MVI)}(\mu_1), \text{Sol(MVI)}(\mu_2)) \leq \left(\frac{n_2}{h_2}\right)^{\frac{1}{\beta_2 - 1}} d^{\frac{\gamma_2}{\beta_2 - 1}}(\mu_1, \mu_2), \textit{resp.}).$$

Proof We present a proof for problem (SVI), and the other one is similar. In order to apply Theorem 4.21, we check that all its conditions are implied by the assumptions of the above corollary. The first assumption of Theorem 4.21 is satisfied since, for $y_1, y_2 \in K$ and $t \in (0, 1)$, by (a) there is $z \in K$ such that, for all $\mu \in U$ and $x \in K$,

$$\langle T(x, \mu), z \rangle \leq (1 - t)\langle T(x, \mu), y_1 \rangle + t\langle T(x, \mu), y_2 \rangle - h_2 t(1 - t)d^{\beta_2}(y_1, y_2),$$

or equivalently

$$\Phi(x, z, \mu) \leq (1 - t)\Phi(x, y_1, \mu) + t\Phi(x, y_2, \mu) - h_2 t(1 - t)d^{\beta_2}(y_1, y_2).$$

The second condition of Theorem 4.21 is fulfilled because, for $x, y \in K$ and $\mu \in U$, one has

$$\Phi(x, y, \mu) + \Phi(y, x, \mu) = \langle T(x, \mu) - T(y, \mu), y - x \rangle \leq 0,$$

as $T(\cdot, \mu)$ is monotone in K.

The last condition of Theorem 4.21 is verified as, for $\mu_1, \mu_2 \in U$ and $x, y \in K$,

$$|\Phi(x, y, \mu_1) - \Phi(x, y, \mu_2)| = |\langle T(x, \mu_1) - T(x, \mu_2), y - x\rangle|$$
$$\leq n_2\|\mu_1 - \mu_2\|^{\gamma_2}\|x - y\|.$$

This brings the proof to the end. □

To present Hölder conditions for (SVI) and (MVI), we need some preliminaries. Let X, Y, Z, and W be Banach spaces. A map $T : X \to X^*$ is said to be STRONGLY MONOTONE (STRONGLY ANTIMONOTONE, resp.) if there is $\alpha > 0$ such that for all $x, y \in X$,

$$\langle T(x) - T(y), x - y\rangle \geq \alpha\|x - y\|^2$$

$$(\langle T(x) - T(y), x - y\rangle \leq \alpha\|x - y\|^2, \text{ resp.}).$$

The map T is said to be MAXIMALLY MONOTONE (MAXIMALLY ANTIMONOTONE, resp.) if it is monotone (antimonotone, resp.), i.e., if T is monotone (antimonotone, resp.) and for any $(x, u) \in \operatorname{gph} T$,

$$\langle u - v, x - y\rangle \geq 0$$

$$(\langle u - v, x - y\rangle \geq 0, \text{ resp}),$$

then $(y, v) \in \operatorname{gph} T$ and no enlargement of gph T is possible without destroying the monotonicity (antimonotonicity, resp.). A map $\tau : X \to Z$ is said to be LOCALLY BOUNDED at $x \in X$ if there is a neighborhood U of x such that $\tau(U)$ is bounded on U. τ is said to be locally bounded on a subset K of X, if τ is locally bounded at each point of K.

We next employ conditions related to monotonicity properties to provide Hölder property results for the underlying problems.

Corollary 4.61 (Hölder continuity of the solution maps of Stampacchia and Minty variational inequalities via conditions related to maximal monotonicity) *For the problem* (SVI), ((MVI), *resp.*) *assume the existence of a neighborhood* $N \times U$ *of* $(\lambda_0, \mu_0) \in \Lambda \times M$ *such that:*

(a) *There is a neighborhood* P *of the solution* $x(\lambda_0, \mu_0)$ *such that for every* $\lambda, \lambda' \in N$,

$$\mathcal{D}(\lambda) \cap P \subset \mathcal{D}(\lambda') + \ell B\left(0, d^{\alpha}(\lambda, \lambda')\right),$$

(i.e., $\mathcal{D}$ *is* (ℓ, α)*-pseudo-Hölder at* λ_0*).*

(b) *There is a real number* $\beta > 1$ *such that for every* $\mu \in U$, *and for every* $x, y \in \mathscr{D}(N)$ *with* $x \neq y$,

$$h\|x - y\|^{\beta} \leq d\left(\langle T(x, \mu), y - x\rangle, \mathbb{R}_{+}\right) + d\left(\langle T(y, \mu), x - y\rangle, \mathbb{R}_{+}\right)$$

$$(h\|x - y\|^{\beta} \leq d\left(\langle T(x, \mu), y - x\rangle, \mathbb{R}_{-}\right) + d\left(\langle T(y, \mu), x - y\rangle, \mathbb{R}_{-}\right), \; resp.).$$

(c) *For every* $\lambda \in N$, *and every* $x \in \mathscr{D}(\lambda)$, $T(x, \cdot)$ *is* (h, ξ)*-Hölder continuous at* μ_0.

(d) *For each* $\mu \in U$, *the map* $T(\cdot, \mu)$ *is maximally monotone (maximally antimonotone, resp.) and locally bounded on* $\mathscr{D}(N)$.

Then, on the neighborhood $N \times U$ *of* (λ_0, μ_0), *the solution map of* (SVI) *((MVI), resp.) satisfies the following Hölder condition:*

$$H(\mathrm{Sol(SVI)}(\lambda_1, \mu_1), \mathrm{Sol(SVI)}(\lambda_2, \mu_2)) \leq k_1 \|\lambda_1 - \lambda_2\|^{\alpha/\beta} + k_2 \|\eta_1 - \eta_2\|^{\xi/(\beta-1)},$$

$$(H(\mathrm{Sol(MVI)}(\lambda_1, \mu_1), \mathrm{Sol(MVI)}(\lambda_2, \mu_2)) \leq k_1 \|\lambda_1 - \lambda_2\|^{\alpha/\beta} + k_2 \|\eta_1 - \eta_2\|^{\xi/(\beta-1)}, \; resp.)$$

where k_1, k_2 *depend on* $\ell, h, \alpha, \beta, \ldots$

Proof Obviously, all the conditions of Theorem 4.21 are satisfied, except the Hölder property of the set-valued map $\mathscr{D}$, and so for applying Theorem 4.21, we need to show that in this case the assumed pseudo-Hölder property in (a) is enough to replace the Hölder property, or equivalently there are neighborhoods $\bar{U}$ of $x(\lambda_0, \mu_0)$, $B(\lambda_0, r)$ and $\widehat{U}$ of μ_0 such that for every $(\lambda, \mu) \in B(\lambda_0, r) \times \widehat{U}$, we have $x(\lambda, \mu) \in \bar{U}$.

Set $x_0 = x(\lambda_0, \mu_0)$ and choose positive real numbers r_1 and r such that $B(x_0, r_1) \subset P$, $B(\lambda_0, r) \subset N$ and $\ell r^{\alpha} \leq r_1$. Then, by condition (a), for every $\lambda \in B(\lambda_0, r)$, there exists $x_\lambda \in \mathscr{D}(\lambda)$ such that

$$\|x_0 - x_\lambda\| \leq \ell d^{\alpha}(\lambda_0, \lambda) \leq \ell r^{\alpha} \leq r_1.$$

Therefore, $\mathscr{D}(\lambda) \cap B(x_0, r_1)$ is closed, convex, and nonempty. For each $\mu \in U$, we consider the set-valued map:

$$x \mapsto T(x, \eta) + N_{\mathscr{D}(\lambda) \cap B(x_0, r_1)}(x), \tag{4.27}$$

where $N_S(\cdot)$ is the normal cone to a convex set S at $(\cdot)$. Because $N_S(.)$ is maximally monotone, and so is $T(\cdot, \mu)$, the set-valued map defined by (4.27) is maximally monotone with bounded domain $\mathscr{D}(\lambda) \cap B(x_0, r_1)$, and so it is onto. Hence, there exists some vector $\hat{x}(\lambda, \mu) \in \mathscr{D}(\lambda) \cap B(x_0, r_1)$ such that

$$0 \in A(\hat{x}(\lambda, \eta), \eta) + N_{K(\lambda) \cap B(x_0, r_1)}(\hat{x}(\lambda, \eta)). \tag{4.28}$$

We prove that

$$N_{K(\lambda) \cap B(x_0, r_1)}(\hat{x}(\lambda, \eta)) = N_{K(\lambda)}(\hat{x}(\lambda, \eta)). \tag{4.29}$$

To verify this, we claim first that, for (λ, μ) close enough to (λ_0, μ_0), $\hat{x}(\lambda, \eta) \in \operatorname{int} B(x_0, r_1)$.

Because $\hat{x}(\lambda, \eta) \in B(x_0, r_1) \subset P$, by condition (a), there exists some vector $\hat{x}_0 \in \mathscr{D}(\lambda_0)$ such that

$$\|\hat{x}(\lambda, \mu) - \hat{x}_0\| \leq \ell d^{\alpha}(\lambda, \lambda_0).$$

Similarly, we can get some vector $x_\lambda \in \mathscr{D}(\lambda) \cap B(x_0, r_1)$ satisfying

$$d(x_0, x_\lambda) \leq \ell d^{\alpha}(\lambda, \lambda_0).$$

On the other hand, since $T(\cdot, \mu)$ is locally bounded, there is a real number $q > 0$ such that, for (λ, μ) in some neighborhood of (λ_0, μ_0),

$$\|T(\hat{x}(\lambda, \mu), \mu)\| \leq q, \quad \|T(x_0, \mu_0)\| \leq q.$$

Because x_0 is a solution of (SVI) and $\hat{x}(\lambda, \eta)$ is a solution of (SVI) restricted to $K(\lambda) \cap B(x_0, r_1)$, the following inequalities hold:

$$\begin{aligned} &\langle T(\hat{x}(\lambda, \mu), \mu), x_\lambda - \hat{x}(\lambda, \mu)\rangle \geq 0, \\ &\langle T(x_0, \mu_0), \hat{x}_0 - x_0\rangle \geq 0. \end{aligned}$$

In view of condition (b), we have

$$\begin{aligned} &h\left\|\hat{x}(\lambda, \eta) - x_0\right\|^{\beta} \\ &\leq d(\langle T(\hat{x}(\lambda, \mu), \mu), x_0 - \hat{x}(\lambda, \mu)\rangle, \langle T(\hat{x}(\lambda, \mu), \mu), x_\lambda - \hat{x}(\lambda, \mu)\rangle) \\ &\quad + d(\langle T(x_0, \mu_0), \hat{x}(\lambda, \mu) - x_0\rangle, \langle T(x_0, \mu_0), \hat{x}_0 - x_0\rangle) \\ &\leq \|T(\hat{x}(\lambda, \mu), \mu)\| \, \|x_0 - x_\lambda\| + \|T(x_0, \mu_0)\| \, \left\|\hat{x}_0 - \hat{x}(\lambda, \mu)\right\| \\ &\leq 2q\ell d^{\alpha}(\lambda, \lambda_0). \end{aligned}$$

Therefore, if (λ, η) is close enough to (λ_0, η_0), then we have $\hat{x}(\lambda, \mu) \in \operatorname{int} B(x_0, r_1)$. Returning back to (4.29), we see that the inclusion $\supset$ is clear.

For the opposite inclusion, take arbitrary v on the left-hand side and $z \in \mathscr{D}(\lambda)$. Since $\hat{x}(\lambda, \mu) \in \operatorname{int} B(x_0, r_1)$, there exists $\varepsilon > 0$ such that

$$z_\alpha = \hat{x}(\lambda, \mu) + \varepsilon(z - x) \in \mathscr{D}(\lambda) \cap B(x_0, r_1),$$

and so

$$\langle v, z - \hat{x}(\lambda, \mu)\rangle = \left\langle v, \frac{1}{\varepsilon}\left(z_\alpha - \hat{x}(\lambda, \mu)\right)\right\rangle \leq 0,$$

or equivalently $v \in N_{\mathscr{D}(\lambda)}(\hat{x}(\lambda, \mu))$. Hence, (4.29) follows, and (4.28) shows that $\hat{x}(\lambda, \mu)$ is a solution of (MVI). By assumption (b), the solution $x(\lambda, \mu)$ of (MVI) is locally unique, and so $\hat{x}(\lambda, \mu) = x(\lambda, \mu)$ and we have the desired assertion with $\bar{U} = B(x_0, r_1)$. Therefore, the proof is complete. □

We next discuss Hölder properties of the solution maps of the underlying problems by using classical conditions.

Corollary 4.62 (Hölder continuity of the solution maps of Stampacchia and Minty variational inequalities via conditions related to strong monotonicity) *Assume that* $x_0 := x(\lambda_0, \mu_0)$ *is a solution of* (SVI) ((MVI), *resp.) at* (λ_0, μ_0) *and that there is a neighborhood* $P \times N \times U$ *of* (x_0, λ_0, μ_0) *such that:*

(a) $\mathscr{D}$ *is pseudo-Lipschitz on* N.
(a) $T(\cdot, \mu)$ *is strongly monotone (strongly antimonotone, resp.) for each* $\mu \in U$.
(a) T *is Lipschitz on* $P \times U$.

Then, on some neighborhood of (λ_0, μ_0)*, the solutions of* (SVI) ((MVI), *resp.) satisfy the Hölder condition:*

$$H(\mathrm{Sol(SVI)}(\lambda_1, \mu_1), \mathrm{Sol(SVI)}(\lambda_2, \mu_2)) \leq k_1 d^{1/2} d(\lambda_1, \lambda_2) + k_2 d(\mu_1, \mu_2),$$

$$(H(\mathrm{Sol(MVI)}(\lambda_1, \mu_1), \mathrm{Sol(MVI)}(\lambda_2, \mu_2))$$
$$\leq k_1 d^{1/2} d(\lambda_1, \lambda_2) + k_2 d(\mu_1, \mu_2),\ resp.).$$

Proof We check the assumptions of Corollary 4.61. Assumption (a) is satisfied with $\alpha = 1$ and (b) is satisfied with $\beta = 2$. Finally, by the Lipschitz continuity of T, in order to apply Corollary 4.61, we only need to confirm the maximal monotonicity of the map $T(\cdot, \mu)$, while the desired assertion is obtained by the monotonicity and the demicontinuity of the map $T(\cdot, \mu)$. □

Finally, we provide Hölder conditions for the approximate solution maps of the reference problems.

Corollary 4.63 (Hölder continuity of the approximate solution maps of Stampacchia and Minty variational inequalities) *For problem* (SVI) ((MVI), *resp.*), *suppose that the following conditions hold:*

(a) $\mathscr{D}$ *is* (ℓ_2, α_2)*-Hölder continuous at* λ_0*, i.e., there is a neighborhood* N *of* λ_0 *such that, for all* $\lambda_1, \lambda_2 \in N$,

$$\mathscr{D}(\lambda_1) \subset \mathscr{D}(\lambda_2) + \ell_2 B(0, ||\lambda_1 - \lambda_2||^{\alpha_2}).$$

(b) *There is a neighborhood* U *of* μ_0 *such that, for each* $\mu \in U$*, the map* $T(\cdot, \mu)$ *is bounded (i.e., there exists* $q_2 > 0$ *such that* $\|T(x, \mu)\| \leq q_2, \ \forall x \in \mathscr{D}(N)$*), monotone (antimonotone, resp.), and affine on* $\mathscr{D}(N)$.
(c) *For* $x \in \mathscr{D}(N)$, $T(x, \cdot)$ *is* (h_2, β_2)*-Hölder continuous on* U.

Then, for each $\bar{\varepsilon} > 0$, $\widetilde{\text{Sol}}$(SVI) ($\widetilde{\text{Sol}}$(MVI), *resp.*) *satisfies the following Hölder condition on* $[\bar{\varepsilon}, +\infty) \times N \times U$:

$$H(\widetilde{\text{Sol}}(\text{SVI})(\varepsilon_1, \lambda_1, \mu_1), \widetilde{\text{Sol}}(\text{SVI})(\varepsilon_2, \lambda_2, \mu_2))$$
$$\leq k_1|\varepsilon_1 - \varepsilon_2| + k_2\|\mu_1 - \mu_2\|^{\beta_2} + k_3\|\lambda_1 - \lambda_2\|^{\alpha_2},$$

$$H(\widetilde{\text{Sol}}(\text{MVI})(\varepsilon_1, \lambda_1, \mu_1), \widetilde{\text{Sol}}(\text{MVI})(\varepsilon_2, \lambda_2, \mu_2))$$
$$\leq k_1|\varepsilon_1 - \varepsilon_2| + k_2\|\mu_1 - \mu_2\|^{\beta_2} + k_3\|\lambda_1 - \lambda_2\|^{\alpha_2},$$

resp., where $k_1, k_2, k_3 > 0$ *and depend on* $\bar{\varepsilon}$, ℓ_2*, etc.*

Proof For (SVI), we only need to check the conditions of Theorem 4.25 with the function Φ defined as above. The first condition of the mentioned theorem is similar to that of this corollary. For the second one of Theorem 4.25, taking arbitrary $x_1, x_2 \in \mathscr{D}(N)$, $t \in [0, 1]$, and $y \in \mathscr{D}(N)$, by condition (b), we have

$$\Phi(tx_1 + (1-t)x_2, y, \mu) - t\Phi(x_1, y, \mu) - (1-t)\Phi(x_2, y, \mu)$$

$$= \langle tT(x_1, \mu) + (1-t)T(x_2, \mu), y - tx_1 - (1-t)x_2 \rangle$$

$$-t\langle T(x_1, \mu), y - x_1 \rangle - (1-t)\langle T(x_2, \mu), y - x_2 \rangle$$

$$= t\langle T(x_1, \mu), (1-t)x_1 - (1-t)x_2 \rangle + (1-t)\langle T(x_2, \mu), -tx_1 + tx_2 \rangle$$

$$= t(1-t)\langle T(x_1, \mu) - T(x_2, \mu), x_1 - x_2 \rangle \geq 0.$$

Therefore, the second condition of Theorem 4.25 is satisfied. Turning to the third one of this statement, for arbitrary $\mu_1, \mu_2 \in U$ and $x, y \in \mathscr{D}(N)$, we have

$$\begin{aligned}|\Phi(x, y, \mu_1) - \Phi(x, y, \mu_2)| &= |\langle T(x, \mu_1) - T(x, \mu_2), y - x\rangle| \\ &\leq \|T(x, \mu_1) - T(x, \mu_2)\| \|x - y\| \\ &\leq h_2\rho_2 \|\mu_1 - \mu_2\|^{\beta_2},\end{aligned}$$

where $\rho_2 = \text{diam}\mathscr{D}(N)$. Hence, the mentioned condition is verified with $h = h_2\rho_2$ and $\beta = \beta_2$.

The last condition (iv) of Theorem 4.25 is also satisfied with $q = q_2$ and $\delta = 1$, since for $y_1, y_2 \in \mathscr{D}(N)$, $x \in \mathscr{D}(N)$ and $\mu \in U$, we have

$$\begin{aligned}|\Phi(x, y_1, \mu) - \Phi(x, y_2, \mu)| &= |\langle T(x, \mu), y_1 - x\rangle - \langle T(x, \mu), y_2 - x\rangle| \\ &= |\langle T(x, \mu), y_1 - y_2\rangle| \\ &\leq \|T(x, \mu)\| \|y_1 - y_2\| \\ &\leq q_2 \|y_1 - y_2\|.\end{aligned}$$

Hence, by Theorem 4.25, we get the first conclusion of Corollary 4.63. The second one of this corollary is confirmed similarly. □

We would like to note here that, although $T(\cdot, \mu)$ is assumed to be monotone, solutions may not be unique, as illustrated by the following.

Example 4.64 (Monotone variational inequalities with nonunique solutions) Let $X = \mathbb{R}$, $\Lambda \equiv M = [0, 1]$, $A = [0, 1]$, $\mathscr{D}(\lambda) = [0, 1]$, and $T(x, \lambda) \equiv 0$. Then, all the assumptions of Corollary 4.63 for the problem (SVI) are satisfied with $\ell_2 = 0$, $\alpha_2 = 1$, $h_2 = 0$, and $\beta_2 = 1$. Therefore, Corollary 4.63 implies the Hölder continuity of $\widetilde{\text{Sol}}$(SVI). By direct calculations we have, for all $(\varepsilon, \lambda) \in \mathbb{R}_+ \times \Lambda$, $\widetilde{\text{Sol}}(\text{SVI})(\varepsilon, \lambda) = [0, 1]$, which is not a singleton.

4.4.3 Fixed-Point Problems

Let X be a locally Hausdorff topological vector space, Y a vector space, and $A \subset X$, $M \subset Y$ nonempty. Let $\tau : A \times M \to A$. For $\mu \in M$, we consider the following fixed-point problem:

(FP) find $\bar{x} \in A$ such that $\tau(\bar{x}, \mu) = \bar{x}$.

Setting

$$\Phi(x, y, \mu) := \langle x - \tau(x, \mu), y - x\rangle,$$

we also consider the auxiliary problem:

(EP) find $\bar{x} \in A$ such that for all $y \in A$, $\Phi(\bar{x}, y, \mu) \geq 0$.

Proposition 4.65 (Relationships between fixed-point problems and equilibrium problems) *$\bar{x}$ is a solution of* (FP) *if and only if it is a solution of* (EP).

Proof If $\bar{x}$ is a solution of (FP), then

$$\langle \bar{x} - \tau(\bar{x}, \mu), y - \bar{x} \rangle = 0$$

for all $y \in A$, or equivalently $\bar{x}$ solves (EP).

Conversely, let $\bar{x}$ be a solution of (EP), i.e.,

$$\langle \bar{x} - \tau(\bar{x}, \mu), y - \bar{x} \rangle \geq 0$$

for all $y \in A$. By taking $y = \tau(\bar{x}, \mu)$, we have

$$\langle \bar{x} - \tau(\bar{x}, \mu), \tau(\bar{x}, \mu) - \bar{x} \rangle \geq 0,$$

and so we must have an equality, i.e., $\bar{x}$ is a solution of (FP). □

For each $\varepsilon \in \mathbb{R}_+$ and $\mu \in M$, we denote the ε-solution set of (EP) ((FP), resp.) by $\widetilde{\text{Sol}}(\text{EP})(\varepsilon, \mu)$ ($\widetilde{\text{Sol}}(\text{FP})(\varepsilon, \mu)$, resp.), i.e.,

$$\widetilde{\text{Sol}}(\text{EP})(\varepsilon, \mu) = \{\bar{x} \in A \mid \Phi(\bar{x}, y, \mu) + \varepsilon \geq 0, \forall y \in A\}$$

$$(\widetilde{\text{Sol}}(\text{FP})(\varepsilon, \mu) = \{\bar{x} \in A \mid \|\bar{x} - \tau(\bar{x}, \mu)\| \leq \varepsilon\}, \text{resp.}).$$

Then, we have the following relation between the approximate solutions of (EP) and (FP).

Proposition 4.66 (Relationships between approximate solutions of fixed-point problems and equilibrium problems) *If $\bar{x}$ is an ε-solution of problem* (EP), *then $\bar{x}$ is a $\sqrt{\varepsilon}$-solution of problem* (FP).

Proof Putting $y = \tau(\bar{x}, \mu)$ into

$$\langle \bar{x} - \tau(\bar{x}, \mu), y - \bar{x} \rangle + \varepsilon \geq 0,$$

one has

$$\langle \bar{x} - \tau(\bar{x}, \mu), \bar{x} - \tau(\bar{x}, \mu) \rangle \leq \varepsilon.$$

Thus,

$$\|\bar{x} - \tau(\bar{x}, \mu)\| \leq \sqrt{\varepsilon}.$$

Consequently, $\bar{x}$ is a $\sqrt{\varepsilon}$-solution of (FP). □

We now discuss the Hausdorff continuity property of the approximate solution maps of (FP).

Corollary 4.67 (Hausdorff continuity of the approximate solution maps of fixed-point problems) *Assume that the following conditions hold for* (FP)*:*

(a) *There is a neighborhood U of μ_0 such that, for each $\mu \in U$, $\tau(\cdot, \mu)$ is affine and $x \mapsto x - \tau(x, \mu)$ is monotone on A.*
(b) *For every $x \in A$, $\tau(x, \cdot)$ is continuous at μ_0.*

Then, $\widetilde{\text{Sol}}$(FP) is Hausdorff continuous at $(\sqrt{\varepsilon_0}, \mu_0)$.

Proof Obviously, in order to apply Theorem 4.15, we need to check only its second condition. For arbitrary $x_1, x_2 \in A$, $t \in [0, 1]$, and $y \in A$,, we have

$$\begin{aligned}
&\Phi(tx_1 + (1-t)x_2, y, \mu_0) - t\Phi(x_1, y, \mu_0) - (1-t)\Phi(x_2, y, \mu_0)\\
&\quad = \langle tx_1 + (1-t)x_2 - \tau(tx_1 + (1-t)x_2, \mu_0), y - tx_1 - (1-t)x_2\rangle\\
&\qquad -t\langle x_1 - \tau(x_1, \mu_0), y - x_1\rangle - (1-t)\langle x_2 - \tau(x_2, \mu_0), y - x_2\rangle\\
&\quad = t\langle x_1 - \tau(x_1, \mu_0), (1-t)x_1 - (1-t)x_2\rangle\\
&\qquad +(1-t)\langle x_2 - \tau(x_2, \mu_0), -tx_1 + tx_2\rangle\\
&\quad = t(1-t)[\langle x_1 - x_2, x_1 - x_2\rangle - \langle \tau(x_1, \mu_0) - \tau(x_2, \mu_0), x_1 - x_2\rangle]\\
&\quad = t(1-t)\langle (x_1 - x_2) - (\tau(x_1, \mu_0) - \tau(x_2, \mu_0)), x_1 - x_2\rangle \geq 0,
\end{aligned}$$

since $-\Phi$ is monotone. Consequently, all the conditions of Theorem 4.15 are satisfied, and so the proof is complete. □

The following result follows from Corollary 4.61.

Corollary 4.68 (Hölder continuity of the solution maps of fixed-point problems) *For problem* (FP)*, assume the existence of a neighborhood U of μ_0 such that fixed points of $\tau(\cdot, \mu)$ exist in U and that:*

(a) *For every $\mu \in U$, and $x, y \in X : x \neq y$,*

$$h\|x - y\|^\beta \leq d(\langle x - \tau(x, \mu), y - x\rangle, \mathbb{R}_+) + d(\langle y - \tau(y, \mu), x - y\rangle, \mathbb{R}_+)$$

for some $h > 0$ and $\beta > 1$.

(b) *For every* $x \in X$, $\tau(x, \cdot)$ *is* (p, ξ)*-Hölder at* μ_0 *for* $p > 0$ *and* $\xi > 0$.
(c) *The map* $(x, \mu) \mapsto x - \tau(x, \mu)$ *is locally bounded on* $X \times U$.

Then, in a neighborhood of μ_0, *the fixed point* $x(\eta)$ *of* $\tau(\cdot, \mu)$ *is unique and satisfies the Hölder condition:*

$$\|x(\eta_1) - x(\eta_2)\| \le k d^{\xi/(\beta-1)}(\eta_1, \eta_2),$$

for some $k > 0$.

The following result is a consequence of Theorem 4.21.

Corollary 4.69 (Hölder continuity of the solution maps of fixed-point problems via strong monotonicity) *Assume that* (FP) *has solutions in a neighborhood of* μ_0 *and the following conditions are satisfied:*

(a$_3$) $\tau(x, \cdot)$ *is* $(\mathcal{N}_1, \delta_1)$*-Hölder at* μ_0 *on* A *and* $(x, \mu) \mapsto x - \tau(x, \mu)$ *is bounded on* A : $\|x - \tau(x, \mu)\| \le \mathcal{N}_2, \forall x \in A$ *and* $\mu \in V(\mu_0)$ *(a neighborhood of* μ_0*)*.
(b$_3$) $\forall \mu \in V(\mu_0)$, $(x, \mu) \mapsto x - \tau(x, \mu)$ *is* (h, β)*-Hölder strongly monotone on* A *for some* $h > 0$ *and* $\beta > 1$.

Then, for each μ, *in a neighborhood of* μ_0, (FP) *has a unique solution* $x(\mu)$, *which satisfies the Hölder condition, for some* $k > 0$,

$$\|x(\mu_1) - x(\mu_2)\| \le k\|\mu_1 - \mu_2\|^{\delta_1/(\beta-1)}.$$

Proof Setting $f(x, y, \mu) := \langle x - B(x, \mu), y - x\rangle$ and $K(x, \lambda) := A$, we check the assumptions of Theorem 4.21. For the Hölder conditions, we have

$$|f(x, y, \mu_1) - f(x, y, \mu_2)| = |\langle x - B(x, \mu_1), y - x\rangle - \langle x - B(x, \mu_2), y - x\rangle|$$

$$= |\langle B(x, \mu_2) - B(x, \mu_1), y - x\rangle| \le |B(x, \mu_2) - B(x, \mu_1)|\|y - x\|$$

$$\le \mathcal{N}_1\|\mu_1 - \mu_2\|^{\delta_1}\|y - x\|,$$

$$|f(x, y_1, \mu) - f(x, y_2, \mu)| = |\langle x - B(x, \mu), y_1 - x\rangle - \langle x - B(x, \mu), y_2 - x\rangle|$$

$$= |\langle x - B(x, \mu), y_1 - y_2\rangle| \le \|x - B(x, \mu)\|\|y_1 - y_2\| \le \mathcal{N}_2\|y_1 - y_2\|.$$

Obviously, the other conditions of Theorem 4.21 are also satisfied, and so the proof follows. □

Corollary 4.70 (Hölder continuity of the solution maps of fixed-point problems via strong pseudomonotonicity) *Corollary 4.69 remains valid if assumption* (b$_3$) *is replaced by*

(b$'_3$) $\forall \mu \in V(\mu_0)$, $(x, \mu) \mapsto x - B(x, \mu)$ *is quasimonotone on* A, *and* $(x, \mu) \mapsto x - B(x, \mu)$ *is* (h, β)*-Hölder-strongly pseudomonotone on* A.

Now, we focus on Hölder conditions for the approximate solution sets of the underlying problems.

Corollary 4.71 (Hölder continuity of the approximate solution maps of fixed-point problems) *Assume that solutions of* (FP) *exist in a neighborhood of the reference point $\mu_0 \in M$. Assume further that:*

(a) *There is a neighborhood U of μ_0 such that, for each $\mu \in U$, $\tau(\cdot, \mu)$ is affine and $x \mapsto x - \tau(x, \mu)$ is monotone on A.*
(b) *For every $x \in A$, $\tau(x, \cdot)$ is (h_3, β_3)-Hölder continuous on A.*

Then, for each $\bar{\varepsilon}$, $\widetilde{\mathrm{Sol}}$(FP) satisfies the following Hölder condition on $[\sqrt{\bar{\varepsilon}}, +\infty) \times U$:

$$H(\widetilde{\mathrm{Sol}}(\mathrm{FP})(\varepsilon_1, \mu_1), \widetilde{\mathrm{Sol}}(\mathrm{FP})(\varepsilon_2, \mu_2)) \leq k_1|\varepsilon_1 - \varepsilon_2| + k_2\|\mu_1 - \mu_2\|^{\beta_3},$$

where $k_1, k_2 > 0$ and depend on $\bar{\varepsilon}$, h_3, and β_3.

Proof In order to use Corollary 4.31, we show that both its assumptions are satisfied. For the first one, taking arbitrary $x_1, x_2 \in A$, $t \in [0, 1]$, $y \in A$, and $\mu \in M$, we have

$$\begin{aligned}
&\Phi(tx_1 + (1-t)x_2, y, \mu) - t\Phi(x_1, y, \mu) - (1-t)\Phi(x_2, y, \mu) \\
&\quad = \langle tx_1 + (1-t)x_2 - \tau(tx_1 + (1-t)x_2, \mu), y - tx_1 - (1-t)x_2\rangle \\
&\qquad -t\langle x_1 - \tau(x_1, \mu), y - x_1\rangle - (1-t)\langle x_2 - \tau(x_2, \mu), y - x_2\rangle \\
&\quad = t\langle x_1 - \tau(x_1, \mu), (1-t)x_1 - (1-t)x_2\rangle \\
&\qquad +(1-t)\langle x_2 - \tau(x_2, \mu), -tx_1 + tx_2\rangle \\
&\quad = t(1-t)[\langle x_1 - x_2, x_1 - x_2\rangle - \langle \tau(x_1, \mu) - \tau(x_2, \mu), x_1 - x_2\rangle] \\
&\quad = t(1-t)\langle (x_1 - \tau(x_1, \mu)) - (x_2 - \tau(x_2, \mu)), x_1 - x_2\rangle \geq 0,
\end{aligned}$$

since $x - \tau(x, \mu)$ is monotone. Hence, the first assumption of Corollary 4.31 is satisfied. Considering the second one, for any $\mu_1, \mu_2 \in U$ and $x, y \in A$, we have

$$\begin{aligned}
|\Phi(x, y, \mu_1) - \Phi(x, y, \mu_2)| &= |\langle x - \tau(x, \mu_1), y - x\rangle - \langle x - \tau(x, \mu_2), y - x\rangle| \\
&= |\langle \tau(x, \mu_1) - \tau(x, \mu_2), y - x\rangle| \\
&\leq \|\tau(x, \mu_1) - \tau(x, \mu_2)\| \|x - y\| \\
&= h_3 \rho_3 \|\mu_1 - \mu_2\|^{\beta_3},
\end{aligned}$$

where $\rho_3 := \text{diam}A$. Hence, the last assumption of Corollary 4.31 is satisfied with $h := h_3\rho_3$ and $\beta := \beta_3$. □

Remark 4.72 (Monotonicity conditions of the map $x \mapsto x - \tau(x)$**)** Let $\tau : A \to A$. Then, the following assertions hold:

(a) If $-\tau(x)$ is monotone on A, then $x \mapsto x - \tau(x)$ is also monotone on A. Indeed, for arbitrary vectors $x_1, x_2 \in A$, we have

$$\langle (x_1 - \tau(x_1)) - (x_2 - \tau(x_2)), x_1 - x_2\rangle = \langle -\tau(x_1) + \tau(x_2), x_1 - x_2\rangle \geq 0,$$

as the map $-\tau(x)$ is monotone. But, in general, the converse is not true as shown by the following example.

Let $\tau : \mathbb{R} \to \mathbb{R}$ be defined as $\tau(x) = ax + b$ where $a \in (0, 1)$ and $b \in \mathbb{R}$. Then, for all $x_1, x_2 \in \mathbb{R}$, we have

$$\langle (-\tau(x_1)) - (-\tau(x_2)), x_1 - x_2\rangle = -a\langle x_1 - x_2, x_1 - x_2\rangle \leq 0.$$

This means that the map $-\tau$ is antimonotone while

$$\langle (x_1 - \tau(x_1)) - (x_2 - \tau(x_2)), x_1 - x_2\rangle = (1 - a)\langle x_1 - x_2, x_1 - x_2\rangle \geq 0),$$

i.e., the map $x \mapsto x - \tau(x) = \frac{1}{2}x - 1$ is monotone in $\mathbb{R}$.

(b) If for the gradient τ' we have $\|\tau'(x)\| \leq 1$ for every $x \in A$, then $x \mapsto x - \tau(x)$ is monotone. Indeed, since for any $u \in A$, $\tau(u) = C$ for some bounded linear map C, for each $x_1, x_2 \in A$, we have

$$\begin{aligned}\langle x_1 - \tau(x_1) - (x_2 - \tau(x_2)), x_1 - x_2\rangle &= \|x_1 - x_2\|^2 - \langle C(x_1 - x_2), x_1 - x_2\rangle \\ &\geq \|x_1 - x_2\|^2 - \|C\|^2\|x_1 - x_2\|^2 \geq 0.\end{aligned}$$

4.5 Exercises

Exercise 4.1 Let $f : E \times E \times M \to \mathbb{R}$ where E is a nonempty convex compact subset of X, $M \subset W$, and X, W are locally convex Hausdorff topological vector spaces. We deal with the parametric scalar equilibrium problem (for short (EP)) of finding $\bar{x} \in E$ such that $f(\bar{x}, y, \mu) \geq 0$ for all $y \in E$. For $\varepsilon \geq 0$, we denote the approximate solution set of (EP) by

$$S(\varepsilon, \mu) := \{\bar{x} \in E \mid f(\bar{x}, y, \mu) + \varepsilon \geq 0, \forall y \in E\}.$$

Prove that, for a given point μ_0, if for every $y \in E$, $f(\cdot, y, \mu_0)$ is continuous on E, then the approximate solution set $S(\varepsilon, \mu_0)$ of (EP) is compact for all $\varepsilon \geq 0$.

Exercise 4.2 Let problem (EP) be defined as in Exercise 4.1 and $S(\varepsilon, \mu)$ be nonempty in a neighborhood of a fixed point (ε_0, μ_0). Prove that if for each $x, y \in E$, $f(x, y, \cdot)$ is continuous at μ_0, then there exists a neighborhood $N \times U$ of (ε_0, μ_0) such that the approximate solution mapping S of (EP) satisfies the following condition:

$$S(\varepsilon_1, \mu_1) \subset S(\varepsilon_2, \mu_2), \forall \mu_1, \mu_2 \in U, \forall \varepsilon_1, \varepsilon_2 \in N, \varepsilon_1 \leq \varepsilon_2.$$

Exercise 4.3 Let problem (EP) be defined as in Exercise 4.1 and $S(\varepsilon, \mu)$ be nonempty in a neighborhood of a fixed point (ε_0, μ_0). Prove that if for each $y \in E$, $f(\cdot, y, \mu_0)$ is a concave function, then the approximate solution set $S(\varepsilon, \mu_0)$ is convex for each $\varepsilon \in [\varepsilon_0 - \delta, \varepsilon_0 + \delta]$ for some $\delta > 0$.

Exercise 4.4 Let problem (EP) be defined as in Exercise 4.1, and $S(\varepsilon, \mu)$ be nonempty in a neighborhood of a fixed point (ε_0, μ_0). Prove that if for $x, y \in E$, $f(\cdot, y, \mu_0)$ is continuous on E, $f(x, y, \cdot)$ is continuous at μ_0, and $f(\cdot, y, \mu_0)$ is a concave function, then the approximate solution map S of (EP) is H-continuous at (ε_0, μ_0).

Exercise 4.5 Consider a function $f : X \times X \times M \to \mathbb{R}$, a map $\mathscr{K} : \Lambda \rightrightarrows X$, $\Lambda \subset Y$, and $M \subset W$, where X, Y, W are metric spaces. For every $\lambda \in \Lambda$ and $\mu \in M$, we study the equilibrium problem (for short (EP)) of finding $\bar{x} \in K(\lambda)$ such that $f(\bar{x}, y, \mu) \geq 0$ for all $y \in K(\lambda)$, and denote its solution set by $S(\lambda, \mu)$. For a given point $\lambda_0 \in \Lambda$, prove that if $K(\lambda_0)$ is nonempty-convex-compact, $f(\cdot, \cdot, \mu)$ is properly quasiconvex, and $f(\cdot, y, \mu)$ is upper semicontinuous for every $y \in K(\lambda_0)$ and $\mu \in M$, then $S(\lambda_0, \mu)$ is nonempty and closed for all μ, and further that it is convex if $f(\cdot, y, \mu)$ is quasiconcave.

Exercise 4.6 Let problem (EP) be defined as in Exercise 4.5. For a given point λ_0, prove that if $K(\lambda_0)$ is nonempty-convex-compact and $f(\cdot, y, \cdot)$ is upper semicontinuous for every $y \in K(\lambda_0)$, then $\nu(x, \mu) = \inf_{y \in K(\lambda_0)} f(x; y; \mu)$ is upper semicontinuous on $K(\lambda_0) \times M$ and in particular, the map $S(\lambda_0, \cdot)$ is closed and upper hemicontinuous.

Exercise 4.7 Let problem (EP) be defined as in Exercise 4.5. For a given point λ_0, prove that if $K(\lambda_0)$ is nonempty-convex-compact, $f(x, x, \mu) = 0$ for all $x \in K(\lambda_0)$, $f(\cdot, \cdot, \mu)$ is pseudomonotone for every $\mu \in K(\lambda_0)$, $f(x, \cdot, \cdot)$ is lower semicontinuous, $f(\cdot, y, \mu)$ is upper semicontinuous on the segments for every $y \in K(\lambda_0)$, $\mu \in M$ (i.e., $\limsup_{t \to 0^+} f((1-t)x_1 + tx_2, y, \mu) \leq f(x_1, y, \mu), \forall x_1, x_2 \in K(\lambda_0)$), $f(x, \cdot, \mu)$ is quasiconvex, and $f(x, \cdot, \mu)$ is sign-preserving for every $x \in K(\lambda)$, $\mu \in M$ (i.e., $[f(x, y_1, \mu) < 0, f(x, y_2; \mu) = 0] \Rightarrow [f(x, (1-t)y_1 + ty_2, \mu) < 0, \forall y_1, y_2 \in K(\lambda_0), \forall t \in (0, 1)]$, then $S(\lambda_0, \cdot)$ is upper hemicontinuous.

Exercise 4.8 Let problem (EP) be defined as in Exercise 4.5 and X, Y be metric spaces. Prove that if for a given point λ_0, $K(\lambda_0)$ is nonempty-convex-compact and

$f(\cdot, y, \cdot)$ is quasiconcave and upper semicontinuous for every $y \in \mathscr{D}(\lambda_0)$, then $S(\lambda_0, \cdot)$ is locally Lipschitz on $\text{int}R(T^{-1})$.

Exercise 4.9 Let (EP), X, Y, and λ_0 be as in Exercise 4.5. Prove that if $K(\lambda_0)$ is nonempty-convex-compact, $M \subset \text{dom}(T)$, $f(\cdot, \cdot, \mu)$ is strongly pseudomonotone for every $\mu \in M$ (i.e., $[f(x, y, \mu) \geq 0] \Rightarrow [f(y, x, \mu) \leq -\alpha d^{\eta}(x, y)$ for some positive numbers α, η), and $f(x, y, \cdot)$ is Hölder continuous uniformly with respect to (x, y) (i.e., $|f(x, y, \mu_1) - f(x, y, \mu_2)| \leq \beta d^{\gamma}(\mu_1, \mu_2)$ for some positive numbers β, γ), then for every $\mu \in M$, the solution of (EP) is unique and this map is Hölder continuous on M.

Exercise 4.10 Let X and Y be two Banach spaces and G be a multivalued mapping from X into Y satisfying the following:

(a) G is convex and closed.
(b) $G(X) = Y$.
(c) $G^{-1}(0)$ is bounded.

Prove that for each bounded neighborhood V_0, there is a bounded closed set $B \subset X$ such that $G^{-1}(e) \subset B$ for all $v \in V_0$. Moreover, if X is reflexive, then G^{-1} is usc in V_0 with respect to $Y \times X_\sigma$. (X_σ denotes the space X endowed with its weak topology).

Exercise 4.11 Let $f : X \times X \to \mathbb{R}$. Assume the conditions specified in Exercise 4.10. Prove that if the following conditions hold:

(a) X is reflexive.
(b) f is lower semicontinuous on $X \times X_\sigma$ and convex in its second argument.
(c) f is hemicontinuous and monotone, i.e.,

$$f(x, x) \geq 0, f(x, y) + f(y, x) \leq 0, \forall x, y \in X,$$

then for each $v \in V_0$, the set $S(e) = \{x \in G^{-1}(e) \mid f(x, y) \geq 0, \forall y \in G^{-1}(e)\}$ is nonempty and the mapping $S(\cdot)$ is upper semicontinuous in the topology of $Y \times X_\sigma$.

Exercise 4.12 Prove that under the same assumptions as in Exercise 4.11, the mapping $S_\epsilon(\cdot)$ is lower semicontinuous at 0 for each $\varepsilon > 0$, where S_ε is given by

$$S_\varepsilon(e) = \begin{cases} S(0) & \text{if} \quad e = 0, \\ \{x \in G^{-1}(e) \mid f(x, y) + \varepsilon \geq 0, \forall y \in G^{-1}(e)\} & \text{otherwise.} \end{cases}$$

4.6 Commentary

Section 4.1

The studies of the stability of parametric scalar equilibrium problems began relatively late. On semicontinuity and continuity of solution maps, we can mention the contributions of [30, 99] followed by [27, 101, 138, 254, 296, 326]. The first works provided extensions of known results on the same topic for variational inequalities. Further developments were more significant and were for various models related to equilibrium problems, including special cases such as variational inequalities, constrained minimization, complementarity problems, Nash equilibria, and traffic networks; see, for example, [24, 141, 322] and also generalizations of equilibrium problems such as inclusion problems and variational relation problems [29, 36, 283]. This section is an effort to provide a complete and comprehensive presentation of the results obtained so far. It also contains some new facts or refinements of the known results.

Section 4.2

Hölder and Lipschitz continuity of solution maps of parametric scalar equilibrium problems has been investigated from the beginning, along with some types of (semi)continuity. Unlike the case of (semi)continuity, to get Hölder continuity of solution maps, besides assumptions related to Hölder continuity of the data of the problem under consideration, additional conditions of different nature should also be imposed. In the literature, these additional conditions have been of three types. The strong monotonicity condition was employed in [26, 33, 99, 101, 360, 461], while a strong convexity assumption was used in [39]. In these cases, solution maps are Hölder continuous, but also locally single-valued. In practical problems, solution sets are usually not singletons and so such results are not applicable. To avoid this drawback, some coercivity conditions were proposed in [323, 324]. The third type of additional conditions can be found in [38]. Namely, this paper dealt with Hölder continuity of approximate solutions (with arbitrary small tolerance), while Hölder calmness, instead of Hölder continuity, was investigated in [42, 115, 144, 146, 260, 320, 472].

The section reflects comprehensively the results in the aforementioned contributions of the three types. The results presented here could be regarded as an endeavor to unify the corresponding contributions in the literature.

Section 4.3

Well-posedness is one of the most important topics in both theory and numerical methods in optimization, and hence it has attracted the attention of several generations of mathematicians. In 1966, Tikhonov introduced the concept of well-posedness for unconstrained optimization problems, which is now known as Tikhonov well-posedness. A minimization problem is said to be Tikhonov well-posed if it has a unique solution to which every minimizing sequence of the problem converges (see [429]). Since then, the study of Tikhonov well-posedness and its extensions has been among the most interesting and important topics in optimization theory. A generalization of Tikhonov well-posedness has been discussed for constrained optimization problems (see, e.g., [311, 319, 391]). Another generalization of the concept given for optimization problems with more than one solution requires the existence and convergence of a subsequence of each minimizing sequence to a solution [184]. Another fundamental generalization of Tikhonov well-posedness, which was first introduced for scalar optimization problems in [96, 474, 475, 477, 478], is well-posedness under perturbations. This generalized notion incorporates both the idea of Tikhonov well-posedness and the continuous dependence of the solution on the data (the latter is called Hadamard well-posedness in optimization [232, 392, 476]). In this section, rather than repeating any given paper, we try to unify the results in the literature. However, we make no attempt to cover all the results to date. Besides the results on well-posedness conditions presented in this section, we also aim to introduce readers to notable works on this topic for important models in optimization, such as in game theory [111, 362], quadratic programming problems [259], variational inequalities [328, 354], and optimization problems [199, 355, 372, 373].

Chapter 5
Stability of Scalar Quasiequilibrium Problems

Abstract By the great importance of quasiequilibrium models in practical applications, stability and well-posedness conditions for them have motivated and inspired studies of many mathematicians. In most works devoted to this topic, topological properties of constraint sets are imposed directly on the fixed points of the set-valued maps, and hence, it is difficult to check them when sufficient conditions for stability and well-posedness are in use. To avoid this drawback, in this chapter, we first study separately many important properties of solution maps of fixed-point problems of set-valued maps and then employ these results to the aforementioned stability and well-posedness conditions directly in terms of the problem data. Furthermore, we also give an overview of sufficient conditions for Hölder and Lipschitz properties of solution maps of quasiequilibrium problems via assumptions related to monotonicity and convexity. Furthermore, under relaxed conditions, the calmness property of such maps is also derived in this chapter. Note that for solution maps of quasiequilibrium problems, the difference between stability and well-posedness in the sense of semicontinuity and continuity and in the sense of Hölder and Lipschitz properties is so big that many approaches and techniques for dealing with equilibrium problems cannot be employed. As a result, several established results for equilibrium problems have not yet been extended to quasiequilibrium problems, and hence, deriving Hölder and Lispchitz conditions for the solution maps of such models is still an open problem.

5.1 Continuity of Solution Mappings

Let A, Λ, M be nonempty sets, $\Phi : A\times A\times M \to \mathbb{R}$ a function, and $\mathscr{D} : A\times\Lambda \rightrightarrows A$ surjective. For each $(\lambda, \mu) \in \Lambda \times M$, we consider the following quasiequilibrium problem:

(QEP)(λ, μ) find $\bar{x} \in \mathscr{D}(\bar{x}, \lambda)$ such that $\Phi(\bar{x}, y, \mu) \geq 0$ for all $y \in \mathscr{D}(\bar{x}, \lambda)$.

The solution set Sol(QEP)(λ, μ) depends on two parameters λ and μ, and hence, the set-valued map $(\lambda, \mu) \mapsto$ Sol(QEP)(λ, μ) is called the SOLUTION MAP of the

L. Q. Anh et al., *Equilibrium Problems*, Forum for Interdisciplinary Mathematics, https://doi.org/10.1007/978-3-031-89003-1_5

parametric quasiequilibrium problem $\{(\text{QEP})_{\Lambda\times M}\} := \{(\text{QEP})(\lambda, \mu) \mid \lambda \in \Lambda, \mu \in M\}$, which is a family of quasiequilibrium problems. In this chapter, we study the properties of this map.

Setting $\mathscr{E}(\lambda) := \{x \in A \mid x \in \mathscr{D}(x, \lambda)\}$, we have $\text{Sol}(\text{QEP})(\lambda, \mu) = \mathscr{E}(\lambda) \cap \text{Sol}(\text{QEP}^1)(\lambda, \mu)$, where $\text{Sol}(\text{QEP}^1)(\lambda, \mu)$ is the solution set of the following problem:

$(\text{QEP}^1)(\lambda, \mu)$ find $\bar{x} \in A$ such that $\Phi(\bar{x}, y, \mu) \geq 0$ for all $y \in \mathscr{D}(\bar{x}, \lambda)$.

In view of the aforementioned remark, we first study continuity properties of the map $\mathscr{E}$ and then combine them with the corresponding properties of $\text{Sol}(\text{QEP}^1)$ to establish continuity properties of the map Sol(QEP).

In this section, we discuss (semi)continuity of the solution maps of quasiequilibrium problems, and hence in the rest of this section, if not otherwise specified, we suppose that X is a Hausdorff topological space, and A, Λ, M are nonempty closed subsets of X.

5.1.1 Continuity of Solution Maps of Fixed-Point Problems

Theorem 5.1 (Upper semicontinuity and closedness of fixed-point maps of multivalued maps)

(a) *If $\mathscr{D}$ is closed on $A \times \Lambda$, then $\mathscr{E}$ is closed on Λ.*
(b) *If $\mathscr{D}$ is closed and A is compact, then $\mathscr{E}$ is both usc and closed on Λ.*

Proof For (a), let $(\lambda^k, x^k) \in \text{gph}\mathscr{E}$ converge to $(\bar{\lambda}, \bar{x})$. Because $x^k \in \mathscr{E}(\lambda^k)$ for all k, i.e., $x^k \in \mathscr{D}(x^k, \lambda^k)$ and $\mathscr{D}$ is closed, we conclude that $\bar{x} \in \mathscr{D}(\bar{x}, \bar{\lambda})$, and hence, $\mathscr{E}$ is closed.

For (b), Because $\mathscr{D}$ is usc and A is compact, $\mathscr{D}$ is closed, and so $\mathscr{E}$ is closed. Suppose on the contrary that there is a $\bar{\lambda} \in \Lambda$ such that $\mathscr{E}$ is not usc at $\bar{\lambda}$. Then, we can find an open neighborhood U of $\mathscr{E}(\bar{\lambda})$, a net $\{\lambda^k\}_k, \lambda^k \to \bar{\lambda}$ and $x^k \in \mathscr{E}(\lambda^k)$ with $x^k \notin U$ for all k. Since A is compact, we can assume that $x^k \to \bar{x}$ for some $\bar{x} \in A$. Then, $\bar{x} \in \mathscr{E}$ as $\mathscr{D}$ is closed, which leads to a contradiction as $x^k \notin U$ for all k. Hence, $\mathscr{E}$ is usc on Λ. □

Lemma 5.2 (A property of convex subsets of a reflexive Banach space) *Let X be a reflexive Banach space, B the closed unit ball of X, and A a nonempty closed convex subset of X. For given $\delta > 0$, if $a + \delta B \subset A + \delta B$, then $a \in A$.*

Proof If a does not belong to A, then since A is closed, one has

$$d(a, A) = \inf_{y \in A} \|a - y\| > 0.$$

Because X is a reflexive Banach space and $A \subset X$ is nonempty closed convex, there exists $b \in A$ such that

$$d(a, A) = \inf_{y \in A} \|a - y\| = \|a - b\| > 0. \tag{5.1}$$

Setting $\lambda := \frac{\|a-b\|}{\delta+\|a-b\|}$, we get $x \in X$ such that $a = \lambda x + (1 - \lambda)b$. Equivalently,

$$x = \frac{a}{\lambda} + b - \frac{b}{\lambda}.$$

Next we show that

$$\|x - y\| \geq \|x - b\|, \forall y \in A.$$

It follows from (5.1) that

$$\|a - y\| \geq \|a - b\|, \forall y \in A. \tag{5.2}$$

Obviously, for any $y \in A$, we have $\lambda y + (1 - \lambda)b \in A$ as A is convex, and hence, (5.2) implies that

$$\|a - (\lambda y + (1 - \lambda)b)\| \geq \|a - b\|.$$

Consequently,

$$\|x - y\| \geq \|x - b\|, \forall y \in A. \tag{5.3}$$

On the other hand,

$$\|x - a\| = \left\| \frac{a}{\lambda} + b - \frac{b}{\lambda} - a \right\| = \left(\frac{1}{\lambda} - 1 \right) \|a - b\| = \delta,$$

and hence, $x \in a + \delta B$. Combining (5.3) with the fact that

$$\|x - b\| = \left\| \frac{a}{\lambda} + b - \frac{b}{\lambda} - b \right\| = \frac{1}{\lambda} \|a - b\| = \delta + \|a - b\| > \delta,$$

we obtain that

$$x \notin A + \delta B,$$

and hence

$$a + \delta B \not\subset A + \delta B,$$

which is a contradiction. The proof is complete. □

Definition 5.3 (Rotund maps) Let X and Y be two topological vector spaces. A set-valued mapping $\mathscr{K} : X \rightrightarrows Y$ is said to be ROTUND if, $\text{gph}(\mathscr{K})$ is convex and for any $(x_1, y_1), (x_2, y_2) \in \text{gph}(\mathscr{K})$ with $x_1 \neq x_2, y_1 \neq y_2$, we have

$$((x_1, y_1), (x_2, y_2)) \cap (\text{bd}(\text{gph}(\mathscr{K})))^c \neq \emptyset,$$

where $((x_1, y_1), (x_2, y_2)) = \{\lambda (x_1, y_1) + (1 - \lambda)(x_2, y_2) \mid \lambda \in (0, 1)\}$.

Theorem 5.4 (Lower semicontinuity of fixed-point maps of multivalued maps) *Let A be a nonempty compact convex subset of a reflexive Banach space X. For each $\lambda \in \Lambda$, assume that*

(a) $x \mapsto \mathscr{D}(x, \lambda)$ *is rotund;*
(b) $\mathscr{D}$ *is continuous on* $A \times \{\lambda\}$ *with nonempty closed convex values.*

Then, $\mathscr{E}$ *is lower semicontinuous on* Λ.

Proof Suppose to the contrary that Sol(QEP) is not lsc at λ_0. Then, there exist a point $x_0 \in \text{Sol(QEP)}(\lambda_0)$, a neighborhood W_0 of $0 \in X$, and a sequence $\{\lambda^k\}_k \subset \Lambda$ with $\lambda^k \to \lambda_0$ such that

$$(x_0 + W_0) \cap \text{Sol(QEP)}(\lambda^k) = \emptyset, \forall k. \tag{5.4}$$

We consider two cases.
Case 1. $\text{Sol(QEP)}(\lambda_0)$ is a singleton.
For $x^k \in \text{Sol(QEP)}(\lambda^k)$, one has

$$x^k \in T(x^k, \lambda^k). \tag{5.5}$$

Since $x^k \in A$ and A is compact, without loss of generality, we can assume that x^k converge to some $\bar{x} \in A$. Noting that T is usc at $(\bar{x}, \lambda_0)$, it follows from (5.5) that there exist $x' \in T(\bar{x}, \lambda_0)$ and a subsequence $\{x^\beta\}$ of $\{x^k\}$ such that $x^\beta \to x'$. It is clear that $\bar{x} = x'$ and so $\bar{x} \in T(\bar{x}, \lambda_0)$. This means that $\bar{x} \in \text{Sol(QEP)}(\lambda_0) = \{x_0\}$. Thus, $x^k \in x_0 + W_0$ for k large enough. This contradicts (5.4).
Case 2. $\text{Sol(QEP)}(\lambda_0)$ is not a singleton.
We can find $x^* \in \text{Sol(QEP)}(\lambda_0)$ with $x^* \neq x_0$. Then, $(x^*, x^*), (x_0, x_0) \in \text{gph}T(\cdot, \lambda_0)$. With $x(t) = tx^* + (1 - t)x_0$, $x(t) \in A$ for any $t \in [0, 1]$. Because $T(\cdot, \lambda_0)$ is rotund, we can find $t_0 \in (0, 1)$ such that

$$x(t_0) \in x_0 + W_0 \text{ and } (x(t_0), x(t_0)) \in \text{int}(\text{gph}T(\cdot, \lambda_0)). \tag{5.6}$$

This implies that there is a constant $\delta > 0$ such that

$$(x(t_0), x(t_0)) + \delta B \times \delta B \subset \text{int}(\text{gph}T(\cdot, \lambda_0)),$$

where B is the closed unit ball in X. This shows that

$$x(t_0) + \delta B \subset T(x(t_0), \lambda_0). \tag{5.7}$$

Since $T(x(t_0), \cdot)$ is H-lsc at λ_0 and the sequence $\{\lambda^k\}_k$ converges to λ_0, for δB, there exists k_0 large enough such that

$$T(x(t_0), \lambda_0) \subset T(x(t_0), \lambda^{k_0}) + \delta B.$$

This together with (5.7) implies that

$$x(t_0) + \delta B \subset T(x(t_0), \lambda^{k_0}) + \delta B. \tag{5.8}$$

By the convexity and closedness of $T(x(t_0), \lambda^{k_0})$, (5.8) yields that $x(t_0) \in T(x(t_0), \lambda^{k_0})$, or equivalently $x(t_0) \in \mathrm{Sol(QEP)}(\lambda^{k_0})$. This contradicts (5.4). The proof is complete. □

5.1.2 Continuity of Solution Maps of Parametric Quasiequilibrium Problems

We observe that, in almost all the results on semicontinuity properties of solution maps of quasiequilibrium problems, the properties of the fixed-point map $\mathscr{E}$ are assumed without presenting any sufficient conditions, while it is still difficult to check the properties of $\mathscr{E}$. It motivates us to derive a different approach to studying semicontinuity properties of solution maps. Firstly, we recall some well-known results.

Lemma 5.5 (Some preservation properties of intersection maps) *Let X, Y be Hausdorff topological spaces, $A \subset X$, and $F, G : A \rightrightarrows Y$ satisfy $F(x) \cap G(x) \neq \emptyset$ for all $x \in A$. Then,*

(a) *$F \cap G$ is usc and compact-valued on A if F and G are usc and compact-valued on A;*
(b) *$F \cap G$ is closed A if F and G are closed on A;*
(c) *$F \cap G$ is lsc on A if F and G are lsc on A.*

Theorem 5.6 (Upper semicontinuity and closedness of solution maps of quasiequilibrium problems) *Assume that*

(a) *A is compact;*
(b) *$\mathscr{D}$ is both closed and lsc on $A \times M$;*
(c) *Φ is 0-upper level closed on $A \times A \times M$, i.e., the set $\{(x, y, \mu) \in A \times A \times M \mid \Phi(x, y, \mu) \geq 0\}$ is closed.*

Then, Sol(QEP) *is usc, closed, and compact-valued on $\Lambda \times M$.*

Proof It follows from Theorem 5.1 that the map $\mathscr{E}$ is usc, closed, and compact-valued on Λ. We prove that so is the map $\mathrm{Sol(QEP^1)}$ on $\Lambda \times M$. Then, the conclusion of the theorem will follow from Lemma 5.5.

Suppose to the contrary that $\mathrm{Sol(QEP^1)}$ is not usc at some $(\bar{\lambda}, \bar{\mu}) \in \Lambda \times M$. Then, there are an open set U containing $\mathrm{Sol(QEP^1)}(\bar{\lambda}, \bar{\mu})$ and some sequence $\{(\lambda^k, \mu^k)\}_k$ converging to $(\bar{\lambda}, \bar{\mu})$ such that for each k, there is $x^k \in \mathrm{Sol(QEP^1)}(\lambda^k, \mu^k)$ but $x^k \notin U$. Because A is compact, we can assume that x^k converge to some $\bar{x} \in A$. Let $\bar{y}$ be arbitrary and belong to $\mathscr{D}(\bar{x}, \bar{\lambda})$. There are $\bar{x}^k \in \mathscr{D}(x^k, \lambda^k)$ converging to $\bar{y}$ as $\mathscr{D}$ is lsc at $(\bar{x}, \bar{\lambda})$. Because the vector x^k belongs to $\mathrm{Sol(QEP^1)}(\lambda^k, \mu^k)$ and Φ is 0-upper level closed, we conclude that for each $\bar{y} \in \mathscr{D}(\bar{x}, \bar{\lambda})$, we have $\Phi(\bar{x}, \bar{y}, \bar{\mu}) \geq 0$, or equivalently $\bar{x}$ is an element of $\mathrm{Sol(QEP^1)}(\bar{\lambda}, \bar{\mu})$. This contradicts the fact that $x^k \notin U$ for all k. Therefore, $\mathrm{Sol(QEP^1)}$ is usc on $\Lambda \times M$.

Using suitable arguments similar to the earlier discussion, we also obtain that $\mathrm{Sol(QEP^1)}$ is closed and compact-valued on $\Lambda \times M$. □

Theorem 5.7 (Lower semicontinuity of solution maps of quasiequilibrium problems) *Let X be a reflexive Banach space and the set*

$$\mathrm{Sol(QEP^{1+})}(\lambda, \mu) := \{x \in A \mid \Phi(x, y, \mu) > 0, \forall y \in \mathscr{D}(x, \lambda)\}$$

be nonempty for all $(\lambda, \mu) \in \Lambda \times M$. Assume further that

(a) *A is nonempty, compact, and convex;*
(b) *$\mathscr{D}$ is continuous, nonempty-closed-convex-valued on $A \times \Lambda$, and rotund in the first component;*
(c) *$-\Phi(\cdot, \cdot, \mu)$ is sign-preserving in the first component for each $\mu \in M$ and Φ is 0-lower level closed, i.e., the set $\{(x, y, \mu) \in A \times A \times M \mid \Phi(x, y, \mu) \leq 0\}$ is closed.*

Then, Sol(QEP) *is lsc on $\Lambda \times M$.*

Proof Due to Theorem 5.4, the map $\mathscr{E}$ is lsc on Λ. We first prove that the solution map $\mathrm{Sol(QEP^1)}$ is usc on $\Lambda \times M$. Then, Lemma 5.5 implies the lower semicontinuity of the solution map Sol(QEP). Using the same arguments as in the proof of Proposition 4.9, we get

$$\mathrm{Sol(QEP^1)}(\lambda, \mu) \subset \mathrm{clSol(QEP^{1+})}(\lambda, \mu). \tag{5.9}$$

Let us show that $\mathrm{Sol(QEP^{1+})}$ is a lower semicontinuous map. Suppose to the contrary that there are $\bar{x} \in \mathrm{Sol(QEP^{1+})}(\bar{\lambda}, \bar{\mu})$ and $\{(\lambda^k, \mu^k)\}_k \subset \Lambda \times M$ converging to $(\bar{\lambda}, \bar{\mu})$ such that all sequences $\{x^k\}_k$ with $x^k \in \mathrm{Sol(QEP^{+})}(\lambda^k, \mu^k)$ cannot converge to $\bar{x}$. Since $\mathscr{E}$ is lower semicontinuous, there is $\bar{x}^k \in \mathscr{E}(\lambda^k)$ such that $\bar{x}^k \to \bar{x}$. Due to the aforementioned contradiction assumption, we can assume that $\bar{x}^k \notin \mathrm{Sol(QEP^{1+})}(\lambda^k, \mu^k)$ for all k. Then, for each k, there exists some vector y^k belonging to $\mathscr{D}(x^k, \lambda^k)$ such that $\Phi(\bar{x}^k, y^k, \mu^k) \leq 0$. Since $\mathscr{D}$ is upper semicontinuous and compact-valued, we can assume that $y^k \to \bar{y}$ for some

$\bar{y} \in \mathscr{D}(\bar{x}, \bar{\lambda})$. Then, by the 0-lower level closedness of Φ, $\Phi(\bar{x}, \bar{y}, \bar{\mu}) \leq 0$, which contradicts the fact that $\bar{x}$ is a solution of Sol(QEP^1).

Consequently, for each net $\{(\lambda^k, \mu^k)\}_k \subset \Lambda \times M$ converging to $(\bar{\lambda}, \bar{\mu}) \in \Lambda \times M$, we have

$$\begin{aligned}\mathrm{Sol(QEP^1)}(\bar{\lambda}, \bar{\mu}) \subset \mathrm{cl}(\mathrm{Sol(QEP^{1+})}(\bar{\lambda}, \bar{\mu})) &\subset \liminf_k \mathrm{Sol(QEP^{1+})}(\lambda^k, \mu^k)\\ &\subset \liminf_k \mathrm{Sol(QEP^1)}(\lambda^k, \mu^k).\end{aligned}$$

Thus, Sol(QEP^1) is lower semicontinuous on $\Lambda \times M$. □

Remark 5.8 (A property of the fixed-point maps of multivalued maps in a reflexive Banach space) Unlike the results related to the upper semicontinuity property of the solution mappings, for the lower semicontinuity property, we have to study the problem in a reflexive Banach space so that sufficient conditions of the corresponding property of fixed-point mapping $\mathscr{E}$ can be established.

5.2 Hölder Continuity of Solution Maps

In this section, we study sufficient conditions for the Hölder continuity of solution maps, and hence, we suppose that X is a metric space and A, Λ, M are nonempty closed subsets of X.

5.2.1 Hölder Continuity of Solution Maps of (QEPs)

Similar to the previous section, for $(\lambda, \mu) \in \Lambda \times M$, we also consider the map

$$\mathscr{E}(\lambda) := \left\{x \in X \mid x \in \mathscr{D}(x, \lambda)\right\}$$

of fixed points and the solution map Sol(QEP)(λ, μ) of (QEP$_{\lambda,\mu}$).

We first prove the following result, which is used in what follows.

Proposition 5.9 (A property of semistrongly monotone functions) *Let $\Omega : A \times A \to \mathbb{R}$ be a real-valued function. If Ω is (h, β)-semistrongly monotone, i.e., for each $x, y \in A, x \neq y$,*

$$\min\{\Omega(x, y), \Omega(y, x)\} + \frac{h}{2} d^{\beta}(x, y) \leq 0, \tag{5.10}$$

then the following property holds true

(A) *for* $x, y \in X, x \neq y,$

$$d(\Omega(x, y), \mathbb{R}_+) + d(\Omega(y, x), \mathbb{R}_+) \geq \frac{h}{2} d^\beta(x, y).$$

Proof For $x, y \in X, x \neq y$, we consider the following three cases.

If $\Omega(y, x) \geq 0$, (5.10) implies that $\Omega(x, y) < 0$, and hence

$$\min\{\Omega(x, y), \Omega(y, x)\} = \Omega(x, y).$$

It follows from (5.10) that

$$\Omega(x, y) + \frac{h}{2} d^\beta(x, y) \leq 0.$$

Consequently,

$$d(\Omega(x, y), R_+) + d(\Omega(y, x), R_+) = -\Omega(x, y) \geq \frac{h}{2} d^\beta(x, y),$$

or equivalently, the condition (A) is true.

If $\Omega(x, y) \geq 0$, similar to the aforementioned case, condition (A) is also satisfied.

Hence, we now move on to the case that both of the considered values are negative, and suppose that

$$\Omega(x, y) \leq \Omega(y, x) < 0.$$

Then, inequality (5.10) leads to

$$\Omega(x, y) + \frac{h}{2} d^\beta(x, y) \leq 0.$$

Therefore,

$$d(\Omega(x, y), R_+) + d(\Omega(y, x), R_+) = -\Omega(x, y) - \Omega(y, x) \geq \frac{h}{2} d^\beta(x, y),$$

which means that (A) is satisfied. □

Theorem 5.10 (Hölder continuity of solution maps of quasiequilibrium problems via a semistrong monotonicity condition) *For problem* (QEP)(λ, μ), *assume that solutions exist in* $\Lambda \times M$ *and the following conditions hold true*

(a) $\mathscr{D}$ *is* $((\ell_1, \alpha_1), (\ell_2, \alpha_2))$*-Hölder in* $A \times \Lambda$;
(b) *for each* $\mu \in M$, Φ *is* (ρ, γ)*-Hölder continuous at* μ *with respect to* M, θ*-uniformly over* A *and* (σ, δ)*-Hölder continuous on* A *in the second component;*

(c) *for each* $\mu \in M$, $\Phi(\cdot, \cdot, \mu)$ *is* $(2h, \beta)$*-semistrongly monotone on* A;
(d) $\alpha_1 \delta = \beta$, $h > 2\sigma \ell_1^\delta$, *and* $\beta > \theta$.

Then, the solution $x(\lambda, \mu, \eta)$ *of* (QEP)(λ, μ) *is unique and satisfies the following condition*

$$d\big(x(\lambda_1, \mu_1), x(\lambda_2, \mu_2)\big) \le k_1 d^{\alpha_2\delta/\beta}(\lambda_1, \lambda_2) + k_2 d^{\gamma/(\beta-\theta)}(\mu_1, \mu_2),$$

where k_1, k_2, *and* k_3 *are positive constants depending on* $h, \beta, \sigma, n_3, \rho, \theta, \ell_1, \ell_2, \ldots$

Proof Let $\lambda_1, \lambda_2 \in \Lambda$ and $\mu_1, \mu_2 \in M$.
Step 1. We prove that, for all $x(\lambda_1, \mu_1) \in$ Sol(QEP)(λ_1, μ_1) and $x(\lambda_1, \mu_2) \in$ Sol(QEP)(λ_1, μ_2),

$$d_1 := d\big(x(\lambda_1, \mu_1), x(\lambda_1, \mu_2)\big) \le \left(\frac{\rho}{h - 2\sigma \ell_1^\delta}\right)^{1/(\beta-\theta)} d^{\gamma/(\beta-\theta)}(\mu_1, \mu_2).$$

Let $x(\lambda_1, \mu_1) \ne x(\lambda_1, \mu_2)$ (if the equality holds then, of course, the aforementioned inequality is true). Because $x(\lambda_1, \mu_1) \in \mathscr{D}(x(\lambda_1, \mu_1), \lambda_1)$, $x(\lambda_1, \mu_2) \in \mathscr{D}(x(\lambda_1, \mu_2), \lambda_1)$, and $\mathscr{D}$ is Hölder continuous, there are $x_1 \in \mathscr{D}(x(\lambda_1, \mu_1), \lambda_1)$ and $x_2 \in \mathscr{D}(x(\lambda_1, \mu_2), \lambda_1)$ such that

$$d\big(x(\lambda_1, \mu_1), x_2\big) \le \ell_1 d^{\alpha_1}\big(x(\lambda_1, \mu_1), x(\lambda_1, \mu_2)\big), \tag{5.11}$$

$$d\big(x(\lambda_1, \mu_2), x_1\big) \le \ell_1 d^{\alpha_1}\big(x(\lambda_1, \mu_1), x(\lambda_1, \mu_2)\big). \tag{5.12}$$

Because $x(\lambda_1, \mu_1)$ and $x(\lambda_1, \mu_2)$ are solutions of $(\mathrm{P}_{\mathrm{r}_1\varphi_2})$, we have

$$\Phi(x(\lambda_1, \mu_1), x_1, \mu_1) \ge 0, \tag{5.13}$$

$$\Phi(x(\lambda_1, \mu_2), x_2, \mu_2) \ge 0. \tag{5.14}$$

Due to assumption (c), we have

$$d(\Phi(x(\lambda_1, \mu_1), x(\lambda_1, \mu_2), \mu_1), R_+) + d(\Phi(x(\lambda_1, \mu_2), x(\lambda_1, \mu_1), \mu_1), R_+) \ge h d_1^\beta.$$

It follows from (5.13) and (5.14) that

$$d(\Phi(x(\lambda_1, \mu_1), x(\lambda_1, \mu_2), \mu_1), \Phi(x(\lambda_1, \mu_1), x_1, \mu_1))$$
$$+ d(\Phi(x(\lambda_1, \mu_2), x(\lambda_1, \mu_1), \mu_1), \Phi(x(\lambda_1, \mu_2), x_2, \mu_2)) \ge h d_1^\beta.$$

Consequently,

$$\begin{aligned} & d\big(\Phi(x(\lambda_1,\mu_1),x_1,\mu_1),\Phi(x(\lambda_1,\mu_1),x(\lambda_1,\mu_2),\mu_1)\big) \\ &\quad + d\big(\Phi(x(\lambda_1,\mu_2),x_2,\mu_2),\Phi(x(\lambda_1,\mu_2),x(\lambda_1,\mu_1),\mu_2)\big) \\ &\quad + d\big(\Phi(x(\lambda_1,\mu_2),x(\lambda_1,\mu_1),\mu_2),\Phi(x(\lambda_1,\mu_2),x(\lambda_1,\mu_1),\mu_1)\big) \geq hd_1^{\beta}. \end{aligned}$$

By assumptions (b) and (c), one has

$$\sigma d^{\delta}(x_1,x(\lambda_1,\mu_2)) + \sigma d^{\delta}(x_2,x(\lambda_1,\mu_1)) + \rho d_1^{\theta} d^{\gamma}(\mu_1,\mu_2) \geq hd_1^{\beta}.$$

Now (5.11) and (5.12) imply that

$$\sigma \ell_1^{\delta} d_1^{\alpha_1\delta} + \sigma \ell_1^{\delta} d_1^{\alpha_1\delta} + \rho d_1^{\theta} d^{\gamma}(\mu_1,\mu_2) \geq hd_1^{\beta}.$$

Then, it follows from assumption (d) that

$$d_1^{\beta-\theta} \leq \left(\frac{\rho}{h-2\sigma\ell_1^{\delta}}\right) d^{\gamma}(\mu_1,\mu_2).$$

Setting $k_1 = \left(\frac{\rho}{h-2\sigma\ell_1^{\delta}}\right)^{\frac{1}{\beta-\theta}}$, we have

$$d_1 \leq k_1 d^{\frac{\gamma}{\beta-\theta}}(\mu_1,\mu_2).$$

Step 2. Now we show that for each $x(\lambda_1,\mu_2)$ and $x(\lambda_2,\mu_2)$ belonging to the solution sets $\mathrm{Sol(QEP)}(\lambda_1,\mu_2)$ and $\mathrm{Sol(QEP)}(\lambda_2,\mu_2)$, resp., we always have

$$d_2 := d\big(x(\lambda_1,\mu_2),x(\lambda_2,\mu_2)\big) \leq \left(\frac{2\sigma\ell_2^{\delta}}{h-2\sigma\ell_1^{\delta}}\right)^{1/\beta} d^{\alpha_2\delta/\beta}(\lambda_1,\lambda_2).$$

As before, we can assume that $x(\lambda_1,\mu_2) \neq x(\lambda_2,\mu_2)$. Thanks to (a), we can get vectors

$$x_1' \in \mathscr{D}(x(\lambda_2,\mu_2),\lambda_1) \quad \text{and} \quad x_2' \in \mathscr{D}(x(\lambda_1,\mu_2),\lambda_2)$$

satisfying the following inequalities

$$d\big(x(\lambda_1,\mu_2),x_2'\big) \leq \ell_2 d^{\alpha_2}(\lambda_1,\lambda_2), \tag{5.15}$$

$$d\big(x(\lambda_2,\mu_2),x_1'\big) \leq \ell_2 d^{\alpha_2}(\lambda_1,\lambda_2). \tag{5.16}$$

Again, due to the Hölder continuity of $\mathscr{D}$, we can pick

$$x_1'' \in \mathscr{D}(x(\lambda_1, \mu_2), \lambda_1) \quad \text{and} \quad x_2'' \in \mathscr{D}(x(\lambda_2, \mu_2), \lambda_2)$$

such that

$$d(x_1', x_1'') \leq \ell_1 d^{\alpha_1}\big(x(\lambda_1, \mu_2), x(\lambda_2, \mu_2)\big), \tag{5.17}$$

$$d(x_2', x_2'') \leq \ell_1 d^{\alpha_1}\big(x(\lambda_1, \mu_2), x(\lambda_2, \mu_2)\big). \tag{5.18}$$

By the definition of the considered problems, we have

$$\Phi(x(\lambda_1, \mu_2), x_1'', \mu_2) \geq 0, \tag{5.19}$$

$$\Phi(x(\lambda_2, \mu_2), x_2'', \mu_2) \geq 0. \tag{5.20}$$

It follows from assumption (c), (5.19), and (5.20) that

$$d\big(\Phi(x(\lambda_1, \mu_2), x_1'', \mu_2), \Phi(x(\lambda_1, \mu_2), x(\lambda_2, \mu_2), \mu_2)\big)$$
$$+ d\big(\Phi(x(\lambda_2, \mu_2), x_2'', \mu_2), \Phi(x(\lambda_2, \mu_2), x(\lambda_1, \mu_2), \mu_2)\big) \geq h d_2^{\beta},$$

and hence,

$$d\big(\Phi(x(\lambda_1, \mu_2), x_1'', \mu_2), \Phi(x(\lambda_1, \mu_2), x_1', \mu_2)\big)$$
$$+ d\big(\Phi(x(\lambda_1, \mu_2), x_1', \mu_2), \Phi(x(\lambda_1, \mu_2), x(\lambda_2, \mu_2), \mu_2)\big)$$
$$+ d\big(\Phi(x(\lambda_2, \mu_2), x_2'', \mu_2), \Phi(x(\lambda_2, \mu_2), x_2', \mu_2)\big)$$
$$+ d\big(\Phi(x(\lambda_2, \mu_2), x_2', \mu_2), \Phi(x(\lambda_2, \mu_2), x(\lambda_1, \mu_2), \mu_2)\big) \geq h d_2^{\beta}.$$

By the Hölder continuity of Φ assumed in (b), we obtain

$$\sigma d^{\delta}(x_1'', x_1') + \sigma d^{\delta}(x_1', x(\lambda_2, \mu_2)) + \sigma d^{\delta}(x_2'', x_2')$$
$$+ \sigma d^{\delta}(x_2', x(\lambda_1, \mu_2)) \geq h d_2^{\beta}.$$

From (5.15), (5.16), (5.17), and (5.18), we have

$$\sigma \ell_1^{\delta} d_2^{\alpha_1 \delta} + \sigma \ell_2^{\delta} d^{\alpha_2 \delta}(\lambda_1, \lambda_2) + \sigma \ell_1^{\delta} d_2^{\alpha_1 \delta} + \sigma \ell_2^{\delta} d^{\alpha_2 \delta}(\lambda_1, \lambda_2) \geq h d_2^{\beta}.$$

It follows from assumption (d) that

$$d_2^\beta \leq \left(\frac{2\sigma \ell_2^\delta}{h - 2\sigma \ell_1^\delta}\right) d^{\alpha_2\delta}(\lambda_1, \lambda_2).$$

Taking $k_2 = \left(\frac{2\sigma \ell_2^\delta}{h-2\sigma \ell_1^\delta}\right)^{\frac{1}{\beta}}$, one has

$$d_2 \leq k_2 d^{\frac{\alpha_2\delta}{\beta}}(\lambda_1, \lambda_2).$$

Step 3. Finally since, for all $x(\lambda_1, \mu_1) \in \mathrm{Sol(QEP)}(\lambda_1, \mu_1)$ and $x(\lambda_2, \mu_2) \in \mathrm{Sol(QEP)}\ (\lambda_2, \mu_2)$,

$$d\big(x(\lambda_1, \mu_1, \eta_1), x(\lambda_2, \mu_2, \eta_2)\big) \leq d_1 + d_2,$$

and consequently,

$$\mathbb{D}\big(\mathrm{Sol(QEP)}(\lambda_1, \mu_1), \mathrm{Sol(QEP)}(\lambda_2, \mu_2)\big) \leq d_1 + d_2.$$

Putting $(\lambda_1, \mu_1) = (\lambda_2, \mu_2)$ in this inequality, one sees that $\mathrm{Sol(QEP)}(\lambda_1, \mu_1)$ is a singleton. Similarly, $\mathrm{Sol(QEP)}(\lambda_2, \mu_2)$ is also a singleton. Thus, the problem has a unique solution, and then the Hölder condition claimed in the theorem is obtained. □

Using Proposition 5.9, we have the following results.

Corollary 5.11 (Hölder continuity of the solution maps of quasiequilibrium problems via conditions related to strong monotonicity and strong convexity) *The conclusion of Theorem 5.10 remains valid if the assumption (A) is replaced by one of the following conditions*

(a) $\Phi(\cdot, \cdot, \mu)$ *is Hölder strongly monotone on* $\mathscr{E}(\Lambda)$ *for all* $\mu \in M$;
(b) $-\Phi(\cdot, \cdot, \mu)$ *is quasimonotone on* $\mathscr{E}(\Lambda)$ *and* $\Phi(\cdot, \cdot, \mu)$ *is Hölder strongly pseudomonotone on* $\mathscr{E}(\Lambda)$ *for all* $\mu \in M$;
(c) $\mathscr{E}(\Lambda)$ *is convex,* Φ *is monotone, and* Φ *is both Hölder continuous and strongly convex in the second component on* $\mathscr{E}(\Lambda)$.

5.2.2 *Hölder Calmness of Solution Maps*

Theorem 5.12 (Hölder calmness of the solution maps of quasiequilibrium problems) *For the problem* $\mathrm{(QEP)}(\bar\lambda, \bar\mu)$, *assume that*

(a) *there are neighborhoods* $U(\bar\lambda)$ *of* $\bar\lambda$ *and* $V(\bar\mu)$ *of* $\bar\mu$ *such that* Φ *is* (ρ, γ)*-Hölder calm at* $\bar\mu$*,* θ*-uniformly in* $\mathscr{E}(U(\bar\lambda))$ *and, for all* $x \in \mathscr{E}(U(\bar\lambda))$ *and* $\mu \in V(\bar\mu)$, $\Phi(x, \cdot, \mu)$ *is* (σ, δ)*-Hölder in* $\mathscr{E}(U(\bar\lambda))$;

(b) *assumption (A) is satisfied;*
(c) *for all* $\lambda \in U(\bar{\lambda})$ *and* $x \in \mathscr{E}(U(\bar{\lambda}))$, $\mathscr{D}(\cdot, \lambda)$ *is* (ℓ_1, α_1)*-Hölder in* $\mathscr{E}(U(\bar{\lambda}))$ *and* $\mathscr{D}(x, \cdot)$ *is* (ℓ_2, α_2)*-Hölder calm at* $\bar{\lambda}$*;*
(d) $\alpha_1\delta = \beta > \theta$ *and* $h > 2\sigma l_1^{\delta}$.

Then, for each (λ, μ) *in a neighborhood of* $(\bar{\lambda}, \bar{\mu})$*, the solution map of (QEP)*(λ, μ) *satisfies the following Hölder calmness condition*

$$\mathbb{D}\big(\mathrm{Sol(QEP)}(\bar{\lambda}, \bar{\mu}), \mathrm{Sol(QEP)}(\lambda, \mu)\big) \leq k_1 d^{\alpha_2\delta/\beta}(\bar{\lambda}, \lambda) + k_2 d^{\gamma/(\beta-\theta)}(\bar{\mu}, \mu),$$

where k_1 *and* k_2 *are positive constants depending on* $h, \beta, \ell_1, \ell_2, \rho, \sigma, \gamma, \delta, \theta,$*, etc.*

Proof Let $(\lambda, \mu) \in U(\bar{\lambda}) \times V(\bar{\mu})$. We split the proof into three steps.
Step 1. For each $x(\bar{\lambda}, \bar{\mu}) \in \mathrm{Sol(QEP)}(\bar{\lambda}, \bar{\mu})$ and $x(\lambda, \bar{\mu}) \in \mathrm{Sol(QEP)}(\lambda, \bar{\mu})$, let us prove the inequality

$$d_1 := d\big(x(\bar{\lambda}, \bar{\mu}), x(\lambda, \bar{\mu})\big) \leq \left(\frac{2\sigma \ell_2^{\delta}}{h - 2\sigma \ell_1^{\delta}}\right)^{1/\beta} d^{\alpha_2\delta/\beta}(\bar{\lambda}, \lambda). \tag{5.21}$$

If $x(\bar{\lambda}, \bar{\mu}) = x(\lambda, \bar{\mu})$, then (5.21) is true. Consider the case $x(\bar{\lambda}, \bar{\mu}) \neq x(\lambda, \bar{\mu})$. Thanks to assumption (c), we can find $\bar{x} \in \mathscr{D}(x(\lambda, \bar{\mu}), \bar{\lambda})$ and $x \in \mathscr{D}(x(\bar{\lambda}, \bar{\mu}), \lambda)$ such that

$$\max\{d\big(x(\bar{\lambda}, \bar{\mu}), x\big), d\big(x(\lambda, \bar{\mu}), \bar{x}\big)\} \leq \ell_2 d^{\alpha_2}(\bar{\lambda}, \lambda). \tag{5.22}$$

From the Hölder continuity of $\mathscr{D}$ given in assumption (c), there are $\bar{x}' \in \mathscr{D}(x(\bar{\lambda}, \bar{\mu}), \bar{\lambda})$ and $x' \in \mathscr{D}(x(\lambda, \bar{\mu}), \lambda)$ such that

$$\max\{d(\bar{x}, \bar{x}'), d(x, x')\} \leq \ell_1 d^{\alpha_1}\big(x(\bar{\lambda}, \bar{\mu}), x(\lambda, \bar{\mu})\big). \tag{5.23}$$

Being solutions of (QEP)$(\bar{\lambda}, \bar{\mu})$ and (QEP)$(\lambda, \bar{\mu})$, resp., the vectors $x(\bar{\lambda}, \bar{\mu})$ and $x(\lambda, \bar{\mu})$ satisfy the following inequalities

$$\Phi(x(\bar{\lambda}, \bar{\mu}), \bar{x}', \bar{\mu}) \geq 0, \quad \Phi(x(\lambda, \bar{\mu}), x', \bar{\mu}) \geq 0. \tag{5.24}$$

By assumption (b),

$$d(\Phi(x(\bar{\lambda}, \bar{\mu}), x(\lambda, \bar{\mu}), \bar{\mu}), R_+) + d(\Phi(x(\lambda, \bar{\mu}), x(\bar{\lambda}, \bar{\mu}), \bar{\mu}), R_+) \geq h d_1^{\beta}. \tag{5.25}$$

From (5.24) and (5.25), we obtain

$$\begin{aligned} & d(\Phi(x(\bar{\lambda}, \bar{\mu}), x(\lambda, \bar{\mu}), \bar{\mu}), \Phi(x(\bar{\lambda}, \bar{\mu}), \bar{x}, \bar{\mu})) \\ & \quad + d(\Phi(x(\bar{\lambda}, \bar{\mu}), \bar{x}, \bar{\mu}), \Phi(x(\bar{\lambda}, \bar{\mu}), \bar{x}', \bar{\mu})) \end{aligned}$$

$$+ d(\Phi(x(\lambda,\bar\mu), x(\bar\lambda,\bar\mu),\bar\mu), \Phi(x(\lambda,\bar\mu), x, \bar\mu))$$
$$+ d(\Phi(x(\lambda,\bar\mu), x, \bar\mu), \Phi(x(\lambda,\bar\mu), x', \bar\mu)) \geq h d_1^\beta.$$

Consequently, the Hölder continuity assumed in (a) for Φ implies that

$$\sigma d^\delta(x(\lambda,\bar\mu),\bar x) + \sigma d^\delta(\bar x,\bar x') + \sigma d^\delta(x(\bar\lambda,\bar\mu), x) + \sigma d^\delta(x, x') \geq h d_1^\beta.$$

Using (5.22) and (5.23), we get

$$\sigma \ell_2^\delta d^{\alpha_2\delta}(\bar\lambda,\lambda) + \sigma\ell_1^\delta d_1^{\alpha_1\delta} + \sigma\ell_2^\delta d^{\alpha_2\delta}(\bar\lambda,\lambda) + \sigma\ell_1^\delta d_1^{\alpha_1\delta} \geq h d_1^\beta.$$

This means that (5.21) follows due to

$$d_1^\beta \leq \left(\frac{2\sigma\ell_2^\delta}{h - 2\sigma\ell_1^\delta}\right) d^{\alpha_2\delta}(\bar\lambda,\lambda).$$

Step 2. We now claim that, for each vector

$$x(\lambda,\bar\mu) \in \mathrm{Sol(QEP)}(\lambda,\bar\mu) \quad \text{and} \quad x(\lambda,\mu) \in \mathrm{Sol(QEP)}(\lambda,\mu),$$

the following inequality holds true:

$$d_2 := d\big(x(\lambda,\bar\mu), x(\lambda,\mu)\big) \leq \left(\frac{\rho}{h - 2\sigma\ell_1^\delta}\right)^{1/(\beta-\theta)} d^{\gamma/(\beta-\theta)}(\bar\mu,\mu). \tag{5.26}$$

Obviously, we only need to prove that the inequality (5.26) is valid when $x(\lambda,\bar\mu) \neq x(\lambda,\mu)$. Since $x(\lambda,\bar\mu) \in \mathscr{D}(x(\lambda,\bar\mu),\lambda)$ and $x(\lambda,\mu) \in \mathscr{D}(x(\lambda,\mu),\lambda)$, by the Hölder continuity of $\mathscr{D}(\cdot,\lambda)$, there exist $\bar x'' \in \mathscr{D}(x(\lambda,\bar\mu),\lambda)$ and $x'' \in \mathscr{D}(x(\lambda,\mu),\lambda)$ such that

$$\max\left\{d\big(x(\lambda,\bar\mu), x''\big), d\big(x(\lambda,\mu),\bar x''\big)\right\} \leq \ell_1 d^{\alpha_1}\big(x(\lambda,\bar\mu), x(\lambda,\mu)\big). \tag{5.27}$$

As $x(\lambda,\bar\mu)$ and $x(\lambda,\mu)$ are solutions of $\mathrm{(QEP)}(\lambda,\bar\mu)$ and $\mathrm{(QEP)}(\lambda,\mu)$, resp.,

$$\Phi(x(\lambda,\bar\mu),\bar x'',\bar\mu) \geq 0, \quad \Phi(x(\lambda,\mu), x'',\mu) \geq 0. \tag{5.28}$$

On the other hand, assumption (b) implies that

$$d(\Phi(x(\lambda,\bar\mu), x(\lambda,\mu),\bar\mu), R_+) + d(\Phi(x(\lambda,\mu), x(\lambda,\bar\mu),\bar\mu), R_+) \geq h d_2^\beta.$$

Then, inequality (5.28) leads to

$$d(\Phi(x(\lambda,\bar{\mu}),x(\lambda,\mu),\bar{\mu}),\Phi(x(\lambda,\bar{\mu}),\bar{x}'',\bar{\mu}))$$
$$+\,d(\Phi(x(\lambda,\mu),x(\lambda,\bar{\mu}),\bar{\mu}),\Phi(x(\lambda,\mu),x(\lambda,\bar{\mu}),\mu))$$
$$+\,d(\Phi(x(\lambda,\mu),x(\lambda,\bar{\mu}),\mu),\Phi(x(\lambda,\mu),x'',\mu))\geq hd_2^{\beta}.$$

Therefore, from assumption (a), we deduce that

$$\sigma d^{\delta}(x(\lambda,\mu),\bar{x}'')+\rho d_2^{\theta}d^{\gamma}(\bar{\mu},\mu)+\sigma d^{\delta}(x(\lambda,\bar{\mu}),x'')\geq hd_2^{\beta}.$$

This inequality together with (5.27) implies that

$$\sigma\ell_1^{\delta}d_2^{\alpha_1\delta}+\rho d_2^{\theta}d^{\gamma}(\bar{\mu},\mu)+\sigma\ell_1^{\delta}d_1^{\alpha_1\delta}\geq hd_2^{\beta}.$$

Thus, by assumption (d), (5.26) is obtained since

$$d_2^{\beta-\theta}\leq\left(\frac{\rho}{h-2\sigma\ell_1^{\delta}}\right)d^{\gamma}(\bar{\mu},\mu).$$

Step 3. Now, we are ready to complete the proof. For all $x(\bar{\lambda},\bar{\mu})\in\mathrm{Sol(QEP)}(\bar{\lambda},\bar{\mu})$ and $x(\lambda,\mu)\in\mathrm{Sol(QEP)}(\lambda,\mu)$, combining the results of Steps 1 and 2, we obtain

$$d\big(x(\bar{\lambda},\bar{\mu}),x(\lambda,\mu)\big)\leq d_1+d_2.$$

Using (5.26) and (5.21) we get, with $k_1=\left(\frac{2\sigma\ell_2^{\delta}}{h-2\sigma\ell_1^{\delta}}\right)^{\frac{1}{\beta}}$ and $k_2=\left(\frac{\rho}{h-2\sigma\ell_1^{\delta}}\right)^{\frac{1}{\beta-\theta}}$,

$$\mathbb{D}\rho\big(S^s(\bar{\lambda},\bar{\mu}),S^s(\lambda,\mu)\big)\leq k_1d^{\alpha_2\delta/\beta}(\bar{\lambda},\lambda)+k_2d^{\gamma/(\beta-\theta)}(\bar{\mu},\mu).$$

The proof is thereby complete. □

Using Proposition 5.9, we have the following results.

Corollary 5.13 (Hölder calmness of the solution maps of quasiequilibrium problems via conditions related to strong monotonicity and strong convexity) *The conclusion of Theorem 5.12 is still valid if assumption (A) is replaced by one of the following conditions:*

(a) *$\Phi(\cdot,\cdot,\mu)$ is $h.\beta$-strongly monotone on $\mathscr{E}(\Lambda)$ for all $\mu\in M$;*
(b) *$-\varphi(\cdot,\cdot,\mu)$ is quasimonotone on $\mathscr{E}(\Lambda)$ and $\varphi(\cdot,\cdot,\mu)$ is $h.\beta$-strongly pseudomonotone on $\mathscr{E}(\Lambda)$ for all $\mu\in M$;*
(c) *$\mathscr{E}(\Lambda)$ is convex, Φ is monotone, and Φ is both Hölder continuous and strongly convex in the second component on $\mathscr{E}(\Lambda)$.*

5.3 Well-Posedness of (QEPs)

Let X be a metric space and A, Λ, M nonempty closed subsets of X. We first recall well-posedness notions.

Definition 5.14 (Approximating sequences for quasiequilibrium problems) Let $\{(\lambda^k, \mu^k)\}_k$ converge to $(\bar{\lambda}, \bar{\mu})$. For $x^k \in \mathscr{D}(x^k, \lambda^k)$, the sequence $\{x^k\}_k$ is said to be an APPROXIMATING SEQUENCE for (QEP) corresponding to $\{(\lambda^k, \mu^k)\}_k$ if there exists a sequence of nonnegative real numbers $\{\varepsilon^k\}_k$ converging to 0 such that, for all $y \in \mathscr{D}(x^k, \lambda^k)$,

$$\Phi(x^k, y, \mu^k) + \varepsilon^k \geq 0.$$

Definition 5.15 (Well-posedness for quasiequilibrium problems) Problem (QEP) is said to be WELL-POSED at $(\bar{\lambda}, \bar{\mu})$ if

(a) the solution set $\mathrm{Sol(QEP)}(\bar{\lambda}, \bar{\mu})$ of $\mathrm{(QEP)}(\bar{\lambda}, \bar{\mu})$ is nonempty;
(b) for any sequence $\{(\lambda^k, \mu^k)\}_k$ converging to $(\bar{\lambda}, \bar{\mu})$, every corresponding approximating sequence for (QEP) has a subsequence converging to some point of $\mathrm{Sol(QEP)}(\bar{\lambda}, \bar{\mu})$.

(QEP) is said to be UNIQUELY WELL-POSED at $(\bar{\lambda}, \bar{\mu})$ if $\mathrm{Sol(QEP)}(\bar{\lambda}, \bar{\mu}) = \{\bar{x}\}$, a singleton, and every approximating sequence converges to $\bar{x}$. (QEP) (or any other problem) is called (uniquely) well-posed if it is (uniquely) well-posed at each $(\lambda, \mu) \in \Lambda \times M$.

5.3.1 *Generalized Level Closedness and Pseudocontinuity of Functions*

Definition 5.16 (Generalized level closedness of functions) Let X and Y be topological spaces, $\Phi : X \to \bar{R}$, and $\Psi : Y \to \bar{R}$.

(a) Φ is called (SEQUENTIALLY) UPPER 0-LEVEL CLOSED with respect to Ψ at $(x^0, y^0) \in X \times Y$ if, for any sequence $\{(x^k, y^k)\}_k$ converging to (x^0, y^0),

$$[\Phi(x^k) + \Psi(y^k) \geq 0, \forall k] \Rightarrow [\Phi(x^0) + \Psi(y^0) \geq 0].$$

(b) Φ is called (SEQUENTIALLY) LOWER 0-LEVEL CLOSED wrt Ψ at (x^0, y^0) if, for any sequence $\{(x^k, y^k)\}_k$ converging to (x^0, y^0),

$$[\Phi(x^k) + \Psi(y^k) \leq 0, \forall k] \Rightarrow [\Phi(x^0) + \Psi(y^0) \leq 0].$$

If we have Φ in the place of $\Phi + \Psi$ in the aforementioned inequalities, we say that Φ is UPPER (or LOWER) 0-LEVEL CLOSED at x_0. While if we have $b \in \mathbb{R}$ instead of 0, then of course "0-level" is replaced by "b-level."

Remark 5.17 (A special case of generalized level-closed functions) If Φ and Ψ are usc (lsc, resp.) at x^0 and y^0, resp., then Φ is upper (lower, resp.) 0-level closed wrt Ψ at (x_0, y_0). Indeed, if $(x^k, y^k) \to (x^0, y^0)$ and

$$\Phi(x^k) + \Psi(y^k) \geq 0,$$

for all k, one has

$$\begin{aligned}\Phi(x_0) + \Psi(y_0) &\geq \limsup \Phi(x^k) + \limsup \Psi(y^k)\\ &\geq \limsup[\Phi(x^k) + \Psi(y^k)] \geq 0.\end{aligned}$$

From now on, we use id to denote the identity map on $\mathbb{R}_+$. The following example shows that the converse of the aforementioned remark is not true.

Example 5.18 (A difference between the class of generalized level-closed functions and that of semicontinuous functions) Let $\Phi : \mathbb{R} \to \mathbb{R}$ be defined by

$$\Phi(x) := \begin{cases} 0 & \text{if } x \in \mathbb{Q}, \\ 1 & \text{if } x \in \mathbb{R} \setminus \mathbb{Q}, \end{cases}$$

where $\mathbb{Q}$ is the set of all rational numbers. Then, Φ is upper 0-level closed wrt id at every $(x, y) \in \mathbb{R} \times \mathbb{R}_+$, but Φ is neither usc at any $x \in \mathbb{Q}$ nor lsc at any $x \in \mathbb{R} \setminus \mathbb{Q}$.

Definition 5.19 (Pseudocontinuous functions) Let X be a topological space and $\Phi : X \to \mathbb{R} \cup \{\infty\}$.

(a) Φ is said to be (SEQUENTIALLY) UPPER PSEUDOCONTINUOUS at $x_0 \in X$ if

$$[\Phi(x) > \Phi(x_0)] \Rightarrow [\text{for any } \{x^k\} \to x_0, \Phi(x) > \limsup \Phi(x^k)].$$

(b) Φ is said to be LOWER PSEUDOCONTINUOUS at $x_0 \in X$ if

$$[\Phi(x) < f(x_0)] \Rightarrow [\text{for any } \{x^k\} \to x_0, \Phi(x) < \liminf \Phi(x^k)].$$

(c) Φ is said to be PSEUDOCONTINUOUS at $x_0 \in X$ if it is both lower and upper pseudocontinuous at this point.

The class of the upper pseudocontinuous functions strictly contains that of the usc functions. We include here a new simple illustrative example.

Example 5.20 (The class of the upper pseudocontinuous functions strictly contains that of the upper semicontinuous functions) Let $\Phi : \mathbb{R} \to \mathbb{R}$ be defined by

$$\Phi(x) := \begin{cases} x+1 & \text{if} \quad x > 0, \\ 0 & \text{if} \quad x = 0, \\ x-1 & \text{if} \quad x < 0. \end{cases}$$

Then, Φ is pseudocontinuous at 0, but neither usc nor lsc at 0.

We observe further that if Φ and Ψ are lsc (or usc) at x_0, then $\Phi + \Psi$ is lsc (usc, resp.) at x_0. Unfortunately, this property does not hold for pseudocontinuous functions as shown by the following.

Example 5.21 (The sum of two pseudocontinuous functions) Let $\Phi_1, \Psi_1 : \mathbb{R} \to \mathbb{R}$ be defined as follows

$$\Phi_1(x) := \begin{cases} 1 & \text{if} \quad x \geq 0, \\ \frac{x}{2} & \text{if} \quad x < 0, \end{cases} \qquad \text{and} \quad \Psi_1(x) = -x.$$

Then, Φ_1 is lower pseudocontinuous at 0 and Ψ_1 is continuous at 0. But,

$$(\Phi_1 + \Psi_1)(x) = \begin{cases} -x+1 & \text{if} \quad x \geq 0, \\ -\frac{x}{2} & \text{if} \quad x < 0 \end{cases}$$

is not lower pseudocontinuous at 0.

To show the same situation for upper pseudocontinuity, let

$$\Phi_2(x) := \begin{cases} -1 & \text{if} \quad x \geq 0, \\ -\frac{x}{2} & \text{if} \quad x < 0, \end{cases} \qquad \text{and} \quad \Psi_2(x) = x.$$

Then, at 0, Φ_2 is upper pseudocontinuous and Ψ_2 is continuous. However,

$$(\Phi_2 + \Psi_2)(x) = \begin{cases} x-1 & \text{if} \quad x \geq 0, \\ \frac{x}{2} & \text{if} \quad x < 0 \end{cases}$$

is not upper pseudocontinuous at 0.

Lemma 5.22 (A characterization of pseudocontinuity of functions in terms of inequality preservation) *Let X be a topological space. Then, $\Phi : X \to \mathbb{R} \cup \{\infty\}$ is pseudocontinuous in X if and only if, for all sequences $\{x^k\}$ and $\{y^k\}$ in X converging to x and y, resp.,*

$$[\Phi(x) < \Phi(y)] \Rightarrow [\limsup \Phi(x^k) < \liminf \Phi(y^k)]. \tag{5.29}$$

Proof Assume first that Φ is pseudocontinuous on X. Let $\Phi(x) < \Phi(y)$, $x^k \to x$, and $y^k \to y$. If there exists a valuc $\Phi(z)$ satisfying $\Phi(x) < \Phi(z) < \Phi(y)$, then we have

$$\limsup \Phi(x^k) < \Phi(z) < \liminf \Phi(y^k).$$

Otherwise, we have

$$(\Phi(x)), \Phi(y)) \cap \text{rang}(\Phi) = \emptyset,$$

where $\text{im}\Phi := \{\Phi(z) \mid z \in X\}$. Since Φ is upper pseudocontinuous at x, we have

$$\limsup \Phi(x^k) < f(y).$$

Now, if $\Phi(x) < \limsup \Phi(x^k)$, then

$$(\Phi(x), \Phi(y)) \cap \text{rang}(\Phi) \neq \emptyset,$$

which is in conflict with the aforementioned assumption. Thus,

$$\limsup \Phi(x^k) \leq \Phi(x) < \Phi(y).$$

Similarly, if Φ is lower pseudocontinuous at y, we get

$$\Phi(y) \leq \liminf \Phi(y^k),$$

and then

$$\limsup \Phi(x^k) \leq \Phi(x) < \Phi(y) \leq \liminf \Phi(y^k),$$

and consequently, (5.29) holds true.

Conversely, we assume that (5.29) is satisfied, $\Phi(x) < \Phi(y)$, $x^k \to x$, and $(y^k)_k$ with $y^k = y$ for all k. Then,

$$\limsup \Phi(x^k) < \Phi(y),$$

or equivalently Φ is upper pseudocontinuous at x. Similarly, one can prove that (5.29) also implies the lower pseudocontinuity of Φ at y. □

5.3.2 Well-Posedness of (QEPs)

In this subsection, we provide well-posedness conditions and discuss metric characterizations. Assume that A, Λ, M are nonempty subsets of a metric space X. For positive ε, the ε-solution set of (QEP)(λ, μ) is defined by

$$\widetilde{\mathrm{Sol}}(\mathrm{QEP})(\varepsilon, \lambda, \mu) := \{x \in \mathscr{D}(x, \lambda) \mid \Phi(x, y, \mu) + \varepsilon \geq 0, \forall y \in \mathscr{D}(x, \lambda)\}.$$

When X and Λ are metric spaces, for positive δ and ε, we define the following set of approximate solutions of the family (QEP), allowing also the parameters to vary around the considered point,

$$\Pi(\delta, \varepsilon, \bar{\lambda}, \bar{\mu}) := \bigcup_{\lambda \in B(\bar{\lambda}, \delta), \mu \in B(\bar{\mu}, \delta)} \widetilde{\mathrm{Sol}}(\mathrm{QEP})(\varepsilon, \lambda, \mu),$$

where $B(\cdot, \delta)$ is the closed ball centered at $(\cdot)$ with radius δ.

Theorem 5.23 *[Well-posedness conditions] Assume that*

(a) *A is compact and $\mathscr{D}$ is closed and lsc on $A \times \Lambda$;*
(b) *Φ is upper 0-level closed wrt id on $A \times A \times M$.*

Then, (QEP) is well-posed. Furthermore, if Sol(QEP) *is a single-valued map, then this problem is uniquely well-posed.*

Proof We first check that $\widetilde{\mathrm{Sol}}(\mathrm{QEP})$ is usc at $(0, \bar{\lambda}, \bar{\mu})$ for each $(\bar{\lambda}, \bar{\mu}) \in \Lambda \times M$. Suppose to the contrary that there exist an open neighborhood U of $\widetilde{\mathrm{Sol}}(\mathrm{QEP})(0, \bar{\lambda}, \bar{\mu})$, a sequence $\{(\lambda^k, \mu^k)\}_k \subset \Lambda \times M$, and nonnegative real numbers $\{\varepsilon^k\}$ converging to $(\bar{\lambda}, \bar{\mu})$ and 0, resp., such that for each k, there is $x^k \in \widetilde{\mathrm{Sol}}(\mathrm{QEP})(\varepsilon^k, \lambda^k, \mu^k)$ and $x^k \notin U$. By the compactness of A, we can assume that x^k converge to some point $\bar{x}$ in A.

Thanks to Theorem 5.1, the map $\mathscr{E}$ is closed at $\bar{\lambda}$, and hence, $\bar{x} \in \mathscr{E}(\bar{\lambda})$. If

$$\bar{x} \notin \widetilde{\mathrm{Sol}}(\mathrm{QEP})(0, \bar{\lambda}, \bar{\mu}) = S(\bar{\lambda}, \bar{\mu}),$$

then there is $\bar{y} \in \mathscr{D}(\bar{x}, \bar{\lambda})$ such that $\Phi(\bar{x}, \bar{y}, \bar{\mu}) < 0$. The lower semicontinuity of $\mathscr{D}$ in turn shows the existence of points $y_n \in \mathscr{D}(x^k, \lambda^k)$ such that y^k converge to $\bar{y}$. On the other hand, as x^k belongs to $\widetilde{\mathrm{Sol}}(\mathrm{QEP})(\varepsilon^k, \lambda^k, \mu^k)$, we have

$$\Phi(x^k, y^k, \mu^k) + \varepsilon^k \geq 0.$$

By the upper 0-level closedness of Φ wrt the map id, we have $\Phi(\bar{x}, \bar{y}, \bar{\mu}) \geq 0,$, which is a contradiction. Thus, $\bar{x} \in \widetilde{\mathrm{Sol}}(\mathrm{QEP})(0, \bar{\lambda}, \bar{\mu}) \subset U$, which is another contradiction as $x^k \notin U$ for all k. Hence, $\widetilde{\mathrm{Sol}}(\mathrm{QEP})$ is usc at $(0, \bar{\lambda}, \bar{\mu})$.

Now we prove that $\mathrm{Sol}(\mathrm{QEP})(\bar{\lambda}, \bar{\mu})$ is compact by checking its closedness. Let $\hat{x}^k \in \mathrm{Sol}(\mathrm{QEP})(\bar{\lambda}, \bar{\mu})$ converge to some point $\hat{x} \in X$. If $\hat{x} \notin \mathrm{Sol}(\mathrm{QEP})(\bar{\lambda}, \bar{\mu})$, then there exists $\hat{y} \in \mathscr{D}(\hat{x}, \bar{\lambda})$ such that $\Phi(\hat{x}, \hat{y}, \bar{\mu}) < 0$. In light of the lower semicontinuity of $\mathscr{D}$, there is $\hat{y}^k \in \mathscr{D}(\hat{x}^k, \bar{\lambda})$ such that $\hat{y}^k$ converge to $\hat{y}$. Then, for all k, we always have $\Phi(\hat{x}^k, \hat{y}^k, \bar{\lambda}) \geq 0$ as $\hat{x}^k \in \mathrm{Sol}(\mathrm{QEP})(\bar{\lambda}, \bar{\mu})$. By the second assumption, we get $\Phi(\hat{x}, \hat{y}, \bar{\lambda}) \geq 0$, which is impossible. Therefore, $\hat{x} \in \mathrm{Sol(QEP)}(\bar{\lambda}, \bar{\mu})$, and hence, $\mathrm{Sol(QEP)}(\bar{\lambda}, \bar{\mu})$ is compact. □

The assumptions of Theorem 5.23 are essential as indicated in the following examples.

Example 5.24 (The compactness of A cannot be dropped) Let $X = A = \mathbb{R}$, $\Lambda = M = \mathbb{R}_+$, $\mathbb{D}(x, \lambda) = \mathbb{R}$, $\bar{\lambda} = \bar{\mu} = 0$, and $\Phi(x, y, \mu) = 2^{x-y} + \mu$.

It is clear that on $A \times \Lambda$, $\mathscr{D}$ is closed lsc and (b) holds as Φ is continuous on $A \times A \times M$. But, $\mathrm{Sol(QEP)}(\lambda, \mu) = \mathbb{R}$ for all $\lambda \in \Lambda, \mu \in M$. Hence, (QEP) is not well-posed at 0. Indeed, by letting $\lambda^k = \frac{1}{k}$ and $\mu^k = \frac{1}{k}$, we see that $x^k := k \in \mathrm{Sol(QEP)}(\bar{\lambda}^k, \mu^k)$ for all k. However, $\{x^k\}_k$ has no convergent subsequence. The reason is that A is not compact.

Example 5.25 (the closedness of $\mathscr{D}$ is essential). Let $X = \mathbb{R}$, $A = [-2, 1]$, $\Lambda = M = [0, 1]$, $\mathscr{D}(x, \lambda) = (-2\lambda, 1]$, $\bar{\lambda} = \bar{\mu} = 0$, and $\Phi(x, y, \mu) = x(x - y)2^{\mu}$.

It is not hard to see that A is compact, $\mathscr{D}$ is lsc on $A \times \Lambda$, and (b) is fulfilled. But,

$$\mathrm{Sol(QEP)}(0, 0) = \{1\} \quad \text{and} \quad \mathrm{Sol(QEP)}\left(\frac{1}{k}, \frac{1}{k}\right) = \{0, 1\},$$

for all k. Therefore, (QEP) is not well-posed at $(0, 0)$. The reason is that $\mathscr{D}$ is not closed. Indeed, by letting $x^k = \lambda^k = \frac{1}{k}$ and $z^k = -\frac{1}{k} \in \mathscr{D}(x^k, \lambda^k) = (-\frac{2}{k}, 1]$, we see that the sequence $\{z^k\}_k$ converges to $0 \notin \mathscr{D}(0, 0)$.

Example 5.26 (The lower semicontinuity of $\mathscr{D}$ cannot be dispensed) Let $X = \mathbb{R}$, $A = [-1, 1]$, $\Lambda = M = [0, 1]$, $\Phi(x, y, \mu) = (x + y)2^{\mu}$, $\bar{\lambda} = \bar{\mu} = 0$, and

$$\mathscr{D}(x, \lambda) = \begin{cases} \{-1, 0, 1\} & \text{if } \lambda = 0, \\ \{0, 1\} & \text{otherwise.} \end{cases}$$

Then, A is compact, $\mathscr{D}$ is closed on $A \times \Lambda$, and (b) holds. But,

$$\mathrm{Sol(QEP)}(0, 0) = \{1\} \quad \text{and} \quad \mathrm{Sol(QEP)}\left(\frac{1}{k}, \frac{1}{k}\right) = \{0, 1\},$$

for all k. Thus, (QEP) is not well-posed at 0. The reason is that $\mathscr{D}$ is not lsc.

Example 5.27 (the generalized level closedness of the objective function cannot be dropped). Let $X = \mathbb{R}$, $A = [0, 1]$, $\Lambda = M = [0, 1]$, $\mathscr{D} = [0, 1]$, and

$$\Phi(x, y, \mu) = \begin{cases} x - y & \text{if } \mu = 0, \\ y - x & \text{otherwise.} \end{cases}$$

It is clear that assumption (a) is satisfied and $\mathrm{Sol(QEP)}(0, 0) = \{1\}$. Let

$$\lambda^k = \mu^k = \varepsilon^k = \frac{1}{k}.$$

The points $x^k \equiv 0 \in \widetilde{\mathrm{Sol}}(\mathrm{QEP})(\varepsilon^k, \lambda^k, \mu^k)$. Therefore, the sequence $\{x^k\}_k$ is an approximating sequence for (QEP) corresponding to $\{\lambda^k, \mu^k\}$. But, $x^k \to 0 \notin \mathrm{Sol(QEP)}(0, 0)$, and hence the reference problem is not well-posed at $\bar{\lambda} = \bar{\mu} = 0$. The reason is that assumption (b) is violated. Indeed, taking $x^k \equiv 0$, $y^k \equiv 1$, $\mu^k \equiv \frac{1}{k}$, and $\varepsilon^k \equiv 0$, one has $\{(x^k, y^k, \lambda^k, \varepsilon^k)\} \to (0, 1, 0, 0)$ and

$$\Phi(x^k, y^k, \lambda^k) + \varepsilon^k = \Phi\left(0, 1, \frac{1}{k}\right) = 1 > 0,$$

but $\Phi(0, 1, 0) = -1 < 0$.

Next we present some metric characterizations of well-posedness.

Theorem 5.28 (Characterizations of unique well-posedness) *For (QEP), the following statements hold true.*

(a) *If (QEP) is uniquely well-posed, then*

$$\mathrm{diam}\Pi(\delta, \varepsilon, \lambda, \mu) \to 0^+ \quad \text{as} \quad (\delta, \varepsilon) \to (0^+, 0^+)$$

for each $(\lambda, \mu) \in \Lambda \times M$.

(b) *Conversely, if X is complete and the following conditions hold*

- $\mathscr{D}$ *is closed and lsc on* $A \times \Lambda$;
- Φ *is upper* 0*-level closed with respect to id on* $A \times A \times M$,

then (QEP) is uniquely well-posed, provided that

$$\mathrm{diam}\Pi(\delta, \varepsilon, \lambda, \mu) \to 0^+ \quad \text{as} \quad (\delta, \varepsilon) \to (0^+, 0^+)$$

for all $(\lambda, \mu) \in \Lambda \times M$.

Proof

(a) Suppose (QEP) is uniquely well-posed and $(\bar{\lambda}, \bar{\mu}) \in \Lambda \times M$, but there is some sequence $\{(\delta^k, \varepsilon^k)\}_k \subset \mathbb{R}_+ \times \mathbb{R}_+$ converging to $(0, 0)$ such that there are $k_0 \in \mathbb{N}$ and $r > 0$ such that, for all $k \geq k_0$,

$$\mathrm{diam}\Pi(\delta^k, \varepsilon^k, \bar{\lambda}, \bar{\mu}) > r.$$

Then, there are $x_1^k, x_2^k \in \Pi(\delta^k, \varepsilon^k, \bar{\lambda}, \bar{\mu})$ such that $d(x_1^k, x_2^k) > \frac{r}{2}$. Consequently, there exist $\lambda_1^k, \lambda_2^k \in B(\bar{\lambda}, \delta^k)$ and $\mu_1^k, \mu_2^k \in B(\bar{\mu}, \delta^k)$ such that

$$\Phi\left(x_1^k, y, \mu_1^k\right) + \varepsilon^k \geq 0, \forall y \in K\left(x_1^k, \lambda_1^k\right)$$

and

$$\Phi\left(x_2^k, y, \mu_2^k\right) + \varepsilon^k \geq 0, \forall y \in K\left(x_2^k, \lambda_2^k\right).$$

That is $\{x_1^k\}_k$ and $\{x_2^k\}_k$ are approximating sequences for (QEP) corresponding to $\{(\lambda_1^k, \mu_1^k)\}_k$ and $\{(\lambda_2^k, \mu_2^k)\}_k$, resp. Hence, x_1^k and x_2^k converge to the unique solution of (QEP)$(\bar{\lambda}, \bar{\mu})$, contradicting the fact that $d(x_1^k, x_2^k) > \frac{r}{2} > 0$ for all k.

(b) Let $\{(\lambda^k, \mu^k)\} \to (\bar{\lambda}, \bar{\mu})$ and $\{x^k\}_k$ be an approximating sequence for (QEP) corresponding to $\{(\lambda^k, \mu^k)\}_k$. Then, there are $\varepsilon^k \to 0^+$ such that, for all $y \in \mathcal{D}(x^k, \lambda^k)$ and all $k \in \mathbb{N}$,

$$\Phi(x^k, y, \mu^k) + \varepsilon^k \geq 0.$$

Consequently, x^k belongs to $\Pi(\delta^k, \varepsilon^k, \bar{\lambda}, \bar{\mu})$ with

$$\delta^k := d((\lambda^k, \mu^k), (\bar{\lambda}, \bar{\mu})) := \max\{d(\lambda^k, \bar{\lambda}), d(\mu^k, \bar{\mu})\},$$

which converge to 0^+ as $k \to +\infty$. Because $\operatorname{diam}\Pi(\delta^k \varepsilon^k, \bar{\lambda}, \bar{\mu}) \to 0^+$, $\{x^k\}_k$ is a Cauchy sequence and hence converges to some $\bar{x}$. By the closedness of the map $\mathcal{D}$ at $(\bar{x}, \bar{\lambda})$, we have $\bar{x} \in \mathcal{D}(\bar{x}, \bar{\lambda})$. Using the same argument as that in the proof of Theorem 5.23, we deduce that $\bar{x}$ belongs to Sol(QEP)$(\bar{\lambda}, \bar{\mu})$.

To complete the proof, we show that (QEP)$(\bar{\lambda}, \bar{\mu})$ has a unique solution. If Sol(QEP)$(\bar{\lambda}, \bar{\mu})$ has two distinct points $\bar{x}_1$ and $\bar{x}_2$, it is not hard to see that $\bar{x}_1$ and $\bar{x}_2$ belong to $\Pi(\delta, \varepsilon, \bar{\lambda}, \bar{\mu})$ for all positive δ and ε. It follows that

$$0 < d(\bar{x}_1, \bar{x}_2) \leq \operatorname{diam}\Pi(\delta, \varepsilon, \bar{\lambda}, \bar{\mu}),$$

which is impossible. □

We will use γ in the sequel to denote either the Hausdorff measure or the Kuratowski measure of noncompactness.

Theorem 5.29 (Characterizations of well-posedness in terms of measures of noncompactness) *For (QEP), the following statements hold true.*

(a) *If (QEP) is well-posed, then*

$$\gamma(\Pi(\delta, \varepsilon, \bar{\lambda}, \bar{\mu})) \to 0^+ \text{ as } (\delta, \varepsilon) \to (0^+, 0^+)$$

for all $(\bar{\lambda}, \bar{\mu}) \in \Lambda \times M$.

(b) *Conversely, if X is complete, Λ, M are compact (or finite dimensional if they are normed spaces), and the following conditions hold*

(a) $\mathscr{D}$ *is closed and lsc on* $A \times \Lambda$;
(b) Φ *is upper b-level closed on* $A \times A \times M$, *for all* $b < 0$, *i.e., the set* $\{(x, y, \mu) \in A \times A \times M \mid \Phi(x, y, \mu) \geq b\}$ *is closed for all* $b < 0$.

Then, (QEP) is well-posed, provided that

$$\gamma(\Pi(\delta, \varepsilon, \bar{\lambda}, \bar{\mu})) \to 0^+ \text{ as } (\delta, \varepsilon) \to (0^+, 0^+)$$

for all $(\bar{\lambda}, \bar{\mu}) \in \Lambda \times M$.

Proof Let γ be the Hausdorff measure η (for the Kuratowski measure case, the argument is similar).

(a) Assume that (QEP) is well-posed, $(\bar{\lambda}, \bar{\mu}) \in \Lambda \times M$, and $(\delta, \varepsilon) \to (0^+, 0^+)$. Since

$$\text{Sol(QEP)}(\bar{\lambda}, \bar{\mu}) \subset \Pi(\delta, \varepsilon, \bar{\lambda}, \bar{\mu})$$

for all $\delta, \varepsilon > 0$, we have

$$H(\Pi(\delta, \varepsilon, \bar{\lambda}, \bar{\mu}), \text{Sol(QEP)}(\bar{\lambda}, \bar{\mu})) = \text{ex}(\Pi(\delta, \varepsilon, \bar{\lambda}, \bar{\mu}), \text{Sol(QEP)}(\bar{\lambda}, \bar{\mu})),$$

where $\text{ex}(A, B) := \sup_{a \in A} d(a, B)$ and $d(a, B) := \inf_{b \in B} d(a, b)$. Let $\{x^k\}_k$ be any sequence in Sol(QEP)$(\bar{\lambda}, \bar{\mu})$. Since $\{x^k\}_k$ is an approximating sequence for (QEP), there is a subsequence converging to some point of Sol(QEP)$(\bar{\lambda}, \bar{\mu})$. So, the solution set Sol(QEP) $(\bar{\lambda}, \bar{\mu})$ is compact.
If Sol(QEP)$(\bar{\lambda}, \bar{\mu}) \subset \bigcup_{k=1}^{n} B(z^k, \varepsilon)$, then

$$\Pi(\delta, \varepsilon, \bar{\lambda}, \bar{\mu}) \subset \bigcup_{k=1}^{n} B\big(z^k, \varepsilon + H(\Pi(\delta, \varepsilon, \bar{\lambda}, \bar{\mu}), \text{Sol(QEP)}(\bar{\lambda}, \bar{\mu}))\big),$$

and hence,

$$\eta(\Pi(\delta, \varepsilon, \bar{\lambda}, \bar{\mu})) \leq H(\Pi(\delta, \varepsilon, \bar{\lambda}, \bar{\mu}), \text{Sol(QEP)}(\bar{\lambda}, \bar{\mu})) + \gamma\big(\text{Sol(QEP)}(\bar{\lambda}, \bar{\mu})\big).$$

Because Sol(QEP)$(\bar{\lambda}, \bar{\mu})$ is compact, $\eta(\text{Sol(QEP)}(\bar{\lambda}, \bar{\mu})) = 0$. So, we have

$$\eta(\Pi(\delta, \varepsilon, \bar{\lambda}, \bar{\mu})) \leq H(\Pi(\delta, \varepsilon, \bar{\lambda}, \bar{\mu}), \text{Sol(QEP)}(\bar{\lambda}, \bar{\mu})).$$

Now we claim that

$$H(\Pi(\delta, \varepsilon, \bar{\lambda}, \bar{\mu}), \text{Sol(QEP)}(\bar{\lambda}, \bar{\mu})) \to 0^+ \text{ as } (\delta, \varepsilon) \to (0^+, 0^+).$$

Indeed, suppose to the contrary that there are a positive real number $r > 0$, a sequence $\{(\delta^k, \varepsilon^k)\}_k$ tending to $(0^+, 0^+)$, and points $x^k \in \Pi(\delta^k, \varepsilon^k, \bar{\lambda}, \bar{\mu})$ such that, for all $k \in \mathbb{N}$,

$$d(x^k, \mathrm{Sol(QEP)}(\bar{\lambda}, \bar{\mu})) \geq r.$$

Since $\{x^k\}_k$ is an approximating sequence for (QEP), there is a subsequence converging to some point of $\mathrm{Sol(QEP)}(\bar{\lambda}, \bar{\mu})$, a contradiction.

(b) Assume that $\eta(\Pi(\delta, \varepsilon, \bar{\lambda}, \bar{\mu})) \to 0^+$ as $(\delta, \varepsilon) \to (0^+, 0^+)$. We first prove that $\Pi(\delta, \varepsilon, \bar{\lambda}, \bar{\mu})$ is closed for all positive δ and ε. Let $x^k \in \Pi(\delta, \varepsilon, \bar{\lambda}, \bar{\mu})$ be such that x^k converge to some point x. Then, for each $k \in \mathbb{N}$, there are $\lambda^k \in B(\bar{\lambda}, \delta)$ and $\mu^k \in B(\bar{\mu}, \delta)$ such that for all $y \in \mathscr{D}(x^k, \lambda^k)$,

$$\Phi(x^k, y, \mu^k) + \varepsilon \geq 0.$$

Since $B(\bar{\lambda}, \delta)$ and $B(\bar{\mu}, \delta)$ are compact, we can assume that $\{(\lambda^k, \mu^k)\} \to (\lambda, \mu)$ for some $(\lambda, \mu) \in B(\bar{\lambda}, \delta) \times B(\bar{\mu}, \delta)$. By the closedness of the map $\mathscr{D}$ at (x, λ), $y \in \mathscr{D}(x, \lambda)$. We claim that, for all $y \in \mathscr{D}(x, \lambda)$,

$$\Phi(x, y, \mu) + \varepsilon \geq 0.$$

Indeed, if there is $y \in \mathscr{D}(x, \lambda)$ such that

$$\Phi(x, y, \mu) + \varepsilon < 0,$$

then there is $y^k \in \mathscr{D}(x^k, \lambda^k)$ such that y^k converge to y as $\mathscr{D}$ is lsc at (x, λ). By the upper $-\varepsilon$-level closedness of the map Φ at (x, y, μ), there is $k_0 \in \mathbb{N}$ such that, for all $k \geq k_0$,

$$\Phi(x^k, y^k, \mu^k) < -\varepsilon,$$

a contradiction. Since $(\lambda, \mu) \in B(\bar{\lambda}, \delta) \times B(\bar{\mu}, \delta)$, we have $x \in \Pi(\delta, \varepsilon, \bar{\lambda}, \bar{\mu})$. Hence, $\Pi(\delta, \varepsilon, \bar{\lambda}, \bar{\mu})$ is closed.

Now we show that

$$\mathrm{Sol(QEP)}(\bar{\lambda}, \bar{\mu}) = \bigcap_{\delta>0, \varepsilon>0} \Pi(\delta, \varepsilon, \bar{\lambda}, \bar{\mu}).$$

We first prove that

$$\bigcap_{\delta>0} \Pi(\delta, \varepsilon, \bar{\lambda}, \bar{\mu}) = \widetilde{\mathrm{Sol}}(\mathrm{QEP})(\varepsilon, \bar{\lambda}, \bar{\mu}).$$

It is easy to see that

$$\bigcap_{\delta>0} \Pi(\delta, \varepsilon, \bar{\lambda}, \bar{\mu}) \supset \widetilde{\mathrm{Sol}}(\mathrm{QEP})(\varepsilon, \bar{\lambda}, \bar{\mu}).$$

Conversely, let $x \in \bigcap_{\delta>0} \Pi(\delta, \varepsilon, \bar{\lambda}, \bar{\mu})$ be arbitrary. Then, there are $(\lambda^k, \mu^k) \in B(\bar{\lambda}, \delta) \times B(\bar{\mu}, \delta)$ such that for all $y \in \mathscr{D}(x, \lambda^k)$,

$$\Phi(x, y, \mu^k) + \varepsilon \geq 0.$$

Since $y \in \mathscr{D}(x, \lambda^k)$, $\lambda^k \to \bar{\lambda}$, and $\mathscr{D}$ is closed, one can see that $y \in \mathscr{D}(x, \bar{\lambda})$. Now we verify that $x \in \widetilde{\mathrm{Sol}}(\mathrm{QEP})(\varepsilon, \bar{\lambda}, \bar{\mu})$. Indeed, for each $y \in \mathscr{D}(x, \bar{\lambda})$, since $\mathscr{D}$ is lsc at $(x, \bar{\lambda})$, there exists $y^k \in \mathscr{D}(x, \lambda^k)$ with $y^k \to y$. Because $x \in \widetilde{\mathrm{Sol}}(\mathrm{QEP})(\varepsilon, \bar{\lambda}, \bar{\mu})$,

$$\Phi(x, y^k, \mu^k) + \varepsilon \geq 0.$$

By the upper $-\varepsilon$-level closedness of the map Φ, one has

$$\Phi(x, y, \bar{\mu}) + \varepsilon \geq 0,$$

and hence $\bigcap_{\delta>0} \Pi(\delta, \varepsilon, \bar{\lambda}, \bar{\mu}) \subset \widetilde{\mathrm{Sol}}(\mathrm{QEP})(\varepsilon, \bar{\lambda}, \bar{\mu})$. Therefore,

$$\bigcap_{\delta>0} \Pi(\delta, \varepsilon, \bar{\lambda}, \bar{\mu}) = \widetilde{\mathrm{Sol}}(\mathrm{QEP})(\varepsilon, \bar{\lambda}, \bar{\mu}).$$

Next, we have

$$\mathrm{Sol}(\mathrm{QEP})(\bar{\lambda}, \bar{\mu}) = \bigcap_{\varepsilon>0} \widetilde{\mathrm{Sol}}(\mathrm{QEP})(\varepsilon, \bar{\lambda}, \bar{\mu}) = \bigcap_{\delta>0, \varepsilon>0} \Pi(\delta, \varepsilon, \bar{\lambda}, \bar{\mu}).$$

Because $\eta(\Pi(\delta, \varepsilon, \bar{\lambda}, \bar{\mu})) \to 0^+$ as $(\delta, \varepsilon) \to (0^+, 0^+)$, the regular measure properties of η imply that $\mathrm{Sol}(\mathrm{QEP})(\bar{\lambda}, \bar{\mu})$ is compact and

$$H(\Pi(\delta, \varepsilon, \bar{\lambda}, \bar{\mu}), \mathrm{Sol}(\mathrm{QEP})(\bar{\lambda}, \bar{\mu})) \to 0^+,$$

as $(\delta, \varepsilon) \to (0^+, 0^+)$.

Let $\{x^k\}_k$ be an approximating sequence for (QEP) corresponding to $\{(\lambda^k, \mu^k)\}_k$, where $(\lambda^k, \mu^k) \to (\bar{\lambda}, \bar{\mu})$. Then, there are $\varepsilon^k \to 0^+$ such that for all $y \in \mathscr{D}(x^k, \lambda^k)$ and all $k \in \mathbb{N}$,

$$\Phi(x^k, y, \mu^k) + \varepsilon^k \geq 0.$$

This means that $x^k \in \Pi(\delta^k, \varepsilon^k, \bar{\lambda}, \bar{\mu})$ with $\delta^k := d((\bar{\lambda}, \bar{\mu}), (\lambda^k, \mu^k))$. We see that

$$d(x^k, \mathrm{Sol}(\mathrm{QEP})(\bar{\lambda}, \bar{\mu})) \leq H(\Pi(\delta, \varepsilon, \bar{\lambda}, \bar{\mu}), \mathrm{Sol}(\mathrm{QEP})(\bar{\lambda}, \bar{\mu})) \to 0^+.$$

Hence, there are points $\bar{x}^k \in \text{Sol(QEP)}(\bar{\lambda}, \bar{\mu})$ such that $d(x^k, \bar{x}^k) \to 0$ as $k \to \infty$. By the compactness of the solution set $\text{Sol(QEP)}(\bar{\lambda}, \bar{\mu})$, there is a subsequence $\{\bar{x}^{k_j}\}_j$ of $\{\bar{x}^k\}_k$ converging to some point $\bar{x}$ of $\text{Sol(QEP)}(\bar{\lambda}, \bar{\mu})$. Therefore, the corresponding subsequence $\{x^{k_j}\}_j$ of $\{x^k\}_k$ tends to $\bar{x}$. Hence, QEP is well-posed at $\bar{\lambda}$. □

5.4 Applications

Since quasiequilibrium problems contain many models as special cases, including variational inequalities, optimization problems, fixed-point and coincidence point problems, complementarity problems, Nash equilibrium problems, etc., we can derive from the results of the previous sections the consequences for such special cases. In this section, we discuss only some corollaries for quasivariational inequalities and traffic network problems as examples.

5.4.1 Quasivariational Inequalities

Let X, Y, W and Λ be Hausdorff topological vector spaces. Let $A \subset X$, $\Lambda \subset Y$, and $M \subset W$ be nonempty. Let $\mathscr{D} : A \times \Lambda \rightrightarrows A$ be a set-valued map, X^* be the dual space of X, and $T : A \times M \to X^*$. We consider the following parametric quasivariational inequalities, for each $\lambda \in \Lambda$,

$$\text{(QSVI) find } \bar{x} \in \mathscr{D}(\bar{x}, \lambda) \text{ such that for every } y \in \mathscr{D}(\bar{x}, \lambda),$$

$$\langle T(\bar{x}, \mu), y - \bar{x} \rangle \geq 0,$$

$$\text{(QMVI) find } \bar{x} \in \mathscr{D}(\bar{x}, \lambda) \text{ such that for every } y \in \mathscr{D}(\bar{x}, \lambda),$$

$$\langle T(y, \mu), \bar{x} - y \rangle \leq 0,$$

where $\langle \cdot, \cdot \rangle$ denotes the pairing between X and X^*.

To convert (QSVI) and (QMVI) to special cases of (QEP), set $Y = \mathbb{R}$, $C = \mathbb{R}_+$, and

$$\Phi(x, y, \mu) = \langle T(x, \mu), y - x \rangle,$$

$$\Phi(x, y, \mu) = -\langle T(y, \mu), x - y \rangle,$$

resp. Let $\mathscr{E}(\lambda) := \{x \in A \mid x \in D(x, \lambda)\}$. The following result is obtained from Theorems 5.1 and 5.6.

Corollary 5.30 (Upper semicontinuity and closedness of the solution maps of Stampacchia and Minty quasivariational inequalities) *Assume for (QSVI) (QMVI, resp.) that*

(a) *for every* $\mu \in M$, *the set*

$$\{(x, y) \in A \times A \mid \langle T(x, \mu), y - x \rangle \geq 0\}$$

$$(\{(x, y) \in A \times A \mid \langle T(y, \mu), x - y \rangle \leq 0\}, \text{ resp.})$$

is closed in $A \times A$;

(b) *for every* $x, y \in A$, *the function*

$$\mu \mapsto \langle T(x, \mu), y - x \rangle$$

$$(\mu \mapsto \langle T(y, \mu), x - y \rangle, \text{ resp.})$$

is $(-\infty, 0)$*-upper semicontinuous* ($(0, +\infty)$*-lower semicontinuous, resp.*);

(c) *either of the following conditions holds:*

(c_1) $\mathscr{E}$ *is upper semicontinuous and compact-valued and* $\mathscr{D}$ *is lower semicontinuous;*

(c_2) A *is compact and* $\mathscr{D}$ *is continuous.*

Then, the solution map Sol(QSVI) (Sol(QMVI), resp.) is usc, closed, and compact-valued.

Remark 5.31 (The generalized level closedness concept in dual spaces) Corollary 5.30 remains valid if we replace assumptions (a) and (b) by

(a') the set

$$\{(x, y, \mu) \mid \langle T(x, \mu), y - x \rangle \geq 0\}$$

$$(\{(x, y, \mu) \mid \langle T(y, \mu), x - y \rangle \leq 0\}, \text{ resp.})$$

is closed in $A \times A \times M$.

Now we are going to discuss Hölder conditions of solution maps. Assume that X, Y, W are normed spaces and recall here some concepts used in the sequel. Let $A \subset X$ be nonempty and $T : A \to X^*$.

(a) T is said to be quasimonotone (quasi-antimonotone, resp.) in A if for every $x, y \in A$,

$$[\langle T(x), y - x \rangle > 0] \Rightarrow [\langle T(y), x - y \rangle \leq 0],$$

$$([\langle T(y), x - y \rangle < 0] \Rightarrow [\langle b(x), y - x \rangle \geq 0], \text{ resp.}).$$

(b) T is said to be (h, α)-Hölder strongly pseudomonotone ((h, α)-strongly pseudo-antimonotone, resp.) in A if for every $x, y \in A$,

$$[\langle T(x), y - x\rangle \geq 0] \Rightarrow [\langle T(y), x - y\rangle + h\|x - y\|^\alpha \leq 0]$$

$$([\langle T(y), x - y\rangle \leq 0] \Rightarrow [\langle T(x), y - x\rangle + h\|x - y\|^\alpha \geq 0], \text{ resp.}).$$

(c) T is said to be (h, α)-Hölder strongly monotone ((h, α)-Hölder strongly antimonotone, resp.) in A if, for all $x, y \in A$,

$$\langle T(x) - T(y), y - x\rangle + n\|x - y\|^\alpha \leq 0,$$

$$(\langle T(y) - T(x), x - y\rangle + n\|x - y\|^\alpha \geq 0, \text{ resp.}).$$

Corollary 5.32 (Hölder continuity of the solution maps of Stampacchia and Minty quasivariational inequalities) *For problem (QSVI) (QMVI), resp.), assume that solutions exist in a neighborhood of* (λ_0, μ_0) *and that*

(a) *there are neighborhoods* N *of* λ_0 *and* U *of* μ_0 *such that for every* $x \in \mathscr{E}(N)$, $T(x, \cdot)$ *is* (ρ', γ')*-Hölder at* μ_0 *and* $T(\cdot, \cdot)$ *is bounded with constant* p *on* $\mathscr{E}(N) \times U$, *and* $\mathscr{E}(N)$ *is bounded with constant* q;
(b) *for every* $\mu \in U$, $T(\cdot, \mu)$ *is* (h, β)*-Hölder strongly monotone* ((h, β)*-Hölder strongly antimonotone, resp.) on* $\mathscr{E}(N)$;
(c) $\mathscr{D}$ *is* $((\ell_1, \alpha_1), (\ell_2, \alpha_2))$*-Hölder on* $\mathscr{E}(N) \times N$;
(d) $\alpha_1 = \beta$ *and* $h > 2p\ell_1$.

Then, on some neighborhood of (λ_0, μ_0), *the solution* $x(\lambda, \mu)$ *of (QSVI) ((QMVI), resp.) is unique and satisfies the following Hölder condition*

$$d\big(x(\lambda_1, \mu_1), x(\lambda_2, \mu_2)\big) \leq k_1 d^{\alpha_2/\beta}(\lambda_1, \lambda_2) + k_2 d^{\gamma/\beta}(\mu_1, \mu_2),$$

where k_1 *and* k_2 *are positive constants depending on* ρ', γ', β, *etc.*

Proof We verify the assumptions of Theorem 5.10 (a) with $\rho = 2q\rho', \gamma = \gamma', \theta = 0, \sigma = p$, and $\delta = 1$.

By (b),

$$0 \geq \langle T(x, \mu) - T(y, \mu), y - x\rangle + h\|y - x\|^\beta$$

$$= f(x, y, \mu) + f(y, x, \mu) + h\|y - x\|^\beta.$$

Other conditions are easy to verify. □

The following example gives a case where all the assumptions and the conclusion of Corollary 5.32 hold with T not being globally bounded.

Example 5.33 (Hölder continuity of the solution maps of Stampacchia and Minty quasivariational inequalities without the global boundedness assumption) Let $X = Y = W = \mathbb{R}$, $\Lambda \equiv M = [0, 1]$, $A = [-2, 2]$, $\mathscr{D}(x, \lambda) = [\lambda, 1]$, $\lambda_0 = 0$, and

$$T(x, \lambda) = \frac{(1 + \lambda)x}{1 + x}.$$

Then, T is locally bounded on $[0, 1] \times [0, 1]$, (b) holds with $h = \frac{1}{4}$ and $\beta = 2$, (c) is fulfilled with $\ell_1 = 0, \ell_2 = \alpha_2 = 1$, and arbitrary α_1, and hence (d) is also satisfied. So, Corollary 5.32 implies the Hölder continuity of the solution at 0 (in fact, $x(\lambda) = \lambda$). But, T is not globally bounded.

Similarly, we also have the following result.

Corollary 5.34 (Hölder continuity of the solution maps of Stampacchia and Minty quasivariational inequalities via pseudostrong monotonicity) *Corollary 5.32 remains valid if assumption (b) is replaced by*

(b') *for every $\mu \in U$, $-T(\cdot, \mu)$ is quasimonotone (quasi-antimonotone, resp.) on $\mathscr{D}(N)$ and $T(\cdot, \mu)$ is (h, β)-Hölder strongly pseudomonotone ((h, β)-Hölder strongly pseudo-antimonotone, resp.) on $\mathscr{D}(N)$.*

The following example gives a case where Corollary 5.34 can be applied but Corollary 5.32 cannot.

Example 5.35 (Incomparability of strong monotonicity and strong pseudomonotonicity together with quasimonotonicity) Let $X, Y, W, \Lambda, M, \lambda_0$ be as Example 5.33, $A = [0, 1]$,

$$\mathscr{D}(x, \lambda) = \left[\frac{x + \lambda}{16}, \frac{2x + \lambda}{16}\right],$$

and

$$T(x, \lambda) = \frac{\lambda + 1}{1 + x}.$$

Then, all the assumptions of Corollary 5.34 are fulfilled with $\rho' = \frac{1}{14}, \gamma' = 1, \theta = 0, p = 2, h = \frac{14}{15}, \beta = 1, \ell_1 = \ell_2 = \frac{1}{16}$, and $\alpha_1 = \alpha_2 = 1$. Hence, Corollary 5.34 implies the Lipschitz continuity of the unique solution (in fact, $x(\lambda) = \frac{\lambda}{15}$). While Corollary 5.32 cannot be employed since T is not strongly monotone.

5.4.2 Traffic Network Problems

The notion of equilibrium flows for transportation network problems was introduced by JG Wardrop in 1952 together with a basic traffic network principle. Since then, these problems have raised a great interest and much developed in both theory and methodology view points. Variational approaches to such problems were initiated by MJ Smith in 1979, who proved that the Wardrop equilibrium can be expressed in terms of variational inequalities. Therefore, the possibility of exploiting the powerful tools and approaches of variational analysis has been applied to study a large variety of models, reaching valuable theoretical results and providing applications in practical situations. In this subsection, we consider a class of equilibrium problems, which can be discussed within the framework of quasivariational inequalities. Namely, we study models with elastic travel demands.

Let us denote the graph of a network by $G = [N, L]$, where N is the set of NODES (e.g., crossroads, airports, and railway stations) and L is the set of directed LINKS (called also ARCS, including stretches of streets,...). Let ℓ be a link of the network, which connects a pair of nodes, and let p be a path consisting of a sequence of links, which connects an ORIGIN-DESTINATION (O/D) PAIR of nodes. We will assume that in the underlying network, there are n links and m paths. Let W denote the set of all O/D pairs with typical O/D pair w. We assume further that the number of elements of W is n with $m > n$.

Let P_w be the set of paths connecting the O/D pair w, and the set of all the paths in the network is denoted by P. The traffic flow on link ℓ is denoted by F_ℓ while V_p is the nonnegative traffic flow on path p. Denote by $F = (F_1, \dots, F_n)^T$ the link flow vector, and $V = (v_1, \dots, v_m)^T$ by the path flow vector. The relationship between the link flows and the path flows is provided by

$$F_\ell = \sum_{p \in P} \gamma_{\ell p} V_p,$$

or equivalently $F = \Gamma V$, where $\Gamma = \left(\gamma_{\ell p}\right)_{\ell \in L, p \in P}$ is the LINK-PATH INCIDENCE MATRIX, where $\gamma_{\ell p}$ is 1 if link ℓ is contained in path p and 0 otherwise. This condition ensures that the flow on a particular link is defined by the sum of the flows on the paths that contain that link. We suppose that the travel demand associated with the users traveling between O/D pairs is not fixed, but depends on the user equilibrium flow denoted by H.

Actually, TRAVEL DEMANDS are affected by users' prediction of the flow amount through the network. Let A be a nonempty, compact, and convex subset of $\mathbb{R}^m_+$ and let $g : E \to \mathbb{R}^n_+$ be the travel demand. Then, the path vector flows meeting the travel demands are called the FEASIBLE PATH VECTOR FLOWS and form the constraint set-valued map $\mathbb{D} : A \rightrightarrows \mathbb{R}^m_+$ given by

$$\mathscr{D}(H) = \{F \in A \mid \Phi F = g(H)\},$$

where $\Phi = (\varphi_{wp})_{w \in W, p \in P}$ is the LINK-ROUTE INCIDENCE MATRIX O/D pairs-paths, where ϕ_{wp} is 1 if path p connects the pair w and 0 otherwise. The condition $\Phi F = g(H)$ means that flows and hence travelers are not lost or generated in the network. Let $t_\ell(F)$ denote the USER TRAVEL COST associated with the link ℓ, and we group the link costs into the vector $t(F) = (t_1(F), \ldots, t_n(F))^T$. Only the case of asymmetric cost, in which a link's cost is affected by all of the network's links' flows, will be considered. Let $\tau_p(V)$ denote the user travel cost on path p and let $\tau(V) = (\tau_1(V), ..., \tau_m(V))^T$ be the path flow vector. Then,

$$\tau_p(V) = \sum_{\ell \in L} \gamma_{\ell p} t_\ell(F),$$

or equivalently

$$\tau(V) = \Gamma^T t(F) = \Gamma^T t(\Gamma V).$$

This relationship ensures that the cost of a path is equal to the sum of the costs on links that form the path.

Finally, we recall the definition of user equilibrium flow and its equivalent quasivariational inequality formulation.

Definition 5.36 (A user traffic equilibrium flow) A flow $H \in \mathscr{D}(H)$ is called a USER TRAFFIC EQUILIBRIUM FLOW if for every $w \in W$ and $p, s \in P_w$,

$$\tau_p(H) > \tau_s(H) \Rightarrow H_p = 0.$$

Lemma 5.37 (Relationships between traffic network problems and quasivariational inequalities) *A flow $H \in \mathscr{D}(H)$ is a user traffic equilibrium flow if and only if it satisfies the following quasivariational inequality*

$$\langle \tau(H), F - H \rangle \geq 0, \quad \forall F \in \mathscr{D}(H).$$

Remark 5.38 (The positive values condition of the travel demand) It is worth noting that if there exists some $H \in A$ such that $g(H) \leq 0$, then the network is empty, and furthermore for the clear physical meaning, an unlimited demand requirement would be unrealistic. Hence, it is reasonable for us to assume that $g(H) > 0$ for every $H \in A$, and flows are nontrivial and norm-bounded from below.

The quasivariational inequality that we are addressing entails finding $H \in \mathscr{D}(H)$ such that

$$\text{(QVI)}\ \langle \tau(H), F - H \rangle \geq 0, \quad \forall F \in \mathscr{D}(H),$$

where $\mathscr{D}(H) = \{F \in A \mid \Phi F = g(H)\}$, A is a convex and compact subset of $\mathbb{R}^m_+$, Φ is an appropriate $n \times m-$ matrix (with $m > n$), and $g : A \to \mathbb{R}^n_+$ the travel demand map. Because flows are nontrivial in this model, we assume that $0 \notin A$. Hence, since the norm is lower semicontinuous on A, feasible flows are bounded from below in norm. Here we focus on the stability of the reference model, and hence, we also assume that the solution set of (QVI) is nonempty and nontrivial. Then, we denote by $\bar{H}$ a solution to (QVI) belonging to some neighborhood $X \subset A$. In order to consider a parametric quasivariational inequality, we suppose that τ is subject to change, as shown by including a parameter μ in a general perturbation form, where μ belongs to a subset M of a finite dimensional normed space. As a result, for each $\mu \in M$, we consider a cost operator $\tau(\cdot, \mu) : A \to \mathbb{R}^m_+$. The constraint set will be perturbed with regard to the map g, whereas the link-route incidence matrix Φ will be fixed. Therefore, we assume that there exists a parameter λ belonging to a subset of Λ, which will act on the travel demand map g.

For each $\lambda \in \Lambda$ and $\mu \in M$, our traffic network problem is equivalent to a quasivariational inequality as follows.

Lemma 5.39 (Relationships between parametric traffic network problems and parametric quasivariational inequalities) *A path vector flow $H \in \mathscr{D}(H, \lambda)$ is an equilibrium flow if and only if it is a solution of the following quasivariational inequality*

(TNP) *find $H \in \mathscr{D}(H, \lambda)$ such that for all $F \in \mathscr{D}(H, \lambda)$,*

$$\langle \tau(H, \mu), F - H \rangle \geq 0.$$

Next we aim to determine how the constraints behave whenever the parameter acting on g and the point H are moving. The following result can be seen as a particular case of the famous Hoffman's lemma.

Lemma 5.40 (A version of Hoffman's lemma) *Let Φ be an $n \times m$ matrix, a_1 and a_2 be given vectors in R^n. The solution set of the linear equality $\Phi X = a_i$, for $i =$ 1, 2, is denoted by S_i. Then, there exists $\delta = \delta(\Phi) > 0$ such that for each $X_1 \in S_1$ there exists $X_2 \in S_2$ satisfying*

$$\|X_1 - X_2\| \leq \delta \|a_1 - a_2\|.$$

Lemma 5.41 (Properties of the feasible maps of parametric traffic network problems) *Assume that g is continuous at (H_0, λ_0). Then, $\mathscr{D}$ is continuous at (H_0, λ_0) and convex, compact-valued.*

Proof Consider the system

$$\Phi F_0 = g(H_0, \lambda_0),$$

$$\Phi F = g(H, \lambda).$$

By applying Lemma 5.40, we can get some real number $\delta > 0$ such that for each element $F_0 \in \mathscr{D}(H_0, \lambda_0)$, there exists $F \in \mathscr{D}(H, \lambda)$ satisfying

$$\|F - F_0\| \leq \delta \|g(H, \lambda) - g(H_0, \lambda_0)\|.$$

Since g is continuous at (H_0, c_0), $\mathscr{D}$ is lower semicontinuous at (H_0, λ_0). If $\mathscr{D}$ is not upper semicontinuous at (H_0, λ_0), then there exist an open neighborhood U of $\mathscr{D}(H_0, \lambda_0)$ and a sequence $(H_k, \lambda_k) \to (H_0, \lambda_0)$ such that for each k, there exists $F_k \in \mathscr{D}(H_k, \lambda_k)$ but $F_k \notin U$. Due to the compactness of A, we can assume that $F_k \to F_0$. Then, by Lemma 5.40, there exists $F_n^0 \in \mathscr{D}(H_0, \lambda_0)$ such that

$$\|F_k - F_n^0\| \leq \delta \|g(H_k, \lambda_k) - g(H_0, \lambda_0)\|.$$

Consequently, $F_k^0 \to F_0$. Since $\Phi F_k^0 = g(H_0, \lambda_0)$, we have $\Phi F_0 = g(H_0, \lambda_0)$, or equivalently $F_0 \in \mathscr{D}(H_0, \lambda_0) \subset U$, which is a contradiction. □

Passing to the Hölder property of the constraint map, we have the following important lemma.

Lemma 5.42 (Hölder continuity of feasible maps of parametric traffic network problems) *Assume that g is $((L_1, \alpha_1), (L_2, \alpha_2))$-Hölder continuous at (x_0, λ_0), i.e., there exist neighborhoods U of x_0 and V of λ_0 such that for every $x_1, x_2 \in U$ and $\lambda_1, \lambda_2 \in V$, we have*

$$\|g(x_1, \lambda_1) - g(x_2, \lambda_2)\| \leq L_1 \|x_1 - x_2\|^{\alpha_1} + L_2 \|\lambda_1 - \lambda_2\|^{\alpha_2}.$$

Then, there exist $\ell_1, \ell_2 > 0$ such that $\mathscr{D}$ is $((\ell_1, \alpha_1), (\ell_2, \alpha_2))$-Hölder continuous at (x_0, λ_0).

Proof Let $\lambda_1, \lambda_2 \in V$ and $H_1, H_2 \in A$. We consider the systems

$$\Phi F = g(H_1, \lambda_1);\ \Phi F' = g(H_1, \lambda_2).$$

From Lemma 5.40, there exists $\delta = \delta(\Phi)$ such that for each $F \in \mathscr{D}(H_1, \lambda_1)$, there exists some $F' \in \mathscr{D}(H_2, \lambda_2)$ such that

$$\left\|F - F'\right\| \leq \delta \left\|g(H_1, \lambda_1) - g(H_2, \lambda_2)\right\|.$$

Due to the Hölder condition on the map g, we see that the map $\mathscr{D}$ is $(\ell_1\alpha_1, \ell_2\alpha_2)$-Hölder continuous with $\ell_i = \delta L_i$ for $i = 1, 2$. The proof is complete. □

Set $X = R^m$ and

$$\Psi(x, y, \mu) = \langle \tau(x, \mu), y - x \rangle.$$

Then, problem (TNP) becomes a special case of (QEP).

The following results are easy to get by applying the aforementioned lemmas and the corresponding results of the previous sections.

Corollary 5.43 (Upper semicontinuity and closedness of the solution maps of parametric traffic network problems) *For problem* (TNP), *assume that*

(a) *g is continuous on $A \times \Lambda$;*
(b) *the set $\{(H, F, \mu) \mid \langle \tau(H, \mu), F - H \rangle \geq 0\}$ is closed in $A \times A \times M$.*

Then, the solution map of problem (TNP) is upper semicontinuous, closed, and compact-valued on $\Lambda \times M$.

Corollary 5.44 (Continuity of the solution maps of parametric traffic network problems via quasimonotonicity conditions) *Impose the assumptions of Corollary 5.43 and the following conditions*

(a) *τ is quasimonotone in A;*
(b) *for every $(\lambda, \mu) \in \Lambda \times M$, and for every $H, H' \in \mathrm{Sol(TNP)}(\lambda, \mu)$, $H \neq H'$,*

$$\langle \tau(H, b_0), H' - H \rangle > 0.$$

Then, the solution map Sol(TNP) is continuous on $\Lambda \times M$.

The second condition of the aforementioned corollary can be replaced by some suitable assumptions given in the following result.

Corollary 5.45 (Continuity of the solution maps of parametric traffic network problems via pseudomonotonicity conditions) *Let the assumptions of Corollary 5.43 and the following conditions be satisfied*

(a) *τ is pseudomonotone on A;*
(b) *for every $(\lambda, \mu) \in \Lambda \times M$ and $H_1, H_2 \in \mathscr{E}(\lambda)$,*

$$[\langle \tau(H_1, \mu), H_2 - H_1 \rangle = 0] \Rightarrow [H_2 = H_1];$$

(c) *for every $(\lambda, \mu) \in \Lambda \times M$ and $H_1, H_2 \in \mathrm{Sol(TNP)}(\lambda, \mu)$,*

$$\langle \tau(H_1, \mu), H_2 - H_1 \rangle \geq 0.$$

Then, the solution map Sol(TNP) is continuous on $\Lambda \times M$.

For each $\lambda \in \Lambda$, we denote by $\mathscr{E}(\lambda)$ the fixed-point sets of the map $\mathscr{D}$, i.e.,

$$\mathscr{E}(\lambda) = \{H \in A \mid H \in \mathscr{D}(H, \lambda)\}.$$

Let us present some typical results obtained from Corollaries 5.32 and 5.34.

Corollary 5.46 (Hölder continuity of solution maps of parametric traffic network problems via strong monotonicity) *For problem (TNP), assume that the following conditions are satisfied.*

(a) *there are neighborhoods N of λ_0 and U of μ_0 such that for every $x \in \mathscr{E}(N)$, $\tau(x, \cdot)$ is (ρ', γ')-Hölder continuous at μ_0, $\tau(\cdot, \cdot)$ is bounded with constant p in $\mathscr{E}(N) \times U$, and $\mathscr{E}(N)$ is bounded with constant q;*
(b) *for every $\mu \in U$, $\tau(\cdot, \mu)$ is (h, β)-Hölder strongly monotone ((h, β)-Hölder strongly antimonotone, resp.) in $\mathscr{E}(N)$;*
(c) *g is $((L_1, \alpha_1), (L_2, \alpha_2))$-Hölder continuous in $E\big(U(\lambda_0)\big) \times \{\lambda_0\}$;*
(d) *$\alpha_1 = \beta$ and $h > 2p\ell_1$.*

Then, in some neighborhood of (λ_0, μ_0), the solution $x(\lambda, \mu)$ of (TNP) is unique and satisfies the following Hölder condition

$$d\big(x(\lambda_1, \mu_1), x(\lambda_2, \mu_2)\big) \leq k_1 d^{\alpha_2/\beta}(\lambda_1, \lambda_2) + k_2 d^{\gamma/\beta}(\mu_1, \mu_2),$$

where k_1 and k_2 are positive constants depending on ρ', γ', β, etc.

Corollary 5.47 (Hölder continuity of solution maps of parametric traffic network problems via pseudostrong monotonicity) *Corollary 5.46 remains valid if assumption (b) is replaced by*

(b') *$\forall \mu \in U, \; -\tau(\cdot, \mu)$ is quasimonotone on $\mathscr{E}(N)$ and $\tau(\cdot, \mu)$ is (h, β)-Hölder-strongly pseudomonotone on $\mathscr{E}(N)$.*

5.5 Exercises

Exercise 5.1 Let $T : \mathbb{R} \rightrightarrows \mathbb{R}$ be defined by

$$T(x) := \left\{ y \in \mathbb{R} \mid 1 - \sqrt{x^2 + 2x} \leq y \leq 1 + \sqrt{x^2 + 2x} \right\}.$$

Prove that T is rotund.

Exercise 5.2 Let E be a compact subset of a Banach space and $\mathscr{P}(E)$ be the collection of all nonempty closed convex subsets of E. Denote the collection of all continuous set-valued mappings $K : E \to \mathscr{P}(E)$ by $\mathscr{D}(E, \mathscr{P}(E))$ with the distance defined as

$$d(K, K') = \max_{z \in E} H(K(z), K'(z)), \forall K, K' \in \mathscr{D}(E, \mathscr{P}(E)),$$

where H is the Hausdorff distance. Prove that $(\mathscr{P}(E), d)$ is a complete metric space.

Exercise 5.3 Let $\mathscr{D}(E, \mathscr{P}(E))$ with the distance d be as defined in Exercise 5.2 Prove that for any $K, K' \in \mathscr{D}(E, \mathscr{P}(E))$, the function $g : E \to \mathbb{R}$ defined by

$$g(x) := d(K(x), K'(x)), \forall x \in E$$

is continuous.

Exercise 5.4 Let $\mathscr{D}(E, \mathscr{P}(E))$ be as in Exercise 5.2 Prove that the set-valued mapping $G : E \times \mathscr{D}(E, \mathscr{P}(E)) \to \mathscr{P}(E)$ defined by

$$G(x, K) := K(x), \forall x \in E, \forall K \in \mathscr{D}(E, \mathscr{P}(E))$$

is continuous on $E \times \mathscr{D}(E, \mathscr{P}(E))$.

Exercise 5.5 Let E be a compact subset of a Banach space and let $\mathscr{D}(E, \mathscr{P}(E))$ be as in Exercise 5.2. Denote the set of all continuous functions $f : E \times E \to \mathbb{R}$ by $\mathscr{C}(E \times E, \mathbb{R})$ with the metric defined as

$$\rho(f, f') := \max_{z \in E \times E} |f(z) - f'(z)|, \forall f, f' \in \mathscr{C}(E \times E, \mathbb{R}).$$

Let Ω be the collection of all $(f, K) \in \mathscr{C}(E \times E, \mathbb{R}) \times \mathscr{D}(E, \mathscr{P}(E))$ such that the solution set of a quasiequilibrium problem

$$S(f, K) := \{x \in K(x) \mid f(x, y) \geq 0, \forall y \in K(x)\}$$

is nonempty. Prove that (Ω, η) is a complete metric space, where the metric η is defined by

$$\eta(\omega, \omega') := \max\{\rho(f, f'), d(K, K')\}, \forall \omega = (f, K), \omega' = (f', K') \in \Omega.$$

Exercise 5.6 Let E be a compact subset of a Banach space X and $\mathscr{D}(E, \mathscr{P}(E))$ be as in Exercise 5.2. Consider the set-valued map $\mathscr{E} : \mathscr{D}(E, \mathscr{P}(E)) \rightrightarrows E$ defined by

$$\mathscr{E}(K) := \{x \in E \mid x \in K(x)\}, \forall K \in \mathscr{D}(E, \mathscr{P}(E)).$$

Prove that $\mathscr{E}$ is upper semicontinuous and compact-valued on $\mathscr{D}(E, \mathscr{P}(E))$.

Exercise 5.7 Let the map $\mathscr{E}$ be defined as in Exercise 5.6. Prove that if $K \in \mathscr{D}(E, \mathscr{P}(E))$ is strongly lower semicontinuous on E (i.e., for every $x \in E$, $y \in K(x)$, and $\varepsilon > 0$, there exists a real number $\delta > \varepsilon$ such that $K(x) \cap B(y, \varepsilon) \neq \emptyset$ for all $x \in \text{bd}B(x, \delta)$), then $\mathscr{E}$ is lower semicontinuous at K.

Exercise 5.8 Let Ω and S be defined as in Exercise 5.5. Prove that the solution map S is upper semicontinuous and compact-valued on Ω.

Chapter 6
Stability of Vector Equilibrium Problems

Abstract In this chapter, we consider sufficient conditions for (semi)continuity, Hölder/ Lipschitz continuity, and well-posedness of parametric vector equilibrium problems (VEPs), including their weak and strong variants. Two approaches are presented: the first is direct on the vector formulation and the second is via scalarization techniques. Regarding the first approach, for stability in the sense of semicontinuity, we apply cone continuity/semicontinuity properties to get upper semicontinuity results for solution maps of VEPs, while to obtain lower semicontinuity ones, generalized convexity assumptions are additionally needed. Finally, for continuity of these solution maps, we also aim at some results independent from the above ones about semicontinuity. The techniques used here are not very far from those in Chap. 5, but the tools employed here are new. For stability in the Hölder/Lipschitz continuity or calmness sense, we resort to relaxed (quasi/pseudo) strong monotonicity and convexity of vector maps. When studying approximate solution maps, these properties are omitted or mitigated. Besides stability, a unified form of the Hadamard and Tikhonov well-posedness concepts in terms of semicontinuity, which is popular in the literature, is investigated as well in the first approach in this chapter. Moreover, we also establish relationships between well-posedness and stability of approximate solutions. Regarding the second approach (using scalarization), for weak VEPs we can use the linear scalarization by adding quasiconvexity hypotheses. But, for strong VEPs, we define many extensions of the Gerstewitz/Tammer nonlinear scalarization functions, and see as a byproduct that they can even be used also for weak VEPs. Furthermore, through these nonlinear scalarization techniques, we find interesting relationships between VEPs and set optimization problems, and then discuss stability for set optimization problems.

6.1 Vector Equilibrium Problems

In this section, we consider vector equilibrium problems. Picking up the main ideas from the previous chapters, we also employ some conditions related to monotonicity or convexity of vector mappings to study sufficient conditions for Hölder continuity of solution mappings of our problems. Let X be a topological space, Y a topological

L. Q. Anh et al., *Equilibrium Problems*, Forum for Interdisciplinary Mathematics, https://doi.org/10.1007/978-3-031-89003-1_6

vector space, $A \subset X$ a nonempty subset, $C \subset Y$ a solid closed convex cone, and Λ, M nonempty subsets of topological spaces. Let $\mathscr{D} : \Lambda \rightrightarrows A$ be surjective and $f : A \times A \times M \to Y$ a vector-valued map. For $x, y \in Y$, we adopt the notations:

$$x \geq y \Leftrightarrow x - y \in C,$$

$$x > y \Leftrightarrow x - y \in \text{int}C,$$

$$x \ngeq y \Leftrightarrow x - y \notin C,$$

and similarly for $\leq, <, \nleq$, etc., where int$C$ stands for the interior of C.

For $(\lambda, \mu) \in \Lambda \times M$, we consider the following parametric equilibrium problems:

(WEP)(λ, μ) find $\bar{x} \in \mathscr{D}(\lambda)$ such that, for all $y \in \mathscr{D}(\lambda)$,

$$f(\bar{x}, y, \mu) \nless 0.$$

(SEP)(λ, μ) find $\bar{x} \in \mathscr{D}(\lambda)$ such that, for all $y \in \mathscr{D}(\lambda)$,

$$f(\bar{x}, y, \mu) \geq 0.$$

For each $(\lambda, \mu) \in \Lambda \times M$, the solution sets of (WEP)(λ, μ) and (SEP)(λ, μ) are denoted by Sol(WEP)(λ, μ) and Sol(SEP)(λ, μ), resp., i.e.,

$$\text{Sol(WEP)}(\lambda, \mu) := \{x \in \mathscr{D}(\lambda) \mid f(x, y, \mu) \nless 0, \forall y \in \mathscr{D}(\lambda)\},$$

$$\text{Sol(SEP)}(\lambda, \mu) := \{x \in \mathscr{D}(\lambda) \mid f(x, y, \mu) \geq 0, \forall y \in \mathscr{D}(\lambda)\}.$$

Then, Sol(SEP) and Sol(WEP) are mappings from $\Lambda \times M$ into A, and we always assume that these mappings have nonempty values on a neighborhood of the reference point.

6.2 Continuity of Solution Mappings of Vector Equilibrium Problems

6.2.1 Upper Semicontinuity of Solution Mappings of Vector Equilibrium Problems

There have been several notions of semicontinuity of vector functions in the literature, and they may appear under slightly different terms. In this section, we call $h : A \to Y$ C-LOWER SEMICONTINUOUS (C-lsc) at x_0 if for any neighborhood U of $h(x_0)$, there is a neighborhood N of x_0 such that $h(N) \subset U + C$. h is called C-UPPER SEMICONTINUOUS (C-usc) at x_0 if for any neighborhood U of $h(x_0)$, there is a neighborhood N of x_0 such that $h(N) \subset U - C$.

We note the following simple relations with semicontinuity of scalar functions. Let $c \in C$, $f : A \to \mathbb{R}$, and $h : A \to Y$ be defined by $h(x) := f(x)c$ for $x \in A$. If f is lsc (usc, resp.) at x_0, then h is C-lsc (C-usc, resp.) at x_0. Furthermore, relations with level sets are also easily obtained as follows. For $b \in Y$, we use the following notations for level sets of h, for different orderings $\leq$ (by the context, no confusion occurs):

$$\mathrm{L}_{\leq b}h := \{x \in A \mid h(x) \leq b\},$$

$$\mathrm{L}_{\not> b}h := \{x \in A \mid h(x) \not> b\},$$

and similarly for other level sets $\mathrm{L}_{\not< b}h$, $\mathrm{L}_{\geq b}h$, $\mathrm{L}_{\not\leq b}h$, $\mathrm{L}_{> b}h$, etc. Then, it is not hard to verify the following equivalences.

Proposition 6.1 (Relationships between semicontinuity and level closedness of maps) *Let* $h : A \to Y$:

(a) h *is* C*-lsc at* x_0 *if and only if* $\mathrm{L}_{\not> b}h$ *is closed at* x_0 *for all* $b \in Y$.
(b) h *is* C*-usc at* x_0 *if and only if* $\mathrm{L}_{\not< b}h$ *is closed at* x_0 *for all* $b \in Y$.

These equivalences suggest the following natural properties weaker than C-semicontinuities, by imposing requirements only at a given level. Here we propose also similar definitions for other preorders $\leq$ and $\geq$.

Definition 6.2 (Generalized cone-semicontinuity) Let $h : A \to Y$ and $b \in Y$ be given.

(a) h is said to be b-LEVEL C-LOWER SEMICONTINUOUS ((b, C)-lsc) at x_0 if $\mathrm{L}_{\not> b}h$ is closed at x_0.
(b) h is called b-LEVEL C-UPPER SEMICONTINUOUS ((b, C)-usc) at x_0 if $\mathrm{L}_{\not< b}h$ is closed at x_0.
(c) h is termed b-LEVEL C-QUASILOWER SEMICONTINUOUS ((b, C)-qlsc) at x_0 if $\mathrm{L}_{\leq b}h$ is closed at x_0.
(d) h is said to be b-LEVEL C-QUASIUPPER SEMICONTINUOUS ((b, C)-qusc) at x_0 if $\mathrm{L}_{\geq b}h$ is closed at x_0.

If a statement among (a)–(d) holds for all $b \in Y$, we omit b in the corresponding terminology.

To make these definitions clearer, we present equivalent statements in terms of neighborhoods as follows.

Proposition 6.3 (Characterizations of generalized cone-semicontinuity) *Let* $h : A \to Y$ *and* $b \in Y$.

(a) h *is* (b, C)*-lsc at* x_0 *if and only if there is a neighborhood* N *of* x_0 *with* $h(N) \subset b + \mathrm{int}C$ *whenever* $h(x_0) \in b + \mathrm{int}C$.
(b) h *is* (b, C)*-usc at* x_0 *if and only if there is a neighborhood* N *of* x_0 *with* $h(N) \subset b - \mathrm{int}C$ *whenever* $h(x_0) \in b - \mathrm{int}C$.

(c) *h is (b, C)-qlsc at x_0 if and only if there exists a neighborhood N of x_0 such that $h(N) \cap (b - C) = \emptyset$ whenever $h(x_0) \notin b - C$.*
(d) *h is (b, C)-qusc at x_0 if and only if there exists a neighborhood N of x_0 such that $h(N) \cap (b + C) = \emptyset$ whenever $h(x_0) \notin b + C$.*

Proof By the similarity, we verify only (c) as an example.

"Only if." Suppose ad absurdum that h is (b, C)-qlsc at x_0 and $h(x_0) \not\leq b$, but for any neighborhood N of x_0, there is $x \in N$ such that, $h(x) \leq b$, i.e., there is a sequence $x^k \to x_0$ with $h(x^k) \leq b$. Then, as h is (b, C)-qlsc at x_0, we get the contradiction that $h(x_0) \leq b$.

"If." Let $x^k \to x_0$ with $h(x^k) \leq b$. We show that $h(x_0) \leq b$. If $h(x_0) \not\leq b$, by the assumption there is a neighborhood N of x_0 such that $h(x) \not\leq b$ for all $x \in N$. Since $x^k \to x_0$, one can assume that $x^k \in N$, i.e., $h(x^k) \not\leq b$, a contradiction. □

The following relationships between C-semicontinuity and C-quasisemicontinuity resemble those for the scalar counterparts, though not very clearly from the above definitions and discussions.

Proposition 6.4 (Relationships between cone-semicontinuity and cone-quasisemicontinuity) *Let $h : A \to Y$:*

(a) *If h is C-lsc at x_0, then h is C-qlsc at x_0.*
(b) *If h is C-usc at x_0, then h is C-qusc at x_0.*

Proof

(a) Suppose that h is C-lsc at x_0 and $x^k \to x_0$ with $h(x^k) \leq b$ for some $b \in Y$, but $h(x_0) \in Y \setminus (b - C)$. Since $Y \setminus (b - C)$ is open, one can assume that

$$h(x^k) \in Y \setminus (b - C) + C.$$

So, there are $t_\alpha \in Y \setminus (b - C)$ and $c_\alpha \in C$ such that $h(x^k) = t^k + c^k$. Because $h(x^k) \in b - C$, there is $\bar{c}^k \in C$ such that $h(x^k) = b - \bar{c}^k$. Therefore, $t^k = b - \bar{c}^k - c^k \in b - C$, which is impossible as $t_\alpha \in Y \setminus (b - C)$.
(b) It may be checked similarly.

□

The following example shows that the converse of Proposition 6.4 is not true.

Example 6.5 (The proper inclusion of a class of cone-semicontinuous maps in that of cone-quasisemicontinuous maps) Let $X = \mathbb{R}, Y = \mathbb{R}^2, A = \mathbb{R}, C = \mathbb{R}^2_+, x_0 = 0$, and

$$h(x) := \begin{cases} (x, 1) & \text{if } x \leq 0, \\ (-\frac{1}{x}, \frac{1}{x}) & \text{otherwise.} \end{cases}$$

Then, h is both C-qlsc and C-qusc at 0. Indeed, by the similarity we check only the latter property. For any $x^k \to 0$ such that $h(x^k) \geq b = (b_1, b_2)$ for some $b \in \mathbb{R}$ and all k, there is a subsequence x^{k_j} such that $x^{k_j} \le 0$ for all k_j, because otherwise one would have

$$h(x^{k_j}) \to (-\infty, +\infty) \not\geq b.$$

It follows from $h(x^{k_j}) \ge b$ that $x^{k_j} \ge b_1$ and $1 \ge b_2$, and hence $0 \ge b_1$. Therefore,

$$h(x_0) = (0, 1) \ge b,$$

or equivalently h is (b, C)-qusc at 0. Because b is arbitrary, the map h is C-qusc at 0.

However, h is neither C-lsc nor C-usc at 0. Indeed, we check only the latter. By letting

$$b = (2, 2), x^k = \frac{1}{k},$$

one has: $h(x^k) \not< b$ for large n. But, $h(0) = (0, 1) < b$. Consequently, h is not (b, C)-usc at 0.

Example 6.6 (Incomparability of generalized cone-semicontinuity and generalized cone-quasi-semicontinuity) Let X, Y, A, C, and $x_0 = 0$ be as in Example 6.5 and

$$h(x) := \begin{cases} (-1, 1) & \text{if } \ x = 0, \\ (-1, -1) & \text{otherwise.} \end{cases}$$

Then, for all $x \in \mathbb{R}$, $h(x) \not> 0$ and hence h is $(0, C)$-lsc at 0 since $\mathrm{L}_{\not>0}h$ is closed. But, h is not $(0, C)$-qlsc at 0 since $\mathrm{L}_{\le 0}h$ is not closed. Indeed, let $x^k = \frac{1}{k} \to 0$. Then, $h(x^k) = (-1, -1) \le 0$, but $h(0) = (-1, 1) \not\le 0$. So, $\mathrm{L}_{\le 0}h$ is not closed.

Now we modify h as

$$h(x) := \begin{cases} (-1, 1) & \text{if } \ x = 0, \\ (1, 1) & \text{otherwise.} \end{cases}$$

Then, for all $x \in \mathbb{R}$, $h(x) \not< 0$, whence h is $(0, C)$-usc at 0 since $\mathrm{L}_{\not<0}h$ is closed. But, h is not $(0, C)$-qusc at 0 since $\mathrm{L}_{\ge 0}h$ is not closed. Indeed, let $x^k = \frac{1}{k}$. Then, $h(x^k) = (1, 1) \ge 0$, but $h(0) = (-1, 1) \not\geq 0$. Hence, $\mathrm{L}_{\ge 0}h$ is not closed.

For considerations in what follows, we need to develop sum rules for generalized semicontinuous and quasisemicontinuous maps.

Proposition 6.7 (A sum rule of two cone-semicontinuous maps) *Let X, W be topological spaces, $A \subset X$ and $B \subset W$ nonempty open sets, $C \subset Y$ a solid closed convex cone, $f : A \to Y, g : B \to Y, A \subset X$, and $B \subset W$. If f and g are C-lsc (or C-usc) at x_0 and y_0, resp., then $h = f + g$ is C-lsc (or C-usc) at (x_0, y_0).*

Proof For reasons of similarity and symmetry, we discuss only the C-upper semicontinuity. Let U be any open superset of $f(x_0) + g(y_0)$. There are open neighborhoods U_1 and U_2 of $f(x_0)$ and $g(x_0)$, resp., such that $U_1 + U_2 \subset U$. Since f and g are C-lsc, there are open neighborhoods N_1 and N_2 of x_0 and y_0, resp., such that

$$f(x) \in U_1 - C \text{ and } g(y) \in U_2 - C$$

for all $(x, y) \in N_1 \times N_2$. Hence,

$$h(x, y) = f(x) + g(y) \in U_1 + U_2 - C \subset U - C$$

for all $(x, y) \in N_1 \times N_2$, and so the map h is C-usc at (x_0, y_0). □

Unfortunately, Proposition 6.7 cannot be extended for C-quasisemicontinuous maps as shown by the following example.

Example 6.8 (A counterexample for nonpreserving of cone-quasisemicontinuity under a sum of maps) Let $X, Y, A, C,$ and x_0 be as in Example 6.5. Let $W = B = X$, $y_0 = x_0$, and f be defined as h in Example 6.5. Let

$$g(x) := \begin{cases} (x, 1) & \text{if } x \leq 0, \\ (\frac{1}{x}, -\frac{1}{x}) & \text{otherwise.} \end{cases}$$

By arguments similar to those in Example 6.5, we see that f and g are C-qlsc at 0. Furthermore,

$$(f + g)(x) = \begin{cases} (2x, 2) & \text{if } x \leq 0, \\ (0, 0) & \text{otherwise.} \end{cases}$$

For $x^k = \frac{1}{k} \to 0$, one has

$$(f + g)(x^k) = (0, 0) \leq (1, 1),$$

but

$$(f + g)(0) = (0, 2) \not\leq (1, 1).$$

So, $f + g$ is not C-qlsc at 0.

Let us now modify f and g by setting

$$f(x) := \begin{cases} (x, -1) & \text{if } x \leq 0, \\ (-\frac{1}{x}, \frac{1}{x}) & \text{otherwise,} \end{cases}$$

$$g(x) := \begin{cases} (x, -1) & \text{if } x \leq 0, \\ (\frac{1}{x}, -\frac{1}{x}) & \text{otherwise.} \end{cases}$$

Then, f and g are C-qusc at 0. But,

$$(f+g)(x) = \begin{cases} (2x, -2) & \text{if } x \leq 0, \\ (0, 0) & \text{otherwise} \end{cases}$$

is not C-qusc at 0 because for $x^k = \frac{1}{k} \to 0$, we have:

$$(f+g)(x^k) = (0, 0) \geq (-1, -1),$$

but,

$$(f+g)(0) = (0, -2) \ngeq (-1, -1).$$

Proposition 6.9 (Sum of cone-quasisemicontinuity and cone-semicontinuity maps) *Assume that X, Y, W, A, B, C, f, and g are defined as in Proposition 6.7:*

(a) *If f is C-qlsc at x_0 and g is C-lsc at y_0, then $h = f + g$ is C-qlsc at (x_0, y_0).*
(b) *If f is C-qusc at x_0 and g is C-usc at y_0, then $h = f + g$ is C-qusc at (x_0, y_0).*

Proof By the similarity and symmetry, we demonstrate only (a). For arbitrary $b \in Y$, assume that $(x^k, y^k) \to (x_0, y_0)$ and $h(x^k, y^k) \leq b$. For each $e \in \text{int}C$, as g is C-lsc at y_0, there is a subsequence y^{k_j} such that $g(y^{k_j}) > g(y_0) - e$ for all k_j. Hence,

$$f(x^{k_j}) + g(y_0) - e \leq b.$$

Because f is C-qlsc at x_0

$$f(x_0) + g(y_0) - e \leq b,$$

Consequently, $f(x_0) + g(y_0) \leq b$. Thus, h is (b, C)-qlsc at (x_0, y_0). By the arbitrariness of b, h is C-qlsc at (x_0, y_0). □

The following examples prevent us from stating results similar to Propositions 6.7 and 6.9 for a given fixed level b.

Example 6.10 (Unpreserving of generalized level closedness under a sum) Let $A = \mathbb{R}, Y = \mathbb{R}^2, C = \mathbb{R}^2_+, x_0 = 0$, and $f, g : \mathbb{R} \to \mathbb{R}^2$ be defined by $g(x) := (2, -1)$ for all x and

$$f(x) := \begin{cases} (-1, 2) & \text{if } \quad x = 0, \\ (-3, -3) & \text{otherwise.} \end{cases}$$

Then, f is $(0, C)$-lsc at 0, since $\mathrm{L}_{\not>0} f$ is closed. g is continuous, whence $(0, C)$-lsc at 0. But,

$$(f+g)(x) = \begin{cases} (1, 1) & \text{if} \quad x = 0, \\ (-1, -4) & \text{otherwise} \end{cases}$$

is not $(0, C)$-lsc at 0, since $\mathrm{L}_{\not>0} h$ is not closed. Indeed, let $x^k = \frac{1}{k}$. Then,

$$h(x^k) = (-1, -4) \not> 0,$$

but

$$h(0) = (1, 1) > 0.$$

So, $\mathrm{L}_{\not>0} h$ is not closed. Therefore, from the fact that f is (b, C)-lsc at x_0 and g is (b', C) lsc at x_0, we cannot deduce that $f + g$ is $(b + b', C)$-lsc at x_0. If $b = b'$, f is (b, C)-lsc at x_0 and g is even continuous at x_0, we do not have that $f + g$ is (b, C)-lsc at x_0.

To get a counterexample for (b, C)-upper semicontinuity, we modify f and g as follows. Put $g(x) := (1, -2)$ and

$$f(x) := \begin{cases} (-2, 1) & \text{if} \quad x = 0, \\ (3, 3) & \text{otherwise.} \end{cases}$$

Then, f is $(0, C)$-usc at 0 and g is continuous (and hence $(0, C)$-usc) at 0. But,

$$(f+g)(x) = \begin{cases} (-1, -1) & \text{if} \quad x = 0, \\ (4, 1) & \text{otherwise} \end{cases}$$

is not $(0, C)$-usc at 0 since $\mathrm{L}_{\not<0} h$ is not closed. Indeed, let $x^k = \frac{1}{k} \to 0$. Then,

$$h(x^k) = (4, 1) \not< 0,$$

but $h(0) = (-1, -1) < 0$, whence $\mathrm{L}_{\not<0} h$ is not closed. This example shows that from f being (b, C)-usc and g being even continuous at x_0, it does not follow that $f + g$ is (b, C)-usc at x_0,

Moving on to discussions of C-quasisemicontinuity, we have similar counterexamples as follows.

Example 6.11 (Counterexample for unpreserving of cone-quasisemicontinuity under a sum of a continuity map and a cone-quasisemicontinuity map) Let Y, A, C, and $x_0 = 0$ be as in Example 6.5. Let $f, g : \mathbb{R} \to \mathbb{R}^2$ be defined by

$g(x) := (2, 2)$ for all x and

$$f(x) := \begin{cases} (-1, -1) & \text{if } \; x = 0, \\ (-3, -3) & \text{otherwise.} \end{cases}$$

Then, f is $(0, C)$-qlsc and g is continuous (and hence $(0, C)$-lsc) at 0. But,

$$(f + g)(x) = \begin{cases} (1, 1) & \text{if } \; x = 0, \\ (-1, -1) & \text{otherwise} \end{cases}$$

is not $(0, C)$-qlsc at 0 (to see this let $x^k = \frac{1}{k}$).

A similar assertion for (b, C)-quasi upper semicontinuity is seen by setting $g(x) := (-2, -2)$ and

$$f(x) := \begin{cases} (1, 1) & \text{if } \; x = 0, \\ (3, 3) & \text{otherwise.} \end{cases}$$

The following example indicates that, for a fixed level b, the sum of a continuous function and a (b, C)-lsc function may not be (b, C)-qlsc.

Example 6.12 (Counterexample for unpreserving of generalized cone-semicontinuity under the sum a continuity map and a generalized cone-semicontinuity map) Let us modify the previous example: only two new items $f(x) := (2, 2)$ and

$$g(x) := \begin{cases} (-1, -1) & \text{if } \; x = 0, \\ (-3, -3) & \text{otherwise.} \end{cases}$$

Then, f is continuous and g is $(0, C)$-lsc at 0. But,

$$(f + g)(x) = \begin{cases} (1, 1) & \text{if } \; x = 0, \\ (-1, -1) & \text{otherwise} \end{cases}$$

is not $(0, C)$-qlsc at 0.

Now let $f(x) := (-2, -2)$, $x_0 = 0$, and

$$g(x) := \begin{cases} (1, 1) & \text{if } \; x = 0, \\ (3, 3) & \text{otherwise.} \end{cases}$$

Then, f is continuous and g is $(0, C)$-usc at 0. But,

$$(f+g)(x) = \begin{cases} (-1, -1) & \text{if} \quad x = 0, \\ (1, 1) & \text{otherwise} \end{cases}$$

is not $(0, C)$-qusc at 0.

Since the existence of solutions of quasiequilibrium problems was intensively studied in the previous chapters, we focus on the stability study, assuming always the needed existence. In the subsections, we assume additionally that the properties of X, Y, Λ, and M involved in our problems satisfy the requires of considering.

Theorem 6.13 (Upper semicontinuity of solution maps of weak equilibrium problems) Sol(WEP) *is usc at* $(\bar{\lambda}, \bar{\mu})$ *if:*

(a) $\mathscr{D}$ *is continuous and compact-valued at* $\bar{\lambda}$;
(b) f *is* $(0, C)$*-usc in* $\mathscr{D}(\Lambda) \times \mathscr{D}(\Lambda) \times \{\bar{\mu}\}$.

Proof Suppose that Sol(WEP) is not usc at $(\bar{\lambda}, \bar{\mu})$, i.e., there is an open neighborhood U of Sol(WEP)$(\bar{\lambda}, \bar{\mu})$ such that there are sequences $\{(\lambda^k, \mu^k)\}_k$ tending to $(\bar{\lambda}, \bar{\mu})$ and $x^k \in$ Sol(WEP)(λ^k, μ^k) with $x^k \notin U$ for all k. By the upper semicontinuity of $\mathscr{D}$ and the compactness of $\mathscr{D}(\bar{\lambda})$, one can assume that $x^k \to \bar{x}$ for some $\bar{x} \in \mathscr{D}(\bar{\lambda})$. We show that $\bar{x}$ belongs to Sol(WEP)(λ^k, μ^k). Suppose to the contrary that there is $\bar{y} \in \mathscr{D}(\bar{\lambda})$ such that

$$f(\bar{x}, \bar{y}, \bar{\mu}) < 0.$$

The lower semicontinuity of $\mathscr{D}$ in turn shows the existence of $y^k \in \mathscr{D}(\lambda^k)$ such that $y^k \to \bar{y}$. Condition (a) allows one to assume that

$$f(x^k, y^k, \mu^k) < 0,$$

which is impossible as x^k is an element of Sol(WEP)(λ^k, μ^k) for all k. Therefore, $\bar{x}$ is an element of Sol(WEP)$(\bar{\lambda}, \bar{\mu}) \subset U$, which is again a contradiction since $x^k \notin U$ for each k. □

Similarly, for (SEP), we have the following result.

Theorem 6.14 (Upper semicontinuity of solution maps of strong equilibrium problems) *Impose assumption* (a) *of Theorem 6.13 and assume further that:*

(b) f *is* $(0, C)$*-qusc in* $\mathscr{D}(\Lambda) \times \mathscr{D}(\Lambda) \times \{\bar{\mu}\}$.

Then, Sol(SEP) *is usc at* $(\bar{\lambda}, \bar{\mu})$.

6.2.2 *Lower Semicontinuity of Solution Mappings of Vector Equilibrium Problems*

Now we discuss several generalized quasiconvexity notions for vector maps and develop their properties in order to use them in discussing the lower semicontinuity of solution maps. In this section, let X be a vector space, Y a topological vector space partially ordered by a solid pointed closed convex cone C, $A \subset X$ a convex set, and $h : A \to Y$. The following definition is known. (The relations $\leq$, $<$, etc., are defined by C.)

Definition 6.15 (Types of cone convexity of maps)

(a) h is said to be CONVEX if, for every $x_1, x_2 \in A$ and $t \in [0, 1]$,

$$h(tx_1 + (1-t)x_2) \leq th(x_1) + (1-t)h(x_2).$$

(b) h is said to be STRICTLY CONVEX if, for every $x_1, x_2 \in A$ and $t \in (0, 1)$,

$$h(tx_1 + (1-t)x_2) < th(x_1) + (1-t)h(x_2).$$

(c) h is said to be QUASICONVEX if, for every $x_1, x_2 \in A$ and $t \in [0, 1]$, either $h(tx_1 + (1-t)x_2) \leq h(x_1)$ or $h(tx_1 + (1-t)x_2) \leq h(x_2)$.
(d) h is said to be STRICTLY QUASICONVEX if, for every $x_1, x_2 \in A$ and $t \in (0, 1)$, either $h(tx_1 + (1-t)x_2) < h(x_1)$ or $h(tx_1 + (1-t)x_2) < h(x_2)$.

For these definitions and for all later convexity-related definitions, by inverting the inequalities, we get the corresponding concavity notions. In other words, h is concave (strictly concave, quasiconcave, etc.) if and only if -h is convex (strictly convex, quasiconvex, etc., resp.).

Remark 6.16 (Irrelationships between convexity properties) If $Y = C \cup -C$, like for scalar functions, convexity (strict convexity, resp.) implies quasiconvexity (strict quasiconvexity). However, this is not true in general as shown by the following example.

Example 6.17 (Class of cone-convex maps is not contained in that of cone-quasiconvex maps) Let $X = A = \mathbb{R}, Y = \mathbb{R}^2, C = \mathbb{R}^2_+$, and $h(x) := (-1-2x, -3+3x)$. Then, h is convex as it is linear. But, h is not quasiconvex. Indeed, for $x_1 = 0, x_2 = 1$, and $t = \frac{1}{2}$, one has:

$$h(x_1) = (-1, -3), h(x_2) = (-3, 0),$$

but

$$h\left(\frac{1}{2}x_1 + \frac{1}{2}x_2\right) = \left(-2, -\frac{3}{2}\right) \not\leq h(x_1), \text{ and } \not\leq h(x_2).$$

Note that, for the scalar case, convexity property is stronger than the quasiconvexity. The following example shows that this assertion is not adequate for the vector case.

Example 6.18 (Class of cone-strict convex maps is not contained in that of cone-quasiconvex maps) Let $X = A = \mathbb{R}$, $Y = \mathbb{R}^2$, $C = \mathbb{R}^2_+$, and

$$h(x) := (4x^2 + 2x - 1, 4x^2 - 10x + 5) =: (h_1(x), h_2(x)).$$

Observe that both h_1 and h_2 are scalar strictly convex functions. Hence, for all $x_1, x_2 \in \mathbb{R}$, all $t \in (0, 1)$, and for $x_t = tx_1 + (1-t)x_2$, one has:

$$h(x_t) < th(x_1) + (1-t)h(x_2),$$

i.e., h is strictly convex. But, h is not quasiconvex. Indeed, for $x_1 = 0, x_2 = 1$, and $t = \frac{1}{2}$, one has:

$$h(x_1) = (-1, 5), h(x_2) = (5, -1),$$

and

$$h\left(\frac{1}{2}x_1 + \frac{1}{2}x_2\right) = (1, 1) \not\leq h(x_1), \quad h\left(\frac{1}{2}x_1 + \frac{1}{2}x_2\right) = (1, 1) \not\leq h(x_2).$$

Definition 6.19 (Types of level convexity) Let $b \in Y$, $x_1, x_2 \in A$, $t \in (0, 1)$, and $x_t = (1-t)x_1 + tx_2$:

(a) If $h(x_1) \not< b$ and $h(x_2) \not\leq b$ imply $h(x_t) \not\leq b$, then h is said to be $\text{lev}_{\not\leq b}$-CONVEX.
(b) If $h(x_1) \geq b$ and $h(x_2) > b$ imply $h(x_t) > b$, then h is called $\text{lev}_{>b}$-CONVEX.

Note that, if $Y = C \cup -C$, these two properties coincide. Furthermore, if h is $\text{lev}_{\not\leq b}$-convex ($\text{lev}_{>b}$-convex, resp.), then $L_{\not\leq b}h$ ($L_{>b}h$, resp.) is convex, but not vice versa as shown by the following example.

Example 6.20 (Difference of types of level convexity) Let $A = [-2, 2]$, $Y = \mathbb{R}$, $C = \mathbb{R}_+$, and $h(x) := \max\{0, x\}$ for $x \in A$.

Then, $\text{L}_{\not\leq 0}h = (0, 2]$ is convex, but h is not $\text{lev}_{\not\leq 0}$-convex. Indeed, let $x_1 = -1, x_2 = 1$, and $t = \frac{1}{2}$. Then, $h(x_1) = 0 \not< 0, h(x_2) = 1 \not\leq 0$, but

$$h(x_t) = h\left(\frac{1}{2}x_1 + \frac{1}{2}x_2\right) = h(0) = 0 \leq 0.$$

Remark 6.21 (A relationship between cone-strict concavity and level convexity) If h is strictly quasiconcave, then h is $\text{lev}_{\not\leq b}$-convex for all $b \in Y$. Indeed, let $x_1, x_2 \in A$ be such that $h(x_1) \not< b$ and $h(x_2) \not\leq b$. Suppose the existence of $c_1 \in C$ such that $h(x_t) - b = -c_1$. Since h is strictly quasiconcave, $h(x_t) > h(x_1)$ or $h(x_t) > h(x_2)$ for all $t \in (0, 1)$. If $h(x_t) > h(x_1)$, there is $c_2 \in \text{int}C$ with

$h(x_t) = h(x_1) + c_2$. Hence, $h(x_1) - b = -c_1 - c_2 \in -\text{int}C$, which is impossible as $h(x_1) \not< b$. The case $h(x_t) > h(x_2)$ is similar.

Example 6.18 shows that $\text{lev}_{\not\leq b}$-convexity does not follow neither from strict concavity nor from strict convexity. Indeed, we know that the map

$$h(x) := (4x^2 + 2x - 1, 4x^2 - 10x + 5)$$

is strictly convex (and $-h$ is strictly concave). For $b = 0$, $x_1 = 0$, $x_2 = 1$, and $t = \frac{1}{2}$, we have $h(x_1) = (-1, 5) \not< 0$ and $h(x_2) = (5, -1) \not\leq 0$, but $h(x_{\frac{1}{2}}) = (1, 1) \leq 0$. Similarly, $-h$ is not $\text{lev}_{\not\leq b}$-convex either. The next example indicates that neither quasiconcavity nor quasiconvexity implies $\text{lev}_{\not\leq b}$-convexity.

Example 6.22 (Irrelationships between level-convexity and cone-concavity/convexity properties) Let $X = Y = \mathbb{R}$, $A = \mathbb{R}$, $C = \mathbb{R}_+$, and

$$h(x) := \begin{cases} 0 & \text{if} \quad x \leq 1, \\ x - 1 & \text{otherwise.} \end{cases}$$

Then, h is both quasiconvex and quasiconcave. But, h is not $\text{lev}_{\not\leq 0}$-convex since for $x_1 = 0$, $x_2 = 2$, one has $h(x_1) = 0 \not\leq 1$, $h(x_2) = 1 \not< 1$, but $h(x_{\frac{1}{2}}) = 0 \not\leq 1$.

Remark 6.23 (Relationships between level-convexity and cone-strict concavity/quasiconcavity maps) If h is strictly quasiconcave or concave, then h is $\text{lev}_{>b}$-convex for all $b \in Y$. Indeed, assume that h is strictly quasiconcave and $x_1, x_2 \in A$ such that

$$h(x_1) \geq b, h(x_2) > b.$$

For any $t \in (0, 1)$, if $h(x_t) > h(x_1)$, there is $c_1 \in \text{int}C$ with $h(x_t) - h(x_1) = c_1$. Since $h(x_1) \geq b$, there is $c_2 \in C$ with $h(x_1) = b + c_2$. Consequently,

$$h(x_t) - b = c_1 + c_2 \in \text{int}C,$$

i.e., $h(x_t) > b$. Similarly, $h(x_t) > b$ if $h(x_t) > h(x_2)$. Thus, h is $\text{lev}_{>b}$-convex. Now assume that h is concave and $x_1, x_2 \in A$ such that $h(x_1) \geq b, h(x_2) > b$. Then, there exist $c_1 \in C$ and $c_2 \in \text{int}C$ such that

$$h(x_1) = b + c_1, h(x_2) = b + c_2.$$

Since h is concave, there exists $c_3 \in C$ such that

$$h(x_t) - th(x_1) - (1 - t)h(x_2) = c_3.$$

Consequently, $h(x_t) - b = tc_1 + (1 - t)c_2 + c_3 \in \text{int}C$. Thus, $h(x_t) > b$.

Example 6.22 shows also that neither quasiconcavity nor quasiconvexity yields $\text{lev}_{>b}$-convexity since in this case $\text{lev}_{\not\leq b}$-convexity coincides with $\text{lev}_{>b}$-convexity.

In the rest of this subsection, we assume additionally that X is a Hausdorff topological vector space and A is convex. We consider the following problems (WEP_1) and (SEP_1) as auxiliary problems of (WEP) and (SEP), resp:

(WEP_1) find $\bar{x} \in \mathscr{D}(\lambda)$ such that, for all $y \in \mathscr{D}(\lambda)$,

$$f(\bar{x}, y, \mu) \not\leq 0;$$

(SEP_1) find $\bar{x} \in \mathscr{D}(\lambda)$ such that, for all $y \in \mathscr{D}(\lambda)$,

$$f(\bar{x}, y, \mu) > 0.$$

Let $\text{Sol(WEP)}_1(\lambda, \mu)$ and $\text{Sol(SEP)}_1(\lambda, \mu)$ be the solution sets of (WEP_1) and (SEP_1), resp., corresponding to (λ, μ).

Theorem 6.24 (Lower semicontinuity of solution maps of auxiliary problems of weak equilibrium problems) *The map* Sol(WEP)_1 *of* (WEP) *is lsc at* $(\bar{\lambda}, \bar{\mu})$ *if:*

(a) *$\mathscr{D}$ is continuous and convex, compact at $\bar{\lambda}$;*
(b) *f is $(0, C)$-qlsc on $A \times A \times \{\bar{\mu}\}$;*
(c) *for each $y \in \mathscr{D}(\bar{\lambda})$, $f(\cdot, y, \bar{\mu})$ is $\text{lev}_{\not\leq 0}$-convex in $\mathscr{D}(\bar{\lambda})$.*

Proof We start by proving that Sol(WEP)_1 is lsc at $(\bar{\lambda}, \bar{\mu})$. Suppose to the contrary that Sol(WEP)_1 is not lsc at $(\bar{\lambda}, \bar{\mu})$, i.e., there are $\bar{x} \in \text{Sol(WEP)}_1(\bar{\lambda}, \bar{\mu})$ and sequence $\{(\lambda^k, \mu^k)\}_k$ tending to $(\bar{\lambda}, \bar{\mu})$ such that, for any choice of $x^k \in \text{Sol(WEP)}_1(\lambda^k, \mu^k)$, $\{x^k\}_k$ does not converge to $\bar{x}$. Since $\mathscr{D}$ is lsc at $\bar{\lambda}$, there is $\bar{x}^k \in \mathscr{D}(\lambda^k)$ with $\bar{x}^k \to \bar{x}$. By our assumption, there must be a subsequence $\bar{x}^{k_j}$ such that $\bar{x}^{k_j} \notin \text{Sol(WEP)}_1(\lambda^{k_j}, \mu^{k_j})$ for all k_j, i.e., for some $y^{k_j} \in \mathscr{D}(\lambda^{k_j})$,

$$f(\bar{x}^{k_j}, y_{k_j}, \mu^{k_j}) \leq 0. \tag{6.1}$$

As $\mathscr{D}$ is usc at $\bar{\lambda}$ and $\mathscr{D}(\bar{\lambda})$ is compact, one has $\bar{y} \in \mathscr{D}(\bar{\lambda})$ such that $y^{k_j} \to \bar{y}$ (taking a subsequence if necessary). By assumption (b), (6.1) yields that $f(\bar{x}, \bar{y}, \bar{\mu}) \leq 0$, which is impossible since $\bar{x} \in \text{Sol(WEP)}_1(\bar{\lambda}, \bar{\mu})$.

Now let us check that

$$\text{Sol(WEP)}(\bar{\lambda}, \bar{\mu}) \subset \text{clSol(WEP)}_1(\bar{\lambda}, \bar{\mu}). \tag{6.2}$$

Let $\bar{x} \in \text{Sol(WEP)}(\bar{\lambda}, \bar{\mu})$, $\bar{x}^1 \in \text{Sol(WEP)}_1(\bar{\lambda}_0, \bar{\mu})$ and $t \in (0, 1)$. Then, $x_t = (1-t)\bar{x} + t\bar{x}^1$ belongs to $\mathscr{D}(\bar{\lambda})$ as $\mathscr{D}(\bar{\lambda})$ is convex. Since $f(\cdot, \cdot, \bar{\mu})$ is $\text{lev}_{\not\leq 0}$-convex, for any $y \in \mathscr{D}(\bar{\lambda})$,

$$f(x_t, y, \bar{\mu}) \not\leq 0,$$

i.e., $x_t \in \mathrm{Sol(WEP)}_1(\bar{\lambda}, \bar{\mu})$. Therefore, (6.2) holds. By the lower semicontinuity of $\mathrm{Sol(WEP)}_1$ at $(\bar{\lambda}, \bar{\mu})$, we have:

$$\begin{aligned}\mathrm{Sol(WEP)}(\bar{\lambda}, \bar{\mu}) \subset \mathrm{clSol(WEP)}_1(\bar{\lambda}, \bar{\mu}) &\subset \liminf \mathrm{Sol(WEP)}_1(\bar{\lambda}^k, \mu^k)\\ &\subset \liminf \mathrm{Sol(WEP)}(\lambda^k, \mu^k).\end{aligned}$$

Equivalently, Sol(WEP) is lsc at $(\bar{\lambda}, \bar{\mu})$ and the proof is complete. □

Example 6.25 (Essentialness of the level-convexity condition) Let $X = Y = A = \mathbb{R}$, $\Lambda = [0, 1]$, $C = \mathbb{R}_+$, $\mathscr{D}(\lambda) = [\lambda, \lambda + 3]$, $\bar{\lambda} = 0$, and

$$f(x, y, \lambda) := x^2 - (\lambda + 1)x.$$

Then, it is easy to see that assumptions (a) and (b) of Theorem 6.24 are satisfied. But, $\mathrm{Sol(WEP)}(0) = \{0\} \cup [1, 3]$ and $\mathrm{Sol(WEP)}(\lambda) = [\lambda+1, \lambda+3]$ for all $\lambda \in (0, 1]$, and hence Sol(WEP) is not lsc at $\bar{\lambda}$. The reason is that (c) is violated. Indeed, let $x^1 = 0, x^2 = \frac{3}{2} \in \mathscr{D}(\bar{\lambda}) = [0, 3]$. Then, for any $y \in \mathscr{D}(\bar{\lambda}) = [0, 3]$, we have:

$$f(x^1, y, \bar{\lambda}) = 0,\ f(x^2, y, \bar{\lambda}) = \frac{3}{4},$$

but

$$f\left(\frac{1}{2}x^1 + \frac{1}{2}x^2, y, \bar{\lambda}\right) = -\frac{3}{16}.$$

Considering now problem (SEP), we obtain a result similar to Theorem 6.24.

Theorem 6.26 (Lower semicontinuity of solution maps of strong equilibrium problems) *The solution map* Sol(SEP) *is lsc at* $(\bar{\lambda}, \bar{\mu})$ *if:*

(a) $\mathscr{D}$ *is continuous and convex, compact at* $\bar{\lambda}$;
(b) f *is* $(0, C)$*-lsc on* $A \times A \times \{\bar{\mu}\}$;
(c) *for each* $y \in \mathscr{D}(\bar{\lambda})$, $f(\cdot, y, \bar{\mu})$ *is* $\mathrm{lev}_{>0}$*-convex in* $\mathscr{D}(\bar{\lambda})$.

Finally, we present sufficient conditions for Hausdorff semicontinuity of solution maps.

Theorem 6.27 (Hausdorff lower semicontinuity of solution maps of weak equilibrium problems) *Impose the assumptions of Theorem 6.24 and*

(d) *for each* $y \in A$, $f(\cdot, y, \bar{\mu})$ *is* $(0, C)$*-usc on* A.

Then, Sol(WEP) *is Hausdorff lower semicontinuous at* $(\bar{\lambda}, \bar{\mu})$.

Proof We first show that $\mathrm{Sol(WEP)}(\bar{\lambda}, \bar{\mu})$ is closed. Assume that $x^k \in \mathrm{Sol(WEP)}(\bar{\lambda}, \bar{\mu})$ and $x^k \to \bar{x}$. Since $\mathscr{D}(\bar{\lambda})$ is compact, $\bar{x} \in \mathscr{D}(\bar{\lambda})$. Suppose there exists $\bar{y} \in \mathscr{D}(\bar{\lambda})$ such that

$$f(\bar{x}, \bar{y}, \bar{\mu}) < 0. \tag{6.3}$$

Because $x^k \in \mathrm{Sol(WEP)}(\bar{\lambda}, \bar{\mu})$, for any $y \in \mathscr{D}(\bar{\lambda})$,

$$f(x^k, y, \bar{\mu}) \nless 0. \tag{6.4}$$

Assumption (d) shows a contradiction between (6.3) and (6.4). Thus, Sol(WEP) $(\bar{\lambda}, \bar{\mu})$ is closed and then it is compact.

Applying Theorem 6.24, we get the lower semicontinuity of Sol(WEP). From this and the above conclusion, we conclude that Sol(WEP) is Hlsc at $(\bar{\lambda}, \bar{\mu})$. □

The following example confirms the essentialness of the added assumption.

Example 6.28 (Cruciality of the generalized semicontinuity condition) Let $X = A = \mathbb{R}^2$, $Y = \mathbb{R}$, $\Lambda = [0, 1]$, $C = \mathbb{R}_+$, $\bar{\lambda} = 0$, $\mathscr{D}(\lambda) = \{(x_1, x_2) \mid x_2 = \lambda x_1\}$, and $f(x, y, \lambda) = 1 + \lambda$. Then, the assumptions of Theorem 6.27, but the compactness of $\mathscr{D}(\bar{\lambda})$, are satisfied. Direct computations give

$$\mathrm{Sol(WEP)}(\lambda) = \{(x_1, x_2) \in \mathbb{R}^2 \mid x_2 = \lambda x_1\},$$

and then Sol(WEP) is not Hlsc at $\bar{\lambda}$ (although Sol(WEP) is lsc at $\bar{\lambda}$).

Similarly, for problem (SEP), we have the following.

Theorem 6.29 (Hausdorff lower semicontinuity of solution maps of strong equilibrium problems) *Impose the assumptions of Theorem 6.27 and the following additional condition:*

(d) *for each $y \in A$, $f(\cdot, y, \bar{\mu})$ is $(0, C)$-qusc on A.*

Then, Sol(SEP) *is Hausdorff lower semicontinuous at $(\bar{\lambda}, \bar{\mu})$.*

6.2.3 *Continuity of Solution Mappings of Vector Equilibrium Problems*

In this section, we avoid concavity assumptions and use relaxed monotonicity conditions. The following monotonicity notions are extensions of the counterparts for the scalar case.

Definition 6.30 (Cone-quasimonotonicity and cone-pseudomonotonicity) Let X be a set, Y a linear space, and $C \subset Y$ a convex cone. Let $y_1 \leq y_2$ mean $y_2 - y_1 \in C$. Let $g : X \times X \to Y$:

(a) g is said to be QUASIMONOTONE of type 1 in X if, for all $x, z \in X$ and $x \neq z$,

$$[g(x, z) \nleq 0] \Rightarrow [g(z, x) \leq 0].$$

(b) g is said to be QUASIMONOTONE of type 2 in X if, for all $x, z \in X$ and $x \neq z$,

$$[g(x, z) > 0] \Rightarrow [g(z, x) \not> 0].$$

(c) g is said to be PSEUDOMONOTONE of type 1 in X if, for all $x, z \in X$ and $x \neq z$,

$$[g(x, z) \not< 0] \Rightarrow [g(z, x) \leq 0].$$

(d) g is said to be PSEUDOMONOTONE of type 2 in X if, for all $x, z \in X$ and $x \neq z$,

$$[g(x, z) \geq 0] \Rightarrow [g(z, x) \not> 0].$$

Remark 6.31 (Relationships between cone-quasimonotonicity and cone-pseudomonotonicity) Note that if g is quasimonotone of type 1 (pseudomonotone of type 1), then g is quasimonotone of type 2 (pseudomonotone of type 2, resp.). Indeed, when g is quasimonotone of type 1, if $g(x, z) > 0$ (i.e., $g(x, z) \not\leq 0$), then $g(z, x) \leq 0$, i.e., $g(z, x) \not> 0$. So, g is quasimonotone of type 2. The relationship between the pseudomonotonicities of type 1 and type 2 is checked similarly.

The following example ensures that the converse is not true.

Example 6.32 (A counterexample for relationships between cone-quasimonotonicity and cone-pseudomonotonicity) Let $X = \mathbb{R}, Y = \mathbb{R}^2, C = \mathbb{R}^2_+$, and $g : \mathbb{R} \times \mathbb{R} \to \mathbb{R} \times \mathbb{R}$ be defined by

$$g(x, z) := \begin{cases} (1, 1) & \text{if } \ x \leq z, \\ (-1, 1) & \text{if } \ x > z. \end{cases}$$

Then, g is both quasimonotone of type 2 and pseudomonotone of type 2 in $\mathbb{R}$, but neither quasimonotone of type 1 nor pseudomonotone of type 1.

Theorem 6.33 (Continuity of solution maps of Sol(WEP)**)** *Impose for* (WEP) *the assumptions of Theorem 6.13 and:*

(a) $f(\cdot, \cdot, \bar{\mu})$ *is quasimonotone of type 1 in* $\mathscr{D}(\bar{\lambda}) \times \mathscr{D}(\bar{\lambda})$;
(b) *for all* $x \in$ Sol(WEP)$(\bar{\lambda}, \bar{\mu})$ *and* $y \in$ Sol(WEP)$(\bar{\lambda}, \bar{\mu}) \setminus \{x\}$, *one has* $f(x, y, \bar{\mu}) \not\leq 0$.

Then, Sol(WEP) *is continuous at* $(\bar{\lambda}, \bar{\mu})$.

Proof It suffices to prove that Sol(WEP) is lsc at $(\bar{\lambda}, \bar{\mu})$. Suppose to the contrary that there are a sequence $\{(\lambda^k, \mu^k)\}_k$ tending to $(\bar{\lambda}, \bar{\mu})$ and a point $\bar{x} \in$ Sol(WEP)$(\bar{\lambda}, \bar{\mu})$ such that, for any choice of $x^k \in$ Sol(WEP)(λ^k, μ^k), the sequence $\{x^k\}_k$ does not converge to $\bar{x}$. Since $\mathscr{D}$ is usc and $\mathscr{D}(\bar{\lambda})$ is compact, we can assume that $x^k \to \bar{x}_0$ for some $\bar{x}_0 \in \mathscr{D}(\bar{\lambda})$.

From the proof of Theorem 6.13, we see that $\bar{x}_0 \in \text{Sol(WEP)}(\bar{\lambda}, \bar{\mu})$. By the contradiction assumption, one has $\bar{x}_0 \neq x_0$. Due to assumption (b), one has:

$$f(\bar{x}_0, \bar{x}, \bar{\mu}) \not\leq 0 \quad \text{and} \quad f(\bar{x}, \bar{x}_0, \bar{\mu}) \not\leq 0,$$

which is impossible since $f(\cdot, \cdot, \bar{\mu})$ is quasimonotone of type 1. □

Similarly, we have the following.

Theorem 6.34 (Continuity of solution map of strong equilibrium problems) *Impose the assumptions of Theorem 6.14. Assume further that:*

(a) $f(\cdot, \cdot, \bar{\mu})$ *is quasimonotone of type 2 in* $\mathscr{D}(\bar{\lambda}) \times \mathscr{D}(\bar{\lambda})$;
(b) *for all* $x \in \text{Sol(SEP)}(\bar{\lambda}, \bar{\mu})$ *and* $y \in \text{Sol(SEP)}(\bar{\lambda}, \bar{\mu}) \setminus \{x\}$, *one has* $f(x, y, \bar{\mu}) > 0$.

Then, Sol(SEP) *is continuous at* $(\bar{\lambda}, \bar{\mu})$.

We can modify the assumptions and use pseudomonotonicity as follows.

Theorem 6.35 (Continuity of solution map of weak equilibrium problems) *Impose the assumptions of Theorem 6.13 and:*

(a') $f(\cdot, \cdot, \bar{\mu})$ *is pseudomonotone of type 1 in* $\mathscr{D}(\bar{\lambda}) \times \mathscr{D}(\bar{\lambda})$;
(b') *if* $f(x, y, \bar{\mu}) \in \text{bd}(-C)$, *then* $x = y$ (bd(.) *denotes the boundary);*
(c') *for all* $\bar{x}$ *and* $\bar{x}_0$ *in* $\text{Sol(WEP)}(\bar{\lambda}, \bar{\mu})$, $f(\bar{x}, \bar{x}_0, \bar{\mu}) \nless 0$.

Then, Sol(WEP) *is continuous at* $(\bar{\lambda}, \bar{\mu})$.

Proof We retain the first part of the proof of Theorem 6.13 to obtain $\bar{x}_0 \neq x_0$. By (c') one has $f(\bar{x}, \bar{x}_0, \bar{\mu}) \nless 0$ and $f(\bar{x}_0, \bar{x}, \bar{\mu}) \nless 0$. Clearly, (a') implies that $f(\bar{x}_0, \bar{x}, \bar{\mu}) \leq 0$. Hence, $f(\bar{x}_0, \bar{x}, \bar{\mu}) \in \text{bd}(-C)$. Assumption (b') now yields a contradiction that $\bar{x}_0 = \bar{x}$. □

By the relationships between the solutions, we have the next result.

Theorem 6.36 (Continuity of solution maps of strong equilibrium problems) *Impose the assumptions in Theorem 6.14 and that:*

(a') $f(\cdot, \cdot, \bar{\mu})$ *is pseudomonotone of type 2 in* $\mathscr{D}(\bar{\lambda}) \times \mathscr{D}(\bar{\lambda})$;
(b') *if* $f(x, y, \bar{\mu}) \in \text{bd}(C)$, *then* $x = y$;
(c') *for all* $\bar{x}$ *and* $\bar{x}_0$ *in* $\text{Sol(SEP)}(\bar{\lambda}, \bar{\mu})$, $f(\bar{x}, \bar{x}_0, \bar{\mu}) \geq 0$.

Then, Sol(SEP) *is continuous at* $(\bar{\lambda}, \bar{\mu})$.

6.3 Hölder Continuity

Throughout this section, let X, Λ and M be metric spaces and Y a metric linear space. For convenience, we use $d(\cdot, \cdot)$ to denote the metric in any of these spaces (the context makes it clear which space is considered). We will need the following two assumptions.

(W) For the reference point $(\bar{\lambda}, \bar{\mu}) \in \Lambda \times M$, there are neighborhoods $U(\bar{\lambda})$ and $V(\bar{\mu})$ of $\bar{\lambda}$ and $\bar{\mu}$ such that, for $h > 0$, $\beta > \theta > 0$, $\mu \in V(\bar{\mu})$, and $x \neq y$ in $\mathscr{D}\big(U(\bar{\lambda})\big)$,

$$hd^{\beta}(x, y) \leq d(f(x, y, \mu), Y \setminus -\text{int}C) + d(f(y, x, \mu), Y \setminus -\text{int}C). \tag{6.5}$$

(S) It is similar to (W), but with (6.5) replaced by

$$hd^{\beta}(x, y) \leq d(f(x, y, \mu), C) + d(f(y, x, \mu), C).$$

Remark 6.37 (A special case of conditions (W) and (S)) These assumptions look complicated. But, they are not hard to be checked. To make their meanings clearer, we consider the simplest case that $f : X \times X \to \mathbb{R}$. Then, (W) and (S) coincide and collapse to the following: for all $x \neq y$ in S

$$hd^{\beta}(x, y) \leq d\big(f(x, y), \mathbb{R}_+\big) + d\big(f(y, x), \mathbb{R}_+\big). \tag{6.6}$$

6.3.1 Hölder Continuity of Solution Maps

Theorem 6.38 (Hölder continuity of solution maps of weak and strong equilibrium problems) *Assume that:*

(a) *$\mathscr{D}$ is (ℓ, α)-Hölder at in $\bar{\lambda}$;*
(b) *there are neighborhoods $U(\bar{\lambda})$ of $\bar{\lambda}$ and $V(\bar{\mu})$ of $\bar{\mu}$ such that f is (n_1, δ_1)-Hölder at $(\bar{\mu}, \theta)$-relative to $\mathscr{D}(U(\bar{\lambda}))$ and, for all $x \in \mathscr{D}(U(\bar{\lambda}))$ and $\mu \in V(\bar{\mu})$, $f(x, \cdot, \mu)$ is (n_2, δ_2)-Hölder in $\mathscr{D}(U(\bar{\lambda}))$;*
(c) *for* (WEP) *(for (SEP)), assumption (W) (assumption (S), resp.) is satisfied.*

Then, for each (λ, μ) in a neighborhood of $(\bar{\lambda}, \bar{\mu})$, (WEP) ((SEP), resp.) has a unique solution $x(\lambda, \mu)$ which satisfies the Hölder condition:

$$d\big(x(\lambda_1, \mu_1), x(\lambda_2, \mu_2)\big) \leq k_1 d^{\alpha_2 \delta_2/\beta}(\lambda_1, \lambda_2) + k_2 d^{\delta_1/(\beta-\theta)}(\mu_1, \mu_2),$$

where k_1 and k_2 are positive constants depending on $h, \beta, n_1, n_2, \delta_1, \delta_2, \theta$, etc.

Proof We demonstrate the assertion only for (SEP). Let $\lambda_1, \lambda_2 \in U(\bar{\lambda})$ and $\mu_1, \mu_2 \in V(\bar{\mu})$.

Step 1. We prove that, for all $x(\lambda_1, \mu_1) \in \text{Sol(SEP)}(\lambda_1, \mu_1)$ and $x(\lambda_1, \mu_2) \in \text{Sol(SEP)}(\lambda_1, \mu_2)$,

$$d_1 := d\big(x(\lambda_1, \mu_1), x(\lambda_1, \mu_2)\big) \leq \left(\frac{n_1}{h}\right)^{1/(\beta-\theta)} d^{\delta_1/(\beta-\theta)}(\mu_1, \mu_2). \tag{6.7}$$

Let $x(\lambda_1, \mu_1) \neq x(\lambda_1, \mu_2)$ (if the equality holds, then the proof is complete). Since $x(\lambda_1, \mu_1)$ and $x(\lambda_1, \mu_2)$ are solutions of (SEP), we have:

$$f(x(\lambda_1, \mu_1), x(\lambda_1, \mu_2), \mu_1) \geq 0, \tag{6.8}$$

$$f(x(\lambda_1, \mu_2), x(\lambda_1, \mu_1), \mu_2) \geq 0. \tag{6.9}$$

Then, combining (6.8) and assumption (c), we get:

$$d(f(x(\lambda_1, \mu_2), x(\lambda_1, \mu_1), \mu_1), C) \geq hd_1^{\beta}.$$

Applying (6.9), we have:

$$d(f(x(\lambda_1, \mu_2), x(\lambda_1, \mu_1), \mu_1), f(x(\lambda_1, \mu_2), x(\lambda_1, \mu_1), \mu_2)) \geq hd_1^{\beta}.$$

This together with assumption (b) yields $n_1 d_1^{\theta} d^{\delta_1}(\mu_1, \mu_2) \geq hd_1^{\beta}$, and consequently

$$d_1^{\beta-\theta} \leq \left(\frac{n_1}{h}\right) d^{\delta_1}(\mu_1, \mu_2).$$

Step 2. We show that, for each $x(\lambda_1, \mu_2) \in \mathrm{Sol(SEP)}(\lambda_1, \mu_2)$ and $x(\lambda_2, \mu_2) \in \mathrm{Sol(SEP)}(\lambda_2, \mu_2)$,

$$d_2 := d\big(x(\lambda_1, \mu_2), x(\lambda_2, \mu_2)\big) \leq \left(\frac{2n_2\ell^{\delta_2}}{h}\right)^{1/\beta} d^{\alpha\delta_2/\beta}(\lambda_1, \lambda_2). \tag{6.10}$$

Let $x(\lambda_1, \mu_2) \neq x(\lambda_2, \mu_2)$. Thanks to (a), we have $x_1' \in \mathscr{D}(\lambda_1)$ and $x_2' \in \mathscr{D}(\lambda_2)$ such that

$$d\big(x(\lambda_1, \mu_2), x_2'\big) \leq \ell d^{\alpha}(\lambda_1, \lambda_2), \tag{6.11}$$

$$d\big(x(\lambda_2, \mu_2), x_1'\big) \leq \ell d^{\alpha}(\lambda_1, \lambda_2). \tag{6.12}$$

Due to the definition of (SEP), we have:

$$f(x(\lambda_1, \mu_2), x_1', \mu_2) \geq 0, \tag{6.13}$$

$$f(x(\lambda_2, \mu_2), x_2', \mu_2) \geq 0. \tag{6.14}$$

It follows from assumption (c) that

$$d(f(x(\lambda_1, \mu_2), x(\lambda_2, \mu_2), \mu_2), C) + d(f(x(\lambda_2, \mu_2), x(\lambda_1, \mu_2), \mu_2), C) \geq hd_2^{\beta}.$$

From (6.13) and (6.14), one has:

$$d(f(x(\lambda_1,\mu_2),x(\lambda_2,\mu_2),\mu_2),\ f(x(\lambda_1,\mu_2),x_1',\mu_2))$$

$$+\,d(f(x(\lambda_2,\mu_2),x(\lambda_1,\mu_2),\mu_2),\ f(x(\lambda_2,\mu_2),x_2',\mu_2)\ge hd_2^{\beta}.$$

Hence, the Hölder continuity assumptions in (b) of f imply that

$$n_2d^{\delta_2}(x(\lambda_2,\mu_2),x_1')+n_2d^{\delta_2}(x(\lambda_1,\mu_2),x_2')\ge hd_2^{\beta}.$$

Employing (6.11) and (6.12), we obtain:

$$n_2\ell^{\delta_2}d^{\alpha\delta_2}(\lambda_1,\lambda_2)+n_2\ell^{\delta_2}d^{\alpha\delta_2}(\lambda_1,\lambda_2)\ge hd_2^{\beta}.$$

Equivalently,

$$d_2^{\beta}\le\left(\frac{2n_2\ell^{\delta_2}}{h}\right)d^{\alpha\delta_2}(\lambda_1,\lambda_2).$$

Step 3. Finally, for all $x(\lambda_1,\mu_1)\in\mathrm{Sol(SEP)}(\lambda_1,\mu_1)$ and all $x(\lambda_2,\mu_2)\in\mathrm{Sol(SEP)}(\lambda_2,\mu_2)$, from

$$d\big(x(\lambda_1,\mu_1),x(\lambda_2,\mu_2)\big)\le d_1+d_2,$$

it follows from (6.86) and (6.10) that

$$\mathbb{D}\big(\mathrm{Sol(SEP)}(\lambda_1,\mu_1),\mathrm{Sol(SEP)}(\lambda_2,\mu_2)\big)\le k_1d^{\alpha\delta_2/\beta}(\lambda_1,\lambda_2)+k_2d^{\delta_1/(\beta-\theta)}(\mu_1,\mu_2),$$

where $k_1=\left(\frac{2n_2\ell^{\delta_2}}{h}\right)^{\frac{1}{\beta}}$ and $k_2=\left(\frac{n_1}{h}\right)^{\frac{1}{\beta-\theta}}$.

Taking $\lambda_2=\lambda_1$ and $\mu_2=\mu_1$, we see that the diameter of $\mathrm{Sol(SEP)}(\lambda_1,\mu_1)$ is 0, i.e., this set is a singleton $\{x(\lambda_1,\mu_1)\}$. Similarly, $\mathrm{Sol(SEP)}(\lambda_2,\mu_2)$ is also a singleton. Thus, the solution is unique and the required Hölder condition is checked. □

Definition 6.39 (Hölder calm maps) Let $h:X\to Y$ and ℓ,α be positive real numbers. h is said to be (ℓ,α)-HÖLDER CALM at $\bar{x}$ if there is a neighborhood N of $\bar{x}$ such that, for all $x\in N$,

$$\|(h(x)-h(\bar{x})\|\le\ell d^{\alpha}(x,\bar{x}).$$

Remark 6.40 (A motivation of the assumption that the Hölder degree is less than 1) For $\alpha>1$, the Hölder condition given in Definition 6.39 is very strong. If X and Y are normed spaces, it implies that the Fréchet derivative of h equals 0 in N and hence h is a constant in N (the Fréchet derivative $h'(\bar{x})=0$, resp.). Hence, in this section, we always consider Hölder condition of orders not greater than 1.

Definition 6.41 (Hölder calm vector maps) Let ℓ, α, m, β be positive and $\theta \geq 0$:

(a) A set-valued map $Q : \Lambda \rightrightarrows X$ is said to be (ℓ, α)-HÖLDER CALM at $\bar{\lambda}$ if there exists a neighborhood U of $\bar{\lambda}$ such that, for all $\lambda \in U$,

$$H(Q(\lambda), Q(\bar{\lambda})) \leq \ell d^{\alpha}(\lambda, \bar{\lambda}).$$

(b) A vector-valued map $h : X \times X \times M \to Y$ is termed (m, β)-HÖLDER CALM at $\bar{\mu} \in M$ θ-UNIFORMLY in $A \subset X$ if there is a neighborhood V of $\bar{\mu}$ such that, for all $\mu \in V$ and $x \neq y$ in A,

$$\|h(x, y, \mu) - h(x, y, \bar{\mu})\| \leq md^{\beta}(\mu, \bar{\mu})d^{\theta}(x, y).$$

If $\theta = 0$, one says that h is (m, β)-Hölder calm at $\bar{\mu}$ uniformly in A.

Employing the techniques used in the proof of Theorem 6.38, we arrive at the following result.

Theorem 6.42 (Hölder calmness of solution maps of weak and strong equilibrium problems) *Assume that:*

(a) *$\mathscr{D}$ is (ℓ, α)-Hölder calm at $\bar{\lambda}$;*
(b) *there are neighborhoods $U(\bar{\lambda})$ of $\bar{\lambda}$ and $V(\bar{\mu})$ of $\bar{\mu}$ such that f is (n_1, δ_1)-Hölder calm at $(\bar{\mu}, \theta)$-uniformly in $\mathscr{D}(U(\bar{\lambda}))$ and, for all $x \in \mathscr{D}(U(\bar{\lambda}))$ and $\mu \in V(\bar{\mu})$, $f(x, \cdot, \mu)$ is (n_2, δ_2)-Hölder in $\mathscr{D}(U(\bar{\lambda}))$;*
(c) *for (WEP)* ((SEP)) *assumption (W) (assumption* (S)*, resp.) is satisfied, resp.*

Then, for each (λ, μ) in a neighborhood of $(\bar{\lambda}, \bar{\mu})$, the solution map of (WEP) *(*(SEP)*, resp.) satisfies the following Hölder calmness condition:*

$$\mathbb{D}\big(\mathrm{Sol(WEP)}S\bar{\lambda}, \bar{\mu}), \mathrm{Sol(WEP)}(\lambda, \mu)\big) \leq k_1 d^{\alpha_2\delta_2/\beta}(\bar{\lambda}, \lambda) + k_2 d^{\delta_1/(\beta-\theta)}(\bar{\mu}, \mu),$$

$$(\mathbb{D}\big(\mathrm{Sol(SEP)}(\bar{\lambda}, \bar{\mu}), \mathrm{Sol(SEP)}(\lambda, \mu)\big) \leq k_1 d^{\alpha_2\delta_2/\beta}(\bar{\lambda}, \lambda) + k_2 d^{\delta_1/(\beta-\theta)}(\bar{\mu}, \mu),\ resp,)$$

where k_1 and k_2 are positive constants depending on $h, \beta, \ell, n_1, n_2, \delta_1, \delta_2, \alpha, \theta$, etc.

The following examples show the essentialness of the assumptions in Theorem 6.42.

Example 6.43 (The Hölder calmness condition of the constraint map is essential) Let $X = Y = A = \mathbb{R}$, $M = \Lambda = [0, 1]$, $C = \mathbb{R}_+ f(x, y, \lambda) = x - y$, $\bar{\lambda} = 0$, and

$$\mathscr{D}(\lambda) = \begin{cases} \{0\} & \text{if } \lambda = 0, \\ \{-1, 0, 1\} & \text{otherwise.} \end{cases}$$

Then, apart from the Hölder continuity of K, all the other assumptions of Theorem 6.42 are fulfilled. Direct calculations give:

$$\text{Sol(WEP)}(\lambda) = \text{Sol(SEP)}(\lambda) = \begin{cases} \{0\} & \text{if } \lambda = 0, \\ \{1\} & \text{otherwise.} \end{cases}$$

Obviously, Sol(WEP) and Sol(SEP) are not Hölder calm at $\bar{\lambda} = 0$. The reason is that condition (a) of Theorem 6.42 is not fulfilled.

Example 6.44 (The Hölder calmness of the objective map cannot be dropped) Let $X = Y = A = \mathbb{R}$, $M = \Lambda = [0; 1]$, $C = \mathbb{R}_+\mathscr{D}(\lambda) = [0; 1]$, $\bar{\lambda} = 0$, and

$$f(x, y, \lambda) = \begin{cases} x - y & \text{if } \lambda = 0, \\ y - x & \text{otherwise.} \end{cases}$$

Then, except (b), all the other assumptions of Theorem 6.42 are satisfied. Indeed, for each fixed $x \neq y$, $f(x, y, \cdot)$ is not Hölder calm at $\bar{\lambda} = 0$. Hence, f cannot be Hölder calm θ-uniformly for any θ. Direct calculations yield:

$$\text{Sol(WEP)}(\lambda) = \text{Sol(SEP)}(\lambda) = \begin{cases} \{1\} & \text{if } \lambda = 0, \\ \{0\} & \text{otherwise.} \end{cases}$$

This solution map is not Hölder calm at $\bar{\lambda} = 0$, which is just due to the violation of assumption (b).

Example 6.45 (The conditions related to monotonicity are crucial) Let $X = Y = A = \mathbb{R}$, $M = \Lambda = [0; 1]$, $C = \mathbb{R}_+\mathscr{D}(\lambda) \equiv [-1; 1]$, $f(x, y, \lambda) = \lambda(x + y)$, and $\bar{\lambda} = 0$. We see that, except for (c), all the other assumptions of Theorem 6.42 are satisfied. By direct computations, one has the solution map:

$$\text{Sol(WEP)}(\lambda) = \text{Sol(SEP)}(\lambda) = \begin{cases} [-1, 1] & \text{if } \lambda = 0, \\ \{1\} & \text{otherwise,} \end{cases}$$

which is not Hölder calm at $\bar{\lambda} = 0$. This happens because assumption (c) is violated. Indeed, taking $x = 1$ and $y = 0$, for all $h, \beta > 0$, we have:

$$d\big(f(1, 0, 0), \mathbb{R}_+\big) + d\big(f(0, 1, 0), \mathbb{R}_+\big) = 0 < h|1 - 0|^\beta = h.$$

6.3.2 *Hölder Continuity of Approximate Solution Maps*

Let $e \in \text{int}C$ and $\bar{\varepsilon} > 0$. From now on in this section, we denote $\Delta = [-\bar{\varepsilon}e, 0]$. For each $(\varepsilon, \lambda, \mu) \in (0, \bar{\varepsilon}) \times \Lambda \times M$, we denote the ε-solution sets of (WEP) and (SEP) corresponding to $(\varepsilon, \lambda, \mu)$ by $\widetilde{\text{Sol}}(\text{WEP})(\varepsilon, \lambda, \mu)$ and $\widetilde{\text{Sol}}(\text{SEP})(\varepsilon, \lambda, \mu)$, resp., i.e.,

$$\widetilde{\text{Sol}}(\text{WEP})(\varepsilon, \lambda, \mu) := \{x \in \mathscr{D}(\lambda) \mid f(x, y, \mu) + \varepsilon e \not< 0, \forall y \in \mathscr{D}(\lambda)\},$$

$$\widetilde{\text{Sol}}(\text{SEP})(\varepsilon, \lambda, \mu) := \{x \in \mathscr{D}(\lambda) \mid f(x, y, \mu) + \varepsilon e \geq 0, \forall y \in \mathscr{D}(\lambda)\}.$$

Definition 6.46 (Cone-Hölder continuous maps) A vector-valued map $f : X \to Y$ is termed (ℓ, α)-HÖLDER CONTINUOUS wrt $e \in \text{int}C$ at $x_0 \in X$ if there is a neighborhood V of x_0 such that, for all $x_1, x_2 \in V$,

$$f(x_1) \in f(x_2) + \ell \|x_1 - x_2\|^{\alpha}[-e, e],$$

where $[-e, e] = (-e + C) \cap (e - C)$. If $\alpha = 1$, then Hölder continuity is called Lipschitz continuity.

Since the existence conditions of exact and approximate solutions of equilibrium problems were intensively studied in previous chapters, in this paper we only focus on sensitivity analysis. We always assume that all kinds of solution sets of these problems are nonempty in a neighborhood of the reference point $(\bar{\varepsilon}, \bar{\lambda}, \bar{\mu})$.

Lemma 6.47 (Hölder continuity of approximate solution maps of weak equilibrium problems) *Assume that:*

(a) *$\mathscr{D}$ is (ℓ, α)-Hölder continuous at $\bar{\lambda}$, i.e., there exists a neighborhood U of $\bar{\lambda}$ such that for all $\lambda_1, \lambda_2 \in U$,*

$$\mathscr{D}(\lambda_1) \subset \mathscr{D}(\lambda_2) + \ell \|\lambda_1 - \lambda_2\|^{\alpha} B(0, 1);$$

(b) *there exists a neighborhood V of $\bar{\mu}$ such that $f(\cdot, y, \mu)$ is Δ-concave of the second type in $\mathscr{D}(U)$ for all $(y, \mu) \in \mathscr{D}(U) \times V$.*

Then, for all $\varepsilon_1, \varepsilon_2 \in (0, \bar{\varepsilon})$ with $\varepsilon_1 < \varepsilon_2$ and $(\lambda, \mu) \in U \times V$,

$$H(\widetilde{\text{Sol}}(\text{WEP})(\varepsilon_1, \lambda, \mu), \widetilde{\text{Sol}}(\text{WEP})(\varepsilon_2, \lambda, \mu)) \leq \frac{\eta}{\varepsilon_2}|\varepsilon_1 - \varepsilon_2|, \tag{6.15}$$

where $\eta := \text{diam}\mathscr{D}(\bar{\lambda}) + 2\ell(\text{diam}(U))^{\alpha}$.

Proof We first claim that $\widetilde{\text{Sol}}(\text{WEP})(\varepsilon_1, \lambda, \mu) \subset \widetilde{\text{Sol}}(\text{WEP})(\varepsilon_2, \lambda, \mu)$. Indeed, suppose that there exists $\bar{x} \in \widetilde{\text{Sol}}(\text{WEP})(\varepsilon_1, \lambda, \mu) \setminus \widetilde{\text{Sol}}(\text{WEP})(\varepsilon_2, \lambda, \mu)$. Then, there is $\bar{y} \in \mathscr{D}(\lambda)$ such that

$$f(\bar{x}, \bar{y}, \mu) + (\varepsilon_1 e - \varepsilon_1 e) + \varepsilon_2 e \in -\text{int}C.$$

This implies that

$$f(\bar{x}, \bar{y}, \mu) + \varepsilon_1 e := (\varepsilon_1 - \varepsilon_2)e + c_1,$$

where $c_1 \in -\text{int}C$. Since C is a convex cone and $(\varepsilon_1 - \varepsilon_2)e + c_1 \in -\text{int}C$,

$$f(\bar{x}, \bar{y}, \mu) + \varepsilon_1 e \in -\text{int}C.$$

This is impossible because $\bar{x}$ belongs to $\widetilde{\text{Sol}}(\text{WEP})(\varepsilon_1, \lambda, \mu)$. Hence,

$$\text{ex}(\widetilde{\text{Sol}}(\text{WEP})(\varepsilon_1, \lambda, \mu), \widetilde{\text{Sol}}(\text{WEP})(\varepsilon_2, \lambda, \mu)) = 0. \tag{6.16}$$

Take any $x_2 \in \widetilde{\text{Sol}}(\text{WEP})(\varepsilon_2, \lambda, \mu)$ and $x_0 \in \widetilde{\text{Sol}}(\text{WEP})(0, \lambda, \mu)$. Then, for all $y \in \mathscr{D}(\lambda)$, we have:

$$f(x_2, y, \mu) + \varepsilon_2 e \notin -\text{int}C \quad \text{and} \quad f(x_0, y, \mu) \in C.$$

By using the Δ-concavity of the second type of f, for $y \in \mathscr{D}(\lambda)$,

$$f\left(\frac{\varepsilon_1}{\varepsilon_2} x_2 + \frac{\varepsilon_2 - \varepsilon_1}{\varepsilon_2} x_0, y, \mu\right) + \varepsilon_1 e \notin -\text{int}C.$$

This comes down to having $x_1 := \frac{\varepsilon_1}{\varepsilon_2} x_2 + \frac{\varepsilon_2 - \varepsilon_1}{\varepsilon_2} x_0 \in \Pi^w(\varepsilon_1, \lambda, \mu)$.

By (a), one has $\mathscr{D}(\lambda) \subset \mathscr{D}(\bar{\lambda}) + \ell B(0, \|\lambda - \lambda_0\|^\alpha) \forall \lambda \in U$ and hence $\text{diam}\mathscr{D}(\lambda) \leq \eta$. Then,

$$\|x_2 - x_1\| = \frac{|\varepsilon_1 - \varepsilon_2|}{\varepsilon_2} ||x_2 - x_0|| \leq \frac{\eta}{\varepsilon_2} |\varepsilon_1 - \varepsilon_2|,$$

and thus

$$\text{ex}(\widetilde{\text{Sol}}(\text{WEP})(\varepsilon_2, \lambda, \mu), \widetilde{\text{Sol}}(\text{WEP})(\varepsilon_1, \lambda, \mu)) \leq \frac{\eta}{\varepsilon_2} |\varepsilon_1 - \varepsilon_2|. \tag{6.17}$$

Utilizing (6.16) and (6.17), we obtain (6.15). □

Lemma 6.48 (Hölder continuity of approximate solution maps of strong equilibrium problems) *Assume that:*

(a) *$\mathscr{D}$ is (ℓ, α)-Hölder continuous at $\bar{\lambda}$, i.e., there exists a neighborhood U of $\bar{\lambda}$ such that for all $\lambda_1, \lambda_2 \in U$,*

$$\mathscr{D}(\lambda_1) \subset \mathscr{D}(\lambda_2) + \ell \|\lambda_1 - \lambda_2\|^\alpha B(0, 1);$$

(b) *there exists a neighborhood V of $\bar{\mu}$ such that $f(\cdot, y, \mu)$ is Δ-concave of the first type in $\mathscr{D}(U)$ for all $(y, \mu) \in \mathscr{D}(U) \times V$.*

Then, for all $\varepsilon_1, \varepsilon_2 \in (0, \bar{\varepsilon})$ *with* $\varepsilon_1 < \varepsilon_2$ *and* $(\lambda, \mu) \in U \times V$, *we have:*

$$H(\widetilde{\mathrm{Sol}}(\mathrm{SEP})(\varepsilon_1, \lambda, \mu), \widetilde{\mathrm{Sol}}(\mathrm{SEP})(\varepsilon_2, \lambda, \mu)) \leq \frac{\eta}{\varepsilon_2}|\varepsilon_1 - \varepsilon_2|,$$

where $\eta := \mathrm{diam}\mathscr{D}(\bar{\lambda}) + 2\ell(\mathrm{diam}(U))^\alpha$.

We omit the proof since it is similar to the proof of Lemma 6.47.

Theorem 6.49 (Hölder continuity of approximate solution maps of parametric weak and strong equilibrium problems) *Assume that the following conditions hold:*

(a) $\mathscr{D}$ *is* (ℓ, α)*-Hölder continuous at* $\bar{\lambda}$, *i.e., there exists a neighborhood* U *of* $\bar{\lambda}$ *such that for all* $\lambda_1, \lambda_2 \in U$,

$$\mathscr{D}(\lambda_1) \subset \mathscr{D}(\lambda_2) + \ell\|\lambda_1 - \lambda_2\|^\alpha B(0, 1);$$

(b) *there is a neighborhood* V *of* $\bar{\mu}$ *such that for each* $(y, \mu) \in \mathscr{D}(U) \times V$, *for (WEP),* $f(\cdot, y, \mu)$ *is* Δ*-concave of the second type in* $\mathscr{D}(U)$, *(for (SEP),* $f(\cdot, y, \mu)$ *is* Δ*-concave of the first type in* $\mathscr{D}(U)$, *resp.);*
(c) *for all* $x, y \in \mathscr{D}(U)$, $f(x, y, \cdot)$ *is* (h, β)*-Hölder continuous wrt e in* V;
(d) *for* $\mu \in V$ *and* $x \in \mathscr{D}(U)$, $f(x, \cdot, \mu)$ *is* (q, δ)*-Hölder continuous wrt e in* $\mathscr{D}(U)$.

Then, for any $\hat{\varepsilon} \in (0, \bar{\varepsilon})$, $\widetilde{\mathrm{Sol}}(\mathrm{WEP})$ *(*$\widetilde{\mathrm{Sol}}(\mathrm{SEP})$*, resp.) is Hölder continuous in* $(\hat{\varepsilon}, \bar{\varepsilon}) \times U \times V$.

Proof We only detail the proof for problem (WEP) since the other one is similarly discussed. We will show that for all $(\varepsilon_1, \lambda_1, \mu_1), (\varepsilon_2, \lambda_2, \mu_2) \in (\hat{\varepsilon}, \bar{\varepsilon}) \times U \times V$,

$$\begin{aligned} &H(\widetilde{\mathrm{Sol}}(\mathrm{WEP})(\varepsilon_1, \lambda_1, \mu_1), \widetilde{\mathrm{Sol}}(\mathrm{WEP})(\varepsilon_2, \lambda_2, \mu_2)) \\ &\quad \leq k_1|\varepsilon_1 - \varepsilon_2| + k_2||\mu_1 - \mu_2||^\beta + k_3||\lambda_1 - \lambda_2||^{\alpha\delta}, \end{aligned}$$

where k_1, k_2, k_3 are positive, depending on $\hat{\varepsilon}, \ell, \alpha, h$, etc. Without loss of generality, we can suppose that $\varepsilon_1 < \varepsilon_2$.
Step 1 We first provide an estimation of $H(\widetilde{\mathrm{Sol}}(\mathrm{WEP})(\varepsilon_1, \lambda_1, \mu_1), \widetilde{\mathrm{Sol}}(\mathrm{WEP})(\varepsilon_2, \lambda_1, \mu_1))$. In view of Lemma 6.47, for all $(\lambda, \mu) \in U \times V$, we have:

$$H(\widetilde{\mathrm{Sol}}(\mathrm{WEP})(\varepsilon_1, \lambda, \mu), \widetilde{\mathrm{Sol}}(\mathrm{WEP})(\varepsilon_2, \lambda, \mu)) \leq \frac{\eta}{\varepsilon_2}|\varepsilon_1 - \varepsilon_2|,$$

which derives

$$\begin{aligned} H(\widetilde{\mathrm{Sol}}(\mathrm{WEP})(\varepsilon_1, \lambda_1, \mu_1), \widetilde{\mathrm{Sol}}(\mathrm{WEP})(\varepsilon_2, \lambda_1, \mu_1)) &\leq \frac{\eta}{\hat{\varepsilon}}|\varepsilon_1 - \varepsilon_2| \\ &=: k_1|\varepsilon_1 - \varepsilon_2|. \end{aligned}$$

Step 2 Now we estimate $H(\widetilde{\text{Sol}}(\text{WEP})(\varepsilon_2, \lambda_1, \mu_1), \widetilde{\text{Sol}}(\text{WEP})(\varepsilon_2, \lambda_1, \mu_2))$ with $\mu_1 \neq \mu_2$. Two cases have to be analyzed now separately.
Case 1. $h\|\mu_1 - \mu_2\|^\beta \leq \varepsilon_2$. Let $r = h\|\mu_1 - \mu_2\|^\beta$, then $0 < r \leq \varepsilon_2$. We conclude that

$$\widetilde{\text{Sol}}(\text{WEP})(\varepsilon_2 - r, \lambda_1, \mu_1) \subset \widetilde{\text{Sol}}(\text{WEP})(\varepsilon_2, \lambda_1, \mu_2).$$

Suppose that there is $\bar{x} \in \widetilde{\text{Sol}}(\text{WEP})(\varepsilon_2 - r, \lambda_1, \mu_1) \setminus \widetilde{\text{Sol}}(\text{WEP})(\varepsilon_2, \lambda_1, \mu_2)$. Then, there exists $\bar{y} \in \mathscr{D}(\lambda_1)$ such that

$$f(\bar{x}, \bar{y}, \mu_1) + f(\bar{x}, \bar{y}, \mu_2) - f(\bar{x}, \bar{y}, \mu_1) + \varepsilon_2 e \in -\text{int}C,$$

which implies that

$$f(\bar{x}, \bar{y}, \mu_1) + (\varepsilon_2 - r)e = f(\bar{x}, \bar{y}, \mu_1) - f(\bar{x}, \bar{y}, \mu_2) - re + \hat{c},$$

where $\hat{c} \in -\text{int}C$. Thanks to (b), we get:

$$f(\bar{x}, y, \mu_1) - f(\bar{x}, y, \mu_2) \in h\|\mu_1 - \mu_2\|^\beta[-e, e] = r[-e, e].$$

Therefore,

$$f(\bar{x}, y, \mu_1) + (\varepsilon_2 - r)e = rs - re + \bar{c},$$

where $s \in [-e, e]$. This leads to $f(\bar{x}, y, \mu_1) + (\varepsilon_2 - r)e \in -\text{int}C$, a contradiction since $\bar{x} \in \widetilde{\text{Sol}}(\text{WEP})(\varepsilon_2 - r, \lambda_1, \mu_1)$. It follows from the result of Step 1 that

$$\begin{aligned}
&\text{ex}(\widetilde{\text{Sol}}(\text{WEP})(\varepsilon_2, \lambda_1, \mu_1), \widetilde{\text{Sol}}(\text{WEP})(\varepsilon_2, \lambda_1, \mu_2)) \\
&\qquad \leq \text{ex}(\widetilde{\text{Sol}}(\text{WEP})(\varepsilon_2, \lambda_1, \mu_1), \widetilde{\text{Sol}}(\text{WEP})(\varepsilon_2 - r, \lambda_1, \mu_1)) \\
&\qquad \leq H(\widetilde{\text{Sol}}(\text{WEP})(\varepsilon_2, \lambda_1, \mu_1), \widetilde{\text{Sol}}(\text{WEP})(\varepsilon_2 - r, \lambda_1, \mu_1)) \\
&\qquad \leq \frac{\eta r}{\varepsilon_2} \\
&\qquad \leq \frac{\eta h}{\hat{\varepsilon}}\|\mu_1 - \mu_2\|^\beta.
\end{aligned}$$

By the same manner, we get:

$$\text{ex}(\widetilde{\text{Sol}}(\text{WEP})(\varepsilon_2, \lambda_1, \mu_2), \widetilde{\text{Sol}}(\text{WEP})(\varepsilon_2, \lambda_1, \mu_1)) \leq \frac{\eta h}{\hat{\varepsilon}}\|\mu_1 - \mu_2\|^\beta.$$

Hence,

$$H(\widetilde{\text{Sol}}(\text{WEP})(\varepsilon_2, \lambda_1, \mu_1), \widetilde{\text{Sol}}(\text{WEP})(\varepsilon_2, \lambda_1, \mu_2)) \leq \frac{\eta h}{\hat{\varepsilon}}\|\mu_1 - \mu_2\|^\beta.$$

Case 2. $h\|\mu_1 - \mu_2\|^\beta > \varepsilon_2$. Let $\eta_1 = \operatorname{diam} V < +\infty$. Observe that there is a natural number n such that $\dfrac{\eta_1}{n} \le \left(\dfrac{\hat{\varepsilon}}{h}\right)^{1/\beta}$. Partitioning the segment $[\mu_1, \mu_2]$ by the nodes $v_1, v_2, ..., v_{n+1}$ satisfying $v_1 = \mu_1, v_{n+1} = \mu_2$, and

$$\|v_i - v_{i+1}\| = \frac{\|\mu_1 - \mu_2\|}{n}, \quad i = 1, ..., n,$$

we have $\|v_i - v_{i+1}\|^\beta = \left(\dfrac{\|\mu_1 - \mu_2\|}{n}\right)^\beta \le \left(\dfrac{\eta_1}{n}\right)^\beta \le \dfrac{\hat{\varepsilon}}{h} \le \dfrac{\varepsilon_2}{h}$. Applying Case 1, one gets:

$$\begin{aligned}
&H(\widetilde{\mathrm{Sol}}(\mathrm{WEP})(\varepsilon_2, \lambda_1, \mu_1), \widetilde{\mathrm{Sol}}(\mathrm{WEP})(\varepsilon_2, \lambda_1, \mu_2)) \\
&\le \sum_{i=1}^{n} H(\widetilde{\mathrm{Sol}}(\mathrm{WEP})(\varepsilon_2, \lambda_1, v_i), \widetilde{\mathrm{Sol}}(\mathrm{WEP})(\varepsilon_2, \lambda_1, v_{i+1})) \\
&\le \frac{\eta h}{\hat{\varepsilon}} \sum_{i=1}^{n} \|v_i - v_{i+1}\|^\beta \\
&\le \frac{n\eta h}{\hat{\varepsilon}} \|\mu_1 - \mu_2\|^\beta \\
&=: k_2 \|\mu_1 - \mu_2\|^\beta.
\end{aligned}$$

Step 3 We now evaluate $H(\widetilde{\mathrm{Sol}}(\mathrm{WEP})(\varepsilon_2, \lambda_1, \mu_2), \widetilde{\mathrm{Sol}}(\mathrm{WEP})(\varepsilon_2, \lambda_2, \mu_2))$ for $\lambda_1 \ne \lambda_2$. Again, we consider two cases.

Case 1. $\|\lambda_1 - \lambda_2\|^\alpha \le \dfrac{1}{\ell}\left(\dfrac{\varepsilon_2}{q}\right)^{1/\delta}$. Let $r' = \ell^\delta q||\lambda_1 - \lambda_2||^{\alpha\delta}$. Then, $0 < r' \le \varepsilon_2$. It is easy to see that for all $y_2 \in K(\lambda_2)$, there is $y_1 \in \mathscr{D}(\lambda_1)$ such that

$$||y_1 - y_2|| \le \ell||\lambda_1 - \lambda_2||^\alpha.$$

Pick any $x' \in \widetilde{\mathrm{Sol}}(\mathrm{WEP})(\varepsilon_2 - r', \lambda_1, \mu_2)$ and $y_2 \in \mathscr{D}(\lambda_2)$, we have:

$$f(x', y_2, \mu_2) + f(x', y_1, \mu_2) - f(x', y_2, \mu_2) + (\varepsilon_2 - r')e \notin -\mathrm{int}C.$$

Combining this and (c), one has

$$f(x', y_2, \mu_2) + \varepsilon_2 e \notin -\mathrm{int}C, \ \forall\ y_2 \in \mathscr{D}(\lambda_2),$$

i.e., $x' \in \widetilde{\mathrm{Sol}}(\mathrm{WEP})(\varepsilon_2, \lambda_2, \mu_2)$. This allows us to conclude that

$$\widetilde{\mathrm{Sol}}(\mathrm{WEP})(\varepsilon_2 - r', \lambda_1, \mu_2) \subset \widetilde{\mathrm{Sol}}(\mathrm{WEP})(\varepsilon_2, \lambda_2, \mu_2).$$

By using Step 1, we obtain the following estimates:

$$\begin{aligned}
&\mathrm{ex}(\widetilde{\mathrm{Sol}}(\mathrm{WEP})(\varepsilon_2, \lambda_1, \mu_2), \widetilde{\mathrm{Sol}}(\mathrm{WEP})(\varepsilon_2, \lambda_2, \mu_2)) \\
&\quad \leq H(\widetilde{\mathrm{Sol}}(\mathrm{WEP})(\varepsilon_2, \lambda_1, \mu_2), \widetilde{\mathrm{Sol}}(\mathrm{WEP})(\varepsilon_2 - r', \lambda_1, \mu_2)) \\
&\quad \leq \frac{\eta q \ell^\delta}{\varepsilon_2} \|\lambda_1 - \lambda_2\|^{\alpha\delta} \\
&\quad \leq \frac{\eta q \ell^\delta}{\hat{\varepsilon}} \|\lambda_1 - \lambda_2\|^{\alpha\delta}.
\end{aligned}$$

In a similar way, we also have the following conclusion:

$$\mathrm{ex}(\widetilde{\mathrm{Sol}}(\mathrm{WEP})(\varepsilon_2, \lambda_2, \mu_2), \widetilde{\mathrm{Sol}}(\mathrm{WEP})(\varepsilon_2, \lambda_1, \mu_2)) \leq \frac{\eta q \ell^\delta}{\hat{\varepsilon}} ||\lambda_1 - \lambda_2||^{\alpha\delta}.$$

Therefore,

$$H(\widetilde{\mathrm{Sol}}(\mathrm{WEP})(\varepsilon_2, \lambda_1, \mu_2), \widetilde{\mathrm{Sol}}(\mathrm{WEP})(\varepsilon_2, \lambda_2, \mu_2)) \leq \frac{\eta q \ell^\delta}{\hat{\varepsilon}} \|\lambda_1 - \lambda_2\|^{\alpha\delta}.$$

Case 2. $\|\lambda_1 - \lambda_2\|^\alpha > \frac{1}{\ell}\left(\frac{\varepsilon_2}{q}\right)^{1/\delta}$. Let $\eta_2 = \mathrm{diam}(U) < +\infty$. We also have a natural number $\mathscr{N}$ satisfying $\frac{\eta_2}{\mathscr{N}} \leq \left[\frac{1}{\ell}\left(\frac{\hat{\varepsilon}}{q}\right)^{1/\delta}\right]^{1/\alpha}$. We partition the segment $[\lambda_1, \lambda_2]$ by $\mathscr{N} + 1$ nodes $u_1, ..., u_{\mathscr{N}+1}$ such that

$$u_1 = \lambda_1, \qquad u_{\mathscr{N}+1} = \lambda_2, \quad \|u_j - u_{j+1}\| = \frac{\|\lambda_1 - \lambda_2\|}{\mathscr{N}}.$$

Hence,

$$\|u_j - u_{j+1}\|^\alpha \leq \left(\frac{\eta_2}{\mathscr{N}}\right)^\alpha \leq \frac{1}{\ell}\left(\frac{\hat{\varepsilon}}{q}\right)^{1/\delta} \leq \frac{1}{\ell}\left(\frac{\varepsilon_2}{q}\right)^{1/\delta}.$$

According to Case 1, we derive:

$$H(\widetilde{\mathrm{Sol}}(\mathrm{WEP})(\varepsilon_2, \lambda_1, \mu_2), \widetilde{\mathrm{Sol}}(\mathrm{WEP})(\varepsilon_2, \lambda_2, \mu_2)) \leq \frac{\eta q \ell^\delta}{\hat{\varepsilon}} \sum_{j=1}^{\mathscr{N}} ||u_j - u_{j+1}||^{\alpha\delta}$$

$$\leq \frac{\mathscr{N} \eta q \ell^\delta}{\hat{\varepsilon}} \left[\frac{1}{\ell}\left(\frac{\varepsilon_2}{q}\right)^{1/\delta}\right]^\delta \leq \frac{\mathscr{N} \eta q \ell^\delta}{\hat{\varepsilon}} \|\lambda_1 - \lambda_2\|^{\alpha\delta} =: k_3 \|\lambda_1 - \lambda_2\|^{\alpha\delta}.$$

Step 4 Finally, we utilize the results of the preceding three steps to obtain the desired conclusion for $\widetilde{\text{Sol}}(\text{WEP})$. □

We now provide an example to illustrate the essentialness of the Δ-concavity condition imposed in Theorem 6.49.

Example 6.50 (Cruciality of generalized concavity conditions) Let $\hat{\varepsilon} = \frac{1}{3}, \bar{\varepsilon} = 3, X = A = \mathbb{R}, \Lambda = M = [0, 1], \mathcal{D}(\lambda) = [-3, 3], Y = \mathbb{R}^2, C = \mathbb{R}^2_+, e = (1, 1) \in \text{int}C$, and

$$f(x, y, \mu) := \begin{cases} (0, 0) & \text{if} \quad x = 0, \\ (-1, -1) & \text{if} \quad x \neq 0. \end{cases}$$

Then, we see that (a), (b), and (c) of Theorem 6.49 are satisfied. Some direct computations give the approximate solution set:

$$\widetilde{\text{Sol}}(\text{WEP})(\varepsilon) \equiv \widetilde{\text{Sol}}(\text{SEP})(\varepsilon) = \begin{cases} [-3, 3] & \text{if} \quad \varepsilon \geq 1, \\ \{0\} & \text{if} \quad \frac{1}{3} \leq \varepsilon < 1, \end{cases}$$

which is not Hölder continuous at $\varepsilon_0 = 1$. The reason is that f is neither Δ-concavity of the first type nor Δ-concavity of the second type in $\Delta = [-3e, 0]$ (to see this, take $z = (-2, -2) \in \Delta, x_1 = 1, x_2 = 0$, and $t = \frac{1}{3}$).

6.4 Well-Posedness for (WEP) and (SEP)

In this section, if not otherwise specified, let X, Λ, M be three metric spaces and Y a Hausdorff topological vector space. Let $e : M \to Y$ be a continuous mapping such that, for each $\mu \in M$ $e(\mu) \in \text{int}C$.

Definition 6.51 (Approximating sequences for weak and strong equilibrium problems) Let $\lambda \in \Lambda$, $\mu \in M$, and $\{(\lambda^k, \mu^k\}_k \subset \Lambda \times M$ be a sequence converging to (λ, μ). A sequence $\{x^k\}_k$, $x^k \in \mathcal{D}(\lambda^k)$, is said to be an APPROXIMATING SEQUENCE for (WEP)(λ, μ) ((SEP)(λ, μ), resp.) corresponding to $\{(\lambda^k, \mu^k\}_k$ if there exists a sequence $\{\varepsilon^k\} \downarrow 0$ such that, for each $k \in \mathbb{N}$ and $y \in \mathcal{D}(\lambda^k)$,

$$f(x^k, y, \mu^k) + \varepsilon^k e(\mu^k) \not< 0,$$

$$(f(x^k, y, \mu^k) + \varepsilon^k e(\mu^k) \geq 0, \text{resp.}).$$

Definition 6.52 (Well-posedness under perturbations of weak and strong equilibrium problems) Problem (WEP) (resp., (SEP)) is said to be WELL-POSED UNDER PERTURBATIONS (well-posed, briefly), if for each $\lambda \in \Lambda$ and $\mu \in M$:

(a) (WEP)(λ, μ) ((SEP)(λ, μ), resp.) has solutions;

(b) for any sequence $\{(\lambda^k, \mu^k)\}_k \subset \Lambda \times M$ converging to (λ, μ), every approximating sequence $\{x^k\}_k$ for (WEP)(λ, μ) ((SEP)(λ, μ), resp.) corresponding to $\{(\lambda^k, \mu^k)\}_k$ must have a subsequence converging to an element in Sol(WEP)(λ, μ) (Sol(SEP)(λ, μ), resp.).

Problem (WEP)(λ, μ) ((SEP)(λ, μ), resp.) is said to be UNIQUELY WELL-POSED UNDER PERTURBATIONS (uniquely well-posed, briefly) if, for each $\lambda \in \Lambda$, $\mu \in M$, the solution set Sol(WEP)(λ, μ) (Sol(SEP)(λ, μ), resp.) is a singleton, and every approximating sequence for (WEP)(λ, μ) ((SEP) (λ, μ), resp.) converges to the unique solution.

Definition 6.53 (Generalized Hadamard well-posedness for weak and strong equilibrium problems) Problem (WEP) ((SEP), resp.) is said to be GENERALIZED HADAMARD WELL-POSED if, for each $\lambda \in \Lambda$ and $\mu \in M$:

(a) (WEP)(λ, μ) ((SEP)(λ, μ), resp.) has solutions;
(b) for any sequence $\{\lambda^k, \mu^k\}_k \subset \Lambda \times M$ converging to (λ, μ) and $x^k \in$ Sol(WEP)(λ^k, μ^k) ($x^k \in$ Sol(SEP)(λ^k, μ^k), resp.), $\{x^k\}_k$ has a subsequence converging to some point of Sol(WEP)(λ, μ) (Sol(SEP)(λ, μ), resp.).

Problem (WEP) ((SEP), resp.) is called HADAMARD WELL-POSED if, for each $\lambda \in \Lambda$, $\mu \in M$, the solution set Sol(WEP)(λ, μ) (Sol(SEP)(λ, μ), resp.) is a singleton, and for $x^k \in$ Sol(WEP)(λ^k, μ^k) ($x^k \in$ Sol(WEP)(λ, μ), resp.), $\{x^k\}_k$ converges to the unique solution, resp.

For each $\lambda \in \Lambda$, $\varepsilon \geq 0$, we denote the ε-solution sets of (WEP)(λ, μ) and (SEP)(λ, μ) by $\widetilde{\text{Sol}}$(WEP)$(\varepsilon, \lambda, \mu)$ and $\widetilde{\text{Sol}}$(SEP)$(\varepsilon, \lambda, \mu)$, resp., which are defined as follows:

$$\widetilde{\text{Sol}}(\text{WEP})(\varepsilon, \lambda, \mu) := \{x \in \mathscr{D}(\lambda) \mid f(x, y, \mu) + \varepsilon e(\mu) \not< 0, \forall y \in \mathscr{D}(\lambda)\},$$

$$\widetilde{\text{Sol}}(\text{SEP})(\varepsilon, \lambda, \mu) := \{x \in \mathscr{D}(\lambda) \mid f(x, y, \mu) + \varepsilon e(\mu) \geq 0, , \forall y \in \mathscr{D}(\lambda)\}.$$

Lemma 6.54 (Convergence characterizations of upper semicontinuity) *Let $S : X \rightrightarrows Y$ be a set-valued mapping between two topological spaces. Then, the following assertions hold:*

(a) *If $S(\bar{x})$ is compact, then S is usc at $\bar{x}$ if and only if for any sequence $\{x^k\}_k$ convergent to $\bar{x}$ and $y^k \in S(x^k)$, there is a subsequence $\{y^{k_j}\}$ convergent to some $y \in S(\bar{x})$.*
(b) *If, in addition, $S(\bar{x}) = \{\bar{y}\}$ is a singleton, then the above limit point y must be $\bar{y}$ and the whole $\{y^k\}_k$ converges to $\bar{y}$.*

Since the existence theory of the equilibrium problems has been intensively studied, in this section we only focus on the well-posedness of such problems, and hence we always assume that the solutions of the considered problems exist in a neighborhood of the reference point.

Theorem 6.55 (Closedness of solution sets of weak and strong equilibrium problems) *Assume that, for given* $\lambda \in \Lambda, \mu \in M$*, the following assumptions hold:*

(a) $\mathscr{D}(\lambda)$ *is closed;*
(b) *for each* $y \in \mathscr{D}(\lambda)$, $f(\cdot, y, \mu)$ *is* C*-upper semicontinuous.*

Then, the solution sets $\text{Sol(WEP)}(\lambda, \mu)$ *and* $\text{Sol(SEP)}(\lambda, \mu)$ *are closed in* X.

Proof Since the proof techniques are similar, we discuss only the closedness of $\text{Sol(WEP)}(\lambda, \mu)$. Let sequence $\{x^k\}_k \subset \text{Sol(WEP)}(\lambda, \mu)$ be such that $x^k \to \bar{x} \in X$, we will show that $\bar{x}$ belongs to $\text{Sol(WEP)}(\lambda, \mu)$. Since x^k is a solution of $\text{(WEP)}(\lambda, \mu)$, for each $n \in \mathbb{N}$, $x^k \in \mathscr{D}(\lambda)$, and

$$f(x^k, y, \mu) \nless 0, \forall y \in \mathscr{D}(\lambda). \tag{6.18}$$

Since $\mathscr{D}(\lambda)$ is closed, $\bar{x} \in \mathscr{D}(\lambda)$. If $\bar{x} \notin \text{Sol(WEP)}(\lambda, \mu)$, then there exists $\bar{y} \in \mathscr{D}(\lambda)$ such that

$$f(\bar{x}, \bar{y}, \mu) < 0.$$

By the openness of $-\text{int}C$, there is a neighborhood V of θ_Y such that $f(\bar{x}, \bar{y}, \lambda) + V \subset -\text{int}C$. Since $f(\cdot, \bar{y}, \mu)$ is C-usc at $\bar{x}$ and $x^k \to \bar{x}$, there exists $k_0 \in \mathbb{N}$ such that for all $k \geqslant k_0$, we have:

$$f(x^k, \bar{y}, \mu) \in f(\bar{x}, \bar{y}, \mu) + V - C \subset -\text{int}C - C = -\text{int}C,$$

which is in contradiction with (6.18). Therefore, $\bar{x} \in \text{Sol(WEP)}(\lambda, \mu)$. □

Now we investigate semicontinuity of approximating solution sets, which plays an important role in well-posedness.

Theorem 6.56 (Upper semicontinuity of approximate solution maps of parametric weak and strong equilibrium problems) *Assume that* f *is* C*-upper semicontinuous and* $\mathscr{D}$ *is continuous and compact-valued. Then,* $\widetilde{\text{Sol}}\text{(WEP)}$ *and* $\widetilde{\text{Sol}}\text{(SEP)}$ *are upper semicontinuous and compact-valued on* $\mathbb{R}_+ \times \Lambda \times M$.

Proof As an example, we present only the proof for problem (SEP). We first prove the upper semicontinuity and compact-valuedness of $\text{Sol(SEP)}(\cdot, \cdot) = \widetilde{\text{Sol}}\text{(SEP)}(0, \cdot, \cdot)$ on $\Lambda \times M$. For each $\lambda \in \Lambda$ and $\mu \in M$, Theorem 6.55 implies the fact that $\text{Sol(SEP)}(\lambda, \mu)$ is a closed subset of $\mathscr{D}(\lambda)$, and so it is compact. We show that Sol(SEP) is usc on $\Lambda \times M$. Suppose that there exist $\lambda \in \Lambda, \mu \in M$ such that Sol(SEP) is not usc at (λ, μ). Then, there are open neighborhood V of $\text{Sol(SEP)}(\lambda, \mu)$ and a sequence $\{(\lambda^k, \mu^k)\}_k \subset \Lambda \times M$ converging (λ, μ) such that, for each $k \in \mathbb{N}$, there exists $x^k \in \text{Sol(SEP)}(\lambda^k, \mu^k) \setminus V$. Since $x^k \in \text{Sol(SEP)}(\lambda^k, \mu^k)$, $x^k \in \mathscr{D}(\lambda^k)$ and

$$f(x^k, y, \mu^k) \geq 0, \forall y \in \mathscr{D}(\lambda^k).$$

Because $\mathscr{D}$ is usc and compact-valued at λ, we can assume that x^k converge to $\bar{x}$ for some $\bar{x} \in \mathscr{D}(\lambda)$. Since $x^k \notin V$ for all k, $\bar{x}$ does not belong to $\text{Sol(SEP)}(\lambda, \mu) \subset V$, i.e., there exists $\bar{y} \in \mathscr{D}(\lambda)$ such that

$$f(\bar{x}, \bar{y}, \mu) \not\geq 0,$$

and hence, there is a neighborhood B of θ_Y such that

$$(f(\bar{x}, \bar{y}, \mu) + B) \subset (Y \setminus C).$$

As f is C-usc at $(\bar{x}, \bar{y}, \mu)$, there is a neighborhood U of $\bar{y}$ and $k_0 \in \mathbb{N}$ such that, for all $y \in U$ and $k \geq k_0$

$$f(x^k, y, \mu^k) \in f(\bar{x}, \bar{y}, \lambda) + B - C \subset (Y \setminus C) - C \subset (Y \setminus C). \tag{6.19}$$

Due to $\bar{y} \in \mathscr{D}(\lambda)$, $\mathscr{D}(\lambda) \cap U \neq \emptyset$. By the lower semicontinuity of $\mathscr{D}$ at λ, there exists $k_1 \in \mathbb{N}$ such that, for all $k \geq k_1$,

$$\mathscr{D}(\lambda^k) \cap U \neq \emptyset.$$

Combining this with (6.19), we conclude that, for all $k \geq k_2 = \max\{k_0, k_1\}$,

$$f(x^k, y, \mu^k) \not\geq 0,$$

for some $y \in \mathscr{D}(\lambda^k)$, which is impossible as $x^k \in \text{Sol(SEP)}(\lambda^k, \mu^k)$. Hence, Sol(SEP) is usc at (λ, μ).

Now we show that $\widetilde{\text{Sol}}\text{(SEP)}$ is usc and compact-valued on $\mathbb{R}_+ \times \Lambda \times M$. Let $g : X \times X \times \mathbb{R}_+ \times M \to Y$ be defined by

$$g(x, y, \varepsilon, \mu) := f(x, y, \mu) + \varepsilon e(\mu).$$

Since f is C-usc and e is continuous, we see that g is C-usc. Using the techniques in the proof for Sol(SEP), where f is replaced by g, we conclude that $\widetilde{\text{Sol}}\text{(SEP)}$ is upper semicontinuous and compact-valued on $\mathbb{R}_+ \times \Lambda \times M$. □

Now let us present important relationships between well-posedness and stability.

Theorem 6.57 (Relationships between well-posedness and stability)

(a) *Problems* (WEP) *and* (SEP) *are well-posed under perturbations if and only if, for each* $\lambda \in \Lambda$ *and* $\mu \in M$, $\widetilde{\text{Sol}}\text{(WEP)}$ *and* $\widetilde{\text{Sol}}\text{(SEP)}$ *are upper semicontinuous with compact values at* $(0, \lambda, \mu)$.

Lemma 6.63 (A version of Eidelheit's separation theorem) *Let S and T be nonempty convex subsets of a real topological linear space X with* $\operatorname{int}S \neq \emptyset$. *Then,* $\operatorname{int}S \cap T \neq \emptyset$ *if and only if there are a continuous linear functional* $\ell \in X^*\backslash\{0_{X^*}\}$ *and a real number α with*

$$\ell(s) \leq \alpha \leq \ell(t) \;\textit{ for all } s \in S \textit{ and } t \in T$$

and

$$\ell(s) < \alpha \;\textit{ for all } s \in \operatorname{int}S.$$

For $\xi \in C^* \setminus \{0\}$, $\lambda \in \Lambda$, and $\mu \in M$, the ξ-solution sets and the ξ-approximate solution sets of problem (WEP) are denoted by

$$\operatorname{Sol(WEP)}_{\xi}(\lambda, \mu) := \{x \in \mathscr{D}(\lambda) \mid \langle \xi, f(x, y, \mu)\rangle \geq 0, \forall y \in \mathscr{D}(\lambda)\},$$

$$\widetilde{\operatorname{Sol}}\operatorname{(WEP)}_{\xi}(\varepsilon, \lambda, \mu) := \{x \in \mathscr{D}(\lambda) \mid \langle \xi, f(x, y, \mu)\rangle + \varepsilon \geq 0, \forall y \in \mathscr{D}(\lambda)\},$$

resp.

Now, we are in a position to state and prove the main result of this subsection.

Theorem 6.64 (Linear scalarization for weak equilibrium problems) *Assume that for each $x \in \mathscr{D}(\Lambda)$ and $(\lambda, \mu) \in \Lambda \times M$, $f(x, \cdot, \mu)$ is convex-like in $\mathscr{D}(\lambda)$, then*

$$\operatorname{Sol(WEP)}(\lambda, \mu) = \bigcup_{\xi \in C^*\backslash\{0\}} \operatorname{Sol(WEP)}_{\xi}(\lambda, \mu) = \bigcup_{\xi \in B_e^*} \operatorname{Sol(WEP)}_{\xi}(\lambda, \mu).$$

Proof

(a) We first prove that

$$\operatorname{Sol(WEP)}(\lambda, \mu) = \bigcup_{\xi \in C^*\backslash\{0\}} \operatorname{Sol(WEP)}_{\xi}(\lambda, \mu).$$

For $x \in \cup_{\xi \in C^*\backslash\{0\}} \operatorname{Sol(WEP)}_{\xi}(\lambda, \mu)$, there is $\bar{\xi} \in C^* \setminus \{0\}$ such that $x \in \operatorname{Sol(WEP)}_{\bar{\xi}}(\lambda, \mu)$. By the definition of $\operatorname{Sol(WEP)}_{\bar{\xi}}(\lambda, \mu)$, for each $y \in \mathscr{D}(\lambda)$,

$$\langle \bar{\xi}, f(x, y, \mu)\rangle \geq 0.$$

So, for all $y \in \mathscr{D}(\lambda)$, $f(x, y, \mu) \notin -\operatorname{int}C$. Indeed, if $f(x, y, \mu) \in -\operatorname{int}C$, Lemma 6.62 implies that $\langle \bar{\xi}, f(x, y, \mu)\rangle < 0$, which is impossible. Hence, $x \in \operatorname{Sol(WEP)}(\lambda, \mu)$.

Conversely, let $x \in \operatorname{Sol(WEP)}(\lambda, \mu)$. By the fact that $f(x, y, \mu) \notin -\operatorname{int}C$ for all $y \in \mathscr{D}(\lambda)$

$$f(x, \mathscr{D}(\lambda), \mu) \cap (-\operatorname{int}C) = \emptyset.$$

Since C is a convex cone, we get:

$$(f(x, \mathscr{D}(\lambda), \mu) + C) \cap (-\text{int}C) = \emptyset.$$

Indeed, if there is $w \in (f(x, \mathscr{D}(\lambda), \mu)+C)\cap(-\text{int}C)$, i.e., $w = z+c \in -\text{int}C$, where $z \in f(x, \mathscr{D}(\lambda), \mu)$ and $c \in C$, one has: $z = -c - w \in -\text{int}C$, which is a contradiction with the above discussions. Next we show that $f(x, \mathscr{D}(\lambda), \mu)+C$ is a convex subset of Y. Let $z_1, z_2 \in f(x, \mathscr{D}(\lambda), \mu) + C$, i.e., there are $y_1, y_2 \in \mathscr{D}(\lambda)$ and $c_1, c_2 \in C$ such that $z_i = f(x, y_i, \mu) + c_i, i = 1, 2$. Since $f(x, \cdot, \mu)$ is convex-like in $\mathscr{D}(\lambda)$, for all $t \in [0, 1]$, there exist $s \in \mathscr{D}(\lambda)$ and $c \in C$ such that

$$tf(x, y_1, \mu) + (1-t)f(x, y_2, \mu) = f(x, s, \mu) + c,$$

i.e., $tz_1+(1-t)z_2 = f(x, s, \mu)+tc_1+(1-t)c_2+c \in f(x, \mathscr{D}(\lambda), \mu)+C$ as C is a convex cone, and hence $f(x, \mathscr{D}(\lambda), \mu)+C$ is convex. Applying Lemma 6.63, there exist a continuous linear functional $\xi \in Y^* \setminus \{0\}$ and a number $\gamma \in \mathbb{R}$ such that

$$\langle \xi, \bar{c})\rangle < \gamma < \langle \xi, z + c\rangle, \forall \bar{c} \in -\text{int}C, \forall z \in f(x, \mathscr{D}(\lambda), \mu), \forall c \in C.$$

Since C is a cone, it follows that, for all $\bar{c} \in -\text{int}C$ and $\alpha > 0$, $\alpha\bar{c} \in -\text{int}C$. So, $\langle \xi, \bar{c}\rangle \leq 0$ for all $\bar{c} \in -\text{int}C$. Hence, $\langle \xi, c\rangle \geq 0$ for all $c \in C$, i.e., $\xi \in C^* \setminus \{0\}$. Since $\bar{c} \in -\text{int}C$ can be chosen arbitrarily close to 0, $\gamma \geq 0$. Similarly, since $c \in C$ can also be taken arbitrarily close to 0 and ξ is a continuous linear function, we derive that $\langle \xi, z\rangle \geq 0$ for all $z \in f(x, \mathscr{D}(\lambda), \mu)$, i.e.,

$$\langle \xi, f(x, y, \mu)\rangle \geq 0, \forall y \in \mathscr{D}(\lambda),$$

and hence $x \in \text{Sol(WEP)}_\xi(\lambda, \mu)$. So,

$$\text{Sol(WEP)}(\lambda, \mu) = \cup_{\xi \in C^* \setminus \{0\}} \text{Sol(WEP)}_\xi(\lambda, \mu).$$

(b) Next we show that

$$\bigcup_{\xi \in C^* \setminus \{0\}} \text{Sol(WEP)}_\xi(\lambda, \mu) = \bigcup_{\xi \in B_e^*} \text{Sol(WEP)}_\xi(\lambda, \mu).$$

Since $B_e^* \subset (C^* \setminus \{0\})$, one has

$$\bigcup_{\xi \in C^* \setminus \{0\}} \text{Sol(WEP)}_\xi(\lambda, \mu) \supset \bigcup_{\xi \in B_e^*} \text{Sol(WEP)}_\xi(\lambda, \mu).$$

Conversely, for each $\xi \in C^* \setminus \{0\}$, from Lemma 6.62, we have $\langle \xi, e\rangle > 0$ as $e \in \text{int}C$. Putting $\bar{\xi} = \frac{1}{\langle \xi, e\rangle}\xi$, then $\bar{\xi} \in B_e^*$ and $\text{Sol(WEP)}_{\xi}(\lambda, \mu) = \text{Sol(WEP)}_{\bar{\xi}}(\lambda, \mu)$ for all $(\lambda, \mu) \in \Lambda \times M$, and hence $\text{Sol(WEP)}_{\xi} \subset \cup_{\xi \in B_e^*} \text{Sol(WEP)}_{\xi}(\lambda, \mu)$ for all $\xi \in C^* \setminus \{0\}$. So,

$$\bigcup_{\xi \in C^* \setminus \{0\}} \text{Sol(WEP)}_{\xi}(\lambda, \mu) \subset \bigcup_{\xi \in B_e^*} \text{Sol(WEP)}_{\xi}(\lambda, \mu).$$

The proof follows.

□

Finally, we present properties of the linear functions via the corresponding ones of the original objective map, which have been applied to formulate qualitative properties of solutions to (WEPs).

Definition 6.65 (Cone-strong monotonicity, cone-strong pseudomonotonicity) Let X be a normed space, $A \subset X$ be nonempty, $h > 0$, and $e \in \text{int}C$. A map $g : X \times X \to Y$ is said to be:

(a) (h, β)-Hölder strongly monotone wrt e in A if for each $x, y \in A$ with $x \neq y$,

$$g(x, y) + g(y, x) + hd^{\beta}(x, y)e \in -C;$$

(b) (h, β)-Hölder strongly pseudomonotone with respect to e in A if for each $x, y \in A$ with $x \neq y$,

$$g(x, y) \notin -\text{int}C \Rightarrow g(y, x) + hd^{\beta}(x, y)e \in -C;$$

(c) quasimonotone in A if for each $x, y \in A$ with $x \neq y$,

$$g(x, y) \in -\operatorname{int} C \Rightarrow f(y, x) \notin -\operatorname{int} C.$$

Definition 6.66 (Cone-Hölder continuity of a vector-valued map) Let $g : X \times X \to Y$, $A \subset X$ nonempty, $h > 0$, and $e \in \text{int}C$ be given. g is said to be (h, β)-Hölder continuous wrt e around $\bar{x}$ if there exists a neighborhood U of $\bar{x}$ such that for any $x_1, x_2 \in U$

$$g(x_1) \in g(x_2) + hd^{\beta}(x_1, x_2)[-e, e],$$

where $[-e, e] := (e - C) \cap (-e + C)$.

Lemma 6.67 (Topological preserving properties via linear scalarization functions) *Suppose that $N(\lambda_0) \times N(\mu_0)$ is a neighborhood of (λ_0, μ_0). Let $e \in \text{int}C$, $\xi \in B_e^*$, and $h, n, m, \beta, \delta, \gamma > 0$. Then, the following assertions hold:*

(a) *If for each $\mu \in N(\mu_0)$, $f(\cdot, \cdot, \mu)$ is (h, β)-Hölder strongly monotone wrt e in $\mathscr{D}(N(\lambda_0))$, then $\langle \xi, f(\cdot, \cdot, \mu)\rangle$ is (h, β)-Hölder strongly monotone in $\mathscr{D}(N(\lambda_0))$.*

(b) *If for each* $\mu \in N(\mu_0)$, $f(\cdot, \cdot, \mu)$ *is* (h, β)*-Hölder strongly pseudomonotone wrt* e *in* $\mathscr{D}(N(\lambda_0))$*, then* $\langle \xi, f(\cdot, \cdot, \mu)\rangle$ *is* (h, β)*-Hölder strongly pseudomonotone in* $\mathscr{D}(N(\lambda_0))$.

(c) *If for each* $\mu \in N(\mu_0)$, $f(\cdot, \cdot, \mu)$ *is quasimonotone in* $\mathscr{D}(N(\lambda_0))$*, then* $\langle \xi, f(\cdot, \cdot, \mu)\rangle$ *is quasimonotone in* $\mathscr{D}(N(\lambda_0))$.

(d) *If for each* $\mu \in N(\mu_0)$ *and* $x \in \mathscr{D}(N(\lambda_0))$, $f(x, \cdot, \mu)$ *is* (n, δ)*-Hölder continuous wrt* e *in* $\mathscr{D}(N(\lambda_0))$*, then* $\langle \xi, f(x, \cdot, \mu)\rangle$ *is* (n, δ)*-Hölder continuous in* $\mathscr{D}(N(\lambda_0))$.

(e) *If for each* $\mu \in N(\mu_0)$ *and* $x \in \mathscr{D}(N(\lambda_0))$, $f(x, \cdot, \mu)$ *is* (n, δ)*-Hölder continuous wrt* e *in* $\mathscr{D}(N(\lambda_0))$*, then* $\langle \xi, f(x, \cdot, \mu)\rangle$ *is* (n, δ)*-Hölder continuous in* $\mathscr{D}(N(\lambda_0))$.

(f) *If for each* $x, y \in \mathscr{D}(N(\lambda_0))$, $f(x, y, \cdot)$ *is* (n, γ)*-Hölder continuous wrt* e *in* $\mathscr{D}(N(\lambda_0))$*, then* $\langle \xi, f(x, y, \cdot)\rangle$ *is* (m, γ)*-Hölder continuous in* $\mathscr{D}(N(\mu_0))$.

Proof By the similarity, we verify only (a) as an example.

For each $\xi \in B_e^* \subset C^*$, one has:

$$\langle \xi, f(x, y, \mu) + f(y, x, \mu) + hd^{\beta}(x, y)e\rangle \leq 0.$$

By the linearity of ξ and $\langle \xi, e\rangle = 1$, we conclude that

$$\langle \xi, f(x, y, \mu)\rangle + \langle \xi, f(y, x, \mu)\rangle + hd^{\beta}(x, y) \leq 0.$$

□

Finally, we consider a very important property, which is the preservation of Hölder continuity under linear homogenization. This property plays a pivotal role in studying stability in the sense of Hölder/Lipschitz continuity, as will be presented in the sequel.

Lemma 6.68 (Cone-Hölder continuity via linear scalarization functions) *Suppose that* $N(\lambda_0) \times N(\mu_0)$ *is a neighborhood of* (λ_0, μ_0) *and* $m, \gamma > 0$*. Assume further that:*

(a) *for each* $x, y \in \mathscr{D}(N(\lambda_0))$, $f(x, y, \cdot)$ *is* (m, γ) *-Hölder continuous wrt* $e \in \mathrm{int}C$ *at* μ_0;

(b) *the function* $f(\cdot, \cdot, \mu_0)$ *is bounded on* $\mathscr{D}(N(\lambda_0)) \times \mathscr{D}(N(\lambda_0))$.

Then, there is a positive real number ℓ_0*, such that for each* $\xi_1, \xi_2 \in B_e^*$*, we have:*

$$|\langle \xi_1, f(x, y, \mu_1)\rangle - \langle \xi_2, f(x, y, \mu_2)\rangle| \leq md^{\gamma}(\mu_1, \mu_2) + \ell_0\|\xi_1 - \xi_2\|.$$

Proof By assumption (b), there is

$$\ell = \sup_{(x,y)\in\mathscr{D}(N(\lambda_0))\times\mathscr{D}(N(\lambda_0))} \|f(x, y, \mu_0)\| < +\infty.$$

Since $f(x, y, \cdot)$ is Hölder continuous at μ_0, there is ℓ_0 such that

$$\|f(x, y, \mu)\| \leq \ell_0, \forall x, y \in \mathscr{D}(N(\lambda_0)), \forall \mu \in N(\mu_0). \tag{6.20}$$

Now we verify the conclusion. Arguing as in the proof of Lemma 6.67, for any $\xi \in B_e^*$, we have:

$$-md^\gamma(\mu_1, \mu_2) \leq \langle \xi, f(x, y, \mu_1)\rangle - \langle \xi, f(x, y, \mu_2)\rangle \leq md^\gamma(\mu_1, \mu_2),$$

and so

$$|\langle \xi, f(x, y, \mu_1)\rangle - \langle \xi, f(x, y, \mu_2)\rangle| \leq md^\gamma(\mu_1, \mu_2).$$

Thus, for each $\xi_1, \xi_2 \in B_e^*$, it follows from (6.20) that

$$|\langle \xi_1, f(x, y, \mu_1)\rangle - \langle \xi_2, f(x, y, \mu_2)\rangle|$$

$$\leq |\langle \xi_1, f(x, y, \mu_1)\rangle - \langle \xi_1, f(x, y, \mu_2)\rangle| + |\langle \xi_1, f(x, y, \mu_2)\rangle - \langle \xi_2, f(x, y, \mu_2)\rangle|$$

$$\leq md^\gamma(\mu_1, \mu_2) + \|f(x, y, \mu_2)\|\|\xi_1 - \xi_2\| \leq md^\gamma(\mu_1, \mu_2) + \ell_0\|\xi_1 - \xi_2\|.$$

□

Stability and Sensitivity Analysis for (WEP)

Theorem 6.69 (Hölder continuity of solution maps of linear scalarized equilibrium problems) *Assume that for each $\xi \in B_e^*$, the ξ-solution set* Sol(WEP)(ξ, λ, μ) *for* (WEP) *exists in a neighborhood $N(\lambda_0) \times N(\mu_0)$ of the considered point (λ_0, μ_0). Furthermore, assume that the following conditions hold*

- (a) *$\mathscr{D}$ is (ℓ, α)-Hölder continuous at $\lambda_0 \in \Lambda$;*
- (b) *for each $\mu \in N(\mu_0)$, $f(\cdot, \cdot, \mu)$ is (h, β)-Hölder strongly pseudomonotone wrt $e \in$ int$C$ in $\mathscr{D}(N(\lambda_0))$ and $-f(\cdot, \cdot, \mu)$ is quasimonotone in $\mathscr{D}(N(\lambda_0))$;*
- (c) *for each $\mu \in N(\mu_0)$ and $x \in \mathscr{D}(N(\lambda_0))$, $f(x, \cdot, \mu)$ is (n, δ)-Hölder continuous wrt $e \in$ int$C$ in $\mathscr{D}(N(\lambda_0))$;*
- (d) *for each $x, y \in \mathscr{D}(N(\lambda_0))$, $f(x, y, \cdot)$ is (m, γ)-Hölder continuous wrt $e \in$ int$C$ at μ_0;*
- (e) *the function $f(\cdot, \cdot, \mu_0)$ is bounded on $\mathscr{D}(N(\lambda_0)) \times \mathscr{D}(N(\lambda_0))$.*

Then, for any $\bar{\xi} \in B_e^*$, there exist open neighborhoods $N'(\bar{\xi})$ of $\bar{\xi}$, $N'_{\bar{\xi}}(\lambda_0)$ of λ_0, and $N_{\bar{\xi}}(\mu_0)$ of μ_0 such that the ξ-solution set on $N'(\bar{\xi}) \times N'_{\bar{\xi}}(\lambda_0) \times N_{\bar{\xi}}(\mu_0)$ is a singleton and satisfies the following condition, for all $(\xi_1, \lambda_1, \mu_1)$ and $(\xi_2, \lambda_2, \mu_2) \in N'(\bar{\xi}) \times N'_{\bar{\xi}}(\lambda_0) \times N_{\bar{\xi}}(\mu_0)$:

$$d(x(\xi_1, \lambda_1, \mu_1), x(\xi_2, \lambda_2, \mu_2))$$

$$\leq \left(mhd^{\gamma}(\mu_1, \mu_2) + \ell h\|\xi_1 - \xi_2\|\right)^{\frac{1}{\alpha}} + \left(\frac{2\ell^{\delta} n}{h}\right)^{\frac{1}{\alpha}} d^{\frac{\beta\delta}{\alpha}}(\lambda_1, \lambda_2), \tag{6.21}$$

where $x(\xi_i, \lambda_i, \mu_i) \in \mathrm{Sol(WEP)}(\xi_i, \lambda_i, \mu_i),\ i = 1, 2.$

Proof For any $\bar{\xi} \in B_e^*$, let $N'(\bar{\xi}) \times N'_{\bar{\xi}}(\lambda_0) \times N_{\bar{\xi}}(\mu_0) \subset B_e^* \times N(\lambda_0) \times N(\mu_0)$ be open (where $N'_{\bar{\xi}}(\lambda_0), N_{\bar{\xi}}(\mu_0)$ depend on $\bar{\xi}$). Obviously, for each $(\xi, \lambda, \mu) \in N'(\bar{\xi}) \times N'_{\bar{\xi}}(\lambda_0) \times N_{\bar{\xi}}(\mu_0)$, the solution set $\mathrm{Sol(WEP)}_{\xi}(\lambda, \mu)$ is nonempty. Fix $(\xi, \lambda, \mu) \in N'(\bar{\xi}) \times N'_{\bar{\xi}}(\lambda_0) \times N_{\bar{\xi}}(\mu_0)$.

If $x_0 \in \mathrm{Sol(WEP)}_{\xi}(\lambda, \mu)$, then

$$\langle \xi, f(x_0, y, \mu)\rangle \geq 0, \forall y \in \mathscr{D}(\lambda). \tag{6.22}$$

By virtue of Lemma 6.67, we have:

$$\langle \xi, f(y, x_0, \mu)\rangle + hd^{\alpha}(x_0, y) \leq 0, \forall y \in \mathscr{D}(\lambda) \setminus \{x_0\}. \tag{6.23}$$

Hence, (6.22) and (6.23) together yield that

$$\langle \xi, f(y, x_0, \mu)\rangle < 0, \forall y \in \mathscr{D}(\lambda) \setminus \{x_0\}.$$

Therefore, $y \in \mathscr{D}(\lambda) \setminus \{x_0\}$ is not a ξ-solution of (WEP) and the ξ-solution uniqueness follows. Arguing as in the proof of Theorem 6.42, we have (6.21) by Lemmas 6.67 and 6.68. □

Theorem 6.70 (Hölder continuity of solution maps of weak equilibrium problems via linear scalarizations for a strongly pseudomonotone map) *Assume that for each* $\xi \in B_e^*$, *the* ξ*-solution set* $\mathrm{Sol(WEP)}_{\xi}(\lambda, \mu)$ *for (WEP) exists in a neighborhood* $N(\lambda_0) \times N(\mu_0)$ *of the considered point* (λ_0, μ_0). *Furthermore, assume that:*

(a) $\mathscr{D}$ *is* (ℓ, α)*-Hölder continuous at* $\lambda_0 \in \Lambda$;
(b) *for each* $\mu \in N(\mu_0)$, $f(\cdot, \cdot, \mu)$ *is* (h, β)*-Hölder strongly pseudomonotone wrt* $e \in \mathrm{int}C$ *in* $\mathscr{D}(N(\lambda_0))$ *and* $-f(\cdot, \cdot, \mu)$ *is quasimonotone in* $\mathscr{D}(N(\lambda_0))$;
(c) *for each* $\mu \in N(\mu_0)$ *and* $x \in \mathscr{D}(N(\lambda_0))$, $f(x, \cdot, \mu)$ *is* (n, δ)*-Hölder continuous wrt* $e \in \mathrm{int}C$ *in* $\mathscr{D}(N(\lambda_0))$;
(d) *for each* $x, y \in \mathscr{D}(N(\lambda_0))$, $f(x, y, \cdot)$ *is* (m, γ)*-Hölder continuous wrt* $e \in \mathrm{int}C$ *at* μ_0;
(e) *the function* $f(\cdot, \cdot, \mu_0)$ *is bounded on* $\mathscr{D}(N(\lambda_0)) \times \mathscr{D}(N(\lambda_0))$.

If for each $x \in \mathscr{D}(N(\lambda_0))$ *and* $\mu \in N(\mu_0)$, $f(x, \cdot, \mu)$ *is* C*-convex like on* $\mathscr{D}(N(\lambda_0))$, *then there exist neighborhoods* $\widetilde{N}(\lambda_0)$ *of* λ_0 *and* $\widetilde{N}(\mu_0)$ *of* μ_0 *such that the weak solution set* Sol(WEP) *on* $\widetilde{N}(\lambda_0) \times \widetilde{N}(\mu_0)$ *is nonempty and satisfies the*

Taking $\bar{x} = \frac{1}{2}(x_{11} + x_{21})$, it is clear that $\bar{x} \in \operatorname{conv}(\mathscr{D}(N(\lambda_0)))$. Since $f(x_{11}, \cdot, \mu_1)$ is $h\beta$-strong convexity wrt e, we have

$$f(x_{11}, \bar{x}, \mu_1) - \frac{1}{2} f(x_{11}, x_{11}, \mu_1) - \frac{1}{2} f(x_{11}, x_{21}, \mu_1) + \frac{h}{4} d_1^{\beta} e \in -C.$$

Then, it follows from the definition of B_e^* that

$$\begin{aligned}0 &\geq \left\langle \xi, f(x_{11}, \bar{x}, \mu_1) - \frac{1}{2} f(x_{11}, x_{11}, \mu_1) - \frac{1}{2} f(x_{11}, x_{21}, \mu_1) + \frac{h}{4} d_1^{\beta} e \right\rangle \\ &= \langle \xi, f(x_{11}, \bar{x}, \mu_1) \rangle - \left\langle \xi, \frac{1}{2} f(x_{11}, x_{11}, \mu_1) \right\rangle - \left\langle \xi, \frac{1}{2} f(x_{11}, x_{21}, \mu_1) \right\rangle \\ &\quad + \left\langle \xi, \frac{h}{4} d_1^{\beta} e \right\rangle,\end{aligned}$$

which arrives that

$$\frac{h}{4} d_1^{\beta} \leq \left\langle \xi, \frac{1}{2} f(x_{11}, x_{11}, \mu_1) \right\rangle + \left\langle \xi, \frac{1}{2} f(x_{11}, x_{21}, \mu_1) \right\rangle - \langle \xi, f(x_{11}, \bar{x}, \mu_1) \rangle. \tag{6.31}$$

Since $f(\cdot, \cdot, \mu_1)$ is monotone and $x_{11} \in \operatorname{Sol(WEP)}_{\xi}(\lambda_1, \mu_1)$, we have:

$$\left\langle \xi, \frac{1}{2} f(x_{11}, x_{11}, \mu_1) \right\rangle = 0 \tag{6.32}$$

and

$$\left\langle \xi, f(x_{11}, x_{21}, \mu_1) \right\rangle \leq -\langle \xi, f(x_{21}, x_{11}, \mu_1) \rangle. \tag{6.33}$$

Using (6.31), (6.32), and (6.33), we derive that

$$\frac{h}{4} d_1^{\beta} \leq -\left\langle \xi, \frac{1}{2} f(x_{21}, x_{11}, \mu_1) \right\rangle - \left\langle \xi, f\ Big(x_{11}, \frac{x_{11} + x_{21}}{2}, \mu_1 \right) \right\rangle. \tag{6.34}$$

Taking $z = x_2$ and $y = \frac{1}{2}(x_{11} + x_1)$ in (6.29), we can get that

$$\min \left\{ \left\langle \xi, f\left(x_{11}, \frac{1}{2}(x_{11} + x_1), \mu_1\right) \right\rangle, \langle \xi, f(x_{21}, x_2, \mu_1) \rangle \right\} \geq 0. \tag{6.35}$$

Then, (6.34) and (6.35) together yield that

$$\frac{h}{4} d_1^{\beta} \leq -\left\langle \xi, \frac{1}{2} f(x_{21}, x_{11}, \mu_1) \right\rangle - \left\langle \xi, f\left(x_{11}, \frac{x_{11} + x_{21}}{2}, \mu_1\right) \right\rangle$$

$$
\begin{aligned}
&+\Big\langle \xi, f\Big(x_{11}, \frac{x_{11}+x_1}{2}, \mu_1\Big)\Big\rangle + \Big\langle \xi, \frac{1}{2} f(x_{21}, x_2, \mu_1)\Big\rangle \\
&\le \Big\langle \xi, f\Big(x_{11}, \frac{x_{11}+x_1}{2}, \mu_1\Big)\Big\rangle - \Big\langle \xi, f\Big(x_{11}, \frac{x_{11}+x_{21}}{2}, \mu_1\Big)\Big\rangle \\
&+\Big\langle \xi, \frac{1}{2} f(x_{21}, x_2, \mu_1)\Big\rangle - \Big\langle \xi, \frac{1}{2} f(x_{21}, x_{11}, \mu_1)\Big\rangle \\
&\le \Big|\Big\langle \xi, f\Big(x_{11}, \frac{x_{11}+x_1}{2}, \mu_1\Big)\Big\rangle - \Big\langle \xi, f\Big(x_{11}, \frac{x_{11}+x_{21}}{2}, \mu_1\Big)\Big\rangle\Big| \\
&+\Big|\Big\langle \xi, \frac{1}{2} f(x_{21}, x_2, \mu_1)\Big\rangle - \Big\langle \xi, \frac{1}{2} f(x_{21}, x_{11}, \mu_1)\Big\rangle\Big|.
\end{aligned}
$$

Hence, using (6.30) and the m-Lipschitz continuity wrt e of $f(x_{11}, \cdot, \mu_1)$ and $f(x_{21}, \cdot, \mu_1)$ given in (b), we have:

$$
\begin{aligned}
\frac{h}{4} d_1^\beta &\le \frac{1}{2} m d(x_{11}, x_2) + \frac{1}{2} m d(x_{21}, x_1) \\
&\le \frac{1}{2} m d^\alpha(\lambda_1, \lambda_2) + \frac{1}{2} m d^\alpha(\lambda_1, \lambda_2) \\
&\le m\ell d^\alpha(\lambda_1, \lambda_2),
\end{aligned}
$$

which gives that

$$
d_1 \le \Big(\frac{4m\ell}{h}\Big)^{\frac{1}{\beta}} d^{\frac{\alpha}{\beta}}(\lambda_1, \lambda_2).
$$

Step II. We claim that for any two points $x_{21} \in \mathrm{Sol(WEP)}_\xi(\lambda_2, \mu_1)$ and $x_{22} \in \mathrm{Sol(WEP)}_\xi(\lambda_2, \mu_2)$,

$$
d_2 := d(x_{21}, x_{22}) \le \Big(\frac{n}{h}\Big)^{\frac{1}{\beta}} d^{\frac{\gamma}{\beta-\theta}}(\mu_1, \mu_2).
$$

By the definition of ξ-solutions, for all $y, z \in \mathscr{D}(\lambda_2)$,

$$
\min\Big\{\langle \xi, f(x_{21}, y, \mu_1)\rangle, \langle \xi, f(x_{22}, z, \mu_2)\rangle\Big\} \ge 0. \tag{6.36}
$$

Putting $y = \frac{1}{2}(x_{21} + x_{22})$ in (6.36), it is clear that $y \in \mathscr{D}(\lambda_2)$, and hence

$$
\langle \xi, f(x_{21}, y, \mu_1)\rangle \ge 0.
$$

By virtue of the $h\beta$-strong convexity wrt e of $f(x_{21}, \cdot, \mu_1)$, one has:

$$
f\Big(x_{21}, \frac{x_{21}+x_{22}}{2}, \mu_1\Big) - \frac{1}{2} f(x_{21}, x_{22}, \mu_1) - \frac{1}{2} f(x_{21}, x_{21}, \mu_1) + \frac{h}{4} d_2^\beta e \in -C.
$$

Therefore,

$$0 \geq \left\langle \xi, f\left(x_{21}, \frac{x_{21}+x_{22}}{2}, \mu_1\right) - \frac{1}{2} f(x_{21}, x_{22}, \mu_1) - \frac{1}{2} f(x_{21}, x_{21}, \mu_1) + \frac{h}{4} d_2^{\beta} e \right\rangle$$
$$\geq \left\langle \xi, -\frac{1}{2} f(x_{21}, x_{22}, \mu_1) - \frac{1}{2} f(x_{21}, x_{21}, \mu_1) + \frac{h}{4} d_2^{\beta} e \right\rangle.$$

Since $\xi \in B_e^*$,

$$\langle \xi, f(x_{21}, x_{22}, \mu_1) \rangle + \langle \xi, f(x_{21}, x_{21}, \mu_1) \rangle - \frac{h}{2} d_2^{\beta} \geq 0. \tag{6.37}$$

By the monotonicity of $f(\cdot, \cdot, \mu_1)$ and the belonging $x_{21} \in \mathrm{Sol(WEP)}_{\xi}(\lambda_2, \mu_1)$, we obtain:

$$f(x_{21}, x_{21}, \mu_1) = 0 \tag{6.38}$$

and

$$\langle \xi, f(x_{21}, x_{21}, \mu_1) \rangle \leq -\langle \xi, f(x_{22}, x_{21}, \mu_1) \rangle. \tag{6.39}$$

From (6.37), (6.38), and (6.39), we get:

$$\frac{h}{2} d_2^{\beta} \leq -\langle \xi, f(x_{22}, x_{21}, \mu_1) \rangle. \tag{6.40}$$

Now, replacing z in (6.36) by $\frac{1}{2}(x_{22} + x_{21})$, we have:

$$\left\langle \xi, f\left(x_{22}, \frac{x_{22} + x_{21}}{2}, \mu_2\right) \right\rangle \geq 0. \tag{6.41}$$

By the strong convexity wrt e of $f(x_{22}, \cdot, \mu_2)$, we also obtain the inequality similar to (6.37)

$$\langle \xi, f(x_{22}, x_{21}, \mu_2) \rangle + \langle \xi, f(x_{22}, x_{22}, \mu_2) \rangle - \frac{h}{2} d_2^{\beta} \geq 0.$$

As $f(x_{22}, x_{22}, \mu_2) = 0$,

$$\frac{h}{2} d_2^{\beta} \leq \langle \xi, f(x_{22}, x_{21}, \mu_2) \rangle. \tag{6.42}$$

Summing (6.40) and (6.42) and combining with assumption (d), we have:

$$h d_2^{\beta} \leq \langle \xi, f(x_{22}, x_{21}, \mu_2) \rangle - \langle \xi, f(x_{22}, x_{21}, \mu_1) \rangle$$

$$\le |\langle \xi, f(x_{22}, x_{21}, \mu_2 - f(x_{22}, x_{21}, \mu_1)\rangle|$$
$$\le nd^{\gamma}(\mu_1, \mu_2)d_2^{\theta},$$

and thus

$$d_2 \le \left(\frac{n}{h}\right)^{\frac{1}{\beta}} d^{\frac{\gamma}{\beta-\theta}}(\mu_1, \mu_2).$$

Step III. Finally, applying Steps I and II, we conclude that for each $x_{11} \in \text{Sol(WEP)}_{\xi}(\lambda_1, \mu_1)$ and $x_{22} \in \text{Sol(WEP)}_{\xi}(\lambda_2, \mu_2)$

$$d(x_{11}, x_{22}) \le d_1 + d_2 \le \left(\frac{4m\ell}{h}\right)^{\frac{1}{\beta}} d^{\frac{\alpha}{\beta}}(\lambda_1, \lambda_2) + \left(\frac{n}{h}\right)^{\frac{1}{\beta}} d^{\frac{\gamma}{\beta-\theta}}(\mu_1, \mu_2).$$

Hence, we get:

$$\mathbb{D}(\text{Sol(WEP)}_{\xi}(\lambda_1, \mu_1), \text{Sol(WEP)}_{\xi}(\lambda_2, \mu_2))$$
$$\le \left(\frac{4m\ell}{h}\right)^{\frac{1}{\beta}} d^{\frac{\alpha}{\beta}}(\lambda_1, \lambda_2) + \left(\frac{n}{h}\right)^{\frac{1}{\beta}} d^{\frac{\gamma}{\beta-\theta}}(\mu_1, \mu_2). \tag{6.43}$$

(b) Now we prove (6.27). We see that B_e^* is a weak* compact set and hence it can be covered by a finite number of open sets:

$$B_e^* \subset \bigcup_{i=1}^{n} N(\xi_i), \tag{6.44}$$

where $\xi_i \in B_e^*$ and $N(\xi_i)$ is a neighborhood of ξ_i, defined in (a). For $N_{\xi_i}(\lambda_0)$ and $N_{\xi_i}(\mu_0)$ determined as in (a), let $N'(\lambda_0) = \bigcap_{i=1}^n N_{\xi_i}(\lambda_0)$ and $N'(\mu_0) = \bigcap_{i=1}^n N_{\xi_i}(\mu_0)$. For each $(\lambda, \mu) \in N'(\lambda_0) \times N'(\mu_0)$ and $\xi \in B_e^*$, (6.44) implies the existence of $i_0 \in \{1, \ldots, n\}$ such that $\xi \in N(\xi_{i_0})$ and obviously, $(\lambda, \mu) \in N_{\xi_{i_0}}(\lambda_0) \times N_{\xi_{i_0}}(\mu_0)$. Combining the convexity of $f(x, ., \mu)$ assumed in (b) with Theorem 6.64, we have

$$\text{Sol(WEP)}(\lambda, \mu) = \bigcup_{\xi \in B_e^*} \text{Sol(WEP)}_{\xi}(\lambda, \mu).$$

For each $(\lambda_1, \mu_1), (\lambda_2, \mu_2) \in N'(\lambda_0) \times N'(\mu_0)$, we now show that

$$H(\text{Sol(WEP)}(\lambda_1, \mu_1), \text{Sol(WEP)}(\lambda_2, \mu_2))$$
$$\le \left(\frac{4m\ell}{h}\right)^{\frac{1}{\beta}} d^{\frac{\alpha}{\beta}}(\lambda_1, \lambda_2) + \left(\frac{n}{h}\right)^{\frac{1}{\beta}} d^{\frac{\gamma}{\beta-\theta}}(\mu_1, \mu_2). \tag{6.45}$$

Indeed, for each $x_{11} \in \text{Sol(WEP)}(\lambda_1, \mu_1) = \cup_{\xi \in B_e^*} \text{Sol(WEP)}_\xi(\lambda_1, \mu_1)$, there is $\hat{\xi} \in B_e^*$ such that $x_{11} \in S_{\hat{\xi}}(\lambda_1, \mu_1)$. As $\text{Sol(WEP)}_{\hat{\xi}}(\lambda_2, \mu_2) \subset \text{Sol(WEP)}(\lambda_2, \mu_2)$, applying (a) one has:

$$\begin{aligned} d(x_1, \text{Sol(WEP)}(\lambda_2, \mu_2)) &\le d(x_1, \text{Sol(WEP)}_{\hat{\xi}}(\lambda_2, \mu_2)) \\ &\le \text{ex}(\text{Sol(WEP)}_{\hat{\xi}}(\lambda_1, \mu_1), \text{Sol(WEP)}_{\hat{\xi}}(\lambda_2, \mu_2)) \\ &\le \mathbb{D}(\text{Sol(WEP)}_{\hat{\xi}}(\lambda_1, \mu_1), \text{Sol(WEP)}_{\hat{\xi}}(\lambda_2, \mu_2)) \\ &\le \left(\frac{4m\ell}{h}\right)^{\frac{1}{\beta}} d^{\frac{\alpha}{\beta}}(\lambda_1, \lambda_2) + \left(\frac{n}{h}\right)^{\frac{1}{\beta}} d^{\frac{\gamma}{\beta-\theta}}(\mu_1, \mu_2). \end{aligned}$$

Therefore,

$$\begin{aligned} &\text{ex}(\text{Sol(WEP)}(\lambda_1, \mu_1), \text{Sol(WEP)}(\lambda_2, \mu_2)) \\ &\le \left(\frac{4ml}{h}\right)^{\frac{1}{\beta}} d^{\frac{\alpha}{\beta}}(\lambda_1, \lambda_2) + \left(\frac{n}{h}\right)^{\frac{1}{\beta}} d^{\frac{\gamma}{\beta-\theta}}(\mu_1, \mu_2). \end{aligned} \tag{6.46}$$

Similarly, we also have:

$$\begin{aligned} &\text{ex}(\text{Sol(WEP)}(\lambda_2, \mu_2), \text{Sol(WEP)}(\lambda_1, \mu_1)) \\ &\le \left(\frac{4m\ell}{h}\right)^{\frac{1}{\beta}} d^{\frac{\alpha}{\beta}}(\lambda_1, \lambda_2) + \left(\frac{n}{h}\right)^{\frac{1}{\beta}} d^{\frac{\gamma}{\beta-\theta}}(\mu_1, \mu_2). \end{aligned} \tag{6.47}$$

Combining (6.46) with (6.47), we obtain (6.27).

□

To illustrate the essentialness of the imposed assumptions, which are not related to Hölder continuity, in Theorem 6.71, we bring out some examples.

Example 6.72 (The strong convexity is essential) Let $X = Y = \mathbb{R}^2$, $\Lambda \equiv M = [0, 1]$, $C = \mathbb{R}^2_+$, $e = (1, 1)$, $\mathcal{D}(\lambda) = [\lambda, 1]$, $\bar{\lambda} = 0$, and $f(x, y, \lambda) := (\lambda(x_1^2 - y_1^2), \lambda(x_2^2 - y_2^2))$, where $x = (x_1, x_2)$ and $y = (y_1, y_2) \in X$. Then, all the assumptions of Theorem 6.71 are fulfilled, except the strong convexity. By direct calculations, we have

$$\text{Sol(WEP)}(\lambda) = \begin{cases} [0, 1] \times [0, 1] & \lambda = 0, \\ \{(1, 1)\} & \lambda \neq 0, \end{cases}$$

which is not lower semicontinuous at $\bar{\lambda} = 0$.

Example 6.73 (The monotonicity is crucial) Let $X = Y = \mathbb{R}^2$, $\Lambda \equiv M = [0, 1]$, $C = \mathbb{R}^2_+$, $e = (1, 1)$, $\mathcal{D}(\lambda) \equiv [0, 1]$, $\bar{\lambda} = 0$, and $f(x, y, \lambda) := (y_1^2 - \lambda x_1, y_2^2 - \lambda x_2)$, where $x = (x_1, x_2), y = (y_1, y_2) \in X$. Then, it is not hard to see that the assumptions of Theorem 6.71 arc satisfied, except the monotonicity of f ($f(x, y, 0) + f(y, x, 0) = (y_1^2 + x_1^2, y_2^2 + x_2^2) > 0$ for many $x, y \in [0, 1] \times [0, 1]$). Direct computations give us the solution set:

$$\mathrm{Sol(WEP)}(\lambda) = \begin{cases} [0, 1] \times [0, 1] & \lambda = 0, \\ \{(0, 0)\} & \lambda \neq 0, \end{cases}$$

which is even not lower semicontinuous at $\bar{\lambda} = 0$.

Stability and Sensitivity Analysis for Approximate Solution Maps of (WEPs)

Theorem 6.74 (Hölder continuity of approximate solution maps of linear scalarized weak equilibrium problems) *For (WEP), assume that for each $\xi \in B_e^*$, ξ-approximate solutions exist in a neighborhood of the considered point $(\lambda_0, \mu_0) \in \Lambda \times M$. Furthermore, impose the following conditions:*

(a) *$\mathscr{D}$ is (ℓ, α)-Hölder continuous around λ_0, i.e., there exists a neighborhood U of λ_0 such that for all $\lambda_1 \lambda_2 \in U$,*

$$\mathscr{D}(\lambda_1) \subset \mathscr{D}(\lambda_2) + \ell \|\lambda_1 - \lambda_2\|^\alpha B(0, 1);$$

(b) *there is a neighborhood V of μ_0 such that for each $y \in \mathscr{D}(U)$ and $\mu \in V$, $f(\cdot, y, \mu)$ is C-concave in $\mathscr{D}(U)$;*
(c) *for $x, y \in \mathscr{D}(U)$, $f(x, y, \cdot)$ is (h, β)-Hölder continuous in V;*
(d) *for $\mu \in V$ and $x \in \mathscr{D}(U)$, $f(x, \cdot, \mu)$ is (q, δ)-Hölder in $\mathscr{D}(U)$.*

Then, for any $\bar{\varepsilon} > 0$ and $\bar{\xi} \in B_e^*$, there exist open neighborhoods $N(\bar{\xi})$ of $\bar{\xi}$, $N_{\bar{\xi}}(\lambda_0)$ of λ_0, and $N_{\bar{\xi}}(\mu_0)$ of μ_0 such that the ξ-approximate solution mapping $\widetilde{\mathrm{Sol}}(\mathrm{WEP})_\xi(\cdot, \cdot, \cdot)$ satisfies the following Hölder property in $[\bar{\varepsilon}, +\infty) \times N_{\bar{\xi}}(\lambda_0) \times N_{\bar{\xi}}(\mu_0)$

$$\begin{aligned} &H(\widetilde{\mathrm{Sol}}(\mathrm{WEP})_\xi(\varepsilon_1, \lambda_1, \mu_1), \widetilde{\mathrm{Sol}}(\mathrm{WEP})_\xi(\varepsilon_2, \lambda_2, \mu_2)) \\ &\leq k_1|\varepsilon_1 - \varepsilon_2| + k_2||\mu_1 - \mu_2||^\beta + k_3||\lambda_1 - \lambda_2||^{\alpha\delta}, \end{aligned}$$

where $\xi \in N(\bar{\xi}), (\varepsilon_i, \lambda_i, \mu_i) \in [\bar{\varepsilon}, +\infty) \times N_{\bar{\xi}}(\lambda_0) \times N_{\bar{\xi}}(\mu_0)$, $i = 1, 2$, and k_1, k_2, k_3 are positive and depend on $\bar{\varepsilon}, \ell, \alpha, h, \beta$, etc.

Proof Let $\bar{\xi} \in B_e^*$ and $N(\bar{\xi}) \times N_{\bar{\xi}}(\lambda_0) \times N_{\bar{\xi}}(\mu_0) \subset B_e^* \times U \times V$ be open. Let $(\varepsilon_1, \lambda_1, \mu_1), (\varepsilon_2, \lambda_2, \mu_2) \in [\bar{\varepsilon}, +\infty) \times N_{\bar{\xi}}(\lambda_0) \times N_{\bar{\xi}}(\mu_0)$ be arbitrarily given. Without loss generality, we always assume that $\varepsilon_1 < \varepsilon_2$.
Step 1. For $0 < \varepsilon_1 < \varepsilon_2$ with $\bar{\varepsilon} < \varepsilon_2$, $(\lambda_1, \mu_1) \in N_{\bar{\xi}}(\lambda_0) \times N_{\bar{\xi}}(\mu_0)$, and $\xi \in N(\bar{\xi})$, we provide an estimation of $H(\widetilde{\mathrm{Sol}}(\mathrm{WEP})_\xi(\varepsilon_1, \lambda_1, \mu_1), \widetilde{\mathrm{Sol}}(\mathrm{WEP})_\xi(\varepsilon_2, \lambda_1, \mu_1))$.

Therefore, for every $\xi_1, \xi_2 \in N(\bar{\xi})$

$$\begin{aligned}|\xi_1(f(x,y,\mu_1)) - \xi_2(f(x,y,\mu_2))| &\le |\xi_1(f(x,y,\mu_1)) - \xi_1(f(x,y,\mu_2))| \\ &\quad + |\xi_1(f(x,y,\mu_2)) - \xi_2(f(x,y,\mu_2))| \\ &\le h\|\mu_1 - \mu_2\|^\beta + \|f(x,y,\mu_2)\|\|\xi_1 - \xi_2\| \\ &\le h\|\mu_1 - \mu_2\|^\beta + M_0\|\xi_1 - \xi_2\|.\end{aligned}$$

Thus, we get (6.53). Now, we divide Step 2 into two cases.
Case 1. $h\|\mu_1 - \mu_2\|^\beta \le \varepsilon_2$. Let $r = h\|\mu_1 - \mu_2\|^\beta$. Then, $0 < r \le \varepsilon_2$. We conclude that $\widetilde{\text{Sol}}(\text{WEP})_\xi(\varepsilon_2 - r, \lambda_1, \mu_1) \subset \widetilde{\text{Sol}}(\text{WEP})_\xi(\varepsilon_2, \lambda_1, \mu_2)$. Indeed, let $\bar{x} \in \widetilde{\text{Sol}}(\text{WEP})_\xi(\varepsilon_2 - r, \lambda_1, \mu_1)$. Then, for all $y \in \mathscr{D}(\lambda_1)$,

$$\xi\,(f(\bar{x}, y, \mu_2)) + \xi\,(f(\bar{x}, y, \mu_1)) - \xi\,(f(\bar{x}, y, \mu_2)) + \varepsilon_2 - r \ge 0.$$

This inequality together with (6.53) implies that $\xi(f(\bar{x}, y, \mu_2)) + \varepsilon_2 \ge 0$ for all $y \in \mathscr{D}(\lambda_1)$, i.e, $\bar{x} \in \widetilde{\text{Sol}}(\text{WEP})_\xi(\varepsilon_2, \lambda_1, \mu_2)$. Hence, the result of Step 1 results in the following estimate:

$$\begin{aligned}&\text{ex}(\widetilde{\text{Sol}}(\text{WEP})_\xi(\varepsilon_2, \lambda_1, \mu_1), \widetilde{\text{Sol}}(\text{WEP})_\xi(\varepsilon_2, \lambda_1, \mu_2)) \\ &\quad \le \text{ex}(\widetilde{\text{Sol}}(\text{WEP})_\xi(\varepsilon_2, \lambda_1, \mu_1), \widetilde{\text{Sol}}(\text{WEP})_\xi(\varepsilon_2 - r, \lambda_1, \mu_1)) \\ &\quad \le H(\widetilde{\text{Sol}}(\text{WEP})_\xi(\varepsilon_2, \lambda_1, \mu_1), \widetilde{\text{Sol}}(\text{WEP})_\xi(\varepsilon_2 - r, \lambda_1, \mu_1)) \\ &\quad \le \frac{\rho r}{\varepsilon_2} \\ &\quad \le \frac{\rho h}{\bar{\varepsilon}}\|\mu_1 - \mu_2\|^\beta.\end{aligned}$$

In the same way, we get

$$\text{ex}(\widetilde{\text{Sol}}(\text{WEP})_\xi(\varepsilon_2, \lambda_1, \mu_2), \widetilde{\text{Sol}}(\text{WEP})_\xi(\varepsilon_2, \lambda_1, \mu_1)) \le \frac{\rho h}{\bar{\varepsilon}}\|\mu_1 - \mu_2\|^\beta.$$

Of course, $\text{ex}(\widetilde{\text{Sol}}(\text{WEP})_{\xi_2}(\varepsilon_2, \lambda_1, \mu_2), \widetilde{\text{Sol}}(\text{WEP})_{\xi_1}(\varepsilon_2, \lambda_1, \mu_1))$ has the same upper bound. Hence,

$$\begin{aligned}&H(\widetilde{\text{Sol}}(\text{WEP})_\xi(\varepsilon_2, \lambda_1, \mu_1), \widetilde{\text{Sol}}(\text{WEP})_\xi(\varepsilon_2, \lambda_1, \mu_2)) \le \frac{\rho h}{\bar{\varepsilon}}\|\mu_1 - \mu_2\|^\beta. \\ &=: k_2\|\mu_1 - \mu_2\|^\beta + k_3\|\xi_1 - \xi_2\|.\end{aligned}$$

Case 2. $h\|\mu_1 - \mu_2\|^\beta > \varepsilon_2$. Let $\rho_1 = \text{diam}(V)$. Then, there exists a natural number n such that $\frac{\rho_1}{n} \le \left(\frac{\bar{\varepsilon}}{h}\right)^{1/\beta}$. Let $P_1 = \{v_1 = \mu_1, v_2, \ldots, v_{n+1} = \mu_2\}$ be a partition of segment $[\mu_1, \mu_2]$ such that

$$\|v_i - v_{i+1}\| = \frac{\|\mu_1 - \mu_2\|}{n}.$$

Then, we have $\|v_i - v_{i+1}\|^\beta = \left(\frac{\|\mu_1-\mu_2\|}{n}\right)^\beta \leq \left(\frac{\rho_1}{n}\right)^\beta \leq \frac{\bar{\varepsilon}}{h} \leq \frac{\varepsilon_2}{h}$. Hence, using the result of Case 1, one gets:

$$\begin{aligned}
&H(\widetilde{\text{Sol}}(\text{WEP})_\xi(\varepsilon_2, \lambda_1, \mu_1), \widetilde{\text{Sol}}(\text{WEP})_\xi(\varepsilon_2, \lambda_1, \mu_2)) \\
&\leq \sum_{i=1}^{n} H(\widetilde{\text{Sol}}(\text{WEP})_\xi(\varepsilon_2, \lambda_1, v_i), \widetilde{\text{Sol}}(\text{WEP})_\xi(\varepsilon_2, \lambda_1, v_{i+1})) \\
&\leq \frac{\rho_1 h}{\bar{\varepsilon}} \sum_{i=1}^{n} \|v_i - v_{i+1}\|^\beta \\
&\leq \frac{n\rho_1 h}{\bar{\varepsilon}} \|\mu_1 - \mu_2\|^\beta \\
&=: k_2 \|\mu_1 - \mu_2\|^\beta.
\end{aligned}$$

Step 3. We estimate $H(\widetilde{\text{Sol}}(\text{WEP})_\xi(\varepsilon_2, \lambda_1, \mu_2), \widetilde{\text{Sol}}(\text{WEP})_\xi(\varepsilon_2, \lambda_2, \mu_2))$ for $\lambda_1 \neq \lambda_2$. We have also two cases.

Case 1. $\|\lambda_1 - \lambda_2\|^\alpha \leq \frac{1}{\ell}(\frac{\varepsilon_2}{q})^{1/\delta}$. Let $r' = \ell^\delta q||\lambda_1 - \lambda_2||^{\alpha\delta}$. Then, $0 < r' \leq \varepsilon_2$. We first claim that $\widetilde{\text{Sol}}(\text{WEP})_\xi(\varepsilon_2 - r', \lambda_1, \mu_2) \subset \widetilde{\text{Sol}}(\text{WEP})_\xi(\varepsilon_2, \lambda_2, \mu_2)$. Indeed, for any $x' \in \widetilde{\text{Sol}}(\text{WEP})_\xi(\varepsilon_2 - r', \lambda_1, \mu_2)$ and $y_2 \in \mathscr{D}(\lambda_2)$, there is $y_1 \in \mathscr{D}(\lambda_1)$ such that

$$||y_1 - y_2|| \leq l||\lambda_1 - \lambda_2||^\alpha,$$

and

$$\xi(f(x', y_2, \mu_2)) + \xi(f(x', y_1, \mu_2)) - \xi(f(x', y_2, \mu_2)) + \varepsilon_2 - r' \geq 0.$$

It follows from (d) that

$$f(x', y_1, \mu_2)) \in f(x', y_2, \mu_2)) + q\|y_1 - y_2\|^\delta B(0, 1).$$

Thus,

$$\begin{aligned}
|\xi(f(x', y_1, \mu_2)) - \xi(f(x', y_2, \mu_2))| &\leq q\|y_1 - y_2\|^\delta \sup\{\xi(e') \mid e' \in B(0, 1)\} \\
&\leq q\|y_1 - y_2\|^\delta \\
&\leq q\ell^\delta \|\lambda_1 - \lambda_2\|^{\alpha\delta} \\
&\leq r'.
\end{aligned}$$

Hence, for all $y_2 \in \mathscr{D}(\lambda_2)$,

$$\xi(f(x', y_2, \mu_2)) + \varepsilon_2 \geq 0,$$

i.e., $x' \in \widetilde{\text{Sol}}(\text{WEP})_\xi(\varepsilon_2, \lambda_2, \mu_2)$, resulting in

$$\widetilde{\text{Sol}}(\text{WEP})_\xi(\varepsilon_2 - r', \lambda_1, \mu_2) \subset \widetilde{\text{Sol}}(\text{WEP})_\xi(\varepsilon_2, \lambda_2, \mu_2).$$

Applying this and the result of Step 1, one has the following estimate:

$$\begin{aligned}
&\text{ex}(\widetilde{\text{Sol}}(\text{WEP})_\xi(\varepsilon_2, \lambda_1, \mu_2), \widetilde{\text{Sol}}(\text{WEP})_\xi(\varepsilon_2, \lambda_2, \mu_2)) \\
&\leq H(\widetilde{\text{Sol}}(\text{WEP})_\xi(\varepsilon_2, \lambda_1, \mu_2), \widetilde{\text{Sol}}(\text{WEP})_\xi(\varepsilon_2 - r', \lambda_1, \mu_2)) \\
&\leq \frac{\rho q \ell^\delta}{\varepsilon_2} \|\lambda_1 - \lambda_2\|^{\alpha\delta} \\
&\leq \frac{\rho q \ell^\delta}{\bar{\varepsilon}} \|\lambda_1 - \lambda_2\|^{\alpha\delta}.
\end{aligned}$$

Similarly,

$$\text{ex}(\widetilde{\text{Sol}}(\text{WEP})_\xi(\varepsilon_2, \lambda_2, \mu_2), \widetilde{\text{Sol}}(\text{WEP})_\xi(\varepsilon_2, \lambda_1, \mu_2)) \leq \frac{\rho q \ell^\delta}{\bar{\varepsilon}} ||\lambda_1 - \lambda_2||^{\alpha\delta}.$$

Thus,

$$H(\widetilde{\text{Sol}}(\text{WEP})_\xi(\varepsilon_2, \lambda_1, \mu_2), \widetilde{\text{Sol}}(\text{WEP})_\xi(\varepsilon_2, \lambda_2, \mu_2)) \leq \frac{\rho q \ell^\delta}{\bar{\varepsilon}} \|\lambda_1 - \lambda_2\|^{\alpha\delta}.$$

Case 2. $\|\lambda_1 - \lambda_2\|^\alpha > \frac{1}{\ell}\left(\frac{\varepsilon_2}{q}\right)^{1/\delta}$. Let $\varepsilon_0 = \left[\frac{1}{\ell}\left(\frac{\bar{\varepsilon}}{q}\right)^{1/\delta}\right]^{1/\alpha}$ and $\wp = \text{diam}(U) < +\infty$. There exists a natural number $\mathscr{N}$ such that $\frac{\wp}{\mathscr{N}} \leq \varepsilon_0$. Let P_2 be a partition of segment $[\lambda_1, \lambda_2]$ with $\mathscr{N} + 1$ nodes $u_1, ..., u_{\mathscr{N}+1}$ such that

$$u_1 = \lambda_1, \qquad u_{\mathscr{N}+1} = \lambda_2, \quad \|u_j - u_{j+1}\| = \frac{\|\lambda_1 - \lambda_2\|}{\mathscr{N}}.$$

Hence,

$$\|u_j - u_{j+1}\|^\alpha \leq \left(\frac{\wp}{\mathscr{N}}\right)^\alpha \leq \frac{1}{\ell}\left(\frac{\bar{\varepsilon}}{q}\right)^{1/\delta} \leq \frac{1}{\ell}\left(\frac{\varepsilon_2}{q}\right)^{1/\delta}.$$

Applying the result of Case 1, one has

$$H(\widetilde{\mathrm{Sol}}(\mathrm{WEP})_\xi(\varepsilon_2, \lambda_1, \mu_2), \widetilde{\mathrm{Sol}}(\mathrm{WEP})_\xi(\varepsilon_2, \lambda_2, \mu_2)) \le \frac{\rho q \ell^\delta}{\bar{\varepsilon}} \sum_{j=1}^{\mathcal{N}} ||u_j - u_{j+1}||^{\alpha\delta}$$

$$\le \frac{\mathcal{N} \rho q \ell^\delta}{\bar{\varepsilon}} \left[\frac{1}{\ell} \left(\frac{\varepsilon_2}{q} \right)^{1/\delta} \right]^\delta \le \frac{\mathcal{N} \rho q \ell^\delta}{\bar{\varepsilon}} \|\lambda_1 - \lambda_2\|^{\alpha\delta} =: k_3 \|\lambda_1 - \lambda_2\|^{\alpha\delta}.$$

Step 4. Now, we are ready to complete the proof. Combining the results of the preceding three steps, we obtain:

$$\begin{aligned}
&H(\widetilde{\mathrm{Sol}}(\mathrm{WEP})_\xi(\varepsilon_1, \lambda_1, \mu_1), \widetilde{\mathrm{Sol}}(\mathrm{WEP})_\xi(\varepsilon_2, \lambda_2, \mu_2)) \\
&\quad \le H(\widetilde{\mathrm{Sol}}(\mathrm{WEP})_\xi(\varepsilon_1, \lambda_1, \mu_1), \widetilde{\mathrm{Sol}}(\mathrm{WEP})_\xi(\varepsilon_2, \lambda_1, \mu_1)) \\
&\quad + H(\widetilde{\mathrm{Sol}}(\mathrm{WEP})_\xi(\varepsilon_2, \lambda_1, \mu_1), \widetilde{\mathrm{Sol}}(\mathrm{WEP})_\xi(\varepsilon_2, \lambda_1, \mu_2)) \\
&\quad + H(\widetilde{\mathrm{Sol}}(\mathrm{WEP})_\xi(\varepsilon_2, \lambda_1, \mu_2), \widetilde{\mathrm{Sol}}(\mathrm{WEP})_\xi(\varepsilon_2, \lambda_2, \mu_2)) \\
&\quad \le k_1|\varepsilon_1 - \varepsilon_2| + k_2\|\mu_1 - \mu_2\|^\beta + k_3\|\lambda_1 - \lambda_2\|^{\alpha\delta}.
\end{aligned}$$

□

By using Theorem 6.74 and a suitable technique, we obtain the following Hölder continuity of approximate solutions maps of (WEP).

Theorem 6.75 (Hölder continuity of approximate solution maps of weak equilibrium problems via linear scalarization functions) *For (WEP), assume that for each $\xi \in B_e^*$, ξ-approximate solutions exist in a neighborhood of the considered point $(\lambda_0, \mu_0) \in \Lambda \times M$. Furthermore, assume that the following conditions hold:*

(a) *$\mathcal{D}$ is (ℓ, α)-Hölder continuous around λ_0, i.e., there exists a neighborhood U of λ_0 such that for all $\lambda_1, \lambda_2 \in U$,*

$$\mathcal{D}(\lambda_1) \subset \mathcal{D}(\lambda_2) + \ell\|\lambda_1 - \lambda_2\|^\alpha B(0, 1);$$

(b) *there is a neighborhood V of μ_0 such that for each $y \in \mathcal{D}(U)$ and $\mu \in V$, $f(\cdot, y, \mu)$ is C-concave on $\mathcal{D}(U)$;*
(c) *for $x, y \in \mathcal{D}(U)$, $f(x, y, \cdot)$ is (h, β)-Hölder continuous on V;*
(d) *for $\mu \in V$ and $x \in \mathcal{D}(U)$, $f(x, \cdot, \mu)$ is C-convex like as well as (q, δ)-Hölder on $\mathcal{D}(U)$.*

Then, for any $\bar{\varepsilon} > 0$, there exist open neighborhoods $N'(\lambda_0)$ of λ_0 and $N'(\mu_0)$ of μ_0 such that the approximate solution map $\widetilde{\mathrm{Sol}}(\mathrm{WEP})$ satisfies the following Hölder property in $[\bar{\varepsilon}, +\infty) \times N'(\lambda_0) \times N'(\mu_0)$:

$$H(\widetilde{\mathrm{Sol}}(\mathrm{WEP})(\varepsilon_1, \lambda_1, \mu_1), \widetilde{\mathrm{Sol}}(\mathrm{WEP})(\varepsilon_2, \lambda_2, \mu_2))$$

$$\leq k_1|\varepsilon_1 - \varepsilon_2| + k_2||\mu_1 - \mu_2||^{\beta} + k_3||\lambda_1 - \lambda_2||^{\alpha\delta},$$

where $(\varepsilon_i, \lambda_i, \mu_i) \in [\bar{\varepsilon}, +\infty) \times N'(\lambda_0) \times N'(\mu_0)$, $i = 1, 2$, *and* k_1, k_2, k_3 *are positive and depend on* $\bar{\varepsilon}, \ell, \alpha, h, \beta$, *etc.*

Proof We see that B_e^* is a weak* compact set and, in Theorem 6.74, the system $\{N(\bar{\xi})\}_{\bar{\xi}\in B_e^*}$ is an open covering of B_e^*. Consequently, there is a finite number of points $\xi_i \in B_e^* (i = 1, ..., n)$ satisfying

$$B_e^* \subset \bigcup_{i=1}^{n} N(\xi_i). \tag{6.54}$$

Let $N'(\mu_0) = \bigcap_{i=1}^n N_{\xi_i}(\mu_0)$ and $N'(\lambda_0) = \bigcap_{i=1}^n N_{\xi_i}(\lambda_0)$. Take arbitrarily $(\lambda, \mu) \in N'(\lambda_0) \times N'(\mu_0)$. Thanks to (6.44), for any $\xi \in B_e^*$, there is $i_0 \in \{1, ..., n\}$ such that $\xi \in N(\xi_{i_0})$. From the construction of the neighborhoods $N'(\lambda_0)$ and $N'(\mu_0)$, we have $(\lambda, \mu) \in N_{\xi_{i_0}}(\lambda_0) \times N_{\xi_{i_0}}(\mu_0)$. So, $N'(\lambda_0)$ and $N'(\mu_0)$ are desired neighborhoods of λ_0 and μ_0, resp.

Due to assumption (d) and in view of Lemma 6.64, one has:

$$\widetilde{\text{Sol}}(\text{WEP})(\varepsilon, \lambda, \mu) = \bigcup_{\xi\in B_e^*} \widetilde{\text{Sol}}(\text{WEP})_{\xi}(\varepsilon, \lambda, \mu).$$

For all $(\lambda_1, \mu_1), (\lambda_2, \mu_2) \in N'(\lambda_0) \times N'(\mu_0)$, we now claim that

$$\begin{aligned} &H(\widetilde{\text{Sol}}(\text{WEP})(\varepsilon_1, \lambda_1, \mu_1), \widetilde{\text{Sol}}(\text{WEP})(\varepsilon_2, \lambda_2, \mu_2)) \\ &\quad \leq k_1|\varepsilon_1 - \varepsilon_2| + k_2\|\mu_1 - \mu_2||^{\beta} + k_3||\lambda_1 - \lambda_2||^{\alpha\delta}. \end{aligned} \tag{6.55}$$

Indeed, for each $x_1 \in \widetilde{\text{Sol}}(\text{WEP})(\varepsilon_1, \lambda_1, \mu_1) = \bigcup_{\xi\in B_e^*} \widetilde{\text{Sol}}(\text{WEP})_{\xi}(\varepsilon_1, \lambda_1, \mu_1)$, there is $\hat{\xi} \in B_e^*$ such that $x_1 \in \widetilde{\text{Sol}}(\text{WEP})_{\hat{\xi}}(\varepsilon_1, \lambda_1, \mu_1)$. As $\widetilde{\text{Sol}}(\text{WEP})_{\hat{\xi}}(\varepsilon_2, \lambda_2, \mu_2) \subset \widetilde{\text{Sol}}(\text{WEP})(\varepsilon_2, \lambda_2, \mu_2)$, applying Theorem 6.74, one has:

$$\begin{aligned} &d(x_1, \widetilde{\text{Sol}}(\text{WEP})(\varepsilon_2, \lambda_2, \mu_2)) \\ &\quad \leq d(x_1, \widetilde{\text{Sol}}(\text{WEP})_{\hat{\xi}}(\varepsilon_2, \lambda_2, \mu_2)) \\ &\quad \leq \text{ex}(\widetilde{\text{Sol}}(\text{WEP})_{\hat{\xi}}(\varepsilon_1, \lambda_1, \mu_1), \widetilde{\text{Sol}}(\text{WEP})_{\hat{\xi}}(\varepsilon_2, \lambda_2, \mu_2)) \\ &\quad \leq H(\widetilde{\text{Sol}}(\text{WEP})_{\hat{\xi}}(\varepsilon_1, \lambda_1, \mu_1), \widetilde{\text{Sol}}(\text{WEP})_{\hat{\xi}}(\varepsilon_2, \lambda_2, \mu_2)). \\ &\quad \leq k_1|\varepsilon_1 - \varepsilon_2| + k_2\|\mu_1 - \mu_2||^{\beta} + k_3\|\lambda_1 - \lambda_2\|^{\alpha\delta}, \end{aligned}$$

where k_1, k_2, k_3 are positive and depend on $\bar{\varepsilon}, l, \alpha, h, \beta$, etc.

Therefore,

$$\begin{aligned}&\text{ex}(\widetilde{\text{Sol}}(\text{WEP})(\varepsilon_1,\lambda_1,\mu_1),\widetilde{\text{Sol}}(\text{WEP})(\varepsilon_2,\lambda_2,\mu_2))\\&\quad\le k_1|\varepsilon_1-\varepsilon_2|+k_2\|\mu_1-\mu_2\|^\beta+k_3\|\lambda_1-\lambda_2\|^{\alpha\delta}.\end{aligned} \tag{6.56}$$

Similarly, we also obtain that

$$\begin{aligned}&\text{ex}(\widetilde{\text{Sol}}(\text{WEP})(\varepsilon_2,\lambda_2,\mu_2),\widetilde{\text{Sol}}(\text{WEP})(\varepsilon_1,\lambda_1,\mu_1))\\&\quad\le k_1|\varepsilon_1-\varepsilon_2|+k_2\|\mu_1-\mu_2\|^\beta+k_3\|\lambda_1-\lambda_2\|^{\alpha\delta}.\end{aligned} \tag{6.57}$$

From (6.56) and (6.57), we have (6.55). Hence, the proof is complete. □

We now provide some examples to illustrate the essentialness of the assumptions in Theorem 6.75.

Example 6.76 (Essentialness of the concavity property of the objective map) Let $\bar{\varepsilon}=\frac{1}{3}$, $X=A=\mathbb{R}$, $\Lambda=M=[0,1]$, $\mathscr{D}(\lambda)=[-3,3]$, $Y=\mathbb{R}^2$, $C=\mathbb{R}^2_+$, $e=(1,1)\in\text{int}C$, and

$$f(x,y,\mu):=\begin{cases}(0,0) & \text{if } x=0,\\ (-1,-1) & \text{if } x\neq 0.\end{cases}$$

Then, we see that (a), (c), (d), and (e) of Theorem 6.75 are satisfied. Some direct computations give the approximate solution set:

$$\widetilde{\text{Sol}}(\text{WEP})(\varepsilon)=\begin{cases}[-3,3] & \text{if } \varepsilon\ge 1,\\ \{0\} & \text{if } \frac{1}{3}\le\varepsilon<1,\end{cases}$$

which is not Hölder continuous at $\varepsilon_0=1$. The cause if that the concavity in the first variable of objective function f is violated (to see this, for instance, take $x_1=0, x_2=1$ and $t=\frac{1}{2}$).

Example 6.77 (The convexity of the constrained map cannot be dispensed) Let $X=A=\mathbb{R}$, $\Lambda\equiv M=[0,4]$, $\mathscr{D}(\lambda)=[1,3]\cup\{4\}$, $Y=\mathbb{R}^2$, $C=\mathbb{R}^2_+$, $\bar{\varepsilon}=10^{-1}$, $e=(1,1)\in\text{int}C$, and $f(x,y,\mu):=(\mu-x,\mu-x)$. It is not hard to check that all the assumptions of Theorem 6.75 are fulfilled. Obviously, $\mathscr{D}$ is not convex. By direct computations, we have $\widetilde{\text{Sol}}(\text{WEP})(\varepsilon,\mu)=(-\infty,\mu+\varepsilon]\cap([1,3]\cup\{4\})$. We see that $\widetilde{\text{Sol}}(\text{WEP})$ is not Hölder continuous at $\mu_0=3$ for $\varepsilon_0=1$.

Example 6.78 (Cruciality of the boundedness of the constraint map) Let $X=A=\mathbb{R}$, $\Lambda\equiv M=[0,1]$, $\mathscr{D}(\lambda)=[0,+\infty)$, $Y=\mathbb{R}^2$, $C=\mathbb{R}^2_+$, $\bar{\varepsilon}=10^{-2}$, $e=(1,1)\in\text{int}C$, and $f(x,y,\mu):=(-\mu x,-\mu x)$. Then, all the assumptions of Theorem 6.75 hold. Direct calculations give

$$\widetilde{\text{Sol}}(\text{WEP})(\varepsilon,0)=[0,+\infty),$$

Consider $\eta = \eta_0 < \dfrac{\varepsilon_0}{\rho + 1}$ and $V' = [\varepsilon_0 - \eta_0, \varepsilon_0 + \eta_0]$. Substituting $\varepsilon_2 = \varepsilon_0$ and $\varepsilon_1 = \varepsilon_0 - \eta_0$ in (6.61). Choosing $\theta = \rho + 1$, one has $1 < \theta < \dfrac{\varepsilon_2}{\varepsilon_2 - \varepsilon_1}$. From (6.59), one has further

$$\widetilde{\text{Sol}}(\text{WEP})_\xi(\varepsilon_0, \lambda_0) \subset \widetilde{\text{Sol}}(\text{WEP})_\xi(\varepsilon_0 - \eta_0, \lambda_0) + B$$
$$\subset \widetilde{\text{Sol}}(\text{WEP})_\xi(\varepsilon, \lambda_0) + B,$$

i.e., $\widetilde{\text{Sol}}(\text{WEP})_\xi(\cdot, \lambda_0)$ is H-lsc at ε_0.

Now substituting $\varepsilon_2 = \varepsilon_0 + \eta_0$ and $\varepsilon_1 = \varepsilon_0$ in (6.61) and choosing $\theta = \rho + 1$, we get:

$$\widetilde{\text{Sol}}(\text{WEP})_\xi(\varepsilon_0 + \eta_0, \lambda_0) \subset \widetilde{\text{Sol}}(\text{WEP})_\xi(\varepsilon_0, \lambda_0) + B.$$

Combining this with (6.59), we have $\widetilde{\text{Sol}}(\text{WEP})_\xi(\varepsilon, \lambda_0) \subset \widetilde{\text{Sol}}(\text{WEP})_\xi(\varepsilon_0, \lambda_0) + B$, and hence $\widetilde{\text{Sol}}(\text{WEP})_\xi(\cdot, \lambda_0)$ is H-usc at ε_0. The proof of Step 1 is complete.

Step 2. We prove that, for any $\varepsilon \in V'$, $\widetilde{\text{Sol}}(\text{WEP})_\xi(\varepsilon, \cdot)$ is Hausdorff continuous at λ_0.

We show $\widetilde{\text{Sol}}(\text{WEP})_\xi(\varepsilon, \cdot)$ is H-usc at λ_0 by proving $\widetilde{\text{Sol}}(\text{WEP})_\xi(\varepsilon, \cdot)$ is usc at λ_0. Suppose to the contrary that $\widetilde{\text{Sol}}(\text{WEP})_\xi(\varepsilon, \cdot)$ is not usc at λ_0. Then, we can find an open set U with $U \supset \widetilde{\text{Sol}}(\text{WEP})_\xi(\varepsilon, \lambda_0)$ and a sequence $\lambda^k \to \lambda_0$ such that there is

$$x^k \in \widetilde{\text{Sol}}(\text{WEP})_\xi(\varepsilon, \lambda^k), x^k \notin U, \ \forall k.$$

The upper semicontinuity of $\mathscr{D}$ at λ_0 and the compactness of $\mathscr{D}(\lambda_0)$ give us that $x^k \to x_0 \in \mathscr{D}(\lambda_0)$. If $x_0 \notin \widetilde{\text{Sol}}(\text{WEP})_\xi(\varepsilon, \lambda_0)$, then there is $y_0 \in \mathscr{D}(\lambda_0)$ such that

$$\langle \xi, f(x_0, y_0, \lambda_0)\rangle + \varepsilon < 0. \tag{6.62}$$

The lower semicontinuity of $\mathscr{D}$ at λ_0 shows the existence of $y^k \in \mathscr{D}(\lambda^k)$ such that $y^k \to y_0$. Since $x^k \in \widetilde{\text{Sol}}(\text{WEP})_\xi(\varepsilon, \lambda^k)$, $\langle \xi, f(x^k, y^k, \lambda^k)\rangle + \varepsilon \geq 0$, which is absurd inasmuch as f is continuous. Hence, $x_0 \in \widetilde{\text{Sol}}(\text{WEP})_\xi(\varepsilon, \lambda_0)$, which again leads to a contradiction as $x^k \notin U$ for all k.

For the Hausdorff lower semicontinuity of $\widetilde{\text{Sol}}(\text{WEP})_\xi(\varepsilon, \cdot)$ at λ_0, it suffices to check $\widetilde{\text{Sol}}(\text{WEP})_\xi(\varepsilon, \cdot)$ is lsc at λ_0 and $\widetilde{\text{Sol}}(\text{WEP})_\xi(\varepsilon, \lambda_0)$ is compact. We start by proving the map $\widetilde{\text{Sol}}(\widehat{\text{WEP}})_\xi(\varepsilon, \cdot)$ is lsc at λ_0. Suppose that $\widetilde{\text{Sol}}(\widehat{\text{WEP}})_\xi(\varepsilon, \cdot)$ is not lsc at λ_0, i.e., there are $x_0 \in \widetilde{\text{Sol}}(\widehat{\text{WEP}})_\xi(\varepsilon, \lambda_0)$ and sequence $\lambda^k \to \lambda_0$, for all $x^k \in \widetilde{\text{Sol}}(\widehat{\text{WEP}})_\xi(\varepsilon, \lambda^k)$, $x^k \not\to x_0$. Thanks to the lower semicontinuity of K at λ_0, there exists $\bar{x}^k \in \mathscr{D}(\lambda^k)$ satisfying $\bar{x}^k \to x_0$. Due to the above contradiction assumption, there must be a subsequence $\{\bar{x}^{k_j}\}$ such that, for all k_j: $\bar{x}^{k_j} \notin \widetilde{\text{Sol}}(\widehat{\text{WEP}})_\xi(\varepsilon, \lambda^{k_j})$, i.e., for some $y^{k_j} \in \mathscr{D}(\lambda^{k_j})$,

$$\langle \xi, f(\bar{x}^{k_j}, y^{k_j}, \lambda^{k_j}) + \varepsilon \leq 0.$$

Since $\mathscr{D}$ is usc at λ_0 and $\mathscr{D}(\lambda_0)$ is compact, one has $y_0 \in \mathscr{D}(\lambda_0)$ such that $y^{k_j} \to y_0$ (take a subsequence if necessary). From the continuity of f, we have:

$$\langle \xi, f(x_0, y_0, \lambda_0) + \varepsilon \leq 0,$$

which is impossible as x_0 belongs to $\widetilde{\text{Sol}}(\widehat{\text{WEP}})_\xi(\varepsilon, \lambda_0)$.

Now, let us show that

$$\widetilde{\text{Sol}}(\text{WEP})_\xi(\varepsilon, \lambda_0) \subset \text{cl}\widetilde{\text{Sol}}(\widehat{\text{WEP}})_\xi(\varepsilon, \lambda_0). \tag{6.63}$$

Take any $\bar{x} \in \widetilde{\text{Sol}}(\text{WEP})_\xi(\varepsilon, \lambda_0)$. Let $x_1 \in \widetilde{\text{Sol}}(\widehat{\text{WEP}})_\xi(\varepsilon, \lambda_0)$ and $x_t = (1-t)\bar{x}+tx_1$ with $t \in (0, 1)$. By the concavity of f, for all $y \in \mathscr{D}(\lambda_0)$, we have:

$$\begin{aligned}
\langle \xi, f(x_t, y, \lambda_0)\rangle + \varepsilon &\geq (1-t)\langle \xi, f(\bar{x}, y, \lambda_0)\rangle + t\langle \xi, f(x_1, y, \lambda_0)\rangle + \varepsilon \\
&= (1-t)(\langle \xi, f(\bar{x}, y, \lambda_0)\rangle + \varepsilon) + t(\langle \xi, f(x_1, y, \lambda_0)\rangle + \varepsilon) \\
&> 0.
\end{aligned}$$

This means that $x_t \in \widetilde{\text{Sol}}(\text{WEP})_\xi(\varepsilon, \lambda_0)$. As $x_t \to \bar{x}$ when $t \to 0$, $\bar{x}$ belongs to $\text{cl}\widetilde{\text{Sol}}(\text{WEP})_\xi(\varepsilon, \lambda_0)$. Therefore, (6.63) holds. By the lower semicontinuity of $\widetilde{\text{Sol}}(\text{WEP})_\xi(\varepsilon, \cdot)$ at λ_0, we have:

$$\begin{aligned}
\widetilde{\text{Sol}}(\text{WEP})_\xi(\varepsilon, \lambda_0) \subset \text{cl}\widetilde{\text{Sol}}(\text{WEP})_\xi(\varepsilon, \lambda_0) &\subset \liminf \widetilde{\text{Sol}}(\text{WEP})_\xi(\varepsilon, \lambda^k) \\
&\subset \liminf \widetilde{\text{Sol}}(\text{WEP})_\xi(\varepsilon, \lambda^k),
\end{aligned}$$

i.e., $\widetilde{\text{Sol}}(\text{WEP})_\xi(\varepsilon, \cdot)$ is lsc at λ_0.

For the compactness of $\widetilde{\text{Sol}}(\text{WEP})_\xi(\varepsilon, \lambda_0)$, it is enough to show that $\widetilde{\text{Sol}}(\text{WEP})_\xi(\varepsilon, \lambda_0)$ is closed in $\mathscr{D}(\lambda_0)$. For any $x^k \in \widetilde{\text{Sol}}(\text{WEP})_\xi(\varepsilon, \lambda_0)$, $x^k \to x_0$, then $x^k \in \mathscr{D}(\lambda_0)$ and $x_0 \in \mathscr{D}(\lambda_0)$. For all $y \in \mathscr{D}(\lambda_0)$, we have $\langle \xi, f(x^k, y, \lambda_0)\rangle + \varepsilon \geq 0$. By the continuity of f, we get $\langle \xi, f(x_0, y, \lambda_0)\rangle + \varepsilon \geq 0$, i.e., $x_0 \in \widetilde{\text{Sol}}(\text{WEP})_\xi(\varepsilon, \lambda_0)$.

Step 3. Utilizing the results of Steps 1 and 2, we obtain the Hausdorff continuity of $\widetilde{\text{Sol}}(\text{WEP})_\xi$ at $(\varepsilon_0, \lambda_0)$. □

Applying the above theorem, we derive the following result.

Theorem 6.80 (Hausdorff continuity of approximate solution maps of weak equilibrium problems via linear scalarization functions) *Consider (WEP). Suppose that for each $\xi \in B_e^*$, ξ-efficient approximate solutions exist in a neighborhood of $(\varepsilon_0, \lambda_0)$. Furthermore, assume that the following conditions hold:*

(a) *$\mathscr{D}$ is continuous at λ_0 and $\mathscr{D}(\lambda_0)$ is compact;*

(b) *there exists a neighborhood N of λ_0 such that f is continuous on $\mathscr{D}(N) \times \mathscr{D}(N) \times \{\lambda_0\}$;*
(c) *for all $y \in \mathscr{D}(N)$, $f(\cdot, y, \lambda_0)$ is C-concave on $\mathscr{D}(N)$;*
(d) *for each $x \in \mathscr{D}(N)$, $f(x, \cdot, \lambda_0)$ is C-convex-like on $\mathscr{D}(N)$.*

Then, $\widetilde{\text{Sol}}(\text{WEP})$ is Hausdorff continuous at $(\varepsilon_0, \lambda_0)$.

Proof At first, we show that $\widetilde{\text{Sol}}(\text{WEP})$ is H-lsc. In view of Lemma 6.64, we have:

$$\widetilde{\text{Sol}}(\text{WEP})(\varepsilon, \lambda) = \bigcup_{\xi \in B_e^*} \widetilde{\text{Sol}}(\text{WEP})_\xi(\varepsilon, \lambda).$$

From Theorem 6.79, one has for each $\xi \in B_e^*$, $\widetilde{\text{Sol}}(\text{WEP})_\xi$ is lsc at $(\varepsilon_0, \lambda_0)$. By Lemma 6.61, $\widetilde{\text{Sol}}(\text{WEP})$ is lsc at $(\varepsilon_0, \lambda_0)$. On the other hand, take $x^k \in \widetilde{\text{Sol}}(\text{WEP})(\varepsilon_0, \lambda_0)$ satisfying $x^k \to \bar{x}$. One has, for all $y \in \mathscr{D}(\lambda_0)$, $f(x^k, y, \lambda_0) + \varepsilon_0 e \notin -\text{int}C$. Thanks to the continuity of f, $f(\bar{x}, y, \lambda_0) + \varepsilon_0 e \notin -\text{int}C$, i.e., $\bar{x} \in \widetilde{\text{Sol}}(\text{WEP})(\varepsilon_0, \lambda_0)$. Hence, $\widetilde{\text{Sol}}(\text{WEP})(\varepsilon_0, \lambda_0)$ is closed in $\mathscr{D}(\lambda_0)$, which is compact. Therefore, $\widetilde{\text{Sol}}(\text{WEP})$ is H-lsc at $(\varepsilon_0, \lambda_0)$.

Next, we show that $\widetilde{\text{Sol}}(\text{WEP})$ is H-usc at $(\varepsilon_0, \lambda_0)$ by proving that $\widetilde{\text{Sol}}(\text{WEP})$ is usc at $(\varepsilon_0, \lambda_0)$. Suppose to the contrary that $\widetilde{\text{Sol}}(\text{WEP})$ is not usc at $(\varepsilon_0, \lambda_0)$. Then, there exists a neighborhood U_0 of $\widetilde{\text{Sol}}(\text{WEP})(\varepsilon_0, \lambda_0)$ and a sequence $\{(\varepsilon^k, \lambda^k)\}_k$ converging $(\varepsilon_0, \lambda_0)$ such that, for each α, there exists $x^k \in \widetilde{\text{Sol}}(\text{WEP})(\varepsilon^k, \lambda^k)$ satisfying $x^k \notin U_0$. Since $\mathscr{D}(\lambda_0)$ is compact, we can assume that $x^k \to x_0 \in \mathscr{D}(\lambda_0)$. If $x_0 \notin \widetilde{\text{Sol}}(\text{WEP})(\varepsilon_0, \lambda_0)$, then there exists $y_0 \in \mathscr{D}(\lambda_0)$ such that

$$f(x_0, y_0, \lambda_0) + \varepsilon_0 e \in -\text{int}C.$$

The lower semicontinuity of $\mathscr{D}$ at λ_0 shows the existence of $y^k \in \mathscr{D}(\lambda^k)$ with $y^k \to y_0$. As $x^k \in \widetilde{\text{Sol}}(\text{WEP})(\varepsilon^k, \lambda^k)$, $f(x^k, y^k, \lambda^k) + \varepsilon^k e \notin -\text{int}C$. This is absurd since f is continuous. Hence, $x_0 \in \widetilde{\text{Sol}}(\text{WEP})(\varepsilon_0, \lambda_0)$, which contradicts the fact that $x^k \notin U_0$ for all k. The proof is complete. □

Example 6.81 (The cone concavity of the objective map is crucial) Let $X = A = \mathbb{R}$, $\Lambda = [0, 1]$, $\mathscr{D}(\lambda) = [\lambda, 2]$, $Y = \mathbb{R}^2$, $C = \mathbb{R}^2_+$, $e = (1, 1) \in \text{int}C$, $\varepsilon_0 = 10^{-2}$, $\lambda_0 = 0$, and $f(x, y, \lambda) = (x^2 - 2x - 3\lambda - 10^{-2}, x^2 - 2x - 3\lambda - 10^{-2})$. We see that assumptions (a), (b), and (d) of Theorem 6.80 are fulfilled. Take $x_1 = 0$, $x_2 = 1$, and $t = \frac{1}{2}$, we have $f\left(\frac{1}{2}x_1 + \frac{1}{2}x_2, y, \lambda_0\right) = (-\frac{3}{4} - 10^{-2}, -\frac{3}{4} - 10^{-2})$, but $\frac{1}{2}f(x_1, y, \lambda_0) + \frac{1}{2}f(x_2, y, \lambda_0) = (-\frac{1}{2} - 10^{-2}, -\frac{1}{2} - 10^{-2})$. So, the C-concavity of f is violated. We will show that approximate solution set $\widetilde{\text{Sol}}(\text{WEP})$ is not Hausdorff continuous at $(10^{-2}, 0)$. One has $0 \in \widetilde{\text{Sol}}(\text{WEP})(10^{-2}, 0) = \{0, 2\}$ and $(\varepsilon^k, \lambda^k) := (10^{-2} + \frac{3}{k}, \frac{1}{k}) \to (\varepsilon_0, \lambda_0) = (10^{-2}, 0)$. But, for any $x^k \in \widetilde{\text{Sol}}(\text{WEP})(\varepsilon^k, \lambda^k) = [\{2\}, x^k \nrightarrow 0$, and hence $\widetilde{\text{Sol}}(\text{WEP})$ is not lower semicontinuous at $(10^{-2}, 0)$. Therefore, $\widetilde{\text{Sol}}(\text{WEP})$ is also not Hausdorff continuous at $(10^{-2}, 0)$.

Let $K \subset A$ be a nonempty compact convex set. By using the same technique as that for Theorem 6.80, we derive the following result.

Theorem 6.82 (Hausdorff continuity of approximate solution maps of weak equilibrium problems with fixed constrained set via linear scalarization functions) *Consider (WEP), where $\mathscr{D}(\lambda) \equiv K$ for all $\lambda \in \Lambda$ is a nonempty convex compact subset. Suppose that for each $\xi \in B_e^*$, ξ-efficient approximate solutions exist in a neighborhood of $(\varepsilon_0, \lambda_0)$. Furthermore, assume that the following conditions hold:*

(a) *for $x, y \in \mathscr{D}$, $f(x, y, \cdot)$ is continuous at λ_0;*
(b) *for all $y \in \mathscr{D}$, $f(\cdot, y, \lambda_0)$ is C-concave on $\mathscr{D}$;*
(c) *for each $x \in \mathscr{D}$, $f(x, \cdot, \lambda_0)$ is C-convex-like on $\mathscr{D}$.*

Then, $\widetilde{\mathrm{Sol}}(\mathrm{WEP})$ is Hausdorff continuous at $(\varepsilon_0, \lambda_0)$.

Remark 6.83 (Convex-like maps in totally ordered spaces) Note that if $Y = C \cup -C$, i.e., Y is a totally ordered space, then each $f : X \to Y$ is convex-like. So, when applying our results to scalar equilibrium problems, convex-likeness assumptions is evidently satisfied.

The following example illustrates applicability of Theorem 6.82.

Example 6.84 (Hausdorff continuity of solution maps of equilibrium problems without assuming continuity conditions) Let $X = A = \mathbb{R}$, $\Lambda = [0, 1]$, $\mathscr{D}(\lambda) = [0, 1]$, $Y = \mathbb{R}^2$, $C = \mathbb{R}^2_+$, $e = (1, 1) \in \mathrm{int}C$, $\varepsilon_0 = 10^{-2}$, $\lambda_0 = 0$, and

$$f(x, y, \lambda) := \begin{cases} (0, 0) & \text{if } x = 0, \\ (1, 1) & \text{if } x \neq 0. \end{cases}$$

Obviously, all the assumptions of Theorem 6.82 are satisfied. Then, the approximate solution map is $\widetilde{\mathrm{Sol}}(\mathrm{WEP})(\varepsilon, \lambda) = [0, 1]$, which is Hausdorff continuous. Note that in this case, the objective function is not continuous at $x = 0$.

6.5.2 *Nonlinear Scalarization Functions*

Sensitivity Analysis for (WEP)

In this section, we discuss the nonlinear scalarization function commonly known as the Gerstewitz function (also called the Tammer function since these two names belong to the same mathematician) in the theory of vector optimization. Let X and Y be normed spaces, and let $C \subset Y$ be a pointed closed and convex cone with nonempty interior int$C$. Let Y^* be the topological dual space of Y, $\langle \cdot, \cdot \rangle$ the natural pairing between Y and Y^*, and C^* the dual cone of C, i.e.,

$$C^* := \left\{ \lambda \in Y^* \mid \langle \lambda, k \rangle \geq 0, \forall k \in C \right\}.$$

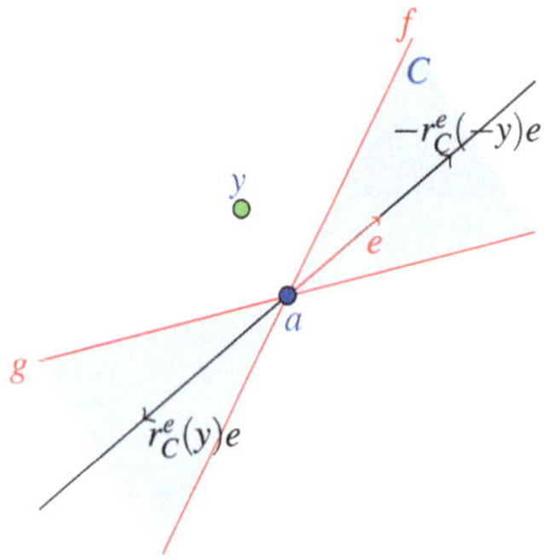

Fig. 6.1 Illustrations of computations for the Gerstewitz function r_C^e

Definition 6.85 (The Gerstewitz function) Given a fixed point $e \in \operatorname{int} C$, the nonlinear scalarization function $r_C^e : Y \to \mathbb{R}$ defined by

$$r_C^e(y) := \inf\{t \in \mathbb{R} \mid y \in te - C\}$$

is called the Gerstewitz (or Tammer) scalarization function. In Fig. 6.1, we provide a visual illustration of the method for determining the value of the function r_C^e at the point y.

It is well known that r_C^e is a continuous, positively homogeneous, subadditive, and convex function on Y, and it is monotone, i.e.,

$$y^2 - y^1 \in C \Rightarrow r_C^e\left(y^1\right) \leq r_C^e\left(y^2\right)$$

and strictly monotone, i.e.,

$$y^2 - y^1 \in \operatorname{int} C \Rightarrow r_C^e\left(y^1\right) < r_C^e\left(y^2\right).$$

In the special case with $Y = \mathbb{R}^n, C = \mathbb{R}^n_+$, and $e = (1, 1, \ldots, 1) \in \operatorname{int} \mathbb{R}^n_+$, this nonlinear scalarization function can be rewritten in the following equivalent form:

$$r_C^e(y) = \max_{1 \leq i \leq n} \{y_i\}, \quad \forall y = (y_1, y_2, \ldots, y_n) \in \mathbb{R}^n.$$

Based on the definition of the function r_C^e, the following properties are easy to prove.

Proposition 6.86 (Equivalence between vector relations and their Gerstewitz function values relations) *For any fixed $e \in \operatorname{int}C$, $y \in Y$ and $r \in \mathbb{R}$:*

(a) $r_C^e(y) < r \Leftrightarrow y \in re - \operatorname{int}C$ *(equivalently,* $r_C^e(y) \geq r \Leftrightarrow y \notin re - \operatorname{int}C$*);*
(b) $r_C^e(y) \leq r \Leftrightarrow y \in re - C$ *(equivalently,* $r_C^e(y) > r \Leftrightarrow y \notin re - C$*);*
(c) $r_C^e(y) = r \Leftrightarrow y \in re - \partial C$*, where* ∂C *denotes the boundary of* C*;*
(d) $r_C^e(re) = r$.

Property (a) of Proposition 6.86 will play an essential role in scalarizations. Property (d) can be strengthened as

$$r_C^e(y + re) = r_C^e(y) + r, \quad \forall y \in Y, r \in \mathbb{R}.$$

Let $e \in \operatorname{int} C$ be a fixed point. Let

$$C^e := \left\{ \lambda \in C^* \mid \langle \lambda, e \rangle = 1 \right\}.$$

Clearly, C^e is a weak*-compact base of C^*, i.e., C^e is convex and weak*-compact such that $0 \notin C^e$ and $C^* = \bigcup_{t \geq 0} tC^e$.

Proposition 6.87 (A formula of the Gerstewitz function) *Let $e \in \operatorname{int} C$. Then, for $y \in Y$,*

$$r_C^e(y) = \max_{\lambda \in C^e} \langle \lambda, y \rangle.$$

Proof Due to $y \in r_C^e(y)e - C$, by the definition of r_C^e, we have:

$$\left\langle \lambda, y - r_C^e(y)e \right\rangle \leq 0, \quad \forall \lambda \in C^e.$$

Therefore,

$$\langle \lambda, y \rangle - r_C^e(y) = \langle \lambda, y \rangle - r_C^e(y)\langle \lambda, e \rangle \leq 0, \quad \forall \lambda \in C^e,$$

and hence, $r_C^e(y) \geq \langle \lambda, y \rangle$ for all $\lambda \in C^e$. Consequently,

$$r_C^e(y) \geq \sup_{\lambda \in C^e} \langle \lambda, y \rangle.$$

On the other hand, let

$$t_e = \sup_{\lambda \in C^e} \langle \lambda, y \rangle.$$

Then,

$$t_e \geq \langle \lambda, y \rangle, \quad \forall \lambda \in C^e.$$

Hence, we get $t_e \langle \lambda, e \rangle \geq \langle \lambda, y \rangle$ as $\langle \lambda, e \rangle = 1$. Therefore,

$$\langle \lambda, y - t_e e \rangle \leq 0, \quad \forall \lambda \in C^e.$$

Because of $C^* = \bigcup_{\gamma \geq 0} \gamma C^e$, we have

$$\langle \lambda, y - t_e e \rangle \leq 0, \quad \forall \lambda \in C^*.$$

Thus, for any $y = (y_1, y_2) \in \mathbb{R}^2$, we get:

$$\begin{aligned} r_C^e(y) &= \max_{(\lambda_1,\lambda_2)\in C^e} \lambda_1 y_1 + \lambda_2 y_2 \\ &= \max_{\lambda_1\in[-0.1,2]} \left(y_1 - \frac{2}{3}y_2\right)\lambda_1 + \frac{1}{3}y_2 \\ &= \begin{cases} 2y_1 - y_2 & \text{if } y_1 \geq \frac{2}{3}y_2, \\ 0.4y_2 - 0.1y_1 & \text{if } y_1 \leq \frac{2}{3}y_2. \end{cases} \end{aligned}$$

Next we verify the Lipschitz property of r_C^e on $\mathbb{R}^2$. Let $y = (y_1, y_2)$ and $y' = (y_1', y_2') \in \mathbb{R}^2$. Four cases should be calculated. Notice that

$$|au + bv| \leq \sqrt{a^2 + b^2}\sqrt{u^2 + v^2}, \forall a, b, u, v \in \mathbb{R}.$$

Case 1. If $y_1 \geq \frac{2}{3}y_2$ and $y_1' \geq \frac{2}{3}y_2'$, then

$$\begin{aligned} \left|r_C^e(y) - \xi_q\left(y'\right)\right| &= \left|2y_1 - y_2 - \left(2y_1' - y_2'\right)\right| = \left|2\left(y_1 - y_1'\right) - \left(y_2 - y_2'\right)\right| \\ &\leq \sqrt{5}\left\|y - y'\right\|. \end{aligned}$$

Case 2. If $y_1 \leq \frac{2}{3}y_2$ and $y_1' \leq \frac{2}{3}y_2'$, then

$$\begin{aligned} \left|r_C^e(y) - r_C^e\left(y'\right)\right| &= \left|0.4y_2 - 0.1y_1 - \left(0.4y_2' - 0.1y_1'\right)\right| \\ &= \left|0.4\left(y_2 - y_2'\right) - 0.1\left(y_1 - y_1'\right)\right| \leq \frac{\sqrt{17}}{10}\left\|y - y'\right\|. \end{aligned}$$

Case 3. If $y_1 \geq \frac{2}{3}y_2$ and $y_1' \leq \frac{2}{3}y_2'$, then

$$\left|r_C^e(y) - r_C^e\left(y'\right)\right| = \left|2y_1 - y_2 - \left(0.4y_2' - 0.1y_1'\right)\right|.$$

If

$$2y_1 - y_2 - \left(0.4y_2' - 0.1y_1'\right) \geq 0,$$

then

$$2\left(y_1 - y_1'\right) - \left(y_2 - y_2'\right) + \left(2.1y_1' - 1.4y_2'\right) = 2y_1 - y_2 - \left(0.4y_2' - 0.1y_1'\right) \geq 0.$$

Thus,

$$2\left(y_1 - y_1'\right) \geq \left(y_2 - y_2'\right) - \left(2.1y_1' - 1.4y_2'\right) \geq y_2 - y_2'.$$

as $2.1y_1' - 1.4y_2' \leq 0$. Whence,

$$\left|r_C^e(y) - r_C^e\left(y'\right)\right| \leq \left|2\left(y_1 - y_1'\right) - \left(y_2 - y_2'\right)\right| \leq \sqrt{5}\left\|y - y'\right\|.$$

If

$$2y_1 - y_2 - \left(0.4y_2' - 0.1y_1'\right) \leq 0,$$

then

$$0.4\left(y_2 - y_2'\right) - 0.1\left(y_1 - y_1'\right) + (2.1y_1) - 1.4y_2) = 2y_1 - y_2 - \left(0.4y_2' - 0.1y_1'\right) \leq 0.$$

Thus,

$$0.4\left(y_2 - y_2'\right) \leq 0.1\left(y_1 - y_1'\right) - (2.1y_1 - 1.4y_2) \leq 0.1\left(y_1 - y_1'\right),$$

as $2.1y_1 - 1.4y_2 \geq 0$. Whence,

$$\left|r_C^e(y) - r_C^e\left(y'\right)\right| \leq \left|0.4\left(y_2 - y_2'\right) - 0.1\left(y_1 - y_1'\right)\right| \leq \frac{\sqrt{17}}{10}\left\|y - y'\right\|.$$

Case 4. If $y_1 \leq \frac{2}{3}y_2$ and $y_1' \geq \frac{2}{3}y_2'$, then

$$\left|r_C^e(y) - r_C^e\left(y'\right)\right| = \left|0.4y_2 - 0.1y_1 - \left(2y_1' - y_2'\right)\right| = \left|2y_1' - y_2' - (0.4y_2 - 0.1y_1)\right|.$$

Then, the calculation is similar to Case 3. In a word, for all $y = (y_1, y_2), y' = \left(y_1', y_2'\right) \in \mathbb{R}^2$,

$$\left|r_C^e(y) - r_C^e\left(y'\right)\right| \leq \sqrt{5}\|y - y'\|.$$

Indeed,

$$L = \sup_{\lambda \in C^e} \|\lambda\| = \|(2, -1)\| = \sqrt{5}.$$

Next we consider a generalized vector equilibrium problem and discuss Hölder conditions of its solution map. Let (X, d_X) be a metric space, Y a linear normed space, Λ and Ω nonempty subsets of metric spaces, and $C \subset Y$ a pointed closed and convex cone with $\text{int}C \neq \emptyset$. Let $\mathcal{D} : X \times \Lambda \rightrightarrows X$ be a set-valued mapping with nonempty values and $f : X \times X \times M \to Y$.

For parameters $\lambda \in \Lambda$ and $\mu \in M$, our problem is

(VQEP) find $\bar{x} \in \mathcal{D}(\bar{x}, \lambda)$ such that $f(\bar{x}, y, \mu) \notin -\text{int}\, C, \quad \forall y \in \mathcal{D}(\bar{x}, \lambda)$.

Let $\mathscr{E}(\lambda) = \{x \in X \mid x \in K(x, \lambda)\}$ and Sol(VQEP)(λ, μ) be the solution set of (VQEP). For the reference point $(\bar{\lambda}, \bar{\mu}) \in \Lambda \times \Omega$, we assume that Sol(VQEP)$(\lambda, \mu) \neq \emptyset$ for every $\lambda \in U(\bar{\lambda})$ and $\mu \in U(\bar{\mu})$, where $U(v)$ is a neighborhood of v. It follows from Proposition 6.86(a) that for any fixed $e \in \operatorname{int} C$

$$\text{Sol(VQEP)}(\lambda, \mu) = \left\{x \in \mathscr{E}(\lambda) \mid r_C^e(f(x, y, \mu)) \geq 0, \forall y \in \mathscr{D}(x, \lambda)\right\}.$$

Definition 6.91 (Hölder continuity of a set-valued map via the Hausdorff distance) A set-valued map $G : X \times \Lambda \rightrightarrows X$ is said to be $((\ell_1, \alpha_1), (\ell_2, \alpha_2))$-HÖLDER CONTINUOUS at (x_0, λ_0) if there exist neighborhoods $U(x_0)$ of x_0 and $U(\lambda_0)$ of λ_0 such that for all $x_1, x_2 \in U(x_0)$, for all $\lambda_1, \lambda_2 \in U(\lambda_0)$, we have:

$$\begin{aligned} &G(x_1, \lambda_1) \\ &\quad \subset \left\{x \in X \mid \exists z \in G(x_2, \lambda_2), d_X(x, z) \leq \ell_1 d^{\alpha_1}(x_1, x_2) + \ell_2 d^{\alpha_2}(\lambda_1, \lambda_2)\right\}, \end{aligned}$$

where $\ell_1, \ell_2 \geq 0$ and $\alpha_1, \alpha_2 > 0$.

Theorem 6.92 (Hölder continuity of solution sets of quasiequilibrium problems via a nonlinear scalarization of a strong monotonicity) *Suppose that the following conditions hold:*

(a) *$\mathscr{D}(\cdot, \cdot)$ is $((\ell_1, \alpha_1), (\ell_2, \alpha_2))$-Hölder continuous on $\mathscr{E}(U(\bar{\lambda})) \times \{\bar{\lambda}\}$;*
(b) *$f(\cdot, \cdot, \mu)$ is $(\hbar, \beta)$-Hölder strongly monotone wrt r_C^e, i.e., there exist $q \in \operatorname{int} C$ and $\hbar > 0$, $\beta > 0$ such that for every $x, y \in E(U(\lambda))$, $x \neq y$*

$$r_C^e(f(x, y, \mu)) + r_C^e(f(y, x, \mu)) + \hbar d_X^\beta(x, y) \leq 0,$$

where $e \in \operatorname{int} C$;
(c) *for each $x \in \mathscr{E}(U(\bar{\lambda}))$, $f(x, \cdot, \mu)$ is (n, δ)-Hölder continuous on $\mathscr{E}(U(\bar{\lambda}))$;*
(d) *$f(x, y, \cdot)$ is (m, γ)-Hölder continuous at $\bar{\mu}$ with θ-relative to $\mathscr{E}(U(\bar{\lambda}))$, i.e., for every $\mu_1, \mu_2 \in U(\bar{\mu})$ and $x, y \in E(U(\bar{\lambda}))$ with $x \neq y$, we have:*

$$\|f(x, y, \mu_1) - f(x, y, \mu_2)\| \leq m d_X^\theta(x, y) d^\gamma(\mu_1, \mu_2);$$

(e) *$\beta = \alpha_1 \delta > \theta$, $\hbar > 2n\rho\ell_1^\delta$, where $\rho := \sup_{\lambda \in C^q} \|\lambda\| \in \left[\frac{1}{\|e\|}, +\infty\right)$ is the Lipschitz constant of r_C^e on Y.*

Then, for every $(\lambda, \mu) \in U(\bar{\lambda}) \times U(\bar{\mu})$, $(VQEP)$ has a unique solution $x(\lambda, \mu)$ and the function $x(\cdot, \cdot)$ satisfies the Hölder condition:

$$d_X\left(x\left(\lambda_1,\mu_1\right),x\left(\lambda_2,\mu_2\right)\right)\le\left(\frac{2n\rho\ell_2^{\delta}}{\hbar-2n\rho\ell_1^{\delta}}\right)^{\frac{1}{\beta}}d^{\frac{\alpha_2\delta}{\beta}}\left(\lambda_1,\lambda_2\right)$$
$$+\left(\frac{m\rho}{\hbar-2n\rho\ell_1^{\delta}}\right)^{\frac{1}{\beta-\theta}}d^{\frac{\gamma}{\beta-\phi}}\left(\mu_1,\mu_2\right).$$

Proof Let $\left(\lambda_1,\mu_1\right),\left(\lambda_2,\mu_2\right)\in U(\bar{\lambda})\times U(\bar{\mu})$. The proof of the theorem is divided into three steps, based on the fact that

$$d_X\left(x\left(\lambda_1,\mu_1\right),x\left(\lambda_2,\mu_2\right)\right)\le d_X\left(x\left(\lambda_1,\mu_1\right),x\left(\lambda_1,\mu_2\right)\right)$$
$$+\,d_X\left(x\left(\lambda_1,\mu_2\right),x\left(\lambda_2,\mu_2\right)\right).$$

Step 1. We prove that, for $x\left(\lambda_1,\mu_1\right)\in\mathrm{Sol(VQEP)}\left(\lambda_1,\mu_1\right)$ and $x\left(\lambda_1,\mu_2\right)\in\mathrm{Sol(VQEP)}\left(\lambda_1,\mu_2\right)$,

$$d_1:=d_X\left(x\left(\lambda_1,\mu_1\right),x\left(\lambda_1,\mu_2\right)\right)\le\left(\frac{m\rho}{\hbar-2n\rho\ell_1^{\delta}}\right)^{\frac{1}{\beta-\theta}}d^{\frac{\gamma}{\beta-\theta}}\left(\mu_1,\mu_2\right). \tag{6.64}$$

If $x\left(\lambda_1,\mu_1\right)=x\left(\lambda_1,\mu_2\right)$, then the estimate (6.64) holds trivially. So, we suppose $x\left(\lambda_1,\mu_1\right)\neq x\left(\lambda_1,\mu_2\right)$. Since $x\left(\lambda_1,\mu_1\right)\in\mathscr{D}\left(x\left(\lambda_1,\mu_1\right),\lambda_1\right)$ and $x\left(\lambda_1,\mu_2\right)\in\mathscr{D}\left(x\left(\lambda_1,\mu_2\right),\lambda_1\right)$, by the Hölder continuity of $\mathscr{D}\left(\cdot,\lambda_1\right)$, there exist $x_1\in\mathscr{D}\left(x\left(\lambda_1,\mu_1\right),\lambda_1\right)$ and $x_2\in\mathscr{D}\left(x\left(\lambda_1,\mu_2\right),\lambda_1\right)$ such that

$$\begin{aligned}d_X\left(x\left(\lambda_1,\mu_1\right),x_2\right)&\le\ell_1d_X^{\alpha_1}\left(x\left(\lambda_1,\mu_1\right),x\left(\lambda_1,\mu_2\right)\right)=\ell_1d_1^{\alpha_1},\\ d_X\left(x\left(\lambda_1,\mu_2\right),x_1\right)&\le\ell_1d_X^{\alpha_1}\left(x\left(\lambda_1,\mu_1\right),x\left(\lambda_1,\mu_2\right)\right)=\ell_1d_1^{\alpha_1}.\end{aligned}\tag{6.65}$$

Because $x\left(\lambda_1,\mu_1\right)\in\mathrm{Sol(VQEP)}\left(\lambda_1,\mu_1\right)$ and $x\left(\lambda_1,\mu_2\right)\in\mathrm{Sol(VQEP)}\left(\lambda_1,\mu_2\right)$, we have:

$$\begin{aligned}&r_C^e\left(f\left(x\left(\lambda_1,\mu_1\right),x_1,\mu_1\right)\right)\ge0,\\ &r_C^e\left(f\left(x\left(\lambda_1,\mu_2\right),x_2,\mu_2\right)\right)\ge0.\end{aligned}\tag{6.66}$$

By virtue of assumption (b) and (6.66), we get:

$$\begin{aligned}\hbar d_1^{\beta}\le&-r_C^e\left(f\left(x\left(\lambda_1,\mu_1\right),x\left(\lambda_1,\mu_2\right),\mu_1\right)\right)-r_C^e\left(f\left(x\left(\lambda_1,\mu_2\right),x\left(\lambda_1,\mu_1\right),\mu_1\right)\right)\\ \le&\left|r_C^e\left(f\left(x\left(\lambda_1,\mu_1\right),x_1,\mu_1\right)\right)-r_C^e\left(f\left(x\left(\lambda_1,\mu_1\right),x\left(\lambda_1,\mu_2\right),\mu_1\right)\right)\right|\\ &+\left|r_C^e\left(f\left(x\left(\lambda_1,\mu_2\right),x_2,\mu_2\right)\right)-r_C^e\left(f\left(x\left(\lambda_1,\mu_2\right),x\left(\lambda_1,\mu_1\right),\mu_2\right)\right)\right|\\ &+\left|r_C^e\left(f\left(x\left(\lambda_1,\mu_2\right),x\left(\lambda_1,\mu_1\right),\mu_2\right)\right)\right.\\ &\left.-r_C^e\left(f\left(x\left(\lambda_1,\mu_2\right),x\left(\lambda_1,\mu_1\right),\mu_1\right)\right)\right|.\end{aligned}$$

From the Lipschitz property of r_C^e (Proposition 2.3) and assumptions (c) and (d), together with (6.65), we can deduce:

$$\begin{aligned}\hbar d_1^{\beta} \leq & \rho \left\| f\left(x\left(\lambda_1, \mu_1\right), x_1, \mu_1\right) - f\left(x\left(\lambda_1, \mu_1\right), x\left(\lambda_1, \mu_2\right), \mu_1\right)\right\| \\ & + \rho \left\| f\left(x\left(\lambda_1, \mu_2\right), x_2, \mu_2\right) - f\left(x\left(\lambda_1, \mu_2\right), x\left(\lambda_1, \mu_1\right), \mu_2\right)\right\| \\ & + \rho \left\| f\left(x\left(\lambda_1, \mu_2\right), x\left(\lambda_1, \mu_1\right), \mu_2\right) - f\left(x\left(\lambda_1, \mu_2\right), x\left(\lambda_1, \mu_1\right), \mu_1\right)\right\| \\ \leq & n\rho d_X^{\delta}\left(x_1, x\left(\lambda_1, \mu_2\right)\right) + n\rho d_X^{\delta}\left(x_2, x\left(\lambda_1, \mu_1\right)\right) + m\rho d_1^{\theta} d^{\gamma}\left(\mu_1, \mu_2\right) \\ \leq & 2n\rho \ell_1^{\delta} d_1^{\alpha_1 \delta} + m\rho d_1^{\theta} d^{\gamma}\left(\mu_1, \mu_2\right).\end{aligned}$$

Whence, assumption (e) yields that

$$d_1^{\beta-\theta} \leq \frac{m\rho}{\hbar - 2n\rho \ell_1^{\delta}} d^{\gamma}\left(\mu_1, \mu_2\right),$$

and that the estimate (6.64) holds.
Step 2. We prove that, for $x\left(\lambda_1, \mu_2\right) \in$ Sol(VQEP) $\left(\lambda_1, \mu_2\right)$ and $x\left(\lambda_2, \mu_2\right) \in$ Sol(VQEP) $\left(\lambda_2, \mu_2\right)$,

$$d_2 := d_X\left(x\left(\lambda_1, \mu_2\right), x\left(\lambda_2, \mu_2\right)\right) \leq \left(\frac{2n\rho \ell_2^{\delta}}{\hbar - 2n\rho \ell_1^{\delta}}\right)^{\frac{1}{\beta}} d^{\frac{\alpha_2 \delta}{\beta}}\left(\lambda_1, \lambda_2\right). \tag{6.67}$$

Let $x\left(\lambda_1, \mu_2\right) \neq x\left(\lambda_2, \mu_2\right)$ (the case with $x\left(\lambda_1, \mu_2\right) = x\left(\lambda_2, \mu_2\right)$ is trivial). Due to $x\left(\lambda_1, \mu_2\right) \in \mathscr{D}\left(x\left(\lambda_1, \mu_2\right), \lambda_1\right)$ and $x\left(\lambda_2, \mu_2\right) \in \mathscr{D}\left(x\left(\lambda_2, \mu_2\right), \lambda_2\right)$, it follows from assumption (a) that there are $x_1' \in \mathscr{D}\left(x\left(\lambda_2, \mu_2\right), \lambda_1\right)$ and $x_2' \in \mathscr{D}\left(x\left(\lambda_1, \mu_2\right), \lambda_2\right)$ such that

$$\begin{aligned} d_X\left(x\left(\lambda_1, \mu_2\right), x_2'\right) &\leq \ell_2 d^{\alpha_2}\left(\lambda_1, \lambda_2\right), \\ d_X\left(x\left(\lambda_2, \mu_2\right), x_1'\right) &\leq \ell_2 d^{\alpha_2}\left(\lambda_1, \lambda_2\right). \end{aligned} \tag{6.68}$$

By the Hölder continuity of $\mathscr{D}(\cdot, \cdot)$, there are $x_1'' \in \mathscr{D}\left(x\left(\lambda_1, \mu_2\right), \lambda_1\right)$ and $x_2'' \in \mathscr{D}\left(x\left(\lambda_2, \mu_2\right), \lambda_2\right)$ such that

$$\begin{aligned} d_X\left(x_1', x_1''\right) &\leq \ell_1 d_X^{\alpha_1}\left(x\left(\lambda_1, \mu_2\right), x\left(\lambda_2, \mu_2\right)\right) = \ell_1 d_2^{\alpha_1}, \\ d_X\left(x_2', x_2''\right) &\leq \ell_1 d_X^{\alpha_1}\left(x\left(\lambda_1, \mu_2\right), x\left(\lambda_2, \mu_2\right)\right) = \ell_1 d_2^{\alpha_1}. \end{aligned} \tag{6.69}$$

Since $x\left(\lambda_1, \mu_2\right) \in$ Sol(VQEP) $\left(\lambda_1, \mu_2\right)$ and $x\left(\lambda_2, \mu_2\right) \in$ Sol(VQEP) $\left(\lambda_2, \mu_2\right)$, we obtain:

$$\begin{aligned} r_C^e\left(f\left(x\left(\lambda_1, \mu_2\right), x_1'', \mu_2\right)\right) &\geq 0, \\ r_C^e\left(f\left(x\left(\lambda_2, \mu_2\right), x_2'', \mu_2\right)\right) &\geq 0. \end{aligned} \tag{6.70}$$

By assumption (b) and (6.70), we get:

$$\begin{aligned}\hbar d_2^{\beta} \leq & -r_C^e \left(f\left(x\left(\lambda_1,\mu_2\right),x\left(\lambda_2,\mu_2\right),\mu_2\right)\right) - r_C^e \left(f\left(x\left(\lambda_2,\mu_2\right),x\left(\lambda_1,\mu_2\right),\mu_2\right)\right) \\ \leq & \left|r_C^e\left(f\left(x\left(\lambda_1,\mu_2\right),x_1'',\mu_2\right)\right) - r_C^e\left(f\left(x\left(\lambda_1,\mu_2\right),x_1',\mu_2\right)\right)\right| \\ & + \left|r_C^e\left(f\left(x\left(\lambda_1,\mu_2\right),x_1',\mu_2\right)\right) - r_C^e\left(f\left(x\left(\lambda_1,\mu_2\right),x\left(\lambda_2,\mu_2\right),\mu_2\right)\right)\right| \\ & + \left|r_C^e\left(f\left(x\left(\lambda_2,\mu_2\right),x_2'',\mu_2\right)\right) - r_C^e\left(f\left(x\left(\lambda_2,\mu_2\right),x_2',\mu_2\right)\right)\right| \\ & + \left|r_C^e\left(f\left(x\left(\lambda_2,\mu_2\right),x_2',\mu_2\right)\right) - r_C^e\left(f\left(x\left(\lambda_2,\mu_2\right),x\left(\lambda_1,\mu_2\right),\mu_2\right)\right)\right|.\end{aligned}$$

From the Lipschitz property of r_C^e and assumption (c), together with (6.68) and (6.69), we can deduce:

$$\begin{aligned}\hbar d_2^{\beta} \leq & \rho \left\| f\left(x\left(\lambda_1,\mu_2\right),x_1'',\mu_2\right) - f\left(x\left(\lambda_1,\mu_2\right),x_1',\mu_2\right)\right\| \\ & + \rho \left\| f\left(x\left(\lambda_1,\mu_2\right),x_1',\mu_2\right) - f\left(x\left(\lambda_1,\mu_2\right),x\left(\lambda_2,\mu_2\right),\mu_2\right)\right\| \\ & + \rho \left\| f\left(x\left(\lambda_2,\mu_2\right),x_2'',\mu_2\right) - f\left(x\left(\lambda_2,\mu_2\right),x_2',\mu_2\right)\right\| \\ & + \rho \left\| f\left(x\left(\lambda_2,\mu_2\right),x_2',\mu_2\right) - f\left(x\left(\lambda_2,\mu_2\right),x\left(\lambda_1,\mu_2\right),\mu_2\right)\right\| \\ \leq & n\rho d_X^{\delta}\left(x_1',x_1''\right) + n\rho d_X^{\delta}\left(x_1',x\left(\lambda_2,\mu_2\right)\right) + n\rho d_X^{\delta}\left(x_2',x_2''\right) \\ & + n\rho d_X^{\delta}\left(x_2',x\left(\lambda_1,\mu_2\right)\right) \\ \leq & 2n\rho \ell_1^{\delta} d_2^{\alpha_1\delta} + 2n\rho \ell_2^{\delta} d^{\alpha_2\delta}\left(\lambda_1,\lambda_2\right).\end{aligned}$$

It follows from assumption (e) that

$$d_2^{\beta} \leq \frac{2n\rho\ell_2^{\delta}}{\hbar - 2n\rho\ell_1^{\delta}} d^{\alpha_2\delta}\left(\lambda_1,\lambda_2\right).$$

This implies that the estimate (6.67) holds.
Step 3. For $x\left(\lambda_1,\mu_1\right) \in \mathrm{Sol(VQEP)}\left(\lambda_1,\mu_1\right)$ and $x\left(\lambda_2,\mu_2\right) \in \mathrm{Sol(VQEP)}\left(\lambda_2,\mu_2\right)$, we have:

$$d_X\left(x\left(\lambda_1,\mu_1\right),x\left(\lambda_2,\mu_2\right)\right) \leq d_1 + d_2.$$

Thus,

$$\begin{aligned}& \mathbb{D}\left(\mathrm{Sol(VQEP)}\left(\lambda_1,\mu_1\right),\mathrm{Sol(VQEP)}\left(\lambda_2,\mu_2\right)\right) \\ & = \sup_{x(\lambda_1,\mu_1)\in \mathrm{Sol(VQEP)}(\lambda_1,\mu_1),\, x(\lambda_2,\mu_2)\in \mathrm{Sol(VQEP)}(\lambda_2,\mu_2)} d_X\left(x\left(\lambda_1,\mu_1\right),x\left(\lambda_2,\mu_2\right)\right) \\ & \leq \left(\frac{2n\rho\ell_2^{\delta}}{\hbar - 2n\rho\ell_1^{\delta}}\right)^{\frac{1}{\beta}} d^{\frac{\alpha_2\delta}{\beta}}\left(\lambda_1,\lambda_2\right) + \left(\frac{m\rho}{\hbar - 2n\rho\ell_1^{\delta}}\right)^{\frac{1}{\beta-\theta}} d^{\frac{\gamma}{\beta-\theta}}\left(\mu_1,\mu_2\right).\end{aligned}$$

Taking $\lambda_2 = \lambda_1$ and $\mu_2 = \mu_1$, we see that the diameter of Sol(VQEP) (λ_1, μ_1) is 0, that is, this set is a singleton $\{x(\lambda_1, \mu_1)\}$. The set Sol(VQEP) (λ_2, μ_2) is similar. Thus, problem (VQEP) has a unique solution in a neighborhood of $(\bar{\lambda}, \bar{\mu})$, and the Hölder condition is satisfied. □

Remark 6.93 (A replacement the technical assumption $\beta > \theta$) If in assumption (d) of Theorem 6.92 $\mathscr{E}(U(\bar{\lambda}))$ is bounded, then without loss of generality, we can take $\theta = 0$ since $d_X(x, y) \le w$ for some $w > 0$ and for all $x, y \in \mathscr{E}(U(\bar{\lambda}))$. Thus, the condition $\beta > \theta$ in Theorem 6.92 can be omitted.

Theorem 6.94 (Hölder continuity of solution sets of quasiequilibrium problems via a nonlinear scalarization of a strong pseudomonotonicity) *Assume that assumption (b) in Theorem 6.92 is replaced by*

(b') $f(\cdot,\cdot,\mu)$ *is* $(\hbar, \beta)$*-Hölder strongly pseudomonotone wrt* $q \in \operatorname{int}C$*, i.e., there exist* $q \in \operatorname{int}C$ *and* $\hbar, \beta > 0$ *such that for every* $x, y \in E(U(\bar{\lambda}))$ *with* $x \neq y$

$$f(x, y, \mu) \notin -\operatorname{int}C \Rightarrow f(y, x, \mu) + \hbar d_X^{\beta}(x, y)q \in -C,$$

and $-f(\cdot,\cdot,\mu)$ *is quasimonotone on* $E(U(\bar{\lambda}))$*, i.e., for all* $x, y \in E(U(\bar{\lambda}))$ *with* $x \neq y$*,*

$$f(x, y, \mu) \in -\operatorname{int}C \Longrightarrow f(y, x, \mu) \notin -\operatorname{int}C.$$

Then, the following Hölder condition holds:

$$d_X\left(x(\lambda_1, \mu_1), x(\lambda_2, \mu_2)\right) \le \left(\frac{n\rho\ell_2^{\delta}}{\hbar - n\rho\ell_1^{\delta}}\right)^{\frac{1}{\beta}} d^{\frac{\alpha_2\delta}{\beta}}(\lambda_1, \lambda_2)$$

$$+\left(\frac{m\rho}{\hbar - n\rho\ell_1^{\delta}}\right)^{\frac{1}{\beta-\theta}} d^{\frac{\gamma}{\beta-\theta}}(\mu_1, \mu_2).$$

Proof Let $(\lambda_1, \mu_1), (\lambda_2, \mu_2) \in U(\bar{\lambda}) \times U(\bar{\mu})$.
Step 1. We prove that, for $x(\lambda_1, \mu_1) \in$ Sol(VQEP) (λ_1, μ_1) and $x(\lambda_1, \mu_2) \in$ Sol(VQEP) (λ_1, μ_2)

$$d_1 := d_X\left(x(\lambda_1, \mu_1), x(\lambda_1, \mu_2)\right) \le \left(\frac{m\rho}{\hbar - n\rho\ell_1^{\delta}}\right)^{\frac{1}{\beta-\theta}} d^{\frac{\gamma}{\beta-\theta}}(\mu_1, \mu_2). \qquad (6.71)$$

Let $x(\lambda_1, \mu_1) \neq x(\lambda_1, \mu_2)$. Obviously, (6.65) and (6.66) still hold as they are not related to assumption (b). If $f\left(x(\lambda_1, \mu_1), x(\lambda_1, \mu_2), \mu_1\right) \notin -\operatorname{int}C$, then condition (b') implies that

$$f(x(\lambda_1,\mu_2),x(\lambda_1,\mu_1),\mu_1)+\hbar d_X^{\beta}(x(\lambda_1,\mu_1),x(\lambda_1,\mu_2))q\in -C.$$

Because of Proposition 6.86(b), we have:

$$r_C^e(f(x(\lambda_1,\mu_2),x(\lambda_1,\mu_1),\mu_1))\leq -\hbar d_X^{\beta}(x(\lambda_1,\mu_1),x(\lambda_1,\mu_2)).$$

Hence, noting the second inequality of (6.66), we obtain:

$$\begin{aligned}\hbar d_1^{\beta}\leq & -r_C^e(f(x(\lambda_1,\mu_2),x(\lambda_1,\mu_1),\mu_1))\\ \leq & \left|r_C^e(f(x(\lambda_1,\mu_2),x(\lambda_1,\mu_1),\mu_2))-r_C^e(f(x(\lambda_1,\mu_2),x(\lambda_1,\mu_1),\mu_1))\right|\\ & +\left|r_C^e(f(x(\lambda_1,\mu_2),x_2,\mu_2))-r_C^e(f(x(\lambda_1,\mu_2),x(\lambda_1,\mu_1),\mu_2))\right|.\end{aligned}$$

From the Lipschitz property of r_C^e and assumptions (c) and (d), together with the first inequality of (6.65), we deduce that

$$\begin{aligned}\hbar d_1^{\beta}\leq & \rho\left\|f(x(\lambda_1,\mu_2),x(\lambda_1,\mu_1),\mu_2)-f(x(\lambda_1,\mu_2),x(\lambda_1,\mu_1),\mu_1)\right\|\\ & +\rho\left\|f(x(\lambda_1,\mu_2),x_2,\mu_2)-f(x(\lambda_1,\mu_2),x(\lambda_1,\mu_1),\mu_2)\right\|\\ \leq & m\rho d_1^{\theta}d^{\gamma}(\mu_1,\mu_2)+n\rho d_X^{\delta}(x(\lambda_1,\mu_1),x_2)\\ \leq & m\rho d_1^{\theta}d^{\gamma}(\mu_1,\mu_2)+n\rho\ell_1^{\delta}d_1^{\alpha_1\delta}.\end{aligned}$$

Thanks to (e), we get the estimate (6.71).

If $f(x(\lambda_1,\mu_1),x(\lambda_1,\mu_2),\mu_1)\in -\operatorname{int}C$, then condition (b') implies that

$$f(x(\lambda_1,\mu_2),x(\lambda_1,\mu_1),\mu_1)\notin -\operatorname{int}C,$$

$$f(x(\lambda_1,\mu_1),x(\lambda_1,\mu_2),\mu_1)+\hbar d_X^{\beta}(x(\lambda_1,\mu_1),x(\lambda_1,\mu_2))q\in -C.$$

Then, similarly as above, using the first inequality of (6.66), the Lipschitz property of r_C^e, assumption (c), and the second inequality of (6.65), we obtain:

$$\begin{aligned}\hbar d_1^{\beta}\leq & -r_C^e(f(x(\lambda_1,\mu_1),x(\lambda_1,\mu_2),\mu_1))\\ \leq & \left|r_C^e(f(x(\lambda_1,\mu_1),x_1,\mu_1))-r_C^e(f(x(\lambda_1,\mu_1),x(\lambda_1,\mu_2),\mu_1))\right|\\ \leq & \rho\left\|f(x(\lambda_1,\mu_1),x_1,\mu_1)-f(x(\lambda_1,\mu_1),x(\lambda_1,\mu_2),\mu_1)\right\|\\ \leq & n\rho d_X^{\delta}(x_1,x(\lambda_1,\mu_2))\\ \leq & n\rho\ell_1^{\delta}d_1^{\alpha_1\delta}.\end{aligned}$$

So, it follows from assumption (e) that $d_1=0$. Hence, the estimate (6.71) holds.

Step 2. We prove that, for $x(\lambda_1,\mu_2)\in \mathrm{Sol(VQEP)}(\lambda_1,\mu_2)$ and $x(\lambda_2,\mu_2)\in \mathrm{Sol(VQEP)}(\lambda_2,\mu_2)$,

$$d_2 := d\left(x\left(\lambda_1, \mu_2\right), x\left(\lambda_2, \mu_2\right)\right) \leq \left(\frac{n\rho\ell_2^{\delta}}{\hbar - n\rho\ell_1^{\delta}}\right)^{\frac{1}{\beta}} d^{\frac{\alpha_2\delta}{\beta}}\left(\lambda_1, \lambda_2\right). \tag{6.72}$$

Let $x\left(\lambda_1, \mu_2\right) \neq x\left(\lambda_2, \mu_2\right)$. Obviously, (6.68)–(6.70) still hold. If $f\left(x\left(\lambda_1, \mu_2\right), x\left(\lambda_2, \mu_2\right), \mu_2\right) \notin -\operatorname{int} C$, then condition (b') implies that

$$\hbar d_2^{\beta} \leq -r_C^e\left(f\left(x\left(\lambda_2, \mu_2\right), x\left(\lambda_1, \mu_2\right), \mu_2\right)\right).$$

Using the second inequalities of (6.69) and (6.70), the first inequality of (6.68), the Lipschitz property of r_C^e, and assumption (c), we obtain:

$$\begin{aligned}
\hbar d_2^{\beta} \leq & \left|r_C^e\left(f\left(x\left(\lambda_2, \mu_2\right), x_2'', \mu_2\right)\right) - r_C^e\left(f\left(x\left(\lambda_2, \mu_2\right), x_2', \mu_2\right)\right)\right| \\
& + \left|r_C^e\left(f\left(x\left(\lambda_2, \mu_2\right), x_2', \mu_2\right)\right) - r_C^e\left(f\left(x\left(\lambda_2, \mu_2\right), x\left(\lambda_1, \mu_2\right), \mu_2\right)\right)\right| \\
\leq & \rho\left\|f\left(x\left(\lambda_2, \mu_2\right), x_2'', \mu_2\right) - f\left(x\left(\lambda_2, \mu_2\right), x_2', \mu_2\right)\right\| \\
& + \rho\left\|f\left(x\left(\lambda_2, \mu_2\right), x_2', \mu_2\right) - f\left(x\left(\lambda_2, \mu_2\right), x\left(\lambda_1, \mu_2\right), \mu_2\right)\right\| \\
\leq & n\rho d_X^{\delta}\left(x_2', x_2''\right) + n\rho d_X^{\delta}\left(x_2', x\left(\lambda_1, \mu_2\right)\right) \\
\leq & n\rho\ell_1^{\delta} d_2^{\alpha_1\delta} + n\rho\ell_2^{\delta} d^{\alpha_2\delta}\left(\lambda_1, \lambda_2\right).
\end{aligned}$$

Thanks to (e), we get the estimate (6.72). If $f\left(x\left(\lambda_1, \mu_2\right), x\left(\lambda_2, \mu_2\right), \mu_2\right) \in -\operatorname{int}C$, then condition (b') implies that

$$\begin{aligned}
& f\left(x\left(\lambda_2, \mu_2\right), x\left(\lambda_1, \mu_2\right), \mu_2\right) \notin -\operatorname{int}C, \\
& f\left(x\left(\lambda_1, \mu_2\right), x\left(\lambda_2, \mu_2\right), \mu_2\right) + \hbar d_X^{\beta}\left(x\left(\lambda_1, \mu_2\right), x\left(\lambda_2, \mu_2\right)\right) q \in -C.
\end{aligned}$$

Then, combining the first inequalities of (6.69) and (6.70), the second inequality of (6.68), the Lipschitz property of r_C^e, and assumption (c), we obtain:

$$\begin{aligned}
\hbar d_2^{\beta} \leq & -r_C^e\left(f\left(x\left(\lambda_1, \mu_2\right), x\left(\lambda_2, \mu_2\right), \mu_2\right)\right) \\
\leq & \left|r_C^e\left(f\left(x\left(\lambda_1, \mu_2\right), x_1'', \mu_2\right)\right) - r_C^e\left(f\left(x\left(\lambda_1, \mu_2\right), x_1', \mu_2\right)\right)\right| \\
& + \left|r_C^e\left(f\left(x\left(\lambda_1, \mu_2\right), x_1', \mu_2\right)\right) - r_C^e\left(f\left(x\left(\lambda_1, \mu_2\right), x\left(\lambda_2, \mu_2\right), \mu_2\right)\right)\right| \\
\leq & \rho\left\|f\left(x\left(\lambda_1, \mu_2\right), x_1'', \mu_2\right) - f\left(x\left(\lambda_1, \mu_2\right), x_1', \mu_2\right)\right\| \\
& + \rho\left\|f\left(x\left(\lambda_1, \mu_2\right), x_1', \mu_2\right) - f\left(x\left(\lambda_1, \mu_2\right), x\left(\lambda_2, \mu_2\right), \mu_2\right)\right\| \\
\leq & n\rho d_X^{\delta}\left(x_1', x_1''\right) + n\rho d_X^{\delta}\left(x_1', x\left(\lambda_2, \mu_2\right)\right) \\
\leq & n\rho\ell_1^{\delta} d_2^{\alpha_1\delta} + n\rho\ell_2^{\delta} d^{\alpha_2\delta}\left(\lambda_1, \lambda_2\right).
\end{aligned}$$

It follows from assumption (e) that the estimate (6.72) is true.

Step 3 is the same as for Theorem 6.92. □

Remark 6.95 (Irrelationships between strong monotonicity of f and quasi-monotonicity of $-f$) From Proposition 6.86, we see that assumption (b) of Theorem 6.92 implies the first part of condition (b') of Theorem 6.94. The converse does not hold generally, even in the special case with $X = Y = \mathbb{R}, C = \mathbb{R}_+$. Assumption (b) and the second part of condition (b') do not follow each other. Whence, Theorem 6.94 is different from Theorem 6.92.

Motivated by the Hölder-related assumptions introduced in the previous chapter, we propose a new condition (b") as follows.

Theorem 6.96 (Hölder continuity of solution sets of quasiequilibrium problems via a nonlinear scalarization function) *Assume condition (b) in Theorem 6.92 is replaced by*

(b") $f(\cdot,\cdot,\mu)$ *is* $(\hbar,\beta)$*-Hölder-related weakly monotone wrt* r^e_C*, i.e., there exist* $q \in \operatorname{int} C$ *and* $\hbar, \beta > 0$ *such that, for every* $x, y \in E(U(\bar{\lambda}))$ *with* $x \neq y$,

$$\hbar d^{\beta}_X(x,y) \le \left|r^e_C(f(x,y,\mu)), \mathbb{R}_+\right| + \left|r^e_C(f(y,x,\mu)), \mathbb{R}_+\right|,$$

where $|a,\mathbb{R}_+| = \inf_{b\in\mathbb{R}_+} |a-b|$ *for* $a \in \mathbb{R}$.

Then, the following Hölder condition holds:

$$\begin{aligned} d_X\left(x\left(\lambda_1,\mu_1\right), x\left(\lambda_2,\mu_2\right)\right) \le &\left(\frac{2n\rho\ell_2^{\delta}}{\hbar - 2n\rho\ell_1^{\delta}}\right)^{\frac{1}{\beta}} d^{\frac{\alpha_2\delta}{\beta}}\left(\lambda_1,\lambda_2\right) \\ &+\left(\frac{m\rho}{\hbar - 2n\rho\ell_1^{\delta}}\right)^{\frac{1}{\beta-\theta}} d^{\frac{\gamma}{\beta-\theta}}\left(\mu_1,\mu_2\right). \end{aligned}$$

Proof Let $(\lambda_1,\mu_1), (\lambda_2,\mu_2) \in U(\bar{\lambda}) \times U(\bar{\mu})$.
Step 1. By virtue of condition (b") and (6.66), we get:

$$\begin{aligned} \hbar d_1^{\beta} &= \hbar d_X^{\beta}\left(x\left(\lambda_1,\mu_1\right), x\left(\lambda_1,\mu_2\right)\right) \\ &\le \left|r^e_C\left(f\left(x\left(\lambda_1,\mu_1\right), x\left(\lambda_1,\mu_2\right),\mu_1\right)\right), \mathbb{R}_+\right| \\ &\quad + \left|r^e_C\left(f\left(x\left(\lambda_1,\mu_2\right), x\left(\lambda_1,\mu_1\right),\mu_1\right)\right), \mathbb{R}_+\right| \\ &\le \left|r^e_C\left(f\left(x\left(\lambda_1,\mu_1\right), x_1,\mu_1\right)\right) - r^e_C\left(f\left(x\left(\lambda_1,\mu_1\right), x\left(\lambda_1,\mu_2\right),\mu_1\right)\right)\right| \\ &\quad + \left|r^e_C\left(f\left(x\left(\lambda_1,\mu_2\right), x_2,\mu_2\right)\right) - r^e_C\left(f\left(x\left(\lambda_1,\mu_2\right), x\left(\lambda_1,\mu_1\right),\mu_1\right)\right)\right| \\ &\le \left|r^e_C\left(f\left(x\left(\lambda_1,\mu_1\right), x_1,\mu_1\right)\right) - r^e_C\left(f\left(x\left(\lambda_1,\mu_1\right), x\left(\lambda_1,\mu_2\right),\mu_1\right)\right)\right| \\ &\quad + \left|r^e_C\left(f\left(x\left(\lambda_1,\mu_2\right), x_2,\mu_2\right)\right) - r^e_C\left(f\left(x\left(\lambda_1,\mu_2\right), x\left(\lambda_1,\mu_1\right),\mu_2\right)\right)\right| \end{aligned}$$

$$+ \mid r_C^e \left(f\left(x\left(\lambda_1, \mu_2\right), x\left(\lambda_1, \mu_1\right), \mu_2\right)\right)$$
$$-r_C^e \left(f\left(x\left(\lambda_1, \mu_2\right), x\left(\lambda_1, \mu_1\right), \mu_1\right)\right) \mid .$$

Then, similar arguments as in Step 1 for Theorem 6.92 complete Step 1.
Step 2. By virtue of condition (b") and (6.70), we get:

$$\begin{aligned}
\hbar d_2^\beta =& \hbar d_X^\beta \left(x\left(\lambda_1, \mu_2\right), x\left(\lambda_2, \mu_2\right)\right) \\
\leq& \left|r_C^e \left(f\left(x\left(\lambda_1, \mu_2\right), x\left(\lambda_2, \mu_2\right), \mu_2\right)\right), \mathbb{R}_+\right| \\
&+ \left|r_C^e \left(f\left(x\left(\lambda_2, \mu_2\right), x\left(\lambda_1, \mu_2\right), \mu_2\right)\right), \mathbb{R}_+\right| \\
\leq& \left|r_C^e \left(f\left(x\left(\lambda_1, \mu_2\right), x_1'', \mu_2\right)\right) - r_C^e \left(f\left(x\left(\lambda_1, \mu_2\right), x\left(\lambda_2, \mu_2\right), \mu_2\right)\right)\right| \\
&+ \left|r_C^e \left(f\left(x\left(\lambda_2, \mu_2\right), x_2'', \mu_2\right)\right) - r_C^e \left(f\left(x\left(\lambda_2, \mu_2\right), x\left(\lambda_1, \mu_2\right), \mu_2\right)\right)\right| \\
\leq& \left|r_C^e \left(f\left(x\left(\lambda_1, \mu_2\right), x_1'', \mu_2\right)\right) - r_C^e \left(f\left(x\left(\lambda_1, \mu_2\right), x_1', \mu_2\right)\right)\right| \\
&+ \left|r_C^e \left(f\left(x\left(\lambda_1, \mu_2\right), x_1', \mu_2\right)\right) - r_C^e \left(f\left(x\left(\lambda_1, \mu_2\right), x\left(\lambda_2, \mu_2\right), \mu_2\right)\right)\right| \\
&+ \left|r_C^e \left(f\left(x\left(\lambda_2, \mu_2\right), x_2'', \mu_2\right)\right) - r_C^e \left(f\left(x\left(\lambda_2, \mu_2\right), x_2', \mu_2\right)\right)\right| \\
&+ \left|r_C^e \left(f\left(x\left(\lambda_2, \mu_2\right), x_2', \mu_2\right)\right) - r_C^e \left(f\left(x\left(\lambda_2, \mu_2\right), x\left(\lambda_1, \mu_2\right), \mu_2\right)\right)\right| .
\end{aligned}$$

Then, the rest of Step 2 is similar to Step 2 for Theorem 6.92.
Step 3 is the same as for Theorem 6.92. □

Proposition 6.97 (Relationships between the conditions related to monotonicity)

(a) *Condition (b) ⇒ condition (b").*
(b) *Condition* (b') ⇒ *condition* (b'').
(c) *Condition* (b'') ⇒ *the first part of condition* (b').

Proof

(a) By condition (b), one has for every $x, y \in \mathscr{E}(U(\bar{\lambda}))$ with $x \neq y$:

$$\begin{aligned}
\hbar d_X^\beta(x, y) &\leq -r_C^e(f(x, y, \mu)) - r_C^e(f(y, x, \mu)) \\
&\leq \left|r_C^e(f(x, y, \mu)), \mathbb{R}_+\right| + \left|r_C^e(f(y, x, \mu)), \mathbb{R}_+\right| .
\end{aligned}$$

(b) If $f(x, y, \mu) \notin -\operatorname{int} C$, then $f(y, x, \mu) + \hbar d_X^\beta(x, y)q \in -C$. By Proposition 6.86(b), we have $r_C^e(f(y, x, \mu)) \leq -\hbar d_X^\beta(x, y)$. Thus,

$$\hbar d_X^\beta(x, y) \leq -r_C^e(f(y, x, \mu)) \leq \left|r_C^e(f(x, y, \mu)), \mathbb{R}_+\right| + \left|r_C^e(f(y, x, \mu)), \mathbb{R}_+\right| .$$

If $f(x, y, \mu) \in -\operatorname{int} C$, then $f(y, x, \mu) \notin -\operatorname{int} C$. Similarly, we also get:

$$\hbar d_X^\beta(x, y) \leq -r_C^e(f(x, y, \mu)) \leq \left|r_C^e(f(x, y, \mu)), \mathbb{R}_+\right| + \left|r_C^e(f(y, x, \mu)), \mathbb{R}_+\right|.$$

(c) Assume that $f(x, y, \mu) \notin -\operatorname{int} C$. Then, it follows from Proposition 6.86(a) that

$$\left|r_C^e(f(x, y, \mu)), \mathbb{R}_+\right| = 0.$$

Condition (b") implies that

$$\left|r_C^e(f(y, x, \mu)), \mathbb{R}_+\right| \geq \hbar d_X^\beta(x, y),$$

whence, we deduce that $r_C^e(f(y, x, \mu)) \leq -\hbar d_X^\beta(x, y)$. Proposition 6.86(b) implies that $f(y, x, \mu) + \hbar d_X^\beta(x, y)e \in -C$.

□

The converse of Proposition 6.97 is not true generally, even in the special case with $X = Y = \mathbb{R}$ and $C = \mathbb{R}_+$. Hence, Theorem 6.96 relaxes the monotonicity assumption in Theorems 6.92 and 6.94, resp., and thus improves Theorems 6.92 and 6.94.

At last, we give the relationship between condition (b") and the Hölder-related assumptions introduced in the previous chapter.

Proposition 6.98 (A generalized monotonicity condition via the Gerstewitz function) *Condition (b") implies that, for* $x, y \in E(U(\bar{\lambda}))$ *with* $x \neq y$,

$$\hbar' d_X^\beta(x, y) \leq d(f(x, y, \mu), Y \setminus -\operatorname{int} C) + d(f(y, x, \mu), Y \setminus -\operatorname{int} C),$$

where $\hbar' = \hbar d(-q, Y \setminus -\operatorname{int} C) > 0$.

Proof Assume that condition (b") holds. Then, for any $x, y \in E(U(\bar{\lambda}))$ with $x \neq y$, we have four cases.

Case 1. If $f(x, y, \mu) \notin -\operatorname{int} C$ and $f(y, x, \mu) \notin -\operatorname{int} C$, then it follows from Proposition 6.86(a) that

$$r_C^e(f(x, y, \mu)) \geq 0 \text{ and } r_C^e(f(y, x, \mu)) \geq 0.$$

Thus, for any $\hbar, \beta > 0$, $\hbar d_X^\beta(x, y) \leq 0$. This yields a contradiction, and hence this case does not occur.

Case 2. If $f(x, y, \mu) \notin -\operatorname{int} C$ and $f(y, x, \mu) \in -\operatorname{int} C$, then it follows from Proposition 6.86(a) that

$$r_C^e(f(x, y, \mu)) \geq 0 \text{ and } r_C^e(f(y, x, \mu)) < 0.$$

By condition (b"), we get that $\left|r_C^e(f(y, x, \mu)), \mathbb{R}_+\right| \geq \hbar d_X^\beta(x, y)$, whence

$$r_C^e(f(y, x, \mu)) \leq -\hbar d_X^\beta(x, y).$$

Then, by Proposition 6.86(b) and $q \in \operatorname{int} C$, one has:

$$f(y, x, \mu) \in -\hbar d_X^\beta(x, y)q - C \subset -\operatorname{int} C.$$

Thus,

$$\begin{aligned} d(f(y, x, \mu), Y \backslash -\operatorname{int} C) &\geq d\left(-\hbar d_X^\beta(x, y)q, Y \backslash -\operatorname{int} C\right) \\ &= \hbar d_X^\beta(x, y) d(-q, Y \backslash -\operatorname{int} C). \end{aligned}$$

Consequently,

$$d(f(x, y, \mu), Y \backslash -\operatorname{int} C) + d(f(y, x, \mu), Y \backslash -\operatorname{int} C) \geq \hbar' d_X^\beta(x, y).$$

Case 3. If $f(x, y, \mu) \in -\operatorname{int} C$ and $f(y, x, \mu) \notin -\operatorname{int} C$, then the proof is similar to Case 2.
Case 4. If $f(x, y, \mu) \in -\operatorname{int} C$ and $f(y, x, \mu) \in -\operatorname{int} C$, then it follows from Proposition 6.86(a) that

$$r_C^e(f(x, y, \mu)) < 0 \text{ and } r_C^e(f(y, x, \mu)) < 0.$$

Moreover, it follows from the definition of r_C^e that

$$\begin{aligned} &f(x, y, \mu) \in r_C^e(f(x, y, \mu))q - C \subset -\mathrm{int} C \\ &\text{and } f(y, x, \mu) \in r_C^e(f(y, x, \mu))q - C \subset -\mathrm{int} C. \end{aligned}$$

Thus,

$$\begin{aligned} d(f(x, y, \mu), Y \backslash -\operatorname{int} C) &\geq d\left(r_C^e(f(x, y, \mu))q, Y \backslash -\operatorname{int} C\right) \\ &= -r_C^e(f(x, y, \mu)) d(-q, Y \backslash -\operatorname{int} C) \end{aligned}$$

and

$$\begin{aligned} d(f(y, x, \mu), Y \backslash -\operatorname{int} C) &\geq d\left(r_C^e(f(y, x, \mu))q, Y \backslash -\operatorname{int} C\right) \\ &= -r_C^e(f(y, x, \mu)) d(-q, Y \backslash -\operatorname{int} C). \end{aligned}$$

Consequently,

$$\begin{aligned} &d(f(x, y, \mu), Y \backslash -\operatorname{int} C) + d(f(y, x, \mu), Y \backslash -\operatorname{int} C) \\ &\geq \left[-r_C^e(f(x, y, \mu)) - r_C^e(f(y, x, \mu))\right] d(-q, Y \backslash -\operatorname{int} C) \\ &= \left[\left|r_C^e(f(x, y, \mu)), \mathbb{R}_+\right| + \left|r_C^e(f(y, x, \mu)), \mathbb{R}_+\right|\right] d(-q, Y \backslash -\operatorname{int} C). \end{aligned}$$

By condition (b'), one has that

$$d(f(x, y, \mu), Y \setminus - \operatorname{int} C) + d(f(y, x, \mu), Y \setminus - \operatorname{int} C) \geq \hbar' d_X^{\beta}(x, y).$$

The proof is complete. □

Sensitivity Analysis for (SEP)

In this section, if not otherwise specified, let X, Y, W be normed spaces, A a nonempty subset of X, and Λ, M nonempty subsets of W. We use $\|\cdot\|$ for the norm in any normed space, int$A$ for the interior of A, and $\mathbb{B}[x, r]$ for the closed ball of radius $r \geq 0$ centered at x.

Let $C \subset Y$ be a pointed closed and convex cone with int$C \neq \emptyset$, $\mathcal{D} : \Lambda \rightrightarrows A$ a set-valued map with nonempty bounded convex values, and $f : A \times A \times M \to Y$ a vector-valued map. For each $(\lambda, \mu) \in \Lambda \times M$, we consider the following parametric strong vector equilibrium problem:

(SEP) find $\bar{x} \in \mathcal{D}(\lambda)$ such that for all $y \in \mathcal{D}(\lambda)$, $f(\bar{x}, y, \mu) \geq 0$.

For $(\varepsilon, \lambda, \mu) \in \mathbb{R}_+ \times \Lambda \times M$ and $e \in$ int$C$, we denote the ε-approximate solution set of (SEP) by

$$\widetilde{\text{Sol}}(\text{SEP})(\varepsilon, \lambda, \mu) = \{x \in \mathcal{D}(\lambda) \mid f(x, y, \mu) + \varepsilon e \geq 0, \ \forall y \in \mathcal{D}(\lambda)\}.$$

Because existence conditions of solutions have been studied widely in the previous chapters, in this work we only focus on stability conditions and always assume that the solution sets of the reference problems are nonempty.

Now, we recall some notions which are needed in the sequel.

Definition 6.99 (Types of Lipschitz continuity conditions)

(a) A function $\varphi : X \to \mathbb{R}$ is said to be ℓ-LIPSCHITZ CONTINUOUS at $x_0 \in X$ if there exists a neighborhood $\mathcal{V}$ of x_0 such that for all $x_1, x_2 \in \mathcal{V}$

$$|\varphi(x_1) - \varphi(x_2)| \leq \ell \|x_1 - x_2\|.$$

(b) A vector-valued map $g : X \to Y$ is said to be C-ℓ-LIPSCHITZ CONTINUOUS at $x_0 \in X$ wrt e if there exists a neighborhood $\mathcal{V}$ of x_0 such that for all $x_1, x_2 \in \mathcal{V}$

$$g(x_1) + \ell \|x_1 - x_2\| e \in g(x_2) + C.$$

(c) A set-valued map $Q : X \rightrightarrows Y$ is said to be ℓ-LIPSCHITZ CONTINUOUS at $x_0 \in X$ if there exists a neighborhood $\mathcal{V}$ of x_0 such that for all $x_1, x_2 \in \mathcal{V}$,

$$Q(x_1) \subset Q(x_2) + \mathbb{B}\left[0, \ell \|x_1 - x_2\|\right].$$

We say that a certain property is satisfied on a subset $\mathscr{D} \subset X$ if so is it at every point of $\mathscr{D}$.

Definition 6.100 (Cone-convex maps) A vector-valued map $g : X \to Y$ is said to be C-CONVEX ON A CONVEX SUBSET $\mathscr{D} \subset X$ if for any $x_1, x_2 \in \mathscr{D}$ and $t \in [0, 1]$,

$$tg(x_2) + (1 - t)g(x_1) \in g(tx_2 + (1 - t)x_1) + C, \tag{6.73}$$

and g is said to be C-concave if (6.73) is replaced by

$$g(tx_2 + (1 - t)x_1) \in tg(x_2) + (1 - t)g(x_1) + C.$$

Definition 6.101 (Generalized concave maps) Let $\Gamma \subset \mathbb{R}, \varphi : X \to \mathbb{R}$ be a function, $g : X \to Y$ a vector-valued map, and $\mathscr{D}$ a convex subset of X:

(a) φ is said to be Γ-CONCAVE on $\mathscr{D}$ if for any $x_1, x_2 \in \mathscr{D}$ and $p \in \Gamma$ such that for all $t \in [0, 1]$,

$$\varphi(x_1) \geq 0, \varphi(x_2) \geq p \Rightarrow \varphi(tx_2 + (1 - t)x_1) \geq tp.$$

(b) g is said to be Γ-CONCAVE WRT e on $\mathscr{D}$ if for any $x_1, x_2 \in \mathscr{D}$ and $p \in \Gamma$ such that for all $t \in [0, 1]$,

$$g(x_1) \in C, g(x_2) \in pe + C \Rightarrow g(tx_2 + (1 - t)x_1) \in tpe + C.$$

The following example depicts the above definition.

Example 6.102 (A special case of generalized concavity in an Euclidean space) Let $X = Y = \mathbb{R}$, $\mathscr{D} = \mathbb{R}$, $C = \mathbb{R}_+$, $\Gamma = -\mathbb{R}_+$, and

$$\varphi(x) := \begin{cases} x & \text{if } x < 0, \\ x^4 & \text{if } x \geq 0. \end{cases}$$

We claim that $\varphi(x)$ is Γ-concave on $\mathscr{D}$. Indeed, for any $x_1, x_2 \in \mathscr{D}$, $t \in [0, 1]$, $\varphi(x_1) = x_1^4 \geq 0$, and $\varphi(x_2) - p = x_2 - p \geq 0$ for all $p \in \Gamma$, we need to show that $\varphi(x_t) - tp \geq 0$ with $x_t := tx_2 + (1 - t)x_1$. We consider two cases:
Case 1: If $x_1 \geq 0$ and $x_2 \geq 0$, then $x_t \geq 0$, and so $\varphi(x_t) - tp = x_t^4 - tp \geq 0$.
Case 2: If $x_1 \geq 0$ and $x_2 < 0$, then either $x_t < 0$ or $x_t \geq 0$. In the case with $x_t \geq 0$, similar to Case 1, we obtain $\varphi(x_t) - tp \geq 0$. If $x_t < 0$, then $\varphi(x_t) - tp = x_t - tp = tx_2 + (1 - t)x_1 - tp = t(x_2 - p) + (1 - t)x_1 \geq 0$.

Lemma 6.103 (Relationships between cone concavity and generalized concavity) *Let g and $\mathscr{D}$ be as in Definition 6.101. If g is C-concave on $\mathscr{D}$, then g is $\mathbb{R}$-concave wrt e on $\mathscr{D}$.*

Proof Let $x_1, x_2 \in \mathscr{D}$, $p \in \mathbb{R}$, $g(x_1) \in C$, and $g(x_2) \in pe + C$. Then, for any $t \in [0, 1]$, the C-concavity of g guarantees that

$$\begin{aligned} g(tx_2 + (1-t)x_1) &\in tg(x_2) + (1-t)g(x_1) + C \\ &\in t(pe + C) + (1-t)C + C \\ &\in tpe + tC + (1-t)C + C \\ &\in tpe + C. \end{aligned}$$

Therefore, g is $\mathbb{R}$-concave wrt e on $\mathscr{D}$. □

The following example shows that the converse of Lemma 6.103 is not true.

Example 6.104 (A map which is generalized concave but not cone-concave) Let $X = \mathbb{R}$, $\mathscr{D} = \mathbb{R}$, $Y = \mathbb{R}^2$, $C = \mathbb{R}^2_+$, $e = (1, 1)$, and

$$g(x) := \begin{cases} (2x, 2x+1) & \text{if } x \geq 0, \\ (x, x+1) & \text{if } x < 0. \end{cases}$$

For any $x_1, x_2 \in \mathscr{D}$, $t \in [0, 1]$, and $p \in \mathbb{R}$, if $g(x_1) \in \mathbb{R}^2_+$ and $g(x_2) \in (p, p) + \mathbb{R}^2_+$, then

$$g(x_t) \in (tp, tp) + \mathbb{R}^2_+,$$

with $x_t := tx_2 + (1-t)x_1$. Indeed, we consider two cases.

Case 1: If $x_1 \geq 0$ and $x_2 \geq 0$, then $x_t \geq 0$. Hence,

$$2x_t - tp \geq 2(tx_2 + (1-t)x_1) - tp \geq t(2x_2 - p) + 2(1-t)x_1 \geq 0,$$

which implies that $g(x_t) \in (tp, tp) + \mathbb{R}^2_+$.

Case 2: If $x_1 \geq 0$ and $x_2 < 0$, then either $x_t \geq 0$ or $x_t < 0$. We only discuss the case $x_t < 0$ (the case $x_t \geq 0$ is trivial as $0 > x_2 \geq p$). We have:

$$x_t - tp = tx_2 + (1-t)x_1 - tp = t(x_2 - p) + (1-t)x_1 \geq 0$$

as $x_1 \geq 0$ and $x_2 - p \geq 0$. Consequently, $g(x_t) \in (tp, tp) + \mathbb{R}^2_+$. Therefore, g is $\mathbb{R}$-concave wrt e on $\mathscr{D}$. However, for $x_1 = -1$, $x_2 = 1$, and $t = \frac{1}{2}$, $g(tx_2 + (1-t)x_1) = g(0) = (0, 1) \notin \frac{1}{2}g(-1) + \frac{1}{2}g(1) + \mathbb{R}^2_+ = [\frac{1}{2}, +\infty) \times [\frac{3}{2}, +\infty)$, which implies that g is not C-concave on $\mathscr{D}$.

We are now in a position to present the main result of this section.

Lemma 6.105 (The approximate solution set of (VEP) in terms of nonlinear scalarized problems) *For* $(\varepsilon, \lambda, \mu) \in \mathbb{R}_+ \times \Lambda \times M$, *we have:*

$$\widetilde{\text{Sol}}(\text{SEP})(\varepsilon, \lambda, \mu) = \left\{x \in \mathscr{D}(\lambda) \mid r^{-e}_{-C}(f(x, y, \mu)) \leq \varepsilon,\ \forall y \in \mathscr{D}(\lambda)\right\}.$$

Proof

($\subset$) Taking arbitrarily $x \in \widetilde{\mathrm{Sol}}(\mathrm{SEP})(\varepsilon, \lambda, \mu)$, one has $x \in \mathscr{D}(\lambda)$ and $f(x, y, \mu) + \varepsilon e \in C$ for all $y \in \mathscr{D}(\lambda)$. It follows from Proposition 6.86 that $r_{-C}^{-e}(f(x, y, \mu)) \leq \varepsilon$ for all $y \in \mathscr{D}(\lambda)$, which implies that

$$\widetilde{\mathrm{Sol}}(\mathrm{SEP})(\varepsilon, \lambda, \mu) \subset \left\{x \in \mathscr{D}(\lambda) \mid r_{-C}^{-e}(f(x, y, \mu)) \leq \varepsilon,\ \forall y \in \mathscr{D}(\lambda)\right\}.$$

($\supset$) Let $x \in \left\{x \in \mathscr{D}(\lambda) \mid r_{-C}^{-e}(f(x, y, \mu)) \leq \varepsilon,\ \forall y \in \mathscr{D}(\lambda)\right\}$. Then, $r_{-C}^{-e}(f(x, y, \mu)) \leq \varepsilon$ for all $y \in \mathscr{D}(\lambda)$. This together with Proposition 6.86(a) implies that $f(x, y, \mu) + \varepsilon e \in C$ for all $y \in \mathscr{D}(\lambda)$. Hence,

$$\left\{x \in \mathscr{D}(\lambda) \mid r_{-C}^{-e}(f(x, y, \mu)) \leq \varepsilon,\ \forall y \in \mathscr{D}(\lambda)\right\} \subset \widetilde{\mathrm{Sol}}(\mathrm{SEP})(\varepsilon, \lambda, \mu).$$

The proof is complete.

□

To establish sufficient conditions for the Lipschitz continuity of ε-solution maps of (SEP), we start with the following result.

Lemma 6.106 (Lipschitz conditions for approximate solution maps of strong equilibrium problems via a nonlinear scalarization function) *Let $\varepsilon_0 \in (0, +\infty)$ and $(\lambda_0, \mu_0) \in \Lambda \times M$. Assume that:*

(a) *there exists a neighborhood $\mathscr{V}$ of λ_0 such that for all $\lambda_1, \lambda_2 \in \mathscr{V}$,*

$$\mathscr{D}(\lambda_1) \subset \mathscr{D}(\lambda_2) + \mathbb{B}\left[0, \ell_1 \|\lambda_1 - \lambda_2\|\right];$$

(b) *f is $-\mathbb{R}_+$-concave wrt e in the first component on $\mathscr{D}(\lambda)$ for all $\lambda \in \mathscr{V}$.*

Then, $\widetilde{\mathrm{Sol}}(\mathrm{SEP})$ is $\frac{\rho}{\varepsilon_0}$-Lipschitz continuous in the first component on $(\varepsilon_0, +\infty)$, where $\rho := \mathrm{diam}(\mathscr{D}(\lambda_0)) + 2\ell_1 \mathrm{diam}\mathscr{V}$.

Proof Let $\bar{\varepsilon} \in (\varepsilon_0, +\infty)$ be given. Then, $(\varepsilon_0, 2\bar{\varepsilon} - \varepsilon_0)$ is a neighborhood of $\bar{\varepsilon}$. Taking any $\varepsilon_1, \varepsilon_2 \in (\varepsilon_0, 2\bar{\varepsilon} - \varepsilon_0)$ with $\varepsilon_1 < \varepsilon_2$, one has $\widetilde{\mathrm{Sol}}(\mathrm{SEP})(\varepsilon_1, \lambda, \mu) \subset \widetilde{\mathrm{Sol}}(\mathrm{SEP})(\varepsilon_2, \lambda, \mu)$ for all $(\lambda, \mu) \in \mathscr{V} \times M$, and so

$$\widetilde{\mathrm{Sol}}(\mathrm{SEP})(\varepsilon_1, \lambda, \mu) \subset \widetilde{\mathrm{Sol}}(\mathrm{SEP})(\varepsilon_2, \lambda, \mu) + \mathbb{B}\left[0, \frac{\rho}{\varepsilon_0}|\varepsilon_1 - \varepsilon_2|\right], \tag{6.74}$$

where $\rho := \mathrm{diam}\mathscr{D}(\lambda_0) + 2\ell_1 \mathrm{diam}\mathscr{V}$. For each $x_2 \in \widetilde{\mathrm{Sol}}(\mathrm{SEP})(\varepsilon_2, \lambda, \mu)$, $x_0 \in \widetilde{\mathrm{Sol}}(\mathrm{SEP})(0, \lambda, \mu)$, and $y \in \mathscr{D}(\lambda)$, we have:

$$f(x_0, y, \mu) \in C, \quad f(x_2, y, \mu) \in -\varepsilon_2 e + C.$$

Thanks to the convexity of $\mathscr{D}(\lambda)$, we get $x_1 := \frac{\varepsilon_1}{\varepsilon_2} x_2 + \frac{\varepsilon_2 - \varepsilon_1}{\varepsilon_2} x_0 \in \mathscr{D}(\lambda)$. Since f is $-\mathbb{R}_+$-concave wrt e on $\mathscr{D}(\lambda)$, we obtain:

$$f\left(\frac{\varepsilon_1}{\varepsilon_2}x_2 + \frac{\varepsilon_2 - \varepsilon_1}{\varepsilon_2}x_0, y, \mu\right) \in -\varepsilon_1 e + C.$$

It follows that $r_{-C}^{-e}(f(x_1, y, \mu)) \leq \varepsilon_1$ or equivalently, $x_1 \in \widetilde{\text{Sol}}(\text{SEP})(\varepsilon_1, \lambda, \mu)$. Clearly

$$\|x_2 - x_1\| = \frac{|\varepsilon_1 - \varepsilon_2|}{\varepsilon_2}\|x_2 - x_0\|.$$

By assumption (a), we have:

$$\mathscr{D}(\lambda) \subset \mathscr{D}(\lambda_0) + \mathbb{B}[0, \ell_1\|\lambda - \lambda_0\|],$$

which implies that $\text{diam}\mathscr{D}(\lambda) \leq \text{diam}\mathscr{D}(\lambda_0) + 2\ell_1\text{diam}\mathscr{V}$, and hence

$$\|x_2 - x_1\| \leq \frac{\rho}{\varepsilon_0}|\varepsilon_1 - \varepsilon_2|.$$

Therefore,

$$x_2 \in x_1 + \mathbb{B}\left[0, \frac{\rho}{\varepsilon_0}|\varepsilon_1 - \varepsilon_2|\right],$$

consequently,

$$\widetilde{\text{Sol}}(\text{SEP})(\varepsilon_2, \lambda, \mu) \subset \widetilde{\text{Sol}}(\text{SEP})(\varepsilon_1, \lambda, \mu) + \mathbb{B}\left[0, \frac{\rho}{\varepsilon_0}|\varepsilon_1 - \varepsilon_2|\right]. \tag{6.75}$$

From (6.74) and (6.75), we obtain that $\widetilde{\text{Sol}}(\text{SEP})$ is $\frac{\rho}{\varepsilon_0}$-Lipschitz continuous in the first component at $\bar{\varepsilon}$. By the arbitrariness of $\bar{\varepsilon}$, we conclude that $\widetilde{\text{Sol}}(\text{SEP})$ is $\frac{\rho}{\varepsilon_0}$-Lipschitz continuous in the first component on $(\varepsilon_0, +\infty)$. □

Applying Lemma 6.106, we investigate the Lipschitz property of the solution map.

Theorem 6.107 (Lipschitz conditions for approximate solution maps of parametric strong equilibrium problems via a nonlinear scalarization function) *Let $\varepsilon_0 \in (0, +\infty)$ and $(\lambda_0, \mu_0) \in \Lambda \times M$ be given. Assume that:*

(a) *there exists a neighborhood $\mathscr{V}$ of λ_0 such that for all $\lambda_1, \lambda_2 \in \mathscr{V}$,*

$$\mathscr{D}(\lambda_1) \subset \mathscr{D}(\lambda_2) + \mathbb{B}[0, \ell_1\|\lambda_1 - \lambda_2\|];$$

(b) *f is $-\mathbb{R}_+$-concave wrt e in the first component on $\mathscr{D}(\lambda)$ for all $\lambda \in \mathscr{V}$;*

(c) *there is a neighborhood* $\mathscr{U}$ *of* μ_0 *such that for all* (x_1, y_1, μ_1) *and* (x_2, y_2, μ_2) *in* $\mathscr{D}(\mathscr{V}) \times \mathscr{D}(\mathscr{V}) \times \mathscr{U}$,

$$f(x_1, y_1, \mu_1) + \ell_2 \|(x_1, y_1, \mu_1) - (x_2, y_2, \mu_2)\| e \in f(x_2, y_2, \mu_2) + C,$$

where $\|(x_1, y_1, \mu_1) - (x_2, y_2, \mu_2)\| = \|x_1 - x_2\| + \|y_1 - y_2\| + \|\mu_1 - \mu_2\|$.

Then, $\widetilde{\text{Sol}}(\text{SEP})$ *satisfies the following Lipschitz condition :*

$$\widetilde{\text{Sol}}(\text{SEP})(\varepsilon_1, \lambda_1, \mu_1) \subset \widetilde{\text{Sol}}(\text{SEP})(\varepsilon_2, \lambda_2, \mu_2) + \mathbb{B}\,[0, r_0]\,,$$

for all $(\varepsilon_1, \lambda_1, \mu_1), (\varepsilon_2, \lambda_2, \mu_2) \in (\varepsilon_0, +\infty) \times \mathscr{V} \times \mathscr{U}$, *where*

$$r_0 := \frac{\rho}{\varepsilon_0}|\varepsilon_1 - \varepsilon_2| + \left(\ell_1 + \frac{2\rho\ell_1\ell_2}{\varepsilon_0}\right)\|\lambda_1 - \lambda_2\| + \frac{\rho\ell_2}{\varepsilon_0}\|\mu_1 - \mu_2\|,$$

with $\rho := \text{diam}\mathscr{D}(\lambda_0) + 2\ell_1 \text{diam}\mathscr{V}$.

Proof For $(\varepsilon_1, \lambda_1, \mu_1), (\varepsilon_2, \lambda_2, \mu_2) \in (\varepsilon_0, +\infty) \times \mathscr{V} \times \mathscr{U}$, we set

$$r := 2\ell_1\ell_2\|\lambda_1 - \lambda_2\|, \quad s := \ell_2\|\mu_1 - \mu_2\| \text{ and } \theta := \varepsilon_2 - \varepsilon_0,$$

and consider two cases.

Case 1. If $r + s \le \theta$, then $\varepsilon_2 - r - s \ge \varepsilon_2 - \theta = \varepsilon_0$. Let $\bar{x} \in \widetilde{\text{Sol}}(\text{SEP})(\varepsilon_2 - r - s, \lambda_1, \mu_1)$ be arbitrary. Since $\bar{x} \in \mathscr{D}(\lambda_1)$, assumption (a) implies that there exists $x_{22} \in \mathscr{D}(\lambda_2)$ such that

$$\|\bar{x} - x_{22}\| \le \ell_1\|\lambda_1 - \lambda_2\|. \tag{6.76}$$

We claim that $x_{22} \in \widetilde{\text{Sol}}(\text{SEP})(\varepsilon_2, \lambda_2, \mu_2)$. For $y_2 \in \mathscr{D}(\lambda_2)$, there exists $y_1 \in \mathscr{D}(\lambda_1)$ satisfying

$$\|y_1 - y_2\| \le \ell_1\|\lambda_1 - \lambda_2\|. \tag{6.77}$$

It follows from $\bar{x} \in \widetilde{\text{Sol}}(\text{SEP})(\varepsilon_2 - r - s, \lambda_1, \mu_1)$ and $y_1 \in \mathscr{D}(\lambda_1)$ that

$$r_{-C}^{-e}(f(\bar{x}, y_1, \mu_1)) \le (\varepsilon_2 - r - s),$$

and so

$$r_{-C}^{-e}(f(x_{22}, y_2, \mu_2)) + r_{-C}^{-e}(f(\bar{x}, y_1, \mu_1)) - r_{-C}^{-e}(f(x_{22}, y_2, \mu_2)) \le (\varepsilon_2 - r - s). \tag{6.78}$$

Thanks to (c), one has:

$$f(\bar{x}, y_1, \mu_1) + \ell_2(\|\bar{x} - x_{22}\| + \|y_1 - y_2\| + \|\mu_1 - \mu_2\|)e \in f(x_{22}, y_2, \mu_2) + C.$$

It follows that

$$f(\bar{x}, y_1, \mu_1) - [f(x_{22}, y_2, \mu_2) - \ell_2(\|\bar{x} - x_{22}\| + \|y_1 - y_2\| + \|\mu_1 - \mu_2\|)e] \in C.$$

By Proposition 6.86(b) and (d), we obtain:

$$r_{-C}^{-e}(f(\bar{x}, y_1, \mu_1)) \le r_{-C}^{-e}(f(x_{22}, y_2, \mu_2)) + \ell_2(\|\bar{x} - x_{22}\| + \|y_1 - y_2\| + \|\mu_1 - \mu_2\|).$$

Consequently,

$$\begin{aligned} -r_{-C}^{-e}(f(x_{22}, y_2, \mu_2)) \le & -r_{-C}^{-e}(f(\bar{x}, y_1, \mu_1)) \\ & + \ell_2(\|\bar{x} - x_{22}\| + \|y_1 - y_2\| + \|\mu_1 - \mu_2\|) \end{aligned}$$

or equivalently,

$$\begin{aligned} r_{-C}^{-e}(f(\bar{x}, y_1, \mu_1)) \le & \, r_{-C}^{-e}(f(x_{22}, y_2, \mu_2)) \\ & + \ell_2(\|\bar{x} - x_{22}\| + \|y_1 - y_2\| + \|\mu_1 - \mu_2\|). \end{aligned}$$

Employing similar arguments, we also get:

$$\begin{aligned} r_{-C}^{-e}(f(x_{22}, y_2, \mu_2)) \le & \, r_{-C}^{-e}(f(\bar{x}, y_1, \mu_1)) \\ & + \ell_2(\|\bar{x} - x_{22}\| + \|y_1 - y_2\| + \|\mu_1 - \mu_2\|). \end{aligned}$$

Therefore,

$$|r_{-C}^{-e}(f(\bar{x}, y_1, \mu_1)) - r_{-C}^{-e}(f(x_{22}, y_2, \mu_2))| \le \ell_2(\|\bar{x} - x_{22}\| + \|y_1 - y_2\| + \|\mu_1 - \mu_2\|).$$

Combining this with (6.76) and (6.77), we arrive at

$$\begin{aligned} & |r_{-C}^{-e}(f(\bar{x}, y_1, \mu_1)) - r_{-C}^{-e}(f(x_{22}, y_2, \mu_2))| \\ & \le \ell_2(\|\bar{x} - x_{22}\| + \|y_1 - y_2\| + \|\mu_1 - \mu_2\|) \\ & \le \ell_1\ell_2\|\lambda_1 - \lambda_2\| + \ell_1\ell_2\|\lambda_1 - \lambda_2\| + \ell_2\|\mu_1 - \mu_2\| \\ & \le 2\ell_1\ell_2\|\lambda_1 - \lambda_2\| + \ell_2\|\mu_1 - \mu_2\| \\ & \le r + s. \end{aligned}$$

This together with (6.78) gives that $r_{-C}^{-e}(f(x_{22}, y_2, \mu_2)) \le \varepsilon_2$, and hence $x_{22} \in \widetilde{\text{Sol}}(\text{SEP})(\varepsilon_2, \lambda_2, \mu_2)$.

By Lemma 6.106, we have:

$$\widetilde{\text{Sol}}(\text{SEP})(\varepsilon_2, \lambda_1, \mu_1) \subset \widetilde{\text{Sol}}(\text{SEP})(\varepsilon_2 - r - s, \lambda_1, \mu_1) + \mathbb{B}\left[0, \frac{\rho}{\varepsilon_0}|r + s|\right]. \quad (6.79)$$

By (6.76),

$$\bar{x} \in x_{22} + \mathbb{B}\left[0, \ell_1\|\lambda_1 - \lambda_2\|\right].$$

Thus,

$$\widetilde{\mathrm{Sol}}(\mathrm{SEP})(\varepsilon_2 - r - s, \lambda_1, \mu_1) \subset \widetilde{\mathrm{Sol}}(\mathrm{SEP})(\varepsilon_2, \lambda_2, \mu_2) + \mathbb{B}\left[0, \ell_1\|\lambda_1 - \lambda_2\|\right]. \tag{6.80}$$

Combining (6.79) with (6.80), one has:

$$\begin{aligned}
&\widetilde{\mathrm{Sol}}(\mathrm{SEP})(\varepsilon_2, \lambda_1, \mu_1) \\
&\quad \subset \widetilde{\mathrm{Sol}}(\mathrm{SEP})(\varepsilon_2, \lambda_2, \mu_2) + \mathbb{B}\left[0, \ell_1\|\lambda_1 - \lambda_2\| + \frac{\rho}{\varepsilon_0}|r+s|\right] \\
&\quad \subset \widetilde{\mathrm{Sol}}(\mathrm{SEP})(\varepsilon_2, \lambda_2, \mu_2) + \\
&\quad\quad + \mathbb{B}\left[0, \ell_1\|\lambda_1 - \lambda_2\| + \frac{\rho}{\varepsilon_0}(2\ell_1\ell_2\|\lambda_1 - \lambda_2\| + \ell_2\|\mu_1 - \mu_2\|)\right] \\
&\quad \subset \widetilde{\mathrm{Sol}}(\mathrm{SEP})(\varepsilon_2, \lambda_2, \mu_2) + \\
&\quad\quad + \mathbb{B}\left[0, \left(\ell_1 + \frac{2\rho\ell_1\ell_2}{\varepsilon_0}\right)\|\lambda_1 - \lambda_2\| + \frac{\rho\ell_2}{\varepsilon_0}\|\mu_1 - \mu_2\|\right].
\end{aligned}$$

Similarly, we also obtain:

$$\begin{aligned}
\widetilde{\mathrm{Sol}}(\mathrm{SEP})(\varepsilon_2, \lambda_2, \mu_2) &\subset \widetilde{\mathrm{Sol}}(\mathrm{SEP})(\varepsilon_2, \lambda_1, \mu_1) \\
&\quad + \mathbb{B}\left[0, \left(\ell_1 + \frac{2\rho\ell_1\ell_2}{\varepsilon_0}\right)\|\lambda_1 - \lambda_2\| + \frac{\rho\ell_2}{\varepsilon_0}\|\mu_1 - \mu_2\|\right].
\end{aligned}$$

Case 2. If $r + s > \theta$, then there is a natural number n satisfying $\frac{1}{n} \le \min\{\frac{\theta}{2r}, \frac{\theta}{2s}\}$. Let $\mathbb{P}$ be a partition of segment $[\lambda_1, \lambda_2]$ with $n+1$ nodes $u_1, u_2, \ldots, u_{n+1}$ such that $u_1 = \lambda_1, u_{n+1} = \lambda_2$, and $\|u_i - u_{i+1}\| = \frac{\|\lambda_1 - \lambda_2\|}{n}$. Then,

$$\|u_i - u_{i+1}\| = \frac{\|\lambda_1 - \lambda_2\|}{n} \le \frac{\theta}{4\ell_1\ell_2},$$

that is,

$$2\ell_1\ell_2\|u_i - u_{i+1}\| \le \frac{\theta}{2}. \tag{6.81}$$

In addition, let $\mathbb{V}$ be a partition of segment $[\mu_1, \mu_2]$ with $n+1$ nodes $v_1, v_2, \ldots, v_{n+1}$, $v_1 = \mu_1, v_{n+1} = \mu_2$, and $\|v_i - v_{i+1}\| = \frac{\|\mu_1 - \mu_2\|}{n}$. It is clear that

$$\|v_i - v_{i+1}\| = \frac{\|\mu_1 - \mu_2\|}{n} \le \frac{\theta}{2\ell_2},$$

and hence

$$\ell_2\|v_i - v_{i+1}\| \le \frac{\theta}{2}. \tag{6.82}$$

From inequalities (6.81) and (6.82), one has:

$$2\ell_1\ell_2\|u_i - u_{i+1}\| + \ell_2\|v_i - v_{i+1}\| \le \theta.$$

Applying Case 1, we get:

$$\begin{aligned}
\widetilde{\mathrm{Sol}}(\mathrm{SEP})(\varepsilon_2, u_i, v_i) &\subset \widetilde{\mathrm{Sol}}(\mathrm{SEP})(\varepsilon_2, u_{i+1}, v_{i+1}) + \\
&+ \mathbb{B}\left[0, \left(\ell_1 + \frac{2\rho\ell_1\ell_2}{\varepsilon_0}\right)\|u_i - u_{i+1}\| + \frac{\rho\ell_2}{\varepsilon_0}\|v_i - v_{i+1}\|\right] \\
&\subset \widetilde{\mathrm{Sol}}(\mathrm{SEP})(\varepsilon_2, u_{i+1}, v_{i+1}) + \\
&+ \mathbb{B}\left[0, \left(\ell_1 + \frac{2\rho\ell_1\ell_2}{\varepsilon_0}\right)\frac{1}{n}\|\lambda_1 - \lambda_2\| + \frac{\rho\ell_2}{\varepsilon_0}\frac{1}{n}\|\mu_1 - \mu_2\|\right].
\end{aligned}$$

Consequently,

$$\begin{aligned}
&\mathrm{Sol}(\mathrm{SEP})(\varepsilon_2, \lambda_1, \mu_1) \subset \mathrm{Sol}(\mathrm{SEP})(\varepsilon_2, \lambda_2, \mu_2) \\
&+ \mathbb{B}\left[0, \left(\ell_1 + \frac{2\rho\ell_1\ell_2}{\varepsilon_0}\right)\|\lambda_1 - \lambda_2\| + \frac{\rho\ell_2}{\varepsilon_0}\|\mu_1 - \mu_2\|\right].
\end{aligned} \tag{6.83}$$

Similarly, we also have:

$$\begin{aligned}
&\widetilde{\mathrm{Sol}}(\mathrm{SEP})(\varepsilon_2, \lambda_2, \mu_2) \subset \widetilde{\mathrm{Sol}}(\mathrm{SEP})(\varepsilon_2, \lambda_1, \mu_1) \\
&+ \mathbb{B}\left[0, \left(\ell_1 + \frac{2\rho\ell_1\ell_2}{\varepsilon_0}\right)\|\lambda_1 - \lambda_2\| + \frac{\rho\ell_2}{\varepsilon_0}\|\mu_1 - \mu_2\|\right].
\end{aligned} \tag{6.84}$$

By (6.83), (6.84), and Lemma 6.106, we conclude that

$$\begin{aligned}
\widetilde{\mathrm{Sol}}(\mathrm{SEP})(\varepsilon_1, \lambda_1, \mu_1) &\subset \widetilde{\mathrm{Sol}}(\mathrm{SEP})(\varepsilon_2, \lambda_1, \mu_1) + \mathbb{B}\left[0, \frac{\rho}{\varepsilon_0}|\varepsilon_1 - \varepsilon_2|\right] \\
&\subset \widetilde{\mathrm{Sol}}(\mathrm{SEP})(\varepsilon_2, \lambda_2, \mu_2) + \mathbb{B}\left[0, \frac{\rho}{\varepsilon_0}|\varepsilon_1 - \varepsilon_2|\right] \\
&+ \mathbb{B}\left[0, \left(\ell_1 + \frac{2\rho\ell_1\ell_2}{\varepsilon_0}\right)\|\lambda_1 - \lambda_2\| + \frac{\rho\ell_2}{\varepsilon_0}\|\mu_1 - \mu_2\|\right].
\end{aligned}$$

This results in

$$\widetilde{\text{Sol}}(\text{SEP})(\varepsilon_1, \lambda_1, \mu_1) \subset \widetilde{\text{Sol}}(\text{SEP})(\varepsilon_2, \lambda_2, \mu_2) + \mathbb{B}\left[0, \frac{\rho}{\varepsilon_0}|\varepsilon_1 - \varepsilon_2| + \left(\ell_1 + \frac{2\rho\ell_1\ell_2}{\varepsilon_0}\right)\|\lambda_1 - \lambda_2\| + \frac{\rho\ell_2}{\varepsilon_0}\|\mu_1 - \mu_2\|\right].$$

In the same way, one has:

$$\widetilde{\text{Sol}}(\text{SEP})(\varepsilon_2, \lambda_2, \mu_2) \subset \widetilde{\text{Sol}}(\text{SEP})(\varepsilon_1, \lambda_1, \mu_1) + \mathbb{B}\left[0, \frac{\rho}{\varepsilon_0}|\varepsilon_1 - \varepsilon_2| + \left(\ell_1 + \frac{2\rho\ell_1\ell_2}{\varepsilon_0}\right)\|\lambda_1 - \lambda_2\| + \frac{\rho\ell_2}{\varepsilon_0}\|\mu_1 - \mu_2\|\right].$$

The proof is complete. □

To illustrate the applicability of Theorem 6.107, we have the following example.

Example 6.108 (Nonlinear scalarizations for strong equilibrium problems in Euclidean spaces) Let $X = W = \mathbb{R}$, $A = \mathbb{R}$, $Y = \mathbb{R}^2$, $C = \mathbb{R}^2_+$, $e = (1, 1) \in \operatorname{int}C$, $\Lambda = [-\frac{2}{5}, \frac{1}{2}]$, $M = [-1, 1]$, $\mathscr{D}(\lambda) = [-\frac{1}{2}, \lambda]$, and

$$f(x, y, \mu) := \begin{cases} (x^2 + |y| + |\mu|, |y| + |\mu| + 2) & x < 0, \\ (-x + |y| + |\mu|, |y| + |\mu| + 2) & x \geq 0. \end{cases} \tag{6.85}$$

Let $\varepsilon_0 = \frac{1}{3}$, $(\lambda_0, \mu_0) = (0, 0)$, and $\mathscr{V} = [-\frac{1}{6}, \frac{1}{6}]$. It is clear that assumption (a) of Theorem 6.107 holds true with $\ell_1 = 1$. We now claim that f is $-\mathbb{R}_+$-concave. Indeed, let $x_1, x_2, y \in [-\frac{1}{2}, \lambda]$, $t \in [0, 1]$, $p \in -\mathbb{R}_+$, and $\mu \in [-1, 1]$ be such that

$$f(x_1, y, \mu) \in \mathbb{R}^2_+ \tag{6.86}$$

and

$$f(x_2, y, \mu) \in (p, p) + \mathbb{R}^2_+. \tag{6.87}$$

We need to show that

$$f(x_t, y, \mu) \in (tp, tp) + \mathbb{R}^2_+, \tag{6.88}$$

where $x_t := tx_2 + (1 - t)x_1$. We consider four cases.

Case 1. If $x_1 < 0$ and $x_2 < 0$, then $x_t < 0$, and hence $x_t^2 + |y| + |\mu| - tp \geq 0$, i.e., (6.88) holds true.

Case 2. If $x_1 \geq 0$ and $x_2 \geq 0$, then $x_t \geq 0$. Combining (6.85), (6.86), and (6.87), we obtain $-x_1 + |y| + |\mu| \geq 0$ and $-x_2 + |y| + |\mu| \geq p$, and hence

$$\begin{aligned}-x_t+|y|+|\mu|-tp&=-tx_2-(1-t)x_1+|y|+|\mu|-tp\\&=t(-x_2+|y|+|\mu|-p)+(1-t)(-x_1+|y|+|\mu|)\\&\geq 0,\end{aligned}$$

which leads to (6.88).

Case 3. If $x_1 < 0 \leq x_2$, then either $x_t < 0$ or $x_t \geq 0$. It follows from (6.85) and (6.87) that $-x_2+|y|+|\mu| \geq p$. If $x_t < 0$, similar to Case 1, (6.88) follows. In the case of $x_t \geq 0$, one has:

$$\begin{aligned}-x_t+|y|+|\mu|-tp&=-tx_2-(1-t)x_1+|y|+|\mu|-tp\\&=t(-x_2+|y|+|\mu|-p)+(1-t)(|y|+|\mu|)\\&\quad-(1-t)x_1\geq 0,\end{aligned}$$

and so we obtain (6.88).

Case 4. If $x_2 < 0 \leq x_1$, we only present the case $x_t \geq 0$ since the case $x_t < 0$ is similar to Case 1. Whenever $x_t \geq 0$, by the definition of F, we have

$$-x_t+|y|+|\mu|-tp \geq -x_1+|y|+|\mu|-tp \geq 0,$$

which guarantees that (6.88) is true.

Moving on to the Lipschitz continuity of f, for any $(x_1, y_1, \mu_1), (x_2, y_2, \mu_2) \in [-\frac{1}{2}, \frac{1}{6}] \times [-\frac{1}{2}, \frac{1}{6}] \times [-1, 1]$, we have the following four cases.

Case 1. If $x_1 < 0$ and $x_2 < 0$, we have

$$\begin{aligned}|x_1^2-x_2^2+|y_1|-|y_2|+|\mu_1|-|\mu_2||&\leq |x_1^2-x_2^2|+||y_1|-|y_2||+||\mu_1|-|\mu_2||\\&\leq |x_1+x_2||x_1-x_2|+|y_1-y_2|+|\mu_1-\mu_2|\\&\leq |x_1-x_2|+|y_1-y_2|+|\mu_1-\mu_2|\\&\leq \|(x_1,y_1,\mu_1)-(x_2,y_2,\mu_2)\|.\end{aligned}$$

Case 2. If $x_1 \geq 0$ and $x_2 \geq 0$, then

$$\begin{aligned}|-x_1+x_2+|y_1|-|y_2|+|\mu_1|-|\mu_2||&\leq |x_1-x_2|+||y_1|-|y_2||+||\mu_1|-|\mu_2||\\&\leq |x_1-x_2|+|y_1-y_2|+|\mu_1-\mu_2|\\&\leq |x_1-x_2|+|y_1-y_2|+|\mu_1-\mu_2|\\&\leq \|(x_1,y_1,\mu_1)-(x_2,y_2,\mu_2)\|.\end{aligned}$$

Case 3. If $x_1 < 0 \leq x_2$, we obtain:

$$|x_1^2+x_2+|y_1|-|y_2|+|\mu_1|-|\mu_2|| \leq |x_1^2+x_2|+||y_1|-|y_2||+||\mu_1|-|\mu_2||$$

$$\leq |-x_1+x_2|+|y_1-y_2|+|\mu_1-\mu_2|$$
$$\leq \|(x_1,y_1,\mu_1)-(x_2,y_2,\mu_2)\|.$$

Case 4. If $x_2 < 0 \leq x_1$, then

$$\begin{aligned}||x_2^2+x_1+|y_1|-|y_2|+|\mu_1|-|\mu_2|| &\leq |x_2^2+x_1|+||y_1|-|y_2||+||\mu_1|-|\mu_2|| \\ &\leq |-x_2+x_1|+|y_1-y_2|+|\mu_1-\mu_2| \\ &\leq \|(x_1,y_1,\mu_1)-(x_2,y_2,\mu_2)\|.\end{aligned}$$

Therefore, f satisfies the Lipschitz condition with $\ell_2 = 1$. Then, Theorem 6.107 says that the map $\widetilde{\text{Sol}}(\text{SEP})$ is Lipschitz continuous for all $(\varepsilon,\lambda,\mu) \in (\varepsilon_0,+\infty)\times \Lambda \times M$. More precisely, by direct computations, we have:

$$\begin{aligned}&\widetilde{\text{Sol}}(\text{SEP})(\varepsilon,\lambda,\mu)\\ &=\left\{x\in\left[-\frac{1}{2},\lambda\right] \middle| \begin{cases}x^2+|y|+|\mu|+\varepsilon\geq 0, \ x<0\\ -x+|y|+|\mu|+\varepsilon\geq 0, \ x\geq 0,\end{cases} \quad \forall y\in\left[-\frac{1}{2}\lambda\right]\right\}\\ &=\left\{x\in\left[-\frac{1}{2},\lambda\right] \middle| \begin{cases}x^2+|\mu|+\varepsilon\geq 0\, x<0\\ -x+|\mu|+\varepsilon\geq 0\, x\geq 0\end{cases}\right\}\\ &=\left[-\frac{1}{2},\min\{\lambda,\varepsilon+|\mu|\}\right]\\ &=\left[-\frac{1}{2},\frac{\lambda+\varepsilon+|\mu|}{2}-\frac{|\lambda-\varepsilon-|\mu||}{2}\right].\end{aligned}$$

Then, for all $(\varepsilon_1,\lambda_1,\mu_1),(\varepsilon_2,\lambda_2,\mu_2)\in(\varepsilon_0,+\infty)\times[-\frac{1}{6},\frac{1}{6}]\times[-1,1]$, one has:

$$\begin{aligned}d &:= \left|\frac{\lambda_1+\varepsilon_1+|\mu_1|}{2}-\frac{\lambda_2+\varepsilon_2+|\mu_2|}{2}+\frac{|\lambda_2-\varepsilon_2-|\mu_2||}{2}-\frac{|\lambda_1-\varepsilon_1-|\mu_1||}{2}\right|\\ &\leq \left|\frac{\lambda_1+\varepsilon_1+|\mu_1|}{2}-\frac{\lambda_2+\varepsilon_2+|\mu_2|}{2}\right|+\left|\frac{|\lambda_2-\varepsilon_2-|\mu_2||}{2}-\frac{|\lambda_1-\varepsilon_1-|\mu_1||}{2}\right|\\ &\leq \frac{1}{2}|\varepsilon_1-\varepsilon_2+\lambda_1-\lambda_2+|\mu_1|-|\mu_2||+\frac{1}{2}|\lambda_2-\varepsilon_2-|\mu_2|-(\lambda_1-\varepsilon_1-|\mu_1|)|\\ &\leq \frac{1}{2}(|\varepsilon_1-\varepsilon_2|+|\lambda_1-\lambda_2|+|\mu_1-\mu_2|)+\frac{1}{2}(|\varepsilon_1-\varepsilon_2|+|\lambda_1-\lambda_2|+|\mu_1-\mu_2|)\\ &\leq |\varepsilon_1-\varepsilon_2|+|\lambda_1-\lambda_2|+|\mu_1-\mu_2|.\end{aligned}$$

However, in Example 6.108, f is not C-concave in the first component on $\mathscr{D}(\lambda)$. To see this, take $(x_1,y_1,\mu_1)=\left(-\frac{1}{2},0,1\right)$, $(x_2,y_2,\mu_2)=(0,0,1)$, and $t=\frac{1}{2}$. Then,

$$f\left(-\frac{1}{4},0,1\right)=\left[\frac{17}{16},3\right]\notin\left[\frac{9}{8},+\infty\right)=\frac{1}{2}f\left(-\frac{1}{2},0,1\right)+\frac{1}{2}f\,(0,0,1)+\mathbb{R}^2_+.$$

Now we investigate a special case of (SEP) where $\mathscr{D}(\lambda)\equiv\mathscr{D}$ is a nonempty convex subset of A with diam$\mathscr{D}\le\rho$. This case plays an important role in studying optimization-related problems with equilibrium constraints. In this special case, not only the Lipschitz continuity of $\mathscr{D}$ but also the Lipschitz continuity of f in the first and second components can be lacked. Therefore, the following result cannot be derived from Theorem 6.107.

Theorem 6.109 (Lipschitz conditions for parametric strong equilibrium problems with fixed constraint sets via a nonlinear scalarization function) *Assume that* $\varepsilon_0>0$ *and*

(a) f *is* $-\mathbb{R}_+$*-concave in the first component on* $\mathscr{D}$;
(b) f *is* C-ℓ*-Lipschitz continuous in the third component at* $\mu_0\in M$.

Then, there exists a neighborhood $\mathscr{U}$ *of* μ_0 *such that* $\widetilde{\text{Sol}}$(SEP) *satisfies the following Lipschitz condition :*

$$\widetilde{\text{Sol}}(\text{SEP})(\varepsilon_1,\mu_1)\subset\widetilde{\text{Sol}}(\text{SEP})(\varepsilon_2,\mu_2)+\mathbb{B}\left[0,\frac{\rho}{\varepsilon_0}|\varepsilon_1-\varepsilon_2|+\frac{\rho\ell}{\varepsilon_0}\|\mu_1-\mu_2\|\right]$$

for all $(\varepsilon_1,\mu_1),(\varepsilon_2,\mu_2)\in(\varepsilon_0,+\infty)\times\mathscr{U}$.

Proof For $(\varepsilon_1,\mu_1),(\varepsilon_2,\mu_2)\in(\varepsilon_0,+\infty)\times\mathscr{U}$, by Lemma 6.106, we have:

$$\widetilde{\text{Sol}}(\text{SEP})(\varepsilon_1,\mu_1)\subset\widetilde{\text{Sol}}(\text{SEP})(\varepsilon_2,\mu_1)+\mathbb{B}\left[0,\frac{\rho}{\varepsilon_0}|\varepsilon_1-\varepsilon_2|\right],\tag{6.89}$$

$$\widetilde{\text{Sol}}(\text{SEP})(\varepsilon_2,\mu_1)\subset\widetilde{\text{Sol}}(\text{SEP})(\varepsilon_1,\mu_1)+\mathbb{B}\left[0,\frac{\rho}{\varepsilon_0}|\varepsilon_1-\varepsilon_2|\right].\tag{6.90}$$

We set $r:=\ell\|\mu_1-\mu_2\|$ and discuss two cases.
Case 1. If $\|\mu_1-\mu_2\|\le\frac{\varepsilon_2}{\ell}$, then $0<r\le\varepsilon_2$. Let $\bar{x}\in\text{Sol(SEP)}(\varepsilon_2-r,\mu_1)$ be arbitrary and $y\in\mathscr{D}$. We have:

$$r^{-e}_{-C}(f(\bar{x},y,\mu_1))\le(\varepsilon_2-r),$$

and so

$$r^{-e}_{-C}(f(\bar{x},y,\mu_2))+r^{-e}_{-C}(f(\bar{x},y_1,\mu_1))-r^{-e}_{-C}(f(\bar{x},y,\mu_2))\le(\varepsilon_2-r).\tag{6.91}$$

It follows from assumption (b) that

$$f(\bar{x},y,\mu_1)+\ell\|\mu_1-\mu_2\|e\in f(\bar{x},y,\mu_2)+C,$$

or equivalently,

$$f(\bar{x}, y, \mu_1) - (f(\bar{x}, y, \mu_2) - \ell\|\mu_1 - \mu_2\|e) \in C.$$

By Proposition 6.86(b) and (d), one has:

$$r_{-C}^{-e}(f(\bar{x}, y, \mu_1)) \leq r_{-C}^{-e}(f(\bar{x}, y, \mu_2)) + \ell\|\mu_1 - \mu_2\|.$$

Analogously, we also get:

$$r_{-C}^{-e}(f(\bar{x}, y, \mu_2)) \leq r_{-C}^{-e}(f(\bar{x}, y, \mu_1)) + \ell\|\mu_1 - \mu_2\|.$$

Therefore,

$$|r_{-C}^{-e}(f(\bar{x}, y, \mu_1)) - r_{-C}^{-e}(f(\bar{x}, y, \mu_2))| \leq \ell\|\mu_1 - \mu_2\| = r. \tag{6.92}$$

From (6.92) and (6.91), we obtain:

$$r_{-C}^{-e}(f(\bar{x}, y, \mu_2)) \leq \varepsilon_2,$$

and hence $\bar{x} \in \widetilde{\text{Sol}}(\text{SEP})(\varepsilon_2, \mu_2)$, which yields that $\text{Sol}(\text{SEP})(\varepsilon_2 - r, \mu_1) \subset \widetilde{\text{Sol}}(\text{SEP})(\varepsilon_2, \mu_2)$.

By Lemma 6.106, one has:

$$\widetilde{\text{Sol}}(\text{SEP})(\varepsilon_2, \mu_1) \subset \widetilde{\text{Sol}}(\text{SEP})(\varepsilon_2 - r, \mu_1) + \mathbb{B}\left[0, \frac{\rho}{\varepsilon_0} r\right].$$

Thus,

$$\widetilde{\text{Sol}}(\text{SEP})(\varepsilon_2, \mu_1) \subset \widetilde{\text{Sol}}(\text{SEP})(\varepsilon_2, \mu_2) + \mathbb{B}\left[0, \frac{\rho}{\varepsilon_0}\ell\|\mu_1 - \mu_2\|\right].$$

Utilizing similar arguments, we also have:

$$\widetilde{\text{Sol}}(\text{SEP})(\varepsilon_2, \mu_2) \subset \widetilde{\text{Sol}}(\text{SEP})(\varepsilon_2, \mu_1) + \mathbb{B}\left[0, \frac{\rho}{\varepsilon_0}\ell\|\mu_1 - \mu_2\|\right].$$

Case 2. $\|\mu_1 - \mu_2\| > \frac{\varepsilon_2}{\ell}$. Let $\delta = \frac{\varepsilon_0}{\ell}$ and $\gamma = \text{diam}\,\mathscr{U} < +\infty$, then there exists a natural number n satisfying $\frac{\gamma}{n} \leq \delta$. Let $\mathbb{V}$ be a partition of segment $[\mu_1, \mu_2]$ with $n+1$ nodes $v_1, v_2, \ldots, v_{n+1}$ with $v_1 = \mu_1$, $v_{n+1} = \mu_2$, and $\|v_i - v_{i+1}\| = \frac{\|\mu_1 - \mu_2\|}{n}$. It is easy to see that

$$\|v_i - v_{i+1}\| = \frac{\|\mu_1 - \mu_2\|}{n} \leq \delta,$$

and hence

$$\|v_i - v_{i+1}\| \leq \frac{\varepsilon_0}{\ell} \leq \frac{\varepsilon_2}{\ell}.$$

Using the result of Case 1, we obtain:

$$\begin{aligned}\widetilde{\mathrm{Sol}}(\mathrm{SEP})(\varepsilon_2, v_i) &\subset \widetilde{\mathrm{Sol}}(\mathrm{SEP})(\varepsilon_2, v_{i+1}) + \mathbb{B}\left[0, \frac{\rho\ell}{\varepsilon_0}\|v_i - v_{i+1}\|\right] \\ &\subset \widetilde{\mathrm{Sol}}(\mathrm{SEP})(\varepsilon_2, v_{i+1}) + \mathbb{B}\left[0, \frac{\rho\ell}{\varepsilon_0}\frac{1}{n}\|\mu_1 - \mu_2\|\right].\end{aligned}$$

Therefore,

$$\widetilde{\mathrm{Sol}}(\mathrm{SEP})(\varepsilon_2, \mu_1) \subset \widetilde{\mathrm{Sol}}(\mathrm{SEP})(\varepsilon_2, \mu_2) + \mathbb{B}\left[0, \frac{\rho}{\varepsilon_0}\ell\|\mu_1 - \mu_2\|\right]. \tag{6.93}$$

Similarly, we arrive at

$$\widetilde{\mathrm{Sol}}(\mathrm{SEP})(\varepsilon_2, \mu_2) \subset \widetilde{\mathrm{Sol}}(\mathrm{SEP})(\varepsilon_2, \mu_1) + \mathbb{B}\left[0, \frac{\rho}{\varepsilon_0}\ell\|\mu_1 - \mu_2\|\right]. \tag{6.94}$$

By (6.89) and (6.93), we conclude that

$$\begin{aligned}&\widetilde{\mathrm{Sol}}(\mathrm{SEP})(\varepsilon_1, \mu_1) \\ &\quad\subset \widetilde{\mathrm{Sol}}(\mathrm{SEP})(\varepsilon_2, \mu_1) + \mathbb{B}\left[0, \frac{\rho}{\varepsilon_0}|\varepsilon_1 - \varepsilon_2|\right] \\ &\quad\subset \widetilde{\mathrm{Sol}}(\mathrm{SEP})(\varepsilon_2, \mu_2) + \mathbb{B}\left[0, \frac{\rho}{\varepsilon_0}|\varepsilon_1 - \varepsilon_2|\right] + \mathbb{B}\left[0, \frac{\rho\ell}{\varepsilon_0}\|\mu_1 - \mu_2\|\right].\end{aligned}$$

Hence,

$$\widetilde{\mathrm{Sol}}(\mathrm{SEP})(\varepsilon_1, \mu_1) \subset \widetilde{\mathrm{Sol}}(\mathrm{SEP})(\varepsilon_2, \mu_2) + \mathbb{B}\left[0, \frac{\rho}{\varepsilon_0}|\varepsilon_1 - \varepsilon_2| + \frac{\rho\ell}{\varepsilon_0}\|\mu_1 - \mu_2\|\right].$$

In the same way, combining (6.90) with (6.94), we get:

$$\widetilde{\mathrm{Sol}}(\mathrm{SEP})(\varepsilon_2, \mu_2) \subset \widetilde{\mathrm{Sol}}(\mathrm{SEP})(\varepsilon_1, \mu_1) + \mathbb{B}\left[0, \frac{\rho}{\varepsilon_0}|\varepsilon_1 - \varepsilon_2| + \frac{\rho\ell}{\varepsilon_0}\|\mu_1 - \mu_2\|\right].$$

The proof is complete. □

In the rest of this section, we present a nonlinear function which can be used to scalarize problem (SEP). We will start by recalling the oriented distance function

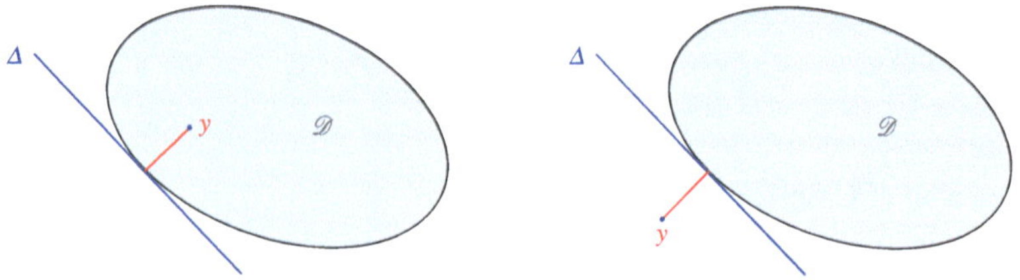

(a) Illustrations of the computation for $\delta_{\mathscr{D}}(y) = -d(y,\Delta)$ (b) Illustrations of the computation for $\delta_{\mathscr{D}}(y) = d(y,\Delta)$

Fig. 6.2 Illustrations of computations for the oriented distance function $\delta_{\mathscr{D}}$

introduced by Hiriart-Urruty and list of its main properties, which play an important role in our analysis.

Definition 6.110 (The oriented distance function) For $\mathscr{D} \subset Y$, the oriented distance function $\delta_{\mathscr{D}} : Y \to \mathbb{R}$ is defined by

$$\delta_{\mathscr{D}}(y) := d(y, \mathscr{D}) - d(y, Y \setminus \mathscr{D}) = \begin{cases} d(y, \mathscr{D}) & \text{if } y \in Y \setminus \mathscr{D}, \\ -d(y, Y \setminus \mathscr{D}) & \text{if } y \in \mathscr{D}, \end{cases}$$

with $d(y, \emptyset) = +\infty$.

In Fig. 6.2, we illustrate the computation of the oriented distance function $\delta_{\mathscr{D}}$ for two cases: when the point y belongs to the set D and when the point y does not belong to the set D.

Let us derive some examples of how function δ_C looks for different choices of the space Y, its norm, and the ordering cone C.

Example 6.111 (Calculating the oriented distance function)

(a) Let $Y = \mathbb{R}^n$ with the Euclidean norm $\|\cdot\|_2$ and $C = \mathbb{R}^n_+$. Then, it holds that

$$d_{-C}(y) = \|y^+\|,$$

where $y_i^+ = \max\{y_i, 0\}$, $i = 1, \ldots, n$, and

$$d_{Y\setminus -C}(y) = \begin{cases} 0 & \text{if } \; y_i \geq 0 \text{ for some } i, \\ -\max y_i & \text{if } \; y_i < 0 \text{ for all } i. \end{cases}$$

Consequently,

$$\delta_{-C}(y) = \begin{cases} \|y^+\| & \text{if } \; y \notin -C, \\ \max y_i & \text{if } \; y \in -C. \end{cases}$$

(b) Let $Y = \mathbb{R}^n$ with the norm $\|y\|_\infty = \max |y_i|$ and $C = \mathbb{R}^n_+$. We have:

$$d_{-C}(y) = \begin{cases} \max y_i & \text{if } y \notin -C, \\ 0, & \text{if } y \in -C, \end{cases}$$

and

$$d_{Y\setminus -C}(y) = \begin{cases} 0 & \text{if } y \notin -C, \\ -\max y_i, & \text{if } y \in -C. \end{cases}$$

This implies that for all $y \in Y$,

$$\delta_{-C}(y) = \max y_i.$$

(c) Let T be a compact subset of a metric space, Y the space of all continuous functions from T into $\mathbb{R}$ with the supremum norm, and the ordering cone C consist of all nonnegative functions: $C = \{y \in Y \mid y(t) \geq 0 \ \ \forall t \in T\}$. Let $y \in Y$ be arbitrary. If there is some $t \in T$ satisfying $y(t) \geq 0$, then

$$\delta_{-C}(y) = d_{-C}(y) = \max_{t \in T} y(t),$$

and so

$$\delta_{-C}(y) = -d_{Y\setminus -C}(y) = \max_{t \in T} y(t).$$

Thus, it holds that for all $y \in Y$,

$$\delta_{-C}(y) = \max_{t \in T} y(t).$$

(d) Let I be an interval of $\mathbb{R}$ and $Y = L^{\infty}(I)$ the space of all essentially bounded functions on I with the usual norm, and the ordering cone C consist of the nonnegative functions. It holds that

$$\delta_{-C}(y) = \sup_{\omega \in I} y(\omega),$$

where sup means essential supremum.

(e) Consider the spaces L^p, $1 \leq p < \infty$, with the usual norm $\|\cdot\|_p$. The nonnegative orthant C has an empty interior. Consequently,

$$\delta_{-C}(y) = d_{-C}(y) = \|y^+\|,$$

where $y^+ = \sup\{y, 0\}$ has the usual meaning of the lattice theory.

Now we collect important properties of the Hiriart-Urruty oriented distance function in the following statement.

Lemma 6.112 (Characterizations of a set via the oriented distance function) *For a subset $\mathscr{D} \subset Y$, $\mathrm{int}\mathscr{D}$ and $\mathrm{bd}\mathscr{D}$ stand for the interior and the boundary of $\mathscr{D}$, resp. Then, the following assertions hold:*

(a) $\delta_{\mathscr{D}}$ *is 1-Lipschitz continuous on Y.*
(b) $\mathrm{int}\mathscr{D} = \{y \in Y \mid \delta_{\mathscr{D}}(y) < 0\}$.
(c) $\mathrm{bd}\mathscr{D} = \{y \in Y \mid \delta_{\mathscr{D}}(y) = 0\}$.
(d) $Y \setminus \mathrm{cl}\mathscr{D} = \{y \in Y \mid \delta_{\mathscr{D}}(y) > 0\}$.
(e) *If $\mathscr{D}$ is convex, then $\delta_{\mathscr{D}}$ is convex on a convex subset of Y.*
(f) *If $\mathscr{D}$ is a closed convex cone, then $\delta_{-\mathscr{D}}$ is decreasing wrt the ordering relation induced by $\mathscr{D}$ on Y, that is, for every $y_1, y_2 \in Y$,*

$$y_2 - y_1 \in \mathscr{D} \Rightarrow \delta_{-\mathscr{D}}(y_2) \leq \delta_{-\mathscr{D}}(y_1).$$

Moreover, if $\mathrm{int}\mathscr{D} \neq \emptyset$, *then*

$$y_2 - y_1 \in \mathrm{int}\mathscr{D} \Rightarrow \delta_{-\mathscr{D}}(y_2) < \delta_{-\mathscr{D}}(y_1).$$

(g) $\delta_{Y\setminus\mathscr{D}}(y) = -\delta_{\mathscr{D}}(y)$ *for all $y \in Y$.*
(h) *If $\mathscr{D}$ is a cone, then $\delta_{\mathscr{D}}$ is positively homogeneous, i.e., for all $\ell > 0$ and $y \in Y$,* $\delta_{\mathscr{D}}(\ell y) = \ell\delta_{\mathscr{D}}(y)$.
(i) $\delta_{-\mathscr{D}}(u + v) \leq \delta_{-\mathbb{D}}(u) + \delta_{-\mathscr{D}}(v)$ *for all $u, v \in Y$.*
(k) $\delta_{\mathscr{D}}(-y) = \delta_{-\mathscr{D}}(y)$ *for all $y \in Y$.*

Lemma 6.113 (Estimations of the oriented distance function) *Assume that Y is a reflexive Banach space and r is a nonnegative real number. Then,*

$$[y \in rB_Y + C] \Leftrightarrow [\delta_{Y\setminus C}(y) + r \geq 0].$$

Proof If $y \in rB_Y + C$, then there are two cases to consider.

Case 1. If $y \in C$, then $\delta_{Y\setminus C}(y) = d(y, Y \setminus C) \geq 0$, and so $\delta_{Y\setminus C}(y) + r \geq 0$.
Case 2. If $y \notin C$, then $\delta_Y(y) = -d(y, C) \geq -r$ as $y \in rB_Y + C$, and hence $\delta_{Y\setminus C}(y) + r \geq 0$.

Conversely, thanks to Lemma 6.112(b)–(d), if $\delta_{Y\setminus C}(y) \geq 0$, then $y \in C \subset rB_Y + C$ for all $r \in \mathbb{R}_+$. If $-r \leq \delta_{Y\setminus C}(y) < 0$, then Lemma 6.112(d) implies that $0 < \delta_C(y) \leq r$, and so

$$\delta_C(y) = d(y, C) = \operatorname*{int}_{c\in C} \|y - c\| \leq r.$$

Because Y is a reflexive Banach space and C is a pointed closed convex cone in Y, there exists $c' \in C$ such that $\|y - c'\| \leq r$, and hence $y \in rB_Y + C$. □

Employing Lemma 6.113, we are able to present the main result of this subsection.

Theorem 6.114 (Scalarizations for strong equilibrium problems via the oriented distance function) *An element $\bar{x}$ is a solution of* (SEP)(λ, μ) *if and only if it it solves the following problem:*

(EP)(λ, μ) *find $x \in \mathscr{D}(\lambda)$ such that $\delta_{Y \setminus C}(f(x, y, \mu)) \geq 0$ for all $y \in \mathscr{D}(\lambda)$.*

6.6 Tikhonov Regularization for Vector Equilibrium Problems

Let (X, τ) and (Y, σ) be Hausdorff topological linear spaces, $C \subset Y$ be a convex proper cone with a nonempty interior and $\langle \cdot, \cdot \rangle$ be the canonical pair between Y and its dual space Y^*. We use $\mathscr{L}(X, Y)$ to stand for the set of continuous mappings from X into Y. We denote $\| \cdot \|$ the norm in all Banach spaces since the context will make it clear which space is considered. Let $C^* := \{\varphi \in Y^* \colon \langle \varphi, y \rangle \geq 0 \text{ for all } y \in C\}$ be the positive polar cone of C. It is known that C can take the form $C = \{y \in Y : \langle \varphi, y \rangle \geq 0 \text{ for all } \varphi \in C^*\}$ whenever Y is locally convex and C is closed. Moreover, $\operatorname{int} C = \{y \in Y \colon \langle \varphi, y \rangle > 0 \text{ for all } \varphi \in C^* \setminus \{0\}\}$.

For nonempty subsets A and B of a real topological linear space, the *algebraic sum* and *algebraic difference* of A and B are defined respectively as

$$A + B := \{a + b : a \in A \text{ and } b \in B\}, \text{ and } A - B := \{a - b : a \in A \text{ and } b \in B\}.$$

For every $t \in \mathbb{R}$, we define $tA := \{ta : a \in A\}$. In particular, $-A := \{-a : a \in A\}$. Let $\mathbb{R}^p_+$ stand for the nonnegative orthant of $\mathbb{R}^p$, and $\mathbb{R}_+ := [0, +\infty[$.

We leave the proof of the following lemma as an exercise.

Lemma 6.115 (Continuity of the scalar multiplication) *If y is an element in a topological linear space (Y, σ), α is a scalar, and V is a neighborhood of αy, then there exist a positive real number r and a neighborhood V_1 of y such that $\beta V_1 \subset V$ whenever $|\beta - \alpha| < r$.*

We propose the following result, which plays a crucial role in the further discussion.

Proposition 6.116 (Continuity of the function multiplication) *Let p be a mapping from (X, τ) into (Y, σ), (Z, δ) be a Hausdorff topological linear space, and $\xi : (Z, \delta) \to \mathbb{R}_+$ be a continuous function. Let π be the product topology on $Z \times X$. The mapping $\xi p : (Z \times X, \pi) \to (Y, \sigma)$, defined by $(z, x) \mapsto \xi(z)p(x)$, is C-upper semicontinuous (C-lower semicontinuous) on $Z \times X$ whenever p is C-upper semicontinuous (C-lower semicontinuous, respectively) on X.*

Proof For a given $(\bar{z}, \bar{x}) \in Z \times X$, let V be an arbitrarily open neighborhood of $\xi(\bar{z})p(\bar{x})$. Since Y is a topological linear space, according to Lemma 6.115, there exist $r > 0$ and a neighborhood V_1 of $p(\bar{x})$ such that

$$\beta V_1 \subset V \text{ whenever } |\beta - \xi(\bar{z})| < r. \tag{6.95}$$

It follows from the continuity of ξ at $\bar{z}$ that there is an open neighborhood Δ of $\bar{z}$ satisfying

$$|\xi(z) - \xi(\bar{z})| < \frac{r}{2} \quad \forall z \in \Delta.$$

Combining this with (6.95), we arrive at $\xi(z)V_1 \subset V$ for all $z \in \Delta$. Because p is C-upper semicontinuous at $\bar{x}$, for the such V_1, there exists an open neighborhood U of $\bar{x}$ such that

$$p(x) \in V_1 - C \quad \forall x \in U.$$

Consequently,

$$\xi(z)p(x) \in \xi(z)V_1 - C \subset V - C \quad \forall (z, x) \in \Delta \times U.$$

Therefore, ξp is C-upper semicontinuous at $(\bar{z}, \bar{x})$. The C-lower semicontinuity of ξp can be shown analogously. □

The following materials will be frequently employed in the remainder of this paper.

Lemma 6.117 (Generalized Archimedean axiom and generalized sign preserving) *Let C be a convex cone with a nonempty interior in a real topological linear space Y:*

(a) (Generalized Archimedean axiom) *For any $x \in \operatorname{int} C$ and $y \in Y$, there exists $n_0 \in \mathbb{N}$ such that $y - nx \in -\operatorname{int} C$ for all $n \geq n_0$.*
(b) (Generalized sign preserving) *If $a \in C$ and $b \notin -\operatorname{int} C$, then $a + b \notin -\operatorname{int} C$.*

Proof

(i) We first show that

$$Y = \bigcup \{tx - \operatorname{int} C : t \in \mathbb{R}_+ \setminus \{0\}\} \tag{6.96}$$

for each $x \in \operatorname{int} C$. Let $U := x - \operatorname{int} C$. Obviously, U is an open neighborhood of the origin in Y. Hence, $Y = \bigcup \{tU : t \in \mathbb{R}_+ \setminus \{0\}\}$. Because C is a cone, one has $tU \subset tx - \operatorname{int} C$ for each $t \in \mathbb{R}_+ \setminus \{0\}$. So,

$$Y = \bigcup \{tU : t \in \mathbb{R}_+ \setminus \{0\}\} \subset \bigcup \{tx - \operatorname{int} C : t \in \mathbb{R}_+ \setminus \{0\}\} \subset Y.$$

Therefore, (6.96) holds true. Now, for any $y \in Y$, it yields from (6.96) that there exists $t \in \mathbb{R}_+ \setminus \{0\}$ satisfying $y \in tx - \operatorname{int} C$. Thus, there is $c \in \operatorname{int} C$ such that $y = tx - c$. By choosing $n \geq \lceil t \rceil + 1$, where $\lceil t \rceil$ is the ceiling function giving the smallest integer greater than or equal to t, we get $y = nx - (n - t)x - c \in nx - \operatorname{int} C$. As a result, $y - nx \in -\operatorname{int} C$.

(ii) Suppose that there are $a \in C$ and $b \notin -\operatorname{int} C$ satisfying $a + b \in -\operatorname{int} C$. Then, there exists $c \in -\operatorname{int} C$ such that $a + b = c$. Taking into account the convexity of C, we obtain $b = -a + c \in -C - \operatorname{int} C = -\operatorname{int} C$, which furnishes a contradiction. Therefore, $a + b \notin -\operatorname{int} C$.

□

We present concepts of monotonicity involving cones for vector-valued maps.

Definition 6.118 (Monotonicity involving cones for vector-valued maps) Let K be a nonempty subset of a Banach space X, and $e \in \operatorname{int} C$. A vector-valued mapping $h : X \times X \to Y$ is said to be

(a) *C-strongly monotone* on K with respect to e if there exists $\gamma > 0$ such that

$$h(x, y) + h(y, x) + \gamma \|x - y\|^2 e \in -C \text{ for all } x, y \in K;$$

(b) *C-monotone* on K if

$$h(x, y) + h(y, x) \in -C \text{ for all } x, y \in K;$$

(c) *C-pseudomonotone* on K if

$$h(x, y) \notin -\operatorname{int} C \Longrightarrow h(y, x) \notin \operatorname{int} C \text{ for all } x, y \in K;$$

(d) *explicitly C-pseudomonotone* on K if

$$h(x, y) \notin -\operatorname{int} C \Longrightarrow h(y, x) \in -C \text{ for all } x, y \in K.$$

Remark 6.119 (Relations between cone-monotonicity properties)

(a) For $e, e' \in \operatorname{int} C$, h is C-strongly monotone with respect to e if and only if it is C-strongly monotone with respect to e'. Indeed, by Lemma 6.117(a), there exists $n \in \mathbb{N} \setminus \{0\}$ such that $e' - ne \in -\operatorname{int} C \subset -C$, and hence $e = n^{-1}(e' + c)$ for some $c \in C$. For all $x, y \in K$, $h(x, y) + h(y, x) + \gamma \|x - y\|^2 e \in -C$, so

$$h(x, y) + h(y, x) + \frac{\gamma}{n} \|x - y\|^2 (e' + c) \in -C.$$

Set $\gamma' = n^{-1}\gamma$, we get:

$$h(x, y) + h(y, x) + \gamma' \|x - y\|^2 e' \in -\gamma' \|x - y\|^2 c - C \subset -C.$$

Hence, h is C-strongly monotone with respect to $e' \in \operatorname{int} C$.

(b) Obviously, (ii) $\Rightarrow$ (iii) and (iv) $\Rightarrow$ (iii). If C is a total odering cone, then (iii) coincides with (iv) and they are weaker than (ii). In general, however, there is no relationship between (ii) and (iv). The following example shows that the bifunction f is C-monotone on K, but it is not explicitly C-pseudomonotone.

Example 6.120 (An example of cone-monotonicity map) Let $X = \mathbb{R}, Y = \mathbb{R}^2, K = \mathbb{R}_+, C = \mathbb{R}^2_+$ and $f : K \times K \to \mathbb{R}^2$ be a mapping defined as

$$f(x, y) = (x^2(y - x), x - y).$$

For all $x, y \in K$, we have:

$$\begin{aligned} f(x, y) + f(y, x) &= (x^2(y - x), x - y) + (y^2(x - y), y - x) \\ &= (-(x - y)^2(y + x), 0) \in -C. \end{aligned}$$

It follows that f is C-monotone on K. However, the mapping f is not explicitly C-pseudomonotone. Indeed, taking $(x, y) = (1, 2) \in K \times K$, we have $f(x, y) = (1, -1) \notin -\operatorname{int} C$ but $f(y, x) = (-4, 1) \notin -C$.

6.6.1 *Tikhonov Regularization for Weak Vector Equilibrium Problems*

In what follows, we always suppose that X is a real reflexive Banach space endowed with the weak topology, and K is a closed convex subset of X. For the bifunction $\mu : K \times K \to \mathbb{R}$ satisfying $\mu(x, x) = 0$ for all $x \in K$, we define a scalar equilibrium problem as

$$\text{find } \bar{x} \in K : \ \mu(\bar{x}, y) \geq 0 \ \forall y \in K. \qquad \text{EP}(\mu, K)$$

For each $y \in K$, we denote:

$$\sigma(y) := \{x \in K : \mu(x, y) \geq 0\}.$$

Obviously, $\bar{x}$ solves the EP(μ, K) if and only if $\bar{x} \in \bigcap_{y \in K} \sigma(y)$. Thus, $\bigcap_{y \in K} \sigma(y) = \text{Sol(EP)}(\mu, K)$, where $\text{Sol(EP)}(\mu, K)$ is the solution sets of EP(μ, K).

A function $\theta : X \to \mathbb{R}$ is said to be *weakly coercive* with respect to the set K, if there exists a number r such that

$$M_r = \{x \in K : \theta(x) \leq r\}$$

is a nonempty and bounded set.

We first recall the following result that extended the well-known Ky-Fan's theorem.

Proposition 6.121 *Suppose that the following conditions hold:*

(i) *$\mu(\cdot, y)$ is upper semicontinuous for each $y \in K$;*
(ii) *$\mu(x, \cdot)$ is quasiconvex for each $x \in K$;*

(iii) *there exists a nonempty convex compact D of K and an element $y \in D$ such that $\mu(x, y) < 0$ for all $x \in K \setminus D$.*

Then, $\mathrm{Sol(EP)}(\mu, K)$ *is nonempty and compact.*

The following assumptions for a bifunction $\mu : K \times K \to \mathbb{R}$ will be used in the rest of this paper:

(A1) For each $y \in K$, the function $x \mapsto \mu(x, y)$ is weakly upper semicontinuous.
(A2) For each $x \in K$, the function $y \mapsto \mu(x, y)$ is explicitly quasiconvex, that is, it is quasiconvex and holds that for all $y_1, y_2 \in K$ with $\mu(x, y_1) \neq \mu(x, y_2)$, and for all $t \in]0, 1[$,

$$\mu(x, ty_1 + (1-t)y_2) < \max\{\mu(x, y_1), \mu(x, y_2)\}.$$

(A3) There exist a function $\theta : X \to \mathbb{R}$, which is lower semicontinuous, convex, and weakly coercive with respect to the set K, and a number r such that for any $x \in K \setminus \{z \in X : \theta(z) \leq r\}$, there is an element $y \in K$ with

$$\min\{\mu(x, y), \theta(y) - \theta(x)\} < 0 \text{ and } \max\{\mu(x, y), \theta(y) - \theta(x)\} \leq 0.$$

We recall an interesting existence result for the problem $\mathrm{EP}(\mu, K)$.

Proposition 6.122 $\mathrm{Sol(EP)}(\mu, K)$ *is nonempty whenever Assumptions* (A1)–(A3) *are satisfied.*

The solution set of $\mathrm{WEP}(f, K)$ is denoted by $\mathrm{Sol(WEP)}(f, K)$. For each $\varphi \in C^* \setminus \{0\}$, we denote the φ-solution set of $\mathrm{WEP}(f, K)$ by taking

$$\mathrm{Sol(WEP)}_\varphi(f, K) := \{x \in K \colon \langle \varphi, f(x, y)\rangle \geq 0 \text{ for all } y \in K\}.$$

According to Theorem 6.64, we have $\mathrm{Sol(WEP)}(f, K) = \bigcup_{\varphi \in C^* \setminus \{0\}} \mathrm{Sol(WEP)}_\varphi (f, K)$.

Inspired by the coercivity conditions proposed by Konnov [308], we introduce a coercivity condition for a vector-valued bifunction $h : X \times X \to Y$.

(H1) There exist a function $\theta : X \to \mathbb{R}$, which is lower semicontinuous, convex, and weakly coercive with respect to the set K, and a number r such that for any $x \in K \setminus \{z \in X : \theta(z) \leq r\}$, there is $y \in K$ such that $\theta(y) < \theta(x)$ and $h(x, y) \notin \operatorname{int} C$.

We are now in a position to present the existence result for $\mathrm{WEP}(f, K)$.

Theorem 6.123 *Assume that the following conditions are satisfied: for each x and y in K :*

(i) *$f(\cdot, y)$ is weakly C-upper semicontinuous on K;*
(ii) *$f(x, \cdot)$ is C-convex on K;*
(iii) *f satisfies Condition (H1).*

Then, Sol(WEP)(f, K) *is nonempty and weakly closed.*

Proof Since f satisfies Condition (H1), there exist a convex lower semicontinuous function θ being weakly coercive with respect to the set K, and a number r such that for any $x \in K \setminus \{x \in X : \theta(x) \le r\}$, there is $y \in K$ such that $\theta(y) < \theta(x)$ and $f(x, y) \notin \operatorname{int} C$. Then, there is $\varphi \in C^* \setminus \{0\}$ such that $\langle \varphi, f(x, y)\rangle \le 0$.

We apply Proposition 6.122 by setting $\psi(x, y) := \langle \varphi, f(x, y)\rangle$.

- ◇ It is clear that ψ holds Assumption (A3).
- ◇ For each $x \in K$, let $y_1, y_2 \in K$ be arbitrary points. Since $f(x, \cdot)$ is C-convex,

$$tf(x, y_1) + (1-t)f(x, y_2) - f(x, ty_1 + (1-t)y_2) \in C, \quad \forall t \in [0, 1].$$

Consequently,

$$\langle \varphi, tf(x, y_1) + (1-t)f(x, y_2)\rangle \ge \langle \varphi, f(x, ty_1 + (1-t)y_2)\rangle,$$

which implies that $\psi(x, \cdot)$ is convex, and hence Assumption (A2) is satisfied.

- ◇ For all $y \in K$, since $f(\cdot, y)$ is weakly C-upper semicontinuous on K, $\psi(\cdot, y)$ is weakly upper semicontinuous on K. Therefore, ψ holds Assumption (A1).

Applying Proposition 6.122, EP(ψ, K) admits a solution. By Lemma 6.174, we conclude that Sol(WEP)(f, K) is a nonempty set.

Let $\{x_n\}$ be an arbitrary sequence in Sol(WEP)(f, K) weakly converging to x_0. Since K is a closed convex set in the Banach space X, we conclude that $x_0 \in K$. We show that $x_0 \in$ Sol(WEP)(f, K). For any $y \in K$, we have:

$$f(x_n, y) \notin -\operatorname{int} C, \ \forall n \in \mathbb{N}, \tag{6.97}$$

inasmuch as $x_n \in$ Sol(WEP)(f, K). For each neighborhood V of the origin 0_Y in Y, there is a balanced neighborhood V_1 of 0_Y such that $V_1 \subset V$. Because $f(\cdot, y)$ is weakly C-upper semicontinuous at x_0, there is a weak neighborhood U of x_0 such that $f(x, y) \in f(x_0, y) + V_1 - C$ for all $x \in U$. Since $\{x_n\}$ weakly converges to x_0, there exists $n_0 \in \mathbb{N}$ such that for each $n \ge n_0$, $x_n \in U$. Thus, for all $n \ge n_0$, one has $f(x_n, y) \in f(x_0, y) + V_1 - C$, or equivalently, $f(x_0, y) \in f(x_n, y) - V_1 + C$. This together with (6.97) and the balance of V_1 implies that

$$f(x_0, y) \in V_1 + Y \setminus (-\operatorname{int} C) + C.$$

It follows from Lemma 6.117(b) that

$$f(x_0, y) \in V_1 + Y \setminus (-\operatorname{int} C) + C \subset V + Y \setminus (-\operatorname{int} C).$$

Since V is arbitrary and $Y \setminus (-\operatorname{int} C)$ is closed, we have $f(x_0, y) \in Y \setminus (-\operatorname{int} C)$, and hence $x_0 \in$ Sol(WEP)(f, K). Therefore, Sol(WEP)(f, K) is weakly closed.

□

Now, we present an example to illustrate the applicability of Theorem 6.123.

Example 6.124 Let $X = Y = \mathbb{R}^2$, $K = \{(x_1, x_2) \in X : x_1 + 2x_2 \geq 0, x_1 + 4x_2 \leq 0\}$, $C = \mathbb{R}^2_+$, and $f : K \times K \to \mathbb{R}^2$ be defined by $f(x, y) := \langle G(x), y - x \rangle$, where

$$G(x) := \|x\| \begin{pmatrix} 1 & 0 \\ 0 & 1 \end{pmatrix} \text{ for } x = (x_1, x_2), \text{ and } \|x\| := (x_1^2 + x_2^2)^{\frac{1}{2}}.$$

It is clear that f satisfies all conditions in Theorem 6.123 with $\theta(x) = \|x\|$ and $r = 1$. Applying this theorem we conclude that the problem WEP(f, K) is solvable (by direct computations, $\mathrm{Sol(WEP)}(f, K) = \{(0, 0)\}$).

Remark 6.125 There are many works devoted to the existence of solutions for vector equilibrium problems in the literature, so we instead study ones being suitable for application to Tikhonov regularization method. Fortunately, the obtained result in Theorem 6.123 somehow contributes to the development of solution existence theory for vector equilibrium problems. Namely, different from most existing results obtained under assumptions related to monotonicity properties of the objective mapping (Theorems 9.1 and 9.2 in [66]) or an auxiliary mapping (Theorem 2.1 in [129]), Theorem 6.123 gives sufficient conditions for the existence of solutions to vector equilibrium problems without imposing any assumption related to monotonicity properties, for instance, the case presented in Example 6.124. Moreover, the coercivity condition (H1) is weaker than the corresponding ones in Theorems 9.1–9.3 of [66], Theorem 2.1 of [129], and Theorem 3.3 of [317]. To clarify this statement, let's recall the coercivity conditions and discuss their relationships:

(C1) [66, 129] There exist a nonempty, compact, convex subset D of K and an element $y \in D$ such that $f(x, y) \in -\operatorname{int} C$ for all $x \in K \setminus D$.

(C2) [66, 129] There exists a nonempty, compact, and convex subset D of K such that for any $x \in K \setminus D$, there is an element $y \in D$ satisfying $f(x, y) \in -\operatorname{int} C$.

(C3) [317] There exists a number $r > 0$ such that for any $x \in K \setminus B_r$ there is an element $y \in K$, $\|y\| < \|x\|$ satisfying $f(x, y) \in -\operatorname{int} C \cup \{0_Y\}$, where $B_r := \{x \in X : \|x\| \leq r\}$.

It is clear that [(C1) $\Longrightarrow$ (C2)] is fulfilled. We claim that [(C2) $\Longrightarrow$ (C3)] holds true. For the nonempty compact D, there exists a number $r > 0$ such that $D \subset B_r := \{x \in X : \|x\| \leq r\}$. Then, it follows from (C2) that for any $x \in K \setminus B_r \subset K \setminus D$, there exists an element $y \in D \subset B_r$ such that $f(x, y) \in -\operatorname{int} C$, and hence [(C2) $\Longrightarrow$ (C3)] holds true. If we choose $\theta(x) = \|x\|$, then (C3) $\Longrightarrow$ (H1). For the case of X being a Hausdorff topological vector space without any norm, although Condition (C3) is not useful, the implication [(C2) $\Longrightarrow$ (H1)] is still valid by choosing $\theta(x) = \iota_D(x)$, where ι_D is the indicator function of D defined by

$$\iota_D(x) := \begin{cases} 0, & \text{if } x \in D, \\ +\infty, & \text{if } x \in X \setminus D. \end{cases}$$

We now consider an example to show the preeminence of Condition (H1).

Example 6.126 Let $X = Y = \mathbb{R}^2$, $K = \{(x_1, x_2) \in X : x_1 \geq 0, x_1 + 4x_2 \leq 0\}$, $C = \mathbb{R}^2_+$, and $f : K \times K \to \mathbb{R}^2$ be defined by $f(x, y) := \langle G(x), x - y \rangle$, where

$$G(x) := (x_1 - x_2) \begin{pmatrix} -1 & 0 \\ 0 & 1 \end{pmatrix} \text{ for } x = (x_1, x_2).$$

Then, for each $x = (x_1, x_2)$ and $y = (y_1, y_2)$, one has:

$$f(x, y) = (x_1 - x_2) \begin{pmatrix} -1 & 0 \\ 0 & 1 \end{pmatrix} \begin{pmatrix} x_1 - y_1 \\ x_2 - y_2 \end{pmatrix} = (x_1 - x_2)\,(y_1 - x_1, x_2 - y_2).$$

Obviously, f satisfies Assumptions (i) and (ii) in Theorem 6.123. Let $\theta(x) = 2^{-1}\|x\|^2$, $r = 1$, and $x = (x_1, x_2)$ be an arbitrary element in K with $x_1^2 + x_2^2 > r$. By choosing $y = (0, 0)$, we have $\theta(y) < \theta(x)$ and

$$f(x, y) = (x_1 - x_2)\,(-x_1, x_2) \notin \operatorname{int} C,$$

which yields that f satisfies Condition (H1). Therefore, the solution set of WEP(f, K) is nonempty.

We now show that Condition (C3) is not valid. For any $r > 0$, let $\hat{x} = (0, x_2) \in K$, with $\|\hat{x}\| > r$. For any $y = (y_1, y_2) \in K$ satisfying $\|y\| < \|\hat{x}\|$, we have $y_2^2 < x_2^2$. Consequently, $y_2 > x_2$ as they are nonpositive numbers, and hence $x_2 - y_2 < 0$. Combining this with $y_1 \geq 0$ and $x_2 \neq 0$, we arrive at

$$f(\hat{x}, y) = -x_2(y_1, x_2 - y_2) \notin -\operatorname{int} C \cup \{(0, 0)\}.$$

Therefore, Condition (C3) is not valid, and so Conditions (C1) and (C2) are also not satisfied.

Remark 6.127 To weaken the weak C-upper semicontinuity in the first argument of f, we need the following stronger coercivity condition:

(H1') There exists a nonempty, compact, and convex subset D of K and an element $y \in D$ such that $h(x, y) \notin C$ for all $x \in K \setminus D$.

Employing Proposition 6.121, we obtain the following result.

Corollary 6.128 *Assume that the following conditions are satisfied: For each x and y in K:*

(i) *$f(\cdot, y)$ is C-upper semicontinuous on K;*
(ii) *$f(x, \cdot)$ is C-convex on K;*
(iii) *f satisfies Condition (H1').*

Then, Sol(WEP)(f, K) *is nonempty and closed.*

Proof Since f satisfies Condition (H1'), there exist a nonempty, compact, and convex subset D of K and an element $y \in D$ such that $f(x, y) \notin C$ for any $x \in K \setminus D$. Then, there is $\varphi \in C^* \setminus \{0\}$ such that $\langle \varphi, f(x, y) \rangle < 0$.

Let $\psi(x, y) := \langle \varphi, f(x, y) \rangle$. The conclusion follows by using the same arguments given in Theorem 6.123 and taking into account Proposition 6.121. □

We now study Tikhonov regularization method to monotone vector equilibrium problems. Let $g : K \times K \to Y$ be C-upper semicontinuous in the first argument, C-convex and C-lower semicontinuous in the second argument, and C-strongly monotone w.r.t. some $e \in \operatorname{int} C$ such that $g(x, x) = 0_Y$ for all $x \in K$. For each $\varepsilon > 0$, we consider the following regularized problem for WEP(f, K)

$$\text{find } x(\varepsilon) \in K : f_\varepsilon(x(\varepsilon), y) := f(x(\varepsilon), y) + \varepsilon g(x(\varepsilon), y) \notin -\operatorname{int} C \ \forall y \in K, \qquad \text{RWEP}(f_\varepsilon, K)$$

g and ε are called the *regularization vector-valued mapping* and the *regularization parameter*, respectively. A curve $\{x(\varepsilon) : \varepsilon > 0\}$, where $x(\varepsilon)$ is a solution to the regularized problem RWEP(f_ε, K), is called a *Tikhonov trajectory*.

Theorem 6.129 *Assume that for each $\varepsilon > 0$, the Problem* RWEP(f_ε, K) *admits a solution $x(\varepsilon)$, and for each y in K, $f(\cdot, y)$ is C-upper semicontinuous on K. Then, every cluster point of any Tikhonov trajectory $\{x(\varepsilon)\}_{\varepsilon>0}$ is a solution to the original problem as $\varepsilon \to 0$.*

Proof For every sequence $\{\varepsilon_n\} \subset \mathbb{R}_+$ converging to 0, let $x_n := x(\varepsilon_n)$ be a solution to RWEP(f_{ε_n}, K) and $\bar{x}$ be a cluster point of the sequence $\{x_n\}$. We can assume, without loss of generality, that $\{x_n\}$ converges to $\bar{x}$. For any $y \in K$, because $x_n \in$ Sol(RWEP)(f_{ε_n}, K), we get

$$f(x_n, y) + \varepsilon_n g(x_n, y) \notin -\operatorname{int} C. \tag{6.98}$$

For each neighborhood V of the origin 0_Y in Y, there exist balanced neighborhoods V_1, V_2 of 0_Y such that $V_1 \subset V$ and $V_2 + V_2 \subset V_1$. Since Y is a topological linear space and $0g(\bar{x}, y) = 0_Y \in V_2$, according to Lemma 6.115, there exist $r > 0$ and a neighborhood B of $g(\bar{x}, y)$ such that

$$\beta B \subset V_2 \text{ whenever } |\beta| < r. \tag{6.99}$$

Because $\varepsilon_n \to 0$, there is $N_1 \in \mathbb{N}$ such that

$$\varepsilon_n < \frac{r}{2} \quad \forall n > N_1.$$

Combining this with (6.99), we arrive at $\varepsilon_n B \subset V_2$ for all $n > N_1$. Because $g(\cdot, y)$ is C-upper semicontinuous at $\bar{x}$, for the neighborhood B, there exists $N_2 \in \mathbb{N}$ such that

$$g(x_n, y) \in B - C \quad \forall n > N_2.$$

Consequently,

$$\varepsilon_n g(x_n, y) \in \varepsilon_n B - C \subset V_2 - C \quad \forall n > \max\{N_1, N_2\}. \tag{6.100}$$

Similarly, due to the C-upper semicontinuity of $f(\cdot, y)$ at $\bar{x}$, we can pick up $N_3 \in \mathbb{N}$ satisfying

$$f(x_n, y) \in f(\bar{x}, y) + V_2 - C \quad \forall n > N_3.$$

This together with (6.100) leads to

$$f(x_n, y) + \varepsilon_n g(x_n, y) \in f(\bar{x}, y) + V_2 + V_2 - C \subset f(\bar{x}, y) + V_1 - C,$$

for all $n > \max\{N_1, N_2, N_3\}$. Equivalently, $f(\bar{x}, y) \in f(x_n, y) + \varepsilon_n g(x_n, y) - V_1 + C$. Combining this with (6.98) and taking into account the balance of V_1, we arrive at

$$f(\bar{x}, y) \in V_1 + Y \setminus (-\operatorname{int} C) + C.$$

According to Lemma 6.117(b), we obtain:

$$f(\bar{x}, y) \in V_1 + Y \setminus (-\operatorname{int} C) \subset V + Y \setminus (-\operatorname{int} C).$$

By the arbitrary choice of V and the closedness of $Y \setminus (-\operatorname{int} C)$, $f(\bar{x}, y) \in Y \setminus (-\operatorname{int} C)$, so $\bar{x} \in \mathrm{Sol(WEP)}(f, K)$. The proof is complete. □

Remark 6.130 Recently, Chadli et al. [129] have established an interesting result on the Tikhonov regularization for vector equilibrium problems presented in Theorem 3.1. In this article, Theorem 6.129 does not require any assumptions related to the monotonicity properties as well as the C-convexity and C-lower semicontinuity in the second argument of the objective bifunction. However, we need the C-upper semicontinuity in the first argument instead of the C-upper hemicontinuity of the objective bifunction as in Theorem 3.1 of [129]. Therefore, Theorem 6.129 is different from Theorem 3.1 in [129].

We now study the existence solutions of the regularized vector equilibrium problems. In this regard, we have the following important result.

Lemma 6.131 *Let $h : K \times K \to Y$ be C-strongly monotone w.r.t. some $e \in \operatorname{int} C$ such that $h(x, x) = 0_Y$ for all $x \in K$. If h is C-convex and C-lower semicontinuous in the second argument, then there exists a nonempty, compact, and convex subset D of K and an element $y \in D$ such that $h(x, y) \notin C$ for all $x \in K \setminus D$.*

Proof Suppose, on the contrary, that for each nonempty, compact, and convex subset D of K, and for all $y \in D$, there exists $x \in K \setminus D$ such that $h(x, y) \in C$. Let E be a finite-dimensional subspace of X such that $E \cap K \neq \emptyset$. Let y_0 be in $E \cap K$. Set $B(y_0, r) := \{x \in X : \|x - y_0\| \leq r\}$. Choose $D_r := B(y_0, r) \cap E \cap K, r > 1$, be a closed ball of E, and hence D_r is compact for all r. By the contradicting assumption, for $y = y_0$ in D_r, we can pick up a point $x_r \in K \setminus D_r$ such that $h(x_r, y_0) \in C$.

Since $h(y_0, \cdot)$ is C-lower semicontinuous at y_0, for all neighborhood V of the origin $0_Y = h(y_0, y_0)$ of Y, there exists a neighborhood U of y_0 such that

$$h(y_0, x) \in V + C, \forall x \in U. \tag{6.101}$$

Let x be in the intersection of U and $[x_r, y_0]$ such that $\|x - y_0\| = \bar{r} < 1$, where $[x_r, y_0]$ is the line segment joining two points x_r and y_0. Then, $x_r - y_0 = \lambda_r(x - y_0)$ for some $\lambda_r > r$. Since h is C-strongly monotone w.r.t. $e \in \operatorname{int} C$, there is $\gamma > 0$ such that

$$h(y_0, x_r) \in -h(x_r, y_0) - \gamma \|x_r - y_0\|^2 e - C.$$

Because $h(x_r, y_0) \in C$, we arrive at

$$h(y_0, x_r) \in -C - \gamma\lambda_r^2 \|x - y_0\|^2 e - C \subset -\gamma(\lambda_r \bar{r})^2 e - C. \tag{6.102}$$

Now we rewrite the element x as $x = \lambda_r^{-1} x_r + \lambda_r^{-1}(\lambda_r - 1) y_0$. Due to the C-convexity of $h(y_0, \cdot)$ and (6.102), one has:

$$\begin{aligned} h(y_0, x) &\in \frac{1}{\lambda_r} h(y_0, x_r) + \frac{\lambda_r - 1}{\lambda_r} h(y_0, y_0) - C \\ &\subset -\bar{r}^2 \gamma \lambda_r e - C \subset -\bar{r}^2 \gamma r e - C. \end{aligned}$$

This together with (6.101) implies that $(V + C) \cap \left(-\bar{r}^2 \gamma r e - C\right) \neq \emptyset$, which is impossible as $r \to +\infty$. □

Applying Lemma 6.131, Corollary 6.128 and Theorem 6.129, we obtain conditions for the existence of solutions to the regularized problem and the convergence of the regularization method.

Theorem 6.132 *Assume that the following conditions are satisfied:*

(i) *for each $x \in K$, $f(x, \cdot)$ is C-lower semicontinuous, and C-convex on K;*
(ii) *for each $y \in K$, $f(\cdot, y)$ is C-upper semicontinuous on K;*
(iii) *f is C-monotone.*

Then, RWEP(f_ε, K) *admits a solution for each $\varepsilon > 0$. Furthermore, every cluster point of any Tikhonov trajectory $\{x(\varepsilon)\}_{\varepsilon > 0}$ is a solution to the original problem as $\varepsilon \to 0$.*

Proof For each $\varepsilon > 0$, because f is C-monotone and g is C-strongly monotone, f_ε is C-strongly monotone. Since $\varepsilon > 0$, and f, g are C-convex, C-lower semiconituous with respect to the second argument, so is f_ε. According to Lemma 6.131, the function f_ε satisfies Condition (H1'). Similarly, we also conclude that $f_\varepsilon(\cdot, y)$ is C-upper semicontinuous for each $y \in K$. Applying Corollary 6.128, we conclude that RWEP(f_ε, K) has a solution $x(\varepsilon)$. The second assertion follows immediately from Theorem 6.129. The proof is complete. □

Next, we study Tikhonov regularization for nonmonotone vector equilibrium problems in a reflexive Banach space endowed with the weak topology. Inspired by Konnov [308] and Konnov and Dyabilkin [309], we consider the following coercivity condition:

(C) There exist a function $\theta : X \to \mathbb{R}$, which is lower semicontinuous, convex, and weakly coercive with respect to the set K, a vector-valued bifunction $g : K \times K \to Y$, which is weakly C-upper semicontinuous in the first argument and C-convex in the second argument, and a number r such that for any $x \in K \setminus \{z \in X : \theta(z) \leq r\}$, there is $y \in K$ such that $\theta(y) < \theta(x)$, and

$$\{f(x, y), g(x, y)\} \cap (-C) \neq \emptyset \text{ and } \{f(x, y), g(x, y)\} \subset (Y \setminus \operatorname{int} C)\,. \tag{6.103}$$

Under Condition (C), we obtain conditions for the existence of solutions to the regularized problem and the convergence result for the Tikhonov regularization of the original problem.

Theorem 6.133 *Assume that Condition (C) is fulfilled and the following conditions are satisfied: for each x and y in K :*

(i) *$f(\cdot, y)$ is weakly C-upper semicontinuous on K;*
(ii) *$f(x, \cdot)$ is C-convex on K.*

Then, RWEP(f_ε, K) *admits a solution for each $\varepsilon \geq 0$. Moreover, every weak cluster point of any Tikhonov trajectory $\{x(\varepsilon)\}_{\varepsilon>0}$ is a solution to the original problem as $\varepsilon \to 0$.*

Proof For each $\varepsilon \geq 0$, we first show that (H1) is satisfied for f_ε. Choose an arbitrary element $x \in K \setminus M_r$, where $M_r = \{z \in K : \theta(z) \leq r\}$. Then, by Condition (C), there exists $y \in K$ with $\theta(y) < \theta(x)$ such that (6.103) is fulfilled. We show that

$$f_\varepsilon(x, y) = f(x, y) + \varepsilon g(x, y) \notin \operatorname{int} C. \tag{6.104}$$

It follows from Condition (C) that

$$\{f(x, y), g(x, y)\} \cap (-C) \neq \emptyset, \tag{6.105}$$

$$\{f(x, y), g(x, y)\} \subset (Y \setminus \operatorname{int} C)\,. \tag{6.106}$$

Following (6.105) and (6.106), there are two situations must be considered:
Case 1. If $f(x, y) \in -C$, then the inclusion (6.104) is obtained by applying Lemma 6.117(b) for the cone $(-C)$ and choosing $a := f(x, y) \in -C$ and $b := \varepsilon g(x, y) \notin \operatorname{int} C$.

Case 2. If $g(x, y) \in -C$, then (6.104) is also satisfied by applying Lemma 6.117(b) for the cone $(-C)$ choosing $a := \varepsilon g(x, y) \in -C$ and $b := f(x, y) \notin \operatorname{int} C$.

Therefore, f_ε satisfies (H1). By Theorem 6.123, we conclude that the Problem RWEP(f_ε, K) has a solution $x(\varepsilon)$. For every sequence $\{\varepsilon_n\} \subset \mathbb{R}_+$ converging to 0, let $x_n := x(\varepsilon_n)$ be a solution to RWEP(f_{ε_n}, K) and $\bar{x}$ be a weak cluster point of the sequence $\{x_n\}$. It can be assumed without loss of generality that $\{x_n\}$ weakly converges to $\bar{x}$. Since $\{x_n\} \subset K$ and K is a closed convex set in the Banach space X, we obtain that $\bar{x} \in K$. We now demonstrate that $\bar{x}$ belongs to Sol(WEP)(f, K). Since $x_n \in$ Sol(RWEP)(f_{ε_n}, K), for all $y \in K$ we have:

$$f(x_n, y) + \varepsilon_n g(x_n, y) \notin -\operatorname{int} C. \tag{6.107}$$

Let V be an arbitrary neighborhood of the origin 0_Y in Y, then there are balanced neighborhoods V_1, V_2 of 0_Y satisfying $V_1 \subset V$ and $V_2 + V_2 \subset V_1$. Using the same arguments given in the proof of Theorem 6.129, we can also pick up a neighborhood B of $g(\bar{x}, y)$ and a number $N_1 \in \mathbb{N}$ such that $\varepsilon_n B \subset V_2$ for all $n > N_1$. Due to the weak C-upper semicontinuity of $g(\cdot, y)$ at $\bar{x}$, there is a weak neighborhood U of $\bar{x}$ satisfying $g(x, y) \in B - C$ for all $x \in U$. Because $\{x_n\}$ weakly converges to $\bar{x}$, there is some number $N_2 \in \mathbb{N}$ such that $x_n \in U$ for all $n > N_2$, and then

$$g(x_n, y) \in B - C \quad \forall n > N_2.$$

Thus, for all $n > \max\{N_1, N_2\}$, we have:

$$\varepsilon_n g(x_n, y) \in \varepsilon_n B - C \subset V_2 - C. \tag{6.108}$$

Employing the weak C-upper semicontinuity of $f(\cdot, y)$ at $\bar{x}$, there is a number $N_3 \in \mathbb{N}$ such that

$$f(x_n, y) \in f(\bar{x}, y) + V_2 - C \quad \forall n > N_3.$$

Combining this and (6.108), we arrive at

$$f(x_n, y) + \varepsilon_n g(x_n, y) \in f(\bar{x}, y) + V_2 + V_2 - C \subset f(\bar{x}, y) + V_1 - C,$$

for all $n > \max\{N_1, N_2, N_3\}$. Consequently, $f(\bar{x}, y) \in f(x_n, y) + \varepsilon_n g(x_n, y) - V_1 + C$. Due to the balance of V_1, this together with (6.107) leads to

$$f(\bar{x}, y) \in V_1 + Y \setminus (-\operatorname{int} C) + C.$$

It follows from Lemma 6.117(b) that

$$f(\bar{x}, y) \in V_1 + Y \setminus (-\operatorname{int} C) \subset V + Y \setminus (-\operatorname{int} C).$$

Because V is arbitrary and the set $Y \setminus (-\operatorname{int} C)$ is closed, we conclude that $f(\bar{x}, y) \in Y \setminus (-\operatorname{int} C)$, and so $\bar{x} \in \text{Sol(WEP)}(f, K)$. The proof follows. □

Remark 6.134

(a) From the proof of Theorem 6.133, we see that Condition (C) can be replaced by a slight modification version as follows:

(C') There exist a function $\theta : X \to \mathbb{R}$, which is lower semicontinuous, convex, and weakly coercive with respect to the set K, a vector-valued bifunction $g : K \times K \to Y$, which is weakly C-upper semicontinuous in the first argument and C-convex in the second argument, and a number r such that for any $x \in K \setminus \{z \in X : \theta(z) \leq r\}$, there is $y \in K$ such that $\theta(y) < \theta(x)$ and

$$f(x, y) + \varepsilon g(x, y) \notin \operatorname{int} C, \quad \text{for } \varepsilon > 0 \text{ sufficiently small.}$$

Condition (C') guarantees the existence of solutions for the regularized problem RWEP(f_ε, K), and does not require the coercivity of the objective mapping of the original problem. Hence, Condition (C') is widely used in regularization methods. For instance, to establish sufficient conditions for existence solutions to regularized problems, Chadli et al. [129] employed Condition ($\mathscr{H}$) in Theorem 3.2 and Condition ($\mathscr{C}$) in Theorem 3.3. Note that Condition ($\mathscr{C}$) $\Longrightarrow$ ($\mathscr{H}$) as shown in the proof of Theorem 3.3 of [129]. Under suitable choice on θ, for instance, $\theta(x) = 2^{-1}\|x\|^2$, Condition ($\mathscr{H}$) implies Condition (C').

(b) It is worth noting that Condition (C) is proposed with various advantages, for instance, it ensures not only the convergence of the Tikhonov regularization method but also the existence of solutions to the original and regularized problems. For the scalar case where $Y = \mathbb{R}$, and $C = \mathbb{R}_+$, Condition (C) reduces to various effective conditions used in many papers [305–309]. For instance, a weak coercive condition, say Condition (**C2**), was proposed in [308] to study the regularization method for the penalized equilibrium problem in a Banach space. We now recall Condition (**C2**) in [308].

(**C2**) There exist a lower semicontinuous and convex function $\theta : X \to \mathbb{R}$, which is coercive with respect to the set K, and an equilibrium bifunction $\Psi : K \times K \to \mathbb{R}$ such that $\Psi(\cdot, y)$ is weakly upper semicontinuous for each $y \in K$, $\Psi(x, \cdot)$ is convex for each $x \in K$, and a number r such that for any $\tau > 0$ and for any element $x \in K \setminus \{z \in X : \theta(z) \leq r\}$, there exists an element $y \in K$ such that $\theta(y) \leq \theta(x)$ and

$$\min\{\Phi_\tau(x, y), \Psi(x, y)\} < 0 \text{ and } \max\{\Phi_\tau(x, y), \Psi(x, y)\} \leq 0, \tag{6.109}$$

where $\Phi_\tau : K \times K \to \mathbb{R}$ is the objective bifunction of the penalized equilibrium problem.

It is easy to see that (6.109) is equivalent to

$$\{\Phi_\tau(x, y), \Psi(x, y)\} \cap (-\operatorname{int}\mathbb{R}_+) \neq \emptyset \text{ and } \{\Phi_\tau(x, y), \Psi(x, y)\} \subset -\mathbb{R}_+.$$

Therefore, in the case of $Y = \mathbb{R}$, and $C = \mathbb{R}_+$, Condition (C) modifies slightly Condition (**C2**) in [308], where f and g in Condition (C) serve as Φ_τ and Ψ in Condition (**C2**), respectively. Moreover, if we take

$$g(x, y) = \theta(y) - \theta(x), \quad \forall x, y \in K,$$

then (C) is a slight modification of Condition (**C1**) in [308].

For the vector case, recently Kassay and Miholca [275] have proposed Conditions, say (G) and (H), to study existence results for vector equilibrium problems given by a sum of two bifunctions. Latter, Chadli et al. [129] have extended these conditions to $(\mathscr{H}_1)$ and $(\mathscr{H}_2)$, and employed them to obtain qualitative properties of solutions to regularized problems presented in Theorem 3.4, which are similar to that of Theorem 6.133. If we note, Condition $(\mathscr{H}_2)$ is strictly stronger than Condition (C). Let's recall Condition $(\mathscr{H}_2)$ in [129].

$(\mathscr{H}_2)$ there exist $a \in \operatorname{int} C$ and $x_0 \in K$ satisfying $\forall t > 0, \exists R = R(t) > 0$ such that for $x \in K$ with $\|x - x_0\| > R$, $-t\|x - x_0\|a - f(x, x_0) \in C$.

It is obvious that Condition $(\mathscr{H}_2)$ implies the following condition:

$(\mathscr{H}_2')$ There is $x_0 \in K$ such that the set

$$D := \{x \in K : f(x, x_0) \notin -C\}$$

is bounded.

Indeed, suppose to the contrary that D is not bounded, then there exists a sequence $\{x_n\}$ in D such that $\|x_n\| \to +\infty$. From $(\mathscr{H}_2)$, we get:

$$f(x_n, x_0) \in -t\|x_n - x_0\|a - C \in -C,$$

which is a contradiction.

We now show that $(\mathscr{H}_2') \Longrightarrow$ (C). For such x_0, let us consider strongly convex functions $\theta, \gamma : X \to \mathbb{R}$ defined by $\theta(x) := M\|x - x_0\|^2$, and $\gamma(x) := N\|x - x_0\|^2$, where M and N are positive real numbers, then $\{x_0\} = \arg\min_{x\in K} \theta(x) = \arg\min_{x\in K} \gamma(x)$. We can find a positive real number r such that $D \subset \{x \in X : \theta(x) \leq r\}$. For any $x \in K \setminus \{x \in X : \theta(x) \leq r\}$, we have $\theta(x) > \theta(x_0)$ and $\gamma(x) > \gamma(x_0)$. By setting $g(x, y) = (\gamma(y) - \gamma(x))\,a$, for $a \in \operatorname{int} C$ defined in Condition $(\mathscr{H}_2)$, we get $g(x, x_0) \in -\operatorname{int} C$ and $f(x, x_0) \in -C$. Therefore, Condition (C) holds true, in which x_0 serves as y.

From above discussions, Condition (C) is a unified form of interesting coercivity conditions used in the literature; see, e.g., [129, 275, 305–309].

In summary, under Conditions (C) and (C'), results related to qualitative properties of solutions to RWEP(f_ε, K) are established without imposing monotonicity and C-lower semicontinuity in the second argument of the objective mapping of the original problem. Therefore, they are different from the existing ones in the literature.

Before presenting relationships for qualitative properties between solutions to the original problem and its regularized one, we consider the following condition.

(H2) For each $x \in K$ and sequence $\{y_n\} \subset K$ with $\|y_n\| \to +\infty$, and for each neighborhood U of the origin of Y, there exist $t > 0$ and $n_0 \in \mathbb{N}$ such that $\|y_n - x\|^{-1}h(x, y_n) \in tU$ for all $n \geq n_0$.

Theorem 6.135 *Assume that for each $y \in K$, $f(\cdot, y)$ and $g(\cdot, y)$ are weakly C-upper semicontinuous on K. Assume further that f is C-explicitly pseudomonotone on K, g is C-strongly monotone w.r.t some $e \in \operatorname{int} C$ and satisfies Condition (H2). If $x(\varepsilon_n)$ is a solution of Problem RWEP(f_{ε_n}, K), where $\{\varepsilon_n\} \subset \mathbb{R}_+ \setminus \{0\}$ converges to 0, then the following assertions are equivalent:*

(a) *The sequence $\{x(\varepsilon_n)\}$ is bounded.*
(b) Sol(WEP)(f, K) *is a nonempty set.*

Proof (a) $\Rightarrow$ (b): Assume that the sequence $\{x(\varepsilon_n)\}$ is bounded. Then, there is a subsequence $\{x_{n_k}\}$ of $\{x_n\}$ which weakly converges to some point $\bar{x}$ in K where $x_n := x(\varepsilon_n)$. Thus,

$$f(x_{n_k}, y) + \varepsilon_{n_k} g(x_{n_k}, y) \notin -\operatorname{int} C \ \forall y \in K.$$

Using the same argument given in the proof of Theorem 6.133, we obtain that $\bar{x} \in$ Sol(WEP)(f, K). Therefore, Sol(WEP)(f, K) is nonempty.
(b) $\Rightarrow$ (a): We suppose that Sol(WEP)(f, K) is nonempty. Let $\bar{x}$ be a solution to WEP(f, K) and ε_n be a sequence converging to 0. Let $x_n := x(\varepsilon_n) \in$ Sol(RWEP)(f_{ε_n}, K), then

$$f(\bar{x}, x_n) \notin -\operatorname{int} C \text{ and } f(x_n, \bar{x}) + \varepsilon_n g(x_n, \bar{x}) \notin -\operatorname{int} C. \tag{6.110}$$

Since f is explicitly C-pseudomonotone, $f(x_n, \bar{x}) \in -C$. Now, we suppose that $\varepsilon_n g(x_n, \bar{x}) \in -\operatorname{int} C$. Then, $f(x_n, \bar{x}) + \varepsilon_n g(x_n, \bar{x}) \in -C - \operatorname{int} C = -\operatorname{int} C$ which contradicts (6.110). Therefore,

$$g(x_n, \bar{x}) \notin -\operatorname{int} C. \tag{6.111}$$

On the other hand, by the C-strong monotonicity of g with respect to e, we have:

$$g(x_n, \bar{x}) + g(\bar{x}, x_n) + \gamma \|x_n - \bar{x}\|^2 e \in -C,$$

or equivalently,

$$-g(x_n, \bar{x}) - g(\bar{x}, x_n) - \gamma \|x_n - \bar{x}\|^2 e \in C. \tag{6.112}$$

By (6.111) and (6.112), we obtain from Lemma 6.117(b) that $-g(\bar{x}, x_n) - \gamma \|x_n - \bar{x}\|^2 e \notin -\operatorname{int} C$. Consequently,

$$-\frac{g(\bar{x}, x_n)}{\|x_n - \bar{x}\|} - \gamma \|x_n - \bar{x}\| e \notin -\operatorname{int} C. \tag{6.113}$$

Suppose, to the contrary that, the sequence $\{x_n\}$ is not bounded. Let $y \in \operatorname{int} C$ and choose $U = (-y + C) \cap (y - C)$. Then, U is a neighborhood of the origin in Y. Since g satisfies Condition (H2), there exist $t > 0$ and $n_0 \in \mathbb{N}$ such that

$$u = \frac{g(\bar{x}, x_n)}{\|x_n - \bar{x}\|} \in tU \quad \forall n \geq n_0.$$

Because $\|x_n - \bar{x}\|$ approaches infinity and $e \in \operatorname{int} C$, according to Lemma 6.117(a), for n sufficiently large, we get $ty - \gamma \|x_n - \bar{x}\| e \in -\operatorname{int} C$. Employing the convexity of C, we have:

$$(ty - tC) - \gamma \|x_n - \bar{x}\| e \subset -\operatorname{int} C - tC = -\operatorname{int} C - C = -\operatorname{int} C.$$

Thus, $tU - \gamma \|x_n - \bar{x}\| e \subset -\operatorname{int} C$, which contradicts (6.113). Therefore, the sequence $\{x_n\}$ is bounded. The proof is complete. □

The following example illustrates the applicability of Theorem 6.135.

Example 6.136 Let $X = \mathbb{R}$, $Y = \mathbb{R}^2$, $K =]-\infty, 1]$, $C = \mathbb{R}^2_+$, $f : K \times K \to \mathbb{R}^2$ be defined by

$$f(x, y) := \Big((x-1)(y-x), (x^3-1)(y-x)\Big) \text{ and}$$

$$g(x, y) := (x(y-x), x(y-x)).$$

Obviously, all the assumptions of Theorem 6.135 are satisfied. A direct computation shows that Sol(WEP)$(f, K) = \{1\}$ and that

$$\text{Sol(RWEP)}(f_\varepsilon, K) = \left\{1, \frac{1}{1+\varepsilon}, \frac{\sqrt[3]{\sqrt{12\varepsilon^3+81}+9}}{\sqrt[3]{18}} - \frac{\sqrt[3]{2/3}\varepsilon}{\sqrt[3]{\sqrt{12\varepsilon^3+81}+9}}\right\}.$$

The set-valued mapping $J_p : X \rightrightarrows X^*$ with $p > 1$, defined by

$$J_p(x) := \{x^* \in X^* : \langle x^*, x\rangle = \|x\| \|x^*\|, \|x^*\| = \|x\|^{p-1}\},$$

is called *duality mapping* of X. Motivated by [191], consider the mapping $g : K \times K \to Y$ defined by

$$g(x, y) := \langle j_p(x), y - x\rangle e, \tag{6.114}$$

where $p > 1$ and $j_p : X \to X^*$ is a single-valued selection of duality mapping (that means $j_p(x) \in J_p(x)$ for all $x \in X$) and $e \in \operatorname{int} C$ is given. We obtain the following results on the convergence of the Tikhonov trajectory to the least-norm solution of the original WEP(f, K).

Theorem 6.137 *Let g be defined as* (6.114). *Assume that f is explicitly C-pseudomonotone on K, and all the assumptions of Theorem 6.123 are satisfied. Then, every weak cluster point of any Tikhonov trajectory is a solution with least-norm of Problem WEP(f, K).*

Proof According to Theorem 6.123, we conclude that Sol(WEP)(f, K) is nonempty and weakly closed. Setting $d = \inf\{\|x\| : x \in \text{Sol(WEP)}(f, K)\}$, there exists a sequence $\{x_n\}$ in Sol(WEP)(f, K) such that $\lim_{n\to\infty} \|x_n\| = d$, and hence this sequence is bounded. Due to the reflexivity of X, there is some subsequence $\{x_{n_k}\}$ of the bounded sequence $\{x_n\}$ weakly converging to some $\hat{x}$. Because Sol(WEP)(f, K) is weakly closed, $\hat{x} \in \text{Sol(WEP)}(f, K)$, so

$$d \le \|\hat{x}\| \le \liminf_{k\to\infty} \|x_{n_k}\| = d.$$

Therefore, Problem WEP(f, K) admits at least a solution with least norm. Let $\bar{x}$ be a solution with least norm of Problem WEP(f, K), and let $x_n := x(\varepsilon_n) \in \text{Sol(RWEP)}(f_{\varepsilon_n}, K)$ be an arbitrary Tikhonov trajectory associated with the sequence $\{\varepsilon_n\}$ of positive real numbers tending to 0. Utilizing the same technique given in the second part of the proof for Theorem 6.135, we conclude that

$$\varepsilon_n \langle j_p(x_n), \bar{x} - x_n \rangle e \notin -\operatorname{int} C.$$

Hence, $\langle j_p(x_n), \bar{x} - x_n \rangle \ge 0$, which implies that $\langle j_p(x_n), \bar{x} \rangle \ge \langle j_p(x_n), x_n \rangle$. Applying Cauchy's inequality and taking into account the definition of $j_p(\cdot)$, we obtain $\|\bar{x}\| \ge \|x_n\|$. Consequently, the sequence $\{x_n\}$ is bounded, and so it has weak cluster points. Let x_0 be a weak cluster point of $\{x_n\}$. Employing the same argument as in the first part of the proof for Theorem 6.135, we see that x_0 is a solution of Problem WEP(f, K). Obviously,

$$\|x_0\| \le \liminf_{k\to\infty} \|x_{n_k}\| \le \|\bar{x}\|.$$

This together with $\|\bar{x}\| \le \|x_0\|$ yields that $\|\bar{x}\| = \|x_0\|$. This brings the proof to the end. $\square$

6.6.2 Tikhonov Regularization for Strong Vector Equilibrium Problems

In this section we consider a regularization method for the strong form of vector equilibrium problems. Similar to the previous sections, we choose a regularization equilibrium map $g : X \times X \to Y$ being C-strongly monotone (with respect to some $e \in \operatorname{int} C$) with $g(x, x) = 0_Y$ for all $x \in K$. For each regularization parameter $\varepsilon > 0$, we consider the following regularized problem defined as

$$\text{RSEP}(f_\varepsilon, K) \qquad \text{find } x(\varepsilon) \in X : f_\varepsilon(x(\varepsilon), y) \in C \quad \forall y \in K,$$

where $f_\varepsilon(x, y) = f(x, y) + \varepsilon g(x, y)$. The solution set of Problem RSEP(f_ε, K) is denoted by Sol(RSEP)(f_ε, K).

Theorem 6.138 *Assume that the following conditions are satisfied:*

(i) *$f(\cdot, y)$ and $g(\cdot, y)$ are C-upper semicontinuous for all $y \in K$;*
(ii) *f is C-momonotone on K;*
(iii) *g satisfies Condition (H2);*
(iv) *for each $\varepsilon > 0$, Sol(RSEP)(f_ε, K) is nonempty.*

Then, the following assertions are equivalent:

(a) Sol(SEP)(f, K) *is nonempty;*
(b) $\lim_{\varepsilon \to 0} \sup \|x(\varepsilon)\| < \infty$, *where $x(\varepsilon)$ is an arbitrary solution to Problem* RSEP(f_ε, K).

Proof (a) $\Rightarrow$ (b) For any sequence $\{\varepsilon_n\}$ converging to 0 and $x \in$ Sol(SEP)(f, K), let a sequence $\{x_n\}$ be such that for each $n \in \mathbb{N}$, $x_n = x(\varepsilon_n)$ with $x(\varepsilon_n) \in$ Sol(SEP)(f_{ε_n}, K). Thus, for each $n \in N$

$$f(x_n, x) + \varepsilon_n g(x_n, x) \in C. \tag{6.115}$$

Because x is a solution to Problem SEP(f, K), we have $f(x, x_n) \in C$. This together with assumption (ii) gives us $f(x_n, x) \in -C$, or equivalently $-f(x_n, x) \in C$. Plugging this inclusion into (6.115), we arrive at $\varepsilon_n g(x_n, x) \in C$, and hence

$$g(x_n, x) \in C. \tag{6.116}$$

Since g is C-strongly monotone with respect to e, we have:

$$g(x_n, x) + g(x, x_n) + \gamma \|x_n - x\|^2 e \in -C. \tag{6.117}$$

Plugging (6.116) into (6.117), we obtain $g(x, x_n) + \gamma \|x_n - x\|^2 e \in -C$, which implies that

$$\frac{1}{\|x_n - x\|} g(x, x_n) + \gamma \|x_n - x\| e \in -C. \tag{6.118}$$

Let us assume that the sequence $\{x_n\}$ is not bounded. For each $c \in \operatorname{int} C$, setting $N = (C - c) \cap (c - C)$, we see that N is a neighborhood of 0_Y. Because of satisfying Condition (H2), there exists $\lambda > 0$ such that for n sufficiently large

$$\frac{1}{\|x_n - x\|} g(x, x_n) \in \lambda N.$$

On the other hand, since $e \in \operatorname{int} C$ and $\|x_n - x\| \to \infty$, from Lemma 6.117(a), it holds eventually that

$$\gamma \|x_n - x\| e - \lambda c \in \operatorname{int} C.$$

Hence, due to the convexity of C, we have

$$\gamma \|x_n - x\| e - \lambda c + \lambda C \subset \operatorname{int} C + C = \operatorname{int} C.$$

Consequently, $\gamma \|x_n - x\| e + \lambda N \subset \operatorname{int} C$, which is a contradiction to (6.118). Therefore, the sequence $\{x_n\}$ is bounded.

(b) $\Rightarrow$ (a) For each sequence of positive real numbers $\{\varepsilon_n\}$, $\varepsilon_n \to 0$, let $x_n := x(\varepsilon_n)$ be a solution to RSEP(f_{ε_n}, K). Since $\lim_{\varepsilon \to 0} \sup \|x(\varepsilon)\| < \infty$, we can assume that the sequence $\{x_n\}$ has a subsequence (will be still denoted by $\{x_n\}$ for simplify) converging to some $\bar{x} \in K$. We show that $\bar{x} \in$ Sol(SEP)(f, K). It follows from $x_n \in$ Sol(SEP)(f_{ε_n}, K) that for all $y \in K$,

$$f(x_n, y) + \varepsilon_n g(x_n, y) \in C. \tag{6.119}$$

For each neighborhood U of the origin in Y, there exits a balanced neighborhood U_1 of the origin in Y such that $U_1 \subset U$. Because $f(\cdot, y)$ and $g(\cdot, y)$ are C-upper semicontinuous at $\bar{x}$, we know from Proposition 6.116 that the map $(\varepsilon_n, x) \mapsto f(x, y) + \varepsilon_n g(x, y)$ is C-upper semicontinuous at $(0, \bar{x})$. Thus, there is $n_0 \in \mathbb{N}$ such that for each $n \geq n_0$, one has:

$$f(x_n, y) + \varepsilon_n g(x_n, y) \in f(\bar{x}, y) + U_1 - C,$$

or equivalently, $f(\bar{x}, y) \in f(x_n, y) + \varepsilon_n g(x_n, y) - U_1 + C$. Combining this with (6.119) and taking into account the balance of U_1, we arrive at

$$f(\bar{x}, y) \in U_1 + C \subset U + C.$$

Since U is arbitrary and C is closed, we have $f(\bar{x}, y) \in C$. Therefore, $\bar{x} \in$ Sol(SEP)(f, K), that is Sol(SEP)$(f, K) \neq \emptyset$. The proof is complete. □

In Theorem 6.138, the non-emptiness of the solution set to Problem RSEP(f_ε, K) plays an important role, so we will study existence conditions for this problem. We first recall an existence result for the strong vector equilibrium problem SEP(h, K), where $h : X \times X \to Y$ is an objective map.

Proposition 6.139 *Assume that K is convex, and for each $x, y \in K$ the following conditions hold:*

(i) *$h(\cdot, y)$ is C-upper semicontinuous for each $y \in K$;*
(ii) *$\{y \in X : h(x, y) \notin C\}$ is convex for each $x \in K$;*
(iii) *h satisfies Condition (H1'): There exists a compact convex set D of K and $y \in D$ such that $h(x, y) \notin C$ for each $x \in K \setminus D$.*

Then, Sol(SEP)(h, K) *is nonempty.*

By the virtue of Proposition 6.139, we provide a convergence analysis for the strong form of monotone vector equilibrium problems.

Theorem 6.140 *Assume that K is convex, and for each $x, y \in K$, the following conditions are satisfied:*

(i) *$f(x, \cdot)$ and $g(x, \cdot)$ are C-lower semicontinuous and C-convex;*
(ii) *$f(\cdot, y)$ and $g(\cdot, y)$ are C-upper semicontinuous;*
(iii) *for each $\varepsilon > 0$, the set $\{y \in X : f_\varepsilon(x, y) \notin C\}$ is convex for each $x \in K$;*
(iv) *f is C-monotone on K.*

Then, the regularized problem RSEP(f_ε, K) has a solution $x(\varepsilon)$ for each $\varepsilon > 0$, and any cluster point of the sequence $\{x(\varepsilon)\}_{\varepsilon>0}$ is a solution to Problem SEP(f, K).

Proof For each $\varepsilon > 0$, it is obvious that $f_\varepsilon(\cdot, y)$ is C-upper semicontinuous for all $y \in K$, and f_ε is C-strongly monotone as f is C-monotone and g is C-strongly monotone. Combining this with the fact that f_ε is C-convex, and C-lower semicontinuous with respect to the second argument, we conclude by applying Lemma 6.131 that the condition (H1') holds true for f_ε. According to Proposition 6.139, the regularized problem RSEP(f_ε, K) admits a solution $x(\varepsilon)$. Using the same arguments given as in the second part of Theorem 6.138, we obtain the remainder statement. □

We now give an example to illustrate Theorem 6.140.

Example 6.141 Let $X = \mathbb{R}, Y = \mathbb{R}^2$, $K = [0, \infty[, C = \mathbb{R}^2_+$ and a bifunction $f, g : K \times K \to \mathbb{R}^2$ be defined for all $x, y \in K$ by

$$f(x, y) := \left(\frac{(y - x)(x - 1)}{x + 2}, \frac{(y - x)(x - 1)}{x + 3} \right), \text{ and}$$

$$g(x, y) := \left(\frac{x(y - x)}{x + 2}, \frac{x(y - x)}{x + 3} \right).$$

Then, all conditions in Theorem 6.140 are satisfied, and hence its conclusion is valid. In fact, by direct computations one has Sol(SEP)$(f, 1) = \{1\}$, Sol(RSEP)$(f_\varepsilon, K) = \left\{(1 + \varepsilon)^{-1}\right\}$.

In the rest of this section, we address a regularization method for the strong form of nonmonotone and nonconvex vector equilibrium problems. In this way, we give an approach to study the solvability of strong vector equilibrium problems by the virtue of techniques appeared in vector optimization. For the problem SEP(h, K), we consider the following condition:

(D1) There exist a map $p : X \to Y$ such that $h(x, y) \in p(y) - p(x) + C$ for all $x, y \in K$.

We consider the following vector optimization problem:

VOP(p, K) Minimize $p(x)$ subject to $x \in K$.

A feasible solution $\bar{x} \in K$ is an *efficient solution* to VOP(p, K) (denoted by $\bar{x} \in \text{Eff}(p, K)$) if there is no any $x \in K$ such that $p(x) \in p(\bar{x}) - C \setminus \{0_Y\}$ (or equivalently, $p(x) \in p(\bar{x}) - C$ for some $x \in K$ implies $p(x) = p(\bar{x})$), an *ideal solution* to VOP(p, K) (denoted by $\bar{x} \in \text{IEff}(p, K)$) if $p(\bar{x}) \in p(x) - C$ for all $x \in K$.

It is easy to see that $\text{IEff}(p, K) \subset \text{Eff}(p, K)$. Motivated by the concepts of generalized convexity proposed in [188], which were employed to study the relationships between weak and efficient solutions, we introduce a concept of generalized pseudo-convex-likeness and use it to establish the converse inclusion.

Definition 6.142 Let K be a nonempty subset of X, and $g : X \to Y$ be a vector-valued map. The map g is said to be EXPLICITLY PSEUDO C-CONVEX-LIKE on K if for every $x, y \in K, x \neq y$, there exists $z \in K$ and $\lambda \in [0, 1]$ such that

$$\lambda g(x) + (1 - \lambda) g(y) \in g(z) + C \setminus \{0\}.$$

Proposition 6.143 *If p is C-lower semicontinuous on a compact set K, then* $\text{Eff}(p, K)$ *is nonempty. Furthermore,* $\text{IEff}(p, K) = \text{Eff}(p, K)$ *whenever p is explicitly pseudo C-convexlike on K.*

Proof The proof for the non-emptiness of $\text{Eff}(p, K)$ can be found in [343, Theorem 4.20]. For the second assertion, it suffices to show that $\text{Eff}(p, K) \subset \text{IEff}(p, K)$. Let $\bar{x} \in \text{Eff}(p, K)$, suppose that there exists $x \in X$ satisfying $p(\bar{x}) \in p(x) + C$. Since $\bar{x} \in \text{Eff}(p, K)$, one has $p(\bar{x}) = p(x)$. Because p is explicitly pseudo C-convexlike, there exists $y \in K$ such that $p(\bar{x}) \in p(y) + C \setminus \{0_Y\}$, which is impossible as $\bar{x} \in \text{Eff}(p, K)$. Therefore, $\bar{x} \in \text{IEff}(p, K)$.

□

Theorem 6.144 *Assume that K is compact and there exists a C-lower semicontinuous and explicitly pseudo C-convexlike map $p : X \to Y$ such that condition (D1) holds true for h. Then,* $\text{Sol(SEP)}(h, K)$ *is nonempty.*

Proof In the light of Proposition 6.143, $\text{IEff}(p, K)$ is nonempty, and the solvability of SEP(h, K) follows from $\text{IEff}(p, K) \subset \text{Sol(SEP)}(h, K)$. □

Remark 6.145 Clearly, if $h(x, y) \in C$ for all $x, y \in K$, condition (D1) is satisfied whenever p is a constant function. Moreover, if h satisfies the triangle inequality property with respect to $\leq_C$ (i.e., $h(x, y)+h(y, z) \leq_C h(x, z)$ for all $x, y, z \in K$), then condition (D1) also holds true. For example, for a fixed but arbitrary element $z \in K$, we can choose $p_z = h(z, \cdot)$, or $p_z = -h(\cdot, z)$, and in this manner the C-lower semicontinuity of p is valid if we assume that h is C-upper semicontinuous in the first argument. We would like to note here that condition (D1) has a potential application to discuss vector Ekeland's variational principle, and hence a class of maps satisfying condition (D1) should be studied more in the future.

In the case of which the feasible is noncompact, we consider the following condition.

(D2) There exists a compact set D of K such that for all $x \in K \setminus D$ there exists $y \in K$ satisfying $\|y\| < \|x\|$ and $p(x) \in p(y) + C$.

Theorem 6.146 *The statement of Theorem 6.144 is still valid if the compactness of K is replaced by condition (D2).*

Proof According to Proposition 6.143, VOP(p, D) admits an ideal solution $\bar{x}$. We show that $\bar{x}$ is also an ideal solution to VOP(p, K). Suppose to the contrary that there is $x \in K \setminus D$ such that $p(x) \notin p(\bar{x}) + C$, or equivalently

$$p(x) - p(\bar{x}) \in Y \setminus C. \tag{6.120}$$

By setting $E = \{z \in X : p(x) \in p(z) + C\}$ and $L = B(0_X, \|x\|) \cap E$, where $B(a, r)$ is a closed ball centered at a with radius r, we obtain that L is nonempty and weakly compact. Let $\hat{x}$ be an element in L with the least norm (this is possible due to the weak lower semicontinuity of the norm). There are two cases:
Case 1. $\hat{x} \in D$. Since $\hat{x} \in E$, one has $p(\hat{x}) \in p(x) - C$. This together with (6.120) gives:

$$p(\hat{x}) - p(\bar{x}) \in p(x) - p(\bar{x}) - C \subset Y \setminus C - C = Y \setminus C,$$

which is impossible because $\bar{x} \in$ IEff(p, D).
Case 2. $\hat{x} \notin D$. It follows from condition (D2) that there is $y \in K$ with $\|y\| < \|\hat{x}\| \leq \|x\|$ satisfying $p(\hat{x}) \in p(y) + C$. Combining this with the fact that $\hat{x} \in E$ and the convexity of C, one has $p(x) \in p(y) + C$. Thus, $y \in L$, but this is a contradiction because $\hat{x}$ is an element with the least norm.

Therefore, $\bar{x} \in$ IEff(p, K), and we conclude that Sol(SEP)(h, K) is nonempty. □

We are now ready to establish a basic convergence result of the regularization method for the strong form of nonmonotone vector equilibrium problems.

Theorem 6.147 *Assume that the following conditions are satisfied:*

(i) *$f(\cdot, y)$ and $g(\cdot, y)$ are C-upper semicontinuous for all $y \in K$;*
(ii) *for each $\varepsilon > 0$, there exists a map $p_\varepsilon : K \to Y$ such that condition (D1) holds true for f_ε;*

(iii) p_ε *is* C*-lower semicontinuous and explicitly pseudo* C*-convexlike on* K;
(iv) p_ε *holds condition (D2).*

Then, the regularized problem RSEP(f_ε, K) *has a solution* $x(\varepsilon)$ *for each* $\varepsilon > 0$, *and any cluster point of the sequence* $\{x(\varepsilon)\}_{\varepsilon>0}$ *is a solution to Problem* SEP(f, K).

Proof It follows from Theorem 6.146 that for each $\varepsilon > 0$, the regularized problem RSEP(f_ε, K) has a solution $x(\varepsilon)$. The rest assertion is obtained by repeating the same arguments which already presented in the second part proof of Theorem 6.138. □

To finalize this section, we give an example to illustrate Theorem 6.147.

Example 6.148 Let $X = \mathbb{R}$, $Y = \mathbb{R}^2$, $K = \{0\} \cup [1, \infty[$, $C = \mathbb{R}^2_+$ and a bifunction $f, g : K \times K \to \mathbb{R}^2$ be defined for all $x, y \in K$ by

$$f(x, y) := \left(x^2 + y - x + (x - y)^2, x^4 + y - x + (y - x)^2\right), \text{ and}$$

$$g(x, y) := \left(y - x - (x - y)^2, y - x - (x - y)^2\right).$$

For all $\varepsilon \in [0, 1]$, condition (D1) holds true for $p_\varepsilon(x) = (1 + \varepsilon)(x, x)$. It is obvious that all conditions in Theorem 6.147 are satisfied, and hence for each $\varepsilon \in [0, 1]$, the problem RSEP(f_ε, K) has a solution.

6.7 Relationships Between Vector Set Optimization Problems and Equilibrium Problems

From now on in this chapter, unless otherwise specified, let X, Y and W be normed spaces, Λ a nonempty subset of W, $C \subset Y$ a pointed closed convex cone with int$C \neq \emptyset$. Let $\mathscr{P}_0(Y)$ be the family of nonempty subsets of Y. For $A, B \in \mathscr{P}_0(Y)$, we define l-type and u-type set less order relations wrt C as follows:

$$A \preceq^l_C B \Leftrightarrow B \subset A + C \quad \text{and} \quad A \prec^l_C B \Leftrightarrow B \subset A + \text{int}C,$$

$$A \preceq^u_C B \Leftrightarrow A \subset B - C \quad \text{and} \quad A \prec^u_C B \Leftrightarrow A \subset B - \text{int}C.$$

Let $\mathscr{P}_{0C}(Y)$ and $\mathscr{P}_{0(-C)}(Y)$ be the family of all nonempty, C-proper and $-C$-proper subsets of Y, resp. Let $e \in -\text{int}C$, $C^* := \{\xi \in Y^* \mid \xi(c) \geq 0 \ \forall c \in C\}$, and $B^* := \{\xi \in C^* \mid \xi(-e) = 1\}$. We see that B^* is a weak* compact base of C^*.

The following notions are needed in what follows.

Definition 6.149 (Cone-proper, cone-closed, cone-bounded, and cone-compact sets) Let A be a nonempty subset of Y:

(a) A is said to be C-PROPER if $A + C \neq Y$.
(b) A is said to be C-CLOSED if $A + C$ is closed.
(c) A is said to be C-BOUNDED if for each neighborhood U of zero in Y, there is $t > 0$ satisfying $A \subset tU + C$.
(d) A is said to be C-COMPACT if any cover of A of the form $\{U_\alpha + C \mid U_\alpha \text{ are open}\}$ admits a finite subcover.

A function $k : X \times X \times X \to \mathbb{R}$ is said to be $(\ell_1\alpha_1, \ell_2\alpha_2, \ell_3\alpha_3)$-HÖLDER CONTINUOUS at $(x_0, y_0, z_0) \in X \times X \times X$ if there exists a neighborhood $N_1 \times N_2 \times N_3$ of (x_0, y_0, z_0) such that for all $(x_1, y_1, z_1), (x_2, y_2, z_2) \in N_1 \times N_2 \times N_3$

$$|k(x_1, y_1, z_1) - k(x_2, y_2, z_2)| \leq \ell_1\|x_1 - x_2\|^{\alpha_1} + \ell_2\|y_1 - y_2\|^{\alpha_2} + \ell_3\|z_1 - z_2\|^{\alpha_3}.$$

We say that a function satisfies a certain property on a subset D if it satisfies that property at every point in D.

Let $F : X \rightrightarrows Y$ and $A \subset X$. F is said to be:

(a) C-CONVEX on A if for any $x_1, x_2 \in A$ and $t \in [0, 1]$,

$$tF(x_1) + (1 - t)F(x_2) \subset F(tx_1 + (1 - t)x_2) + C;$$

(b) C-CONCAVE on A if for any $x_1, x_2 \in A$ and $t \in [0, 1]$,

$$F(tx_1 + (1 - t)x_2) \subset tF(x_1) + (1 - t)F(x_2) + C;$$

(c) C-QUASICONVEX on A if for any $x_1, x_2 \in A$ and $t \in [0, 1]$, either

$$F(x_1) \subset F(tx_1 + (1 - t)x_2) + C,$$

or

$$F(x_2) \subset F(tx_1 + (1 - t)x_2) + C;$$

(d) C-QUASICONCAVE on A if for any $x_1, x_2 \in A$ and $t \in [0, 1]$, either

$$F(tx_1 + (1 - t)x_2) \subset F(x_1) + C,$$

or

$$F(tx_1 + (1 - t)x_2) \subset F(x_2) + C.$$

It is easy to verify that if F is C-convex (C-concave) on A, then it is also C-quasiconvex (resp., C-quasiconcave) on A.

In the case that F is single-valued, $Y = \mathbb{R}$, and $C = \mathbb{R}_+$, the above concepts related to convexity reduce to the classical ones.

Definition 6.150 (Level quasiconcave maps) Let $A \subset X$ be nonempty and convex and $\varphi : A \to \mathbb{R}$. φ is said to be 0-LOWER-LEVEL QUASICONCAVE on A if for each $x_1, x_2 \in A$, $\lambda \in [0, 1]$, and $\varphi(\lambda x_1 + (1 - \lambda)x_2) < 0$, then either $\varphi(x_1) < 0$ or $\varphi(x_2) < 0$.

Now we provide some conditions equivalent to the 0-lower-level quasiconcavity.

Lemma 6.151 (Equivalent statements of 0-level quasiconcave conditions) *Let φ and A be as in Definition 6.150. The following statements are equivalent:*

(a) *φ is 0-lower-level quasiconcave on A.*
(b) *For each finite subset $\{x_1, x_2, \ldots, x_n\} \subset A$ and $x \in \text{conv}\{x_1, x_2, \ldots, x_n\}$ with $\varphi(x) < 0$, there is $i \in \{1, 2, \ldots, n\}$ such that $\varphi(x_i) < 0$, where $\text{conv}\{x_1, x_2, \ldots, x_n\}$ stands for the convex hull of $\{x_1, x_2, \ldots, x_n\}$, i.e., $\text{conv}\{x_1, x_2, \ldots, x_n\} = \{x \in X \mid x = \sum_{i=1}^{n} \lambda_i x_i, \lambda_i \geq 0, \sum_{i=1}^{n} \lambda_i = 1\}$.*
(c) *The set $\mathscr{A} := \{x \in A \mid \varphi(x) \geq 0\}$ is convex.*

Proof (a) $\Rightarrow$(b). It follows from the definition of φ.

(b) $\Rightarrow$ (c). For all $x_1, x_2 \in \mathscr{A}$ and $t \in [0, 1]$, one has $\varphi(x_1) \geq 0$ and $\varphi(x_2) \geq 0$. If

$$\varphi(tx_1 + (1 - t)x_2) < 0,$$

then either $\varphi(x_1) < 0$ or $\varphi(x_2) < 0$, which contradicts the preceding relation. Thus,

$$\varphi(tx_1 + (1 - t)x_2) \geq 0,$$

and hence $\mathscr{A}$ is convex.

(c) $\Rightarrow$ (a) Suppose that φ is not 0-lower level quasiconcave on A. Then, there are $x_1, x_2 \in \mathscr{A}$ and $t \in [0, 1]$ with

$$\varphi(tx_1 + (1 - t)x_2) < 0.$$

Since $\mathscr{A}$ is convex, we get

$$\varphi(tx_1 + (1 - t)x_2) \geq 0,$$

which is a contradiction. Therefore, φ is 0-lower-level quasiconcave on A. □

Remark 6.152 (Relationships between level quasiconcavity and quasiconcavity properties) If φ is quasiconcave in A, i.e.,

$$\varphi(\lambda x_1 + (1 - \lambda)x_2) \geq \min\{\varphi(x_1), \varphi(x_2)\},$$

for all $x_1, x_2 \in A$ and $\lambda \in [0, 1]$, then it is 0-lower level quasiconcave on A.

The following example shows that the converse of Remark 6.152 is not true.

Example 6.153 (A case of level quasiconcavity but not quasiconcavity) Let $A = \mathbb{R}$ and $\varphi : A \to \mathbb{R}$ be defined by $\varphi(x) := -x^4 + 2x^2$ for all $x \in A$. Then,

$$\{x \in A \mid \varphi(x) \geq 0\} \equiv [-\sqrt{2}, \sqrt{2}],$$

and hence it is convex. Lemma 6.151 says that φ is 0-lower-level quasiconcave on A. However, for $x_1 = -1, x_2 = 1 \in A$, and $\lambda = \frac{1}{2}$, we have

$$\min\{\varphi(x_1), \varphi(x_2)\} = 1,$$

while

$$\varphi(\lambda x_1 + (1 - \lambda)x_2) = 0.$$

So, φ is not quasiconcave on A.

Remark 6.154 (Level-quasiconvexity and weak quasiconvexity) As usual, we say that φ is a 0-lower-level quasiconvex function if $-\varphi$ is 0-lower level quasiconcave. The following examples show that, in general, the 0-lower level quasiconvexity and the WEAK QUASICONVEXITY (in the sense that there exists $y_0 \in \text{dom}\varphi$ such that the set $\{x \in X \mid \varphi(x) \leq \lambda\}$ is convex for all $\lambda \geq \varphi(y_0)$) are incomparable to each other.

Example 6.155 (Level-quasiconvexity and weak convexity are incomparable)

(a) Let $A = (-\infty, 0]$ and $f(x) : A \to \mathbb{R}$ be defined by $f(x) := x \cos x - 2x$ for all $x \in A$. Clearly, f is 0-lower-level quasiconvex on A. Then, for all $y \in \text{dom}f$, there exists $\lambda_0 > f(y)$ such that the level set $\{x \in A \mid f(x) \leq \lambda_0\}$ consists of the intervals whose intersection is empty, and hence f is not weakly quasiconvex on A.
(b) The function $g(x) = x^3 - 3x^2 + 2$ for all $x \in \mathbb{R}$ is weakly quasiconvex on $\mathbb{R}$, but not 0-lower-level quasiconvex on $\mathbb{R}$. Indeed, if $x_1 = -1, x_2 = 2$, and $t = \frac{1}{2}$, then $g(\frac{1}{2}) = \frac{11}{8}$ while $g(-1) = g(2) = -2$.

For $A, B \subset X$, in what follows we will use the Hausdorff distance $H(A, B)$ of A and B. In order to be convenient in considering set optimization problems involving set less order relations via the nonlinear scalarization method, we will recall and modify many versions of the Gerstewitz functions.

Definition 6.156 (Generalized Gerstewitz functions) Let $a \in Y$, $A \subset Y$ be nonempty, and $e' \in \text{int}C$. Then, we consider the following functions:

(a) $g^u_{e',a} : Y \to \mathbb{R}$ defined by $g^u_{e',a}(y) := \inf\{t \in \mathbb{R} \mid y \in te' + a - C\}$ $\forall y \in Y$;
(b) $G^u_{e',A} : Y \to \mathbb{R}$ defined by $G^u_{e',A}(y) := \inf\{t \in \mathbb{R} \mid y \in te' + A - C\}$ $\forall y \in Y$;
(c) $\mathcal{G}^u_{e'} : Y \times Y \to \mathbb{R} \cup \{\infty\}$ defined by $\mathcal{G}^u_{e'}(P, Q) := \sup_{p \in P} G^u_{e',Q}(p)$ $\forall P, Q \in Y$.

Motivated by these definitions, we define the Gerstewitz and generalized Gerstewitz functions for the l-type set less order relation as follows.

Definition 6.157 (Dual generalized Gerstewitz functions) Let $a \in Y$, $A \subset Y$ be nonempty, and $e \in -\text{int}C$. The Gerstewitz function and generalized Gerstewitz function are the following:

(a) $g^l_{e,a} : Y \to \mathbb{R}$ is defined by $g^l_{e,a}(y) := \inf\{t \in \mathbb{R} \mid y \in te + a + C\}\ \forall y \in Y$.
(b) $G^l_{e,A} : Y \to \mathbb{R}$ is defined by $G^l_{e,A}(y) := \inf\{t \in \mathbb{R} \mid y \in te + A + C\}\ \forall y \in Y$.
(c) $\mathscr{G}^l_e : Y \times Y \to \mathbb{R}\cup\{\infty\}$ is defined by $\mathscr{G}^l_e(P, Q) := \sup_{q\in Q} G^l_{e,P}(q)\ \forall P, Q \in Y$.

Clearly, $G^l_{e,A}(y) = \inf_{a\in A} g^l_{e,a}(y)$ and $G^u_{e',A}(y) = \inf_{a\in A} g^u_{e',a}(y)$ for any given $y \in Y$.

Proposition 6.158 (Estimations of types of Gerstewitz functions)

(a) *Let* $A, B \in \mathscr{P}_{0(-C)}(Y)$ *and* $e \in \text{int}C$. *If* B *is* $-C$*-closed, then*

$$G^u_e(A, B) \le r \Leftrightarrow A \subset re + B - C.$$

(b) *Let* $A, B \in \mathscr{P}_{0(C)}(Y)$ *and* $e' \in -\text{int}C$. *If* A *is* C*-closed, then*

$$G^l_{e'}(A, B) \le r \Leftrightarrow B \subset re' + A + C.$$

Proof

(a) We see that $G^u_e(A, B) \le r$ if and only if $\varphi^u_{e,B}(a) \le r$ for all $a \in A$. Hence, we have:

$$a \in re + B + \text{cl}(B - C).$$

Due to the $-C$-closedness of B, we get:

$$a \in re + B + B - C.$$

Therefore, $A \subset re + B - C$.
(b) This assertion is proved similarly as part (a).

□

Remark 6.159 (Characterizations of nonpositive values of the generalized Gerstewitz function) In the special case with $r = 0$, from Proposition 6.158, we have

$$G^u_e(A, B) \le 0 \Leftrightarrow A \preceq^u_C B,$$

$$G^l_{e'}(A, B) \le 0 \Leftrightarrow A \preceq^l_C B.$$

Let $F : X \rightrightarrows Y$ be nonempty-valued. For $a \in Y$, $e \in -\text{int}C$, and $e' \in \text{int}C$, we consider the functions $\tau^l_a, \tau^u_a : X \to \mathbb{R}$ (called the Gerstewitz functions):

$$\tau_a^l(x) := G^l_{e,F(x)}(a) = \inf\{t \in \mathbb{R} \mid a \in te + F(x) + C\} \; \forall x \in X,$$

$$\tau_a^u(x) := G^u_{e',F(x)}(a) = \inf\{t \in \mathbb{R} \mid a \in te' + F(x) - C\} \; \forall x \in X.$$

The functions $\mathcal{T}^l, \mathcal{T}^u : X \times X \to \mathbb{R} \cup \{\infty\}$ (also called generalized Gerstewitz functions) are defined as follows, for $x, y \in X$,

$$\mathcal{T}^l(x, y) := \mathcal{G}^l_e(F(x), F(y)) = \sup_{a \in F(y)} \tau_a^l(x) :$$

$$\mathcal{T}^u(x, y) := \mathcal{G}^u_{e'}(F(x), F(y)) = \sup_{a \in F(x)} \tau_a^u(y).$$

The following results are easy to verify.

Lemma 6.160 (Finite values conditions for generalized Gerstewitz functions)

(a) *If F has C-proper and C-bounded values, then*

$$-\infty < \mathcal{T}^l(x, y) < +\infty$$

for all $(x, y) \in X \times X$.

(b) *If F has $-C$-proper and $-C$-bounded values, then*

$$-\infty < \mathcal{T}^u(x, y) < +\infty$$

for all $(x, y) \in X \times X$.

In what follows, we discuss some properties of the aforementioned functions. These properties play important roles for investigating the stability analysis of solutions of (SOPs).

Lemma 6.161 (Calculating Gerstewitz functions) *For $x, y \in X$, the following statements are true:*

(a) $\tau_a^u(x) = \inf_{b \in F(x)} \sup_{\xi \in B^*} \xi(a - b)$ *for each $a \in Y$.*
(b) $\mathcal{T}^l(x, y) = \sup_{b \in F(y)} \inf_{a \in F(x)} \sup_{\xi \in B^*} \xi(a - b)$.

Proposition 6.162 (Convexity and quasiconvexity of Gerstewitz functions) *For each $a \in Y$, the following assertions hold:*

(a) τ_a^l *is convex (resp., quasiconvex) on X provided that F is C-convex (resp., C-quasiconvex) on X.*
(b) τ_a^u *is convex (resp., quasiconvex) on X provided that F is $-C$-convex (resp., $-C$-quasiconvex) on X.*

Proof We only provide the proof for τ_a^u (for (a), the same techniques are used).
* [*F is $-C$-convex*] $\Rightarrow$ [*τ_a^u is convex*]

For all $y_1, y_2 \in X$ and $t \in [0, 1]$, the $-C$-convexity of F gives us:

$$tF(y_1) + (1 - t)F(y_2) \subset F(ty_1 + (1 - t)y_2) - C.$$

For any $b_1 \in F(y_1)$ and $b_2 \in F(y_2)$, there exist $b \in F(ty_1 + (1-t)y_2)$ and $c \in C$ such that

$$b = tb_1 + (1-t)b_2 + c.$$

For every $\xi \in B^*$, one has:

$$\xi(a-b) \le t\xi(a-b_1) + (1-t)\xi(a-b_2).$$

Therefore,

$$\sup_{\xi \in B^*} \xi(a-b) \le t \sup_{\xi \in B^*} \xi(a-b_1) + (1-t) \sup_{\xi \in B^*} \xi(a-b_2),$$

which implies that

$$\inf_{\bar{b} \in F(ty_1+(1-t)y_2)} \sup_{\xi \in B^*} \xi(a-\bar{b}) \le t \sup_{\xi \in B^*} \xi(a-b_1) + (1-t) \sup_{\xi \in B^*} \xi(a-b_2).$$

Consequently,

$$\inf_{\bar{b} \in F(ty_1+(1-t)y_2)} \sup_{\xi \in B^*} \xi(a-\bar{b})$$

$$\le t \inf_{\bar{b}_1 \in F(y_1)} \sup_{\xi \in B^*} \xi(a-\bar{b}_1) + (1-t) \inf_{\bar{b}_2 \in F(y_2)} \sup_{\xi \in B^*} \xi(a-\bar{b}_2).$$

By Lemma 6.161(a), we obtain:

$$\tau_a^u(ty_1 + (1-t)y_2) \le t\tau_a^u(y_1) + (1-t)\tau_a^u(y_2).$$

* [*F is $-C$-quasiconvex*] $\Rightarrow$ [*τ_a^u is quasiconvex*]

For all $y_1, y_2 \in X$ and $t \in [0,1]$, since F is $-C$-quasiconvex, one has two cases.
Case 1. $F(y_1) \subset F(ty_1 + (1-t)y_2) - C$. For each $\xi \in B^*$ and $b_1 \in F(y_1)$, there are $b \in F(ty_1 + (1-t)y_2)$ and $c \in C$ such that $b_1 = b - c$. For all $a \in F(y)$, one has:

$$\xi(a-b_1) = \xi(a-b) + \xi(c)$$

with $\xi(c) \ge 0$, that is $\xi(a-b) \le \xi(a-b_1)$. Therefore,

$$\sup_{\xi \in B^*} \xi(a-b) \le \sup_{\xi \in B^*} \xi(a-b_1),$$

which implies that

$$\inf_{\bar{b}\in F(ty_1+(1-t)y_2)} \sup_{\xi\in B^*} \xi(a-\bar{b}) \le \sup_{\xi\in B^*} \xi(a-b_1).$$

Hence,

$$\inf_{\bar{b}\in F(ty_1+(1-t)y_2)} \sup_{\xi\in B^*} \xi(a-\bar{b}) \le \inf_{\bar{b}_1\in F(y_1)} \sup_{\xi\in B^*} \xi(a-\bar{b}_1),$$

i.e.,

$$\tau_a^u(ty_1+(1-t)y_2) \le \tau_a^u(y_1). \tag{6.121}$$

Case 2. $F(y_2) \subset F(ty_1+(1-t)y_2) - C$. Similar to Case 1, we also have:

$$\tau_a^u(ty_1+(1-t)y_2) \le \tau_a^u(y_2). \tag{6.122}$$

Combining (6.121) and (6.122), one has $\tau_a^u(ty_1+(1-t)y_2) \le \max\{\tau_a^u(y_1), \tau_a^u(y_2)\}$.
□

Proposition 6.163 (Convexity and quasiconvexity of generalized Gerstewitz functions)

(a) $\mathscr{T}^l$ *is convex (resp., quasiconvex) in the first component on* X *provided that* F *is* C*-convex (resp.,* C*-quasiconvex) on* X.
(b) $\mathscr{T}^u$ *is convex (resp., quasiconvex) in the first component on* X *provided that* F *is* $-C$*-concave (resp.,* $-C$*-quasiconcave) on* X.

Proof We only discuss assertion (a) as (b) can be proved similarly.
* [*F is C-convex*] $\Rightarrow$ [*$\mathscr{T}^l$ is convex*]
Thanks to the C-convexity of F, for any $x_1, x_2 \in X$ and $t \in [0, 1]$, we have:

$$tF(x_1) + (1-t)F(x_2) \subset F(tx_1+(1-t)x_2) + C. \tag{6.123}$$

For each $a_1 \in F(x_1)$ and $a_2 \in F(x_2)$, there exist $a \in F(tx_1+(1-t)x_2)$ and $c \in C$ such that

$$a = ta_1 + (1-t)a_2 - c.$$

For every $b \in F(y)$ and $\xi \in B^*$, we have:

$$\xi(a-b) \le t\xi(a_1-b) + (1-t)\xi(a_2-b).$$

Hence,

$$\sup_{\xi\in B^*} \xi(a-b) \le t \sup_{\xi\in B^*} \xi(a_1-b) + (1-t) \sup_{\xi\in B^*} \xi(a_2-b), \tag{6.124}$$

which implies that

$$\inf_{a\in F(tx_1+(1-t)x_2)} \sup_{\xi\in B^*} \xi(a-b) \le t \sup_{\xi\in B^*} \xi(a_1-b)+(1-t) \sup_{\xi\in B^*} \xi(a_2-b). \tag{6.125}$$

Therefore,

$$\begin{aligned}&\inf_{a\in F(tx_1+(1-t)x_2)} \sup_{\xi\in B^*} \xi(a-b)\\ &\quad\le t \inf_{a_1\in F(x_1)} \sup_{\xi\in B^*} \xi(a_1-b)+(1-t) \inf_{a_2\in F(x_2)} \sup_{\xi\in B^*} \xi(a_2-b).\end{aligned} \tag{6.126}$$

By Lemma 6.161(b) and (6.126), one has:

$$\mathcal{G}_e^l(F(tx_1+(1-t)x_2), F(y)) \le t\mathcal{G}_e^l(F(x_1), F(y)) + (1-t)\mathcal{G}_e^l(F(x_2), F(y)),$$

i.e.,

$$\mathcal{T}^l(tx_1+(1-t)x_2, y) \le t\mathcal{T}^l(x_1, y) + (1-t)\mathcal{T}^l(x_2, y).$$

* [*F is C-quasiconvex*] $\Rightarrow$ [$\mathcal{T}^l$ *is quasiconvex*]

For all $x_1, x_2 \in X$ and $t \in [0, 1]$, since F is C-quasiconvex, one has two cases. *Case 1.* $F(x_1) \subset F(tx_1 + (1-t)x_2) + C$. For every $\xi \in B^*$, $a_1 \in F(x_1)$, and $b \in F(y)$, there exist $a \in F(tx_1+(1-t)x_2)$ and $c \in C$ with $\xi(a-b) \le \xi(a_1-b)$. Therefore,

$$\sup_{\xi\in B^*} \xi(a-b) \le \sup_{\xi\in B^*} \xi(a_1-b).$$

Consequently,

$$\inf_{\bar a\in F(tx_1+(1-t)x_2)} \sup_{\xi\in B^*} \xi(\bar a-b) \le \sup_{\xi\in B^*} \xi(a_1-b).$$

This leads to

$$\inf_{\bar a\in F(tx_1+(1-t)x_2)} \sup_{\xi\in B^*} \xi(\bar a-b) \le \inf_{\bar a_1\in F(x_1)} \sup_{\xi\in B^*} \xi(\bar a_1-b),$$

which implies that

$$\sup_{b\in F(y)} \inf_{\bar a\in F(tx_1+(1-t)x_2)} \sup_{\xi\in B^*} \xi(\bar a-b) \le \sup_{b\in F(y)} \inf_{\bar a_1\in F(x_1)} \sup_{\xi\in B^*} \xi(\bar a_1-b).$$

This inequality means that

$$\mathcal{T}^l(tx_1+(1-t)x_2, y) \le \mathcal{T}^l(x_1, y). \tag{6.127}$$

Case 2. $F(x_2) \subset F(tx_1 + (1-t)x_2) + C$. Similar to Case 1, we also have:

$$\mathcal{T}^l(tx_1 + (1-t)x_2, y) \leq \mathcal{T}^l(x_1, y). \tag{6.128}$$

Combining (6.127) and (6.128), one has:

$$\mathcal{T}^l(tx_1 + (1-t)x_2, y) \leq \max\{\mathcal{T}^l(x_1, y), \mathcal{T}^l(x_2, y)\}.$$

The proof is complete. □

Proposition 6.164 (Continuity of generalized Gerstewitz functions) *If F is continuous with nonempty and compact values on X, then $\mathcal{T}^u$ and $\mathcal{T}^l$ are continuous on $X \times X$.*

Proof Because of the similarity, we only prove the continuity of $\mathcal{T}^u$. To see the upper semicontinuity of $\mathcal{T}^u$, we show that for all $\alpha_0 \in \mathbb{R}$, the subset

$$P := \{(x, y) \in X \times X \mid \mathcal{T}^u(x, y) \geq \alpha_0\}$$

is closed. Let $(x^k, y^k) \in P$ with $(x^k, y^k) \to (x_0, y_0)$. We show that $(x_0, y_0) \in P$. If $\mathcal{T}^u(x_0, y_0) < \alpha_0$, then

$$\mathcal{T}^u(x_0, y_0) = G_e^u(F(x_0), F(y_0)) = \sup_{a \in F(x_0)} \{\varphi^u_{e, F(y_0)}(a)\} < \alpha_0.$$

Hence, $\varphi^u_{e, F(y_0)}(a) < \alpha_0$ for all $a \in F(x_0)$. This yields that

$$F(x_0) \subset \alpha_0 e + F(y_0) - \text{int} C. \tag{6.129}$$

We claim that

$$F(x^k) \subset \alpha_0 e + F(y^k) - \text{int} C,$$

for k large enough. Suppose that we can find a subsequence $\{x^{k_j}\}$ satisfying

$$F(x^{k_j}) \not\subset \alpha_0 e + F(y^{k_j}) - \text{int} C,$$

for all k_j. Then, there exist $v^{k_j} \in F(x^{k_j})$ such that

$$v^{k_j} \notin \alpha_0 e + F(y^{k_j}) - \text{int} C. \tag{6.130}$$

Since F is usc with compact values, we can assume that $v^{k_j} \to v_0 \in F(x_0)$. From (6.129), there exists $u_0 \in F(y_0)$ such that

$$v_0 \in \alpha_0 e + u_0 - \text{int} C. \tag{6.131}$$

Since F is lsc, there exists $u^{k_j} \in F(y^{k_j})$ such that $u^{k_j} \to u_0$. It follows from (6.131) that

$$v^{k_j} \in \alpha_0 e + u^{k_j} - \mathrm{int}C,$$

for k_j large enough, which contradicts (6.130). Hence, there is $k_0 \in \mathbb{N}$ such that

$$F(x^k) \subset \alpha_0 e + F(y^k) - \mathrm{int}C,$$

for all $k \geq k_0$. Therefore, we get:

$$\mathscr{T}^u(x^k, y^k) < \alpha_0,$$

which is impossible as $(x^k, y^k) \in P$.

Moving on to the lower semicontinuity of $\mathscr{T}^u$, we show that, for all $\alpha_0 \in \mathbb{R}$, the subset

$$Q := \{(x, y) \in X \times X \mid \mathscr{T}^u(x, y) \leq \alpha_0\}$$

is closed. Let $(x^k, y^k) \in Q$ converge to (x_0, y_0), we prove that $\mathscr{T}^u(x_0, y_0) \leq \alpha_0$. If $\mathscr{T}^u(x_0, y_0) > \alpha_0$, then there exists $a_0 \in F(x_0)$ such that $\varphi^u_{e,F(y_0)}(a_0) > \alpha_0$. Consequently, one has:

$$a_0 \notin \alpha_0 e + \mathrm{cl}(F(y_0) - C). \tag{6.132}$$

Applying Proposition 6.158, we can pick $a^k \in F(x^k)$ and $b^k \in F(y^k)$ satisfying

$$a^k \in \alpha_0 e + b^k - C.$$

Since F is usc and compact-valued, we can assume that $b^k \to b_0 \in F(y_0)$. From (6.132), we obtain:

$$a_0 \notin \alpha_0 e + b_0 - C.$$

By the closedness of $\alpha_0 e - C$, we have

$$a^k \notin \alpha_0 e + b^k - C$$

for k large enough, a contradiction. □

Proposition 6.165 (Hölder continuity of generalized Gerstewitz functions) *For any $a \in Y$, if F is Hölder continuous on X, then so are τ_a^l, τ_a^u, $\mathscr{T}^l$, and $\mathscr{T}^u$.*

Proof For reasons of the similarity, we only present the proof for τ_a^l and $\mathscr{T}^l$. Assume that F is $m\alpha$-Hölder continuous on X. Then, for each $x_0 \in X$, there exists a neighborhood U of x_0 such that for all $x_1, x_2 \in U$

$$F(x_2) \subset F(x_1) + m\mathbb{B}(0, \|x_1 - x_2\|^\alpha). \tag{6.133}$$

For any $\varepsilon > 0$, there is $t_0 \in \mathbb{R}$ satisfying $\tau_a^l(x_2) \leq t_0 < \tau_a^l(x_2) + \varepsilon$ and

$$a \in t_0 e + F(x_2) + C. \tag{6.134}$$

Thanks to $e \in -\text{int}C$, there exists $\gamma > 0$ with $\mathbb{B}(0, \gamma) \subset e + C$, and hence

$$\mathbb{B}(0, \|x_1 - x_2\|^\alpha) \subset \frac{1}{\gamma}\|x_1 - x_2\|^\alpha e + C.$$

This together with (6.133) and (6.134) yields that

$$a \in t_0 e + F(x_1) + m\mathbb{B}(0, \|x_1 - x_2\|^\alpha) + C \subset \left(t_0 + \frac{m}{\gamma}\|x_1 - x_2\|^\alpha\right) e + F(x_1) + C,$$

which implies that

$$\tau_a^l(x_1) \leq t_0 + \frac{m}{\gamma}\|x_1 - x_2\|^\alpha \leq \tau_a^l(x_2) + \varepsilon + \frac{m}{\gamma}\|x_1 - x_2\|^\alpha.$$

As ε is arbitrary, we get:

$$\tau_a^l(x_1) \leq \tau_a^l(x_2) + \frac{m}{\gamma}\|x_1 - x_2\|^\alpha.$$

Similarly, we have:

$$\tau_a^l(x_2) \leq \tau_a^l(x_1) + \frac{m}{\gamma}\|x_1 - x_2\|^\alpha.$$

Therefore, τ_a^l is Hölder continuous.

For all $x_0, y_0 \in X$, there are neighborhoods U of x_0 and V of y_0 such that

$$F(x_2) \subset F(x_1) + m\mathbb{B}(0, \|x_1 - x_2\|^\alpha)\ \forall x_1, x_2 \in U, \tag{6.135}$$

$$F(y_1) \subset F(y_2) + m\mathbb{B}(0, \|y_1 - y_2\|^\alpha)\ \forall y_1, y_2 \in V. \tag{6.136}$$

For $x_1, x_2 \in U$ and $y_1, y_2 \in V$, we can find a positive number η satisfying

$$\mathbb{B}(0, m\|x_1 - x_2\|^\alpha + m\|y_1 - y_2\|^\alpha) \subset \left(\frac{m}{\eta}\|x_1 - x_2\|^\alpha + \frac{m}{\eta}\|y_1 - y_2\|^\alpha\right) e + C. \tag{6.137}$$

Setting

$$p := \frac{m}{\eta}\|x_1 - x_2\|^\alpha + \frac{m}{\eta}\|y_1 - y_2\|^\alpha,$$

we now prove that

$$\mathcal{T}^l(x_1, y_1) \leq \mathcal{T}^l(x_2, y_2) + p. \tag{6.138}$$

Suppose that $\mathcal{T}^l(x_1, y_1) > \mathcal{T}^l(x_2, y_2) + p$. By the definition of $\mathcal{T}^l(x_1, y_1)$, there is $a_1 \in F(y_1)$ such that

$$\mathcal{T}^l(x_1, y_1) \geq \tau^l_{a_1}(x_1) > \mathcal{T}^l(x_2, y_2) + p.$$

It follows from (6.136) that there exists $a_2 \in F(y_2)$ satisfying

$$a_1 \in a_2 + m\mathbb{B}(0, \|y_1 - y_2\|^\alpha). \tag{6.139}$$

Since

$$\tau^l_{a_1}(x_1) > \mathcal{T}^l(x_2, y_2) + p \geq \tau^l_{a_2}(x_2) + p,$$

one has:

$$\tau^l_{a_1}(x_1) - p > \tau^l_{a_2}(x_2).$$

By the definition of $\tau^l_{a_2}(x_2)$, there exists $t \in \mathbb{R}$ with

$$\tau^l_{a_1}(x_1) - p > t \geq \tau^l_{a_2}(x_2),$$

and hence

$$a_2 \in te + F(x_2) + C. \tag{6.140}$$

Combining (6.135), (6.137), (6.139) with (6.140), we get

$$a_1 \in a_2 + m\mathbb{B}(0, \|y_1 - y_2\|^\alpha),$$

and hence

$$
\begin{aligned}
a_1 &\in te + F(x_2) + C + m\mathbb{B}(0, \|y_1 - y_2\|^{\alpha}) \\
&\subset te + F(x_1) + \mathbb{B}\left(0, m\|x_1 - x_2\|^{\alpha} + m\|y_1 - y_2\|^{\alpha}\right) + C \\
&\subset te + F(x_1) + \frac{1}{\eta}\left(m\|x_1 - x_2\|^{\alpha} + m\|y_1 - y_2\|^{\alpha}\right)e + C \\
&\subset \left(t + \frac{m}{\eta}\|x_1 - x_2\|^{\alpha} + \frac{m}{\eta}\|y_1 - y_2\|^{\alpha}\right)e + F(x_1) + C \\
&\subset (t + p)e + F(x_1) + C.
\end{aligned}
$$

Since $\tau^l_{a_1}(x_1) > t + p$, we have:

$$a_1 \notin (t + p)e + F(x_1) + C,$$

a contradiction. Thus, (6.138) is proved. Analogously, we have

$$\mathcal{T}^l(x_2, y_2) \leq \mathcal{T}^l(x_1, y_1) + p. \tag{6.141}$$

From (6.138) and (6.141), we obtain the Hölder continuity for $\mathcal{T}^l$. □

6.7.1 Existence Conditions for (SOP)

In this subsection, let e, e' be defined as in the previous subsection, X, Y, W normed spaces, $D \subset X$ a nonempty and convex set, and Λ, M nonempty subsets of W. Let $\mathcal{D} : \Lambda \rightrightarrows D$ be a set-valued map with bounded convex values and $F : X \times M \rightrightarrows Y$ be a set-valued map with nonempty values.

For $(\lambda, \mu) \in \Lambda \times M$, we consider the following parametric set optimization problem:

$$\min F(x, \mu) \text{ subject to } x \in \mathcal{D}(\lambda). \tag{SOP}$$

For each $(\varepsilon, \lambda, \mu) \in \mathbb{R}_+ \times \Lambda \times M$, we denote the ε-solution sets of (SOP) corresponding to $\alpha \in \{l, u\}$ by $\widetilde{\text{Sol}}(\text{SOP})^{\alpha}(\varepsilon, \lambda, \mu)$, namely,

$$\widetilde{\text{Sol}}(\text{SOP})^{l}(\varepsilon, \lambda, \mu) := \left\{x \in \mathcal{D}(\lambda) \mid F(x, \mu) \preceq^l_C F(y, \mu) - \varepsilon e \ \forall y \in \mathcal{D}(\lambda)\right\},$$

$$\widetilde{\text{Sol}}(\text{SOP})^{u}(\varepsilon, \lambda, \mu) := \left\{x \in \mathcal{D}(\lambda) \mid F(x, \mu) \preceq^u_C F(y, \mu) + \varepsilon e' \ \forall y \in \mathcal{D}(\lambda)\right\}.$$

We now recall the notion of KKM map and a version of the well-known KKM-Fan lemma discussed in the previous chapters. Let E be a nonempty subset of X.

Definition 6.166 (KKM maps) A set-valued map $H : X \rightrightarrows Y$ is called a KKM map on E if for any finite subset $\{x_1, ..., x_n\}$ of E,

$$\operatorname{conv}\{x_1, ..., x_n\} \subset \bigcup_{i=1}^{n} H(x_i).$$

Lemma 6.167 (The KKM-Fan lemma) *If $H : X \rightrightarrows Y$ is a KKM map on E with closed values and $H(x_0)$ is compact for at least one $x_0 \in E$, then*

$$\cap_{x \in E} H(x) \neq \emptyset.$$

Lemma 6.168 (An extended version of KKM-Fan lemma) *Let $E \subset X$ be nonempty and $H : X \rightrightarrows Y$ a KKM map on E with closed values. Assume that there exists a nonempty compact convex subset B of E such that $\cap_{x \in B} H(x)$ is compact. Then,*

$$\cap_{x \in E} H(x) \neq \emptyset.$$

Lemma 6.169 (KKM maps in terms of set less order relations) *Let $\lambda \in \Lambda$, $\mu \in M$, and $\alpha \in \{l, u\}$ be given. If for each $x \in \mathscr{D}(\lambda)$, the set*

$$\{b \in \mathscr{D}(\lambda) \mid F(x, \mu) \not\preceq_C^{\alpha} F(b, \mu)\}$$

is convex, then the set-valued map $H : \mathscr{D}(\lambda) \rightrightarrows \mathscr{D}(\lambda)$ defined by

$$H(b) := \{\bar{x} \in \mathscr{D}(\lambda) \mid F(\bar{x}, \mu) \preceq_C^{\alpha} F(b, \mu)\}$$

is a KKM map.

Proof We prove this lemma for the case of $\alpha = u$. Suppose that H is not a KKM-map, i.e., there exists a finite subset $\{x_1, ..., x_n\} \subset \mathscr{D}(\lambda)$ such that a point $x \in \operatorname{conv}\{x_1, ..., x_n\}$ outside

$$\bigcup_{i=1}^{n} H(x_i).$$

Then, $x \notin H(x_i)$ for all i, which implies that

$$F(x, \mu) \not\preceq_C^{u} F(x_i, \mu).$$

That is,

$$x_i \in \{b \in \mathscr{D}(\lambda) \mid F(x, \mu) \not\preceq_C^{u} F(b, \mu)\},$$

and hence

$$x \in \{b \in \mathscr{D}(\lambda) \mid F(x,\mu) \not\preceq^u_C F(b,\mu)\},$$

which is impossible. □

Lemma 6.170 *Characterizations of cone-quasiconvexity and cone-quasiconcavity via set less order relations For $\lambda \in \Lambda$ and $\mu \in M$, the following assertions hold:*

(a) *If $F(\cdot,\mu)$ is C-quasiconvex on $\mathscr{D}(\lambda)$, then the set*

$$P = \{b \in \mathscr{D}(\lambda) \mid F(x,\mu) \not\preceq^l_C F(b,\mu)\}$$

is convex for all $x \in \mathscr{D}(\lambda)$.

(b) *If $F(\cdot,\mu)$ is $-C$-quasiconcave on $\mathscr{D}(\lambda)$, then the set*

$$P = \{b \in \mathscr{D}(\lambda) \mid F(x,\mu) \not\preceq^u_C F(b,\mu)\}$$

is convex for all $x \in \mathscr{D}(\lambda)$.

Proof

(a) For $b_1, b_2 \in P$, we show that for all $t \in [0,1]$

$$b = tb_1 + (1-t)b_2 \in P.$$

Suppose that $b \notin P$, i.e.,

$$F(x,\mu) \preceq^l_C F(b,\mu),$$

and hence

$$F(b,\mu) \subset F(x,\mu) + C.$$

Because $b_1, b_2 \in P$, there are $z_1 \in F(b_1,\mu)$ and $z_2 \in F(b_2,\mu)$ such that

$$z_1, z_2 \notin F(x,\mu) + C.$$

The C-quasiconvexity of F gives us either

$$F(b_1,\mu) \subset F(b,\mu) + C$$

or

$$F(b_2,\mu) \subset F(b,\mu) + C.$$

If $F(b_1, \mu) \subset F(b, \mu) + C$, then for any $z_1 \in F(b_1, \mu)$, one has:

$$z_1 \in F(b, \mu) + C,$$

which is a contradiction. Similarly, the case of

$$F(b_2, \mu) \subset F(b, \mu) + C$$

also leads to a contradiction.

(b) The proof is analogous to (a).

□

Theorem 6.171 (Existence conditions for approximate solutions of set optimization problems in terms of set less order relations) *For $\lambda \in \Lambda$, $\mu \in M$, $\varepsilon \in \mathbb{R}_+$, and $\alpha \in \{u, l\}$, assume that:*

(a) *for each $b \in \mathscr{D}(\lambda)$, the set $\{x \in \mathscr{D}(\lambda) \mid F(x, \mu) \preceq_C^\alpha F(b, \mu)\}$ is closed;*
(b) *for every $x \in \mathscr{D}(\lambda)$, the set $\{b \in \mathscr{D}(\lambda) \mid F(x, \mu) \not\preceq_C^\alpha F(b, \mu)\}$ is convex;*
(c) *there exists a nonempty compact subset $A \subset \mathscr{D}(\lambda)$ such that for each finite subset $B = \{b_1, b_2, ..., b_n\} \subset \mathscr{D}(\lambda)$, there is a nonempty convex compact subset $N \subset \mathscr{D}(\lambda)$ with $B \subset N$, and for all $x \in N \setminus A$, there is $b \in B$ satisfying $F(x, \mu) \not\preceq_C^\alpha F(b, \mu)$.*

Then, $\widetilde{\mathrm{Sol}}(\mathrm{SOP})^\alpha(\varepsilon, \lambda, \mu)$ is nonempty.

Proof The proof here is presented for the case $\alpha = l$. The case $\alpha = u$ is proved similarly. Since

$$\widetilde{\mathrm{Sol}}(\mathrm{SOP})^l(0, \lambda, \mu) \subset \widetilde{\mathrm{Sol}}(\mathrm{SOP})^l(\varepsilon, \lambda, \mu)$$

for all $\varepsilon \in \mathbb{R}_+$, it suffices to show that $\widetilde{\mathrm{Sol}}(\mathrm{SOP})^l(0, \lambda, \mu)$ is nonempty. Define the set-valued map $H : \mathscr{D}(\lambda) \rightrightarrows \mathscr{D}(\lambda)$ by

$$H(b) = \{x \in \mathscr{D}(\lambda) \mid F(x, \mu) \preceq_C^l F(b, \mu)\}.$$

Let Q be a restriction of H on N. Clearly, $b \in Q(b)$ and hence it is nonempty for each $b \in N$. Let $\{x_1, ..., x_n\} \subset N \subset \mathscr{D}(\lambda)$ and $x \in \mathrm{conv}\{x_1, ..., x_n\}$. Then, $x \in N$ as N is convex. By (b) and Lemma 6.169, H is a KKM map on $\mathscr{D}(\lambda)$. Thus, there exists $j \in \{1, \dots, n\}$ satisfying $x \in H(x_j)$, so $x \in Q(x_j)$. Hence, Q is a KKM map on N.

Furthermore, $Q(b) = N \cap H(b)$ is compact since N is compact and $H(b)$ is closed for each $b \in \mathscr{D}(\lambda)$. Applying Lemma 6.167, we get:

$$\cap_{b \in N} Q(b) \neq \emptyset,$$

i.e., there exists $\bar{x} \in \cap_{b\in N} Q(b)$ which implies that $\bar{x} \in N$ and $\bar{x} \in H(b)$, and hence

$$F(\bar{x}, \mu) \preceq_C^l F(b, \mu)$$

for all $b \in N$. This together with (c) yields that $\bar{x} \in A$.

Let $\{b_1, b_2, \cdots, b_n\} \subset N$ be arbitrary. We have:

$$\bar{x} \in A \cap \big(\cap_{i=1}^n Q(b_i)\big),$$

and so

$$A \cap \big(\cap_{i=1}^n H(b_i)\big) \neq \emptyset.$$

By (a) and the compactness of A, one has:

$$A \cap \big(\cap_{b\in N} H(b)\big) \neq \emptyset.$$

Thanks to Lemma 6.168, we obtain:

$$A \cap \big(\cap_{b\in \mathscr{D}(\lambda)} H(b)\big) \neq \emptyset,$$

and thus

$$\cap_{b\in \mathscr{D}(\lambda)} H(b) \neq \emptyset.$$

Therefore, $\widetilde{\mathrm{Sol}}(\mathrm{SOP})^l(0, \lambda, \mu)$ is nonempty. □

6.7.2 Relationships Between (SOP) and (EP)

In this subsection, we establish a bridge between (SOP) and (EP) by using generalized nonlinear scalarization functions in the previous subsection. Namely, we provide conditions for the coincidence of the solution set of (SOP) with the one of (EP). This allows us to work on stability analysis for (SOP) via (EP), which will be discussed in detail in the next section. We now assume that equilibrium problem (EP) suffers perturbations, which are expressed in terms of perturbing parameters $\lambda \in \Lambda$ and $\mu \in M$ where Λ and M are subsets of a normed space W (parameter space). Then, the parametric equilibrium problem (EP) under our consideration is of

$$\text{finding } \bar{x} \in \mathscr{D}(\lambda) \text{ such that } f(\bar{x}, y, \mu) \geq 0 \text{ for all } y \in \mathscr{D}(\lambda),$$

where $\mathscr{D} : \Lambda \rightrightarrows D$ is a set-valued map with bounded convex values and $f : X \times X \times M \to \mathbb{R}$ is a trifunction satisfying $f(x, x, \mu) = 0$ for all $x \in X$ and

$\mu \in M$. For $(\varepsilon, \lambda, \mu) \in \mathbb{R}_+ \times \Lambda \times M$, we denote the ε-solution set of (EP) by $\widetilde{\text{Sol}}(\text{EP})(\varepsilon, \lambda, \mu)$, i.e.,

$$\widetilde{\text{Sol}}(\text{EP})(\varepsilon, \lambda, \mu) := \{x \in \mathscr{D}(\lambda) \mid f(x, y, \mu) + \varepsilon \geq 0, \forall y \in \mathscr{D}(\lambda)\}.$$

Lemma 6.172 (Translations for generalized Gerstewitz functions)

(a) *Let* $\gamma \in \mathbb{R}$, *B be* $-C$*-closed, and* $e \in \text{int}C$. *If* $\mathscr{G}_e^u(A, B) < \infty$, *then*

$$\mathscr{G}_e^u(A, B + \gamma e) = \mathscr{G}_e^u(A, B) - \gamma.$$

(b) *Let* $\gamma \in \mathbb{R}$, *A be C-closed, and* $e' \in -\text{int}C$. *If* $\mathscr{G}_{e'}^l(A, B) < \infty$, *then*

$$\mathscr{G}_{e'}^l(A, B + \gamma q) = \mathscr{G}_{e'}^l(A, B) - \gamma.$$

Proof

(a) Let $t = G_e^u(A, B)$ and $s = G_e^u(A, B + \gamma e)$. It follows from Proposition 6.158 that

$$A \subset te + B - C,$$

which yields that

$$A \subset (t - \gamma)e + B + \gamma e - C.$$

Thanks to Proposition 6.158, one has:

$$\mathscr{G}_e^u(A, B + \gamma e) \leq t - \gamma.$$

On the other hand, by Proposition 6.158

$$A \subset se + B + \gamma e - C = (s + \gamma)e + B - C.$$

As a result,

$$t = \mathscr{G}_e^u(A, B) \leq s + \gamma = \mathscr{G}_e^u(A, B + \gamma e) + \gamma,$$

and then

$$\mathscr{G}_e^u(A, B + \gamma e) = G_e^u(A, B) - \gamma.$$

(b) Let $p = G_{e'}^l(A, B)$ and $m = G_{e'}^l(A, B + \gamma q)$. Thanks to Proposition 6.158, we have:

$$B \subset pe' + A + C,$$

which implies that

$$B + \gamma q \subset pe' + \gamma q + A + C = (p - \gamma)e' + A + C.$$

By Proposition 6.158, one has:

$$\mathcal{G}^l_{e'}(A, B + \gamma q) \leq p - \gamma.$$

Furthermore, using Proposition 6.158, we get:

$$B + \gamma q \subset me' + A + C,$$

which means that

$$B \subset (m + \gamma)e' + A + C,$$

and hence

$$\mathcal{G}^l_{e'}(A, B) \leq m + \gamma,$$

which implies

$$\mathcal{G}^l_{e'}(A, B + \gamma q) \geq p - \gamma.$$

Therefore,

$$\mathcal{G}^l_{e'}(A, B + \gamma q) = \mathcal{G}^l_{e'}(A, B) - \gamma.$$

□

Let $F : X \times M \rightrightarrows Y$ have nonempty values.

Lemma 6.173 (Nonlinear scalarizations for set optimization problems) *For $\varepsilon \in \mathbb{R}_+$, the following statements are true:*

(a) *Let F have C-proper, C-closed, and C-bounded values. Then, $x \in \widetilde{\text{Sol}}(\text{SOP})^l(\varepsilon, \lambda, \mu)$ if and only if*

$$\mathcal{G}^l_e(F(x, \mu)F(y, \mu)) - \varepsilon \leq 0$$

for all $y \in \mathcal{D}(\lambda)$.

(b) *Let F have $-C$-proper, $-C$-closed and $-C$-bounded values. Then, $x \in \widetilde{\text{Sol}}(\text{SOP})^u(\varepsilon, \lambda, \mu)$ if and only if*

$$\mathcal{G}^u_{e'}(F(x, \lambda), F(y, \lambda)) - \varepsilon \leq 0$$

for all $y \in \mathcal{D}(\lambda)$.

Proof

(a) Let $q = -e \in \text{int}C$. Then,

$$F(x, \mu) \preceq^l_C F(y, \mu) + \varepsilon q$$

if and only if $\mathcal{G}^l_e(F(x, \mu), F(y, \mu) + \varepsilon q) \leq 0$. On the other hand, one has:

$$\mathcal{G}^l_e(F(x, \mu), F(y, \mu) + \varepsilon q) = \mathcal{G}^l_e(F(x, \mu), F(y, \mu)) - \varepsilon.$$

(b) It can be proved in a similar way.

□

We now consider the functions $f^l, f^u : X \times X \times M \to \mathbb{R} \cup \{\infty\}$ defined by

$$f^l(x, y, \mu) := -\mathcal{G}^l_e(F(x, \mu), F(y, \mu)) \; \forall x, y \in X, \forall \mu \in M,$$

$$f^u(x, y, \mu) := -\mathcal{G}^u_{e'}(F(x, \mu), F(y, \mu)) \; \forall x, y \in X, \forall \mu \in M.$$

We are in a position to present the main result of this subsection by employing Lemma 6.173.

Proposition 6.174 (Relationships between equilibrium problems and set optimization problems) *For $\alpha \in \{l, u\}$ and $\varepsilon \in \mathbb{R}_+$, assume that F has nonempty, Ω-proper, Ω-closed, and Ω-bounded values. Then, x is an ε-solution of (SOP) corresponding to α if and only if it is an ε-solution of the following equilibrium problem:*

$$\text{find } \bar{x} \in \mathcal{D}(\lambda) \text{ such that } f^\alpha(\bar{x}, y, \mu) \geq 0 \text{ for all } y \in \mathcal{D}(\lambda),$$

where $\Omega = C$ if $\alpha = l$ and $\Omega = -C$ if $\alpha = u$.

6.7.3 *Hausdorff Continuity of Solution Maps of Set Optimization Problems*

For investigating the Hausdorff continuity of solution maps of (EP$_\alpha$), as an auxiliary approximate equilibrium problems, we consider the following one:

(EP$^1_\alpha$) find $\bar{x} \in \mathcal{D}(\lambda)$ such that $f^\alpha(\bar{x}, y, \lambda) + \varepsilon > 0$ for all $y \in \mathcal{D}(\lambda)$,

where $\varepsilon \in \mathbb{R}_+$ and $\alpha \in \{u, l\}$. We denote the solution set of (EP$^1_\alpha$) by

$$\widetilde{\text{Sol}}(\text{EP})^\alpha_1(\varepsilon, \lambda) := \{x \in \mathcal{D}(\lambda) \mid f^\alpha(x, y, \lambda) + \varepsilon > 0, \; \forall y \in \mathcal{D}(\lambda)\},$$

and always assume that: $\widetilde{\text{Sol}}(\text{EP})^\alpha_1(\varepsilon, \lambda) \neq \emptyset$ for all $(\varepsilon, \lambda) \in \mathbb{R}_+ \times \Lambda$.

Lemma 6.175 (Hausdorff continuity of approximate solution maps of equilibrium problems via auxiliary problems) *For* (EP_α), *assume that*

(a) $\mathcal{D}$ *is continuous and compact-valued and* $\mathcal{D}(\Lambda)$ *is bounded;*
(b) f^α *is continuous on* $\mathcal{D}(\Lambda) \times \mathcal{D}(\Lambda) \times \Lambda$;
(c) *for each* $y \in \mathcal{D}(\Lambda)$ *and* $\lambda \in \Lambda$, $f^\alpha(\cdot, y, \lambda)$ *is concave in* $\mathcal{D}(\Lambda)$.

Then, $\widetilde{\mathrm{Sol}}(\mathrm{EP})^\alpha$ *is Hausdorff continuous on* $(0, +\infty) \times \Lambda$.

Proof The proof that $\widetilde{\mathrm{Sol}}(\mathrm{EP})^\alpha$ is Hausdorff continuous at an arbitrary $(\varepsilon_0, \lambda_0) \in (0, +\infty) \times \Lambda$ is divided into three steps.
Step 1. We first show that for any neighborhood $\mathbb{B}$ of the origin in X, there exists a neighborhood V of ε_0 such that for all $\lambda \in \Lambda$ and $\varepsilon \in V$

$$\widetilde{\mathrm{Sol}}(\mathrm{EP})^\alpha(\varepsilon, \lambda) \subset \widetilde{\mathrm{Sol}}(\mathrm{EP})^\alpha(\varepsilon_0, \lambda) + \mathbb{B} \text{ and } \widetilde{\mathrm{Sol}}(\mathrm{EP})^\alpha(\varepsilon_0, \lambda) \subset \widetilde{\mathrm{Sol}}(\mathrm{EP})^\alpha(\varepsilon, \lambda) + \mathbb{B}. \tag{6.142}$$

For each δ satisfying $0 < \delta < \varepsilon_0$, let $V' = [\varepsilon_0 - \delta, \varepsilon_0 + \delta]$ be a given neighborhood of ε_0. It is easy to see that for any $\varepsilon_1, \varepsilon_2 \in V'$ with $\varepsilon_1 < \varepsilon_2$ and $\lambda \in \Lambda$, one has:

$$\widetilde{\mathrm{Sol}}(\mathrm{EP})^\alpha(\varepsilon_1, \lambda) \subset \widetilde{\mathrm{Sol}}(\mathrm{EP})^\alpha(\varepsilon_2, \lambda).$$

Applying this for $\varepsilon \in V'$, we have

$$\widetilde{\mathrm{Sol}}(\mathrm{EP})^\alpha(\varepsilon_0 - \delta, \lambda) \subset \widetilde{\mathrm{Sol}}(\mathrm{EP})^\alpha(\varepsilon, \lambda) \subset \widetilde{\mathrm{Sol}}(\mathrm{EP})^\alpha(\varepsilon_0 + \delta, \lambda). \tag{6.143}$$

Next, for each real number θ satisfying

$$1 < \theta < \frac{\varepsilon_2}{\varepsilon_2 - \varepsilon_1}$$

and

$$\gamma = \varepsilon_2 + \theta(\varepsilon_1 - \varepsilon_2),$$

we claim that

$$\frac{1}{\theta}\widetilde{\mathrm{Sol}}(\mathrm{EP})^\alpha(\gamma, \lambda) + \left(1 - \frac{1}{\theta}\right)\widetilde{\mathrm{Sol}}(\mathrm{EP})^\alpha(\varepsilon_2, \lambda) \subset \widetilde{\mathrm{Sol}}(\mathrm{EP})^\alpha(\varepsilon_1, \lambda). \tag{6.144}$$

Indeed, take any $\bar{x} \in \widetilde{\mathrm{Sol}}(\mathrm{EP})^\alpha(\gamma, \lambda)$ and $\bar{x}' \in \widetilde{\mathrm{Sol}}(\mathrm{EP})^\alpha(\varepsilon_2, \lambda)$. By virtue of the convexity of $\mathcal{D}(\lambda)$, one has

$$\frac{1}{\theta}\bar{x} + \left(1 - \frac{1}{\theta}\right)\bar{x}' \in \mathcal{D}(\lambda).$$

For all $y \in \mathcal{D}(\lambda)$,

$$\min\{f^\alpha(\bar{x}, y, \lambda) + \gamma,\ f^\alpha(\bar{x}', y, \lambda) + \varepsilon_2\} \geq 0.$$

Therefore,

$$\frac{1}{\theta}f^{\alpha}(\bar{x},y,\lambda)+\frac{1}{\theta}\gamma+\left(1-\frac{1}{\theta}\right)f^{\alpha}(\bar{x}',y,\lambda)+\left(1-\frac{1}{\theta}\right)\varepsilon_2\geq 0.$$

The concavity of f^{α} gives us

$$f^{\alpha}\left(\frac{1}{\theta}\bar{x}+\left(1-\frac{1}{\theta}\right)\bar{x}',y,\lambda\right)+\varepsilon_1\geq 0,$$

i.e.,

$$\frac{1}{\theta}\bar{x}+\left(1-\frac{1}{\theta}\right)\bar{x}'\in\widetilde{\mathrm{Sol}}(\mathrm{EP})^{\alpha}(\varepsilon_1,\lambda),$$

and hence (6.144) is proved.

Inclusion (6.144) can be rewritten as

$$\left(1-\frac{1}{\theta}\right)\widetilde{\mathrm{Sol}}(\mathrm{EP})^{\alpha}(\varepsilon_2,\lambda)\subset\widetilde{\mathrm{Sol}}(\mathrm{EP})^{\alpha}(\varepsilon_1,\lambda)-\frac{1}{\theta}\widetilde{\mathrm{Sol}}(\mathrm{EP})^{\alpha}(\gamma,\lambda),$$

which implies that

$$\begin{aligned}\widetilde{\mathrm{Sol}}(\mathrm{EP})^{\alpha}(\varepsilon_2,\lambda)&\subset\widetilde{\mathrm{Sol}}(\mathrm{EP})^{\alpha}(\varepsilon_1,\lambda)+\frac{1}{\theta-1}\left[\widetilde{\mathrm{Sol}}(\mathrm{EP})^{\alpha}(\varepsilon_1,\lambda)-\widetilde{\mathrm{Sol}}(\mathrm{EP})^{\alpha}(\gamma,\lambda)\right]\\&\subset\widetilde{\mathrm{Sol}}(\mathrm{EP})^{\alpha}(\varepsilon_1,\lambda)+\frac{1}{\theta-1}[\mathscr{D}(\Lambda)-\mathscr{D}(\Lambda)].\end{aligned}$$

By the boundedness of $\mathscr{D}(\Lambda)$, for each closed convex neighborhood $\mathbb{B}$ of the origin in X, there is $\rho>0$ such that

$$\mathscr{D}(\Lambda)-\mathscr{D}(\Lambda)\subset\rho\mathbb{B}.$$

Consequently,

$$\widetilde{\mathrm{Sol}}(\mathrm{EP})^{\alpha}(\varepsilon_2,\lambda)\subset\widetilde{\mathrm{Sol}}(\mathrm{EP})^{\alpha}(\varepsilon_1,\lambda)+\frac{\rho}{\theta-1}\mathbb{B}\ \forall\lambda\in\Lambda.\tag{6.145}$$

Let $0<\delta_0<\dfrac{\varepsilon_0}{\rho+1}$ and $V=[\varepsilon_0-\delta_0,\varepsilon_0+\delta_0]$ be a neighborhood of ε_0. From (6.145), taking $\varepsilon_2=\varepsilon_0$, $\varepsilon_1=\varepsilon_0-\delta_0$, and $\theta=\rho+1$, one has

$$1<\theta<\frac{\varepsilon_2}{\varepsilon_2-\varepsilon_1}.$$

Combining this with (6.143), one gets:

$$\widetilde{\text{Sol}}(\text{EP})^\alpha(\varepsilon_0, \lambda) \subset \widetilde{\text{Sol}}(\text{EP})^\alpha(\varepsilon_0 - \delta_0, \lambda) + \mathbb{B} \subset \widetilde{\text{Sol}}(\text{EP})^\alpha(\varepsilon, \lambda) + \mathbb{B}.$$

Now substituting $\varepsilon_2 = \varepsilon_0 + \delta_0$, $\varepsilon_1 = \varepsilon_0$ in (6.145), and $\theta = \rho + 1$, we have:

$$\widetilde{\text{Sol}}(\text{EP})^\alpha(\varepsilon_0 + \delta_0, \lambda) \subset \widetilde{\text{Sol}}(\text{EP})^\alpha(\varepsilon_0, \lambda) + \mathbb{B}.$$

This inclusion together with (6.143) implies that

$$\widetilde{\text{Sol}}(\text{EP})^\alpha(\varepsilon, \lambda) \subset \widetilde{\text{Sol}}(\text{EP})^\alpha(\varepsilon_0, \lambda) + \mathbb{B}$$

for all $\lambda \in \Lambda$. Therefore, (6.142) is proved.

Step 2. For each $\varepsilon \in V$, we prove that $\widetilde{\text{Sol}}(\text{EP})^\alpha(\varepsilon, \cdot)$ is Hausdorff continuous at λ_0.

We show that $\widetilde{\text{Sol}}(\text{EP})^\alpha(\varepsilon, \cdot)$ is H-usc at λ_0 by proving that $\widetilde{\text{Sol}}(\text{EP})^\alpha(\varepsilon, \cdot)$ is usc at λ_0. Suppose $\widetilde{\text{Sol}}(\text{EP})^\alpha(\varepsilon, \cdot)$ is not usc at λ_0, i.e., there are open set U with $\widetilde{\text{Sol}}(\text{EP})^\alpha(\varepsilon, \lambda_0) \subset U$ and sequence $\lambda^k \to \lambda_0$ such that

$$x^k \in \widetilde{\text{Sol}}(\text{EP})^\alpha(\varepsilon, \lambda^k), x^k \notin U,$$

for all k. The upper semicontinuity of $\mathscr{D}$ at λ_0 and the compactness of $\mathscr{D}(\lambda_0)$ allow us to assume that $x^k \to x_0$ for some $x_0 \in \mathscr{D}(\lambda_0)$. If $x_0 \notin \widetilde{\text{Sol}}(\text{EP})^\alpha(\varepsilon, \lambda_0)$, then there is $y_0 \in \mathscr{D}(\lambda_0)$ such that

$$f^\alpha(x_0, y_0, \lambda_0) + \varepsilon < 0. \tag{6.146}$$

The lower semicontinuity of $\mathscr{D}$ at λ_0 in turn shows the existence of $y^k \in \mathscr{D}(\lambda^k)$ satisfying $y^k \to y_0$. Since $x^k \in \widetilde{\text{Sol}}(\text{EP})^\alpha(\varepsilon, \lambda^k)$,

$$f^\alpha(x^k, y^k, \lambda^k) + \varepsilon \geq 0$$

which is impossible because of the continuity of f. Hence,

$$x_0 \in \widetilde{\text{Sol}}(\text{EP})^\alpha(\varepsilon, \lambda_0),$$

which implies a contradiction since $x^k \notin U$ for all n.

To get the Hausdorff lower semicontinuity of $\widetilde{\text{Sol}}(\text{EP})^\alpha(\varepsilon, \cdot)$ at λ_0, we show that $\widetilde{\text{Sol}}(\text{EP})^\alpha(\varepsilon, \cdot)$ is lsc at λ_0 and $\widetilde{\text{Sol}}(\text{EP})^\alpha(\varepsilon, \lambda_0)$ is compact.

First, we verify that $\widetilde{\text{Sol}}(\text{EP})_1^\alpha(\varepsilon, \cdot)$ is lsc at λ_0. Suppose that $\widetilde{\text{Sol}}(\text{EP})_1^\alpha(\varepsilon, \cdot)$ is not lsc at λ_0, i.e., there are $\hat{x} \in \widetilde{\text{Sol}}(\text{EP})_1^\alpha(\varepsilon, \lambda_0)$ and a sequence $\lambda^k \to \lambda_0$ such that for all $x^k \in \widetilde{\text{Sol}}(\text{EP})_1^\alpha(\varepsilon, \lambda^k)$, $x^k \nrightarrow \hat{x}$.

Since $\mathscr{D}$ is lsc at λ_0, there is $\bar{x}^k \in \mathscr{D}(\lambda^k)$ with $\bar{x}^k \to \hat{x}$. By the above contradiction assumption, there must be a subsequence $\{\bar{x}^{k_j}\}$ such that, for all k_j

$$\bar{x}^{k_j} \notin \widetilde{\text{Sol}}(\text{EP})_1^\alpha(\varepsilon, \lambda^{k_j}),$$

i.e., for some $y^{k_j} \in \mathscr{D}(\lambda^{k_j})$,

$$f^{\alpha}(\bar{x}^{k_j}, y^{k_j}, \lambda^{k_j}) + \varepsilon \leq 0.$$

Since $\mathscr{D}$ is usc at λ_0 and $\mathscr{D}(\lambda_0)$ is compact, one has $\hat{y} \in \mathscr{D}(\lambda_0)$ such that $y^{k_j} \to \hat{y}$ (take a subsequence if necessary). By the continuity of f^{α}, we have:

$$f^{\alpha}(\hat{x}, \hat{y}, \lambda_0) + \varepsilon \leq 0,$$

which is impossible as $\hat{x} \in \widetilde{\mathrm{Sol}}(\mathrm{EP})_1^{\alpha}(\varepsilon, \lambda_0)$.

Now, let us check that

$$\widetilde{\mathrm{Sol}}(\mathrm{EP})^{\alpha}(\varepsilon, \lambda_0) \subset \mathrm{cl}\widetilde{\mathrm{Sol}}(\mathrm{EP})_1^{\alpha}(\varepsilon, \lambda_0). \tag{6.147}$$

Taking any $\bar{x} \in \widetilde{\mathrm{Sol}}(\mathrm{EP})^{\alpha}(\varepsilon, \lambda_0)$ and $x_1 \in \widetilde{\mathrm{Sol}}(\mathrm{EP})_1^{\alpha}(\varepsilon, \lambda_0)$. Let $x_t = (1-t)\bar{x} + tx_1$ with $t \in (0, 1)$. For all $y \in \mathscr{D}(\lambda_0)$, the concavity of $f^{\alpha}(\cdot, y, \lambda_0)$ gives us:

$$\begin{aligned} f^{\alpha}(x_t, y, \lambda_0) + \varepsilon &\geq (1-t) f^{\alpha}(\bar{x}, y, \lambda_0) + t f^{\alpha}(x_1, y, \lambda_0) + \varepsilon \\ &\geq (1-t)(f^{\alpha}(\bar{x}, y, \lambda_0) + \varepsilon) + t(f^{\alpha}(x_1, y, \lambda_0) + \varepsilon) > 0. \end{aligned}$$

This means that $x_t \in \widetilde{\mathrm{Sol}}(\mathrm{EP})_1^{\alpha}(\varepsilon, \lambda_0)$. As $x_t \to \bar{x}$ when $t \to 0$, $\bar{x}$ belongs to $\mathrm{cl}\widetilde{\mathrm{Sol}}(\mathrm{EP})_1^{\alpha}(\varepsilon, \lambda_0)$. Therefore, (6.147) holds. By the lower semicontinuity of $\widetilde{\mathrm{Sol}}(\mathrm{EP})_1^{\alpha}(\varepsilon, \cdot)$ at λ_0, we have:

$$\begin{aligned} \widetilde{\mathrm{Sol}}(\mathrm{EP})^{\alpha}(\varepsilon, \lambda_0) \subset \mathrm{cl}\widetilde{\mathrm{Sol}}(\mathrm{EP})_1^{\alpha}(\varepsilon, \lambda_0) &\subset \mathrm{Liminf}\widetilde{Sol}(EP)_1^{\alpha}(\varepsilon, \lambda^k) \\ &\subset \mathrm{Liminf}\widetilde{Sol}(EP)^{\alpha}(\varepsilon, \lambda^k), \end{aligned}$$

i.e., $\widetilde{\mathrm{Sol}}(\mathrm{EP})^{\alpha}(\varepsilon, \cdot)$ is lsc at λ_0.

For the compactness of $\widetilde{\mathrm{Sol}}(\mathrm{EP})^{\alpha}(\varepsilon, \lambda_0)$, it suffices to show that $\widetilde{\mathrm{Sol}}(\mathrm{EP})^{\alpha}(\varepsilon, \lambda_0)$ is a closed subset of $\mathscr{D}(\lambda_0)$. For any $x^k \in \widetilde{\mathrm{Sol}}(\mathrm{EP})^{\alpha}(\varepsilon, \lambda_0)$, $x^k \to \widetilde{x}$, then $x^k \in \mathscr{D}(\lambda_0)$ and $\widetilde{x} \in \mathscr{D}(\lambda_0)$. For all $y \in \mathscr{D}(\lambda_0)$, we have

$$f^{\alpha}(x^k, y, \lambda_0) + \varepsilon \geq 0.$$

By the continuity of f^{α}, we get:

$$f^{\alpha}(\widetilde{x}, y, \lambda_0) + \varepsilon \geq 0,$$

i.e., $\widetilde{x} \in \widetilde{\mathrm{Sol}}(\mathrm{EP})^{\alpha}(\varepsilon, \lambda_0)$.

Step 3. We prove that $\widetilde{\mathrm{Sol}}(\mathrm{EP})^{\alpha}$ is Hausdorff continuous at $(\varepsilon_0, \lambda_0)$.

For each neighborhood $\mathbb{B}$ of the origin in X, there is a neighborhood $\mathbb{B}_1$ of the origin in X such that

$$\mathbb{B}_1 + \mathbb{B}_1 \subset \mathbb{B}. \tag{6.148}$$

By Step 1, for any neighborhood $\mathbb{B}_1$ of the origin in X, there is a vicinity V_1 of ε_0 such that for all $\lambda \in \Lambda$,

$$\widetilde{\mathrm{Sol}}(\mathrm{EP})^\alpha(\varepsilon, \lambda) \subset \widetilde{\mathrm{Sol}}(\mathrm{EP})^\alpha(\varepsilon_0, \lambda) + \mathbb{B}_1 \text{ and } \widetilde{\mathrm{Sol}}(\mathrm{EP})^\alpha(\varepsilon_0, \lambda) \subset \widetilde{\mathrm{Sol}}(\mathrm{EP})^\alpha(\varepsilon, \lambda) + \mathbb{B}_1. \tag{6.149}$$

For each $\varepsilon \in V_1$, due to Step 2, $\widetilde{\mathrm{Sol}}(\mathrm{EP})^\alpha(\varepsilon, \cdot)$ is H-continuous at λ_0, and consequently there is a neighborhood N_1 of λ_0 such that for all $\lambda \in N_1$,

$$\begin{aligned} \widetilde{\mathrm{Sol}}(\mathrm{EP})^\alpha(\varepsilon_0, \lambda) \subset \widetilde{\mathrm{Sol}}(\mathrm{EP})^\alpha(\varepsilon_0, \lambda_0) + \mathbb{B}_1 \text{ and } \widetilde{\mathrm{Sol}}(\mathrm{EP})^\alpha(\varepsilon_0, \lambda_0) \\ \subset \widetilde{\mathrm{Sol}}(\mathrm{EP})^\alpha(\varepsilon_0, \lambda) + \mathbb{B}_1. \end{aligned} \tag{6.150}$$

Obviously, $V_1 \times N_1$ is a vicinity of $(\varepsilon_0, \lambda_0)$. Combining this with (6.148), (6.149), and (6.150), for all $(\varepsilon, \lambda) \in V_1 \times N_1$, we get:

$$\widetilde{\mathrm{Sol}}(\mathrm{EP})^\alpha(\varepsilon, \lambda) \subset \widetilde{\mathrm{Sol}}(\mathrm{EP})^\alpha(\varepsilon_0, \lambda_0) + \mathbb{B} \text{ and } \widetilde{\mathrm{Sol}}(\mathrm{EP})^\alpha(\varepsilon_0, \lambda_0) \subset \widetilde{\mathrm{Sol}}(\mathrm{EP})^\alpha(\varepsilon, \lambda) + \mathbb{B}.$$

Therefore, $\widetilde{\mathrm{Sol}}(\mathrm{EP})^\alpha$ is Hausdorff continuous on $(0, +\infty) \times \Lambda$. □

Now we come to the main result of this section.

Theorem 6.176 (Hausdorff continuity conditions for set optimization problems) *For (SOP), assume that $\mathscr{D}$, F are continuous and compact-valued and $\mathscr{D}(\Lambda)$ is bounded. Then,*

(a) $\widetilde{\mathrm{Sol}}(\mathrm{SOP})^u$ *is Hausdorff continuous on* $(0, +\infty) \times \Lambda$ *if F is upper C-convex in the first component on $\mathscr{D}(\Lambda)$ and has $-C$-proper-closed-bounded values*;
(b) $\widetilde{\mathrm{Sol}}(\mathrm{SOP})^l$ *is Hausdorff continuous on* $(0, +\infty) \times \Lambda$ *if F is lower C-convex in the first component on $\mathscr{D}(\Lambda)$ and has C-proper-closed-bounded values.*

Proof

(a) Since F has nonempty $-C$-proper-closed-bounded values, in view of Theorem 6.174, we conclude that for all $(\varepsilon, \lambda) \in (0, +\infty) \times \Lambda$,

$$\widetilde{\mathrm{Sol}}(\mathrm{SOP})^u(\varepsilon, \lambda) = \widetilde{\mathrm{Sol}}(\mathrm{EP})^u(\varepsilon, \lambda).$$

Therefore, it suffices to check that all the assumptions of Lemma 6.175 are satisfied. Nothing is to do for checking (a). Assumption (b) is derived by the continuity and compactness of F together with Proposition 6.164. Combining the upper C-convexity of F with Theorem 6.163, the concavity condition in (c) is fulfilled.
(b) Since the proof is similar to that of (a), we omit it.

□

Example 6.177 (An illustration of Theorem 6.176) Let $X = \Omega = \mathbb{R}$, $Y = \mathbb{R}^2$, $\Lambda = [0, 1]$, $C = \mathbb{R}^2_+$, $\mathscr{D}(\lambda) = [0, \lambda]$, $e = (1, 1) \in \mathrm{int}C$, and

$$F(x, \lambda) = [0, 1] \times [0, x^2].$$

It is not hard to verify that the assumptions in case (a) are satisfied. For every $\varepsilon \geq 0$, we compute directly the solution set

$$\widetilde{\mathrm{Sol}}(\mathrm{SOP})^u(\varepsilon, \lambda) = [0, \lambda] \cap [0, \sqrt{\varepsilon}],$$

which is Hausdorff continuous on $(0, +\infty) \times \Lambda$.

6.7.4 Hölder Continuity of Solution Maps of Set Optimization Problems

Until now, our stability analysis of solution maps of parametric (SOPs) is only for the upper and lower semicontinuity. Because of difficult techniques, there is no work which studies the Hölder continuity of the solution maps of (SOPs) directly. Fortunately, thanks to the idea in Sect. 6.4, we consider stability in this sense by employing generalized scalarization functions. More precisely, sufficient conditions for Hölder continuity of solution maps of (EPs) are first established. Then, these results are applied to derive the same property for solution maps of (SOPs) by utilizing Proposition 6.174.

For $\alpha \in \{l, u\}$, we consider the approximate solution map $\widetilde{\mathrm{Sol}}(\mathrm{EP})^\alpha : \mathbb{R}_+ \times M \times \Lambda \rightrightarrows X$ defined by

$$\widetilde{\mathrm{Sol}}(\mathrm{EP})^\alpha(\varepsilon, \mu, \lambda) := \{x \in \mathscr{D}(\lambda) \mid f^\alpha(x, y, \mu) + \varepsilon \geq 0, \forall y \in \mathscr{D}(\lambda)\},$$

for $(\varepsilon, \lambda, \mu) \in \mathbb{R}_+ \times \Lambda \times M$.

Lemma 6.178 (Höder conditions for equilibrium problems) *For $\alpha \in \{l, u\}$, assume that:*

(a) *$\mathscr{D}$ is (m, γ)-Hölder continuous on Λ;*
(b) *for each $(\lambda, \mu) \in \Lambda \times M$ and $y \in \mathscr{D}(\lambda)$, the function*

$$(x, \varepsilon) \mapsto f^\alpha(x, y, \mu) + \varepsilon$$

is 0-lower-level quasiconcave on $\mathscr{D}(\lambda) \times (0, +\infty)$;
(c) *f^α is $((\ell_1, \gamma_1), (\ell_2, \gamma_2), (h, \beta))$-Hölder continuous on $\mathscr{D}(\Lambda) \times \mathscr{D}(\Lambda) \times M$.*

Then, the approximate solution map $\widetilde{\mathrm{Sol}}(\mathrm{EP})^\alpha$ is Hölder continuous on $(0, +\infty) \times M \times \Lambda$.

Proof Let $(\varepsilon_0, \mu_0, \lambda_0) \in (0, +\infty) \times M \times \Lambda$ and $\bar{\varepsilon}_0$ be arbitrary with $0 < \bar{\varepsilon}_0 < \varepsilon_0$.
Step 1. By (a), there is a neighborhood N of λ_0 satisfying, for $\lambda \in N$

$$\mathscr{D}(\lambda) \subset \mathscr{D}(\lambda_0) + m\mathbb{B}(0, \|\lambda - \lambda_0\|^{\gamma}),$$

and hence

$$\operatorname{diam}\mathscr{D}(\lambda) \leq \rho := \operatorname{diam}\mathscr{D}(\lambda_0) + 2m(\operatorname{diam}(N))^{\gamma}.$$

We now estimate $H(\widetilde{\mathrm{Sol}}(\mathrm{EP})^{\alpha}(\varepsilon_1, \lambda_1, \mu_1), \widetilde{\mathrm{Sol}}(\mathrm{EP})^{\alpha}(\varepsilon_2, \lambda_1, \mu_1))$, where $\varepsilon_1, \varepsilon_2 \in [\bar{\varepsilon}_0, +\infty)$, $\varepsilon_1 < \varepsilon_2$, and $(\lambda_1, \mu_1) \in \Lambda \times M$. Thanks to

$$\widetilde{\mathrm{Sol}}(\mathrm{EP})^{\alpha}(\varepsilon_1, \lambda_1, \mu_1) \subset \widetilde{\mathrm{Sol}}(\mathrm{EP})^{\alpha}(\varepsilon_2, \lambda_1, \mu_1),$$

one has:

$$\operatorname{ex}(\widetilde{\mathrm{Sol}}(\mathrm{EP})^{\alpha}(\varepsilon_1, \lambda_1, \mu_1), \widetilde{\mathrm{Sol}}(\mathrm{EP})^{\alpha}(\varepsilon_2, \lambda_1, \mu_1)) = 0. \tag{6.151}$$

For each

$$x_2 \in \widetilde{\mathrm{Sol}}(\mathrm{EP})^{\alpha}(\varepsilon_2, \lambda_1, \mu_1), \quad x_0 \in \widetilde{\mathrm{Sol}}(\mathrm{EP})^{\alpha}(0, \lambda_1, \mu_1),$$

and $y \in \mathscr{D}(\lambda_1)$, we have:

$$f^{\alpha}(x_2, y, \mu_1) + \varepsilon_2 \geq 0, \quad f^{\alpha}(x_0, y, \mu_1) \geq 0,$$

i.e., $(x_2, \varepsilon_2), (x_0, 0) \in L := \{(x, \varepsilon) \in \mathscr{D}(\lambda_1) \times (0, +\infty) \mid f^{\alpha}(x, y, \mu_1) + \varepsilon \geq 0\}$.
By (b) and Lemma 6.151, L is convex, which implies that

$$(x_1, \varepsilon_1) := \frac{\varepsilon_2 - \varepsilon_1}{\varepsilon_2}(x_0, 0) + \frac{\varepsilon_1}{\varepsilon_2}(x_2, \varepsilon_2) \in L.$$

This means that

$$f^{\alpha}(x_1, y, \mu_1) + \varepsilon_1 \geq 0,$$

i.e.,

$$x_1 \in \widetilde{\mathrm{Sol}}(\mathrm{EP})^{\alpha}(\varepsilon_1, \lambda_1, \mu_1).$$

Then,

$$\|x_2 - x_1\| = \frac{|\varepsilon_1 - \varepsilon_2|}{\varepsilon_2}\|x_2 - x_0\| \leq \frac{\rho}{\varepsilon_2}|\varepsilon_1 - \varepsilon_2|.$$

Therefore,

$$\mathrm{ex}(\widetilde{\mathrm{Sol}}(\mathrm{EP})^{\alpha}(\varepsilon_2, \lambda_1, \mu_1), \widetilde{\mathrm{Sol}}(\mathrm{EP})^{\alpha}(\varepsilon_1, \lambda_1, \mu_1)) \leq \frac{\rho}{\bar{\varepsilon}_0}|\varepsilon_1 - \varepsilon_2| := \kappa_1|\varepsilon_1 - \varepsilon_2|. \tag{6.152}$$

It follows from (6.151) and (6.152) that

$$H(\widetilde{\mathrm{Sol}}(\mathrm{EP})^{\alpha}(\varepsilon_1, \lambda_1, \mu_1), \widetilde{\mathrm{Sol}}(\mathrm{EP})^{\alpha}(\varepsilon_2, \lambda_1, \mu_1)) \leq \kappa_1|\varepsilon_1 - \varepsilon_2|.$$

Step 2 Condition (c) shows the existence of a neighborhood U of μ_0 such that for all $x, y \in \mathscr{D}(N)$ and $\mu_1, \mu_2 \in U$:

$$|f^{\alpha}(x, y, \mu_1) - f^{\alpha}(\bar{x}, y, \mu_2)| \leq h\|\mu_1 - \mu_2\|^{\beta}. \tag{6.153}$$

We give an estimation for $H(\widetilde{\mathrm{Sol}}(\mathrm{EP})^{\alpha}(\varepsilon_2, \lambda_1, \mu_1), \widetilde{\mathrm{Sol}}(\mathrm{EP})^{\alpha}(\varepsilon_2, \lambda_1, \mu_2))$ with $\varepsilon_2 \in [\bar{\varepsilon}_0, +\infty)$, $\lambda_1 \in N$ and $\mu_1, \mu_2 \in U$, $\mu_1 \neq \mu_2$. There are two cases:
Case 1. If

$$\|\mu_1 - \mu_2\|^{\beta} \leq \frac{\varepsilon_2}{h},$$

then

$$0 < s := h\|\mu_1 - \mu_2\|^{\beta} \leq \varepsilon_2.$$

We conclude that

$$\widetilde{\mathrm{Sol}}(\mathrm{EP})^{\alpha}(\varepsilon_2 - s, \lambda_1, \mu_1) \subset \widetilde{\mathrm{Sol}}(\mathrm{EP})^{\alpha}(\varepsilon_2, \lambda_1, \mu_2).$$

Indeed, let $\bar{x} \in \widetilde{\mathrm{Sol}}(\mathrm{EP})^{\alpha}(\varepsilon_2 - s, \lambda_1, \mu_1)$, then for all $y \in \mathscr{D}(\lambda_1)$,

$$f^{\alpha}(\bar{x}, y, \mu_2) + f^{\alpha}(\bar{x}, y, \mu_1) - f^{\alpha}(\bar{x}, y, \mu_2) + \varepsilon_2 - s \geq 0.$$

Combining this with (6.153), one has:

$$f^{\alpha}(\bar{x}, y, \mu_2) + \varepsilon_2 \geq 0,$$

for all $y \in \mathscr{D}(\lambda_1)$, i.e.,

$$\bar{x} \in \widetilde{\mathrm{Sol}}(\mathrm{EP})^{\alpha}(\varepsilon_2, \lambda_1, \mu_2).$$

Using the result of Step 1, one has:

$$\begin{aligned}
&\mathrm{ex}(\widetilde{\mathrm{Sol}}(\mathrm{EP})^{\alpha}(\varepsilon_2, \lambda_1, \mu_1), \widetilde{\mathrm{Sol}}(\mathrm{EP})^{\alpha}(\varepsilon_2, \lambda_1, \mu_2)) \\
&\quad \leq \mathrm{ex}(\widetilde{\mathrm{Sol}}(\mathrm{EP})^{\alpha}(\varepsilon_2, \lambda_1, \mu_1), \widetilde{\mathrm{Sol}}(\mathrm{EP})^{\alpha}(\varepsilon_2 - s, \lambda_1, \mu_1))
\end{aligned}$$

$$\leq H(\widetilde{\mathrm{Sol}}(\mathrm{EP})^{\alpha}(\varepsilon_2, \lambda_1, \mu_1), \widetilde{\mathrm{Sol}}(\mathrm{EP})^{\alpha}(\varepsilon_2 - s, \lambda_1, \mu_1))$$
$$\leq \frac{\rho h}{\bar{\varepsilon}_0}||\mu_1 - \mu_2||^{\beta}.$$

Similarly,

$$\mathrm{ex}(\widetilde{\mathrm{Sol}}(\mathrm{EP})^{\alpha}(\varepsilon_2, \lambda_1, \mu_2), \widetilde{\mathrm{Sol}}(\mathrm{EP})^{\alpha}(\varepsilon_2, \lambda_1, \mu_1)) \leq \frac{\rho h}{\bar{\varepsilon}_0}||\mu_1 - \mu_2||^{\beta}.$$

Hence,

$$H(\widetilde{\mathrm{Sol}}(\mathrm{EP})^{\alpha}(\varepsilon_2, \lambda_1, \mu_1), \widetilde{\mathrm{Sol}}(\mathrm{EP})^{\alpha}(\varepsilon_2, \lambda_1, \mu_2)) \leq \frac{\rho h}{\bar{\varepsilon}_0}||\mu_1 - \mu_2||^{\beta} := \kappa_2||\mu_1 - \mu_2||^{\beta}. \tag{6.154}$$

Case 2. If $\|\mu_1 - \mu_2\|^{\beta} > \frac{\varepsilon_2}{h}$, then there is a natural number n_0 satisfying

$$\frac{\mathrm{diam}U}{n_0} \leq \left(\frac{\bar{\varepsilon}_0}{h}\right)^{1/\beta}.$$

Let $\mathbb{P}$ be a partition of the segment $[\mu_1, \mu_2]$ with $n_0 + 1$ nodes $z_1, ..., z_{n_0+1}$ such that

$$z_1 = \mu_1, z_{n_0+1} = \mu_2, \|z_i - z_{i+1}\| = \frac{\|\mu_1 - \mu_2\|}{n_0}.$$

Then,

$$\|z_i - z_{i+1}\| \leq \frac{\rho}{n_0} \leq \left(\frac{\bar{\varepsilon}_0}{h}\right)^{1/\beta},$$

with $\rho := \mathrm{diam}U$. Thus,

$$\|z_i - z_{i+1}\|^{\beta} \leq \frac{\bar{\varepsilon}_0}{h} \leq \frac{\varepsilon_2}{h}.$$

From (6.154), one has:

$$H(\widetilde{\mathrm{Sol}}(\mathrm{EP})^{\alpha}(\varepsilon_2, \lambda_1, \mu_1), \widetilde{\mathrm{Sol}}(\mathrm{EP})^{\alpha}(\varepsilon_2, \lambda_1, \mu_2)) \leq \frac{\rho h}{\bar{\varepsilon}_0}\sum_{i=1}^{n_0}\|z_i - z_{i+1}\|^{\beta}$$
$$\leq \frac{n_0 \rho h}{\bar{\varepsilon}_0}\frac{\varepsilon_2}{h} \leq \frac{n_0 \rho h}{\bar{\varepsilon}_0}\|\mu_1 - \mu_2\|^{\beta}$$
$$=: \kappa_2\|\mu_1 - \mu_2\|^{\beta}.$$

Step 3. Let $\varepsilon_2 \in [\bar{\varepsilon}_0, +\infty)$, $\mu_2 \in U$, and $\lambda_1, \lambda_2 \in N'$, where $N' \subset N$ is a neighborhood of λ_0 with $\text{diam}N' \leq \frac{1}{2}$. We will evaluate $H(\widetilde{\text{Sol}}(\text{EP})^\alpha(\varepsilon_2, \lambda_1, \mu_2), \widetilde{\text{Sol}}(\text{EP})^\alpha(\varepsilon_2, \lambda_2, \mu_2))$ with $\lambda_1 \neq \lambda_2$. Putting

$$\eta := \varepsilon_2 - \bar{\varepsilon}_0 \text{ and } \delta := (\ell_1 m^{\gamma_1} + \ell_2 m^{\gamma_2})\|\lambda_1 - \lambda_2\|^{\gamma_3},$$

where $\gamma_3 := \min\{\gamma, \gamma\gamma_1, \gamma\gamma_2\}$, we have two cases.
Case 1. If $\delta \leq \eta$, then $\varepsilon_2 - \delta \geq \varepsilon_2 - \eta = \bar{\varepsilon}$. For all $x_1 \in \widetilde{\text{Sol}}(\text{EP})^\alpha(\varepsilon_2 - \delta, \lambda_1, \mu_2)$, $x_1 \in \mathscr{D}(\lambda_1)$ and

$$f^\alpha(x_1, y, \mu_2) + \varepsilon_2 - \delta \geq 0$$

for all $y \in \mathscr{D}(\lambda_1)$. By (a), there exists $x_2 \in \mathscr{D}(\lambda_2)$ such that

$$\|x_1 - x_2\| \leq m\|\lambda_1 - \lambda_2\|^\gamma.$$

For each $y_2 \in \mathscr{D}(\lambda_2)$, also by (a), there is $y_1 \in \mathscr{D}(\lambda_1)$ satisfying

$$\|y_1 - y_2\| \leq m\|\lambda_1 - \lambda_2\|^\gamma.$$

Since $y_1 \in \mathscr{D}(\lambda_1)$, one has:

$$f^\alpha(x_2, y_2, \mu_2) + f^\alpha(x_1, y_1, \mu_2) - f^\alpha(x_2, y_2, \mu_2) + \varepsilon_2 - \delta \geq 0.$$

By (c), we get:

$$\begin{aligned}
|f^\alpha(x_1, y_1, \mu_2) - f^\alpha(x_2, y_2, \mu_2)| &\leq \ell_1\|x_1 - x_2\|^{\gamma_1} + \ell_2\|y_1 - y_2\|^{\gamma_2} \\
&\leq \ell_1 m^{\gamma_1}\|\lambda_1 - \lambda_2\|^{\gamma\gamma_1} + \ell_2 m^{\gamma_2}\|\lambda_1 - \lambda_2\|^{\gamma\gamma_2} \\
&\leq \delta.
\end{aligned}$$

Therefore,

$$f^\alpha(x_2, y_2, \mu_2) + \varepsilon_2 \geq 0,$$

i.e., $x_2 \in \widetilde{\text{Sol}}(\text{EP})^\alpha(\varepsilon_2, \lambda_2, \mu_2)$.

Now, for each $x_0 \in \widetilde{\text{Sol}}(\text{EP})^\alpha(\varepsilon_2, \lambda_1, \mu_2)$ and $x_1 \in \widetilde{\text{Sol}}(\text{EP})^\alpha(\varepsilon_2 - \delta, \lambda_1, \mu_2)$, we have:

$$\begin{aligned}
d(x_0, \widetilde{\text{Sol}}(\text{EP})^\alpha(\varepsilon_2, \lambda_2, \mu_2)) &\leq \|x_2 - x_0\| \leq \|x_2 - x_1\| + \|x_1 - x_0\| \\
&\leq m\|\lambda_1 - \lambda_2\|^\gamma + \|x_1 - x_0\|.
\end{aligned}$$

Because x_1 is arbitrary, one gets:

$$d(x_0, \widetilde{\text{Sol}}(\text{EP})^\alpha(\varepsilon_2, \lambda_2, \mu_2)) \leq m\|\lambda_1 - \lambda_2\|^\gamma + d(x_0, \widetilde{\text{Sol}}(\text{EP})^\alpha(\varepsilon_2 - \delta, \lambda_1, \mu_2)),$$

which implies

$$\begin{aligned}
&\text{ex}(\widetilde{\text{Sol}}(\text{EP})^\alpha(\varepsilon_2, \lambda_1, \mu_2), \widetilde{\text{Sol}}(\text{EP})^\alpha(\varepsilon_2, \lambda_2, \mu_2)) \\
&\quad \leq m\|\lambda_1 - \lambda_2\|^\gamma + \text{ex}(\widetilde{\text{Sol}}(\text{EP})^\alpha(\varepsilon_2, \lambda_1, \mu_2), \widetilde{\text{Sol}}(\text{EP})^\alpha(\varepsilon_2 - \delta, \lambda_1, \mu_2)) \\
&\quad \leq m\|\lambda_1 - \lambda_2\|^\gamma + H(\widetilde{S}^\alpha(\varepsilon_2, \lambda_1, \mu_2), \widetilde{\text{Sol}}(\text{EP})^\alpha(\varepsilon_2 - \delta, \lambda_1, \mu_2)) \\
&\quad \leq m\|\lambda_1 - \lambda_2\|^\gamma + \frac{\rho}{\bar{\varepsilon}_0}|\delta| \\
&\quad \leq m\|\lambda_1 - \lambda_2\|^\gamma + \frac{\rho(\ell_1 m^{\gamma_1} + \ell_2 m^{\gamma_2})}{\bar{\varepsilon}_0}\|\lambda_1 - \lambda_2\|^{\gamma_3} \\
&\quad \leq \left(m + \frac{\rho(\ell_1 m^{\gamma_1} + \ell_2 m^{\gamma_2})}{\bar{\varepsilon}_0}\right)\|\lambda_1 - \lambda_2\|^{\gamma_3}.
\end{aligned}$$

Case 2. If $\delta > \eta$, then there is a natural number n satisfying $\frac{\delta}{n} \leq \eta$. Let $\mathbb{P}'$ be a partition of the segment $[\lambda_1, \lambda_2]$ with $n + 1$ nodes $u_1, u_2, \ldots, u_{n+1}$ with

$$u_1 = \lambda_1, u_{n+1} = \lambda_2, \|u_i - u_{i+1}\| = \frac{\|\lambda_1 - \lambda_2\|}{n}.$$

Then,

$$\|u_i - u_{i+1}\|^{\gamma_3} \leq \frac{\|\lambda_1 - \lambda_2\|^{\gamma_3}}{n} \leq \frac{\eta}{\ell_1 m^{\gamma_1} + \ell_2 m^{\gamma_2}},$$

which implies that

$$(\ell_1 m^{\gamma_1} + \ell_2 m^{\gamma_2})\|u_i - u_{i+1}\|^{\gamma_3} \leq \eta.$$

Applying Case 1, we get:

$$\begin{aligned}
&H(\widetilde{\text{Sol}}(\text{EP})^\alpha(\varepsilon_2, \lambda_1, \mu_2), \widetilde{\text{Sol}}(\text{EP})^\alpha(\varepsilon_2, \lambda_2, \mu_2)) \\
&\quad \leq \sum_{i=1}^{n} H(\widetilde{\text{Sol}}(\text{EP})^\alpha(\varepsilon_2, u_i, \mu_2), \widetilde{\text{Sol}}(\text{EP})^\alpha(\varepsilon_2, u_{i+1}, \mu_2)) \\
&\quad \leq n\left(m + \frac{\rho(\ell_1 m^{\gamma_1} + \ell_2 m^{\gamma_2})}{\bar{\varepsilon}_0}\right)\|u_i - u_{i+1}\|^{\gamma_3} \\
&\quad \leq \left(m + \frac{\rho(\ell_1 m^{\gamma_1} + \ell_2 m^{\gamma_2})}{\bar{\varepsilon}_0}\right)n^{1-\alpha_3}\|\lambda_1 - \lambda_2\|^{\gamma_3}
\end{aligned}$$

$$=: \kappa_3 \|\lambda_1 - \lambda_2\|^{\gamma_3}.$$

Step 4 We have:

$$H(\widetilde{\mathrm{Sol}}(\mathrm{EP})^{\alpha}(\varepsilon_1, \lambda_1, \mu_1), \widetilde{\mathrm{Sol}}(\mathrm{EP})^{\alpha}(\varepsilon_2, \lambda_2, \mu_2))$$

$$\begin{aligned} &\leq H(\widetilde{\mathrm{Sol}}(\mathrm{EP})^{\alpha}(\varepsilon_1, \lambda_1, \mu_1), \widetilde{\mathrm{Sol}}(\mathrm{EP})^{\alpha}(\varepsilon_2, \lambda_1, \mu_1)) \\ &\quad + H(\widetilde{\mathrm{Sol}}(\mathrm{EP})^{\alpha}(\varepsilon_2, \lambda_1, \mu_1), \widetilde{\mathrm{Sol}}(\mathrm{EP})^{\alpha}(\varepsilon_2, \lambda_1, \mu_2)) \\ &\quad + H(\widetilde{\mathrm{Sol}}(\mathrm{EP})^{\alpha}(\varepsilon_2, \lambda_1, \mu_2), \widetilde{\mathrm{Sol}}(\mathrm{EP})^{\alpha}(\varepsilon_2, \lambda_2, \mu_2)) \\ &\leq \kappa_1 |\varepsilon_1 - \varepsilon_2| + \kappa_2 \|\mu_1 - \mu_2\|^{\beta} + \kappa_3 \|\lambda_1 - \lambda_2\|^{\gamma_3}. \end{aligned}$$

We complete the proof. □

The following example illustrates the applicability of Lemma 6.178.

Example 6.179 (A motivated example for Hölder conditions for equilibrium problems) Let $D = \mathbb{R}$, $\Lambda = M = [0, 1]$, $\mathscr{D}(\lambda) = [0, \lambda]$, and $f(x, y, \lambda) = x^2 + \lambda$. Then, all the assumptions of Lemma 6.178 are satisfied. The approximate solution set is $\widetilde{\mathrm{Sol}(EP)}(\lambda) = [0, \lambda]$, which is Hölder continuous.

The following lemma concerning with the Lipschitz continuity of the solution map of (EP) is derived from Lemma 6.178.

Lemma 6.180 (Lipschitz properties of approximate solution maps of equilibrium problems) *For $\alpha \in \{u, l\}$, assume that:*

(i) *$\mathscr{D}$ is Lipschitz continuous on Λ;*
(ii) *for each $(\lambda, \mu) \in \Lambda \times M$ and $y \in \mathscr{D}(\lambda)$, $(x, \varepsilon) \mapsto f^{\alpha}(x, y, \mu) + \varepsilon$ is 0-lower level quasiconcave on $\mathscr{D}(\lambda) \times [\varepsilon_0, +\infty)$;*
(iii) *f^{α} is Lipschitz continuous on $\mathscr{D}(\Lambda) \times \mathscr{D}(\Lambda) \times M$.*

Then, the approximate solution map $\widetilde{\mathrm{Sol}}(\mathrm{EP})^{\alpha}$ is Lipschitz continuous on $(0, +\infty) \times M \times \Lambda$.

Theorem 6.181 (Hölder properties of solution sets of set optimization problems) *For (SOP) assume that:*

(i) $\mathscr{D}$ is Hölder continuous and compact-valued on Λ;
(ii) F is Hölder continuous in each component separately on $\mathscr{D}(\Lambda) \times M$;
(iii) F has nonempty, Ω-proper, Ω-closed, Ω-bounded, and Ω-compact values.

Then, the solution set of (SOP) corresponding to $\alpha \in \{l, u\}$ is nonempty and Hölder continuous on $(0, +\infty) \times M \times \Lambda$, provided that:

(a) *for each $\lambda \in \Lambda$, F is C-quasiconvex in the first component on $\mathscr{D}(\lambda)$ with $\alpha = l$ and $\Omega = C$;*

(b) *for each* $\lambda \in \Lambda$, F *is* $-C$*-quasiconcave in the first component on* $\mathscr{D}(\lambda)$ *with* $\alpha = u$ *and* $\Omega = -C$.

Proof Because of similarity, we only discuss the case $\alpha = l$. We first show that the solution set of (SOP) is nonempty by applying Theorem 6.171. Let

$$x^k \in P := \{x \in \mathscr{D}(\lambda) \mid F(x, \mu) \preceq_C^l F(b, \mu)\},$$

for all $b \in \mathscr{D}(\lambda)$ and $x^k \to x_0$, one has:

$$F(x^k, \mu) \preceq_C^l F(b, \mu),$$

i.e.,

$$F(b, \mu) \subset F(x^k, \mu) + C.$$

This together with the continuity of F in (b) and the compactness of $F(x, \mu)$ in (c) yields that

$$F(b, \mu) \subset \mathrm{bd}(F(x_0, \mu) + C) = F(x_0, \mu) + C,$$

as $F(x_0, \mu)$ is compact and C is closed. Thus, $x_0 \in P$, which shows that (a) of Theorem 6.171 holds.

Assumption (b) of Theorem 6.171 is obtained from (a) and Lemma 6.170. Assumption (c) of Theorem 6.171 is evidently fulfilled since $\mathscr{D}(\lambda)$ is compact. Therefore, the solution set of (SOP) is nonempty.

We now verify the Hölder continuity of $\widetilde{\mathrm{Sol}}(\mathrm{SOP})^l$. By Proposition 6.174, it suffices to check that all the assumptions of Lemma 6.178 are satisfied. Clearly, (a) of this lemma is fulfilled. For (b), thanks to Proposition 6.163, $\mathscr{G}_e^l(F(\cdot, \mu), F(y, \mu))$ is quasiconvex for all $y \in \mathscr{D}(\lambda)$ and $\mu \in M$. This implies that the 0-lower-level quasiconcave condition in (b) of Lemma 6.178 is satisfied by Remark 6.152. Combining (c) with Proposition 6.165, the Hölder conditions in (c) of Lemma 6.178 are fulfilled. Therefore, the proof is complete. □

With the same arguments, we have the following result dealing with Lipschitz continuity of solution maps of (SOPs).

Corollary 6.182 (Lipschitz properties of set optimization problems) *For (SOP), assume that:*

(a) $\mathscr{D}$ *is Lipschitz continuous and compact-valued on* Λ;
(b) F *is Lipschitz continuous in each component separately on* $\mathscr{D}(\Lambda) \times M$;
(c) F *has nonempty,* Ω*-proper,* Ω*-closed,* Ω*-bounded, and* Ω*-compact values.*

Then, the approximate solution set of (SOP) corresponding to $\alpha \in \{l, u\}$ *is nonempty and Lipschitz continuous on* $(0, +\infty) \times M \times \Lambda$*, provided that:*

(a) *for each* $\lambda \in \Lambda$, F *is* C*-quasiconvex in the first component on* $\mathscr{D}(\lambda)$ *with* $\alpha = l$ *and* $\Omega = C$;
(b) *for each* $\lambda \in \Lambda$, F *is* $-C$*-quasiconcave in the first component on* $\mathscr{D}(\lambda)$ *with* $\alpha = u$ *and* $\Omega = -C$.

6.7.5 Well-Posedness of Set Optimization Problems

Throughout this subsection, unless stated otherwise, let X and Y be Banach spaces, $\mathrm{cl}\, A$, $\mathrm{bd}\, A$, and $\mathrm{int}\, A$ the closure, boundary, and topological interior, resp., of a subset A, $C \subset Y$ a nonempty pointed, closed, convex, and solid cone, and $\mathscr{P}(Y)$ the family of all subsets of Y. We say that a nonempty set $A \in \mathscr{P}(Y)$ is C-proper if $A + C \neq Y$; C-closed if $A + C$ is a closed set; and C-bounded if for each neighborhood U of the origin 0_Y in Y, $A \subset tU + C$ for some $t > 0$. For brevity, we denote the families of C-proper, C-closed, and C-bounded by $\mathscr{P}_{\mathrm{pro}\, C}(Y)$, $\mathscr{P}_{\mathrm{cl}\, C}(Y)$, and $\mathscr{P}_{\mathrm{bd}\, C}(Y)$, resp.

We first recall the cone-continuity concepts of a set-valued map.

Definition 6.183 (Cone-semicontinuity of set-valued maps) For $\bar{x} \in X$, a set-valued map F from X to Y is said to be:

(a) C-UPPER SEMICONTINUOUS (C-usc) at $\bar{x}$, if for any open set U containing $F(\bar{x})$, there exists a neighborhood N of $\bar{x}$ such that $F(x) \subset U + C$ for all $x \in N$;
(b) C-LOWER SEMICONTINUOUS (C-lsc) at $\bar{x}$, if for any open set U of Y with $F(\bar{x}) \cap U \neq \emptyset$, there exists a neighborhood N of $\bar{x}$ such that $F(x) \cap (U - C) \neq \emptyset$ for all $x \in N$;
(c) C-CONTINUOUS at $\bar{x}$ if it is both C-usc and C-lsc at $\bar{x}$.

F is called C-usc (C-lsc, C-continuous, resp.) on a subset M of X if it is C-usc (C-lsc, C-continuous, resp.) at every $x \in M$.

Definition 6.184 (Upper and lower pseudocontinuity) Let ϕ be an extended real-valued function defined on X and $\bar{x} \in X$. ϕ is said to be:

(a) LOWER PSEUDOCONTINUOUS at $\bar{x}$ if

$$[\forall x \in X, \phi(x) < \phi(\bar{x})] \Rightarrow [\forall x^k \to \bar{x}, \phi(x) < \lim\ \inf_{n\to\infty} \phi(x^k)];$$

(b) UPPER PSEUDOCONTINUOUS at $\bar{x}$ if $(-\phi)$ is lower pseudocontinuous at $\bar{x}$.

With $A, B \in \mathscr{P}(Y)$, for convenience, we recall binary relations on $\mathscr{P}(Y)$ wrt C defined at the beginning of Sect. 6.6:

$$A \preceq_C^l B \Leftrightarrow B \subset A + C \quad \text{and} \quad A \preceq_C^u B \Leftrightarrow A \subset B - C.$$

$$A \prec_C^l B \Leftrightarrow A \subset B - \mathrm{int}\, C \quad \text{and} \quad A \prec_C^u B \Leftrightarrow A \subset B - \mathrm{int}\, C.$$

For each $\alpha \in \{l, u\}$, we consider an equivalence relation on $\mathscr{P}(Y)$ defined as follows: for $A, B \in \mathscr{P}(Y)$,

$$A \sim^{\alpha} B \quad \text{if, and only if,} \quad A \preceq_C^{\alpha} B \quad \text{and} \quad B \preceq_C^{\alpha} A.$$

Motivated by the lattice approach, we present an equivalent relation in the decision space via that in the image space. For all $x, y \in M$, it is said that $x \sim_F^{\alpha} y$ if, and only if, $F(x) \sim^{\alpha} F(y)$. By $[x]_F^{\alpha}$ we denote the class of all decisions which are equivalent to x:

$$[x]_F^{\alpha} = \{y \in M \mid y \sim_F^{\alpha} x\}.$$

Let M be a nonempty closed subset of X, and $F : M \to \mathscr{P}(Y)$ be a set-valued mapping. For each $\alpha \in \{l, u\}$, we consider the following set optimization problem:

α -SOP $\qquad$ α-minimize $F(x)$ subject to $x \in M$,

where the "α-minimize" is understood as to minimize using the relation $\preceq_C^{\alpha}$.

Definition 6.185 (Efficient points of set optimization problems) For each $\alpha \in \{l, u\}$, a feasible solution $\bar{x} \in M$ is said to be an α-PARETO EFFICIENT SOLUTION of (α -SOP) if, for all $x \in M$, $F(x) \preceq_C^{\alpha} F(\bar{x})$ implies $F(\bar{x}) \preceq_C^{\alpha} F(x)$.

The set of all Pareto efficient solutions of problem (α-SOP) is denoted by α -Eff.

We now introduce concepts of well-posedness for problem (α-SOP). In this way, for each $e \in \operatorname{int} C$, we present two kinds of approximate sequences wrt the vector e as follows.

Definition 6.186 (Generalized minimizing sequences for set optimization problems) For $\alpha \in \{l, u\}$, a sequence $\{x_n\} \subset M$ is said to be a GENERALIZED e-MINIMIZING SEQUENCE for problem (α -SOP) if there exist a sequence of nonnegative real numbers $\{\varepsilon_n\}$ converging to 0 and a sequence $\{v_n\} \subset \alpha$ -Eff such that

$$F(x_n) \preceq_C^{\alpha} F(v_n) + \varepsilon_n e.$$

If $v_n = v$ for all n, then the generalized e-minimizing sequence $\{x_n\}$ is called an e-MINIMIZING SEQUENCE corresponding to v.

The following result shows that the notion defined in Definition 6.186 does not depend on the choice of e.

Proposition 6.187 (Characterizations of generalized minimizing sequences for set optimization problems) *Let $e \in \operatorname{int} C$ be fixed and $\alpha \in \{l, u\}$. The following statements are true:*

(a) *a sequence $\{x_n\} \subset M$ is an e-minimizing sequence corresponding to a vector v belonging to α -Eff for problem (α -SOP) if and only if there exists a sequence*

$\{c_n\} \subset C$ converging to 0_Y such that

$$F(x_n) \preceq^{\alpha}_{C} F(v) + c_n;$$

(b) *a sequence $\{x_n\} \subset M$ is a generalized e-minimizing sequence for problem (α-SOP) if and only if there exist sequences $\{c_n\} \subset C$ converging to 0_Y and $\{v_n\}$ in α-Eff such that*

$$F(x_n) \preceq^{\alpha}_{C} F(v_n) + c_n.$$

Proof We need only to prove (b) as (a) can be discussed similarly. If $\{x_n\}$ is a generalized e-minimizing sequence for problem (α-SOP), obviously we have the above result with $c_n = \epsilon_n e$. For the converse, we present the proof of $\alpha = u$. Let $\{c_n\} \subset C$ converging to 0_Y, $\{x_n\} \subset M$, $\{v_n\} \subset u$-Eff satisfying $F(x_n) \preceq^{u}_{C} F(v_n) + c_n$ for all n, then we have:

$$F(x_n) \subset F(v_n) + c_n - C. \tag{6.155}$$

Since $(e - \operatorname{int} C)$ is a neighborhood of the origin in Y, there exists $\delta > 0$ such that $\delta B(0_Y, 1) \subset (e - \operatorname{int} C)$, where $B(0_Y, 1)$ is the closed unit ball. For each $n \in \mathbb{N}$, we have $c_n \in \|c_n\| B(0_Y, 1) \subset \|c_n\|\delta^{-1}(e - C) = \|c_n\|\delta^{-1}e - C$. For $\varepsilon_n = \|c_n\|\delta^{-1}$, $\{\varepsilon_n\}$ converges to 0 and $c_n \in \varepsilon_n e - C$. Plugging this into (6.155), we obtain:

$$F(x_n) \subset F(v_n) + \varepsilon_n e - C,$$

i.e., $F(x_n) \preceq^{u}_{C} F(v_n) + \varepsilon_n e$. So, $\{x_n\}$ is a generalized α-e-maximizing sequence for problem (u-SOP). □

According to Proposition 6.187, we simply write "(generalized) α-minimizing sequence" instead of "(generalized) α-e-minimizing sequence."

Definition 6.188 (Well-posedness concepts for set optimization problems) Problem (α-SOP) is said to be:

(a) α-WELL-POSED at α-Pareto efficient solution v if for each α-minimizing sequence $\{x^k\}$ corresponding to v, it holds that $x^k \to v$;
(b) GENERALIZED α-WELL-POSED at α-Pareto efficient solution v if every α-minimizing sequence corresponding to v has a subsequence converging to some α-Pareto efficient solution x^* such that $x^* \sim^{\alpha}_{F} v$;
(c) GLOBALLY α-WELL-POSED if every generalized α-minimizing sequence has a subsequence converging to some α-Pareto efficient solution x^*.

Remark 6.189 (Relationships between well-posedness types) Later in this section, we will consider well-posedness for α-set optimization problems via Gerstewitz functions and use assumptions related to the non-emptiness of the interior of ordering cones. When $\alpha = l$, Definition 6.186(a) collapses to pointwise well-posedness and Definitions 6.186(b), (c) collapse to L-well-posedness. When $\alpha = u$, Definitions 6.186(a), (b) become the concepts of DH-well-posedness and L-well-posedness, resp. Obviously, if problem (u-SOP) is globally u-well-posed in the sense of Definition 6.186(c), then it is generalized B-well-posedness considered by many authors. The following example shows that Definition 6.186(c) is stronger than the generalized B-well-posedness.

Example 6.190 (A case with generalized B-well-posedness but not global L-well-posedness) Let $X = \mathbb{R}, Y = \mathbb{R}^2, M = [-1, 1], C = \mathbb{R}^2_+$, and $F : M \to \mathscr{P}(Y)$ be defined by

$$F(x) := \begin{cases} \{(y_1, y_2) \in Y \mid (y_1 - x)^2 + (y_2 - x^2)^2 \leq 1\} & \text{if } x \neq 0, \\ \{(y_1, y_2) \in Y \mid y_1^2 + (y_2 - 1)^2 \leq 1\} & \text{if } x = 0. \end{cases}$$

It is easy to check that u-Eff $= [-1, 0)$ and problem (u-SOP) is generalized B-well-posed, but it is not globally L-well-posed for any $e \in \operatorname{int} C$.

In the rest of this section, we consider Tikhonov well-posedness of scalar optimization problems.

Definition 6.191 (Tikhonov well-posedness for scalar optimization problems) For $\phi : X \to \mathbb{R}$, consider the following scalar optimization problem:

$$\text{OP}(M, \phi) \quad \text{minimize } \phi(x) \quad \text{subject to} \quad x \in M.$$

(a) $\text{OP}(M, \phi)$ is said to be TIKHONOV WELL-POSED if ϕ has a unique minimizer $\bar{x}$ on M and every sequence $\{x^k\} \subset M$ such that $\phi(x^k) \to \inf \phi(M)$ (called a MINIMIZING SEQUENCE) converges to $\bar{x}$;
(b) $\text{OP}(M, \phi)$ is said to be GENERALIZED TIKHONOV WELL-POSED if the set of minimizers of $\text{OP}(M, \phi)$ is nonempty and every minimizing sequence $\{x^k\} \subset M$ has some subsequence $\{x^{k_j}\}$ converging to a minimizer of $\text{OP}(M, \phi)$.

Lemma 6.192 (Tikhonov well-posedness of scalar optimization problems) *Let M be a compact set and ϕ lower pseudocontinuous on M. Then, problem* $\text{OP}(M, \phi)$ *is generalized Tikhonov well-posed. Moreover, if ϕ has a unique minimizer $\bar{x}$ on M, then* $\text{OP}(M, \phi)$ *is Tikhonov well-posed.*

Well-Posedness of Set Optimization Problems Involving Uniform Set Less Order Relations

In this section, we consider well-posedness properties of set optimization problems wrt the two important cases of $\preceq^{\alpha}_C$, namely, $\alpha \in \{l, u\}$, via nonlinear scalarization functions.

Let $a \in Y$ and $e \in \operatorname{int} C$. For each $\alpha \in \{l, u\}$, we consider the nonlinear scalarization function $\tau_{e,a}^{(\alpha)} : Y \to \mathbb{R} \cup \{-\infty\}$ defined for all $y \in Y$ by

$$\tau_{e,a}^{(\alpha)}(y) := \begin{cases} \inf\{t \in \mathbb{R} \mid y \in -te + a + C\} & \text{if } \alpha = l, \\ \inf\{t \in \mathbb{R} \mid y \in te + a - C\} & \text{if } \alpha = u. \end{cases}$$

If the vector a is replaced by a set $A \in \mathscr{P}(Y)$, we define the function $T_{e,A}^{(\alpha)} : Y \to \mathbb{R} \cup \{-\infty\}$ for all $y \in Y$ as follows:

$$T_{e,A}^{(\alpha)}(y) := \begin{cases} \inf\{t \in \mathbb{R} \mid y \in -te + A + C\} & \text{if } \alpha = l, \\ \inf\{t \in \mathbb{R} \mid y \in te + A - C\} & \text{if } \alpha = u \end{cases}$$

is also called a GERSTEWITZ FUNCTION. Then, we have the following important relation.

Proposition 6.193 (A characterization of Gerstewitz functions) *For each $\alpha \in \{l, u\}$, the following relation*

$$T_{e,A}^{(\alpha)}(y) = \inf_{a \in A}\{\tau_{e,a}^{(\alpha)}(y)\}$$

is valid for all $y \in Y$.

Proof We give the proof for $\alpha = u$ as the other case is similar. Let $y \in Y$ be given. For each $a \in A$ and $\varepsilon > 0$, we can pick up a real number $\bar{t}$ such that

$$\bar{t} < \tau_{e,a}^{(u)}(y) + \varepsilon \text{ and } y \in \bar{t}e + a - C,$$

and consequently

$$y \in \bar{t}e + A - C.$$

This together with the definition of $T_{e,A}^{(\alpha)}$ implies that

$$T_{e,A}^{(u)}(y) \leq \bar{t} < \tau_{e,a}^{(u)}(y) + \varepsilon, \forall \varepsilon > 0.$$

Since ε is arbitrary, $T_{e,A}^{(u)}(y) \leq \tau_{e,a}^{(u)}(y)$ for all $a \in A$. Therefore,

$$T_{e,A}^{(u)}(y) \leq \inf_{a \in A}\{\tau_{e,a}^{(u)}(y)\}.$$

If $\inf_{a \in A}\{\tau_{e,a}^{(u)}(y)\} > T_{e,A}^{(u)}(y)$, then there would be $\bar{t}$ satisfying

$$T_{e,A}^{(u)}(y) < \bar{t} < \inf_{a \in A}\{\tau_{e,a}^{(u)}(y)\}$$

and $y \in \bar{t}e + A - C$. Consequently, there exists an element $\bar{a} \in A$ such that $y \in \bar{t}e + \bar{a} - C$, which leads to the following contradiction:

$$\tau^{(u)}_{e,\bar{a}}(y) \le \bar{t} < \inf_{a\in A}\{\tau^{(u)}_{e,a}(y)\} \le \tau^{(u)}_{e,\bar{a}}(y).$$

Therefore, $T^{(u)}_{e,A}(y) = \inf_{a\in A}\{\tau^{(u)}_{e,a}(y)\}$ for all $y \in Y$. □

Next, for the sake of simplification, we use the notations $\mathscr{P}^{(\alpha)}_{\text{pro}}(Y)$, $\mathscr{P}^{(\alpha)}_{\text{cl}}(Y)$ to denote the families of C-proper, C-closed if $\alpha = l$ and of $(-C)$-proper, $(-C)$-closed if $\alpha = u$, resp. For $(A_1, A_2) \in \mathscr{P}^{(\alpha)}_{\text{pro}}(Y)^2$, we define the GENERALIZED GERSTEWITZ FUNCTIONS $G^{(\alpha)}_e$ acting from $\mathscr{P}^{(\alpha)}_{\text{pro}}(Y)^2$ into $\mathbb{R} \cup \{\infty\}$ as follows:

$$G^{(\alpha)}_e(A_1, A_2) = \begin{cases} \sup_{a\in A_2}\left\{T^{(\alpha)}_{e,A_1}(a)\right\} & \text{if } \alpha = l, \\ \sup_{a\in A_1}\left\{T^{(\alpha)}_{e,A_2}(a)\right\} & \text{if } \alpha = u. \end{cases}$$

Lemma 6.194 (Estimations for generalized Gerstewitz functions) *For each $\alpha \in \{l, u\}$ and $r \in \mathbb{R}$, the following assertions are true for all $A_1, A_2 \in \mathscr{P}^{(\alpha)}_{\text{cl}}(Y) \cap \mathscr{P}^{(\alpha)}_{\text{pro}}(Y)$:*

(a) *$G^{(\alpha)}_e(A_1, A_2) \le r$ if and only if $A_1 \preceq^{\alpha}_C re + A_2$.*
(b) *If $G^{(\alpha)}_e(A_1, A_2) < r$, then $A_1 \prec^{\alpha}_C re + A_2$.*

Proof

(a) We present the proof for $\alpha = l$. Let $G^{(l)}_e(A_1, A_2) \le r$, i.e., for all $u \in A_2$:

$$T^{(\alpha)}_{e,A_1}(u) \le r. \tag{6.156}$$

By the definition of $T^{(l)}_{e,A_1}$, (6.156) is equivalent to

$$u \in -re + A_1 + C.$$

Therefore,

$$A_2 \subset -re + A_1 + C,$$

and consequently $A_1 \preceq^{l}_C A_2 + re$.

(b) We only discuss the case $\alpha = u$. Since $G^{(u)}_e(A_1, A_2) < r$, for all $x \in A_1$ we have that $T_{e,A_2}(x) < r$. Then, there is at least one $\bar{r} < r$ satisfying

$$x \in \bar{r}e + A_2 - C.$$

Noting that $\bar{r}e = re - k$ for some $k \in \operatorname{int} C$, and hence

$$x \in re - k + A_2 - C \subset re + A_2 - \operatorname{int} C.$$

Since x is an arbitrary element in A_1, we have:

$$A_1 \subset re + A_2 - \operatorname{int} C,$$

or equivalently $A_1 \prec^u_C A_2 + re$.

□

We now present some characterizations of α-Pareto efficient solutions of problem (α-SOP) via the generalized α-Gerstewitz function.

Lemma 6.195 (A characterization of efficient points of set optimization problems via generalized Gerstewitz functions) *For each $\alpha \in \{l, u\}$, suppose that $F : M \to \mathscr{P}^{(\alpha)}_{\text{cl}}(Y) \cap \mathscr{P}^{(\alpha)}_{\text{pro}}(Y)$. If v is an α-Pareto efficient solution of problem* (α -SOP), *then the following statements hold true:*

(a) $G^{(\alpha)}_e(F(x), F(v)) \geq 0$ *for all* $x \in M$;
(b) *for each* $x \in M$, $G^{(\alpha)}_e(F(x), F(v)) = 0$ *if and only if* $x \sim^{\alpha}_F v$.

Proof

(a) By the similarity, we give only a proof for $\alpha = u$. Suppose to the contrary that there exist a vector $x \in M$ and a real number $r < 0$ such that $G^{(u)}_e(F(x), F(v)) < r$. According to Lemma 6.194(b), we then obtain $F(x) \prec^u_C F(v) + re$, i.e.,

$$F(x) \subset F(v) + re - \operatorname{int} C.$$

Consequently,

$$F(x) \subset F(v) - \operatorname{int} C. \tag{6.157}$$

Since v is an u-Pareto efficient solution of (u -SOP)

$$F(v) \subset F(x) - C.$$

Combining this with (6.157), we obtain $F(x) - C \subset F(x) - \operatorname{int} C$. Thus, $F(x) - C = F(x) - \operatorname{int} C$ which is impossible as $F(z) - C \neq Y$ and $F(z) - C$ is closed for all $z \in M$.

(b) Suppose that $G^{(\alpha)}_e(F(x), F(v)) = 0$. Then, by Lemma 6.194(a), we obtain $F(x) \preceq^{\alpha}_C F(v)$. Since $v \in \alpha$ -Eff, we have that $F(v) \preceq^{\alpha}_C F(x)$.
Conversely, suppose that $F(x) \preceq^{\alpha}_C F(v)$ and $F(v) \preceq^{\alpha}_C F(x)$. It follows from $F(x) \preceq^{\alpha}_C F(v)$ that $G^{(\alpha)}_e(F(x), F(v)) \leq 0$. Due to (i), $G^{(\alpha)}_e(F(x), F(v)) = 0$.

□

Remark 6.196 (Relationships between efficient points of set optimization problems and that of scalar optimization problems) From the definition of function $G_e^{(\alpha)}$, we have

$$G_e^{(\alpha)}(F(x), F(x)) = 0$$

for all $x \in M$. Therefore, according to Lemma 6.195(a), we conclude that

$$\min_{x \in M} G_e^{(\alpha)}(F(x), F(v)) = 0$$

for every $v \in \alpha$-Eff.

Now we use the generalized α-Gerstewitz function to scalarize the well-posedness of problem (α-SOP).

Theorem 6.197 (Well-posedness for set optimization problems via nonlinear scalarization functions) *For each $\alpha \in \{l, u\}$, suppose that $F : M \to \mathscr{P}_{\text{cl}}^{(\alpha)}(Y) \cap \mathscr{P}_{\text{pro}}^{(\alpha)}(Y)$ and $v \in \alpha$-Eff. Then, problem (α-SOP) is*

(a) *generalized α-well-posed at v if and only if problem* OP$(M, G_e^{(\alpha)}(F(\cdot), F(v)))$ *is generalized Tykhonov well-posed;*
(b) *α-well-posed at v if and only if problem* OP$(M, G_e^{(\alpha)}(F(\cdot), F(v)))$ *is Tykhonov well-posed.*

Proof

(a) Let $\{x_n\}$ be a minimizing sequence of OP$(M, G_e^{(\alpha)}(F(\cdot), F(v)))$. Then, by Remark 6.196,

$$G_e^{(\alpha)}(F(x_n), F(v)) \to \min_{x \in M} G_e^{(\alpha)}(F(x), F(v)) = 0.$$

Let $\varepsilon_n = G_e^{(\alpha)}(F(x_n), F(v))$. Then, according to Lemma 6.195(a), we obtain $\varepsilon_n \geq 0$ as $v \in \alpha$-Eff. By using Lemma 6.194(a), we conclude that $F(x_n) \preceq_C^{\alpha} F(v) + \varepsilon_n e$, and hence $\{x_n\}$ is a minimizing sequence corresponding to v when $\alpha = l$, and a maximizing sequence corresponding to v when $\alpha = u$. Since (α-SOP) is generalized α-well-posed at v, $\{x_n\}$ has a subsequence $\{x_{n_k}\}$ converging to some α-Pareto efficient solution $\bar{x}$ with $\bar{x} \sim_F^{\alpha} v$. Applying Lemma 6.195(b), we come to $G_e^{(\alpha)}(F(\bar{x}), F(v)) = 0$, and so by Remark 6.196, the vector $\bar{x}$ is a minimizer of problem OP$(M, G_e^{(\alpha)}(F(\cdot), F(v)))$.
Conversely, let $\{x_n\}$ be an α-minimizing sequence corresponding to v, then there is a sequence $\{\varepsilon_n\} \subset \mathbb{R}_+$ converging to 0 such that

$$F(x_n) \preceq_C^{\alpha} \varepsilon_n e + F(v).$$

It follows from Remark 6.196 and Lemma 6.194(a) that

$$0 = \min_{x \in M} G_e^{(\alpha)}(F(x), F(v)) \leq G_e^{(\alpha)}(F(x_n), F(v)) \leq \varepsilon_n.$$

Because problem $\mathrm{OP}(M, G_e^{(\alpha)}(F(\cdot), F(v)))$ is generalized Tykhonov well-posed at v, $\{x_n\}$ has a subsequence converging to some minimizer x^* of $\mathrm{OP}(M, G_e^{(\alpha)}(F(\cdot), F(v)))$. Therefore, by Remark 6.196 and Lemma 6.195(b), we have $x^* \sim_F^\alpha v$.

Similarly, we have also proven the statement (b).

□

For a nonempty subset $A \subset M$, we consider an α-MARGINAL NONLINEAR SCALARIZATION FUNCTION $\gamma_A^{(\alpha)} : M \to \mathbb{R}$ defined as

$$\gamma_A^{(\alpha)}(x) = \inf_{a \in A} G_e^{(\alpha)}(F(x), F(a)) \text{ for all } x \in M.$$

By employing this function, we obtain the following result which plays an important role as a bridge between global well-posedness of set optimization problems and scalar optimization problems.

Theorem 6.198 (Global well-posedness for set optimization problems via nonlinear scalarization functions) *For each* $\alpha \in \{l, u\}$, *suppose that* $F : M \to \mathscr{P}_{\mathrm{cl}}^{(\alpha)}(Y) \cap \mathscr{P}_{\mathrm{pro}}^{(\alpha)}(Y)$. *Then, problem* ($\alpha$ -SOP) *is globally* α*-well-posed if and only if problem* $\mathrm{OP}(M, \gamma_{\alpha\text{-Eff}}^{(\alpha)})$ *is generalized Tykhonov well-posed.*

Proof Let $\{x_n\}$ be a sequence such that

$$\gamma_{\alpha\text{-Eff}}^{(\alpha)}(x_n) \to \inf_{x \in M} \gamma_{\alpha\text{-Eff}}^{(\alpha)}(x) = 0.$$

Then, there are sequences $\{\delta_n\} \subset \mathbb{R}_+$ converging to 0 and $\{v_n\} \subset \alpha$ -Eff such that

$$0 \leq G_e^{(\alpha)}(F(x_n), F(v_n)) < \delta_n.$$

By setting $G_e^{(\alpha)}(F(x_n), F(v_n)) = \varepsilon_n$, one has $\varepsilon_n \in [0, \delta_n[$ and $F(x_n) \preceq_C^\alpha F(v_n) + \varepsilon_n e$. Thus, $\{x_n\}$ is a generalized minimizing sequence for problem (α-SOP), and hence it has a subsequence converging to some solution $\bar{x} \in \alpha$ -Eff, as problem (α -SOP) is globally α-well-posed. Therefore, $\mathrm{OP}(M, \gamma_{\alpha\text{-Eff}}^{(\alpha)})$ is generalized Tykhonov well-posed, because the vector $\bar{x}$ is also a minimizer of $\mathrm{OP}(M, \gamma_{\alpha\text{-Eff}}^{(\alpha)})$.

Conversely, let $\{x_n\}$ be a generalized minimizing sequence for (α -SOP). Then, there are a sequence of nonnegative real numbers $\{\epsilon_n\}$ converging to 0 and a sequence $\{v_n\} \subset \alpha$ -Eff such that $F(x_n) \preceq_C^\alpha F(v_n) + \epsilon_n e$. Then,

$$0 = \min_{x \in M} \gamma_{\alpha\text{-Eff}}^{(\alpha)}(x) \leq \gamma_{\alpha\text{-Eff}}^{(\alpha)}(x_n) \leq G_e^{(\alpha)}(F(x_n), F(v_n)) = \varepsilon_n.$$

Hence, $\{x_n\}$ is a minimizing sequence for $\mathrm{OP}(M, \gamma^{(\alpha)}_{\alpha\text{-Eff}})$. Because problem $\mathrm{OP}(M, \gamma^{(\alpha)}_{\alpha\text{-Eff}})$ is generalized Tykhonov well-posed, the sequence $\{x_n\}$ has a subsequence approaching some minimizer $\bar{x}$ of $\mathrm{OP}(M, \gamma^{(\alpha)}_{\alpha\text{-Eff}})$. Consequently, there exists $\bar{v} \in \alpha$ -Eff such that

$$\gamma^{(\alpha)}_{\alpha\text{-Eff}}(\bar{x}) = \inf_{v \in \alpha\text{-Eff}} G^{(\alpha)}_e(F(\bar{x}), F(v)) = G^{(\alpha)}_e(F(\bar{x}), F(\bar{v})) = 0.$$

This together with Lemma 6.195(b) implies that $\bar{x}$ is also an α-Pareto efficient solution of (α-SOP), and so problem (α -SOP) is globally α-well-posed. This completes the proof. □

To unify the presentation, we propose the following concept: the mapping F is said to be C-SEMICONTINUOUS (resp., C-CONTINUOUS) WRT α at $\bar{x} \in X$ if it is C-usc (resp., C-continuous) at $\bar{x}$ when $\alpha = l$ and $(-C)$-lsc (resp., $(-C)$-continuous) at $\bar{x}$ when $\alpha = u$.

Lemma 6.199 (Lower semicontinuity of nonlinear scalarization functions) *For each $\alpha \in \{l, u\}$ and $y \in M$, if the mapping $F : M \to \mathscr{P}^{(\alpha)}_{\mathrm{cl}}(Y) \cap \mathscr{P}^{(\alpha)}_{\mathrm{pro}}(Y)$ is C-semicontinuous wrt α, then $G^{(\alpha)}_e(F(\cdot), F(y))$ is lower semicontinuous on M. Furthermore, if F is C-continuous wrt α on M, then $G^{(\alpha)}_e(F(\cdot), F(\cdot))$ is lower semicontinuous on M.*

Proof By the similarity, we only deal with $\alpha = u$.

For $y \in M$, we need to prove that $G^{(u)}_e(F(\cdot), F(y))$ is lower semicontinuous or, equivalently, for every $r \in \mathbb{R}$, the lower-level set $L_y = \{x \in M \mid G^{(u)}_e(F(x), F(y)) \le r\}$ is closed. According to Lemma 6.194(a), we have:

$$L_y = \{x \in M \mid F(x) \preceq^u_C re + F(y)\} = \{x \in M \mid F(x) \subset re + F(y) - C\}.$$

For any sequence $\{x_n\} \subset L_y$ converging to x, one has:

$$F(x_n) \subset re + F(y) - C. \tag{6.158}$$

We have that $x \in L_y$. Otherwise, we would have $F(x) \not\subset re + F(y) - C$, or equivalently $F(x) \cap Y \setminus (re + F(y) - C) \neq \emptyset$. Since $F \in \mathscr{P}^{(u)}_{\mathrm{cl}}(Y)$, one has $Y \setminus (re + F(y) - C)$ is an open set. This together with the $(-C)$-lower semicontinuity of F at x implies that for n sufficiently large

$$F(x_n) \cap [Y \setminus (re + F(y) - C) + C] \neq \emptyset.$$

Note that $Y \setminus (re + F(y) - C) + C = Y \setminus (re + F(y) - C)$, and hence

$$F(x_n) \cap [Y \setminus (re + F(y) - C)] \neq \emptyset,$$

which contradicts (6.158). Therefore, $G_e^{(u)}(F(\cdot), F(y))$ is lower semicontinuous on M.

For the second statement, it suffices to prove that for each $r \in \mathbb{R}$ the lower-level set

$$L = \left\{(x, y) \in M \times M \mid G_e^{(u)}(F(x), F(y)) \leq r\right\}$$
$$= \{(x, y) \in M \times \mid F(x) \subset re + F(y) - C\}$$

is closed, i.e., for any sequence $\{(x_n, y_n)\} \subset L$ converging to (x, y), $(x, y) \in L$. Let $z \in F(x)$ be arbitrary. For each neighborhood V of the origin in Y, there exists a balanced neighborhood V_1 such that $V_1 + V_1 \subset V$ and $F(x) \cap (z + V_1) \neq \emptyset$. Since F is $(-C)$-lsc,

$$F(x_n) \cap (z + V_1 + C) \neq \emptyset \text{ for } n \text{ sufficiently large.}$$

Then, there exists $w_n \in F(x_n)$ such that $w_n \in z + V_1 + C$, or equivalently

$$z \in w_n + V_1 - C. \tag{6.159}$$

Note that $F(y) \subset F(y) + V_1$, and because F is $(-C)$-usc at y, $F(y_n) \subset F(y) + V_1 - C$ for n sufficiently large. Since $F(x_n) \subset re + F(y_n) - C$, one has $F(x_n) \subset re + F(y) + V_1 - C$. Combining this with (6.159) and taking into account the convexity of C, we arrive at

$$z \in re + F(y) + V_1 + V_1 - C - C \subset re + F(y) + V - C.$$

Because V is arbitrarily chosen and $F(y) - C$ is closed, $z \in re + F(y) - C$, which implies that $F(x) \subset re + F(y) - C$, i.e., $(x, y) \in L$. Therefore, $G_e^{(u)}(F(\cdot), F(\cdot))$ is lower semicontinuous on M. □

Lemma 6.200 (Lower semicontinuity of the marginal nonlinear scalarization function) *For each $\alpha \in \{l, u\}$, let $F : M \to \mathscr{P}_{\text{cl}}^{(\alpha)}(Y) \cap \mathscr{P}_{\text{pro}}^{(\alpha)}(Y)$ be a mapping. If F is C-continuous wrt α, then for any nonempty compact subset A of M, the function $\gamma_A^{(\alpha)}(x)$ is lower pseudocontinuous on M.*

Proof Let $\bar{x}$ and w be in M satisfying $\gamma_A^{(\alpha)}(w) < \gamma_A^{(\alpha)}(\bar{x})$, and let a sequence $\{x_n\}$ converge to $\bar{x}$. Let's consider the following two cases.
Case 1. There is $z \in M$ such that

$$\gamma_A^{(\alpha)}(w) < \gamma_A^{(\alpha)}(z) < \gamma_A^{(\alpha)}(\bar{x}). \tag{6.160}$$

Then, there is an element $y^* \in A$ satisfying for all $y \in A$,

$$G_e^{(\alpha)}(F(z), F(y^*)) < G_e^{(\alpha)}(F(\bar{x}), F(y)). \tag{6.161}$$

We show that there is an index $n_0 \in \mathbb{N}$ such that

$$\gamma_A^{(\alpha)}(x_n) > G_e^{(\alpha)}(F(z), F(y^*)) \text{ for all } n \geq n_0.$$

Otherwise, for each n_0, there is $n \geq n_0$ satisfying $\gamma_A^{(\alpha)}(x_n) \leq G_e^{(\alpha)}(F(z), F(y^*))$. Then, for any sequence of positive real numbers $\{\varepsilon_n\}$ converging to 0, due to the definition of the function $\gamma_A^{(\alpha)}$, there exist a sequence $\{y_n\} \subset A$ such that

$$G_e^{(\alpha)}(F(x_n), F(y_n)) - \varepsilon_n < G_e^{(\alpha)}(F(z), F(y^*)) \text{ for all } n. \tag{6.162}$$

Because of the compactness of A, we can assume that the sequence $\{y_n\}$ converges to some $\bar{y} \in A$. Combining (6.162) with the lower semicontinuity of the function $G_e^{(\alpha)}(F(\cdot), F(\cdot))$, we have

$$G_e^{(\alpha)}(F(\bar{x}), F(\bar{y})) \leq \liminf_{n\to\infty}[G_e^{(\alpha)}(F(x_n), F(y_n)) - \varepsilon_n] \leq G_e^{(\alpha)}(F(z), F(y^*)),$$

which contradicts (6.161). Thus, for n sufficiently large,

$$\gamma_A^{(\alpha)}(x_n) > G_e^{(\alpha)}(F(z), F(y^*)) \geq \gamma_A^{(\alpha)}(z). \tag{6.163}$$

This together with (6.160) implies that

$$\liminf_{n\to\infty} \gamma_A^{(\alpha)}(x_n) \geq \gamma_A^{(\alpha)}(z) > \gamma_A^{(\alpha)}(w).$$

Case 2. There does not exist any vector $z \in M$ such that

$$\gamma_A^{(\alpha)}(w) < \gamma_A^{(\alpha)}(z) < \gamma_A^{(\alpha)}(\bar{x}).$$

By the definition of γ_A, there is an element $y^* \in A$ satisfying for all $y \in A$,

$$G_e^{(\alpha)}(F(w), F(y^*)) < G_e^{(\alpha)}(F(\bar{x}), F(y)). \tag{6.164}$$

By repeating the above arguments after (6.161) (w serves as z), we have:

$$\gamma_A^{(\alpha)}(x_n) > G_e^{(\alpha)}(F(w), F(y^*)) \geq \gamma_A^{(\alpha)}(w) \text{ for } n \text{ sufficiently large.}$$

Since there is not $z \in M$ such that

$$\gamma_A^{(\alpha)}(w) < \gamma_A^{(\alpha)}(z) < \gamma_A^{(\alpha)}(\bar{x}),$$

then

$$\gamma_A^{(\alpha)}(\bar{x}) \leq \gamma_A^{(\alpha)}(x_n).$$

Consequently,

$$\liminf_{n\to\infty} \gamma_A^{(\alpha)}(x_n) \geq \gamma_A^{(\alpha)}(\bar{x}) > \gamma_A^{(\alpha)}(w).$$

This completes the proof. □

Definition 6.201 (**α-converse property**) A set-valued mapping $F : M \to \mathscr{P}(Y)$ is said to have the α-CONVERSE PROPERTY at $\bar{x} \in M$ wrt $x^* \in M \setminus \{\bar{x}\}$ if either $F(x^*) \not\preceq_C^\alpha F(\bar{x})$ or for each sequence $\{x^k\} \subset M$ converging to $\bar{x}$, one has $F(x^*) \preceq_C^\alpha F(x^{k_0})$ for some $k_0 \in \mathbb{N}$.

Now, we use the above concept to discuss the closedness of an α-Eff set.

Lemma 6.202 (Closedness conditions of solution sets of set optimization problems) *For each $\alpha \in \{l, u\}$, assume that $F : M \to \mathscr{P}_{\text{cl}}^{(\alpha)}(Y)$ has the α-converse property at every $x \in M$ wrt each $y \in M\setminus\{x\}$. If F is C-semicontinuous wrt α on M, then α-Eff is a closed set.*

Proof Let $v_n \in \alpha$-Eff converge to $\bar{v}$. Since M is closed, $\bar{v} \in M$. We assume that there exists $\bar{x} \in M$ such that $F(\bar{x}) \preceq_C^\alpha F(\bar{v})$. Now we need to show that $F(\bar{v}) \preceq_C^\alpha F(\bar{x})$. It is trivial if $\bar{x} = \bar{v}$. So, we consider the situation $\bar{x} \neq \bar{v}$.
Case 1. $\alpha = l$. Because F has the l-converse property at $\bar{v}$ wrt $\bar{x}$, there exists a subsequence $\{v_{n_k}\}$ of the sequence $\{v_n\}$ such that $F(\bar{x}) \preceq_C^l F(v_{n_k})$ for all k. Due to $v_{n_k} \in l$-Eff, we have $F(v_{n_k}) \preceq_C^l F(\bar{x})$, i.e.,

$$F(\bar{x}) \subset F(v_{n_k}) + C. \tag{6.165}$$

Let V be an arbitrary balanced neighborhood of the origin in Y. We have, of course, that $F(\bar{v}) \subset F(\bar{v}) + V$. By the C-upper semicontinuity of F at $\bar{v}$, and the convergence $v_{n_k} \to \bar{v}$, one has $F(v_{n_k}) \subset F(\bar{v}) + V + C$ for k sufficiently large. This together with (6.165) gives $F(\bar{x}) \subset F(\bar{v}) + V + C$. Since V is arbitrary and $F(\bar{v}) + C$ is closed, we have $F(\bar{x}) \subset F(\bar{v}) + C$ which implies that $F(\bar{v}) \preceq_C^l F(\bar{x})$. Therefore, $\bar{v} \in l$-Eff.
Case 2. $\alpha = u$. For any $z \in F(\bar{v})$, let V be an arbitrary balanced neighborhood of the origin in Y. We have, of course, that $F(\bar{v}) \cap (z + V) \neq \emptyset$. By the $(-C)$-lower semicontinuity of F at $\bar{v}$, there exists a neighborhood U of $\bar{z}$ satisfying $F(v) \cap (z + V + C) \neq \emptyset$ for all $v \in U$. Since $v_n \to \bar{v}$, there is $n_0 \in \mathbb{N}$ such that $F(v_n) \cap (z + V + C) \neq \emptyset$ for all $n \geq n_0$, and hence we can pick up a sequence of vectors z_n with $z_n \in F(v_n) \cap (z + V + C)$, which yields that

$$z \in z_n + V - C. \tag{6.166}$$

On the other hand, because F has the u-converse property at $\bar{v}$ wrt $\bar{x}$, there exists a subsequence $\{v_{n_k}\}$ of the sequence $\{v_n\}$ such that $F(\bar{x}) \preceq_C^u F(v_{n_k})$ for all k. Due to $v_{n_k} \in u$-Eff, we have $F(v_{n_k}) \preceq F(\bar{x})$. This together with (6.166) gives $z \in F(\bar{x}) + V - C$. Since V is arbitrary and $F(\bar{x}) - C$ is closed, we have $z \in F(\bar{x}) - C$.

Consequently, $F(\bar{v}) \subset F(\bar{x}) - C$, that is, $F(\bar{v}) \preceq^u_C F(\bar{x})$. Therefore, $\bar{v} \in u$-Eff. This completes the proof. □

Next, we employ the above results to study well-posedness properties for problem (α-SOP).

Theorem 6.203 (Well-posedness conditions for set optimization problems) *For $\alpha \in \{l, u\}$ and $v \in \alpha$-Eff, assume that*

(i) *M is compact;*
(ii) *$F : M \to \mathscr{P}^{(\alpha)}_{\text{cl}}(Y) \cap \mathscr{P}^{(\alpha)}_{\text{pro}}(Y)$ is C-semicontinuous wrt α.*

Then, problem (α-SOP) is generalized α-well-posed at v. Moreover, if $[v]^{\alpha}_F = \{v\}$ then (α-SOP) is α-well-posed at v.

Proof We give the proof for $\alpha = u$, the other case is analogous.

We first verify that problem (u-SOP) is solvable. Let $\{F(x_i)\}_{i \in I}$ be an arbitrary chain. Since M is compact, without loss of generality, we can assume that $\{x_i\}$ converges to a vector $x \in M$. We show that $F(x)$ is a lower bound of the chain $\{F(x_i)\}_{i \in I}$. Suppose that there is an index $i_0 \in I$ such that $F(x) \npreceq^u_C F(x_{i_0})$, i.e., $F(x) \not\subset F(x_{i_0}) - C$. Since F is $(-C)$-lsc at x and $Y \setminus (F(x_{i_0}) - C)$ is open, there exists an open neighborhood $U \subset M$ of x such that for all $z \in U$

$$F(z) \cap (Y \setminus (F(x_{i_0}) - C) + C) \neq \emptyset,$$

and hence

$$F(z) \cap (Y \setminus (F(x_{i_0}) - C)) \neq \emptyset.$$

Consequently,

$$F(z) \npreceq^u_C F(x_{i_0}), \forall z \in U. \tag{6.167}$$

Because $\{x_i\}$ converges to x, x_i will belong to the neighborhood U for i sufficiently large. Then, by (6.167), we conclude that $F(x_i) \npreceq^u_C F(x_{i_0})$ for i sufficiently large, which is impossible as $\{F(x_i)\}_{i \in I}$ is a chain. Therefore, u-Eff is nonempty.

We now prove the u-well-posedness of problem (u-SOP). Firstly, applying Lemma 6.199, we derive that $G^{(u)}_e(F(\cdot), F(y))$ is lower semicontinuous on M. Next, using Lemma 6.192, we conclude that problem OP($M, G^{(u)}_e(F(\cdot), F(v))$) is generalized Tykhonov well-posed. The final of the proof follows from Theorem 6.197. □

Now, we discuss the global α-well-posedness of problem (α-SOP), herein the converse property will play an important role.

Theorem 6.204 (Global well-posedness conditions for set optimization problems) *For each $\alpha \in \{l, u\}$, assume that:*

(i) *M is compact;*
(ii) $F : M \to \mathscr{P}^{(\alpha)}_{\text{cl}}(Y) \cap \mathscr{P}^{(\alpha)}_{\text{pro}}(Y)$ *is C-continuous wrt* α;
(iii) *F has the* α*-converse property at every* $x \in M$ *wrt each* $y \in M \backslash \{x\}$.

Then, problem (α -SOP) *is* α*-globally well-posed.*

Proof Firstly, from the first part of the proof of Theorem 6.203 and Lemma 6.202, we see that α -Eff is a nonempty closed set. Secondly, applying Lemma 6.200, we obtain that the function $\gamma^{(\alpha)}_{\alpha\text{-Eff}}$ is lower pseudocontinuous on M. According to Lemma 6.192, problem OP$(M, \gamma^{(\alpha)}_{\alpha\text{-Eff}})$ is generalized Tykhonov well-posed. Finally, from the conclusion of Theorem 6.198, (α -SOP) is globally α-well-posed. □

Remark 6.205 (Cruciality of the closedness of the solution set of a set optimization problem) Usually, on studying global well-posedness for vector optimization problems, the closedness property of efficient solution sets is a key assumption. Here, we propose sufficient conditions for this important property. Note further that the closedness of efficient solution sets plays an important role not only in considering well-posedness but also in discussing the stability and convergence for problems related to optimization, and so the result presented in Lemma 6.202 has a potential application in studying the mentioned topics. Moreover, we would like to point out here that most of works in the literature dealing with well-posedness of set optimization problems usually impose the non-emptiness of efficient solution sets, and hence existence conditions given in Theorems 6.203 and 6.204 are meaningful.

We finish this section with an example to illustrate the applicability of Theorem 6.204.

Example 6.206 (A special case of globally well-posed set optimization problems) Let $X = Y = \mathbb{R}^2$, $C = \mathbb{R}^2_+$

$$M := \{(x_1, x_2) \in [0, 2]^2 \mid x_1 + x_2 \geq 1, x_1 - 8x_2 \geq -8\},$$

and $F : M \to \mathscr{P}(Y)$ be defined by

$$F(x_1, x_2) := \{(y_1, y_2) \mid (y_1 - x_1)^2 + (y_2 - x_2)^2 \leq 1\}, \forall (x_1, x_2) \in M.$$

Then, for each $\alpha \in \{0, 1\}$, all the assumptions of Theorem 6.204 are satisfied, and hence Theorem 6.204 ensures that problem (α-SOP) is globally α-well-posed. In fact, α -Eff $= \{(x_1, x_2) \in M \mid x_1 \in [0, 1], x_1 + x_2 = 1\}$. It is worth noting that the weakly efficient solution set, i.e., the case that C is replaced by $\operatorname{int} C$, of (α-SOP), is

$$\{(x_1, x_2) \in M \mid (x_1 \in [0, 1], x_1 + x_2 = 1) \vee (x_1 \in]1, 2], x_2 = 0)\},$$

which is different from α -Eff.

6.8 Applications

Since the equilibrium problem contains many optimization-related models, we can derive consequences of the results of the previous sections for them. In this section, we discuss some important particular cases as examples.

6.8.1 *Browder Variational Inclusions*

Let X, Y, W be Hausdorff vector spaces, $A \subset X$, $\Lambda \subset W$, $M \subset W$, $C \subset Y$ a solid cones, and $\mathscr{K} : \Lambda \rightrightarrows A$. The Browder variational inclusion has the form:

(BVI) find $\bar{x} \in \mathscr{K}(\lambda)$ such that $\langle \Omega(\mu), \bar{x} \rangle \subset C$,

where $\Omega : M \rightrightarrows \mathscr{L}(X, Y)$ is a set-valued map, $\mathscr{L}(X, Y)$ is the space of all linear continuous operators from X into Y, $\langle \cdot, \cdot \rangle$ is the pairing between $\mathscr{L}(X, Y)$ and X, and

$$\langle \Omega(\mu), x \rangle := \{\langle x^*, x \rangle \mid x^* \in \Omega(\mu)\}.$$

Usually, problem (BVI) is considered as a set-variational inequality. Here, in order to motivate applicability of the underlying problem in this book, we discuss (BVI) via an equilibrium problem. For $\theta = (\lambda, \mu)$, $\Theta := \Lambda \times M$ and $\mathscr{D} : \Theta \rightrightarrows A \times \mathscr{L}(X, Y)$, $f : (A \times \mathscr{L}(X, Y))^2 \to Y$ are defined by

$$\mathscr{D}(\theta) := \mathscr{K}(\lambda) \times \Omega(\mu),$$

$$f(\bar{z}, z) := \langle z_2, \bar{z}_1 \rangle$$

for all $\bar{\mathbf{z}} = (\bar{z}_1, \bar{z}_2)$ and $\mathbf{z} = (z_1, z_2) \in \mathscr{D}(\theta)$. Then, for each $\theta = (\lambda, \mu)$ problem (BVI) becomes the following equilibrium problem:

(EP) find $\bar{z} \in \mathscr{D}(\theta)$ such that $f(\bar{z}, z) \geq 0, \forall z \in \mathscr{D}(\theta)$.

Recall that A set-valued mapping $G : X \rightrightarrows Y$ is convex on a convex subset A of X if gph$G$ is a convex set in $X \times Y$.

Remark 6.207

(a) G is convex on A if and only if for all $x_1, x_2 \in A$ and $t \in [0, 1]$,

$$tG(x_1) + (1 - t)G(x_2) \subset G(tx_1 + (1 - t)x_2).$$

It follows from this that G is said to be concave on A if for all $x_1, x_2 \in A$ and $t \in [0, 1]$,

$$G(tx_1 + (1 - t)x_2) \subset tG(x_1) + (1 - t)G(x_2).$$

(b) G is convex if and only if it is C-convex for any pointed cone C.

For $e \in \text{int}C$, the following results are obtained from Theorems 6.49 and 6.80.

Corollary 6.208 (Hölder conditions for the Browder variational inclusion) *Assume that:*

(a) $\mathscr{K}$ *is Hölder continuous on* Λ;
(b) Ω *is Hölder continuous and concave on* M.

Then, the approximate solution mapping

$$(\varepsilon, \lambda) \mapsto \widetilde{\text{Sol}}(\text{BVI})(\varepsilon, \lambda) := \{x \in \mathscr{K}(\lambda) \mid \langle \Omega(\mu), \bar{x} \rangle + \varepsilon e \subset C\}$$

is Hölder continuous on $(0, +\infty) \times \Lambda \times M$.

Corollary 6.209 (Hausdorff conditions for the Browder variational inclusion) *Assume that:*

(a) $\mathscr{K}$ *is continuous, compact-valued on* Λ, *and* $\mathscr{K}(\Lambda)$ *is bounded*;
(b) Ω *is continuous, concave, and compact-valued on* M.

Then, the approximate solution mapping

$$(\varepsilon, \lambda) \mapsto \widetilde{\text{Sol}}(\text{BVI})(\varepsilon, \lambda) := \{x \in \mathscr{K}(\lambda) \mid \langle \Omega(\mu), \bar{x} \rangle + \varepsilon e \subset C\}$$

is Hausdorff continuous on $(0, +\infty) \times \Lambda \times M$.

6.8.2 *Vector Optimization Problems*

Let X be a locally convex Hausdorff topological vector space; Y, W vector topological spaces; $A \subset X$, $\Lambda \subset W$, $M \subset W$, Y equipped with a pointed, closed, and convex cone C with nonempty interior; and $A \subset X$ a nonempty convex subset. Let $\mathscr{D} : \Lambda \rightrightarrows A$ be a set-valued map with convex compact values and $g : A \times M \to Y$. For $(\lambda, \mu) \in \Lambda \times M$, we consider the vector optimization problem:

(VOP) $\quad C\text{-}\min g(x, \mu)$ subject to $x \in \mathscr{D}(\lambda)$.

Definition 6.210 (Strong, efficient, weak solutions of vector optimization problems) For $\lambda \in \Lambda$, $\mu \in M$, and $\bar{x} \in \mathscr{D}(\lambda)$:

(a) $\bar{x}$ is called a STRONGLY EFFICIENT SOLUTION of (VOP) if for any $x \in \mathscr{D}(\lambda)$

$$g(\bar{x}, \mu) \leq g(x, \mu);$$

(b) $\bar{x}$ is called an EFFICIENT SOLUTION of (VOP) if for any $x \in \mathscr{D}(\lambda)$,

$$[g(x,\mu) \leq g(\bar{x},\mu)] \Rightarrow [g(\bar{x},\mu) \leq g(x,\mu)];$$

(c) $\bar{x}$ is called a WEAKLY EFFICIENT SOLUTION of (VOP) if for any $x \in \mathscr{D}(\lambda)$,

$$[g(x,\mu) < g(\bar{x},\mu)] \Rightarrow [g(\bar{x},\mu) < g(x,\mu)],$$

or equivalently, there is no $x \in \mathscr{D}(\lambda)$ such that $g(x,\mu) < g(\bar{x},\mu)$.

For each $\lambda \in \Lambda$ and $\mu \in M$, we denote by s-Eff(λ,μ), Eff(λ,μ), and w-Eff(λ,μ) the set of all strongly efficient solutions, efficient solutions, and weakly efficient solutions, resp., and assume that they are nonempty in a neighborhood of the reference points.

By applying Theorems 6.13 and 6.14, we obtain the following result.

Corollary 6.211 (Upper semicontinuity and closedness of strong and weak solution maps of vector optimization problems) *Assume that:*

(a) *$\mathscr{D}$ is continuous and compact-valued on Λ;*
(b) *g is continuous of $A \times M$;*
(c) *g is C-convex in the first component on A.*

Then, s-Eff, *and w-Eff are upper semicontinuous, closed, and compact-valued on $\Lambda \times M$.*

Moving on to Hölder conditions of these solution sets, we have the following result.

Corollary 6.212 (Hölder continuity of strong and weak solution maps of vector optimization problems) *For (VOP), assume that:*

(a) *$\mathscr{D}$ is (ℓ_1,α_1)-Hölder continuous on a neighborhood N of $\bar{\lambda}$;*
(b) *there is a neighborhood U of $\bar{\mu}$ such that, for each $\mu \in U$, $g(\cdot,\mu)$ is C-(h_1,β_1)-strongly quasiconvex ($Y\setminus(-\text{int}C)$-(h_1,β_1)- strongly quasiconvex, resp.) as well as (m_1,δ_1)-Hölder continuous wrt e on* conv *$\mathscr{D}(N)$;*
(c) *for every $\mu \in U$, $g(\cdot,\mu)$ is C-continuous on $\mathscr{D}(N)$;*
(d) *for each $x \in \mathscr{D}(N)$, $g(x,\cdot)$ is (n_1,γ_1)-Hölder continuous wrt e on U.*

Then, s-Eff (w-Eff, resp.) is nonempty on $N \times U$ and satisfies the Hölder condition on $N \times U$, i.e., for all $(\lambda_1,\mu_1), (\lambda_2,\mu_2) \in N \times U$, and $\alpha \in \{s, w\}$:

$$\mathbb{D}(\alpha - \text{Eff}(\lambda_1,\mu_1), \alpha - \text{Eff}(\lambda_2,\mu_2)) \leq \left(\frac{4m_1 l_1^{\delta_1}(1+2^{-\delta_1})}{h_1}\right)^{\frac{1}{\beta_1}} \|\lambda_1 - \lambda_2\|^{\frac{\alpha_1\delta_1}{\beta_1}} + \left(\frac{8n_1}{h_1}\right)^{\frac{1}{\beta_1}} \|\mu_1 - \mu_2\|^{\frac{\gamma_1}{\beta_1}}.$$

Proof Setting $f(x, y, \mu) := g(y, \mu) - g(x, \mu)$, (VOP) becomes a special case of (VEP). For $\alpha = s$, to prove this corollary, it is sufficient to check that f satisfies all the assumptions of Theorem 6.181. It is easy to check that assumptions (a), (c), and (d) of Theorem 6.181 are fulfilled. For the strong quasiconvexity in (b), for all $x \in \mathscr{D}(N)$, $\mu \in U$, $y_1, y_2 \in \text{conv}\mathscr{D}(N)$, and $t \in (0, 1)$, by the strong quasiconvexity of g, we have two cases to consider. The first one is

$$g(y_1, \mu) - g((1-t)y_1 + ty_2, \mu) - h_1 t(1-t)\|y_1 - y_2\|^{\beta_1} e \in C.$$

This leads to

$$[g(y_1, \mu) - g(x, \mu)] - [g((1-t)y_1 + ty_2, \mu) - g(x, \mu)] - h_1 t(1-t)\|y_1 - y_2\|^{\beta_1} e \in C,$$

or equivalently,

$$f(y_1, \mu) - f((1-t)y_1 + ty_2, \mu) - h_1 t(1-t)\|y_1 - y_2\|^{\beta_1} e \in C.$$

For the second case, $g(y_2, \mu) - g((1-t)y_1 + ty_2, \mu) - h_1 t(1-t)\|y_1 - y_2\|^{\beta_1} e \in C$, by the same arguments, we also obtain

$$f(y_1, \mu) - f((1-t)y_1 + ty_2, \mu) - h_1 t(1-t)\|y_1 - y_2\|^{\beta_1} e \in C.$$

Hence, the strong quasiconvexity of f in (b) of Theorem 6.181 is satisfied with $h = h_1, \beta = \beta_1$. It is not hard to verify that the Hölder continuity of f in (b) of Theorem 6.181 holds with $m = m_1, \delta = \delta_1$.

For (e) of Theorem 6.181, taking arbitrarily $\mu_1, \mu_2 \in U$ and $x, y \in \mathscr{D}(N)$, by the Hölder continuity of g, we have:

$$g(y, \mu_1) - g(y, \mu_2) + n_1\|\mu_1 - \mu_2\|^{\delta_1} e \in C, \tag{6.168}$$

$$g(x, \mu_1) - g(x, \mu_2) + n_1\|\mu_1 - \mu_2\|^{\delta_1} e \in C. \tag{6.169}$$

From (6.168), (6.169), and the convexity of the cone C, we get:

$$[g(y, \mu_1) - g(x, \mu_1)] - [g(y, \mu_2) - g(x, \mu_2)] + 2n_1\|\mu_1 - \mu_2\|^{\delta_1} e \in C.$$

This means that

$$f(x, y, \mu_1) - f(x, y, \mu_2) + 2n_1\|\mu_1 - \mu_2\|^{\delta_1} e \in C.$$

So, (e) of Theorem 6.181 is satisfied with $n = 2n_1$,$\gamma = \gamma_1$, and $\theta = 0$. The case of $\alpha = w$ is similarly proved. □

6.8.3 Vector Variational Inequalities

Let X be a locally convex Hausdorff topological vector space, Y, W be vector topological spaces, $K \subset X$ be a nonempty convex compact subset, $M \subset W$ be a nonempty subset, $C \subset Y$ be a pointed closed convex cone with int$C \neq \emptyset$ (int$C$ is the interior of C), and $T : K \times \Lambda \to \mathscr{L}(X, Y)$. For $\mu \in M$, the Stampacchia vector variational inequality problem under our consideration is

(SVVI) find $\bar{x} \in K$ such that, for all $y \in K$,

$$\langle T(\bar{x}, \mu), y - \bar{x} \rangle \in C.$$

To convert (SVVI) to a special case of (SVEP), we simply put $f(x, y, \mu) = \langle T(x, \mu), y - x \rangle$.

The dual of (SVVI) is the Minty vector variational inequality:

(MVVI) find $\bar{x} \in K$ such that, for all $y \in K$, $\langle T(y, \mu), \bar{x} - y \rangle \in -C$.

Similarly for (SVVI), let $f(y, x, \mu) = \langle T(y, \mu), \bar{x} - y \rangle$ to see that (MVVI) becomes a special case of (DSVEP).

For $(\varepsilon, \lambda) \in \mathbb{R}_+ \times M$, we denote the approximate solution set of (SVVI) and (MVVI) by $\widetilde{\text{Sol}}$(SVVI)(ε, μ) and $\widetilde{\text{Sol}}$(MVVI)(ε, λ), resp.

Corollary 6.213 (Hausdorff continuity of approximate solution maps of Stampacchia and Minty vector variational inequalities) *Assume that the following conditions hold:*

(a) *for each $x \in K$, $T(x, \cdot)$ is uniformly continuous on M;*
(b) *for each $\mu \in M$, $T(\cdot, \mu)$ is monotone and affine in K.*

Then, $\widetilde{\text{Sol}}$(SVVI) *and* $\widetilde{\text{Sol}}$(MVVI) *are Hausdorff continuous on* $\mathbb{R}_+ \times M$.

Proof In order to apply Theorem 6.79, we have to check only the concavity of f. For all $x_1, x_2 \in K, t \in [0, 1]$ and $y \in K$, we have:

$$f(tx_1 + (1-t)x_2, y, \lambda_0) - tf(x_1, y, \lambda_0) - (1-t)f(x_2, y, \lambda_0)$$

$$= \langle tT(x_1, \lambda_0) + (1-t)T(x_2, \lambda_0), y - tx_1 - (1-t)x_2 \rangle$$

$$-t\langle T(x_1, \lambda_0), y - x_1 \rangle - (1-t)\langle T(x_2, \lambda_0), y - x_2 \rangle$$

$$= t\langle T(x_1, \lambda_0), (1-t)x_1 - (1-t)x_2 \rangle + (1-t)\langle T(x_2, \lambda_0), -tx_1 + tx_2 \rangle$$

$$= t(1-t)\langle T(x_1, \lambda_0) - T(x_2, \lambda_0), x_1 - x_2 \rangle \in C.$$

The proof is complete. □

To consider the Hölder property of the underlying problems, we assume that X, Y, W are normed spaces.

Corollary 6.214 (Hölder continuity of the solution maps of the Stampacchia vector variational inequality) *For (SVVI), assume that the following conditions hold:*

(a) *there is a neighborhood U of $\bar{\mu}$ such that for all $x \in K$ and $\mu \in U$, $\langle T(x, \mu), \cdot\rangle$ is (h_2, β_2)-strongly convex-like on K with $\beta_2 > 1$;*
(b) *for each $\mu \in U$, $T(\cdot, \mu)$ is monotone on K, i.e, for all $x_1, x_2 \in K$, $\langle T(x_1, \mu) - T(x_2, \mu), x_1 - x_2\rangle \in C$;*
(c) *for $x \in K$, $T(x, \cdot)$ is (n_2, γ_2)-Hölder continuous on U.*

Then, the solution map to (SVVI) is single-valued and satisfies the following Hölder condition on U: for $\mu_1, \mu_2 \in U$,

$$\mathbb{D}(\mathrm{Sol(SVVI)}(\mu_1), \mathrm{Sol(SVVI)}(\mu_2)) \leq \left(\frac{n_2}{h_2}\right)^{\frac{1}{\beta_2 - 1}} d^{\frac{\gamma_2}{\beta_2 - 1}}(\mu_1, \mu_2).$$

Proof To apply Theorem 6.181 to this special case, we need to check that all the assumptions of this theorem are fulfilled with $f(x, y, \mu) = \langle h(x, \mu), y - x\rangle$. Assumption (a) is satisfied since for all $y_1, y_2 \in K$ and $t \in (0, 1)$, by (a), there is $z \in K$ such that, for all $\mu \in U$ and $x \in K$,

$$-\langle h(x, \mu), z\rangle + (1-t)\langle h(x, \mu), y_1\rangle + t\langle h(x, \mu), y_2\rangle - h_2 t(1-t)d^{\beta_2}(y_1, y_2)e \in C,$$

which implies that

$$- f(x, z, \mu) + (1 - t)\langle f(x, y_1, \mu) + tf(x, y_2, \mu) - h_2 t(1 - t)d^{\beta_2}(y_1, y_2)e \in C.$$

For $x, y \in K$ and $\mu \in U$, one has:

$$f(x, y, \mu) + f(y, x, \mu) = \langle h(x, \mu) - h(y, \mu), y - x\rangle \in -C,$$

as $h(\cdot, \mu)$ is monotone on K, and hence Assumption (b) is fulfilled.

Assumption (c) is checked as, for $\mu_1, \mu_2 \in U$ and $x, y \in K$,

$$\begin{aligned} &f(x, y, \mu_1) - f(x, y, \mu_2) + n_2 d^{\gamma_2}(\mu_1, \mu_2)d(x, y)e \\ &= \langle h(x, \mu_1) - h(x, \mu_2), y - x\rangle + n_2 d^{\gamma_2}(\mu_1, \mu_2)d(x, y)e \in C. \end{aligned}$$

Therefore, the proof is complete. □

We have the following result for the Hölder continuity of solution maps of (MVVI).

Corollary 6.215 (Hölder continuity of the solution map of the Minty vector variational inequality) *For (MVVI), assume (c), (b) of Corollary 6.214 with f replaced by $-f$ and assume further that* ($\mathrm{a_d}$) *there is a neighborhood U of $\bar{\mu}$ such that for all $x \in K$ and $\mu \in U$, $\langle h(x,\mu), \cdot\rangle$ is $h_2\beta_2$-strongly concave-like on K with $\beta_2 > 1$.*

Then, the same conclusion as that of Corollary 6.214 on the solution map holds for (MVVI).

6.8.4 Vector Traffic Equilibrium Problems

We now consider a traffic network $\mathbb{G} = (\mathbb{N}, \mathbb{A}, \mathbb{C}, \Omega)$, where $\mathbb{N}$ and $\mathbb{A}$ are the sets of nodes and directed arcs, resp., and $\mathbb{C} = (c_a)_{a\in\mathbb{A}}$ denotes the capacity vector where c_a is the capacity of arc $a \in \mathbb{A}$ and Ω stands for the set of the origin/destination (O/D) pairs.

Denote the set of all demand vectors by

$$\mathscr{D} := \{d = (d_\omega)_{\omega\in\Omega} \mid d_\omega \in \mathbb{R}_+\},$$

where d_ω is the demand of traffic flow wrt $\omega \in \Omega$. In this model, we consider the case of d_ω being bounded, and forms the demand set $\mathscr{D}$. For each $a \in \mathbb{A}$ and $w \in \Omega$, we denote the capacity constraint of the arc flow f_a and the set of available paths joining O/D pair ω by $[0, c_a]$ and P_ω, resp. Given a path $k \in P_\omega$, h_k stands for the traffic flow on this path. Then,

$$f_a = \sum_{\omega\in\Omega}\sum_{k\in P_\omega} \delta_{ak} h_k,$$

where $\delta_{ak} = 1$ if arc a is contained in path k and $\delta_{ak} = 0$ otherwise.

Let $h = (h_1, h_2, \ldots, h_m) \in \mathbb{R}^m_+$ be the path flow with $m = \sum_{\omega\in\Omega} |P_\omega|$. We say that a path flow h satisfies the demand constraints if

$$\sum_{k\in P_\omega} h_k = d_\omega.$$

If h satisfies the demand constraints and capacity constraints, then it is called a feasible path flow. Hence, the set of the feasible path flows depends on demand vectors and it is defined by

$$\mathscr{H}(d) := \left\{ h \in \mathbb{R}^m_+ \mid \forall\omega \in \Omega, \sum_{k\in P_\omega} h_k = d_\omega \text{ and } \forall a \in \mathbb{A}, c_a \geq f_a \geq 0 \right\}$$

$$= \left\{ h \in \mathbb{R}^m_+ \mid \forall \omega \in \Omega, \sum_{k \in P_\omega} h_k = d_\omega \text{ and } \forall a \in \mathbb{A}, c_a \geq \sum_{\omega \in \Omega} \sum_{k \in P_\omega} \delta_{ak} h_k \geq 0 \right\}.$$

It is clear that for each d, $\mathcal{H}(d)$ is nonempty, convex, and compact. Moreover, $\mathcal{H}(\mathcal{D})$ is bounded due to the boundedness of $\mathcal{D}$. Assume that a vector-valued cost function τ_a given on each arc $a \in \mathbb{A}$ takes values in a finite-dimensional space $\mathbb{R}^r_+$ with $r \geq 2$. In the literature, one often considers the case of $r = 2$, which corresponds to two criteria: the travel time and the travel cost. Let $\mathcal{T}_k : \mathbb{R}^m_+ \to \mathbb{R}^r_+$ be a vector-valued cost function along the path k. Then, the vector-valued cost on the path k is equal to the sum of all costs of the flow h_k through arcs, which belong to the path k, that is,

$$\mathcal{T}_k(h) = \Sigma_{a \in \mathbb{A}} \delta_{ak} \tau_a(h_k).$$

Let $\mathcal{T}(h) = (\mathcal{T}_1(h), \mathcal{T}_2(h), \ldots, \mathcal{T}_m(h)) \in \mathbb{R}^{r \times m}$.

Definition 6.216 (Saturated arc) For every $d \in \mathbb{R}^{|\Omega|}_+$ and $h \in \mathcal{H}(d)$:

(a) for $a \in \mathbb{A}$, if $f_a = c_a$, then a is called a SATURATED ARC of flow h, otherwise a nonsaturated arc of flow h;
(b) for $k \in \cup_{\omega \in \Omega} P_\omega$, if there exists a SATURATED ARC a of flow h such that a belongs to path k, then k is called a saturated path of flow h, otherwise a nonsaturated path of flow h.

We now generalize (weak) vector equilibrium principle to the case of capacity constraints of arcs and propose the following equilibrium principle with capacity constraints of arcs.

Definition 6.217 (Weak vector equilibrium flow) For each $d \in \mathbb{R}^{|\Omega|}_+$, $h \in \mathcal{H}(d)$ is called a WEAK VECTOR EQUILIBRIUM FLOW if for all $\omega \in \Omega$ and $k, j \in P_\omega$, the following condition is statisfied:

$$\mathcal{T}_k(h) - \mathcal{T}_j(h) \in \text{int}\mathbb{R}^r_+ \Rightarrow h_k = 0 \text{ or path } j \text{ is a saturated path of flow } h.$$

Lemma 6.218 (Relationships between vector traffic equilibrium problems and vector variational inequalities) *For $d \in \mathbb{R}^{|\Omega|}_+$, $h \in \mathcal{H}(d)$ is a weak vector equilibrium flow if it satisfies the following vector variational inequality problem:*
(VVI) find $h \in \mathcal{H}(d)$ such that for all $v \in \mathcal{H}(d)$,

$$\langle \mathcal{T}(h), v - h \rangle \notin -\text{int}\mathbb{R}^r_+.$$

To convert (VVI) to a special case of (VEP), we put $f(h, v, d) = \langle \mathcal{T}(h), v - h \rangle$. For $\varepsilon \in \mathbb{R}_+$ and $d \in \mathbb{R}^{|\Omega|}_+$, we denote the approximate solution set of (VVI) by $S_2(\varepsilon, d)$. Now we provide sufficient conditions for the Hausdorff continuity of the solution map of (VVI) in the following result.

Corollary 6.219 (Hausdorff conditions for vector traffic equilibrium problems) *For (VVI) and $\varepsilon_0 > 0$, assume that:*

(a) $\mathscr{H}$ *is continuous at* d_0;
(b) $\mathscr{T}$ *is monotone and affine.*

Then, the solution map of problem (VVI) is Hausdorff continuous at (ε_0, d_0).

6.8.5 Some Practical Situations

Fuel-Optimal Frictionless Horizontal Motion of a Mass Point

In this subsection, we discuss in detail a practical situation which plays an important role in engineering applications of technology. To be more precise, we study the nonempty value and the Hölder property of a solution map of a parametric fuel-optimal frictionless horizontal motion of a mass point problem.

State variables:

$$x_1 = \text{ position}, x_2 = \text{ velocity},$$

control variable:

$$v = \text{ acceleration},$$

subject to the constraint

$$v(t) \in [-a_{\max}, a_{\max}].$$

Find an acceleration $v : [t_0, t_1] \to \mathbb{R}$ such that the dynamic system

$$\begin{cases} \dot{x}_1(t) = x_2(t) + \lambda(t), \\ \dot{x}_2(t) = v(t) + \lambda(t), \\ v(t) \in [-a_{\max}, a_{\max}] \end{cases}$$

is transferred from the initial state

$$x_1(t_0) = x_{01}, x_2(t_0) = x_{02},$$

to an arbitrary final state $x_1(t_1), x_2(t_1) \in \mathbb{R}$, and such that the cost functional

$$\int_{t_0}^{t_1} (|v(t)| + \mu(t))\, dt$$

is minimized, where $a_{\max}, x_{01}, x_{02}, t_0, t_1$ are given real numbers and $\mu \in \mathcal{L}^\infty([t_0, t_1], \mathbb{R})$, $\lambda \in \mathcal{L}^r([t_0, t_1], \Theta) =: \Lambda$ are parameters and Θ is a bounded subset of $\mathbb{R}$. Here, $\mathcal{L}^p([t_0, t_1], \mathbb{R}^n)$ $(1 \le p \le \infty)$ stands for the Lebesgue space of the integrable functions defined on $[t_0, t_1]$.

Let $X := \mathcal{L}^1([t_0, t_1], \mathbb{R}^2)$ and $V \subset \mathcal{L}^1([t_0, t_1], \mathbb{R})$ be the solutions set of linear differential equations, which is defined by

$$\begin{aligned} V := \{ & v_b \in \mathcal{L}^p([t_0, t_1], \mathbb{R}) \mid \dot{v}_b(t) + a_0 v_b(t) = bt, v_b(t_0) = 0, \\ & b \in I \subset \mathbb{R}, t \in [t_0, t_1]\}. \end{aligned} \tag{6.170}$$

Let us set

$$A(t) := \begin{pmatrix} 0 & 1 \\ 0 & 0 \end{pmatrix}, B(t) := \begin{pmatrix} 0 \\ 1 \end{pmatrix}, C(t) := \begin{pmatrix} 1 \\ 1 \end{pmatrix},$$

$$x := \begin{pmatrix} x_1 \\ x_2 \end{pmatrix}, x_0 := \begin{pmatrix} x_1(t_0) \\ x_2(t_0) \end{pmatrix},$$

$$f_2(t, v(t), \mu(t)) := |v(t)| + \mu(t), \quad f_1(v, \mu) := \int_{t_0}^{t_1} (|v(t)| + \mu(t))\, dt,$$

and

$$\begin{aligned} \mathcal{D}(\lambda) = \{ v \in V \mid \exists x \in X, \dot{x}(t) = A(t)x(t) + B(t)v(t) + C(t)\lambda(t), x(t_0) = x_0, \\ v(t) \in [-a_{\max}, a_{\max}]\}. \end{aligned}$$

Definition 6.220 (Equicontinuous function) A family Ξ of functions ξ defined on $[t_0, t_1]$ is said to be EQUICONTINUOUS on $[t_0, t_1]$ if for every $\varepsilon > 0$, there exists a $\delta > 0$ such that

$$|\xi(t) - \xi(t')| < \varepsilon,$$

whenever $|t - t'| < \delta$, $t, t' \in [t_0, t_1]$, and $\xi \in \Xi$.

The following results are not hard to check, and so we leave them to the reader.

Lemma 6.221 (Integrand inequality) *Suppose* $1 \le p \le q \le \infty$. *If* $x \in \mathcal{L}^q([t_0, t_1], \mathbb{R}^n)$, *then* $x \in \mathcal{L}^p([t_0, t_1], \mathbb{R}^n)$ *and*

$$\left(\int_{t_0}^{t_1} |x(t)|^p dt \right)^{\frac{1}{p}} \le (t_1 - t_0)^{\frac{1}{p} - \frac{1}{q}} \left(\int_{t_0}^{t_1} |x(t)|^q dt \right)^{\frac{1}{q}}.$$

Lemma 6.222 (Lipschitz conditions of state equations) *The set-valued map $\mathscr{D}$ is Lipschitz continuous on Λ.*

Lemma 6.223 (Compactness and convexity conditions of state equations) *If V is equicontinuous on $[t_0, t_1]$, then $\mathscr{D}(\lambda)$ is nonempty compact convex valued for all $\lambda \in \Lambda$.*

Proof It follows from the solution existence theorem for linear differential equations that $K(\lambda) \neq \emptyset$ for all $\lambda \in \Lambda$, and it is obvious that for all $\lambda \in \Lambda$, $\mathscr{D}(\lambda)$ is a convex set.

We prove that $\mathscr{D}(\lambda)$ is compact for all $\lambda \in \Lambda$. It follows from the equicontinuity of V that $\mathscr{D}(\lambda)$ is equicontinuous. On the other hand, for all $v \in \mathscr{D}(\Lambda)$ and $t \in [t_0, t_1]$, one has $v(t) \in [-a_{\max}, a_{\max}]$, and thus $K(\Lambda)$ is bounded. Particularly, $\mathscr{D}(\lambda)$ is uniformly bounded and, consequently, by Arzela-Ascoli theorem, $\mathscr{D}(\lambda)$ is a relatively compact set for all $\lambda \in \Lambda$.

Let $v_i \in \mathscr{D}(\lambda)$ with $v_i \to v$ in $\mathscr{L}^1([t_0, t_1], \mathbb{R})$, then

$$\begin{cases} \dot{x}_1(t) = x_2(t) + \lambda(t), \\ \dot{x}_2(t) = v_i(t) + \lambda(t), \\ v_i(t) \in [-a_{\max}, a_{\max}]. \end{cases}$$

By letting $i \to \infty$, we obtain:

$$\begin{cases} \dot{x}_1(t) = x_2(t) + \lambda(t), \\ \dot{x}_2(t) = v(t) + \lambda(t), \\ v(t) \in [-a_{\max}, a_{\max}]. \end{cases}$$

Therefore, $v \in \mathscr{D}(\lambda)$ and so $\mathscr{D}(\lambda)$ is closed. That is, $\mathscr{D}(\lambda)$ is compact. □

We now reformulate the above problem in the following form:

(OCP): min $f_1(v, \mu)$ subject to $v \in \mathscr{D}(\lambda)$.

We denote the solution set of (OCP) at (λ, μ) by $S_0(\lambda, \mu)$, i.e.,

$$S_0(\lambda, \mu) = \{v \in \mathscr{D}(\lambda) \mid f_1(v, \mu) \leq f_1(w, \mu), \forall w \in \mathscr{D}(\lambda)\}.$$

Corollary 6.224 (Hölder conditions for fuel-optimal frictionless horizontal motion of a mass point problem) *The solution map S_0 is nonempty-valued and Hölder continuous on $\mathscr{L}^r([t_0, t_1], \Theta) \times \mathscr{L}^\infty([t_0, t_1], \mathbb{R})$, provided that $a_0 \in (\frac{1}{t_0}, +\infty)$ and I is a subset of*

$$\left[\frac{-a_{\max} a_0}{t_1 - \frac{1}{a_0} - e^{a_0(t_0 - t_1)}(t_0 - \frac{1}{a_0})}, \frac{a_{\max} a_0}{t_1 - \frac{1}{a_0} - e^{a_0(t_0 - t_1)}(t_0 - \frac{1}{a_0})}\right].$$

Proof We will prove this lemma by applying Corollary 6.212. Assumptions (a) of Corollary 6.212 is satisfied due to Lemma 6.222. It is not hard to check the Hölder continuity and the continuity in (b), (c), and (d) of Corollary 6.212. It follows from (6.170) that

$$V = \left\{ v_b \mid v_b(t) = \frac{b}{a_0}\left(t - \frac{1}{a_0} - e^{a_0(t_0-t)}\left(t_0 - \frac{1}{a_0}\right)\right), b \in I, t \in [t_0, t_1] \right\},$$

and thus V is equicontinuous on $[t_0, t_1]$. Therefore, by Lemma 6.223, we derive that $\mathscr{D}(\lambda)$ is nonempty compact convex valued for all λ.

Due to $b \in I$, we have:

$$\frac{-a_{\max}a_0}{t_1 - \frac{1}{a_0} - e^{a_0(t_0-t_1)}\left(t_0 - \frac{1}{a_0}\right)} \le b \le \frac{a_{\max}a_0}{t_1 - \frac{1}{a_0} - e^{a_0(t_0-t_1)}\left(t_0 - \frac{1}{a_0}\right)},$$

which implies that

$$-a_{\max} \le \frac{b}{a_0}\left(t - \frac{1}{a_0} - e^{a_0(t_0-t)}\left(t_0 - \frac{1}{a_0}\right)\right) \le a_{\max},$$

for all $t \in [t_0, t_1]$, and so $v_b(t) \in [-a_{\max}, a_{\max}]$.

We now focus on proving the strong quasiconvexity of f_1. It is not hard to check that f_2 is $2\hat{h}$-strongly quasiconvex in the second component. Then, for all $\theta \in [0, 1]$, $v_{b_1}, v_{b_2} \in V$, one gets:

$$\begin{aligned} f_2(t, \theta v_{b_1}(t) + (1-\theta)v_{b_2}(t), \mu(t)) \le \max\{f_2(t, v_{b_1}(t), \mu(t)), f_2(t, v_{b_2}(t), \mu(t))\} \\ - \hat{h}\theta(1-\theta)|v_{b_1}(t) - v_{b_2}(t)|^2. \end{aligned}$$

From $v_{b_1}, v_{b_2} \in V$, there are $b_1, b_2 \in I$ such that

$$v_{b_1}(t) = \frac{b_1}{a_0}\left(t - \frac{1}{a_0} - e^{a_0(t_0-t)}\left(t_0 - \frac{1}{a_0}\right)\right),$$

$$v_{b_2}(t) = \frac{b_2}{a_0}\left(t - \frac{1}{a_0} - e^{a_0(t_0-t)}\left(t_0 - \frac{1}{a_0}\right)\right).$$

Since the graphs of functions $v_{b_1}(t)$ and $v_{b_2}(t)$ either coincide or intersect exactly once at $t = t_0$ on $[t_0, t_1]$, the following estimations hold true:

$$\begin{aligned} & f_1(\theta v_{b_1}(t) + (1-\theta)v_{b_2}(t), \mu(t)) \\ &= \int_{t_0}^{t_1} f_2(t, \theta v_{b_1}(t) + (1-\theta)v_{b_2}(t), \mu(t))dt \end{aligned}$$

$$
\begin{aligned}
&\leq \int_{t_0}^{t_1} \max\{f_2(t, v_{b_1}(t), \mu(t)), f_2(t, v_{b_2}(t), \mu(t))\}dt \\
&\quad - \hat{h}\theta(1-\theta) \int_{t_0}^{t_1} |v_{b_1}(t) - v_{b_2}(t)|^2 dt \\
&= \int_{t_0}^{t_1} \max\{|v_{b_1}(t)| + \mu(t), |v_{b_2}(t)| + \mu(t)\}dt - \hat{h}\theta(1-\theta) \\
&\quad \times \int_{t_0}^{t_1} |v_{b_1}(t) - v_{b_2}(t)|^2 dt \\
&= \max\left\{\int_{t_0}^{t_1} \left(|v_{b_1}(t)| + \mu(t)\right) dt, \int_{t_0}^{t_1} \left(|v_{b_2}(t)| + \mu(t)\right) dt\right\} \\
&\quad - \hat{h}\theta(1-\theta) \int_{t_0}^{t_1} |v_{b_1}(t) - v_{b_2}(t)|^2 dt.
\end{aligned}
$$

This together with Lemma 6.221 implies that

$$
\begin{aligned}
f_1(\theta v_{b_1}(t) + (1-\theta)v_{b_2}(t), \mu(t)) &\leq \max\left\{f_1(v_{b_1}(t), \mu(t)), f_1(v_{b_2}(t), \mu(t))\right\} \\
&\quad -\hat{h}(t_1 - t_0)^{1-\beta_2}\theta(1-\theta)\left(\int_{t_0}^{t_1} |v_{b_1}(t) - v_{b_2}(t)| dt\right)^2,
\end{aligned}
$$

or equivalently

$$
\begin{aligned}
f_1(\theta v_{b_1}(t) + (1-\theta)v_{b_2}(t), \mu(t)) &\leq \max\left\{f_1(v_{b_1}(t), \mu(t)), f_1(v_{b_2}(t), \mu(t))\right\} \\
&\quad -\hat{h}(t_1 - t_0)^{1-\beta_2}\theta(1-\theta)|v_{b_1}(t) - v_{b_2}(t)|^2,
\end{aligned}
$$

that is, f_1 is a strongly quasiconvex in the second component. □

A Parametric Quadratic Programming

Now we present an application of our results to a practical situation. Namely, we study qualitative properties of solution maps of parametric quadratic programming, a special case of optimization problems, which has well-known applications in the view of practice. A good sample is the model of mean-variance portfolio based on investor's utility maximization. This problem can be formulated as a parametric quadratic optimization problem. Consider a universe of n assets with the known data $c = (c_1, ..., c_n)^T$ and $Q = [\sigma_{ij}]$, where c_i is the expected return for asset i and σ_{ij} is the covariance of returns for assets i and j. So, c is the vector of expected returns and Q is the $n \times n$ variance-covariance matrix of asset returns. Herein, Q is a positive definite matrix which follows from the properties of variance-covariance matrices. Denote the vector of asset holdings by $x = (x_1, ..., x_k)^T$. Then, the expected return of the portfolio x is $c^T x$ and its variance is $\sigma^2 = x^T Qx$. A portfolio is said to be efficient if for some fixed level of expected return, no other portfolio gives smaller variance.In other words, an efficient portfolio can be considered as the one for which at some fixed level of variance no other portfolio gives larger expected return. The determination of the efficient portfolio frontier in this mean-variance model is equivalent to solving the following parametric problem:

$$\min \frac{1}{2}x^T Qx - \mu c^T x \quad \text{s.t.} \quad Ax = b,\ x \in \mathbb{R}^n_+, \tag{6.171}$$

where $\mu \geq 0$ is an investor's variance aversion parameter. The linear constraints $Ax = b$ can represent budget constraints, bounds on asset holdings, etc. Nonnegativity constraints $x \geq 0$ are short-sale constraints (nonnegative asset holdings). If μ is allowed to vary, (6.171) becomes a parametric optimization problem. For $\mu \geq 0$, we denote the solution set of (6.171) by $S^{QO}(\mu)$.

Lemma 6.225 (Sum of strong convexity and convexity functions) *If $u(x)$ is $h\beta$-strongly convex and $v(x)$ is convex on a convex set $A \subset \mathbb{R}^n$, then $g(x) = u(x)+v(x)$ is $h\beta$-strongly convex on A.*

Proof For all $x_1, x_2 \in A$ and $t \in [0, 1]$, we have

$$\begin{aligned}
&g(tx_1 + (1-t)x_2)\\
&\quad = u(tx_1 + (1-t)x_2) + v(tx_1 + (1-t)x_2)\\
&\quad \leq tu(x_1) + (1-t)u(x_2) - ht(1-t)d^{\beta}(x_1, x_2) + tv(x_1) + (1-t)v(x_2)\\
&\quad \leq t(u(x_1) + v(x_1)) + (1-t)(u(x_2) + v(x_2)) - ht(1-t)d^{\beta}(x_1, x_2)\\
&\quad \leq tg(x_1) + (1-t)g(x_2) - ht(1-t)d^{\beta}(x_1, x_2).
\end{aligned}$$

Hence, g is a strongly convex function. □

Lemma 6.226 (Strong convexity of a quadratic function given by a positive definite matrix) *The function $g(x) = x^T Qx$ is strongly convex wrt the Euclidean norm in $\mathbb{R}^n$, where Q is a positive definite matrix.*

Proof Since Q is a positive definite matrix, we can consider the norm defined by $\|x\| := \sqrt{x^T Qx}$ on $\mathbb{R}^n$. We first prove that the function $g(x) = x^T Qx$ is strongly convex wrt this norm. Indeed, for all $x_1, x_2 \in \mathbb{R}^n$ and $t \in [0, 1]$, one has:

$$\begin{aligned}
&g(tx_1 + (1-t)x_2)\\
&\quad = (tx_1 + (1-t)x_2)^T Q(tx_1 + (1-t)x_2)\\
&\quad = t^2 x_1^T Qx_1 + t(1-t)x_1^T Qx_2 + t(1-t)x_2^T Qx_1 + (1-t)^2 x_2^T Qx_2
\end{aligned}$$

$$= t^2 x_1^T Qx_1 + (1-t)^2 x_2^T Qx_2 + 2t(1-t)x_1^T Qx_2$$
$$= tx_1^T Qx_1 + (1-t)x_2^T Qx_2 - t(1-t)(x_1^T Qx_1 + x_2^T Qx_2 - 2x_1^T Qx_2)$$
$$= tx_1^T Qx_1 + (1-t)x_2^T Qx_2 - t(1-t)\|x_1 - x_2\|^2$$
$$= tg(x_1) + (1-t)g(x_2) - t(1-t)\|x_1 - x_2\|^2.$$

We now show that $g(x) = x^T Qx$ is strongly convex wrt the Euclidean norm $\|\cdot\|_E$. Since any two norms on a finite-dimensional normed space are equivalent, there are $a, b > 0$ such that $a\|\cdot\|_E \le \|\cdot\| \le b\|\cdot\|_E$. Therefore, for all $x_1, x_2 \in \mathbb{R}^n$ and $t \in [0, 1]$, we get:

$$g(tx_1 + (1-t)x_2) \le tg(x_1) + (1-t)g(x_2) - a^2 t(1-t)\|x_1 - x_2\|_E^2.$$

The proof is complete. □

From Lemmas 6.225 and 6.226, we have the following result.

Corollary 6.227 (Hölder conditions for a parametric quadratic programming) *If Q is a positive definite matrix, then the solution map $S^{QO}(\cdot)$ is Hölder continuous on $[0, +\infty)$.*

6.9 Exercises

Exercise 6.1 Let X, Y be normed spaces; $C \subset Y$ a pointed, convex, closed, and solid cone; $K \subset X$ a nonempty compact convex subset; and M an open subset of a metric space. Consider a function $f : K \times K \times M \to Y$ satisfying the following assumptions:

(a) $f(\cdot, y, \bar{\mu})$ is upper semicontinuous for every $y \in Y$;
(b) $f(x, y, \cdot)$ is $(-C)$-continuous at $\bar{\mu}$ uniformly wrt $x, y \in K$, i.e., for every neighborhood U of $f(x, y, \bar{\mu})$, there exists a neighborhood N of $\bar{\mu}$ such that

$$f(x, y, \mu) \in U - C, \forall \mu \in N.$$

Prove that the solution map $S : M \rightrightarrows K$ defined by

$$S(\mu) := \{x \in K \mid f(x, y, \mu) \nless 0, \forall y \in K\}$$

is upper hemicontinuous at $\bar{\mu}$.

Exercise 6.2 Let X, Y be normed spaces; $C \subset Y$ a pointed, convex, closed, and solid cone; $K \subset X$ a nonempty compact convex subset; and M an open subset of a metric space. Consider a function $f : K \times K \times M \to Y$ satisfying the following assumptions:

(a) $f(\cdot, \cdot, \mu)$ is pseudomonotone for every $\mu \in M$, i.e., $f(x, y, \mu) \nless 0 \Rightarrow f(y, x, \mu) \ngtr 0$;
(b) $f(\cdot, \cdot, \cdot)$ is (C)-continuous on $K \times K \times M$;
(c) $f(x, \cdot, \mu)$ is quasiconvex for every $x \in K$, $\mu \in M$;
(d) $f(\cdot, \cdot, \mu)$ is sign-preserving for every $\mu \in M$ and $f(x, x, \mu) \geq 0$ for every $x \in K$ and $\mu \in M$.

Prove that the solution map $S : M \rightrightarrows K$ defined by

$$S(\mu) := \{x \in K \mid f(x, y, \mu) \nless 0, \forall y \in K\}$$

is upper hemicontinuous M.

Exercise 6.3 Let X, Y be normed spaces; $C \subset Y$ a pointed, convex, closed, and solid cone; $K \subset X$ a nonempty compact convex subset; and M an open subset of a metric space. Consider a function $f : K \times K \times M \to Y$ satisfying the following assumptions:

(a) $f(\cdot, \cdot, \mu)$ is strongly pseudomonotone on $K \times K$ uniformly wrt μ (i.e., there exist $e \in \text{int}C$ and positive number a such that for every $x, y \in K$, $[f(x, y, \mu) \nless 0] \Rightarrow [f(y, x, \mu) \leq -d^a(x, y)e]$);
(b) there exist $u \in \text{int}C$ and a positive number b such that, for every $x, y \in K$,

$$-\|\mu - \mu'\|^b u \leq f(x, y, \mu) - f(x, y, \mu') \leq \|\mu - \mu'\|^b u,$$

when $\|\mu - \mu'\|$ is small enough.

Prove that the solution map $S : M \rightrightarrows K$ defined by

$$S(\mu) := \{x \in K \mid f(x, y, \mu) \nless 0, \forall y \in K\}$$

is locally Hölder continuous.

Exercise 6.4 Let X, Y be normed spaces, $C \subset Y$ a pointed, convex, closed, and solid cone, $K \subset X$ a nonempty compact convex subset, Λ and M open subsets of a metric space. Consider a set-valued map $K : \Lambda \rightrightarrows X$ and a function $f : X \times X \times M \to Y$ satisfying the following assumptions:

(a) $f(\cdot, \cdot, \mu)$ is strongly pseudomonotone on $K \times K$ uniformly wrt μ (i.e., there exist $e \in \text{int}C$ and positive number a such that for every $x, y \in K$, $[f(x, y, \mu) \nless 0] \Rightarrow [f(y, x, \mu) \leq -d^a(x, y)e]$) and $-f(\cdot, \cdot, \mu)$ is weakly quasimonotone (i.e., $[-f(x, y, \mu) > 0] \Rightarrow [-f(y, x, \mu) \ngtr 0]$);

(b) there exist $u \in \text{int}C$ and a positive number b such that, for every $x, y \in K$,

$$-\|\mu - \mu'\|^b u \leq f(x, y, \mu) - f(x, y, \mu') \leq \|\mu - \mu'\|^b u,$$

when $\|\mu - \mu'\|$ is small enough;

(c) there exist $v \in \text{int}C$ and a positive number g such that for every $x, y, y' \in X$, $\mu \in M$,

$$-d^g(y, y')^v \leq f(x, y, \mu) - f(x, y', \mu) \leq d^g(y, y')v;$$

(d) the map K is convex-compact-valued and is Hölder continuous.

Prove that the solution map $S : \Lambda \times M \rightrightarrows X$ defined by

$$S(\lambda, \mu) := \{x \in K(\lambda) \mid f(x, y, \mu) \not< 0, \forall y \in K(\lambda)\}$$

is locally Hölder continuous.

Exercise 6.5 Let X, Y be Banach spaces, $K \subset X$ be nonempty compact convex, $C \subset Y$ be a closed-convex-solid cone and $f : K \times K \to Y$ belong to $\mathscr{C}(K \times K, Y)$ the set of all continuous maps from $K \times K$ into Y with the norm defined by $\|f\| := \max_{(x,y)\in K\times K} \|f(x, y)\|$. Let Φ be the collection of all $f \in \mathscr{C}(K \times K, Y)$ such that

$$S(f) = \{x \in K \mid f(x, y) \not< 0, \forall y \in K\} \neq \emptyset.$$

Prove that the subspace Φ of $\mathscr{C}(K \times K, Y)$ is a complete metric space.

Exercise 6.6 Let X, Y be Banach spaces, $K \subset X$ nonempty compact convex, and $C \subset Y$ a closed-convex-solid cone. Let $\mathscr{C}(K \times K, Y)$ be the set of all continuous maps from $K \times K$ into Y with the norm defined by $\|f\| := \max_{(x,y)\in K\times K} \|f(x, y)\|$. Let Φ be the collection of all $f \in \mathscr{C}(K \times K, Y)$ such that

$$S(f) := \{x \in K \mid f(x, y) \not< 0, \forall y \in K\} \neq \emptyset.$$

Prove that the solution map S is upper semicontinuous and compact-valued on Φ.

Exercise 6.7 Let X, Y be Banach spaces, $K \subset X$ nonempty compact convex, and $C \subset Y$ a closed-convex-solid cone. Let $\mathscr{C}(K \times K, Y)$ be the set of all continuous maps from $K \times K$ into Y with the norm defined by $\|f\| := \max_{(x,y)\in K\times K} \|f(x, y)\|$. Let Φ be the collection of all $f \in \mathscr{C}(K \times K, Y)$ such that

$$S(f) := \{x \in K \mid f(x, y) \not< 0, \forall y \in K\} \neq \emptyset.$$

Let $\Psi = \{f \in \Phi \mid \text{the solution map } S \text{ is continuous at } f\}$. Prove that Ψ contains a dense G_δ subset of Φ, where G_δ is a set that equals a countable intersection of open sets of X.

Exercise 6.8 Let X, Y be Banach spaces, $K \subset X$ nonempty compact convex, and $C \subset Y$ a closed-convex-solid cone. Let $\mathscr{C}(K \times K, Y)$ be the set of all continuous maps from $K \times K$ into Y with the norm defined by $\|f\| := \max_{(x,y)\in K\times K} \|f(x, y)\|$. Let Φ be the collection of all $f \in \mathscr{C}(K \times K, Y)$ such that

$$S(f) := \{x \in K \mid f(x, y) \not< 0, \forall y \in K\} \neq \emptyset.$$

Prove that if $f \in \Phi$ and $x \mapsto f(x, y)$ is strictly C-weakly quasiconcave for each $y \in K$, then the solution map S is lower semicontinuous at f.

Exercise 6.9 Let X, Y be Banach spaces, $K \subset X$ nonempty compact convex, and $C \subset Y$ a closed-convex-solid cone. Let $\mathscr{C}(K \times K, Y)$ be the set of all continuous maps from $K \times K$ into Y with the norm defined by $\|f\| := \max_{(x,y)\in K\times K} \|f(x, y)\|$. Let Φ be the collection of all $f \in \mathscr{C}(K \times K, Y)$ such that

$$S(f) := \{x \in K \mid f(x, y) \not< 0, \forall y \in K\} \neq \emptyset.$$

Prove that if $f \in \Phi$ and f is strongly pseudomonotone, then the solution set $S(f)$ is a singleton and Hölder continuous at f.

Exercise 6.10 Let X, Y, W be normed spaces, $C \subset Y$ a closed convex solid cone, and M nonempty subset of W. Let $\mathscr{D} : M \rightrightarrows X$ be continuous with nonempty compact convex values and $f : \mathscr{D}(M) \times \mathscr{D}(M) \times M \to Y$. Suppose that the following conditions are satisfied:

(a) $\mathscr{D}$ is continuous with compact values on M;
(b) f is continuous on $\mathscr{D}(M) \times \mathscr{D}(M) \times M$;
(c) for each $\mu \in M$ and $x \in \mathscr{D}(\mu) \setminus S(\mu)$, there is $y \in S(\mu)$ such that

$$d(x, y) \leq d(f(x, y, \mu), Y \setminus -\text{int}C).$$

Then, prove that the mapping S given by

$$S(\mu) := \{x \in \mathscr{D}(\mu) \mid f(x, y, \mu) \notin -C \setminus \{0\}, \forall y \in \mathscr{D}(\mu)\},$$

is lower semicontinuous on M.

Exercise 6.11 Let X, Y, W be normed spaces, $A \subset X$ a convex subset, and Λ and M nonempty subsets of W. Let $\mathscr{D} : \Lambda \rightrightarrows A$ be continuous with nonempty compact convex values, $C : A \times M \rightrightarrows Y$ closed-convex-solid-cone-valued and $f : A \times A \times M \to Y$ and $g : A \to A$. Prove that if the following conditions hold:

(a) $\mathcal{D}$ is continuous with compact convex values;
(b) f and g are continuous;
(c) for any $\mu \in M$ and $x \in A$, $f(x, x, \mu) \in C(x, \mu)$, and $f(\cdot, \cdot, \mu)$ is C-convex wrt its second argument;
(d) $K : A \times M \rightrightarrows Y$ defined by $K(x, \mu) := Y \setminus -\text{int}C(x, \mu)$ is closed,

then the map $S : \Lambda \times M \rightrightarrows A$ given by

$$S(\lambda, \mu) := \{x \in \mathcal{D}(\lambda) \mid f(g(x), y, \mu) \notin -\text{int}C(x, \mu), \forall y \in \mathcal{D}(\lambda)\}$$

is usc on $\Lambda \times M$.

Exercise 6.12 Let X, Y, W be normed spaces, $A \subset X$ a convex subset, and Λ and M nonempty subsets of W. Let $\mathcal{D} : \Lambda \rightrightarrows A$ be continuous with nonempty compact convex values, $C : A \times M \rightrightarrows Y$ closed-convex-solid-cone-valued and $f : A \times A \times M \to Y$ and $g : A \to A$. Suppose that:

(a) $\mathcal{D}$ is lsc with compact convex values;
(b) f and g are continuous;
(c) for any $\mu \in M$ and $x \in A$, $f(x, x, \mu) \in C(x, \mu)$, and $f(\cdot, \cdot, \mu)$ is C-convex wrt its second argument;
(d) $K : X \times M \rightrightarrows Y$ defined by $K(x, \mu) := Y \setminus -\text{int}C(x, \mu)$ is closed;
(e) for a given $(\lambda_0, \mu_0) \in \Lambda \times M$, there exist neighborhoods $U(\lambda_0)$ of λ_0 and $V(\mu_0)$ of μ_0 such that for any $\lambda \in U(\lambda_0)$, $\mu \in V(\mu_0)$, and $x \in \mathcal{D}(\lambda)$, $f(\cdot, \cdot, \mu)$ is locally nonpositive at x wrt g.

Then, prove that the map $S : \Lambda \times M \rightrightarrows A$ given by

$$S(\lambda, \mu) := \{x \in \mathcal{D}(\lambda) \mid f(g(x), y, \mu) \notin -\text{int}C(x, \mu), \forall y \in \mathcal{D}(\lambda)\}$$

is lsc on $\Lambda \times M$.

Exercise 6.13 Let X be a nonempty subset of a topological space and Y a real topological vector space. Let $f : X \to Y$, $C : X \rightrightarrows Y$ have proper solid convex cone values and $K : X \rightrightarrows Y$ be defined by $K(x) := Y \setminus -\text{int}C(x)$. If K is usc on X and for each $x \in X$, f is $(-C(x))$-continuous at x, then prove that the set $\mathfrak{L} := \{x \in X \mid f(x) \in -\text{int}C(x)\}$ is open.

Exercise 6.14 Let X be a nonempty subset of a topological space and Y a real topological vector space. Let $C : X \rightrightarrows Y$ have proper solid convex cone values and $\widehat{C} : X \rightrightarrows Y$ an intersectional mapping of C with solid convex cone values. Suppose $f : X \to Y$ and $K : X \to Y$, defined by $K(x) := Y \setminus -\text{int}C(x)$. Prove that if K is closed and for each $x \in X$, f is $(-\widehat{C}(x))$-continuous at x, then the set $\mathfrak{L} := \{x \in X \mid f(x) \in -\text{int}C(x)\}$ is open.

Exercise 6.15 Let X be a nonempty subset of a real topological vector space and Y a real topological vector space, $C : X \times \Lambda \rightrightarrows Y$ with proper solid convex cone values, and Λ a subset of some topological space. Let $\mathscr{D} : X \times \Lambda \rightrightarrows X \setminus \{\emptyset\}$ and $f : X \times X \times \Lambda \to Y$. Prove that if the following conditions are satisfied:

(a) $S(\lambda) := \{x \in \mathscr{D}(x, \lambda) \mid f(x, y, \lambda) \notin -\text{int}C(x, \lambda), \forall y \in \mathscr{D}(x, \lambda)\}$ is non empty for each $\lambda \in \Lambda$;
(b) X is compact;
(c) $\mathscr{D}$ is usc with compact values on $X \times \Lambda$;
(d) the set $\{(x, y, \lambda) \in X \times X \times \Lambda \mid f(x, y, \lambda) \in -\text{int}C(x, \lambda)\}$ is open,

then the map S is usc on Λ.

Exercise 6.16 Let X and Λ be two nonempty subsets of two real topological vector spaces, resp., Y a real topological vector space, and $C : X \times \Lambda \rightrightarrows Y$ have proper solid convex cone values. Let $\mathscr{D} : X \times \Lambda \rightrightarrows X \setminus \{\emptyset\}$ and $f : X \times X \times \Lambda \to W$. Prove that if f is $C(x, \lambda)$-continuous at (x, y, λ) for every $x, y \in X$, $\lambda \in \Lambda$, and C is usc on $X \times \Lambda$, then the set

$$\mathfrak{U} := \{(x, y, \lambda) \in X \times X \times \Lambda \mid f(x, y, \lambda) \notin -\text{cl}C(x, \lambda)\}$$

is open.

Exercise 6.17 Let X and Λ be two nonempty subsets of two real topological vector spaces, resp., Y a real topological vector space and $C : X \times \Lambda \rightrightarrows Y$ with proper solid convex cone values, and let $K : X \times \Lambda \rightrightarrows X \setminus \{\emptyset\}$. Suppose that $f : X \times X \times \Lambda \to Y$ is a vector-valued map and that $D : X \times \Lambda \rightrightarrows Y$ is a intersectional mapping of C with solid convex cone values. Prove that if f is $D(x, \lambda)$-continuous at (x, y, λ) for every $x, y \in X$, $\lambda \in \Lambda$, and C is closed, then the set

$$\mathfrak{U} := \{(x, y, \lambda) \in X \times X \times \Lambda \mid f(x, y, \lambda) \notin -\text{cl}C(x, \lambda)\}$$

is open.

Exercise 6.18 Let X and Λ be two nonempty subsets of two real topological vector spaces, resp., Y a real topological vector space, and $C : X \times \Lambda \rightrightarrows Y$ a set-valued map with proper solid convex cone values. Let $\mathscr{D} : X \times \Lambda \rightrightarrows X \setminus \{\emptyset\}$ and $f : X \times X \times \Lambda \to W$. Prove that if the following conditions hold true

(a) the set $\{x \in \mathscr{D}(\lambda) \mid f(x, y, \lambda) \notin -\text{int}C(x, \lambda)\} \neq \emptyset$ for each $\lambda \in \Lambda$;
(b) $\mathscr{D}$ is usc with compact convex values on X;
(c) for each $\lambda \in \Lambda$, $\mathscr{D}(\cdot, \lambda)$ is concave on X;
(d) $\mathscr{E} : \Lambda \rightrightarrows X$, defined by $\mathscr{E}(\lambda) := \{x \in X \mid x \in \mathscr{D}(x, \lambda)\}$, is lsc on Λ and $\mathscr{E}(\lambda)$ is convex for each $\lambda \in \Lambda$;
(e) the set $\{(x, y, \lambda) \in X \times X \times \Lambda \mid f(x, y, \lambda) \notin -\text{cl}C(x, \lambda)\}$ is open;
(f) f is C-weakly quasiconcave on $X \times X \times \Lambda$,

then the set-valued map $S : \Lambda \rightrightarrows X$ given by

$$S(\lambda) := \{x \in \mathscr{D}(x,\lambda) \mid f(x,y,\lambda) \notin -\text{int}C(x,\lambda)\}$$

is lsc on Λ.

Exercise 6.19 (Generalized Archimedean axiom) Let C be a convex cone with a nonempty interior in a real topological linear space Y. Prove that for any $x \in \text{int}C$ and $y \in Y$, there exists $n_0 \in \mathbb{N}$ such that

$$y - nx \in -\text{int}C, \forall n \geq n_0.$$

Exercise 6.20 (Generalized sign preserving) Let C be a convex cone with a nonempty interior in a real topological linear space Y. Prove that if $a \in C$ and $b \notin -\text{int}C$, then

$$a + b \notin -\text{int}C.$$

6.10 Commentary

Section 6.2
In this section, we present various kinds of relaxed semicontinuity and generalized convexity of vector maps via ordering cones, and discuss their properties as well as their relationships. Then, they are applied to studying stability in the sense of (semi)continuity of solution maps of weak and strong vector equilibrium problems. The techniques used in this section are not very far from those in Chap. 5, but the tools employed herein are new, and, under suitable adjustments, they can also be applied to studying qualitative properties of solution sets of many models in vector optimization. For more details on the relationships between the techniques and approaches in this section and the existing ones, we refer the reader to some typical works [7, 12, 17, 19, 21, 29, 32, 34, 37, 44, 45, 51, 57, 136, 214, 296, 325, 325].

Section 6.3
To support the study of Hölder and Lipschitz conditions of solution maps of weak and strong equilibrium problems, we provide many concepts related to Hölder/Lipschitz continuity, (quasi/pseudo) strongly monotonicity, and convexity of vector maps. Proprties and relationships of these concepts are also discussed in this section. They are taken mainly from [13, 25, 26, 31, 33, 38, 42, 43, 46, 49, 55, 89, 101, 240, 323, 324]. Under suitable techniques, these concepts have become key conditions in formulating (calm) Hölder/Lipschitz conditions of solution maps and approximate solution maps of the reference problems. Note that due to the complexity of techniques used in this study, the reader should have a close look on the proofs of these results to see the main contribution of each key conditions. We hope that these approaches can be applied to studying the corresponding properties of solution maps of problems in vector optimization.

Section 6.4

The main idea of studying well-posedness for any problem is to provide conditions for convergence of approximating sequences for the reference problem, and therefore the concepts of approximating sequences play a key role in this issue, and based on them, many kinds of well-posedness are proposed and discussed. In this section, we introduce concepts of approximate sequences for the weak and strong quasiequilibirum problems, and then by using the concepts of relaxed semicontinuity and generalized convexity of vector maps presented in Sect. 6.2, we formulate well-posedness conditions as well as their characterizations for such problems. Due to the generalization of the well-posedness considered in this section, many kinds of well-posedness such as the Tikhonov well-posedness, the Hadamard well-posedness, and Levitin-Polyak well-posedness are also established under suitable settings. For more details on the relationships between the techniques and approaches in this section and the existing ones, we refer the reader to some typical works [7, 8, 20, 22, 35, 40, 41, 59–61, 141, 155–157, 157, 166, 258, 297, 297, 349].

Section 6.5

In this section, we provide a scalarization method for vector equilibrium problems. For the weak equilibrium problem, we see that the linear scalarization method is an efficient approach in studying qualitative properties of solution maps, while the nonlinear scalarization method is the best one in considering the strong equlibrium problem. Moreover, the results on scalarization approaches presented in this section also give us motivations for investigating scalarization functions for vector models ralated to optimization. For more details on the applications of scalarization methods in studying qualitative properties of solutions to vector optimization models, we refer the reader to several notable works [10, 11, 23, 47, 48, 50, 137, 227, 230, 231, 244, 267, 425].

Section 6.6

In most methods designed for solving optimization-related problems, it is implicitly assumed that the underlying reference problem is well-posed, meaning that its solution uniquely exists and continuously depends on perturbed input data; see, for instance, [18, 147, 184, 187, 474]. In practice, a problem is, however, typically ill-posed due to at least one of the following reasons: it has no solution or more than one solution; or its solutions do not continuously depend on the input data. We refer the readers to [207, 359] for further discussions of ill-posed problems. In general, it is difficult to solve an ill-posed problem because a very small error in numerical computations may lead to a completely erroneous solution. One of the most widespread and applicable methods to handle ill-posed problems is the Tikhonov regularization. It was originally introduced by Tikhonov [428] and first applied to variational inequality problems by Browder [113]. Tikhonov regularization together with its extensions has since then been extensively studied for inverse problems, complementarity problems, variational inequalities, and optimal control problems [163, 191, 417, 464]. The Tikhonov regularization for scalar equilibrium problems was originally introduced by Gwinner in [228], where the author called such problems variational inequalities and obtained basic convergence results for this method under certain monotonicity conditions. Chadli et al. [131] initially used Tikhonov regularization procedure for monotone equilibrium problems to give a "Viscosity Principle" for the considered problems. Later, Konnov [307] provided sufficient conditions for the existence of solutions to equilibrium problems in finite-dimensional spaces, and applied them to show the well-definiteness and convergence of the Tikhonov trajectory without monotonicity assumptions. Then, Konnov and Dyabilkin [309] introduced a general coercivity condition, an extension of the one in [100, 307], for the existence of solutions and also showed that this condition could create a broad family of regularization methods that preserve well-definiteness and convergence of the iteration sequence without additional monotonicity assumptions. Tikhonov regularization was also extended to pseudomonotone equilibrium problems in real Euclidean spaces by Hung and Muu [255]. The authors proved that the regularized problems admits a solution if and only if the original one does. The authors showed that any Tikhonov trajectory tends to the same limit, which is a solution to the original problem. From the above observations, it can be concluded that Tikhonov regularization method for scalar equilibrium problems has been a significant topic and attracted attention of various researchers, and hence it is worthy to study this topic for vector models. In line with the aim of this section, we present only the typical results on Tikhonov regularization methods for vector equilibrium problems. To provide a comprehensive view of contributions to this topic, along with the closely related penalty function methods, we would like to introduce readers to several highly relevant works for further reference [14, 14–16, 53, 58, 140, 229, 246, 247, 255, 279, 384, 434–438, 438, 439]. We hope that these resources may not only help readers deepen their understanding of the coercivity conditions used in this field but also inspire new ideas for applying regularization methods to generalized optimization models.

This section focuses on the Tikhonov regularization method for weak and strong models of vector equilibrium problems. For weak models, a coercive condition is introduced to study basic types of convergence of the regularization method in both monotone and nonmonotone cases. The results and approaches are new and different from the existing in the literature and were first considered in [58]. For strong models, assumptions related to generalized convexity are employed to study the regularization method for convex and monotone problems. For nonconvex and nonmonotone problems, this method is applied using a coercivity property concerning an auxiliary vector optimization problem, which has potential applications in studying vector Ekeland variational principle. The results obtained for the second model are sourced from the paper [15]. The most noteworthy contribution of this section is the original regularization results it establishes for strong vector equilibrium problems. Overall, this section provides a comprehensive investigation of the Tikhonov regularization and introduces innovative approaches to address both weak and strong models of vector equilibrium problems.

Section 6.7

This section is dedicated to relationships between set optimization problems and equilibrium models. Set-valued optimization is an interesting and important branch of applied mathematics, which is designed to solve optimization problems, in which either the objective mapping or the constraint mappings are set-valued mappings acting between abstract spaces. Concerning set-valued optimization, there are two main approaches, depending on what the notion of minimality is considered. The classical one is vector approach, where the definition of a minimizer considers only an efficient point of the union of all images of the set-valued objective mapping and identifies the image set containing this minimal element as the "best" available [265, 351]. However, a serious disadvantage of this approach is that, in general, only one element does not necessarily imply that the whole image set is in a certain sense minimal with respect to all image sets. In an effort to overcome this serious drawback, a more appropriate order relation for comparing sets, which has been introduced to optimization by Kuroiwa [314], called a set order relation. With the aim of investigating preferences over sets, it seems that the set optimization approach is more natural and appropriate than the vector one. Until now, set optimization approach has received much attention by many authors; see, for instance, [139, 154, 189, 221, 245, 264, 266, 280, 294] and the references therein. Regarding this approach, there are two main set order relations considered in set-valued optimization problems: upper type set less order relation and lower type set less order relation. We refer the reader to [139, 266, 280] and the references therein for more details. By blending these two set order relations, we obtain another one, say "set less order relation," which has been independently utilized in algebra by Young [463], in fixed-point theory by Nishnianidze [382], and in computer science and interval analysis by Chiriaev and Walster [143]. For a given set optimization problem, the study of the behavior of minimal sets under perturbations is one of the significant topics in set-valued optimization. The classical way concerning stability in set optimization is to consider the semicontinuity/continuity and Hölder/Lipschitz properties of the optimal multifunction for parametric set optimization problems with the parameter perturbed in a space of parameters; see, for instance, [9, 56, 241, 242, 293, 390, 442, 443, 458]. Recently, many works have focused on an alternative way to study stability of set optimization problems with respect to the lower and upper types set less order relation in the image space in terms of the convergence (in the sense of Painlevé-Kuratowski or Hausdorff) for a sequence of solutions of perturbed problems with convergent sequences of constraint sets to a solution of the given problem. Because these perturbed problems are different from such parametric problems with the parameter perturbed in a parameter space, the study of convergence of the solution sets in [17, 54, 225] is interesting and deserving attention.

Among many tools for studying vector problems in optimization, the Gerstewitz function is an effective one. In this section, we investigate the stability of set optimization problems by using generalized Gerstewitz functions. More precisely, we first introduce the generalized Gerstewitz scalarization functions and study their continuity and convexity properties. Next, we use these functions as objective functions of equilibrium problems to study stabililty of set optimization problems. Since the equilibrium problem has very close relations with many optimization-related problems such as variational inclusion problems and implicit multifunction problems, the current approach has the evident potential to explore other continuity properties, e.g., the inner and outer semicontinuity, Lipschitz-like, and Hölder-like properties of these problems (see, e.g., [42, 145, 146]).

Chapter 7
Variational Convergence of Bifunctions and Approximations of Equilibrium Problems

Abstract In Sects. 7.1 and 7.2, the traditional types of variational convergence, epi-convergence for functions and epi/hypo- and lopsided convergence for bifunctions, and their variants, inside, weak, and weak inside types, are examined in detail, with a focus on characterizations and variational properties together with relationships between the convergence types, especially with those related to their suitability for the study of variational properties. Applications of the obtained results to approximations of equilibrium and strong equilibrium problems are considered in Sect. 7.3, including discussions of the two kinds of approximate solutions and also their convergence for pairs of dual problems. The subsequent five sections are devoted to important special cases: multiobjective optimization, approximations of optimization problems in terms of optimality functions, Nash equilibria, Walras economic equilibria, and consistency of statistical estimators of solutions of stochastic optimization models. In each section, convergence results for optimal values and solutions/approximate solutions in terms of appropriate types of variational convergence are discussed.

In this chapter, we begin a new topic about approximations of equilibrium problems, which is different enough from the ones studied in the preceding chapters. Hence, let us start with motivations and brief descriptions of the topic. The term "approximation" appears in all areas of mathematics because when facing a complicated object or problem, the first and popular idea is to study simpler replacements approximating the original one. There is an extensive literature on local approximation, see, e.g., the nice monographs [81, 109, 368, 396]. The literature on global approximation is remarkably more sparse. This chapter and the next one are concerned with global approximation in the following sense. Approximating problems naturally mean those that converge in a certain sense to the original one. Most optimization-related problems are crucially related to variational properties such as being their extremum values, minimizers, maximizers, minsup-values, minsup-points, and saddle points. Hence, the most meaningful and interesting types of convergence are those that preserve variational properties, i.e., such variational properties are guaranteed to be preserved under such convergence. The term "variational

L. Q. Anh et al., *Equilibrium Problems*, Forum for Interdisciplinary Mathematics, https://doi.org/10.1007/978-3-031-89003-1_7

convergence" is not a particular concept, but a piece of common terminology for the convergence types that preserve variational properties. Both the variational convergence and preserved properties are global (i.e., satisfied in the whole domains under consideration). Approximations of problems in terms of sequences that variationally converge to the original problem may be called variational approximations. They are necessary in a multitude of situations arising in applications, including demands of computation and/or modeling. We must start by getting used to variational convergence whose main types are convergence of bivariate functions (bifunctions) (bifunctions/two-component functions/two-variable functions/two-argument functions/two-input functions) because the main ingredient of an equilibrium problem is a bifunction. But the first section presents the basic variational convergence of univariate/uni-component functions (unifunctions, or simpler, functions, hereafter), followed by a section about variational convergence of bifunctions. The first reason for the necessity of this long preparation to the main study of variational approximations of equilibrium problems is that this study relies heavily on types of variational convergence. Hence, such types of convergence are emerging as important tools in the study of approximation for mathematical programs, variational inequalities, complementarity problems, Nash and Walras equilibrium problems, fixed points, robust optimizations, etc. But these tools are not available in the previous material of the book. We will see that the properties (widely used in mathematics, in general) such as types of continuity, convexity, and monotonicity are much less crucial for variational approximations, unlike in the preceding chapters. The second reason is that variational convergence is rather different from the other types of convergence widely employed in various areas of mathematics. So, a detailed study of variational convergence is necessary and inevitable. The third reason is that there is a single book by Attouch [74] devoted to variational convergence, published 40 years ago. It focuses on epi-convergence of extended-real-valued unifunctions on the whole spaces (one finds only a short additional part about epi/hypo-convergence of bifunctions defined on the whole product spaces). It is good that very recently, when we were about to submit this monograph, we discovered the interesting book [403] with Chapters 8 and 9 strongly related to lop-convergence. Chapters for epi-convergence of extended-real-valued unifunctions on the whole spaces are also included in the famous books by Aubin and Frankowska [82] and Rockafellar and Wets [396]. A key result on the epi-convergence says that any sequence of functions on a separable metric space admits an epi-convergent subsequence (see Theorem 2.6 in [78]), ensuring the applicability of this basic variational convergence in practice. Theorem 6.1 in that paper asserts the corresponding result for epi/hypo-convergence, the first basic variational convergence of bifunctions: any sequence of bifunctions on the product of two separable metric spaces admits an epi/hypo-convergent subsequence. Therefore, we will see that comprehensive and detailed studies of variational convergence in Sects. 7.1 and 7.2 are really justified by their applications in considering approximations of equilibrium problems we study in detail in the subsequent sections of this chapter.

7.1 Epi-Convergence of Functions

This first section is devoted to several types of epi-convergence (for unifunctions). Epi-convergence is a fundamental concept in variational analysis and, particularly an inevitable base for variational convergence of bifunctions, the main ingredient in equilibrium problems. Note that in the literature, epi-convergence is also called Γ-convergence. If not otherwise specified, in this section, let X be a metric space, $A^k, A \subset X$ nonempty, $\varphi^k, \varphi : X \to \mathbb{R} \cup \{+\infty\}$, φ^k and φ be finite on A^k and A, resp. With no fear of confusions, we also say that φ^k and φ are defined on the entire A^k and A, resp., and write $\varphi^k : A^k \to \mathbb{R}$ and $\varphi : A \to \mathbb{R}$. We also often write $\varphi^k, \varphi \in$ fv-fcn(X), which means finite-valued functions on the considered subsets of X). The basic concepts of variational convergence for functions we use are presented next.

7.1.1 *Types of Epi-Convergence*

Consider X and φ^k, φ as above.

Definition 7.1 (Epi-convergence, inside epi-convergence) φ^k EPI-CONVERGES to φ, written $\varphi^k \overset{e}{\to} \varphi$ or $\varphi = \text{e-lim}_k \varphi^k$, when:

(a) For every subsequence $A^k \ni x^k \to x$, $\liminf_j \varphi^{k_j}(x^{k_j}) \geq \varphi(x)$ if $x \in A$ and $\varphi^{k_j}(x^{k_j}) \to +\infty$ if $x \notin A$
(b) For every $x \in A$, there exists $A^k \ni x^k \to x$ such that $\limsup_k \varphi^k(x^k) \leq \varphi(x)$

Omitting the case $x \notin A$ together with the infinity condition, i.e., only consider $x \in A$, one has INSIDE EPI-CONVERGENCE, which is denoted by $\varphi^k \overset{i-e}{\to} \varphi$ or $\varphi =$ i-e-$\lim_k \varphi^k$.

Definition 7.2 (Weak epi-convergence, weak inside epi-convergence) φ^k WEAKLY EPI-CONVERGES to φ, written $\varphi^k \overset{w-e}{\to} \varphi$ or $\varphi =$ w-e-$\lim_k \varphi^k$, when the following conditions are satisfied:

(a) For every sequence $A^k \ni x^k \to x$, $\liminf_k \varphi^k(x^k) \geq \varphi(x)$ if $x \in A$ and $\varphi^k(x^k) \to +\infty$ if $x \notin A$
(b) For every $x \in A$, there exists $A^k \ni x^k \to x$ such that $\limsup_k \varphi^k(x^k) \leq \varphi(x)$

Omitting the case $x \notin A$ with the infinity condition, one has WEAK INSIDE EPI-CONVERGENCE, which is denoted by $\varphi^k \overset{w-i-e}{\to} \varphi$ or $\varphi =$ w-i-e-$\lim_k \varphi^k$.

Among the above four types, epi-convergence is a well-known concept and appears in many publications on various topics. The other three modified and weaker notions aim at applications in variational approximations of optimization-related problems. We will see that each of them can be useful in particular situations. Unlike the concepts of convergence of functions appearing in the preceding chapters, the above types of variational convergence involve direct function-variable relationships. This is one of the main construction features for them to guarantee preservation of variational properties. In the literature, the case $A^k = A = X$ for all k and $\mathbb{R}$ replaced by $\mathbb{R} \cup \{+\infty\}$ is more traditional and has been extensively studied. Definitions 7.1 and 7.2 as well as the facts related to epi-convergence in Chaps. 7 and 8 are still in force for this case (of course with the infinity condition and the inside convergence being dropped).

The basic variational objects of $\varphi : A \to \mathbb{R}$ are the infimum value $\inf_A \varphi$ of φ on A; a minimizer (minimum point), i.e., a point $\bar{x} \in A$ where the function achieves the infimum value ($\varphi(\bar{x}) = \inf_A \varphi$); and the whole set of minimizers, denoted by $\text{argmin}_A \varphi := \{x \in A \mid \varphi(x) = \inf_A \varphi\}$. Approximates of such $\bar{x}$ are also important as they exist more often in practice. The definition is for $\varepsilon \geq 0$, $\bar{x}_\varepsilon \in A$ is an ε-MINIMUM POINT or an ε-MINIMIZER of φ on A if $\varphi(\bar{x}_\varepsilon) \leq \inf_A \varphi + \varepsilon$. We denote the set of ε-minimum points by ε-$\text{argmin}_A \varphi$.

In Chaps. 7 and 8, we often use basic concepts and facts of set-valued analysis. The ones already presented at the beginning of Chap. 4 are not recalled here. We provide some others about limits of a sequence of sets in a metric space.

Let X be a metric space and $C^k \subset X$. The (KURATOWSKI-PAINLEVÉ) UPPER/OUTER LIMIT $\text{Limsup}_{k\to+\infty} C^k$ of the sequence of sets C^k is defined as

$$\text{Limsup}_{k\to\infty} C^k := \{x \in X \mid \text{liminf}_{k\to\infty} d(x, C^k) = 0\},$$

which is equivalent to

$$\text{Limsup}_{k\to\infty} C^k := \{x \in X \mid x = \lim_{j\to\infty} x^{k_j}, x^k \in C^k\},$$

or

$$\text{Limsup}_{k\to\infty} C^k := \cap_{K>0} \text{cl} \cup_{k\geq K} C^k = \cap_{\varepsilon>0} \cap_{K>0} \cup_{k\geq K} B(C^k, \varepsilon)$$

(recall that $B(C^k, \varepsilon)$ is the ε enlargement of C^k). The (KURATOWSKI-PAINLEVÉ) LOWER/INNER LIMIT is similarly defined with the above formulas replaced by, correspondingly,

$$\text{Liminf}_{k\to\infty} C^k := \{x \in X \mid \lim_{k\to\infty} d(x, C^k) = 0\},$$

which is equivalent to

$$\text{Liminf}_{k\to\infty} C^k := \{x \in X \mid x = \lim_{k\to\infty} x^k, x^k \in C^k\},$$

or

$$\text{Liminf}_{k\to\infty} C^k := \cap_{\varepsilon>0} \cup_{K>0} \cap_{k\geq K} B(C^k, \varepsilon).$$

The (full) (Kuratowski-Painlevé) limit is defined as follows. If

$$\mathrm{Limsup}_{k\to\infty}C^k = \mathrm{Liminf}_{k\to\infty}C^k,$$

this common (set) value is called the (full) (KURATOWSKI-PAINLEVÉ) LIMIT of C^k and denoted by $\lim_{k\to\infty}C^k$.

From now on, we often use the abbreviations li, ls, Li, and Ls for liminf, limsup, Liminf, and Limsup, resp.

Proposition 7.3 (Basic variational properties of epi-convergence)

(i) *If* $\varphi^k \overset{w-i-e}{\to} \varphi$, *then* $\mathrm{ls}_k(\inf_{A^k}\varphi^k) \le \inf_A\varphi$.
(ii) *If* $\varphi^k \overset{e}{\to} \varphi$, *then for any* $\varepsilon, \varepsilon^k \ge 0$ *and* $\varepsilon^k \to \varepsilon$, $\mathrm{Ls}_k(\varepsilon^k\text{-argmin}_{A^k}\varphi^k) \subset \varepsilon\text{-argmin}_A\varphi$.
(iii) *If* $\varphi^k \overset{w-e}{\to} \varphi$, $\varepsilon^k \to 0^+$, *and there exists* $x^k \in \varepsilon^k\text{-argmin}_{A^k}\varphi^k$ *converging to* x, *then* $x \in \mathrm{argmin}_A\varphi$ *and* $\lim_k(\inf_{A^k}\varphi^k) = \inf_A\varphi$.

Proof

(i) Assume that $\{x^k\}_k \subset A$ is an infimal sequence of φ (i.e., $\varphi(x^k) \to \inf_A\varphi$). By condition (b) of weak inside epi-convergence, for each k, there is $A^k_j \ni x^k_j \to x^k$ such that $\mathrm{ls}_j\varphi^j(x^k_j) \le \varphi(x^k)$. Then, for all k, $\mathrm{ls}_j(\inf_{A^j}\varphi^j) \le \mathrm{ls}_j\varphi^j(x^k_j) \le \varphi(x^k)$. Taking limit as $k \to +\infty$ of both sides yields $\mathrm{ls}_j(\inf_{A^j}\varphi^j) \le \inf_A\varphi$. (Note that $\inf_A\varphi$ may be $-\infty$ in this assertion.)
(ii) Assume that $x \in \mathrm{Ls}_k(\varepsilon^k\text{-argmin}_{A^k}\varphi^k)$. Then, there exists $x^{k_j} \in \varepsilon^{k_j}\text{-argmin}_{A^{k_j}}\varphi^{k_j}$ tending to x. So,

$$\mathrm{ls}_j\varphi^{k_j}(x^{k_j}) \le \mathrm{ls}_j(\inf_{A^{k_j}}\varphi^{k_j} + \varepsilon^{k_j}) \le \inf_A\varphi + \varepsilon < +\infty.$$

By condition (a) of epi-convergence, $x \in A$ and so $\varphi(x) \le \mathrm{li}_j\varphi^{k_j}(x^{k_j}) \le \inf_A\varphi + \varepsilon$. Therefore, $x \in \varepsilon\text{-argmin}_A\varphi$.
(iii) Let $x^k \in \varepsilon^k\text{-argmin}_{A^k}\varphi^k$ and $x^k \to x$. In view of part (i),

$$\mathrm{ls}_k\varphi^k(x^k) \le \mathrm{ls}_k(\inf_{A^k}\varphi^k + \varepsilon^k) \le \inf_A\varphi < +\infty.$$

Hence, $x \in A$ by condition (a) of the weak epi-convergence. Also, for any subsequence φ^{k_j},

$$\begin{aligned}\varphi(x) &\le \mathrm{li}_k\varphi^k(x^k) \le \mathrm{li}_j\varphi^{k_j}(x^{k_j}) \le \mathrm{ls}_j\varphi^{k_j}(x^{k_j}) \le \mathrm{ls}_k\varphi^k(x^k)\\ &\le \mathrm{ls}_k(\inf_{A^k}\varphi^k + \varepsilon^k) \le \inf_A\varphi\end{aligned}$$

(the last inequality follows from assertion (i)). Therefore, $x \in \mathrm{argmin}_A\varphi$ and

$$\varphi(x) \le \mathrm{li}_k\varphi^k(x^k) \le \mathrm{li}_k(\inf_{A^k}\varphi^k + \varepsilon^k) \le \inf_A\varphi,$$

i.e., $\mathrm{li}_k \inf_{A^k} \varphi^k = \inf_A \varphi$. From the above reasoning, clearly $\lim_k(\inf_{A^k} \varphi^k) = \inf_A \varphi$.

□

Example 7.4 (Failure of some types of epi-convergence, while other types hold)

(a) (epi-convergence fails, the three weaker types hold) Let A^k be equal to $\{0, 1\}$ for the odd k's and to $\{0\}$ for the even k's, $A = \{0\}$, $\varphi^k(x) = 0$ for all $k \in \mathbb{N}$ and $x \in A^k$, and $\varphi(x) = 0$ for $x \in A$; see Fig. 7.1. Then, φ^k does not epi-converge to φ (see Exercise 7.1). But the three other types of epi-convergence hold (see Exercise 7.2). We also have that for any $\varepsilon^k \to 0^+$, $\mathrm{Ls}_k(\varepsilon^k\text{-argmin}_{A^k} \varphi^k)$ is not contained in $\mathrm{argmin}_A \varphi$ because $\mathrm{argmin}_A \varphi = \{0\}$, but (see Exercise 7.3)

$$\varepsilon^k\text{–}\mathrm{argmin}_{A^k} \varphi^k = \begin{cases} \{0, 1\} & \text{for odd } k's, \\ \{0\} & \text{for even } k's. \end{cases}$$

This example illustrates Proposition 7.3(i),(iii) for the two types that are weaker than epi-convergence. This is clear from direct computations: $\lim_k(\inf_{A^k} \varphi^k) = 0 = \inf_A \varphi$; for $\varepsilon^k \equiv 0$, $0 \in \varepsilon^k\text{-argmin}_{A^k} \varphi^k$ tends to $0 \in \mathrm{argmin}_A \varphi$.

(b) (weak epi-convergence fails, the inside types take place) To see the reasonableness of the inside convergence, we consider the following functions. Let $\alpha \in (0, 2)$ and $\varphi : [\alpha, 2] \to \mathbb{R}$, $\varphi^k : [0, 2] \to \mathbb{R}$ be defined by $\varphi(x) \equiv 0$ and

$$\varphi^k(x) = \begin{cases} -k^2 x + k & \text{for } 0 \le x \le k^{-1}, \\ (2k-1)^{-1}(x - k^{-1}) & \text{for } k^{-1} \le x \le 2. \end{cases}$$

Then, φ^k does not weakly epi-converge to φ since for $\beta \in (0, \alpha)$ (outside the domain of φ) and $A^k \ni x^k \to \beta$, $\lim_k \varphi^k(x^k) = 0$, not $+\infty$ as condition (a) of weak epi-convergence requires. However, φ^k inside epi-converge since for every $A^{k_j} \ni x^{k_j} \to x \in [\alpha, 2]$, $\lim_j \varphi^{k_j}(x^{k_j}) = 0 = \varphi(x)$; see Fig. 7.2. This example shows a case when assertion (i) of Proposition 7.3 really holds for a weak inside epi-convergent sequence: $\lim_k(\inf_{A^k} \varphi^k) = 0 = \inf_A \varphi$.

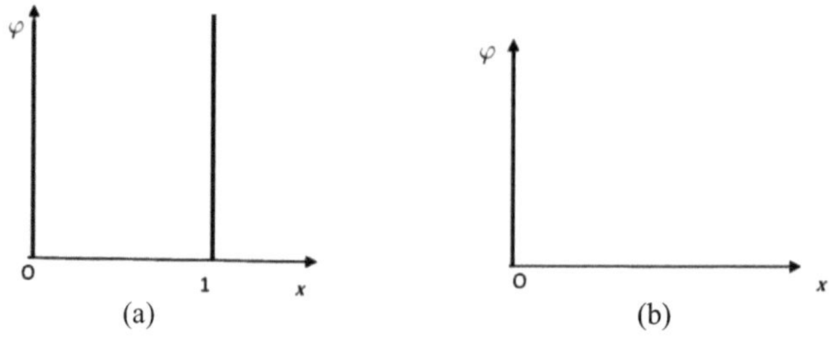

Fig. 7.1 φ^k i-e-, w-e-, w-i-e-converges to φ, but the epi-convergence fails. (**a**) $\mathrm{epi}\varphi^k$ for odd k. (**b**) $\mathrm{epi}\varphi^k = \mathrm{epi}\varphi$ for even k

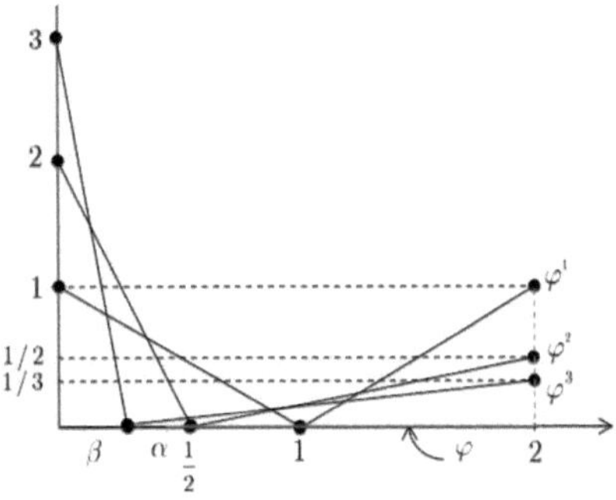

Fig. 7.2 φ^k inside epi-converges but not weakly epi-converges

Epi-convergence has a geometric reformulation in terms of epigraphs as follows.

Definition 7.5 (Epigraph)

(a) EPIGRAPH of $\varphi : X \to \mathbb{R} \cup \{+\infty\}$ is the following set:

$$\text{epi}\varphi := \{(x, r) \in X \times \mathbb{R} \mid r \geq \varphi(x)\}.$$

(b) For $\varphi \in$ fv-fcn(X) (namely, $\varphi : A \to \mathbb{R}$), define $\zeta\varphi : X \to \mathbb{R} \cup \{+\infty\}$ by

$$\zeta\varphi(x) := \begin{cases} \varphi(x) \text{ if } x \in A, \\ +\infty \text{ otherwise} \end{cases}$$

and regard point $(x, +\infty)$ as belonging to epi$\zeta\varphi$ when $x \notin A$. Then, define epi$\varphi :=$ epi$\zeta\varphi$.

Proposition 7.6 (Equivalence between the variational and geometric formulations of epi-convergence) *Condition* (a) *in Definition 7.1 means that* $\text{Ls}_k\text{epi}\varphi^k \subset \text{epi}\varphi$ *and condition* (b) *is just* $\text{Li}_k\text{epi}\varphi^k \supset \text{epi}\varphi$. *Therefore,* $\varphi^k \overset{e}{\to} \varphi$ *if and only if* $\text{epi}\varphi^k \overset{P-K}{\to} \text{epi}\varphi$, *i.e.,* $\lim_k \text{epi}\varphi^k = \text{epi}\varphi$.

Proof From $\text{Ls}_k\text{epi}\varphi^k \subset \text{epi}\varphi$, for any $x^{k_j} \in A^{k_j}$ with $(x^{k_j}, \varphi^{k_j}(x^{k_j})) \to (x, \alpha)$, either $x \in A$ and $\alpha \geq \varphi(x)$ (in particular $x \in A$ and $\text{li}_j\varphi^{k_j}(x^{k_j}) \geq \varphi(x)$) or $x \notin A$ and $\alpha = +\infty$. That is, condition (a) is satisfied. Conversely, by condition (a), $A^{k_j} \ni x^{k_j} \to x \in A$ and $\lim_j \varphi^{k_j}(x^{k_j}) = \alpha$ imply $\alpha \geq \varphi(x)$. If $(x^{k_j}, \alpha^{k_j}) \in \text{epi}\varphi^{k_j}$ tends to (x, α'), then $\alpha' \geq \alpha \geq \varphi(x)$ when $x \in A$ and (x^{k_j}, α^{k_j}) tends to $(x, +\infty)$ when $x \notin A$. So, $\text{Ls}_k\text{epi}\varphi^k \subset \text{epi}\varphi$. Now consider condition (b). $\text{Epi}\varphi \subset \text{Li}_k\text{epi}\varphi^k$ means that $\forall(x, \alpha) \in \text{epi}\varphi$, $\exists(x^k, \alpha^k) \in \text{epi}\varphi^k$, which tends to (x, α). Particularly, for $(x, \varphi(x))$, there exists $(x^k, \alpha^k) \to (x, \varphi(x))$ with $\varphi^k(x^k) \leq \alpha^k$ for all k. Then, $\text{ls}_k\varphi^k(x^k) \leq \text{ls}_k\alpha^k = \varphi(x)$. Conversely, for all $(x, \alpha) \in \text{epi}\varphi$ with $x \in A$, condition (b) entails that $\alpha \geq \varphi(x) \geq \text{ls}_k\varphi^k(x^k)$ for some $A^k \ni x^k \to x$. We take $\alpha^k \geq \varphi^k(x^k)$ such that $\alpha^k \to \alpha$. Then, $(x^k, \alpha^k) \in \text{epi}\varphi^k$ tends to (x, α). If $(x, \alpha) \in \text{epi}\varphi = \text{epi}\zeta\varphi$ with $x \notin A$, then $\alpha = +\infty$. We take arbitrarily $(x^k, +\infty) \in \text{epi}\zeta\varphi^k = \text{epi}\varphi^k$ tending to (x, α). So, $\text{epi}\varphi \subset \text{Li}_k\text{epi}\varphi^k$. □

We see that epi-convergence is characterized by the convergence of epigraphs in the Painlevé-Kuratowski sense. (From this, in particular, epi-limits are always lower semicontinuous, i.e., their epigraphs are closed.) This geometric explanation of the convergence is crucial for many research topics in optimization, especially in studies of optimality conditions and related areas. But, for the preceding parts of our book, epigraphs are not used, while they are important for the purpose of variational approximations. However, the three weaker types of convergence are also sufficient in many results, although none of them guarantees the convergence of epigraphs, i.e., these types of convergence are properly weaker than epi-convergence as shown by Example 7.4.

To ensure variational properties stronger than those given in Proposition 7.3, the following relaxed uniform compactness condition is needed.

Definition 7.7 (Tightness of the types of epi-convergence) We say that $\{\varphi^k\}_k$ EPI-CONVERGES (I-EPI- or W-EPI- or W-I-EPI-CONVERGES, resp.) TIGHTLY to φ, if φ^k epi-converges (i-epi- or w-epi- or w-i-epi-converges, resp.) to φ and for each $\varepsilon > 0$, there exist a compact subset B_ε and an upper index k_ε such that for all $k \geq k_\varepsilon$,

$$\inf_{B_\varepsilon \cap A^k} \varphi^k \leq \inf_{A^k} \varphi^k + \varepsilon.$$

Definition 7.8 (Lower semicontinuous regularization) Let X be a metric space, $\xi : X \to \mathbb{R} \cup \{+\infty\}$. The LOWER SEMICONTINUOUS REGULARIZATION $\mathrm{lsc}\xi$ (or the LOWER CLOSURE $\mathrm{cl}\xi$) of ξ is defined by the formula

$$\mathrm{lsc}\xi(x) := \begin{cases} \mathrm{li}_{\mathrm{dom}\xi \ni y^k \to y} \xi(y^k) & \text{if } x \in \mathrm{dom}\xi, \\ +\infty & \text{otherwise.} \end{cases}$$

Proposition 7.9 (Variational properties under tight convergence)

(i) *If φ^k epi-converges to φ tightly, then* $\inf_A \varphi > -\infty$ *and* $\inf_{A^k} \varphi^k \to \inf_A \varphi$. *If φ^k weakly epi-converges or epi-converges to φ,* $\inf_A \varphi > -\infty$, *and* $\inf_{A^k} \varphi^k \to \inf_A \varphi$, *then the convergence is tight.*

(ii) *If $\varphi^k \xrightarrow{e} \varphi$ tightly, then for all $\varepsilon > 0$,* $\mathrm{argmin}_A \varphi \subset \mathrm{Li}_k(\varepsilon\text{-}\mathrm{argmin}_{A^k} \varphi^k)$.

(iii) *Let φ^k epi-converge to φ. If there exists $\varepsilon^k \to 0^+$ such that* ε^k-$\mathrm{argmin}_{A^k} \varphi^k$ *converges in the Painlevé-Kuratowski sense to* $\mathrm{argmin}_A \varphi$, *then the epi-convergence is tight. Conversely, if $\varphi^k \xrightarrow{e} \varphi$ tightly, then there exists $\varepsilon^k \to 0^+$ such that* $\mathrm{argmin}_A \varphi = \mathrm{Lim}_k(\varepsilon^k\text{-}\mathrm{argmin}_{A^k} \varphi^k)$.

Proof

(i) To prove the first assertion, assume that φ^k epi-converges to φ tightly. First, suppose that $\inf_A \varphi = -\infty$. Then, (i) of Proposition 7.3 implies that $\inf_{A^k} \varphi^k \to -\infty$. By the tightness, there exists a compact subset $B \subset X$ such that $\inf_{B \cap A^k} \varphi^k \to -\infty$ and hence also $x^k \in B \cap A^k$ such that $\varphi^k(x^k) \to -\infty$. The compactness of B implies that $\lim_j x^{k_j} = x$ for some subsequence $\{x^{k_j}\}_j$ and some $x \in B$. In view of condition (a) of epi-convergence, $\mathrm{li}_j \varphi^{k_j}(x^{k_j}) \geq$

$\varphi(x) \in \mathbb{R}$ or $\varphi^{k_j}(x^{k_j}) \to +\infty$, each of which contradicts $\varphi^k(x^k) \to -\infty$. Hence, $\inf_A \varphi > -\infty$.

We show that for any $\varepsilon > 0$ and B_ε given by the assumed tightness, $\{\inf_{B_\varepsilon \cap A^k} \varphi^k \mid k \in \mathbb{N}\}$, except possibly a finite number of k, is bounded from $-\infty$, by contradiction as well. Suppose for any $r > 0$, one finds a subsequence $\{k_j\}_j$ such that $\inf_{B_\varepsilon \cap A^{k_j}} \varphi^{k_j} < -r$ for all j. Then, there exists $x^{k_j} \in B_\varepsilon \cap A^{k_j}$ such that $\varphi^{k_j}(x^{k_j}) < -r$ for all j. As B_ε is compact, there exists a cluster point $\bar{x}$ of $\{x^{k_j}\}_j$. As $\varphi^{k_j}(x^{k_j})$ cannot tend to $+\infty$, by virtue of condition (a) of epi-convergence, $\bar{x}$ must be in A and $\mathrm{li}_j \varphi^{k_j}(x^{k_j}) \geq \varphi(\bar{x}) \geq \inf_A \varphi$, which is a contradiction. Hence, the boundedness from $-\infty$ of $\{\inf_{B_\varepsilon \cap A^k} \varphi^k \mid k \in \mathbb{N}\}$ is verified. This boundedness yields $x^k \in \varepsilon\text{-argmin}_{B_\varepsilon \cap A^k} \varphi^k$ for all k. Let $\bar{x}$ be a cluster point in B_ε of $\{x^k\}_k$. Again by condition (a) of epi-convergence, $\varphi(\bar{x}) \leq \mathrm{li}_k \varphi(x^k)$. Hence,

$$\inf_A \varphi - \varepsilon \leq \varphi(\bar{x}) - \varepsilon \leq \mathrm{li}_k \varphi(x^k) - \varepsilon \leq \mathrm{li}_k \inf_{B_{\varepsilon \cap A^k}} \varphi^k.$$

On the other hand, by Proposition 7.3(i),

$$\mathrm{li}_k \inf_{B_\varepsilon \cap A^k} \varphi^k \leq \mathrm{li}_k \inf_{A^k} \varphi^k + \varepsilon \leq \mathrm{ls}_k \inf_{A^k} \varphi^k + \varepsilon \leq \inf_A \varphi + \varepsilon \leq +\infty.$$

The arbitrariness of ε ensures that equalities hold throughout the above two chains of inequalities and so $\inf_{A^k} \varphi^k \to \inf_A \varphi$.
Now we verify the second assertion only for the case of weak epi-convergence since the other case is similar (as condition (b) of the convergence is common). With a given $\varepsilon > 0$, there exists $x \in A$ such that $\varphi(x) \leq \inf_A \varphi + \varepsilon/3$, and by assumption, $\inf_A \varphi \leq \inf_{A^k} \varphi^k + \varepsilon/3$ for large k. By condition (b) of weak epi-convergence, there exists $A^k \ni x^k \to x$ such that $\varphi^k(x^k) \leq \varphi(x) + \varepsilon/3$ for large k. Moreover, one can find a compact subset $B_\varepsilon \subset X$ containing $\{x^k\}_k$ (for instance, B_ε is the same set $\{x^k\}_k$). Hence, for k large enough,

$$\inf_{B_\varepsilon \cap A^k} \varphi^k \leq \varphi^k(x^k) \leq \inf_{A^k} \varphi^k + \varepsilon.$$

Therefore, the weak epi-convergence is tight.

(ii) In view of condition (b) of epi-convergence, for $x \in \mathrm{argmin}\varphi$, there exists $A^k \ni x^k \to x$ such that $\mathrm{ls}_k \varphi^k(x^k) \leq \varphi(x)$ and so one can find k_1 such that $\varphi^k(x^k) \leq \varphi(x) + \varepsilon/2$ for all $k \geq k_1$. The above assertion (i) gives $k_2 \geq k_1$ satisfying $\inf_A \varphi \leq \inf_{A^k} \varphi^k + \varepsilon/2$ for all $k \geq k_2$. So, for all $k \geq k_2$,

$$\varphi^k(x^k) \leq \varphi(x) + \varepsilon/2 = \inf_A \varphi + \varepsilon/2 \leq \inf_{A^k} \varphi^k + \varepsilon,$$

i.e., $x \in \mathrm{Li}_k(\varepsilon\text{-argmin}_{A^k} \varphi^k)$.

(iii) For $x \in \text{argmin}_A\varphi$, by assumption, there exists $x^k \in \varepsilon^k\text{-argmin}_{A^k}\varphi^k$ tending to x. Condition (a) of epi-convergence implies that

$$\inf_A\varphi = \varphi(x) \le \text{li}_k\varphi(x^k) \le \text{li}_k(\inf_{A^k}\varphi^k + \varepsilon^k) = \text{li}_k(\inf_{A^k}\varphi^k).$$

This in conjunction with (i) of Proposition 7.3 ensures that $\inf_{A^k}\varphi^k \to \inf_A\varphi$. Consequently, assertion (i) yields the tightness of the epi-convergence.

To see the converse, let $x \in \text{argmin}\varphi$ and $\varepsilon^j \to 0^+$. By (ii), there exists $x_j^k \in \varepsilon^j\text{-argmin}_{A^k}\varphi^k$ tending to x for each fixed j. Applying the diagonalization argument, we get $x_{j(k)}^k \in \varepsilon^{j(k)}\text{-argmin}_{A^k}\varphi^k$ tending to x. We set $x^k := x_{j(k)}^k$ and $\varepsilon^k := \varepsilon^{j(k)}$ to have $x^k \in \varepsilon^k\text{-argmin}_{A^k}\varphi^k$ and so $x \in \text{Li}_k(\varepsilon^k\text{-argmin}_{A^k}\varphi^k)$. This and Proposition 7.3(ii) imply that $\text{argmin}\varphi = \text{Lim}_k(\varepsilon^k\text{-argmin}_{A^k}\varphi^k)$. □

We see from the above results that the main construction feature of variational convergence to guarantee preservation of variational properties is that they involve function-variable relationships. However, there are also other types of convergence in mathematical analysis, which involve such relationships. We clarify differences between epi-convergence and some typical ones among such types of convergence. This sheds light onto core properties not only of variational convergence but also of convergence notions in general, while these notions are both conceptional and basic technical tools in continuous mathematics. First, we present other equivalent formulations of the types of epi-convergence complementing the one given in Proposition 7.6. These formulations and also relationships between the aforementioned types of convergence are based on the following definition.

Definition 7.10 (Types of lower and upper epi-limits) Let $\varphi^k \in \text{fv-fcn}X$ and $x \in X$.

(a) The LOWER EPI-LIMIT of $\{\varphi^k\}_k$ at x is $eli\varphi^k(x) := \inf_{\{A^{k_j} \ni x^{k_j} \to x\}} \text{li}_j\varphi^{k_j}(x^{k_j})$, where the notation $\{A^{k_j} \ni x^{k_j} \to x\}$ means that we consider all possible subsequences $\{k_j\}_j$ such that $x^{k_j} \to x$. The UPPER EPI-LIMIT of $\{\varphi^k\}_k$ at x is $els\varphi^k(x) := \inf_{\{A^{k_j} \ni x^{k_j} \to x\}} \text{ls}_j\varphi^{k_j}(x^{k_j})$.

(b) When considering the convergence of φ^k to $\varphi : A \subset X \to \mathbb{R}$, the two functions defined as above but restricted to $x \in A$ instead of $x \in X$, we obtain notions weaker than the ones in (a) with the additional term "inside" in the names and the prefix "i-" in the notation.

(c) If subsequence $\{A^{k_j} \ni x^{k_j} \to x\}$ in (a) and (b) is replaced by the whole sequence $\{A^k \ni x^k \to x\}$ (and correspondingly li_j, ls_j by li_k, ls_k), then we have the corresponding weak types of convergence with the prefixes "w-" or "w-i" suitably.

Observe from the above definition that $\text{eli}\varphi^k$ and $\text{els}\varphi^k$ are the functions associated with a given sequence $\{\varphi^k\}_k$. Their domains are obtained from the formulas for their values and are incomparable, when we are interested in a given φ (cf. Example 7.11).

Example 7.11 (Types of lower and upper epi-limits) We revisit the functions φ^k and φ in Example 7.4(a) and compute its lower and upper epi-limits:

$$\mathrm{eli}\varphi^k(0) = \inf_{\{A^{k_j} \ni x^{k_j} \to 0\}} \mathrm{li}_j \varphi^{k_j}(x^{k_j}) = 0,$$

$$\mathrm{els}\varphi^k(0) = \inf_{\{A^{k_j} \ni x^{k_j} \to 0\}} \mathrm{ls}_j \varphi^{k_j}(x^{k_j}) = 0,$$

$$\mathrm{eli}\varphi^k(1) = \inf_{\{A^{k_j} \ni x^{k_j} \to 1\}} \mathrm{li}_j \varphi^{k_j}(x^{k_j}) = 0,$$

$$\mathrm{els}\varphi^k(1) = \inf_{\{A^{k_j} \ni x^{k_j} \to 1\}} \mathrm{ls}_j \varphi^{k_j}(x^{k_j}) = 0.$$

Moreover, w-$\mathrm{eli}\varphi^k(0) = \inf_{\{A^{k_j} \ni x^{k_j} \to 0\}} \mathrm{li}_k \varphi^k(x^k) = +\infty$. Using the same convention that the infimum over an empty set of any numerical function is $+\infty$, we have w-$\mathrm{els}\varphi^k(0) =$ w-$\mathrm{eli}\varphi^k(1) =$ w-$\mathrm{els}\varphi^k(1) = +\infty$. For all $x \in \mathbb{R} \setminus \{0, 1\}$, the set $\{A^{k_j} \ni x^{k_j} \to x\}$ is empty, and hence all the values corresponding to the above ones with 0 or 1 replaced by x are $+\infty$.

The types of lower and upper epi-limits are used in particular to characterize the corresponding types of epi-limits in Proposition 7.13.

Lemma 7.12 (Nonstandard diagonalization) *Let $\{a^{k,j} \mid k, j \in \mathbb{N}\}$ be a two-index family in $\overline{\mathbb{R}} := [-\infty, +\infty]$. Then, there exists a map $j \mapsto j(k)$ such that*

$$\mathrm{ls}_k a^{k,j(k)} \le \mathrm{ls}_j(\mathrm{ls}_k a^{k,j}).$$

If interested in the proof of this auxiliary result, the reader is referred to Corollary A3 in [78].

Proposition 7.13 (Characterizations of the types of epi-convergence)

(i) $\mathrm{eli}\varphi^k(x) = \min_{\{A^{k_j} \ni x^{k_j} \to x\}} \mathrm{li}_j \varphi^{k_j}(x^{k_j})$ *and* $\mathrm{els}\varphi^k(x) = \min_{\{A^{k_j} \ni x^{k_j} \to x\}} \mathrm{ls}_j \varphi^{k_j}(x^{k_j})$ *for $x \in X$. The corresponding characterizations of the other three types of epi-convergence can be similarly stated, replacing* inf *in the definitions by* min.

(ii) *Condition* (a) *of epi-convergence means that* $\mathrm{els}\varphi^k(x) \le \varphi(x)$ *and* (b) *is* $\mathrm{eli}\varphi^k(x) \ge \varphi(x)$. *For the weaker three types of epi-convergence, we have similar equivalences.*

(iii) $\varphi^k \xrightarrow{e} \varphi$ *if and only if* $\mathrm{eli}\varphi^k(x) = \mathrm{els}\varphi^k(x) = \varphi(x)$ *for $x \in A$ and* $\mathrm{eli}\varphi^k(x) = +\infty$ *for $x \notin A$. The corresponding characterizations of the other three types of epi-convergence are similarly formulated.* $\mathrm{eli}\varphi^k$ *and* $\mathrm{els}\varphi^k$ *have the geometric reformulation:* $\mathrm{epi}(\mathrm{eli}\varphi^k) = \mathrm{Ls}_k(\mathrm{epi}\varphi^k)$ *and* $\mathrm{epi}(\mathrm{els}\varphi^k) = \mathrm{Li}_k(\mathrm{epi}\varphi^k)$.

(iv) $\varphi^k \xrightarrow{e} \varphi$ *if and only if* $\mathrm{epi}\varphi^k \to \mathrm{epi}\varphi$ *in the sense of Painlevé-Kuratowski.*

Proof

(i) By similarity, we only prove the second formula for weak epi-convergence. If w-$\mathrm{els}\varphi^k(x) < +\infty$, then for any positive γ^j, there exists $x_j^k \in A^k$

tending to x such that $\text{ls}_k\varphi^k(x_j^k) \leq \text{w-els}\varphi^k(x) + \gamma^j$ (including the case w-els$\varphi^k(x) = -\infty$). Letting $\gamma^j \to 0^+$, one has $\text{ls}_j\text{ls}_k\varphi^k(x_j^k) \leq \text{w-els}\varphi^k(x)$. By Lemma 7.12, for each k, there is $j(k)$ such that $\text{ls}_k\varphi^k(x_{j(k)}^k) \leq \text{w-els}\varphi^k(x)$. Of course, $\{x_{j(k)}^k\}_k$ is a minimizer for the expression defining w-els$\varphi^k(x)$. If w-els$\varphi^k(x) = +\infty$, any sequence $\{x^k \in A^k\}_k$ tending to x is a minimizer.

(ii) This assertion is evident from the definitions.

(iii) We only discuss the required epi-convergence. The first equivalence is straightforward from (i) and Definition 7.1 because one always has eli$\varphi^k \leq$ elsφ^k (this relation is understood pointwise in the considered domain). For the second equivalence (geometric reformulation), we only prove the first equality. It follows from (a) that (x, α) lying in the left-hand side means the existence of $A^{k_j} \ni x^{k_j} \to x$ such that $\alpha \geq \text{eli}\varphi^k(x) = \text{li}_j\varphi^{k_j}(x^{k_j}) = \lim_j\varphi^{k_j}(x^{k_j})$ for some subsequence $\{k_j\}_j$. This is equivalent to the fact that, for any $\varepsilon^{k_j} > 0$ such that $\alpha^{k_j} := \varphi^{k_j}(x^{k_j}) + \varepsilon^{k_j} \to \alpha$, we have $(x^{k_j}, \alpha^{k_j}) \in \text{epi}\varphi^{k_j}$, and it tends to (x, α). The last sentence is nothing else but (x, α) belonging to $\text{Ls}_k(\text{epi}\varphi^k)$.

(iv) This equivalence is already proved in Proposition 7.6. Here, it follows immediately from (ii): by the first equivalence in (ii), $\varphi^k \xrightarrow{e} \varphi$ if and only if epi(eli φ^k) = epi(els φ^k) = epiφ when considered on A and eli$\varphi^k(x)$ = els$\varphi^k(x) = +\infty$ for $x \notin A$. By the second equivalence, this is $\text{Ls}_k(\text{epi}\varphi^k) = \text{Li}_k(\text{epi}\varphi^k) = \text{epi}\varphi$.

□

Example 7.14 (Applications of characterizations of epi-convergence) We apply the characterizations given in Proposition 7.13 to check the types of epi-convergence of the functions given in Examples 7.4(a) and 7.11 to regain the conclusion already obtained in Example 7.4(a) (by Definitions 7.1 and 7.2 and a geometric argument). As for $x = 1 \notin A$, eli$\varphi^k(1) = 0$, not $+\infty$, the epi-convergence does not hold. But we have the inside epi-convergence. Also, since all the weak lower and upper epi-limits (at 0 and 1) are $+\infty$, the weak epi-convergence is verified and of course so is the weak inside epi-convergence.

We need also the following notion, which is symmetric to the types of epi-convergence (and largely concerning maximization). A sequence $\{\varphi^k\}_k$ is said to HYPO-CONVERGE (INSIDE HYPO-CONVERGE, WEAK HYPO-CONVERGE, AND WEAK INSIDE HYPO-CONVERGE, resp.) to φ, denoted by $\varphi^k \xrightarrow{h} \varphi$ or $\varphi = \text{h-lim}_k\varphi^k$ (for the other types, use i-h, w-h, and w-i-h, resp., instead of h), if $-\varphi^k$ epi-converges (converges in the other senses, resp.) to $-\varphi$. Similar to epi-convergence, $\varphi^k \xrightarrow{h} \varphi$ if and only if hypo$\varphi^k \xrightarrow{P-K}$ hypoφ, but the other (weaker) types of convergence do not possess geometric reformulations. Here, hypoφ is the hypograph of φ defined symmetrically as in Definition 7.5:

Definition 7.15 (Hypograph)

(a) HYPOGRAPH of $\varphi : X \to \mathbb{R} \cup \{-\infty\}$ is the following set:

$$\text{hypo}\varphi := \{(x, r) \in X \times \mathbb{R} \mid r \leq \varphi(x)\}.$$

(b) For $\varphi \in \text{fv-fcn}(X)$, define $\zeta\varphi : X \to \mathbb{R} \cup \{-\infty\}$ by

$$\zeta\varphi(x) := \begin{cases} \varphi(x) & \text{if } x \in A, \\ -\infty & \text{otherwise,} \end{cases}$$

and regard a point $(x, -\infty)$ as belonging to epi$\zeta\varphi$ when $x \notin A$. Then, define hypo$\zeta\varphi$ =: hypoφ.

The definition of the tightness concept for the types of hypo-convergence is symmetrically induced by the ones in Definition 7.7. Similarly, for the types of hypo-convergence, we clearly have the statements symmetric to the above ones for epi-convergence. We only show explicitly the formulas for the lower and upper hypo-limits for our convenient use in what follows:

$$\text{hli}\varphi^k(x) := \max_{\{A^{k_j} \ni x^{k_j} \to x\}} \text{li}_j \varphi^{k_j}(x^{k_j}), \ \ \text{hls}\varphi^k(x) = \max_{\{A^{k_j} \ni x^{k_j} \to x\}} \text{ls}_j \varphi^{k_j}(x^{k_j}).$$

We provide a geometric interpretation of the lower and upper epi-limits, which is clearer than that in Proposition 7.13(ii). For this, the following lemma about the Painlevé-Kuratowski set limits is needed.

Lemma 7.16 (Geometry of set convergence, [74], Definition 1.31 and Proposition 1.34) *For a sequence of sets C^k in a metric space X, one has the formulas for its lower and upper limits:*

$$\text{Li}_k C^k = \text{cl} \cup_{k_0 \in \mathbb{N}} \cap_{k \geq k_0} \text{cl} C^k,$$

$$\text{Ls}_k C^k = \cap_{k_0 \in \mathbb{N}} \text{cl} \cup_{k \geq k_0} C^k.$$

Proposition 7.17 (Geometric formulas for lower and upper epi-limits) *Let $x \in X$.*

(a) $\text{epi}(\text{eli}\varphi^k) = \cap_{k_0 \in \mathbb{N}} \text{cl} \cup_{k \geq k_0} \text{epi}\varphi^k$.
(b) $\text{epi}(\text{els}\varphi^k) = \text{cl} \cup_{k_0 \in \mathbb{N}} \cap_{k \geq k_0} \text{cl}(\text{epi}\varphi^k)$.

Proof This is a direct consequence of Proposition 7.13(ii) and Lemma 7.16. (Here, epiφ^k is understood in the sense given in Definition 7.5(b) for functions in fv-fcn(X).) □

To have "more symmetric" and analogous formulas for lower and upper limits, we use the following notation. The set $\mathcal{N}_\infty$ is the so-called FILTER of the neighborhoods of $+\infty$ that are implicit in the notation $k \to +\infty$. We can view

$\mathscr{N}_\infty$ as the collection of all tails of the sequence $k \to +\infty$, i.e., the collection of the sets $\{k\}_{k\in N}$ with $N \in \mathscr{N}_\infty$. The set $\mathscr{N}_\infty^\sharp$ is its associated GRILL, which can be roughly understood as the collection of all the subsequences of $k \to +\infty$, i.e., the collection of the sets $\{k\}_{k\in N}$ with $N \in \mathscr{N}_\infty^\sharp$. Then, we have a natural duality between $\mathscr{N}_\infty$ and $\mathscr{N}_\infty^\sharp$

$$\mathscr{N}_\infty^\sharp = \{N \subset \mathbb{N} \mid \forall N' \in \mathscr{N}_\infty, N \cap N' \neq \emptyset\},$$

$$\mathscr{N}_\infty = \{N \subset \mathbb{N} \mid \forall N' \in \mathscr{N}_\infty^\sharp, N \cap N' \neq \emptyset\}.$$

Lemma 7.18 (Another geometry of the set convergence, [74], Proposition 1.32) *For a sequence of sets C^k in a metric space X,*

$$\mathrm{Li}_k C^k = \cap_{N\in\mathscr{N}_\infty^\sharp} \mathrm{cl} \cup_{k\in N} C^k,$$

$$\mathrm{Ls}_k C^k = \cap_{N\in\mathscr{N}_\infty} \mathrm{cl} \cup_{k\in N} C^k.$$

Proposition 7.13(ii) and Lemma 7.18 yield the following.

Proposition 7.19 (Symmetric geometric formulas for lower and upper epi-limits) *Let $x \in X$.*

(a) $\mathrm{epi}(\mathrm{eli}\varphi^k) = \cap_{N\in\mathscr{N}_\infty} \mathrm{cl} \cup_{k\in N} \mathrm{epi}\varphi^k$.
(b) $\mathrm{epi}(\mathrm{els}\varphi^k) = \cap_{N\in\mathscr{N}_\infty^\sharp} \mathrm{cl} \cup_{k\in N} \mathrm{epi}\varphi^k$.

7.1.2 Comparison of Epi-Convergence with Other Types of Convergence

We now discuss relationships between epi- and hypo-convergence and the following classical concepts of convergence. They show that epi-convergence can be expected to be a useful notion.

Definition 7.20 (Classical types of convergence) Let $\varphi^k : A^k \to \mathbb{R}$ and $\varphi : A \to \mathbb{R}$.

(a) When $A^k \equiv A$, $\{\varphi^k\}_k$ is called POINTWISE CONVERGENT to φ, written $\varphi^k \xrightarrow{p} \varphi$, if $\varphi^k(x) \to \varphi(x)$ for all $x \in A$.
(b) $\{\varphi^k\}_k$ is said to CONVERGE GRAPHICALLY to φ, written $\varphi^k \xrightarrow{g} \varphi$, if we have the convergence of the graphs: $\mathrm{gph}\varphi^k \xrightarrow{P-K} \mathrm{gph}\varphi$.
(c) $\{\varphi^k\}_k$ is said to CONVERGE CONTINUOUSLY to φ relative to $A^k \xrightarrow{P-K} A$, written $\varphi^k \xrightarrow{c} \varphi$, if for any $A^k \ni x^k \to x$, $\varphi^k(x^k) \to \varphi(x)$.

(d) When $A^k \equiv A$, $\{\varphi^k\}_k$ is called UNIFORMLY CONVERGENT to φ if, for any positive ε and compact subset $D \subset A$, there is a subscript $k_0(\varepsilon)$ such that $|\varphi^k(x) - \varphi(x)| \le \varepsilon$ for all $x \in D$ and $k \ge k_0(\varepsilon)$.
(e) When $A^k \equiv A$, φ^k is called ALMOST EQUICONTINUOUS if, for any $x \in A$ and positive ε, there exists an upper index $k_0(x, \varepsilon)$ such that $|\varphi^k(y) - \varphi^k(x)| \le \varepsilon$ for all $y \in B(x, \varepsilon) \cap A$ and $k \ge k_0(x, \varepsilon)$.

Proposition 7.21 (Relationships between types of convergence) *Let $A^k, A \subset X$ be nonempty, $\varphi^k : A^k \to \mathbb{R}$, and $\varphi : A \to \mathbb{R}$.*

(a) *The following assertions are equivalent:*
  (a_1) *φ^k converge graphically to φ.*
  (a_2) *φ^k both epi- and hypo-converge to φ.*
  (a_3) *φ^k both w-epi- and w-hypo-converge to φ.*
  (a_4) *φ^k converge continuously to φ relative to $A^k \overset{P-K}{\longrightarrow} A$.*
  (a_5) $\text{hls}\varphi^k \le \varphi \le \text{eli}\varphi^k$.
  (a_6) $\text{w-hls}\varphi^k \le \varphi \le \text{w-eli}\varphi^k$.
(b) *If $A^k \equiv A$, then each of the following is equivalent to the ones in* (a)*:*
  (b_1) *φ^k are uniformly convergent to φ and φ is continuous.*
  (b_2) *φ^k are almost equicontinuous and pointwise convergent to φ.*

Proof We demonstrate (a) in the following steps:

(a_1) $\Rightarrow$ (a_2). Let $\varphi^k \overset{g}{\longrightarrow} \varphi$. Then, clearly $A^k \overset{P-K}{\to} A$. To have condition (a) of epi-convergence, suppose to the contrary the existence of $A^k \ni x^k \to x$ with $\text{li}_k\varphi^k(x^k) = \beta < \varphi(x)$. Let x^{k_j} satisfy $\lim_j \varphi^{k_j}(x^{k_j}) = \beta$. Then, $(x^{k_j}, \varphi^{k_j}(x^{k_j})) \in \text{gph}\varphi^{k_j}$ tends to $(x, \beta) \notin \text{gph}\varphi$, contradicting the assumed graphical convergence. Condition (b) of epi-convergence is clear as for any $x \in A$, there exists $(x^k, \varphi^k(x^k)) \to (x, \varphi(x))$ by the graphic convergence. For hypo-convergence, as $-\varphi^k \overset{g}{\longrightarrow} -\varphi$, one has

$$\text{hypo}\varphi^k = -\text{epi}(-\varphi^k) \overset{P-K}{\longrightarrow} -\text{epi}(-\varphi) = \text{hypo}\varphi.$$

(a_2) $\Rightarrow$ (a_3). This is evident.

(a_3) $\Rightarrow$ (a_4). This is a straightforward consequence of the definitions. (Note that from the simultaneous w-epi- and w-hypo-convergence of φ^k to φ, it follows that $A^k \overset{P-K}{\longrightarrow} A$.)

(a_4) $\Rightarrow$ (a_5). The continuous convergence implies that $\text{eli}\varphi^k = \text{els}\varphi^k = \varphi = \text{hli}\varphi^k = \text{hls}\varphi^k$. Hence, the assertion about inequalities is clear.

(a_5) $\Rightarrow$ (a_6). This is obvious.

Moreover, (a_4) $\Rightarrow$ (a_1) because for every $x^k \in A^k$ tending to x (x must be in A), $\varphi^k(x^k) \to \varphi(x)$. Then, $(x^k, \varphi^k(x^k)) \in \text{gph}\varphi^k$ tends to $(x, \varphi(x)) \in \text{gph}\varphi$. Finally, it is evident that (a_6) $\Rightarrow$ (a_3). Therefore, all six assertions are equivalent.

(b) We verify the string of implications (b_1) $\Rightarrow$ (a_4) $\Rightarrow$ (b_2) $\Rightarrow$ (b_1).

(b_1) $\Rightarrow$ (a_4). For any $x \in A$ and $x^k \in A$ tending to x, $D := \{x, x^k \mid k \in \mathbb{N}\}$ is a compact set containing itself. As φ is continuous, it is uniformly continuous on D, i.e., $\forall\varepsilon > 0, \exists\delta > 0, \forall x, y \in D$: $d(x, y) < \delta$, $|\varphi(x) - \varphi(y)| < \varepsilon$. As $x^k \in D$ for all k, by the uniform convergence of φ^k on D, $\forall\varepsilon > 0, \exists k_\varepsilon, \forall k > k_\varepsilon$, $|\varphi^k(x^k) - \varphi(x^k)| < \varepsilon$ and so, as $x^k \to x$ implies the existence of $\hat{k}_\varepsilon$ with $d(x^k, x) < \delta$ for all $k > \hat{k}_\varepsilon$, for all $k > \max\{k_\varepsilon, \hat{k}_\varepsilon\}$,

$$|\varphi(x) - \varphi^k(x^k)| \le |\varphi(x) - \varphi(x^k)| + |\varphi(x^k) - \varphi^k(x^k)| < 2\varepsilon,$$

i.e., $\varphi^k(x^k) \to \varphi(x)$.

(a_4) $\Rightarrow$ (b_2). For any $x \in A$, choosing $x^k \equiv x$ yields $\varphi^k(x) \to \varphi(x)$, i.e., the pointwise convergence holds. Suppose the almost equicontinuity does not hold, i.e., there is $x \in A$ such that $\forall k_0$, $\exists y^k \in B(x, 1/k) \cap A$, $\exists i(k) > k_0$: increasing with k, $|\varphi^{i(k)}(y^k) - \varphi^{i(k)}(x)| \ge 1/k$ for large k. This contradicts the continuous convergence as $y^k \to x$.

(b_2) $\Rightarrow$ (b_1). For any x in a compact subset $D \subset A$ and $\varepsilon > 0$, (b_2) yields an index $k_0(x, \varepsilon)$ such that $|\varphi^k(x) - \varphi^k(y)| \le \varepsilon$ for all $y \in B(x, \varepsilon) \cap A$ and $k \ge k_0(x, \varepsilon)$. So, by the pointwise convergence, $|\varphi(y) - \varphi(x)| \le \varepsilon$ for all $y \in B(x, \varepsilon) \cap A$, i.e., φ is continuous. Since D is compact, there exists a finite subset $\{x_1, \dots, x_m\}$ in D such that $D \subset \cup_{i=1}^m B(x_i, \varepsilon) \cap A$. Hence, if $y \in D$, one has $y \in B(x_i, \varepsilon) \cap A$ for some $x_i \in \{x_1, \dots, x_m\}$ and therefore

$$|\varphi(y) - \varphi^k(y)| \le |\varphi(y) - \varphi(x_i)| + |\varphi(x_i) - \varphi^k(x_i)| + |\varphi^k(x_i) - \varphi^k(y)|.$$

Here, on the right-hand side, the first term is not greater than ε by the continuity of φ; the second one is less than ε for all $k \ge k_i$ for some k_i by the pointwise convergence of φ^k; and the third term is less than ε for all $k \ge k_0(x_i, \varepsilon)$ by the equicontinuity of φ^k. Therefore, for all $k \ge \max_{1 \le i \le m}\{k_0(x_i, \varepsilon), k_i\}$, $|\varphi(y) - \varphi^k(y)| < 3\varepsilon$. Thus, φ^k uniformly converges to φ on D. □

Remark 7.22 (Relationship between types of convergence)

(a) In all the assertions of Propositions 7.21, A^k must tend to A.

(b) Proposition 7.21 indicates that the involved types of convergence are equivalent to epi- and hypo-convergence holding together and hence possess the variational properties of both epi- and hypo-convergence (i.e., properties concerning both minimization and maximization). But these types of convergence are too narrow and difficult to satisfy in practice. They are stronger than the pointwise convergence (p-convergence for short) as well. However, epi-convergence (and hypo-convergence too) is incomparable with (independent from) the pointwise convergence as we argue now; see Fig. 7.3. First, taking any function φ, which is not lower semicontinuous, we see that $\varphi^k := \varphi$, for all k, pointwise converge, but do not epi-converge to φ because any epi-limit must be lower semicontinuous, i.e., pointwise convergence does not imply epi-convergence (it does not imply any of the three weaker types of epi-convergence either). A converse situation is given next.

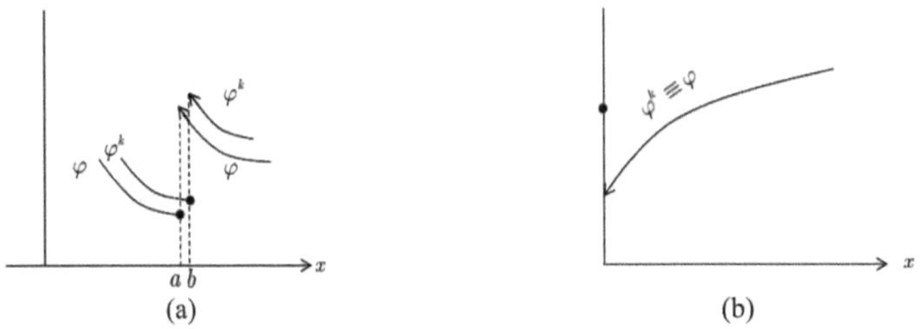

Fig. 7.3 E-convergence and p-convergence are incomparable. (**a**) φ^k and φ are close epigraphically but not pointwise. (**b**) e-convergence fails p-convergence holds

Example 7.23 (Epi-convergence holds; pointwise convergence fails) Let $\varphi^k : [0, 1] \to \mathbb{R}$ be defined by

$$\varphi^{2m}(x) := \begin{cases} 1/m & \text{if } x \text{ is rational,} \\ 1 - 1/m & \text{else,} \end{cases}$$

$$\varphi^{2m+1}(x) := \begin{cases} 1 - 1/m & \text{if } x \text{ is rational,} \\ 1/m & \text{else.} \end{cases}$$

Then, at any $x \in [0, 1]$, one has two convergent subsequences:

$$\varphi^{2m}(x) \text{ converging to } g(x) = \begin{cases} 0 & \text{if } x \text{ is rational,} \\ 1 & \text{else,} \end{cases}$$

$$\varphi^{2m+1}(x) \text{ converging to } h(x) = \begin{cases} 1 & \text{if } x \text{ is rational,} \\ 0 & \text{else.} \end{cases}$$

So, φ^k do not pointwise converge. Condition (a) of the epi-convergence of this sequence to $\varphi = 0$ is verified as follows. For any subsequence $x^{k_j} \to x$, one has for the even indices k_j:

$$\varphi^{k_j}(x^{k_j}) = \begin{cases} \frac{2}{k_j} & \text{if } x^{k_j} \text{ is rational,} \\ 1 - \frac{2}{k_j} & \text{else.} \end{cases}$$

Similarly, one has for the odd indices k_j:

$$\varphi^{k_j}(x^{k_j}) = \begin{cases} 1 - \frac{2}{k_j - 1} & \text{if } x^{k_j} \text{ is rational,} \\ \frac{2}{k_j - 1} & \text{else.} \end{cases}$$

Hence, any subsequence φ^{k_j} has a subsequence, which converges to 1 or to 0, and it cannot have a cluster point less than $0 = \varphi(x)$. Therefore, $\mathrm{li}_j \varphi^{k_j}(x^{k_j}) \geq \varphi(x)$, i.e., (a) of epi-convergence is fulfilled.

Condition (b) of the epi-convergence is easier to check as follows. For any $x \in [0, 1]$, one can choose the following sequence $x^k \to x$. Take rational numbers x^k for the even k's and irrational numbers x^k for the odd k's. Then,

$$\varphi^k(x^k) = \begin{cases} \frac{2}{k} & \text{if } k \text{ is even,} \\ \frac{2}{k-1} & \text{else} \end{cases}$$

tends to $0 = \varphi(x)$, and clearly $\varphi^k(x^k)$ cannot have a cluster point greater than 0. Hence, $\mathrm{ls}_k \varphi^k(x^k) \geq \varphi(x)$, i.e., (b) of the epi-convergence is checked.

Example 7.24 (Pointwise convergence does not preserve variational properties) Let $\varphi^k : [-1, 1] \to \mathbb{R}$ be defined by $\varphi^k(x) := \min\{1, 4.2^{2(k-1)}x^2 + 2^{-1}, 1 - 2^{k-1}(2^k - 1)^{-1}(x - 2^{-k})\}$. Then, φ^k pointwise converges to φ defined by $\varphi(x) := \min\{1, 1 - x\}$ for $x \in [-1, 1]$. It is also not hard to see that φ^k is continuous and attains its minimal value $1/2$ uniquely at $x = 0$, but φ attains the minimal value 0 at $x = 1$. So, one does not have the variational property $\mathrm{ls}_k(\inf_{A^k}\varphi^k) \leq \inf_A \varphi$ or $\mathrm{Limsup}_k(\varepsilon^k\text{-}\mathrm{argmin}_{A^k}\varphi^k) \subset \mathrm{argmin}_A \varphi$ for any $\varepsilon^k \to 0^+$ as stated in Proposition 7.3 for the epi-convergence. This also justifies that the epi-convergence is a basic type of variational convergence of unifunctions.

In order to have a more complete picture of types of convergence, let us note some additional insights into converse relations between the pointwise convergence and the stronger notions in Definition 7.20, under usual regularity assumptions. (These relations are known, but apart from our study here and hence have not been presented in detail.) If $A^k = A$ and convex function φ^k pointwise converges to φ, then φ must be convex and in fact φ^k converges continuously. Furthermore, if φ^k is lower semicontinuous, monotonously increasing with respect to k, and pointwise converges to a continuous function φ, then φ^k converges uniformly.

The next example shows that the strong types of convergence given in Definition 7.20 in general do not guarantee that $\inf_{A^k}\varphi^k \to \inf_A \varphi$ as the tight epi-convergence does (see Proposition 7.9).

Example 7.25 (The absence of convergence of infima under uniform convergence) Let $A^k = A = [0, +\infty)$, $\varphi(x) = x^2$, and

$$\varphi^k(x) := \begin{cases} x^2 & \text{if } 0 \leq x \leq k, \\ (x - 3k)^2 - k^2 & \text{if } k < x. \end{cases}$$

Then, φ^k uniformly converges to the continuous function φ. We see that $\inf_A \varphi = 0$ but $\lim_k (\inf_{A^k}\varphi^k) = -\infty$ as $\inf_{A^k}\varphi^k = -k^2$; and 0 is the minimizer of φ, but the minimizer of φ^k is $3k$ and tends to $+\infty$. For the detailed computations, see Exercise 7.4.

7.2 Epi/Hypo-Convergence and Lopsided Convergence of Bifunctions

The main ingredient of the equilibrium problem (EP) under consideration in this chapter is a bifunction $\Phi : A \times B \to \mathbb{R}$, where A and B are nonempty subsets of (possibly different) metric spaces X and Y, resp., unlike the bifunctions defined on $A \times A$ investigated in the previous chapters. (Note that corresponding to the case of unifunctions in Sect. 7.1, we write $\Phi : A \times B \to \mathbb{R}$, but in fact, $\Phi : X \times Y \to \bar{\mathbb{R}} := [-\infty, +\infty] := \mathbb{R} \cup \{-\infty, +\infty\}$ with finite values on $A \times B$.) The reason is that in many practical EPs in economics, control, management, etc., $x \in A$ is the state variable (location in dynamic problems, goods in economics, etc.), while $y \in B$ plays the role of a dual variable (control in dynamic problems, price in economics, etc.). As explained at the beginning of the chapter, variational approximations of EPs are based on types of variational convergence of bifunctions, and so we have to discuss such convergence first. In this section if not otherwise specified, X and Y are metric spaces, $A, A^k \subset X$ and $B, B^k \subset Y$ are nonempty, $\Phi : A \times B \to \mathbb{R}$, and $\Phi^k : A^k \times B^k \to \mathbb{R}$. The set of all bifunctions of this type is denoted by rec$(X \times Y)$ as $A \times B$ is a kind of a rectangle (in a general sense) in $X \times Y$.

7.2.1 *Epi/Hypo-Convergence*

Definitions and Characterizations

Epi/hypo-convergence (briefly, e/h-convergence) is a direct extension of the epi-convergence to the case of bifunctions and basic variational convergence for bifunctions. In Sect. 7.1, we have already seen that continuous convergence, graph convergence, and some other notions are types of variational convergence stronger than epi-convergence, but they are very difficult to satisfy and are rarely applied. Similarly, for bifunctions, continuous convergence, graph convergence, and some others also preserve variational properties, but they are too narrow/strict and useful only in a limited number of cases.

Definition 7.26 (Types of epi/hypo-convergence) Let $\{\Phi, \Phi^k\}_k \subset \text{rec}(X \times Y)$. Φ^k INSIDE EPI/HYPO-CONVERGES (I-E/H-CONVERGES) to Φ if:

(a) For every subsequence $A^{k_j} \ni x^{k_j} \to x \in A$ and $y \in B$, there exists $B^{k_j} \ni y^{k_j} \to y$ such that $\text{li}_j \Phi^{k_j}(x^{k_j}, y^{k_j}) \geq \Phi(x, y)$
(b) For every subsequence $B^{k_j} \ni y^{k_j} \to y \in B$ and $x \in A$, there exists $A^{k_j} \ni x^{k_j} \to x$ such that $\text{ls}_j \Phi^{k_j}(x^{k_j}, y^{k_j}) \leq \Phi(x, y)$

Φ^k EPI/HYPO-CONVERGES (E/H-CONVERGES) to Φ if the following additional infinity conditions are satisfied: for all $A^{k_j} \ni x^{k_j} \to x \notin A$, there exists $y^{k_j} \in B^{k_j}$ such that $\Phi^{k_j}(x^{k_j}, y^{k_j}) \to +\infty$ in condition (a), and for all $B^{k_j} \ni y^{k_j} \to y \notin B$, there exists $x^{k_j} \in A^{k_j}$ such that $\Phi^{k_j}(x^{k_j}, y^{k_j}) \to -\infty$ in condition (b). e/h-convergence is denoted by $\Phi = \text{e/h-lim}_k \Phi^k$ or $\Phi^k \xrightarrow{e/h} \Phi$ and for the inside variant, a prefix "i-" is added.

Definition 7.27 (Weak epi/hypo-convergence, weak inside epi/hypo-convergence) We say that Φ^k WEAKLY INSIDE EPI/HYPO-CONVERGES (W-I-E/H-CONVERGES) to Φ if:

(a) For every sequence $A^k \ni x^k \to x \in A$ and $y \in B$, there exists $B^k \ni y^k \to y$ such that $\text{li}_k \Phi^k(x^k, y^k) \geq \Phi(x, y)$
(b) For every sequence $B^k \ni y^k \to y \in B$ and $x \in A$, there exists $A^k \ni x^k \to x$ such that $\text{ls}_k \Phi^k(x^k, y^k) \leq \Phi(x, y)$

Φ^k WEAKLY E/H-CONVERGES (W-E/H-CONVERGES) to Φ if, in addition, the following infinity conditions are satisfied: for all $A^k \ni x^k \to x \notin A$, there exists $y^k \in B^k$ such that $\Phi^k(x^k, y^k) \to +\infty$ in condition (a), and for all $B^k \ni y^k \to y \notin B$, there exists $x^k \in A^k$ such that $\Phi^k(x^k, y^k) \to -\infty$ in condition (b).

Clearly, e/h-convergence implies the other three types. i-e/h-convergence implies w-i-e/h-convergence. All four types of e/h-convergence have applications, depending on situations, so weaker types have a priority to be chosen if possible.

Example 7.28 (Four types of e/h-convergence) Let $A^k = B^k = [1/k, 1]$, $A = B = [0, 1]$, and

$$\Phi^k(x, y) = \begin{cases} 1 & \text{if } (x, y) \in A^k \times B^k \text{ and } x \neq y, \\ 0 & \text{if } (x, y) \in A^k \times B^k \text{ and } x = y. \end{cases}$$

We verify that Φ^k e/h-converges to

$$\Phi(x, y) = \begin{cases} 1 & \text{if } (x, y) \in A \times B \text{ and } x \neq y, \\ 0 & \text{if } (x, y) \in A \times B \text{ and } x = y. \end{cases}$$

Condition (a). For all $y \in B$ and $A^{k_j} \ni x^{k_j} \to x$, x must belong to A. If $x \neq y$, there exists a neighborhood $U \times V \subset A \times B$ of (x, y), which is disjoint from the diagonal. Take $y^{k_j} \neq x^{k_j}$ tending to y with $(x^{k_j}, y^{k_j}) \in U \times V$ for large j. Then, $\Phi^{k_j}(x^{k_j}, y^{k_j}) \equiv 1$ tends to $1 = \Phi(x, y)$. If $x = y$, take $y^{k_j} = x^{k_j}$. Then, $y^{k_j} \to y$ and $\Phi^{k_j}(x^{k_j}, y^{k_j}) \equiv 0$ tends to $\Phi(x, y)$.

Condition (b) is symmetric to (a) with the role of x and y interchanged. So, the argument in this case is the same. Hence, Φ^k e/h-converges to Φ, and so the other three weaker types of e/h-convergence also hold.

Example 7.29 (i-e/h-convergence holds; w-e/h-convergence fails) Consider the same Φ^k and Φ as in Example 7.28, but modify the domains as $A^k = B^k = (1/k, 1)$, $A = B = (0, 1)$. For i-e/h-convergence, we check only condition (a) in the same way as in the above example to see that $\Phi^{k_j}(x^{k_j}, y^{k_j}) \to \Phi(x, y)$. Hence, Φ^k i-e/h-converges to Φ. However, we do not have w-e/h-convergence because, for

$x = 0$ or $x = 1$ (not in A) and any $A^{k_j} \ni x^{k_j} \to x$, we cannot find $y^k \in B^k$ such that $\Phi^{k_j}(x^{k_j}, y^{k_j}) \to +\infty$.

Example 7.30 (e/h-convergence does not hold; the other three weaker types do) Let A^k be equal to $\{0, 1\}$ for the odd k's and to $\{0\}$ for the even k's, $A = \{0\}$, $B^k = B = \{1\}$ for all k, $\Phi^k(x, y) = \Phi(x, y) = 0$ for all k, $(x, y) \in A^k \times B^k$, and $(x, y) \in A \times B$. Then, Φ^k do not e/h-converge to Φ as the infinity condition in (a) of the definition is violated (see Exercise 7.7). But we have all the other three weaker types of e/h-convergence (see Exercise 7.8).

Remark 7.31 (Comments on e/h-convergence)

(a) Each type of e/h-convergence is symmetric in the sense that conditions (a) and (b) are symmetric (wrt x and y, liminf and limsup, etc.). We discuss further for e/h-convergence as a representative. If one defines the hypo/epi-convergence of Φ^k to Φ by the condition that $-\Phi^k$ e/h-converges to $-\Phi$, then h/e-convergence corresponds to $\max_x \min_y$, i.e., maximization wrt x and minimization wrt y, and is different from e/h-convergence. If one insists on $\min_x \max_y$ and is interested only in changing the order of the two operations, e/h-convergence is symmetric in the sense that $(x, y) \mapsto \Phi^k(x, y)$ e/h-converges to $(x, y) \mapsto \Phi(x, y)$ if and only if $(x, y) \mapsto \Phi^k(y, x)$ h/e-converges to $(x, y) \mapsto \Phi(y, x)$. So, h/e-convergence may be considered to coincide with e/h-convergence (with (b) coming before (a)). In this book, we accept the latter setting and need to consider only e/h-convergence. For the other three types of e/h-convergence, we have the same situation.

(b) The symmetry feature of the types of e/h-convergence may be regarded as the first interest for bifunctions since the object of the first and traditional interest for bifunctions is saddle points, which are symmetric. Many symmetry issues such as duality properties are also needed in applications.

(c) In the case $A^k = A$ and $B^k = B$ for all k, if $\Phi^k(\cdot, y) \xrightarrow{e} \Phi(\cdot, y)$ for all $y \in B$ and $\Phi^k(x, \cdot) \xrightarrow{h} \Phi(x, \cdot)$ for all $x \in A$, then $\Phi^k \xrightarrow{e/h} \Phi$ (see Exercise 7.4). But the converse does not hold, i.e., e/h-convergence is weaker than the above epi- and hypo-convergence together. However, the following related assertion holds: if Φ, Φ^k are defined on $A \times B$ and $A^k \times B$, resp., but do not depend on y, i.e., $\Phi(x, y) = \Phi(x, y')$ for all $x \in A$ and $y, y' \in B$ and similarly for Φ^k, then $\Phi^k \xrightarrow{e/h} \Phi$ is equivalent to $\Phi^k(\cdot, y) \xrightarrow{e} \Phi(\cdot, y)$ for all $y \in B$. The corresponding assertion for the other (weaker) types of e/h- and epi-convergence is also valid. By the symmetry, the similar statements for hypo-convergence instead of epi-convergence are easy to be verified.

(d) Definitions 7.26 and 7.27 and the facts related to e/h-convergence in the book are also in force for the case that $A^k = A = X$, $B^k = B = Y$, and $\Phi^k, \Phi : X \times Y \to \mathbb{R} \cup \{+\infty, -\infty\}$ if the infinity conditions and the inside convergence are removed.

Proposition 7.32 (Characterizations of e/h-convergence in terms of epi- and hypo-convergence of slices)

(i) *Condition* (a) *of weak e/h-convergence* $\Leftrightarrow$ *for every* $A^k \ni x^k \to x \in A$, *one has condition* (b) *of* $\Phi^k(x^k, \cdot) \overset{w-h}{\to} \Phi(x, \cdot)$ *and for every* $A^k \ni x^k \to x \notin A$, *there exists* $y^k \in B^k$ *with* $\Phi^k(x^k, y^k) \to +\infty$.

(ii) *Condition* (b) *of weak e/h-convergence* $\Leftrightarrow$ *for every* $B^k \ni y^k \to y \in B$, *one has condition* (b) *of* $\Phi^k(\cdot, y^k) \overset{w-e}{\to} \Phi(\cdot, y)$ *and for every* $B^k \ni y^k \to y \notin B$, *there exists* $x^k \in A^k$ *such that* $\Phi^k(x^k, y^k) \to -\infty$.

The corresponding assertions for the other three types of e/h-convergence (and the other three types of epi- and hypo-convergence) also hold true.

Proof

(i) From the definition of weak hypo-convergence, the first part of condition (a) in the definition of weak e/h-convergence (about the inequality for liminf) clearly means condition (b) of the weak hypo-convergence of $\Phi^k(x^k, \cdot)$, while the second part related to minus infinity is the same in both equivalent assertions under question.

(ii) This is immediate since in the definition of the involved convergence, condition (b) is symmetric to condition (a).

The proofs of the corresponding assertions for the other types of e/h-convergence are similar.

□

We introduce the following four e/h-limit bifunctions, associated with a given sequence of bifunctions $\{\Phi^k\}_k$ and Φ to have other characterizations.

Definition 7.33 (Upper and lower e/h-limits and weak e/h-limits) Let Φ^k be in $\text{rec}(X \times Y)$ for $k \in \mathbb{N}$ and $A \times B \subset X \times Y$:

(a) For $x \in A$ and $y \in B$, the UPPER E/H-LIMIT is

$$\text{ehls}\Phi^k(x, y) := \inf_{\{A^{k_j} \ni x^{k_j} \to x\}} \sup_{\{B^{k_j} \ni y^{k_j} \to y\}} \text{li}_j \Phi^{k_j}(x^{k_j}, y^{k_j});$$

and for $x \notin A$ (and ignoring y), it is

$$\text{ehls}\Phi^k(x) := \inf_{\{A^{k_j} \ni x^{k_j} \to x\}} \sup_{\{y^{k_j} \in B^{k_j}\}} \text{li}_j \Phi^{k_j}(x^{k_j}, y^{k_j}).$$

(b) For $y \in B$ and $x \in A$, the LOWER E/H-LIMIT is

$$\text{ehli}\Phi^k(x, y) := \sup_{\{B^{k_j} \ni y^{k_j} \to y\}} \inf_{\{A^{k_j} \ni x^{k_j} \to x\}} \text{ls}_j \Phi^{k_j}(x^{k_j}, y^{k_j});$$

and for $y \notin B$ (and not depending on x), it is

$$\text{ehli}\Phi^k(y) := \sup_{\{B^{k_j} \ni y^{k_j} \to y\}} \inf_{\{x^{k_j} \in A^{k_j}\}} \text{ls}_j \Phi^{k_j}(x^{k_j}, y^{k_j}).$$

(c) For $(x, y) \in A \times B$, the WEAK UPPER E/H-LIMIT of Φ^k is

$$\text{w-ehls}\Phi^k(x, y) := \inf_{\{A^k \ni x^k \to x\}} \sup_{\{B^k \ni y^k \to y\}} \text{li}_k \Phi^k(x^k, y^k).$$

The other three types: WEAK UPPER E/H-LIMIT for $x \notin A$, WEAK LOWER E/H-LIMIT, and WEAK UPPER E/H-LIMIT for $y \notin B$ are defined similarly. We adopt the notation $\text{w-ehls}\Phi^k(x)$, $\text{w-ehli}\Phi^k(x, y)$, and $\text{w-ehli}\Phi^k(y)$, resp.

In Sect. 7.2.2 about lopsided convergence, we will see that these four lower and upper e/h-limits are also crucial for characterizing the types of lopsided convergence.

Lemma 7.34 (Nonstandard diagonalization) *Let* $\{a^{k,j} \mid k, j \in \mathbb{N}\}$ *be a two-index family in* $\overline{\mathbb{R}}$. *There exists a map* $j \mapsto j(k)$ *such that*

$$\mathrm{li}_k a^{k,j(k)} \geq \mathrm{li}_j(\mathrm{li}_k a^{k,j}).$$

If interested in the proof of this auxiliary result, the reader is referred to Corollary A1 in [78].

Theorem 7.35 (Formulas for the upper and lower e/h-limits)

(a) (upper e/h-limits) *For* $x \in A$ *and* $y \in B$,
$$\mathrm{ehls}\Phi^k(x, y) = \inf_{\{A^{k_j} \ni x^{k_j} \to x\}} \max_{\{B^{k_j} \ni y^{k_j} \to y\}} \mathrm{li}_j \Phi^{k_j}(x^{k_j}, y^{k_j});$$
and for $x \notin A$,
$$\mathrm{ehls}\Phi^k(x) = \inf_{\{A^{k_j} \ni x^{k_j} \to x\}} \max_{\{y^{k_j} \in B^{k_j}\}} \mathrm{li}_j \Phi^{k_j}(x^{k_j}, y^{k_j});$$

(b) (lower e/h-limits) *For* $y \in B$ *and* $x \in A$,
$$\mathrm{ehli}\Phi^k(x, y) = \sup_{\{B^{k_j} \ni y^{k_j} \to y\}} \min_{\{A^{k_j} \ni x^{k_j} \to x\}} \mathrm{ls}_j \Phi^{k_j}(x^{k_j}, y^{k_j});$$
and for $y \notin B$,
$$\mathrm{ehli}\Phi^k(y) = \sup_{\{B^{k_j} \ni y^{k_j} \to y\}} \min_{\{x^{k_j} \in A^{k_j}\}} \mathrm{ls}_j \Phi^k(x^{k_j}, y^{k_j}).$$

Proof Because of similarity, we only prove (a) in detail. To verify the first equality, fix $A^{k_j} \ni x^{k_j} \to x$ and denote

$$s := \sup_{\{B^{k_j} \ni y^{k_j} \to y\}} \mathrm{li}_j \Phi^{k_j}(x^{k_j}, y^{k_j}).$$

We check the attainment of the involved supremum. If $s = -\infty$, any $B^{k_j} \ni y^{k_j} \to y$ gives the maximum. If $-\infty < s < +\infty$, for any $\varepsilon_i \to 0^+$ and each fixed i, there exists $B^{k_j} \ni y_i^{k_j} \to y$ as $j \to +\infty$ with $\mathrm{li}_j \Phi^{k_j}(x^{k_j}, y_i^{k_j}) \geq s - \varepsilon_i$. Now varying i to take $\liminf_i$ as $i \to +\infty$, we have $\mathrm{li}_i \mathrm{li}_j \Phi^{k_j}(x^{k_j}, y_i^{k_j}) \geq s$. In view of Lemma 7.34, there exists $i(j)$ for each j such that $\mathrm{li}_j \Phi^{k_j}(x^{k_j}, y_{i(j)}^{k_j}) \geq s$. Of course, the sequence $\{y_{i(j)}^{k_j}\}_j$ gives the aforementioned maximum. By the arbitrariness of $\{x^{k_j}\}_j$, we obtain the required equality. If $s = +\infty$, then for any $\theta_i \to +\infty$ and each fixed i, one finds $y_i^{k_j} \in B^{k_j}$ tending to y as $j \to +\infty$ such that $\mathrm{li}_j \Phi^{k_j}(x^{k_j}, y_i^{k_j}) \geq \theta_i$. Taking $\liminf_i$ as $i \to +\infty$, one arrives at $\mathrm{li}_i \mathrm{li}_j \Phi^{k_j}(x^{k_j}, y_i^{k_j}) = +\infty$. Again by Lemma 7.34, one gets $i(j)$ for each j such that $\mathrm{li}_j \Phi^{k_j}(x^{k_j}, y_{i(j)}^{k_j}) = +\infty$. Hence, $\{y_{i(j)}^{k_j}\}_j$ is a maximizer for the sup-expression in the definition of s.

To verify the second formula, we can repeat the above argument, writing $y_i^{k_j} \in B^{k_j}$ instead of $B^{k_j} \ni y_i^{k_j} \to y$. In fact, $\mathrm{ehli}\Phi^k = -\mathrm{ehls}(-\Phi^k)$ as can be checked straightforwardly. So, the similarity of the proof of (a) and (b) follows naturally. □

The formulas for the corresponding weak upper and lower e/h-limits are similar, simply replacing all the subsequences (e.g., $A^{k_j} \ni x^{k_j} \to x$) by the corresponding sequences. Note that it can be checked directly that, by definition, $\text{ehli}\Phi^k = -\text{ehls}(-\Phi^k)$. So, the similarity in the proof of (a) and (b) of Theorem 7.35 follows naturally.

The task of the lower and upper e/h-limits in characterizing the types of e/h-convergence is now easy to be accomplished as follows.

Theorem 7.36 (Characterizations of e/h- and i-e/h-convergence) *Let Φ and Φ^k, $k \in \mathbb{N}$, be in* $\text{rec}(X \times Y)$. *Then, Φ^k e/h-converges to Φ if and only if the following two assertions hold:*

(i) $\Phi(x, y) \leq \text{ehls}\Phi^k(x, y)$ *when* $(x, y) \in A \times B$, *and* $\text{ehls}\Phi^k(x) = +\infty$ *when* $x \notin A$.
(ii) $\text{ehli}\Phi^k(x, y) \leq \Phi(x, y)$ *when* $(x, y) \in A \times B$, *and* $\text{ehli}\Phi^k(y) = -\infty$ *when* $y \notin B$.

Omitting the infinity conditions, we have a characterization of i-e/h-convergence.

Proof From the definition of e/h-convergence and the formulas given in Theorem 7.35, it is obvious that conditions (a) and (b) of e/h-convergence coincide with assertions (i) and (ii), resp., of Theorem 7.36. □

Remark 7.37 (About characterizations of all the types of e/h-convergence)

(a) Given a sequence $\Phi^k \in \text{rec}(X \times Y)$, which one of $\text{ehli}\Phi^k(x, y)$ and $\text{ehls}\Phi^k(x, y)$ is smaller changes from point to point in general. Theorem 7.36 stipulates that if $\text{ehli}\Phi^k(x, y) \leq \text{ehls}\Phi^k(x, y)$ for all $(x, y) \in A \times B$, any Φ satisfying the inequality $\text{ehli}\Phi^k \leq \Phi \leq \text{ehls}\Phi^k$ on $A \times B$ is an i-e/h-limit of Φ^k, and if the infinity conditions are also fulfilled, it is an e/h-limit. (Later on, for an equality or inequality of a function or bifunction which holds at all points in a domain, we write it without variable like $\text{ehli}\Phi^k \leq \Phi \leq \text{ehls}\Phi^k$.) The bifunction interval $[\text{ehli}\Phi^k, \text{ehls}\Phi^k]$ is called the inside e/h-limit interval (of Φ^k). So, this interval being a singleton means that the inside e/h-limit is unique. By Theorem 7.36, e/h-convergence occurs if and only if one has the interval $[\text{ehli}\Phi^k, \text{ehls}\Phi^k]$ together with the two infinity conditions. Then, the same interval becomes the e/h-limit interval.
(b) The corresponding characterization of w-e/h- and w-i-e/h-convergence is similar: Φ^k w-e/h-converges to Φ if and only if the following two assertions hold:

- $\Phi(x, y) \leq \text{w-ehls}\Phi^k(x, y)$ if $(x, y) \in A \times B$, and $\text{w-ehls}\Phi^k(x) = +\infty$ if $x \notin A$.
- $\text{w-ehli}\Phi^k(x, y) \leq \Phi(x, y)$ if $(x, y) \in A \times B$, and $\text{w-ehli}\Phi^k(y) = -\infty$ if $y \notin B$.

Omitting the infinity conditions, we have a characterization of w-i-e/h-convergence.

Example 7.38 (Applications of the characterizations of the types of e/h-convergence)

(a) (e/h-limits) Let $A^k = A = (0, e]$, $B^k = B = (0, 1]$, and $\Phi^k(x, y) = \ln(y/x)$ for all k.
As $\{\Phi^k\}_k$ is a constant sequence (i.e., all its entries are the same bifunction), e/h-convergence coincides with weak e/h-convergence, to apply Theorem 7.36, one can argue with sequences instead of subsequences. First, for $(x, y) \in A \times B$, one computes

$$\text{ehls}\Phi^k(x, y) = \inf_{\{A^k \ni x^k \to x\}} \max_{\{B^k \ni y^k \to y\}} \text{li}_k \ln(y^k/x^k) = \ln(y/x)$$

by the continuity of the logarithmic functions. Similarly, $\text{ehli}\Phi^k(x, y) = \ln(y/x)$. Hence, the two inequalities in Theorem 7.36 are fulfilled as equalities. Now, considering the infinity condition for $x = 0 \notin A$ in the theorem, one calculates $\text{ehls}\Phi^k(0) = \inf_{\{A^k \ni x^k \to 0\}} \max_{\{y^k \in B^k\}} \text{li}_k \ln(y^k/x^k)$. Taking $y^k \equiv y$ for any $y \in (0, 1]$, one can see that $\text{ehls}\Phi^k(0) = +\infty$. Similarly, for $y = 0 \notin B$, $\text{ehli}\Phi^k(0) = -\infty$. Therefore, Theorem 7.36 says that the unique e/h-limit is $\ln(y/x)$. Figure 7.4 gives the graph of this bifunction.

(b) (weak e/h-limits) Revisit Φ^k in Example 7.30, where we saw that $\Phi(x, y) = 0$ for all $(x, y) \in A \times B = \{0\} \times \{1\}$ is a limit in the sense of the three types of convergence weaker than epi/hypo-convergence. Now, we apply the characterizations given in Theorem 7.36 and Remark 7.37 to find the whole limit intervals. For instance, we only compute the interval of weak epi-limits by verifying (i) and (ii) in Theorem 7.36 using upper and lower weak epi/hypo-limits.

  (i) For $(x, y) \in A \times B = (0, 1)$, $\text{w-ehls}\Phi^k(x, y) = 0$ since one has only $x^k \equiv 0$ tending to x and $y^k \equiv 1$ approaching y and so $\Phi^k(x^k, y^k) \equiv 0$; there is no $A^k \ni x^k \to x \notin A$ and so $\text{w-ehls}\Phi^k(x) = +\infty$.
  (ii) Similarly, for $(x, y) = (0, 1) \in A \times B$, $\text{w-ehli}\Phi^k(x, y) = 0$ and $\text{w-ehls}\Phi^k(y) = -\infty$ for $y \notin B$.

Hence, we know from Example 7.30 that the bifunction $\Phi = 0$ is just the unique weak e/h-limit.

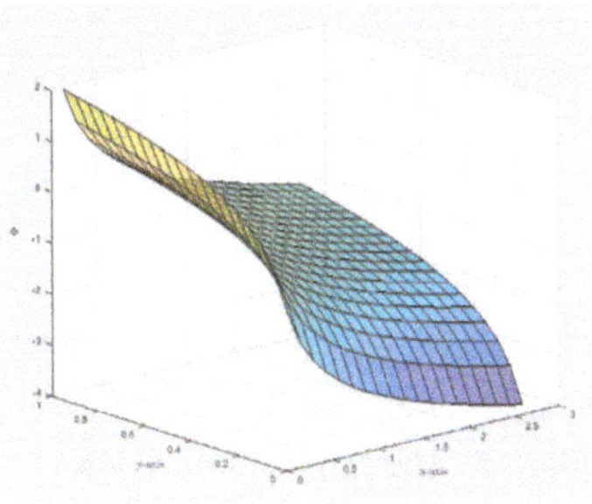

Fig. 7.4 $\Phi^k(x, y) = \Phi(x, y) = \ln(y/x)$ on $(0, e] \times (0, 1]$

The following example shows that even for a constant sequence on a fixed square, the e/h-limits may form a whole interval of bifunctions and may also be an empty set. We see that some properties of bifunctions (generally popular in mathematics) may vary from one e/h-limit to another (very different from an epi-limit!).

Example 7.39 (e/h-limits form a whole interval, not a singleton, with varying properties) Let $X = Y = \mathbb{R}$, $A = A^k = B = B^k = [0, 1]$, and $\Phi^k(x, y) = \Phi(x, y) = y^x$ for all $k \in \mathbb{N}$, with the convention that $0^0 = 1$. Considering e/h-convergence, we need to check only $(0, 0)$, since for all other (x, y) of $[0, 1]^2$, for every $A^{k_j} \ni x^{k_j} \to x$ and $B^{k_j} \ni y^{k_j} \to y$, $\Phi^{k_j}(x^{k_j}, y^{k_j}) \to y^x$. We have $\mathrm{ehli}\Phi^k(0, 0) = 0 \le \mathrm{ehls}\Phi^k(0, 0) = 1$.

$$\Phi^k(x, y) = \Phi(x, y) = y^x \text{ on } [0, 1]^2,$$

$$\Phi_a(x, y) = \begin{cases} y^x & \text{if } (x, y) \in [0, 1]^2 \setminus \{(0, 0)\}, \\ a & \text{if } (x, y) = (0, 0), \forall a \in [0, 1] \end{cases}$$

Therefore, the equivalence class of e/h-limits is the whole class:

$$\Phi_a(x, y) = \begin{cases} y^x & \text{if } (x, y) \in [0, 1]^2 \setminus \{(0, 0)\}, \\ a & \text{if } (x, y) = (0, 0), \end{cases}$$

for a being any number in $[0, 1]$. The graphs of Φ^k and Φ_a are shown in Fig. 7.5.

Observe that all Φ^k and Φ_a, $0 \le a \le 1$, are convex-concave. But, regarding semicontinuity, we see that some of these bifunctions are not lsc-usc on $[0, 1]^2$.

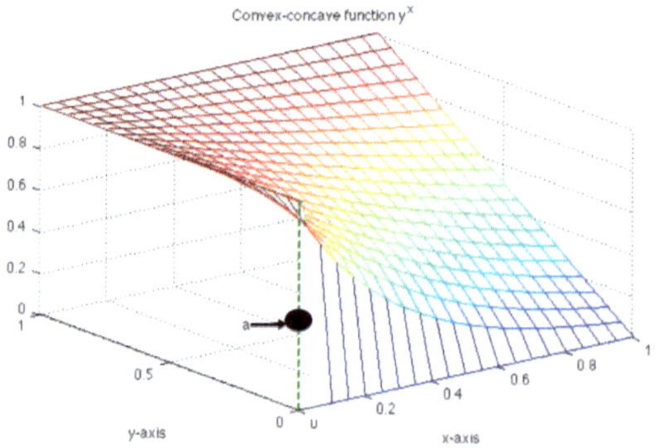

$$\Phi^k(x,y) = \Phi(x,y) = y^x \text{ on } [0,1]^2, \quad \Phi_a(x,y) = \begin{cases} y^x & \text{if } (x,y) \in [0,1]^2 \setminus \{(0,0)\}, \\ a & \text{if } (x,y) = (0,0), \forall a \in [0,1] \end{cases}$$

Fig. 7.5 The graph of bifunction y^x

(Recall that a finite-valued bifunction is said to be lsc-usc (or, in other words, lower-upper closed) if it is lsc in x-variable for each y and usc in y-variable for each x). In fact, Φ^k is usc in y at every points but not lsc in x at $(0, 0)$. We see that, in this example, all e/h-limits are convex-concave too (i.e., the convexity-concavity is preserved under e/h-convergence). However, for many values of a, the e/h-limit Φ_a does not inherit the semicontinuity. Namely, $\Phi_a(0, \cdot)$ is not usc at $(0, 0)$ if $a < 1$ and $\Phi_a(\cdot, 0)$ is not lsc at $(0, 0)$ if $a > 0$.

Now, we consider a modification with the same data, except the only change that now $A = A^k = B = B^k = (0, 1]$. Then, the only difference is that instead of ehls$\Phi^k(0, 0)$, now we need to compute ehls$\Phi^k(0) = 1$, and then it is not $+\infty$ as required. Hence, there is no e/h-limit. But one has the unique i-e/h-limit Φ.

Epi/hypo-convergence does not have a direct geometric interpretation (unlike epi- or hypo-convergence). But lower and upper e/h-limits are intimately tied to geometry of epi- and hypo-limits as follows.

Theorem 7.40 (Geometry of the lower and upper e/h-limits (and w-e/h-limits) being the end points of the e/h-equivalence class) *For $(x, y) \in A \times B$, the following formulas are valid:*

epi(ehli$\Phi^k(\cdot, y)) = \bigcap_{\{y^{k_j} \in B^{k_j} \to y\}} \mathrm{Li}_j(\mathrm{epi}\Phi^{k_j}(\cdot, y^{k_j}))$ for ehli$\Phi^k(\cdot, y))$ on A,

hypo(ehls$\Phi^k(x, \cdot)) = \bigcap_{\{x^{k_j} \in A^{k_j} \to x\}} \mathrm{Li}_j(\mathrm{hypo}\Phi^{k_j}(x^{k_j}, \cdot))$ for ehls$\Phi^k(x, \cdot)$ on B,

epi(w-ehli$\Phi^k(\cdot, y)) = \bigcap_{\{y^k \in B^k \to y\}} \mathrm{Li}_k(\mathrm{epi}\Phi^k(\cdot, y^k))$ for w-ehli$\Phi^k(\cdot, y)$ on A,

hypo(w-ehls$\Phi^k(x, \cdot)) = \bigcap_{\{x^k \in A^k \to x\}} \mathrm{Li}_k(\mathrm{hypo}\Phi^k(x^k, \cdot))$ for w-ehls$\Phi^k(x, \cdot)$ on B.

Proof Because of similarity, we only prove the first formula (for e/h-limits). (x, α) belonging to the right-hand side means that for all $y^{k_j} \in B^{k_j} \to y$ and $\varepsilon > 0$, there exists $j_0 \in \mathbb{N}$ such that $\forall j \geq j_0$,

$$[(A^{k_j} \cap B(x, \varepsilon)) \times (-\infty, \alpha + \varepsilon)] \cap \mathrm{epi}\Phi^{k_j}(., y^{k_j}) \neq \emptyset. \tag{7.1}$$

This relation means the existence of $x^{k_j} \in A^{k_j} \cap B(x, \varepsilon)$ such that $\Phi^{k_j}(x^{k_j}, y^{k_j}) < \alpha + \varepsilon$ for large j. Hence, the above statement is equivalent to

$$\sup_{\{B^{k_j} \ni y^{k_j} \to y\}} \min_{\{A^{k_j} \ni x^{k_j} \to x\}} \mathrm{ls}_j \Phi^{k_j}(x^{k_j}, y^{k_j}) \leq \alpha,$$

which means that ehli$\Phi^k(x, y) \leq \alpha$, i.e., $(x, \alpha) \in$ epi(ehli$\Phi^k(., y)$).

□

From the above theorem, it follows in particular that ehli$\Phi^k(\cdot, y)$ is always lower semicontinuous (i.e., its epigraph is always closed), and similarly ehls$\Phi^k(x, \cdot)$ is always upper semicontinuous (i.e., its hypograph is always closed).

Similar to the epi-convergence case, using filters and grills in a suitable way, in the special case that $A^k \to A$ and $B^k \to B$, we obtain another geometric formula for lower and upper e/h-limits as follows. For $y \in B$, denote the grill of the filter

$\mathscr{B}_y$ of the neighborhoods $\{B^k \cap B(y, 1/k) \mid k \in \mathbb{N}\}$ of $y \in B$ by $\mathscr{B}_y^\sharp := \{B^{k_j} \cap B(y, 1/k_j) \mid k \in \mathbb{N}\}$, and set the filter $\mathscr{U}_y := \mathscr{N}_\infty \times \mathscr{B}_y$ and so its grill is $\mathscr{U}_y^\sharp := \mathscr{N}_\infty^\sharp \times \mathscr{B}_y^\sharp$. For $x \in A$, the corresponding filters and grills are $\mathscr{B}_x$, $\mathscr{B}_x^\sharp := \{A^{k_j} \cap B(x, 1/k_j) \mid k \in \mathbb{N}\}$, $\mathscr{U}_x := \mathscr{N}_\infty \times \mathscr{B}_x$, and $\mathscr{U}_x^\sharp := \mathscr{N}_\infty^\sharp \times \mathscr{B}_x^\sharp$. Then, we have the following statement, which shows clearly the analogy and symmetry of lower and upper e/h-limits.

Theorem 7.41 (Another geometric formula for the lower and upper e/h-limits being the end points of the e/h-equivalence class) *Let $A^k \to A$ and $B^k \to B$ (in the sense of Painlevé-Kuratowski). For $(x, y) \in A \times B$, it holds that*

$$\text{epi}(\text{ehli}\Phi^k(\cdot, y)) = \cap_{U \in \mathscr{U}_y^\sharp} \text{cl}(\cup_{(k,y') \in U} \text{epi}\Phi^k(\cdot, y') \text{ for } \text{ehli}\Phi^k(\cdot, y)) \text{ on } A,$$
$$\text{hypo}(\text{ehls}\Phi^k(x, \cdot)) = \cap_{U \in \mathscr{U}_x^\sharp} \text{cl}(\cup_{(k,x') \in U} \text{hypo}\Phi^k(x', \cdot) \text{ for } \text{ehls}\Phi^k(x, \cdot) \text{ on } B.$$

Proof We only prove the first formula. By looking from the end backward to the beginning of the proof of Theorem 7.40, we see that (x, α) in $\text{epi}(\text{ehli}\Phi^k(\cdot, y))$ means that (corresponding to (7.1)) for any $U \in \mathscr{U}_y^\sharp$ and any $\varepsilon > 0$, there exists $(k, y') \in U$ such that

$$[(A^{k_j} \cap B(x, \varepsilon)) \times (-\infty, \alpha + \varepsilon)] \cap \text{epi}\Phi^k(., y') \neq \emptyset.$$

This is equivalent to the fact that (x, α) lies in the right-hand side of the claimed formula.

□

Remark 7.42 (Similarity between the formulas for e/h- and epi-limits) Observe that Theorem 7.41 for lower and upper e/h-limits is quite similar to Proposition 7.19 for lower and upper epi-limits. Namely, for all $y \in B$, $\text{epi}(\text{ehli}\Phi^k(\cdot, y))$ is indeed the set limit, wrt both k and y of epigraphs (instead of only wrt k in the case of $\text{epi}(\text{eli}\varphi)$). The situation for $\text{hypo}(\text{ehls}\Phi^k(x, \cdot))$ is similar and symmetric: the corresponding items are $x \in A$, k and x, and hypographs.

Main Variational Objects of Bifunctions

The main variational objects of $\Phi \in \text{rec}(X \times Y)$ are contained in the next definition.

Definition 7.43 (minsup-point, maxinf-point, saddle point, infsup-inequality, and infsup-equality)

(a) A point $\bar{x} \in A$ is called a MINSUP-POINT ($\bar{y} \in B$ is a MAXINF-POINT, resp.) of Φ, written $\bar{x} \in \text{mis}\Phi$ ($\bar{y} \in \text{mai}\Phi$, resp.), if

$$\sup_{y \in B} \Phi(\bar{x}, y) = \min_{x \in A} \sup_{y \in B} \Phi(x, y)$$

$$(\inf_{x \in A} \Phi(x, \bar{y}) = \max_{y \in B} \inf_{x \in A} \Phi(x, y), \text{ resp.}).$$

(b) A point $(\bar{x}, \bar{y}) \in A \times B$ is called a SADDLE POINT of Φ, written $(\bar{x}, \bar{y}) \in \text{sdl}\Phi$, if $\Phi(\bar{x}, y) \le \Phi(\bar{x}, \bar{y}) \le \Phi(x, \bar{y})$ for all $x \in A$ and $y \in B$.

(c) The always-fulfilled relation $\sup_B \inf_A \Phi \leq \inf_A \sup_B \Phi$ is called thc INFSUP-INEQUALITY. If here we have equality instead of $\leq$, then it is called the INFSUP-EQUALITY. (Note that in the literature, usually the word "minimax" is used instead of "infsup" for these relations, but to emphasize that the extremal values are not necessarily attained, we make use of "infsup.")

Proposition 7.44 (Characterizations of saddle points) *Let $\Phi \in \text{rec}(X \times Y)$. The following assertions are equivalent:*

(a) $(\bar{x}, \bar{y}) \in \text{sdl}\Phi$;
(b) $\bar{x} \in \text{mis}\Phi$, $\bar{y} \in \text{argmax}_{y \in B} \Phi(\bar{x}, y)$, $\bar{y} \in \text{mai}\Phi$, *and* $\bar{x} \in \text{argmin}_{x \in A} \Phi(x, \bar{y})$;
(c) $\bar{x} \in \text{mis}\Phi$, $\bar{y} \in \text{mai}\Phi$, *and the infsup-equality holds.*

Proof (a) $\Rightarrow$ (b). We show first that $\bar{x} \in \text{mis}\Phi$. Suppose there exists $x_1 \in A$ with $\sup_B \Phi(x_1, \cdot) < \sup_B \Phi(\bar{x}, \cdot)$. Then, $\Phi(x_1, y) < \Phi(\bar{x}, \bar{y}) \leq \Phi(x, \bar{y})$ for all $x \in A$ and $y \in B$, leading to the contradiction $\Phi(x_1, \bar{y}) < \Phi(x_1, \bar{y})$. Similarly, $\bar{y} \in \text{mai}\Phi$.

Next, we show that $\bar{y} \in \text{argmax}_{y \in B} \Phi(\bar{x}, y)$. Having already $\bar{x} \in \text{mis}\Phi$, suppose there is $y_1 \in B$ such that $\Phi(\bar{x}, y_1) > \Phi(\bar{x}, \bar{y})$. But this contradicts the left saddle inequality. $\bar{x} \in \text{argmin}_{x \in A} \Phi(x, \bar{y})$ is verified similarly.

(b) $\Rightarrow$ (c). We need to ensure the infsup-equality. Due to $\bar{x} \in \text{mis}\Phi$ and $\bar{y} \in \text{argmax}_{y \in B} \Phi(\bar{x}, y)$,

$$\inf_{x \in A} \sup_{y \in B} \Phi(x, y) = \sup_{y \in B} \Phi(\bar{x}, y) = \Phi(\bar{x}, \bar{y}),$$

Similarly, from $\bar{y} \in \text{mai}\Phi$ and $\bar{x} \in \text{argmin}_{x \in A} \Phi(x, \bar{y})$, it follows that

$$\sup_{y \in B} \inf_{x \in A} \Phi(x, y) = \Phi(\bar{x}, \bar{y}).$$

These equalities imply the infsup-equality.

(c) $\Rightarrow$ (a). This is the "converse case" in Proposition 1.1 (in Chap. 1). □

Example 7.45 (The essentialness of the assumptions in Proposition 7.44)

(a) (infsup-equality in (c) is crucial; a case when a bilinear bifunction does not have saddle points) Let $A := \{(0, 0), (0, 4), (4, 4), (4, 0), (1, 2), (2, 1)\} \subset \mathbb{R}^2$, $B = \mathbb{R}$, and $\Phi(x, y) = -2x_1 + x_2 + y(x_1 + x_2 - 3)$. Direct computations give $\bar{x} = (2, 1) \in \text{mis}\Phi$ and $\min_A \sup_B \Phi = -3$. For Proposition 7.44(c), computing $\text{mai}\Phi$, we first have

$$\inf_A \Phi(\cdot, y) = \begin{cases} -4 + 5y & \text{if } y < -1 \text{ and attained at } x = (4, 4), \\ -8 + y & \text{if } y < -1 \text{ and attained at } x = (4, 0), \\ -9 & \text{if } y = -1 \text{ and attained at } x = (4, 0), (4, 4), \\ -8 + y & \text{if } -1 < y \leq 2 \text{ and attained at several } x, \\ -3y & \text{if } y > 2 \text{ and attained at } x = (0, 0). \end{cases}$$

The maximizer of $\inf_A \Phi(\cdot, y)$ is $\bar{y} = 2 \in \text{mai}\Phi$ and $\max_B \inf_A \Phi = -6$. The infsup-equality does not hold and so Proposition 7.44(c) is violated. One can also compute directly that $\text{sdl}\Phi = \emptyset$.

(b) ($\bar{y} \in \text{mai}\Phi$ in (c) cannot be dropped) Let $A = B = [0, +\infty)$ and

$$\Phi(x, y) = \begin{cases} \frac{1}{x} + 1 - \frac{1}{y+1} & \text{if } x \neq 0, \\ 1 & \text{if } x = 0. \end{cases}$$

To verify the condition $\bar{x} \in \text{mis}\Phi$, consider $x = 0$. We have for all $y \in B$,

$$\sup_{y\in B} \Phi(0, y) = \max_{y\in B} \Phi(0, y) = 1.$$

For $x \neq 0$, we have

$$\begin{aligned} \sup_{y\in B} \Phi(x, y) &= \sup_{y\in B} \left(\frac{1}{x} + 1 - \frac{1}{y+1}\right) = \frac{1}{x} + \sup_{y\in B} \left(1 - \frac{1}{y+1}\right) \\ &= \frac{1}{x} + 1 > 1. \end{aligned}$$

Hence, $\bar{x} = 0 \in \text{mis}\Phi$ and

$$\inf_{x\in A} \sup_{y\in B} \Phi(x, y) = 1.$$

Now considering the condition $\bar{y} \in \text{mai}\Phi$, we have for $y \in B$,

$$\begin{aligned} \inf_{x\in A} \Phi(x, y) &= \min \left\{1, \inf_{x\in(0,+\infty)} \left(\frac{1}{x} + 1 - \frac{1}{y+1}\right)\right\} \\ &= \min \left\{1, 1 - \frac{1}{y+1}\right\} = 1 - \frac{1}{y+1} \end{aligned}$$

which is obtained as $x \to +\infty$. Hence,

$$\sup_{y\in B} \inf_{x\in A} \Phi(x, y) = \sup_{y\in B} \left(1 - \frac{1}{y+1}\right) = 1.$$

So, the infsup-equality holds but $\text{mai}\Phi = \emptyset$, not satisfying the aforementioned condition. We see directly that for any $y \in B$, $(0, y)$ is not a saddle point since for all $x > y + 1$,

$$\Phi(0, y) = 1 > \frac{1}{x} + 1 - \frac{1}{y+1} = \Phi(x, y).$$

The reason is that $\text{mai}\Phi = \emptyset$.

Similar to ε-minimum points of a function in Sect. 7.1, approximates of the above points are important for bifunctions. An ε-MINSUP POINT $\bar{x}_\varepsilon$ of Φ, denoted by $\bar{x}_\varepsilon \in \varepsilon\text{-mis}\Phi$, is defined by the condition

$$\sup_B \Phi(\bar{x}_\varepsilon, \cdot) \leq \inf_A \sup_B \Phi + \varepsilon.$$

The definition of an ε-MAXINF POINT $\bar{y}_\varepsilon$ is similar. For such a point, we write $\bar{y}_\varepsilon \in \varepsilon\text{-mai}\Phi$. Observe that $\bar{x}_\varepsilon \in \varepsilon\text{-mis}\Phi$ and $\bar{y}_\varepsilon \in \varepsilon\text{-mai}\Phi$ at the same time imply that

$$\sup_B \inf_A \Phi - \varepsilon \leq \Phi(\bar{x}_\varepsilon, \bar{y}_\varepsilon) \leq \inf_A \sup_B \Phi + \varepsilon. \tag{7.2}$$

Having two components, approximate saddle points can be understood in several ways. We are interested in some definitions. The first natural concept is as follows: $(\bar{x}_\varepsilon, \bar{y}_\varepsilon)$ is called an ε-SADDLE POINT of Φ, written $(\bar{x}_\varepsilon, \bar{y}_\varepsilon) \in \varepsilon\text{-sdl}\Phi$, if one has for all $x \in A$ and $y \in B$,

$$\Phi(\bar{x}_\varepsilon, y) - \varepsilon \leq \Phi(\bar{x}_\varepsilon, \bar{y}_\varepsilon) \leq \Phi(x, \bar{y}_\varepsilon) + \varepsilon.$$

Note that these inequalities are stronger than (7.2), but incomparable with the fact that $\bar{x}_\varepsilon \in \varepsilon\text{-mis}\Phi$ and $\bar{y}_\varepsilon \in \varepsilon\text{-mai}\Phi$ simultaneously. (Recall that for the corresponding "exact" objects, $(\bar{x}, \bar{y}) \in \text{sdl}\Phi$ implies that $\bar{x} \in \text{mis}\Phi$ together with $\bar{y} \in \text{mai}\Phi$.) A better approximation type of saddle points is defined as follows: $(\bar{x}_\varepsilon, \bar{y}_\varepsilon)$ is called a STRONG ε-SADDLE POINT of Φ, written $(\bar{x}_\varepsilon, \bar{y}_\varepsilon) \in \text{str-}\varepsilon\text{-sdl}\Phi$, if $(\bar{x}_\varepsilon, \bar{y}_\varepsilon) \in \varepsilon\text{-sdl}\Phi$, $\bar{x}_\varepsilon \in \text{mis}\Phi$, and $\bar{y}_\varepsilon \in \text{mai}\Phi$.

Variational Properties of the Types of E/H-Convergence The question is what variational objects of bifunctions are preserved under e/h-convergence? The answer is almost all, with a saddle point playing the central role. The preservation properties (possibly under some additional assumptions) of e/h-convergence are called variational properties. Now, we investigate variational properties of an arbitrary e/h-limit, possibly under some additional conditions. We will see that all e/h-limits in an equivalence class share many common properties. This fact should be highlighted, since in many applications, it helps to avoid dealing with whole equivalence classes of e/h-limits.

A function $\xi : A \to \mathbb{R}$ ($\eta : B \to \mathbb{R}$, resp.) is called the SUP-PROJECTION (INF-PROJECTION, resp.) of $\Phi \in \text{rec}(X \times Y)$ if, for $x \in A$ ($y \in B$, resp.),

$$\xi(x) := \sup_{y \in B} \Phi(x, y) \; (\eta(y) := \inf_{x \in A} \Phi(x, y), \; \text{resp.}).$$

Observe that $\text{dom}\xi \subset A$ and it may be empty. For instance, if $A = [0, 1]$, $B = \mathbb{R}_+$, and $\Phi(x, y) = y(x + 1)$, then $\sup_{y \in B} \Phi(x, y) = +\infty$ for each $x \in A$. Note also that $\bar{x} \in \text{argmin}\xi$ if and only if $\bar{x} \in \text{mis}\Phi$, and also $\bar{x} \in \varepsilon\text{-argmin}\xi$ (i.e., $\varphi(\bar{x}) \leq \inf_A \varphi$) if and only if $\bar{x} \in \varepsilon\text{-mis}\Phi$ (i.e., $\sup_B \Phi(\bar{x}, \cdot) \leq \inf_A \sup_B \Phi + \varepsilon$.

Theorem 7.46 (Condition (a) of a type of e/h-convergence implies (a) of the corresponding type of epi-convergence of sup-projections)

(i) *If* dom$\xi^k \neq \emptyset$, *then condition (a) of the e/h-convergence of* Φ^k *to* Φ *implies that* dom$\xi \neq \emptyset$ *and (a) of the epi-convergence of* ξ^k *to* ξ.

(ii) *The corresponding fact holds also for the other types of e/h-convergence.*

Proof

(i) We apply the geometric equivalence of condition (a) of epi-convergence given in Proposition 7.6. Assume that $(x, \alpha) \in \mathrm{Ls}_k(\mathrm{epi}\xi^k)$, i.e., there exist $x^{k_j} \in \mathrm{dom}\xi^{k_j}$ and $\alpha^{k_j} \geq \xi^{k_j}(x^{k_j})$ such that $(x^{k_j}, \alpha^{k_j}) \to (x, \alpha)$. If $x \notin A$, then by assumption, there exists $y^{k_j} \in B^{k_j}$ such that $\Phi^{k_j}(x^{k_j}, y^{k_j}) \to +\infty$. Since $\alpha^{k_j} \geq \xi^{k_j}(x^{k_j}) \geq \Phi^{k_j}(x^{k_j}, y^{k_j})$, $\alpha^{k_j} \to +\infty$, which is a contradiction as $\alpha^{k_j} \to \alpha$. So, $x \in A$. If $x \notin \mathrm{dom}\xi$, then one exhibits $y \in B$ with $\Phi(x, y) \geq \alpha + 1$. By assumption, there is $y^{k_j} \to y$ such that $\mathrm{li}_j \Phi^{k_j}(x^{k_j}, y^{k_j}) \geq \Phi(x, y)$. Hence,

$$\alpha = \lim_j \alpha^{k_j} \geq \mathrm{li}_j \xi^{k_j}(x^{k_j}) \geq \mathrm{li}_j \Phi^{k_j}(x^{k_j}, y^{k_j}) \geq \Phi(x, y) \geq \alpha + 1,$$

which is impossible. Consequently, x must be in domξ. For any positive ε, take $y_\varepsilon \in B$ such that $\Phi(x, y_\varepsilon) \geq \sup_B \Phi(x, \cdot) - \varepsilon$. Again by condition (a) of e/h-convergence, one has $B^{k_j} \ni y^{k_j} \to y_\varepsilon \in B$ such that

$$\mathrm{li}_j \xi^{k_j}(x^{k_j}) \geq \mathrm{li}_j \Phi^{k_j}(x^{k_j}, y^{k_j}) \geq \Phi(x, y_\varepsilon) \geq \xi(x) - \varepsilon.$$

Therefore,

$$\alpha = \lim \alpha^{k_j} \geq \mathrm{li}_j \xi^{k_j}(x^{k_j}) \geq \xi(x) - \varepsilon$$

and so $(x, \alpha) \in \mathrm{epi}\xi$ as ε is arbitrary. Now, Proposition 7.6 applies so that condition (a) of the epi-convergence of ξ^k is satisfied.

(ii) Because of similarity, we only discuss the case of weak convergence. Condition (a) of w-e/h-convergence means that for all sequences $A^k \ni x^k \to x$ and all $y \in B$, there exists $y^k \to y$ such that $\mathrm{li}_k \Phi^k(x^k, y^k) \geq \Phi(x, y)$ if $x \in A$ and $\Phi^k(x^k, y^k) \to +\infty$ if $x \notin A$. Hence, when $x \in A$, $\mathrm{li}_k \sup_{\{B^k \ni y^k \to y\}} \Phi^k(x^k, y^k) \geq \Phi(x, y)$ for all $y \in B$ and so $\mathrm{li}_k \sup_{\{y^k \in B^k\}} \Phi^k(x^k, y^k) \geq \sup_B \Phi(x, \cdot)$, which means that $\mathrm{li}_k \xi^k(x^k) \geq \xi(x)$. If $x \notin A$, then for all $y \in B$, $\sup_{B^k} \Phi^k(x^k, \cdot) \geq \sup_{\{B^k \ni y^k \to y\}} \Phi^k(x^k, y^k)$ $\to +\infty$, i.e., $\xi^k(x^k) \to +\infty$. Therefore, condition (a) of w-epi-convergence of ξ^k is obtained.

□

Theorem 7.47 (Convergence of approximate saddle points) *Let* $\varepsilon, \varepsilon^k \geq 0$ *and* $\varepsilon^k \to \varepsilon$.

(a) *If* $\Phi^k \overset{w-i-e/h}{\to} \Phi$, $(\bar{x}^k, \bar{y}^k) \in \varepsilon^k$-sdl$\Phi^k$ *for all k, and* $(\bar{x}, \bar{y}) = \lim_k(\bar{x}^k, \bar{y}^k)$ *lies in* $A \times B$, *then* $(\bar{x}, \bar{y}) \in \varepsilon$- sdl$\Phi$.

(b) *If* $\Phi^k \overset{i-e/h}{\to} \Phi$, *then* $\mathrm{Ls}_k(\varepsilon^k\text{-sdl}\Phi^k) \cap (A \times B) \subset \varepsilon\text{-sdl}\Phi$. *If specifically* $\varepsilon = 0$ *and* $\lim_j(\bar{x}^{k_j}, \bar{y}^{k_j}) = (\bar{x}, \bar{y})$, *we have*

$$\Phi(\bar{x}, \bar{y}) = \lim_j \Phi^{k_j}(\bar{x}^{k_j}, \bar{y}^{k_j}).$$

(c) *If* $\Phi^k \overset{e/h}{\to} \Phi$, *then* $\mathrm{Ls}_k(\varepsilon^k\text{-sdl}\Phi^k) \subset \varepsilon\text{-sdl}\Phi$.

Proof

(a) Consider arbitrary $(x, y) \in A \times B$. As $(\bar{x}^k, \bar{y}^k) \in \varepsilon^k$- sdl$\Phi^k$, for any $(x^k, y^k) \in A^k \times B^k$, it holds

$$\Phi^k(\bar{x}^k, y^k) - \varepsilon^k \le \Phi^k(\bar{x}^k, \bar{y}^k) \le \Phi^k(x^k, \bar{y}^k) + \varepsilon^k. \tag{7.3}$$

Assume that $(\bar{x}^k, \bar{y}^k) \to (\bar{x}, \bar{y}) \in A \times B$. By (7.3), we have

$$\sup_{\{B^k \ni y^k \to y\}} \mathrm{li}_k(\Phi^k(\bar{x}^k, y^k) - \varepsilon^k) \le \mathrm{li}_k \Phi^k(\bar{x}^k, \bar{y}^k) \le \mathrm{ls}_k \Phi^k(\bar{x}^k, \bar{y}^k)$$
$$\le \inf_{\{A^k \ni x^k \to x\}} \mathrm{ls}_k(\Phi^k(x^k, \bar{y}^k) + \varepsilon^k).$$

By the definition of w-i-e/h-convergence, one has

$$\Phi(\bar{x}, y) - \varepsilon \le \sup_{\{B^k \ni y^k \to y\}} \mathrm{li}_k(\Phi^k(\bar{x}^k, y^k) - \varepsilon^k)$$
$$\le \inf_{\{A^k \ni x^k \to x\}} \mathrm{ls}_k(\Phi^k(x^k, \bar{y}^k) + \varepsilon^k) \le \Phi(x, \bar{y}) + \varepsilon.$$

These inequalities mean that $(\bar{x}, \bar{y})$ is an ε-saddle point of Φ.

(b) When $\Phi^k \overset{i-e/h}{\to} \Phi$, $(\bar{x}^{k_j}, \bar{y}^{k_j}) \in \varepsilon^{k_j}$-sdl$\Phi^{k_j}$, and $(\bar{x}, \bar{y}) = \lim_j(\bar{x}^{k_j}, \bar{y}^{k_j}) \in A \times B$, the above argument can be written for the inequalities in terms of subsequences given by conditions (a) and (b) of i-e/h-convergence, and we also obtain the ε-saddle inequalities for $(\bar{x}, \bar{y})$. To see that $\Phi(\bar{x}, \bar{y}) = \lim_j \Phi^{k_j}(\bar{x}^{k_j}, \bar{y}^{k_j})$ for the case $\varepsilon = 0$, simply observe that condition (a) of i-e/h-convergence, (7.3), and $\bar{x}^{k_j} \to \bar{x}$ ensure the existence of $B^{k_j} \ni y^{k_j} \to \bar{y}$ satisfying

$$\Phi(\bar{x}, \bar{y}) \le \mathrm{li}_j(\Phi^{k_j}(\bar{x}^{k_j}, y^{k_j}) - \varepsilon^{k_j} \le \mathrm{li}_j \Phi^{k_j}(\bar{x}^{k_j}, \bar{y}^{k_j}).$$

With the role played by x and y reversed, a similar reasoning gives $\Phi(\bar{x}, \bar{y}) \ge \mathrm{ls}_j \Phi^{k_j}(\bar{x}^{k_j}, \bar{y}^{k_j})$.

(c) Due to assertion (b), we only need to show that no subsequence $(\bar{x}^{k_j}, \bar{y}^{k_j}) \in \varepsilon^{k_j}\text{-sdl}\Phi^{k_j}$ can tend to a point outside $A \times B$. For the sake of contradiction, suppose that there exists $(\bar{x}^{k_j}, \bar{y}^{k_j}) \to (x, y) \notin A \times B$, say $x \notin A$ (the case $y \notin B$ is of course similar since both e/h-convergence and saddle points are symmetric wrt x and y). By condition (a) of e/h-convergence, there is $y^{k_j} \in B^{k_j}$ with $\Phi^{k_j}(\bar{x}^{k_j}, y^{k_j}) \to +\infty$. Then, (7.3) implies that for any $x^{k_j} \in A^{k_j}$,

$$\Phi^{k_j}(x^{k_j}, \bar{y}^{k_j}) + \varepsilon^{k_j} \geq \Phi^{k_j}(\bar{x}^{k_j}, y^{k_j}) - \varepsilon^{k_j} \to +\infty. \tag{7.4}$$

We have two cases: $y \in B$ or $y \notin B$. If $y \in B$, then condition (b) of e/h-convergence implies that for any $\hat{x} \in A$, there exists $A^{k_j} \ni \hat{x}^{k_j} \to \hat{x}$ such that $\text{ls}_j \Phi^{k_j}(\hat{x}^{k_j}, \bar{y}^{k_j}) \leq \Phi(\hat{x}, y)$, which contradicts (7.4). So only $y \notin B$ remains possible. But, then again by the above condition (b), there exists $\hat{x}^{k_j} \in A^{k_j}$ with $\Phi^{k_j}(\hat{x}^{k_j}, \bar{y}^{k_j}) \to -\infty$ contradicting (7.4) as well. The proof is complete.

□

Note that the assumptions in the above theorem are only about types of e/h-convergence, without any usually imposed assumptions such as (relaxed) closedness, compactness, semicontinuity, or convexity-concavity. Furthermore, in fact when considering Φ, one is concerned only with points $(x, y) \in A \times B$, and so the case $x^k \in A^k \to x \notin A$ is not of interest in many situations. The inclusion $(x, y) \in A \times B$ is in general easily verifiable. That is why one needs only types of inside e/h-convergence in many situations.

Theorem 7.47 gives a lower bound of ε-sdlΦ in terms of $\text{Ls}_k(\varepsilon^k\text{-sdl}\Phi^k)$. With a stronger assumption, later we will have an upper bound of this set in terms of $\text{Li}_k(\varepsilon^k\text{-sdl}\Phi^k)$.

Example 7.48 (Convergence of approximate saddle points) Consider $\Phi^k(x, y) = \ln x \ln y$ defined on

$$A^k \times B^k = ([1/2, 1 - k^{-1}] \cup [1 + k^{-1}, 3/2]) \times ([1/2, 1 - k^{-1}] \cup [1 + k^{-1}, 3/2]).$$

We can check directly that Φ^k e/h-converges to $\Phi(x, y) = \ln x \ln y$, whose graph is provided in Fig. 7.6, defined on $([1/2, 1] \times [1/2, 1]) \cup ([1, 3/2] \times [1, 3/2])$, and Φ^k does not have saddle points. But it does have approximate saddle points. Namely,

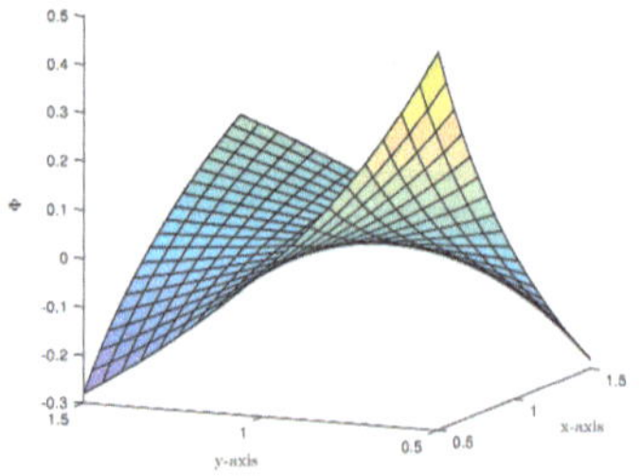

Fig. 7.6 $\Phi(x, y) = \ln x \ln y$ defined on $([1/2, 1] \times [1/2, 1]) \cup ([1, 3/2] \times [1, 3/2])$

with

$$\varepsilon^k = \max\{\ln(1-k^{-1})\ln 1/2 - \ln(1-k^{-1})\ln 3/2, \ln(1+k^{-1})\ln 3/2 - \ln(1+k^{-1})\ln 1/2\},$$

ε^k-sdlΦ^k is the set consisting of two horizontal intervals $([1/2, 1-k^{-1}] \times \{1-k^{-1}\}) \cup ([1+k^{-1}, 3/2]) \times \{1+k^{-1}\})$ and two vertical intervals $(\{2/3\} \times [1/2, 1-k^{-1}]) \cup (\{1+k^{-1}\} \times [1+k^{-1}, 3/2])$. Hence, ε^k-sdlΦ^k really set-converges to

$$\text{sdl}\Phi = ([1/2, 3/2] \times \{1\}) \cup (\{1\} \times [1/2, 3/2]).$$

Example 7.49 (A saddle-point set completely preserved) We revisit Example 7.39: $X = Y = \mathbb{R}$, $A = A^k = B = B^k = [0, 1]$, and $\Phi^k(x, y) = y^x$ for all k. That example already shows that Φ^k e/h-converges and the equivalence class of e/h-limits is the whole interval of bifunctions:

$$\Phi_a(x, y) = \begin{cases} y^x & \text{if } (x, y) \in [0, 1]^2 \setminus \{(0, 0)\}, \\ a & \text{if } (x, y) = (0, 0), \end{cases}$$

for all numbers a in $[0, 1]$. Hence, Theorem 7.47 ensures that for $\varepsilon^k \to 0^+$, $\text{Ls}_k(\varepsilon^k$-sdl$\Phi^k) \subset$ sdlΦ. However, we can compute directly that sdlΦ^k = sdl$\Phi_a = \{(x, 1) \mid x \le 1\}$, i.e., in this particular case, we have the stronger assertion that sdlΦ^k tend to sdlΦ.

To obtain more variational properties, we need the tightness conditions in the following definition.

Definition 7.50 (Tightness properties)

(i) Φ^k is said to WEAKLY I-E/H-CONVERGE to Φ x-ANCILLARY TIGHTLY if condition (a) of weak i-e/h-convergence and the following condition are satisfied
(b-m) condition (b) of weak i-e/h-convergence is satisfied together, additionally, with $\text{ls}_k \xi^k(x^k) \le \xi(x)$, where $x^k \to x$ is given in condition (b).

(ii) y-ANCILLARY TIGHT WEAK I-E/H-CONVERGENCE is defined correspondingly, with condition (b) of weak i-e/h-convergence and
(a-m) condition (a) of weak i-e/h-convergence is fulfilled together, additionally, with, $\text{li}_k \eta^k(y^k) \ge \eta(y)$ ($y^k \to y$ given in condition (a)).

(iii) If we replace in (i) "weak i-e/h-convergence" by "i-e/h-convergence" and $\text{ls}_k \xi^k(x^k) \le \xi(x)$ in condition (b-m) by $\text{ls}_j \xi^{k_j}(x^{k_j}) \le \xi(x)$, then Φ^k is said to I-E/H-CONVERGE TO Φ x-ANCILLARY TIGHTLY. If we replace in (ii) "weak i-e/h-convergence" by "i-e/h-convergence" and $\text{li}_k \eta^k(y^k) \ge \eta(y)$ in condition (a-m) by $\text{li}_j \eta^{k_j}(y^{k_j}) \ge \eta(y)$, then Φ^k I-E/H-CONVERGES TO Φ y-ANCILLARY TIGHTLY.

(iv) Φ^k E/H-CONVERGES TO Φ x-ANCILLARY TIGHTLY if condition (a) of e/h-convergence, condition (b-m) in (iii), and the infinity condition in (b) of

e/h-convergence are fulfilled. y-ANCILLARY TIGHT E/H-CONVERGENCE is defined similarly.
When both types of tightness in condition (iv) are satisfied, Φ^k is said to E/H-CONVERGE TO Φ (FULLY) TIGHTLY.

Theorem 7.51 (Epi-convergence of sup-projections)

(i) *If* $\Phi^k \overset{e/h}{\to} \Phi$ *x-ancillary tightly and* $\mathrm{dom}\xi^k \neq \emptyset$, *then* $\mathrm{dom}\xi \neq \emptyset$ *and* e-$\lim_k \xi^k = \xi$.
(ii) *Let* Φ^k *i-e/h-converge to* Φ *(weakly e/h-converge and weakly i-e/h-converge, resp.) x-ancillary tightly. Then,* $\mathrm{dom}\xi \neq \emptyset$ *and* ξ^k *i-e-converges (weakly e-converges and weakly i-e-converges, resp.) to* ξ.

Proof

(i) Theorem 7.46 implies that $\mathrm{epi}\xi \supset \mathrm{Ls}_k\mathrm{epi}\xi^k$. By condition (b-m) in the property x-tightness, any $(x, \alpha) \in \mathrm{epi}\xi$ corresponds to a sequence $A^k \ni x^k \to x$ such that $\mathrm{ls}_k\xi^k(x^k) \leq \xi(x)$, which is condition (b) of the epi-convergence of ξ^k. Then, in view of Proposition 7.6, $\mathrm{epi}\xi \subset \mathrm{Li}_k\mathrm{epi}\xi^k$.
(ii) Because of similarity, we only prove that $\xi^k \overset{w-i-e}{\to} \xi$. We use the analytic formulation as for the "weak types" no geometric equivalence statement holds. As for the "inside types" if $\mathrm{dom}\xi = \emptyset$, we have nothing to verify. Let $x \in \mathrm{dom}\xi$ and $\varepsilon > 0$. There exists $y_\varepsilon \in B$ such that $\Phi(x, y_\varepsilon) > \xi(x) - \varepsilon$. By the weak i-e/h-convergence of Φ^k, for every $A^k \ni x^k \to x$, there exists $B^k \ni y_\varepsilon^k \to y_\varepsilon$ such that $\mathrm{li}_k\Phi^k(x^k, y_\varepsilon^k) \geq \Phi(x, y_\varepsilon) > \xi(x) - \varepsilon$. We have

$$\mathrm{li}_k\xi^k(x^k) \geq \mathrm{li}_k\Phi^k(x^k, y_\varepsilon^k) > \xi(x) - \varepsilon.$$

After recalling that ε is arbitrary, we have that condition (a) of weak i-e-convergence is fulfilled. By condition (b-m), for all $x \in A$, we find $A^k \ni x^k \to x$ such that $\mathrm{ls}_k\xi^k(x^k) \leq \xi(x)$, i.e., condition (b) of weak i-e-convergence is satisfied.

□

A direct consequence of Theorem 7.51(i) is the lower semicontinuity of ξ, being an epi-limit, when $\Phi^k \overset{e/h}{\to} \Phi$ x-ancillary tightly. Moreover, assuming additionally that X is a linear space and $\Phi(\cdot, y)$ is convex for all $y \in B^k$, ξ is convex. Indeed, $\xi^k(\cdot) = \sup_{y \in B^k}\Phi(\cdot, y)$ is convex for all k. Hence, its epi-limit ξ is also convex.

As a set limit is always a closed set, from the above theorem, we see that if $\Phi^k \overset{e/h}{\to} \Phi$ x-ancillary tightly and A is compact, then there exist minsup-points of Φ. Furthermore, if for each k, A^k is compact and $\Phi^k(\cdot, y)$ is lower semicontinuous for each $y \in B^k$, then there exists a minsup-point x^k of Φ^k for all k and each cluster point of $\{x^k\}_k$ belongs to $\mathrm{mis}\Phi$.

Theorem 7.52 (Types of x-ancillary tight e/h-convergence subsume convergence of minsup-points)

(a) *If Φ^k w-i-e/h-converges to Φ x-ancillary tightly, then*

$$\text{ls}_k[\inf_{x\in A^k}\sup_{y\in B^k}\Phi^k(x,y)] \le \inf_{x\in A}\sup_{y\in B}\Phi(x,y).$$

(b) *If Φ^k e/h-converges to Φ x-ancillary tightly, $\varepsilon, \varepsilon^k \ge 0$, and $\varepsilon^k \to \varepsilon$, then* $\text{Ls}_k(\varepsilon^k\text{-mis}\Phi^k) \subset \varepsilon\text{-mis}\Phi$.

(c) *If Φ^k w-e/h-converges to Φ x-ancillary tightly, $\varepsilon^k \to 0^+$, and there exists $x^k \in$ $\varepsilon^k\text{-mis}\Phi^k$ tending to $\bar{x}$, then $\bar{x} \in \text{mis}\Phi$ and the minsup-value of Φ^k converges to the minsup-value of Φ.*

Proof

(a) To prove this assertion, apply Proposition 7.3(a) and Theorem 7.51(ii).
(b) This follows from Proposition 7.3(b) and Theorem 7.51(i).
(c) This follows from Proposition 7.3(c) and Theorem 7.51(ii).

□

Example 7.53 (An illustration of Theorem 7.52) Let $A^k = A = (0, e]$, $B^k = B = (0, 1]$, and $\Phi^k(x, y) = \Phi(x, y) = \ln(y/x)$. By the continuity of the logarithmic function, to apply Theorem 7.51, we need to consider only the point $(0, 0)$ since, for all $(x^k, y^k) \ne (0, 0)$ tending to $(x, y) \ne (0, 0)$, $\ln(y^k/x^k)$ converges to $\ln(y/x)$. To show condition (a) of e/h-convergence, for $(0, e] \ni x^k \to 0$, take $y^k = (x^k)^{1/2}$. Then, $\Phi(x^k, y^k) = \ln((x^k)^{-1/2}) \to +\infty$ as required. For condition (b) and any $(0, 1] \ni y^k \to 0$, choose $x^k = (y^k)^{1/2}$ to show that $\ln(y^k/x^k) = \ln((y^k)^{1/2}) \to -\infty$, i.e., (b) is satisfied. Finally, for any $x \in (0, e]$, clearly any $(0, e] \ni x^k \to x$ satisfies the x-ancillary tightness property because $\text{ls}_k\xi^k(x^k) = \ln((x^k)^{-1}) \to \ln(x^{-1}) = \xi(x)$. Hence, Theorem 7.52 says that, for all $\varepsilon^k \to 0^+$, $\text{Ls}_k(\varepsilon^k\text{- mis}\Phi^k) \subset \text{mis}\Phi = \{e\}$ (Fig. 7.4 shows that $\text{mis}\Phi = \{e\}$). This conclusion can be checked by direct calculations. In fact, one has a stronger result in this case: $\varepsilon^k\text{-mis}\Phi^k = [e^{1-\varepsilon^k}, e]$ tends to $\{e\} = \text{mis}\Phi$.

Observe that Theorem 7.52 does not guarantee the existence of minsup-points of Φ, though it provides some convergence of approximate minsup-points of Φ^k to such points of Φ. This existence is obtained, for instance, as follows. Besides $\Phi^k \overset{e/h}{\to} \Phi$ x-ancillary tightly, assume that Φ^k has ξ^k with nonempty domain being lsc, and there is a compact subset C^k containing $\text{dom}\xi^k$. Then, there exists $x^k \in \text{mis}\Phi^k$ and any of its cluster points is a minsup-points of Φ.

Theorem 7.54 (Complete convergence of approximate saddle points) *Let $\varepsilon, \varepsilon^k \ge 0$ and $\varepsilon^k \to \varepsilon$.*

(i) *If Φ^k inside epi/hypo-converges tightly to Φ, then*

$$\varepsilon\text{-sdl}\Phi \subset \cap_{\delta>\varepsilon}\text{Li}_k(\delta\text{-sdl}\Phi^k)$$

and hence, $\varepsilon\text{-sdl}\Phi \subset \text{Li}_k(\varepsilon^k\text{-sdl}\Phi^k)$.

(ii) *If Φ^k epi/hypo-converges tightly to Φ, then*

$$\mathrm{Lim}_k(\varepsilon^k\text{-sdl}\Phi^k) = \varepsilon\text{-sdl}\Phi,$$
$$\mathrm{Ls}_k(\text{str-}\varepsilon^k\text{-sdl}\Phi^k) \subset \text{str-}\varepsilon\text{-sdl}\Phi.$$

Proof

(i) Let $(\bar{x}, \bar{y}) \in \varepsilon\text{-sdl}\Phi$. By conditions (a) and (b), resp., of the i-e/h-convergence of Φ^k, we have, resp.,

$$\begin{cases} \text{for every } A^{k_j} \ni x^{k_j} \to \bar{x}, \text{ there exists } B^{k_j} \ni y^{k_j} \to \bar{y} \text{ such that} \\ \quad \mathrm{li}_j \Phi^{k_j}(x^{k_j}, y^{k_j}) \geq \Phi(\bar{x}, \bar{y}), \\ \text{for every } B^{k_j} \ni y^{k_j} \to \bar{y}, \text{ there exists } A^{k_j} \ni x^{k_j} \to \bar{x}, \text{ such that} \\ \quad \mathrm{ls}_j \Phi^{k_j}(x^{k_j}, y^{k_j}) \leq \Phi(\bar{x}, \bar{y}). \end{cases} \tag{7.5}$$

The assumed full tightness implies further that

$$\begin{cases} \text{for } 3x^{k_j} \text{ in the above second relation, } \mathrm{ls}_j \xi^{k_j}(x^{k_j}) \leq \xi(\bar{x}), \\ \text{for } y^{k_j} \text{ in the above first relation, } \mathrm{li}_j \eta^{k_j}(y^{k_j}) \geq \eta(\bar{y}). \end{cases} \tag{7.6}$$

To demonstrate the first inclusion in (i), It suffices to show that for (x^{k_j}, y^{k_j}) in (7.6), for all $\delta > \varepsilon$, there exists k_δ such that for all $k_j \geq k_\delta$, (x^{k_j}, y^{k_j}) is a δ-weak saddle point of Φ^{k_j}.
Suppose to the contrary that there exists $\delta_0 > \varepsilon$ and a subsequence of k_j, denoted also by k_j, such that for all j, at least one of the two inequalities in the definition of $(x^{k_j}, y^{k_j}) \in \delta_0\text{-wsdl}\Phi^{k_j}$ is violated. First, suppose the right-hand one fails, i.e.,

$$\Phi^{k_j}(x^{k_j}, y^{k_j}) > \inf_{x \in A^{k_j}} \Phi^{k_j}(x, y^{k_j}) + \delta_0 = \eta^{k_j}(y^{k_j}) + \delta_0.$$

Taking limsup of both sides, we get

$$\Phi(\bar{x}, \bar{y}) \geq \mathrm{ls}_j \Phi^{k_j}(x^{k_j}, y^{k_j}) \geq \mathrm{ls}_j \eta^{k_j}(y^{k_j}) + \delta_0 \geq \mathrm{li}_j \eta^{k_j}(y^{k_j}) + \delta_0 > \eta(\bar{y}) + \varepsilon,$$

where the first inequality follows from the second inequality of (7.5) and the last one from the second inequality of (7.6). This contradicts the right-hand side inequality in the definition of the ε-weak saddle point $(\bar{x}, \bar{y})$.
For the case that the left-hand-side inequality is violated:

$$\Phi^{k_j}(x^{k_j}, y^{k_j}) < \sup_{y \in B^{k_j}} \Phi^{k_j}(x^{k_j}, y) - \delta_0 = \xi^{k_j}(x^{k_j}) - \delta_0,$$

taking liminf of both sides and using the first inequality of (7.5) and the first inequality of (7.6), we similarly arrive at a contradiction. Therefore, we obtain the first inclusion in (i).

For the second inclusion, let $(\bar{x}, \bar{y}) \in \varepsilon\text{-sdl}\Phi$. We find a sequence $(x^k, y^k) \in \varepsilon^k\text{-sdl}\Phi^k$ converging to $(\bar{x}, \bar{y})$ as follows. For any $\ell > \varepsilon$, by the above first inclusion, one has $(x^k_\ell, y^k_\ell) \in \varepsilon^\ell\text{-sdl}\Phi^k$ converging to $(\bar{x}, \bar{y})$. Pass $\varepsilon^\ell \to \varepsilon$ and use the diagonalization argument to obtain $(x^k_{\ell(k)}, y^k_{\ell(k)}) \in \varepsilon^{\ell(k)}\text{-sdl}\Phi^k$ converging to $(\bar{x}, \bar{y})$. Then, $(x^k, y^k) := (x^k_{\ell(k)}, y^k_{\ell(k)})$ is just a desired sequence.

(ii) The (first) equality follows from (i) and Theorem 7.47(c). For the (second) inclusion, by assertion (b) of Theorem 7.52 and its counterpart for maxinf-points, we have that $\text{Ls}_k\text{mis}\Phi^k \subset \text{mis}\Phi$ and $\text{Ls}_k\text{mai}\Phi^k \subset \text{mai}\Phi$. This fact together with the preceding equality completes the proof.

□

In this subsection, we have seen that the types of e/h-convergence preserve many variational properties of bifunctions (this preservation helps to get variational approximations of equilibrium problems as we will see clearly later). The subsection has also shown that these direct extensions of the types of epi-convergence are especially suitable for (approximate) saddle points. The main reason is that both objects are symmetric and the types of e/h-convergence play crucial roles in preservation of symmetric properties and in duality studies. In the next subsection, we consider the types of lopsided convergence (lop-convergence), the types of variational convergence of bifunctions, which are not symmetric and suitable mainly for studies of nonsymmetric/one-sided properties in variational approximations.

7.2.2 Lopsided Convergence

Definitions and Characterizations

Besides the types of e/h-convergence, which are symmetric and suitable for studies of symmetric objects such as saddle points, minimax problems, duality properties, etc., the other variants of variational convergence are the types of lopsided convergence (lop-convergence), which are one-sided and also play crucial roles in applications. The reason is that, besides saddle points, one is often interested in the convergence of either minsup-points or maxinf-points but not both, and so less concerned about saddle points or duality relations. This subsection is devoted to lop-convergence, which is appropriate for such aims. We investigate sequences as in Sect. 7.2.1: $A, A^k \subset X$, $B, B^k \subset Y$, and bifunctions Φ^k and Φ on $A^k \times B^k$ and $A \times B$, resp.

Definition 7.55 (Minsup-lopsided convergence, inside minsup-lopsided convergence) Φ^k is said to MINSUP-LOPSIDEDLY (MIS-LOP) CONVERGE to Φ if:

(a) For every $A^{k_j} \ni x^{k_j} \to x \in A$ and $y \in B$, there exists $B^{k_j} \ni y^{k_j} \to y$ such that $\text{li}_j \Phi^{k_j}(x^{k_j}, y^{k_j}) \geq \Phi(x, y)$, and for every $A^{k_j} \ni x^{k_j} \to x \notin A$, there exists $y^{k_j} \in B^{k_j}$ such that $\Phi^{k_j}(x^{k_j}, y^{k_j}) \to +\infty$
(b) For every $x \in A$, there exists $A^{k_j} \ni x^{k_j} \to x$ such that for every $B^{k_j} \ni y^{k_j} \to y$, it holds $\text{ls}_j \Phi^{k_j}(x^{k_j}, y^{k_j}) \leq \Phi(x, y)$ if $y \in B$ and $\Phi^{k_j}(x^{k_j}, y^{k_j}) \to -\infty$ if $y \notin B$

We use the notation $\Phi^k \overset{mis-lop}{\to} \Phi$ or $\Phi = \text{mis-lop-lim}_k \Phi^k$ for this mis-lop convergence. If we consider only $x \in A$ in condition (a) and $y \in B$ in condition (b) (and omit the infinity conditions), Φ^k is said to INSIDE MINSUP-LOPSIDEDLY CONVERGE (I-MIS-LOP CONVERGE). This convergence is denoted by $\Phi^k \overset{i-mis-lop}{\to} \Phi$ or $\Phi = \text{i-mis-lop-lim}_k \Phi^k$.

Definition 7.56 (Weak minsup-lopsided convergence, weak inside minsup-lopsided convergence) We say that Φ^k WEAKLY MINSUP-LOPSIDEDLY (MIS-LOP) CONVERGES to Φ if:

(a) For every $A^k \ni x^k \to x \in A$ and $y \in B$, there exists $B^k \ni y^k \to y$ such that $\text{li}_k \Phi^k(x^k, y^k) \geq \Phi(x, y)$, and for every $A^k \ni x^k \to x \notin A$, there exists $y^k \in B^k$ such that $\Phi^k(x^k, y^k) \to +\infty$
(b) For every $x \in A$, there exists $A^k \ni x^k \to x$ such that for every $B^k \ni y^k \to y$, it holds $\text{ls}_k \Phi^k(x^k, y^k) \leq \Phi(x, y)$ if $y \in B$ and $\Phi^k(x^k, y^k) \to -\infty$ if $y \notin B$.

Then, we use the notation $\Phi^k \overset{w-mis-lop}{\to} \Phi$ or $\Phi = \text{w-mis-lop-lim}_k \Phi^k$. If we consider only $x \in A$ in condition (a) and $y \in B$ in condition (b) (and omit the infinity conditions), Φ^k is said to WEAKLY INSIDE MINSUP-LOPSIDEDLY CONVERGE (W-I-MIS-LOP CONVERGE) to Φ. Correspondingly, the notation for this convergence involves "w-i-mis-lop."

Observe that these definitions are not symmetric (between (a) and (b), x and y, etc.), and so we have the following other types of lop-convergence, correspondingly.

Definition 7.57 (Maxinf-lopsided convergence, inside maxinf-lopsided convergence) Φ^k MAXINF-LOPSIDEDLY (MAI-LOP) CONVERGES to Φ if:

(a) For every $B^{k_j} \ni y^{k_j} \to y$ and $x \in A$, there exists $A^{k_j} \ni x^{k_j} \to x$ such that $\text{ls}_j \Phi^{k_j}(x^{k_j}, y^{k_j}) \leq \Phi(x, y)$; for every $B^{k_j} \ni y^{k_j} \to y \notin B$, there exists $x^{k_j} \in A^{k_j}$ such that $\Phi^{k_j}(x^{k_j}, y^{k_j}) \to -\infty$
(b) For every $y \in B$, there exists $B^{k_j} \ni y^{k_j} \to y$ such that for every $A^{k_j} \ni x^{k_j} \to x$, we have $\text{li}_j \Phi^{k_j}(x^{k_j}, y^{k_j}) \geq \Phi(x, y)$ if $x \in A$ and $\Phi^{k_j}(x^{k_j}, y^{k_j}) \to +\infty$ if $x \notin A$

We denote this convergence by $\Phi^k \overset{mai-lop}{\to} \Phi$ or $\Phi = \text{mai-lop-lim}_k \Phi^k$. The INSIDE MAI-LOP CONVERGENCE is defined and denoted naturally similarly.

Definition 7.58 (Weak maxinf-lopsided convergence, weak inside maxinf-lopsided convergence) Φ^k WEAKLY MAXINF-LOPSIDEDLY (W-MAI-LOP) CONVERGES to Φ if:

(a) for every $B^k \ni y^k \to y$ and $x \in A$, there exists $A^k \ni x^k \to x$ such that $\text{ls}_k \Phi^k(x^k, y^k) \leq \Phi(x, y)$; for every $B^k \ni y^k \to y \notin B$, there exists $x^k \in A^k$ with $\Phi^k(x^k, y^k) \to -\infty$;
(b) for every $y \in B$, there exists $B^k \ni y^k \to y$ such that for every $A^k \ni x^k \to x$, it holds $\text{li}_k \Phi^k(x^k, y^k) \geq \Phi(x, y)$ if $x \in A$ and $\Phi^k(x^k, y^k) \to +\infty$ if $x \notin A$.

This convergence is denoted by $\Phi^k \overset{w-mai-lop}{\to} \Phi$ or $\Phi = \text{w-mai-lop-lim}_k \Phi^k$. The WEAK INSIDE MAI-LOP CONVERGENCE is defined and denoted naturally similarly, with additional "inside" and "i," resp.

Remark 7.59 (Comments on the concepts of lop-convergence)

(a) The types of mis-lop and mai-lop convergence in Definitions 7.55–7.58 are not in the sense that the types of mai-lop convergence of Φ^k to Φ mean the corresponding types of mis-lop convergence of $-\Phi^k$ to $-\Phi$. The latter way of definition corresponds to $\max_x \min_y$, while in all the above definitions of the types of lop-convergence, we keep always $\min_x \max_y$ and change only the order of the two operations. In what follows, our main attention is paid to "minsup framework" due to the similarity.
(b) The weak types of lop-convergence are really weaker than the corresponding ("strong") types. This can be justified similarly as for the types of e/h-convergence. The introduction of the inside types is also justified by their applications in variational approximation of equilibrium problems in many situations.
(c) Definitions 7.55–7.58 are also applicable to the case $A^k = A = X$, $B^k = B = Y$, and $\Phi^k, \Phi : X \times Y \to \mathbb{R} \cup \{+\infty, -\infty\}$ if the infinity conditions and the inside convergence are removed.
(d) Clearly, each mis-lop or mai-lop convergence is stronger than e/h-convergence. We have also similar assertions for the types of inside and weak convergence.
(e) The last (second) part of Remark 7.31(c) about the equivalence of some e/h-convergence and epi-convergence is also valid for mis-lop convergence. The corresponding fact about hypo-convergence is clearly valid for mai-lop convergence.

Later on, by "lop-convergence," we refer to any of the types of convergence described in the above definitions, when there is no need to specify which one is spoken about.

Example 7.60 (All the types of lop-convergence hold) Revisit Example 7.38(a): $A^k = A = (0, e]$, $B^k = B = (0, 1]$, and $\Phi^k(x, y) = \Phi(x, y) = \ln(y/x)$. As in this case lop-convergence coincides with weak lop-convergence, we apply Definition 7.56 for ("usual") lop-convergence. For the mis-lop convergence, as condition (a) is the same as (a) of e/h-convergence, Example 7.38(a) verified that condition (a) is fulfilled. For condition (b), consider any $x \in A$. We take $x^k \equiv x$ to get that, for every $B^k \ni y^k \to y \in B$, $\ln(y^k/x^k) \to \ln(y/x)$, and $\ln(y^k/x^k) \to -\infty$ if $y^k \in B^k \to 0 \notin B$. Hence, the mis-lop convergence is verified.

For the mai-lop convergence, similar to the above argument, we need to consider only the point $(x, y) = (0, 0)$. For condition (a), for any $B^k \ni y^k \to 0 \notin B$, we take $x^k = 1 \in B^k$ for all k to see that $\ln(y^k/x^k) \to -\infty$. For condition (b), we pick $y^k = 1 \in B^k$ for all k to have $\ln(y^k/x^k) \to +\infty$ for any $A^k \ni x^k \to 0$. So, the mai-lop convergence holds.

Example 7.61 (Failure of the mis-lop and mai-lop convergence, the weaker types of lop-convergence hold) Let A^k be equal to $\{0, 1\}$ for the odd k's and to $\{0\}$ for the even k's, $A = \{0\}$, $B^k = B = \{1\}$ for all k, $\Phi^k(x, y) = 0$ for all k and $(x, y) \in A^k \times B^k$, and $\Phi(x, y) = 0$ for all $(x, y) \in A \times B$. Then, both the mis-lop and mai-lop convergence do not hold (see Exercise 8.13). But we have all the other (six) weaker types of lop-convergence (see Exercises 8.13 and 8.14).

Proposition 7.62 (Characterizations of the weak mis-lop convergence in terms of slices) *We have the following equivalent assertions:*

(i) *Condition* (a) *of w-mis-lop convergence* $\Leftrightarrow$ *for every* $A^k \ni x^k \to x \in A$, *it holds* $\text{w-hli}\Phi^k(x^k, \cdot) \geq \Phi(x, \cdot)$ *with* $\Phi^k(x^k, \cdot)$ *defined on* B^k *and* $\Phi(x, \cdot)$ *on* B, *and for every* $A^k \ni x^k \to x \notin A$, *there exists* $y^k \in B^k$ *with* $\Phi^k(x^k, y^k) \to +\infty$ $\Leftrightarrow$ *for every* $A^k \ni x^k \to x \in A$, *one has condition (b) of* $\Phi^k(x^k, \cdot) \overset{w-h}{\to} \Phi(x, \cdot)$ *with the domains as above, and for every* $A^k \ni x^k \to x \notin A$, *there exists* $y^k \in B^k$ *with* $\Phi^k(x^k, y^k) \to +\infty$.

(ii) *Condition* (b) *of w-mis-lop convergence* $\Leftrightarrow$ *for every* $x \in A$, *there exists* $A^k \ni x^k \to x$ *such that for every* $y \in B$, *we have* $\text{w-hls}\Phi^k(x^k, y) \leq \Phi(x, y)$ *and for every* $y \notin B$, *we have* $\max_{\{y^k \in B^k \to y\}} \text{ls}_k \Phi^k(x^k, y^k) = -\infty$ $\Leftrightarrow$ *for every* $x \in A$, *there exists* $A^k \ni x^k \to x$ *such that condition* (a) *of* $\Phi^k(x^k, \cdot) \overset{w-h}{\to} \Phi(x, \cdot)$ *(with the domains as above) holds.*

(iii) *If* Φ^k *do not depend on* y *(on* x, *resp.), then* $\Phi^k \overset{w-mis-lop}{\to} \Phi$ *implies that* $\Phi^k \overset{w-e}{\to} \Phi$ *(*$\Phi^k \overset{w-h}{\to} \Phi$, *resp.).*

Proof

(i) This is just condition (a) for the weak e/h-convergence in Proposition 7.32(i).
(ii) The first equivalence can be proved similarly to that in (i) and the second one is clear by definition.
(iii) This is clear.

□

Similar characterizations hold true for all the other types of lop convergence.

To have more characterizations of the types lop-convergence, we use the four "semi" e/h-limits in Definition 7.33 for each type of e/h-convergence and introduce the following two "semi" lop-limits (for each type of lop-convergence) of a given sequence $\{\Phi^k\}_k$. It should be noted (unlike the four types of e/h-convergence) that these two "semi" lop-limits are defined for points (x, y) both inside $A \times B$ and outside it in a "critical" way that only x or y is outside.

Definition 7.63 (Weak lower and weak upper lop-limits) For Φ^k in rec$(X \times Y)$ and $A \times B \subset X \times Y$, the WEAK LOWER AND WEAK UPPER LOP-LIMITS are defined by, resp.,

w-lli$\Phi^k(x, y) := \sup_{\{B^k \ni y^k \to y\}} \inf_{\{A^k \ni x^k \to x\}} \mathrm{li}_k \Phi^k(x^k, y^k)$ for $y \in B$ and $x \in X$,

w-lls$\Phi^k(x, y) := \inf_{\{A^k \ni x^k \to x\}} \sup_{\{B^k \ni y^k \to y\}} \mathrm{ls}_k \Phi^k(x^k, y^k)$ for $x \in A$ and $y \in Y$.

Remark 7.64 (Comparison of "semi" e/h- and "semi" lop-limits)

(a) From Definitions 7.33 and 7.63, we have the important relationships:

$$\text{w-lls}\Phi^k(x, y) \geq \begin{cases} \text{w} - \text{lli}\Phi^k(x, y) & \text{for } (x, y) \in A \times B, \\ \text{w} - \text{ehls}\Phi^k(x, y) & \text{for } (x, y) \in A \times B, \\ \text{w} - \text{ehli}\Phi^k(x, y) & \text{for } (x, y) \in A \times B, \\ \text{w} - \text{ehli}\Phi^k(y) & \text{for } x \in A \text{ and } y \notin B, \end{cases}$$

$$\text{w-lli}\Phi^k(x, y) \leq \begin{cases} \text{w} - \text{lls}\Phi^k(x, y) & \text{for } (x, y) \in A \times B, \\ \text{w} - \text{ehls}\Phi^k(x, y) & \text{for } (x, y) \in A \times B, \\ \text{w} - \text{ehls}\Phi^k(x) & \text{for } y \in B \text{ and } x \notin A, \\ \text{w} - \text{ehli}\Phi^k(x, y) & \text{for } (x, y) \in A \times B. \end{cases}$$

The above inequalities show that thanks to the two "semi" lop-limits, all the six "semi" limits are connected by crucial relations, while the four "semi" e/h-limits are incomparable. It should be emphasized that in Chap. 8, we will see that the above relations no longer hold for bifunctions on nonrectangular domains and so the relationship between e/h-convergence and lop-convergence is more complicated.

(b) We can write similarly the corresponding definition of the "strong" lower and upper lop-limits, e.g.,

lls$\Phi^k(x, y) := \min_{\{A^{k_j} \ni x^{k_j} \to x\}} \sup_{\{B^{k_j} \ni y^{k_j} \to y\}} \mathrm{ls}_j \Phi^{k_j}(x^{k_j}, y^{k_j})$ for $x \in A$ and $y \in Y$.

Of course, the relations in (a) also hold true for these limits.

Theorem 7.65 (Formulas for the lower and upper weak lop-limits) *For* $(x, y) \in X \times Y$,

w-lli$\Phi^k(x, y) = \max_{\{B^k \ni y^k \to y\}} \inf_{\{A^k \ni x^k \to x\}} \mathrm{li}_k \Phi^k(x^k, y^k)$ *for* $y \in B$ *and* $x \in X$, w-lls$\Phi^k(x, y)$:= $\min_{\{A^k \ni x^k \to x\}} \sup_{\{B^k \ni y^k \to y\}} \text{ls}_k \Phi^k(x^k, y^k)$ for $x \in A$ *and* $y \in Y$.

Proof Because of similarity, we discuss only the second formula. Denote $u :=$ w-lls$\Phi^k(x, y)$. If $u = +\infty$, any $A^k \ni x^k \to x$ is a minimizer for the considered expression. If $-\infty < u < +\infty$, for any $\varepsilon_j \to 0^+$ and each fixed j, there exists $A^k \ni x_j^k \to x$ (as $k \to +\infty$) such that for all $B^k \ni y_j^k \to y$, $\text{ls}_k \Phi^k(x_j^k, y_j^k) \leq u + \varepsilon_j$. Now we vary j and take $\limsup_j$ as $j \to +\infty$ to get $\text{ls}_j \text{ls}_k \Phi^k(x_j^k, y_j^k) \leq u$. Due to Lemma 7.12, for each k, there is $j(k)$ such that $\text{ls}_k \Phi^k(x_{j(k)}^k, y_{j(k)}^k) \leq u$. Hence, the sequence $\{x_{j(k)}^k\}_k$ is a minimizer. If $u = -\infty$, for any $\theta_j \to +\infty$ and each fixed j, one has $A^k \ni x_j^k \to x$ (as $k \to +\infty$) such that for every $B^k \ni y_j^k \to y$, it holds $\text{ls}_k \Phi^k(x_j^k, y_j^k) \leq -\theta_j$. Similarly as above, by using Lemma 7.12, we obtain a sequence $\{x_{j(k)}^k\}_k$, which is a minimizer. □

Similar formulas hold for the "strong" lower and upper lop-limits.

Theorem 7.66 (Characterizations of weak lop-convergence)

(i) Φ^k *mis-lop converges weakly to* Φ *if and only if for* $x \in A$, w-lls$\Phi^k(x, y) = \Phi(x, y) =$ w-ehls$\Phi^k(x, y)$ *if* $y \in B$ *and* w-lls$\Phi^k(x, y) = -\infty$ *if* $y \notin B$*; and for* $x \notin A$, w-ehls$\Phi^k(x) = +\infty$. *Hence, if the weak mis-lop limit exists, it is always unique and is just the right end of the interval of the weak e/h-limits.*

(ii) Φ^k *mai-lop converges weakly to* Φ *if and only if for* $y \in B$, w-ehli$\Phi^k(x, y) = \Phi(x, y) =$ w-lli$\Phi^{(}x, y)$ *if* $x \in A$ *and* w-lli$\Phi^k(x, y) = +\infty$ *if* $x \notin A$*, and for* $y \notin B$, w-ehli$\Phi^k(y) = -\infty$. *So, if the weak mai-lop limit exists, it is always unique and is just the left end of the interval of the weak e/h-limits.*

Proof

(i) Condition (a) of the weak mis-lop convergence is the same as (a) of the weak e/h-convergence and means that $\Phi(x, y) \leq$ w-ehls$\Phi^k(x, y)$ if $x \in A$, $y \in B$ and w-ehls$\Phi^k(x) = +\infty$ if $x \notin A$. So, we get a half of the required equivalence in assertion (i). To consider its remaining part and condition (b) of the weak mis-lop convergence, we use the following abbreviation. For $x \in A$, $y \in Y$, $\varepsilon^j \to 0^+$, and $\gamma^j \to +\infty$, set

$$\theta^j(x, y) := \begin{cases} \Phi(x, y) + \varepsilon^j & \text{if } y \in B, \\ -\gamma^j & \text{if } y \notin B. \end{cases}$$

The aforementioned condition (b) means that

$$\min_{\{A^k \ni x^k \to x\}} \sup_{\{B^k \ni y^k \to y\}} \text{ls}_k \Phi^k(x^k, y^k) \leq \begin{cases} \Phi(x, y) & \text{if } y \in B, \\ -\infty & \text{if } y \notin B. \end{cases} \tag{7.7}$$

This inequality says that for each j, there exists $A^k \ni x_j^k \to x$ such that for every $B^k \ni y_j^k \to y$, $\mathrm{ls}_k \Phi^k(x_j^k, y_j^k)) \le \theta^j(x, y)$. Hence,

$$\mathrm{ls}_j \mathrm{ls}_k \Phi^k(x_j^k, y_j^k) \le \begin{cases} \Phi(x, y) & \text{if } y \in B, \\ -\infty & \text{if } y \notin B. \end{cases}$$

Lemma 7.12 gives an index $j(k)$ for each k such that

$$\mathrm{ls}_k \Phi^k(x_{j(k)}^k, y_{j(k)}^k) \le \begin{cases} \Phi(x, y) & \text{if } y \in B, \\ -\infty & \text{if } y \notin B, \end{cases}$$

which in turn ensures that the sequence $x^k := x_{j(k)}^k$ is a minimizer for (7.7). So, following Theorem 7.65, (7.7) means that w-lls$\Phi^k(x, y) \le \Phi(x, y)$ if $(x, y) \in A \times B$ and w-lls$\Phi^k(x, y) = -\infty$ if $y \notin B$. Then, Remark 7.64(a) implies that w-lls$\Phi^k(x, y) = \Phi(x, y) =$ w-ehls$\Phi^k(x, y)$. Hence, the required equivalence is proved. Therefore, if the weak mis-lop limit exists, it is unique and equals to the mentioned right end point.

(ii) The proof is similar to (i), but applying Lemma 7.34.

□

Remark 7.67 (Relationship between the types of variational convergence of bifunctions)

(a) Clear modifications need to be made to Theorems 7.65 and 7.66 to have characterizations of the weak inside lop-convergence and the types of ("strong") lop-convergence.
(b) It should be emphasized again that a weak type of variational convergence is properly weaker than the corresponding "usual" (strong) type. The origin of this important fact is already pointed out in Example 7.4 for the types of epi-convergence.
(c) By the inequalities given in Remark 7.64, the above theorem also reconfirms the important relationship between the types of variational convergence (which was observed after the definitions of the types of lop-convergence at the beginning of Sect. 7.2.2) that e/h- (or weak-e/h-) convergence holds if either of mis-lop (or weak-mis-lop, resp.) or mai-lop (or weak-mai-lop, resp.) convergence holds, i.e., lop-convergence is stronger than e/h-convergence.
The uniqueness case should be discussed in detail. From Theorem 7.66 and Remark 7.64, if the i-e/h-limit is unique, we only have for $(x, y) \in A \times B$,

$$\mathrm{lli}\Phi^k \le \mathrm{ehli}\Phi^k = \mathrm{ehls}\Phi^k \le \mathrm{lls}\Phi^k.$$

Hence, one or both inequalities may still be strict and, correspondingly, one or both i-mis-lop and i-mai-lop-convergence fail. The complete uniqueness occurs when the i-e/h-limit is unique and both i-mis-lop and i-mai-lop limits exist. Then, all the three limits coincide.

In Example 7.60, we apply the definition of the lop-convergence to verify whether or not a given Φ is a lop-limit of given Φ^k. In case we are not given a candidate Φ, it may be convenient to apply Theorem 7.66 as in the following example.

Example 7.68 (Finding lop-limits by their characterizations)

(i) Consider again Example 7.60, applying Theorem 7.66. For the mis-lop convergence (in the case of constant sequences, the weak and "strong" concepts are the same), direct computations (using the continuity of the logarithmic function) yield $\text{lls}\Phi^k(x, y) = \text{ehls}\Phi^k(x, y) = \Phi(x, y)$ for $x \in A$ and $y \in B$. Checking the four infinity conditions when $x = 0$ or $y = 0$ (outside A and B, resp.) is similar to Example 7.60, for condition (a) of lop-convergence, one has $\text{lls}\Phi^k(x, 0) = -\infty$ for $x \neq 0$ and $\text{ehls}\Phi^k(0, y) = +\infty$ for $y \neq 0$. Similarly, for condition (b), one has $\text{lli}\Phi^k(0, y) = +\infty$ for $y \neq 0$ and $\text{ehli}\Phi^k(x, 0) = -\infty$ for $x \neq 0$ as required. Hence, by Theorem 7.66, we have the unique mis-lop-$\lim_k \Phi^k$ that is Φ. Similar computations show that the mai-lop-$\lim_k \Phi^k$ is the unique Φ as well.

(ii) Revisit Example 7.39. We know that for $\Phi^k(x, y) = y^x$ on $[0, 1]^2$ for all k, by the continuity of Φ for $(x, y) \neq (0, 0)$, we need to consider only the point $(0, 0)$. We already have $\text{ehls}\Phi^k(0, 0) = 1$ and $\text{ehli}\Phi^k(0, 0) = 0$. Now, direct computations give $\text{lls}\Phi^k(0, 0) = 1$ and $\text{lli}\Phi^k(0, 0) = 0$. Hence, by Theorem 7.66, mis-lop-$\lim_k \Phi^k$ and mai-lop-$\lim_k \Phi^k$ are

$$\Phi_a(x, y) = \begin{cases} y^x & \text{if } (x, y) \in [0, 1]^2 \setminus \{(0, 0)\}, \\ a & \text{if } (x, y) = (0, 0) \end{cases}$$

with $a = 0$ and $a = 1$, resp.

Example 7.69 (Finding lop-limits when e/h-limits exist and are known) Let $A^k = A = (0, e]$, $B^k = B = (0, 1]$, and $\Phi^k(x, y) = \ln(y/x)$ for all k. Example 7.38 shows that the unique e/h-limit is $\ln(y/x)$. To find the two lop-limits, by the continuity of Φ^k, for (x, y) in the domain, all the four limits involved in Theorem 7.66 are equal to $\ln(y/x)$. Hence, we only need to calculate $\text{lls}\Phi^k(x, y)$ for $y = 0 \notin B$ and all $x \in A$ and $\text{lli}\Phi^k(x, y)$ for $x = 0 \notin A$ and all $y \in B$. To compute the above $\text{lls}\Phi^k(x, 0)$, take any $y^k \to 0^+$, $x \in A$, and $A^k \ni x^k \to x$. Then, for large k, $x^k > \sqrt{y^k}$ and $\ln\frac{y^k}{x^k} < \ln\sqrt{y^k} \to -\infty$. Hence, $\text{lls}\Phi^k(x, 0) = \inf_{\{A^k \ni x^k \to x\}} \sup_{\{B^k \ni y^k \to 0\}} \text{ls}_k \Phi^k(x^k, y^k) = -\infty$. So, the mis-lop-limit is $\ln(y/x)$ by Theorem 7.66. Similarly, to get the above $\text{lli}\Phi^k(0, y)$, take any $x^k \to 0^+$, $y \in B$, and $B^k \ni y^k \to y$. Then, for large k, $y^k > \sqrt{x^k}$ and $\ln\frac{y^k}{x^k} > \ln\frac{1}{\sqrt{x^k}} \to +\infty$. Hence, $\text{lli}\Phi^k(0, y) = \sup_{\{A^k \ni x^k \to 0\}} \inf_{\{B^k \ni y^k \to y\}} \text{li}_k \Phi^k(x^k, y^k) = +\infty$ and so the mai-lop-limit is $\ln(y/x)$. That is, all the e/h-, mis-lop-, and mai-lop-limits are unique and equal to $\ln(y/x)$.

The above characterizations of the lop-convergence are important, but the assumptions are relatively complicated, involving lower and upper epi-, hypo-, and lop-limits. In practice, sufficient conditions (in general not necessary) with simpler assumptions may be more convenient to use. We provide here one of such results for illustration only, since the focus of this chapter is approximation. Of course, this is also a sufficient condition for e/h-convergence.

Theorem 7.70 (Sufficiency for mis-lop convergence) *Let A be closed, $A^k \equiv A$, and $B^k \overset{P-K}{\to} B$. Then, we have the mis-lop convergence of Φ^k to Φ provided that there exists a lower semicontinuous (lsc) bifunction $\bar{\Phi} : X \times Y \to \mathbb{R}$ such that $\bar{\Phi}(x, \cdot)$ is upper semicontinuous (usc) for all $x \in A$, $\Phi = \bar{\Phi}$ on $A \times B$, and $\Phi^k = \bar{\Phi}$ on $A \times B^k$.*

Proof We verify first condition (a) of the mis-lop convergence. For all $y \in B$, we have $B^k \ni y^k \to y$ by the set convergence of B^k. As $\bar{\Phi}$ is lsc, for all $A^k \equiv A \ni x^k \to x$, $\mathrm{li}_k \bar{\Phi}(x^k, y^k) \geq \bar{\Phi}(x, y)$. Hence, for all $(x^{k_j}, y^{k_j}) \to (x, y)$, $\mathrm{li}_j \bar{\Phi}(x^{k_j}, y^{k_j}) \geq \bar{\Phi}(x, y)$. Following the expression of Φ and Φ^k in terms of $\bar{\Phi}$, for all $A^{k_j} \ni x^{k_j} \to x \in A$, there exists $B^{k_j} \ni y^{k_j} \to y$ such that $\mathrm{li}_j \Phi^{k_j}(x^{k_j}, y^{k_j}) \geq \Phi(x, y)$ as condition (a) requires.

To check condition (b) of mis-lop convergence, for every $x \in A$, take $x^k = x$ for all k. For every $B^k \ni y^k \to y$, y must be in B. As $\bar{\Phi}(x, \cdot)$ is usc, $\mathrm{ls}_k \bar{\Phi}(x^k, y^k) \leq \bar{\Phi}(x, y)$. Hence, for $x^{k_j} \equiv x$ and any $B^{k_j} \ni y^{k_j} \to y$, $\mathrm{ls}_j \Phi^{k_j}(x^{k_j}, y^{k_j}) \leq \Phi(x, y)$ as condition (b) requires. (Note that the infinity conditions in (a) and (b) are not involved in the assumptions of the theorem.) □

Note that Chapter 9 in [403] contains other sufficient conditions for mis-lop convergence.

Variational Properties of the Types of Lop-Convergence

Following Theorem 7.66 and Remark 7.67, if a type of either mis-lop or mai-lop convergence holds, then the corresponding type of e/h-convergence also holds. Hence, each type of lop-convergence has all the variational properties of the corresponding type of e/h-convergence (when tightness conditions are not involved), which we discussed in Sect. 7.2.1. Now, we study tight lop-convergence and their variational properties.

Definition 7.71 (Types of tight lop-convergence)

(i) (x-ancillary tight weak minsup-lop convergence) $\{\Phi^k\}_k$ x-ANCILLARY TIGHTLY WEAKLY MINSUP-LOP CONVERGES to Φ if one has condition (a) of the weak minsup-lop convergence and condition (b) strenghthened as (b-m) condition (b) holds, $\mathrm{dom}\xi^k \neq \emptyset$, and for the sequence $A^k \ni x^k \to x$ selected in (b), $\mathrm{ls}_k \xi^k(x^k) \leq \xi(x)$.

(ii) (x-tight weak minsup-lop convergence) $\{\Phi^k\}_k$ x-TIGHTLY WEAKLY MINSUP-LOP CONVERGES to Φ if it x-ancillary tightly minsup-lop converges and condition (a) is strengthened as
(a-m) condition (a) holds and for all $\varepsilon > 0$, one has a compact subset $A_\varepsilon \subset X$ such that for large k,

$$\inf_{x\in A_\varepsilon\cap A^k}\sup_{y\in B^k}\Phi^k(x, y) \leq \inf_{x\in A^k}\sup_{y\in B^k}\Phi^k(x, y) + \varepsilon.$$

(iii) The definitions of the corresponding types of tight weak maxinf-lop convergence are similar/symmetric, where one has "y-ancillary" instead of "x-ancillary" in condition (i), and "y-tightly" instead of "x-tightly" in condition (ii). If both the x-tight weak mis-lop and the y-tight weak mai-lop convergence occur, we say that the weak lop-convergence is (fully) tight.
The corresponding types of the weak inside convergence are naturally similar. So are also the types of the "strong" lop-convergence.

Unlike the types of tightness for e/h-convergence, the above definitions are not symmetric.

Proposition 7.72 (A type of epi-convergence of sup-projections in terms of the corresponding type of x-ancillary tight mis-lop convergence) *Assume* $\mathrm{dom}\xi^k \neq \emptyset$ *and a type of* Φ^k *mis-lop converges* x*-ancillary tightly to* Φ*. Then,* $\mathrm{dom}\xi \neq \emptyset$ *and the corresponding type of epi-convergence of* ξ^k *to* ξ *holds.*

Proof Under the assumption of a type of x-ancillary tight mis-lop convergence and the nonemptiness of $\mathrm{dom}\xi^k$, $\mathrm{dom}\xi \neq \emptyset$ and condition (a) of the corresponding type of epi-convergence are given in Theorem 7.46 because condition (a) is common for any type of e/h- and mis-lop convergence. So, it remains to check condition (b) in the type of epi-convergence. We write this for weak epi-convergence as an example. Condition (b-m) in the definition of x-ancillary tight weak mis-lop convergence gives $A^k \ni x^k \to x \in A$ such that $\mathrm{ls}_k\xi^k(x^k) \leq \xi(x)$, which is just the required condition (b). □

Note that since the tightness conditions for the e/h- and lop-convergence are incomparable, Theorem 7.51 and Proposition 7.72 are different. The same observation is valid for the subsequent results about the types of tight lop-convergence.

Theorem 7.73 (Convergence of minsup-points in terms of mis-lop-convergence)

(a) *Let* Φ^k *weakly inside mis-lop converge* x*-ancillary tightly to* Φ*. Then,*

$$\mathrm{ls}_k[\inf_{x\in A^k}\sup_{y\in B^k}\Phi^k(x, y) \leq \inf_{x\in A}\sup_{y\in B}\Phi(x, y).$$

(b) *Let* $\Phi^k \overset{mis-lop}{\to} \Phi$ x*-ancillary tightly. Then, for any* $\varepsilon, \varepsilon^k \geq 0$ *with* $\varepsilon^k \to \varepsilon$*,* $\mathrm{Ls}_k(\varepsilon^k\text{-mis}\Phi^k) \subset \varepsilon\text{-mis}\Phi$.

(c) *If* $\Phi^k \overset{w-mis-lop}{\to} \Phi$ x*-ancillary tightly,* $\varepsilon^k \to 0^+$*, and there exists* $x^k \in \varepsilon^k$-$\mathrm{mis}\Phi^k$ *converging to* x*, then* $x \in \mathrm{mis}\Phi$ *and* $\min_{A^k}\sup_{B^k}\Phi^k \to \min_A\sup_B\Phi$.

Proof Apply Propositions 7.72 and then 7.3. □

With some additional compactness and lower semicontinuity assumptions like those mentioned after Example 7.53, the existence of minsup-points of Φ is also ensured by Theorem 7.73.

Of course, we have also the version corresponding to the above theorem for maxinf-points in terms of y-ancillary tight mai-lop convergence. Note that Proposition 7.72 and Theorem 7.73 provide sufficient conditions, which are not necessary.

Theorem 7.74 (Complete convergence of minsup-points under x-tight mis-lop convergence)

(a) *Assume the x-tightness of one of the types of mis-lop convergence of Φ^k to Φ. Then, we have the tightness of the corresponding types of epi-convergence of ξ^k to ξ.*

(b) *If Φ^k mis-lop converges x-tightly to Φ, then*

$$\inf_{x\in A^k}\sup_{y\in B^k}\Phi^k(x, y) \to \inf_{x\in A}\sup_{y\in B}\Phi(x, y) \text{ finitely.}$$

(c) *If Φ^k mis-lop converges x-tightly to Φ, then there exists $\varepsilon^k \to 0^+$ such that $\varepsilon^k\text{-mis}\Phi^k \overset{P-K}{\to} \text{mis}\Phi$.*

Proof

(a) The corresponding types of epi-convergence of ξ^k to ξ are asserted in Proposition 7.72. By the x-tightness defined in Definition 7.71(ii), the tightness of the mentioned types of epi-convergence is guaranteed.

(b) Apply the above assertion (a) and then Proposition 7.9(a), with the observation that $\Phi^k \overset{mis-lop}{\to} \Phi$ x-tightly implies that $\xi^k \overset{e}{\to} \xi$ tightly, to see that the limit is finite.

(c) The proof is similar to (a) with the use of (c) of Proposition 7.9 instead of (a). (Note here that $\varepsilon\text{-mis}\Phi = \varepsilon\text{-argmin}\xi$.)

□

Theorem 7.74 shows an advantage of the lop-convergence over the e/h-convergence: the convergence of infsup-values in (b) and the complete convergence of approximate minsup-points in (c) are specialties of the x-tight mis-lop convergence. This advantage stems from the tight epi-convergence of sup-projections as mentioned in (a). We also have a "complete" counterpart for lop-convergence of Theorem 7.54 (for e/h-convergence) in Theorem 7.75. A very recent examination of convergence of minsup-points and minsup-values can be found in [403].

Theorem 7.75 (Complete convergence of approximate saddle points under tight lop-convergence) *Let $\varepsilon, \varepsilon^k \geq 0$ and $\varepsilon^k \to \varepsilon$.*

(a) *If Φ^k inside mis-lop converges both x- and y-ancillary-tightly or inside mai-lop converges both x- and y-ancillary-tightly to Φ, then* ε-sdl$\Phi \subset \bigcap_{\delta>\varepsilon} \text{Li}_k(\delta\text{-sdl}\Phi^k)$ *and so* ε-sdl$\Phi \subset \text{Li}_k(\varepsilon^k\text{-sdl}\Phi^k)$.

(b) *If one removes the restriction "inside" in assertion* (a), *then*

$$\text{Lim}_k(\varepsilon^k\text{-sdl}\Phi^k) = \varepsilon\text{-sdl}\Phi,$$
$$\text{Ls}_k(\text{str-}\varepsilon^k\text{-sdl}\Phi^k) \subset \text{str-}\varepsilon\text{-sdl}\Phi.$$

Proof

(a) The arguments are similar to those in the proof of assertion (i) of Theorem 7.54.
(b) Similar to assertion (ii) of Theorem 7.54, here we apply Theorem 7.47(c) (which is of course valid for the lop-convergence), assertion (a) of Theorems 7.75, and 7.73 instead of Theorem 7.52.

□

Here, we should emphasize again that epi/hypo-convergence is a better means of considering saddle points since simultaneous x-tight mis-lop and y-tight mai-lop convergence is a more restrictive condition than the tight e/h-convergence.

7.3 Approximations of Equilibrium Problems

Consider the following EQUILIBRIUM PROBLEM

$$\text{(EP)} \qquad \text{find } \bar{x} \in A \text{ such that } \Phi(\bar{x}, y) \leq 0 \text{ for all } y \in B,$$

where X and Y are metric spaces, $A \times B \subset X \times Y$, and $\Phi : A \times B \to \mathbb{R}$. (As explained at the beginning of Sect. 7.2, though we write $\Phi : A \times B \to \mathbb{R}$ or $\Phi \in \text{rec}(X \times Y)$, Φ is finite on $A \times B$ but in fact, Φ is defined on $X \times Y$.) This equilibrium problem is also called a KY FAN INEQUALITY. As mentioned at the beginning of this chapter, the model (EP) here is in terms of subsets A and B in different spaces, not equal sets $A = B$ as in the preceding chapters, to meet the general situations in practice. This (EP) model has been shown to be a genuine generalization of many optimization-related problems. In the literature, the economical and social meaning of "equilibrium" has been confirmed more clearly than its physical/technical meaning, by many considerations of practical problems such as traffic networks, Nash games (noncooperative games), and economic models. Observe also that in consideration of approximations of (EP) in terms of variational convergence, we let (EP) to involve the inequality $\Phi(\bar{x}, y) \leq 0$, which is reverse to the corresponding inequality in the preceding chapters, in order to suit better e/h-convergence and minsup framework (not h/e-convergence and maxinf framework). Any $\bar{x} \in A$ with $\sup_{y \in B} \Phi(\bar{x}, y) \leq 0$ belongs to Sol(EP), the solution set of (EP), including of course

Fig. 7.7 Sol(EP) $= [0, 1]$, mis$\Phi = \{1\}$

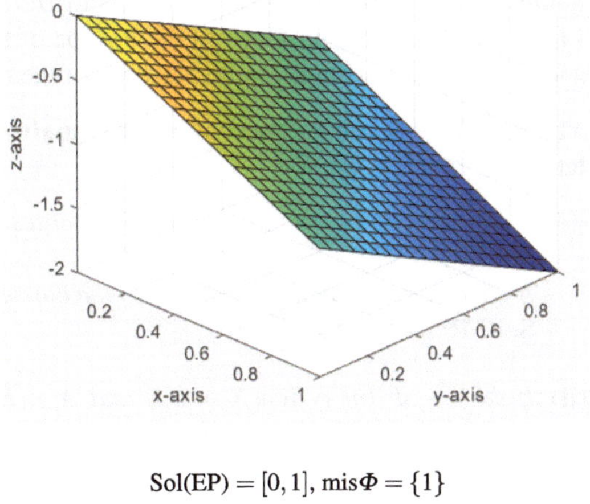

the minsup-points of Φ which satisfy this inequality. But a solution of (EP) may not be a minsup-point of Φ (simply think of $\Phi(x, y) = -x - y$ and $A = B = [0, 1]$; then Sol(EP) $= [0, 1]$ but mis$\Phi = \{1\}$); see Fig. 7.7. Therefore, we also investigate the following STRONG EQUILIBRIUM PROBLEM :

$$\text{(sEP)} \qquad \text{find } \bar{x} \in A \text{ solving } \min_{x\in A}\sup_{y\in B}\Phi(x, y) \leq 0.$$

OBSERVATION. The above simple example shows that *(EP) is quite different from (sEP)* since Sol(EP) $= [0, 1]$ and Sol(sEP) $= \{1\}$. Any solution of (sEP) clearly also solves (EP), but the converse is not true. This should be noted here since some contributions in the literature may confuse readers on this issue. The second important fact is that (sEP) has been defined very recently. Usually, in the scalar case, only (EP) is considered.

The solution set of (sEP) is denoted by Sol(sEP). The definition of approximate solutions is natural: for $\varepsilon \geq 0$, $\bar{x}_\varepsilon \in A$ is an ε-SOLUTION of (EP) or of (sEP), denoted by $\bar{x}_\varepsilon \in \varepsilon$-Sol(EP) or $\bar{x}_\varepsilon \in \varepsilon$-Sol(sEP), if

$$\Phi(\bar{x}_\varepsilon, y) \leq \varepsilon \text{ for all } y \in B,$$

$$\bar{x}_\varepsilon \in \text{mis}\Phi \text{ such that } \min_A\sup_B\Phi \leq \varepsilon, \text{ resp.}$$

Assume for (EP) that one wants to have a sequence of good (in a certain sense) approximating problems (EP^k) defined (with an identical setting as (EP)) by $A^k \subset X$, $B^k \subset Y$, and $\Phi^k : A^k \times B^k \to \mathbb{R}$. For instance, in statistics, the original problem (EP) may be known, by sampling, only inexactly via a sequence of probability distributions and one needs its estimators (EP^k) whose variational objects converge to those of (EP). That is, one needs (EP^k) to converge variationally to (EP). A similar situation for (sEP) and (sEP^k) may be met in practice as well.

For considering approximations of (EP), in some cases, we need the following notion. $A \times B$ is said to be STRONGLY INCLUDED in $\mathrm{Li}_k(A^k \times B^k)$, written $A \times B \sqsubset \mathrm{Li}_k(A^k \times B^k)$, if the presence of either type of the convergence, $A^{k_j} \ni x^{k_j} \to x$ and $B^{k_j} \ni y^{k_j} \to y$, implies the presence of the other.

Theorem 7.76 (Convergence of approximate solutions of equilibrium problems) *Let* $\varepsilon, \varepsilon^k \geq 0$ *and* $\varepsilon^k \to \varepsilon$.

(i) *If* Φ^k *e/h- (or mis-lop or mai-lop) converges to* Φ, *then* $\mathrm{Ls}_k(\varepsilon^k\text{-Sol}(\mathrm{EP}^k)) \subset \varepsilon\text{-Sol(EP)}$.
If the convergence of Φ^k *reduces to the corresponding inside type, then* $\mathrm{Ls}_k(\varepsilon^k\text{-Sol}(\mathrm{EP}^k)) \cap A \subset \varepsilon\text{-Sol(EP)}$.

(ii) *If* $\Phi^k \xrightarrow{e} \Phi$ *(in* fv-fcn$(X \times Y)$*) and* $A \times B \sqsubset \mathrm{Li}_k(A^k \times B^k)$, *then* $\mathrm{Ls}_k(\varepsilon^k\text{-Sol}(\mathrm{EP}^k)) \subset \varepsilon\text{-Sol(EP)}$. *If in the above assertion, epi-convergence reduces to the inside one, then* $\mathrm{Ls}_k(\varepsilon^k\text{-Sol}(\mathrm{EP}^k)) \cap A \subset \varepsilon\text{-Sol(EP)}$.

Proof

(i) Let $\bar{x} \in \mathrm{Ls}_k(\varepsilon^k\text{-Sol}(\mathrm{EP}^k))$, i.e., there exists x^{k_j} in $\varepsilon^{k_j}\text{-Sol}(\mathrm{EP}^{k_j})$ converging to $\bar{x}$ as $j \to +\infty$. If $\bar{x} \in A$, then for every $y \in B$, by condition (a) of e/h-convergence, there is $B^{k_j} \ni y^{k_j} \to y$ such that $\mathrm{li}_j \Phi^{k_j}(x^{k_j}, y^{k_j}) \geq \Phi(\bar{x}, y)$. As $x^{k_j} \in \varepsilon^{k_j}\text{-Sol(EP)}^{k_j}$, $\Phi(\bar{x}, y) \leq \varepsilon$, i.e., $\bar{x}$ lies in ε-Sol(EP). Suppose that $\bar{x} \notin A$. Again by condition (a), one has $y^{k_j} \in B^{k_j}$ such that $\Phi^{k_j}(x^{k_j}, y^{k_j}) \to +\infty$, contradicting the fact that $\Phi^{k_j}(x^{k_j}, y) \leq \varepsilon^{k_j}$ for all $y \in B^{k_j}$.
The argument for the case Φ^k i-e/h-converges to Φ is similar. For the assertions about lop-convergence, apply Remarks 7.59(d) and 7.67(c) about the relationship between e/h- and lop-convergence.

(ii) We only give the proof for the case Φ^k epi-converges to Φ. Let x^{k_j} in $\varepsilon^{k_j}\text{-Sol}(\mathrm{EP}^{k_j})$ tend to $\bar{x}$. If $\bar{x} \notin A$, then for every $B^{k_j} \ni y^{k_j} \to y$, $(\bar{x}, y) \notin A \times B$. Hence, condition (a) of the epi-convergence implies that $\Phi^{k_j}(x^{k_j}, y^{k_j}) \to +\infty$, which is impossible as $\Phi^{k_j}(x^{k_j}, y^{k_j}) \leq \varepsilon^{k_j}$ for all $y^{k_j} \in B^{k_j}$. So, $\bar{x} \in A$. It remains to show that $\Phi(\bar{x}, y) \leq \varepsilon$ for each $y \in B$. The assumed strong inclusion implies that for each $y \in B$, there exists $B^{k_j} \ni y^{k_j} \to y$. Then, the above condition (a) implies that $\mathrm{li}_j \Phi^{k_j}(x^{k_j}, y^{k_j}) \geq \Phi(\bar{x}, y)$. Hence, $\Phi(\bar{x}, y) \leq \varepsilon$ as $\Phi^{k_j}(x^{k_j}, y^{k_j}) \leq \varepsilon^{k_j}$.

□

Theorem 7.77 (Convergence of approximate solutions of strong equilibrium problems) *Let* $\varepsilon, \varepsilon^k \geq 0$ *and* $\varepsilon^k \to \varepsilon$.

(a) *If* Φ^k *e/h-converges x-ancillary tightly or mis-converges x-ancillary tightly to* Φ, *then* $\mathrm{Ls}_k(\varepsilon^k\text{- Sol}(\mathrm{sEP}^k)) \subset \varepsilon\text{-Sol(sEP)}$.
If the above types of convergence reduce to the inside types, then $\mathrm{Ls}_k(\varepsilon^k\text{-Sol}(\mathrm{sEP}^k)) \cap A \subset \varepsilon\text{- Sol(sEP)}$.

(b) *If* $\Phi^k \overset{w-e/h}{\to} \Phi$ *x-ancillary tightly or* $\Phi^k \overset{w-mis-lop}{\to} \Phi$ *x-ancillary tightly,* $\varepsilon = 0$*, and there exists* $x^k \in \varepsilon^k$-Sol(sEPΦ^k) *converging to* x*, then* $x \in$ Sol(sEPΦ) *and* $\min_{A^k}\sup_{B^k}\Phi^k \to \min_A\sup_B\Phi$.

Proof

(a) Consider the case of e/h-convergence. By similarity, we only discuss the first assertion. Let $\bar{x} \in \mathrm{Ls}_k(\varepsilon^k$- Sol(sEP^k)). If $\bar{x} \in A$, from part (i) in the proof of Theorem 7.76, we have $\Phi(\bar{x}, y) \le \varepsilon$ for all $y \in B$. Next, we employ Theorem 7.52(b) (with $\varepsilon^k \equiv \varepsilon$ in that theorem) to see that $\bar{x} \in \mathrm{mis}\Phi$ with $\min_A\sup_B\Phi \le \varepsilon$. Hence, $\bar{x} \in \varepsilon$-Sol(sEP). For the case $\bar{x} \notin A$, the argument is the same as for Theorem 7.76(a).
The argument for the case of mis-lop convergence is the same as above, with Theorem 7.73(b) replacing Theorem 7.52(b).

(b) The proof resembles (a), but applying both Theorems 7.52(c) and 7.73(c).

□

For the case $\varepsilon^k \equiv 0$, the preceding two results can be stated in terms of outer semicontinuity of solution maps as follows. For instance, look only at Theorem 7.76. Consider the (set-valued) solution map $\mathscr{S}$: fv-biv$(X \times Y) \rightrightarrows X$ defined by $\mathscr{S}(\Phi) :=$ Sol(EP). Then, the Limsup inclusion in Theorem 7.76 can be restated as the outer semicontinuity of $\mathscr{S}$.

We consider another type of approximate solutions to show the possible variety of mathematical definitions to meet practical applications when necessary. For the case $A = B$ is a subset of a normed space X, let us also investigate variational approximations of equilibrium problems with the following concept of approximate solutions. Here, the tolerance of the approximation is not controlled by ε for all $y \in B$ as for the first-kind approximate solution, but depends on the difference $\bar{x}_{\varepsilon\theta} - y$ where y varies in B.

Definition 7.78 (A second kind of approximate solutions) Let $X = Y$ be a normed space, $A = B$, and $\theta : X \to \mathbb{R}$ be such that $\theta(x) > 0$ if $x \neq 0$. For $\varepsilon \ge 0$, $\bar{x}_{\varepsilon\theta} \in A$ is called an $\varepsilon\theta$-SOLUTION of (EP) if $\Phi(\bar{x}_{\varepsilon\theta}, y) \le \varepsilon\theta(\bar{x}_{\varepsilon\theta} - y)$ for all $y \in A$. Denote the set of the $\varepsilon\theta$-solutions of (EP) by $\varepsilon\theta$-Sol(EP).

Note that no corresponding notion is defined for strong equilibrium problems. The proof of the following statement is a modification of that of Theorem 7.76, and so is omitted. The reader can refer to the proof of Proposition 8.54 in the next chapter, a more general result.

Theorem 7.79 (Convergence of second-kind approximate solutions) *Let* $\varepsilon, \varepsilon^k \ge 0$ *and* $\varepsilon^k \to \varepsilon$.

(a) *If* Φ^k *i-e/h- (or i-mis-lop or i-mai-lop) converges to* Φ *and* θ *is continuous, then*

$$\mathrm{Ls}_k(\varepsilon^k\theta\text{-Sol(EP}^k)) \cap A \subset \varepsilon\theta\text{-Sol(EP)}.$$

If Φ^k e/h- (or mis-lop) converges to Φ, θ is bounded from above, and $\varepsilon = 0$, then

$$\mathrm{Ls}_k(\varepsilon^k\theta\text{-Sol}(\mathrm{EP}^k)) \subset \mathrm{Sol(EP)}.$$

(b) *If $\Phi^k \xrightarrow{e} \Phi$ as unifunctions (on $A^k \times B^k$ and $A \times B$, resp.), instead of the convergence of Φ^k as bifunctions in assertion* (a)*, then we have the same convergence of second-kind approximate solutions as in assertion* (a) *under the additional assumption $A \times B \sqsubset \mathrm{Li}_k(A^k \times B^k)$.*

We consider also the following DUAL EQUILIBRIUM PROBLEMS:

(DEP) find $\bar{y} \in B$ such that $\Phi(x, \bar{y}) \geq 0$ for all $x \in A$,

(DsEP) find $\bar{y} \in \mathrm{mai}\Phi$ such that $\max_B \inf_A \Phi \geq 0$.

(EP) is also usually called a STAMPACCHIA EQUILIBRIUM PROBLEM and (DEP) is a MINTY EQUILIBRIUM PROBLEM. Observe that (EP) and (DEP) are dual to each other, i.e., the dual of (DEP) is just (EP). Indeed, by setting $\overline{\Phi}(\bar{y}, x) = -\Phi(x, \bar{y})$, (DEP) becomes $(\overline{\mathrm{EP}})$ for $\overline{\Phi}$ on $B \times A$. Then, it is easy to check that the dual to $(\overline{\mathrm{EP}})$ is just (EP). Similarly, the dual of (DsEP) is again (sEP).

The two (primal-dual) pairs (EP)-(DEP) and (sEP)-(DsEP) have different properties. Following Proposition 7.44, if $(\bar{x}, \bar{y})$ is a saddle point of Φ in $A \times B$, then $\bar{x}$ is a minsup-point of Φ and $\bar{y}$ is a maxinf-point of Φ. However, the reverse implications do not hold in general: $\bar{x}$ is a minsup-point of Φ and $\bar{y}$ is a maxinf-point of Φ do not imply that $(\bar{x}, \bar{y})$ is a saddle point of Φ. This reverse implication holds in the following case.

Theorem 7.80 (Saddle components are primal and dual solutions) *We have the equivalent assertions:*

(a) *$(\bar{x}, \bar{y})$ is a saddle point of Φ and $\Phi(\bar{x}, \bar{y}) = 0$;*
(b) *$\bar{x}$ is a solution of* (EP) *and $\bar{y}$ is that of* (DEP)*;*
(c) *$\bar{x}$ is a solution of* (sEP) *and $\bar{y}$ is that of* (DsEP).

Proof (a) $\Leftrightarrow$ (b). The "only if" assertion is evident. For the "if" part, observe first that $\Phi(\bar{x}, \bar{y})$ must be zero. Then, $\Phi(\bar{x}, y) \leq 0 = \Phi(\bar{x}, \bar{y}) \leq \Phi(x, \bar{y})$ for all $(x, y) \in A \times B$.

(a) $\Leftrightarrow$ (c). The "only if" assertion is obvious. The "if" part is immediate from (b) $\Rightarrow$ (a) as a solution of a strong problem is of course a solution of a "usual" problem. □

Note that solutions of any one of the aforementioned four problems are not unique in general. To see this, simply think of the case $\Phi = 0$ (the zero bifunction) and arbitrary A and B, where any point in $A \times B$ is a solution pair of the strong problems (sEP) and (DsEP). Observe from Theorem 7.80 that we have a zero duality gap for any couple of solutions $\bar{x}$, $\bar{y}$ of (EP) and (DEP), resp.

APPROXIMATE SOLUTIONS of the dual problems (DEP) and (DsEP) are defined as follows: for $\varepsilon \geq 0$, $\bar{y}_\varepsilon \in B$ is an ε-solution of (DEP) or $\bar{y}_\varepsilon \in B$ is an ε-solution of (DsEP) if

$$\Phi(x, \bar{y}_\varepsilon) \geq -\varepsilon \text{ for all } x \in A,$$

$$\bar{y}_\varepsilon \in \text{mai}\Phi \text{ such that } \max_B \inf_A \Phi \geq -\varepsilon, \text{ resp.}$$

Together with the condition $\Phi(\bar{x}_\varepsilon, \bar{y}_\varepsilon) = 0$, we obtain the following relationship between the two types of approximate saddle points and approximate solutions of (EP) and (sEP) and their dual problems, which corresponds to Theorem 7.80 for exact solutions. The subtle difference between the two theorems shows the need for the two types of problems and of approximate saddle points (defined prior to Theorem 7.46).

Theorem 7.81 (Approximate sdl-components are approximate primal and dual solutions)

(a) $\bar{x}_\varepsilon$ *is an* ε*-solution of* (EP), $\bar{y}_\varepsilon$ *is that of* (DEP), *and* $\Phi(\bar{x}_\varepsilon, \bar{y}_\varepsilon) = 0$ *if and only if* $(\bar{x}_\varepsilon, \bar{y}_\varepsilon)$ *is an* ε*-saddle point of* Φ *with* $\Phi(\bar{x}_\varepsilon, \bar{y}_\varepsilon) = 0$.
(b) $\bar{x}_\varepsilon$ *is an* ε*-solution of* (sEP), $\bar{y}_\varepsilon$ *is that of* (DsEP), *and* $\Phi(\bar{x}_\varepsilon, \bar{y}_\varepsilon) = 0$ *if and only if* $(\bar{x}_\varepsilon, \bar{y}_\varepsilon)$ *is a strong* ε*-saddle point of* Φ *with* $\Phi(\bar{x}_\varepsilon, \bar{y}_\varepsilon) = 0$.

Proof

(a) $\bar{x}_\varepsilon \in \varepsilon$-Sol(EP) means that $\sup_B \Phi(\bar{x}_\varepsilon, \cdot) \leq \varepsilon$, i.e., $\inf_A \sup_B \Phi(\cdot, \cdot) \leq \xi(\bar{x}_\varepsilon) \leq \varepsilon$. This together with $\bar{y}_\varepsilon \in \varepsilon$-Sol(DEP) is equivalent to

$$-\varepsilon \leq \eta(\bar{y}_\varepsilon) \leq \sup_B \inf_A \Phi(\cdot, \cdot) \leq \inf_A \sup_B \Phi(\cdot, \cdot) \leq \xi(\bar{x}_\varepsilon) \leq \varepsilon. \tag{7.8}$$

These inequalities together with $\Phi(\bar{x}_\varepsilon, \bar{y}_\varepsilon) = 0$ imply that

$$\xi(\bar{x}_\varepsilon) - \varepsilon \leq 0 = \Phi(\bar{x}_\varepsilon, \bar{y}_\varepsilon) \leq \eta(\bar{y}_\varepsilon) + \varepsilon,$$

i.e., $(\bar{x}_\varepsilon, \bar{y}_\varepsilon) \in \varepsilon$-sdl$\Phi$.
Conversely, if $(\bar{x}_\varepsilon, \bar{y}_\varepsilon) \in \varepsilon$-sdl$\Phi$ and $\Phi(\bar{x}_\varepsilon, \bar{y}_\varepsilon) = 0$, then $0 = \Phi(\bar{x}_\varepsilon, \bar{y}_\varepsilon) \leq \Phi(x, \bar{y}_\varepsilon) + \varepsilon$ for all $x \in A$, i.e., $\bar{y}_\varepsilon \in \varepsilon$-Sol(DEP). And similarly one has also $\bar{x}_\varepsilon \in \varepsilon$-Sol(EP).
(b) $\bar{x}_\varepsilon \in \varepsilon$-Sol(sEP) and $\bar{y}_\varepsilon \in \varepsilon$-Sol(DsEP) mean that

$$-\varepsilon \leq \eta(\bar{y}_\varepsilon) = \sup_B \inf_A \Phi(\cdot, \cdot) \leq \inf_A \sup_B \Phi(\cdot, \cdot) = \xi(\bar{x}_\varepsilon) \leq \varepsilon,$$

which is stronger than (7.8) due to the two equalities. So, it is also true that $(\bar{x}_\varepsilon, \bar{y}_\varepsilon) \in \varepsilon$-sdl$\Phi$. The two equalities mean that $\bar{y}_\varepsilon \in \text{mai}\Phi$ and $\bar{x}_\varepsilon \in \text{mis}\Phi$. Hence, $(\bar{x}_\varepsilon, \bar{y}_\varepsilon) \in$ str-ε-sdlΦ.

□

For dual problems, let us record only the following statement corresponding to Theorem 7.76 for primal problems since the two proofs are quite similar.

Theorem 7.82 (Convergence of approximate solutions of dual equilibrium problems) *We have all the assertions for the dual problems corresponding to those of Theorem 7.76, with the replacement of epi-convergence by hypo-convergence in assertion* (ii).

Theorem 7.77 is accompanied by the following statement for dual strong problems, whose proof is analogous and so omitted.

Theorem 7.83 (Convergence of approximate solutions of dual strong equilibrium problems) *Let* $\varepsilon, \varepsilon^k \geq 0$ *and* $\varepsilon^k \to \varepsilon$.

(a) *If* Φ^k *e/h-converges y-ancillary tightly or mai-lop converges y-ancillary tightly to* Φ, *then* $\mathrm{Ls}_k(\varepsilon^k\text{- Sol(DsEP}^k)) \subset \varepsilon\text{-Sol(DsEP)}$.
If the above types of convergence are only inside, then $\mathrm{Ls}_k(\varepsilon^k\text{-Sol(DsEP}^k)) \cap B \subset \varepsilon\text{-Sol(DsEP)}$.
(b) *If* $\Phi^k \overset{w-e/h}{\to} \Phi$ *y-ancillary tightly or* $\Phi^k \overset{w-mai-lop}{\to} \Phi$ *y-ancillary tightly to* Φ, $\varepsilon = 0$, *and* $y^k \in \varepsilon^k\text{-Sol(DsEP}^k)$ *converges to* y, *then* $y \in \text{Sol(DsEP)}$ *and* $\mathrm{maxinf}\Phi^k \to \mathrm{maxinf}\Phi$.

Denote

$$\mathscr{C}\Phi^k := \{(x, y) \in A \times B \mid \Phi^k(x, y) = 0\},\ \mathscr{C}\Phi := \{(x, y) \in A \times B \mid \Phi(x, y) = 0\}.$$

Theorem 7.84 (Convergence of approximate solutions of pairs of primal and dual problems and approximate saddle points) *Let* $\varepsilon, \varepsilon^k \geq 0$ *and* $\varepsilon^k \to \varepsilon$.

(a) *If* $\Phi^k \overset{i-e/h}{\to} \Phi$ *or* $\Phi^k \overset{i-mis-lop}{\to} \Phi$ *or* $\Phi^k \overset{i-mai-lop}{\to} \Phi$ *and* $\mathrm{Ls}_k\mathscr{C}\Phi^k \subset \mathrm{cl}\mathscr{C}\Phi$, *then*

$$\begin{aligned}&\mathrm{Ls}_k(\mathscr{C}\Phi^k \cap \varepsilon^k - \mathrm{sdl}\Phi^k) \cap (A \times B)\\ &\quad= \mathrm{Ls}_k[\mathscr{C}\Phi^k \cap (\varepsilon^k - \mathrm{Sol(EP}^k) \times \varepsilon^k - \mathrm{Sol(DEP}^k))] \cap (A \times B)\\ &\subset \mathrm{cl}\mathscr{C}\Phi \cap (\varepsilon - \mathrm{Sol(EP)} \times \varepsilon - \mathrm{Sol(DEP)}) = \mathrm{cl}\mathscr{C}\Phi \cap \varepsilon - \mathrm{sdl}\Phi.\end{aligned}$$

If the above restriction "inside" is removed, then "$\cap (A \times B)$*" is also removed from the above relation.*
(b) *If* Φ^k *i-e/h- tightly or both x- and y-ancillary-tightly i-mis-lop or both x- and y-ancillary-tightly i-mai-lop converges to* Φ *and* $\mathrm{cl}\mathscr{C}\Phi \subset \mathrm{Li}_k\mathscr{C}\Phi^k$, *then*
$$\begin{aligned}\varepsilon\text{-sdl}\Phi \cap \mathrm{cl}\mathscr{C}\Phi &= (\varepsilon\text{-Sol(EP)} \times \varepsilon\text{-Sol(DEP)}) \cap \mathrm{cl}\mathscr{C}\Phi\\ &\subset \cap_{\delta>\varepsilon}\mathrm{Li}_k(\delta\text{-Sol(EP}^k) \times \delta\text{-Sol(DEP}^k)) \cap \mathrm{Li}_k\mathscr{C}\Phi^k\\ &= \cap_{\delta>\varepsilon}\mathrm{Li}_k\delta\text{-sdl}\Phi^k \cap \mathrm{Li}_k\mathscr{C}\Phi^k.\end{aligned}$$

Hence,

$$\varepsilon\text{-sdl}\Phi \cap \text{cl}\mathscr{C}\Phi = (\varepsilon\text{-Sol(EP)} \times \varepsilon\text{-Sol(DEP)}) \cap \text{cl}\mathscr{C}\Phi$$
$$\subset \text{Li}_k(\varepsilon^k\text{-Sol(EP}^k) \times \varepsilon^k\text{-Sol(DEP}^k)) \cap \text{Li}_k\mathscr{C}\Phi^k = \text{Li}_k\varepsilon^k\text{-sdl}\Phi^k \cap \text{Li}_k\mathscr{C}\Phi^k.$$

(c) *If* $\text{Ls}_k\mathscr{C}\Phi^k \subset \text{cl}\mathscr{C}\Phi$, $\Phi^k \overset{w-e/h}{\rightarrow} \Phi$ *tightly or both x-ancillary-tightly w-mislop and y-ancillary-tightly w-mai-lop converges simultaneously,* $\varepsilon = 0$, *and* $(x^k, y^k) \in (\varepsilon^k\text{-Sol(sEP}^k) \times \varepsilon^k\text{- Sol(DsEP}^k) \cap \mathscr{C}\Phi^k$ *tends to* (x, y), *then* $(x, y) \in (\text{Sol(sEP)}) \times (\text{Sol(DsEP)}) \cap \text{sdl}\Phi \cap \text{cl}\mathscr{C}\Phi$.

(d) *Assume that* $\text{Lim}_k\mathscr{C}\Phi^k = \text{cl}\mathscr{C}\Phi$, *the above restriction "inside" in assertion* (b) *is removed, and the involved types of lop-convergence are imposed together (i.e., the word "or" is replaced by "and"). Then,*

$$\varepsilon\text{-sdl}\Phi \cap \text{cl}\mathscr{C}\Phi = (\varepsilon\text{-Sol(EP)} \times \varepsilon\text{-Sol(DEP)}) \cap \text{cl}\mathscr{C}\Phi$$
$$= \text{Lim}_k(\varepsilon^k\text{-Sol(EP}^k) \times \varepsilon^k\text{-Sol(DEP}^k)) \cap \text{Lim}_k\mathscr{C}\Phi^k$$
$$= \text{Lim}_k\varepsilon^k\text{-sdl}\Phi^k \cap \text{Lim}_k\mathscr{C}\Phi^k.$$

Proof

(a) This assertion follows from Theorems 7.76, 7.80, 7.81, and 7.82 (taking into account that the Limsup of the intersection of two sets is included in the intersection of the Limsups of these sets and that the Limsup of the product of sets is equal to the product of the Limsups of these sets).

(b) We see that the two required equalities follow from Theorems 7.80 and 7.81. It is known that the Liminf of the intersection of two sets is included in the intersection of the Liminfs of these sets and that the Liminf of the product of sets is equal to the product of the Liminfs of these sets. Hence, the conclusion follows from Theorems 7.54(i) and 7.75(a).

(c) The proof is a simple application of Theorems 7.77, 7.80, 7.81, and 7.83.

(d) We need to show the equality in the chain of relations of this assertion. By the second chain of relations in assertion (b), we only need to check $\text{Ls}_k(\varepsilon^k\text{-wsdl}\Phi^k) \subset \varepsilon\text{-wsdl}\Phi$. By assertion (a), we have to show only that any subsequence $(\bar{x}^{k_j}, \bar{y}^{k_j}) \in \varepsilon^{k_j}\text{-wsdl}\Phi^{k_j}$ cannot tend to a point outside $A \times B$. Suppose to the contrary the existence of $(\bar{x}^{k_j}, \bar{y}^{k_j}) \to (x, y) \notin A \times B$, say $x \notin A$. We argue only the case of tight e/h-convergence. By condition (a) of this convergence, there is $y^{k_j} \in B^{k_j}$ with $\Phi^{k_j}(\bar{x}^{k_j}, y^{k_j}) \to +\infty$. Then, for any $x^{k_j} \in A^{k_j}$, $\Phi^{k_j}(x^{k_j}, \bar{y}^{k_j}) + \varepsilon^{k_j} \geq \Phi^{k_j}(\bar{x}^{k_j}, y^{k_j}) - \varepsilon^{k_j} \to +\infty$. We have two cases: $y \in B$ or $y \notin B$. If $y \in B$, then condition (b) in the definition of e/h-convergence implies that for any $\hat{x} \in A$, there exists $A^{k_j} \ni \hat{x}^{k_j} \to \hat{x}$ such that $\text{ls}_j\Phi^{k_j}(\hat{x}^{k_j}, \bar{y}^{k_j}) \leq \Phi(\hat{x}, y)$, a contradiction. So only $y \notin B$ remains possible. Then, again by the above assertion (b), there exists $\hat{x}^{k_j} \in A^{k_j}$ with $\Phi^{k_j}(\hat{x}^{k_j}, \bar{y}^{k_j}) \to -\infty$, which is impossible as well.

□

From Theorems 7.54, 7.75, 7.77, 7.80, 7.81, and 7.83, we have the following statement for the pair (sEP)-(DsEP), corresponding to Theorem 7.84(a) for (EP)-(DEP). We need to notice the absence of the counterparts of assertions (b) and (c) in that theorem here. This difference stems from Theorems 7.54 and 7.75.

Theorem 7.85 (Convergence of pairs of approximate solutions of strong problems and strong approximate saddle points) *Let $\varepsilon, \varepsilon^k \geq 0$ and $\varepsilon^k \to \varepsilon$. If Φ^k i-e/h-converges tightly or both i-mis-lop converges x- and y-ancillary-tightly and i-mai-lop converges x-and y-ancillary-tightly to Φ, and $\mathrm{Ls}_k \mathscr{C}\Phi^k \subset \mathrm{cl}\mathscr{C}\Phi$, then*

$$\mathrm{Ls}_k \mathscr{C}\Phi^k \cap (\mathrm{str} - \varepsilon^k - \mathrm{sdl}\Phi^k) \cap (A \times B)$$

$$= \mathrm{Ls}_k[\mathscr{C}\Phi^k \cap (\varepsilon^k - \mathrm{Sol(sEP)}^k \times \varepsilon^k - \mathrm{Sol(DsEP}^k))] \cap (A \times B)$$

$$\subset \mathrm{cl}\mathscr{C}\Phi \cap (\varepsilon - \mathrm{Sol(sEP)} \times \varepsilon - \mathrm{Sol(DsEP)}) = \mathrm{cl}\mathscr{C}\Phi \cap (\mathrm{str} - \varepsilon - \mathrm{sdl}\Phi).$$

If the above restriction "inside" is removed, then "$\cap (A \times B)$" can also be removed from the above relation.

Remark 7.86 (Comparisons of the obtained statements about convergence of approximate solutions of equilibrium problems and approximate saddle points) Let us make some comments on the statements about convergence of approximate solutions of equilibrium problems and strong equilibrium problems in relation with convergence of approximate saddle points we have obtained so far.

(a) In all the results, without tightness conditions, only inclusions can be proved, not equalities. Results about saddle points of bifunctions, which are still not relevant to equilibrium problems, are proved for both e/h-convergence and mis-lop together with mai-lop convergence, without tightness, in Theorems 7.47. Theorem 7.54 for tight e/h-convergence and Theorem 7.74 for x-tight mis-lop and y-tight mai-lop convergence simultaneous establish a Painlevé-Kuratowski convergence of approximate saddle points. But, for Limsups of strong approximate saddle points, only an inclusion in Theorem 7.85 is proved.

(b) When considering separately an equilibrium problem or a strong equilibrium problem, we only have convergence of (approximate) solutions but not yet related to (approximate) saddle points; see Theorems 7.76, 7.77, 7.82, and 7.83. These theorems say that the aforementioned types of convergence (not necessarily tight) ensure inclusions for Limsups of approximate solutions of both equilibrium problems and dual equilibrium problems. While for strong equilibrium problems and their dual models, we need the corresponding ancillary tight convergence to confirm similar inclusions for Limsups.

(c) When discussing pairs of dual problems (EP)-(DEP) or (sEP)-(DsEP) in Theorems 7.84 and 7.85, under the tight convergence used in Theorems 7.54 and 7.74, we have equalities of the Painlevé-Kuratowski limits of approximate saddle points and of pair of approximate solutions of primal and dual equilibrium problems (associated with limits of $\mathscr{C}\Phi^k$ and $\mathscr{C}\Phi$, the sets of "zero points" of Φ^k and Φ, resp.). While for pairs of dual strong equilibrium problems, these types of tight convergence only guarantee inclusions for Limsups of approximate solutions and strong approximate saddle points.

Turn now to the second kind of approximate solutions. A concept dual to Definition 7.78 is the following.

Definition 7.87 (The second kind of approximate dual solutions) Let $X = Y$ be a normed space, $A = B$, and $\theta : X \to \mathbb{R}$ be such that $\theta(x) > 0$ if $x \neq 0$. For $\varepsilon \geq 0$, $\bar{y}_{\varepsilon\theta} \in A$ is called an $\varepsilon\theta$-SOLUTION of problem (DEP) if $\Phi(x, \bar{y}_{\varepsilon\theta}) \geq -\varepsilon\theta(x - \bar{y}_{\varepsilon\theta})$ for all $x \in A$. Denote the set of all $\varepsilon\theta$-solutions of (DEP) by $\varepsilon\theta$-Sol(DEP).

From the relationship between the types of variational convergence (cf. Remark 7.67), one obviously has the following counterpart of Theorem 7.79 for the second-kind approximate dual solutions.

Theorem 7.88 (Convergence of second-kind approximate dual solutions) *For dual equilibrium problems, one has the assertions similar to those in Theorem 7.79 for equilibrium problems, with the natural modifications: the mis-lop convergence of Φ^k is replaced by the mai-lop one; the boundedness from above of θ, by the boundedness from below; and the epi-convergence of Φ^k, by the hypo-convergence.*

Definition 7.89 (Second-kind approximate saddle points) Let ε and θ be as in Definition 7.78. A point $(\bar{x}_{\varepsilon\theta}, \bar{y}_{\varepsilon\theta}) \in A \times A$ is called an $\varepsilon\theta$-SADDLE POINT of Φ, written $(\bar{x}_{\varepsilon\theta}, \bar{y}_{\varepsilon\theta}) \in \varepsilon\theta\text{-sdl}\Phi$, if for all $(x, y) \in A \times A$,

$$\Phi(\bar{x}_{\varepsilon\theta}, y) - \varepsilon\theta(\bar{x}_{\varepsilon\theta} - y) \leq \Phi(\bar{x}_{\varepsilon\theta}, \bar{y}_{\varepsilon\theta}) \leq \Phi(x, \bar{y}_{\varepsilon\theta}) + \varepsilon\theta(x - \bar{y}_{\varepsilon\theta}).$$

The following relationship for the above kind of approximations is similar to Theorem 7.81(a) and the two proofs are also similar.

Proposition 7.90 (The components of second-kind approximate saddle points are approximate solutions of the two dual problems) $(\bar{x}_{\varepsilon\theta}, \bar{y}_{\varepsilon\theta}) \in \varepsilon\theta\text{-sdl}\Phi$ *and* $\Phi(\bar{x}_{\varepsilon\theta}, \bar{y}_{\varepsilon\theta}) = 0$ *if and only if* $\bar{x}_{\varepsilon\theta} \in \varepsilon\theta$-Sol(EP), $\bar{y}_{\varepsilon\theta} \in \varepsilon\theta$-Sol(DEP), *and* $\Phi(\bar{x}_{\varepsilon\theta}, \bar{y}_{\varepsilon\theta}) = 0$.

From Theorems 7.79 and 7.88 and Proposition 7.90, we have the following.

Theorem 7.91 (Convergence of second-kind approximate saddle points and pairs of approximate primal and dual solutions) *Let* $\varepsilon, \varepsilon^k \geq 0$ *and* $\varepsilon^k \to \varepsilon$.

(a) *If* $\mathrm{Ls}_k \mathscr{C}\Phi^k \subset \mathrm{cl}\mathscr{C}\Phi$, Φ^k *i-e/h- (or i-mis-lop or i-mai-lop) converges to* Φ, *and* θ *is continuous, then*

$$\mathrm{Ls}_k(\mathscr{C}\Phi^k \cap \varepsilon^k\theta - \mathrm{sdl}\Phi^k) \cap (A \times A)$$

$$= \mathrm{Ls}_k[\mathscr{C}\Phi^k \cap (\varepsilon^k\theta - \mathrm{Sol(EP}^k) \times \varepsilon^k\theta - \mathrm{Sol(DEP}^k))] \cap (A \times A)$$

$$\subset \mathrm{cl}\mathscr{C}\Phi \cap (\varepsilon\theta - \mathrm{Sol(EP)} \times \varepsilon\theta - \mathrm{Sol(DEP)}) = \mathrm{cl}\mathscr{C}\Phi \cap \varepsilon\theta - \mathrm{sdl}\Phi.$$

If $\mathrm{Ls}_k \mathscr{C}\Phi^k \subset \mathrm{cl}\mathscr{C}\Phi$, Φ^k *e/h- (or mis-lop or mai-lop) converges to* Φ, θ *is bounded, and* $\varepsilon = 0$, *then*

$$\mathrm{Ls}_k(\mathscr{C}\Phi^k \cap \varepsilon^k\theta - \mathrm{sdl}\Phi^k) = \mathrm{Ls}_k[\mathscr{C}\Phi^k \cap (\varepsilon^k\theta - \mathrm{Sol(EP}^k) \times \varepsilon^k\theta - \mathrm{Sol(DEP}^k))]$$

$$\subset \mathrm{cl}\mathscr{C}\Phi \cap (\mathrm{Sol(EP)} \times \mathrm{Sol(DEP)}) = \mathrm{cl}\mathscr{C}\Phi \cap \mathrm{sdl}\Phi.$$

(b) *If the epi- or i-epi-convergence of* Φ^k *to* Φ *replaces the types of convergence of* Φ^k *assumed in* (a)*, then we have the same convergence of the second-kind approximate solutions as in* (a) *under the additional assumption* $A \times A \sqsubset \mathrm{Li}_k(A^k \times A^k)$.

It should be noted here that one can relate sdlΦ to the saddle points of another bifunction Ψ on the rectangle $(A \times B) \times (A \times B)$ defined by

$$\Psi((x, y), (x', y')) := \Phi(x, y') - \Phi(x', y) \text{ for } (x, y), (x', y') \in A \times B$$

as follows. Denote by (EP$_\Psi$) and (DEP$_\Psi$) the pair of equilibrium problems defined by Ψ.

Proposition 7.92 (Using other bifunctions and equilibrium models) $(\bar{x}, \bar{y}) \in$ sdlΦ *if* $(\bar{x}, \bar{y})$ *is a solution of* (EP$_\Psi$)*. Conversely,* $(\bar{x}, \bar{y}) \in$ sdlΦ *implies that* $((\bar{x}, \bar{y}), (\bar{x}, \bar{y}))$ *is a solution of both* (EP$_\Psi$)*,* (DEP$_\Psi$) *and* $\Psi((\bar{x}, \bar{y}), (\bar{x}, \bar{y})) = 0$. *The converse assertion is equivalent to* $((\bar{x}, \bar{y}), (\bar{x}, \bar{y})) \in$ sdlΨ.

Proof For the first assertion, as $(\bar{x}, \bar{y})$ is a solution of (EP$_\Psi$), one has

$$0 \geq \Psi((\bar{x}, \bar{y}), (x', y')) = \Phi(\bar{x}, y') - \Phi(x', \bar{y})$$

for all $(x', y') \in A \times B$. Taking $x' = \bar{x}$ and $y' = \bar{y}$, one obtains $\Phi(\bar{x}, y') \leq \Phi(\bar{x}, \bar{y})$ and $\Phi(\bar{x}, \bar{y}) \leq \Phi(x', \bar{y})$, resp., for all $(x', y') \in A \times B$, which means that $(\bar{x}, \bar{y}) \in$ sdlΦ.

For the converse, observe that $\Phi(\bar{x}, y') \leq \Phi(x', \bar{y})$ for $(x', \bar{y})$ and $(\bar{x}, y')$ in gph$\mathscr{D}$ implies that $\Psi((\bar{x}, \bar{y}), (x', y')) \leq 0$ and $\Psi((x', y'), (\bar{x}, \bar{y})) \geq 0$ for $(x', y') \in A \times B$, i.e., $((\bar{x}, \bar{y}), (\bar{x}, \bar{y}))$ is a solution of both (EP$_\Psi$), (DEP$_\Psi$) and $\Psi((\bar{x}, \bar{y}), (\bar{x}, \bar{y})) = 0$. The last statement is evidently equivalent to $((\bar{x}, \bar{y}), (\bar{x}, \bar{y})) \in$ sdlΨ. □

Proposition 7.92 shows that bifunctions and equilibrium problems can be applied in many ways for the same given goal. For instance, saddle points of bifunctions Φ and hence also solutions of the pairs of dual equilibrium problems expressed by Φ can be described as the corresponding objects of another bifunction.

7.4 Approximations of Multiobjective Optimization

As argued above, it is well-known that equilibrium models encompass most optimization-related problems. However, in this chapter, we restrict ourselves to the particular equilibrium problem (EP), which is a scalar problem. So, as usual, it contains scalar optimization-related models. But we can apply results in the preceding section for our *scalar* problem (EP) to the following *multiobjective* minimization problem. Let X be a metric space, $A \subset X$ nonempty, $\varphi :=$

$(\varphi_1, \ldots, \varphi_m)$, and $\varphi_i : A \to \mathbb{R}$ for $i = 1, \ldots, m$. (Here, like in the previous sections, in fact, $\varphi_i : X \to \mathbb{R} \cup \{+\infty\}$, but is finite on A.) Our multiobjective minimization problem is of

(OP) finding $\bar{x} \in A$ such that $\varphi(\bar{x}) - \varphi(y) \notin \mathrm{int}\mathbb{R}^m_+$ for all $y \in A$.

Then, $\bar{x}$ is called a WEAK MINIMIZER (or weakly efficient point) of φ on A and also a WEAK SOLUTION of (OP). We can convert (OP) to a special case (EP$_\varphi$) defined by the following Φ_φ on $A \times A$

$$\Phi_\varphi(x, y) = \min_{1 \le i \le m}(\varphi_i(x) - \varphi_i(y)).$$

Indeed, we have the three equivalent assertions, for all $y \in A$,

$$\Phi_\varphi(\bar{x}, y) \le 0 \Leftrightarrow \text{ there exists } i, \varphi_i(\bar{x}) - \varphi_i(y) \le 0 \Leftrightarrow \varphi(\bar{x}) - \varphi(y) \notin \mathrm{int}\mathbb{R}^m_+.$$

The dual to (OP) according to the duality scheme for (EP$_\varphi$), i.e., problem (DEP$_\varphi$) for Φ_φ, is

(DOP) find $\bar{y} \in A$ such that $\varphi_i(x) - \varphi_i(\bar{y}) \ge 0$ for all $x \in A, i = 1, \ldots, m$,

i.e., $\varphi(x) - \varphi(\bar{y}) \in \mathbb{R}^m_+$. Such a solution $\bar{y}$ is called a STRONG/IDEAL MINIMIZER (or a strongly efficient point) of φ on A and also a SOLUTION of (DOP). This duality scheme is different from the well-known ones for multiobjective set-constrained minimization (the Lagrange, Wolfe, and Mond-Weir duality schemes), which involve a primal minimization problem and a dual maximization problem. From the definition of (OP) and (DOP), we see that $\bar{x}, \bar{y}$ are solutions of (OP) and (DOP), resp., if and only if

$$A \subset \{x \in X \mid \varphi(x) \in (\varphi(\bar{x}) + \mathbb{R}^m \setminus (-\mathrm{int}\mathbb{R}^m_+)) \cap (\varphi(\bar{y}) + \mathbb{R}^m_+)\}.$$

By substituting $\bar{x}$ and $\bar{y}$ in this inclusion, we obtain $\varphi_i(\bar{x}) - \varphi_i(\bar{y}) \in \mathrm{bd}\mathbb{R}^m_+$. Furthermore, $\bar{y}$ must be unique, but $\bar{x}$ does not have to. Thus, we have a simple geometric explanation in the objective space $\mathbb{R}^m$ for (OP) and (DOP).

Furthermore, in comparison with (EP), being a special case of (EP), (OP) has the following stronger features.

Proposition 7.93 (Characterizations of solutions of (OP) and (DOP) in terms of Φ_φ)

(a) *$\bar{x}$ is a solution of* (OP) *if and only if $\bar{x}$ is a minsup-point of Φ_φ with the* minsup-*value being* 0*, if and only if* $\sup_{y \in A} \Phi_\varphi(\bar{x}, y) = 0$.
(b) *$\bar{y}$ is a solution of* (DOP) *if and only if $\bar{y}$ is a* maxinf-*point of Φ_φ with the* maxinf-*value being* 0*, if and only if* $\inf_{x \in A} \Phi_\varphi(x, \bar{y}) = 0$.
(c) *Therefore, $\bar{x}, \bar{y}$ are solutions of* (OP) *and* (DOP)*, resp., if and only if $(\bar{x}, \bar{y})$ is a saddle point of Φ_φ with $\Phi_\varphi(\bar{x}, \bar{y}) = 0$.*

Proof For (a), as $\Phi_\varphi(x, x) = 0$, $\sup_{y\in A}\Phi_\varphi(x, y) \geq 0$. $\bar{x}$ being a solution of (OP) means that $\sup_{y\in A}\Phi_\varphi(\bar{x}, y) \leq 0$ and so $\sup_{y\in A}\Phi_\varphi(\bar{x}, y) = 0$. This fact is equivalent to $\bar{x} \in \text{mis}\Phi_\varphi$ with the minsup-value being 0.

(b) This is similar to (a).

(c) This is a straightforward consequence of (a), (b), and Theorem 7.81. □

In fact, we need to solve only (DOP) to have some saddle points of Φ_φ as follows. Because any solution $\bar{y}$ of (DOP) also solves (OP), $(\bar{y}, \bar{y})$ is a saddle point with the saddle value 0.

Now, discuss the following approximations of (OP). Let $A^k \subset X$ and $\varphi_i^k : A^k \to \mathbb{R}$ for $i = 1, \ldots, m$. Our approximating problems of (OP) are

$$(\text{OP}^k) \qquad \text{find } \bar{x} \in A^k \text{ such that } \varphi^k(\bar{x}) - \varphi^k(y) \notin \text{int}\mathbb{R}_+^m \text{ for all } y \in A^k.$$

Similarly as above, each problem (OP^k) is reformulated as (EP$_\varphi^k$) by using the bifunction $\Phi_{\varphi^k}^k : A^k \times A^k \to \mathbb{R}$ defined by

$$\Phi_{\varphi^k}^k(x, y) := \min_{1\leq i\leq m}(\varphi_i^k(x) - \varphi_i^k(y)).$$

Let $e := (1, \ldots, 1) \in \mathbb{R}^m$. For $\varepsilon \geq 0$, a point $\bar{x}_\varepsilon \in A$ is called an ε-SOLUTION of (OP) if for all $y \in A$,

$$\varphi(\bar{x}_\varepsilon) - \varphi(y) - \varepsilon e \notin \text{int}\mathbb{R}_+^m.$$

Clearly, this is equivalent to $\Phi_\varphi(\bar{x}_\varepsilon, y) \leq \varepsilon$. Denote the set of all ε-solutions of (OP) by ε-Sol(OP).

Studying the convergence of the solutions of (OP^k) directly through its data is quite troublesome, so we do not go this way, but use the equivalence over the bifunctions $\Phi_{\varphi^k}^k$ given in Proposition 7.93. By Theorem 7.76 and Proposition 7.93, we have the following.

Proposition 7.94 (Convergence of approximate solutions of (OP^k)) *Let* $\varepsilon, \varepsilon^k \geq 0$ *and* $\varepsilon^k \to \varepsilon$.

(a) *If* $\Phi_{\varphi^k}^k$ *e/h- (or mis-lop or mai-lop) converges to* Φ_φ*, then* $\text{Ls}_k(\varepsilon^k\text{- Sol(OP}^k)) \subset \varepsilon\text{-Sol(OP)}$.
If the convergence of $\Phi_{\varphi^k}^k$ *reduces to the corresponding inside type, then* $\text{Ls}_k(\varepsilon^k\text{-Sol(OP}^k)) \cap A \subset \varepsilon\text{-Sol(OP)}$.

(b) *If* $\Phi_{\varphi^k}^k \xrightarrow{e} \Phi_\varphi$ *(in* fv-fcn$(X \times X)$*) and* $A \times A \sqsubset \text{Li}_k(A^k \times A^k)$*, then* $\text{Ls}_k(\varepsilon^k\text{-Sol(OP}^k)) \subset \varepsilon\text{-Sol(OP)}$. *If the inside convergence replaces the above epi-convergence, then* $\text{Ls}_k(\varepsilon^k\text{-Sol(OP}^k)) \cap A \subset \varepsilon\text{-Sol(OP)}$.

We consider approximating problems (DOP^k) of (DOP) and denote their solution sets by Sol(DOP^k) and Sol(DOP), resp. A point $\bar{y}_\varepsilon$ is called an ε-SOLUTION of

(DOP) if $\varphi(x) - \varphi(\bar{y}_\varepsilon) + \varepsilon e \in \mathbb{R}^m_+$ for all $x \in A$. Evidently, this is equivalent to $\Phi_\varphi(x, \bar{y}_\varepsilon) \geq -\varepsilon$ for all $x \in A$, i.e., $\bar{y}_\varepsilon$ is an ε-solution of (DEP$_\varphi$).

In view of Theorem 7.82, we have the complete counterpart of Proposition 7.94 for the convergence of approximate solutions of (DOP^k). Hence, for the pair (OP^k)-(DOP^k), a special case of the primal-dual pair of equilibrium problems, we define

$$\mathscr{C}\Phi^k_{\varphi^k} := \{(x, y) \in A^k \times B^k \mid \Phi^k_{\varphi^k}(x, y) = 0\},$$
$$\mathscr{C}\Phi_\varphi := \{(x, y) \in A \times B \mid \Phi_\varphi(x, y) = 0\}.$$

Then, we have the following consequence of Theorem 7.84.

Proposition 7.95 (Convergence of approximate solutions of pairs of multiobjective optimization problems and approximate saddle points) *Let $\varepsilon, \varepsilon^k \geq 0$ and $\varepsilon^k \to \varepsilon$.*

(a) If $\Phi^k_{\varphi^k}$ i-e/h- or i-mis-lop or i-mai-lop converges to Φ_φ simultaneously and $\mathrm{Ls}_k\mathscr{C}\Phi^k_{\varphi^k} \subset \mathrm{cl}\mathscr{C}\Phi_\varphi$, then

$$\begin{aligned}&\mathrm{Ls}_k(\mathscr{C}\Phi^k_{\varphi^k} \cap \varepsilon^k\text{-sdl}\Phi^k_{\varphi^k}) \cap (A \times A)\\ &= \mathrm{Ls}_k[\mathscr{C}\Phi^k_{\varphi^k} \cap (\varepsilon^k\text{-Sol(OP)}^k \times \varepsilon^k\text{-Sol(DOP}^k))] \cap (A \times A)\\ &\subset \mathrm{cl}\mathscr{C}\Phi_\varphi \cap (\varepsilon\text{-Sol(OP)} \times \varepsilon\text{-Sol(DOP)}) = \mathrm{cl}\mathscr{C}\Phi_\varphi \cap \varepsilon\text{-sdl}\Phi_\varphi.\end{aligned}$$

If the above restriction "inside" is removed, then in the above relations, "$\cap (A \times A)$" is also superfluous.

(b) If $\Phi^k_{\varphi^k}$ i-e/h-converges tightly or both x- and y-ancillary tightly i-mis-lop or both x- and y-ancillary tightly i-mai-lop converges to Φ_φ, and $\mathrm{cl}\mathscr{C}\Phi_\varphi \subset \mathrm{Li}_k\mathscr{C}\Phi^k_{\varphi^k}$, then

$$\begin{aligned}\varepsilon\text{-sdl}\Phi_\varphi \cap \mathrm{cl}\mathscr{C}\Phi_\varphi &= (\varepsilon\text{-Sol(OP)} \times \varepsilon\text{-Sol(DOP)}) \cap \mathrm{cl}\mathscr{C}\Phi_\varphi\\ &\subset \cap_{\delta>\varepsilon}\mathrm{Li}_k(\delta\text{-Sol(OP}^k) \times \delta\text{-Sol(DOP}^k)) \cap \mathrm{Li}_k\mathscr{C}\Phi^k_{\varphi^k}\\ &= \cap_{\delta>\varepsilon}\mathrm{Li}_k\delta\text{-sdl}\Phi^k_{\varphi^k} \cap \mathrm{Li}_k\mathscr{C}\Phi^k_{\varphi^k}.\end{aligned}$$

Hence,

$$\begin{aligned}\varepsilon\text{-sdl}\Phi_\varphi \cap \mathrm{cl}\mathscr{C}\Phi_\varphi &= (\varepsilon\text{-Sol(OP)} \times \varepsilon\text{-Sol(DOP)}) \cap \mathrm{cl}\mathscr{C}\Phi_\varphi\\ &\subset \mathrm{Li}_k(\varepsilon^k\text{-Sol(OP}^k) \times \varepsilon^k\text{-Sol(DOP}^k)) \cap \mathrm{Li}_k\mathscr{C}\Phi^k_{\varphi^k}\\ &= \mathrm{Li}_k\varepsilon^k\text{-sdl}\Phi^k_{\varphi^k} \cap \mathrm{Li}_k\mathscr{C}\Phi^k_{\varphi^k}.\end{aligned}$$

(c) Assuming that $\mathrm{Lim}_k\mathscr{C}\Phi^k_{\varphi^k} = \mathrm{cl}\mathscr{C}\Phi_\varphi$, the above restriction "inside" in (b) is dropped, and the involved types of lop-convergence are imposed together. Then,

$$\begin{aligned}\varepsilon\text{-sdl}\Phi_\varphi \cap \mathrm{cl}\mathscr{C}\Phi_\varphi &= (\varepsilon\text{-Sol(OP)} \times \varepsilon\text{-Sol(DOP)}) \cap \mathrm{cl}\mathscr{C}\Phi_\varphi\\ &= \mathrm{Lim}_k(\varepsilon^k\text{-Sol(OP}^k) \times \varepsilon^k\text{-Sol(DOP}^k)) \cap \mathrm{Lim}_k\mathscr{C}\Phi^k_{\varphi^k}\\ &= \mathrm{Lim}_k\varepsilon^k\text{-sdl}\Phi^k_{\varphi^k} \cap \mathrm{Lim}_k\mathscr{C}\Phi^k_{\varphi^k}.\end{aligned}$$

(d) *If* $\mathrm{Lim}_k\mathscr{C}\Phi^k_{\varphi^k} \subset \mathrm{cl}\mathscr{C}\Phi_\varphi$, $\Phi^k_{\varphi^k} \overset{w-e/h}{\to} \Phi_\varphi$ *tightly or* $\Phi^k_{\varphi^k} \overset{w-mis-lop}{\to} \Phi_\varphi$ *x-ancillary-tightly and* $\Phi^k_{\varphi^k} \overset{w-mai-lop}{\to} \Phi_\varphi$ *y-ancillary-tightly simultaneously, and* $(x^k, y^k) \in (\varepsilon^k\text{-Sol}(\text{OP}^k) \times \varepsilon^k\text{-(Sol(DOP}^k))$ $\cap \mathscr{C}\Phi^k_{\varphi^k}$ *tends to* (x, y)*, then* $(x, y) \in (\text{Sol(OP)}) \times (\text{Sol(DOP)}) \cap \text{sdl}\Phi_\varphi \cap \text{cl}\mathscr{C}\Phi_\varphi$.

For the second kind of approximate solutions of multiobjective optimization problems, the definition is natural as follows.

Definition 7.96 (Second-kind approximate solutions of optimization problems) Let ε and θ be as in Definition 7.87. $\bar{x}_{\varepsilon\theta}$ ($\bar{y}_{\varepsilon\theta}$, resp.) is an $\varepsilon\theta$-SOLUTION of (OP) ((DOP), resp.) if for $e := (1, \ldots, 1) \in \mathbb{R}^m$,

$$\varphi(\bar{x}_{\varepsilon\theta}) - \varphi(y) - \varepsilon\theta(\bar{x}_{\varepsilon\theta} - y)e \notin \text{int}\mathbb{R}^m_+, \text{ for all } y \in A$$

$$(\varphi(x) - \varphi(\bar{y}_{\varepsilon\theta}) + \varepsilon\theta(x - \bar{y}_{\varepsilon,\theta})e \in \mathbb{R}^m_+ \text{ for all } x \in A, \text{ resp.}).$$

It is easy to verify that $\bar{x}_{\varepsilon\theta}$ and $\bar{y}_{\varepsilon\theta}$ are $\varepsilon\theta$-solutions of (OP) and (DOP), resp., if and only if they are $\varepsilon\theta$-solutions of (EP_φ) and (DEP_φ), resp., and if and only if they are an $\varepsilon\theta$-weak minimizer and an $(\varepsilon\theta)$-strong minimizer, resp., of φ on A. Denote the set of all $\varepsilon\theta$-solutions of (OP) ((DOP), resp.) by $\varepsilon\theta$-Sol(OP) ($\varepsilon\theta$-Sol(DOP), resp.). From Proposition 7.90, Theorem 7.91, and the above relationship between $\varepsilon\theta$-solutions, we obtain the following convergence result.

Proposition 7.97 (Convergence of second-kind approximate solutions of pairs of multiobjective optimization problems) *Let* $\varepsilon, \varepsilon^k \geq 0$ *and* $\varepsilon^k \to \varepsilon$.

(a) *If* $\Phi^k_{\varphi^k}$ *i-e/h- (or i-mis-lop or i-mai-lop) converges to* Φ_φ, $\mathrm{Ls}_k\mathscr{C}\Phi^k_{\varphi^k} \subset \mathrm{cl}\mathscr{C}\Phi_\varphi$, *and* θ *is continuous, then*

$$\mathrm{Ls}_k(\mathscr{C}\Phi^k_{\varphi^k} \cap \varepsilon^k\theta - \text{sdl}\Phi^k_{\varphi^k}) \cap (A \times A)$$

$$= \mathrm{Ls}_k[\mathscr{C}\Phi^k_{\varphi^k} \cap (\varepsilon^k\theta - \text{Sol}(\text{OP}^k) \times \varepsilon^k\theta - \text{Sol}(\text{DOP}^k))] \cap (A \times A)$$

$$\subset \mathrm{cl}\mathscr{C}\Phi_\varphi \cap (\varepsilon\theta - \text{Sol(OP)} \times C\theta - \text{Sol(DOP)}) = \mathrm{cl}\mathscr{C}\Phi_\varphi \cap \varepsilon\theta - \text{sdl}\Phi_\varphi.$$

If $\Phi^k_{\varphi^k}$ *e/h- or mis-lop or mai-lop converges to* Φ_φ, θ *is bounded,* $\mathrm{Ls}_k\mathscr{C}\Phi^k_{\varphi^k} \subset \mathrm{cl}\mathscr{C}\Phi_\varphi$, *and* $\varepsilon = 0$, *then*

$$\mathrm{Ls}_k(\mathscr{C}\Phi^k_{\varphi^k} \cap \varepsilon^k\theta - \text{sdl}\Phi^k_{\varphi^k}) = \mathrm{Ls}_k[\mathscr{C}\Phi^k_{\varphi^k} \cap (\varepsilon^k\theta - \text{Sol}(\text{OP}^k) \times \varepsilon^k\theta - \text{Sol}(\text{DOP}^k))]$$

$$\subset \mathrm{cl}\mathscr{C}\Phi_\varphi \cap \text{Sol(OP)} \times \text{Sol(DOP)} = \mathrm{cl}\mathscr{C}\Phi_\varphi \cap \text{sdl}\Phi_\varphi.$$

(b) *If the e- or i-e-convergence of* $\Phi^k_{\varphi^k}$ *to* Φ_φ *replaces the convergence of* $\Phi^k_{\varphi^k}$ *assumed in* (a)*, then we have the same convergence of the second-kind approximate solutions as in* (a) *under the additional assumption* $A \times A \sqsubset \mathrm{Li}_k(A^k \times A^k)$.

7.5 Approximations of Optimization Problems in Terms of Optimality Functions

We know that equilibrium model (EP) encompasses many optimization-related problems. By applying results for (EP) in Sect. 7.3, the preceding Sect. 7.4 investigates multiobjective constrained minimization problem (OP). There, we consider solutions of (OP). But, in general, one does not know solutions of optimization problems. The usual tests for solutions of an optimization problem are optimality conditions, including necessary conditions and sufficient conditions for optimality. A point satisfying a necessary condition is called a STATIONARY POINT. Hence, a stationary point can be visualized as an approximate solution of the optimization problem under consideration, and one finds or studies stationary points instead of solutions. Stationary points are more convenient than solutions in both theory and computation studies. The following two forms can be considered the main forms of necessary conditions. The first one states that in a neighborhood of a local minimizer/solution, the cost/objective function must increase. This fact is usually expressed as a disjunction and called a necessary condition in the primal form. The second form, called the dual form, is a consequence of the first one and consists of an equation involving derivatives and (Lagrange) multipliers. Usually, when one looks for a method to verify whether necessary conditions in either of the above two forms are satisfied at a point, one is led to the third form that characterizes local solutions as zeros of an easily evaluated OPTIMALITY FUNCTIONS. Furthermore, there can be more than one optimality function associated with a particular necessary optimality condition. One can construct an optimality function as the minimum value of a strictly convex quadratic function, which is a first-order approximation of the cost function.

7.5.1 Examples: Optimality Functions of Basic Simple Problems

To explain the idea of the construction and use of optimality functions, let us begin with the simplest constrained optimization problem:

$$\min_{x \in A} \varphi(x), \tag{7.9}$$

where $X = \mathbb{R}^n$, $A \subset X$ is convex, and we assume that the directional derivative $d\varphi(x; h)$ exists for all $x, h \in X$. Then, a basic necessary condition for $\bar{x}$ to be a local solution of (7.9) is

$$d\varphi(\bar{x}; x - \bar{x}) \geq 0 \text{ for all } x \in A. \tag{7.10}$$

The proof is very simple. Let $\varphi(y) \geq \varphi(\bar{x})$ in $U \cap A$ for a neighborhood U of $\bar{x}$. Then, for small positive t, $\varphi(\bar{x} + t(x - \bar{x})) \geq \varphi(\bar{x})$ for all $x \in A$. Hence, $d\varphi(\bar{x}; x - \bar{x}) := \lim_{t\to 0^+} t^{-1}[\varphi(\bar{x} + t(x - \bar{x})) - \varphi(\bar{x})] \geq 0$. We express (7.10) in another way. Assume that $d\varphi(\bar{x}; \cdot)$ is continuous and convex for all x. Let $\theta_1 : \mathbb{R}^n \to \mathbb{R}$ be defined by

$$\theta_1(x) := \min_{h\in \mathrm{cl} B(0,1)\cap(A-x)} d\varphi(x; h). \tag{7.11}$$

We prove that (7.10) holds if and only if $\theta_1(\bar{x}) = 0$. For the "If" part, by the definition of θ_1, $\theta_1(\bar{x}) = 0 \leq d\varphi(\bar{x}; \lambda(x - \bar{x}))$ if $\|\lambda(x - \bar{x})\| \leq 1$ and $x \in A$. Then, by the positive homogeneity, (7.10) holds. For the "Only if" part, we define the vector

$$h_1(\bar{x}) := \mathrm{argmin}_{h\in \mathrm{cl} B(0,1)\cap(A-\bar{x})} d\varphi(\bar{x}; h).$$

By (7.10), clearly $d\varphi(\bar{x}; h_1(\bar{x})) = 0$ as $d\varphi(\bar{x}; 0) = 0$. Hence, $\theta_1(\bar{x}) = d\varphi(\bar{x}; h_1(\bar{x})) = 0$.

Furthermore, if (7.10) is not satisfied at $\bar{x}$, then $\theta_1(\bar{x}) = \min_{h\in \mathrm{cl} B(0,1)\cap(A-x)} d\varphi (\bar{x}; h) < 0$, i.e., $h_1(\bar{x})$ defines a steepest feasible decent direction for the cost function φ at $\bar{x}$.

The function θ_1 is called the OPTIMALITY FUNCTION for φ (in problem (7.9)). We see, from the above simple example, the advantages of expressing optimality conditions in terms of optimality functions: the latter give a convenient way for verifying whether a point x satisfies a basic necessary optimality condition in cost function value in a ball around x; and when x does not satisfy this condition, the evaluation of optimality functions produces a feasible DESCENT DIRECTION $h_1(x)$, i.e., there is $\lambda_x \in (0, 1]$ such that $\varphi(x + \lambda_x h_1(x)) < \varphi(x)$ and $x + \lambda_x h_1(x) \in A$ for all $\lambda \in (0, \lambda_x]$.

Because we can choose a good one among possible optimality functions for a given φ, we construct an optimality function for problem (7.9) when φ is continuously differentiable of the form

$$\begin{aligned}\theta_2(x) &:= \min_{x'\in A} \langle \varphi'(x), x' - x\rangle + (1/2)\|x' - x\|^2 \\ &= \min_{h\in A-x} \langle \varphi'(x), h\rangle + (1/2)\|h\|^2,\end{aligned} \tag{7.12}$$

and also its associated descent direction function $h_2 : \mathbb{R}^n \to \mathbb{R}^n$ by

$$h_2(x) := \mathrm{argmin}_{h\in A-x} \langle \varphi'(x), h\rangle + (1/2)\|h\|^2. \tag{7.13}$$

Proposition 7.98 (Properties of θ_2 and h_2) *Assume that φ is continuously differentiable and A is a closed convex subset of $\mathbb{R}^n$. Then, θ_2 defined by (7.12) and h_2 by (7.13) have the following properties:*

(a) *$h_2(x)$ is unique for $x \in A$.*
(b) *$\theta_2(x) \le 0$ for all $x \in A$.*
(c) *θ_2 and h_2 are continuous.*
(d) *$d\varphi(x; h_2(x)) \le \theta_2(x)$ for all $x \in A$.*
(e) *For $x \in A$, $\theta_2(x) = 0$ if and only if (7.10) holds (i.e., x is a stationary point of problem (7.9)).*

Proof

(a) This is evident due to the strict convexity of $\langle \varphi'(x), h\rangle + (1/2)\|h\|^2$ as a function of h and the convexity of $A - x$. (This property is one of the main reasons for considering θ_2, h_2 instead of θ_1, h_1.)
(b) This holds because $x' = x$ is an admissible selection in (7.12).
(c) First, we express θ_2 as a minimum over a compact convex set. Let $x^* \in A$ and $\gamma > 0$ be arbitrary. Then, there exists a real number $b < +\infty$ such that $\|\varphi'(x)\| \le b$ for all $x \in A \cap \mathrm{cl}B(x^*, \gamma)$. Hence,

$$\langle \varphi'(x), x' - x\rangle + (1/2)\|x' - x\|^2 > 0$$

for all $x \in A \cap \mathrm{cl}B(x^*, \gamma)$ and $x' \in A$ such that $\|x' - x\| > 2b$. Clearly, $A^* := A \cap \mathrm{cl}B(x^*, \gamma + 2b)$ is compact and convex and $\mathrm{cl}B(x, 2b) \subset \mathrm{cl}B(x^*, \gamma + 2b)$ for all $x \in \mathrm{cl}B(x^*, \gamma)$. So, for all $x \in \mathrm{cl}B(x^*, \gamma)$, we have a desired expression of θ_2:

$$\begin{aligned}\theta_2(x) &= \min\nolimits_{x' \in A^*} \langle \varphi'(x), x' - x\rangle + (1/2)\|x' - x\|^2 \\ &= \min\nolimits_{h \in A^* - x} \langle \varphi'(x), h\rangle + (1/2)\|h\|^2.\end{aligned}$$

Hence, as the minimum over a compact set depending continuously on parameter x of a continuous function of x and h, θ_2 is continuous. For h_2, suppose to the contrary that the (single-valued) function h_2 is not continuous, i.e., there are $x_i \to x$ but $h_2(x_i) \to c \ne h_2(x)$ (by the compactness of $A^* - x$, we can assume that $h_2(x_i)$ converge). Then,

$$\begin{aligned}\theta_2(h_2(x_i)) = \langle \varphi'(x_i), h_2(x_i)\rangle + (1/2)\|h_2(x_i)\|^2 &\to \langle \varphi'(x), c\rangle + (1/2)\|c\|^2 \\ &> \theta_2(h_2(x)),\end{aligned}$$

which contradicts the continuity of θ_2.
(d) Since φ is continuously differentiable, we have $d\varphi(x; h_2(x)) = \langle \varphi'(x), h_2(x)\rangle$. By the definition of $h_2(x)$, it holds $\theta_2(x) = \langle \varphi'(x), h_2(x)\rangle + (1/2)\|h_2(x)\|^2$. Hence, the assertion holds.

(e) For the "if" part, suppose to the contrary that $\theta_2(x) < 0$. Then, $d\varphi(x; h_2(x)) < 0$ by assertion (d), contradicting (7.10). For the "only if" part, suppose $\theta_2(\bar{x}) = 0$ for some $\bar{x} \in A$ but there is $x' \in A$ with $d\varphi(x; x' - x) < 0$. For $\lambda \in [0, 1]$, consider $x(\lambda) := \bar{x} + \lambda(x' - x) \in A$ (so $x(\lambda) - \bar{x} = \lambda(x' - x)$). By the definition of $\theta_2(\bar{x})$, one has

$$\theta_2(\bar{x}) \leq \min_{\lambda \in [0,1]} \langle \varphi'(\bar{x}), x(\lambda) - \bar{x} \rangle + (1/2)\|x(\lambda) - \bar{x}\|^2 < 0$$

because we can find $\lambda^* \in (0, 1]$ such that the function under minimization sign being $\langle \varphi'(\bar{x}), \lambda^* (x' - x) \rangle + (1/2)\|\lambda^* x' - x\|^2 < 0$. This contradicts the definition of $\theta_2(\bar{x})$ and completes the proof.

□

We see from the above discussion that the optimality functions θ_1, θ_2 defined by (7.11) and (7.12), resp., together with their associated descent/search directions h_1, h_2, resp., are useful in both theory and computation studies to replace optimality conditions for the considered optimal solutions for the basic (set-)constrained problem (7.9). But most problems in practice are functional constrained problems, and the basic version for them is the problem with inequality constraints:

$$\begin{cases} \min \varphi^0(x), \\ \varphi^j(x) \leq 0, \ j \in Q := \{1, \ldots, q\}. \end{cases} \tag{7.14}$$

To persuade the reader about the importance of optimality functions, we consider one more example: the min-max problem:

$$\min_{x \in \mathbb{R}^n} \max_{j \in Q} \varphi^j(x). \tag{7.15}$$

We show that this problem is equivalent to the following particular case of problem (7.14) with a particular case of constraints:

$$\begin{cases} \min \varphi^0(x), \\ \varphi^j(x) - \varphi^0(x) \leq 0, \ j \in Q := \{1, \ldots, q\}. \end{cases}$$

Setting $x^{n+1} := \varphi^0(x)$, this problem is reformulated as

$$\begin{cases} \min x^{n+1}, \\ \varphi^j(x) - x^{n+1} \leq 0, \ j \in Q := \{1, \ldots, q\}, \end{cases}$$

which is clearly equivalent to the min-max problem (7.15). It is known that conversely, one can construct a problem with inequality constraints that are locally equivalent to a given min-max problem of the form (7.15). So, min-max problems are not only important in their own right but also provide a very useful bridge to types of inequality constrained problems.

Now, we assume for the min-max problem (7.15) that φ^j is continuously differentiable for $j \in Q$. Following the way leading to introducing optimality functions for problem (7.9), we first discuss optimality conditions. For convenience, set

$$\psi(x) := \max_{j \in Q} \varphi^j(x). \tag{7.16}$$

For the sake of completeness, we prove some properties of $d\psi(\cdot;\cdot)$ for our use.

Lemma 7.99 (Properties of ψ) *Assume that φ^j is continuously differentiable for $j \in Q$. Then, ψ defined in (7.16) possesses the following properties:*

(a) *ψ is continuous.*
(b) *$d\psi(x; h)$ exists for all $x, h \in \mathbb{R}^n$ and is given by*

$$d\psi(x; h) = \max_{j \in Q(x)} \langle (\varphi^j)'(x), h \rangle, \tag{7.17}$$

where $Q(x) := \{j \in Q \mid \varphi^j(x) = \psi(x)\}$.
(c) *$d\psi(\cdot;\cdot)$ is usc.*

Proof

(a) Since Q is finite and φ^j are continuous, for any $\bar{x} \in \mathbb{R}^n$, there exists a neighborhood U of $\bar{x}$ such that for all $x \in U$, there is j_0 such that $\psi(x) := \max_{j \in Q} \varphi^j(x) = \varphi^{j_0}(x)$. Hence, ψ is continuous at $\bar{x}$ because so is φ^{j_0}.
(b) For $x, h \in \mathbb{R}^n$ and $t > 0$,

$$\frac{\psi(x + th) - \psi(x)}{t} = \max_{j \in Q} \frac{\varphi^j(x + th) - \psi(x)}{t}.$$

Observe that, when $t \to 0^+$, $t^{-1}(\varphi^j(x + th) - \psi(x)) \to -\infty$ for all $j \notin Q(x)$ and this quotient tends to $d\varphi^j(x; h)$ for all $j \in Q(x)$. Hence, for $t > 0$ small enough,

$$\frac{\psi(x + th) - \psi(x)}{t} = \max_{j \in Q(x)} \frac{\varphi^j(x + th) - \varphi^j(x)}{t}.$$

Letting $t \to 0^+$, we get the assertion.
(c) Let $x, h \in \mathbb{R}^n$ be arbitrary and $x_i \to x$, $h_i \to h$. Since ψ is continuous (by assertion (a)), there exists i_0 such that for all $i \geq i_0$, we have the implication: $\psi(x_i) = \varphi^j(x_i)$ implies $\psi(x) = \varphi^j(x)$ (i.e., $Q(x_i) \subset Q(x)$). So, for all $i \geq i_0$, assertion (b) implies that

$$d\psi(x_i; h_i) = \max_{j \in Q(x_i)} \langle (\varphi^j)'(x_i), h_i \rangle \leq \max_{j \in Q(x)} \langle (\varphi^j)'(x_i), h_i \rangle.$$

By the continuity of $(\varphi^j)'$ and of $\langle\cdot,\cdot\rangle$, taking the upper limit of the above relations as $i \to +\infty$, we obtain

$$\text{ls}_i d\psi(x_i; h_i) \leq \max_{j\in Q(x)}\langle(\varphi^j)'(x), h\rangle = d\psi(x; h).$$

□

Adapting the form of θ_1 for problem (7.9), we define the optimality function $\theta_3 : \mathbb{R}^n \to \mathbb{R}$ for problem (7.15) by

$$\theta_3(x) := \min_{\|h\|_\infty\leq 1} d\psi(x; h) = \min_{\|h\|_\infty\leq 1}\max_{j\in Q(x)}\langle(\varphi^J)'(x), h\rangle. \tag{7.18}$$

(Note the very good property that this expression of θ_3 can be recast as a linear program.) The associated descent/search direction function $h_3 : \mathbb{R}^n \to \mathbb{R}^n$ is determined by

$$h_3(x) := \text{argmin}_{\|h\|_\infty\leq 1} d\psi(x; h). \tag{7.19}$$

The following statement indicates that θ_3 and h_3 have properties similar to those of the above θ_ℓ and h_ℓ with $\ell = 1, 2$ for problem (7.8).

Proposition 7.100 (Properties of θ_3 and h_3) *Assume for problem (7.15) that φ^j is continuously differentiable for $j \in Q$. Then:*

(a) *$h_3(x)$ exists for each $x \in \mathbb{R}^n$.*
(b) *$\theta_3(x) \leq 0$ for each $x \in \mathbb{R}^n$.*
(c) *$d\psi(x; h(x)) = \theta_3(x)$ for all $x \in \mathbb{R}^n$.*
(d) *θ_3 is usc.*
(e) *$d\psi(x; h) \geq 0$ for all $h \in \mathbb{R}^n$ (i.e., x is a stationary point of problem (7.14) if and only if $\theta_3(x) = 0$.*

Proof

(a) This is due to the definition of $h_3(x)$. Namely, for each $x \in \mathbb{R}^n$, $h_3(x)$ is the minimum of $d\psi(x;\cdot)$ on a compact set and $d\psi(x;\cdot)$ is the maximum of a finite family of continuous linear functions and so is continuous and convex.
(b) For each $x \in \mathbb{R}^n$, $d\psi(x; 0) = 0$. Hence, its minimum with respect to x, i.e., θ_3, must be nonpositive.
(c) This is clear by the definition of θ_3 and h_3.
(d) By Lemma 7.99(c), $d\psi(\cdot;\cdot)$ is usc. Hence, the assertion follows from a well-known fact for parametric minimization (see, e.g., Theorem 1.4.16 in [82]).
(e) The "only if" part is evident since $d\psi(x; h) \geq 0$ for all $h \in \mathbb{R}^n$ and $\theta_3(x) = d\psi(x; 0) = 0$. The "if" part follows directly from the definition of θ_3.

□

Similar to θ_2 and h_2 for problem (7.9), we can design another optimality function θ_4 and its descent direction function h_4 via the following local first-order, quadratic model for φ^j. Given $x \in \mathbb{R}^n$, for any $x' \in \mathbb{R}^n$ and $j \in Q$, we take the approximate of $\varphi^j(x')$

$$\hat{\varphi}^j(x, x') := \varphi^j(x) + \langle (\varphi^j)'(x), x' - x \rangle + (1/2)\|x' - x\|^2.$$

Then, $\hat{\varphi}^j(x, \cdot)$ is strictly convex and a first-order approximation to $\varphi^j(\cdot)$ because it is obvious that there exists $C < +\infty$ such that

$$\text{ls}_{x' \to x} \frac{|\varphi^j(x') - \hat{\varphi}^j(x, x')|}{\|x' - x\|} \leq C.$$

Next, we define the following first-order, piecewise, quadratic local model for ψ

$$\hat{\psi}(x, x') := \max_{j \in Q} \hat{\varphi}^j(x, x'), \tag{7.20}$$

which approximates the value $\psi(x')$. Note that $\hat{\psi}(x, \cdot)$ is convex since $\hat{\varphi}^j(x, \cdot)$ is convex for $j \in Q$.

Now, we can define the aforementioned θ_4 and h_4 by

$$\theta_4(x) := \min_{h \in \mathbb{R}^n} \hat{\psi}(x, x + h) - \psi(x)$$

$$= \min_{h \in \mathbb{R}^n} \max_{j \in Q} \{\varphi^j(x) - \psi(x) + \langle (\varphi^j)'(x), h \rangle + (1/2)\|h\|^2\},$$

$$h_4(x) := \text{argmin}_{h \in \mathbb{R}^n} \max_{j \in Q} \{\varphi^j(x) - \psi(x) + \langle (\varphi^j)'(x), h \rangle + (1/2)\|h\|^2\}.$$

To investigate useful properties of θ_4 and h_4, we state without proof the well-known fundamental von Neumann minimax theorem.

Theorem 7.101 (The von Neumann minimax theorem) *Let $A \subset \mathbb{R}^n$, $B \subset \mathbb{R}^m$ be nonempty compact convex sets, and $\Phi : \mathbb{R}^n \times \mathbb{R}^m \to \mathbb{R}$ satisfy the conditions: Φ is continuous, $\Phi(\cdot, y)$ is convex on A for each $y \in \mathbb{R}^m$, and $\Phi(x, \cdot)$ is concave on B for each $x \in \mathbb{R}^n$. Then,*

$$\min_A \max_B \Phi(x, y) = \max_B \min_A \Phi(x, y)$$

and Φ has saddle points in $A \times B$.

The following statement is an extension of the preceding theorem to the case when either A or B is unbounded.

Corollary 7.102 (Minimax equality for a noncompact case) *Impose the assumptions in Theorem 7.101, except that A is replaced by $\mathbb{R}^n$ together with the additional (coersivity) condition that $\Phi(x, y) \to +\infty$ uniformly in $y \in B$ as $\|x\| \to +\infty$. Then, the same conclusion as in Theorem 7.101 holds with A replaced by $\mathbb{R}^n$.*

Proof Let $\hat{x} \in \mathbb{R}^n$ be arbitrary, $A_y := \{x \in \mathbb{R}^n \mid \Phi(x, y) \leq \Phi(\hat{x}, y)\}$ (a level set) for every $y \in B$, and A the convex hull of $\cup_{y \in B} A_y$. By the assumed uniform convergence to infinity, A is convex and compact. Let $\psi : \mathbb{R}^n \to \mathbb{R}$ be defined by $\psi(x) := \max_{y \in B} \Phi(x, y)$ and $\hat{A} := \{x \in \mathbb{R}^n \mid \psi(x) \leq \psi(\hat{x})\}$ (another level set). Then, if $y_x, \hat{y} \in B$ satisfy $\psi(x) = \Phi(x, y_x)$ and $\psi(\hat{x}) = \Phi(\hat{x}, \hat{y})$, then $\Phi(\hat{x}, \hat{y}) \geq \Phi(x, y_x) \geq \Phi(x, \hat{y})$, which shows that $x \in A_{\hat{y}}$ and so $\hat{A} \subset A$. Hence,

$$\min_{x \in \mathbb{R}^n} \max_{y \in B} \Phi(x, y) = \min_{x \in A} \max_{y \in B} \Phi(x, y).$$

Now, for any $y \in B$, let $x_y \in \mathbb{R}^n$ be such that $\Phi(x_y, y) = \min_{x \in \mathbb{R}^n} \Phi(x, y)$. Then, clearly $x \in A_y$ and so

$$\max_{y \in B} \min_{x \in \mathbb{R}^n} \Phi(x, y) = \max_{y \in B} \min_{x \in A} \Phi(x, y).$$

Therefore, the conclusion follows from the von Neumann theorem.

□

Theorem 7.103 (Properties of θ_4 and h_4) *Assume that φ^j is continuously differentiable for $j \in Q$. Then, θ_4 and h_4 possess the following properties:*

(a) $\theta_4(x) \leq 0$ *for all* $x \in \mathbb{R}^n$.
(b) $d\psi(x; h_4(x)) \leq \theta_4(x) - (1/2)\|h_4(x)\|^2 \leq \theta_4(x)$.
(c) *Alternative expressions for θ_4 and h_4 are*

$$\theta_4(x) = -\min_{\mu \in \Sigma_q} \left\{ \sum_{j=1}^{q} \mu^j (\psi(x) - \varphi^j(x)) + (1/2)\| \sum_{j=1}^{q} \mu^j (\varphi^j)'(x)\|^2 \right\}, \tag{7.21}$$

$$h_4(x) = -\sum_{j=1}^{q} \mu_x^j (\varphi^j)'(x), \tag{7.22}$$

where Σ_q is the unit simplex $\{\mu \in \mathbb{R}^q \mid \mu^j \geq 0, \sum_{j=1}^{q} \mu^j = 1\}$ and μ_x is any solution of (7.22). The following other (equivalent) expressions for θ_4 and h_4 also hold:

$$\theta_4(x) = -\min_{\bar{\xi} \in G\psi(x)} \{\xi^0 + (1/2)\|\xi\|^2\}, \tag{7.23}$$

$$\bar{h}(x) := (h^0(x), h_4(x)) = -\operatorname{argmin}_{\bar{\xi} \in G\psi(x)} \{\xi^0 + (1/2)\|\xi\|^2\},$$

where $G\psi(x) \subset \mathbb{R}^{n+1}$ is the set of elements $\bar{\xi} = (\xi^0, \xi)$ with $\xi^0 \in \mathbb{R}$ and $\xi \in \mathbb{R}^n$ defined by

$$G\psi(x) := \operatorname*{co}_{j \in Q} \left\{ \begin{pmatrix} \psi(x) - \varphi^j(x) \\ (\varphi^j)'(x) \end{pmatrix} \right\}.$$

(d) θ_4 *is continuous.*
(e) h_4 *is a single-valued continuous map from* $\mathbb{R}^n$ *to itself.*

Proof

(a) Since the value of the maximum in the expression of θ_4 is nonpositive at $h = 0$, being the minimum of this maximum, $\theta_4(x) \leq 0$ for any x.
(b) Since $\varphi^j(x) - \psi(x) = 0$ for $j \in Q(x)$, looking at the formula for θ_4, we see that

$$d\psi(x; h_4(x)) = \max_{j \in Q(x)} \langle (\varphi^j)'(x), h_4(x) \rangle \leq \theta_4(x) - (1/2)\|h_4(x)\|^2 \leq \theta_4(x).$$

(c) Employing (7.21) and the fact that the maximum over a set of scalars is equal to the maximum over their convex hull, we have

$$\theta_4(x) = \min_{h \in \mathbb{R}^n} \max_{\mu \in \Sigma_q} \times \left\{ \sum_{j=1}^{q} \mu^j (\varphi^j(x) - \psi(x)) + \sum_{j=1}^{q} \langle (\varphi^j)'(x), h \rangle + (1/2)\|h\|^2 \right\}.$$

Applying Corollary 7.102 to this formula, we obtain

$$\theta_4(x) = \max_{\mu \in \Sigma_q} \min_{h \in \mathbb{R}^n} \times \left\{ \sum_{j=1}^{q} \mu^j (\varphi^j(x) - \psi(x)) + \sum_{j=1}^{q} \langle (\varphi^j)'(x), h \rangle + (1/2)\|h\|^2 \right\}. \tag{7.24}$$

Next, consider the function

$$\nu(\mu) := \min_{h \in \mathbb{R}^n} \times \left\{ \sum_{j=1}^{q} \mu^j (\varphi^j(x) - \psi(x)) + \sum_{j=1}^{q} \mu^j \langle (\varphi^j)'(x), h \rangle + (1/2)\|h\|^2 \right\}.$$

Solving the expression in the right-hand side for h in terms of μ, we find that

$$h = -\sum_{j=1}^{q} \mu^j (\varphi^j)'(x), \tag{7.25}$$

and then, that

$$\nu(\mu) = \sum_{j=1}^{q} \mu^j (\varphi^j(x) - \psi(x)) + (1/2)\left\| \sum_{j=1}^{q} \mu^j (\varphi^j)'(x) \right\|^2.$$

Substituting this back into (7.24), we obtain (7.22).

The expression (7.23) for θ_4 is obtained from (7.22) by direct inspection, while (7.24) follows from (7.25) and Corollary 7.102.

(d) Formula (7.21) expresses $-\theta_4$ as the minimum of a continuous function of x and μ over a compact set of μ. Hence, θ_4 is continuous.

(e) First, we show that $\bar{h}$ is a continuous single-valued map by contradiction. Suppose that $\bar{\xi}_1 \neq \bar{\xi}_2$ are two distinct minimizers for problem (7.23). Then,

$$\xi_1^0 + (1/2)\|\xi_1\|^2 = \xi_2^0 + (1/2)\|\xi_2\|^2. \tag{7.26}$$

Suppose that $\xi_1 \neq \xi_2$. Then, for any $\lambda \in (0,1)$, the vector $\bar{\xi}_\lambda := \lambda\bar{\xi}_1 + (1-\lambda)\bar{\xi}_2$ satisfies

$$\begin{aligned}\xi_\lambda^0 + (1/2)\|\xi_\lambda\|^2 &< \lambda(\xi_1^0 + (1/2)\|\xi_2\|^2) + (1-\lambda)(\xi_2^0 + (1/2)\|\xi_2\|^2)\\ &= \xi_1^0 + (1/2)\|\xi_1\|^2.\end{aligned}$$

Since this is impossible by the strict convexity of $\|\cdot\|^2$, we must assume that $\xi_1 = \xi_2$. Hence, by (7.26), $\xi_1^0 = \xi_2^0$ and so $\bar{\xi}_1 = \bar{\xi}_2$. Thus, $\bar{h}$ is single-valued. Now, we prove that h_4 is continuous. By Corollary 5.3.9 in [389], $G\psi(\cdot)$ is a continuous set-valued map with compact values. Hence, in view of Theorem 5.4.3 in that book (we do not include these auxiliary results here), $\{\bar{h}(\cdot)\}$ is outer semicontinuous. But, as it is single-valued, this means that $\bar{h}(\cdot)$ is continuous. Therefore, h_4 is also continuous.

□

The above three formulations of θ_4 show that it is related to the cost functions not directly as θ_1, θ_2, and θ_3 are. Hence, for θ_4, we do not have exact counterparts of assertion (e) in Propositions 7.98 and 7.100.

7.5.2 Optimality Functions

From the above examples, we are confident to present a theoretical development of the concept of optimality functions. Let X be a metric space, $A \subset X$ be nonempty, and $\varphi : X \to \mathbb{R}$. Consider the optimization problem $\min_{x\in A}\varphi(x)$. A general definition of optimality functions is as follows.

Definition 7.104 (Optimality functions) An upper semicontinuous function $\theta : C \to \mathbb{R}_- \cup \{-\infty\}$ is said to be an OPTIMALITY FUNCTION for φ if $A \subset C$ and

$$x \in A \text{ is locally optimal for } \varphi \;\Rightarrow \theta(x) = 0.$$

The sets of STATIONARY POINTS and QUASI-STATIONARY POINTS (of θ) are $\mathscr{S}_{A,\theta} := \{x \in A \mid \theta(x) = 0\}$ and $\mathscr{Q}_\theta := \{x \in C \mid \theta(x) = 0\}$, resp.

We note that the upper semicontinuity of optimality functions ensures the computationally significant property that if $x^k \to x$ and $\theta(x^k) \to 0^-$, for example, with x^k obtained as approximate solutions of a corresponding optimization problem with gradually smaller tolerance, then $\theta(x) = 0$ and x is quasi-stationary (i.e., $x \in \mathscr{Q}_\theta$). So, we can understand that optimality functions are also instrumental in constructing descent directions for respective optimization problems.

Observe that the optimality functions in the examples considered in the preceding subsection indeed meet the general Definition 7.104.

7.5.3 Consistent Approximations

Definition 7.105 (Consistent approximations) An optimality function pair $\{(\varphi^k : A^k \to \mathbb{R}, \theta^k : C^k \to \mathbb{R}_- \cup \{-\infty\})\}_k$ is called a WEAKLY CONSISTENT APPROXIMATION of a pair $(\varphi : A \to \mathbb{R}, \theta : C \to \mathbb{R}_- \cup \{-\infty\})$ if:

(a) φ^k weakly epi-converges to φ;
(b) θ^k satisfy condition (a) of hypo-convergence to θ (i.e., for every $C^{k_j} \ni x^{k_j} \to x \in X$, $ls_j \theta^{k_j}(x^{k_j}) \leq \theta(x)$ if $x \in C$ and $\theta^{k_j}(x^{k_j}) \to -\infty$ otherwise).

If in addition $\theta^k(x) < 0$ for all $x \in C^k \setminus A^k$ and k, then the pair is called a CONSISTENT APPROXIMATION of $(\varphi : A \to \mathbb{R}, \theta : C \to \mathbb{R}_- \cup \{-\infty\})$.

Example 7.106 (An example of consistent approximations) Let $Z, Z^k \subset \mathbb{R}^p$ be compact subsets such that $Z \subset \mathrm{Li}_k Z^k$ (and so $Z \subset Z^k$ for all k), $\psi : \mathbb{R}^n \times \mathbb{R}^p \to \mathbb{R}$ be such that the derivative ψ'_x is continuous in both components, $\varphi(x) := \max_{z\in Z} \psi(x, z)$, and $\varphi^k(x) := \max_{z\in Z^k} \psi(x, z)$ for $x \in \mathbb{R}^n$. Consider the bifunctions on $\mathbb{R}^n \times \mathbb{R}^n$

$$\Phi(x, y) := \max_{z\in Z}[\psi(x, z) - \varphi(x) + \langle \psi'_x(x, z), y - x\rangle + (1/2)\|y - x\|^2],$$

$$\Phi^k(x, y) := \max_{z\in Z^k}[\psi(x, z) - \varphi^k(x) + \langle \psi'_x(x, z), y - x\rangle + (1/2)\|y - x\|^2].$$

Let us define θ and θ^k by

$$\theta(x) := \min_{x\in\mathbb{R}^n} \Phi(x, y), \quad \theta^k(x) := \min_{x\in\mathbb{R}^n} \Phi^k(x, y).$$

We verify the consistency following Definition 7.105. As $C^k = A^k = \mathbb{R}^n$ for all k, the "constraint qualification" $\theta^k(x) < 0$ for all $x \in C^k \setminus A^k$ is viewed as fulfilled. Hence, the consistency coincides with the weak consistency.

Following Theorem 7.51 rephrased for inf-projections, for condition (b) of consistency, we simply need to check condition (b) of the e/h-convergence of Φ^k to Φ on $\mathbb{R}^n \times \mathbb{R}^n$. Given any $\mathbb{R}^n \ni y^{k_j} \to y \in \mathbb{R}^n$ and $x \in \mathbb{R}^n$, take $x^{k_j} = x$ for all k_j. Then,

$$\begin{aligned}&\mathrm{ls}_j \Phi^{k_j}(x^{k_j}, y^{k_j})\\ &\quad = \mathrm{ls}_j \max_{z \in Z^{k_j}} [\psi(x,z) - \varphi^{k_j}(x) + \langle \psi'_x(x,z), y^{k_j} - x\rangle + (1/2)\|y^{k_j} - x\|^2].\end{aligned}$$

By the continuity assumption on ψ and ψ'_x and the assumption on the sets Z, Z^k, $\varphi^{k_j}(x) \to \varphi(x)$. Hence, $\Phi^{k_j}(x^{k_j}, y^{k_j}) \to \Phi(x, y)$. Thus, condition (b) of e/h-convergence is satisfied.

To verify condition (a) of consistency, i.e., the weak epi-convergence of φ^k to φ, simply observe that for any $x^k \to x$, $\varphi^k(x^k) \to \varphi(x)$ by the aforementioned properties of ψ, Z, and Z^k.

Theorem 7.107 (Convergence of approximate quasi-stationary points) *Assume that the pair $\{(\varphi^k : A^k \to \mathbb{R}, \theta^k : C^k \to \mathbb{R}_- \cup \{-\infty\})\}_k$ is a weakly consistent approximations of $(\varphi : A \to \mathbb{R}, \theta : C \to \mathbb{R}_- \cup \{-\infty\})$, $\varepsilon^k \to 0^+$, and $x^k \in C^k$ is a ε^k-quasi-stationary point of θ^k (i.e., $\theta^k(x^k) \geq -\varepsilon^k$). Then, every cluster point of $\{x^k\}_k$ belongs to $\mathcal{Q}_\theta$.*

If in addition the pair is a consistent approximation, φ^k epi-converges, $\varepsilon^k = 0$ for large k, and $\{\varphi^k(x^k)\}_k$ is bounded from above, then $x \in \mathcal{S}_{A,\theta}$.

Proof Consider a cluster point x of $\{x^k\}_k$, i.e., there exists a subsequence x^{k_j} converging to x. Since $\theta^{k_j}(x^{k_j}) \geq -\varepsilon^{k_j}$ for all j, by condition (b) of weak consistency, x must belong to C. We have

$$0 \leq \mathrm{ls}_j \theta^{k_j}(x^{k_j}) \leq \theta(x) \leq 0,$$

and so $x \in \mathcal{Q}_\theta$.

As $\theta^k(x^k) = 0$ for large k, by the additional condition for consistency (not only weak consistency) $x^k \in A^k$ for such k. By the epi-convergence, $\mathrm{li}_j \varphi^{k_j}(x^{k_j}) \geq \varphi(x)$ if $x \in A$ and $\varphi^{k_j}(x^{k_j}) \to +\infty$. But the latter is impossible due to the assumed boundedness. Hence, $x \in \mathcal{S}_{A,\theta}$.

□

Observe from Sect. 7.5.1 that many optimality functions have the form

$$\theta(x) = \inf_{y \in D} \Phi(x, y)$$

for some bifunction Φ. Observe from the definition of optimality functions that if $\mathcal{Q}_\theta \neq \emptyset$, then $\mathcal{Q}_\theta = \mathrm{argmax}_C \theta$. Consequently, clearly, the consideration of maxinf problems of the form

$$\max_{x \in C} \inf_{y \in D} \Phi(x, y)$$

will provide direct insight on stationary and quasi-stationary points of optimization problems. Therefore, the following theorem is a good result in this direction.

Theorem 7.108 (Convergence of maxinf-points of y-ancillary tightly mai-lop convergent bifunctions to quasi-stationary points) *Let X, Y be metric spaces, $C, C^k \subset X$, $D, D^k \subset Y$, $\Phi^k : C^k \times D^k \to \mathbb{R}$, $\Phi : C \times D \to \mathbb{R}$, and $\theta : C \to \mathbb{R}_- \cap \{-\infty\}$ defined by $\theta(x) = \inf_{y \in D} \Phi(x, y)$. Assume that Φ^k y-ancillary tightly mai-lop converges to Φ.*

Then, θ is upper semicontinuous. Moreover, if additionally $x^k \in \mathrm{mai}\Phi^k$ for all k and $\mathcal{Q}_\theta \neq \emptyset$, then any cluster point of $\{x^k\}_k$ is quasi-stationary, i.e., $x \in \mathcal{Q}_\theta$. Equivalently, this means that $\mathrm{Ls}_k(\mathrm{mai}\Phi^k) \subset \mathcal{Q}_\theta$.

Proof Following Proposition 7.72 reformulated for mai-lop convergence of bifunctions Φ^k to Φ and inf-projections θ^k (of Φ^k) to θ, θ^k hypo-converges to θ, or equivalently $\mathrm{hypo}\theta^k \overset{P-K}{\to} \mathrm{hypo}\theta$. As a set limit, $\mathrm{hypo}\theta$ must be closed, i.e., θ is upper semicontinuous.

If $x^k \in \mathrm{mai}\Phi^k$, in view of Proposition 7.3 rephrased for hypo-convergence of θ^k, $\mathrm{Ls}_k(\mathrm{mai}\theta^k) \subset \mathrm{mai}\theta$. Hence, $\theta(x) = \max_C \theta$, i.e., $\theta(x) = 0$ (since $\theta : C \to \mathbb{R}_- \cap \{-\infty\}$ and $\mathcal{Q}_\theta \neq \emptyset$).

□

Theorem 7.109 (P-K convergence of approximate maxinf-points of y-tightly mai-lop convergent bifunctions to quasi-stationary points) *Consider Φ^k and Φ satisfying the assumptions in the preceding theorem. Assume additionally that Φ^k y-tightly mai-lop converges to Φ. Then, there exists $\varepsilon^k \to 0^+$ such that ε^k-$\mathrm{mai}\Phi^k \overset{P-K}{\to} \mathcal{Q}_\theta$.*

Proof The conclusion is simply an implication of Theorem 7.75(c) (reformulated for mai-lop convergence).

□

7.6 Approximations of Nash Equilibria

In Chaps. 1 and 2, the Nash game (also called Nash equilibrium problem or noncooperative game) was considered but only results on existence were presented. Here, we discuss approximations. Moreover, different models and notions for Nash equilibria are discussed. Hence, we provide investigations from a detailed formulation of the problem. Consider the following Nash game with m players $i \in I := \{1, \ldots, m\}$. Let X_i be a metric space, $A_i \subset X_i$ be the SET OF AVAILABLE STRATEGIES of player i, and $c_i(x_i, x_{-i})$ be the COST FUNCTION (called also the PAYOFF FUNCTION) of player i for the choice of strategy $x_i \in A_i$, where x_{-i} is the vector of the strategies chosen by the remaining players $I \setminus \{i\}$. Then, $c_i : A_i \times \prod_{j \in I \setminus \{i\}} A_j \to \mathbb{R}$ for $i \in I$. The (vector) strategy $\bar{x} := (\bar{x}_i)_{i \in I}$ determines

(c) If the dual Nash game has a solution $\bar{x}$, then we have "full" duality between the pair of Nash games (with zero gap and with the zero value of Nikaido-Isoda bifunction N and $(\bar{x}, \bar{x})$ being a saddle point of N).

From the relationships between Nash equilibria and solutions of (EP$_N$) given in Proposition 7.110, it is natural to define approximate Nash equilibria as follows.

Definition 7.112 (Approximate Nash equilibria) Let $\varepsilon \geq 0$.

(a) A point $\bar{x} \in A$ is said to be a NASH ε-EQUILIBRIUM of $\mathscr{G}$ if $\sup_{x \in A} N(\bar{x}, x) \leq \varepsilon$, and a DUAL NASH ε-EQUILIBRIUM if $\inf_{x \in A} N(x, \bar{x}) \geq -\varepsilon$. The sets of such points are denoted by ε-Ne($\mathscr{G}$) and ε-Ne($D\mathscr{G}$), resp.
(b) A point $\bar{x} \in A$ is said to be a STRONG NASH ε-EQUILIBRIUM of $\mathscr{G}$ if $\bar{x} \in$ mis$N$ with minsup$N \leq \varepsilon$, and a STRONG DUAL NASH ε-EQUILIBRIUM if $\bar{x} \in$ mai$N$ with maxinf$N \geq -\varepsilon$. We denote these facts by $\bar{x} \in \varepsilon$-Ne($s\mathscr{G}$) and $\bar{x} \in \varepsilon$-Ne($Ds\mathscr{G}$), resp.
(c) Let X be a normed space, $\theta : X \to \mathbb{R}$ be such that $\theta(x) > 0$ if $x \neq 0$, and $\varepsilon \geq 0$. A point $\bar{x}_\theta \in A$ is called a NASH $\varepsilon\theta$-EQUILIBRIUM of $\mathscr{G}$ if $N(\bar{x}_\theta, y) \leq \varepsilon\theta(\bar{x}_\theta - y)$ for all $y \in A$. We denote this fact by $\bar{x}_\theta \in \varepsilon\theta$-Ne($\mathscr{G}$).
(d) Let X, $\theta : X \to \mathbb{R}$, and $\varepsilon \geq 0$ be as in (c). $\bar{y}_\theta \in A$ is called a DUAL NASH $\varepsilon\theta$-EQUILIBRIUM of $\mathscr{G}$ if $N(x, \bar{y}_\theta) \geq -\varepsilon\theta(x - \bar{y}_\theta)$ for all $x \in A$. We denote this fact by $\bar{y}_\theta \in \varepsilon\theta$-Ne($D\mathscr{G}$).

Proposition 7.113 (Properties of approximate Nash equilibria)

(a) ε-Ne($s\mathscr{G}$) $\subset$ ε-Ne($\mathscr{G}$) *and* ε-Ne($Ds\mathscr{G}$) $\subset$ ε-Ne($D\mathscr{G}$).
(b) $\bar{x} \in \varepsilon$-Ne($\mathscr{G}$), $\bar{y} \in \varepsilon$-Ne($D\mathscr{G}$), *and* $N(\bar{x}, \bar{y}) = 0$ *if and only if* $(\bar{x}, \bar{y}) \in \varepsilon$-sdl$N$ *and* $N(\bar{x}, \bar{y}) = 0$.
(c) $\bar{x} \in \varepsilon$-Ne($s\mathscr{G}$), $\bar{y} \in \varepsilon$-Ne($Ds\mathscr{G}$), *and* $N(\bar{x}, \bar{y}) = 0$ *if and only if* $(\bar{x}, \bar{y}) \in$ str-ε-sdl$N$ *and* $N(\bar{x}, \bar{y}) = 0$.
(d) $\bar{x}_\theta \in \varepsilon\theta$-Ne($\mathscr{G}$), $\bar{y}_\theta \in \varepsilon\theta$-Ne($D\mathscr{G}$), *and* $N(\bar{x}_\theta, \bar{y}_\theta) = 0$ *if and only if* $(\bar{x}_\theta, \bar{y}_\theta) \in \varepsilon\theta$-sdl$N$ *and* $N(\bar{x}_\theta, \bar{y}_\theta) = 0$.

Proof (a) Let $\bar{x} \in \varepsilon$-Ne($s\mathscr{G}$), i.e., $\bar{x} \in$ mis$N$ and minsup$N \leq \varepsilon$. Then, $\xi_N(\bar{x}) := \sup_A N(\bar{x}, \cdot) =$ minsup$N \leq \varepsilon$, and so $\bar{x} \in \varepsilon$-Ne($\mathscr{G}$). The second inclusion can be checked similarly.

(b)–(d) By Definition 7.112, assertions (b) and (c) are simply a reformulation of Theorem 7.81, and (d) is a rephrased form for $\mathscr{G}$ in Proposition 7.90.

□

Remark 7.114 (Relationships between pairs of approximate Nash equilibria and approximate saddle points)

(a) In Definition 7.112, $\bar{x} \in \varepsilon$-Ne($\mathscr{G}$) implies that $c_i(\bar{x}_i, \bar{x}_{-i}) - c_i(x_i, \bar{x}_{-i}) \leq \varepsilon$ for all $i \in I$ and $c_i(\bar{x}_i, \bar{x}_{-i}) - c_i(x_i, \bar{x}_{-i}) \leq (1/m)\varepsilon$ for at least one i, i.e., each player gets the minimum cost with the tolerance ε, while at least one gets $(1/m)\varepsilon$. Of course, if each player gets the minimum cost with the tolerance $(1/m)\varepsilon$, then $\bar{x} \in \varepsilon$-Ne($\mathscr{G}$).

(b) Clearly the opposite inclusions to those in Proposition 7.113(a) are not valid in general.
(c) For approximate Nash equilibria, we do not have the implications and equivalences similar to those in Remark 7.111(a),(b) for (exact) Nash equilibria.

Theorem 7.115 (Convergence of approximate Nash equilibria in terms of convergence of Nikaido-Isoda bifunctions) *Let* $\varepsilon, \varepsilon^k \geq 0$ *and* $\varepsilon^k \to \varepsilon$.

(a) *If* N^k *i-e/h- (or i-mis-lop or i-mai-lop) converges to* N *or* N^k *i-e-converges to* N *together with* $A \times A \sqsubset \mathrm{Li}_k(A^k \times A^k)$, *then* $\mathrm{Ls}_k(\varepsilon^k\text{-Ne}(\mathcal{G}^k)) \cap A \subset \varepsilon\text{-Ne}(\mathcal{G})$.
If N^k *e/h- (or mis-lop) converges to* N, *or* N^k *epi-converges to* N *together with* $A \times A \subset \mathrm{Li}_k(A^k \times A^k)$, *and* θ *is bounded from above, then* $\mathrm{Ls}_k(\varepsilon^k\theta\text{-Ne}(\mathcal{G}^k)) \subset \varepsilon\theta\text{-Ne}(\mathcal{G})$.
(b) *If* N^k *e/h-converges* x*-tightly or mis-lop converges* x*-ancillary tightly to* N, *then* $\mathrm{Ls}_k(\varepsilon^k\text{- Ne}(s\mathcal{G}^k)) \subset \varepsilon\text{-Ne}(s\mathcal{G})$.
If N^k *i-e/h-converges* x*-ancillary tightly or i-mis-lop converges* x*-ancillary tightly to* N, *then* $\mathrm{Ls}_k(\varepsilon^k\text{- Ne}(s\mathcal{G}^k)) \cap A \subset \varepsilon\text{-Ne}(s\mathcal{G})$.
(c) *If* N^k *w-e/h-converges* x*-ancillary tightly or w-mis-lop converges* x*-ancillary tightly to* N, $\varepsilon = 0$, *and* $x^k \in \varepsilon^k\text{- Ne}(s\mathcal{G}^k)$ *tends to* x, *then* $x \in \mathrm{Ne}(s\mathcal{G})$.
(d) *If the assumptions of assertion* (a) *are satisfied and one has additionally the continuity or boundedness from above of* θ *together with* $\varepsilon = 0$, *then resp.*,

$$\mathrm{Ls}_k(\varepsilon^k\theta\text{-Ne}(\mathcal{G}^k)) \cap A \subset \varepsilon\theta\text{-Ne}(\mathcal{G}),$$
$$\mathrm{Ls}_k(\varepsilon^k\theta\text{-Ne}(\mathcal{G}^k)) \subset \mathrm{Ne}(\mathcal{G}).$$

Proof

(a) By Definition 7.112(a), $\varepsilon^k\text{-Ne}(\mathcal{G}^k) = \varepsilon^k\text{-Sol}(\mathrm{EP}^k_N)$. Hence, by Theorem 7.76, when N^k e/h-converges or epi-converges to N, $\mathrm{Ls}_k(\varepsilon^k\text{-}Ne(\mathcal{G}^k)) \subset \varepsilon\text{-Ne}(\mathcal{G})$.
If the convergence of N^k is relaxed to the inside type, the corresponding assertion also follows from Theorem 7.76.
(b) The proof resembles that of (a), based on Theorem 7.77 instead of Theorem 7.76 and (b) of Definition 7.112 instead of (a).
(c) Apply Theorem 7.77(c).
(d) The proof is similar to that of assertion (a), with the use of Theorem 7.79 instead of Theorem 7.76 and condition (c) of Definition 7.112 instead of (a).

□

Theorem 7.116 (Convergence of approximate dual Nash equilibria in terms of convergence of Nikaido-Isoda bifunctions) *Let* $\varepsilon, \varepsilon^k \geq 0$ *and* $\varepsilon^k \to \varepsilon$.

(a) *If* N^k *e/h- or mis-lop or mai-lop converges to* N, *or hypo-converges to* N *together with* $A \times A \sqsubset \mathrm{Li}_k(A^k \times A^k)$, *then* $\mathrm{Ls}_k(\varepsilon^k\text{-Ne}(D\mathcal{G}^k)) \subset \varepsilon\text{-Ne}(D\mathcal{G})$.
If the above types of convergence of N^k *reduce to the inside types, then* $\mathrm{Ls}_k(\varepsilon^k\text{-}Ne(D\mathcal{G}^k)) \cap A \subset \varepsilon\text{-Ne}(D\mathcal{G})$.
(b) *If* N^k *e/h-converges* y*-ancillary tightly or mai-lop converges* y*-ancillary tightly to* N, *then* $\mathrm{Ls}_k(\varepsilon^k\text{- Ne}(Ds\mathcal{G}^k)) \subset \varepsilon\text{-Ne}(Ds\mathcal{G})$.

If the above types of tight convergence of N^k *reduce to the corresponding tight inside types, then* $\mathrm{Ls}_k(\varepsilon^k\text{- Ne}(Ds\mathscr{G}^k)) \cap A \subset \varepsilon\text{-Ne}(Ds\mathscr{G})$.

(c) *If* N^k *w-e/h-converges y-ancillary tightly or w-mai-lop converges y-ancillary tightly to* N, $\varepsilon = 0$, *and* $y^k \in \varepsilon^k\text{- Ne}(Ds\mathscr{G}^k)$ *converges to* y, *then* $y \in \mathrm{Ne}(Ds\mathscr{G})$.

(d) *If the assumptions of assertion* (a) *are satisfied and one has additionally the continuity or boundedness from above of* θ *together with* $\varepsilon = 0$, *then resp.,*

$$\mathrm{Ls}_k(\varepsilon^k\theta\text{-Ne}(D\mathscr{G}^k)) \cap A \subset \varepsilon\theta\text{-Ne}(D\mathscr{G}),$$

$$\mathrm{Ls}_k(\varepsilon^k\theta\text{-Ne}(D\mathscr{G}^k)) \subset \mathrm{Ne}(D\mathscr{G}).$$

Proof The statement can verified similarly to that of Theorem 7.116, invoking Theorems 7.82, 7.83, and 7.88 about (general) dual equilibrium problems. □

Now we move further on to convergence of pairs of approximate Nash equilibria. From the relationship between these pairs given in Proposition 7.113, we need the sets

$$\mathscr{C}N^k := \{(x, y) \in A\times A \mid N^k(x, y) = 0\}, \ \mathscr{C}N := \{(x, y) \in A\times A \mid N(x, y) = 0\}.$$

Theorem 7.117 (Convergence of approximate primal-dual Nash equilibria pairs) *Let* $\varepsilon, \varepsilon^k \geq 0$ *and* $\varepsilon^k \to \varepsilon$.

(a) *If* $N^k \overset{i-e/h}{\to} N$ *or* $N^k \overset{i-mis-lop}{\to} N$ *and* $N^k \overset{i-mai-lop}{\to} N$ *simultaneously and* $\mathrm{Ls}_k\mathscr{C}N^k \subset \mathrm{cl}\mathscr{C}N$, *then*

$$\mathrm{Ls}_k(\mathscr{C}N^k \cap \varepsilon^k - \mathrm{sdl}N^k) \cap (A \times A)$$
$$= \mathrm{Ls}_k[\mathscr{C}N^k \cap (\varepsilon^k - \mathrm{Ne}(\mathscr{G}^k) \times \varepsilon^k - \mathrm{Ne}(D\mathscr{G}^k))] \cap (A \times A)$$
$$\subset \mathrm{cl}\mathscr{C}N \cap (\varepsilon - \mathrm{Ne}(\mathscr{G}) \times \varepsilon - \mathrm{Ne}(D\mathscr{G})) = \mathrm{cl}\mathscr{C}N \cap \varepsilon - \mathrm{sdl}N.$$

If in the above relation the term "inside" (which is denoted by "i-") is removed, then "$\cap (A \times A)$*" can also be removed.*

(b) *If* $N^k \overset{e/h}{\to} N$ *tightly or both x- and y-ancillary-tightly mis-lop and both x- and y-ancillary-tightly mai-lop converges simultaneously and* $\mathrm{cl}\mathscr{C}N = \mathrm{Lim}_k\mathscr{C}N^k$, *then*

$$\varepsilon\text{-sdl}N \cap \mathrm{cl}\mathscr{C}N = (\varepsilon\text{-Ne}(\mathscr{G}) \times \varepsilon\text{-Ne}(D\mathscr{G})) \cap \mathrm{cl}\mathscr{C}N$$

$$= \mathrm{Lim}_k(\varepsilon^k\text{-Ne}(\mathscr{G})^k) \times \varepsilon^k\text{-Ne}(D\mathscr{G}^k) \cap \mathrm{Lim}_k\mathscr{C}N^k$$
$$= \mathrm{Lim}_k\varepsilon^k\text{-sdl}N^k \cap \mathrm{Lim}_k\mathscr{C}N^k.$$

(c) *If N^k e/h-converges tightly or both mis-lop converges x- and y-ancillary-tightly and mai-lop converges x- and y-ancillary-tightly simultaneously to N and* $\mathrm{Ls}_k \mathscr{C} N^k \subset \mathrm{cl}\mathscr{C} N$*, then*

$$\mathrm{Ls}_k(\mathscr{C}N^k \cap \mathrm{str} - \varepsilon^k - \mathrm{sdl}N^k) = \mathrm{Ls}_k[\mathscr{C}N^k \cap (\varepsilon^k - \mathrm{Ne}(s\mathscr{G}^k) \times \varepsilon^k - \mathrm{Ne}(Ds\mathscr{G}^k))]$$

$$\subset \mathrm{cl}\mathscr{C}N \cap (\varepsilon - \mathrm{Ne}(s\mathscr{G}) \times \varepsilon - \mathrm{Ne}(Ds\mathscr{G})) = \mathrm{cl}\mathscr{C}N \cap \mathrm{str} - \varepsilon - \mathrm{sdl}N.$$

(d) *Impose the assumptions of assertion* (d) *of Theorem 7.116. Then, we have the similar relations written for the second-kind approximate of the objects corresponding to the ones in assertion* (a).

Proof Simply employ Theorems 7.84, 7.85, 7.91, and Proposition 7.113. □

Studying approximations of Nash equilibria, employing the Nikaido-Isoda bifunctions N^k and N as above is convenient and suitable because it allows one to apply many preceding results for (general) equilibrium problems. However, the imposed assumptions are not in terms of the data $\{(A_i^k, c_i^k) \mid i \in I\}$ of $\mathscr{G}^k$. For using directly these data, we need the following concept.

Definition 7.118 (Weak multi-epi-convergence) For $i \in I$ and $k \in \mathbb{N}$, let X_i be a metric space, $X := \prod_{i \in I} X_i$, $\varphi_i : A \subset X \to \mathbb{R}$, and $\varphi_i^k : A^k \subset X \to \mathbb{R}$. The sequence of multicomponent functions $\{(\varphi_1^k, \dots, \varphi_m^k)\}_k$ is said to *weakly multi-epi-converge* to the multicomponent function $(\varphi_1, \dots, \varphi_m)$ if, for all $i \in I$,

(a) for every $A^k \ni x^k \to x$, $\mathrm{li}_k \varphi_i^k(x^k) \geq \varphi_i(x)$ if $x \in A$ and $\varphi_i^k(x^k) \to +\infty$ if $x \notin A$;
(b) for every $x_i \in A_i$ and $A_{-i}^k \ni x_{-i}^k \to x_{-i}$, there exists $A_i^k \ni x_i^k \to x_i$ such that $\mathrm{ls}_k \varphi_i^k(x^k) \leq \varphi_i(x)$.

Now, for the underlying spaces of the Nash game, we consider normed spaces X_i and X. We incorporate the following REGULARIZED NIKAIDO-ISODA BIFUNCTION into our study, for $s > 0$ and $(x, y) \in A \times A$:

$$N_s(x, y) := \Sigma_{i \in I}[c_i(x_i, x_{-i}) - c_i(y_i, x_{-i}) - \frac{s}{2}\langle x_i - y_i, x_i - y_i \rangle].$$

We also consider the sup-projection $\xi_s(x) := \sup_A N_s(x, \cdot)$.

Proposition 7.119 (Properties of ξ_s) *For every $i \in I$, let A_i be closed and convex and $c_i(\cdot, x_{-i})$ be convex and bounded for all $x_{-i} \in A_{-i}$. Then, for any $s > 0$,*

(a) *$\xi_s(\cdot)$ is a nonnegative function and attains its maximum at most one maximizer;*
(b) *$\bar{x}$ is a Nash equilibrium of $\mathscr{G}$ if and only if $\bar{x} \in \mathrm{argmin}_A \xi_s = \mathrm{mis}N_s$ with the min-value and minsup-value being zero;*

(c) *for every $x \in A$ and $i \in I$, the infimum*

$$\inf_{y_i \in A_i} \left[c_i(y_i, x_{-i}) + \frac{s}{2}\langle x_i - y_i, x_i - y_i \rangle \right]$$

can be attained at most one point $y_i(x) \in A_i$.
Furthermore, if $\dim X$ is finite and the above boundedness of $c_i(\cdot, x_{-i})$ is replaced by continuity, then $y_i(\cdot)$ is well-defined and continuous.

Proof

(a) This is evident since $N_s(x, x) = 0$ for all $x \in A$ and $N_s(x, \cdot)$ is strictly concave.
(b) "Only if". If $\bar{x} \in \mathrm{Ne}\mathcal{G}$, then $\bar{x} \in \mathrm{mis}N$ and the minsup-value of N is nonpositive by (c) of Proposition 7.110. Clearly, $N_s(\bar{x}, y) \le N(\bar{x}, y) \le 0$ for all $y \in A$ and $N_s(\bar{x}, \bar{x}) = 0$. Hence, $\bar{x} \in \mathrm{mis}N_s$ and the minsup-value of N_s is zero.
"If". Assume that $\bar{x} \in \mathrm{mis}N_s$ and the minsup-value is zero, i.e., $\xi_s(\bar{x}) = 0$. Fix arbitrary $k \in I$ and $x_k \in A_k$. For any $\lambda \in [0, 1]$, take $y_k = \lambda \bar{x}_k + (1 - \lambda)x_k$ and $y_{-k} = \bar{x}_{-k}$. Then,

$$0 \ge N_s(\bar{x}, y) = c_k(\bar{x}_k, \bar{x}_{-k}) - c_k(\lambda \bar{x}_k + (1 - \lambda)x_k, \bar{x}_{-k}) - \frac{s}{2}(1 - \lambda)^2 \langle \bar{x}_k - x_k, \bar{x}_k - x_k \rangle$$

$$\ge (1 - \lambda)[c_k(\bar{x}_k, \bar{x}_{-k}) - c_k(x_k, \bar{x}_{-k}) - \frac{s}{2}(1 - \lambda)\langle \bar{x}_k - x_k, \bar{x}_k - x_k \rangle].$$

Dividing the inequalities by $1 - \lambda$ and then letting $\lambda \to 1^-$, we obtain $0 \ge c_k(\bar{x}_k, \bar{x}_{-k}) - c_k(x_k, \bar{x}_{-k})$, i.e., $\bar{x} \in \mathrm{Ne}\mathcal{G}$.
(c) This follows from well-known results in parametric optimization as the function $c_i(\cdot, x_{-i}) + \frac{s}{2}\langle x_i - \cdot, x_i - \cdot \rangle$ is strictly convex on A_i.

□

From Proposition 7.119(b), in a natural way, for $\varepsilon \ge 0$, $\bar{x} \in A$ called a NASH ε-EQUILIBRIUM if $\xi_s(\bar{x}) \le \varepsilon$.

Theorem 7.120 (Convergence of approximate Nash equilibria in terms of weak multi-epi-convergence) *For all $i \in I$, let X_i be a normed space, $A_i \subset X_i$ closed and convex, and $c_i(\cdot, x_{-i})$, $c_i^k(\cdot, x_{-i})$ convex and bounded for all $k \in \mathbb{N}$, and $x_{-i} \in A_{-i}$. Let $\varepsilon^k \to 0^+$. Assume that $\{(c_1^k, \ldots, c_m^k)\}_k$ weakly multi-epi-converges to $(c_i, \ldots, c_m)$ and $x^k \in \varepsilon^k$-$\mathrm{Ne}(\mathcal{G}^k)$ tends to $\bar{x}$. Then, $\bar{x}$ is a Nash equilibrium of $\mathcal{G}$.*

Proof By definition,

$$\xi_s^k(x^k) = \sup_{y^k \in A^k} N_s^k(x^k, y^k) = \sum_{i=1}^{m} \left[c_i^k \left(x_i^k, x_{-i}^k \right) - \zeta_i^k \left(x_i^k \right) \right],$$

where we set

$$\zeta_i^k(x_i^k) := \inf_{y_i^k \in A_i^k} \left[c_i^k\left(y_i^k, x_{-i}^k\right) + \frac{s}{2}\langle x_i^k - y_i^k, x_i^k - y_i^k\rangle \right],$$

$$\zeta_i(x_i) := \inf_{y_i \in A_i} \left[c_i(y_i, x_{-i}) + \frac{s}{2}\langle x_i - y_i, x_i - y_i\rangle \right].$$

In view of Proposition 7.119(b), $x^k \in \varepsilon^k\text{-Ne}(Q\mathcal{G}^k) \Leftrightarrow \xi_s^k(x^k) \le \varepsilon^k$ and $\bar{x} \in \text{Ne}(Q\mathcal{G}) \Leftrightarrow \xi_s(\bar{x}) \le 0$ (i.e., $\xi_s(\bar{x}) = 0$ by Proposition 7.119(a)). From the connection between $\xi_s^k(x^k)$ and $\zeta_i^k(x_i^k)$, we now consider the latter. We claim that $\text{ls}_k \zeta_i^k(x_i^k) \le \zeta_i(\bar{x}_i)$ for all $i \in I$ and x^k given in the assumptions. Assume that $\{y_i^\ell\}_\ell$ is a minimal sequence of $c_i(\cdot, \bar{x}_{-i}) + \frac{s}{2}\langle \bar{x}_i - \cdot, \bar{x}_i - \cdot\rangle$, i.e., $c_i(y_i^\ell, \bar{x}_{-i}) + \frac{s}{2}\langle \bar{x}_i - y_i^\ell, \bar{x}_i - y_i^\ell\rangle \to \inf_{y_i \in A_i}[c_i(y_i, \bar{x}_{-i}) + \frac{s}{2}\langle \bar{x}_i - y_i, \bar{x}_i - y_i\rangle] = \zeta_i(\bar{x}_i)$ as $\ell \to +\infty$. The assumed weak multi-epi-convergence implies the weak epi-convergence of $c_i^k(\cdot, x_{-i}^k) + (s/2)\langle x_i^k - \cdot, x_i^k - \cdot\rangle$ (on A_i^k) to $c_i(\cdot, \bar{x}_{-i}) + (s/2)\langle \bar{x}_i - \cdot, \bar{x}_i - \cdot\rangle$ (on A_i) for each i. By condition (b) of this weak epi-convergence, there exists $A_i^k \ni y_i^{\ell,k} \to y_i^\ell$ as $k \to +\infty$ such that

$$\text{ls}_k \left[c_i^k(y_i^{\ell,k}, x_{-i}^k) + \frac{s}{2}\langle x_i^k - y_i^{\ell,k}, x_i^k - y_i^{\ell,k}\rangle \right] \le c_i(y_i^\ell, \bar{x}_{-i}) + \frac{s}{2}\langle \bar{x}_i - y_i^\ell, \bar{x}_i - y_i^\ell\rangle.$$

Then,

$$\begin{aligned} \text{ls}_k \zeta_i^k(x_i^k) &\le \text{ls}_k \left[c_i^k(y_i^{\ell,k}, x_{-i}^k) + \frac{s}{2}\langle x_i^k - y_i^{\ell,k}, x_i^k - y_i^{\ell,k}\rangle \right] \\ &\le c_i(y_i^\ell, \bar{x}_{-i}) + \frac{s}{2}\langle \bar{x}_i - y_i^\ell, \bar{x}_i - y_i^\ell\rangle. \end{aligned}$$

Taking limit as $\ell \to +\infty$ of the sides, we get $\text{ls}_k \zeta_i^k(x_i^k) \le \zeta_i(\bar{x}_i)$ for all $i \in I$ as claimed. Hence, by (a) of the aforementioned weak-epi-convergence, $\bar{x} \in A$. We have

$$\text{li}_k \xi_s^k(x^k) \ge \Sigma_{i \in I} \text{li}_k [c_i^k(x_i^k, x_{-i}^k) - \text{ls}_k \zeta_i^k(x_i^k)] \ge \Sigma_{i \in I}[c(\bar{x}_i) - \zeta_i(\bar{x}_i)] = \xi_s(\bar{x}).$$

As $\varepsilon^k \to 0^+$ and $\xi_s^k(x^k) \le \varepsilon^k$, we have $\xi_s(\bar{x}) \le 0$ and so $\bar{x} \in \text{Ne}(Q\mathcal{G})$.

□

Observe that by applying the regularized Nikaido-Isoda bifunction N_s^k, with additional convexity assumptions, Theorem 7.120 yields a convergence result of approximate solutions of $\mathcal{G}^k$ only in terms of the same convergence of $c_i, i \in I$, directly, not via convergence of bifunctions. However, it seems that looking for a similar result for $D\mathcal{G}^k$ and then for the pair $\mathcal{G}^k$-$D\mathcal{G}^k$ is expected to be a significant challenge since even a formulation of $D\mathcal{G}$ in terms of its data (not in N) is not available for us.

To end this section, we discuss some approximation results that include also existence assertions by employing the Ky Fan function.

Definition 7.121 (Ky Fan function) A bifunction $\Phi : A \times A \to \mathbb{R}$ with A nonempty convex such that:

(a) $\Phi(\cdot, y)$ is lsc on A for all $y \in A$;
(b) $\Phi(x, \cdot)$ is concave on A for all $x \in A$, is called a KY FAN FUNCTION.

Lemma 7.122 (Solutions of the Ky Fan inequality) *Assume that $\Phi : A \times A \to \mathbb{R}$ is a Ky Fan function with A compact and convex. Then, there exists a solution $\bar{x}$ of the Ky Fan inequality:*

$$\text{find } \bar{x} \in A \text{ such that } \sup_{y\in A} \Phi(\bar{x}, y) \leq \sup_{y\in A} \Phi(y, y). \tag{7.27}$$

Hence, if $\Phi(y, y) \leq 0$ for all $y \in A$, then $\bar{x}$ is a minsup-points of Φ, i.e., it is just a solution of the equilibrium problem (EP) *and also of the strong equilibrium problem* (SEP) *(defined in Sect. 7.3).*

Note that this classical result can be derived from the results in Chap. 1, which are its later developments. But it is more convenient for the reader to find it in well-known books, e.g., [81, 82] (for a better insight, look also at Proposition 7.110). Moreover, if for Φ, one is interested in maximizing over $x \in A$ and minimizing over $y \in A$, then in Definition 7.121 of Ky Fan functions, "lsc" in condition (a) and "concave" in condition (b) are replaced by "usc" and "convex," resp.

Remark 7.123 (Results equivalent to the Ky Fan inequality) In some books and papers, the Ky Fan inequality is defined with (7.27) replaced by

$$\text{find } \bar{x} \in A \text{ such that } \sup_{y\in A} \Phi(\bar{x}, y) \leq 0.$$

Then, when $\Phi(y, y) \leq 0$ for all $y \in A$, it is completely equivalent to (EP$_N$) and also to (sEP$_N$). Before 1994, the Ky Fan inequality was deeply investigated in connection to fixed points. From the work [108], its equivalent formulation as (EP$_N$) has been attracting researches in many aspects of optimization-related problems besides the solution existence.

Applying Lemma 7.122 about the Ky Fan inequality, we easily arrive at the following statement about convergence of Nash equilibria.

Theorem 7.124 (Convergence of Nash equilibria in terms of tight e/h- or mis-lop convergence of N^k) *Assume that Nikaido-Isoda bifunction N^k is a Ky Fan function with A^k being compact.*

(a) *If N^k x-ancillary tightly e/h- or x-ancillary tightly mis-lop converges to a bifunction N, then there exists a strategy $\bar{x}^k \in$ Ne($\mathcal{G}^k$) and any cluster point $\bar{x}$ of $\{\bar{x}^k\}_k$ is a Nash equilibrium of $\mathcal{G}$.*

(b) *If* $N^k \overset{mis-lop}{\rightarrow} N$ *x-tightly, then there exists* $\varepsilon^k \to 0^+$ *such that* ε^k-misN^k *Painlevé-Kuratowski converges to* $\mathrm{Ne}(\mathscr{G})$.

Proof

(a) By Lemma 7.122, $\bar{x}^k \in \text{mis}N^k$ (observe that N^k vanishes on the diagonal of $A \times A$), and so $\bar{x}^k \in \mathrm{Ne}(\mathscr{G}^k)$. In view of Theorems 7.52 and 7.73, any cluster point $\bar{x}$ of $\{\bar{x}^k\}_k$ belongs to mis$N$ and then is a Nash equilibrium of $\mathscr{G}$.
(b) This assertion is a consequence of Theorem 7.74 and Proposition 7.110.

□

The reader can find results on the Nash equilibrium problem also in Chapter 9 of the recent book [403].

7.7 Approximations of Walras Economic Equilibria

7.7.1 Problem Setting

Let $(\Omega, \mathscr{A}', P')$ and $(\Theta, \mathscr{A}, P)$ be complete probability spaces, where Θ is a separable metric space and P is the probability law of a random element $\xi : \Omega \to \Theta$. That is, we think of Θ as the set of possible values of the random element ξ defined on the probability space of events $(\Omega, \mathscr{A}', P')$.

7.7.2 Walras Pure Exchange Model

One of the classical economic models is the WALRAS PURE EXCHANGE MODEL (also called the WALRAS BARTER PROBLEM). Actually, a large number of economic models are designed to study conditions for balancing supply and demand of commodities/goods. Hence, they are essentially equilibrium problems. In this Walras model, the economy is described by

$$\mathscr{E} := \{(u_i, e_i) \mid i = 1, \ldots, n\},$$

where we have n consumers/agents; $e_i \in \mathbb{R}^\ell_+$ is the endowment (in trading) of ith consumer, which is a bundle of goods he/she trades; and $u_i : \mathbb{R}^\ell_+ \to \mathbb{R}$ is the ith consumer's utility function.

Trading takes place at a per-unit market price p^j for good j, $j = 1, \ldots, \ell$. So, clearly ith consumer can acquire the bundle of goods x only if the BUDGETARY CONSTRAINT $\langle p, x\rangle \le \langle p, e_i\rangle$ is fulfilled. It is assumed that the consumers act as utility maximizers. Thus, given the price $p \in \mathbb{R}^\ell_+$, ith consumer will end up with his/her consumption (a bundle of goods):

$$c_i(p) = \operatorname{argmax}_{x \in D_i(p)}\{u_i(x) \mid \langle p, x\rangle \le \langle p, e_i\rangle\},$$

where

$$D_i(p) := \operatorname{dom} u_i \cap \{x \in \mathbb{R}^\ell_+ \mid \langle p, x - e_i\rangle \le 0\}. \tag{7.28}$$

The set $D_i(p)$ is called the SURVIVAL SET of ith consumer and assumed to be nonempty. As $c_i(\alpha p) = c_i(p)$ for $\alpha > 0$ (i.e., the consumers' demand functions are homogeneous of degree 0 wrt prices), we may restrict the choice of p to the price unit simplex $\Delta := \{p \in \mathbb{R}^\ell_+ \mid \sum_{j=1}^{\ell} p^j = 1\}$,

The economy $\mathscr{E}$ is operational only if, for a given price p, total supply (of all consumers) exceeds total demand, i.e.,

$$\sum_{i=1}^{n} s_i(p) =: s(p) \ge 0 \text{ with } s_i(p) := e_i - c_i(p).$$

$s : \mathbb{R}^\ell_+ \to \mathbb{R}^\ell$ is called the EXCESS SUPPLY FUNCTION.

Proposition 7.125 (Properties of survival sets) *Assume that* $\operatorname{dom} u_i$ *is closed, convex with nonempty interior and* $e_i \in \operatorname{int}(\operatorname{dom} u_i)$ *for* $i = 1, \dots, n$. *Then,* $D_i : \Delta \rightrightarrows \mathbb{R}^\ell_+$ *is closed-convex-valued,* $\operatorname{int} D_i(p) \ne \emptyset$ *for* $p \in \Delta$, *and* $D_i(\cdot)$ *is continuous.*

Proof Evidently, D_i is closed-convex-valued by (7.28). Since $e_i \in \operatorname{int}(\operatorname{dom} u_i)$, $\operatorname{int} D_i(p) \ne \emptyset$ for all $p \in \Delta$. Consider any $\bar p \in \Delta$ and $\bar z \in \operatorname{int} D_i(\bar p)$. Since for any subsequences $p^{k_j} \to \bar p$ and $z^{k_j} \to \bar z$, $\langle p^{k_j}, z^{k_j} - e_i\rangle \to \langle \bar p, \bar z - e_i\rangle$, $\operatorname{Ls}_{p \in \Delta \to \bar p} D_i(p) \subset D_i(\bar p)$.

Now, we prove that $\operatorname{Li}_{p \in \Delta \to \bar p} D_i(p) \supset D_i(\bar p)$. We claim first that for all $\bar p \in \Delta$ and $\bar z \in \operatorname{int} D_i(\bar p)$, there exist neighborhoods U of $\bar p$ and V of $\bar z$ such that $(U \cap \Delta) \times V \subset \operatorname{gph} D_i(\bar p)$. Indeed, we can find a neighborhood $V \subset D_i(\bar p)$ of $\bar z$ such that $\langle \bar p, z - e_i\rangle < 0$ for all $z \in V$. Thus, the claim is verified. Due to it, for all $p^k \in \Delta \to \bar p$, p^k belongs to $U \cap \Delta$ for all large k. Hence, we can find $z^k \in V \subset D_i(\bar p)$ with $z^k \to \bar z$. Thus, $\operatorname{Li}_{p \in \Delta \to \bar p} D_i(p) \supset \operatorname{int} D_i(\bar p)$. But the closedness of the left-hand side of this containment implies that $\operatorname{Li}_{p \in \Delta \to \bar p} D_i(p) \supset D_i(\bar p)$. So, $D_i(p) \overset{P-K}{\to} D_i(\bar p)$, i.e., D_i is continuous.

□

Definition 7.126 (Equilibrium prices, strict Walras law)

(a) A price $\bar p \in \Delta$ is called a (WALRAS) EQUILIBRIUM PRICE of $\mathscr{E}$ if $s(\bar p) \ge 0$. The set of these points is denoted by $\operatorname{Ep}(\mathscr{E})$.

(b) A price $\bar{p} \in \Delta$ is said to satisfy the STRICT WALRAS LAW if, for each (good) j, either $s^j(\bar{p}) > 0$ or $\bar{p}^j > 0$.

We aim at analyzing the stability/convergence of equilibrium prices to (global) approximations of the economy:

- approximations of the consumers' utilities,
- approximations of the consumers' endowments.

For this purpose, we define the bifunction $W : \Delta \times \Delta \to \mathbb{R}$, called the WALRASIAN, by

$$W(p, y) := \langle y, s(p) \rangle. \tag{7.29}$$

The next statement identifies the equilibrium prices of economy $\mathscr{E}$ with the maxinf-points of the Walrasian W.

Proposition 7.127 (Walras equilibrium prices and maxinf-points) *Every maxinf-point $\bar{p}$ of the Walrasian W such that $W(\bar{p}, \cdot) \geq 0$ on Δ is an equilibrium price. Conversely, every equilibrium price $\bar{p}$ satisfying the strict Walras law is a maxinf-point of W such that $W(\bar{p}, \cdot) \geq 0$ on Δ.*

Proof Assume that $\bar{p} \in \text{mai}W$ with $W(\bar{p}, \cdot) \geq 0$ on Δ. In the place of $\cdot$, we take the unit vector $d^j := (0, \ldots, 1, \ldots, 0)$, with j entry being 1, in Δ to have $W(\bar{p}, d^j) = \langle d^j, s(\bar{p}) \rangle = s^j(\bar{p}) \geq 0$. We carry out this act for all $j = 1, \ldots, \ell$ to obtain that $s(\bar{p}) \geq 0$, i.e., that $\bar{p}$ is an equilibrium price.

Conversely, assume that $\bar{p}$ is an equilibrium price satisfying the strict Walras law. Then, of course $W(\bar{p}, \cdot) \geq 0$ on Δ. The encountered law implies that for any $j = 1, \ldots, \ell$, if $s^j(\bar{p}) = 0$, then $\bar{p}^j > 0$ while if $s^j(\bar{p}) > 0$, then $\bar{p}^j = 0$. Hence, $\langle \bar{p}, s(\bar{p}) \rangle = 0$ and so $\inf_{y \in \Delta} W(\bar{p}, y) = \langle \bar{p}, s(\bar{p}) \rangle = 0$. Now, suppose that $\bar{p} \notin \text{mai}W$ and another p' satisfying the aforementioned law together with $\inf_{y \in \Delta} W(p', y) \geq \inf_{y \in \Delta} W(\bar{p}, y) = 0$. If p' is not an equilibrium price, by the law, there exists $j_0 \in \{1, \ldots, \ell\}$ such that $s^{j_0}(p') < 0$ and $(p')^{j_0} > 0$, and so $\inf_{y \in \Delta} W(p', y) < 0$, a contradiction. Hence, p' is in fact also an equilibrium price satisfying the strict Walras law. Thus, $\inf_{y \in \Delta} W(p', y) = 0 = \inf_{y \in \Delta} W(\bar{p}, y)$. Therefore, $\inf_{y \in \Delta} W(\bar{p}, y) = \text{maxinf}W$, i.e., $\bar{p} \in \text{mai}W$.

□

The study of existence was presented comprehensively in Chaps. 1–3; here we provide only few results while focusing on approximation issues.

Theorem 7.128 (Existence of Walras equilibrium prices) *Assume that the excess supply function $s(\cdot)$ is usc on Δ and the Walrasian satisfies $W(p, p) \geq 0$ for $p \in \Delta$. Then, the economy $\mathscr{E}$ has at least one equilibrium price, say $\bar{p}$. Moreover, $W(\bar{p}, \cdot) \geq 0$ on Δ.*

Proof Since $s(\cdot)$ is usc, so is $W(\cdot, p)$ on Δ for $p \in \Delta$. Being linear, $W(p, \cdot)$ is convex and so W is a Ky Fan function. Then, in view of Lemma 7.122, there exists $\bar{p} \in \text{mai}W$ such that $W(\bar{p}, \cdot) \geq 0$ on Δ. □

Note that the assumption $W(p, p) \geq 0$ for all $p \in \Delta$ is not restrictive. It is satisfied in many cases. For instance, the Nikaido-Isoda bifunction N in (7.30) automatically satisfies this assumption. Now, we present briefly another method of studying the Walras equilibrium problem by expressing it as a special case of Nash equilibria for noncooperative games. To have such an expression, besides the n players being the n consumers, we add to the model $\mathscr{E}$, described at the beginning of this subsection, the so-called WALRAS AUCTIONEER, whose main function is to choose a market price system to secure a market equilibrium. For $i = 1, \ldots, n$, we define the return/payoff function of ith player by

$$r_i(x_i, x_{-i}, p) := u_i(x) \text{ when } \langle p, x_i - e_i \rangle \leq 0, x_i \in D_i(p)$$

and for the Walras auctioneer,

$$r_W(p, x) := \inf_{q \geq 0} \langle q, \sum_{i=1}^{n} (e_i - x_i) \rangle \text{ when } 0 \leq p \neq 0.$$

In this model,

- r_i does not depend on x_{-i}; it is included in the arguments simply for consistency;
- when $0 \leq p \neq 0$, $r_W(p, x) = -\infty$ unless $\sum_{i=1}^{n} (e_i - x_i) \geq 0$, in which case $r_W = 0$, its upper bound.

To suit the framework of Nash equilibria for noncooperative games, in the Walras pure exchange model, if $\bar{p} \neq 0$ is an equilibrium price and $\bar{x} \in c_i(\bar{p})$, we call $(\bar{x}, \bar{p})$ an equilibrium point of the model. Then, like

$$\bar{x}_i \in \operatorname{argmax}_x r_i(x_i, \bar{x}_{-i}, \bar{p}) = \operatorname{argmax}_{x \in D_i(\bar{p})} \{u_i(x) \mid \langle \bar{p}, e_i - x_i \rangle \geq 0\} = c_i(\bar{p}),$$

we have $\bar{p} \in \operatorname{argmax}_{0 \leq p \neq 0} r_W(p, \bar{x})$ and the excess supply $s(\bar{p}) = \sum_{i=1}^{n} (e_i - \bar{x}_i) \geq 0$. The return function $r_W(p, x)$ does not depend explicitly on p, but actually does so implicitly. In fact, once a price system $\bar{p}$ has been taken such that $\sum_{i=1}^{n} (e_i - x_i) \geq 0$, provided that $\bar{x}_i \in \operatorname{argmax}_x r_i(x_i, \bar{x}_{-i}, \bar{p})$, any other (nonnegative) price system p would "maximize" r_W but it would not guarantee that $\bar{x}_i$ would then maximize $r_i(\cdot, \bar{x}_{-i}, \bar{p})$. Hence, $(\bar{x}, \bar{p})$ is a Nash equilibrium for the noncooperative game defined by the return functions $(\{r_i \mid i = 1, \ldots, n\}, r_W)$.

Conversely, if $(\bar{x}, \bar{p})$ is a Nash equilibrium for the noncooperative game with return functions $(\{r_i \mid i = 1, \ldots, n\}, r_W)$, then necessarily for all $i = 1, \ldots, n$, $\bar{x}_i \in \operatorname{argmax}_{x \in D_i(\bar{p})} \{u_i(x) \mid \langle \bar{p}, e_i - x_i \rangle \geq 0\}$ and $\sum_{i=1}^{n} (e_i - \bar{x}_i) \geq 0$ since otherwise $\operatorname{argmax} r_W$ would be empty. This means that $0 \leq p \neq 0$ and $(\bar{x}, \bar{p})$ is an equilibrium point of the Walras equilibrium problem.

We can then write the corresponding Nikaido-Isoda bifunction:

$$N((x, p), (y, q)) := \sum_{i=1}^{n} [(r_i(x_i, x_{-i}, p) - r_i(y_i, x_{-i}, p)) + (r_W(p, x) - r_W(q, x))]. \tag{7.30}$$

However, such an expression, which is relatively explicit in terms of the data (u_i, e_i), $i = 1, \ldots, n$, gets a little unwieldy, since it makes the analysis of convergence properties more involved than it should be. Hence, we will work with the Walrasian bifunction (7.29) in studying approximations below.

7.7.3 Approximations of Walras Equilibrium Models

Now, we apply variational convergence to approximations of the Walras pure exchange problem. Consider approximating economies $\mathscr{E}^k := \{(u_i^k, e_i^k) \mid i = 1, \ldots, n\}$ of the economy $\mathscr{E}$ (with the corresponding survival sets D_i^k).

Theorem 7.129 (Convergence of equilibrium prices under ancillary tight e/h-convergence of Walrasians) *Assume that the Walrasian W^k of the economy $\mathscr{E}^k$ y-ancillary tightly e/h- or y-ancillary tightly mai-lop converges to W and $\varepsilon^k \to 0^+$. If $p^{k_j} \in \varepsilon^{k_j}\text{-mai}W^{k_j}$ converges to $\bar{p}$ and $W(\bar{p}, \cdot) \geq 0$ on Δ, then $\bar{p}$ is a Walras equilibrium price of $\mathscr{E}$.*

Assume additionally that $\varepsilon^k \equiv 0$, $W^k(p^k, \cdot) \geq 0$ on Δ, and p^k, p satisfy the strict Walras law. Then, for any $p^{k_j} \in \mathrm{Ep}(\mathscr{E}^{k_j})$ converging to $\bar{p}$ with $W(\bar{p}, \cdot) \geq 0$ on Δ, $\bar{p} \in \mathrm{Ep}(\mathscr{E})$.

Proof This statement is a consequence of Theorems 7.52 and 7.73 rephrased for maxinf-points and Proposition 7.127. □

We know from Sect. 7.2 that mai-lop convergence is more appropriate for dealing with maxinf-points. This advantage is also useful for our current interest here as follows.

Theorem 7.130 (A complete convergence of equilibrium prices under tight lop-convergence of Walrasians) *If we assume y-tight mai-lop convergence instead of the convergence of W^k in Theorem 7.129, then we have the additional (to the conclusion of that theorem) assertion: there exists $\varepsilon^k \to 0^+$ such that ε^k-$\mathrm{mai}W^k \overset{P-K}{\to} \mathrm{mai}W$.*

Proof Apply Theorem 7.74(c) rewritten for maxinf-points and Proposition 7.127. □

Section 9H in [403] is also devoted to the Walras barter model.

7.8 Consistency of Statistical Estimators of Solutions of Stochastic Optimization Problems

In this section, variational convergence will be applied to approximations of a general stochastic optimization problem. Here, one "statistically" approaches to approximation by independent sampling of the values of a random element ξ in the problem setting to obtain an approximating sequence $\{P^k\}_k$ for the probability law P of the problem, when one has only limited information about P. One can apply variational convergence since the approximate expectations of the random variable $\xi \mapsto f(x, \xi)$ of the approximating problems epi-converge to the original expectation if P^k weakly converge to P (in the probability sense, defined in Sect. 7.8.2). The resulting convergence of approximate solutions to the original one is then expressed as consistency of statistical estimators of solutions, whose meaning is explained now in Sect. 7.8.1, where a general topological problem setting is presented. Section 7.8.2 is devoted to establishing epi-convergence of approximating expectation functions when approximating probabilities P^k weakly converge, also for the topological case. The next subsection, Sect. 7.8.3, applies the above epi-convergence to establish consistency of statistical estimators of solutions and of infimal values of stochastic optimization problems in metric spaces. The last subsection, Sect. 7.8.4, extends the consistency results to stochastic minimax problems and pairs of dual stochastic mathematical programs in metric spaces, by employing epi/hypo-convergence to expectation bifunctions associated with these two types of problems.

7.8.1 Problem Setting

Let $(\Omega, \mathscr{A}', P')$ and $(\Theta, \mathscr{A}, P)$ be complete probability spaces, where Θ is a separable metric space and P is the probability law of a random element $\xi : \Omega \to \Theta$. That is, we think of Θ as the set of possible values of the random element ξ defined on the probability space of events $(\Omega, \mathscr{A}', P')$. Let X be a Hausdorff topological space and $f : X \times \Theta \to \mathbb{R} \cup \{+\infty\}$ have $\mathrm{dom} f := \{(x, \xi) \in X \times \Theta \mid f(x, \xi) < +\infty\} \neq \emptyset$. Our problem consists of

$$\text{finding } \bar{x} \in X \text{ that minimizes } Ef(x), \tag{7.31}$$

where $Ef(x) := \int_\Theta f(x, \xi)P(d\xi)$ is the expectation of the random variable $f(x, \cdot)$ for each $x \in X$. This problem setting contains many models for decision making problems under uncertainty, including Monte Carlo simulations, Bayesian decisions, maximum likelihood estimations, portfolio optimization, etc. Assume that we have only limited information about the law P. Namely, we know a sample space $(Z, \mathscr{F}, \mu)$ with an increasing sequence $\{\mathscr{F}^k\}_k$ of sub-σ-fields of the σ-field $\mathscr{F}$. A sample $\zeta \in Z$, i.e., $\zeta = \{\xi_1, \xi_2, \ldots\}$ is obtained by independent sampling of the values of ξ and gives discrete approximations of P, i.e., a sequence $\{P^k(\cdot, \zeta)\}_k$ of probability measures on $(\Theta, \mathscr{A})$. As only the information gained up to stage k can be used in the choice of P^k, we must assume that, for all $k \in \mathbb{N}$ and $A \in \mathscr{A}$, $\zeta \mapsto P^k(A, \zeta)$ is $\mathscr{F}^k$-measurable. Then, for each $\zeta \in Z$, we have a sequence of approximate problems

$$\text{find } x^k(\zeta) \in X \text{ that minimizes } E^k f(x)(\zeta), \tag{7.32}$$

where

$$E^k f(x)(\zeta) = \int_{\Theta} f(x, \xi) P^k(d\xi, \zeta).$$

An optimal solution $x^k(\zeta)$ of (7.32) is an approximate solution of (7.31) and $x^k(\cdot)$ is called a (STATISTICAL) ESTIMATOR of the solution $\bar{x}$ of (7.31). A sequence $\{x^k(\cdot)\}_k$ of statistical estimators is said to be CONSISTENT if it converges μ-almost surely to $\bar{x}$, i.e., for all ζ in a full μ-measure subset Z_0 of Z, $x^k(\zeta) \to \bar{x}$. Note that though the formulation of problem (7.31) is an unconstrained minimization problem, in fact we can incorporate constraints into the objective by allowing f to take on the value $+\infty$ or by using a Lagrangian (cf. Sect. 7.8.4). Observe also that the integral defining $Ef(x)$ is interpreted as $+\infty$ when $f(x, \cdot) \notin L^1(\Theta, \mathscr{A}, P)$.

We will establish conditions for the above discretization scheme of stochastic optimization problem (7.31) to produce a consistency of the estimators of solutions and also the μ-almost sure convergence of the optimal values of problem (7.32) to those of (7.31). For such a consistency, since (7.31) is a minimization problem, a classical and best (weakest) framework for analyzing approximation is the almost μ-sure epi-convergence of $E^k f(\cdot)$ to the limit $Ef(\cdot)$, i.e., $E^k f(\cdot)$ is EPI-CONSISTENT. Then, in view of variational properties of epi-convergence, we obtain the aforementioned consistency for the optimal values and estimators. Obviously, the epi-convergence of $E^k f(\cdot)$ must be in terms of a type of convergence of the sequence P^k obtained by the aforementioned sampling (together with additional assumptions). We will see this in Theorem 7.133, where the epi-convergence is obtained when the probability P^k converges in the following sense (a basic concept in probability theory).

Definition 7.131 (Weak convergence of probability measures) A sequence P^k is said to WEAKLY CONVERGE to P if for every bounded continuous function $h : \Theta \to \mathbb{R}$, $\int_\Theta h(\xi)P^k(d\xi)$ tends to $\int_\Theta h(\xi)P(d\xi)$. Weak convergence is also known as CONVERGENCE IN DISTRIBUTION or NARROW CONVERGENCE.

Applying epi/hypo-convergence, we also extend these consistency results to stochastic minimax problems and pairs of dual stochastic mathematical programs in metric spaces, by employing epi/hypo-convergence to expectation bifunctions associated with these two types of problems.

7.8.2 Epi-Convergence of Expectation Functions

For a topological space X and functions $\varphi^k : X \to \mathbb{R} \cup \{+\infty\}$, the extension of Definition 7.10 (a) for a metric space X is as follows.

Definition 7.132 (Lower and upper epi-limits) Let X be a topological space and $\varphi^k : X \to \mathbb{R} \cup \{+\infty\}$. The LOWER EPI-LIMIT of φ^k at $x \in X$ is

$$\text{eli}\varphi^k(x) := \sup_{U \in \mathcal{N}(x)} \text{li}_k \inf_{y \in U} \varphi^k(y),$$

and the UPPER EPI-LIMIT is

$$\text{els}\varphi^k(x) := \sup_{U \in \mathcal{N}(x)} \text{ls}_k \inf_{y \in U} \varphi^k(y),$$

where $\mathcal{N}(x)$ designates the collection of all neighborhoods of $x \in X$. If these two limits are equal to $\varphi(x)$, φ^k epi-converges at x. If φ^k epi-converges at every $x \in X$, we omit "at x" and denote as before this convergence by e-$\lim_k \varphi^k = \varphi$ or $\varphi^k \xrightarrow{e} \varphi$.

For $g : X \to \mathbb{R} \cup \{+\infty\}$, f as in (7.31), and $V \in \mathcal{N}(x)$, denote the lower closure (the lower semicontinuous regularization) of g at x by

$$\text{cl}g(x) = \sup_{V \in \mathcal{N}(x)} \inf_{y \in V} g(y),$$

(cf. Definition 7.8) and set

$$f_V(\xi) := \text{cl}[\inf_{y \in V} f(y, \xi)].$$

Observe that, for any usc function $g : X \to \mathbb{R} \cup \{+\infty\}$, cl$g$ is also usc. Indeed, let $\delta > 0$ and $x \in X$ be arbitrary. There exists $V \in \mathcal{N}(x)$ such that $g(x') \leq g(x) + \delta$ for all $x' \in V$. Then,

$$\text{cl}g(x') \leq \text{cl}(g(x) + \delta) \leq \text{cl}g(x) + \delta,$$

i.e., cl$g$ is usc.

Now we proceed to the investigation of epi-convergence of approximating expectation functions of stochastic minimization problem (7.31) with $\mathcal{A}$ being a Borel σ-algebra. We define a TIGHTNESS property (called also ASYMPTOTIC NEGLIGIBILITY) as follows. For $h : \Theta \to \mathbb{R} \cup \{+\infty\}$, a sequence of probability measures $\{P^k\}_k$ on $(\Theta, \mathcal{A})$ is said to be h-TIGHT if, for every positive ε, there exists a compact subset $K_\varepsilon \subset \Theta$ such that, for all k,

$$\int_{\Theta \setminus K_\varepsilon} |h(\xi)| P^k(d\xi) < \varepsilon.$$

We will impose the following assumptions in what follows.

(A1) (continuity of f) For all $\xi \in \Theta$, the domain of $f(\cdot, \xi)$ is a nonempty closed subset $S \subset X$, i.e., dom$f = S \times \Theta$. Note that this condition is satisfied in many practical applications. Furthermore, f is jointly lsc and $f(x, \cdot)$ is usc on Θ for each $x \in S$.

(A2) (weak convergence) For $Z_0 \subset Z$ with full μ-measure and all $\zeta \subset Z_0$, $P^k(\cdot, \zeta)$ converges weakly to $P(\cdot)$. Here, following Definition 7.132, $P^k(\cdot, \zeta)$ weakly converges to $P(\cdot)$ if for each bounded continuous function $h : \Theta \to \mathbb{R}$, $\int_\Theta h(\xi)P^k(d\xi, \zeta)$ tends to $\int_\Theta h(\xi)P(d\xi)$.

(A3) (tightness) For all $x \in S$ and $\zeta \in Z_0$, the sequence $\{P^0(\cdot, \zeta) := P(\cdot), P^k(\cdot, \zeta)\}_k$ is $\{f(x, \cdot), f_V(\cdot)\}_{V \in \mathcal{N}(x)}$-tight.

Theorem 7.133 (Epi- and pointwise convergence of expectation functions) *Let f, P and P^k satisfy conditions (A1), (A2), and (A3). Then, we have for $\zeta \in Z_0$,*

$$\int_\Theta f(\cdot, \xi)P(d\xi) = \text{e-lim}_k \int_\Theta f(\cdot, \xi)P^k(d\xi, \zeta) \text{ on } X,$$

$$Ef(x) = \lim_k E^k f(x)(\zeta) \textit{for all} x \in X,$$

i.e., $E^k f(\cdot)(\zeta)$ converges to $Ef(\cdot)$ in both epigraphic and pointwise sense.

Proof We fix any $\zeta \in Z_0$. For brevity, we omit ζ in all the notations in this proof. It is sufficient to verify the relations

$$\sup_{U \in \mathcal{N}(x)} \text{li}_k \inf_{y \in U} E^k f(y) \geq Ef(x), \tag{7.33}$$

$$\sup_{U \in \mathcal{N}(x)} \text{ls}_k \inf_{y \in U} E^k f(y) \leq Ef(x). \tag{7.34}$$

We verify (7.33) first. If $x \notin S$ (S is closed and given in (A1)), then there exists $U \in \mathcal{N}(x)$ such that $U \cap S = \emptyset$. Hence, $f(y, \xi) = +\infty$ for all $y \in U$ and $\xi \in \Theta$, and the left-hand side of (7.33) equals $+\infty$ and hence this relation is satisfied trivially. Now assume that $x \in S$ and fix $V \in \mathcal{N}(x)$. In view of the tightness assumption, for any $\varepsilon > 0$, K_ε provided by (A3), and all $k = 0, 1, \ldots,$

$$\int_{\Theta \setminus K_\varepsilon} |f_V(\xi)|P^k(d\xi) < \varepsilon.$$

Set

$$h_V^\varepsilon = \max_{\xi \in K_\varepsilon} |f_V(\xi)|,$$

$$\ell_V^\varepsilon(\xi) = \begin{cases} f_V(\xi) \text{ if } |f_V(\xi)| \leq h_V^\varepsilon, \\ h_V^\varepsilon \quad \text{ if } f_V(\xi) > h_V^\varepsilon, \\ -h_V^\varepsilon \text{ if } f_V(\xi) < -h_V^\varepsilon. \end{cases}$$

As $\inf_{y \in V} f(y, \cdot)$ is usc, $f_V(\cdot)$ is continuous and thus its truncation $\ell_V^\varepsilon(\cdot)$ is bounded, continuous, equal to $f_V(\cdot)$ on K_ε, and its absolute values do not exceed those of

$f_V(\cdot)$ on the whole Θ. Now we have the following chain of estimates

$$\text{li}_k \text{inf}_{y\in V} E^k f(y) \geq \text{li}_k \int_{\Theta} \text{inf}_{y\in V} f(y,\xi) P^k(d\xi) \geq \text{li}_k \int_{\Theta} f_V(\xi) P^k(d\xi)$$

$$= \text{li}_k \Big(\int_{\Theta\setminus K_\varepsilon} [f_V(\xi) - \ell_V^\varepsilon(\xi)] P^k(d\xi) + \int_{\Theta} \ell_V^\varepsilon(\xi) P^k(d\xi)\Big)$$

$$\geq -2\varepsilon + \text{li}_k \int_{\Theta} \ell_V^\varepsilon(\xi) P^k(d\xi) = -2\varepsilon + \int_{\Theta} \ell_V^\varepsilon(\xi) P(d\xi)$$

$$= -2\varepsilon + \int_{\Theta\setminus K_\varepsilon} [\ell_V^\varepsilon(\xi) - f_V(\xi)] P(d\xi) + \int_{\Theta} f_V(\xi) P(d\xi) \geq -4\varepsilon + \int_{\Theta} f_V(\xi) P(d\xi),$$

where the fourth and seventh inequalities follow from the tightness and the fifth equality from the weak convergence. As ε and V are arbitrary, this implies that

$$\text{sup}_{V\in\mathcal{N}(x)} \text{li}_k (\text{inf}_{y\in V} E^k f(y)) \geq \text{sup}_{V\in\mathcal{N}(x)} E\{f_V(\xi)\}.$$

By definition, we have

$$\text{sup}_{V\in\mathcal{N}(x)} f_V(\xi)) = \text{sup}_{V\in\mathcal{N}(x)} \text{sup}_{U\in\mathcal{N}(\xi)} \text{inf}_{\theta\in U} \text{inf}_{y\in V} f(y,\theta).$$

It is known from general topology theory that this quantity is equal to

$$\text{sup}_{W\in\mathcal{N}((x,\xi))} \text{inf}_{(y,\theta)\in W} f(y,\theta).$$

This is equal to $f(x,\xi)$ by the joint lower semicontinuity of f. Since, for $V_1, V_2 \in \mathcal{N}(x)$, $V_1 \cap V_2 \in \mathcal{N}(x)$, and $f_{V_1\cap V_2}(\xi) \geq \max\{f_{V_1}(\xi), f_{V_2}(\xi)\}$ for all $\xi \in \Theta$, the family $\{f_V(\cdot)\}_{V\in\mathcal{N}(x)}$ is directed upwards with the upper bound $f(x,\cdot)$. Hence, by Theorem 3.49 in [170],

$$\text{sup}_{V\in\mathcal{N}(x)} E\{f_V(\xi)\} = Ef(x).$$

We move on to proving inequality (7.34) and the pointwise convergence. Clearly, showing the latter is enough. If $x \notin S$, then $f(x,\xi) = +\infty$ for all $\xi \in \Theta$ and thus $E^k f(x) = Ef(x) = +\infty$. Hence, the convergence becomes trivial. Now fix arbitrary $x \in S$ and $V \in \mathcal{N}(x)$. For all $\varepsilon > 0$, let K_ε be given by the tightness such that, for all $k = 0, 1, \ldots$,

$$\int_{\Theta\setminus K_\varepsilon} |f(x,\xi)| P^k(d\xi) < \varepsilon.$$

Set

$$p^{\varepsilon}(x) := \max_{\xi \in K_{\varepsilon}} |f(x,\xi)|,$$

$$f^{\varepsilon}(x,\xi)) := \begin{cases} f(x,\xi) \text{ if } |f(x,\xi)| \leq p^{\varepsilon}(x), \\ p^{\varepsilon}(x) \quad \text{if } f(x,\xi) > p^{\varepsilon}(x), \\ -p^{\varepsilon}(x) \text{ if } f(x,\xi) < -p^{\varepsilon}(x). \end{cases}$$

Then, $f^{\varepsilon}(x,\cdot)$ is bounded and continuous. It is equal to f on K_{ε} and its absolute values do not exceed those of $f(x,\cdot)$ on the whole Θ. Now we have the following chain of estimates

$$\left| \int_{\Theta} f(x,\xi) P^k(d\xi) - \int_{\Theta} f(x,\xi) P(d\xi) \right|$$

$$= \left| \left[\int_{\Theta \setminus K_{\varepsilon}} (f(x,\xi) - f^{\varepsilon}(x,\xi)) P^k(d\xi) + \int_{\Theta} f^{\varepsilon}(x,\xi) P^k(d\xi) \right] \right.$$

$$\left. - \left[\int_{\Theta \setminus K_{\varepsilon}} (f(x,\xi) - f^{\varepsilon}(x,\xi)) P(d\xi) + \int_{\Theta} f^{\varepsilon}(x,\xi) P(d\xi) \right] \right|$$

$$\leq 2\varepsilon + 2\varepsilon + \left| \int_{\Theta} f^{\varepsilon}(x,\xi) P^k(d\xi) - \int_{\Theta} f^{\varepsilon}(x,\xi) P(d\xi) \right|,$$

where the equality is due to the fact that f and f^{ε} are equal on K_{ε} and the inequality to the tightness. Since ε is arbitrary and $f^{\varepsilon}(x,\cdot)$ is bounded and continuous, passing to limit as $k \to +\infty$ completes the proof. □

7.8.3 Consistency of the Infimal Values and Estimators of the Solutions

In this section, we discuss consistency properties for stochastic minimization problems (7.31) and (7.32) by applying Theorem 7.133 on the epi-convergence of the expectation functions of approximate problem (7.32) to this function of problem (7.31) under consideration and the following concept of a random lower semicontinuous function and its properties. We keep the notations as stated in Sect. 7.8.2 for these problems, and assume that X is a metric space. One has a σ-field $\mathscr{B}$ on X, and the sample space $(Z, \mathscr{F}, \mu)$ is a complete probability space. A function h from $X \times Z$ to $\mathbb{R} \cup \{+\infty\}$ is said to be a RANDOM LOWER SEMICONTINUOUS

FUNCTION (called also a NORMAL INTEGRAND), following [394], if h is $\mathscr{B} \otimes \mathscr{F}$-measurable and, for each $\zeta \in Z$, $h(\cdot, \zeta)$ is lsc.

Remark 7.134 (A characterization of a random lsc function) The following assertion stated in [394] for the case dim$X$ is finite, but in fact, it remains true for X being a metric space. Assume that $(Z, \mathscr{F}, \mu)$ is complete. Then, $h : X \times Z \to \mathbb{R} \cup \{+\infty\}$ is a random lower semicontinuous function if and only if the multifunction $\zeta \mapsto \text{epi}h(\cdot, \zeta)$ is nonempty-closed-valued and $\mathscr{F}$-measurable.

Theorem 7.135 (Convergence of approximate solutions, closedness, and measurability of their sets) *Impose (A1), (A2), and (A3). Then, for all $\zeta \in Z_0$, the following assertions hold*

(a)

$$\text{ls}_k \text{inf}_{x \in X} E^k f(x)(\zeta) \leq \text{inf}_{x \in X} Ef(x).$$

Moreover, for all $\zeta \in Z_0$, any cluster point of any sequence $\{x^k(\zeta)\}_k$ with $x^k(\zeta) \in \text{argmin}_{x \in X} E^k f(x)(\zeta)$ belongs to $\text{argmin}_{x \in X} f(x)$.

(b) *For all $k \in \mathbb{N}$ and $\varepsilon \geq 0$, the multifunction $\varepsilon\text{-argmin}_{x \in X} E^k f(x)(\cdot) : Z_0 \rightrightarrows X$ is closed-valued and $\mathscr{F}^k$-measurable.*

Proof

(a) This assertion follows from Proposition 7.3(b) with $\varepsilon = 0$.
(b) The proof of this assertion is based on the following three lemmas.

□

Lemma 7.136 (Measurability of infimal functions, closedness, and measurability of approximate minimizers of normal integrands) *Let $h : X \times Z \to \mathbb{R} \cup \{+\infty\}$ be a random lower semicontinuous function. Then, the infimal function $\text{inf}_{x \in X} h(x, \cdot)$ is $\mathscr{F}$-measurable and, for any $\varepsilon \geq 0$, the ε-optimal solution map $\varepsilon\text{-argmin}_{x \in X} h(x, \cdot)$ is closed-valued and $\mathscr{F}$-measurable.*

Proof To prove the $\mathscr{F}$-measurability of $\text{inf}_{x \in X} h(x, \cdot)$, we have to show that, for $\beta \in \mathbb{R}$, the set $\{\zeta \in Z \mid \text{inf}_{x \in X} h(x, \zeta) < \beta\}$ belongs to $\mathscr{F}$. We define the multifunction $\text{epi}\mathscr{H} : Z \rightrightarrows \mathbb{R} \cup \{+\infty\}$ by

$$(\text{epi}\mathscr{H})(\zeta) := \text{epi}h(\cdot, \zeta).$$

Then,

$$\begin{aligned}(\text{epi}\mathscr{H})^{-1}(X \times (-\infty, \beta)) &= \{\zeta \in Z \mid \text{epi}h(\cdot, \zeta) \cap (X \times (-\infty, \beta)) \neq \emptyset\} \\ &= \{\zeta \in Z \mid \text{inf}_{x \in X} h(\cdot, \zeta) < \beta\}. \qquad (7.35)\end{aligned}$$

epi$\mathscr{H}$ is closed-valued as $h(\cdot, \zeta)$ is lsc, for fixed ζ. It is also measurable since $h(\cdot, \cdot)$ is $\mathscr{B} \times \mathscr{F}$ measurable, where $\mathscr{B}$ is the Borel σ-algebra of subsets of X. Since $X \times (-\infty, \beta)$ is open, $\{\zeta \in Z \mid \inf_{x \in X} h(x, \zeta) < \beta\}$ belongs to $\mathscr{F}$ in view of (7.35).

To prove the second assertion (for multifunction ε-argmin$_{x \in X} h(x, \cdot)$), we define the function $g : X \times Z \to \mathbb{R} \cup \{+\infty\}$ by

$$g(x, \zeta) := h(x, \zeta) - \inf_{x \in X} h(x, \zeta) - \varepsilon.$$

Of course, g is a normal integrand since h is (we use the convention $+\infty - (+\infty) = +\infty$). Then, $\zeta \mapsto \text{epi}g(\cdot, \zeta)$ is a closed-valued measurable multifunction.

It remains to show that the multifunction $\text{lev}_0 g(\cdot, \zeta) = \varepsilon\text{-argmin}_X h(\cdot, \zeta)$ is closed-valued and measurable. For $\alpha \in \mathbb{R}$, we have the multifunction $\text{lev}_\alpha \mathscr{H} : Z \rightrightarrows X$ defined by $(\text{lev}_\alpha \mathscr{H})(\zeta) := \text{lev}_\alpha h(\cdot, \zeta) \subset X$. Then, for any closed set $M \subset X$,

$$(\text{lev}_\alpha \mathscr{H})^{-1}(M) = (\text{epi}\mathscr{H})^{-1}(M \times \{\alpha\}).$$

Since epi$\mathscr{H}$ is measurable, $(\text{lev}_\alpha \mathscr{H})^{-1}$ is measurable ($\in \mathscr{F}$). By the arbitrariness of the closed set M, $\text{lev}_\alpha \mathscr{H}$ is measurable. We claim that $\text{gph}(\text{lev}_\alpha \mathscr{H})$ is a measurable subset of $X \times Z$. Indeed,

$$\text{gph}(\text{lev}_\alpha \mathscr{H}) = \cap_{k=1}^{+\infty} \cap_{i=1}^{+\infty} [B_{ik} \times (\text{lev}_\alpha \mathscr{H})^{-1}(B_{ik})], \tag{7.36}$$

where $\{B_{ik} \mid i, k \in \mathbb{N}\}$ is the collection of all the rational balls with centers indexed by i and having radius k^{-1}. Since $B_{ik} \times (\text{lev}_\alpha \mathscr{H})^{-1}(B_{ik}) \in \mathscr{B} \times \mathscr{F}$, by (7.36), we have $\text{gph}(\text{lev}_\alpha \mathscr{H}) \in \mathscr{B} \times \mathscr{F}$, i.e., our claim is verified. Since this holds for all $\alpha \in \mathbb{R}$, $\text{lev}_\alpha h(\cdot, \zeta)$ is measurable ($\in \mathscr{B}$). As $g(\cdot, \zeta)$ is a vertical translation of $h(\cdot, \zeta)$, $\text{lev}_\alpha g(\cdot, \zeta)$ is closed-valued and measurable. The case $\alpha = 0$ is just what we have to show. □

Lemma 7.137 (The domains of the approximate expectation functions equal the domain of the original integrand) *Let conditions* (A1), (A2), *and* (A3) *hold. Then, for S given in* (A1), $\zeta \in Z_0$, *and* $k = 0, 1, \ldots$,

$$\text{dom}E^k f(\cdot)(\zeta) = S.$$

Furthermore, $E^k f(\cdot)(\zeta)$ *is proper and lsc.*

Proof Fix $\zeta \in Z_0$ and omit ζ in the notations. We denote $P^0 = P$ as before. If $x \notin S$, by (A1), one can see that $E^k f(x) = +\infty$, i.e., $\text{dom}E^k f \subset S$ for all $k = 0, 1 \ldots$. For $x \in S$, any ε, and K_ε given in (A3), one has

$$\int_\Theta f(x, \xi) P^k(d\xi) \leq \max_{\xi \in K_\varepsilon} |f(x, \xi)| \times P^k(K_\varepsilon) + \int_{\Theta \setminus K_\varepsilon} f(x, \xi) P^k(d\xi) < +\infty.$$

Hence, $S \subset \mathrm{dom} E^k f$ and in fact one has equality. Furthermore, for $x \in X$,

$$\int_\Theta f(x,\xi)P^k(d\xi)$$

$$\geq -\max_{\xi \in K_\varepsilon} |f(x,\xi)| \times P^k(K_\varepsilon) - \int_{\Theta \setminus K_\varepsilon} |f(x,\xi)| P^k(d\xi) > -\infty.$$

Thus, $E^k f$ is proper. Now fix $x \in X$ and $k = 0, 1, \ldots$. Then,

$$\sup_{V \in \mathscr{N}(x)} \inf_{x' \in V} E^k f(x') \geq \sup_{V \in \mathscr{N}(x)} \int_\Theta \inf_{x' \in V} f(x',\xi) P^k(d\xi)$$

$$\geq \sup_{V \in \mathscr{N}(x)} \int_\Theta f_V(\xi) P^k(d\xi) = E^k f(x),$$

where the last equality comes from Theorem 3.49 in [170] as argued in the proof of Theorem 7.135. This means that $E^k f$ is lsc at x. □

Lemma 7.138 (Approximate expectation functionals are epi-measurable normal integrands) *Assume* (A1), (A2), *and* (A3). *Then, for* $k \in \mathbb{N}$, *the expectation functional* $E^k f(\cdot)(\cdot)$ *is a random lsc function on* $X \times Z_0$ *such that the multifunction* $\zeta \mapsto \mathrm{epi} E^k f(\cdot)(\zeta)$ *is* $\mathscr{F}^k$*-measurable.*

Proof In view of the preceding lemma, the multifunction $\zeta \mapsto \mathrm{epi} E^k f(.)(\zeta)$ has the nonempty closed values. By Remark 7.134, it remains to prove that $\zeta \mapsto \mathrm{epi} E^k f(\cdot)(\zeta)$ is $\mathscr{F}^k$-measurable for $k \in \mathbb{N}$. Theorem 7.133 says that, for $\zeta \in Z_0$, the multifunction $P^k \mapsto \mathrm{epi} E^k f(.)(\zeta)$ is continuous wrt the topology of weak convergence on the collection $\mathscr{P}$ of the probability measures on $(\Theta, \mathscr{A})$ and the product topology on $X \times \mathbb{R}$. As $\zeta \mapsto P^k(\cdot,\zeta)$ is $\mathscr{F}^k$-measurable on $\mathscr{A}$ by assumption, for any class of closed sets $\{F_i \subset \Theta\}_i$ and real numbers $\{\beta_i\}_i$, the set $\{\zeta \in Z | \; P^k(F_i,\zeta) < \beta_i, i \in \mathbb{N}\}$ belongs to $\mathscr{F}^k$. Therefore, the map $\zeta \mapsto P^k(\cdot,\zeta)$ from Z to $\mathscr{P}$ is $\mathscr{F}^k$-measurable. Consequently, the composition $\zeta \mapsto P^k(\cdot,\zeta) \rightrightarrows \mathrm{epi} E^k f(\cdot)(\zeta)$ of an $\mathscr{F}^k$-measurable and a continuous multifunction is $\mathscr{F}^k$-measurable. □

Proof *Finishing the proof of Theorem 7.135.* As stated before, applying the above three lemmas, we obtain assertion (b) and complete the proof of the theorem. □

To establish complete characterizations of the convergence of the infimal values of problem (7.31) and consistency properties of the estimators of the solutions, we need the following tightness condition (additional to (A3)): A sequence $\{P^k(\cdot,\zeta)\}_k$ is declared INF-f-TIGHT if, for each $\varepsilon > 0$, there exists a compact subset $C_\varepsilon \subset X$ such that, for all $k \in \mathbb{N}$,

$$\inf_{x \in C_\varepsilon} E^k f(x)(\zeta) \leq \inf_{x \in X} E^k f(x)(\zeta) + \varepsilon.$$

Notice that, by Lemma 7.136, this tightness condition is evidently satisfied if the subset S (given in (A1)) is compact.

Theorem 7.139 (Limit properties of approximate solutions) *Let* (A1), (A2), *and* (A3) *hold and* $\zeta \in Z_0$.

(a) *The following assertions are equivalent.*
(a1) $\operatorname{argmin}_{x\in X} Ef(x) \neq \emptyset$ *and*

$$\lim_k[\inf_{x\in X} E^k f(x)(\zeta)] = \inf_{x\in X} Ef(x). \tag{7.37}$$

(a2) *There exist a sequence* $\varepsilon^k \to 0^+$, *a compact set* $D \subset X$, *and a sequence* $\{x^k(\zeta)\}_k$ *such that* $x^k(\zeta) \in (\varepsilon^k\text{-}\operatorname{argmin}_{x\in X} E^k f(x)(\zeta)) \cap D$.
Furthermore, any cluster point of sequence $\{x^k(\zeta)\}_k$ *in* (a2) *lies in* $\operatorname{argmin}_{x\in X} Ef(x)$. *If, in addition,* $\{x^*\} = (\operatorname{argmin} Ef) \cap D$, *then* $\lim_k x^k(\zeta) = x^*$.

(b) *Each assertion in* (a) *implies the* inf-f-*tightness of sequence* $\{P^k(\cdot, \zeta)\}_k$ *for large* k. *Conversely, this* inf-f-*tightness yields (7.37). Under the additional assumption that there is* $\varepsilon^k \downarrow 0$ *such that* $\operatorname{Ls}_k(\varepsilon^k\text{-}\operatorname{argmin}_{x\in X} E^k f(x)(\zeta)) \neq \emptyset$, *this* inf-$f$-*tightness is equivalent to the assertions in* (a).

Proof We omit a fixed ζ in the notations.

(a) "(a1) $\Rightarrow$ (a2)." Let $\bar{x} \in \operatorname{argmin}_{x\in X} Ef(x)$. By the epi-convergence provided by Theorem 7.133 and the closedness of the expectations (cf. Lemma 7.137), one has $x^k \in \operatorname{dom} E^k f$ converging to $\bar{x}$ such that $Ef(\bar{x}) = \lim_k E^k f(x^k)$. We claim that, for each $\varepsilon > 0$, there exists $k = k(\varepsilon)$ such that, for $k \geq k(\varepsilon)$, $x^k \in \varepsilon\text{-}\operatorname{argmin}_{x\in X} E^k f(x)$. Indeed, suppose to the contrary the existence of $\bar{\varepsilon}$ and a subsequence $\{k_i\}_i$ such that, for all i, $E^{k_i} f(x^{k_i}) > \inf_{x\in X} E^{k_i} f(x) + \bar{\varepsilon}$. Then, by (7.37),

$$Ef(\bar{x}) = \lim_i E^{k_i} f(x^{k_i}) \geq \lim_i \inf_{x\in X} E^{k_i} f(x) + \bar{\varepsilon} = \inf_{x\in X} Ef(x) + \bar{\varepsilon},$$

a contradiction. Now, for each $\varepsilon > 0$ and $k \geq k(\varepsilon)$, we rewrite x^k as x^k_ε. However, for $1 \leq k < k(\varepsilon)$, we take x^k_ε arbitrarily in $\varepsilon\text{-}\operatorname{argmin}_{x\in X} E^k f(x)$. Then, for each $\varepsilon > 0$, we have a sequence $\{x^k_\varepsilon\}_k$ converging to $\bar{x}$ as $k \to +\infty$. Take a sequence $\varepsilon^\ell \to 0^+$. For each $\ell \in \mathbb{N}$, we have $\lim_k x^k_{\varepsilon^\ell} = \bar{x}$. Consequently, by the usual diagonalization argument, we get $\varepsilon^k := \varepsilon^{\ell(k)} \to 0^+$ as $k \to +\infty$ such that $x^k_{\varepsilon^k} \in \varepsilon^k$-$\operatorname{argmin}_{x\in X} E^k f(x)$ tends to $\bar{x}$ as $k \to +\infty$ and hence lies in a compact subset D (take, for instance, D being a tail of $\{x^k_{\varepsilon^k}\}$).

"(a2) $\Rightarrow$ (a1)." We have

$$\operatorname{li}_k(\inf_{x\in X} E^k f(x)) \geq \operatorname{li}_k(E^k f(x^k) - \varepsilon^k) \geq Ef(\bar{x}) \geq \inf_{x\in X} Ef(x),$$

where $\bar{x}$ is a cluster point of $\{x^k\}_k$ and the second inequality is due to Theorem 7.133. This together with the inequality in Theorem 7.135(a) yields (a1).

The conclusion about cluster points is true for epi-convergence in general. Now, suppose $(\text{argmin} f) \cap D$ is a singleton $\{x^*\}$, but for some $\zeta \in Z_0$, $\delta > 0$, and subsequence $\{x^{k_i}(\zeta)\}_i$, one has $d(x^*, x^{k_i}(\zeta)) > \delta$. Then, by the compactness of D and what we have shown, there is another subsequence of $\{x^{k_i}(\zeta)\}_i$ converging to x^*, a contradiction.

(b) Using (a1), we can show even more than the inf-f-tightness that, for $\bar{x} \in \text{agrmin}_{x\in X} Ef(x)$, for all $\varepsilon > 0$ and large k, we have this inf-f-tightness with $C_\varepsilon = \{\bar{x}\}$, i.e., $E^k f(\bar{x}) \leq \inf_{x\in X} E^k f(x) + \varepsilon$. Suppose to the contrary that, with some ε' and subsequence $\{k_i\}_i$, $E^{k_i} f(\bar{x}) > \inf_{x\in X} E^{k_i} f(x)+\varepsilon'$. Passing to limit as $i \to +\infty$ and employing Theorem 7.135 to the left-hand side and assumption (a1) to the right-hand side, we get a contradiction: $Ef(\bar{x}) > Ef(\bar{x}) + \varepsilon'$.

For the converse, we first claim the existence of $i_0 \in \mathbb{N}$ such that, for large k, $\inf_{x\in C_\varepsilon} E^k f(x) \geq -i_0$. Indeed, suppose to the contrary that, for every $i \in \mathbb{N}$, there are $k_i \to +\infty$ as $i \to +\infty$ and $x^{k_i} \in C_\varepsilon$ such that $E^{k_i} f(x^{k_i}) < -i$. Let $\bar{x}$ be a cluster point of $\{x^{k_i}\}_i$. Then, by the epi-convergence of $E^k f(\cdot)$,

$$-\infty = \text{li}_i E^{k_i} f(x^{k_i}) \geq Ef(\bar{x}),$$

contradicting the properness of Ef. Now, by this boundedness from below, for large k and all positive ε, there exists $x_\varepsilon^k \in \varepsilon\text{-argmin}_{x\in C_\varepsilon} E^k f(x)$. Let $\bar{x}_\varepsilon$ be a cluster point of $\{x_\varepsilon^k\}_k$. Then,

$$\begin{aligned}\text{li}_k \inf_{x\in X} E^k f(x) &\geq \text{li}_k \inf_{x\in C_\varepsilon} E^k f(x) - \varepsilon \geq \text{li}_k E^k f(x_\varepsilon^k) - 2\varepsilon \\ &\geq Ef(\bar{x}_\varepsilon) - 2\varepsilon \geq \inf_{x\in X} Ef(x) - 2\varepsilon.\end{aligned}$$

Since ε is arbitrary, these inequalities combined with Theorem 7.135(a) yield (7.36). If $\text{Ls}_{k\to\infty}(\varepsilon^k\text{-argmin}_{x\in X} E^k f(x) \neq \emptyset$ then, by (the proof of) Theorem 7.140(a), $\text{argmin}_{x\in X} Ef(x) \neq \emptyset$. □

Moving on to the multifunctions of the optimal solutions $x \mapsto \text{argmin}_{x\in X} E^k f(x)$ (ζ) for $\zeta \in Z_0$, we have the following consistency properties.

Theorem 7.140 (Consistency of approximate statistical estimators) *Let conditions* (A1), (A2), *and* (A3) *be fulfilled. Let* $\zeta \in Z_0$, $\varepsilon \geq 0$, *and* $\epsilon^k \to 0^+$.

(a)

$$\text{Ls}_k(\varepsilon\text{–argmin}_{x\in X} E^k f(x)(\zeta)) \subset \varepsilon\text{–argmin}_{x\in X} Ef(x),$$

$$\text{Ls}_k(\varepsilon^k\text{–argmin}_{x\in X} E^k f(x)(\zeta)) \subset \text{argmin}_{x\in X} Ef(x).$$

Hence, if the left-hand side of the latter inclusion is nonempty, then

$$\text{ls}_k \inf_{x\in X} E^k f(x)(\zeta) = \min_{x\in X} Ef(x).$$

(b) *Assuming that* $\text{argmin}_{x\in X} Ef(x) \neq \emptyset$, *we have*

$$\text{argmin}_{x\in X} Ef(x) = \cap_{\varepsilon>0}\text{Li}_k(\varepsilon-\text{argmin}_{x\in X} E^k f(x)(\zeta))$$

$$= \cap_{\varepsilon>0}\text{Ls}_k(\varepsilon-\text{argmin}_{x\in X} E^k f(x)(\zeta))$$

if and only if

$$\text{lim}_k(\text{inf}_{x\in X} E^k f(x)(\zeta)) = \text{min}_{x\in X} Ef(x).$$

Proof We fix a $\zeta \in Z_0$ and omit it in the notations.

(a) If the left-hand side of the former inclusion is empty, this inclusion is trivially satisfied. While if $\bar{x}$ belongs to this side, i.e., there is a subsequence of $k \in \mathbb{N}$ (we use the same index k) such that $x^k \in \varepsilon\text{-argmin}_{x\in X} E^k f(x)$ and $x^k \to \bar{x}$. Then, by Theorems 7.133 and 7.135,

$$Ef(\bar{x}) \leq \text{li}_k E^k f(x^k) \leq \text{ls}_k E^k f(x^k)$$

$$\leq \text{ls}_k(\text{inf}_{x\in X} E^k f(x) + \varepsilon) \leq \text{inf}_{x\in X} Ef(x) + \varepsilon,$$

i.e., $\bar{x}$ belongs to the right-hand side. Similarly, if there exists $\bar{x}$ in the left-hand side of the latter inclusion in (a), then we have the above string of inequalities with the last two terms replaced by

$$\leq \text{ls}_k(\text{inf}_{x\in X} E^k f(x) + \varepsilon^k) \leq \text{inf}_{x\in X} Ef(x),$$

i.e., $\text{inf}_{x\in X} Ef(x)$ is attained at $\bar{x}$. Then, the desired equality is obtained from Theorem 7.135(a).

(b) "If." By (a), for any $\varepsilon > 0$,

$$\text{Ls}_k(\varepsilon-\text{argmin}_{x\in X} E^k f(x)(\zeta)) \subset \varepsilon-\text{argmin}_{x\in X} Ef(x).$$

Because

$$\text{argmin}_{x\in X} Ef(x) = \cap_{\varepsilon>0}(\varepsilon-\text{argmin}_{x\in X} Ef(x)),$$

it remains only to show that

$$\text{argmin}_{x\in X} Ef(x) \subset \cap_{\varepsilon>0}\text{Li}_k(\varepsilon-\text{argmin}_{x\in X} E^k f(x)(\zeta)). \tag{7.38}$$

As $\text{epi}E^k f \to \text{epi}Ef$ and the considered epigraphs are closed by Lemma 7.137, for any $\bar{x} \in \text{argmin}_{x\in X} Ef(x)$, there exists $(x^k, \alpha^k) \in \text{epi}E^k f$ tending to $(\bar{x}, Ef(\bar{x}))$. If $x^k \in \varepsilon^k\text{-argmin}_{x\in X} E^k f(x)$ for a sequence $\varepsilon^k \to 0^+$, then for every $\varepsilon > 0$, there is k_0 such that $\varepsilon^k < \varepsilon$ for $k > k_0$. Hence,

$$x^k \in \varepsilon\text{−argmin}_{x\in X} E^k f(x).$$

Consequently, for all $\varepsilon > 0$,

$$\bar{x} = \lim_k x^k \in \text{Li}_k(\varepsilon\text{−argmin}_{x\in X} E^k f(x)),$$

i.e., we obtain (7.38). If we do not have $x^k \in \varepsilon^k\text{-argmin}_{x\in X} E^k f(x)$ eventually, there exist $\bar{\varepsilon} > 0$ and a subsequence $\{x^{k_i}\}_i$ with $E^{k_i} f(x^{k_i}) > \inf_{x\in X} E^{k_i} f(x) + \bar{\varepsilon}$. Taking liminf, by the assumption of the "if" part, we arrive at

$$Ef(\bar{x}) \geq \text{li}_i E^{k_i} f(x^{k_i}) \geq \lim_i(\inf_{x\in X} E^{k_i} f(x) + \bar{\varepsilon}) = \min_{x\in X} Ef(x) + \bar{\varepsilon}.$$

This contradiction yields the desired two equalities.

"Only if." We only need to use the first equality. Let

$$\bar{x} \in \cap_{\varepsilon>0}\text{Li}_k(\varepsilon\text{−argmin}_{x\in X} E^k f(x)) = \text{argmin}_{x\in X} Ef(x).$$

Take a sequence $\varepsilon_\ell \to 0^+$. For each ℓ, one can find $x_\ell^k \in \varepsilon_\ell\text{-argmin}_{x\in X} E^k f(x)$ such that $\lim_k x_\ell^k = \bar{x}$. By the usual diagonalization argument, one finds $x^k := x_{\ell(k)}^k$ such that $x^k \in \varepsilon_{\ell(k)}\text{-argmin}_{x\in X} E^k f(x)$ and $x^k \to \bar{x}$. Furthermore,

$$\min_{x\in X} Ef(x) = Ef(\bar{x}) \leq \text{li}_k E^k f(x^k)$$

$$\leq \text{li}_k(\inf_{x\in X} E^k f(x) + \varepsilon_{\ell(k)}) = \text{li}_k(\inf_{x\in X} E^k f(x)).$$

This together with Theorem 7.135(a) means the convergence of the infima of $E^k f$ to that of Ef. □

Theorem 7.140 provides asymptotic properties of the optimal solution sets, but not yet the convergence of argmin$E^k f$. To have such a complete consistency of the optimal solution sets, we need an additional assumption as follows.

Theorem 7.141 (Complete consistency of statistical estimators) *Let* (A1), (A2), (A3), *the* inf-*f*-*tightness hold, and* $\text{argmin}_{x\in X} Ef \neq \emptyset$. *Assume further that for every* $x \in \text{argmin}_X Ef(\cdot)$ *and* $V \in \mathcal{N}(x)$, *there exist* $\delta > 0$, *an open neighborhood* $W \in \mathcal{N}(x)$, *and a subsequence* $\{k_i\}_i$ *such that, for* $\zeta \in Z_0$,

$$[V \cap \text{argmin}_{x\in X} E^{k_i} f(x)(\zeta) = \emptyset] \Rightarrow [\inf_{x\in W} E^{k_i} f(x)(\zeta) \geq \inf_{x\in X} E^{k_i} f(x)(\zeta) + \delta].$$

Then,

$$\text{Lim}_k(\text{argmin}_{x\in X} E^k f(x)(\zeta)) = \text{argmin}_{x\in X} Ef(x).$$

Proof We omit $\zeta \in Z_0$ in the notations. In view of Theorem 7.140(a), we only need to verify the containment:

$$\mathrm{Li}_k(\mathrm{argmin}_{x\in X} E^k f(x)) \supset \mathrm{argmin}_{x\in X} Ef(x). \tag{7.39}$$

Theorem 7.139(b) yields that

$$\alpha := \inf_{x\in X} Ef(x) = \lim_k(\inf_{x\in X} E^k f(x)) =: \lim_k \alpha_k.$$

Suppose $\bar{x}$ belongs to the right-hand side of (7.39), but does not belong to the left-hand side. This means the existence of $V \in \mathcal{N}(\bar{x})$ and a subsequence $\{k_i\}_i$ such that, for all i,

$$V \cap \mathrm{argmin}_{x\in X} E^{k_i} f(x) = \emptyset.$$

Then, with the given $W \in \mathcal{N}(\bar{x})$ and δ provided by assumption,

$$\alpha = \inf_{x\in W} Ef(x) \geq \mathrm{li}_i \inf_{x\in W} E^{k_i} f(x)$$

$$\geq \mathrm{li}_i \inf_{x\in X} E^{k_i} f(x) + \delta = \lim_i \alpha_{k_i} + \delta = \alpha + \delta,$$

a contradiction. (The first inequality in the above string of comparisons follows from our Theorem 7.135 and (3.23) in [397] (which is valid for a metric space X instead of $X = \mathbb{R}^n$).) □

7.8.4 *Consistency for a Primal-Dual Pair of Stochastic Mathematical Programs*

Let the notations be as defined in Sect. 7.8.3 for problems (7.31) and (7.32) and Y be a metric space. Let $\varphi : X \times Y \times \Theta \to \bar{\mathbb{R}}$. We naturally use the following notations, corresponding to the similar ones in Sect. 7.8.2, for $h : Y \to \mathbb{R} \cup \{-\infty\}$, the above φ, $V \in \mathcal{N}(x)$, and $W \in \mathcal{N}(y)$,

$$\mathrm{cl}^y h(y) = \inf_{W\in\mathcal{N}(y)} \sup_{y'\in W} h(y'),$$

$$\varphi_V(y, \xi) = \mathrm{cl}_x[\inf_{x'\in V} \varphi(x', y, \xi)],$$

$$\varphi^W(x, \xi) = \mathrm{cl}^y[\sup_{y'\in W} \varphi(x, y', \xi)],$$

i.e., cl^y stands for an upper closure (or usc regularization) with respect to y (and cl_x is a lower closure).

For φ and $\{P^0 := P, P^k\}_k$, we impose (A2) and

(B1) (continuity of φ). For all $\xi \in \Theta$, $\mathrm{dom}_x\varphi = S_x$ and $\mathrm{dom}_y\varphi = S_y$, where $S_x \subset X$ and $S_y \subset Y$ are nonempty and closed. $\varphi(\cdot, y, \cdot)$ is jointly lsc for each $y \in S_y$ and $\varphi(x, \cdot, \cdot)$ is jointly usc for each $x \in S_x$;

(B3) (tightness). For all $(x, y) \in S_x \times S_y$ and $\zeta \in Z_0$, a full μ-measure subset of Z, the sequence $\{P^0, P^k(\cdot, \zeta)\}_k$ is $\{\varphi(x, y, \cdot), \varphi_V(y, \cdot), \varphi^W(x, \cdot)\}_{V\in\mathcal{N}(x), W\in\mathcal{N}(y)}$-tight.

Applying Theorem 7.133 and Remark 7.31(c), we have

Theorem 7.142 (epi/hypo-convergence of expectation bifunctions) *Let* $\varphi : X \times Y \times \Theta \to \bar{\mathbb{R}}$ *and* $\{P^0, P^k\}_k$ *satisfy* (B1), (A2), *and* (B3). *Then, for all* $\zeta \in Z_0$, $E\varphi(\cdot, \cdot) = \text{e/h-lim}_k E^k\varphi(\cdot, \cdot)(\zeta)$ *on* $X \times Y$ *and* $E\varphi$ *is lsc in x and usc in y.*

The sequence $\{P^k\}_k$ is said to be INF-SUP-φ-TIGHT if, for each $\varepsilon > 0$, there exist compact subsets $C_\varepsilon \subset X$ and $D_\varepsilon \subset Y$ such that, for all $k \in \mathbb{N}$ and $(x, y) \in S_x \times S_y$,

$$\inf_{x'\in C_\varepsilon} E^k\varphi(x', y)(\zeta) \le \inf_{x'\in X} E^k\varphi(x', y)(\zeta) + \varepsilon,$$

$$\sup_{y'\in D_\varepsilon} E^k\varphi(x, y')(\zeta) \ge \sup_{y'\in Y} E^k\varphi(x, y')(\zeta) - \varepsilon.$$

This tightness condition is clearly satisfied if the sets S_x and S_y (given in (B1)) are compact.

Because hypo-convergence is completely symmetric to epi-convergence, we can straightforwardly define a RANDOM UPPER SEMICONTINUOUS FUNCTION and state the results corresponding to Lemmas 7.136–7.138 and Theorem 7.135. Then, applying these facts, we obtain the following consistency result for stochastic minimax problems in a similar way as for Theorem 7.141.

Theorem 7.143 (Consistency for stochastic minimax problems) *Let* (B1), (A2), *and* (B3) *be fulfilled and* $\zeta \in Z_0$.

(a) *The following assertions are equivalent.*

(a1) *There exist saddle points of* $E\varphi$ *and*

$$\inf_{x\in X}\sup_{y\in Y} E^k\varphi(x, y)(\zeta) \to \inf_{x\in X}\sup_{y\in Y} E\varphi(x, y)(\zeta). \tag{7.40}$$

(a2) *There exist* $\varepsilon^k \to 0^+$, *compact sets* $C \subset X$ *and* $D \subset Y$, *and* ε^k*-saddle points* $(x^k(\zeta), y^k(\zeta))$ *of* $E^k\varphi(\cdot, \cdot)(\zeta)$ *in* $C \times D$.

Furthermore, any cluster point of sequence $\{(x^k(\zeta), y^k(\zeta))\}_k$ *in* (a2) *is a saddle point of* $E\varphi$. *If, in addition,* (x^*, y^*) *is a single saddle point of* $E\varphi$ *in* $C \times D$, *then*

$$\lim_k (x^k(\zeta), y^k(\zeta)) = (x^*, y^*).$$

(b) *Each assertion in* (a) *implies the* inf-sup-φ*-tightness of sequence* $\{P^k(\cdot, \zeta)\}_k$ *for large k. Conversely, this* inf-sup-φ*-tightness yields (7.40). If, additionally, there* *is* $\varepsilon^k \to 0^+$ *such that, for* $(x, y) \in S_x \times S_y$,

$$\mathrm{Ls}_k(\varepsilon^k\text{–}\mathrm{argmin}_{x' \in X} E^k\varphi(x', y)(\zeta)) \neq \emptyset,$$

$$\mathrm{Ls}_k(\varepsilon^k\text{–}\mathrm{argmax}_{y' \in Y} E^k\varphi(x, y')(\zeta)) \neq \emptyset.$$

Then, this inf-sup-φ*-tightness is equivalent to assertions* (a1) *and* (a2).

We can obtain also consistency results for stochastic minimax problems similar to Theorems 7.140 and 7.141.

Now, move on to considering a stochastic mathematical program and its dual program (in the sense of the Lagrange duality scheme). Let f and $f_i, i = 1, \ldots, m$, be lsc finite-valued functions and $f(x, \cdot)$ be usc for $x \in X$. Consider the following stochastic mathematical program (P) :

$$\begin{aligned} &\text{minimize} \quad E\{f(x)\} \\ &\text{subject to} \quad f_i(x) \le 0, \quad i = 1, \ldots, s, \\ &\qquad\qquad f_i(x) = 0, \quad i = s+1, \ldots, m. \end{aligned}$$

The Lagrangian of program (P) is

$$L(x, y) := \begin{cases} Ef(x) + \sum_{i=1}^m y_i f_i(x) & \text{if } y_i \ge 0, \text{ for } i = 1, \ldots, s, \\ -\infty & \text{otherwise.} \end{cases}$$

Problem (P) is equivalent to the following minsup-problem for the Lagrangian L

$$\min_{x \in X} \sup_{y \in \mathbb{R}^m} L(x, y).$$

By definition, the (Lagrange) dual problem of (P) is the following maxinf-problem

$$\max_{y \in \mathbb{R}^m} \inf_{x \in X} L(x, y).$$

Then, assumption (B1) is clearly satisfied for the Lagrangian L (playing the role of φ).

Theorem 7.144 (Any cluster point of approximate saddle points of Lagrangians is a solution-pair of primal-dual stochastic programs) *For problem* (P), *assume that* P^k *fulfills (A2) and (B3) for* $k \in \mathbb{N}$ *and* $\zeta \in Z_0$. *Then, if for* $\varepsilon^k \downarrow 0$ *and each* $\zeta \in Z_0$, *we have a sequence of* ε^k*-saddle points of the approximate Lagrangians:*

$$L^k(x, y)(\zeta) = \begin{cases} E^k f(x)(\zeta) + \sum_{i=1}^m y_i f_i(x) & \text{if } y_i \ge 0 \text{ for } i = 1, \ldots, s, \\ -\infty & \text{otherwise} \end{cases}$$

on compact sets $C \subset X$ and $D \subset \mathbb{R}^s_+ \times \mathbb{R}^{m-s}$, then any cluster point of this sequence is a saddle point (x^*, y^*) of $L(.,.)$, and x^* is an optimal solution to (P) *and y^* is an optimal solution to its dual problem.*

Proof This is a direct consequence of Theorem 7.143 and the basic fact in mathematical programming that when (x^*, y^*) is a saddle point of the Lagrangian, then x^* is an optimal solution to the primal program and y^* is an optimal solution to the dual problem. □

7.9 Exercises

Exercise 7.1 (Epi-convergence fails) *Let $A = \{0\}$ and*

$$A^k = \begin{cases} \{0, 1\} \text{ if } k \text{ is odd,} \\ \{0\} \quad \text{otherwise.} \end{cases}$$

Let $\varphi(x) = 0$ for $x \in A$ and $\varphi^k(x) = 0$ for all $k \in \mathbb{N}$ and $x \in A^k$. Prove that φ^k do not epi-converge to φ.

Exercise 7.2 (Types of epi-convergence, which are weaker than the main type, hold) *For φ^k and φ given in Exercise 7.1, prove that i-epi-, w-epi-, and w-i-epi-convergence hold.*

Exercise 7.3 (Computing approximate minimizers) *Consider φ^k and φ given in Exercise 7.1. Prove that $\text{argmin}_A \varphi = \{0\}$ and*

$$\varepsilon^k\text{−argmin}_{A^k}\varphi^k = \begin{cases} \{0, 1\} \text{ for odd } k's, \\ \{0\} \quad \text{for even } k's. \end{cases}$$

Exercise 7.4 (Both non-tight epi-convergence and uniform convergence do not guarantee the convergence of minimizers) *Let $A^k = A = [0, +\infty)$, $\varphi(x) = x^2$, and*

$$\varphi^k(x) := \begin{cases} x^2 & \text{if } 0 \le x \le k, \\ (x - 3k)^2 - k^2 & \text{if } k < x. \end{cases}$$

Prove that φ^k uniformly converges to the continuous function φ, but $\inf_A \varphi = 0$ and $\lim_k (\inf_{A^k} \varphi^k) = \lim_k(-k^2) = -\infty$. Moreover, 0 is the minimizer of φ, but the minimizer of φ^k is $3k$ and tends to $+\infty$. So, the conclusion of Proposition 7.9 is not true and the cause is that φ^k epi-converges but not tightly.

Exercise 7.5 (Epi- and hypo-convergence of slices together imply e/h-convergence) *Assume that $A^k = A$ and $B^k = B$ for all k. Prove that if*

$\Phi^k(\cdot, y) \xrightarrow{e} \Phi(\cdot, y)$ *for all* $y \in B$ *and* $\Phi^k(x, \cdot) \xrightarrow{h} \Phi(x, \cdot)$ *for all* $x \in A$, *then* $\Phi^k \xrightarrow{e/h} \Phi$.

Exercise 7.6 (Failing of the converse of the conclusion in Exercise 7.5) *Find an example showing that the converse implication to the assertion in Exercise 7.5 is not true.*

Exercise 7.7 (e/h-convergence fails) *Let* A^k *be equal to* $\{0, 1\}$ *for the odd* k*'s and to* $\{0\}$ *for the even* k*'s,* $A = \{0\}$*;* $B^k = B = \{1\}$ *for all* k*;* $\Phi^k(x, y) = 0$ *for all* $k \in \mathbb{N}$ *and* $(x, y) \in A^k \times B^k$*; and* $\Phi(x, y) = 0$ *for all* $(x, y) \in A \times B$*. Show that* Φ^k *does not e/h-converge to* Φ*. (Hint: the infinity condition in (a) of the definition is not satisfied.)*

Exercise 7.8 (Weak and inside e/h-convergence hold, while e/h-convergence does not) *Prove that the* Φ^k *in Exercise 7.7 i-e/h-converges and weak e/h-converges to the bifunction* $\Phi(x, y) = 0$ *on* $A \times B = \{(0, 1)\}$.

Exercise 7.9 (y-ancillary tight e/h-convergence implies hypo-convergence of inf-projections) *Prove the following assertions:*

(a) *Let* $\Phi^k \xrightarrow{e/h} \Phi$ *y-ancillary tightly and the domains of* η^k, η *be nonempty. Then,* $\text{h-lim}_k \eta^k = \eta$.
(b) *Let* Φ^k *i-e/h-converge to* Φ *(w-e/h-converge and w-i-e/h-converge, resp.) y-ancillary tightly and the domains of* η^k, η *be nonempty. Then,* η^k *i-h-converges (w-h-converges and w-i-h-converges, resp.) to* η.

Exercise 7.10 (Convergence of maxinf-points under y-ancillary tight e/h-convergence) *State and prove the statement for convergence of maxinf-points corresponding to Theorem 7.52.*

Exercise 7.11 (e/h-convergence preserves saddle points) *Let* $X = Y = \mathbb{R}$, $A = A^k = B = B^k = [0, 1]$, *and* $\Phi^k(x, y) = \Phi(x, y) = y^x$ *for all* k*. Consider the e/h-convergence of* Φ^k *to* Φ *(tight in some sense or not). Compute saddle point sets of* Φ^k *and* Φ *to see a full preservation.*

Exercise 7.12 (Computing lop-limits) *Let* $A^k = A = (0, e]$, $B^k = B = (0, 1]$, *and* $\Phi^k(x, y) = \ln(y/x)$ *for all* k*. Apply the characterization of lop-convergence to find the mai-lop and w-mai-lop limits of* Φ^k.

Exercise 7.13 (mis-lop convergence fails, but its three weaker types hold) *Consider the* Φ^k *and* Φ *in Exercise 7.7. Verify that one does not have mis-lop convergence but does the three weaker types of mis-lop convergence.*

Exercise 7.14 (mai-lop convergence fails, but its three weaker types hold) *You have the same task as in Exercise 7.13, but for mai-lop convergence.*

Exercise 7.15 (Second kind of approximate saddle points corresponds to second-kind approximate solutions of the primal-dual pair of equilibrium problems) *Consider second-kind approximate solutions of a primal-dual pair of* time, there are lots of studies on calculus rules in general with effective applications. We go in a bit more detail on the issue since the reader may be interested in it. We found the following limited sources of calculus rules for epi-convergence. Exercise 7.8 in [396] talks about the epi-convergence of the sums of functions on $\mathbb{R}^n$ of the form $\varphi^k + g$ (g is a fixed function) or $\varphi^k(x) = \lambda^k f^k(x + a^k) + \alpha^k$ (λ^k and α^k are real numbers), and of compositions of the form $\varphi^k \circ F$ with a fixed mapping $F : \mathbb{R}^n \to \mathbb{R}^n$ or $\theta \circ \varphi^k$ for $\theta : \mathbb{R} \to \mathbb{R}$; all the rules are under restrictive assumptions. Subsection 7.4 in [82] discusses a sum combined with a composition of the form $\varphi^k + \psi^k \circ A$, where $\psi^k : \mathbb{R}^m \to \mathbb{R}^n$ and $A \in L(\mathbb{R}^n, \mathbb{R}^m)$. An inequality for the lower epi-limits is obtained and, under strong assumptions including a stability condition concerning epiderivatives, a sum rule at a fixed point is proved. In Section 2 of [77], calculus rules are studied for an epi-sum (defined by $(\varphi\#\xi)(x) := \inf_{x_1+x_2=x}\{\varphi(x_1) + \xi(x_2)\}$) and an epi-multiplication (defined by $(\lambda \star \varphi)(x) := \lambda\varphi(\lambda^{-1}x)$ for $\lambda > 0$), but not for the usual sum and composition. The very recent article [401] establishes a sum rule for the epi-convergence of $\varphi^k + g^k$ with $g^k : \mathbb{R}^n \to (-\infty, +\infty]$ under the assumptions: $-g^k \xrightarrow{e} -g$ and both φ^k and g^k pointwise converge. The following epi-convergence of a composition is included as well. For $\varphi, \varphi^k : \mathbb{R}^n \to [-\infty, +\infty]$ and $h, h^k : \mathbb{R} \to [-\infty, +\infty]$, $h^k \circ \varphi^k \xrightarrow{e} h \circ \varphi$ in either of the following cases:

(a) $h^k \xrightarrow{e} h$, h^k pointwise converges to h, $\mathrm{dom}\varphi = \mathbb{R}$, and φ is continuous.
(b) $\varphi^k \xrightarrow{e} \varphi$ and h is continuous, nondecreasing, and extended to $[-\infty, +\infty]$ with the conventions $h(-\infty) = \inf_{\mathbb{R}} h$ and $h(+\infty) = \sup_{\mathbb{R}} h$.

Section 7.2

Epi/Hypo-convergence and Lopsided Convergence of Bifunctions

Variational convergence of bifunctions is a powerful tool for studying variational approximations of equilibrium problems and all models with the main data being bifunctions. The first and immediate generalization of epi-convergence is epi/hypo-convergence, which was proposed in Attouch and Wets [79] and rigorously treated in [78]. In the paper [76], a variational convergence type that is stronger than epi/hypo-convergence, called lopsided convergence (in short, lop-convergence), was given its airing. All the abovementioned works dealt with extended-real-valued functions or bifunctions. In [219, 270], the lop-convergence was not considered for extended-real-valued bifunctions, but for finite-valued bifunctions defined on what is called a rectangle $C \times D \subset \mathbb{R}^n \times \mathbb{R}^m$. The motivation is that important bifunctions appearing in applications such as Lagrangians in constrained optimization, payoff functions in zero-sum games, or Hamiltonians in the calculus of variations and optimal control can be expressed as finite-valued bifunctions defined on product sets. Another reason for finite-valued bifunctions to be used is that a value of infinity is dramatically different if viewed from the minimization perspective than from the maximization one, and so there is a need for finiteness to be more explicit in concepts under question. The epi/hypo-convergence for such bifunctions was proposed and investigated with applications in approximations in [175, 178], while the lop-convergence for them was developed in [269, 346]. Section 7.2 provides a comprehensive exploration of both the epi/hypo- and lop-convergence of bifunctions on metric spaces, focusing on their characterizations and variational properties, and hence constitutes developments of the aforementioned articles. Note that here we discuss not only the classical types of variational convergence but also the new types called inside, weak, and weak inside convergence. Also, the types of tightness for the lop-convergence are different from those in [270, 404]. (The inside epi- and e/h-convergence were introduced in [179] and the others in [181].) Note that since the purpose of this section is preparing all the types of variational convergence for applications in approximation studies in the subsequent sections, we do not include what is not necessary. Some examples and characterizations of the e/h-convergence were obtained in [75, 178]. When we were about to submit this monograph, we found the recent book [403] with Section 9H about the mis-lop convergence presented mainly in connection with minsup-points and the Nash equilibrium problem. Hence, we managed to refer to this nice book in the related places of our work to draw the attention of the reader to it. For more contributions to epi/hypo- and lop-convergence and applications, see, e.g., [84, 455].

Unfortunately, we have not found any contributions on calculus rules for variational convergence of bifunctions in the literature to include in this book. This topic is very important, especially when optimization problems with functional constraints are targeted. So far applications of variational convergence have been only considered for problems with geometric constraints (set constraints). Studying calculus rules for the mentioned convergence is a big challenge for researchers.

Section 7.3

Approximations of Equilibrium Problems

The first applications of variational convergence of bifunctions naturally aimed at equilibrium problems, whose data consisted of only bifunctions (which were the main ingredients) and domains. This is also because (almost) all aspects of those problems were attracting increasing attention due to their general form covering most optimization-related models. The recent books by Konnov [304], Kassay and Radulescu [276], and Bigi, Castelalni, Pappalardo, and Passcantando [106] present various topics for equilibrium problems and their particular cases, but do not mention variational approximations, which have been dealt with in [175, 269, 346]. In this section, we further develop in many aspects the results of these papers, focusing on pairs of equilibrium problems dual to each other and all the types of variational convergence mentioned in the commentary to Sect. 7.2 above. The types of tightness conditions are also well presented. We discuss in detail two kinds of approximate solutions. The second kind, i.e., the $\varepsilon\theta$-solutions, was probably introduced in [346], considered in [175], and further developed in [181]. Especially, also pairs of strong equilibrium problems have been considered as well.

Section 7.4

Approximations of Multiobjective Optimization

We follow the manner of expressing a multiobjective optimization problem as a special case of equilibrium problems in [346], including the idea of using the epi- (or hypo-) convergence. Its developments can be found in [175]. Here, we significantly develop these results. Like in the preceding section, we also consider convergence of two kinds of approximate solutions for pairs of dual problems.

Section 7.5

Approximations of Optimization Problems in Terms of Optimality Functions

In his excellent comprehensive book [389], Polak develops quantification of necessary optimality conditions in terms of optimality functions with various detailed conceptual algorithms for solving optimization problems. In the book and in many later developments including the recent papers [206, 388, 400, 402, 406], there are numerous applications in nonlinear programming, semi-infinite optimization, optimal control, non-smooth optimization, stochastic programming, and control of uncertain systems. The framework of optimality functions provides a pathway to constructing implementable algorithms consisting only of a finite number of arithmetic operations and function evaluations. Note that after minor adjustments, other familiar functions such as gap functions, merit functions, and residual functions can be viewed as optimality functions. The examples of optimality functions with their properties in Sect. 7.5.1 are based on [389]. The general definitions of optimality functions, stationary, and quasi-stationary points in Sect. 7.5.2 are taken from the recent paper [406]. Results in that paper for the case of metric spaces are also used in the presentation of consistent approximations and convergence of quasi-stationary points under the types of ancillary tight mai-lop convergence in Sect. 7.5.3.

Section 7.6

Approximations of Nash Equilibria

The terminology "Nash equilibrium" can be found in thousands of books and papers. This concept is attributed to John Nash, who developed it in the beginning of the 1950s, while the first insights into the concept had been laid out earlier by Emile Borel and John von Neumann. John Nash was awarded many prestigious prizes, including the John von Neumann Prize in 1978 and the Abel Prize in 2015, for his influential works developing in various ways his results from the 1950s. The fundamental result of Nash about the aforementioned concept states that in a noncooperative game with a finite number of players, each endowed with a finite set of strategies and any kind of payoff function, no player can obtain a higher payoff by switching from his/her equilibrium strategy, i.e., the one in a Nash equilibrium array of strategies, to a different strategy while all the other players keep equilibrium strategies. There are a huge number of contributions to many aspects of existence and properties of Nash equilibria, for various particular problems in a wide range of areas related to mathematics, economics, management, etc. However, the variational approximation of Nash equilibria has received relatively minimal attention in the literature. We observe good related contributions [219, 374].

The first part of this section presents convergence results for approximate Nash and dual Nash equilibria of noncooperative games in terms of various types of epi-, e/h-, and lop-convergence of the approximating Nikaido-Isoda functions. Some of these results are developments of those in [175, 179] and the others are investigated in [181]. There, two kinds of approximate Nash equilibria and different tightness conditions are discussed. The second part considers convergence of these equilibria directly in terms of the data of Nash equilibrium problems. Proposition 7.119 and Theorem 7.120 are taken from [181]; they extend results in [219]. The third part about approximations combined with existence facts by applying the Ky Fan function is taken from [269] with some new developments.

Section 7.7

Approximations of Walras Economic Equilibria

In 1874, Leon Walras presented in his famous "*Elements d'Economie Pure*" the general equilibrium concept as a solution of a system of nonlinear equations. This concept, also developed by Walras' followers, consists in letting price systems play a role by "providing" enough information about the economy for the consumers. Later, neoclassical economists assume that the demands of consumers are derived by maximizing a utility function, in compliance with the main idea of Adam Smith in his seminal publication "*The Wealth of Nations*" in 1776, which is stated in the passage "Every individual endeavors to employ his capital so that its produce may be of greatest value. He generally neither intends to promote the public interest, nor knows how much he is promoting it. He intends only his own security, only his own gain. And he is in this led by an invisible hand to promote an end which was no part of his intention." Now, there is no doubt that Adam Smith was the first scientist to state what we now call decentralization, i.e., the ability of a complex system moved by different actions to pursue different objectives to achieve an allocation of scarce resources, and also to use the phrase "led by an invisible hand." The Walras study, a century later than Adam Smith, made the ideas related to decentralization and invisible hand clearer. But it required one more century for Arrow, Debreu, Gale, Nikaido, and others (see, e.g., [73, 169]) to provide rigorous considerations of equilibrium economic problems. Section 7.7 discusses the Walras pure exchange model following mainly the results in [268, 269], which in turn are based on the pattern first laid out in [81], but include in the model utility functions as the objectives of consumers to maximize, instead of the demand functions being the bundles of goods the consumers buy.

The main contents, i.e., the approximation statements, in Sect. 7.7.2 are improvements of the corresponding ones in [269]. We apply theorems in Sect. 7.2 about convergence of maxinf-points, including those that are based on the tight epi/hypo-convergence, while only the lop-convergence was applied in [269].

Section 7.8

Consistency of Statistical Estimators of Solutions of Stochastic Optimization Problems

This section contains the results of [173], which are extensions to the case of metric spaces of the ones in [186] for finite dimensions. Here, consistency of the statistical estimators and optimal values of stochastic optimization problems is established based on basic properties of epi-convergence provided in Sect. 7.1 and theory of random lower semicontinuous functions initiated in [394]. This result is also relatively naturally extended to consistency for pairs of dual stochastic programming problems by using a connection between the epi-convergence of functions and the epi/hypo-convergence of bifunctions (given in Remark 7.31(c)) to get the epi/hypo-convergence of expectation bifunctions for replacing the epi-convergence of expectation functions of stochastic optimization problems.

Theorem 7.133 in Sect. 7.8.2 about the epi- and pointwise convergence of the (approximating) expectation functions when the approximating probabilities weakly converge is a slight improvement in [181] of the corresponding result in [468]. Unlike the epi-convergence explored in Sect. 7.1 for functions on metric spaces, here the epi-convergence is for functions on topological spaces. Hence, a brief for this general framework of epi-convergence is taken from [181]. Rich properties of random semicontinuous functions employed here are natural extensions to the metric space case of the ones in [360] for finite dimensions. In Sect. 7.8.4, the Lagrange duality scheme for a stochastic optimization problem is presented, and the existence and Painlevé-Kuratowski convergence of saddle points of Lagrangians are established, which correspond to the same type of results for solutions of a stochastic optimization problem and its Lagrange dual. (We know that if a scalar optimization problem is investigated as a special case of an equilibrium problem, then its dual in the duality scheme for equilibrium problems coincides with the primal problem. Here, we develop Lagrangian duality results unlike the duality studies in the previous sections.)

Interesting results related to the ones presented in this subsection can be found in [272, 356, 416]. Recently, the work [201] on epi-convergence under varying probability measures and integrands provides relations between epi-convergence and probability related to the epi-convergence of expectations studied in [173, 186, 186].

of a quasiequilibrium problem are investigated to clarify the differences between equilibrium problems and quasiequilibrium problems.

8.1 Bifunctions on General Domains and Quasiequilibrium Problems

8.1.1 Variational Objects of Bifunctions on General Domains and Types of Quasiequilibrium Problems

In this subsection, we clarify the roles of bifunctions on general domains and their variational objects in connection with types of quasiequilibrium problems. As in Chap. 7, let X and Y be metric spaces. Our MAIN QUASIEQUILIBRIUM PROBLEM is

$$\text{(QEP)} \qquad \text{find } \bar{x} \in A \text{ such that } \Phi(\bar{x}, y) \leq 0 \text{ for all } y \in \mathscr{D}(\bar{x}),$$

where $A \subset X$ and $B \subset Y$ are nonempty, $\mathscr{D} : A \rightrightarrows B$ is defined on the whole A and is onto, and $\Phi : \mathrm{gph}\mathscr{D} \to \mathbb{R}$. We also denote this quasiequilibrium problem by (QEP(Φ, $\mathscr{D}$)) including explicitly the data. First, we make sure that most quasiequilibrium problems in the literature and all those in the previous chapters can be converted to special cases of our (QEP). We consider several typical examples.

(a) Assume that $\hat{A} \subset X$, $\hat{\mathscr{D}} : \hat{A} \rightrightarrows \hat{A}$, and $\hat{\Phi}_1 : \mathrm{gph}\hat{\mathscr{D}} \to \mathbb{R}$. An often-met quasiequilibrium problem is

$$\text{(QEP1)} \qquad \text{find } \bar{x} \in \hat{\mathscr{D}}(\bar{x}) \text{ such that } \hat{\Phi}_1(\bar{x}, y) \leq 0 \text{ for all } y \in \hat{\mathscr{D}}(\bar{x}).$$

By setting $A = \mathrm{Fix}\hat{\mathscr{D}}$, which is the fixed-point set of $\hat{\mathscr{D}}$, $B = \hat{\mathscr{D}}(\mathrm{Fix}\hat{\mathscr{D}})$, $\mathscr{D} : A \rightrightarrows B$ defined by $\mathscr{D}(x) := \hat{\mathscr{D}}(x)$, and $\Phi_1 : \mathrm{gph}\mathscr{D} \to \mathbb{R}$ defined by $\Phi_1(x, y) := \hat{\Phi}_1(x, y)$, we convert (QEP1) into the special case (QEP(Φ_1, $\mathscr{D}$)) of (QEP).

(b) Suppose that $\hat{A} \subset X$, $B \subset Y$, $\mathscr{D}_1 : \hat{A} \rightrightarrows \hat{A}$, $\mathscr{D}_2 : \mathrm{Fix}\mathscr{D}_1 \rightrightarrows B$ such that $\mathscr{D}_2(\mathrm{Fix}\mathscr{D}_1) = B$, and $\hat{\Phi}_2 : \mathrm{gph}\mathscr{D}_2 \to \mathbb{R}$. Another usual quasiequilibrium problem is

$$\text{(QEP2)} \qquad \text{find } \bar{x} \in \mathscr{D}_1(\bar{x}) \text{ such that } \hat{\Phi}_2(\bar{x}, y) \leq 0 \text{ for all } y \in \mathscr{D}_2(\bar{x}).$$

By setting $A = \mathrm{Fix}\mathscr{D}_1$, $\mathscr{D} = \mathscr{D}_2$, and $\Phi_2 : \mathrm{gph}\mathscr{D} \to \mathbb{R}$ defined by $\Phi_2(x, y) := \hat{\Phi}_2(x, y)$, problem (QEP2) becomes the special case (QEP(Φ, $\mathscr{D}$)) of (QEP).

Observe that the domain $\mathrm{gph}\mathscr{D}$ of the (general) bifunction Φ of (QEP) can represent any subset of $X \times Y$. So, the problem setting is general. We denote $A := \mathrm{dom}\mathscr{D}$ and $B := \mathrm{rang}\mathscr{D} = \mathscr{D}(A)$ and regard this set-valued map as a map $\mathscr{D} : A \rightrightarrows B$. So, the case that $\mathscr{D}(x) \equiv B$ aligns with the rectangular case.

Let Sol(QEP) stand for the solution set of (QEP). Observe that any $\bar{x} \in A$ with $\sup_{y \in \mathscr{D}(\bar{x})} \Phi(\bar{x}, y) \leq 0$ belongs to Sol(QEP), including of course the minsup-points of Φ on gph$\mathscr{D}$ satisfying this inequality. But, a solution of (QEP) may not be a minsup-point even in the special rectangular case. (This observation is similar to that for (EP) we had in Chap. 7.) That is why we introduce the second problem besides (QEP), which is the following STRONG QUASIEQUILIBRIUM PROBLEM

$$\text{(sQEP)} \qquad \text{find } \bar{x} \in A \text{ solving } \min_{x \in A} \sup_{y \in \mathscr{D}(x)} \Phi(x, y) \leq 0.$$

As bifunctions on general domains are inevitable for defining and studying types of quasiequilibrium problems, we discuss first variational objects of a bifunction Φ defined on a general domain gph$\mathscr{D}$, which is denoted by $\Phi \in \text{nrec}(X \times Y)$ (nrec signifies "nonrectangular").

Definition 8.1 (Variational objects of bifunctions on general sets)

(a) A point $\bar{x} \in A$ is a MINSUP-POINT ($\bar{y} \in B$ is a MAXINF-POINT, resp.) of Φ (on gph$\mathscr{D}$), written $\bar{x} \in \text{mis}\Phi$ ($\bar{y} \in \text{mai}\Phi$, resp.), if

$$\sup_{y \in \mathscr{D}(\bar{x})} \Phi(\bar{x}, y) = \min_{x \in A} \sup_{y \in \mathscr{D}(x)} \Phi(x, y)$$

$$(\inf_{x \in \mathscr{D}^{-1}(\bar{y})} \Phi(x, \bar{y}) = \max_{y \in B} \inf_{x \in \mathscr{D}^{-1}(y)} \Phi(x, y), \text{ resp.}).$$

(b) A point $(\bar{x}, \bar{y}) \in \text{gph}\mathscr{D}$ is a WEAK SADDLE POINT of Φ, written $(\bar{x}, \bar{y}) \in \text{wsdl}\Phi$, if for all $(\bar{x}, y)$ and $(x, \bar{y})$ in gph$\mathscr{D}$,

$$\Phi(\bar{x}, y) \leq \Phi(\bar{x}, \bar{y}) \leq \Phi(x, \bar{y}),$$

or equivalently, $\Phi(\bar{x}, y) \leq \Phi(x, \bar{y})$ for the above points.

(c) We say that the INFSUP-EQUALITY holds for Φ (on gph$\mathscr{D}$) if

$$\inf_{x \in A} \sup_{y \in \mathscr{D}(x)} \Phi(x, y) = \sup_{y \in B} \inf_{x \in \mathscr{D}^{-1}(y)} \Phi(x, y).$$

(d) If the infsup-equality holds and $(\bar{x}, \bar{y}) \in \text{wsdl}\Phi$, then $(\bar{x}, \bar{y})$ is a SADDLE POINT of Φ (on gph$\mathscr{D}$), written $(\bar{x}, \bar{y}) \in \text{sdl}\Phi$.

Proposition 8.2 (Characterizations and properties of saddle points of general bifunctions)

(a) *The following assertions are equivalent*
  (a_1) $(\bar{x}, \bar{y}) \in \text{sdl}\Phi$;
  (a_2) $\bar{x} \in \text{mis}\Phi$, $\bar{y} \in \text{mai}\Phi$, *and the infsup-equality holds.*

(b) *If* $(\bar{x}, \bar{y}) \in \text{sdl}\Phi$, *then*
  (b_1) $(\bar{x}, \bar{y}) \in \text{wsdl}\Phi$, $\bar{x} \in \text{mis}\Phi$, *and* $\bar{y} \in \text{mai}\Phi$;
  (b_2) $\bar{x} \in \text{mis}\Phi$, $\bar{y} \in \text{argmax}_{y \in \mathscr{D}(\bar{x})} \Phi(\bar{x}, y)$, $\bar{y} \in \text{mai}\Phi$, *and* $\bar{x} \in \text{argmin}_{x \in \mathscr{D}^{-1}(\bar{y})} \Phi(x, \bar{y})$.

(c) *If* $\mathscr{D}(x) \equiv B$, *i.e.*, gph$\mathscr{D}$ *is a rectangle, then* $(\bar{x}, \bar{y}) \in$ sdlΦ *is equivalent to each of the following*

(c_1) $(\bar{x}, \bar{y}) \in$ wsdlΦ*;*

(c_2) $\bar{x} \in$ misΦ, $\bar{y} \in \text{argmax}_{y \in \mathscr{D}(\bar{x})} \Phi(\bar{x}, y)$, $\bar{y} \in$ maiΦ, *and* $\bar{x} \in \text{argmin}_{x \in \mathscr{D}^{-1}(\bar{y})} \Phi(x, \bar{y})$.

Proof

(a) (a_1) $\Rightarrow$ (a_2). Clearly,

$$\inf_{x \in A} \sup_{y \in \mathscr{D}(x)} \Phi(x, y) \le \sup_{y \in \mathscr{D}(\bar{x})} \Phi(\bar{x}, y),$$

$$\inf_{x \in \mathscr{D}^{-1}(\bar{y})} \Phi(x, \bar{y}) \le \sup_{y \in B} \inf_{x \in \mathscr{D}^{-1}(y)} \Phi(x, y).$$

As $(\bar{x}, \bar{y}) \in$ wsdlΦ,

$$\sup_{y \in \mathscr{D}(\bar{x})} \Phi(\bar{x}, y) \le \Phi(\bar{x}, \bar{y}) \le \inf_{x \in \mathscr{D}^{-1}(\bar{y})} \Phi(x, \bar{y}).$$

By the infsup-equality, all the above five quantities are equal. Hence,

$$\sup_{y \in \mathscr{D}(\bar{x})} \Phi(\bar{x}, y) = \min_{x \in A} \sup_{y \in \mathscr{D}(x)} \Phi(x, y), \text{ i.e., } \bar{x} \in \text{mis}\Phi,$$

$$\inf_{x \in \mathscr{D}^{-1}(\bar{y})} \Phi(x, \bar{y}) \le \max_{y \in B} \inf_{x \in \mathscr{D}^{-1}(y)} \Phi(x, y), \text{ i.e., } \bar{y} \in \text{mai}\Phi.$$

(a_2) $\Rightarrow$ (a_1). We need to prove that $(\bar{x}, \bar{y}) \in$ wsdlΦ. As $\bar{x} \in$ misΦ, it holds

$$\min_{x \in A} \sup_{y \in \mathscr{D}(x)} \Phi(x, y) = \sup_{y \in \mathscr{D}(\bar{x})} \Phi(\bar{x}, y) \ge \Phi(\bar{x}, \bar{y}).$$

As $\bar{y} \in$ maiΦ, we have

$$\sup_{y \in B} \inf_{x \in \mathscr{D}^{-1}(y)} \Phi(x, y) = \inf_{x \in \mathscr{D}^{-1}(\bar{y})} \Phi(x, \bar{y}) \le \Phi(\bar{x}, \bar{y}).$$

In view of the infsup-equality, all the above quantities are equal, and hence $(\bar{x}, \bar{y}) \in$ wsdlΦ.

(b) (b_1). We need to show that $\bar{x} \in$ misΦ and $\bar{y} \in$ maiΦ. Since $(\bar{x}, \bar{y}) \in$ wsdlΦ, we have

$$\inf_{x \in A} \sup_{y \in \mathscr{D}(x)} \Phi(x, y) \le \sup_{y \in \mathscr{D}(\bar{x})} \Phi(\bar{x}, y) \le \Phi(\bar{x}, \bar{y}),$$

$$\Phi(\bar{x}, \bar{y}) \le \inf_{x \in \mathscr{D}^{-1}(\bar{y})} \Phi(x, \bar{y}) \le \sup_{y \in B} \inf_{x \in \mathscr{D}^{-1}(y)} \Phi(x, y). \tag{8.1}$$

These inequalities and the infsup-equality imply the equality of all the involved quantities. Hence, (8.1) indicates that $\bar{x} \in$ misΦ and $\bar{y} \in$ maiΦ.

(b_2). Since $(\bar{x}, \bar{y}) \in$ sdlΦ is equivalent to assertion (a_2), we assume both. What we need to verify is $\bar{y} \in \text{argmax}_{y \in \mathscr{D}(\bar{x})} \Phi(\bar{x}, y)$ and $\bar{x} \in$

$\text{argmin}_{x\in\mathscr{D}^{-1}(\bar{y})}\Phi(x,\bar{y})$, which mean that $\sup_{y\in\mathscr{D}(\bar{x})}\Phi(\bar{x},y) = \Phi(\bar{x},\bar{y}) = \min_{x\in\mathscr{D}^{-1}(\bar{y})}\Phi(x,\bar{y})$.

For the first equality, suppose to the contrary that $\Phi(\bar{x},y') > \Phi(\bar{x},\bar{y})$ for some $y' \in \mathscr{D}(\bar{x})$. But, this is in a contradiction with $(\bar{x},\bar{y}) \in \text{wsdl}\Phi$ asserted in (a). The second equality is similarly shown.

(c) (c_1). This follows from the definition of a saddle point and Proposition 1.1.

(c_2). Following assertion (a_2), we have to show that the infsup-equality holds. Due to $\bar{x} \in \text{mis}\Phi$ and $\bar{y} \in \text{argmax}_{y\in\mathscr{D}(\bar{x})}\Phi(\bar{x},y)$, we have

$$\inf_{x\in A}\sup_{y\in\mathscr{D}(x)}\Phi(x,y) = \sup_{y\in\mathscr{D}(\bar{x})}\Phi(\bar{x},y) = \Phi(\bar{x},\bar{y}),$$

Similarly, from $\bar{y} \in \text{mai}\Phi$ and $\bar{x} \in \text{argmin}_{x\in\mathscr{D}^{-1}(\bar{y})}\Phi(x,\bar{y})$, it follows that

$$\sup_{y\in B}\inf_{x\in\mathscr{D}^{-1}(y)}\Phi(x,y) = \Phi(\bar{x},\bar{y}).$$

So, the infsup-equality holds true.

□

Example 8.3 (Examples of variational objects of bifunctions, illustrations of Proposition 8.2) Let $X = Y = \mathbb{R}$ and $\Phi(x,y) = y^x$. In the four cases below, we discuss different types of the domain $\text{gph}\mathscr{D}$ of the above bifunction Φ to underline that different domains cause significantly different (even somehow surprising) situations.

(a) Let $A = B = [0,1]$ and $\mathscr{D}(x) = [0,x]$.

- To compute $\text{mis}\Phi$, we have

$$\sup_{y\in\mathscr{D}(\bar{x})}y^{\bar{x}} = \begin{cases} 1 & \text{if } \bar{x}=0 \text{ (and attained in } \{y=0\}=\mathscr{D}(0)), \\ \bar{x}^{\bar{x}} & \text{if } 0<\bar{x}\le 1 \text{ (and attained at } y=\bar{x}\in\mathscr{D}(\bar{x})). \end{cases} \tag{8.2}$$

  Hence,

$$\min_{\bar{x}\in A}\sup_{y\in\mathscr{D}(\bar{x})}\Phi(\bar{x},y) = \min_{\bar{x}\in A\setminus\{0\}}\{1,\bar{x}^{\bar{x}}\} = e^{-1/e}$$

  and attained at $\bar{x} = e^{-1}$. So, $\text{mis}\Phi = \{e^{-1}\}$.
- Similarly, we have $\text{mai}\Phi = \{1\} =: \bar{y}$.
- $\text{wsdl}\Phi$ consists of the points $(\bar{x},\bar{y}) \in \text{gph}\mathscr{D}$ satisfying $y^{\bar{x}} \le \bar{y}^{\bar{x}}$ for all $y \in \mathscr{D}(\bar{x})$ and $\bar{y}^{\bar{x}} \le \bar{y}^x$ for all $x \in \mathscr{D}^{-1}(\bar{y})$. According to (8.2), the second relation is fulfilled for $(\bar{x},\bar{y}) \in ((0,1]\times\{0\}) \cup (\{1\}\times(0,1]) \cup \{(e^{-1},1)\}$. In this set, the first relation holds at $\{(1,1),(e^{-1},1)\}$. Consequently, $\text{wsdl}\Phi = \{(1,1),(e^{-1},1)\}$.

- Consider sdlΦ. We see that the infsup-equality is violated and *even the infsup-value is strictly smaller than the supinf-value* in this case:

$$\inf_{x\in A}\sup_{y\in\mathscr{D}(x)}\Phi(x,y) < \sup_{y\in B}\inf_{x\in\mathscr{D}^{-1}(y)}\Phi(x,y). \tag{8.3}$$

  This is *striking* (the opposite inequality always holds for bifunctions on rectangles). Hence, there is no saddle point.

(b) Let $A = B = [0, 1]$ and $\mathscr{D}(x) = [x, 1]$. Then, $\min_{x\in A}\sup_{y\in\mathscr{D}(x)}\Phi(x, y) = 1$, mis$\Phi = [0, 1]$, $\max_{y\in B}\inf_{x\in\mathscr{D}^{-1}(y)}\Phi(x, y) = 1$, mai$\Phi = \{0\}$. Calculations yield wsdl$\Phi = \{(0, 0), (1, 1), (0, 1)\}$. Since the infsup-equality holds, sdlΦ = wsdlΦ, see Fig. 8.1.

(c) Let $A = B = [0, 1]$ and $\mathscr{D}(x) = \{x\}$. Then, $\min_{x\in A}\sup_{y\in\mathscr{D}(x)}\Phi(x, y) = e^{-1/e}$, mis$\Phi = \{e^{-1}\}$, $\max_{y\in B}\inf_{x\in\mathscr{D}^{-1}(y)}\Phi(x, y) = 1$ (so the infsup-equality fails), and mai$\Phi = \{0, 1\}$, see Fig. 8.2. Direct verification gives wsdlΦ = gph$\mathscr{D}$, and sdl$\Phi = \emptyset$. Observe that, although the domain of Φ is simply a straight line and the infsup-equality fails, but wsdlΦ equals to the entire domain, totally different from the rectangular case (cf. Proposition 1.1).

(d) Let $A = [0, 1)$, $B = (0, 1)$, and $\mathscr{D}(x) = (x, 1)$. Then, $\min_{x\in A}\sup_{y\in\mathscr{D}(x)}\Phi(x, y) = 1$, mis$\Phi = [0, 1)$, $\sup_{y\in B}\inf_{x\in\mathscr{D}^{-1}(y)}\Phi(x, y) = 1$ (i.e., the infsup-equality holds), but mai$\Phi = \emptyset$, and so sdl$\Phi = \emptyset$ by Proposition 8.2(a_2). By Definition 8.1(d), wsdl$\Phi = \emptyset$.

From Example 8.3, we can deduce the two facts which are essentially different from the corresponding ones for bifunctions on rectangles: One may have inequality (8.3), which is opposite to the popular infsup-inequality (for the rectangular case), and we have to distinguish between a saddle point and a weak saddle point. Note further the two extreme cases: in Example 8.3(c), wsdlΦ= gph$\mathscr{D}$ but sdl$\Phi = \emptyset$ and if gph$\mathscr{D} \subset X \times Y$ is arbitrary and Φ is a constant, then sdlΦ = gph$\mathscr{D}$.

Fig. 8.1 sdlΦ = wsdlΦ = $\{(0, 0), (1, 1), (0, 1)\}$

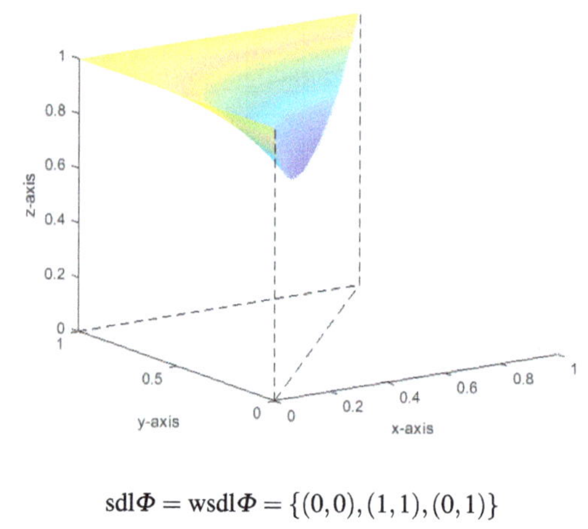

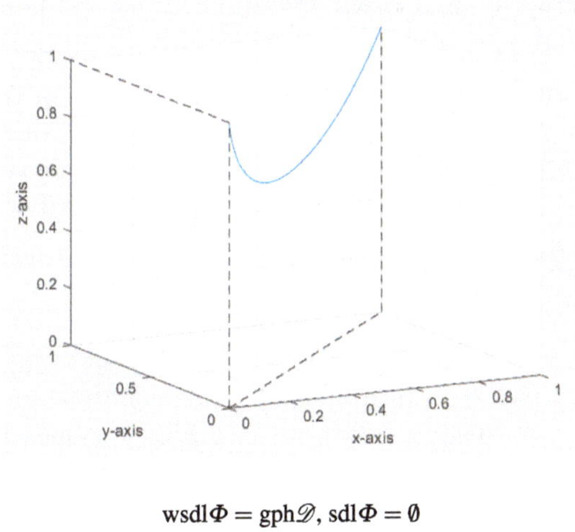

Fig. 8.2 $\text{wsdl}\Phi = \text{gph}\mathcal{D}$, $\text{sdl}\Phi = \emptyset$

Similar to Chap. 7, the main focus of this chapter is on duality properties for (QEP) and (sQEP) and how the variational convergence of general bifunctions preserves variational properties of these problems. Hence, we define now their dual problems.

The DUAL QUASIEQUILIBRIUM PROBLEM is

(DQEP) find $\bar{y} \in B$ such that $\Phi(x, \bar{y}) \geq 0$ for all $x \in \mathcal{D}^{-1}(\bar{y})$.

The DUAL STRONG QUASIEQUILIBRIUM PROBLEM is

(DsQEP) find $\bar{y} \in B$ solving $\max_{y \in B} \inf_{x \in \mathcal{D}^{-1}(y)} \Phi(x, y) \geq 0$.

(QEP) is also often called the STAMPACCHIA QUASIEQUILIBRIUM PROBLEM and (DQEP) is the MINTY QUASIEQUILIBRIUM PROBLEM. Observe that (QEP) and (DQEP) are dual to each other, i.e., the dual of (DQEP) is just (QEP). Indeed, by setting $\hat{\Phi}(\bar{y}, x) = -\Phi(x, \bar{y})$, (DQEP) becomes ($\widehat{\text{QEP}}$) for $\hat{\Phi}$ on $\text{gph}\mathcal{D}^{-1}$. Then, it is easy to check that the dual to ($\widehat{\text{QEP}}$) is just (QEP). Similarly, the dual of (DsQEP) is again (sQEP). Note that solutions of any problem of the aforementioned four types are not unique in general. To see this, simply think of the case $\Phi = 0$ (the zero bifunction) and an arbitrary $\mathcal{D}$, where any point in $\text{gph}\mathcal{D}$ is a solution pair of the strong problems (sQEP) and (DsQEP).

The relationship between solutions of primal-dual pairs of quasiequilibrium problems and saddle points, corresponding to Theorem 7.81 for equilibrium problems, consists of two separate assertions for the two cases as follows.

Proposition 8.4 (Relationship between solutions of primal and dual problems and types of saddle points)

(a) $\bar{x}$ *is a solution of* (QEP), $\bar{y}$ *is a solution of* (DQEP), *and* $(\bar{x}, \bar{y}) \in \mathrm{gph}\mathscr{D}$ *if and only if* $(\bar{x}, \bar{y})$ *is a weak saddle point of* Φ *and* $\Phi(\bar{x}, \bar{y}) = 0$.
(b) $\bar{x}$ *is a solution of* (sQEP), $\bar{y}$ *is a solution of* (DsQEP), *and* $(\bar{x}, \bar{y}) \in \mathrm{gph}\mathscr{D}$ *if and only if* $(\bar{x}, \bar{y})$ *is a saddle point of* Φ *with* $\Phi(\bar{x}, \bar{y}) = 0$.

Hence, if primal and dual solutions exist, the duality gap is always zero.

Proof

(a) "If." $(\bar{x}, \bar{y}) \in \mathrm{gph}\mathscr{D}$ since it is a weak saddle point. $\bar{x} \in \mathrm{Sol(QEP)}$ by the left "wsdl-inequality" and $\bar{y} \in \mathrm{Sol(DQEP)}$ by the right one.
"Only if." Observe first that $\Phi(\bar{x}, \bar{y})$ must be zero. Then, $\inf_{x\in\mathscr{D}^{-1}(\bar{y})}\Phi(x, \bar{y})$ is attained at $\bar{x}$ and $\sup_{y\in\mathscr{D}(\bar{x})}\Phi(\bar{x}, y)$ is attained at $\bar{y}$. Consequently, $(\bar{x}, \bar{y}) \in \mathrm{wsdl}\Phi$.
(b) "If." Already having $\bar{x} \in \mathrm{mis}\Phi$, if $\bar{x} \notin \mathrm{Sol(sQEP)}$, i.e., $\Phi(\bar{x}, y) > 0$ for some $y \in \mathscr{D}(\bar{x})$, then the right "wsdl-inequality" is violated, contradicting $(\bar{x}, \bar{y}) \in \mathrm{sdl}\Phi$.
"Only if." Being the solutions of a pair of strong quasiequilibrium problems, $\bar{x}$ and $\bar{y}$ satisfy

$$0 \geq \min_{x\in A}\sup_{y\in\mathscr{D}(x)}\Phi(x, y) = \sup_{y\in\mathscr{D}(\bar{x})}\Phi(\bar{x}, y) \geq \Phi(\bar{x}, \bar{y}),$$

$$0 \leq \min_{y\in B}\mathrm{x} \in \mathscr{D}^{-1}(\mathrm{y})\Phi(x, y) = \inf_{x\in\mathscr{D}^{-1}(\bar{y})}\Phi(x, \bar{y}) \leq \Phi(\bar{x}, \bar{y}).$$

Therefore, all the involved quantities are equal to 0, and one has the infsup-equality. By assertion (a) of Proposition 8.2, $(\bar{x}, \bar{y})$ is a saddle point.

□

Example 8.5 (Illustration of Proposition 8.4) Let $A \times B = [0, 1]^2$, $\mathscr{D}(x) = \{x\}$, and $\Phi(x, y) = x^2 + y^2$. Then, $\bar{x} = 0$ is the unique solution of both (QEP) and (sQEP) and also the unique minsup-point of Φ, while all the points in $[0, 1]$ are solutions of (DQEP), and $\hat{y} = 1$ is the unique solution of (DsQEP); see Fig. 8.3. Hence, the only point satisfying Proposition 8.4(a) is $(\bar{x}, \bar{y}) = (0, 0)$, and no point satisfies Proposition 8.4(b) (the point $(0, 1)$ does not belong to $\mathrm{gph}\mathscr{D}$). Furthermore, any point (x, x) for $x \in [0, 1]$ is a weak saddle point, and there is no saddle point.

It should be emphasized that results on duality of the usual scheme (QEP)-(DQEP) concern the notion of a weak saddle point, while the saddle point is suitable for the scheme (sQEP)-(DsQEP). For the special rectangular case, the saddle points coincide with weak saddle points (by Proposition 8.2(c)).

Fig. 8.3 $\{\bar{x}\} = \text{Sol(QEP)} = \text{Sol(sQEP)}, \{\hat{y}\} = \text{Sol(DsQEP)}$

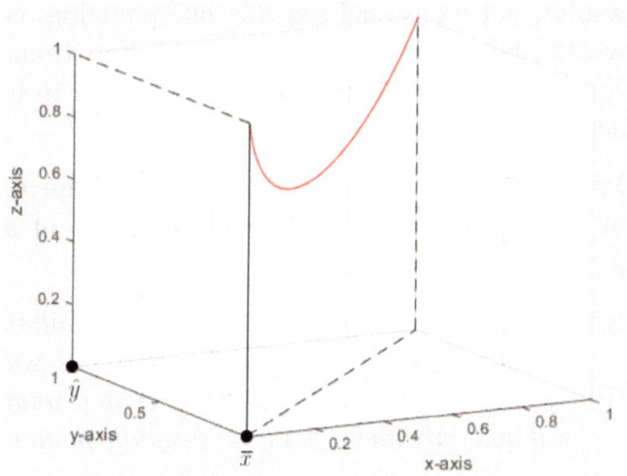

$\{\bar{x}\} = \text{Sol(QEP)} = \text{Sol(sQEP)}, \{\hat{y}\} = \text{Sol(DsQEP)}$

8.1.2 *Types of Approximate Solutions and Approximate Saddle Points*

Naturally, important are concepts and properties of approximate solutions and of approximate saddle points together with their relationships. The definitions of (type 1) approximate solutions of (QEP) and (sQEP) are in the same line as those of (EP) and (sEP). For $\varepsilon \geq 0$, $\bar{x}_\varepsilon \in \varepsilon$-Sol(QEP) or $\bar{x}_\varepsilon \in \varepsilon$-Sol(sQEP), resp., if

$$\Phi(\bar{x}_\varepsilon, y) \leq \varepsilon \text{ for all } y \in \mathscr{D}(\bar{x}_\varepsilon),$$

$$\bar{x}_\varepsilon \in \text{mis}\Phi \text{ such that } \min_{x\in A}\sup_{y\in\mathscr{D}(x)}\Phi(x, y) \leq \varepsilon, \text{ resp.}$$

The corresponding definitions for (DQEP) and (DsQEP) are $\bar{y}_\varepsilon \in \varepsilon$-Sol(DQEP) or $\bar{y}_\varepsilon \in \varepsilon$-Sol(DsQEP), resp., if

$$\Phi(x, \bar{y}_\varepsilon) \geq -\varepsilon \text{ for all } x \in \mathscr{D}^{-1}(\bar{y}_\varepsilon),$$

$$\bar{y}_\varepsilon \in \text{mai}\Phi \text{ such that } \max_{y\in B}\inf_{x\in\mathscr{D}^{-1}(y)}\Phi(x, y) \geq -\varepsilon, \text{ resp.}$$

Correspondingly, we define several types of approximate saddle points as follows. A point $(\bar{x}, \bar{y}) \in \text{gph}\mathscr{D}$ is an ε-WEAK SADDLE POINT of Φ (on gph$\mathscr{D}$), written $(\bar{x}, \bar{y}) \in \varepsilon$-wsdl$\Phi$, if for all $(\bar{x}, y)$ and $(x, \bar{y})$ in gph$\mathscr{D}$, $\Phi(\bar{x}, y) - \varepsilon \leq \Phi(\bar{x}, \bar{y}) \leq \Phi(x, \bar{y}) + \varepsilon$. A point $(\bar{x}, \bar{y}) \in \text{gph}\mathscr{D}$ is an ε-SADDLE POINT of Φ, written $(\bar{x}, \bar{y}) \in \varepsilon$-sdl$\Phi$, if $(\bar{x}, \bar{y}) \in \varepsilon$-wsdl$\Phi$, $\bar{x} \in \text{mis}\Phi$, and $\bar{y} \in \text{mai}\Phi$. Looking at the related definitions for the rectangular case (see a discussion between Example 7.45 and Theorem 7.46), literally, $(\bar{x}, \bar{y}) \in \varepsilon$-sdl$\Phi$ can be understood as $(\bar{x}, \bar{y}) \in$ str-ε-

wsdlΦ, but we do not use this interpretation. Note that for $\varepsilon = 0$, 0-wsdlΦ = wsdlΦ, but 0-sdlΦ is still a type of approximations, not sdlΦ, since the infsup-equality is not ensured. It is not convenient to include an ε-infsup-equality in the definition of an approximate saddle point.

Proposition 8.6 (Relationships between approximate primal and dual solutions of quasiequilibrium problems and types of approximate saddle points) *Let* $\varepsilon \geq 0$.

(a) $\bar{x}_\varepsilon$ *is an* ε*-solution of* (QEP), $\bar{y}_\varepsilon$ *is an* ε*-solution of* (DQEP), *and* $\Phi(\bar{x}_\varepsilon, \bar{y}_\varepsilon) = 0$ *if and only if* $(\bar{x}_\varepsilon, \bar{y}_\varepsilon)$ *is an* ε*-weak saddle point of* Φ *with* $\Phi(\bar{x}_\varepsilon, \bar{y}_\varepsilon) = 0$.
(b) $\bar{x}_\varepsilon$ *is an* ε*-solution of* (sQEP), $\bar{y}_\varepsilon$ *is an* ε*-solution of* (DsQEP), *and* $\Phi(\bar{x}_\varepsilon, \bar{y}_\varepsilon) = 0$ *if and only if* $(\bar{x}_\varepsilon, \bar{y}_\varepsilon)$ *is an* ε*-saddle point of* Φ *with* $\Phi(\bar{x}_\varepsilon, \bar{y}_\varepsilon) = 0$.

Proof Assertion (a) is evident. We prove (b). For the "if," simply substitute $\Phi(\bar{x}_\varepsilon, \bar{y}_\varepsilon) = 0$ into the weak saddle point inequalities. For the "only if," it is clear that $(\bar{x}_\varepsilon, \bar{y}_\varepsilon)$ is an ε-weak saddle point because $\inf_{x \in \mathscr{D}^{-1}(\bar{y}_\varepsilon)} \Phi(x, \bar{y}_\varepsilon) + \varepsilon \geq 0 = \Phi(\bar{x}_\varepsilon, \bar{y}_\varepsilon)$, and similarly $\sup_{y \in \mathscr{D}(\bar{x}_\varepsilon)} \Phi(\bar{x}_\varepsilon, y) - \varepsilon \leq \Phi(\bar{x}_\varepsilon, \bar{y}_\varepsilon)$. $(\bar{x}_\varepsilon, \bar{y}_\varepsilon)$ is an ε-saddle point as $\bar{x}_\varepsilon \in$ misΦ and $\bar{y}_\varepsilon \in$ maiΦ. □

Observe the clear correspondence between the above statement about approximate solutions and types of approximate saddle points and its counterpart in Theorem 7.82. Proposition 8.6 clarifies that a weak saddle point is defined only for the nonrectangular case and a strong approximate saddle point for the rectangular case corresponds to an approximate saddle point for the nonrectangular case.

8.2 Epi/Hypo-Convergence and Lopsided Convergence

8.2.1 Epi/Hypo-Convergence

Epi/hypo-convergence (e/h-convergence) of general bifunctions is an extension of this type of convergence of bifunctions on rectangles studied in the preceding chapter. Hence, the definitions and properties correspond to the known ones in that chapter. For completeness, convenience in applications, and avoiding confusion, most statements are presented completely, but proofs may be only sketches with the focus on the differences.

We consider sequences $A^k, A \subset X$, $B^k, B \subset Y$, $\mathscr{D}^k : A^k \rightrightarrows B^k$, $\mathscr{D} : A \rightrightarrows B$, $\Phi^k : \text{gph}\mathscr{D}^k \to \mathbb{R}$, and $\Phi : \text{gph}\mathscr{D} \to \mathbb{R}$, where $\mathscr{D}^k$ and $\mathscr{D}$ are defined on the whole A^k and A, resp., and are surjective.

Definitions and Characterizations

Definition 8.7 (Epi/hypo-convergence, inside epi/hypo-convergence) Φ^k INSIDE EPI/HYPO-CONVERGES (i-e/h-converges) to Φ (wrt gph$\mathscr{D}^k$ and gph$\mathscr{D}$) if

(a) for every $A^{k_j} \ni x^{k_j} \to x \in A$ and $y \in \mathscr{D}(x)$, there exists $\mathscr{D}^{k_j}(x^{k_j}) \ni y^{k_j} \to y$ such that $\text{li}_j \Phi^{k_j}(x^{k_j}, y^{k_j}) \geq \Phi(x, y)$;
(b) for every $B^{k_j} \ni y^{k_j} \to y \in B$ and $x \in \mathscr{D}^{-1}(y)$, there exists $(\mathscr{D}^{k_j})^{-1}(y^{k_j}) \ni x^{k_j} \to x$ such that $\text{ls}_j \Phi^{k_j}(x^{k_j}, y^{k_j}) \leq \Phi(x, y)$.

Φ^k E/H-CONVERGES to Φ if additionally the following infinity conditions are fulfilled: For every $A^{k_j} \ni x^{k_j} \to x \notin A$, there exists $y^{k_j} \in \mathscr{D}^{k_j}(x^{k_j})$ such that $\Phi^{k_j}(x^{k_j}, y^{k_j}) \to +\infty$ in condition (a), and for every $B^{k_j} \ni y^{k_j} \to y \notin B$, there exists $x^{k_j} \in (\mathscr{D}^{k_j})^{-1}(y^{k_j})$ such that $\Phi^{k_j}(x^{k_j}, y^{k_j}) \to -\infty$ in condition (b).

Similar to the rectangular case, we denote i-e/h-convergence by $\Phi^k \overset{i-e/h}{\to} \Phi$ or $\Phi = \text{i-e/h-lim}_k \Phi^k$, and the abbreviations without "i-" stand for e/h-convergence. We propose the notion of i-e/h-convergence, because in many applications, one is interested only in points in the domain of Φ and direct checking whether a point belongs to gph$\mathscr{D}$ is not hard.

Definition 8.8 (Weak epi/hypo-convergence, weak inside epi/hypo-convergence) We say that Φ^k WEAKLY INSIDE EPI/HYPO-CONVERGES (W-I-E/H-CONVERGES) to Φ if

(a) for every $A^k \ni x^k \to x \in A$ and $y \in \mathscr{D}(x)$, there exists $\mathscr{D}^k(x^k) \ni y^k \to y$ such that $\text{li}_k \Phi^k(x^k, y^k) \geq \Phi(x, y)$;
(b) for every $B^k \ni y^k \to y \in B$ and $x \in \mathscr{D}^{-1}(y)$, there exists $(\mathscr{D}^k)^{-1}(y^k) \ni x^k \to x$ such that $\text{ls}_k \Phi^k(x^k, y^k) \leq \Phi(x, y)$.

Φ^k WEAKLY E/H-CONVERGES (W-E/H-CONVERGES) to Φ if additionally the following conditions are satisfied: for every $A^k \ni x^k \to x \notin A$, there exists $y^k \in \mathscr{D}^k(x^k)$ such that $\Phi^k(x^k, y^k) \to +\infty$, and for every $B^k \ni y^k \to y \notin B$, there exists $x^k \in (\mathscr{D}^k)^{-1}(y^k)$ such that $\Phi^k(x^k, y^k) \to -\infty$.

Evidently, e/h-convergence implies the other three types. We present four types of e/h-convergence for the application purpose.

We note the following fact in the above definitions, only speaking about Definition 8.8. The existence of sequences $\{x^k\}_k$ and $\{y^k\}_k$ mentioned here (still ignoring the conditions on Φ and Φ^k) is ensured by $\text{Li}_k\text{gph}\mathscr{D}^k \supset \text{gph}\mathscr{D}$, which in turn is guaranteed by the graphical convergence of $\mathscr{D}^k$ to $\mathscr{D}$ or by the continuous convergence of $\mathscr{D}^k$ to $\mathscr{D}$ relative to $\text{Lim}_k A^k = A$. But, the reverse does not hold.

Example 8.9 (i-e/h-convergence holds, e/h-convergence fails) For $k \in \mathbb{N}$, let $A^k = A = (0, e-1]$, $B^k = B = (0, 1]$, $\mathscr{D}^k(x) = \mathscr{D}(x) = [\ln(x+1), 1]$, and $\Phi^k(x, y) = \Phi(x, y) = \ln(y/x)$. We check conditions (a) and (b) of Definition 8.7.

As the sequences in the data are constant sequences, we can argue with sequences instead of subsequences.

(a) For every $A^k \ni x^k \to x$ and $y \in \mathscr{D}(x)$, we take in $\mathscr{D}^k(x^k)$

$$y^k = \begin{cases} y & \text{if } 0 < x^k \le x, \\ y + \gamma^k & \text{such that } \gamma^k \to 0^+ \text{ and } \ln(x^k + 1) \\ & \le y + \gamma^k \le 1 \text{ if } x < x^k \le e - 1. \end{cases}$$

Then, $\lim_k \Phi^k(x^k, y^k) = \Phi(x, y)$ by the continuity of $\Phi^k = \Phi$.

For every $A^k \ni x^k \to 0 \notin A$, $y^k \equiv 1 \in \mathscr{D}^k(x^k)$ satisfy $\Phi^k(x^k, y^k) \to +\infty$.

(b) We see that for every $B^k \ni y^k \to y$ and $x \in \mathscr{D}^{-1}(y)$, $x^k := x$ satisfies $\lim_k \Phi^k(x^k, y^k) = \Phi(x, y)$. Therefore, Φ^k i-e/h-converges to Φ. Let $B^k \ni y^k \to 0 \notin B$ and x^k be in $(\mathscr{D}^k)^{-1}(y^k)$. Then, $x^k \le e^{y^k} - 1$ and $\Phi^k(x^k, y^k) \ge \ln(y^k(e^{y^k} - 1)^{-1}) \to 0$, not $-\infty$ as required. So, Φ^k does not weakly e/h-converge to Φ.

Example 8.10 (e/h-convergence holds) We keep the same $\Phi^k(x, y) = \Phi(x, y) = \ln(y/x)$ and change their domains as follows: $A^k = A = B^k = B = (0, 1]$ and $\mathscr{D}^k(x) = \mathscr{D}(x) = [x^2, 1]$ for all k. Then, by a technique similar to that in Example 8.9, we obtain the e/h-convergence. Here, we have only one different computation: for $B^k \ni y^k \to 0 \notin B$, we take in $\mathscr{D}^{-1}(y^k)$ the point $x^k = (y^k)^{1/2}$ to see that $\Phi^k(x^k, y^k) \to -\infty$.

As for the rectangular case, the types of e/h-convergence are symmetric and the types of hypo/epi-convergence coincide with the corresponding types of e/h-convergence.

Proposition 8.11 (Weak e/h-convergence versus weak epi- and hypo-convergence of slices)

(i) *Condition* (a) *of weak e/h-convergence* $\Leftrightarrow$ *for every* $A^k \ni x^k \to x \in A$, w-hli$\Phi^k(x^k, \cdot) \ge \Phi(x, \cdot)$ *with* $\Phi^k(x^k, \cdot)$ *defined on* $\mathscr{D}^k(x^k)$ *and* $\Phi(x, \cdot)$ *on* $\mathscr{D}(x)$, *and for every* $A^k \ni x^k \to x \notin A$, *there exists* $y^k \in \mathscr{D}^k(x^k)$ *with* $\Phi^k(x^k, y^k) \to +\infty$ $\Leftrightarrow$ *for every* $A^k \ni x^k \to x \in A$, *one has condition* (b) *of* $\Phi^k(x^k, \cdot) \overset{w-h}{\to} \Phi(x, \cdot)$ *with the domains as above, and for every* $A^k \ni x^k \to x \notin A$, *there exists* $y^k \in \mathscr{D}^k(x^k)$ *with* $\Phi^k(x^k, y^k) \to +\infty$.

(ii) *Condition* (b) *of weak e/h-convergence* $\Leftrightarrow$ *for every* $B^k \ni y^k \to y \in B$, w-els$\Phi^k(\cdot, y^k) \le \Phi(\cdot, y)$ *with* $\Phi^k(\cdot, y^k)$ *defined on* $(\mathscr{D}^k)^{-1}(y^k)$ *and* $\Phi(\cdot, y)$ *on* $\mathscr{D}^{-1}(y)$, *and for every* $B^k \ni y^k \to y \notin B$, *there exists* $x^k \in (\mathscr{D}^k)^{-1}(y^k)$ *such that* $\Phi^k(x^k, y^k) \to -\infty$ $\Leftrightarrow$ *for every* $B^k \ni y^k \to y \in B$, *one has condition* (b) *of* $\Phi^k(\cdot, y^k) \overset{w-e}{\to} \Phi(\cdot, y)$ *with the above domains, and for every* $B^k \ni y^k \to y \notin B$, *there exists* $x^k \in (\mathscr{D}^k)^{-1}(y^k)$ *such that* $\Phi^k(x^k, y^k) \to -\infty$.

Omitting the infinity conditions in the above statements, we obtain the assertions for weak inside e/h-convergence. The corresponding statements for e/h- and inside e/h-convergence are similar.

Proof By symmetry, we only prove assertion (i). For the first equivalence, the first part of condition (a) of weak e/h-convergence means that w-hli$\Phi^k(x^k, \cdot) \geq \Phi(x, \cdot)$. The remaining part of assertion (i) is the same in the two member assertions to be shown equivalent.

For the second equivalence, the first part of assertion (i) clearly means condition (b) of the weak hypo-convergence of $\Phi^k(x^k, \cdot)$, while the minus infinity part is kept the same in both equivalent assertions in question. □

Definition 8.12 (Lower and upper e/h-limits, weak lower and upper e/h-limits) Consider Φ and Φ^k in nrec$(X \times Y)$ and $(x, y) \in X \times Y$.

(a) For $x \in A$ and $y \in \mathscr{D}(x)$, the UPPER E/H-LIMIT is defined by
$$\text{ehls}\Phi^k(x, y) := \inf_{\{A^{k_j} \ni x^{k_j} \to x\}} \sup_{\{\mathscr{D}^{k_j}(x^{k_j}) \ni y^{k_j} \to y\}} \text{li}_j \Phi^{k_j}(x^{k_j}, y^{k_j});$$
and for $x \notin A$, it is
$$\text{ehls}\Phi^k(x) := \inf_{\{A^{k_j} \ni x^{k_j} \to x\}} \sup_{\{y^{k_j} \in \mathscr{D}^{k_j}(x^{k_j})\}} \text{li}_j \Phi^{k_j}(x^{k_j}, y^{k_j}).$$
(b) For $y \in B$ and $x \in \mathscr{D}^{-1}(y)$, the LOWER E/H-LIMIT is defined by
$$\text{ehli}\Phi^k(x, y) := \sup_{\{B^{k_j} \ni y^{k_j} \to y\}} \inf_{\{(\mathscr{D}^{k_j})^{-1}(y^{k_j}) \ni x^{k_j} \to x\}} \text{ls}_j \Phi^{k_j}(x^{k_j}, y^{k_j});$$
and for $y \notin B$, it is
$$\text{ehli}\Phi^k(y) := \sup_{\{B^k \ni y^{k_j} \to y\}} \inf_{\{x^{k_j} \in (\mathscr{D}^{k_j})^{-1}(y^{k_j})\}} \text{ls}_j \Phi^{k_j}(x^{k_j}, y^{k_j}).$$
(c) For $(x, y) \in \text{gph}\mathscr{D}$, the WEAK UPPER E/H-LIMIT of Φ^k is
$$\text{w-ehls}\Phi^k(x, y) := \inf_{\{A^k \ni x^k \to x\}} \sup_{\{\mathscr{D}^k(x^k) \ni y^k \to y\}} \text{li}_k \Phi^k(x^k, y^k).$$
The other three types of weak e/h-limits are defined and denoted similarly.

Theorem 8.13 (Formulas for the types of upper and lower e/h-limits)

(a) *For $(x, y) \in \text{gph}\mathscr{D}$, we have*
$$\text{ehls}\Phi^k(x, y) = \inf_{\{A^{k_j} \ni x^{k_j} \to x\}} \max_{\{\mathscr{D}^{k_j}(x^{k_j}) \ni y^{k_j} \to y\}} \text{li}_j \Phi^{k_j}(x^{k_j}, y^{k_j});$$
and for $x \notin A$,
$$\text{ehls}\Phi^k(x) = \inf_{\{A^{k_j} \ni x^{k_j} \to x\}} \max_{\{y^{k_j} \in \mathscr{D}^{k_j}(x^{k_j})\}} \text{li}_j \Phi^{k_j}(x^{k_j}, y^{k_j});$$
(b) *For $(y, x) \in \text{gph}\mathscr{D}^{-1}$, we have*
$$\text{ehli}\Phi^k(x, y) = \sup_{\{B^{k_j} \ni y^{k_j} \to y\}} \min_{\{(\mathscr{D}^{k_j})^{-1}(y^{k_j}) \ni x^{k_j} \to x\}} \text{ls}_j \Phi^{k_j}(x^{k_j}, y^{k_j});$$
and for $y \notin B$,
$$\text{ehli}\Phi^k(y) = \sup_{\{B^{k_j} \ni y^{k_j} \to y\}} \min_{\{x^{k_j} \in (\mathscr{D}^{k_j})^{-1}(y^{k_j})\}} \text{ls}_j \Phi^k(x^{k_j}, y^{k_j}).$$

By replacing the subsequences in the above formulas by the corresponding sequences, we have the formulas for the four types of weak convergence.

Proof Because of similarity, we only say for assertion (a) that the proof is similar to that of Theorem 7.35 for the rectangular case with natural modifications, e.g., instead of the expression of s in Theorem 7.35, now the corresponding expression is

$$s := \sup_{\{\mathscr{D}^{k_j}(x^{k_j})\ni y^{k_j}\to y\}} \mathrm{li}_j \Phi^{k_j}(x^{k_j}, y^{k_j}).$$

□

Theorem 8.14 (Characterization of the types of e/h-convergence) *Let Φ and Φ^k for $k \in \mathbb{N}$, be in* nrec$(X \times Y)$. *Then, Φ^k e/h-converging to Φ is equivalent to the following pair of assertions*

(a) $\Phi(x, y) \le \mathrm{ehls}\Phi^k(x, y)$ *when* $x \in A$, $y \in \mathscr{D}(x)$, *and* $\mathrm{ehls}\Phi^k(x) = +\infty$ *when* $x \notin A$;
(b) $\mathrm{ehli}\Phi^k(x, y) \le \Phi(x, y)$ *when* $y \in B$, $x \in \mathscr{D}^{-1}(y)$, *and* $\mathrm{ehli}\Phi^k(y) = -\infty$ *when* $y \notin B$.

Omitting the infinity conditions, we have a characterization of i-e/h-convergence. Replacing ehls *by* w-ehls *and* ehli *by* w-ehli, *we get a characterization of the types of weak e/h-convergence.*

Proof Simply apply Definitions 8.7 and 8.8, and Theorem 8.13.

□

Note that the limits of a given sequence of general bifunctions which converge with respect to a type of e/h-convergence are not unique bifunctions but form a whole interval of bifunctions. In Example 8.9, we use the definition to see that Φ^k i-e/h-converges but does not e/h-converge to a given candidate Φ, while Example 8.10 gives a case when $\Phi^k \equiv \Phi$ e/h-converges to Φ. Theorem 8.14 allows us to obtain all the possible e/h-limits.

Example 8.15 (A case when the e/h-limit is unique) We apply Theorem 8.14 to revisit Example 8.10. Following assertion (a), for $x \in A$ and $y \in \mathscr{D}(x)$, by the continuity of Φ^k, $\mathrm{ehls}\Phi^k(x, y) = \lim_k \Phi^k(x^k, y^k) = \Phi(x, y)$ for all $(x^k, y^k) \in \mathrm{gph}\mathscr{D}^k$ tending to (x, y). For $x = 0 \notin A$, $\mathrm{ehls}\Phi^k(x) = +\infty$ because $\max_{\{y^k\in\mathscr{D}^k(x^k)\}} \mathrm{li}_k\, \Phi^k(x^k, y^k) \ge \lim_k \ln(\frac{1}{x^k}) = +\infty$ for every $A^k \ni x^k \to 0$.

Similarly, with assertion (b) of Theorem 8.14, we have $\mathrm{ehli}\Phi^k(x, y) = \Phi(x, y)$ for $y \in B$, $x \in \mathscr{D}^{-1}(y)$. To calculate $\mathrm{ehli}\Phi^k(y)$ for $y = 0 \notin B$, for all $B^k \ni y^k \to 0$, we take $x^k = (y^k)^{1/2} \in (\mathscr{D}^k)^{-1}(y^k)$ to see that $\mathrm{ehli}\Phi^k(y) \le \mathrm{ls}_k \ln(\frac{y^k}{x^k}) = -\infty$. Therefore, Φ is the unique e/h-limit.

Theorem 8.16 (Geometry of the types of lower and upper e/h-limits ending the equivalence class)

$\mathrm{epi}(\mathrm{ehli}\Phi^k(\cdot, y)) = \bigcap_{\{B^{k_j}\ni y^{k_j}\to y\}} \mathrm{Li}_j(\mathrm{epi}\Phi^{k_j}(\cdot, y^{k_j}))$ for $\mathrm{ehli}\Phi^k(\cdot, y)$ on A,
$\mathrm{hypo}(\mathrm{ehls}\Phi^k(x, \cdot)) = \bigcap_{\{A^{k_j}\ni x^{k_j}\to x\}} \mathrm{Li}_j(\mathrm{hypo}\Phi^{k_j}(x^{k_j}, \cdot))$ for $\mathrm{ehls}\Phi^k(x, \cdot)$ on B,
$\mathrm{epi}(\text{w-ehli}\Phi^k(\cdot, y)) = \bigcap_{\{B^k\ni y^k\to y\}} \mathrm{Li}_k(\mathrm{epi}\Phi^k(\cdot, y^k))$ for w-$\mathrm{ehli}\Phi^k(\cdot, y)$ on A,

$\mathrm{hypo}(\text{w-ehls}\Phi^k(x,\cdot)) = \cap_{\{A^k \ni x^k \to x\}} \mathrm{Li}_k(\mathrm{hypo}\Phi^k(x^k,\cdot))$ for $\text{w-ehls}\Phi^k(x,\cdot)$ on B.

Proof Because of similarity, we only prove the first formula. (x, α) lying inside the right-hand side means that for all $B^{k_j} \ni y^{k_j} \to y$ and $\varepsilon > 0$, there exists $j_0 \in \mathbb{N}$ such that for all $j \geq j_0$,

$$[((\mathscr{D}^{k_j})^{-1}(y^{k_j}) \cap B(x,\varepsilon)) \times (-\infty, \alpha+\varepsilon)] \cap \mathrm{epi}\Phi^{k_j}(.,y^{k_j}) \neq \emptyset.$$

This relation means the existence of $x^{k_j} \in (\mathscr{D}^{k_j})^{-1}(y^{k_j}) \cap B(x,\varepsilon)$ such that $\Phi^{k_j}(x^{k_j}, y^{k_j}) < \alpha + \varepsilon$ for large j. Hence, the above statement is equivalent to

$$\sup_{\{B^{k_j} \ni y^{k_j} \to y\}} \min_{\{(\mathscr{D}^{k_j})^{-1}(y^{k_j}) \ni x^{k_j} \to x\}} \mathrm{ls}_j \Phi^{k_j}(x^{k_j}, y^{k_j}) \leq \alpha,$$

i.e., $\mathrm{ehli}\Phi^k(x,y) \leq \alpha$, or equivalently $(x,\alpha) \in \mathrm{epi}(\mathrm{ehli}\Phi^k(.,y))$. □

Variational properties of the types of e/h-convergence

Naturally, variational properties of the types of variational convergence of (general) bifunctions are about the preservation of their main variational objects given in Definition 8.1 and their approximate objects through the variational convergence types in question.

Theorem 8.17 (Convergence of approximate weak saddle points under e/h-convergence) *Assume that* $\varepsilon, \varepsilon^k \geq 0$, $\varepsilon^k \to \varepsilon$, *and* $(\bar{x}^k, \bar{y}^k) \in \varepsilon^k\text{-wsdl}\Phi^k$ *for all* $k \in \mathbb{N}$.

(a) *If* $\Phi^k \overset{w-i-e/h}{\to} \Phi$ *and* $(\bar{x}, \bar{y}) = \lim_k(\bar{x}^k, \bar{y}^k)$ *is in* $\mathrm{gph}\mathscr{D}$, *then* $(\bar{x}, \bar{y}) \in \varepsilon\text{-wsdl}\Phi$.

(b) *If* $\Phi^k \overset{i-e/h}{\to} \Phi$, *then* $\mathrm{Ls}_k(\varepsilon^k\text{-wsdl}\Phi^k) \cap \mathrm{gph}\mathscr{D} \subset \varepsilon\text{- wsdl}\Phi$. *If specifically* $\varepsilon = 0$ *and* $\lim_j(\bar{x}^{k_j}, \bar{y}^{k_j}) = (\bar{x}, \bar{y})$, *then*

$$\Phi(\bar{x},\bar{y}) = \lim_j \Phi^{k_j}(\bar{x}^{k_j}, \bar{y}^{k_j}).$$

(c) *If* $\Phi^k \overset{e/h}{\to} \Phi$, *then* $\mathrm{Ls}_k(\varepsilon^k\text{-wsdl}\Phi^k) \subset \varepsilon\text{-wsdl}\Phi$.

Proof (a) The proof goes along the same line as that of Theorem 7.47 for the rectangular case. We only mention what is different. Assume that Φ^k w-i-e/h-converges to Φ. Consider arbitrary $(x, \bar{y}), (\bar{x}, y) \in \mathrm{gph}\mathscr{D}$. As $(\bar{x}^k, \bar{y}^k) \in \varepsilon^k$-$\mathrm{wsdl}\Phi^k$, for any $(\mathscr{D}^k)^{-1}(\bar{y}^k) \ni x^k \to x$ and $\mathscr{D}^k(\bar{x}^k) \ni y^k \to y$, we have

$$\Phi^k(\bar{x}^k, y^k) - \varepsilon^k \leq \Phi^k(\bar{x}^k, \bar{y}^k) \leq \Phi^k(x^k, \bar{y}^k) + \varepsilon^k. \tag{8.4}$$

Assume that $(\bar{x}^k, \bar{y}^k) \to (\bar{x}, \bar{y}) \in \mathrm{gph}\mathscr{D}$. By (8.4), the remaining part of the proof as the same as the part after (7.3) in the proof of Theorem 7.47, with only $\mathrm{gph}\mathscr{D}^k$ and $\mathrm{gph}\mathscr{D}$ replacing $A^k \times B^k$ and $A \times B$.

(b) and (c) are proved similarly as those in Theorem 7.47 with (8.4) replacing (7.3). □

A function $\xi : A \to \mathbb{R}$ ($\eta : B \to \mathbb{R}$, resp.) is called the SUP-PROJECTION (INF-PROJECTION, resp.) of a bifunction $\Phi \in \text{nrec}(X \times Y)$ if, for $x \in A$ ($y \in B$, resp.),

$$\xi(x) := \sup_{y\in\mathcal{D}(x)}\Phi(x, y) \; (\eta(y) := \inf_{x\in\mathcal{D}^{-1}(y)}\Phi(x, y), \text{ resp.}).$$

Theorem 8.18 (Condition (a) of a type of e/h-convergence implies (a) of the corresponding type of epi-convergence of sup-projections)

(i) Condition (a) of the e/h-convergence of Φ^k to Φ together with the nonemptiness of $\text{dom}\xi^k$ *implies* $\text{dom}\xi \neq \emptyset$ *and condition* (a) *of the epi-convergence of* ξ^k.
(ii) The corresponding assertion holds also for the other types of e/h-convergence.

This statement extends Theorem 7.46 to the nonrectangular case, and the proof remains almost the same as that of Theorem 7.46 and so is omitted.

Similar to the rectangular case, we need the tightness conditions to have more variational properties. We only formulate and briefly discuss, if necessary, the definitions and the properties which follow now.

Definition 8.19 (Types of tightness conditions)

(i) We say that Φ^k WEAKLY I-E/H-CONVERGES TO Φ x-ANCILLARY TIGHTLY if condition (a) of weak i-e/h-convergence and the following condition are satisfied

(b-m) condition (b) of weak i-e/h-convergence is satisfied together, additionally, with $\text{ls}_k\xi^k(x^k) \le \xi(x)$.

(ii) y-ANCILLARY TIGHT WEAK I-E/H-CONVERGENCE is defined correspondingly, with condition (b) of weak i-e/h-convergence and

(a-m) condition (a) of weak i-e/h-convergence is fulfilled together, additionally, with $\text{li}_k\eta^k(y^k) \ge \eta(y)$.

(iii) If we replace in condition (i) "weak i-e/h-convergence" by "i-e/h-convergence" and $\text{ls}_k\xi^k(x^k) \le \xi(x)$ in condition (b-m) by $\text{ls}_j\xi^{k_j}(x^{k_j}) \le \xi(x)$, then Φ^k I-E/H-CONVERGES TO Φ x-ANCILLARY TIGHTLY. If we replace in condition (ii) "weak i-e/h-convergence" by "i-e/h-convergence" and $\text{li}_k\eta^k(y^k) \ge \eta(y)$ in condition (a-m) by $\text{li}_j\eta^{k_j}(y^{k_j}) \ge \eta(y)$, then Φ^k I-E/H-CONVERGES TO Φ y-ANCILLARY TIGHTLY.

(iv) Φ^k E/H-CONVERGES TO Φ x-ANCILLARY TIGHTLY if condition (a) of e/h-convergence, condition (b-m) in condition (iii), and the infinity requirement in condition (b) of e/h-convergence are fulfilled. y-ANCILLARY TIGHT E/H-CONVERGENCE is defined similarly.

When both types of tightness in assertion (iv) are satisfied, Φ^k E/H-CONVERGES TO Φ (FULLY) TIGHTLY.

Observe that in Definition 8.19, item (i) is symmetric to (ii), item (iii) is symmetric to itself, and all the conditions in (i)-(iv) are not (directly) connected to compactness, while we already know from Chap. 7 that the tightness conditions for minsup-lop convergence represent types of relaxed uniform compactness conditions. The tightness conditions for minsup-lop convergence are appropriate only for minsup-points (maxinf-lop convergence is defined separately and applied to maxinf-points). In the next statements, we show that the types of tight e/h-convergence are really useful not only for considering weak saddle points but also for minsup- and maxinf-points.

Theorem 8.20 (epi-convergence of sup-projections)

(a) *If* $\Phi^k \overset{e/h}{\to} \Phi$ *x-ancillary tightly and* $\mathrm{dom}\xi^k \neq \emptyset$, *then* $\mathrm{dom}\xi \neq \emptyset$ *and* e-$\lim_k \xi^k = \xi$.
(b) *If* Φ^k *i-e/h-converges to* Φ *(or w-e/h converges or w-i-e/h-converges, resp.) x-ancillary tightly, then* $\mathrm{dom}\xi \neq \emptyset$ *and* ξ^k *i-e-converges (or w-e-converges or w-i-e-converges, resp.) to* ξ.

The proof is almost the same as that of Theorem 7.52, with the only change from $A^k \times B^k$ and $A \times B$ to $\mathrm{gph}\mathscr{D}^k$ and $\mathrm{gph}\mathscr{D}$, resp. With a similar argument, we also have the following corresponding statement for inf-projections.

Proposition 8.21 (epi-convergence of inf-projections) *Let the e/h-convergence of* Φ^k *to* Φ *in* $\mathrm{nrec}(X \times Y)$ *be y-ancillary tight and* $\mathrm{dom}\eta^k \neq \emptyset$. *Then,* $\mathrm{dom}\eta \neq \emptyset$ *and* h-$\lim_k \eta^k = \eta$. *The corresponding assertions for the other types of e/h-convergence also hold.*

Theorem 8.22 (Convergence of minsup-points) *Assume that* $\varepsilon, \varepsilon^k \geq 0$ *and* $\varepsilon^k \to \varepsilon$.

(a) *If* Φ^k *w-i-e/h-converges to* Φ *x-ancillary tightly, then*

$$\mathrm{ls}_k[\inf_{x\in A^k}\sup_{y\in\mathscr{D}^k(x)}\Phi^k(x,y)] \leq \inf_{x\in A}\sup_{y\in\mathscr{D}(x)}\Phi(x,y).$$

(b) *If* Φ^k *e/h-converges to* Φ *x-ancillary tightly, then* $\mathrm{Ls}_k(\varepsilon^k\text{-mis}\Phi^k) \subset \varepsilon\text{- mis}\Phi$.
(c) *If* Φ^k *w-e/h-converges to* Φ *x-ancillary tightly,* $\varepsilon^k \to 0^+$, *and there exists* $x^k \in \varepsilon^k\text{-mis}\Phi^k$ *tending to* $\bar{x}$, *then* $\bar{x} \in \mathrm{mis}\Phi$ *and the minsup-value of* Φ^k *converges to the minsup-value of* Φ.

The proof is almost the same as that of Theorem 7.52, but applying Theorem 8.20 instead of Theorem 7.51.

Example 8.23 (The tightness condition in Theorem 8.22 is essential but not necessary) Looking again at Example 8.9, we see that Φ^k only i-e/h-converges but does not e/h-converge to Φ. But, the (unique) minsup-point $\bar{x}^k = e - 1$ of Φ^k tends to the (unique) minsup-point $e - 1$ of Φ.

Theorem 8.24 (Convergence of approximate weak saddle points, a lower bound of approximate saddle points) . *Let $\varepsilon, \varepsilon^k \geq 0$ and $\varepsilon^k \to \varepsilon$.*

(a) *If Φ^k inside epi/hypo-converges tightly to Φ, then*

$$\varepsilon\text{-wsdl}\Phi \subset \cap_{\delta>\varepsilon}\text{Li}_k(\delta\text{-wsdl}\Phi^k)$$

and hence, $\varepsilon\text{-wsdl}\Phi \subset \text{Li}_k(\varepsilon^k\text{-wsdl}\Phi^k)$.

(b) *If Φ^k epi/hypo-converges tightly to Φ, then*

$$\text{Lim}_k(\varepsilon^k\text{-wsdl}\Phi^k) = \varepsilon\text{-wsdl}\Phi,$$

$$\text{Ls}_k(\varepsilon^k\text{-sdl}\Phi^k) \subset \varepsilon\text{-sdl}\Phi.$$

Proof

(a) The proof of this assertion is almost the same as that of assertion (i) of Theorem 7.54 for the rectangular case (applying Proposition 8.11 instead of Proposition 7.32).
(b) The first relation is a consequence of assertion (a) and Theorem 8.17. To obtain the second one, simply apply the first one and Theorem 8.22 together with its counterpart for maxinf-points.

□

Example 8.25 (Illustration of Theorem 8.24) Revisit the bifunctions Φ^k and Φ in Example 8.10. That example ensures the e/h-convergence of Φ^k to Φ. To check condition (b-m), for $y \in (0, 1]$, $x \in \mathscr{D}^{-1}(y)$, and $(0, 1] \ni y^k \to y$, we only need to find $x^k \in (\mathscr{D}^k)^{-1}(y^k)$ tending to x, as $\xi^k(x) = \xi(x) = \sup_{y\in[x^2,1]}\ln(y/x) = \ln(1/x)$ are continuous. It is clear that $x^k = x$ if $y^k \geq y$ and $x^k = x - \gamma^k$ if $y^k < y$, where $\gamma^k \to 0^+$ satisfies $x - \sqrt{y^k} \leq \gamma^k \leq x$, fulfill this requirement. For condition (a-m), with given $x \in (0, 1]$, $y \in \mathscr{D}(x)$, and $(0, 1] \ni x^k \to x$, take $y^k = y$ if $x^k \leq x$ and $y^k = y + \beta^k$ such that $\beta^k \to 0^+$ satisfies $(x^k)^2 \leq \beta^k \leq 1 - y$ if $x^k > x$. Then, the assumptions of Theorem 8.24 are checked. The conclusions of this theorem can also be seen directly.

8.2.2 *Lopsided Convergence*

Naturally similar to the situation of the rectangular case, for bifunctions on general domains, lopsided convergence must go along with epi/hypo-convergence in both theory and application, with similarity and differences between them being similar to some extent to what we know in Chap. 7. So, we discuss it in detail only when necessary.

Definitions and Characterizations

As in Sect. 8.2.1, let $A^k, A \subset X$, $B^k, B \subset Y$, $\mathscr{D}^k : A^k \rightrightarrows B^k$, $\mathscr{D} : A \rightrightarrows B$, $\Phi^k : \text{gph}\mathscr{D}^k \to \mathbb{R}$, and $\Phi : \text{gph}\mathscr{D} \to \mathbb{R}$, where $\mathscr{D}^k$ and $\mathscr{D}$ are defined on the whole A^k and A, resp., and are surjective.

Definition 8.26 (minsup-lopsided convergence, inside minsup-lopsided convergence) Φ^k MINSUP-LOPSIDED (MIS-LOP) CONVERGES to Φ if

(a) for every $A^{k_j} \ni x^{k_j} \to x \in A$ and $y \in \mathscr{D}(x)$, there exists $\mathscr{D}^{k_j}(x^{k_j}) \ni y^{k_j} \to y$ such that $\mathrm{li}_j \Phi^{k_j}(x^{k_j}, y^{k_j}) \geq \Phi(x, y)$, and for every $A^{k_j} \ni x^{k_j} \to x \notin A$, there exists $y^{k_j} \in \mathscr{D}^{k_j}(x^{k_j})$ such that $\Phi^{k_j}(x^{k_j}, y^{k_j}) \to +\infty$;
(b) for every $x \in A$, there exists $A^{k_j} \ni x^{k_j} \to x$ such that for all $y^{k_j} \in \mathscr{D}^{k_j}(x^{k_j})$ tending to y, it holds $\mathrm{ls}_j \Phi^{k_j}(x^{k_j}, y^{k_j}) \leq \Phi(x, y)$ if $y \in \mathscr{D}(x)$ and $\Phi^{k_j}(x^{k_j}, y^{k_j}) \to -\infty$ if $y \notin \mathscr{D}(x)$.

We use the notation $\Phi^k \overset{mis-lop}{\to} \Phi$ or $\Phi = \text{mis-lop-lim}_k \Phi^k$. If we consider only $x \in A$ in condition (a) and $y \in \mathscr{D}(x)$ in condition (b) (and omit the infinity conditions), Φ^k INSIDE MINSUP-LOPSIDED CONVERGES and we have "i-" added to the notation.

Simply replacing the subsequences in the above definition by the corresponding sequences, we have the concepts of WEAK MINSUP-LOPSIDED CONVERGENCE and WEAK INSIDE MINSUP-LOPSIDED CONVERGENCE, resp. Definition 8.26 is not symmetric, and we have also other types of lop-convergence as follows.

Definition 8.27 (maxinf-lopsided convergence, inside maxinf-lopsided convergence) Φ^k MAXINF-LOPSIDED (MAI-LOP) CONVERGES to Φ if

(a) for every $B^{k_j} \ni y^{k_j} \to y$ and $x \in \mathscr{D}^{-1}(y)$, there exists $(\mathscr{D}^{k_j})^{-1}(y^{k_j}) \ni x^{k_j} \to x$ such that $\mathrm{ls}_j \Phi^{k_j}(x^{k_j}, y^{k_j}) \leq \Phi(x, y)$; for every $B^{k_j} \ni y^{k_j} \to y \notin B$, there exists $x^{k_j} \in (\mathscr{D}^{k_j})^{-1}(y^{k_j})$ such that $\Phi^{k_j}(x^{k_j}, y^{k_j}) \to -\infty$;
(b) for every $y \in B$, there exists $B^{k_j} \ni y^{k_j} \to y$ such that for all $(\mathscr{D}^{k_j})^{-1}(y^{k_j}) \ni x^{k_j} \to x$, it holds $\mathrm{li}_j \Phi^{k_j}(x^{k_j}, y^{k_j}) \geq \Phi(x, y)$ if $x \in \mathscr{D}^{-1}(y)$ and $\Phi^{k_j}(x^{k_j}, y^{k_j}) \to +\infty$ if $x \notin \mathscr{D}^{-1}(y)$.

We denote this convergence by $\Phi^k \overset{mai-lop}{\to} \Phi$ or $\Phi = \text{mai-lop-lim}_k \Phi^k$. The INSIDE MAI-LOP CONVERGENCE is defined and denoted naturally similarly, by omitting the infinity conditions.

With the sequences replacing the subsequences in the above sentences, we receive the notions of WEAK MAXINF-LOPSIDED CONVERGENCE and WEAK INSIDE MAXINF-LOPSIDED CONVERGENCE, resp.

The following two examples give cases when one has some types of lop-convergence and does not have the other ones. They also show explicitly differences between these types of lop-convergence and the corresponding ones of e/h-convergence and make clear comparisons between the rectangular and nonrectangular cases.

Example 8.28 (Mis-lop convergence holds; weak mai-lop and e/h-convergence fail) We revisit Example 8.9: $A^k = A = (0, e-1]$, $B^k = B = (0, 1]$, $\mathscr{D}^k(x) = \mathscr{D}(x) = [\ln(x+1), 1]$, and $\Phi^k(x, y) = \Phi(x, y) = \ln(y/x)$. As Φ^k forms a constant sequence, we apply the definition of the weak mis-lop convergence to check sequences instead of subsequences for convenience. We see in Example 8.9

that condition (a) of mis-lop convergence is fulfilled. For condition (b), with any $x \in A$, we take $x^k \equiv x$ to get that, for every $y^k \in \mathscr{D}^k(x^k)$ tending to $y \in \mathscr{D}(x)$, $\mathrm{ls}_k \Phi^k(x^k, y^k) = \Phi(x, y)$. Furthermore, for all $x \in A = (0, e-1]$, taking arbitrary $A^k \ni x^k \to x$, there is no $y^k \in \mathscr{D}^k(x^k) = \mathscr{D}(x)$ tending to a point outside $\mathscr{D}(x)$. Hence, condition (b) is fulfilled, and the mis-lop convergence is ensured. Observe that Example 8.9 already shows that the weak e/h-convergence does not hold. So, *completely unlike the rectangular case, here mis-lop convergence does not imply e/h-convergence.*

Consider now mai-lop convergence. We show that even condition (a) of weak mai-lop convergence is not satisfied. The "inside part" of that condition (a) is checked since for any $B^k \ni y^k \to y$ and $x \in \mathscr{D}^{-1}(y)$, we pick in $(\mathscr{D}^k)^{-1}(y^k)$ the point

$$x^k = \begin{cases} x + \gamma^k & \text{such that } \gamma^k \to 0^+ \text{ and } x + \gamma^k \le e^{y^k} - 1 \text{ if } 0 < y^k \le y, \\ x & \text{if } y < y^k \le 1 \end{cases}$$

to see that $\lim_k \Phi^k(x^k, y^k) = \Phi(x, y)$ (by the continuity). But, the "outside part" (i.e., the infinity condition) is violated since for any $B^k \ni y^k \to 0 \notin B$, taking $x^k = e^{y^k} - 1$ in $(\mathscr{D}^k)^{-1}(y^k)$ to see that $\ln \Phi^k(x^k, y^k) \ge \ln(y^k(e^{y^k} - 1)^{-1}) \to 0$, not $-\infty$ as required. Therefore, Φ^k does not weakly mai-lop converge to Φ. It is easy to see that condition (b) of i-mai-lop convergence is satisfied by the continuity and so one does have the inside mai-lop convergence in this case.

Example 8.29 (All e/h-, mis-lop-, and mai-lop-limits are equal to a singleton) Let us revisit Example 8.10: Keep the same $\Phi^k(x, y) = \Phi(x, y) = \ln(y/x)$ as in Example 8.28, and change their domains to make Φ^k mai-lop converge to Φ: $A^k = A = B^k = B = (0, 1]$ and $\mathscr{D}^k(x) = \mathscr{D}(x) = [x^2, 1]$. Similar to Example 8.28, the mis-lop convergence holds. For condition (a) of mai-lop convergence, we have only one different computation (ensuring (a) to hold): For $B^k \ni y^k \to 0 \notin B$, we take in $\mathscr{D}^{-1}(y^k)$ the point $x^k = (y^k)^{1/2}$ to see that $\Phi^k(x^k, y^k) \to -\infty$. To check condition (b), for any $y \in (0, 1]$, take $y^k \equiv y$. If $(0, y^{1/2}] \ni x^k \to x \neq 0$, then $\mathrm{li}_k \Phi^k(x^k, y^k) = \Phi(x, y)$, while if $x = 0$, then $\lim_k \Phi(y/x^k) = +\infty$, i.e., condition (b) is fulfilled. Therefore, one has the mai-lop convergence. Combining with Example 8.10, one can see that all the e/h-, mis-lop-, and mai-lop-limits are equal to $\ln(y/x)$ defined on $\mathrm{gph}\mathscr{D}$; see Fig. 8.4.

Later on, by "lop-convergence," we mean any of the types of lop-convergence when there is no need to specify which one is spoken about.

Proposition 8.30 (Characterizations of weak mis-lop convergence in terms of slices)

(i) *Condition* (a) *of weak mis-lop convergence* $\Leftrightarrow$ *for every* $A^k \ni x^k \to x \in A$, *it holds* w-hli$\Phi^k(x^k, \cdot) \ge \Phi(x, \cdot)$ *with* $\Phi^k(x^k, \cdot)$ *defined on* $\mathscr{D}^k(x^k)$ *and* $\Phi(x, \cdot)$ *on* $\mathscr{D}(x)$, *and for every* $A^k \ni x^k \to x \notin A$, *there exists* $y^k \in \mathscr{D}^k(x^k)$ *with* $\Phi^k(x^k, y^k) \to +\infty$ $\Leftrightarrow$ *for every* $A^k \ni x^k \to x \in A$, *one has condition (b) of*

Fig. 8.4 e/h-lim$\Phi^k = \{\Phi\} =$ mis-lop-lim$\Phi^k =$ mai-lop-limΦ^k

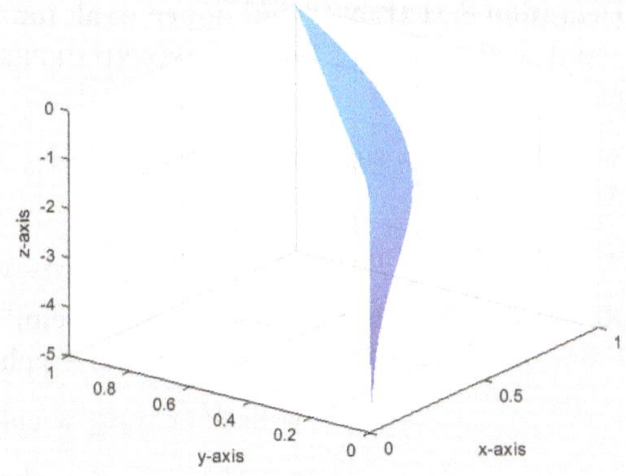

e/h-lim$\Phi^k = \{\Phi\} =$ mis-lop-lim$\Phi^k =$ mai-lop-limΦ^k

$\Phi^k(x^k, \cdot) \overset{w-h}{\to} \Phi(x, \cdot)$ with the domains as above, and for every $A^k \ni x^k \to x \notin A$, there exists $y^k \in \mathscr{D}^k(x^k)$ with $\Phi^k(x^k, y^k) \to +\infty$.

(ii) *Condition* (b) *of weak mis-lop convergence $\Leftrightarrow$ for all $x \in A$, there exists $A^k \ni x^k \to x$ such that for all $y \in \mathscr{D}(x)$, we have* w-hls$\Phi^k(x^k, y) \leq \Phi(x, y)$ *and for $y \notin \mathscr{D}(x)$,* $\max_{\{\mathscr{D}^k(x^k)\ni y^k \to y\}} \text{ls}_k \Phi^k(x^k, y^k)$ $= -\infty$ *$\Leftrightarrow$ for every $x \in A$, there exists $A^k \ni x^k \to x$ such that condition* (a) *of $\Phi^k(x^k, \cdot)$ on $\mathscr{D}^k(x^k)$ hypo-converging weakly to $\Phi(x, \cdot)$ on $\mathscr{D}(x)$ holds.*

(iii) *Hence, if $\Phi^k \overset{w-mis-lop}{\to} \Phi$, then for every $x \in A$, there exists $A^k \ni x^k \to x$ such that $\Phi^k(x^k, \cdot) \overset{w-h}{\to} \Phi(x, \cdot)$, and* $\max_{\{y^k \in \mathscr{D}^k(x^k)\}} \lim_k \Phi^k(x^k, y^k) = +\infty$ *if $x \notin A$.*

(iv) *If Φ^k does not depend on y (on x, resp.), then $\Phi^k \overset{w-mis-lop}{\to} \Phi$ implies that $\Phi^k \overset{w-e}{\to} \Phi$ ($\Phi^k \overset{w-h}{\to} \Phi$, resp.).*

Proof

(i) This is the same assertion (a) for weak e/h-convergence (cf. Proposition 8.11).
(ii) The first equivalence can be proved similarly to that in assertion (i) and the second one is clear by definition.
(iii) This is clear.

□

Similar characterizations in terms of slices of all the other types of lop-convergence hold as well. In what follows, because of similarity, in many situations, we present results only for the weak types of lop-convergence.

Next, to characterize the types of lop-convergence, we use the lower and upper e/h-limits defined in Definition 8.12 together with the following two other ones especially for lop-limits.

Definition 8.31 (Lower and upper weak lop-limits) Let Φ^k be in nrec$(X \times Y)$ and $A \times B \subset X \times Y$. The LOWER AND UPPER WEAK LOP-LIMITS are defined by, resp.,

w-lli$\Phi^k(x, y) := \sup_{\{B^k \ni y^k \to y\}} \inf_{\{(\mathscr{D}^k)^{-1}(y^k) \ni x^k \to x\}} \text{li}_k \Phi^k(x^k, y^k)$ for y $\in$ B and $x \in X$,

w-lls$\Phi^k(x, y) := \inf_{\{A^k \ni x^k \to x\}} \sup_{\{\mathscr{D}^k(x^k) \ni y^k \to y\}} \text{ls}_k \Phi^k(x^k, y^k)$ for $x \in A$ and $y \in Y$.

Remark 8.32 (Relationships between "semi" lop-limits and "semi" e/h-limits) From Definition 8.31, we have for $(x, y) \in \text{gph}\mathscr{D}$,

$$\text{w-lls}\Phi^k(x, y) \geq \text{w-ehls}\Phi^k(x, y),$$

$$\text{w-lli}\Phi^k(x, y) \leq \text{w-ehli}\Phi^k(x, y),$$

the same as the corresponding inequalities in Remark 7.64(a) for the rectangular case. However, the other inequalities corresponding to those in that remark fail to hold.

Theorem 8.33 (Formulas for the lower and upper weak lop-limits)

w-lli$\Phi^k(x, y) = \max_{\{B^k \ni y^k \to y\}} \inf_{\{(\mathscr{D}^k)^{-1}(y^k) \ni x^k \to x\}} \text{li}_k \Phi^k(x^k, y^k)$ for y $\in$ B and $x \in X$,

w-lls$\Phi^k(x, y) = \min_{\{A^k \ni x^k \to x\}} \sup_{\{\mathscr{D}^k(x^k) \ni y^k \to y\}} \text{ls}_k \Phi^k(x^k, y^k)$ for x $\in$ A and $y \in Y$.

The proof is almost the same as that of Theorem 7.65 (with clear modifications of the domains) and is omitted. But, the next result about characterizations contains a significant difference (concerning a relation with e/h-convergence) compared to the counterpart in Chap. 7 (Theorem 7.66).

Theorem 8.34 (Characterizations of weak lop-convergence)

(a) *Φ^k mis-lop converges weakly to Φ if and only if: for $x \in A$,* w-lls$\Phi^k(x, y) = \Phi(x, y) =$ w-ehls$\Phi^k(x, y)$ *if $y \in \mathscr{D}(x)$ and* w-lls$\Phi^k(x, y) = -\infty$ *if $y \notin \mathscr{D}(x)$; and for $x \notin A$,* w-ehls$\Phi^k(x) = +\infty$. *Hence, if exists the weak mis-lop limit is always unique, and if the weak e/h-convergence holds as well, then the weak mis-lop limit is just the right end of the interval consisting of the weak e/h-limits.*

(b) *Φ^k mai-lop converges weakly to Φ if and only if: for $y \in B$,* w-ehli$\Phi^k(x, y) = \Phi(x, y) =$ w-lli$\phi^k(x, y)$ *if $x \in \mathscr{D}^{-1}(y)$ and* w-lli$\Phi^k(x, y) = +\infty$ *if $x \notin \mathscr{D}^{-1}(y)$, and for $y \notin B$,* w-ehli$\Phi^k(y) = -\infty$. *So, if exists the weak mai-lop limit is always unique, and if the weak e/h-convergence also holds, then the weak mai-lop limit is just the left end of the interval of the weak e/h-limits.*

Proof We only consider (a). Apart from the last assertion about a relation with weak e/h-convergence, the arguments are in line with those for Theorem 7.66, with clear changes of the domains of bifunctions and with Remark 8.32 replacing Remark 7.64(a).

Now we justify the last assertion. When the weak mis-lop limit (uniquely) exists, the weak e/h-convergence may still fail. The main cause is that for bifunctions on nonrectangular domains, inequality (8.3) may happen, and so we do not have some inequalities given in Remark 7.64(a) for the rectangular case. Hence, the additional conditions on w-ehli$\Phi^k(x, y)$ and w-ehli$\Phi^k(y) = -\infty$ are needed for the weak e/h-convergence to hold, and then the weak mis-lop limit coincides with the right end of the interval of the weak e/h-limits. (Example 8.28 gives a case when mis-lop convergence holds, but we do not have weak e/h-convergence.) □

Remark 8.35 (Comparisons between e/h- and lop-convergence)

(a) As an example, discuss the weak inside convergence. As shown in Remark 8.32, w-lls$\Phi^k(x, y)$ and w-ehli$\Phi^k(x, y)$ are incomparable. Hence, w-i-mis-lop convergence does not imply w-i-e/h-convergence (different from the rectangular case). The similar facts also occur for the other pairs of the corresponding types of lop- and e/h-convergences.
(b) However, if both mis-lop and e/h-convergence hold, then the mis-lop limit is just the right end of the interval consisting of the e/h-limits. Correspondingly, if both mai-lop and e/h-convergence hold, then the mai-lop limit is just the left end of the interval of the e/h-limits. Figure 8.5 shows that in general lliΦ^k and llsΦ^k lie outside the e/h-limit interval. But, if all mis-lop, mai-lop, and e/h-limits exist, then lliΦ^k coincides with e/h-liΦ^k and llsΦ^k with e/h-liΦ^k.
(c) Even the existence of both mis-lop and mai-lop limits does not imply i-e/h-convergence. Conversely, when the e/h-limit is unique, it may still happen that no inside lop-limit exists. But, when the e/h-limit is unique and both mis-lop and mai-lop limits exist, they must coincide.

Later, we will see that due to these implications, lop-convergence may also ensure convergence properties for (approximate) saddle points and duality like e/h-convergence, but in fact, the assumptions are stronger.

When we do not have a given candidate for a type of limits, characterizations like Theorem 8.34 ensure the uniqueness and also give a tool to compute lop-limits. The above characterizations of lop-convergence are in terms of objects of variational convergence studied in Chaps. 7 and 8, and their formulations are complicated. Now we address one sufficient condition for mis-lop convergence in terms of classical concepts of continuity, which looks simple but the assumptions are restrictive.

Fig. 8.5 Two lop-limits are outside or just the ends of the e/h-limit interval

Theorem 8.36 (A sufficient condition for mis-lop convergence) *Assume that*

(a) $A^k = A$ *are closed for all* k*;*
(b) $\mathscr{D}^k$ *continuously converges to* $\mathscr{D}$ *relative to* A *in the sense that for every subsequence* $A \ni x^{k_j} \to x \in A$, $\mathscr{D}^{k_j}(x^{k_j}) \overset{P-K}{\to} \mathscr{D}(x)$*;*
(c) Φ^k *continuously converges to* Φ *relative to their domains in the sense that for every subsequence* $(x^{k_j}, y^{k_j}) \in \mathrm{gph}\mathscr{D}^{k_j}$ *converging to* $(x, y) \in \mathrm{gph}\mathscr{D}$*, one has* $\Phi^{k_j}(x^{k_j}, y^{k_j}) \to \Phi(x, y)$.

Then, $\Phi^k \overset{mis-lop}{\to} \Phi$.

Proof If $A \ni x^{k_j} \to x$, then $x \in A$ by the closedness of A. Hence, by the continuous convergence of $\mathscr{D}^k$, $\mathscr{D}^{k_j}(x^{k_j}) \overset{P-K}{\to} \mathscr{D}(x)$. Then, for any $y \in \mathscr{D}(x)$, we find $y^{k_j} \in \mathscr{D}^{k_j}(x^{k_j})$ tending to y. In turn, assumption (c) implies that $\lim_j \Phi(x^{k_j}, y^{k_j}) = \Phi(x, y)$ and condition (a) of mis-lop convergence is achieved. For its condition (b), given $x \in A$, we choose $x^{k_j} = x$ for all k_j to have $\mathscr{D}^{k_j}(x^{k_j}) \overset{P-K}{\to} \mathscr{D}(x)$. So, whenever $y^{k_j} \in \mathscr{D}^{k_j}(x)$ tends to y, y must be in $\mathscr{D}(x)$. Therefore, condition (b) is satisfied as $\Phi(x^{k_j}, y^{k_j}) \to \Phi(x, y)$ in view of assumption (c). □

At this point, we caution that certain "natural" conditions are not sufficient ones for lop-convergence, one of which is shown in the following.

Example 8.37 (Failure of weak mis-lop convergence under graphical convergence) Let $A^k = A = \mathbb{R}$, $\mathscr{D}^k(x) = \mathscr{D}(x) = [-1, 1]$ for $x \leq 0$ and $\{0\}$ otherwise, and $\Phi^k = \Phi$ for all k with some bifunction Φ. Then, clearly one has the graphical convergence of set-valued maps $\mathscr{D}^k$ to $\mathscr{D}$ and also of bifunctions Φ^k to Φ. Now, considering condition (a) of mis-lop convergence with $x^k > 0$ tending to $x = 0$ and $y = 1/2 \in \mathscr{D}(x)$, one does not have $\mathscr{D}^k(x^k) \ni y^k \to y$ and so weak mis-lop convergence fails.

Variational properties of the types of lop-convergence

First, observe that although lop-convergence and e/h-convergence in the nonrectangular domain case are incomparable, since condition (a) in the definition is the same for types of e/h- and mis-lop convergence, the properties which hold under only this condition (a) are in force for both e/h- and mis-lop types of convergence. For example, Theorem 8.18 about the implication of condition (a) of the types of epi-convergence of sup-projections from condition (a) of the corresponding types of e/h-convergence is still valid for lop-convergence.

Definition 8.38 (Tightness of the types of mis-lop convergence)

(i) (x-ancillary tight weak mis-lop convergence) $\{\Phi^k\}_k$ x-ANCILLARY TIGHTLY WEAKLY MIS-LOP CONVERGES to Φ if one has condition (a) of weak mis-lop convergence and condition (b) strengthened as

(b-m) condition (b) holds, $\mathrm{dom}\xi^k \neq \emptyset$, and for $A^k \ni x^k \to x \in A$ given in condition (b), $\mathrm{ls}_k \xi^k(x^k) \leq \xi(x)$.

(ii) (x-tight weak mis-lop convergence) $\{\Phi^k\}_k$ x-TIGHTLY WEAKLY MIS-LOP CONVERGES to Φ if it x-ancillary tightly weakly mis-lop converges and additionally with condition (a) replaced by

(a-m) condition (a) holds and for every $\varepsilon > 0$, one has a compact subset $A_\varepsilon \subset X$ and k_ε such that

$$\inf_{x\in A_\varepsilon\cap A^k}\sup_{y\in\mathscr{D}^k(x)}\Phi^k(x,y) \le \inf_{x\in A^k}\sup_{y\in\mathscr{D}^k(x)}\Phi^k(x,y)+\varepsilon\ \forall k\ge k_\varepsilon.$$

(iii) The definitions of y-ancillary tight and y-tight weak mai-lop convergence are similar/symmetric to the above ones. If both above x-tight and y-tight types of convergence hold, we have tight weak lop-convergence. The types of inside convergence are naturally similar.

Proposition 8.39 (A type of epi-convergence of sup-projections in terms of the corresponding type of x-ancillary tight mis-lop convergence) *Assume* $\text{dom}\xi^k \neq \emptyset$ *and a type of* Φ^k *mis-lop converging x-ancillary tightly to* Φ. *Then,* $\text{dom}\xi \neq \emptyset$ *and the corresponding type of epi-convergence of* ξ^k *to* ξ *holds.*

Proof The proof goes along the same line as that of Proposition 7.72 for the rectangular case. □

Theorem 8.40 (Convergence of minsup-points in terms of x-ancillary tight mis-lop convergence)

(a) *Assume that* $\Phi^k \overset{w-i-mis-lop}{\rightarrow} \Phi$ *x-ancillary tightly. Then,*

$$\text{ls}_k[\inf_{x\in A^k}\sup_{y\in\mathscr{D}^k(x)}\Phi^k(x,y)] \le \inf_{x\in A}\sup_{y\in\mathscr{D}(x)}\Phi(x,y).$$

(b) *If* $\Phi^k \overset{mis-lop}{\rightarrow} \Phi$ *x-ancillary tightly, then for any* $\varepsilon, \varepsilon^k \ge 0$ *with* $\varepsilon^k \to \varepsilon$, $\text{Ls}_k(\varepsilon^k\text{-mis}\Phi^k) \subset \varepsilon\text{-mis}\Phi$.

(c) *If* $\Phi^k \overset{w-mis-lop}{\rightarrow} \Phi$ *x-ancillary tightly,* $\varepsilon^k \to 0^+$, *and there exists* $x^k \in \varepsilon^k\text{-mis}\Phi^k$ *converging to x, then* $x \in \text{mis}\Phi$ *and* $\min_{x\in A^k}\sup_{y\in\mathscr{D}^k(x)}\Phi^k(x,y) \to \min_{x\in A}\sup_{y\in\mathscr{D}(x)}\Phi(x,y)$.

Proof Simply apply Propositions 8.39 and 7.3 (the argument is similar to that in the proof of Theorem 7.73). □

Example 8.41 (x-ancillary tightness in Proposition 8.39 and Theorem 8.40 is essential; these statements are not necessary conditions) Let $A = B = A^k = B^k = \mathbb{R}$ for all k, $\Phi(x,y) \equiv 0$, and $\Phi^k(x,y) = 1$ if $x \in \mathbb{R}$ and $y = k$, and zero otherwise. Then, it is easy to see that $\Phi^k \overset{mis-lop}{\rightarrow} \Phi$, $\xi^k(x) \equiv 1$, and $\xi(x) \equiv 0$. So, ξ^k does not epi-converge to ξ. The cause is that the x-ancillary tightness fails. This example also shows that the inclusion asserted in Theorem 8.18 for e/h-convergence, which is valid also for mis-lop convergence (as based only on condition (a) of the convergence), may be strict. Moreover, in this case, one even has $\text{Lim}_k\text{mis}\xi^k = \mathbb{R} = \text{mis}\xi$, which is stronger than what is asserted in Theorem 8.40, and so this theorem does not provide a necessary condition. The following case ensures that both Proposition 8.39 and Theorem 8.40 are sufficient but not necessary conditions. Consider the case when $A = B = A^k = B^k = [0, 1]$ for all k, $\Phi(x, y) \equiv 1$, and $\Phi^k(x, y) = y$ for all $x, y \in [0, 1]$. Then, $\xi^k(x) \equiv \xi(x) \equiv 1$ and so evidently ξ^k epi-converges to ξ. Moreover, $\mathrm{Lim}_k \mathrm{mis}\xi^k = [0, 1] = \mathrm{mis}\xi$ (stronger than the inclusion in Theorem 8.40). However, even condition (a) of both e/h- and mis-lop convergences is not satisfied.

Note that the discussions in this example only involve the rectangular case. So, the observations here remain valid for Proposition 7.72 and Theorem 7.73 for that case.

The following statement corresponds to Theorem 7.74 for the rectangular case.

Theorem 8.42 (Complete convergence of minsup-points under x-tight lop-convergence) *Assume that Φ^k mis-lop converges x-tightly to Φ. Then, the following assertions hold*

(a)

$$\inf_{x\in A^k} \sup_{y\in \mathscr{D}^k(x)} \Phi^k(x, y) \to \inf_{x\in A} \sup_{y\in \mathscr{D}(x)} \Phi(x, y) \text{ finitely;}$$

(b) *there exists* $\varepsilon^k \to 0^+$ *such that* $\varepsilon^k\text{-mis}\Phi^k \overset{P-K}{\to} \text{mis}\Phi$.

Proof

(a) By definition, the x-tightness of mis-lop convergence of Φ^k directly guarantees the tightness of the epi-convergence of their sup-projections ξ^k. Since $\varepsilon\text{-mis}\Phi = \varepsilon\text{-argmin}\xi$ for any bifunction Φ, assertion (a) follows from Proposition 7.9(a).
(b) This is a consequence of Proposition 7.9(c).

□

Theorem 8.43 (Complete convergence of approximate saddle points under tight lop-convergence) *Let* $\varepsilon, \varepsilon^k \geq 0$ *and* $\varepsilon^k \to \varepsilon$.

(a) *If Φ^k inside mis-lop converges both x- and y-ancillary-tightly or inside mai-lop converges both x- and y-ancillary-tightly to Φ, then*

$$\varepsilon\text{-wsdl}\Phi \subset \cap_{\delta>\varepsilon} \mathrm{Li}_k(\delta\text{-wsdl}\Phi^k)$$

and so $\varepsilon\text{-wsdl}\Phi \subset \mathrm{Li}_k(\varepsilon^k\text{-wsdl}\Phi^k)$.
(b) *If the restriction of inside convergence in assertion* (a) *is removed, then*

$$\mathrm{Lim}_k(\varepsilon^k\text{-wsdl}\Phi^k) = \varepsilon\text{-wsdl}\Phi,$$

$$\mathrm{Ls}_k(\varepsilon^k\text{-sdl}\Phi^k) \subset \varepsilon\text{-sdl}\Phi.$$

This theorem is a counterpart of the lop-convergence in Theorem 7.75 in the rectangular case and Theorem 8.24 for e/h-convergence in the general domain case. Hence, the proof is similar.

8.3 Quantification of Variational Convergence

8.3.1 Epi-Distance Between Unifunctions

We define for the class fv-fcn(X) the following subset consisting of lower semicontinuous functions on $A \subset X$

$$\text{lsc-fcn}(X) := \{\varphi : A \to \mathbb{R} \mid \varphi \text{ is lsc}\},$$

where X is a metric space and $A \subset X$ is nonempty. For a fixed point $\bar{x} \in A$, the EPI-DISTANCE between $\varphi, \psi \in \text{lsc-fcn}(X)$ is defined by

$$\mathsf{d}^e(\varphi, \psi) := \int_0^{+\infty} \mathsf{d}^e_\rho(\varphi, \psi) e^{-\rho} d\rho,$$

where $\mathbb{B} := \{x \in X \mid d(\bar{x}, x) \le 1\}$ and $\rho \ge 0$, the ρ-EPI-DISTANCE is given by

$$\mathsf{d}^e_\rho(\varphi, \psi) := \sup\{|\text{dist}((x, \alpha), \text{epi}\varphi)\text{-dist}((x, \alpha), \text{epi}\psi)| \mid d(\bar{x}, x) \le \rho, |\alpha| \le \rho\},$$

and

$$\text{dist}((x, \alpha), \text{epi}\varphi) := \inf\{\max\{d(x, u), |\alpha - \beta|\} \mid (u, \beta) \in \text{epi}\varphi\},$$

with a similar expression for epiψ. Observe that $\bar{x}$ serves as the center of $\mathbb{B}_\rho$ (called the CENTROID) on which d^e_ρ is computed. Hence, the value of the distance depends on the choice of $\bar{x}$, but the topology induced by the epi-distance is the same for any $\bar{x}$. We now apply the following basic and important properties of d^e without proof since they are theoretical issues in topology. If interested in detail, the reader is referred to [90], Sections 3.1, 5.1, and 5.2. d^e is a metric on lsc-fcn(X) and (lsc-fcn(X),d^e) is complete provided that so is (X, d). For $\varphi^k, \varphi \in$ lsc-fcn(X), $\mathsf{d}^e(\varphi^k, \varphi) \to 0$ implies that $\varphi^k \xrightarrow{e} \varphi$. If X is finitely compact, then the converse holds. Recall that a metric space is said to be FINITELY COMPACT (or PROPER) if all its closed balls are compact. In particular, (lsc-fcn($\mathbb{R}^n$, d^e) is a finitely compact complete separable metric space. This is one of the reasons for many applications of d^e in nonparametric statistics, curve fitting, stochastic processes, among others.

Correspondingly/symmetrically, for another subset of fv-fcn(Y) consisting of upper semicontinuous functions on nonempty subset B of a metric space Y

$$\text{usc-fcn}(Y) := \{\varphi : B \to \mathbb{R} \mid \varphi \text{ is usc}\},$$

the HYPO-DISTANCE between two hypographs $\mathsf{d}^h(\varphi, \psi)$ and the ρ-HYPO-DISTANCE $\mathsf{d}^h_\rho(\varphi, \psi)$ are defined by the above expressions with "epi" replaced by "hypo" everywhere.

8.3.2 Variational Distance Between Bifunctions

Parallel to the above metric spaces lsc-fcn(X), for bifunctions with lsc sup-projections

$$\text{lsc-nrec}(X, Y) := \{\Phi : \text{gph}\mathscr{D} \to \mathbb{R} \mid \text{its sup} - \text{projection } \xi \text{ is lsc}\},$$

we define the EBI-DISTANCE between two bifunctions Φ, Φ' as the epi-distance between their sup-projections ξ, ξ':

$$\mathsf{d}^{ebi}(\Phi, \Phi') := \mathsf{d}^{e}(\xi, \xi').$$

Bifunctions with usc inf-projection $\eta(y)$, for $y \in B$, in the class usc-nrec(X, Y) are signified similarly/symmetrically in a natural way and for them the HBI-DISTANCE d^{hbi} is defined as the hypo-distance between the inf-projections. The main reason for these definitions is that by Theorem 8.20 and Proposition 8.39 about epi-convergence of sup-projections under x-ancillary tight e/h- (mis-lop, resp.) convergence and also their counterparts for hypo-convergence of inf-projections under y-ancillary tight e/h- (mai-lop, resp.) convergence, the above distances d^{ebi} and d^{hbi} quantify the types of variational convergence of bifunctions. Notice that in the name "ebi-distance," "e" refers to"epi" (epi-distance) and "bi" to bifunctions (epi-distance of their sup-projections). The ebi-distance and the hbi-distance are pseudometrics, since there are nonidentical bifunctions with the same sup-projection or inf-projection. Hence, we have to deal with equivalence classes. Bifunctions $\Phi, \Phi' \in$ lsc-nrec(X, Y) are called EQUIVALENT, denoted by $\Phi \sim \Phi'$, if their sup-projections exist and are identical. A collection of all equivalent bifunctions is termed an EQUIVALENCE CLASS. The same for $\Phi, \Phi' \in$ usc-nrec(X, Y) with sup-projections replaced by inf-projections, and we use the same name "equivalence class" and notation " $\sim$" without fear of confusion. Observe that the domains of two equivalent bifunctions may not coincide, but the domains and the values of their sup-projections (or inf-projections, resp.) are equal (we see a clear example in Theorem 8.45 below, where bifunctions of interest in nrec$(X \times Y)$ are equivalent to chosen ones in rec$(X \times Y)$). Clearly, d^{ebi} and d^{hbi} are metrics on the quotient set of lsc-nrec(X, Y) and usc-nrec(X, Y), resp.

Lemma 8.44 (A case when e/h-, mis-lop, and epi-convergence are equivalent) *If Φ^k, Φ defined on $A^k \times B$ and $A \times B$, resp., do not depend on y, then*

$$\Phi^k \overset{e/h}{\to} \Phi \Leftrightarrow \Phi^k(\cdot, y) \overset{e}{\to} \Phi(\cdot, y)\ \forall y \in B \Leftrightarrow \Phi^k \overset{mis-lop}{\to} \Phi.$$

Proof We prove the first equivalence. For " $\Rightarrow$" condition (a) of e/h-convergence implies that for all $A^{k_j} \ni x^{k_j} \to x$ with $x \in A$, $\text{li}_j \Phi^{k_j}(x^{k_j}, y) \geq \Phi(x, y)$ for all $y \in B$, which is the corresponding part of condition (a) of the required epi-convergence. If $x \notin A$, then $\Phi^{k_j}(x^{k_j}, y) \to +\infty$ for all $y \in B$, which is the same condition for $\Phi^k(\cdot, y) \overset{e}{\to} \Phi(\cdot, y)$. The argument for condition (b) of the two types of convergence is similar. Also the implication " $\Leftarrow$" is verified analogously.

The second equivalence can be checked similarly. □

Theorem 8.45 (Quantification of variational convergence of bifunctions)

(a) *For* $\Phi^k, \Phi \in \text{lsc-nrec}(X \times Y)$, *if* $\mathsf{d}^{ebi}(\Phi^k, \Phi) \to 0$, *then there exist* $\hat{\Phi}^k, \hat{\Phi} \in \text{lsc-nrec}(X \times Y)$ *such that* $\hat{\Phi} \sim \Phi$, $\hat{\Phi}^k \sim \Phi^k$, $\hat{\Phi}^k \overset{e/h}{\to} \hat{\Phi}$, *and* $\hat{\Phi}^k \overset{mis-lop}{\to} \hat{\Phi}$.

If $\Phi^k \overset{e/h}{\to} \Phi$ *or* $\Phi^k \overset{mis-lop}{\to} \Phi$ *x-ancillary tightly and X is finitely compact, then* $\mathsf{d}^{ebi}(\Phi, \Phi^k) \to 0$.

(b) *For* $\Phi^k, \Phi \in \text{usc-nrec}(X \times Y)$, *if* $\mathsf{d}^{hbi}(\Phi^k, \Phi) \to 0$, *then there exist* $\tilde{\Phi}^k, \tilde{\Phi} \in \text{usc-nrec}(X \times Y)$ *such that* $\tilde{\Phi} \sim \Phi$, $\tilde{\Phi}^k \sim \Phi^k$, $\tilde{\Phi}^k \overset{e/h}{\to} \tilde{\Phi}$, *and* $\tilde{\Phi}^k \overset{mis-lop}{\to} \tilde{\Phi}$.

If $\Phi^k \overset{e/h}{\to} \Phi$ *or* $\Phi^k \overset{mai-lop}{\to} \Phi$ *y-ancillary tightly and X is finitely compact, then* $\mathsf{d}^{hbi}(\Phi, \Phi^k) \to 0$.

Proof

(a) First, we verify $\hat{\Phi}^k \overset{e/h}{\to} \hat{\Phi}$. Take $\hat{A} = \text{dom}\xi$, $\hat{\mathscr{D}}(x) = Y$ for all $x \in A$, and $\hat{\Phi} : \hat{A} \times Y \to \mathbb{R}$ defined by $\hat{\Phi}(x, y) := \xi(x)$ for all $(x, y) \in \hat{A} \times Y$. Then, for any $x \in \hat{A}$,

$$\text{sup}_{y \in Y} \hat{\Phi}(x, y) = \text{sup}_{y \in Y} \xi(x) = \xi(x) < +\infty.$$

So, the sup-projection of $\hat{\Phi}$ is $\hat{\varphi} = \varphi$ with the domain $\hat{A}$, and then $\hat{\Phi} \in$ lsc-nrec$(X \times Y)$ and $\hat{\Phi} \sim \Phi$. The corresponding construction for $\hat{\Phi}^k$ gives $\hat{\xi}^k$ with domain $\hat{A}^k$ and that $\hat{\Phi}^k \in \text{lsc-nrec}(X \times Y)$, $\hat{\Phi}^k \sim \Phi^k$. Therefore, for all $y \in Y$,

$$\mathsf{d}^e(\hat{\Phi}^k(\cdot, y), \hat{\Phi}(\cdot, y)) = \mathsf{d}^e(\xi^k, \xi) = \mathsf{d}^{ebi}(\Phi^k, \Phi) \to 0.$$

Since $\mathsf{d}^e(\hat{\Phi}^k(\cdot, y), \hat{\Phi}(\cdot, y)) \to 0$ implies that $\hat{\Phi}^k(\cdot, y) \overset{e}{\to} \hat{\Phi}(\cdot, y)$, by Lemma 8.44, $\hat{\Phi}^k \overset{e/h}{\to} \hat{\Phi}$.

For the converse, in view of Theorem 8.20, we have $\xi^k \overset{e}{\to} \xi$. As X is finitely compact, this leads to $\mathsf{d}^e(\xi^k, \xi) \to 0$, i.e., $\mathsf{d}^{ebi}(\Phi^k, \Phi) \to 0$.

Now, to complete the proof of assertion (a), by a manner similar to the above argument, we see that the same $\hat{\Phi}^k$ and $\hat{\Phi}$ as above satisfy $\hat{\Phi}^k \overset{mis-lop}{\to} \hat{\Phi}$.

(b) Since all the objects involved in this assertion are similar/symmetric to the corresponding ones in assertion (a), so is the proof. In other words, the proof of assertion (b) is somewhat a dualization of that of assertion (a).

□

Notice that the finite compactness assumed for X in the above theorem holds in many important applications. For instance, this is the case in the optimization problem under stochastic ambiguity investigated in the last subsection of this chapter.

8.4 Approximations of Quasiequilibrium Problems

In this section, we investigate approximations of quasiequilibrium problem (QEP) and strong quasiequilibrium problem (sQEP) defined in Sect. 8.1.1. Approximations of these problems are expressed in terms of (QEP^k) and (sQEP^k) defined by $A^k \subset X$, $B^k \subset Y$, $\mathscr{D}^k : A^k \rightrightarrows B^k$, and $\Phi^k : \text{gph}\mathscr{D}^k \to \mathbb{R}$ such that $\mathscr{D}^k$ are defined on the whole A^k and map onto B^k.

8.4.1 Qualitative Approximations of Quasiequilibrium Problems

A natural variant of the notion of strong inclusion related to set limits for product sets defined prior to Theorem 7.76 for graphs of set-valued maps is as follows. For $\mathscr{D}^k : A^k \rightrightarrows B^k$ and $\mathscr{D} : A \rightrightarrows B$, gph$\mathscr{D}$ is called STRONGLY INCLUDED in $\text{Ls}_k\text{gph}\mathscr{D}$, denoted by $\text{gph}\mathscr{D} \sqsubset \text{Li}_k\text{gph}\mathscr{D}^k$, if for any $(x, y) \in \text{gph}\mathscr{D}$ and a subsequence $\{k_j\}_j$, the presence of either type of the convergence: $A^{k_j} \ni x^{k_j} \to x$ and $\mathscr{D}^{k_j}(x^{k_j}) \ni y^{k_j} \to y$ implies the other.

Proposition 8.46 (Convergence of approximate solutions of quasiequilibrium problems) *Let* $\varepsilon, \varepsilon^k \geq 0$ *and* $\varepsilon^k \to 0^+$.

(i) *If* Φ^k *i-e/h- or i-mis-lop converges to* Φ, *then* $A \cap \text{Ls}_k(\varepsilon^k\text{-Sol(QEP}^k)) \subset \varepsilon\text{-Sol(QEP)}$. *If* Φ^k *e/h- or mis-lop converges to* Φ, *then* $\text{Ls}_k(\varepsilon^k\text{-Sol(QEP}^k)) \subset \varepsilon\text{-Sol(QEP)}$.

(ii) *If* $\Phi^k \overset{e}{\to} \Phi$ *(in* fv-fcn$(X \times Y)$*) and* $\text{gph}\mathscr{D} \sqsubset \text{Li}_k\text{gph}\mathscr{D}^k$, *then*

$$\text{Ls}_k(\varepsilon^k\text{-Sol(QEP}^k)) \subset \varepsilon\text{-Sol(QEP)}.$$

If $\Phi^k \overset{i-e}{\to} \Phi$ *replaces the epi-convergence, then*

$$A \cap \text{Ls}_k(\varepsilon^k\text{-Sol(QEP}^k)) \subset \varepsilon\text{-Sol(QEP)}.$$

(iii) *If* Φ^k *w-e/h-converges or w-mis-lop converges to* Φ *and* $x^k \in \varepsilon^k\text{-Sol(QEP}^k)$ *tends to* x, *then* $x \in \varepsilon\text{-Sol(QEP)}$.

Proof

(i) We only discuss the case $\Phi^k \overset{e/h}{\to} \Phi$. Let $\bar{x} \in \text{Ls}_k(\varepsilon^k\text{-Sol(QEP}^k))$, i.e., there exists a sequence $\{x^{k_j}\}_j$ in $\varepsilon^{k_j}\text{-Sol(QEP}^{k_j})$ converging to $\bar{x}$. If $\bar{x} \in A$, then for all $y \in \mathscr{D}(\bar{x})$, condition (a) of Definition 8.7 yields $\mathscr{D}^{k_j}(x^{k_j}) \ni y^{k_j} \to y$ such that $\text{li}_j \Phi^{k_j}(x^{k_j}, y^{k_j}) \geq \Phi(\bar{x}, y)$. As $x^{k_j} \in \varepsilon^{k_j}\text{-Sol(QEP)}^{k_j}$, $\Phi(\bar{x}, y) \leq \varepsilon$, i.e., $\bar{x}$ lies in ε-Sol(QEP). Suppose that $\bar{x} \notin A$. Again condition (a) gives $y^{k_j} \in \mathscr{D}^{k_j}(x^{k_j})$ such that $\Phi^{k_j}(x^{k_j}, y^{k_j}) \to +\infty$, contradicting the inequality

$\Phi^{k_j}(x^{k_j}, y) \leq \varepsilon^{k_j}$ for all $y \in \mathscr{D}^{k_j}(x^{k_j})$. From the above argument, the case Φ^k i-e/h-converges is clear.

The assertion for mis-lop convergence is automatic, since the above reasoning for e/h-convergence only utilizes the common condition (a) in these two types of convergence.

(ii) Let $\{x^{k_j}\}_j$ in ε^{k_j}-Sol(QEP^{k_j}) tend to $\bar{x}$. Suppose that $\bar{x} \notin A$. Then, for every $y^{k_j} \in \mathscr{D}^{k_j}(x^{k_j})$ tending to y, $(\bar{x}, y) \notin \text{gph}\mathscr{D}$. Then, condition (a) of Definition 7.1 ensures that $\Phi^{k_j}(x^{k_j}, y^{k_j}) \to +\infty$, which contradicts the fact that $\Phi^{k_j}(x^{k_j}, y^{k_j}) \leq \varepsilon^{k_j}$ for all $y^{k_j} \in \mathscr{D}^{k_j}(x^{k_j})$. So, $\bar{x} \in A$. Now, we have to show that $\Phi(\bar{x}, y) \leq \varepsilon$ for each $y \in \mathscr{D}(\bar{x})$. By the assumed strong inclusion, there exists $y^{k_j} \in \mathscr{D}^{k_j}(x^{k_j})$ tending to y. Then, condition (a) of Definition 7.1 implies that $\text{li}_j \Phi^{k_j}(x^{k_j}, y^{k_j}) \geq \Phi(\bar{x}, y)$. As $\Phi^{k_j}(x^{k_j}, y^{k_j}) \leq \varepsilon^{k_j}$, $\Phi(\bar{x}, y) \leq \varepsilon$.

From the above argument, the case $\Phi^k \overset{i-e}{\to} \Phi$ is clear.

(iii) Let $E^k \overset{w-mis-lop}{\to} E$ and $x^k \in \varepsilon^k$-Sol(QEP^k) tend to x. By condition (a) of the convergence of E^k, for every $y \in \mathscr{D}(x)$, there exists $y^k \in \mathscr{D}^k(x^k)$ tending to y such that $\text{li}_k E^k(x^k, y^k) \geq E(x, y)$ if $x \in A$. Since $E^k(x^k, y^k) \leq \varepsilon^k$, $E(x, y) \leq \varepsilon$ for all $y \in \mathscr{D}(x)$, i.e., $x \in \varepsilon$-Sol(QEP). Suppose that $x \notin A$. Again by condition (a), $E^k(x^k, y^k) \to +\infty$ for some $y^k \in \mathscr{D}^k(x^k)$ contradicting the fact that $E^k(x^k, y^k) \leq \varepsilon^k$ for all $y^k \in \mathscr{D}^k(x^k)$.

□

Proposition 8.47 (Convergence of approximate solutions of strong quasiequilibrium problems) *Let $\varepsilon, \varepsilon^k \geq 0$ and $\varepsilon^k \to \varepsilon$.*

(a) *If Φ^k e/h-converges x-ancillary tightly or mis-lop converges x-ancillary tightly to Φ, then* $\text{Ls}_k(\varepsilon^k\text{-Sol(sQEP}^k)) \subset \varepsilon\text{-Sol(sQEP)}$.

(b) *If Φ^k w-e/h-converges x-ancillary or w-mis-lop converges x-ancillary to Φ,* $\text{dom}\xi^k \neq \emptyset$, $\varepsilon = 0$, *and* $x^k \in \varepsilon^k$-Sol(sQEP^k) *tends to x, then* $x \in$ Sol(sQEP).

Proof

(a) The argument resembles that used in the proof of Theorem 7.77, but here we employ Theorems 8.22(b) and 8.40(b) (about minsup-points) instead of Theorems 7.52(b) and 7.73(b), and also Proposition 8.46(i) to replace Theorem 7.76(i).

(b) In view of Theorem 8.20(b) for the case of w-e/h-convergence and Proposition 8.39 for the case of w-mis-convergence, $\xi^k \overset{w-e}{\to} \xi$. On the other hand, as x^k is also an ε^k-solution of (QEP^k), $x \in$ Sol(QEP) by Proposition 8.46(iii). Furthermore, Theorems 8.22(c) and 8.40(c) (for $\varepsilon^k \equiv 0$) yield that $x \in \text{mis}\Phi$. Hence, $x \in$ Sol(sQEP).

□

Proposition 8.48 (Convergence of approximate solutions of dual quasiequilibrium problems) *Let* $\varepsilon, \varepsilon^k \geq 0$ *and* $\varepsilon^k \to \varepsilon$.

(a) *If* Φ^k *i-e/h- or i-mai-lop converges to* Φ, *then*

$$\mathrm{Ls}_k(\varepsilon^k\text{-Sol(DQEP}^k)) \cap B \subset \varepsilon\text{-Sol(DQEP)}.$$

If Φ^k *e/h- or mai-lop converges to* Φ, *then*

$$\mathrm{Ls}_k(\varepsilon^k\text{-Sol(DQEP}^k)) \subset \varepsilon\text{-Sol(DQEP)}.$$

(b) *Assume that* $\mathrm{gph}\mathscr{D} \sqsubset \mathrm{Li}_k\mathrm{gph}\mathscr{D}^k$. *If* $\Phi^k \overset{i-h}{\to} \Phi$, *then* $\mathrm{Ls}_k(\varepsilon^k\text{-Sol(DQEP}^k))\cap B \subset \varepsilon\text{-Sol(DQEP)}$. *If* $\Phi^k \overset{h}{\to} \Phi$, *then* $\mathrm{Ls}_k(\varepsilon^k\text{-Sol(DQEP}^k)) \subset \varepsilon\text{-Sol(DQEP)}$.
(c) *If* Φ^k *w-e/h-converges or w-mai-lop converges to* Φ *and* $y^k \in \varepsilon^k\text{-Sol(DQEP}^k)$ *tends to* x, *then* $y \in \varepsilon\text{-Sol(DQEP)}$.

The proof is similar to that of Proposition 8.46.

Proposition 8.49 (Convergence of approximate solutions of dual strong quasiequilibrium problems) *Let* $\varepsilon, \varepsilon^k \geq 0$ *and* $\varepsilon^k \to \varepsilon$.

(a) *If* Φ^k *e/h-converges y-ancillary tightly or mai-lop converges y-ancillary tightly to* Φ, *then* $\mathrm{Ls}_k(\varepsilon^k\text{-Sol(DsQEP}^k)) \subset \varepsilon\text{-Sol(DsQEP)}$.
(b) *If* Φ^k *w-e/h-converges y-ancillary or w-mai-lop converges y-ancillary to* Φ, $\mathrm{dom}\eta^k \neq \emptyset$, $\varepsilon = 0$, *and* $y^k \in \varepsilon^k\text{-Sol(DsQEP}^k)$ *tends to* y, *then* $y \in$ Sol(DsQEP).

The proof is similar to that of Proposition 8.47.

The following notation is a direct extension of its counterpart in Chap. 7:

$$\mathscr{C}\Phi^k := \{(x, y) \in \mathrm{gph}\mathscr{D}^k \mid \Phi^k(x, y) = 0\},\ \mathscr{C}\Phi := \{(x, y) \in \mathrm{gph}\mathscr{D} \mid \Phi(x, y) = 0\}.$$

Theorem 8.50 (Convergence of approximate primal and dual solutions of quasiequilibrium problems) *Let* $\varepsilon, \varepsilon^k \geq 0$ *and* $\varepsilon^k \to \varepsilon$.

(a) *If* Φ^k *i-e/h- or both i-mis-lop and i-mai lop converges to* Φ *and* $\mathrm{Ls}_k\mathscr{C}\Phi^k \subset \mathrm{cl}\mathscr{C}\Phi$, *then*

$$\begin{aligned}\mathrm{Ls}_k(\mathscr{C}\Phi^k \cap \varepsilon^k\text{-wsdl}\Phi^k) \cap \mathrm{gph}\mathscr{D} &= \mathrm{Ls}_k[\mathscr{C}\Phi^k \cap (\varepsilon^k\text{-Sol(QEP)}^k \\ &\quad \times \varepsilon^k\text{-Sol(DQEP}^k))] \cap \mathrm{gph}\mathscr{D} \\ &\subset \mathrm{cl}\mathscr{C}\Phi \cap (\varepsilon\text{-Sol(QEP)} \times \varepsilon\text{-Sol(DQEP)}) = \mathrm{cl}\mathscr{C}\Phi \cap \varepsilon\text{-wsdl}\Phi.\end{aligned}$$

If the restriction "inside" is removed, then " $\cap\, \mathrm{gph}\mathscr{D}$ *" in the above relation can be dropped.*

(b) *If* Φ^k *i-e/h-converges tightly or i-mis-lop both* x*- and* y*-ancillary-tightly or i-mai-lop both* x*- and* y*-ancillary-tightly converges to* Φ *and* $\mathrm{cl}\mathscr{C}\Phi \subset \mathrm{Li}_k\mathscr{C}\Phi^k$, *then*

$$\begin{aligned}\varepsilon\text{-wsdl}\Phi \cap \mathrm{cl}\mathscr{C}\Phi &= (\varepsilon\text{-Sol(QEP)} \times \varepsilon\text{-Sol(DQEP)}) \cap \mathrm{cl}\mathscr{C}\Phi\\ &\subset \cap_{\delta>\varepsilon}\mathrm{Li}_k(\delta\text{-Sol(QEP}^k) \times \delta\text{-Sol(DQEP}^k)) \cap \mathrm{Li}_k\mathscr{C}\Phi^k\\ &= \cap_{\delta>\varepsilon}\mathrm{Li}_k(\delta\text{-wsdl}\Phi^k) \cap \mathrm{Li}_k\mathscr{C}\Phi^k.\end{aligned}$$

Hence,

$$\begin{aligned}\varepsilon\text{-wsdl}\Phi \cap \mathrm{cl}\mathscr{C}\Phi &= (\varepsilon\text{-Sol(QEP)} \times \varepsilon\text{-Sol(DQEP)}) \cap \mathrm{cl}\mathscr{C}\Phi\\ &\subset \mathrm{Li}_k(\varepsilon^k\text{-Sol(QEP}^k) \times \varepsilon^k\text{-Sol(DQEP}^k)) \cap \mathrm{Li}_k\mathscr{C}\Phi^k\\ &= \mathrm{Li}_k(\varepsilon^k\text{-wsdl}\Phi^k) \cap \mathrm{Li}_k\mathscr{C}\Phi^k.\end{aligned}$$

(c) *Assume that in assertion* (b), *the restriction "inside" in the types of convergence of* Φ^k *is removed, and the involved types of lop-convergence are imposed together (i.e., the word "or" is replaced by "and"). Assume further that* $\mathrm{Lim}_k\mathscr{C}\Phi^k = \mathrm{cl}\mathscr{C}\Phi$. *Then,*

$$\begin{aligned}\varepsilon\text{-wsdl}\Phi \cap \mathrm{cl}\mathscr{C}\Phi &= (\varepsilon\text{-Sol(QEP)} \times \varepsilon\text{-Sol(DQEP)}) \cap \mathrm{cl}\mathscr{C}\Phi\\ &= \mathrm{Lim}_k(\varepsilon^k\text{-Sol(QEP}^k) \times \varepsilon^k\text{-Sol(DQEP}^k)) \cap \mathrm{Lim}_k\mathscr{C}\Phi^k\\ &= \mathrm{Lim}_k(\varepsilon^k\text{-wsdl}\Phi^k) \cap \mathrm{Lim}_k\mathscr{C}\Phi^k.\end{aligned}$$

Proof

(a) This assertion follows from Propositions 8.4, 8.6, 8.46, and 8.48 (with the observation that Limsup of the intersection of two sets is included in the intersection of the Limsups and that Limsup of the product of sets is equal to the product of the Limsups of these sets).

(b) We prove the first chain of relations. The two required equalities follow from Propositions 8.4 and 8.6. We verify now the required inclusion. Assume that $(\bar{x}, \bar{y}) \in \varepsilon\text{-wsdl}\Phi$. Consider first the case of i-e/h-convergence of Φ^k. This convergence means that

$$\begin{cases}\text{for every } A^{k_j} \ni x^{k_j} \to \bar{x}, \text{ there exists } \mathscr{D}^{k_j}(x^{k_j}) \ni y^{k_j} \to \bar{y}\\ \quad \text{such that } \mathrm{li}_j\Phi^{k_j}(x^{k_j}, y^{k_j}) \geq \Phi(\bar{x}, \bar{y}),\\ \text{for every } B^{k_j} \ni y^{k_j} \to \bar{y}, \text{ there exists } (D^{k_j})^{-1}(y^{k_j}) \ni x^{k_j} \to \bar{x}\\ \quad \text{such that } \mathrm{ls}_j\Phi^{k_j}(x^{k_j}, y^{k_j}) \leq \Phi(\bar{x}, \bar{y}).\end{cases} \tag{8.5}$$

The assumed full tightness implies further that

$$\begin{cases} \text{for } x^{k_j} \text{ in the 2nd relation in (8.5)}, \ \mathrm{ls}_j \xi^{k_j}(x^{k_j}) \le \xi(\bar{x}), \\ \text{for } y^{k_j} \text{ in the 1st relation in (8.5)}, \ \mathrm{li}_j \eta^{k_j}(y^{k_j}) \ge \eta(\bar{y}). \end{cases} \tag{8.6}$$

We show that for (x^{k_j}, y^{k_j}) in (8.6), for all $\delta > \varepsilon$, there exists k_δ such that for all $k_j \ge k_\delta$, (x^{k_j}, y^{k_j}) is a δ-weak saddle point of Φ^{k_j}. Suppose to the contrary that there exists $\delta_0 > \varepsilon$ and a subsequence of $\{k_j\}_j$, denoted also by $\{k_j\}_j$, such that for all j, at least one of the two inequalities in the definition of $(x^{k_j}, y^{k_j}) \in \delta_0\text{-wsdl}\Phi^{k_j}$ is violated. First suppose the right-hand one fails, i.e.,

$$\Phi^{k_j}(x^{k_j}, y^{k_j}) > \inf_{x \in (\mathcal{D}^{k_j})^{-1}(y^{k_j})} \Phi^{k_j}(x, y^{k_j}) + \delta_0 = \eta^{k_j}(y^{k_j}) + \delta_0.$$

Taking limsup of both sides, we obtain

$$\Phi(\bar{x}, \bar{y}) \ge \mathrm{ls}_j \Phi^{k_j}(x^{k_j}, y^{k_j}) \ge \mathrm{ls}_j \eta^{k_j}(y^{k_j}) + \delta_0 \ge \mathrm{li}_j \eta^{k_j}(y^{k_j}) + \delta_0 > \eta(\bar{y}) + \varepsilon,$$

where the first inequality follows from the second relation in (8.5) and the last one from that of (8.6). This contradicts the right-hand side inequality in the definition of the ε-weak saddle point $(\bar{x}, \bar{y})$.

For the case of violation of the left-hand-side inequality:

$$\Phi^{k_j}(x^{k_j}, y^{k_j}) < \sup_{y \in \mathcal{D}^{k_j}(x^{k_j})} \Phi^{k_j}(x^{k_j}, y) - \delta_0 = \xi^{k_j}(x^{k_j}) - \delta_0,$$

taking liminf of both sides and using the first relation in (8.5) and (8.6), we similarly arrive at a contradiction.

For the case with the assumptions on the types of lop-convergence, because of similarity, we only address the assumption that $\Phi^k \overset{i-mai-lop}{\to} \Phi$ both x- and y-ancillary tightly. The i-mai-lop convergence means that

$$\begin{cases} \forall B^{k_j} \ni y^{k_j} \to \bar{y}, \forall x \in \mathcal{D}^{-1}(\bar{y}), \exists (\mathcal{D}^{k_j})^{-1}(y^{k_j}) \ni x^{k_j} \to x, \\ \quad \mathrm{ls}_j \Phi^{k_j}(x^{k_j}, y^{k_j}) \le \Phi(x, \bar{y}), \\ \exists B^{k_j} \ni y^{k_j} \to \bar{y}, \forall x \in \mathcal{D}^{-1}(\bar{y}), \forall (\mathcal{D}^{k_j})^{-1}(y^{k_j}) \ni x^{k_j} \to x, \\ \quad \mathrm{li}_j \Phi^{k_j}(x^{k_j}, y^{k_j}) \ge \Phi(x, \bar{y}). \end{cases} \tag{8.7}$$

The assumed tightness yields

$$\begin{cases} \text{for } x^{k_j} \text{ in the 1st relation in (8.7)}, \ \mathrm{ls}_j \xi^{k_j}(x^{k_j}) \le \xi(\bar{x}), \\ \text{for } y^{k_j} \text{ in the 2nd relation in (8.7)}, \ \mathrm{li}_j \eta^{k_j}(y^{k_j}) \ge \eta(\bar{y}). \end{cases} \tag{8.8}$$

For (x^{k_j}, y^{k_j}) in (8.8), we show the same fact as for the above (x^{k_j}, y^{k_j}) in (8.6). Suppose the right-hand side inequality of $(x^{k_j}, y^{k_j}) \in \delta_0\text{-wsdl}\Phi^{k_j}$ does not hold. Taking limsup of this hypothesis yields the same contradiction as the case of (x^{k_j}, y^{k_j}) in (8.6). If instead, the left-hand side inequality fails, taking liminf we also obtain a contradiction as with (8.6).

To get the second chain of relations, we need to verify only that ε-$wsdl\Phi \subset \mathrm{Li}_k(\varepsilon^k$-$wsdl\Phi^k)$. Assume that $(\bar{x}, \bar{y}) \in \varepsilon$-wsdl$\Phi$. We find a sequence $(x^k, y^k) \in \varepsilon^k$-wsdl$\Phi^k$ converging to $(\bar{x}, \bar{y})$ as follows. For any fixed $\ell > \varepsilon$, by the above inclusion in the first chain, one has $(x^k_\ell, y^k_\ell) \in \varepsilon^\ell$-wsdl$\Phi^k$. Pass $\varepsilon^\ell \to \varepsilon$ and use the diagonalization argument to obtain $(x^k_{\ell(k)}, y^k_{\ell(k)}) \in \varepsilon^{\ell(k)}$-wsdl$\Phi^k$ converging to $(\bar{x}, \bar{y})$. Then, $(x^k, y^k) := (x^k_{\ell(k)}, y^k_{\ell(k)})$ is a desired sequence.

(c) It is sufficient to show that the inclusion in the chain of relations is in fact an equality. By the second chain of relations in assertion (b), we only need to check $\mathrm{Ls}_k(\varepsilon^k$-wsdl$\Phi^k) \subset \varepsilon$-wsdl$\Phi$. By assertion (a), we have to show only that no subsequence $(\bar{x}^{k_j}, \bar{y}^{k_j}) \in \varepsilon^{k_j}$-wsdl$\Phi^{k_j}$ can tend to a point outside gph$\mathscr{D}$. Suppose to the contrary the existence of $(\bar{x}^{k_j}, \bar{y}^{k_j}) \to (x, y) \notin \mathrm{gph}\mathscr{D}$, say $x \notin A$. By assertion (a) of e/h-convergence, there is $y^{k_j} \in \mathscr{D}^{k_j}(\bar{x}^{k_j})$ with $\Phi^{k_j}(\bar{x}^{k_j}, y^{k_j}) \to +\infty$. Then, for any $x^{k_j} \in (\mathscr{D}^{k_j})^{-1}(\bar{y}^{k_j})$, $\Phi^{k_j}(x^{k_j}, \bar{y}^{k_j}) + \varepsilon^{k_j} \geq \Phi^{k_j}(\bar{x}^{k_j}, y^{k_j}) - \varepsilon^{k_j} \to +\infty$. We have two cases: $y \in B$ or $y \notin B$. If $y \in B$, then condition (b) of e/h-convergence implies that for any $\hat{x} \in A$, there exists $\hat{x}^{k_j} \in (\mathscr{D}^{k_j})^{-1}(\bar{y}^{k_j})$ tending to $\hat{x}$ such that $\mathrm{ls}_j\Phi^{k_j}(\hat{x}^{k_j}, \bar{y}^{k_j}) \leq \Phi(\hat{x}, y)$, a contradiction. So only $y \notin B$ is possible. Then, again by the above condition (b), there exists $\hat{x}^{k_j} \in (\mathscr{D}^{k_j})^{-1}(\bar{y}^{k_j})$ with $\Phi^{k_j}(\hat{x}^{k_j}, \bar{y}^{k_j}) \to -\infty$ which is impossible either.

□

Similarly, we have the following counterpart of Theorem 7.85 for primal and dual strong quasiequilibrium problems.

Theorem 8.51 (Convergence of approximate primal and dual solutions of strong quasiequilibrium problems and approximate saddle points) *Let $\varepsilon, \varepsilon^k \geq 0$ and $\varepsilon^k \to \varepsilon$. If Φ^k i-e/h-converges tightly or both i-mis-lop converges x and y-ancillary-tightly and i-mai-lop converges x- and y-ancillary-tightly to Φ and $\mathrm{Ls}_k\mathscr{C}\Phi^k \subset \mathrm{cl}\mathscr{C}\Phi$, then*

$$\mathrm{Ls}_k(\mathscr{C}\Phi^k \cap \varepsilon^k\text{-sdl}\Phi^k) \cap \mathrm{gph}\mathscr{D} = \mathrm{Ls}_k[\mathscr{C}\Phi^k \cap (\varepsilon^k\text{-Sol(sEP)}^k$$

$$\times\, \varepsilon^k\text{-Sol(DsEP}^k))] \cap \mathrm{gph}\mathscr{D}$$

$$\subset \mathrm{cl}\mathscr{C}\Phi \cap (\varepsilon\text{-Sol(sEP)} \times \varepsilon\text{-Sol(DsEP)}) = \mathrm{cl}\mathscr{C}\Phi \cap (\varepsilon\text{-sdl}\Phi).$$

If the above restriction "inside" is removed, then "$\cap\,\mathrm{gph}\mathscr{D}$" can also be deleted.

Remark 8.52 (Comparisons between the nonrectangular and rectangular cases about convergence of approximate solutions of the types of equilibrium problems and the types of saddle points) Regarding the differences between the role of e/h- and lop-convergence, from the above chain of results on convergence of variational objects, we see that the same comments as in Remark 7.86 for the rectangular case remain true for the case of general bifunctions. We bring to the fore again that e/h-convergence is more appropriate for saddle points and duality results, while lop-convergence is better for studying minsup- and maxinf-points. Here, we have one new observation that weak saddle points are related to pairs of dual quasiequilibrium problems while saddle points to pairs of strong quasiequilibrium problems. (Recall that for the rectangular case a weak saddle point and a saddle point coincide.)

We now deal with the second kind of approximate solutions by the following definition, which is a direct extension to the nonrectangular case of Definition 7.78.

Definition 8.53 (The second kind of approximate solutions of (QEP)) Assume that $X = Y$ is a normed space, $A = B$, and $\theta : X \to \mathbb{R}$ satisfies $\theta(x) > 0$ for $x \neq 0$. For $\varepsilon \geq 0$, $\bar{x}_{\varepsilon\theta} \in A$ is called an $\varepsilon\theta$-SOLUTION of (QEP) if $\Phi(\bar{x}_{\varepsilon\theta}, y) \leq \varepsilon\theta(\bar{x}_{\varepsilon\theta} - y)$ for all $y \in \mathscr{D}(\bar{x}_{\varepsilon\theta})$. Denote the set of the $\varepsilon\theta$-solutions of (QEP) by $\varepsilon\theta$-Sol(QEP).

Proposition 8.54 (Convergence of second-kind approximate solutions) *Let* $\varepsilon, \varepsilon^k \geq 0$ *and* $\varepsilon^k \to \varepsilon$.

(i) *If* Φ^k *i-e/h- or i-mis-lop converges to* Φ *and* θ *is continuous, then*

$$\mathrm{Ls}_k(\varepsilon^k\theta\text{-Sol}(\mathrm{QEP}^k)) \cap A \subset \varepsilon\theta\text{-Sol}(\mathrm{QEP}).$$

If Φ^k *e/h- or mis-lop converges to* Φ, θ *is bounded from above, and* $\varepsilon = 0$, *then*

$$\mathrm{Ls}_k(\varepsilon^k\theta\text{-Sol}(\mathrm{QEP}^k)) \subset \mathrm{Sol}(\mathrm{QEP}).$$

(ii) *If* $\mathrm{gph}\mathscr{D} \sqsubset \mathrm{Li}_k\mathrm{gph}\mathscr{D}^k$ *and the i-e- or e-convergence of* Φ^k *to* Φ *replaces the types of convergence of* Φ^k *in assertion* (i), *then we have the same convergence of second-kind approximate solutions as in assertion* (i).

Proof

(i) Assume that $\bar{x} \in \mathrm{Ls}_k(\varepsilon^k\theta\text{-Sol}(\mathrm{QEP}^k))$, i.e., there exists $x^{k_j} \in \varepsilon^{k_j}\theta\text{-Sol}(\mathrm{QEP}^{k_j})$ tending to $\bar{x}$. For the first assertion, we only argue for i-e/h-convergence, the weaker assumption. If $\bar{x} \in A$, condition (a) of the assumed inside convergence implies that for all $y \in \mathscr{D}(\bar{x})$, we have $\mathscr{D}(x^{k_j}) \ni y^{k_j} \to y$ such that $\mathrm{li}_j\Phi^{k_j}(x^{k_j}, y^{k_j}) \geq \Phi(\bar{x}, y)$. Since $x^{k_j} \in \varepsilon^{k_j}\theta\text{-Sol}(\mathrm{QEP}^{k_j})$ and θ is continuous, $\Phi(\bar{x}, y) \leq \varepsilon\theta(\bar{x} - y)$ as required.

For the second assertion, using the assumed boundedness of θ and $\varepsilon = 0$, the argument for the case $\bar{x} \in A$ is similar to the one used above. Suppose that $\bar{x} \notin A$. Then, condition (a) of the types of convergence of Φ^k gives $y^{k_j} \in \mathscr{D}(x^{k_j})$ such that $\Phi^{k_j}(x^{k_j}, y^{k_j}) \to +\infty$, which is impossible since $x^{k_j} \in \varepsilon^{k_j}\theta$-Sol($\mathrm{QEP}^{k_j}$) and θ is bounded from above.

(ii) Let $x^{k_j} \in \varepsilon^{k_j}\theta\text{-Sol}(\mathrm{QEP}^{k_j})$ tend to $\bar{x}$. For the first assertion, we have to check that $\Phi(\bar{x}, y) \leq \varepsilon\theta(\bar{x} - y)$ for all $y \in \mathscr{D}(\bar{x})$. As $\mathrm{gph}\mathscr{D} \sqsubset \mathrm{Li}_k\mathrm{gph}\mathscr{D}^k$, there exists $\mathscr{D}(x^{k_j}) \ni y^{k_j} \to y$. Then, condition (a) of the inside epi-convergence ensures that $\mathrm{li}_j\Phi^{k_j}(x^{k_j}, y^{k_j}) \geq \Phi(\bar{x}, y)$. Similar to the argument in assertion (i), this leads to $\bar{x} \in \varepsilon\theta$- Sol(QEP).

For the second assertion, it suffices to eliminate the case $\bar{x} \notin A$. For such an $\bar{x}$, as for any $\mathscr{D}(x^{k_j}) \ni y^{k_j} \to y$, $(\bar{x}, y) \notin \text{gph}\mathscr{D}$ no matter $y \in A$ or not, by condition (a) of the assumed epi-convergence, $\Phi^{k_j}(x^{k_j}, y^{k_j}) \to +\infty$. Similar to the above consideration, this is ruled out.

□

One has an analogous fact for second-kind approximate dual solutions in the following statement. (The definition of the second kind of approximate solutions of (DQEP) is naturally similar to Definitions 7.87 and 8.53.) The proof is analogous to that of the preceding theorem.

Proposition 8.55 (Convergence of second-kind approximate dual solutions) *Let $\varepsilon, \varepsilon^k \geq 0$ and $\varepsilon^k \to \varepsilon$.*

(a) *If Φ^k i-e/h- or i-mai-lop converges to Φ and θ is continuous, then*

$$\text{Ls}_k(\varepsilon^k\theta\text{-Sol}(\text{DQEP}^k)) \cap B \subset \varepsilon\theta\text{-Sol}(\text{DQEP}).$$

If Φ^k e/h- or mai-lop converges to Φ, θ is bounded from below, and $\varepsilon = 0$, then

$$\text{Ls}_k(\varepsilon^k\theta\text{-Sol}(\text{DQEP}^k)) \subset \text{Sol}(\text{DQEP}).$$

(b) *One has the same sentence as in assertion* (b) *of the preceding theorem with hypo-convergence replacing epi-convergence.*

It is easy to see that we have a definition of the second kind of approximate weak saddle points and a relationship between such points and pairs of the second-kind primal and dual approximate solutions completely corresponding to Definition 7.89 and Proposition 7.90 for the rectangular case and Proposition 8.4 for the first kind of approximations in the nonrectangular case, resp. From these facts and Propositions 8.54 and 8.55, we obtain the following.

Theorem 8.56 (Convergence of second-kind approximate weak saddle points and approximate primal and dual solutions) *Let $\varepsilon, \varepsilon^k \geq 0$ and $\varepsilon^k \to \varepsilon$.*

(a) *If Φ^k i-e/h (or i-mis-lop and i-mai-lop simultaneously) converges to Φ, $\text{Ls}_k\mathscr{C}\Phi^k \subset \text{cl}\mathscr{C}\Phi$, and θ is continuous, then*

$$\begin{aligned} &\text{Ls}_k(\mathscr{C}\Phi^k \cap \varepsilon^k\theta\text{-wsdl}\Phi^k) \cap \text{gph}\mathscr{D} \\ &= \text{Ls}_k[\mathscr{C}\Phi^k \cap (\varepsilon^k\theta\text{-Sol}(\text{QEP}^k) \times \varepsilon^k\theta\text{-Sol}(\text{DQEP}^k))] \cap \text{gph}\mathscr{D} \\ &\subset \text{cl}\mathscr{C}\Phi \cap (\varepsilon\theta\text{-Sol}(\text{QEP}) \times \varepsilon\theta\text{-Sol}(\text{DQEP})) = \text{cl}\mathscr{C}\Phi \cap \varepsilon\theta\text{-wsdl}\Phi. \end{aligned}$$

If Φ^k e/h- or both mis-lop and mai-lop converges simultaneously to Φ, $\mathrm{Ls}_k\mathscr{C}\Phi^k \subset \mathrm{cl}\mathscr{C}\Phi$, θ is bounded, and $\varepsilon = 0$, then

$$\begin{aligned}\mathrm{Ls}_k(\mathscr{C}\Phi^k \cap \varepsilon^k\theta\text{-wsdl}\Phi^k) &= \mathrm{Ls}_k[\mathscr{C}\Phi^k \cap (\varepsilon^k\theta\text{-Sol}(\mathrm{QEP}^k) \\ &\qquad \times \varepsilon^k\theta\text{-Sol}(\mathrm{DQEP}^k))] \\ &\subset \mathrm{cl}\mathscr{C}\Phi \cap (\mathrm{Sol}(\mathrm{QEP}) \times \mathrm{Sol}(\mathrm{DQEP})) = \mathrm{cl}\mathscr{C}\Phi \cap \mathrm{wsdl}\Phi.\end{aligned}$$

(b) *If the types of convergence of bifunctions Φ^k in* (a) *are replaced by the corresponding epi- and hypo-convergences of unifunctions Φ^k, resp., and additionally* $\mathrm{gph}\mathscr{D} \sqsubset \mathrm{Li}_k\mathrm{gph}\mathscr{D}^k$ *is satisfied, then one has the same two conclusions in* (a).

8.4.2 Quantitative Approximations of Quasiequilibrium Problems

In this subsection, we will combine the results in Sect. 8.3 about quantification of variational convergence with convergence results for solutions of (sQEP) in the preceding subsection to get estimate properties for solutions of this problem. We limit the scope to the case $E^k, E \in \text{lsc-nrec}(\mathbb{R}^n \times \mathbb{R}^m) \cap \text{usc-nrec}(\mathbb{R}^n \times \mathbb{R}^m)$. Note that for finite dimensional cases, the origin 0 is always taken for the centroid $\bar{x}$. For $\varphi, \psi \in \text{lsc-fcn}\mathbb{R}^n$, we need the auxiliary quantity

$$\hat{\mathsf{d}}^e_\rho(\varphi, \psi) := \inf\{\eta \geq 0 \mid \mathrm{epi}\varphi \cap \rho\mathbb{S} \subset \mathrm{epi}\psi + \eta\mathbb{S}, \mathrm{epi}\psi \cap \rho\mathbb{S} \subset \mathrm{epi}\varphi + \eta\mathbb{S}\},$$

where $\mathbb{S} := \{(x, \gamma) \in \mathbb{R}^n \times \mathbb{R} \mid \max\{\|x\|_{\mathbb{R}^n}, |\gamma|\} \leq 1\}$ (the unit ball). This auxiliary quantity is closely related to d^e_ρ and usually more accessible. Another auxiliary quantity $\hat{\mathsf{d}}^h_\rho(\varphi, \psi)$ is defined similarly/symmetrically for $\varphi, \psi \in \text{usc-fcn}\mathbb{R}^m$, with the epigraphs replaced by the hypographs, and is not given explicitly here.

Proposition 8.57 (Key estimates of variational distances between bifunctions)

(a) *For $\Phi^k, \Phi \in \text{lsc-nrec}(\mathbb{R}^n \times \mathbb{R}^m)$ with their sup-projections ξ^k, ξ and $\rho \geq 0$,*

$$(1 - e^{-\rho})|d^k - d| + e^{-\rho}\hat{\mathsf{d}}^e_\rho(\xi^k, \xi) \leq \mathsf{d}^{ebi}(\Phi^k, \Phi)$$

$$\leq (1 - e^{-\rho})\hat{\mathsf{d}}^e_{\bar{\rho}}(\xi^k, \xi) + e^{-\rho}(\max\{d^k, d\} + \rho + 1),$$

where $d^k := \mathrm{dist}(0, \mathrm{epi}\xi^k)$, $d := \mathrm{dist}(0, \mathrm{epi}\xi)$, and $\bar{\rho} \geq 2\rho + \max\{d^k, d\}$.

(b) *For $\Phi^k, \Phi \in \text{usc-nrec}(\mathbb{R}^n \times \mathbb{R}^m)$ with their inf-projections η^k, η, we have the corresponding inequalities with d^{ebi} is replaced by d^{hbi}, the epigraphs by the hypographs, ξ^k, ξ by η^k, η, and $\hat{\mathsf{d}}^e_{\bar{\rho}}$ by $\hat{\mathsf{d}}^h_{\bar{\rho}}$.*

The proof of this proposition is technically involved and beyond the scope of our book. We note for the interested reader that Proposition 8.57(a) is Proposition 5.1 in [407] and its assertion (b) is simply the symmetric/dual reformulation of (a).

Proposition 8.58 (Estimates of saddle values and saddle points) *Assume that* $\rho \geq 0$ *and* $E^k, E \in$ lsc-nrec$(\mathbb{R}^n \times \mathbb{R}^m) \cap$usc-nrec$(\mathbb{R}^n \times \mathbb{R}^m)$ *have saddle points* sdlE^k, sdl$E$ *and saddle values* sdl-valE^k, sdl-val$E$, *resp., satisfying* ($\mathbb{B}_{\mathbb{R}^n}$ *is the unit ball in* $\mathbb{R}^n$)

$$\text{sdl} - \text{val}E \in [-\rho, \rho], \quad \text{sdl}E \cap \rho(\mathbb{B}_{\mathbb{R}^n} \times \mathbb{B}_{\mathbb{R}^m}) \neq \emptyset,$$

with a similar condition for E^k. *Then,*

$$|\text{sdl} - \text{val}E^k - \text{sdl} - \text{val}E| \leq \min\{\hat{\mathsf{d}}^e_\rho(\xi^k, \xi), \hat{\mathsf{d}}^h_\rho(\eta^k, \eta)\},$$

where ξ^k, ξ *are the sup-projections of* E^k, E, *resp., and* η^k, η *are their inf-projections.*

If, additionally, there exists an increasing and continuous function $\psi : \mathbb{R}_+ \to \mathbb{R}_+$, *with* $\psi(0) = 0$, *such that*

$$\begin{aligned}&\min\{\xi(x), \eta(y)\} - \text{sdl} - \text{val}E \geq \psi(\text{dist}((x, y), \text{sdl} - \text{val}E))\\ &\forall (x, y) \in \rho'(\mathbb{B}_{\mathbb{R}^n} \times \mathbb{B}_{\mathbb{R}^m}),\end{aligned}$$

where $\rho' > \rho + \max\{\hat{\mathsf{d}}^e_\rho(\xi^k, \xi), \hat{\mathsf{d}}^h_\rho(\eta^k, \eta)\}$, *then*

$$\begin{aligned}&\text{sdl}E^k \cap \rho(\mathbb{B}_{\mathbb{R}^n} \times \mathbb{B}_{\mathbb{R}^m})\\ &\subset \text{sdl}E + \min\{\hat{\mathsf{d}}^e_\rho(\xi^k, \xi) + \psi^{-1}(2\hat{\mathsf{d}}^e_\rho(\xi^k, \xi)), \hat{\mathsf{d}}^h_\rho(\eta^k, \eta) + \psi^{-1}\\ &(2\hat{\mathsf{d}}^h_\rho(\eta^k, \eta))\}(\mathbb{B}_{\mathbb{R}^n} \times \mathbb{B}_{\mathbb{R}^m}).\end{aligned}$$

Proposition 8.58 can be proved by employing Proposition 8.57, similar to the proof of Corollary 5.4 (for minsup-values and minsup-points) in [407]. We omit here this technically complicated proof. Following Theorem 8.51, if E^k tightly e/h-converges or both x-tightly mis-lop and y-tightly mai-lop converges to E, then sdlE^k, Sol(sQEP^k), and Sol(DsQEP^k) converge. Moreover, in view of Theorem 8.45, under these types of convergence of E^k, they also converge with respect to both d^{ebi} and d^{hbi}. Therefore, combined with these two theorems, Propositions 8.57 and 8.58 provide estimates for pairs of solutions of (sQEP^k)-(DsQEP^k) to those of (sQEP)-(DsQEP) and at the same time estimates for saddle points. This is an important quantitative duality result for these strong quasiequilibrium problems.

8.5 Approximations of Nash Quasigames

Applying quasiequilibrium models allows us to consider the following extension of the Nash game (noncooperative game) investigated in Sect. 7.6. It is usually called a NASH QUASIGAME or GENERALIZED NONCOOPERATIVE GAME or NASH QUASIQUILIBRIUM PROBREM. In a Nash quasigame $Q\mathscr{G}$ of m players/agents, each player $i \in I := \{1, \ldots, m\}$ chooses a strategy x_i in the set of available strategies $\mathscr{D}_i(x_{-i})$, which depends on the strategies chosen by the other players $x_{-i} := (x_1, \ldots, x_{i-1}, x_{i+1}, \ldots, x_m)$. The extension $\mathscr{D}_i(x_{-i})$ from a fixed set A_i in Sect. 7.6 is just the main factor leading to the need of quasiequilibrium setting with bifunctions on nonrectangular domains. To avoid the boredom and monotony in presentation and to show diversity in the application, we discuss the payoff/objective function $c_i(x_i, x_{-i})$ as the return of agent i, not a cost as in Sect. 7.6. We think of $\mathscr{D}_i(x_{-i})$ as a subset of a metric space X_i. Then, $\mathscr{D}_i : X_{-i} \rightrightarrows X_i$. We set $\mathscr{D} := \prod_{i\in I} \mathscr{D}_i : \prod_{i\in I} X_i \rightrightarrows \prod_{i\in I} X_i =: X$. However, $\mathscr{D}$ can be restricted to the following set, which is the domain of $\mathscr{D}$,

$$A := \{x \in X \mid x_i \in \mathscr{D}_i(x_{-i}),\ \forall i \in I\},$$

since its outside points are irrelevant, i.e., $\mathscr{D} : A \rightrightarrows X$. Obviously, for any $x \in A$, $x \in \mathscr{D}(x)$ and so $(x, x) \in \operatorname{gph}\mathscr{D}$ (this does not hold for general (QEP)). Hence, $\mathscr{D}$ is defined on the whole A and maps onto $\mathscr{D}(A) \supset A$. So, $\bar{x} \in A$ is a NASH QUASIEQUILIBRIUM if for all $i \in I$,

$$\bar{x} \in \operatorname{argmax}_{x_i \in \mathscr{D}_i(\bar{x}_{-i})} c_i(x_i, \bar{x}_{-i})$$

and the problem of finding Nash quasiequilibria is just $Q\mathscr{G}$. Leveraging the Nikaido-Isoda bifunction of $Q\mathscr{G}$: $N : \operatorname{gph}\mathscr{D} \to \mathbb{R}$ defined by

$$N(x, y) := \sum_{i=1}^{m} [c_i(x_i, x_{-i}) - c_i(y_i, x_{-i})],$$

we convert $Q\mathscr{G}$ to the special case (QEP$_N$)

$$\text{find } \bar{x} \in A \text{ such that } N(\bar{x}, y) \geq 0 \text{ for all } y \in \mathscr{D}(\bar{x})$$

of (QEP). Verifying the equivalence between $Q\mathscr{G}$ and (QEP$_N$) is included in the proof of Proposition 8.59 below about properties of Nash quasiequilibria. Following the consideration of (QEP) in Sect. 8.1, the dual problem of $Q\mathscr{G}$ is

$$(\text{DQEP}_N) \quad \text{find } \bar{y} \in \mathscr{D}(A) \text{ such that } N(x, \bar{y}) \leq 0 \text{ for all } x \in \mathscr{D}^{-1}(\bar{y}).$$

Problem (DQEP$_N$) is naturally called the DUAL NASH QUASIGAME or DUAL NASH QUASIEQUILIBRIUM PROBLEM and also DUAL NONCOOPERATIVE GAME, denoted by $DQ\mathscr{G}$ with the DUAL NASH QUASIEQUILIBRIUM SET $\mathrm{NE}(DQ\mathscr{G})$. However, its formulation in the data $\mathscr{D}$ and c is complicated, and so we conveniently use N, i.e., the formulation of (DQEP_N).

We also have the STRONG NASH QUASIGAME or STRONG NASH QUASIEQUILIBRIUM PROBLEM (sQEP_N) and its dual problem, resp., as follows

$$(\mathrm{sQEP}_N) \quad \text{find } \bar{x} \in A \text{ solving } \max_{x\in A}\inf_{y\in\mathscr{D}(x)}N(x,y) \geq 0,$$

$$(\mathrm{DsQEP}_N) \quad \text{find } \bar{y} \in \mathscr{D}(A) \text{ solving } \min_{y\in\mathscr{D}(A)}\sup_{x\in\mathscr{D}^{-1}(y)}N(x,y) \leq 0.$$

By Proposition 8.59, these two problems are also denoted by $(sQ\mathscr{G})$ and $(DsQ\mathscr{G})$, resp. The case $Q\mathscr{G}$ of (QEP) is so special that the above three problems are only a little different from $Q\mathscr{G}$ as follows.

Proposition 8.59 (Properties of Nash quasiequilibria)

(a) *For* $x \in A$, $N(x,x) = 0$, $\inf_{\mathscr{D}(x)}N(x,\cdot) \leq 0$, *and* $\sup_{\mathscr{D}^{-1}(x)}N(\cdot,x) \geq 0$.
(b) *The following assertions are equivalent*
  (b1) $\bar{x} \in \mathrm{Sol}(\mathrm{DQEP}_N)$ *(then* $\sup_{\mathscr{D}^{-1}(\bar{x})}N(\cdot,\bar{x}) = 0$*), i.e.,* $\bar{x} \in \mathrm{Ne}(DQ\mathscr{G})$*;*
  (b2) $\bar{x} \in \mathrm{Sol}(\mathrm{DsQEP}_N)$ *(then* $\bar{x} \in \mathrm{mis}N$ *and the minsup-value of N is zero).*
(c) *The following assertions are equivalent*
  (c1) $\bar{x} \in \mathrm{Sol}(\mathrm{QEP}_N)$ *(then* $\inf_{\mathscr{D}(\bar{x})}N(\bar{x},\cdot) = 0$*);*
  (c2) $\bar{x} \in \mathrm{Sol}(\mathrm{sQEP}_N)$ *(then* $\bar{x} \in \mathrm{mai}N$ *and the maxinf-value of N equals zero);*
  (c3) $\bar{x} \in \mathrm{Ne}(Q\mathscr{G})$.
(d) *Any assertion in* (b) *implies all assertions in* (c) *and also that* $(\bar{x},\bar{x}) \in \mathrm{sdl}N$ *with the saddle value being zero.*

Proof The proof is similar to that of Proposition 7.110 with the order: (a), (c1) $\Leftrightarrow$ (c3), (b1) $\Rightarrow$ (c3), (b1) $\Leftrightarrow$ (b2), (c1) $\Leftrightarrow$ (c2), and (d). Here, we only present several fragments as examples.

(b1) $\Rightarrow$ (c3). $\bar{x} \in \mathrm{Sol}(\mathrm{DQEP}_N)$ signifies that $\sup_{\mathscr{D}^{-1}(\bar{x})}N(\cdot,\bar{x}) \leq 0$, i.e., for all $x \in \mathscr{D}^{-1}(\bar{x})$,

$$\Sigma_{i\in I}[c_i(x_i,x_{-i}) - c_i(\bar{x}_i,x_{-i})] \leq 0.$$

Hence, for any given $k \in I$ and $x_k \in \mathscr{D}_k(\bar{x}_{-k})$, substituting $x = (x_k,\bar{x}_{-k})$ into this inequality gives $c_k(x_k,\bar{x}_{-k}) - c_k(\bar{x}_k,\bar{x}_{-k}) + \Sigma_{j\in I\setminus\{k\}}[c_j(\bar{x}_j,(\bar{x}_{-j,k},x_k)) - c_j(\bar{x}_j,(\bar{x}_{-j,k},x_k))] \leq 0$ ($\bar{x}_{-j,k}$ is the collection of the $m-2$ components of $x = (x_k,\bar{x}_{-k})$, except $\bar{x}_j$ and x_k) or equivalently, $c_i(\bar{x}_i,\bar{x}_{-i}) \geq c_i(x_i,\bar{x}_{-i})$ for all i and $x_i \in \mathscr{D}_i(\bar{x}_{-i})$, i.e., $\bar{x} \in \mathrm{Ne}(Q\mathscr{G})$.

(b1) $\Leftrightarrow$ (b2). $\Leftarrow$ is obvious. To see the converse, having already (b1) $\Rightarrow$ (c1), as $(\bar{x},\bar{x}) \in \mathrm{gph}\mathscr{D}$, by Proposition 8.4, $\bar{x} \in \mathrm{Sol}(\mathrm{sQEP}_N)$ and equivalently $\bar{x} \in \mathrm{Sol}(\mathrm{DsQEP}_N)$. This together with (a) implies that $\mathrm{minsup}N = 0$. □

In view of Proposition 8.59, we see that under the assumptions of this statement, in the formulation of the four problems above, in fact the involved inequalities are equalities and so the duality gap is always zero. Moreover, any solution of $(D Q\mathscr{G})$ solves the other three problems $Q\mathscr{G}$, $(s Q\mathscr{G})$, and $(Ds Q\mathscr{G})$.

Now we discuss approximate solutions and their convergence as usual in accordance with the goal of Chaps. 7 and 8.

Definition 8.60 (Approximate Nash quasiequilibria) Let $\varepsilon \geq 0$.

(a) A point $\bar{x} \in A$ is a NASH ε-QUASIEQUILIBRIUM when $\inf_{x \in \mathscr{D}(\bar{x})} N(\bar{x}, x) \geq -\varepsilon$, and a DUAL NASH ε-QUASIEQUILIBRIUM when $\sup_{x \in \mathscr{D}^{-1}(\bar{x})} N(x, \bar{x}) \leq \varepsilon$. The sets of such points are denoted by ε-Ne$(Q\mathscr{G})$ and ε-Ne$(DQ\mathscr{G})$, resp.
(b) A point $\bar{x} \in A$ is a STRONG NASH ε-QUASIEQUILIBRIUM when $\bar{x} \in \mathrm{mai}N$ with $\mathrm{maxinf}N \geq -\varepsilon$, and a STRONG DUAL NASH ε-QUASIEQUILIBRIUM when $\bar{x} \in \mathrm{mis}N$ with $\mathrm{minsupf}N \leq \varepsilon$. We denote these facts by $\bar{x} \in \varepsilon$-Ne$(sQ\mathscr{G})$ and $\bar{x} \in \varepsilon$-Ne$(DsQ\mathscr{G})$, resp.
(c) Let X be a normed space, $\theta : X \to \mathbb{R}$ be such that $\theta(x) > 0$ if $x \neq 0$, and $\varepsilon \geq 0$. A point $\bar{x}_\theta \in A$ is a NASH $\varepsilon\theta$-QUASIEQUILIBRIUM when $N(\bar{x}_\theta, y) \geq -\varepsilon\theta(\bar{x}_\theta - y)$ for each $y \in \mathscr{D}(\bar{x}_\theta)$. We denote this fact by $\bar{x}_\theta \in \varepsilon\theta$-Ne$(Q\mathscr{G})$.
(d) Let X, $\theta : X \to \mathbb{R}$, and $\varepsilon \geq 0$ be as in condition (c). $\bar{y}_\theta \in A$ is a DUAL NASH $\varepsilon\theta$-QUASIEQUILIBRIUM when $N(x, \bar{y}_\theta) \leq \varepsilon\theta(x - \bar{y}_\theta)$ for every $x \in \mathscr{D}^{-1}(\bar{y}_\theta)$. We denote this fact by $\bar{y}_\theta \in \varepsilon\theta$-Ne$(DQ\mathscr{G})$.

Proposition 8.61 (Properties of approximate Nash quasiequilibria)

(a) ε-Ne$(sQ\mathscr{G}) \subset \varepsilon$-Ne$(Q\mathscr{G})$ *and* ε-Ne$(DsQ\mathscr{G}) \subset \varepsilon$-Ne$(DQ\mathscr{G})$.
(b) $\bar{x} \in \varepsilon$-Ne$(Q\mathscr{G})$ *and* $\bar{y} \in \varepsilon$-Ne$(DQ\mathscr{G})$ *if and only if* $(\bar{x}, \bar{y}) \in \varepsilon$-wsdl$N$ *and* $N(\bar{x}, \bar{y}) = 0$.
(c) $\bar{x} \in \varepsilon$-Ne$(sQ\mathscr{G})$ *and* $\bar{y} \in \varepsilon$-Ne$(DsQ\mathscr{G})$ *if and only if* $(\bar{x}, \bar{y}) \in \varepsilon$-sdl$N$ *and* $N(\bar{x}, \bar{y}) = 0$.
(d) $\bar{x}_\theta \in \varepsilon\theta$-Ne$(Q\mathscr{G})$ *and* $\bar{y}_\theta \in \varepsilon\theta$-Ne$(DQ\mathscr{G})$ *if and only if* $(\bar{x}_\theta, \bar{y}_\theta) \in \varepsilon\theta$-wsdl$N$ *and* $N(\bar{x}_\theta, \bar{y}_\theta) = 0$.

Proof The proof goes along the same line as that of its counterpart for the Nash equilibria case, with reference to Proposition 8.4 and the note prior to Theorem 8.56. □

Note that apart from assertion (a) of the above statement, for approximate quasiequilibria, one does not have relationships similar to those in Proposition 8.55 for exact quasiequilibria. Now assume that we have a sequence of approximating Nash quasigames $Q\mathscr{G}^k$ with the data $\mathscr{D}^k$ and c^k.

The following statement is a specialization for the approximate Nash quasiequilibria of Propositions 8.46, 8.47, and 8.54 for (general) quasiequilibria. It is important to note that, in this Sect. 8.5, the types of e/h- and lop-convergence and the corresponding types of tightness are different (in a symmetric way) from the definitions in Sect. 8.2 since here the problem is maximizing the payoff function.

Theorem 8.62 (Convergence of approximate Nash quasiequilibria) *Let* $\varepsilon, \varepsilon^k \geq 0$ *and* $\varepsilon^k \to \varepsilon$.

(a) *If* N^k *i-e/h- or i-mai-lop converges to* N *or i-hypo-converges to* N *and* $\mathrm{gph}\mathscr{D} \sqsubset \mathrm{Li}_k(\mathrm{gph}\mathscr{D}^k)$, *then* $\mathrm{Ls}_k(\varepsilon^k\text{-Ne}(Q\mathscr{G}^k)) \cap A \subset \varepsilon\text{-Ne}(Q\mathscr{G})$.

The e/h- or mai-lop or hypo-convergence of N^k *with the strong inclusion* $\mathrm{gph}\mathscr{D} \sqsubset \mathrm{Li}_k(\mathrm{gph}\mathscr{D}^k)$ *guarantees that* $\mathrm{Ls}_k(\varepsilon^k\text{-Ne}(Q\mathscr{G}^k)) \subset \varepsilon\text{-Ne}(Q\mathscr{G})$.

(b) *If* N^k *i-e/h-converges* x*-ancillary tightly or i-mai-lop converges* x*-ancillary tightly to* N, *then* $\mathrm{Ls}_k(\varepsilon^k\text{-Ne}(sQ\mathscr{G}^k)) \cap A \subset \varepsilon\text{-Ne}(sQ\mathscr{G})$.

If the term"inside" is removed, then $\mathrm{Ls}_k(\varepsilon^k\text{-Ne}(sQ\mathscr{G}^k)) \subset \varepsilon\text{-Ne}(sQ\mathscr{G})$.

(c) *If* N^k *w-e/h-converges* x*-ancillary tightly or w-mai-lop converges* x*-ancillary tightly to* N, $\varepsilon = 0$, *and* $x^k \in \varepsilon^k\text{-Ne}(sQ\mathscr{G}^k)$ *tends to* x, *then* $x \in \mathrm{Ne}(sQ\mathscr{G})$.

(d) *If the assumptions in assertion* (a) *are satisfied and one has additionally, resp., the continuity or boundedness of* θ *from below with* $\varepsilon = 0$, *then resp.,*

$$\mathrm{Ls}_k(\varepsilon^k\theta\text{-Ne}(Q\mathscr{G}^k)) \cap A \subset \varepsilon\theta\text{-Ne}(Q\mathscr{G}),$$

$$\mathrm{Ls}_k(\varepsilon^k\theta\text{-Ne}(Q\mathscr{G}^k)) \subset \mathrm{Ne}(Q\mathscr{G}).$$

Remark 8.63 (Convergence of approximate dual Nash quasiequilibria) We have a statement for dual Nash quasiequilibrium problems similar to Theorem 8.62 with easily understood modifications: y-ancillary tight i-mai-lop convergence is replaced by x-ancillary tight i-mis-lop one in assertion (b), the similar change in tightness of the weak types of convergence in assertion (c), and boundedness of θ from below by the boundedness from above in assertion (d).

Employing Theorems 8.50 and 8.56 for quasiequilibrium problems to Nash quasiequilibirum problems, we obtain the following.

Theorem 8.64 (Convergence of primal and dual approximate Nash quasiequilibria and approximate weak saddle points) *Let* $\varepsilon, \varepsilon^k \geq 0$ *and* $\varepsilon^k \to \varepsilon$.

(a) *If* N^k *i-e/h- or i-mis-lop and i-mai-lop converges to* N *and* $\mathrm{Ls}_k\mathscr{C}N^k \subset \mathrm{cl}\mathscr{C}N$, *then*

$$\begin{aligned}\mathrm{Ls}_k(\mathscr{C}N^k \cap \varepsilon^k\text{-wsdl}N^k) \cap \mathrm{gph}\mathscr{D} &= \mathrm{Ls}_k[\mathscr{C}N^k \cap (\varepsilon^k\text{-Ne}(Q\mathscr{G}^k) \\ &\quad \times \varepsilon^k\text{-Ne}(DQ\mathscr{G}^k))] \cap \mathrm{gph}\mathscr{D} \\ &\subset \mathrm{cl}\mathscr{C}N \cap (\varepsilon\text{-Ne}(Q\mathscr{G}) \times \varepsilon\text{-Ne}(DQ\mathscr{G}) \cap \mathrm{gph}\mathscr{D} = \mathrm{cl}\mathscr{C}N \cap \varepsilon\text{-wsdl}\Phi.\end{aligned}$$

If N^k *e/h- or simultaneously mis-lop and mai-lop converges simultaneously to* N *and* $\mathrm{Ls}_k\mathscr{C}N^k \subset \mathrm{cl}\mathscr{C}N$, *then we have the above relation with"* $\cap\, \mathrm{gph}\mathscr{D}$*" removed.*

(b) *If* $N^k \stackrel{e/h}{\to} N$ *tightly or both* x*-ancillary-tightly mis-lop and* y*-ancillary-tightly mai-lop converges simultaneously and* $\mathrm{cl}\mathscr{C}N = \mathrm{Lim}_k\mathscr{C}N^k$*, then*

$$\begin{aligned}\varepsilon\text{-wsdl}N \cap \mathrm{cl}\mathscr{C}N &= (\varepsilon\text{-Ne}(Q\mathscr{G}) \times \varepsilon\text{-Ne}(DQ\mathscr{G})) \cap \mathrm{cl}\mathscr{C}N \\ &= \mathrm{Lim}_k(\varepsilon^k\text{-Ne}(Q\mathscr{G}^k) \times \varepsilon^k\text{-Ne}(DQ\mathscr{G}^k)) \cap \mathrm{Lim}_k\mathscr{C}N^k \\ &= \mathrm{Lim}_k\varepsilon^k\text{-wsdl}N^k \cap \mathrm{Lim}_k\mathscr{C}N^k.\end{aligned}$$

(c) *If* N^k *e/h-converges tightly or both mis-lop converges* x*-ancillary-tightly and mai-lop converges* y*-ancillary-tightly simultaneously to* N *and* $\mathrm{Ls}_k\mathscr{C}N^k \subset \mathrm{cl}\mathscr{C}N$*, then*

$$\begin{aligned}\mathrm{Ls}_k(\mathscr{C}N^k \cap \varepsilon^k\text{-sdl}N^k) &= \mathrm{Ls}_k[\mathscr{C}N^k \cap (\varepsilon^k\text{-Ne}(sQ\mathscr{G}^k) \\ &\quad \times\varepsilon^k\text{-Ne}(DsQ\mathscr{G}^k))] \\ &\subset \mathrm{cl}\mathscr{C}N \cap (\varepsilon\text{-Ne}(sQ\mathscr{G}) \times \varepsilon\text{-Ne}(DsQ\mathscr{G})) = \mathrm{cl}\mathscr{C}N \cap \varepsilon\text{-sdl}N.\end{aligned}$$

(d) *If the assumptions in assertion* (a) *are satisfied and one has additionally, resp., the continuity or boundedness of* θ *with* $\varepsilon = 0$*, then we have for the second-kind approximate solutions, resp.,*

$$\begin{aligned}\mathrm{Ls}_k(\mathscr{C}N^k \cap \varepsilon^k\theta\text{-wsdl}N^k) \cap \mathrm{gph}\mathscr{D} &= \mathrm{Ls}_k\mathscr{C}N^k \cap (\varepsilon^k\theta\text{-Ne}(Q\mathscr{G}^k) \\ &\quad \times\varepsilon^k\theta\text{-Ne}(DQ\mathscr{G}^k)) \cap \mathrm{Ls}_k\mathscr{C}N^k \\ &\subset \mathrm{cl}\mathscr{C}N \cap (\varepsilon\theta\text{-Ne}(Q\mathscr{G}) \times \varepsilon\theta\text{-Ne}(DQ\mathscr{G})) = \mathrm{cl}\mathscr{C}N \cap \varepsilon\theta\text{-wsdl}N, \\ \mathrm{Ls}_k(\mathscr{C}N^k \cap \varepsilon^k\theta\text{-wsdl}N^k) &= \mathrm{Ls}_k[\mathscr{C}N^k \cap (\varepsilon^k\theta\text{-Sol}(Q\mathscr{G}^k)) \times \varepsilon^k\theta\text{-Ne}(DQ\mathscr{G}^k)] \\ &\subset \mathrm{cl}\mathscr{C}N \cap (\mathrm{Ne}(Q\mathscr{G}) \times \mathrm{Ne}(DQ\mathscr{G})) = \mathrm{cl}\mathscr{C}N \cap \mathrm{wsdl}N.\end{aligned}$$

In the above study of generalized noncooperative games, the results about convergence of approximate Nash quasiequilibria are in terms of the Nikaido-Isoda bifunctions. The aim of the subsequent consideration is such a convergence directly in terms of the data $c_i, \mathscr{D}_i, i \in I$ of the game. (Such an investigation was not presented for the special case of Nash games in Chap. 7.) For this goal, a natural first step is the following result.

Theorem 8.65 (mai-lop convergence of Nikaido-Isoda bifunctions in terms of the data of $Q\mathscr{G}^k$ **and** $Q\mathscr{G}$**)** *Assume that for all* $i \in I$,

(a) $\mathscr{D}_i^k$ *continuously converges to* $\mathscr{D}_i$ *in the sense that for every* $x^{k_j} \in X \to x \in X$, $\mathscr{D}_i^{k_j}(x_{-i}^{k_j}) \stackrel{P-K}{\to} \mathscr{D}_i(x_{-i})$;
(b) c_i^k *continuously converges to* c_i *in the sense that for every* $x^{k_j} \in X \to x \in X$, $c_i^{k_j}(x_i^{k_j}, x_{-i}^{k_j}) \to c_i(x_i, x_{-i})$;

(c) *either* $\mathscr{D}_i^k(x_{-i}^k) \supset \mathscr{D}_i(x_{-i})$ *for all* $x_{-i} \in X_{-i}$ *and* $k \in \mathbb{N}$ *or the maps* $\mathscr{D}_i^k$ *and* $\mathscr{D}_i$ *are constant for all* $k \in \mathbb{N}$ *and* $i \in I$.

Then, N^k *simultaneously mai-lop, mis-lop, and e/h-converges to* N.

Proof We verify Definition 8.27 of mai-lop convergence. For $A^{k_j} \ni y^{k_j} \to y \in \mathscr{D}(A)$ and $x \in \mathscr{D}^{-1}(y)$, $(\mathscr{D}^{k_j})^{-1}(y^{k_j}) \overset{P-K}{\to} \mathscr{D}^{-1}(y)$ by assumption (a). Hence, there exists $(\mathscr{D}^{k_j})^{-1}(y^{k_j})$ tending to x. The continuous convergence of c_i^k carries over to N^k, which leads to $N^{k_j}(x^{k_j}, y^{k_j}) \to N(x, y)$ and the first part of Definition 8.27(a) is checked. The formula for A^k and assumption (a) imply that

$$\mathrm{Ls}_k A^k \subset \cap_{i \in I} \mathrm{Ls}_k \{y \in X \mid y_i \in \mathscr{D}_i^k(y_{-i})\} \subset \cap_{i \in I} \{y \in X \mid y_i \in \mathscr{D}_i(y_{-i})\} = \mathscr{D}(A).$$

So, each convergent sequence $y^{k_j} \in A^{k_j}$ must have a limit in $\mathscr{D}(A)$, and the second part of Definition 8.27(a) is automatically verified. Next, to consider Definition 8.27(b) for the first case in assumption (c), let $y \in \mathscr{D}(A)$ be arbitrary. Take $y^{k_j} = y \in \mathscr{D}(A)$, which must be in $\mathscr{D}^{k_j}(A^{k_j})$. Assume that $x^{k_j} \in (\mathscr{D}^{k_j})^{-1}(y^{k_j})$ tends to $x \in \mathscr{D}^{-1}(y)$. Assumption (b) again guarantees that $N^{k_j}(x^{k_j}, y^{k_j}) \to N(x, y)$. Since the possibility that $x^{k_j} \in (\mathscr{D}^{k_j})^{-1}(y^{k_j})$ tends to $x \notin \mathscr{D}^{-1}(y)$ is ruled out by assumption (a), Definition 8.27(b) is verified. Now assume that $\mathscr{D}^k$ and $\mathscr{D}$ are constant for all $k \in \mathbb{N}$. It is clear that $A^k \overset{P-K}{\to} A$ and for all $y \in \mathscr{D}(A)$, one has a subsequence $\mathscr{D}^{k_j}(A^{k_j}) \ni y^{k_j} \to y$. Suppose that $(\mathscr{D}^{k_j})^{-1}(y^{k_j}) \ni x^{k_j} \to x \in \mathscr{D}^{-1}(y)$. Again assumption (b) is utilized to see that $N^{k_j}(x^{k_j}, y^{k_j}) \to N(x, y)$. This concludes the proof, since we can rule out the case $x^{k_j} \in (\mathscr{D}^{k_j})^{-1}(y^{k_j})$ tending to $x \notin \mathscr{D}^{-1}(y)$ as argued above.

It is not hard to see that the statements of assumptions (a) and (c) for $\mathscr{D}^k$ and $(\mathscr{D}^k)^{-1}$ are equivalent (remember that $\mathscr{D}^k$ maps the whole A^k onto $\mathscr{D}^k(A^k)$). Hence, the verification of Definition 8.26 of mis-lop convergence is an easy modification of the above part. A glance at Remark 8.35 indicates the e/h-convergence of N^k since as argued above the infinity conditions in the definitions of the types of convergence in question are irrelevant. □

Examining the above proof, we recognize that Theorem 8.65 is a very useful first step in considering convergence of approximate solutions of the approximating games $Q\mathscr{G}^k$ and $DQ\mathscr{G}^k$ to (approximate) solutions of the actual games $Q\mathscr{G}$ and $DQ\mathscr{G}$, resp., and also of pairs of primal and dual approximate solutions. Clearly, the next step is applying Theorems 8.62(a) and (c) and 8.64(a) and (d), which employ the convergences of N^k given in this theorem. It should be noted that only condition (a) of the mentioned types of convergence of N^k is used in these theorems to get convergence of the approximate solutions and in the proof of Theorem 8.65 only its assumptions (a) and (b) are needed for the above condition (a).

We can easily realize that although the assumptions of Theorem 8.65 are directly in terms of the data of the considered games, these assumptions are strong. Now we move on to studying approximations of the game $Q\mathscr{G}$ also in terms of the data $\mathscr{D}_i$ and c_i, $i \in I$, but making use of Definition 7.118 about weak multi-epi-convergence, remarkably weaker than the continuous convergence used in the above theorem. Assuming that X_i are normed spaces, also to consider approximation of Nash quasiequilibria, next we use the following REGULARIZED NIKAIDO-ISODA BIFUNCTION for $s > 0$: $N_s : \text{gph}\mathscr{D} \to \mathbb{R}$ defined by

$$N_s(x, y) := \Sigma_{i\in I}[c_i(x_i, x_{-i}) - c_i(y_i, x_{-i}) + \frac{s}{2}\langle x_i - y_i, x_i - y_i\rangle]$$

and the inf-projection $\eta_s(x) := \inf_{\mathscr{D}(x_-)} N_s(x, \cdot)$, where $\mathscr{D}(x_-) := \mathscr{D}_1(x_{-1}) \times \ldots \times \mathscr{D}_m(x_{-m})$. Note that we also use the following *different* notation

$$\mathscr{D}_{-i}(x_{-(-i)}) := \mathscr{D}_1(x_{-1}) \times \ldots \times \mathscr{D}_{i-1}(x_{-(i-1)}) \times \mathscr{D}_{i+1}(x_{-(i+1)}) \times \ldots \times \mathscr{D}_m(x_{-m}).$$

Proposition 8.66 (Properties of η_s) *For every $i \in I$, let A_i be closed and convex and $c_i(\cdot, x_{-i})$ be concave and bounded for all $x_{-i} \in \mathscr{D}_{-i}(x_{-(-i)})$. Then, for any $s > 0$,*

(a) *for any given $x \in A$ $N_s(x, \cdot)$ attains its minimum at exactly one point and $\eta_s(x) \leq 0$;*
(b) *$\bar{x}$ is a Nash quasiequilibrium of $Q\mathscr{G}$ if and only if $\bar{x} \in \text{argmax}_A \eta_s = \text{mai}N_s$ with the max-value and maxinf-value being zero;*
(c) *for every $i \in I$ and $x \in A$, the supremum*

$$\sup_{y_i\in\mathscr{D}_i(x_i)}[c_i(y_i, x_{-i}) - \frac{s}{2}\langle x_i - y_i, x_i - y_i\rangle]$$

can be attained at at most one point $y_i(x) \in \mathscr{D}_i(x_i)$.

Furthermore, if $\dim X$ *is finite and the above boundedness of $c_i(\cdot, x_{-i})$ is replaced by continuity, then $y_i(\cdot)$ is well-defined and continuous.*

Proof

(a) The first claim is evident, since $N_s(x, \cdot)$ is strictly convex. The second one is also valid as $N_s(x, x) = 0$.
(b) "Only if." When $\bar{x} \in \text{Ne}(Q\mathscr{G})$, then $\bar{x} \in \text{mai}N$ and the maxinf-value of N is zero by assertion (c) of Proposition 8.55. Clearly, $\eta_s(\bar{x}) \geq \eta(\bar{x}) = 0$ and then $\eta(\bar{x}) = 0$ by assertion (a), i.e., $\bar{x} \in \text{argmax}\eta_s$ with the max-value being zero. Hence, $\bar{x} \in \text{mai}N_s$ and the maxinf-value of N_s is zero.

For the "If," the argument is similar to that in the proof of Proposition 7.119 with the assumed concavity of $c_i, i \in I$, instead of convexity, leading to $0 \leq c_k(\bar{x}_k, \bar{x}_{-k}) - c_k(x_k, \bar{x}_{-k})$, i.e., $\bar{x} \in \text{Ne}(Q\mathscr{G})$.
(c) This follows from well-known results in parametric optimization as the function $c_i(\cdot, x_{-i}) + \frac{s}{2}\langle x_i - \cdot, x_i - \cdot\rangle$ is strictly concave on A_i.

□

In view of Proposition 8.66(b), in a natural way, for $\varepsilon \geq 0$, $\bar{x} \in A$ is called a NASH ε-QUASIEQUILIBRIUM if $\eta_s(\bar{x}) \geq -\varepsilon$. Furthermore, we naturally have the concept of weak multi-hypo-convergence by modifying the definition of weak multi-epi-convergence when applying the commonly used relationship between the types of epi- and hypo-convergence. Employing this concept, we arrive at the following statement about convergence of approximate Nash quasiequilibria directly in terms of the data of $Q\mathscr{G}$. Its proof technique is similar to that for Theorem 7.120 (with modifications due to the problem setting being now for quasiequilibria and the objective functions being returns, not costs) and so omitted.

Theorem 8.67 (Convergence of approximate Nash quasiequilibria in terms of weak multi-hypo-convergence) *For $i \in I$, let X_i be a metric space, $X := \prod_{i\in I} X_i$, $A_i^k, A_i \subset X_i$ closed and convex, $c_i(\cdot, x_{-i})$, $c_i^k(\cdot, x_{-i})$ concave and bounded for all $x_{-i} \in \mathscr{D}_{-i}(x_{-(-i)})$, and $\varepsilon^k \to 0^+$. Assume that $\{(c_1^k, \dots, c_m^k)\}_k$ weakly multi-hypo-converges to $(c_1, \dots, c_m)$ and $x^k \in \varepsilon^k$-$\mathrm{Ne}(Q\mathscr{G}^k)$ tends to $\bar{x}$. Then, $\bar{x}$ is a Nash quasiequilibrium of $Q\mathscr{G}$.*

8.6 Approximations of Set-Valued Quasivariational Inequalities and Traffic Networks

8.6.1 Set-Valued Quasivariational Inequalities

Given two nonempty-set-valued maps $\mathscr{T} : \mathbb{R}^m \rightrightarrows (\mathbb{R}^m)^*$ with compact values and $\mathscr{K} : \mathbb{R}^m \rightrightarrows \mathbb{R}^m$, the so-called WEAK SET-VALUED QUASIVARIATIONAL INEQUALITY is

$$\text{(wQVI)} \quad \text{find } \bar{x} \in \mathscr{K}(\bar{x}) \text{ such that } \forall y \in \mathscr{K}(\bar{x}), \exists \bar{t}_y \in \mathscr{T}(\bar{x}), \langle \bar{t}_y, y - \bar{x} \rangle \geq 0.$$

Clearly, $\bar{x} \in \mathscr{K}(\bar{x})$ means that $\bar{x} \in \mathrm{fix}\mathscr{K}$, the fixed-point set of $\mathscr{K}$. Hence, it is natural to always assume that $\mathrm{fix}\mathscr{K} \subset \mathrm{dom}\mathscr{T}$. We also denote this quasivariational inequality by (wQVI($\mathscr{T}, \mathscr{K}$)) to make the data clear, and its solution set by Sol(wQVI). This style of notation is always adopted later for various problems.

To consider approximations of (wQVI($\mathscr{T}, \mathscr{K}$)) and its associated problems, we use the following bifunction on $\mathbb{R}^m \times \mathbb{R}^m$

$$\Phi(x, y) := \sup\nolimits_{t\in\mathscr{T}(x)} \langle t, y - x \rangle,$$

which is easily seen to be equivalent to

$$\Phi(x, y) := \sup\nolimits_{t\in\mathrm{conv}\,\mathscr{T}(x)} \langle t, y - x \rangle. \tag{8.9}$$

Hence, when considering (wQVI) via Φ, we can always assume that $\mathscr{T}$ is convex-valued (but, in the general problem setting, $\mathscr{T}$ may be nonconvex-valued). Moreover, $\Phi(x, \cdot)$ is convex for any $x \in \mathbb{R}^m$. We denote the fixed-point set of $\mathscr{K}$ by $A := \mathrm{fix}\mathscr{K}$, set $B := \mathscr{K}(A)$, and always assume that $A \neq \emptyset$. Hereafter, we restrict $\mathscr{K}$ as the map $\mathscr{K} : A \rightrightarrows B$, and so $\mathscr{K}$ is surjective. Then, (wQVI($\mathscr{T}, \mathscr{K}$))

is equivalent to the following case of quasiequilibrium problem

$$(\text{QEP}(\Phi, \mathscr{K})) \quad \text{find } \bar{x} \in A \text{ such that for all } y \in \mathscr{K}(\bar{x}),\ \Phi(\bar{x}, y) \geq 0.$$

Remark 8.68 (QEP($\Phi, \mathscr{K}$)) is equivalent to (sQEP($\Phi, \mathscr{K}$)) For a general quasiequilibrium problem (QEP($\Psi, \mathscr{D}$)) with $\mathscr{D} : A \rightrightarrows \mathscr{D}(A)$ and $\Psi : \text{gph}\mathscr{D} \to \mathbb{R}$, we know from Sect. 8.4 that the corresponding maxinf-problem

$$(\text{sQEP}(\Psi, \mathscr{D})) \quad \text{find } \bar{x} \in A \text{ solving } \max_{x \in A} \inf_{y \in \mathscr{D}(x)} \Psi(x, y) \geq 0$$

is properly stronger than (QEP($\Psi, \mathscr{D}$)). However, (QEP($\Phi, \mathscr{K}$)) is equivalent to (sQEP($\Phi, \mathscr{K}$)). Indeed, $\bar{x}$ being a solution of (QEP($\Phi, \mathscr{K}$)) means that $\inf_{y \in \mathscr{K}(\bar{x})} \Phi(\bar{x}, y) \geq 0$. On the other hand, as $\Phi(x, x) = 0$, one has $\inf_{y \in \mathscr{K}(x)} \Phi(x, y) \leq 0$ for all $x \in A$. So, $\inf_{y \in \mathscr{K}(\bar{x})} \Phi(\bar{x}, y) \geq 0$ is equivalent to $\bar{x}$ being a maxinf-point of Φ on gph$\mathscr{D}$ (i.e., a solution of (sQEP($\Phi, \mathscr{K}$)) with maxinf-value being zero. That is, the following set-valued quasivariational inequality

$$\text{find } \bar{x} \in \mathscr{K}(\bar{x}) \text{ solving } \max_{x \in \text{fix}\mathscr{K}} \inf_{y \in \mathscr{K}(x)} \sup_{t \in \mathscr{T}(x)} \langle t, y - x \rangle \geq 0,$$

which is equivalent to (sQEP($\Phi, \mathscr{K}$)), is just problem (wQVI($\mathscr{T}, \mathscr{K}$)).

We also consider the following so-called STRONG SET-VALUED QUASIVARIATIONAL INEQUALITY, which is usually considered in the literature,

$$(\text{sQVI}) \quad \text{find } \bar{x} \in \mathscr{K}(\bar{x}) \text{ such that } \exists \bar{t} \in \mathscr{T}(\bar{x}), \forall y \in \mathscr{K}(\bar{x}), \langle \bar{t}, y - \bar{x} \rangle \geq 0.$$

Proposition 8.69 (Relationship between (sQVI) and (wQVI)) Sol(sQVI) $\subset$ Sol(wQVI) *always holds. Moreover, under the condition that $\mathscr{T}$ is convex-valued and $\mathscr{K}$ is closed-convex-valued,* Sol(wQVI) $\subset$ Sol(sQVI), *i.e., the two problems are equivalent.*

Proof By definition, if $\bar{x} \in$ Sol(wQVI), then $\inf_{y \in \mathscr{K}(\bar{x})} \max_{t \in \mathscr{T}(\bar{x})} \langle t, y - \bar{x} \rangle \geq 0$. By the closed-convex-valuedness of $\mathscr{K}$, due to Sion's minimax theorem ([421], Corollary 3.3), inequality is equivalent to $\max_{t \in \mathscr{T}(\bar{x})} \inf_{y \in \mathscr{K}(\bar{x})} \langle t, y - \bar{x} \rangle \geq 0$. Therefore, there exists $\bar{t} \in \mathscr{T}(\bar{x})$ such that $\langle \bar{t}, y - \bar{x} \rangle \geq 0$ for all $y \in \mathscr{K}(x)$, i.e., $\bar{x} \in$ Sol(sQVI). □

The following example shows that the above inclusion may be proper.

Example 8.70 ((sQVI) is properly stronger than (wQVI)) Let $m = 2$, $\mathscr{K}(x) = \{\alpha x \mid \alpha \geq 0\}$, and $\mathscr{T}(x) = \{(-1, -1), (1, 1)\}$ for all $x \in \mathbb{R}^2$. Then, fix$\mathscr{K} = \mathbb{R}^2$. We first consider Sol(sQVI). Let $x = (x_1, x_2) \in \text{fix}\mathscr{K} = \mathbb{R}^2$. For the case $x_1 > x_2$, $t = (-1, -1)$ is not appropriate since $\inf_{y \in \{\alpha x | \alpha \geq 0\}} (1 - \alpha)(x_1 + x_2) = -\infty$. The situation with $t = (1, 1)$ is similar, since the corresponding infimum is less than 0. Hence, points x with $x_1 > x_2$ do not belong to Sol(sQVI). Now it is sufficient to show that such points x belong to Sol(wQVI). We have to check two cases for $y \in \{\alpha x \mid \alpha \geq 0\}$. If $\alpha \in [0, 1]$, one takes $t = (-1, -1)$ to see that the above infimum is nonnegative. If $\alpha > 1$, one takes $t = (1, 1)$ to have that this infimum is positive. Hence, points x with $x_1 > x_2$ do belong to Sol(wQVI).

Example 8.71 ((sQVI) is properly stronger than (wQVI) even with convex-valued $\mathscr{T}$) Assume that $m = 2$, $\mathscr{T}(x) = \{(x_1, x_2) \in \mathbb{R}^2 \mid 0 \le x_1 \le 1, x_2 \in [x_1, 1] \cap [0.1, 1]\}$, and

$$\mathscr{K}(x) = \begin{cases} \{(0,0), (-1, 0.5), (1, -1)\} & \text{if } x \in \mathbb{R}^2_+, \\ \mathbb{R}^2 \setminus \mathbb{R}^2_+ & \text{otherwise.} \end{cases}$$

Then, $\text{fix}\mathscr{K} = \{(0,0)\} \cup (\mathbb{R}^2 \setminus \mathbb{R}^2_+)$. We first consider $x = (0,0) \in \text{fix}\mathscr{K}$ for (sQVI). Clearly, there does not exist $t \in \mathscr{T}(0,0)$ such that $\inf_{y \in \mathscr{K}(0,0)} \langle t, y - (0,0) \rangle \ge 0$. Hence, $x = (0,0) \notin$ Sol(sQVI). However, this point is a solution of (wQVI), since for each $y \in \mathscr{K}(0,0)$, there obviously exists $t \in \mathscr{T}(0,0)$ such that $\langle t, y - (0,0) \rangle \ge 0$. Therefore, Sol(sQVI) is properly smaller than Sol(wQVI).

Example 8.72 (Proposition 8.69 provides a sufficient condition, not a necessary one) Assume that $m = 1$, $\mathscr{T}(x) = \{-1, 1\}$ for all $x \in \mathbb{R}$, and

$$\mathscr{K}(x) = \begin{cases} \{0\} & \text{if } x \in [0, 1), \\ \{1\} & \text{if } x \in [1, +\infty), \\ \{-1\} & \text{if } x \in (-\infty, 0]. \end{cases}$$

Then, $\text{fix}\mathscr{K} = \{-1\} \cup \{0\} \cup \{1\}$. To have Sol(wQVI) $=$ Sol(sQVI)), it suffices to show that Sol(sQVI) $= \text{fix}\mathscr{K}$. For $x = -1$, we take $t = -1$ to obtain $\inf_{y \in \mathscr{K}(-1)} [-(-1 - (-1))] = 0$, and so $x = -1$ is a strong solution. For $x = 0$, we take $t = 1$ to obtain the same infimum value and so $x = 0$ is a strong solution. For $x = 1$, since $\mathscr{K}(1) = \{1\}$, the corresponding minimum is equal to 0 for any $t \in \{-1, 1\}$. Therefore, Sol(sQVI) $= \text{fix}\mathscr{K}$ and is evidently equal to Sol(wQVI).

In many cases, exact solutions of a problem do not exist or are difficult to compute. Hence, approximate solutions play a crucial role.

Definition 8.73 (Approximate solutions) Let $\varepsilon \ge 0$.

(a) A point $\bar{x} \in \text{fix}\mathscr{K}$ is called an ε-SOLUTION OF (WQVI), written $\bar{x} \in \varepsilon$-Sol(wQVI), when $\forall y \in \mathscr{K}(\bar{x}), \exists \bar{t}_y \in \mathscr{T}(\bar{x}), \langle \bar{t}_y, y - \bar{x} \rangle \ge -\varepsilon$.
(b) A point $\bar{x} \in \text{fix}\mathscr{K}$ such that $\exists \bar{t} \in \mathscr{T}(\bar{x}), \forall y \in \mathscr{K}(\bar{x}), \langle \bar{t}, y - \bar{x} \rangle \ge -\varepsilon$ is said to be an ε-SOLUTION OF (SQVI), written $\bar{x} \in \varepsilon$-Sol(sQVI).

Proposition 8.74 (Approximate solutions when $\mathscr{K}$ is closed-convex-valued) *Assume that $\mathscr{K}$ is closed-convex-valued. Then,* ε-Sol(sQVI) $= \varepsilon$-Sol(wQVI).

Proof $\bar{x} \in \varepsilon$-Sol(wQVI)) means that $\inf_{y \in \mathscr{K}(\bar{x})} \max_{t \in \mathscr{T}(\bar{x})} \langle t, y - \bar{x} \rangle \ge -\varepsilon$. In view of Sion's minimax theorem ([421], Corollary 3.3), this is equivalent to

$\max_{t\in\mathscr{T}(\bar{x})}\inf_{y\in\mathscr{K}(\bar{x})}\langle t, y-\bar{x}\rangle \geq -\varepsilon$, i.e., there exists $\bar{t} \in \mathscr{T}(\bar{x})$ such that $\inf_{y\in\mathscr{K}(\bar{x})}\langle t, y-\bar{x}\rangle \geq -\varepsilon$. This means that $\bar{x} \in \varepsilon$-Sol(sQVI). □

We also consider the following second-kind approximate solution $\bar{x}$, where the tolerance depends not only on ε as for the first kind but is controlled by $\varepsilon\theta(\bar{x}-y)$ for y varying in $\mathscr{K}(\bar{x})$.

Definition 8.75 (second-kind approximate solutions) Assume that $\theta : \mathbb{R}^m \to \mathbb{R}$ satisfies $\theta(x) > 0$ for $x \neq 0$ and $\varepsilon \geq 0$. A point $\bar{x} \in \text{fix}\mathscr{K}$ is an $\varepsilon\theta$-SOLUTION OF (wQVI) when $\forall y \in \mathscr{K}(\bar{x})$, $\exists \bar{t}_y \in \mathscr{T}(\bar{x})$, $\langle \bar{t}_y, y-\bar{x}\rangle \geq -\varepsilon\theta(\bar{x}-y)$.

To define the dual of the above-considered problems, we assume that $\mathscr{K}(\text{fix}\mathscr{K}) \subset \text{dom}\mathscr{T}$, which is "symmetric/dual" to $\text{fix}\mathscr{K} \subset \text{dom}\mathscr{T}$ for the primal problem. The DUAL PROBLEM (DwQVI($\mathscr{T},\mathscr{K}$)) of (wQVI($\mathscr{T},\mathscr{K}$)) is defined naturally:

$$\text{(DwQVI}(\mathscr{T},\mathscr{K}))\quad \text{find } \bar{y} \in \mathscr{K}(\text{fix}\mathscr{K}) \text{ such that } \forall x \in \mathscr{K}^{-1}(\bar{y}),$$
$$\exists \bar{t}_x \in \mathscr{T}(\bar{y}), \langle \bar{t}_x, \bar{y}-x\rangle \leq 0.$$

This problem can be reformulated as

$$\text{(DwQVI}(\mathscr{T},\mathscr{K}))\quad \text{find } \bar{y} \in \mathscr{K}(\text{fix}\mathscr{K}) \text{ such that } \forall x \in \mathscr{K}^{-1}(\bar{y}),$$
$$\inf_{t\in\mathscr{T}(\bar{y})}\langle t, \bar{y}-x\rangle \leq 0.$$

This reformulation means that (DwQVI($\mathscr{T},\mathscr{K}$)) is just (DQEP($\hat{\Phi},\mathscr{K}$)) for

$$\hat{\Phi}(x, y) := \inf_{t\in\mathscr{T}(y)}\langle t, y-x\rangle,$$

which is the dual of (QEP($\hat{\Phi},\mathscr{K}$)), not of (QEP($\Phi,\mathscr{K}$)) as in the Stampacchia-Minty duality scheme for quasiequilibrium problem (QEP($\Phi,\mathscr{K}$)). It should be noted that if $\mathscr{T} = t$, a single-valued map, then (DwQVI($t,\mathscr{K}$)) collapses to the classical Minty quasivariational inequality. The mentioned mismath when expressing our dual couple in terms of quasiequilibrium problems (intermediate problems) is due to the set-valuedness of the quasivariational inequalities. This note also holds for the dual strong problem defined as follows.

The DUAL PROBLEM (DsQVI) of (sQVI) is

$$\text{(DsQVI)}\quad \text{find } \bar{y} \in \mathscr{K}(\text{fix}\mathscr{K}) \text{ such that } \exists \bar{t} \in \mathscr{T}(\bar{y}), \forall x \in \mathscr{K}^{-1}(\bar{y}), \langle \bar{t}, \bar{y}-x\rangle \leq 0.$$

Proposition 8.76 (Relationship between convex problems (DsQVI) and (DwQVI)) *It always holds that* Sol(DsQVI) $\subset$ Sol(DwQVI). *Moreover, under the condition that ($\mathscr{T}$ is compact-convex-valued and) $\mathscr{K}^{-1}$ is closed-convex-valued, one has that $\bar{y} \in \text{fix}\mathscr{K}^{-1}$ and $\bar{y} \in$* Sol(DwQVI) *imply that $\bar{y} \in$* Sol(DsQVI).

Proof By definition, $\bar{y} \in \text{Sol(DwQVI)}$ means that $\sup_{x \in \mathscr{K}^{-1}(\bar{y})} \min_{t \in \mathscr{T}(\bar{y})} \langle t, \bar{y} - x\rangle \leq 0$. By Sion's minimax theorem ([421], Corollary 3.3), this implies that $\min_{t \in \mathscr{T}(\bar{y})} \sup_{x \in \mathscr{K}^{-1}(\bar{y})} \langle t, \bar{y} - x\rangle \leq 0$, which means that $\bar{y} \in \text{Sol(DsQVI)}$. □

Definition 8.77 (Approximate dual solutions) Assume that $\varepsilon \geq 0$.

(a) A point $\bar{y} \in B$ is an ε-SOLUTION OF (DwQVI($\mathscr{T}, \mathscr{K}$)), written $\bar{y} \in \varepsilon$-Sol(DwQVI($\mathscr{T}, \mathscr{K}$)), when $\forall x \in \mathscr{K}^{-1}(\bar{y})$, $\exists \bar{t} \in \mathscr{T}(\bar{y})$, $\langle \bar{t}, \bar{y} - x\rangle \leq \varepsilon$.
(b) A point $\bar{y} \in \mathscr{K}(\text{fix}\mathscr{K})$ such that $\exists \bar{t} \in \mathscr{T}(\bar{y})$, $\forall x \in \mathscr{K}^{-1}(\bar{y})$, $\langle \bar{t}, \bar{y} - x\rangle \leq \varepsilon$ is said to be an ε-SOLUTION OF (DsQVI($\mathscr{T}, \mathscr{K}$)), written $\bar{y} \in \varepsilon$-Sol(DsQVI($\mathscr{T}, \mathscr{K}$)).

8.6.2 *Approximations of Set-Valued Quasivariational Inequalities*

Assume that we have a sequence of problems approximating our problems defined by $\mathscr{T}^k : \mathbb{R}^m \rightrightarrows (\mathbb{R}^m)^*$ with compact values, $\mathscr{K}^k : \mathbb{R}^m \rightrightarrows \mathbb{R}^m$, and $\Phi^k : \text{gph}\mathscr{K}^k \to \mathbb{R}$ defined by $\Phi^k(x, y) := \sup_{t \in \mathscr{T}^k(x)} \langle t, y - x\rangle$. Similar to $\mathscr{K}$, we view $\mathscr{K}^k$ as a set-valued map from the whole $A^k := \text{fix}\mathscr{K}^k$ and onto $B^k := \mathscr{K}^k(\text{fix}\mathscr{K}^k)$. So, (QEP($\Phi^k, \mathscr{K}^k$)) is the problem of finding $\bar{x} \in A^k$ such that for all $y \in \mathscr{K}^k(\bar{x})$, it holds $\Phi^k(\bar{x}, y) \geq 0$. The other approximating problems under consideration are stated in the same way.

Proposition 8.78 (Convergence of approximate solutions of (wQVI^k)) *Let* $\varepsilon^k, \varepsilon \geq 0$, *and* $\varepsilon^k \to \varepsilon$.

(i) *If* $-\Phi^k$ *i-e/h- or i-mis-lop converges to* $-\Phi$, *then* $A \cap \text{Ls}_k(\varepsilon^k\text{-Sol(wQVI}^k)) \subset \varepsilon\text{-Sol(wQVI)}$. *If* $-\Phi^k$ *e/h- or mis-lop converges to* $-\Phi$, *then* $\text{Ls}_k(\varepsilon^k\text{-Sol(wQVI}^k)) \subset \varepsilon\text{-Sol(wQVI)}$.
(ii) *If* $-\Phi^k \xrightarrow{e} -\Phi$ *and* $\text{gph}\mathscr{K} \sqsubset \text{Li}_k(\text{gph}\mathscr{K}^k)$, *then*

$$\text{Ls}_k(\varepsilon^k\text{-Sol(wQVI}^k)) \subset \varepsilon\text{-Sol(wQVI)}.$$

If $-\Phi^k \xrightarrow{i-e} -\Phi$ *replaces the epi-convergence, then*

$$A \cap \text{Ls}_k(\varepsilon^k\text{-Sol(wQVI}^k)) \subset \varepsilon\text{-Sol(wQVI)}.$$

(iii) *If* $-\Phi^k$ *w-e/h- or w-mis-lop converges to* $-\Phi$ *and* $x^k \in \varepsilon^k\text{-Sol(wQVI}^k)$ *tends to* x, *then* $x \in \varepsilon\text{- Sol(wQVI)}$.

Proof

(i) Consider the case $-\Phi^k \xrightarrow{e/h} -\Phi$. Let $\bar{x} \in \text{Ls}_k(\varepsilon^k\text{-Sol(wQVI}^k))$, i.e., there exists a subsequence $\{x^{k_j}\}_j$ in $\varepsilon^{k_j}\text{-Sol(wQVI}^{k_j})$ converging to $\bar{x}$. If $\bar{x} \in A$, then for all $y \in \mathscr{K}(\bar{x})$, condition (a) of the assumed e/h-convergence yields

$y^{k_j} \in \mathscr{K}^{k_j}(x^{k_j})$ tending to y such that $\text{ls}_j \Phi^{k_j}(x^{k_j}, y^{k_j}) \le \Phi(\bar{x}, y)$. As $\Phi^{k_j}(x^{k_j}, y^{k_j}) \ge -\varepsilon^{k_j}$ for all $y^{k_j} \in \mathscr{K}^{k_j}(x^{k_j})$, it holds that $\Phi(\bar{x}, y) \ge -\varepsilon$. Suppose that $\bar{x} \notin A$. Again condition (a) gives $y^{k_j} \in \mathscr{K}^{k_j}(x^{k_j})$ such that $\Phi^{k_j}(x^{k_j}, y^{k_j}) \to -\infty$, contradicting the inequality $\Phi^{k_j}(x^{k_j}, y^{k_j}) \ge -\varepsilon^{k_j}$ for all $y^{k_j} \in \mathscr{K}^{k_j}(x^{k_j})$. Hence, $\bar{x} \in \varepsilon$-Sol(wQVI). The case of mis-lop convergence can be proved similarly.

(ii) $\bar{x} \in \text{Ls}_k(\varepsilon^k\text{-Sol(wQVI}^k))$ implies the existence of $\{x^{k_j}\}_j$ in ε^{k_j}-Sol(QEP^{k_j}) tending to $\bar{x}$. Suppose $\bar{x} \notin A$. For every $y^{k_j} \in \mathscr{K}^{k_j}(x^{k_j})$ tending to y, $(\bar{x}, y) \notin \text{gph}\mathscr{K}$ and condition (a) of the epi-convergence gives $\Phi^{k_j}(x^{k_j}, y^{k_j}) \to -\infty$, contradicting the fact that $\Phi^{k_j}(x^{k_j}, y^{k_j}) \ge -\varepsilon^{k_j}$ for all $y^{k_j} \in \mathscr{K}^{k_j}(x^{k_j})$. So, $\bar{x} \in A$. It suffices to verify that $\Phi(\bar{x}, y) \ge -\varepsilon$ for each $y \in \mathscr{K}(\bar{x})$. Since $(\bar{x}, y) \in \text{gph}\mathscr{K}$, by the assumed strong inclusion, there exists $y^{k_j} \in \mathscr{K}^{k_j}(x^{k_j})$ converging to y. Then, the above condition (a) implies that $\text{ls}_j \Phi^{k_j}(x^{k_j}, y^{k_j}) \le \Phi(\bar{x}, y)$. As $\Phi^{k_j}(x^{k_j}, y^{k_j}) \ge -\varepsilon^{k_j}$, $\Phi(\bar{x}, y) \ge -\varepsilon$. The assertion for the remaining case $-\Phi^k \overset{i-e}{\to} -\Phi$ is clear.

(iii) $x^k \in \varepsilon^k$-Sol(wQVI^k) means that $\Phi^k(x^k, y^k) \ge -\varepsilon^k$ for all $y^k \in \mathscr{K}^k(x^k)$. If $x \in A$, then for every $y \in \mathscr{K}(x)$, condition (a) of the w-e/h-convergence yields $y^k \in \mathscr{K}^k(x^k)$ tending to y such that $\text{ls}_k \Phi^k(x^k, y^k) \le \Phi(x, y)$. The rest of the argument is similar to that in (i) with sequences replacing subsequences. The case of the w-mis-lop convergence is similar.

□

Proposition 8.79 (Convergence of approximate solutions of (DwQVI^k)) *Let $\varepsilon^k, \varepsilon \ge 0$, and $\varepsilon^k \to \varepsilon$.*

(i) *If $-\hat{\Phi}^k$ i-e/h- or i-mai-lop converges to $-\hat{\Phi}$, then* $B \cap \text{Ls}_k(\varepsilon^k\text{-Sol(DwQVI}^k)) \subset \varepsilon\text{-Sol(DwQVI)}$. *If $-\hat{\Phi}^k$ e/h- or mai-lop converges to $-\hat{\Phi}$, then* $\text{Ls}_k(\varepsilon^k\text{-Sol(DwQVI}^k)) \subset \varepsilon\text{-Sol(DwQVI)}$.

(ii) *If* $\text{gph}\mathscr{K} \sqsubset \text{Li}_k(\text{gph}\mathscr{K}^k)$ *and the hypo- or i-hypo-convergence of $-\hat{\Phi}^k$ to $-\hat{\Phi}$ replaces the type of convergence of $-\hat{\Phi}^k$ imposed in assertion* (i), *then the conclusion in this assertion still holds.*

(iii) *If $-\hat{\Phi}^k$ w-e/h- or w-mai-lop converges to $-\hat{\Phi}$ and $y^k \in \varepsilon^k$*-Sol(DwQVI^k) *tends to y, then $y \in \varepsilon$*-Sol(DwQVI).

Proof

(i) Consider the case $-\hat{\Phi}^k \overset{mai-lop}{\to} -\hat{\Phi}$. $\bar{y} \in \text{Ls}_k(\varepsilon^k\text{-Sol(DwQVI}^k))$ means that there exists $y^{k_j} \in \varepsilon^{k_j}$-Sol(DwQVI^{k_j}) tending to $\bar{y}$. If $\bar{y} \in B$, then condition (a) of the mai-lop convergence gives that, for all $x \in \mathscr{K}^{-1}(\bar{y})$, there exists $x^{k_j} \in (\mathscr{K}^{k_j})^{-1}(y^{k_j})$ converging to x such that $\text{li}_j \hat{\Phi}^{k_j}(x^{k_j}, y^{k_j}) \ge \hat{\Phi}(x, \bar{y})$. Since $\hat{\Phi}^{k_j}(x^{k_j}, y^{k_j}) \le \varepsilon^{k_j}$ for all $x^{k_j} \in (\mathscr{K}^{k_j})^{-1}(y^{k_j})$, $\hat{\Phi}(x, \bar{y}) \le \varepsilon$. Suppose $\bar{y} \notin B$. Then, for every $x \in \mathscr{K}^{-1}(\bar{y})$, the above condition (a) yields $x^{k_j} \in (\mathscr{K}^{k_j})^{-1}(y^{k_j})$ such that $\hat{\Phi}^{k_j}(x^{k_j}, y^{k_j}) \to +\infty$, which is a contradiction as $\hat{\Phi}^{k_j} \in \varepsilon^{k_j}$-Sol(DwQVI^{k_j}). So, $\bar{y} \in B$ and for all $x \in \mathscr{K}^{-1}(\bar{y})$, $\hat{\Phi}(x, \bar{y}) \le \varepsilon$, i.e., there exists $\bar{t} \in \mathscr{T}(\bar{y})$ such that $\langle \bar{t}, \bar{y} - x \rangle \le \varepsilon$, which means $\bar{y} \in \varepsilon$-

Sol(DwQVI). The proof for the case $-\hat{\Phi}^k \overset{e/h}{\rightarrow} -\hat{\Phi}$ is similar, with the use of condition (b) of the e/h-convergence.

(ii) The proof is similar to (ii) of the preceding proposition with condition (a) of the epi-convergence replaced by condition (a) of the hypo-convergence.

(iii) The proof is similar to (i) (cf. (iii) in the proof of the preceding proposition).

□

Proposition 8.80 (Convergence of second-kind approximate solutions of (wQVI^k)) *Let $\varepsilon^k, \varepsilon \geq 0$, $\varepsilon^k \to \varepsilon$, and θ be as in the definition of the second kind of approximate solutions.*

(a) *If $-\Phi^k$ i-e/h- or i-mis-lop converges to $-\Phi$, then* $A \cap \text{Ls}_k(\varepsilon^k\theta\text{-Sol}(\text{wQVI}^k)) \subset \varepsilon\theta\text{-Sol}(\text{wQVI})$. *If $-\Phi^k$ e/h- or mis-lop converges to $-\Phi$, then* $\text{Ls}_k(\varepsilon^k\theta\text{-Sol}(\text{wQVI}^k)) \subset \varepsilon\theta\text{-Sol}(\text{wQVI})$.

(b) *If* $\text{gph}\mathscr{K} \sqsubset \text{Li}_k(\text{gph}\mathscr{K}^k)$ *and the i-e- or e-convergence of $-\Phi^k$ to $-\Phi$ replaces the types of convergence of $-\Phi^k$ in assertion* (a), *then the conclusion of this assertion still holds.*

(c) *If $-\Phi^k$ w-e/h- or w-mis-lop converges to $-\Phi$ and $x^k \in \varepsilon^k\theta$-Sol(wQVI^k) tends to x, then $x \in \varepsilon\theta$-Sol(wQVI).*

The proof is analogous to that of Proposition 8.78. Furthermore, we also have the convergence of second-kind approximate solutions of (DwQVI^k) similar to Proposition 8.79.

Now move on to discuss approximations for strong set-valued quasi-variational inequalities. Clearly, we are only interested in the case with nonconvex-valued $\mathscr{K}$, since the approximate solution set of a strong inequality with $\mathscr{K}$ closed-convex-valued coincides with that of the corresponding weak inequality as we saw above. (Hence, Propositions 8.78 and 8.79 are also about approximations of strong set-valued quasi-variational inequalities when $\mathscr{K}$ is closed-convex-valued.) For a nonconvex case, an inconvenience is that bifunctions (like Φ in (8.9)) cannot be used, and so variational convergence of bifunctions is still inapplicable. However, we can go deeper into concepts of approximations as follows. Solving our strong problem (sQVI) consists of finding two elements: a solution $\bar{x} \in \text{fix}\mathscr{K}$ and a point $\bar{t} \in \mathscr{T}(\bar{x})$. Hence, one can be interested in the following "second-level" approximations of (sQVI) associated with a sequence $\{x^k\}_k$ of its ε^k-solutions and an ε-solution $\bar{x}$, where $\varepsilon^k, \varepsilon \geq 0$ and $\varepsilon^k \to \varepsilon$.

Assume that $x^k \in \varepsilon^k$-Sol(sQVI), i.e., we have a problem of finding t^k solving $\max_{t\in\mathscr{T}(x^k)}\inf_{y\in\mathscr{K}(x^k)} \langle t, y - x^k\rangle \geq -\varepsilon^k$, and that $\bar{x} \in \varepsilon$-Sol(sQVI), i.e., we have a problem of finding $\bar{t}$ solving $\max_{t\in\mathscr{T}(\bar{x})} \inf_{y\in\mathscr{K}(\bar{x})}\langle t, y - \bar{x}\rangle \geq -\varepsilon$. Setting $A^k := \mathscr{T}(x^k)$, $B^k := \mathscr{K}(x^k)$, and $F^k(t, y) := \langle t, y - x^k\rangle$ for $t \in (\mathbb{R}^m)^*$ and $y \in \mathbb{R}^m$, we have a set-valued strong equilibrium problem (not "quasi-"!) (sEP(F^k, A^k, B^k)) (the following definition of strong equilibrium problems was given in [179]) of finding $t^k \in A^k$ to solve $\max_{t\in A^k}\inf_{y\in B^k}\langle t, y - x^k\rangle \geq -\varepsilon^k$ (hence t^k is necessarily a maxinf-point of F^k on $A^k \times B^k$). Similarly, with $A := \mathscr{T}(\bar{x})$, $B := \mathscr{K}(\bar{x})$, and $F(t, y) := \langle t, y - \bar{x}\rangle$, we have that $\bar{t}$ is a maxinf-point of F on $A \times B$, and it solves problem (sEP(F, A, B)): $\max_{t\in\mathscr{T}(\bar{x})}\inf_{y\in\mathscr{K}(\bar{x})}\langle t, y - \bar{x}\rangle \geq -\varepsilon$. Now we have new bifunctions F^k and F. Then, we can apply variational convergence to investigate global approximations of (sEP(F^k, A^k, B^k)) and (sEP(F, A, B)). Such approximations can be viewed as SECOND-LEVEL APPROXIMATIONS FOR (SQVI). It should be emphasized that in the following study of second-level approximations of strong quasivariational inequalities, we do not impose neither compactness nor convexity assumptions. We only need approximate solutions if our approximating problems exist.

A function $\eta : A \to \mathbb{R}$ defined by $\eta(t) := \inf_{y\in B} F(t, y)$ for $t \in A$ is called the INF-PROJECTION of F. The following notion of tightness modifies the one used before for our current research context.

Definition 8.81 (Ancillary tightness for convergence of $-F^k$) Assume that X is a metric space, $A^k, A, B^k, B \subset X$, $F^k : A^k \times B^k \to \mathbb{R}$, and $F : A \times B \to \mathbb{R}$. If $-F^k$ e/h-converges (mis-lop converges or mai-lop converges, resp.) to $-F$ and for every $t \in A$, there exists $A^k \ni t^k \to t$ such that $\text{li}_k(\eta^k(t^k)) \geq \eta(t)$, then we say that $-F^k$ e/h-converges (mis-lop converges or mai-lop converges, resp.) t-ancillary tightly to $-F$.

Proposition 8.82 (Second-level approximations of (sQVI)) *Assume that $\varepsilon^k, \varepsilon \geq 0$, $\varepsilon^k \to \varepsilon$, $x^k \in \varepsilon^k$-Sol(sQVI), and $\bar{x} \in \varepsilon$-Sol(sQVI). If $-F^k$ e/h- or mis-lop or mai-lop converges t-ancillary tightly to $-F$, then* $\text{Ls}_k(\varepsilon^k\text{-Sol(sEP}(F^k, A^k, B^k)))$ $\subset \varepsilon\text{-Sol(sEP}(F, A, B))$.

Proof Let $\bar{t} \in \text{Ls}_k(\varepsilon^k\text{-Sol(sEP}(F^k, A^k, B^k)))$. Then, there exists $t^{k_j} \in \varepsilon^{k_j}$-Sol(sEP($F^{k_j}, A^{k_j}, B^{k_j}$)) tending to $\bar{t}$ as $j \to +\infty$. Assume the t-ancillary tight e/h-convergence of $-F^k$ to $-F$. If $\bar{t} \in A$, then for all $y \in B$, by condition (a) of the assumed e/h-convergence, there exists $y^{k_j} \in B^{k_j}$ tending to y such that $\text{li}_j(-F^{k_j}(t^{k_j}, y^{k_j})) \geq -F(\bar{t}, y)$. As t^{k_j} is an ε^{k_j}-solution of (sEP($F^{k_j}, A^{k_j}, B^{k_j}$)), $F^{k_j}(t^{k_j}, y^{k_j})) \geq -\varepsilon^{k_j}$. So, $F(\bar{t}, y) \geq -\varepsilon$, i.e., $\bar{t} \in \varepsilon$-Sol(EP(F, A, B)). Suppose $\bar{t} \notin A$. Then, the above condition (a) says that there exists $y^{k_j} \in B^{k_j}$ such that $-F^{k_j}(t^{k_j}, y^{k_j}) \to +\infty$, contradicting the fact that t^{k_j} is an ε^{k_j}-solution as aforementioned.

It suffices now to show that $\bar{t}$ is a maxinf-point of F on $A \times B$. The first step is to ensure that $\text{Ls}_k(\text{hypo}\eta^k) \subset \text{hypo}\eta$. Assume that $(t, \alpha) \in \text{Ls}_k(\text{hypo}\eta^k)$, i.e., there exist $t^{k_j} \in \text{dom}\eta^{k_j}$ and $\alpha^{k_j} \leq \eta^{k_j}(t^{k_j})$ such that $(t^{k_j}, \alpha^{k_j}) \to (t, \alpha)$. If $t \notin A$, then by condition (a) of the assumed e/h-convergence, there exists $y^{k_j} \in B^{k_j}$ such that $-F^{k_j}(t^{k_j}, y^{k_j}) \to +\infty$. Since $\alpha^{k_j} \leq \eta^{k_j}(t^{k_j}) \leq F^{k_j}(t^{k_j}, y^{k_j})$, it holds that $\alpha^{k_j} \to -\infty$, which is a contradiction. So, $t \in A$. If $t \notin \text{dom}\eta$, then one has $y \in B$ with $F(t, y) \leq \alpha - 1$. By the above (a), there is $B^{k_j} \ni y^{k_j} \to y$ such that $\text{li}_j(-F^{k_j}(t^{k_j}, y^{k_j})) \geq -F(t, y)$. Hence,

$$\alpha = \lim_j \alpha^{k_j} \leq \text{ls}_j \eta^{k_j}(t^{k_j}) \leq \text{ls}_j(F^{k_j}(t^{k_j}, y^{k_j})) \leq F(t, y) \leq \alpha - 1,$$

which is impossible. Consequently, t must be in $\text{dom}\eta$. For any positive δ, take $y_\delta \in B$ such that $F(t, y_\delta) \leq \eta(t) + \varepsilon$. Again by condition (a) of the e/h-convergence, one has $B^{k_j} \ni y^{k_j} \to y_\delta \in B$ such that

$$\text{ls}_j \eta^{k_j}(t^{k_j}) \le \text{ls}_j F^{k_j}(t^{k_j}, y^{k_j}) \le F(t, y_\delta) \le \eta(t) + \varepsilon.$$

Therefore,

$$\alpha = \lim \alpha^{k_j} \le \text{ls}_j \eta^{k_j}(t^{k_j}) \le \eta(t) + \varepsilon$$

and so $(t, \alpha) \in \text{hypo}\eta$ as δ is arbitrary.

The second step is to verify that $\text{hypo}\eta \subset \text{Li}_k(\text{hypo}\eta^k)$. By the assumed ancillary tightness, for all $t \in A$, there exists $A^k \ni t^k \to t$ such that $\text{li}_k(\eta^k(t^k)) \ge \eta(t)$, i.e., $\text{ls}_k(-\eta^k(t^k)) \le -\eta(t)$. This is equivalent to condition (b) of the epi-convergence of $-\eta^k$ to $-\eta$, whose geometric formulation is $\text{epi}(-\eta) \subset \text{Li}_k(\text{epi}(-\eta^k))$. The last inclusion is equivalent to $\text{hypo}\eta \subset \text{Li}_k \text{hypo}\eta^k$.

The conclusion for the t-ancillary case of these types of convergence is clearly also verified, because both the mis-lop and mai-lop convergence are stronger than the e/h-convergence of bifunctions defined on rectangles. (Although these types of convergence are incomparable for (general) bifunctions defined on domains of the form $\text{gph}\mathscr{D}$.) □

Now consider dual strong problem (DsQVI). Assume that $y^k \in \varepsilon^k$-Sol(DsQVI), i.e., we have the problem of finding t^k solving problem $\min_{t \in \mathscr{T}(y^k)} \sup_{x \in \mathscr{K}^{-1}(y^k)} \langle t, y^k - x\rangle \le \varepsilon^k$, and $\bar{y} \in \varepsilon$-Sol(DsQVI), i.e., we have the problem of finding $\bar{t}$ solving $\min_{t \in \mathscr{T}(\bar{y})} \sup_{x \in \mathscr{K}^{-1}(\bar{y})} \langle t, \bar{y} - x\rangle \le \varepsilon$. Setting $C^k := \mathscr{T}(y^k)$, $D^k := \mathscr{K}^{-1}(y^k)$, $G^k := \langle t, y^k - x\rangle$, $C := \mathscr{T}(\bar{y})$, $D := \mathscr{K}^{-1}(\bar{y})$, $G := \langle t, \bar{y} - x\rangle$, the above two problems become dual strong set-valued equilibrium problems (DsEP(G^k, C^k, D^k)) and (DsEP(G, C, D)), resp. Global approximations of them in terms of variational convergence of G^k and G can be viewed as second-level global approximations of (DsQVI). For these problems, we need the following notions.

A function $\xi : C \to \mathbb{R}$ defined by $\xi(t) := \sup_{x \in D} G(t, x)$ for $t \in A$ is called the SUP-PROJECTION of G.

Definition 8.83 (Ancillary tightness for convergence of G^k) Assume that X is a metric space, $C^k, C, D^k, D \subset X$, $G^k : C^k \times D^k \to \mathbb{R}$, and $G : C \times D \to \mathbb{R}$. If G^k e/h-converges (mis-lop converges or mai-lop converges, resp.) to G and for all $t \in C$, there exists $C^k \ni t^k \to t$ such that $\text{ls}_k(\xi^k(t^k)) \le \xi(t)$, then we say that G^k e/h-converges (mis-lop converges or mai-lop converges, resp.) t-ancillary tightly to G.

Proposition 8.84 (Second-level approximations of (DsQVI)) *Assume that $\varepsilon^k, \varepsilon \ge 0$, $\varepsilon^k \to \varepsilon$, $y^k \in \varepsilon^k$-Sol(DsQVI), and $\bar{y} \in \varepsilon$-Sol(DsQVI). If G^k e/h- or mis-lop or mai-lop converges t-ancillary tightly to G, then* $\text{Ls}_k(\varepsilon^k$-Sol(DsEP($G^k, C^k, D^k$))) $\subset \varepsilon$-Sol(DsEP(G, C, D)).

The proof is similar to that of Proposition 8.82 and omitted.

In the above several statements, we view $\bar{t}$ in the strong quasivariational inequality (sQVI) as a kind of second-level solution (after solution $\bar{x}$) and consider second-level approximations for this variable. A related idea to consider with "first-level" approximations is as follows. We can modify Definition 8.73 of ε-solution $\bar{x}$ of (sQVI) as follows to give variable t a role of an approximate second-level solution.

Recall from convex analysis that for a convex function $h : \mathbb{R}^m \to \mathbb{R} \cup (+\infty)$, its subdifferential at x is defined by $\partial h(x) := \{t \in (\mathbb{R}^m)^* \mid h(y) - h(x) \geq \langle t, y - x\rangle \ \forall y \in \mathbb{R}^m\}$.

Lemma 8.85 (Subdifferential of Φ in terms of $\mathscr{T}$) *Let $\mathscr{T} : \mathbb{R}^m \rightrightarrows (\mathbb{R}^m)^*$ be nonempty-convex-valued and Φ defined by (8.9). Then, $\partial\Phi(x, \cdot)(x) = \mathscr{T}(x)$ for all $x \in \mathbb{R}^m$.*

Proof Since $\Phi(x, x) = 0$, by the definition of Φ, $\Phi(x, y) - \Phi(x, x) = \Phi(x, y) \geq \langle t, y - x\rangle$ for all $t \in \mathscr{T}(x)$ and $y \in \mathbb{R}^m$. Hence, $\mathscr{T}(x) \subset \partial\Phi(x, \cdot)(x)$. For the reverse inclusion, suppose $t_0 \notin \mathscr{T}(x)$. The separation theorem gives $\bar{y} \in (\mathbb{R}^m)^*$ such that $\max_{t\in\mathscr{T}(x)}\langle t, \bar{y}\rangle < \langle t_0, \bar{y}\rangle$. Setting $\bar{y} = y_0 - x$ yields

$$\Phi(x, y_0) - \Phi(x, x) < \langle t_0, y_0 - x\rangle,$$

and so $t_0 \notin \partial\Phi(x, \cdot)(x)$. It follows that $\partial\Phi(x, \cdot)(x) \subset \mathscr{T}(x)$. Consequently $\partial\Phi(x, \cdot)(x) = \mathscr{T}(x)$. □

Recall further some auxiliary facts. For $\varepsilon \geq 0$, the ε-NORMAL SET of a convex set $M \subset \mathbb{R}^m$ at $x \in M$ is $N_\varepsilon(M, x) := \{\{t \in (\mathbb{R}^m)^* \mid \langle t, y - x\rangle \leq \varepsilon,\ \forall y \in M\}$. For a convex function $h : \mathbb{R}^m \to \bar{\mathbb{R}}$, its ε-SUBDIFFERENTIAL at x is $\partial_\varepsilon h(x) = \{t \in (\mathbb{R}^m)^* \mid h(y) - h(x) \geq \langle t, y - x\rangle - \varepsilon \ \ \forall y \in \mathbb{R}^m\}$. The INDICATOR FUNCTION i of M is defined by $i(M, x)$ being equal to 0 if $x \in M$ and to $+\infty$ if $x \notin M$. The ε-normal set to M at $x \in M$ is equal to $\partial_\varepsilon i(M, x)$. Naturally, in the special case $\varepsilon = 0$, we omit ε in the names and notations.

Lemma 8.85 motivates the following definition.

Definition 8.86 (ε-enlargement of $\mathscr{T}(x)$) For $\varepsilon \geq 0$, bifunction Φ defined in the beginning of Sect. 8.6, and $x \in \mathrm{dom}\mathscr{T}$, ε-$\mathscr{T}(x) := \partial_\varepsilon\Phi(x, \cdot)(x)$ is called an ε-ENLARGEMENT OF $\mathscr{T}(x)$.

Lemma 8.87 (Approximate subdifferential of a sum, [248], Proposition 1.3) *Let h_1 and h_2 be proper lsc convex functions and $\varepsilon \geq 0$. Assume that there exists a point at which h_1 and h_2 are finite, and one of them is continuous. Then, there exist $\varepsilon_1, \varepsilon_2 \geq 0$ with $\varepsilon_1 + \varepsilon_2 = \varepsilon$ such that*

$$\partial_\varepsilon(h_1 + h_2)(x) = \partial_{\varepsilon_1}h_1(x) + \partial_{\varepsilon_2}h_2(x).$$

for all $x \in \mathrm{dom}h_1 \cap \mathrm{dom}h_2$.

Definition 8.88 (Mixed ε-solution of (sQVI)) $\bar{x} \in \mathrm{fix}\mathscr{K}$ such that there exist $\varepsilon_1, \varepsilon_2 \geq 0$ with $\varepsilon_1 + \varepsilon_2 = \varepsilon$ and $\bar{t} \in \varepsilon_2$-$\mathscr{T}(\bar{x})$ such that for all $y \in \mathscr{K}(\bar{x})$,

$\langle \bar{t}, y - \bar{x} \rangle \geq -\varepsilon_1$ is said to be a MIXED ε-SOLUTION OF (SQVI), written $\bar{x} \in$ mix-ε-Sol(sQVI).

For convex (sQVI), it is interesting that we have the following equalities of the approximate solution sets.

Proposition 8.89 (Relationship between approximate solution sets under convexity assumptions) *Assume that $\mathscr{K}$ is closed-convex-valued. Then, all the following assertions are equivalent: $\bar{x} \in \varepsilon$-Sol(QEP($\Phi, \mathscr{K}$)), $\bar{x} \in$ mix-ε-Sol(sQVI($\mathscr{T}, \mathscr{K}$)), $\bar{x} \in \varepsilon$-Sol(sQVI($\mathscr{T}, \mathscr{K}$)), and $\bar{x} \in \varepsilon$-Sol(wQVI($\mathscr{T}, \mathscr{K}$)).*

Proof Impose the assumptions. By Lemma 8.85, for $x \in \text{dom}\mathscr{T}$, there exists $t \in \mathscr{T}(x)$ such that $\Phi(x, y) = \langle t, y - x \rangle$ for all $y \in \mathbb{R}^m$. So, $\text{dom}\Phi(x, \cdot) = \mathbb{R}^m$ and $\Phi(x, \cdot)$ are continuous. It is easy to verify that $\bar{x} \in \varepsilon$-Sol(QEP) implies that $\bar{x} \in \varepsilon\text{-argmin}_{\mathscr{K}(\bar{x})}\Phi(\bar{x}, \cdot)$. The latter fact is equivalent to the first assertion in the following chain (by a well-known characterization of approximate solutions of convex minimization problems)

$$
\begin{aligned}
& 0 \in \partial_\varepsilon[\Phi(\bar{x}, \cdot) + i(\mathscr{K}(\bar{x}), \cdot)](\bar{x}) \\
\Leftrightarrow\ & \text{there exist } \varepsilon_1, \varepsilon_2 \geq 0 \text{ with } \varepsilon_1 + \varepsilon_2 = \varepsilon \text{ such that } 0 \in \partial_{\varepsilon_2}\Phi(\bar{x}, \cdot) \\
& (\bar{x}) + N_{\varepsilon_1}(\mathscr{K}(\bar{x}), \bar{x}) \\
\Leftrightarrow\ & \exists \varepsilon_1, \varepsilon_2 \geq 0 : \varepsilon_1 + \varepsilon_2 = \varepsilon, \exists \bar{t} \in \varepsilon_2 - \mathscr{T}(\bar{x}), \forall y \in \mathscr{K}(\bar{x}), \langle \bar{t}, y - \bar{x} \rangle \geq -\varepsilon_1 \\
\Leftrightarrow\ & \bar{x} \in \text{mix} - \varepsilon\text{-Sol(sQVI)}.
\end{aligned}
$$

Here, the first equivalence follows from Lemma 8.87. Now Proposition 8.74 implies the equivalence stated in Proposition 8.79. □

Similar to Definition 8.88, one can define mixed ε-solution of (DsQVI) and prove an equivalence assertion similar to Proposition 8.89 for convex problems. However, for nonconvex (sQVI) and (DsQVI), studies of the relationship between the mentioned types of approximate solutions are still not available.

8.6.3 Approximations of Traffic Networks with Demands Depending on Equilibrium Flows

Traffic network problems constitute one of the most important special classes of quasivariational inequalities and so also of quasivariational equilibrium models. In Chaps. 1, 2, and 5, traffic networks were studied, but either only with basic facts or in a particular direction. Hence here, we present the theory in detail from the beginning to make things clear. Recently, traffic networks with capacity constraints on arcs have been usually studied. The typical model is as follows. For traffic network $\mathscr{N}$, N is the set of nodes, L is that of directed arcs (called also links), W is the set of origin-destination pairs (O/D pairs) of nodes. Assume further that O/D pair $w \in W$ is connected by set P_w of paths, $P = \cup_{w \in W} P_w$, and $m = |P|$ (the number of paths in the network). Let $F := (F_1, \dots, F_m) = (F_p)_{p=1,\dots,m}$ denote a vector flow (also briefly called flow). Let the capacity constraint on arc $\ell \in L$ be $c_\ell > 0$. So, any arc flow v_ℓ on ℓ needs to satisfy $0 \le v_\ell \le c_\ell$ for $\ell \in L$. A vector flow F traversing arc ℓ is the following v_ℓ^F

$$v_\ell^F = \sum_{w \in W} \sum_{p \in P_w} \delta_{\ell p} F_p, \quad \text{where } \delta_{\ell p} = \begin{cases} 1 \text{ if } \text{ arc } \ell \text{ is involved in path } p, \\ 0 \text{ otherwise.} \end{cases}$$

Then, the set of vector flows satisfying the arc capacity constraints is

$$\{F \in \mathbb{R}_+^m \mid \sum_{w \in W} \sum_{p \in P_w} \delta_{\ell p} F_p \le c_\ell \text{ for all } \ell \in L\}.$$

If the (fixed) demand of O/D pair w is d_w, then the set of feasible vector flows is

$$K = \{F \in \mathbb{R}_+^m \mid \sum_{w \in W} \sum_{p \in P_w} \delta_{\ell p} F_p \le c_\ell \text{ for all } \ell \in L \text{ and}$$

$$\sum_{p \in P_w} F_p = d_w \text{ for all } w \in W\}.$$

Assume that the cost of vector flow F is $t(F) := (t(F_1), \dots, t(F_m))$, i.e, we have the cost map $t : \mathbb{R}_+^m \to (\mathbb{R}^m)^*$. When a given F is clear, we also write simply $t := (t_1, \dots, t_m)$.

Definition 8.90 (Types of approximate saturated arcs and approximate saturated paths) Let $F \in K$.

(a) For $F \in \varepsilon$-K, $\ell \in L$, $w \in W$, and $j, k \in P_w$ containing arc ℓ, if $v_\ell^F \ge c_\ell + \frac{\varepsilon(t_j + t_k)}{t_j t_k}$, then ℓ is said to be a (j, k)-ε-SATURATED ARC of F; if $0 \le v_\ell^F < c_\ell + \frac{\varepsilon(t_j + t_k)}{t_j t_k}$, then ℓ is a (j, k)-ε-NONSATURATED ARC of F (note that $v_\ell^F = F_j = F_k$).

When $\varepsilon = 0$, (j, k)-0-nonsaturatedness is called simply (j, k)-NONSATURATEDNESS and when $v_\ell^F = c_\ell$, ℓ is called a (j, k)-SATURATED ARC of F.

(b) If there exists a (j, k)-ε-saturated arc ℓ in a vector flow $F \in \varepsilon$-K such that ℓ belongs to a path p (i.e., F_p) of F, then p is called a (j, k)-ε-SATURATED PATH of F; otherwise, it is a (j, k)-ε-NONSATURATED PATH of F.

Similar to the above property, if $\varepsilon = 0$, we drop "ε" in the names.

Definition 8.91 (Wardrop approximate equilibrium flows for (TP(t, K))) A vector flow $H \in \varepsilon$-K is called an ε-EQUILIBRIUM FLOW if for all $w \in W$ and

$j, k \in P_w$, $t_k < t_j$ implies that $0 \leq H_j \leq \frac{\varepsilon(t_j+t_k)}{t_j t_k}$ or path k is a (j, k)-ε-saturated path of the vector flow H.

If $\varepsilon = 0$, we drop"ε" in the name.

The TRAFFIC EQUILIBRIUM PROBLEM (briefly, TRAFFIC PROBLEM), denoted by (TP(t, K)), is to find equilibrium flows or ε-equilibrium flows.

It is well-known that H is a vector equilibrium flow of (TP(t, K)) if H is a solution of the variational inequality: find $H \in K$ such that for all $F \in K$, $\langle t(H), F - H\rangle \geq 0$.

In this subsection, our traffic problem is more general than (TP(t, K)) (to suit diverse practical situations) as follows. Assume that the cost $T(F)$ of a vector flow F is a compact subset of $(\mathbb{R}^m_+)^*$, i.e., one has a compact-set-valued map $T : \mathbb{R}^m_+ \rightrightarrows (\mathbb{R}^m_+)^*$. Assume further that each demand d_w depends on equilibrium costs, or more directly, on equilibrium pattern flows H, i.e., we have a map $d_w : \mathbb{R}^m_+ \to \mathbb{R}_+$. Moreover, we are interested in the following case with a tolerance of the demands: $\sum_{p \in P_w} F_p \geq d_w(H) - \gamma$ for some fixed $\gamma \in \mathbb{R}_+$ and all O/D pair $w \in W$.

For our problem with a tolerance ε for capacity restrictions and γ for demands, the set of ε, γ-(APPROXIMATE) FEASIBLE VECTOR FLOWS is

$$\varepsilon, \gamma - K(H) := \{F \in \mathbb{R}^m_+ \mid \sum_{w \in W} \sum_{p \in P_w} \delta_{\ell p} F_p \leq c_\ell + \varepsilon \text{ for all } \ell \in L$$

$$\text{and } \sum_{p \in P_w} F_p \geq d_w(H) - \gamma \text{ for all } w \in W \text{ and } j, k \in P_w\}.$$

Then, ε, γ-$K(H)$ is clearly a closed, compact, and convex set for each H.

Definition 8.90 of approximate saturatedness applies also to our traffic problem with demand d_w depending on equilibrium costs and with tolerances, with the only replacement of ε-K by ε, γ-$K(H)$.

The Wardrop approximate equilibrium concept is extended as follows.

Definition 8.92 (Wardrop concepts for the case of elastic demands depending on equilibrium flows and set-valued costs)

(a) A vector flow $H \in \varepsilon, \gamma$-$K(H)$ such that $\forall w \in W, \forall j, k \in P_w, \exists t \in T(H)$ such that

$$t_k < t_j \;\Rightarrow 0 \leq H_j \leq \frac{\varepsilon(t_j + t_k)}{t_j t_k} \quad \text{or} \quad k \text{ is a (j, k)} - \varepsilon\text{-saturated path}$$

is called a WEAK ε, γ-EQUILIBRIUM FLOW.

(b) For $H \in \varepsilon, \gamma$-$K(H)$, if there exists $t \in T(H)$ such that $\forall w \in W, \forall j, k \in P_w$,

$$t_k < t_j \;\Rightarrow 0 \leq H_j \leq \frac{\varepsilon(t_j + t_k)}{t_j t_k} \quad \text{or} \quad k \text{ is a (j, k)} - \varepsilon\text{-saturated path,}$$

then H is called a STRONG ε, γ-EQUILIBRIUM FLOW.

THE TRAFFIC PROBLEM of finding equilibrium or approximate equilibrium flows in the sense given in Definition 8.92 is denoted by (TP(T, K)). We also use the simple notation (TP) for this problem, since in the rest of this subsection, we always speak about this model.

We now demonstrate a relationship between approximate equilibrium flows of our traffic problem with capacity restrictions on arcs, elastic demands, costs depending on equilibrium flows and approximate solutions of a set-valued quasivariational inequality.

Theorem 8.93 (Relationship between (wQVI) or (sQVI) and (TP))

(a) *A vector flow H is a weak ε, γ-equilibrium flow of* (TP) *if H is an ε, γ-solution of the corresponding weak set-valued quasi-variational inequality* (wQVI$_{TP}$) *in the sense that*

$$H \in \varepsilon, \gamma - K(H) \text{ such that } \forall F \in \varepsilon, \gamma - K(H), \exists t_F \in T(H), \langle t_F, F - H\rangle \geq -\varepsilon$$

and the following condition is satisfied

$$\max_{w \in W; t \in T(H); j,k \in P_w} \frac{\varepsilon(t_j + t_k)}{t_j t_k} \leq \gamma. \tag{8.10}$$

(b) *H is a strong ε, γ-equilibrium flow of* (TP) *if H is an ε, γ-solution of the corresponding strong set-valued quasivariational inequality* (sQVI$_{TP}$) *in the following sense*

$$H \in \varepsilon, \gamma - K(H) \text{ such that } \exists \in T(H), \forall F \in \varepsilon, \gamma - K(H), \langle t, F - H\rangle \geq \varepsilon$$

and condition (8.10) is satisfied.

Proof

(a) Suppose to the contrary that $H \in \varepsilon, \gamma$-Sol(wQVI$_{TP}$), but not a weak ε, γ-equilibrium flow of *(TP)*. The latter means that there exist $w \in W$ and $j, k \in P_w$ such that for all $t \in T(H)$ with $t_k < t_j$, one has $H_j > \frac{\varepsilon(t_j + t_k)}{t_j t_k}$ and k is a (j, k)-ε-nonsaturated path of the flow H.

Consider the following vector flow $F = (F_i)_{i=1,\dots,m}$

$$F_i = \begin{cases} H_i & \text{if } i \neq k, i \neq j, \\ H_j - \frac{\varepsilon}{t_j} & \text{if } i = j, \\ H_k - \frac{\varepsilon}{t_k} & \text{if } i = k. \end{cases}$$

Then, as we can consider for F only the case when arc ℓ belongs to both path j and k due to the definition of (j,k)-ε-nonsaturatedness, we have

$$v_\ell^F = \sum_{w\in W}\sum_{i\in P_w} \delta_{\ell i}F_i = \delta_{\ell i}H_i + \sum_{w\in W}\Big(\sum_{j,k\in P_w;\, j\neq i,k\neq i} (H_j - \frac{\varepsilon}{t_j}) + (H_k - \frac{\varepsilon}{t_k})\Big)$$

$$= v_\ell^H - \frac{\varepsilon}{t_j} - \frac{\varepsilon}{t_k}.$$

As $\frac{\varepsilon}{t_j} + \frac{\varepsilon}{t_k} < H_j = v_\ell^H$, one has $0 \le v_\ell^F \le c_\ell + \varepsilon$.

Moreover, for the chosen F, the supply is $\sum_{p\in P_w} F_p = d_w(H) - \frac{\varepsilon(t_j+t_k)}{t_j t_k} \ge d_w(H) - \gamma$ by condition (8.10). Hence, $F \in \varepsilon, \gamma$-$K(H)$.

We now have

$$\langle t, F-H\rangle = \sum_{i=1}^m t_i(F_i - H_i) = t_k(F_k - H_k) + t_j(F_j - H_j) = -2\varepsilon < -\varepsilon,$$

which is impossible since $H \in \varepsilon, \gamma$-Sol(wQVI$_{TP}$).

(b) The proof is similar to that of (i) (with some clear modifications).

□

Now let us investigate approximations of traffic networks by applying the results of Sects. 8.6.1 and 8.6.2. Hence, we consider the following bifunction

$$\Phi_{TP}(F_1, F_2) := \max_{t\in T(F_1)}\langle t, F_2 - F_1\rangle \quad \text{for } F_1, F_2 \in \mathbb{R}^m_+.$$

We study the traffic networks (TP^k), approximating the original network (TP), defined by sequences of data T^k with compact values and ε^k, γ^k-K^k, and then $\Phi^k_{TP^k}$. Namely, the convergence of weak approximate equilibrium flows of approximating problems (TP^k) to those of (TP) in terms of variational convergence is investigated.

We denote the set of weak ε, γ-equilibrium and strong ε, γ-equilibrium flows, resp., of problem (TP), by ε, γ-weq(TP) and ε, γ-seq(TP).

Corollary 8.94 (Convergence of approximate weak equilibria) *Assume that $\varepsilon^k \to \varepsilon$, $\gamma^k \to \gamma$.*

(a) *If $-\Phi^k_{TP^k}$ e/h- or mis-lop converges to $-\Phi_{TP}$, there is a subsequence $H^{k_j} \in \varepsilon^{k_j}, \gamma^{k_j}$-Sol(wQVI$^{k_j}_{TP^{k_j}}$) $\to H$ as $j \to +\infty$, and condition (8.10) is satisfied, then $H \in \varepsilon, \gamma$-weq(TP).*

(b) *If $-\Phi^k_{TP^k} \xrightarrow{e} -\Phi_{TP}$, there is a subsequence $H^{k_j} \in \varepsilon^{k_j}, \gamma^{k_j}$-Sol(wQVI$^{k_j}_{TP^{k_j}}$) $\to H$ as $j \to +\infty$, condition (8.10) is satisfied, and gph(ε, γ-K) $\sqsubset$ Li$_k$(gph(ε^k, γ^k-K^k)), then $H \in \varepsilon, \gamma$-weq(TP).*

(c) *If* $-\Phi^k_{TP^k}$ *w-e/h- or w-mis-lop converges to* $-\Phi_{TP}$ *and* $H^k \in \varepsilon^k, \gamma^k$-Sol(wQVI$^k_{TP^k}$) *tends to* H*, and condition (8.10) is satisfied, then* $H \in \varepsilon, \gamma$-weq(TP).

Proof

(a) Consider the case of e/h-convergence. By Theorem 8.93(i), ε, γ-Sol(wQVI$_{TP}$) $\subset \varepsilon, \gamma$-weq(TP), hence $H^{k_j} \in \varepsilon^{k_j}, \gamma^{k_j}$-weq(TP$_{TP^{k_j}}$). In virtue of Proposition 8.80(i), $H \in \varepsilon, \gamma$-Sol(wQVI$_{TP}$).
(b) The verification of this assertion is similar to that of (a) (using now Proposition 8.78(ii)).
(c) The proof is also analogous, using (iii) of Proposition 8.78 instead of (ii).

□

Corollary 8.95 (Convergence of approximate strong equilibria) *Assume that* $\varepsilon^k \to \varepsilon$*,* $\gamma^k \to \gamma$*.*

(a) *If* $-\Phi^k_{TP^k}$ *e/h- or mis-lop converges to* $-\Phi_{TP}$*, there is a subsequence* $H^{k_j} \in \varepsilon^{k_j}, \gamma^{k_j}$-Sol(sQVI$^{k_j}_{TP^{k_j}}$) $\to H$ *as* $j \to +\infty$*, and condition (8.10) is satisfied, then* $H \in \varepsilon, \gamma$-seq(TP).
(b) *If* $-\Phi^k_{TP^k} \overset{e}{\to} -\Phi_{TP}$*, there is a subsequence* $H^{k_j} \in \varepsilon^{k_j}, \gamma^{k_j}$-Sol(sQVI$^{k_j}_{TP^{k_j}}$) $\to H$ *as* $j \to +\infty$*, condition (8.10) is satisfied, and* gph(ε, γ-K) $\sqsubset$ Li$_k$(gph(ε^k, γ^k-K^k))*, then* $H \in \varepsilon, \gamma$-seq(TP).
(c) *If* $-\Phi^k_{TP^k}$ *w-e/h- or w-mis-lop converges to* $-\Phi_{TP}$*,* $H^k \in \varepsilon^k, \gamma^k$-Sol(sQVI$^k_{TP^k}$) *tends to* H*, and condition (8.10) is satisfied, then* $H \in \varepsilon, \gamma$-seq(TP).

Proof Apply Propositions 8.74 (since ε, γ-K and ε^k, γ^k-K^k are closed-convex-valued) and Theorem 8.93(b). □

8.7 Approximations of Problems of Optimization Under Stochastic Ambiguity

In this section, we apply variational convergence of bifunctions to approximations of an important class of minsup-problems, which are optimization problems closely related to strong quasiequilibrium problems. However, they are problems of finding minsup-values and minsup-points of bifunctions without connection with the zero value as quasiequilibrium problems. Hence, we rather employ results on variational convergence of bifunctions in Sect. 8.2 than make use of those in Sect. 8.4. Of course, we need also a lot of additional tools, since the theme is rather complex.

8.7.1 Problem Formulation and Examples

Let the set of all probability measures on $(\mathbb{R}^m, \mathscr{B}_{\mathbb{R}^m})$ be denoted by $\mathscr{M}$, where $\mathscr{B}_{\mathbb{R}^m}$ is the Borel σ-algebra on $\mathbb{R}^m$. Given a real-valued cost function φ defined on a subset of $\mathbb{R}^m \times \mathscr{M}$ and $A \subset \mathbb{R}^m$, the PROBLEM OF OPTIMIZATION UNDER STOCHASTIC AMBIGUITY is

$$\min_{x\in A}\sup_{P\in\mathscr{M}_0(x)}\varphi(x, P),$$

where the AMBIGUITY SET $\mathscr{M}_0 \subset \mathscr{M}$ depends on x-decisions in general.

This problem provides practitioners with machinery to model decision-making under uncertainty accounting for possibly uncertain, incomplete, and decision-dependent descriptions of unknown parameters. So, it comes in with a wide variety of forms. Here are two illustrative examples.

(a) *Expectation minimization.* Assume that $\varphi(x, P) = \int \psi(x, \xi)P(d\xi)$ for some $\psi : A \times \mathbb{R}^m \to \mathbb{R}$, which is P-integrable for every P in $\mathscr{M}_0(x) \subset \mathscr{M}$ for all $x \in A$. This case encapsulates stochastic optimization under distributional uncertainty and stochastic programs with decision-dependent probability measure and single-valued $\mathscr{M}_0(x)$ for each x, among others.
(b) *Robust optimization.* For some mapping $\Theta : A \rightrightarrows \mathbb{R}^m$ with domain being the whole A, let $\mathscr{M}_0(x) := \{P \in \mathscr{M} \mid P(\xi) = 1 \text{ for some } \xi \in \Theta(x)\}$. ($\xi$ is a random vector with distribution function P.) Then, optimization under stochastic ambiguity in the form of example (a) leads to the problem $\min_{x\in A}\sup_{\xi\in\Theta(x)}\psi(x, \xi)$, which captures many robust optimization and semi-infinite programming problems.

To permit a unified treatment of the vast number of possibilities, we come back to our formulation of a minsup-problem by considering ambiguity set $\mathscr{M}_0(x)$ as a subset $\mathscr{D}(x)$ of a general metric space Y, which encompasses the above two examples and many more. Thus, we define the PROBLEM OF OPTIMIZATION UNDER STOCHASTIC AMBIGUITY as

$$\text{(AP)} \qquad \min_{x\in A}\sup_{y\in\mathscr{D}(x)}\varphi(x, y),$$

where $\mathscr{D}$ is defined on the whole $A \subset \mathbb{R}^m$ and $\text{dom}\varphi = \text{gph}\mathscr{D}$.

Formulation Using Distribution Functions

We will use the following expression of ambiguity centered on accumulative distribution functions on $\mathbb{R}^m$. To quantify the distance between two distribution functions, a geometrically intuitive metric, the hypo-distance is defined below. It has a convenient property that closed balls in the space are compact and so facilitates results such as those related to existence of solutions. Recall that every probability measure P on $(\mathbb{R}^m, \mathscr{B}_{\mathbb{R}^m})$ defines a DISTRIBUTION FUNCTION $F : \mathbb{R}^m \to [0, 1]$ through $F(\xi) := P(S_\xi)$ for $\xi \in \mathbb{R}^m$, where $S_\xi := \{\zeta \in \mathbb{R}^m \mid \zeta \leq \xi\}$ (vector inequalities are understood componentwise). F has the following properties.

It is nondecreasing and continuous from above. $F(\xi^k) \to 0$ if $\xi_i^k \to -\infty$ for some $i = 1, \ldots, m$ with the other components fixed, and $F(\xi^k) \to 1$ when $\xi_i^k \to +\infty$ for all i. Moreover, $\Delta_A F \geq 0$ for every rectangle A, where $\Delta_A F := \sum_{j=1}^{2^m} (\mathrm{sgn}_A v^j) F(v^j)$, with $v^j, j = 1, \ldots, 2^m$, being the vertices of A and $\mathrm{sgn}_A v^j = 1$ if the number of components v_i^j at a lower bound of A is even and $\mathrm{sgn}_A v^j = -1$ if the number is odd. (For instance, with $m = 2$, $\Delta_A F = F(v^1) - F(v^2) - F(v^3) + F(v^4)$, where v^1 is the upper right vertex, v^2 and v^3 are the upper left and lower right vertices, and v^4 is the lower left vertex.) For every $F : \mathbb{R}^m \to \mathbb{R}$ with these properties, there exists a unique probability measure P on $(\mathbb{R}^m, \mathscr{B}_{\mathbb{R}^m})$ such that $P(A) = \Delta_A F$ for rectangles A and $P(S_\xi) = F(\xi)$ for all $\xi \in \mathbb{R}^m$ (see Theorem 14.1 in [107]). Basic knowledge of probability (mostly known at university knowledge level) is also found in probability textbooks like, e.g., [185] and others. We only explain in detail some results to make things clear and avoid confusion. Let

$$\mathrm{cd-fcns}(\mathbb{R}^m) := \{F : \mathbb{R}^m \to [0, 1] \mid \exists P \in \mathscr{M} \text{ with } F(\xi) := P(S_\xi)\ \forall \xi \in \mathbb{R}^m\}$$

be the set of (cumulative) distribution functions. Then, formulation of (AP) in terms of distribution functions amounts to specifying ambiguity sets that are subsets of cd-fcns$(\mathbb{R}^m)$.

To facilitate analysis of distribution functions, we view them as subsets of the larger class usc-fcns$(\mathbb{R}^m; [0, 1])$ of upper semicontinuous (usc) functions $g : \mathbb{R}^m \to [0, 1]$. If $F \in$ cd-fcns$(\mathbb{R}^m)$, $\xi^k \to \xi$, and $\zeta_i^k = \max\{\xi_i, \xi_i^k\}$ for all i, then $F(\xi^k) \leq F(\zeta^k)$ as F is nondecreasing. By the continuity from above, $F(\zeta^k) \to F(\xi)$ and hence F is usc. So, cd-fcns$(\mathbb{R}^m) \subset$ usc-fcns$(\mathbb{R}^m; [0, 1])$. We equip the space usc-fcns$(\mathbb{R}^m; [0, 1])$ with the hypo-distance d^h. Note that $\mathsf{d}^h(g, g') \leq 1$ for all $g, g' \in$ usc-fcns$(\mathbb{R}^m; [0, 1])$, and (usc-fcns$(\mathbb{R}^m; [0, 1]), \mathsf{d}^h)$ is a complete, separable, and finitely compact metric space (see [405], Corollary 3.6).

8.7.2 *Foundations of Distribution Functions*

We will develop formulations of (AP) with $Y =$ usc-fcns$(\mathbb{R}^m; [0, 1])$ and $d_Y = \mathsf{d}^h$, i.e., ambiguity sets are given in terms of distribution functions and the analysis is carried out using the hypo-distance. Such formulations are referred to as DISTRIBUTION-BASED. So, cd-fcns$(\mathbb{R}^m)$ is always viewed as a subset of the metric space (usc-fcns$(\mathbb{R}^m; [0, 1]), \mathsf{d}^h)$. Apparently, distribution-based formulations of stochastic ambiguity require one to develop the necessary mathematical tools for analyzing the inner maximization over usc functions. This subsection is devoted to discussing distribution functions as subsets of usc functions.

1. Weak Convergence and Other Properties

Even though the space usc-fcns($\mathbb{R}^m$; [0, 1]) is not linear, it is useful to consider the *weighted averages* $(1-\lambda)g + \lambda g'$ for $g, g' \in$ usc-fcns($\mathbb{R}^m$; [0, 1]) and $\lambda \in [0, 1]$ defined by $((1-\lambda)g + \lambda g')(\xi) := (1-\lambda)g(\xi) + \lambda g'(\xi)$ for $\xi \in \mathbb{R}^m$ (pointwise manner).

Definition 8.96 (Convexity on usc-fcns($\mathbb{R}^m$; [0, 1])) A set $S \subset$ usc-fcns($\mathbb{R}^m$; [0, 1]) is called a CONVEX SET if $(1-\lambda)g + \lambda g' \in S$ for every $g, g' \in S$ and $\lambda \in [0, 1]$.

$\psi : S \to \mathbb{R}$ is called a CONVEX FUNCTION if $S \subset$ usc-fcns($\mathbb{R}^m$; [0, 1]) is convex and $\psi((1-\lambda)g + \lambda g') \le (1-\lambda)\psi(g) + \lambda\psi(g')$.

It is not hard to see that cd-fcns($\mathbb{R}^m$) and usc-fcns($\mathbb{R}^m$; [0, 1]) are convex and every local minimizer of a convex function over a convex set is a global minimizer.

Like in Sect. 7.8, weak convergence is a crucial concept for approximating probability measures P^k. For convenience, here we will employ some equivalent formulations of this convergence. Let us recall some basic facts from probability theory for convenience of the reader. For $A, B \in \mathscr{B}_{\mathbb{R}^m}$, $P(A \cup B) = P(A) + P(B) - P(A \cap B)$. Then, by induction, one deduces the so-called inclusion-exclusion formula as follows. Write the preceding formula as $P(\cup_{i=1}^{\ell+1} A_i) = P(\cup_{i=1}^{\ell} A_i) + P(A_{\ell+1}) - P(\cup_{i-1}^{\ell}(A_i \cap A_{i+1}))$, and use it to derive by induction the INCLUSION-EXCLUSION FORMULA $P(\cup_{i=1}^{\ell} A_i) = \sum_{i=1}^{\ell} P(A_i) - \sum_{i<j} P(A_i \cap A_j) + \sum_{i<j<p} P(A_i \cap A_j \cap A_p) + \ldots + (-1)^{\ell+1} P(A_i \cap \ldots \cap A_\ell)$. The indicator function is defined by, unlike the function with the same name in optimization theory,

$$I_A(x) := \begin{cases} 1 \text{ if } x \in A, \\ 0 \text{ if } x \notin A. \end{cases}$$

We also need the notion of a P-continuity set: A set $A \in \mathscr{B}_{\mathbb{R}^m}$ is signified as a P-CONTINUITY SET if $P(\mathrm{bd}A) = 0$.

Proposition 8.97 (Equivalent formulations of weak convergence) *For probabilities P^k and P on $(\mathbb{R}^m, \mathscr{B}_{\mathbb{R}^m})$ and the corresponding distribution functions F^k and F, the following statements are equivalent.*

(a) *For every bounded continuous function $h : \mathbb{R}^m \to \mathbb{R}$, $\int_{\mathbb{R}^m} h(\xi)P^k(d\xi)$ tends to $\int_{\mathbb{R}^m} h(\xi)P(d\xi)$.*
(b) $\mathrm{ls}_k P^k(A) \le P(A)$ *for every closed $A \in \mathscr{B}_{\mathbb{R}^m}$.*
(c) $\mathrm{li}_k P^k(C) \ge P(C)$ *for every open $C \in \mathscr{B}_{\mathbb{R}^m}$.*
(d) $\lim_k P^k(A) = P(A)$ *for every P-continuity set A.*
(e) $\lim_k F^k(\xi) = F(\xi)$ *for every continuity point ξ of F, where F^k and F are the distribution functions of P^k and P, resp. In this case F^k are also said to weakly converge to F.*

Proof (a) $\Rightarrow$ (b). Let

$$\varphi_j(t) = \begin{cases} 1 & \text{if } t \le 0, \\ 1 - jt & \text{if } 0 \le t \le j^{-1}, \\ 0 & \text{if } j^{-1} \le t. \end{cases}$$

Then, $f_j(\xi) := \varphi_j(\text{dist}(\xi, A))$ is continuous and bounded above by 1, and $f_j(\xi) \to I_A(\xi)$ on the right as $j \to +\infty$ for closed $A \in (\mathbb{R}^m, \mathscr{B}_{\mathbb{R}^m})$. If (a) holds, then $\text{ls}_k P^k(A) \le \lim_k \int_{\mathbb{R}^m} f_j(\xi) P^k(d\xi) = \int_{\mathbb{R}^m} f_j(\xi) P(d\xi)$. We have $\int_{\mathbb{R}^m} f_j(\xi) P(d\xi) \to \int_{\mathbb{R}^m} I_A(\xi) P(d\xi) = P(A)$ as $j \to +\infty$, i.e., (b) is derived.

(b) $\Leftrightarrow$ (c). Simply take $A = \mathbb{R}^m \setminus C$.

(b) and (c) $\Rightarrow$ (d). We have

$$P(\text{int}A) \le \text{li}_k P^k(\text{int}A) \le \text{li}_k P^k(A) \le \text{ls}_k P^k(A) \le \text{ls}_k P^k(\text{cl}A) \le P(\text{cl}A).$$

As A is a P-continuity set, $P(\text{cl}A) - P(\text{int}A) = P(\text{bd}A) = 0$. Hence, the above inequalities imply assertion (d).

(d) $\Rightarrow$ (a). Assume that f is continuous and $|f|$ is bounded by K. Given $\varepsilon > 0$, take $\alpha_1 < \ldots < \alpha_\ell$ such that $\alpha_1 < -K < K < \alpha_\ell$, $\alpha_{i+1} - \alpha_i < \varepsilon$, and $P\{\xi \mid f(\xi) = \alpha_i\} = 0$ (the last condition is achieved as $\{\xi \mid f(\xi) = \alpha\}$ are disjoint for different α). Set $A_i := \{\xi \mid \alpha_i < f(\xi) \le \alpha_{i+1})\}$. By the continuity of f, $\text{cl}A_i \subset \{\xi \mid \alpha_i \le f(\xi) \le \alpha_{i+1})\}$ and $A_i \supset \{\xi \mid \alpha_i < f(\xi) < \alpha_{i+1})\}$. Consequently, $\text{bd}A_i \subset \{\xi \mid f(\xi) = \alpha_i\} \cap \{\xi \mid f(\xi) = \alpha_{i+1}\}$, and hence $P(\text{bd}A_i) = 0$. Now $|\int_{\mathbb{R}^m} f_j(\xi) P^k(d\xi) - \sum_{i=1}^{\ell} \alpha_i P^k(A_i)| \le \varepsilon$ and similar for P, and so $\sum_{i=1}^{\ell} \alpha_i P^k(A_i)| \to \sum_{i=1}^{\ell} \alpha_i P(A_i)|$ in view of assertion (d). Because of the arbitrariness of ε, assertion (a) follows.

(d) $\Rightarrow$ (e). We claim first that for $S_\xi := \{\zeta \in \mathbb{R}^m \mid \zeta \le \xi\}$ as before, F is continuous at ξ if and only if $P(\text{bd}S_\xi) = 0$. Since F is nondecreasing, it is continuous at ξ if and only if it is continuous from below in the sense that $\lim F(\xi - h) = F(\xi)$ as $h \to 0^+$. Thus, F is continuous at ξ if and only if $F(\xi) = P(S_\xi) = P(\text{int}S_\xi)$, which holds if and only if $P(\text{bd}S_\xi) = 0$, i.e., the claim is checked. Therefore, $F^k(\xi) = P^(S_\xi) \to P(S_\xi) = F(\xi)$ by assertion (d).

(e) $\Rightarrow$ (c). Since only countably many paralleled closed hyperplanes can have positive P-measure, there exists a dense set D of reals such that $P\{\xi \mid \xi_i = d\} = 0$ for $d \in D$ and $i = 1, \ldots, m$. Let $\mathscr{A}$ be the class of rectangles $A = \{\xi \mid a_i \le \xi_i \le b_i, i = 1, \ldots, m\}$ with all a_i, b_i lying in D. All 2^m vertices of such a rectangle are continuity points of F, and so $\lim_k F^k = F$ (by assertion (e)) implies that $\Delta_A F^k \to \Delta_A F$. Since for rectangle A, $P(A) = \Delta_A F$ and $P^k(A) = \Delta_A F^k$, $P^k(A) \to P(A)$ by the definition of $\Delta_A F$. The inclusion-exclusion formula implies that $P^k(B) \to P(B)$ for finite unions B of elements of $\mathscr{A}$. Now consider an open set $C \subset \mathbb{R}^m$. Because D is dense on the line, C is a countable union of sets $A_q \in \mathscr{A}$. But, $P(\cup_{q \le Q} A_q) = \lim_k P^k(\cup_{q \le Q} A_q) \le \text{li}_k P^k(C)$. Letting $Q \to +\infty$ yields assertion (c). □

Theorem 8.98 (Equivalence of weak convergence and hypo-convergence) *For distribution functions $F^k, F \in \text{cd-fcns}(\mathbb{R}^m)$, $\mathsf{d}^h(F^k, F) \to 0$ if and only if F^k converges weakly to F.*

Proof "If" Assume that F^k converges weakly to F, i.e., for any $\xi \in \mathbb{R}^m$, $\text{ls}_k F^k(\xi) = \text{ls}_k P^k(S_\xi) \le P(S_\xi) = F(\xi)$. We verify condition (a) of hypo-convergence. Let $\xi^k \to \xi$ and $\varepsilon > 0$ be arbitrary. The continuity from above of F implies the existence of $\zeta > 0$ in $\mathbb{R}^m$ such that $F(\xi + \zeta) \le F(\xi) + \varepsilon$. There also is a $\bar{k}$ such that $\xi^k \le \xi + \zeta$ for all $k \ge \bar{k}$. Hence, by the monotonicity of F^k, $F(\xi^k) \le F(\xi + \zeta)$ for that large k. So, for $k \ge \bar{k}$, $\text{ls}_k F^k(\xi^k) \le \text{ls}_k F^k(\xi + \zeta) \le F(\xi + \zeta) \le F(\xi) + \varepsilon$. By the arbitrariness of ε, the required condition (a) holds. To show condition (b), let $\xi \in \mathbb{R}^m$ be arbitrary. Pick $\zeta > 0$ in $\mathbb{R}^m$. There exists an open set A with $S_\xi \subset A \subset S_{\xi+\zeta}$. Since the complement A^c is closed, $\text{ls}_k P^k(A^c) \le P(A^c)$, and so $\text{li}_k P^k(A) \ge P(A)$. The monotonicity of P^k and P implies that

$$\text{li}_k F^k(\xi + \zeta) = \text{li}_k P^k(S_{\xi+\zeta}) \ge \text{li}_k P^k(A) \ge P(A) \ge P(S_\xi) = F(\xi). \tag{8.11}$$

Let $\varepsilon > 0$, $k_0 = 1$, $\zeta^\ell \to 0^+$ with $\zeta^\ell > 0$ for $\ell \in \mathbb{N}$. For each ℓ, there exists by (8.11) a $k_\ell \ge k_{\ell-1}$ such that for all $k \ge k_\ell$, $F^k(\xi + \zeta^\ell) \ge F(\xi) - \varepsilon$. Construct the sequence ξ^k as follows. For $k_1 \le k \le k_2$, set $\xi^k = \xi + \zeta^1$, for $\ell = 2, 3, \ldots$, set $\xi^k = \xi + \zeta^\ell$ for $k_\ell < k \le k_{\ell+1}$. By construction, $F^k(\xi^k) \ge F(\xi) - \varepsilon$. Moreover, $\xi^k \to \xi$, and condition (b) of hypo-convergence is verified.

"Only if" Supposing that $\mathsf{d}^h(F^k, F) \to 0$, it follows by (a) of the definition that $\text{ls}_k F^k(\xi) \le F(\xi)$ for all $\xi \in \mathbb{R}^m$. By (e) of Proposition 8.97, it remains to show that $\text{li}_k F^k(\xi) \ge F(\xi)$ for each continuity point ξ of F. For any $\varepsilon > 0$ and that point ξ, there exists $\delta > 0$ such that $F(\zeta) \ge F(\xi) - \varepsilon$ for all ζ satisfying $\|\zeta - \xi\|_2 \le \delta$. Let ζ be such a point with $\zeta < \xi$. Condition (b) of the hypo-convergence definition implies that there exist $\zeta^k \to \zeta$ such that $\text{li}_k F^k(\zeta^k) \ge F(\zeta)$. Moreover, there exists $\bar{k}$ such that $\zeta^k \le \xi$ for $k \ge \bar{k}$. By the monotonicity of F^k, $F^k(\xi) \ge F^k(\xi^k)$ for all $k \ge \bar{k}$. Consequently,

$$\text{li}_k F^k(\xi) \ge \text{li}_k F^k(\zeta^k) \ge F(\zeta) \ge F(\xi) - \varepsilon.$$

We are in a position to conclude that $F^k(\xi) \to F(\xi)$ for all continuity point ξ of F. □

As empirical distribution functions are crucial in application, the following lemma is important.

Lemma 8.99 (Almost sure convergence of empirical measures, Theorem 11.4.1, [185]) *The empirical measure P^k of a probability measure P on a separable metric space converges almost surely to P.*

By this lemma, an immediate consequence of Theorem 8.98 is the fact that the empirical distribution function under independent sampling hypo-converges almost surely to the distribution function from which the samples are drawn. An obstacle for application is the known fact that the weak limit of a sequence of distribution functions might not be a distribution function. To overcome this, one needs the following tightness concept. A subset $S \subset \text{cd-fcns}(\mathbb{R}^m)$ is said to be TIGHT if for all $\varepsilon > 0$, there exists a rectangle A such that $\Delta_A F \geq 1 - \varepsilon$ for all $F \in S$.

Proposition 8.100 (Convergence to distribution function, Theorem 29.3 [107]) *If the sequence* $\{F^k\}_k \subset \text{cd-fcns}(\mathbb{R}^m)$ *is tight, then*

(a) *the exists a subsequence* $\{F^{k_j}\}_j$ *and an* $F \in \text{cd-fcns}(\mathbb{R}^m)$ *such that* $\mathsf{d}^h(F^{k_j}, F) \to 0$*;*
(b) *If* $\mathsf{d}^h(F^k, F) \to 0$ *for some* $F : \mathbb{R}^m \to \mathbb{R}$*, then* $F \in \text{cd-fcns}(\mathbb{R}^m)$.

Tightness is close to compactness as stated in the following.

Lemma 8.101 (Relationship between tightness and compactness, Proposition 3.4 [407]) *For* $S \subset \text{cd-fcns}(\mathbb{R}^m)$*, one has*

(a) *if* S *is compact, then* S *is tight;*
(b) *if* S *is tight, then* $\text{cl}S$ *is compact and* $S \subset \text{cd-fcns}(\mathbb{R}^m)$.

2. Ambiguity Sets in Distribution-Based Formulations
We illustrate one of the possibilities in distribution-based formulations by the following example.

Example 8.102 (Stochastic dominance) Let $F_0(\cdot, x), G_0(\cdot, x) \in \text{cd-fcns}(\mathbb{R}^m)$, which might depend on $x \in \mathbb{R}^m$, and

$$S(x) := \{F \in \text{cd} - \text{fcns}(\mathbb{R}^m) \mid F_0(\xi, x) \leq F(\xi) \leq G_0(\xi, x) \text{ for all } \xi \in \mathbb{R}^m\}.$$

We say that a random vector ξ DOMINATES IN THE FIRST ORDER a random vector ζ if $F(\xi; x) \leq F(\zeta; x)$ for all $x \in \mathbb{R}^m$. Then, S specifies the investigation of all random vectors that are dominated in the first order with distribution function $G_0(\cdot, x)$ and also by a random vector with distribution function $F_0(\cdot, x)$ in the reoriented sense obtained by switching from the traditional focus on profit/gain to cost/loss. For $\xi \in \mathbb{R}^m$ and $S(x) \ni F^k \to F$, condition (a) of hypo-convergence gives $F_0(\xi, x) \leq F(\xi)$. Condition (b) and the fact that $G_0(\cdot, x)$ is usc yield that for some $\xi^k \to \xi$,

$$G_0(\xi, x) \geq \text{ls}_k G_0(\xi^k, x) \geq \text{li}_k F^k(\xi^k) \geq F(\xi).$$

Therefore, by virtue of the tightness of $S(x)$ and Lemma 8.101, we have the compactness of $S(x)$. From the definition of $S(x)$, we see that it is also convex. Thus, this ambiguity set has good properties for application in distribution-based formulations.

8.7.3 Lop-Convergence in Optimization Under Stochastic Ambiguity

In this subsection, we present some specific results under various assumptions about (AP) and its approximating problems (AP^k). From Proposition 8.39, Theorems 8.40, 8.42, and 8.43, apparently x-ancillary tight mis-lop convergence of φ^k plays a crucial role in the convergence of approximate solutions of the minsup-problems (AP^k) to those of (AP). By Lemma 8.101, $\varphi^k \overset{mis-lop}{\to} \varphi$ implies the x-tightness of this convergence if $\{\mathscr{D}^k(x^k)\}_k$ is a tight set for every $x^k \to x$ selected in condition (b) of mis-lop convergence. In turn, for mis-lop convergence, we can apply besides the definition, Theorems 8.34 (its characterization) or 8.36 (a sufficient condition).

Now we provide an example of distribution-based formulations, where $\varphi^k \overset{mis-lop}{\to} \varphi$ holds.

Example 8.103 (Approximations of stochastic dominance) Let φ^k, φ be bifunctions on $\mathbb{R}^m \times$ usc-fcns($\mathbb{R}^m$; [0, 1]) with φ^k converging continuously to φ and the domains specified as follows. $A^k = A$ are closed for all k and for all $x \in A$, $\{F_0(\cdot, x), G_0(\cdot, x), F_0^k(\cdot, x), G_0^k(\cdot, x)\}_k \subset$ cd-fcns($\mathbb{R}^m$) are reference distribution functions depending on $x \in A$. Moreover, $F_0^k(\cdot, x), G_0^k(\cdot, x)$ continuously converge to $F_0(\cdot, x), G_0(\cdot, x)$, resp. The domains of φ^k, φ are then defined by

$$\mathscr{D}(x) := \{F \in \text{cd} - \text{fcns}(\mathbb{R}^m) \mid F_0(\xi, x) \le F(\xi) \le G_0(\xi, x) \text{ for all } \xi \in \mathbb{R}^m\},$$

$$\mathscr{D}^k(x) := \{F \in \text{cd} - \text{fcns}(\mathbb{R}^m) \mid F_0^k(\xi, x) \le F(\xi) \le G_0^k(\xi, x) \text{ for all } \xi \in \Theta^k\},$$

where $\Theta^k \subset \mathbb{R}^m$ Painlevé-Kuratowski converges to $\mathbb{R}^m$. In this case, the stochastic dominance constraint is approximated in two ways: The reference distribution functions are approximated, and the constraints are enforced on a smaller set, possibly consisting of a finite number of points. (Since $\mathbb{R}^m$ is separable, there are finite sets Θ^k set-converging to $\mathbb{R}^m$.) We claim that $\mathscr{D}^k$ continuously converges to $\mathscr{D}$. Indeed, fix any $A \ni x^k \to x \in A$ and $F \in \text{ls}_k \mathscr{D}^k(x^k)$. Then, there exists $F^\ell \in \mathscr{D}^{k_\ell}(x^{k_\ell})$ tends to F, and so for all $\xi \in \Theta^{k_\ell}$, $F_0^{k_\ell}(\xi, x^{k_\ell}) \le F^\ell(\xi) \le G_0^{k_\ell}(\xi, x^{k_\ell})$. Let $\xi \in \mathbb{R}^m$. There exists $\Theta^k \ni \xi^k \to \xi$ by the set-convergence of Θ^k. Since $F^\ell \to F$,

$$F(\xi) \ge \text{ls}_\ell F^\ell(\xi^{k_\ell}) \ge \text{ls}_\ell F^{k_\ell}(\xi^{k_\ell}, x^{k_\ell}) = F_0(\xi, x).$$

Hence, the lower bound is attained. To consider the upper bound, let $\varepsilon > 0$. By the continuous convergence of G_0^k, there exist $\delta > 0$ and ℓ_1 such that when $\|\eta - \xi\|_2 \le \delta$ and $\ell \ge \ell_1$,

$$G_0^{k_\ell}(\eta, x^{k_\ell}) \le G_0(\xi, x) + \varepsilon/2.$$

Take $\zeta > \xi$ in $\mathbb{R}^m$ such that $0 < \|\zeta - \xi\|_2 < \delta$. Since Θ^k set-converges to $\mathbb{R}^m$, we find $\zeta^\ell \in \Theta^{k_\ell}$ tending to ζ and $\ell_2 \geq \ell_1$ such that $0 < \|\zeta^\ell - \xi\|_2 < \delta$ and $\zeta^\ell > \xi$ for all $\ell \geq \ell_2$. The hypo-convergence of F^ℓ to F guarantees the existence of $\xi^\ell \to \xi$ such that $\mathrm{li}_\ell F^\ell(\xi^\ell) \geq F(\xi)$. Hence, there exists $\ell_3 \geq \ell_2$ such that $F^\ell(\xi^\ell) + \varepsilon/2 \geq F(\xi)$ and $\xi^\ell \leq \zeta^\ell$ for all $\ell \geq \ell_3$. Therefore, now the monotonicity of F^ℓ implies that for all $\ell \geq \ell_3$,

$$F(\xi) \leq F^\ell(\xi^\ell) + \varepsilon/2 \leq F^\ell(\zeta^\ell) + \varepsilon/2 \leq G_0^{k_\ell}(\zeta^\ell, x^{k_\ell}) + \varepsilon/2 \leq G_0(\xi, x) + \varepsilon.$$

As ε is arbitrary, we conclude that F satisfies the upper bound. Being a convergent sequence, $\{F^\ell\}_\ell$ is a compact set and so is a tight set by Lemma 8.101. As a consequence, F is a distribution function (see Proposition 8.100), and thus $F \in \mathscr{D}(x)$ and $\mathrm{Ls}_k \mathscr{D}^k(x^k) \subset \mathscr{D}(x)$. It remains to consider the corresponding liminf. Take $F \in \mathscr{D}(x)$, and construct $F^k : \mathbb{R}^m \to [0, 1]$ by setting for $x \in \mathbb{R}^m$,

$$F^k(\xi) = \max\{F_0^k(\xi, x^k), \min[F(\xi), G_0^k(\xi, x^k)]\},$$

which is a distribution function and belongs to $\mathscr{D}^k(x^k)$. We indicate that $F \xrightarrow{h} F$ by means of definition. For condition (a), let $\xi^k \to \xi$. By the continuous convergence of $F_0^k(\cdot, x)$, $G_0^k(\cdot, x)$ and the upper semicontinuity of F, $\mathrm{ls}_k F^k(\xi^k) \leq \max\{F_0(\xi, x), \min[F(\xi), G_0(\xi, x)]\} = F(\xi)$ as (a) required. For condition (b), let $\xi^k = \xi$ for all k. Then, $\mathrm{li}_k F^k(\xi^k) = \max\{F_0(\xi, x), \min[F(\xi), G_0(\xi, x)]\} = F(\xi)$. Hence, $F^k \xrightarrow{h} F$ and thus $\mathscr{D}(x) \subset \mathrm{Li}_k F^k(x^k)$. Therefore, $\mathscr{D}^k(x^k) \xrightarrow{P-K} \mathscr{D}(x)$ and $\varphi^k \xrightarrow{mis-lop} \varphi$ by Theorem 8.36.

Having this mis-lop convergence, one can apply its variational properties given in Sect. 8.2.2 to obtain approximation results for the optimization problem under stochastic ambiguity in question.

Many distributed-based optimization problems under stochastic ambiguity of the form (AP) have the objective bifunction φ satisfying $\varphi^k \xrightarrow{mis-lop} \varphi$ like in the above example. Then, the following quantification result, which is in fact a part of Proposition 8.58, is important.

Proposition 8.104 (Estimates of minsup-values and minsup-points) *Assume that $\varphi^k, \varphi \in$ lsc-nrec$(\mathbb{R}^n; Y)$, $\rho \geq 0$, minsup$\varphi \in [-\rho, \rho]$, argminsup$\varphi \cap \rho\mathbb{B}_{\mathbb{R}^n} \neq \emptyset$, and φ^k satisfy similar conditions. Then,*

$$|\mathrm{minsup}\varphi^k\text{-}\mathrm{minsup}\varphi| \leq \hat{\mathrm{d}}_\rho^e(\xi^k, \xi).$$

If, additionally, there exists an increasing and continuous function $\psi : \mathbb{R}_+ \to \mathbb{R}_+$ with $\psi(0) = 0$ such that

$$\xi(x)\text{-}\mathrm{minsup}\varphi \geq \psi(\mathrm{dist}(x, \mathrm{argminsup}\varphi)) \quad \forall x \in \rho'\mathbb{B}_{\mathbb{R}^n},$$

where $\rho' > \rho + \hat{\mathsf{d}}^e_\rho(\xi^k, \xi)$, *then*

$$(\text{argminsup}\varphi^k \cap \rho\mathbb{B}_{\mathbb{R}^n}) \subset \text{argminsup}\varphi + (\hat{\mathsf{d}}^e_\rho(\xi^k, \xi) + \psi^{-1}(2\hat{\mathsf{d}}^e_\rho(\xi^k, \xi)))\mathbb{B}_{\mathbb{R}^n}.$$

8.8 Exercises

Exercise 8.1 (Formulae for weak upper and weak lower e/h-limits) Prove the following formulae.

(a) For $x \in A$ and $y \in B$,

$$\text{w} - \text{ehls}\Phi^k(x, y) = \inf_{\{A^k \ni x^k \to x\}} \max_{\{\mathscr{D}^k(x^k) \ni y^k \to y\}} \text{li}_k \Phi^k(x^k, y^k);$$

and for $x \notin A$,

$$\text{w} - \text{ehls}\Phi^k(x) = \inf_{\{A^k \ni x^k \to x\}} \max_{\{y^k \in \mathscr{D}^k(x^k)\}} \text{li}_k \Phi^k(x^k, y^k);$$

(b) For $y \in B$ and $x \in A$,

$$\text{w} - \text{ehli}\Phi^k(x, y) = \sup_{\{B^k \ni y^k \to y\}} \min_{\{(\mathscr{D}^k)^{-1}(y^k) \ni x^k \to x\}} \text{ls}_k \Phi^k(x^k, y^k);$$

and for $y \notin B$,

$$\text{w} - \text{ehli}\Phi^k(y) = \sup_{\{B^k \ni y^k \to y\}} \min_{\{x^k \in (\mathscr{D}^k)^{-1}(y^k)\}} \text{ls}_k \Phi^k(x^k, y^k).$$

Exercise 8.2 (A characterization of weak e/h-convergence) Prove that Φ^k weakly e/h-converging to Φ is equivalent to the following assertions

(a) $\Phi(x, y) \le$ w-ehls$\Phi^k(x, y)$ when $x \in A$, $y \in \mathscr{D}(x)$, and w-ehls$\Phi^k(x) = +\infty$ when $x \notin A$;
(b) w-ehli$\Phi^k(x, y) \le \Phi(x, y)$ when $y \in B$, $x \in \mathscr{D}^{-1}(y)$, and w-ehli$\Phi^k(y) = -\infty$ when $y \notin B$.

Exercise 8.3 (Weak inside epi-convergence of sup-projections) Prove that if $\Phi^k \overset{w-i-e/h}{\to} \Phi$ x-ancillary tightly, then dom$\xi \neq \emptyset$ and w-i-e-lim$_k \xi^k = \xi$, where ξ^k and ξ are the sup-projections of Φ^k and Φ, resp.

Exercise 8.4 (Convergence of maxinf-points under types of e/h-convergence) Assume that the domains of η^k and η are nonempty, $\varepsilon \ge 0$, and $\varepsilon^k \to \varepsilon^+$. Verify the assertions

(a) if Φ^k w-i-e/h-converges to Φ y-ancillary tightly, then

$$\text{li}_k[\sup_{y \in B^k} \inf_{x \in (\mathscr{D}^k)^{-1}(y)} \Phi^k(x, y)] \ge \sup_{y \in B} \inf_{x \in (\mathscr{D})^{-1}(y)} \Phi(x, y);$$

(b) if Φ^k e/h-converges to Φ y-ancillary tightly, $\varepsilon \geq 0$, $\varepsilon^{k_j} \to \varepsilon^+$, and $y^{k_j} \in \varepsilon^{k_j}$-mai$\Phi^{k_j} \to \bar{y}$, then $\bar{y} \in \varepsilon$-maiΦ, i.e., $\mathrm{Ls}_k(\varepsilon^k$- mai$\Phi^k) \subset \varepsilon$-mai$\Phi$, and the ε^{k_j}-maxinf-value of Φ^{k_j} converges to the ε-maxinf-value of Φ.

Exercise 8.5 (Outer limits of epigraphs of sup-projections under condition (a) of e/h-convergence) Prove that condition (a) of the e/h-convergence of Φ^k to Φ implies that $\mathrm{dom}\xi \neq \emptyset$ and

$$\mathrm{Ls}_k(\mathrm{epi}\xi^k) \subset \mathrm{epi}\xi.$$

Hence, $\mathrm{li}_j \xi^{k_j}(x^{k_j}) \geq \xi(x)$ for any $A^{k_j} \ni x^{k_j} \to x$.

Exercise 8.6 (A sufficient condition for mis-lop convergence) Assume that

(a) $A^k = A$ for all k and A is closed;
(b) $\mathscr{D}^k$ continuously converge to $\mathscr{D}$ relative to A in the sense that for every $A \ni x^{k_j} \to x \in A$, $\mathscr{D}^{k_j}(x^{k_j}) \stackrel{P-K}{\to} \mathscr{D}(x)$;
(c) Φ^k continuously converges to Φ relative to their domains in the sense that for every $(x^k, y^k) \in \mathrm{dom}\Phi^k$ tending to $(x, y) \in \mathrm{dom}\Phi$, $\Phi^k(x^k, y^k) \to \Phi(x, y)$.

Prove that Φ^k mis-lop converges to Φ.

Exercise 8.7 (Epi- or hypo-convergence of slices under mis-lop convergence) Show that if Φ^k does not depend on y (on x, resp.), then $\Phi^k \stackrel{mis-lop}{\to} \Phi$ implies that $\Phi^k \stackrel{e}{\to} \Phi$ ($\Phi^k \stackrel{h}{\to} \Phi$, resp.).

Exercise 8.8 (Formulas for the lower and upper lop-limits) Prove the formulas for the lower and upper lop-limits:

$$\mathrm{lli}\Phi^k(x, y) = \max_{\{B^{k_j} \ni y^{k_j} \to y\}} \inf_{\{(\mathscr{D}^{k_j})^{-1}(y^{k_j}) \ni x^{k_j} \to x\}} \mathrm{li}_j \Phi^{k_j}(x^k k_j, y^k k_j) \text{ for}$$

$$\mathrm{y} \in \mathrm{B} \text{ and } x \notin \mathscr{D}^{-1}(y),$$

$$\mathrm{lls}\Phi^k(x, y) = \min_{\{A^{k_j} \ni x^{k_j} \to x\}} \sup_{\{\mathscr{D}^{k_j}(x^{k_j}) \ni y^{k_j} \to y\}} \mathrm{ls}_j \Phi^{k_j}(x^{k_j}, y^{k_j})$$

$$\text{for } x \in A \text{ and } y \notin \mathscr{D}(x).$$

Exercise 8.9 (Sufficient condition for mai-lop convergence) Prove the counterpart for mai-lop convergence of Theorem 8.36.

Exercise 8.10 (Weak epi-convergence of sup-projections under x-ancillary tight weak mis-lop convergence) Prove that if $\Phi^k \stackrel{w-mis-lop}{\to} \Phi$ x-ancillary tightly, then $\xi^k \stackrel{w-e}{\to} \xi$.

Exercise 8.11 (Hypo-convergence of inf-projections under y-ancillary tight mai-lop convergence) Prove that if $\Phi^k \stackrel{mai-lop}{\to} \Phi$ y-ancillary tightly, then $\xi^k \stackrel{h}{\to} \xi$.

Exercise 8.12 (Convergence of approximate maxinf-points under y-ancillary tight mai-lop convergence) Verify that if $\varepsilon, \varepsilon^k \geq 0$, $\varepsilon^k \to \varepsilon$, and $\Phi^k \overset{mai-lop}{\to} \Phi$ y-ancillary tightly, then $\mathrm{Ls}_k(\varepsilon^k\text{-mai}\Phi^k) \subset \varepsilon\text{-mai}\Phi$.

Exercise 8.13 (Convergence of approximate maxinf-points under y-tight mai-lop convergence) Show that if Φ^k y-tightly mai-lop converges to Φ, then there exists $\varepsilon^k \to 0^+$ such that $\varepsilon^k\text{-mai}\Phi^k \overset{P-K}{\to} \text{mai}\Phi$.

Exercise 8.14 (Convergence of approximate dual Nash quasiequilibria under hypo-convergence) Prove that if $\varepsilon, \varepsilon^k \geq 0$, $\varepsilon^k \to \varepsilon$, $\Phi^k \overset{h}{\to} \Phi$, and $\mathrm{gph}\mathscr{D} \subset \mathrm{Li}_k$ $\mathrm{gph}\mathscr{D}^k$, then $\mathrm{Ls}_k\varepsilon^k\text{-(Sol(DQEP}^k)) \subset \varepsilon\text{-(DQEP)}$.

Exercise 8.15 (Convergence of approximate dual strong Nash quasiequilibria) Prove that if Φ^k e/h- or mai-lop converges y-ancillary tightly to Φ, $\varepsilon, \varepsilon^k \geq 0$, and $\varepsilon^k \to \varepsilon$, then

$$\mathrm{Ls}_k(\varepsilon^k\text{-Sol(DsQEP}^k)) \subset \varepsilon\text{-Sol(DsQEP)}.$$

Exercise 8.16 (Convergence of pairs of approximate solutions of strong quasiequilibrium problems and approximate saddle points) Let $\varepsilon, \varepsilon^k \geq 0$ and $\varepsilon^k \to \varepsilon$. Prove that if Φ^k i-e/h-converges tightly to Φ and $\mathrm{Ls}_k\mathscr{C}\Phi^k \subset \mathscr{C}\Phi$, then

$$\begin{aligned}
&\mathrm{Ls}_k(\mathscr{C}\Phi^k \cap \varepsilon^k\text{-sdl}\Phi^k) \cap \mathrm{gph}\mathscr{D} \\
&= \mathrm{Ls}_k[\mathscr{C}\Phi^k \cap (\varepsilon^k\text{-Sol(sEP)}^k \times \varepsilon^k\text{-Sol(DsEP}^k))] \cap \mathrm{gph}\mathscr{D} \\
&\subset \mathscr{C}\Phi \cap (\varepsilon\text{-Sol(sEP)} \times \varepsilon\text{-Sol(DsEP)}) = \mathscr{C}\Phi \cap (\varepsilon\text{-sdl}\Phi).
\end{aligned}$$

Exercise 8.17 (Convergence of the second-kind approximate dual solutions) Let $\varepsilon, \varepsilon^k \geq 0$ and $\varepsilon^k \to \varepsilon$. Verify that if Φ^k i-mis-lop or i-mai-lop converges to Φ and θ is continuous, then

$$\mathrm{Ls}_k(\varepsilon^k\theta\text{-Sol(DQEP}^k)) \cap B \subset \varepsilon\theta\text{-Sol(DQEP)}.$$

Exercise 8.18 (Properties of Nash quasiequilibria) Prove that if $(\bar{x}, \bar{x}) \in \mathrm{gph}\mathscr{D}$, then $\bar{x} \in \mathrm{Sol(QEP}_N)$ if and only if $\bar{x} \in \mathrm{Sol}(Q\mathscr{G})$.

Exercise 8.19 (Properties of the second-kind approximate Nash quasiequilibria) Prove that $\bar{x}_\theta \in \varepsilon\theta\text{-Ne}(Q\mathscr{G})$, $\bar{y}_\theta \in \varepsilon\theta\text{-Ne}(DQ\mathscr{G})$, and $(\bar{x}, \bar{y}) \in \mathrm{gph}\mathscr{D}$ if and only if $(\bar{x}_\theta, \bar{y}_\theta) \in \varepsilon\theta\text{-wsdl}N$ and $N(\bar{x}_\theta, \bar{y}_\theta) = 0$.

Exercise 8.20 (Convergence of approximate Nash quasiequilibria) Let $\varepsilon, \varepsilon^k \geq 0$ and $\varepsilon^k \to \varepsilon$. Verify that if N^k epi-converges to N and $\mathrm{gph}\mathscr{D} \subset \mathrm{Li}_k\mathrm{gph}\mathscr{D}^k$, then $\mathrm{Ls}_k(\varepsilon^k\text{- Ne}(Q\mathscr{G}^k)) \subset \varepsilon\text{- Ne}(Q\mathscr{G})$.

Exercise 8.21 (Convergence of the second-kind approximate Nash quasiequilibria) Let $\varepsilon^k \to 0^+$. Prove that if N^k e/h-converges to N and θ is bounded from above, then $\mathrm{Ls}_k(\varepsilon^k\theta\text{- Ne}(Q\mathcal{G}^k)) \subset \mathrm{Ne}(Q\mathcal{G})$.

Exercise 8.22 (Painlevé-Kuratowski convergence of pairs of dual approximate Nash quasiequilibria and approximate weak saddle points) Establish the following statement. If N^k both x-tightly mis-lop and y-tightly mai-lop converges to N simultaneously and $\mathscr{C}N = \mathrm{Lim}_k \mathscr{C}\Phi^k$, then

$$\begin{aligned}\varepsilon\text{-wsdl}N \cap \mathscr{C}N &= (\varepsilon\text{-Sol}(Q\mathcal{G}) \times \varepsilon\text{-Sol}(DQ\mathcal{G})) \cap \mathscr{C}N \\ &= \mathrm{Lim}_k(\varepsilon^k\text{-Sol}(Q\mathcal{G}^k) \times \varepsilon^k\text{-Sol}(DQ\mathcal{G}^k)) \cap \mathrm{Lim}_k \mathscr{C}N^k \\ &= \mathrm{Lim}_k \varepsilon^k\text{-wsdl}N^k \cap \mathrm{Lim}_k \mathscr{C}N^k.\end{aligned}$$

8.9 Commentary

Section 8.1

Bifunctions on General Domains and Quasiequilibrium Problems
Although general bifunctions and quasiequilibrium problems are already present in Chaps. 3 and 6 when we provide a comprehensive presentation about the existence of solutions and their stability, we devote this section to clarifying a number of important facts which may cause confusion. First we discuss variational objects and approximate variational objects, which have properties quite different from the corresponding ones of bifunctions on rectangular domains. Then we expose differences between a quasiequilibrium problem and a strong quasiequilibrium problem and their dual ones in connection with minsup-points, maxinf-points, weak saddle and saddle points together with their approximate points. These facts are developments of some parts in [179] and also suggest how to arrange the contents of Sects. 8.2 and 8.3 in a reasonable way. In particular, based on this section, we always try to fully research and clearly show the relationship between the above variational objects and problems.

Section 8.2

Epi/Hypo-Convergence and Lopsided Convergence
The study of variational convergence of bifunctions on general domains and approximations to quasiequilibrium problems began very recently. The first contributions are [179, 404, 407]. The definition of e/h- and inside e/h-convergence (for general bifunctions) in this section is taken from [179], while the weak e/h- and inside weak-

e/h-convergence are from [181]. The definition of mis-lop convergence is taken from [404], and the other types of lop-convergence are taken from [181]. A part of the characterizations, properties, and especially variational properties of the above types of e/h- and lop-convergence are taken from [179, 404, 407]. The other part contains results selected from [181], including the concepts of types of tight lop-convergence. Here we also pay attention to clarifying what is common and what is only true for some stronger types of variational convergence and also to exposing the differences between the convergence types, especially between the types of e/h- and lop-convergence and between the rectangular and general (nonrectangular) cases.

Section 8.3

Quantification of Variational Convergence

An introduction to quantification of variational convergence is provided. We begin with a brief description of the concept of Attouch-Wets's epi-distance, which was started in [80] and two further parts of this paper, providing fundamental properties and applications of this distance as a tool for quantification of epi-convergence. (For some important applications of Attouch-Wets's epi-distance, see, e.g., [399, 408].) Then, an extension of this epi-distance to mis-lop distance of two bifunctions is introduced, and its applications in estimates of minsup-values and minsup-points are obtained in [404, 407]. The variational distances ebi-distance and hbi-distance to quantify variational convergence of bifunctions in general, not only for mis-lop convergence, are proposed in [181] based on the lop-distance introduced in [404]. Using these two variational distances, under suitable assumptions, we can get estimates of saddle values and saddle points to quantify the epi/hypo-, mis-lop, and mai-lop convergence.

Section 8.4

Approximations of Quasiequilibrium Problems

The first subsection is a comprehensive exploration of quasiequilibrium and strong quasiequilibrium problems and their dual models. The presentation goes along the same line as that in Sect. 7.3 about the counterparts of these problems in the rectangular case for a good comparison as well as for brevity. Namely, we try to state the results exactly but prove only what is different from Sect. 7.3 and demonstrate these differences relatively in detail. We select a part of the results in [179] and take developments, e.g., the study of second-kind approximate solutions and their convergence under the aforementioned types of variational convergence of bifunctions from [181]. We also make effort to show connections of the results here with the variational properties of the types of e/h- and lop-convergence obtained in Sect. 8.2. In the second subsection, applications of the quantification results in Sect. 8.3 in estimating solutions of a pair of dual strong quasiequilibrium problems and saddle points of their Nikaido-Isoda bifunctions are obtained.

Section 8.5

Approximations of Generalized Noncooperative Games
The presentation here also goes along the line similar to that of Sect. 7.6 about approximations of Nash equilibria when we deal with direct extensions of the results there. Apart from the results taken from [404], some content here is brought from [181]. The first part provides extensions to the quasiequilibrium models stated in Sect. 8.1 of the results on convergence of first- and second-kind approximate solutions in Sect. 7.6, in terms of the Nikaido-Isoda bifunction of the problems in question. Apart from the issues considered in Sect. 7.6, here we also briefly discuss the application of the estimates of solutions in Sect. 8.4.2 to the case of a pair of strong Nash quasiequilibrium problems.

The second part contains some developments of the study in [219] of convergence of the above solutions (in the rectangular case) directly in terms of the data of the problems not via the Nikaido-Isoda bifunction. When applied to the special rectangular case, these results also sharpen the corresponding ones in [219]. An important feature of this section is considering the return of the agents as the payoff/objective function of the considered Nash quasiequilibrium problem not the loss as in Sect. 7.6.

For approximations of Nash equilibria and Nash quasiequilibria in terms of variational convergence, the reader is referred to, e.g., [176, 219, 269, 374, 404].

Section 8.6

Approximations of Set-Valued Quasivariational Inequalities and Traffic Networks
Quasivariational inequalities are one of the most important subclasses of quasiequilibrium problems. This section is devoted to approximations in terms of variational convergence of two set-valued quasivariational inequalities: a weak set-valued quasivariational inequality and a strong set-valued quasivariational inequality together with applications in traffic networks with arc capacity constraints, multivalued costs, and elastic demands depending on equilibrium flows. The content of the section is taken from [177]. Quasivariational inequalities have been intensively studied for two third of a century now, since they also encompass a major part of special cases of quasiequilibrium problems and even more developed than the latter problems outside optimization-related fields such as in partial differential equations. Among

the huge number of works on variational and quasivariational inequalities, we would mention here only some examples [191, 192, 298, 331].

The paper [177] is the first attempt to study approximations of quasivariational inequalities. Here, the relationships between the aforementioned two types of inequalities and also between them and related quasiequilibrium problems are discussed. The types of approximate solutions are new as well. The obtained approximation results are global, since they are in terms of variational convergence. Note that in Proposition 1(iii) of [164], Sion's minimax theorem was applied for a result in the special case of variational inequalities (not "quasivariational," i.e., with $\mathscr{K}(x) = K$ for all $x \in \mathbb{R}^m$). In the proof of Propositions 8.69 and 8.74, we apply an extended version of this theorem given in Corollary 3.3 of [421]. The important feature that approximation results are global is effectively inherited in the application part about approximations of the above traffic networks, both in novel notions and results. Let us say a few words about some contributions on traffic network problems in the literature. The basic optimality/equilibrium concept was introduced by Wardrop [449]. The paper [422] showed that traffic problems can be considered as cases of variational inequalities. Set-valued costs were proposed and studied in [284, 285, 364]. The paper [286] considered well-posedness of a traffic network with elastic demands depending on weak equilibrium flows. The work [342] studied the case with arc capacity constraints. Section 8.6.3 includes all these features of traffic network problems.

Section 8.7

Approximations of Problems of Optimization Under Stochastic Ambiguity
Approximations of problems of optimization under stochastic ambiguity presented here are taken from the very recent articles that pursue this important research direction [404, 407]. However, we do not include all the detailed results because of their technical complexity, and so our aim is to only briefly introduce this new research direction with important practical applications based on the powerful variational convergence tool. The interested reader is referred to the above two recent papers for more details. In addition, if there are difficulties with some basic knowledge of probability, the reader should consider textbooks such as [107, 185].

Chapter 9
Approximations of Vector Quasiequilibrium Problems

Abstract This chapter considers approximations of vector quasiequilibrium problems and their particular case: vector Nash quasigames (vector generalized noncooperative games), which are among the most important models in applied mathematics. Since there have been no studies of variational convergence of vector bifunctions in the literature, our approach consists of first scalarizing the problems by the Tammer and Hiriart-Urruty methods and then applying variational convergence of scalar bifunctions. Sufficient conditions for convergence of the three types of approximate solutions of the problems approximating the original one under appropriate types of variational convergence are established.

In the preceding two chapters, only approximations of scalar equilibrium problems and scalar quasiequilibrium problems are studied in terms of the types of variational convergences of (scalar) bifunctions. Vector equilibrium and quasiequilibrium problems are so important in applied mathematics that numerous contributions have appeared so far, and Chaps. 3 and 6 of this book are devoted to existence and stability aspects of these models. We discuss approximation issues of these vector problems in this chapter. Since only several papers about epi-convergence of vector functions can be spotted in the literature and absolutely no works on this theme for vector bifunctions can be found, we resort to scalarization methods in this chapter. (Note that scalarization techniques are very frequently applied in vector optimization as can be seen partially in Chaps. 3 and 6.)

L. Q. Anh et al., *Equilibrium Problems*, Forum for Interdisciplinary Mathematics, https://doi.org/10.1007/978-3-031-89003-1_9

9.1 Problem Formulation and Concepts of Approximate Solutions

Let, X, Y be metric spaces, Z a normed space, $A \subset X$ and $B \subset Y$ nonempty, $C \subset Z$ a closed convex set, $\mathscr{D} : A \rightrightarrows B$ defined on the whole A and surjective (so $B = \mathscr{D}(A)$), and $f : \mathrm{gph}\mathscr{D} \to Z$. Consider VECTOR STRONG QUASIEQUILIBRIUM PROBLEM

$$\text{(VsQEP)} \quad \text{find } \bar{x} \in A \text{ such that } f(\bar{x}, y) \in -C \text{ for all } y \in \mathscr{D}(\bar{x})$$

and, assuming $\mathrm{int}C \neq \emptyset$, also the VECTOR WEAK QUASIEQUILIBRIUM PROBLEM

$$\text{(VwQEP)} \quad \text{find } \bar{x} \in A \text{ such that } f(\bar{x}, y) \notin \mathrm{int}C \text{ for all } y \in \mathscr{D}(\bar{x}).$$

We should make some notes here. In fact, (VwQEP) can be considered also with $\mathrm{int}C = \emptyset$, and in this case, every $x \in A$ is its solution. Hence, avoiding this trivial situation, we always assume that $\mathrm{int}C \neq \emptyset$ for this problem. Moreover, when $Z = \mathbb{R}$, the above two models collapse to scalar quasiequilibrium problems involving $f(\bar{x}, y) \leq 0$ considered in Chaps. 7 and 8, but not to scalar strong quasiequilibrium problems there. We also investigate the following dual setting of the above two problems: the DUAL VECTOR WEAK QUASIEQUILIBRIUM PROBLEM (when $\mathrm{int}C \neq \emptyset$)

$$\text{(DVwQEP)} \quad \text{find } \bar{y} \in B \ \text{ such that } f(x, \bar{y}) \notin -\mathrm{int}C \text{ for all } x \in \mathscr{D}^{-1}(\bar{y})$$

and the DUAL VECTOR STRONG QUASIEQUILIBRIUM PROBLEM

$$\text{(DVsQEP)} \quad \text{find } \bar{y} \in B \ \text{ such that } f(x, \bar{y}) \in C \text{ for all } x \in \mathscr{D}^{-1}(\bar{y}).$$

Following our notational conventions, we use Sol(VwQEP), Sol(VsQEP), Sol(DVwQEP), and Sol(DVsQEP) to denote the solution sets of the above four models.

Among numerous special cases of vector quasiequilibrium problems, we choose to consider the following vector Nash quasigame (a generalization of the corresponding scalar problem in Chap. 8), since it is among the most important problems in applied mathematics. Let $I := \{1, \ldots, m\}$, x_i be a strategy of player i in the available strategy set $\mathscr{D}_i(x_{-i})$ in metric space X_i, where $x_{-i} := (x_1, \ldots, x_{i-1}, x_{i+1}, \ldots, x_m)$. Then, we have set-valued map $\mathscr{D}_i : X_{-i} \rightrightarrows X_i$. Set $\mathscr{D} := \prod_{i \in I} \mathscr{D}_i : \prod_{i \in I} X_i \rightrightarrows \prod_{i \in I} X_i =: X$ and

$$A := \{x \in X \mid x_i \in \mathscr{D}_i(x_{-i}), \forall i \in I\}.$$

We can ignore points outside A since they are irrelevant and restrict the domain of $\mathscr{D}$ as $\mathscr{D} : A \rightrightarrows X$. As $(x, x) \in \text{gph}\mathscr{D}$ for all $x \in A$, $\mathscr{D}(A) \supset A$. Let Z be a normed space and $c_i(x_i, x_{-i}) \in Z$ be the return of player i. Then, we have a map $c_i : A \to Z$. The preference in Z is defined by a closed convex set C. A traditional approach is of using the following Nikaido-Isoda vector bifunction/bimap $N : \text{gph}\mathscr{D} \to Z$ defined by

$$N(x, y) := \sum_{i=1}^{m} [c(x_i, x_{-i}) - c_i(y_i, x_{-i})].$$

Corresponding to the strong problem (VsQEP), the VECTOR STRONG NASH QUASIGAME is the one of

(VsQ$\mathscr{G}$) finding $\bar{x} \in A$ such that $N(\bar{x}, y) \in C$ for all $y \in \mathscr{D}(\bar{x})$.

Assuming that int$C \neq \emptyset$, the VECTOR WEAK NASH QUASIGAME is of

(VwQ$\mathscr{G}$) finding $\bar{x} \in A$ such that $N(\bar{x}, y) \notin -\text{int}C$ for all $y \in \mathscr{D}(\bar{x})$.

The latter problem is shown in Proposition 9.25 to be equivalent to the following formulation directly in terms of the data

(VwQ$\mathscr{G}$) find $\bar{x} \in A$ such that $c_i(\bar{x}_i, \bar{x}_{-i}) - c_i(x_i, \bar{x}_{-i}) \notin -\text{int}C$,
$\forall i \in I, \forall x \in \mathscr{D}(\bar{x})$.

(Note that because the players' objectives are returns, i.e., of the max type, here the sign of C is opposite to that in the above vector quasiequilibrium problems.) A solution $\bar{x}$ of (VsQ$\mathscr{G}$) is called a STRONG NASH QUASIEQUILIBRIUM VECTOR, written $\bar{x} \in$ Ne(VsQ$\mathscr{G}$), and a solution $\bar{x}$ of (VwQ$\mathscr{G}$) is called a WEAK NASH QUASIEQUILIBRIUM VECTOR, written $\bar{x} \in$ Ne(VwQ$\mathscr{G}$). Naturally, the DUAL VECTOR STRONG NASH QUASIGAME is

(DVsQEP) find $\bar{y} \in B$ such that $N(x, \bar{y}) \in -C$ for all $x \in \mathscr{D}^{-1}(\bar{y})$

and the DUAL VECTOR WEAK NASH QUASIGAME is

(DVwQEP) find $\bar{y} \in B$ such that $N(x, \bar{y}) \notin \text{int}C$ for all $x \in \mathscr{D}^{-1}(\bar{y})$.

Of course, in approximation studies, approximate solutions are common objects of research. Furthermore, in practice, it may be inconvenient to deal with exact solutions; e.g., they may not exist or their existence can be unknown. Even when we are sure that the solutions exist, it can often be very difficult to find them, while finding an approximate solution can be easier. In this chapter, we deal with three types of approximate solutions. The first one is defined as follows. Let $C \subset Z$ be a pointed closed convex cone with nonempty interior, $e \in \text{int}C$, and $\varepsilon \geq 0$. The approximate solution sets of the above four vector quasiequilibrium problems are defined, resp., as follows:

$$\begin{aligned}
&\varepsilon\text{-Sol(VwQEP)} := \{\bar{x} \in A \mid f(\bar{x}, y) \notin \varepsilon e + \text{int}C \text{ for all } y \in \mathscr{D}(\bar{x})\},\\
&\varepsilon\text{-Sol(VsQEP)} := \{\bar{x} \in A \mid f(\bar{x}, y) \in \varepsilon e - C \text{ for all } y \in \mathscr{D}(\bar{x})\},\\
&\varepsilon\text{-Sol(DVwQEP)} := \{\bar{y} \in A \mid f(x, \bar{y}) \notin -\varepsilon e - \text{int}C \text{ for all } x \in \mathscr{D}^{-1}(\bar{y})\},\\
&\varepsilon\text{-Sol(DVsQEP)} := \{\bar{y} \in B \mid f(x, \bar{y}) \in -\varepsilon e + C \text{ for all } x \in \mathscr{D}^{-1}(\bar{y})\}.
\end{aligned}$$

In the next sections, we also discuss the second kind of approximate solutions that directly generalizes the concept under the same name for the scalar problems in Chaps. 7 and 8. However, in this chapter, we also examine the third kind of approximate solutions which is defined for C being simply a closed convex subset of Z, but only for the two strong problems among the above four problems as follows, for $t \geq 0$ and closed unit ball B_Z,

$$\begin{aligned}
&t\text{-Sol(VsQEP)} := \{\bar{x} \in A \mid f(\bar{x}, y) \in tB_Z - C \text{ for all } y \in \mathscr{D}(\bar{x})\},\\
&t\text{-Sol(DVsQEP)} := \{\bar{y} \in B \mid f(x, \bar{y}) \in tB_Z + C \text{ for all } x \in \mathscr{D}^{-1}(\bar{y})\}.
\end{aligned}$$

We have correspondingly the following definitions of the first-kind approximate solutions of vector weak and strong Nash quasigames and their dual models

$$\begin{aligned}
&\varepsilon\text{-Ne(VwQ}\mathscr{G}) := \{\bar{x} \in A \mid N(\bar{x}, y) \notin -\varepsilon e - \text{int}C \text{ for all } y \in \mathscr{D}(\bar{x})\},\\
&\varepsilon\text{-Ne(VsQ}\mathscr{G}) := \{\bar{x} \in A \mid N(\bar{x}, y) \in -\varepsilon e + C \text{ for all } y \in \mathscr{D}(\bar{x})\},\\
&\varepsilon\text{-Ne(DVwQ}\mathscr{G}) := \{\bar{y} \in A \mid N(x, \bar{y}) \notin \varepsilon e + \text{int}C \text{ for all } x \in \mathscr{D}^{-1}(\bar{y})\},\\
&\varepsilon\text{-Ne(DVsQ}\mathscr{G}) := \{\bar{y} \in B \mid N(x, \bar{y}) \in \varepsilon e - C \text{ for all } x \in \mathscr{D}^{-1}(\bar{y})\}.
\end{aligned}$$

The third kind of approximate solutions of vector strong Nash quasigames and their dual models is

$$\begin{aligned}
&t\text{-Ne(VsQ}\mathscr{G}) := \{\bar{x} \in A \mid N(\bar{x}, y) \in tB_Z + C \text{ for all } y \in \mathscr{D}(\bar{x})\},\\
&t\text{-Ne(DVsQ}\mathscr{G}) := \{\bar{y} \in B \mid N(x, \bar{y}) \in tB_Z - C \text{ for all } x \in \mathscr{D}^{-1}(\bar{y})\}.
\end{aligned}$$

For the above problems, we aim to obtain convergence of approximate solutions of approximating problems to approximate or exact solutions of the original problems under suitable types of variational convergence of the problems via the Tammer and Hiriart-Urruty scalarization methods. Observe that these two scalarization methods have probably been the most useful ones, besides the linear scalarization, in vector optimization during the last several decades.

9.2 Approximations of Vector Quasiequilibria via the Tammer Scalarization Function

9.2.1 Approximations of Vector Weak Quasiequilibrium Problems

For the primal vector weak quasiequilibrium problem (VwQEP), we use the GERSTEWITZ/TAMMER SCALARIZATION FUNCTION $r_{-C}^{-e} : Z \to \mathbb{R}$ defined by

$$r_{-C}^{-e}(z) := \inf\{t \in \mathbb{R} \mid z \in -te + C\},$$

which is just a particular case with $q = -e$ and $E = -C$ of r_E^q in Definition 3.8. Figure 9.1 shows the value of r_{-C}^{-e} at a given point z.

We will need the following two among the known good properties of r_{-C}^{-e}, for any $z \in Z$ and $t \in \mathbb{R}$,

(a$_{-e}$) $r_{-C}^{-e}(z) > t \Leftrightarrow z \notin -te + C$ (equivalently, $r_{-C}^{-e}(z) \leq t \Leftrightarrow z \in -te + C$);

(b$_{-e}$) $r_{-C}^{-e}(z) \geq t \Leftrightarrow z \notin -te + \text{int}C$ (equivalently, $r_{-C}^{-e}(z) < t \Leftrightarrow z \in -te + \text{int}C$).

The coincidence shown below of approximate vector weak quasiequilibria and approximate solutions of the corresponding scalar problem follows from property (b$_{-e}$) of r_{-C}^{-e}

$$\varepsilon\text{-Sol(VwQEP)} = \{\bar{x}_\varepsilon \in A \mid r_{-C}^{-e}(f(\bar{x}_\varepsilon, y)) \geq -\varepsilon \text{ for all } y \in \mathscr{D}(\bar{x}_\varepsilon)\}.$$

Definition 9.1 (Variational convergence of vector functions and vector bifunctions via scalarization) Let $r : Z \to \mathbb{R}$.

(a) For $A^k \subset X$, we say that map $\varphi^k : A^k \to Z$ r-EPI-CONVERGES to $\varphi : A \to Z$ if $r(\varphi^k)$ epi-converges to $r(\varphi)$. The other three types: r-inside epi-, r-weak epi-, and r-weak inside epi-convergence are defined correspondingly. The four corresponding types of hypo-convergence are determined analogously.

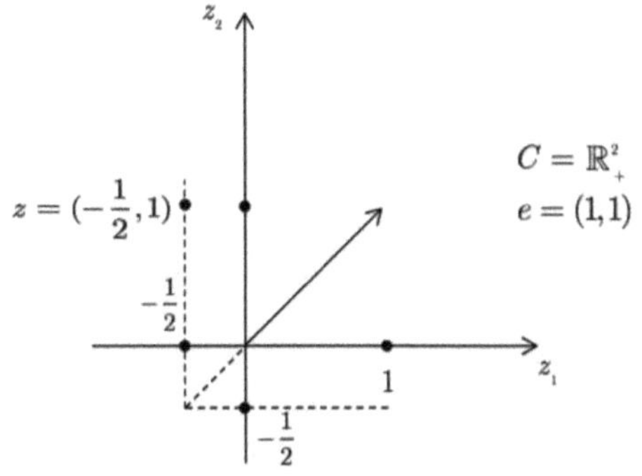

Fig. 9.1 $r_{-C}^{-e}(z) = \frac{\sqrt{2}}{2}$

(b) For $A^k \subset X$ and $\mathscr{D}^k : A^k \rightrightarrows B^k$, vector bifunction/bimap $f^k : \mathrm{gph}\mathscr{D}^k \to Z$ r-EPI/HYPO-CONVERGES to $f : \mathrm{gph}\mathscr{D} \to Z$ if $r(f^k)$ epi/hypo converges to $r(f)$. The three weaker types are naturally similarly defined.
(c) For $A^k \subset X$ and $\mathscr{D}^k : A^k \rightrightarrows B^k$, vector bifunction/bimap $f^k : \mathrm{gph}\mathscr{D}^k \to Z$ r-MIS-LOP CONVERGES to $f : \mathrm{gph}\mathscr{D} \to Z$ if $r(f^k)$ mis-lop converges to $r(f)$. The remaining three types of mis-lop and four types of r-MAI-LOP CONVERGENCE are analogously defined.

Approximating problems of problem (VwQEP) (and other vector quasiequilibrium problems) are defined in the same way as the approximating problems of scalar problems in Chaps. 7 and 8, i.e., by the data A^k, B^k, $\mathscr{D}^k$, f^k, with the natural notation (VwQEP^k). Our first approximation statement is the following.

Proposition 9.2 (Convergence of approximate solutions of vector weak quasiequilibrium problems) *For problems* (VwQEP^k) *and* (VwQEP), *let* $\varepsilon, \varepsilon^k \geq 0$ *and* $\varepsilon^k \to \varepsilon$.

(i) *If* f^k $-\, r^{-e}_{-C}$*-i-e/h- or* $-\, r^{-e}_{-C}$*-i-mis-lop converges to* f, *then* $A \cap \mathrm{Ls}_k(\varepsilon^k$-Sol(VwQEP^k)) $\subset \varepsilon$-Sol(VwQEP). *If* f^k $-\, r^{-e}_{-C}$*-e/h- or* $-\, r^{-e}_{-C}$*-mis-lop converges to* f, *then* $\mathrm{Ls}_k(\varepsilon^k$-Sol(VwQEP^k)) $\subset \varepsilon$-Sol(VwQEP).
(ii) *If* $\mathrm{gph}\mathscr{D} \sqsubset \mathrm{Li}_k \mathrm{gph}\mathscr{D}^k$ *and the* $-\, r^{-e}_{-C}$*-epi- or* $-\, r^{-e}_{-C}$*-i-epi-convergence of* f^k *replaces the variatiobal convergence assumed in* (i), *then the conclusion of assertion* (i) *still holds.*
(iii) *If* f^k $-r^{-e}_{-C}$*-w-e/h-converges or* $-r^{-e}_{-C}$*-w-mis-lop converges to* f *and* $x^k \in \varepsilon^k$-Sol(VwQEP^k) *tends to* $\bar{x}_\varepsilon$, *then* $\bar{x}_\varepsilon \in \varepsilon$-Sol(VwQEP).

Proof

(i) We have to address only the case of $-\, r^{-e}_{-C}$-e/h- and $-\, r^{-e}_{-C}$-mis-lop types of convergence. Let $\bar{x}_\varepsilon \in \mathrm{Ls}_k(\varepsilon^k$-Sol(VwQEP^k)), i.e., there exists a subsequence $\{x^{k_j}\}_j$ in ε^{k_j}-Sol(VwQEP^{k_j}) converging to $\bar{x}_\varepsilon$. If $\bar{x}_\varepsilon \in A$, then for all $y \in \mathscr{D}(\bar{x}_\varepsilon)$, condition (a) of $-\, r^{-e}_{-C}$-e/h-convergence yields $\mathscr{D}^{k_j}(x^{k_j}) \ni y^{k_j} \to y$ such that $\mathrm{li}_j[-r^{-e}_{-C}(f^{k_j}(x^{k_j}, y^{k_j}))] \geq -r^{-e}_{-C}(f(\bar{x}_\varepsilon, y)$, i.e., $\mathrm{ls}_j[r^{-e}_{-C}(f^{k_j}(x^{k_j}, y^{k_j}))] \leq r^{-e}_{-C}(f(\bar{x}_\varepsilon, y))$. As $x^{k_j} \in \varepsilon^{k_j}$-Sol(VwQEP)^{k_j}), $r^{-e}_{-C}(f^{k_j}(x^{k_j}, y^{k_j})) \geq -\varepsilon^{k_j}$. Hence, $r^{-e}_{-C}(f(\bar{x}_\varepsilon, y)) \geq -\varepsilon$, which means that $\bar{x}_\varepsilon$ lies in ε-Sol(VwQEP). Suppose that $\bar{x}_\varepsilon \notin A$. The above condition (a) also gives $y^{k_j} \in \mathscr{D}^{k_j}(x^{k_j})$ such that $-\, r^{-e}_{-C}(f^{k_j}(x^{k_j}, y^{k_j})) \to +\infty$, contradicting the inequality $r^{-e}_{-C}(f^{k_j}(x^{k_j}, y^{k_j})) \geq -\varepsilon^{k_j}$ for all $y \in \mathscr{D}^{k_j}(x^{k_j})$. The rest of the assertion for $-\, r^{-e}_{-C}$-mis-lop convergence is verified automatically, since the used condition (a) is common for the mentioned types of e/h- and mis-lop convergence.
(ii) Let x^{k_j} in ε^{k_j}-Sol(VwQEP^{k_j}) tend to $\bar{x}_\varepsilon$. If $\bar{x}_\varepsilon \notin A$, then any $\mathscr{D}^{k_j}(x^{k_j}) \ni y^{k_j} \to y$ gives $(\bar{x}_\varepsilon, y) \notin \mathrm{gph}\mathscr{D}$. Due to condition (a) of the assumed $-\, r^{-e}_{-C}$-epi-convergence, $-\, r^{-e}_{-C}(f^{k_j}(x^{k_j}, y^{k_j})) \to +\infty$, which is impossible as $x^{k_j} \in \varepsilon^{k_j}$-Sol(VwQEP^{k_j}). So, $\bar{x}_\varepsilon$ must be in A. We have to show that

$r_{-C}^{-e}(f(\bar{x}_\varepsilon, y)) \geq -\varepsilon$ for each $y \in \mathscr{D}(\bar{x}_\varepsilon)$. As $\mathrm{gph}\mathscr{D} \sqsubset \mathrm{Li}_k \mathrm{gph}\mathscr{D}^k$, for all $y \in \mathscr{D}(\bar{x}_\varepsilon)$, one finds $\mathscr{D}^{k_j}(x^{k_j}) \ni y^{k_j} \to y$. Since $x^{k_j} \in \varepsilon^{k_j}$-Sol(VwQEP^{k_j}), $-r_{-C}^{-e}(f^{k_j}(x^{k_j}, y^{k_j})) \leq \varepsilon^{k_j}$. Then, again the above condition (a) implies that $\mathrm{li}_j[-r_{-C}^{-e}(f^{k_j}(x^{k_j}, y^{k_j}))] \geq -r_{-C}^{-e}(f(\bar{x}_\varepsilon, y))$. Hence, $-r_{-C}^{-e}(f(\bar{x}_\varepsilon, y)) \leq \varepsilon$. The case of $-r_{-C}^{-e}$-i-e-convergence is now clear.

(iii) We only address the case of $-r_{-C}^{-e}$-w-mis-lop convergence of f^k. Suppose that convergence and $x^k \in \varepsilon^k$-Sol(VwQEP^k) converging to $\bar{x}_\varepsilon$. By condition (a) of the convergence of f^k, for all $y \in \mathscr{D}(\bar{x}_\varepsilon)$, one finds $\mathscr{D}^k(x^k) \ni y^k \to y$ such that $\mathrm{ls}_k(r_{-C}^{-e}(f^k(x^k, y^k))) \leq r_{-C}^{-e}(f(\bar{x}_\varepsilon, y))$ if $\bar{x}_\varepsilon \in A$. As $r_{-C}^{-e}(f^k(x^k, y^k)) \geq -\varepsilon^k$, $r_{-C}^{-e}(f(\bar{x}_\varepsilon, y)) \geq -\varepsilon$, i.e., $\bar{x}_\varepsilon \in \varepsilon$-Sol(VwQEP). Now assume that $x \notin A$. By the above condition (a), $-r_{-C}^{-e}(f^k(x^k, y^k)) \to +\infty$ for some $y^k \in \mathscr{D}^k(x^k)$ contradicting the fact that $r_{-C}^{-e}(f^k(x^k, y^k)) \geq -\varepsilon^k$ for all $y^k \in \mathscr{D}^k(x^k)$.

□

Example 9.3 (Convergence of approximate solutions of (VwQEP^k)) Pursuing a simple illustration, we focus on the vector character of the problem and consider problem (VwEP) and its approximating problems. Here, (VwEP) denotes the vector weak equilibrium (not "quasi") problem, i.e., the special case of (VwQEP) with $\mathscr{D}(x) = B$ for all $x \in A$. Let its data be $X = Y = \mathbb{R}$, $Z = \mathbb{R}^2$, $C = \mathbb{R}^2_+$, $A = [0, 1]$, $B = [0, 1]$, and $f := (f_1, f_2)$ with $f_1(x, y) = f_2(x, y) = 1$ for $(x, y) \in A \times B$. A direct verification shows that Sol(VwEP) $= \emptyset$. We consider a sequence (VwEP^k) of approximating problems with the data $A^k = A$, $B^k = B$, $\varepsilon^k > \sqrt{2}$ and tending to $\varepsilon = \sqrt{2}$, and $f_1^k = f_2^k = \varepsilon^k/\sqrt{2}$ for $k \in \mathbb{N}$. We deal with approximate solutions with $e = (1, 1)$. It's not hard to see that $r_{-C}^{-e}(f^k(x, y)) = -\varepsilon^k$ for all $(x, y) \in A^k \times B^k$ and $k \in \mathbb{N}$. Hence, clearly f^k $-r_{-C}^{-e}$-e/h-, $-r_{-C}^{-e}$-mis-lop, and $-r_{-C}^{-e}$-epi- converges to f. Also, $\mathrm{gph}\mathscr{D} \sqsubset \mathrm{Li}_k \mathrm{gph}\mathscr{D}^k$. So, all the assumptions in (i)-(iii) of Proposition 9.2 are satisfied. A direct verification by computations yields that ε^k-Sol(VwEP^k) = ε-Sol(VwEP) = [0, 1], reconfirming Proposition 9.2.

For the dual vector weak problem (DVwQEP), we use the following scalarization function

$$r_C^e(z) := \inf\{t \in \mathbb{R} \mid z \in te - C\}.$$

This variant of the Tammer function enjoys properties similar to those of r_{-C}^{-e}, and the following ones will be used, for any $z \in Z$ and $t \in \mathbb{R}$,

(a$_e$) $r_C^e(z) < t \Leftrightarrow z \in te - \mathrm{int}C$ (equivalently, $r_C^e(z) \geq t \Leftrightarrow z \notin te - \mathrm{int}C$);
(b$_e$) $r_C^e(z) \leq t \Leftrightarrow z \in te - C$ (equivalently, $r_C^e(z) > t \Leftrightarrow z \notin te - C$).

We deduce from property (a$_e$) the following relationship between approximate solutions of (DVwQEP) and those of its scalarized problem, for $\varepsilon \geq 0$,

$$\mathrm{Sol(DVwQEP)} = \{\bar{y} \in B \mid r_C^e(f(x, \bar{y})) \geq 0 \text{ for all } x \in \mathscr{D}^{-1}(\bar{y})\},$$

$$\varepsilon\text{-}\mathrm{Sol(DVwQEP)} = \{\bar{y}_\varepsilon \in B \mid r_C^e(f(x, \bar{y}_\varepsilon)) \geq -\varepsilon \text{ for all } x \in \mathscr{D}^{-1}(\bar{y}_\varepsilon)\}.$$

Proposition 9.4 (Convergence of approximate dual vector weak quasiequilibria) *For problems* (DVwQEP^k) *and* (DVwQEP), *let* $\varepsilon, \varepsilon^k \geq 0$ *and* $\varepsilon^k \to \varepsilon$.

(i) *If* f^k r_C^e*-i-e/h- or* r_C^e*-i-mai-lop converges to* f, *then* $B \cap \mathrm{Ls}_k(\varepsilon^k$-Sol(DVwQEP^k)) $\subset \varepsilon$-Sol(DVwQEP). *If* f^k r_C^e*-e/h- or* r_C^e*-mai-lop converges to* f, *then* $\mathrm{Ls}_k(\varepsilon^k$-Sol(DVwQEP^k)) $\subset \varepsilon$-Sol(DVwQEP).

(ii) *If* $\mathrm{gph}\mathscr{D} \sqsubset \mathrm{Li}_k \mathrm{gph}\mathscr{D}^k$ *and the variational convergence of* f^k *in assertion* (i) *is replaced by the* r_C^e*-hypo- or* r_C^e*-i-hypo-convergence, resp., then one still has the conclusion of that assertion.*

(iii) *If* f^k r_C^e*-w-e/h-converges or* r_C^e*-w-mai-lop converges to* f *and* $y^k \in \varepsilon^k$-Sol(DVwQEP^k) *tends to* $\bar{y}_\varepsilon$, *then* $\bar{y}_\varepsilon \in \varepsilon$-Sol(DVwQEP).

Proof

(i) We address the case of r_C^e-mai-lop convergence. Assume that $\bar{y}_\varepsilon \in \mathrm{Ls}_k(\varepsilon^k$-Sol(DVwQEP^k)), i.e., there exists a subsequence $\{y^{k_j}\}_j$ in ε^{k_j}-Sol(DVwQEP^{k_j}) converging to $\bar{y}_\varepsilon$. If $\bar{y}_\varepsilon \in B$, then for all $x \in \mathscr{D}^{-1}(\bar{y}_\varepsilon)$, condition (a) of r_C^e-mai-lop convergence yields $x^{k_j} \in (D^{k_j})^{-1}(y^{k_j})$ tending to x such that $\mathrm{ls}_j[r_C^e(f^{k_j}(x^{k_j}, y^{k_j}))] \leq r_C^e(f(x, \bar{y}_\varepsilon))$. As $y^{k_j} \in \varepsilon^{k_j}$-Sol(DVwQEP)^{k_j}), $r_C^e(f^{k_j}(x^{k_j}, y^{k_j})) \geq -\varepsilon^{k_j}$. Hence, $r_C^e(f(x, \bar{y}_\varepsilon)) \geq -\varepsilon$, i.e., $\bar{y}_\varepsilon \in \varepsilon$-Sol(DVwQEP). Suppose that $\bar{y}_\varepsilon \notin B$. The aforementioned condition (a) gives $x^{k_j} \in (D^{k_j})^{-1}(y^{k_j})$ such that $r_C^e(f^{k_j}(x^{k_j}, y^{k_j})) \to -\infty$, which is impossible since $r_C^e(f^{k_j}(x^{k_j}, y^{k_j})) \geq -\varepsilon^{k_j}$ for all $x^{k_j} \in (D^{k_j})^{-1}(y^{k_j})$.

The rest of the assertions for r_C^e-e/h-convergence is verified automatically, since its condition (a) is also imposed for the mentioned types of e/h- and mis-lop convergence. The case of the inside convergence is a simpler version of the above argument.

Parts (ii) and (iii) can be proved similarly to (ii) and (iii) in the proof of Proposition 9.2. □

Example 9.5 (Convergence of approximate solutions of (DVwQEP^k)) Let the data of (DVwQEP) be $X = Y = \mathbb{R}$, $Z = \mathbb{R}^2$, $C = \mathbb{R}^2_+$, $A = [0, 1]$, $B = [0, 1]$ (i.e., the case of (DVwEP), not "quasi"), and for $(x, y) \in A \times B$, $f(x, y) := (f_1, f_2)(x, y) = (-1, -1)$ for all $(x, y) \in [0, 1] \times [0, 1]$. Then, an easy verification shows that Sol(DVwEP) $= \emptyset$. For a sequence of approximating problems (DVwEP^k) with $k \in \mathbb{N}$, $A^k = A$, $B^k = B$, $\varepsilon^k > \sqrt{2}$ tending to $\varepsilon = \sqrt{2}$, $f_1^k(x, y) = f_2^k(x, y) = -\varepsilon^k/\sqrt{2}$, we employ the scalarization function r_C^e with $e = (1, 1)$. Direct calculations give $r_C^e(f^k(x, y)) = -\varepsilon^k$ for all $(x, y) \in A^k \times B^k$ and $k \in \mathbb{N}$. Hence, clearly f^k r_C^e-e/h-, r_C^e-mai-lop, and r_C^e-hypo- converges to f. We also have $\mathrm{gph}\mathscr{D} \sqsubset \mathrm{Li}_k \mathrm{gph}\mathscr{D}^k$. Therefore, all the assumptions (i)-(iii) of Proposition 9.4 are fulfilled. Direct computations yield that ε^k-Sol(DVwEP^k) = ε-Sol(DVwEP) = [0, 1] in agreement with that proposition.

Now we consider second-kind approximate solutions of our vector problems that are extensions of the ones for the scalar problems in Chaps. 7 and 8.

Definition 9.6 (Second-kind approximate solutions of (VwQEP)) Assume that $X = Y$ is a normed space, $A = B, \theta : X \to \mathbb{R}$ satisfies $\theta(x) > 0$ for $x \neq 0$, and $\varepsilon \geq 0$. $\bar{x}_{\varepsilon\theta} \in A$ is an $\varepsilon\theta$-SOLUTION of (VwQEP), written $\bar{x}_{\varepsilon\theta} \in \varepsilon\theta$-Sol(VwQEP), if $f(\bar{x}_{\varepsilon\theta}, y) \notin \varepsilon\theta(\bar{x}_{\varepsilon\theta} - y)e + \text{int}C$ for all $y \in \mathscr{D}(\bar{x}_{\varepsilon\theta})$.

The second-kind approximate solutions of (DVwQEP) are defined in a similar way.

Proposition 9.7 (Convergence of second-kind approximate solutions of (VwQEP^k)) *Let $\varepsilon^k, \varepsilon \geq 0$ and $\varepsilon^k \to \varepsilon$.*

(i) *If f^k $-r^{-e}_{-C}$-i-e/h-converges or $-r^{-e}_{-C}$-i-mis-lop converges to f and θ is continuous, then*

$$A \cap \text{Ls}_k(\varepsilon^k\theta\text{-Sol(VwQEP}^k)) \subset \varepsilon\theta\text{-Sol(VwQEP)}.$$

If f^k $-r^{-e}_{-C}$-e/h-converges or $-r^{-e}_{-C}$-mis-lop converges to f, θ is bounded from above, and $\varepsilon = 0$, then

$$\text{Ls}_k(\varepsilon^k\theta\text{-Sol(VwQEP}^k)) \subset \text{Sol(VwQEP)}.$$

(ii) *If* gph$\mathscr{D} \sqsubset$ Li$_k$gph$\mathscr{D}^k$, *the $-r^{-e}_{-C}$-epi- or $-r^{-e}_{-C}$-i-epi-convergence of f^k to f replaces the variational convergence imposed in assertion* (i), *and θ fulfills the conditions there, then we have the same convergence of the second-kind approximate solutions as in assertion* (i).

Proof

(i) Assume that $\bar{x}_{\varepsilon\theta} \in \text{Ls}_k(\varepsilon^k\theta\text{-Sol(VwQEP}^k))$, i.e., there exists $x^{k_j} \in \varepsilon^{k_j}\theta$-Sol(VwQEP^{k_j}) converging to $\bar{x}_{\varepsilon\theta}$. For the first assertion, we only address the assumed i-e/h-convergence, since the case of i-mis-lop convergence is similar. Condition (a) of $-r^{-e}_{-C}$-i-e/h-convergence implies that for all $y \in \mathscr{D}(\bar{x}_{\varepsilon\theta})$, there exists $\mathscr{D}^{k_j}(x^{k_j}) \ni y^{k_j} \to y$ such that $\text{li}_j(-r^{-e}_{-C}(f^{k_j}(x^{k_j}, y^{k_j}))) \geq -r^{-e}_{-C}(f(\bar{x}_{\varepsilon\theta}, y))$. Since $x^{k_j} \in \varepsilon^{k_j}\theta$-Sol(VwQEP^{k_j}) and θ is continuous, $-r^{-e}_{-C}(f(\bar{x}_{\varepsilon\theta}, y)) \leq \varepsilon\theta(\bar{x}_{\varepsilon\theta} - y)$, i.e., $\bar{x}_{\varepsilon\theta} \in \varepsilon\theta$-Sol(VwQEP).
For the second assertion, using the assumed boundedness of θ and $\varepsilon = 0$, the argument for the case that $\bar{x}_{\varepsilon\theta} \in A$ is similar to the one used above. Suppose that $\bar{x}_{\varepsilon\theta} \notin A$. Then, condition (a) of the convergence of f^k yields $y^{k_j} \in \mathscr{D}^{k_j}(x^{k_j})$ such that $-r^{-e}_{-C}(f^{k_j}(x^{k_j}, y^{k_j})) \to +\infty$, which is impossible, since $x^{k_j} \in \varepsilon^{k_j}\theta$-Sol(VwQEP^{k_j}) and θ is bounded from above.

(ii) Let $x^{k_j} \in \varepsilon^{k_j}\theta$-Sol(VwQEP^{k_j}) tend to $\bar{x}_{\varepsilon\theta}$. If $\bar{x}_{\varepsilon\theta} \in A$, as gph$\mathscr{D} \sqsubset$ Li$_k$gph$\mathscr{D}^k$, for all $y \in \mathscr{D}(\bar{x}_{\varepsilon\theta})$, there exists $\mathscr{D}^{k_j}(x^{k_j}) \ni y^{k_j} \to y$. Then, condition (a) of the inside epi-convergence ensures that $\text{li}_j(-r^{-e}_{-C}(f^{k_j}(x^{k_j}, y^{k_j}))) \geq -r^{-e}_{-C}(f(\bar{x}_{\varepsilon\theta}, y))$. Similar to the argument for assertion (i), this leads to $\bar{x}_{\varepsilon\theta} \in \varepsilon\theta$-Sol(VwQEP). If $\bar{x}_{\varepsilon\theta} \notin A$, as for any $\mathscr{D}^{k_j}(x^{k_j}) \ni y^{k_j} \to y$, $(\bar{x}_{\varepsilon\theta}, y) \notin$ gph$\mathscr{D}$ no matter $y \in A$ or not. By condition (a) of the epi-convergence, $-r^{-e}_{-C}(f^{k_j}(x^{k_j}, y^{k_j})) \to +\infty$. Similar to the above consideration, this is impossible.

□

Proposition 9.8 (Convergence of second-kind approximate solutions of (DVwQEP^k)) *Let $\varepsilon^k, \varepsilon \geq 0$ and $\varepsilon^k \to \varepsilon$.*

(i) *If f^k r^e_C-i-e/h-converges or r^e_C-i-mai-lop converges to f and θ is continuous, then*

$$B \cap \mathrm{Ls}_k(\varepsilon^k\theta\text{-Sol(DVwQEP}^k)) \subset \varepsilon\theta\text{-Sol(DVwQEP)}.$$

If f^k r^e_C-e/h-converges or r^e_C-mai-lop converges to f, θ is bounded from below, and $\varepsilon = 0$, then

$$\mathrm{Ls}_k(\varepsilon^k\theta\text{-Sol(DVwQEP}^k)) \subset \text{Sol(DVwQEP)}.$$

(ii) *If $\mathrm{gph}\mathscr{D} \sqsubset \mathrm{Li}_k\mathrm{gph}\mathscr{D}^k$, the r^e_C-hypo- or r^e_C-i-hypo-convergence of f^k to f replaces the convergence assumed in* (i), *and θ satisfies the conditions there, then we have the same convergence of the second-kind approximate solutions as in* (i).

Proof

(i) Assume that $\bar{y}_{\varepsilon\theta} \in \mathrm{Ls}_k(\varepsilon^k\theta\text{-Sol(DVwQEP}^k))$, i.e., there exists $y^{k_j} \in \varepsilon^{k_j}\theta$-Sol(DVwQEP^{k_j}) tending to $\bar{y}_{\varepsilon\theta}$. For the first assertion, we only address the assumed i-e/h-convergence, since the case of i-mai-lop convergence is similar. Condition (b) of r^e_C-i-e/h-convergence implies that for all $x \in \mathscr{D}^{-1}(\bar{y}_{\varepsilon\theta})$, there exists $(\mathscr{D}^{k_j})^{-1}(y^{k_j}) \ni x^{k_j} \to x$ such that $\mathrm{ls}_j(r^e_C(f^{k_j}(x^{k_j}, y^{k_j}))) \leq r^e_C(f(x, \bar{y}_{\varepsilon\theta}))$. Since $y^{k_j} \in \varepsilon^{k_j}\theta$-Sol(DVwQEP^{k_j}) and θ is continuous, $r^e_C(f(x, \bar{y}_{\varepsilon\theta})) \geq \varepsilon\theta(\bar{y}_{\varepsilon\theta} - x)$, i.e., $\bar{y}_{\varepsilon\theta} \in \varepsilon\theta$-Sol(DVwQEP).
For the second inclusion, using the assumed boundedness of θ and $\varepsilon = 0$, the argument for the case that $\bar{y}_{\varepsilon\theta} \in B$ is similar to the one used above. Suppose that $\bar{y}_{\varepsilon\theta} \notin B$. Then, condition (b) of the assumed convergence of f^k yields $x^{k_j} \in (\mathscr{D}^{k_j})^{-1}(y^{k_j})$ such that $r^e_C(f^{k_j}(x^{k_j}, y^{k_j})) \to -\infty$, which is impossible, since $y^{k_j} \in \varepsilon^{k_j}\theta$-Sol(DVwQEP^{k_j}) and θ are bounded from below.

(ii) Let $y^{k_j} \in \varepsilon^{k_j}\theta$-Sol(DVwQEP^{k_j}) converge to $\bar{y}_{\varepsilon\theta}$. If $\bar{y}_{\varepsilon\theta} \in B$, as $\mathrm{gph}\mathscr{D} \sqsubset \mathrm{Li}_k\mathrm{gph}\mathscr{D}^k$, for all $x \in \mathscr{D}^{-1}(\bar{y}_{\varepsilon\theta})$, there exists $(\mathscr{D}^{k_j})^{-1}(y^{k_j}) \ni x^{k_j} \to x$. Then, condition (a) of the assumed hypo-convergence ensures that $\mathrm{ls}_j(r^e_C(f^{k_j}(x^{k_j}, y^{k_j}))) \leq r^e_C(f(x, \bar{y}_{\varepsilon\theta}))$. Similar to the argument for assertion (i), this implies that $\bar{y}_{\varepsilon\theta} \in \varepsilon\theta$-Sol(DVwQEP). If $\bar{y}_{\varepsilon\theta} \notin B$, as for any $(\mathscr{D}^{k_j})^{-1}(y^{k_j}) \ni x^{k_j} \to x$, $(x, \bar{y}_{\varepsilon\theta}) \notin \mathrm{gph}\mathscr{D}$. By condition (a) of the hypo-convergence, $r^e_C(f^{k_j}(x^{k_j}, y^{k_j})) \to -\infty$. Similar to the above consideration, this is impossible.

□

9.2.2 *Approximations of Vector Strong Quasiequilibrium Problems*

This section is devoted to the strong models (defined in Sect. 9.1) corresponding to the weak problems investigated in the preceding subsection. We also apply (in certain different ways) the scalarization functions employed in Sect. 9.2.1. From properties (b_e) of r^e_C and (b_{-e}) of r^{-e}_{-C}, one has the following relationship between approximate solutions, resp.,

$$\begin{aligned}\varepsilon\text{-}Sol(VsQEP) &= \{\bar{x}_\varepsilon \in A \mid r^e_C(f(\bar{x}_\varepsilon, y)) \leq \varepsilon \text{ for all } y \in \mathscr{D}(\bar{x}_\varepsilon)\},\\ \varepsilon\text{-}Sol(DVsQEP) &= \{\bar{y}_\varepsilon \in B \mid r^{-e}_{-C}(f(x, \bar{y}_\varepsilon)) \leq \varepsilon \text{ for all } x \in \mathscr{D}^{-1}(\bar{y}_\varepsilon)\}.\end{aligned}$$

Employing these coincidences and arguments for the scalarized problem, we have the following approximation result.

Proposition 9.9 (Convergence of approximate vector strong quasiequilibria) *Let $\varepsilon, \varepsilon^k \geq 0$ and $\varepsilon^k \to \varepsilon$.*

(i) *If f^k r^e_C-i-e/h- or r^e_C-i-mis-lop converges to f, then*

$$\mathrm{Ls}_k(\varepsilon^k\theta\text{-Sol(VsQEP}^k)) \cap A \subset \varepsilon\text{-Sol(VsQEP)}.$$

If the above "inside" (denoted by "-i-") is removed, then $\mathrm{Ls}_k(\varepsilon^k\theta$-Sol(VsQEP^k)) $\subset \varepsilon$-Sol(VsQEP).

(ii) *Assume that* gph$\mathscr{D} \sqsubset \mathrm{Li}_kgph\mathscr{D}^k$. *If the r^e_C-epi- or r^e_C-i-epi-convergence of f^k replaces the variational convergence assumed in assertion* (i), *then the conclusion in that assertion still holds.*

(iii) *If f^k r^e_C-w-e/h- or r^e_C-w-mis-lop converges to f and $x^k \in \varepsilon^k$*-Sol(VsQEP^k) *tends to $\bar{x}_\varepsilon$, then $\bar{x}_\varepsilon \in \varepsilon$*-Sol(VsQEP).

Proof The statement is proved analogously as Proposition 9.2 with $-r^{-e}_{-C}$ replaced by r^e_C. □

Example 9.10 (Convergence of approximate vector strong quasiequilibria) Consider again the case of (VsEP), and its approximating problems with the same data as in Example 9.5. We know from that example: $r^e_C(f^k(x, y)) = 0$ for all $(x, y) \in A^k \times B^k$, f^k r^e_C-e/h- r^e_C-mis-lop, and r^e_C-epi-converges to f, and gph$\mathscr{D} \sqsubset \mathrm{Li}_kgph\mathscr{D}^k$. Hence, all the assumptions (i)–(iii) of Proposition 9.9 are satisfied. By a direct verification, we see the suitable approximate solutions sets Sol(VsEP) = ε^k-Sol(VsEP^k) = [0, 1].

Proposition 9.11 (Convergence of approximate dual vector strong quasiequilibria) *Assume that $\varepsilon, \varepsilon^k \geq 0$ and $\varepsilon^k \to \varepsilon$.*

(i) *If f^k $-r^{-e}_{-C}$-i-e/h- or $-r^{-e}_{-C}$-i-mai-lop converges to f, then $B \cap \mathrm{Ls}_k(\varepsilon^k$*-Sol(DVsQEP^k)) $\subset \varepsilon$-Sol(DVsQEP). *If f^k $-r^{-e}_{-C}$-e/h- or $-r^{-e}_{-C}$-mai-lop converges to f, then* $\mathrm{Ls}_k(\varepsilon^k$-Sol(DVsQEP^k)) $\subset \varepsilon$-Sol(DVsQEP).

(ii) *If* gph$\mathscr{D} \sqsubset \mathrm{Li}_kgph\mathscr{D}^k$ *and the assumed variational convergence in assertion* (i) *are replaced by the $-r^{-e}_{-C}$-hypo- or $-r^{-e}_{-C}$-i-hypo-convergence, resp., then the conclusion in* (i) *still holds.*

(iii) *If f^k $-r^{-e}_{-C}$-w-e/h- or $-r^{-e}_{-C}$-w-mai-lop converges to f and $y^k \in \varepsilon^k$*-Sol(DVsQEP^k) *tends to $\bar{y}_\varepsilon$, then $\bar{y}_\varepsilon \in \varepsilon$*-Sol(DVsQEP).

Proof

(i) Consider the case of $-r^{-e}_{-C}$-e/h-convergence of f^k. Assume that $\bar{y}_\varepsilon \in \mathrm{Ls}_k(\varepsilon^k$-Sol(DVsQEP^k)), which means that there exists $y^{k_j} \in \varepsilon^{k_j}$-Sol(DVsQEP^{k_j}) tending to $\bar{y}_\varepsilon$. If $\bar{y}_\varepsilon \in B$, due to condition (b) of the assumed e/h-convergence, for all $x \in \mathcal{D}^{-1}(\bar{y}_\varepsilon)$, one finds $x^{k_j} \in (\mathcal{D}^{k_j})^{-1}(y^{k_j})$ tending to x with $\mathrm{ls}_j(-r^{-e}_{-C}(f^{k_j}(x^{k_j}, y^{k_j}))) \leq -r^{-e}_{-C}(f(x, \bar{y}))$. As $r^{-e}_{-C}(f^{k_j}(x^{k_j}, y^{k_j})) \leq \varepsilon^{k_j}$, $r^{-e}_{-C}(f(x, \bar{y}_\varepsilon)) \leq \varepsilon$, i.e., $\bar{y}_\varepsilon \in \varepsilon$-Sol(DVsQEP). If $\bar{y}_\varepsilon \notin B$, by the above condition (b), there exists $x^{k_j} \in (\mathcal{D}^{k_j})^{-1}(y^{k_j})$ with $-r^{-e}_{-C}(f^{k_j}(x^{k_j}, y^{k_j})) \to -\infty$ contradicting $y^{k_j} \in \varepsilon^{k_j}$-Sol(DVsQEP^{k_j}). The proof for the case of $-r^{-e}_{-C}$-i-e/h-convergence is simply the part of the above reasoning without considering $\bar{y}_\varepsilon \notin B$. The case of $-r^{-e}_{-C}$-mai-lop convergence is clearly analogous, with using condition (a) of this type of convergence (not (b) as above).

(ii) As in (i), assume that there exists $y^{k_j} \in \varepsilon^{k_j}$-Sol(DVsQEP^{k_j}) tending to $\bar{y}_\varepsilon$. If $\bar{y}_\varepsilon \notin B$, then for every $A^{k_j} \ni x^{k_j} \to x$, $(x, \bar{y}_\varepsilon) \notin \mathrm{gph}\mathcal{D}$. By condition (a) of the $-r^{-e}_{-C}$-hypo-convergence, $-r^{-e}_{-C}(f^{k_j}(x^{k_j}, y^{k_j})) \to -\infty$ contradicting the fact $-r^{-e}_{-C}(f^{k_j}(x^{k_j}, y^{k_j})) \geq -\varepsilon$. Hence, $x \in \mathcal{D}^{-1}(\bar{y}_\varepsilon)$. Since $\mathrm{gph}\mathcal{D} \sqsubset \mathrm{Li}_k \mathrm{gph}\mathcal{D}^k$, one finds $A^{k_j} \ni x^{k_j} \to x$. Again in view of the above condition (a), $\mathrm{ls}_k(-r^{-e}_{-C}(f^{k_j}(x^{k_j}, y^{k_j}))) \leq -r^{-e}_{-C}(f(x, \bar{y}_\varepsilon))$. As $y^{k_j} \in \varepsilon^{k_j}$-Sol(DVsQEP^{k_j}), $-r^{-e}_{-C}(f(x, \bar{y}_\varepsilon)) \geq -\varepsilon$, i.e., $\bar{y}_\varepsilon \in \varepsilon$-Sol(DVsQEP). Now the case of the $-r^{-e}_{-C}$-i-h-convergence becomes evident.

(iii) The proof of this assertion is similar to that of (iii) in Proposition 9.2.

□

Example 9.12 (Convergence of approximate dual vector strong quasiequilibria) Again, consider the rectangular-domain case, i.e., (DVsEP), and its approximating problems with the same data as in Example 9.3, from which we know that $r^{-e}_{-C}(f^k(x, y)) = -\varepsilon^k$ for all $(x, y) \in A^k \times B^k$, f^k $-r^{-e}_{-C}$-e/h-, $-r^{-e}_{-C}$-mai-lop, and $-r^{-e}_{-C}$-hypo- converges to f, and $\mathrm{gph}\mathcal{D} \sqsubset \mathrm{Li}_k \mathrm{gph}\mathcal{D}^k$. Thus, all the assumptions of Proposition 9.11 are verified. Easy calculations yield the approximate solution sets Sol(DVsEP) = ε^k-Sol(DVsEP^k) = [0, 1], meeting the assertions of Proposition 9.11.

Considering convergence of approximate second-kind solutions of the strong quasiequilibrium problems, we have two statements for the primal and dual problems that are completely similar to Propositions 9.7 and 9.8 with the only change of the type of variational convergence of f^k. Namely, we have r^e_C replacing $-r^{-e}_{-C}$ in Proposition 9.7, and $-r^{-e}_{-C}$ replacing r^e_C in Proposition 9.8. We do not provide these statements here.

9.2.3 *Duality Between Vector Weak and Strong Quasiequilibrium Problems*

In this subsection, pairs of vector primal-dual quasiequilibrium problems are examined. We will see that, unlike the scalar case studied in Chaps. 7 and 8, here a pair of problems does not contain the same weak problems or the same strong problems but has a "deviation": "weak primal" goes with "strong dual" and "strong primal" with "weak dual." And correspondingly, we need the following notion of approximate dual weak saddle points. $(\bar{x}_\varepsilon, \bar{y}_\varepsilon) \in \operatorname{gph}\mathscr{D}$ is called an ε-DUAL WEAK SADDLE POINT (of a scalar bifunction Φ on $\operatorname{gph}\mathscr{D}$) if, for all $(\bar{x}_\varepsilon, y)$ and $(x, \bar{y}_\varepsilon)$ in $\operatorname{gph}\mathscr{D}$, $\Phi(x, \bar{y}_\varepsilon) - \varepsilon \le \Phi(\bar{x}_\varepsilon, \bar{y}_\varepsilon) \le \Phi(\bar{x}_\varepsilon, y) + \varepsilon$. We denote the set of all ε-dual weak saddle points by ε-dwsdlΦ.

Proposition 9.13 (Relationship between approximate vector strong primal and weak dual quasiequilibria and approximate weak saddle points of scalarized bimaps) *Let* $\varepsilon \ge 0$.

(a) $\bar{x}_\varepsilon \in \varepsilon\text{-Sol(VsQEP)}$, $\bar{y}_\varepsilon \in \varepsilon\text{-Sol(DVwQEP)}$, *and* $f(\bar{x}_\varepsilon, \bar{y}_\varepsilon) \in \operatorname{bd}(-C)$ *if and only if* $(\bar{x}_\varepsilon, \bar{y}_\varepsilon) \in \varepsilon\text{-wsdl}(r_C^e(f)) \cap \{(x, y) \mid f(x, y) \in \operatorname{bd}(-C)\}$.

(b) $\bar{x}_\varepsilon \in \varepsilon\text{-Sol(VwQEP)}$, $\bar{y}_\varepsilon \in \varepsilon\text{-Sol(DVsQEP)}$, *and* $f(\bar{x}_\varepsilon, \bar{y}_\varepsilon) \in \operatorname{bd}(C)$ *if and only if* $(\bar{x}_\varepsilon, \bar{y}_\varepsilon) \in \varepsilon\text{-dwsdl}(r_{-C}^{-e}(f)) \cap \{(x, y) \mid f(x, y) \in \operatorname{bd}(C)\}$.

Proof

(a) We have the equivalences

$$\begin{aligned}
&\text{for any } (x, y) \in X \times Y, r_C^e(f(x, y)) = 0 \Leftrightarrow f(x, y) \in \operatorname{bd}(-C);\\
&\quad \bar{x}_\varepsilon \in \varepsilon\text{-Sol(VsQEP)} \Leftrightarrow r_C^e(f(\bar{x}_\varepsilon, y)) \le \varepsilon \ \forall y \in \mathscr{D}(\bar{x}_\varepsilon);\\
&\quad \bar{y}_\varepsilon \in \varepsilon\text{-Sol(DVwQEP)} \Leftrightarrow r_C^e(f(x, \bar{y}_\varepsilon)) \ge -\varepsilon \ \forall x \in \mathscr{D}^{-1}(\bar{y}_\varepsilon).
\end{aligned}$$

So, $(\bar{x}_\varepsilon, \bar{y}_\varepsilon) \in \varepsilon\text{-wsdl}(r_C^e(f)) \cap \{(x, y) \mid f(x, y) \in \operatorname{bd}(-C)\}$ means that $r_C^e(f(\bar{x}_\varepsilon, y)) - \varepsilon \le 0 = r_C^e(f(\bar{x}_\varepsilon, \bar{y}_\varepsilon)) \le r_C^e(f(x, \bar{y}_\varepsilon)) + \varepsilon$, which is equivalent to $\bar{x}_\varepsilon \in \varepsilon\text{-Sol(VsQEP)}$, $\bar{y}_\varepsilon \in \varepsilon\text{-Sol(DVwQEP)}$, and $f(\bar{x}_\varepsilon, \bar{y}_\varepsilon) \in \operatorname{bd}(-C)$ by Proposition 8.6.

(b) We give a short proof, though it's similar to (a). In this case, the equivalences we have are

$$\begin{aligned}
&\text{for any } (x, y) \in X \times Y, r_{-C}^{-e}(f(x, y)) = 0 \Leftrightarrow f(x, y) \in \operatorname{bd}(C);\\
&\quad \bar{x}_\varepsilon \in \varepsilon\text{-Sol(VwQEP)} \Leftrightarrow r_{-C}^{-e}(f(\bar{x}_\varepsilon, y)) \ge -\varepsilon \ \forall y \in \mathscr{D}(\bar{x}_\varepsilon);\\
&\quad \bar{y}_\varepsilon \in \varepsilon\text{-Sol(DVsQEP)} \Leftrightarrow r_{-C}^{-e}(f(x, \bar{y}_\varepsilon)) \le -\varepsilon \ \forall x \in \mathscr{D}^{-1}(\bar{y}_\varepsilon).
\end{aligned}$$

Hence, $(\bar{x}_\varepsilon, \bar{y}_\varepsilon) \in \varepsilon\text{-dwsdl}(r_{-C}^{-e}(f)) \cap \{(x, y) \mid f(x, y) \in \operatorname{bd}(C)\}$ means that $r_{-C}^{-e}(f(\bar{x}_\varepsilon, y)) + \varepsilon \ge 0 = r_{-C}^{-e}(f(\bar{x}_\varepsilon, \bar{y}_\varepsilon)) \ge r_{-C}^{-e}(f(x, \bar{y}_\varepsilon)) - \varepsilon$, which is just $\bar{x}_\varepsilon \in \varepsilon\text{- Sol(VwQEP)}$, $\bar{y}_\varepsilon \in \varepsilon\text{-Sol(DVsQEP)}$, and $f(\bar{x}_\varepsilon, \bar{y}_\varepsilon) \in \operatorname{bd}(C)$. □

To go further, set

$$\mathscr{C}_1 f^k = \{(x, y) \in \mathrm{gph}\mathscr{D}^k \mid f^k(x, y) \in \mathrm{bd}(-C)\},$$
$$\mathscr{C}_1 f = \{(x, y) \in \mathrm{gph}\mathscr{D} \mid f(x, y) \in \mathrm{bd}(-C)\},$$

$$\mathscr{C}_2 f^k = \{(x, y) \in \mathrm{gph}\mathscr{D}^k \mid f^k(x, y) \in \mathrm{bd}(C)\},$$
$$\mathscr{C}_2 f = \{(x, y) \in \mathrm{gph}\mathscr{D} \mid f(x, y) \in \mathrm{bd}(C)\}.$$

Theorem 9.14 (Duality convergence of approximate vector strong primal and weak dual quasiequilibria) *Let* $\varepsilon, \varepsilon^k \geq 0$ *and* $\varepsilon^k \to \varepsilon$.

(i) *Assume that* f^k r^e_C*-i-e/h- or both* r^e_C*-i-mis-lop and* r^e_C*-i-mai-lop converges to* f. *Assume further that* $\mathrm{Ls}_k \mathscr{C}_1 f^k \subset \mathrm{cl}\mathscr{C}_1 f$. *Then,*

$$\mathrm{Ls}_k[\mathscr{C}_1 f^k \cap (\varepsilon^k - \mathrm{wsdl}(r^e_C(f^k)))] \cap \mathrm{gph}\mathscr{D} =$$

$$\mathrm{Ls}_k[\mathscr{C}_1 f^k \cap ((\varepsilon^k - \mathrm{Sol(VsQEP^k)}) \times (\varepsilon^k - \mathrm{Sol(DVwQEP^k)}))] \cap \mathrm{gph}\mathscr{D}$$

$$\subset \mathrm{cl}\mathscr{C}_1 f \cap ((\varepsilon - \mathrm{Sol(VsQEP)} \times (\varepsilon - \mathrm{Sol(DVwQEP)})) \cap \mathrm{gph}\mathscr{D}$$
$$= \mathrm{cl}\mathscr{C}_1 f \cap (\varepsilon - \mathrm{wsdl}(r^e_C(f))).$$

If the above restriction "inside" (denoted by "-i-") is removed, then the conclusion holds without the two involved " $\cap\, \mathrm{gph}\mathscr{D}$*."*

(ii) *If* f^k r^e_C*-i-e/h-converges tightly or* r^e_C*-i-mis-lop both* x*- and* y*-ancillary tightly converges or* r^e_C *-i-mai-lop both* x*- and* y*-ancillary tightly converges to* f *and* $\mathrm{cl}\mathscr{C}_1 f \subset \mathrm{Li}_k \mathscr{C}_1 f^k$, *then*

$$\varepsilon\text{-wsdl}(r^e_C(f)) \cap \mathrm{cl}\mathscr{C}_1 f = ((\varepsilon\text{-Sol(VsQEP)}) \times (\varepsilon\text{- Sol(DVwQEP)})) \cap \mathrm{cl}\mathscr{C}_1 f$$

$$\subset \cap_{\delta > \varepsilon} \mathrm{Li}_k(\delta\text{-Sol(VsQEP}^k) \times \delta\text{-Sol(DVwQEP}^k)) \cap \mathrm{Li}_k \mathscr{C}_1 f^k$$

$$= \cap_{\delta > \varepsilon} \mathrm{Li}_k(\delta\text{-wsdl} r^e_C(f^k)) \cap \mathrm{Li}_k \mathscr{C}_1 f^k.$$

Hence,

$$\varepsilon\text{-wsdl}(r^e_C(f)) \cap \mathrm{cl}\mathscr{C}_1 f = (\varepsilon\text{-Sol(VsQEP)} \times \varepsilon\text{- Sol(DVwQEP)}) \cap \mathrm{cl}\mathscr{C}_1 f$$

$$\subset \mathrm{Li}_k((\varepsilon^k\text{-Sol(VsQEP}^k)) \times (\varepsilon^k\text{-Sol(DVwQEP}^k))) \cap \mathrm{Li}_k \mathscr{C}_1 f^k$$

$$= \mathrm{Li}_k(\varepsilon\text{-wsdl}(r^e_C(f))) \cap \mathrm{Li}_k \mathscr{C}_1 f^k.$$

(iii) *Assume that in assertion* (ii), *the restriction "inside" in the types of variational convergence is removed, and the mentioned types of lop-convergence are imposed simultaneously. Assume further that* $\mathrm{Lim}_k \mathscr{C}_1 f^k = \mathrm{cl}\mathscr{C}_1 f$. *Then,*

$$\begin{aligned}\varepsilon\text{-wsdl}(r^e_C(f)) \cap \mathrm{cl}\mathscr{C}_1 f &= (\varepsilon\text{-Sol(VsQEP)} \times \varepsilon\text{-Sol(DVwQEP)}) \cap \mathrm{cl}\mathscr{C}_1 f \\ &= \mathrm{Lim}_k(\varepsilon^k\text{-Sol(VsQEP}^k) \times \varepsilon^k\text{-Sol(DVwQEP}^k)) \cap \mathrm{Lim}_k \mathscr{C}_1 f^k \\ &= \mathrm{Lim}_k(\varepsilon^k\text{-wsdl}(r^e_C(f^k))) \cap \mathrm{Lim}_k \mathscr{C}_1 f^k.\end{aligned}$$

Proof

(i) The two required equalities follow from Proposition 9.13(a). To prove the inclusion, first observe that the upper limit (Limsup) of an intersection of two sets is included in the intersection of the upper limits of these sets, and the upper limit of a product of sets is equal to the product of the upper limits of these sets. Then, the required inclusion is due to Propositions 9.9 and 9.4.

(ii) We address the first chain of relations. The two required equalities follow from Proposition 9.13(a). We check now the required inclusion, making an observation about inner limits similar to that in assertion (i) about upper limits. Assume that $(\bar{x}, \bar{y}) \in \varepsilon\text{-wsdl}(r^e_C(f))$. The assumed r^e_C-i-e/h-convergence means that

$$\begin{cases} \text{for every } A^{k_j} \ni x^{k_j} \to \bar{x}, \text{ there exists } \mathscr{D}^{k_j}(x^{k_j}) \ni y^{k_j} \to \bar{y}, \\ \quad \mathrm{li}_j(r^e_C(f^{k_j}(x^{k_j}, y^{k_j}))) \geq r^e_C(f(\bar{x}, \bar{y})), \\ \text{for every } B^{k_j} \ni y^{k_j} \to \bar{y}, \text{ there exists } (D^{k_j})^{-1}(y^{k_j}) \ni x^{k_j} \to \bar{x}, \\ \quad \mathrm{ls}_j(r^e_C(f^{k_j}(x^{k_j}, y^{k_j}))) \leq r^e_C(f(\bar{x}, \bar{y})). \end{cases} \tag{9.1}$$

The assumed tightness of the r^e_C-i-e/h-convergence implies that

$$\begin{cases} \text{for } x^{k_j} \text{ in the 2nd relation in (9.1)}, \\ \quad \mathrm{ls}_j[\sup_{y \in \mathscr{D}^{k_j}(x^{k_j})}(r^e_C(f^{k_j}(x^{k_j}, y)))] \leq \sup_{y \in \mathscr{D}(\bar{x})}(r^e_C(f(\bar{x}, y))), \\ \text{for } y^{k_j} \text{ in the 1st relation in (9.1)}, \\ \quad \mathrm{li}_j[\inf_{x \in (\mathscr{D}^{k_j})^{-1}(y^{k_j})}(r^e_C(f^{k_j}(x, y^{k_j})))] \geq \inf_{x \in \mathscr{D}^{-1}(\bar{y})}(r^e_C(f(x, \bar{y}))). \end{cases} \tag{9.2}$$

We claim that for (x^{k_j}, y^{k_j}) in (9.2), $\forall \delta > \varepsilon, \exists k_\delta, \forall k_j \geq k_\delta$,

$$(x^{k_j}, y^{k_j}) \in \delta - \mathrm{wsdl}(r^e_C(f^{k_j})). \tag{9.3}$$

Suppose to the contrary that there exists $\delta_0 > \varepsilon$ and a subsequence of $\{k_j\}_j$, denoted also by $\{k_j\}_j$, tending to $+\infty$, such that for all j, (x^{k_j}, y^{k_j}) does not satisfy (9.2) for δ_0. First, assume that (x^{k_j}, y^{k_j}) does not fulfill the right-hand side inequality of $(x^{k_j}, y^{k_j}) \in \delta_0$-wsdl$(r_C^e(f^{k_j}))$, i.e.,

$$r_C^e(f^{k_j}(x^{k_j}, y^{k_j})) > \inf_{x\in(\mathcal{D}^{k_j})^{-1}(y^{k_j})}(r_C^e(f^{k_j}(x, y^{k_j}))) + \delta_0. \tag{9.4}$$

Taking limsup of both sides, we get

$$\begin{aligned} r_C^e(f(\bar{x}, \bar{y})) &\geq \mathrm{ls}_j(r_C^e(f^{k_j}(x^{k_j}, y^{k_j}))) \\ &\geq \mathrm{ls}_j[\inf_{x\in(\mathcal{D}^{k_j})^{-1}(y^{k_j})}(r_C^e(f^{k_j}(x, y^{k_j})))] + \delta_0 \\ &\geq \mathrm{li}_j[\inf_{x\in(\mathcal{D}^{k_j})^{-1}(y^{k_j})}(r_C^e(f^{k_j}(x, y^{k_j})))] + \delta_0 \\ &\geq \inf_{x\in\mathcal{D}^{-1}(\bar{y})}(r_C^e(f(x, \bar{y}))) + \delta_0 \\ &> \inf_{x\in\mathcal{D}^{-1}(\bar{y})}(r_C^e(f(x, \bar{y}))) + \varepsilon, \end{aligned} \tag{9.5}$$

where the first inequality follows from the second relation in (9.1) and the last but one inequality from the second relation in (9.2). This contradicts the right-hand side inequality in the definition of $(\bar{x}, \bar{y}) \in \varepsilon$-wsdl$(r_C^e(f))$. If (x^{k_j}, y^{k_j}) violates the left-hand-side inequality of δ_0-wsdl$(r_C^e(f^{k_j}))$, the argument is similar, but applying the first relation in (9.1) and (9.2). Therefore, the claim is verified.

Now address the cases of the types of lop-convergence. The r_C^e-i-mis-lop convergence is equivalent to

$$\begin{cases} \text{for every } A^{k_j} \ni x^{k_j} \to \bar{x}, \text{ there exists } \mathcal{D}^{k_j}(x^{k_j}) \ni y^{k_j} \to \bar{y}, \\ \quad \mathrm{li}_j(r_C^e(f^{k_j}(x^{k_j}, y^{k_j}))) \geq r_C^e(f(\bar{x}, \bar{y})), \\ \text{there exists } A^{k_j} \ni x^{k_j} \to \bar{x}, \text{ such that for all } \mathcal{D}^{k_j}(x^{k_j}) \ni y^{k_j} \to \bar{y}, \\ \quad \mathrm{ls}_j(r_C^e(f^{k_j}(x^{k_j}, y^{k_j}))) \leq r_C^e(f(\bar{x}, \bar{y})). \end{cases} \tag{9.6}$$

(Note that the first relations in (9.1) and (9.6) are the same, but the second one of (9.6) is stronger than that of (9.1).)

The x-ancillary tightness of the r_C^e-i-mis-lop convergence yields that for the existing x^{k_j} in (9.6),

$$\mathrm{ls}_j[\sup_{y\in\mathcal{D}^{k_j}(x^{k_j})}(r_C^e(f^{k_j}(x^{k_j}, y)))] \leq \sup_{y\in\mathcal{D}(\bar{x})}(r_C^e(f(\bar{x}, y))), \tag{9.7}$$

and the y-ancillary tightness of this lop-convergence gives that for y^{k_j} in (9.6),

$$\mathrm{li}_j[\inf_{x\in(\mathcal{D}^{k_j})^{-1}(y^{k_j})}(r_C^e(f^{k_j}(x, y^{k_j})))] \geq \inf_{x\in\mathcal{D}^{-1}(\bar{y})}(r_C^e(f(x, \bar{y}))). \tag{9.8}$$

Arguing by contradiction, suppose that the aforementioned (x^{k_j}, y^{k_j}) violates the right-hand side inequality for $(x^{k_j}, y^{k_j}) \in \delta_0$- wsdl$(r_C^e(f^{k_j}))$. Then, employing the second relation in (9.6) and (9.8), we get a contradiction similar to the above argument for this violation of the x-ancillary r_C^e-i-e/h-convergence case. If the violated inequality is the left-hand-side one, the employed relations are the first relation in (9.6) and (9.7). The proof for the case of r_C^e-i-mai-lop convergence is also analogous, but instead of (9.6)–(9.8), we make use of the corresponding facts of r_C^e-i-mai-lop convergence.

For the second chain of relations, it suffices to show only that ε-wsdl$(r_C^e(f)) \subset \mathrm{Li}_k(\varepsilon^k$-wsdl$(r_C^e(f^k)))$. Assume that $(\bar{x}, \bar{y}) \in \varepsilon$-wsdl$(r_C^e(f))$. Then, $(x^k, y^k) \in \varepsilon^k$-wsdl$(r_C^e(f^k))$ tending to $(\bar{x}, \bar{y})$ can be found as follows. For any $\ell > \varepsilon$, by the inclusions in the first chain of relations, we have $(x_\ell^k, y_\ell^k) \in \varepsilon^\ell$-wsdl$(r_C^e(f^k))$ tending to $(\bar{x}, \bar{y})$. Passing $\varepsilon^\ell \to \varepsilon$ and using the diagonalization argument, we get $(x_{\ell(k)}^k, y_{\ell(k)}^k) \in \varepsilon^{\ell(k)}$-wsdl$(r_C^e(f^k))$ tending to $(\bar{x}, \bar{y})$. Then, $(x^k, y^k) := (x_{\ell(k)}^k, y_{\ell(k)}^k)$ is just the required sequence.

(iii) By the second chain of relations in assertion (ii), we only need to check $\mathrm{Ls}_k(\varepsilon^k$-wsdl$(r_C^e(f^k))) \subset \varepsilon$-wsdl$(r_C^e(f))$. By assertion (i), we have to show only that no subsequence $(\bar{x}^{k_j}, \bar{y}^{k_j}) \in \varepsilon^{k_j}$-wsdl$(r_C^e(f^{k_j}))$ can tend to a point outside gph$\mathscr{D}$. Suppose to the contrary the existence of $(\bar{x}^{k_j}, \bar{y}^{k_j}) \to (x, y) \notin \mathrm{gph}\mathscr{D}$, say $x \notin A$. By condition (a) of the assumed e/h-convergence, there is $y^{k_j} \in \mathscr{D}^{k_j}(\bar{x}^{k_j})$ with $r_C^e(f^{k_j}(\bar{x}^{k_j}, y^{k_j})) \to +\infty$. Then, for any $x^{k_j} \in (\mathscr{D}^{k_j})^{-1}(\bar{y}^{k_j})$, $r_C^e(f^{k_j}(x^{k_j}, \bar{y}^{k_j})) + \varepsilon^{k_j} \geq r_C^e(f^{k_j}(x^{k_j}, \bar{y}^{k_j})) - \varepsilon^{k_j} \to +\infty$. We have two cases: $y \in B$ or $y \notin B$. If $y \in B$, then condition (b) in the assumed e/h-convergence implies that for any $\hat{x} \in A$, there exists $\hat{x}^{k_j} \in (\mathscr{D}^{k_j})^{-1}(\bar{y}^{k_j})$ tending to $\hat{x}$ such that $\mathrm{ls}_j(r_C^e(f^{k_j}(\hat{x}^{k_j}, \bar{y}^{k_j}))) \leq r_C^e(f(\hat{x}, y))$, a contradiction. So only $y \notin B$ remains possible. Then, again by condition (b) of the e/h-convergence, there exists $\hat{x}^{k_j} \in (\mathscr{D}^{k_j})^{-1}(\bar{y}^{k_j})$ with $r_C^e(f^{k_j}(\hat{x}^{k_j}, \bar{y}^{k_j})) \to -\infty$ which is impossible either.

□

Example 9.15 (Duality convergence of approximate strong primal solutions and weak dual solutions) Like in the previous examples, we focus on illustrating the main "vector features" of Theorem 9.14, i.e., the problem setting, the concept of approximate solutions, the scalarization, and only using simple data. Let $X = Y = \mathbb{R}$, $Z = \mathbb{R}^2$, $C = \mathbb{R}_+^2$, $A = B = [0, 1]$, and $f_1(x, y) = 0$, $f_2(x, y) = -1$ for every $(x, y) \in [0, 1]^2$. The approximating data are: $\varepsilon^k > \sqrt{2} \to \varepsilon = \sqrt{2}$, $A^k = B^k = [0, 1]$, and $f_1^k(x, y) = 0$, $f_2^k(x, y) = -\frac{\varepsilon^k}{\sqrt{2}}$ for all $(x, y) \in [0, 1]^2$ and k. Then, $\mathscr{C}_1 f^k = \mathscr{C}_1 f = [0, 1]^2$ for all k, so they satisfy the assumption. $r_C^e(f^k(x, y)) = r_C^e(f(x, y)) = 0$ for all $(x, y) \in [0, 1]^2$ and k, so the types of variational convergence assumed in the theorem hold. Moreover, $\xi^k(r_C^e(f^k))(x) = \sup_{y \in B^k} r_C^e(f^k(x, y)) = 0$ and, similarly, $\xi(r_C^e(f))(x) = 0$ for all $x \in [0, 1]$ and k. $\eta^k(r_C^e(f^k))(y) = \inf_{x \in A^k} r_C^e(f^k(x, y)) = 0$ and $\eta(r_C^e(f))(y) = 0$ for all $y \in [0, 1]$

and k. Hence, the imposed tightness conditions, the other assumptions, are also fulfilled.

Direct computation yields that ε^k-wsdl$(r^e_C(f^k)) = \varepsilon$-wsdl$(r^e_C(f)) = [0, 1]^2$ for all k and all the involved approximate solution sets are equal to [0, 1]. They meet the conclusion of Theorem 9.14.

Theorem 9.16 (Duality convergence of approximate solutions of vector primal weak and dual strong quasiequilibrium problems) *Assume that* $\varepsilon, \varepsilon^k \geq 0$ *and* $\varepsilon^k \to \varepsilon$.

(a) *Assume that* f^k $-r^{-e}_{-C}$*-i-e/h- or both* $-r^{-e}_{-C}$*-i-mis-lop and* $-r^{-e}_{-C}$*-i-mai-lop converges to* f. *Assume further that* $\mathrm{Ls}_k\mathscr{C}_2 f^k \subset \mathrm{cl}\mathscr{C}_2 f$. *Then,*

$$\mathrm{Ls}_k[\mathscr{C}_2 f^k \cap (\varepsilon^k - \mathrm{dwsdl}(r^{-e}_{-C}(f^k)))] \cap \mathrm{gph}\mathscr{D}$$

$$= \mathrm{Ls}_k[\mathscr{C}_2 f^k \cap ((\varepsilon^k - \mathrm{Sol(VwQEP^k)}) \times (\varepsilon^k - \mathrm{Sol(DVsQEP^k)}))] \cap \mathrm{gph}\mathscr{D}$$

$$\subset \mathrm{cl}\mathscr{C}_2 f \cap ((\varepsilon - \mathrm{Sol(VwQEP)} \times (\varepsilon - \mathrm{Sol(DVsQEP)})) \cap \mathrm{gph}\mathscr{D}$$

$$= \mathrm{cl}\mathscr{C}_2 f \cap (\varepsilon - \mathrm{dwsdl}(r^{-e}_{-C}(f))).$$

If the above restriction "inside" is removed, then the conclusion holds without the two involved " $\cap\, \mathrm{gph}\mathscr{D}$*."*

(b) *If* f^k $-r^{-e}_{-C}$*-i-e/h-converges tightly or* $-r^{-e}_{-C}$*-i-mis-lop both* x*- and* y*-ancillary tightly converges or* $-r^{-e}_{-C}$*-i-mai-lop both* x*- and* y*-ancillary tightly converges to* f *and* $\mathrm{cl}\mathscr{C}_2 f \subset \mathrm{Li}_k\mathscr{C}_2 f^k$, *then*

$$\varepsilon\text{-dwsdl}(r^{-e}_{-C}(f)) \cap \mathrm{cl}\mathscr{C}_2 f = ((\varepsilon\text{-Sol(VwQEP)}) \times (\varepsilon\text{- Sol(DVsQEP)})) \cap \mathrm{cl}\mathscr{C}_2 f$$

$$\subset \cap_{\delta>\varepsilon}\mathrm{Li}_k(\delta\text{-Sol}(\mathrm{VwQEP}^k) \times \delta\text{-Sol}(\mathrm{DVsQEP}^k)) \cap \mathrm{Li}_k\mathscr{C}_2 f^k$$

$$= \cap_{\delta>\varepsilon}\mathrm{Li}_k(\delta\text{-dwsdl}r^{-e}_{-C}(f^k)) \cap \mathrm{Li}_k\mathscr{C}_2 f^k.$$

Hence,

$$\varepsilon\text{-dwsdl}(r^{-e}_{-C}(f)) \cap \mathrm{cl}\mathscr{C}_2 f = (\varepsilon\text{-Sol(VwQEP)} \times \varepsilon\text{- Sol(DVsQEP)}) \cap \mathrm{cl}\mathscr{C}_2 f$$

$$\subset \mathrm{Li}_k((\varepsilon^k\text{-Sol}(\mathrm{VwQEP}^k)) \times (\varepsilon^k\text{-Sol}(\mathrm{DVsQEP}^k))) \cap \mathrm{Li}_k\mathscr{C}_2 f^k$$

$$= \mathrm{Li}_k(\varepsilon\text{-dwsdl}(r^{-e}_{-C}(f))) \cap \mathrm{Li}_k\mathscr{C}_2 f^k.$$

(c) *Assume that in assertion* (b), *the restriction "inside" in the type of variational convergence is removed, and the mentioned types of lop-convergence are imposed simultaneously. Assume further that* $\mathrm{Lim}_k \mathscr{C}_2 f^k = \mathrm{cl}\mathscr{C}_2 f$. *Then,*

$$\begin{aligned}\varepsilon\text{-dwsdl}(r^{-e}_{-C}(f)) \cap \mathrm{cl}\mathscr{C}_2 f &= (\varepsilon\text{-Sol(VwQEP)} \times \varepsilon\text{- Sol(DVsQEP)}) \cap \mathrm{cl}\mathscr{C}_2 f \\ &= \mathrm{Lim}_k(\varepsilon^k\text{-Sol(VwQEP}^k) \times \varepsilon^k\text{-Sol(DVsQEP}^k)) \cap \mathrm{Lim}_k \mathscr{C}_2 f^k \\ &= \mathrm{Lim}_k(\varepsilon^k\text{-dwsdl}(r^{-e}_{-C}(f^k))) \cap \mathrm{Lim}_k \mathscr{C}_2 f^k.\end{aligned}$$

Proof

(a) The equalities are due to Proposition 9.13(b), and the inclusion is due to Propositions 9.2 and 9.9.

(b) We discuss the first chain of relations. The two required equalities follow from Proposition 9.13(b). We verify the required inclusion. Assume that $(\bar{x}, \bar{y}) \in \varepsilon$-dwsdl$(r^{-e}_{-C}(f))$. The assumed $-r^{-e}_{-C}$-i-e/h-convergence gives

$$\begin{cases} \text{for every } A^{k_j} \ni x^{k_j} \to \bar{x}, \text{ there exists } \mathscr{D}^{k_j}(x^{k_j}) \ni y^{k_j} \to \bar{y}, \\ \quad \mathrm{ls}_j(r^{-e}_{-C}(f^{k_j}(x^{k_j}, y^{k_j}))) \leq r^{-e}_{-C}(f(\bar{x}, \bar{y})), \\ \text{for every } B^{k_j} \ni y^{k_j} \to \bar{y}, \text{ there exists } (D^{k_j})^{-1}(y^{k_j}) \ni x^{k_j} \to \bar{x}, \\ \quad \mathrm{li}_j(r^{-e}_{-C}(f^{k_j}(x^{k_j}, y^{k_j}))) \geq r^{-e}_{-C}(f(\bar{x}, \bar{y})). \end{cases} \tag{9.9}$$

From the tightness in the assumptions of the $-r^{-e}_{-C}$-i-e/h-convergence, it follows that

$$\begin{cases} \text{for } y^{k_j} \text{ in the 1st relation in (9.9)}, \\ \quad \mathrm{ls}_j[\inf_{x \in (\mathscr{D}^{k_j})^{-1}(y^{k_j})}(r^{-e}_{-C}(f^{k_j}(x, y^{k_j})))] \leq \inf_{x \in \mathscr{D}^{-1}(\bar{y})}(r^{-e}_{-C}(f(x, \bar{y}))), \\ \text{for } x^{k_j} \text{ in the 2nd relation in (9.9)}, \\ \quad \mathrm{li}_j[\sup_{y \in \mathscr{D}^{k_j}(x^{k_j})}(r^{-e}_{-C}(f^{k_j}(x^{k_j}, y)))] \geq \inf_{y \in \mathscr{D}(\bar{x})}(r^{-e}_{-C}(f(\bar{x}, y))). \end{cases} \tag{9.10}$$

The next step is to ensure that for (x^{k_j}, y^{k_j}) in (9.10), $\forall \delta > \varepsilon, \exists k_\delta, \forall k_j \geq k_\delta$,

$$(x^{k_j}, y^{k_j}) \in \delta - \mathrm{dwsdl}(r^{-e}_{-C}(f^{k_j})). \tag{9.11}$$

Suppose the existence of $\delta_0 > \varepsilon$ and a subsequence of $\{k_j\}_j$ with the same index notation $\{k_j\}_j$ such that for all j, (x^{k_j}, y^{k_j}) does not satisfy (9.11) for δ_0. Assume that (x^{k_j}, y^{k_j}) does not fulfill the right-hand side inequality of $(x^{k_j}, y^{k_j}) \in \delta_0$-*dwsdl*$(r^{-e}_{-C}(f^{k_j}))$, i.e.,

$$r_{-C}^{-e}(f^{k_j}(x^{k_j}, y^{k_j})) > \inf_{y \in \mathscr{D}^{k_j}(x^{k_j})}(r_{-C}^{-e}(f^{k_j}(x^{k_j}, y))) + \delta_0. \tag{9.12}$$

Taking limsup of both sides, we get

$$r_{-C}^{-e}(f(\bar{x}, \bar{y})) > \inf_{y \in \mathscr{D}(\bar{x})}(r_{-C}^{-e}(f(\bar{x}, y))) + \varepsilon, \tag{9.13}$$

which is impossible due to the right-hand side inequality of the ε-dwsd-point $(\bar{x}, \bar{y})$.
The reasoning for the case of the violation of the left-hand side inequality is analogous. And the proof for the case of the type of lop-convergence is also similar. We easily see that the second chain of relations in assertion (b) is established in the same manner as that in Theorem 9.14.

(c) The proof goes along the same lines as that for assertion (c) in Theorem 9.14.

□

Example 9.17 (Duality convergence of approximate weak primal solutions and strong dual solutions) Let $X = Y = \mathbb{R}$, $Z = \mathbb{R}^2$, $C = \mathbb{R}^2_+$, $A = B = [0, 1]$, and $f_1(x, y) = 0$, $f_2(x, y) = 1$ for every $(x, y) \in [0, 1]^2$. The approximating data are: $\varepsilon^k > \sqrt{2} \to \varepsilon = \sqrt{2}$, $A^k = B^k = [0, 1]$, and $f_1^k(x, y) = 0$, $f_2^k(x, y) = \frac{\varepsilon^k}{\sqrt{2}}$ for all $(x, y) \in [0, 1]^2$ and k. Then, $\mathscr{C}_2 f^k = \mathscr{C}_2 f = [0, 1]^2$ for all k, so they satisfy the assumption. $r_{-C}^{-e}(f^k(x, y)) = r_{-C}^{-e}(f(x, y)) = 0$ for all $(x, y) \in [0, 1]^2$ and k, so the types of variational convergence assumed in Theorem 9.16 hold. Moreover, $\xi^k(r_{-C}^{-e}(f^k))(x) = \sup_{y \in B^k} r_{-C}^{-e}(f^k(x, y)) = 0$ for all $x \in [0, 1]$ and k and, similarly, $\xi(r_{-C}^{-e}(f))(x) = 0$ for all $x \in [0, 1]$. $\eta^k(r_{-C}^{-e}(f^k))(y) = \inf_{x \in A^k} r_{-C}^{-e}(f^k(x, y)) = 0$ and $\eta(r_{-C}^{-e}(f))(y) = 0$ for all $y \in [0, 1]$ and k. Hence, the imposed tightness conditions, and the other assumptions, are also fulfilled.

Direct computation yields that ε^k-dwsdl$(r_{-C}^{-e}(f^k)) = \varepsilon$-dwsdl$(r_{-C}^{-e}(f)) = [0, 1]^2$ for all k and all the involved approximate solution sets are equal to $[0, 1]$. They meet the conclusion of Theorem 9.16.

9.3 Approximations of Vector Quasiequilibrium Problems via the Hiriart-Urruty Scalarization

In this section, we employ the Hiriart-Urruty scalarization. The scalarization function is defined as follows. Let Z be a normed space, $H \subset Z$, and C a closed convex set in Z. The HIRIART-URRUTY SCALARIZATION FUNCTION (also called the ORIENTED DISTANCE FUNCTION or the SIGNED DISTANCE FUNCTION) $\delta_H : Z \to \mathbb{R} \cup \{-\infty, +\infty\}$ is defined as

$$\delta_H(z) := d(z, H) - d(z, Z \setminus H) = \begin{cases} d(z, H) & \text{if } z \in Z \setminus H, \\ -d(z, Z \setminus H) & \text{if } z \in H \end{cases}$$

Fig. 9.2 Oriented distance $\delta_{Z\setminus C}(z) = -t,\ \delta_{Z\setminus C}(z') = t$

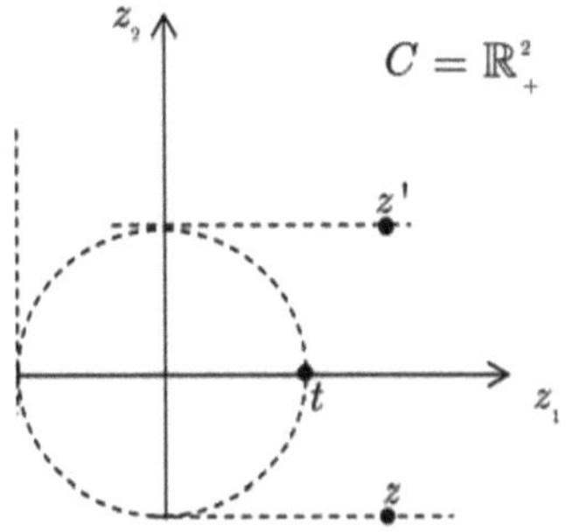

with the convention that $d(z, \emptyset) = +\infty$. Note that if H is a proper subset of Z, i.e., H is nonempty and $H \neq Z$, then δ_H is finite (see, e.g., [462]). (Later we will use the evident property $\delta_{Z\setminus H}(z) = -\delta_H(z)$ for all $z \in Z$.) Figure 9.2 shows the value of $\delta_{Z\setminus C}$ at two given points z and z'.

Proposition 9.18 (Relationship between approximates of C and those of its scalarized sets) *Assume that Z is reflexive, $t \geq 0$, and B_Z is the closed unit ball in Z. Then,*

(a) $z \in tB_Z + C$ *if and only if* $\delta_{Z\setminus C}(z) \geq -t$;
(b) $z \in tB_Z - C$ *if and only if* $\delta_{Z\setminus(-C)}(z) \geq -t$.

Proof

(a) Assume that $z \in tB_Z + C$. If $z \in C$, then $\delta_{Z\setminus C}(z) \geq 0 \geq -t$. If $z \notin C$, then $\delta_{Z\setminus C}(z) = -\delta_C(z) \geq -t$. Conversely, if $\delta_{Z\setminus C}(z) \geq -t$, then $\delta_C(z) = -\delta_{Z\setminus C}(z) \leq t$. Since Z is reflexive and C is closed and convex, $tB_Z + C$ is closed and so $z \in tB_Z + C$.
(b) We argue similarly as for assertion (a), but with $-C$ instead of C.

□

Now consider our two vector primal and dual strong quasiequilibrium problems, but with the following THIRD KIND OF APPROXIMATE SOLUTIONS

$$t\text{-Sol(VsQEP)} := \{\bar{x} \in A \mid f(\bar{x}, y) \in tB_Z - C \text{ for all } y \in \mathcal{D}(\bar{x})\},$$

$$t\text{-Sol(DVsQEP)} := \{\bar{y} \in B \mid f(x, \bar{y}) \in tB_Z + C \text{ for all } x \in \mathcal{D}^{-1}(\bar{y})\}.$$

Following this definition, Proposition 9.18 deduces the next coincidences of the approximate solutions of our vector problems and those of their scalarized problems. These scalarized problems meet the following new definition. The so-called SCALAR REVERSE QUASIEQUILIBRIUM PROBLEMS are

$$(\text{QEP}_Y) \qquad \text{find } \bar{y} \in B \text{ such that } \Phi(x, \bar{y}) \leq 0 \text{ for all } x \in \mathcal{D}^{-1}(\bar{y}),$$

$$(\text{DQEP}_Y) \qquad \text{find } \bar{x} \in A \text{ such that } \Phi(\bar{x}, y) \geq 0 \text{ for all } y \in \mathcal{D}(\bar{x}).$$

(Observe that the word "reverse" here signifies that we address infimum wrt y and supremum wrt x, in a reverse way to our conventional problems.) The sets of approximate solutions naturally are

$$t\text{-Sol}(\text{QEP}_Y) := \{\bar{y} \in B \mid \Phi(x, \bar{y}) \leq t \text{ for all } x \in \mathscr{D}^{-1}(\bar{y})\},$$

$$t\text{-Sol}(\text{DQEP}_Y) := \{\bar{x} \in A \mid \Phi(\bar{x}, y) \geq -t \text{ for all } y \in \mathscr{D}(\bar{x})\}.$$

Proposition 9.19 (Coincidence between the approximate solutions of the vector strong problems and those of their scalarized problems) *Assume that Z is reflexive, C is closed and convex, and $t \geq 0$. Then,*

(a) $t\text{-Sol(VsQEP)} = \{\bar{x} \in A \mid \delta_{Z\setminus(-C)}(f(\bar{x}, y)) \geq -t \ \forall y \in \mathscr{D}(\bar{x})\}$;

(b) $t\text{-Sol(DVsQEP)} = \{\bar{y} \in B \mid \delta_{Z\setminus C}(f(x, \bar{y})) \geq -t \ \forall x \in \mathscr{D}^{-1}(\bar{y})\}$.

Proof

(a) $\bar{x} \in t$-Sol(VsQEP) means that $f(\bar{x}, y) \in tB_Z - C$ for all $y \in \mathscr{D}(\bar{x})\}$, which is equivalent to $\delta_{Z\setminus(-C)}(f(\bar{x}, y)) \geq -t$ for such y by Proposition 9.18(b).
(b) The proof of this assertion is similar to that of assertion (a), but applying assertion (a) of the preceding proposition (instead of (b)).

□

Remark 9.20 (Comparison between the results using the Tammer and Hiriart-Urruty scalarizations of vector quasiequilibrium problems)

(a) It is clear from the above investigation that when scalarizing by the Hiriart-Urruty scalarization function, C is only assumed to be a closed and convex set, but we need additionally that Z is reflexive. Moreover, scalarizing results are obtained for the other kind of approximate solutions and only for the two vector strong problems. Furthermore, the two scalarizing functions are different (to each other), and then we cannot get duality results. The corresponding facts are clearly different when applying the Tammer scalarization.
(b) Proposition 9.19 gives a scalarizing result that is different from the ones obtained by the Tammer scalarization method. Namely, the scalarized problem of (VsQEP) is a scalar dual reverse quasiequilibrium problem and that of (DVsQEP) is a scalar dual quasiequilibrium problem (not reverse).
(c) Because of the assumption that C is closed and convex, the two vector weak problems cannot be involved in Proposition 9.19.

Next, we consider convergence of approximate vector strong quasiequilibria. Let our strong problems have approximating problems defined by A^k, B^k, $\mathscr{D}^k$ and f^k as usual.

Proposition 9.21 (Convergence of approximate vector primal strong quasiequilibria in terms of the Hiriart-Urruty's scalarization) *Assume that Z is reflexive, C is closed and convex, $t, t^k \geq 0$, and $t^k \to t$.*

(a) *If* f^k $-\delta_{Z\setminus(-C)}$*-i-e/h- or* $-\delta_{Z\setminus(-C)}$*-i-mis-lop converges to* f*, then* $A \cap \mathrm{Ls}_k(t^k$-Sol(VsQEP^k)) $\subset t$-Sol(VsQEP). *If* f^k $-\delta_{Z\setminus(-C)}$*-e/h- or* $-\delta_{Z\setminus(-C)}$*-mis-lop converges to* f*, then* $\mathrm{Ls}_k(t^k$-Sol(VsQEP^k)) $\subset t$-Sol(VsQEP).

(b) *If* $\mathrm{gph}\mathscr{D} \sqsubset \mathrm{Li}_k \mathrm{gph}\mathscr{D}^k$ *and the types of varriational convergence assumed in* (a) *are replaced by the* $-\delta_{Z\setminus(-C)}$*-epi- or* $-\delta_{Z\setminus(-C)}$*-i-epi-convergence, resp., then the conclusion in* (a) *is still valid.*

(c) *If* f^k $-\delta_{Z\setminus(-C)}$*-w-e/h-converges or* $-\delta_{Z\setminus(-C)}$*-w-mis-lop converges to* f *and* $x^k \in t^k$-Sol(VsQEP^k) *tends to* $\bar{x}$*, then* $\bar{x} \in t$-Sol(VsQEP).

The proof is analogous to that of Proposition 9.2 with the replacement of $-r_{-C}^{-e}$ by $-\delta_{Z\setminus(-C)}$.

Example 9.22 (Convergence of approximate vector primal strong quasiequilibria in terms of Hiriart-Urruty's scalarization) Let us consider convergence of approximate strong equilibria in terms of the Hiriart-Urruty scalarization. Assume that $X = Y = \mathbb{R}$, $Z = \mathbb{R}^2$, $A = (0, e]$ (note: in this section e is the Euler number, not a vector in Z as in the preceding part of the chapter), $B = (0, 1]$, $C = \mathbb{R}^2_+$, and $f(x, y) = (\ln(y/x), 0)$ for $(x, y) \in A \times B$. It is not hard to directly compute

$$\delta_{Z\setminus(-C)} f(x, y) = \begin{cases} 0 & \text{if } x \geq y, \\ -\ln(y/x) & \text{if } x \leq y. \end{cases}$$

Assume that we have approximating strong equilibrium problems with $A^k = A$, $B^k = B$, and $f^k(x, y) = (\ln(y/x), t^k)$ for $t^k \geq 0$. Then, similar to $\delta_{Z\setminus(-C)} f(x, y)$, we have

$$\delta_{Z\setminus(-C)} f^k(x, y) = \begin{cases} t^k & \text{if } x \geq y, \\ -\sqrt{(\ln(y/x))^2 + (t^k)^2} & \text{if } x \leq y. \end{cases}$$

We easily see that when $t^k \to 0^+$, $-\delta_{Z\setminus(-C)} f^k$ e/h-, mis-lop, and epi- converges to $-\delta_{Z\setminus(-C)} f$, and also $gph\mathscr{D} \sqsubset \mathrm{Li}_k \mathrm{gph}\mathscr{D}^k$. Therefore, all the assumptions imposed in Proposition 9.21 are satisfied. Direct calculations yield that t^k-Sol(VsEP^k) $= [1, e]$ Painlevé-Kuratowski converges to Sol(VsEP) $= [1, e]$.

Proposition 9.23 (Convergence of approximate vector dual strong quasiequilibria in terms of Hiriart-Urruty's scalarization) *Assume that* Z *is reflexive,* C *is closed and convex,* $t, t^k \geq 0$*, and* $t^k \to t$*. Then,*

(a) *if* f^k $\delta_{Z\setminus C}$*-i-e/h- or* $\delta_{Z\setminus C}$*-i-mai-lop converges to* f*, then* $B \cap \mathrm{Ls}_k(t^k$-Sol(DVsQEP^k)) $\subset t$-Sol(DVsQEP). *If* f^k $\delta_{Z\setminus C}$*-e/h- or* $\delta_{Z\setminus C}$*-mai-lop converges to* f*, then* $\mathrm{Ls}_k(t^k$-Sol(DVsQEP^k)) $\subset t$-Sol(DVsQEP)*;*

(b) *if* $\mathrm{gph}\mathscr{D} \sqsubset \mathrm{Li}_k \mathrm{gph}\mathscr{D}^k$ *and the* $\delta_{Z\setminus C}$*-hypo- or* $\delta_{Z\setminus C}$*-i-hypo-convergence of* f^k *replaces the variational convergence assumed in assertion* (a)*, then one also has the same conclusion as in that assertion.*

(c) *If* f^k $\delta_{Z\setminus C}$*-w-e/h-converges or* $\delta_{Z\setminus C}$*-w-mai-lop converges to* f *and* $y^k \in t^k$-Sol(DVsQEP^k) *tends to* $\bar{y}$*, then* $\bar{y} \in t$-Sol(DVsQEP).

The proof is similar to that of Proposition 9.4 with r_C^e replaced by $\delta_{Z\setminus C}$.

Example 9.24 (Convergence of approximate vector dual strong quasiequilibria) We consider approximate dual strong equilibria in terms of the Hiriart-Urruty scalarization. Assume that $X = Y = \mathbb{R}$, $Z = \mathbb{R}^2$, $A = [1, e]$, $B = (0, e]$, $C = \mathbb{R}^2_+$, and $f(x, y) = (-\ln(y/x), 0)$ for $(x, y) \in A \times B$. It is not hard to directly compute

$$\delta_{Z\setminus C} f(x, y) = \begin{cases} 0 & \text{if } x \geq y, \\ -\ln(y/x) & \text{if } x \leq y. \end{cases}$$

Assume that we have approximating dual strong equilibrium problems with $A^k = A$, $B^k = B$, and $f^k(x, y) = (-\ln(y/x), -t^k)$ for $t^k \geq 0$. Then, similar to $\delta_{Z\setminus C} f(x, y)$, we have

$$\delta_{Z\setminus C} f^k(x, y) = \begin{cases} -t^k & \text{if } x \geq y, \\ -\sqrt{(\ln(y/x))^2 + (t^k)^2} & \text{if } x \leq y. \end{cases}$$

We easily see that when $t^k \to 0^+$, $\delta_{Z\setminus C} f^k$ e/h-, mai-lop, and hypo- converges to $\delta_{Z\setminus C} f$, and also gph$\mathscr{D} \sqsubset$ Li$_k$gph$\mathscr{D}^k$. Hence, the assumptions imposed in Proposition 9.23 are fulfilled. Direct computations give that t^k-Sol(DVsEP^k) $=$ $(0, 1]$ Painlevé-Kuratowski converges to Sol(DVsEP) $= (0, 1]$.

9.4 Approximations of Vector Nash Quasigames

In this section, we study primal-dual pairs of both vector weak and strong Nash quasigames (vector generalized noncooperative games), which are already defined in Sect. 9.1. Besides suitable applications of the results in Sects. 9.2 and 9.3, many involved arguments based directly on the structure of the games need to participate in the consideration.

First, we discuss relationships between formulations.

Proposition 9.25 (Relationship between some formulations)

(a) $\bar{x} \in$ Ne(VwQ$\mathscr{G}$) *if and only if* $c_i(\bar{x}_i, \bar{x}_{-i}) - c_i(x_i, \bar{x}_{-i}) \notin -intC$ *for all* $i \in I$ *and* $x \in \mathscr{D}(\bar{x})$.

(b) Sol(DVwQEP$_N$) $\subset$ Ne(VwQ$\mathscr{G}$).

Proof

(a) Assume that $\bar{x} \in \text{Ne}(\text{VwQ}\mathscr{G})$, i.e., for all $x \in \mathscr{D}(\bar{x})$,

$$\sum_{i \in I} [c_i(\bar{x}_i, \bar{x}_{-i}) - c(x_i, \bar{x}_{-i})] \notin -\text{int}C. \tag{9.14}$$

For any fixed $k \in I$ and $x_k \in A_k$, substituting $x = (x_k, \bar{x}_{-k})$ into (9.14) gives

$$c_k(\bar{x}_k, \bar{x}_{-k}) - c_k(x_k, \bar{x}_{-k}) + \sum_{j \in I \setminus \{k\}} [c_j(\bar{x}_j, (\bar{x}_{j,k}, \bar{x}_k)) - c_j(\bar{x}_j, (\bar{x}_{j,k}, \bar{x}_k))] \notin -\text{int}C$$

(where $\bar{x}_{j,k}$ denotes the $m-2$ components of $x = (\bar{x}_i, \bar{x}_{-i})$ except $\bar{x}_j$ and $\bar{x}_k$), i.e., $c_k(\bar{x}_k, \bar{x}_{-k}) - c_k(x_k, \bar{x}_{-k}) \notin -\text{int}C$ for all k.

Conversely, let $c_i(\bar{x}_i, \bar{x}_{-i}) - c_i(x_i, \bar{x}_{-i}) \notin -\text{int}C$ for all i and all $x \in \mathscr{D}(\bar{x})$. Suppose that $\bar{x} \notin \text{Ne}(\text{VwQ}\mathscr{G})$, i.e., there exists $x \in \mathscr{D}(\bar{x}) = \{x \in X \mid x_i \in \mathscr{D}_i(\bar{x}_{-i}) \ \forall i \in I\}$ with $N(\bar{x}, x) \in -\text{int}C$. Assume that $x = (x_k, \bar{x}_{-k})$ (i.e., its k-component is out of the quasiequilibrium and the others are not). Substituting this x into $N(\bar{x}, x) \in -\text{int}C$, we get

$$c_k(\bar{x}_k, \bar{x}_{-k}) - c_k(x_k, \bar{x}_{-k}) + \sum_{j \in I \setminus \{k\}} [c_j(\bar{x}_j, (\bar{x}_{j,k}, \bar{x}_k)) - c_j(\bar{x}_j, (\bar{x}_{j,k}, \bar{x}_k))] \in -\text{int}C$$

(where $\bar{x}_{j,k}$ is the notation used above), i.e., $c_k(\bar{x}_k, \bar{x}_{-k}) - c_k(x_k, \bar{x}_{-k}) \in -\text{int}C$. This is a contradiction.

(b) Let $\bar{x} \in \text{Ne}(\text{DVwQ}\mathscr{G})$, that is for all $x \in \mathscr{D}^{-1}(\bar{x})$,

$$N(x, \bar{x}) = \sum_{i \in I} [c_i(x_i, x_{-i}) - c(\bar{x}_i, x_{-i})] \notin \text{int}C.$$

For each fixed $k \in I$, taking any $x_k \in A_k$ and substituting $x = (x_k, \bar{x}_{-k})$ into this relation, one has

$$c_k(x_k, \bar{x}_{-k}) - c_k(\bar{x}_k, \bar{x}_{-k}) + \sum_{j \in I \setminus \{k\}} [c_j(\bar{x}_j, (\bar{x}_{j,k}, x_k)) - c_j(\bar{x}_j, (\bar{x}_{j,k}, x_k))] \notin \text{int}C,$$

where $\bar{x}_{j,k}$ is the vector of $m-2$ components (after excluding $\bar{x}_j$ and x_k) consisting of $m-2$ enters $\bar{x}_i$ for all i. Therefore, $c_k(x_k, \bar{x}_{-k}) - c_k(\bar{x}_k, \bar{x}_{-k}) \notin \text{int}C$ which is $c_k(\bar{x}_k, \bar{x}_{-k}) - c_k(x_k, \bar{x}_{-k}) \notin -\text{int}C$ for all k or (the same) $\bar{x} \in \text{Ne}(\text{VwQ}\mathscr{G})$.

□

9.4.1 Approximations of Vector Nash Quasigames via the Tammer Scalarization

Similar to the vector quasiequilibrium models in the previous Sects. 9.1 and 9.2, for vector weak Nash quasigames, assume that we have approximate problems defined by A^k, $\mathscr{D}^k$, c^k, and N^k, which are as usual denoted by (VwQ$\mathscr{G}^k$), (DVwQ$\mathscr{G}^k$), etc. We aim at convergence of their approximate solutions to those of the original problems in terms of variational convergence. Apply the Tammer scalarization function $r_C^e : Z \to \mathbb{R}$ defined (in Sect. 9.1) by $r_C^e(z) := \inf\{t \in \mathbb{R} \mid z \in te - C\}$. Due to its property (a$_e$), the approximate solutions of (VwQ$\mathscr{G}$) are expressed as those of its scalarized problem as follows

$$\varepsilon\text{-Ne(VwQ}\mathscr{G}) = \{\bar{x}_\varepsilon \in A \mid r_C^e(N(\bar{x}, x)) \geq -\varepsilon \text{ for all } x \in \mathscr{D}(\bar{x}_\varepsilon)\}.$$

Our first convergence result for approximate solutions of (VwQ$\mathscr{G}^k$) is

Proposition 9.26 (Convergence of approximate solutions of vector Nash quasiequilibria) *For problems* (VwQ$\mathscr{G}^k$) *and* (VwQ$\mathscr{G}$), *let* $\varepsilon, \varepsilon^k \geq 0$ *and* $\varepsilon^k \to \varepsilon$.

(i) *If* N^k r_C^e*-i-e/h- or* r_C^e*-i-mis-lop converges to* N, *then* $A \cap \text{Ls}_k(\varepsilon^k\text{-Ne(VwQ}\mathscr{G}^k) \subset \varepsilon\text{-Ne(VwQ}\mathscr{G})$. *If* N^k r_C^e*-e/h- or* r_C^e*-mis-lop converge to* N, *then* $\text{Ls}_k(\varepsilon^k\text{-Ne(VwQ}\mathscr{G}^k) \subset \varepsilon\text{-Ne(VwQ}\mathscr{G})$.

(ii) *If* gph$\mathscr{D} \sqsubset \text{Li}_k\text{gph}\mathscr{D}^k$ *and the* r_C^e*-epi- or* r_C^e*-i-epi-convergence replaces the e/h- and mis-lop convergence in assertion* (i), *then the conclusion in that assertion also holds.*

(iii) *If* N^k r_C^e*-w-e/h-converges or* r_C^e*-w-mis-lop converges to* N *and* $x^k \in \varepsilon^k$-Ne(VwQ$\mathscr{G}^k$) *tends to* $\bar{x}_\varepsilon$, *then* $\bar{x}_\varepsilon \in \varepsilon$-Ne(VwQ$\mathscr{G}$).

Proof

(i) Address first the case of r_C^e-mis-lop convergence. Let $\bar{x}_\varepsilon \in \text{Ls}_k(\varepsilon^k$-Ne(VwQ$\mathscr{G}^k$)); i.e., there exists $x^{k_j} \in \varepsilon^{k_j}$-Ne(VwQ$\mathscr{G}^{k_j}$) tending to $\bar{x}_\varepsilon$. If $\bar{x}_\varepsilon \in A$, then by condition (b) of r_C^e-mis-lop convergence, for every $y \in \mathscr{D}(\bar{x}_\varepsilon)$, there exists $\mathscr{D}^{k_j}(x^{k_j}) \ni y^{k_j} \to y$ such that $\text{ls}_j[r_C^e(N^{k_j}(x^{k_j}, y^{k_j}))] \leq r_C^e(N(\bar{x}, y))$. As $r_C^e(N^{k_j}(x^{k_j}, y^{k_j})) \geq -\varepsilon^{k_j}$, $r_C^e(N(\bar{x}, y)) \geq -\varepsilon$, i.e., $\bar{x}_\varepsilon \in \varepsilon$-Ne(VwQ$\mathscr{G}$). Suppose that $\bar{x}_\varepsilon \notin A$. The above condition (b) gives $y^{k_j} \in \mathscr{D}^{k_j}(x^{k_j})$ such that $r_C^e(N^{k_j}(x^{k_j}, y^{k_j})) \to -\infty$, that contradicts the inequality $r_C^e(N^{k_j}(x^{k_j}, y^{k_j})) \geq -\varepsilon^{k_j}$ for all $y^{k_j} \in \mathscr{D}^{k_j}(x^{k_j})$.
For the case of r_C^e-e/h-convergence, we argue similarly. The proof of the case of inside convergence becomes now evident.

(ii) For this assertion, the reasoning is similar to that for assertion (i) with condition (a) of r_C^e-epi-convergence replacing condition (b) used in assertion (i).

(iii) Under the imposed hypotheses, by condition (b) of the w-e/h-convergence, for every $y \in \mathscr{D}(\bar{x}_\varepsilon)$, there exists $\mathscr{D}^k(x^k) \ni y^k \to y$ such that $\text{ls}_k(r_C^e(N^k(x^k, y^k)) \leq r_C^e(N(\bar{x}_\varepsilon, y))$. If $\bar{x}_\varepsilon \in A$, then $r_C^e(N(\bar{x}_\varepsilon, y)) \geq -\varepsilon$. If

$\bar{x}_\varepsilon \notin A$, the above (b) yields $\mathscr{D}^k(x^k) \ni y^k \to y$ with $r^e_C(N^k(x^k, y^k)) \to -\infty$, which is a contradiction.

□

Move on now to the dual vector weak Nash quasigame. Property (b_{-e}) of r^{-e}_{-C} ensures the following equivalence of approximate solution sets

$$\varepsilon\text{-Ne(DVwQ}\mathscr{G}) = \{\bar{y}_\varepsilon \in \mathscr{D}(A) \mid r^{-e}_{-C}(N(x, \bar{y}_\varepsilon)) \geq -\varepsilon \text{ for all } x \in \mathscr{D}^{-1}(\bar{y}_\varepsilon)\}.$$

By a technique analogous to the one for Ne(VwQ$\mathscr{G}$), we arrive at the corresponding assertion for Ne(DVwQ$\mathscr{G}$):

Proposition 9.27 (Convergence of approximate dual vector Nash quasiequilibria) *For problems* (DVwQ$\mathscr{G}^k$) *and* (DVwQ$\mathscr{G}$), *let* $\varepsilon, \varepsilon^k \geq 0$ *and* $\varepsilon^k \to \varepsilon$.

(a) *If* N^k $-r^{-e}_{-C}$*-i-e/h- or* $-r^{-e}_{-C}$*-i-mai-lop converges to* N, *then* $\mathscr{D}(A) \cap \text{Ls}_k(\varepsilon^k$-Ne(DVwQ$\mathscr{G}^k$)) $\subset$ ε-Ne(DVwQ$\mathscr{G}$). *If* N^k $-r^{-e}_{-C}$*-e/h- or* $-r^{-e}_{-C}$*-mai-lop converges to* N, *then* $\text{Ls}_k(\varepsilon^k$-Ne(DVwQ$\mathscr{G}^k$)) $\subset$ ε-Ne(DVwQ$\mathscr{G}$).
(b) *If* gph$\mathscr{D} \sqsubset \text{Li}_kgph\mathscr{D}^k$ *and the* $-r^{-e}_{-C}$*-hypo- or* $-r^{-e}_{-C}$*-i-hypo-convergence of* N^k *replaces the types of variational convergence assumed in assertion* (a), *then one has the same assertion as in* (a).
(c) *If* N^k $-r^{-e}_{-C}$*-w-e/h-converges or* $-r^{-e}_{-C}$*-w-mai-lop converges to* N *and* $y^k \in \varepsilon^k$-Ne(DVwQ$\mathscr{G}^k$) *tends to* $\bar{y}_\varepsilon$, *then* $\bar{y}_\varepsilon \in \varepsilon$-Ne(DVwQ$\mathscr{G}$).

Since we investigate vector weak Nash quasigames as special cases of vector weak quasiequilibrium problems and also via the Nikaido-Isoda bimap and the Tammer scalarization function, with reasoning similar to that in Sect. 9.1, we have the following counterparts for strong problems.

The sets of approximate vector strong Nash quasiequilibria are expressed in terms of their scalarized quasiequilibria as follows:

$$\varepsilon\text{-Ne(VsQ}\mathscr{G}) = \{\bar{x}_\varepsilon \in A \mid r^{-e}_{-C}(N(\bar{x}_\varepsilon, y)) \leq \varepsilon \text{ for all } y \in \mathscr{D}(\bar{x}_\varepsilon)\},$$

$$\varepsilon\text{-Ne(DVsQ}\mathscr{G}) = \{\bar{y}_\varepsilon \in \mathscr{D}(A) \mid r^{e}_{C}(N(x, \bar{y}_\varepsilon)) \leq \varepsilon \text{ for all } x \in \mathscr{D}^{-1}(\bar{y}_\varepsilon)\}.$$

The convergence results we have are the following two statements.

Proposition 9.28 (Convergence of approximate vector strong Nash quasiequilibria) *For* (VsQ$\mathscr{G}^k$) *and* (VsQ$\mathscr{G}$), *let* $\varepsilon, \varepsilon^k \geq 0$ *and* $\varepsilon^k \to \varepsilon$.

(a) *If* N^k $-r^{-e}_{-C}$*-i-e/h- or* $-r^{-e}_{-C}$*-i-mis-lop converges to* N, *then* $A \cap \text{Ls}_k(\varepsilon^k$-Ne(VsQ$\mathscr{G}^k$)) $\subset$ ε-Nel(VsQ$\mathscr{G}$). *If* N^k $-r^{-e}_{-C}$*-e/h- or* $-r^{-e}_{-C}$*-mis-lop converges to* N, *then* $\text{Ls}_k(\varepsilon^k$-Ne(VsQ$\mathscr{G}^k$)) $\subset$ ε-Nel(VsQ$\mathscr{G}$).
(b) *If* gph$\mathscr{D} \sqsubset \text{Li}_kgph\mathscr{D}^k$ *and the* $-r^{-e}_{-C}$*-epi- or* $-r^{-e}_{-C}$*-i-epi-convergence, resp., replaces the types of variational convergence in assertion* (a), *then the same conclusion as in* (a) *holds.*
(c) *If* N^k $-r^{-e}_{-C}$*-w-e/h- or* $-r^{-e}_{-C}$*-w-mis-lop converges to* N *and* $x^k \in \varepsilon^k$-Ne(VsQ$\mathscr{G}^k$) *tends to* $\bar{x}_\varepsilon$, *then* $\bar{x}_\varepsilon \in \varepsilon$-Nel(VsQ$\mathscr{G}$).

Proposition 9.29 (Convergence of approximate dual vector strong Nash quasiequilibria) *Let* $\varepsilon, \varepsilon^k \geq 0$ *and* $\varepsilon^k \to \varepsilon$.

(a) *If* N^k r^e_C*-i-e/h- or* r^e_C*-i-mai-lop converges to* N, *then*

$$\mathscr{D}(A) \cap \mathrm{Ls}_k(\varepsilon^k\text{-Ne(DVsQ}\mathscr{G}^k)) \subset \varepsilon\text{-Ne(DVsQ}\mathscr{G}).$$

If the above restriction "inside" (i.e., the abbreviation "-i-") is removed, then $\mathrm{Ls}_k(\varepsilon^k\text{-Ne(DVsQ}\mathscr{G}^k)) \subset \varepsilon\text{-Ne(DVsQ}\mathscr{G})$.

(b) *Assume that* $\mathrm{gph}\mathscr{D} \sqsubset \mathrm{Li}_k \mathrm{gph}\mathscr{D}^k$. *If the* r^e_C*-hypo- or* r^e_C*-i-hypo-convergence replaces the types of variational convergence in assertion* (a), *then the conclusion in this assertion still holds.*

(c) *If* N^k r^e_C*-w-e/h- or* r^e_C*-w-mai-lop converges to* N *and* $y^k \in \varepsilon^k\text{-Ne(DVsQ}\mathscr{G}^k)$ *tends to* $\bar{y}_\varepsilon$, *then* $\bar{y}_\varepsilon \in \varepsilon\text{-Nel(DVsQ}\mathscr{G})$.

Duality between Vector Weak Nash Quasiequilibria and their Strong Counterparts via the Tammer Scalarization

Observe first that we clearly have a specialization for vector weak Nash quasiequilibria of Proposition 9.13 (for general vector weak quasiequilibria). Corresponding to $\mathscr{C}_1 f$ and $\mathscr{C}_2 f$ (defined prior to Theorem 9.14), we have the following for vector weak Nash quasiequilibrium problems:

$$\mathscr{C}_1 N^k = \{(x, y) \in \mathrm{gph}\mathscr{D}^k \mid N^k(x, y) \in \mathrm{bd}(-C)\},$$
$$\mathscr{C}_1 N = \{(x, y) \in \mathrm{gph}\mathscr{D} \mid N(x, y) \in \mathrm{bd}(-C)\},$$

$$\mathscr{C}_2 N^k = \{(x, y) \in \mathrm{gph}\mathscr{D}^k \mid N^k(x, y) \in \mathrm{bd}(C)\},$$
$$\mathscr{C}_2 N = \{(x, y) \in \mathrm{gph}\mathscr{D} \mid N(x, y) \in \mathrm{bd}(C)\}.$$

With a proof that modifies the one of Theorem 9.14, we obtain

Proposition 9.30 (Duality in convergence of approximate vector weak Nash quasiequilibria and their dual strong counterparts) *Assume that* $\varepsilon, \varepsilon^k \geq 0$ *and* $\varepsilon^k \to \varepsilon$.

(a) *Assume that* N^k r^e_C*-i-e/h- or both* r^e_C*-i-mis-lop and* r^e_C*-i-mai-lop converges to* N *and* $\mathrm{Ls}_k \mathscr{C}_1 N^k \subset \mathrm{cl}\mathscr{C}_1 N$. *Then,*

$$\begin{aligned}
&\mathrm{Ls}_k[\mathscr{C}_1 N^k \cap (\varepsilon^k - \mathrm{wsdl}(r^e_C(N^k)))] \cap \mathrm{gph}\mathscr{D} \\
&= \mathrm{Ls}_k[\mathscr{C}_1 N^k \cap ((\varepsilon^k - \mathrm{Ne}(VwQ\mathscr{G}^k)) \times (\varepsilon^k - \mathrm{Ne}(DVsQ\mathscr{G}^k)))] \\
&\cap \mathrm{gph}\mathscr{D} \subset \mathrm{cl}\mathscr{C}_1 N \cap ((\varepsilon - \mathrm{Ne}(VwQ\mathscr{G}) \times (\varepsilon - \mathrm{Ne}(DVsQ\mathscr{G}))) \cap \mathrm{gph}\mathscr{D} \\
&= \mathrm{cl}\mathscr{C}_1 N \cap (\varepsilon - \mathrm{wsdl}(r^e_C(N))).
\end{aligned}$$

If the above restriction "inside" (denoted by "-i-") is removed, then the conclusion holds without the two involved " $\cap\, \mathrm{gph}\mathscr{D}$*."*

(b) *If* f^k r^e_C*-i-e/h-converges tightly or* r^e_C*-i-mis-lop both* x*- and* y*-ancillary tightly converges or* r^e_C*-i-mai-lop both* x*- and* y*-ancillary tightly converges to* N *and* $\text{cl}\mathscr{C}_1 N \subset \text{Li}_k \mathscr{C}_1 N^k$*, then*

$$\varepsilon\text{-wsdl}(r^e_C(N)) \cap \text{cl}\mathscr{C}_1 N = ((\varepsilon\text{-Ne}(\text{VwQ}\mathscr{G})) \times (\varepsilon\text{-Ne}(\text{DVsQ}\mathscr{G}))) \cap \text{cl}\mathscr{C}_1 N$$

$$\subset \cap_{\delta>\varepsilon}\text{Li}_k(\delta\text{-Ne}(\text{VwQ}\mathscr{G}^k) \times \delta\text{-Ne}(\text{DVsQ}\mathscr{G}^k)) \cap \text{Li}_k \mathscr{C}_1 N^k$$
$$= \cap_{\delta>\varepsilon}\text{Li}_k(\delta\text{-wsdl}r^e_C(N^k)) \cap \text{Li}_k \mathscr{C}_1 N^k .$$

Hence,

$$\varepsilon\text{-wsdl}(r^e_C(N)) \cap \text{cl}\mathscr{C}_1 N = (\varepsilon\text{-Ne}(\text{VwQ}\mathscr{G}) \times \varepsilon\text{-Ne}(\text{DVsQ}\mathscr{G})) \cap \text{cl}\mathscr{C}_1 N$$

$$\subset \text{Li}_k((\varepsilon^k\text{-Ne}(\text{VwQ}\mathscr{G}^k)) \times (\varepsilon^k\text{-Ne}(\text{DVsQ}\mathscr{G}^k))) \cap \text{Li}_k \mathscr{C}_1 N^k$$
$$= \text{Li}_k(\varepsilon\text{-wsdl}(r^e_C(N))) \cap \text{Li}_k \mathscr{C}_1 N^k.$$

(c) *Assume that in assertion* (b)*, the restriction "inside" in the variational convergence is removed, and the mentioned types of lop-convergence are imposed simultaneously. Assume further that* $\text{Lim}_k \mathscr{C}_1 N^k = \text{cl}\mathscr{C}_1 N$*. Then,*

$$\varepsilon\text{-wsdl}(r^e_C(N)) \cap \text{cl}\mathscr{C}_1 N = (\varepsilon\text{-Ne}(\text{VwQ}\mathscr{G}) \times \varepsilon\text{-Ne}(\text{Sol}(\text{DVsQ}\mathscr{G})) \cap \text{cl}\mathscr{C}_1 N$$

$$= \text{Lim}_k(\varepsilon^k\text{-Ne}(\text{VwQ}\mathscr{G}^k) \times \varepsilon^k\text{-Ne}(\text{DVsQ}\mathscr{G}^k)) \cap \text{Lim}_k \mathscr{C}_1 N^k$$

$$= \text{Lim}_k(\varepsilon^k\text{-wsdl}(r^e_C(N^k))) \cap \text{Lim}_k \mathscr{C}_1 N^k.$$

Naturally, with reasoning similar to that for Theorem 9.16, we also have the following statement which is somewhat symmetric to that theorem.

Proposition 9.31 (Duality in convergence of approximate vector strong Nash quasiequilibria and their dual weak counterparts) *Assume that* $\varepsilon, \varepsilon^k \geq 0$ *and* $\varepsilon^k \to \varepsilon$*.*

(a) *Assume that* N^k $-r^{-e}_{-C}$*-i-e/h- or both* $-r^{-e}_{-C}$*-i-mis-lop and* $-r^{-e}_{-C}$*-i-mai-lop converges to* N *and* $\text{Ls}_k \mathscr{C}_2 N^k \subset \text{cl}\mathscr{C}_2 N$*. Then,*

$$\text{Ls}_k[\mathscr{C}_2 N^k \cap (\varepsilon^k - \text{dwsdl}(r^{-e}_{-C}(N^k)))] \cap \text{gph}\mathscr{D}\}$$

$$= \text{Ls}_k[\mathscr{C}_2 N^k \cap ((\varepsilon^k - \text{Ne}(\text{VsQ}\mathscr{G}^k)) \times (\varepsilon^k - \text{Ne}(\text{DVwQ}\mathscr{G}^k)))] \cap \text{gph}\mathscr{D}$$

$$\subset \text{cl}\mathscr{C}_2 N \cap ((\varepsilon - \text{Ne}(\text{VsQ}\mathscr{G})) \times (\varepsilon - \text{Ne}(\text{DVwQ}\mathscr{G}))) \cap \text{gph}\mathscr{D}$$

$$= \text{cl}\mathscr{C}_2 N \cap (\varepsilon - \text{dwsdl}(r^{-e}_{-C}(N))).$$

If the above restriction "inside" is removed, then the conclusion holds without " $\cap\, \text{gph}\mathscr{D}$*."*

(b) *If* N^k $-r^{-e}_{-C}$*-i-e/h-converges tightly or* $-r^{-e}_{-C}$*-i-mis-lop both* x*- and* y*-ancillary tightly converges or* $-r^{-e}_{-C}$*-i-mai-lop both* x*- and* y*-ancillary tightly converges to* N *and* $\mathrm{cl}\mathscr{C}_2 N \subset \mathrm{Li}_k \mathscr{C}_2 N^k$*, then*

$$\varepsilon\text{-dwsdl}(r^{-e}_{-C}(N)) \cap \mathrm{cl}\mathscr{C}_2 N = ((\varepsilon\text{-Ne}(\mathrm{VsQ}\mathscr{G})) \times (\varepsilon\text{- Ne}(\mathrm{DVwQ}\mathscr{G}))) \cap \mathrm{cl}\mathscr{C}_2 N$$

$$\subset \cap_{\delta>\varepsilon} \mathrm{Li}_k (\delta\text{-Ne}(\mathrm{VsQ}\mathscr{G}^k) \times \delta\text{-Ne}(\mathrm{DVwQ}\mathscr{G}^k)) \cap \mathrm{Li}_k \mathscr{C}_2 N^k$$

$$= \cap_{\delta>\varepsilon} \mathrm{Li}_k (\delta\text{-dwsdl} r^{-e}_{-C}(N^k)) \cap \mathrm{Li}_k \mathscr{C}_2 N^k.$$

Hence,

$$\varepsilon\text{-dwsdl}(r^{-e}_{-C}(N)) \cap \mathrm{cl}\mathscr{C}_2 N = (\varepsilon\text{-Ne}(\mathrm{VsQ}\mathscr{G}) \times \varepsilon\text{-Ne}(\mathrm{DVwQ}\mathscr{G})) \cap \mathrm{cl}\mathscr{C}_2 N$$

$$\subset \mathrm{Li}_k ((\varepsilon^k\text{-Ne}(\mathrm{VsQ}\mathscr{G}^k)) \times (\varepsilon^k\text{-Ne}(\mathrm{DVwQ}\mathscr{G}^k))) \cap \mathrm{Li}_k \mathscr{C}_2 N^k$$
$$= \mathrm{Li}_k (\varepsilon\text{-dwsdl}(r^{-e}_{-C}(N))) \cap \mathrm{Li}_k \mathscr{C}_2 N^k.$$

(c) *Assume that in assertion* (b)*, the restriction "inside" in the variational convergence is removed, and the mentioned types of lop-convergence are imposed simultaneously. Assume further that* $\mathrm{Lim}_k \mathscr{C}_2 N^k = \mathrm{cl}\mathscr{C}_2 N$*. Then,*

$$\varepsilon\text{-dwsdl}(r^{-e}_{-C}(N)) \cap \mathrm{cl}\mathscr{C}_2 N = (\varepsilon\text{-Ne}(\mathrm{VsQ}\mathscr{G}) \times \varepsilon\text{- Ne}(\mathrm{DVwQ}\mathscr{G})) \cap \mathrm{cl}\mathscr{C}_2 N$$

$$= \mathrm{Lim}_k (\varepsilon^k\text{-Ne}(\mathrm{VsQ}\mathscr{G}^k) \times \varepsilon^k\text{-Ne}(\mathrm{DVwQ}\mathscr{G}^k)) \cap \mathrm{Lim}_k \mathscr{C}_2 N^k$$

$$= \mathrm{Lim}_k (\varepsilon^k\text{-dwsdl}(r^{-e}_{-C}(N^k))) \cap \mathrm{Lim}_k \mathscr{C}_2 N^k.$$

9.4.2 Approximations of Vector Nash Quasigames via the Hiriart-Urruty Scalarization

In this subsection, we employ the Hiriart-Urruty/signed distance function for scalarizing vector Nash quasigames. For a subset H in a normed space Z and a nonempty closed and convex subset $C \subset Z$, this function $\delta_H : Z \to \mathbb{R} \cup \{+\infty\}$ has already been defined at the beginning of Sect. 9.2. For $t \geq 0$, we consider the following types of approximate strong Nash quasiequilibria

$$t\text{-Ne}(\mathrm{VsQ}\mathscr{G}) := \{\bar{x} \in A \mid N(\bar{x}, x) \in tB_Z + C \text{ for all } x \in \mathscr{D}(\bar{x})\},$$

$$t\text{-Ne}(\mathrm{DVsQ}\mathscr{G}) := \{\bar{x} \in \mathscr{D}(A) \mid N(x, \bar{x}) \in tB_Z - C \text{ for all } x \in \mathscr{D}^{-1}(\bar{x})\}.$$

We always assume in this subsection that Z is reflexive. In virtue of Proposition 9.13, the following expressions of the above approximate strong Nash quasiequilibria in terms of the Hiriart-Urruty scalarization are easily verified

(a) $t\text{-Ne}(\mathrm{VsQ}\mathscr{G}) = \{\bar{x} \in A \mid \delta_{Z\setminus C}(N(\bar{x}, x)) \geq -t,\ \forall x \in \mathscr{D}(\bar{x})\}$;
(b) $t\text{-Ne}(\mathrm{DVsQ}\mathscr{G}) = \{\bar{x} \in \mathscr{D}(A) \mid \delta_{Z\setminus(-C)}(N(x, \bar{x})) \geq -t,\ \forall x \in \mathscr{D}^{-1}(\bar{x})\}$.

These scalarized expressions allow us to obtain the following convergence results for approximate vector strong Nash quasiequilibria.

Proposition 9.32 (Convergence of approximate vector strong Nash quasiequilibria in terms of the Hiriart-Urruty scalarization) *Assume that Z is reflexive, C is closed and convex, $t, t^k \geq 0$, and $t^k \to t$. Then,*

(a) *If N^k $\delta_{Z\setminus C}$-i-e/h- or $\delta_{Z\setminus C}$-i-mis-lop converges to N, then $A \cap \mathrm{Ls}_k(t^k$-$\mathrm{Ne}(\mathrm{VsQ}\mathscr{G}^k)) \subset t$-$\mathrm{Ne}(\mathrm{VsQ}\mathscr{G})$. If N^k $\delta_{Z\setminus C}$-e/h- or $\delta_{Z\setminus C}$-mis-lop converges to N, then $\mathrm{Ls}_k(t^k$-$\mathrm{Ne}(\mathrm{VsQ}\mathscr{G}^k)) \subset t$-$\mathrm{Ne}(\mathrm{VsQ}\mathscr{G})$.*

(b) *If $\mathrm{gph}\mathscr{D} \sqsubset \mathrm{Li}_k\mathrm{gph}\mathscr{D}^k$ and the $\delta_{Z\setminus C}$-epi- or $\delta_{Z\setminus C}$-i-epi-convergence replaces the variational convergence imposed in assertion* (a), *then the concluded inclusion in this assertion still holds.*

(c) *If N^k $\delta_{Z\setminus C}$-w-e/h-converges or $\delta_{Z\setminus C}$-w-mis-lop converges to N and $x^k \in t^k$-$\mathrm{Ne}(\mathrm{VsQ}\mathscr{G}^k)$ tends to $\bar{x}$, then $\bar{x} \in t$-$\mathrm{Ne}(\mathrm{VsQ}\mathscr{G})$.*

Proposition 9.33 (Convergence of approximate dual vector strong Nash quasiequilibria in terms of the Hiriart-Urruty scalarization) *Assume that Z is reflexive, C is closed and convex, $t, t^k \geq 0$, and $t^k \to t$. Then,*

(a) *If N^k $-\delta_{Z\setminus(-C)}$-i-e/h- or $-\delta_{Z\setminus(-C)}$-i-mai-lop converges to N, then $\mathscr{D}(A) \cap \mathrm{Ls}_k(t^k$-$\mathrm{Ne}(\mathrm{DVsQ}\mathscr{G}^k)) \subset t$-$\mathrm{Ne}(\mathrm{DVsQ}\mathscr{G})$. If N^k $-\delta_{Z\setminus(-C)}$-e/h- or $-\delta_{Z\setminus(-C)}$-mai-lop converges to N, then $\mathrm{Ls}_k(t^k$-$\mathrm{Ne}(\mathrm{DVsQ}\mathscr{G}^k)) \subset t$-$\mathrm{Ne}(\mathrm{DVsQ}\mathscr{G})$.*

(b) *If $\mathrm{gph}\mathscr{D} \sqsubset \mathrm{Li}_k\mathrm{gph}\mathscr{D}^k$ and the $-\delta_{Z\setminus(-C)}$-hypo- or $-\delta_{Z\setminus(-C)}$-i-hypo-convergence, resp., replaces the variational convergence in assertion* (a), *then in conclusion one has the same inclusion as in* (a).

(c) *If N^k $-\delta_{Z\setminus(-C)}$-w-e/h-converges or $-\delta_{Z\setminus(-C)}$-w-mai-lop converges to N and $x^k \in t^k$-$\mathrm{Ne}(\mathrm{DVsQ}\mathscr{G}^k)$ tends to $\bar{x}$, then $\bar{x} \in t$-$\mathrm{Ne}(\mathrm{DVsQ}\mathscr{G})$.*

The proofs of the preceding two propositions are easy combinations of those of Propositions 9.23 and 9.21 with the above facts about scalarizing strong Nash quasigames.

9.5 Exercises

Exercise 9.1 Consider problem (VwQEP) with the data $X = Y = \mathbb{R}$, $Z = \mathbb{R}^2$, $C = \mathbb{R}^2_+$, $A = [-1, 1]$, $B = [0, 1]$, and for $(x, y) \in A \times B$,

$$f_1(x, y) = \begin{cases} y^x & \text{if } y > 0, y \neq 1, \\ 1 & \text{if } y = 1, x \neq 0 \text{ or } y \in [0, 1],\ x = 0, \\ 0 & \text{if } y = 0, x \neq 0, \end{cases}$$

$f_2(x, y) = 1$ for all $(x, y) \in [-1, 1] \times [0, 1]$. Compute Sol(VwQEP) and ε-Sol(VwQEP) for any $\varepsilon > 0$.

Exercise 9.2 Consider problem (DVwQEP), (DVwQEP^k) (defined in Sect. 9.2), $\varepsilon, \varepsilon^k \geq 0$, and $\varepsilon^k \to \varepsilon$. Assume that f^k r^e_C-w-e/h-converges or r^e_C-w-mai-lop converges to f and $y^k \in \varepsilon^k$-Sol(DVwQEP^k) tends to $\bar{y}_\varepsilon$. Verify that $\bar{y}_\varepsilon \in \varepsilon$-Sol(DVwQEP).

Exercise 9.3 Let $X = Y = \mathbb{R}$, $Z = \mathbb{R}^2$, $C = \mathbb{R}^2_+$, $A^k = [-1, 1]$, $B^k = [0, 1]$, and for $(x, y) \in A^k \times B^k$,

$$f_1^k(x, y) = \begin{cases} -y^x & \text{if } y > 0, y \neq 1, \\ -1 & \text{if } y = 1, x \neq 0 \text{ or } y \in [0, 1], x = 0, \\ 0 & \text{if } y = 0, x \neq 0, \end{cases}$$

and $f_2^k(x, y) = -\varepsilon^k/\sqrt{2}$. Find the r_C^e-e/h-, r_C^e-mai-lop, and r_C^e-hypo-limits of f^k.

Exercise 9.4 Consider the dual vector strong quasiequilibrium problem (DVsQEP) with f given in Exercise 9.1 and (DVsQEP^k) with $A^k = [-1, 1]$, $B^k = [0, 1]$, $f_1^k = f_1$, and $f_2^k(x, y) = \varepsilon^k/\sqrt{2}$. Compute Sol(DVsQEP) and Sol(DVsQEP^k) and show that the assumptions and conclusions of Proposition 9.11 hold.

Exercise 9.5 Consult Proposition 9.7 to state, and prove the convergence of the second-kind approximate solutions of vector strong quasiequilibrium problems under suitable variational convergence of their scalarized problems.

Exercise 9.6 Study the vector strong primal and weak dual quasiequilibrium problems with the data given in Exercise 9.3 and $A = [-1, 1]$, $B = [0, 1]$, $f_1 = f_1^k$, and $f_2 = -1$. Verify that the assumptions and conclusions of Theorem 9.14 hold and compute directly the sides of the relations in assertion (c) of that theorem.

Exercise 9.7 Consider the vector primal strong quasiequilibrium problems (VsQEP) and (VsQEP^k) with the data: $X = Y = \mathbb{R}$, $Z = \mathbb{R}^2$, $A = A^k = (0, e]$ (where e is the Euler number), $B = B^k = (0, 1]$, $C = \mathbb{R}^2_+$, $f(x, y) = (\ln(y/x), 0)$, and $f^k(x, y) = (\ln(y/x), t^k)$ for $t^k \geq 0$ and $(x, y) \in A \times B$. It is not hard to directly compute $\delta_{Z\setminus(-C)} f(x, y)$ and $\delta_{Z\setminus(-C)} f^k(x, y)$. Verify that when $t^k \to 0^+$, it holds that $-\delta_{Z\setminus(-C)} f^k$ e/h-, mis-lop, and epi- converges to $-\delta_{Z\setminus(-C)} f$, and also $gph\mathscr{D} \sqsubset \mathrm{Li}_k \mathrm{gph}\mathscr{D}^k$. Compute directly the solution sets t^k-Sol(VsQEP^k) and Sol(VsQEP).

9.6 Commentary

Section 9.1

Introduction To make the topics of this chapter clear, we formulate all the considered objects: the vector primal weak and strong quasiequilibrium problems, their dual settings, the vector weak Nash quasigames, and the three kinds of approximate solutions of these models. We also state that convergence of these approximate solutions under suitable types of variational convergence of the scalarized considered problems by the Tammer and Hiriart-Urruty scalarization methods are examined in this chapter. The reason for approaching these vector problems via scalarization techniques is that there are too few results on variational convergence of vector functions; see [243, 347, 467], and none for vector bifunctions, and it even seems that the developments of variational convergence for vector bifunctions are not promising.

The content of this chapter is based on the very recent work [180].

Section 9.2

Approximations of Vector Quasiequilibria via the Tammer Scalarization Function Vector weak and strong quasiequilibrium problems and their dual models are scalarized by two appropriate variants of the Tammer scalarization functions. For these functions, see books [218, 280, 424], paper [227], and the references therein. (Note that there are also other scalarization methods; see, e.g., [282, 351], but, recently these two scalarization functions have become most popular.) Then, sufficient conditions for the convergence of approximate solutions of our vector problems are established by applying suitable types of variational convergence of the scalarized problems. Like in Chaps. 7 and 8, here we try to use various types and as weak as possible to relax the assumptions in these sufficient conditions.

Pairs of vector primal and dual quasiequilibrium problems are also considered. The convergence of approximate solution pairs of these problems together with types of approximate weak saddle points under suitable types of variational convergence of the scalarized problems are obtained. However, unlike the case of scalar problems studied in Chaps. 7 and 8, here a pair of problems does not contain the same weak problems or the same strong problems, but has a "deviation": "weak primal" goes with "strong dual" and "strong primal" with "weak dual." (In any case, the meanings of "strong" in the scalar and vector cases are quite different.)

The contents of this chapter are mainly taken from the very recent paper [180].

Section 9.3

Approximations of Vector Quasiequilibrium Problems via the Hiriart-Urruty Scalarization Unlike the preceding two sections with the use of the Tammer scalarization function for considering convergence of the first and second kind of approximate solutions to both vector weak and strong quasiequilibrium problems, in this section, employing the Hiriart-Urruty scalarization function (introduced in [249]; see also [457, 465]), we only establish convergence results for vector primal and dual strong quasiequilibrium problems. Moreover, the third-kind approximate solutions are investigated, and this scalarization function is not adequate for investigating the first and second ones. In the very recent work [462], the Hiriart-

Urruty scalarization is applied successfully in vector optimization with variable coradiant sets (these sets are generalizations of cones).

Section 9.4

Approximations of Vector Nash Quasigames This section is devoted to discussing the topics of the preceding two sections, but for vector Nash quasigames. Observe that for vector Nash games or quasigames, most of the contributions focus on existence studies; see, e.g., [5, 6, 148]. We have not been able to find results on approximations of these problems except for the very recent manuscript [180], on which this chapter is based. Namely, convergence of approximate primal and dual vector Nash quasiequilibria under types of variational convergence of the Tammer-scalarized problems is addressed first. Then, duality in convergence of primal-dual pairs of problems with a "deviation" similar to that in Sect. 9.2 is obtained. Finally, convergence of approximate vector strong Nash quasiequilibria in terms of the Hiriart-Urruty scalarization method is established.

Chapter 10
Variational Inclusion Problems and Variational Relation Problems

Abstract In this chapter, we delve into two generalized models of the equilibrium problem: the variational inclusion problem and variational relation problems. In fact, under appropriate conditions, these problems are equivalent. However, with applications in mind, we present results about properties of the solutions to these problems independently. Similar to previous chapters, we will explore the existence of solutions and the stability of these problems. First we prove several existence results for the variational relation problem. Then, applying these results, we obtain some existence conditions for the variational inclusion problems. We employ semicontinuity properties of set-valued maps to establish sufficient conditions for the (semi)continuity of solution maps and prove well-posedness conditions for the considered variational inclusion problem. As for the variational relation problem, we also provide stability results related to (semi)continuity, convergence, and Holder calmness of solution maps. The results presented in this chapter encompass all the findings for these models, and we acknowledge that there remain many topics worthy of further study.

10.1 Problems

Much effort has been made in the last several decades to propose general problem settings related to optimization, beginning with equilibrium problems. Recently, inclusion problems have been investigated as a generalization of equilibrium problems, in order to include a wide class of problems in diverse fields such as variational inequalities, vector optimization, game theory, fixed point and coincidence point problems, Nash equilibrium problems, complementarity problems, traffic equilibria, etc. It should be noted here that the term "variational inclusion" is understood in different ways in several recent works; for instance, it can be considered as multivalued variational inequalities or as problems of finding zeroes of maximal monotone mappings, which do not include some general equilibrium problems or are not convenient for studying existence of solutions (see (c) below). Therefore, in this section, we propose four variants of general inclusion problems to amend

L. Q. Anh et al., *Equilibrium Problems*, Forum for Interdisciplinary Mathematics, https://doi.org/10.1007/978-3-031-89003-1_10

existing problem settings and establish sufficient conditions for the existence of solutions.

In this chapter, we consider the following model.

(VRP) Find $\bar{a} \in A$ such that $\bar{a} \in \mathscr{D}_1(\bar{a})$ and $R(\bar{a}, b)$ holds for all $b \in \mathscr{D}_2(\bar{a})$,

where A, B are nonempty sets, $\mathscr{D}_1 : A \rightrightarrows A$, $\mathscr{D}_2 : A \rightrightarrows B$ and R is a relation linking $a \in A$ and $b \in B$ such that the images of $\mathscr{D}_1$ and $\mathscr{D}_2$ are nonempty.

The problem (VRP) is called a VARIATIONAL RELATION PROBLEM. In this problem, $\mathscr{D}_1$ and $\mathscr{D}_2$ are constraint mappings, and the relation R describes the relationship between the decision variable and the dependent variable. In the literature, the variational relation problem is also studied in implicit form as follows. Let Y be another nonempty set, $\mathscr{P} : A \times B \rightrightarrows Y$ and $\mathscr{R}(a, b, y)$ be a relation linking $a \in A$, $b \in B$ and $y \in Y$. Let the notation $\alpha_1(u, U)$ mean $\forall u \in U$, $\alpha_2(u, U)$ mean $\exists u \in U$, $\overline{\alpha_1} = \alpha_2$ and $\overline{\alpha_2} = \alpha_1$. For $\alpha \in \{\alpha_1, \alpha_2\}$, consider the problem

$$(\mathrm{VRP}_\alpha) \qquad \begin{cases} \text{find } \bar{a} \in A \text{ such that } \bar{a} \in \mathscr{D}_1(\bar{a}), \text{ and for all } y \in \mathscr{D}_2(\bar{a}), \\ \alpha(y, \mathscr{P}(\bar{a}, b)), \ \mathscr{R}(\bar{a}, b, y) \text{ holds.} \end{cases}$$

Then, (VRP_α) is equivalent to (VRP). To see this equivalence simply, define a relation R linking $a \in A$ and $b \in B$ as follows:

$$R(a, b) \text{ holds if and only if } \alpha(y, \mathscr{P}(a, b)), \mathscr{R}(a, b, y) \text{ holds.} \tag{10.1}$$

A variational relation problem can represent various situations, depending on the relationship between variables. All problems considered in previous chapters are special cases of (VRP). The following VARIATIONAL INCLUSION PROBLEMS are particular cases of (VRP) which have been studied extensively in the literature. Let $A, B, Y, \mathscr{D}_1, \mathscr{D}_2$ and $\mathscr{P}$ be as above, Z be another nonempty set, $F : Y \times B \times A \rightrightarrows Z$ and $G : Y \times A \rightrightarrows Z$ be set-valued mappings. For subsets U and V and points x, y under consideration, we adopt the notations

$$r_1(U, V) \quad \text{means} \quad U \subseteq V; \ \bar{r}_1(U, V) \quad \text{means} \quad U \nsubseteq V;$$

$$r_2(U, V) \quad \text{means} \quad U \cap V \neq \emptyset; \ \bar{r}_2(U, V) \quad \text{means} \quad U \cap V = \emptyset.$$

For each $r \in \{r_1, r_2\}$, $\alpha \in \{\alpha_1, \alpha_2\}$, we consider the following quasivariational inclusion problem

$$(\mathrm{IP}_{r\alpha}) \qquad \begin{cases} \text{find } \bar{a} \in A \text{ such that } \bar{a} \in \mathscr{D}_1(\bar{a}), \text{ and for all } b \in \mathscr{D}_2(\bar{a}), \\ \alpha(y, \mathscr{P}(\bar{a}, b)), \ r(F(y, b, \bar{a}), G(y, \bar{a})). \end{cases}$$

If we define the relation $\mathscr{R}$ linking $a \in A$, $b \in B$ and $y \in Y$ by:

$$\mathscr{R}(a, b, y) = 1 \text{ if and only if } r(F(y, b, a), G(y, a)), \tag{10.2}$$

then (VRP_α) becomes ($\text{IP}_{r\alpha}$).

To motivate the problem setting, let us look at several special cases of variational relation problems and variational inclusion problems. Assume that Z is a vector space and $C : A \rightrightarrows Z$ is a set-valued mapping such that $C(a)$ is a cone for all $a \in A$.

(a) If $A \equiv B$, $C(a) \equiv C$, $G(y, a) = F(y, a, a) + C$, where $C \subset Z$ is a closed cone, ($\text{IP}_{r_1\alpha_1}$) becomes the variational inclusion problem with constraints (VIP), namely,
(VIP) Find $\bar{a} \in \mathscr{D}_1(\bar{a})$ such that, for all $b \in \mathscr{D}_2(\bar{a})$, and for all $y \in \mathscr{P}(\bar{a}, b)$,

$$F(y, b, \bar{a}) \subset F(y, \bar{a}, \bar{a}) + C.$$

If we replace $\mathscr{P}$ by the mapping $(a, b) \mapsto \mathscr{P}(a, a)$, we get a variational inclusion problem of Minty type. If $\mathscr{P}$ is replaced by the mapping $(a, b) \mapsto \mathscr{P}(b, b)$, one has a variational inclusion problem of Stampacchia type.
If $\mathscr{P}(a, b)$ is replaced by $(a, b) \mapsto \mathscr{P}(a, a) := \mathscr{P}(a)$, and $\mathscr{D}_1(a) = \mathscr{D}_2(a) := \mathscr{D}(a)$, then (VIP) coincides with the upper variational inclusion problem:
(UIP) Find $\bar{a} \in \mathscr{D}(\bar{a})$ such that, for all $b \in \mathscr{D}(\bar{a})$, and for all $y \in \mathscr{P}(\bar{a})$,

$$F(y, b, \bar{a}) \subset F(y, \bar{a}, \bar{a}) + C.$$

(b) Note that (VIP) and (UIP) do not include the following general equilibrium problem (without severe assumptions on F):
(GEP) Find $\bar{a} \in \mathscr{D}_1(\bar{a})$ such that, for all $b \in \mathscr{D}_1(\bar{a})$, and for all $y \in \mathscr{P}(\bar{a})$,

$$F(y, b, \bar{a}) \subset C.$$

But our problem ($\text{IP}_{r_1\alpha_1}$) clearly does.

(c) If $A \equiv B$, $\mathscr{D}_1(a) = \mathscr{D}_2(a) \equiv A$, $\mathscr{P}$ is replaced by $(a, b) \mapsto \mathscr{P}(a, a) := \mathscr{P}(a)$, $Y = \mathscr{L}(A, Z)$ (the space of linear continuous mappings of A into Z), F is single-valued and $G(y, a) = Z \setminus -\operatorname{int} C(a)$, then ($\text{IP}_{r_2\alpha_2}$) collapses to the implicit vector variational inequality:
(IVI) Find $\bar{a} \in A$ such that, for all $b \in A$, there exists $y \in \mathscr{P}(\bar{a})$,

$$F(y, b, \bar{a}) \notin -\operatorname{int} C(\bar{a}).$$

(d) If A and B are topological vector spaces, $\mathscr{D}_1(a) = \mathscr{D}_2(a) := \mathscr{D}(a)$, $Y = \mathscr{L}(A, Z)$, $F(y, b, a) = (y, a - b)$ and $G(y, a) = Z\setminus$-int $C(a)$, where (y, a) denotes the value of a linear mapping y at a, then ($\text{IP}_{r_2\alpha_2}$) reduces to the vector quasivariational inequality problem:

(QVI) Find $\bar{a} \in \mathscr{D}(\bar{a})$ such that, for all $b \in \mathscr{D}(\bar{a})$, there exists $y \in \mathscr{P}(\bar{a})$,

$$(y, b - \bar{a}) \notin -\operatorname{int} C(\bar{a}).$$

(e) Let $A \equiv B$, $\mathscr{D}_1(a) = \mathscr{D}_2(a) := \mathscr{D}(a)$, $C(a) \equiv C \subset Z$, $f : A \times A \to Z$ and R linking $a \in A$ and $b \in A$ be defined by: $R(a, b) := 1$ if and only if $f(a, b) \notin -C$. Then, (VRP) collapses to the vector quasiequilibrium problem: (VQEP) Find $\bar{a} \in A$ such that $\bar{a} \in \mathscr{D}(\bar{a})$ and $f(\bar{a}, b) \notin -C$ for all $b \in \mathscr{D}(\bar{a})$.

10.2 Existence of Solutions

For nonempty sets A, B and the relation R linking $a \in A$ and $b \in B$, to study (VRP), we define a map $R : A \times B \to \{0, 1\}$ by $R(a, b) = 1$ if a links b, i.e., $R(a, b)$ holds, and $R(a, b) = 0$ if $R(a, b)$ does not hold. Though R stands for both the relation and this map, no confusion may occur. Moreover, for $a \in A$ and $b \in B$, we use $(R^a)^{-1}$ and R_b^{-1} to denote the inverse maps of $R(a, \cdot) : B \longrightarrow \{0, 1\}$ and $R(\cdot, b) : A \longrightarrow \{0, 1\}$, respectively. Similarly, we have a map $\mathscr{R} : A \times B \times Y \to \{0, 1\}$ by $\mathscr{R}(a, b, y) = 1$ if $\mathscr{R}(a, b, y)$ holds, and $\mathscr{R}(a, b, y) = 0$ if $\mathscr{R}(a, b, y)$ does not hold.

For a bifunction $\Phi : A \times B \to \mathbb{R}$, and $a \in A$, $b \in B$, we set

- $\operatorname{lev}_{\geq} \Phi(\cdot, b) = \{a \in A \mid \Phi(a, b) \geq 0\}$;
- $\operatorname{lev}_{<} \Phi(\cdot, b) = \{a \in A \mid \Phi(a, b) < 0\}$;
- $\operatorname{lev}_{\geq} \Phi(a, \cdot) = \{b \in B \mid \Phi(a, b) \geq 0\}$;
- $\operatorname{lev}_{<} \Phi(a, \cdot) = \{b \in B \mid \Phi(a, b) < 0\}$.

For relation R linking $a \in A$ and $b \in B$, we define $\Phi_R : A \times B \to \mathbb{R}$, by

$$\Phi_R(a, b) := \begin{cases} 1 \text{ if } R(a, b) = 1, \\ -1 \text{ if } R(a, b) = 0. \end{cases} \quad \text{for all } a \in A \text{ and } b \in B. \tag{10.3}$$

Then, it is clear that R is completely defined if and only if Φ_R is defined. Thus, the problem (VRP) is equivalent to GENERALIZED QUASIEQUILIBRIUM PROBLEM:

(GQEP) find $\bar{a} \in A$ such that $\bar{a} \in \mathscr{D}_1(\bar{a})$ and $\Phi_R(\bar{a}, b) \geq 0$ for all $b \in \mathscr{D}_2(\bar{a})$.

The problem (GQEP) was considered in Exercise 2.12 (Chap. 2) for the case where $A \equiv B$.

First we prove the following lemma, which is an extension of Lemma 1.31 (Chap. 1).

Lemma 10.1 *Let A be a nonempty compact subset of a topological vector space, B be a nonempty set, $\Phi : A \times B \to \mathbb{R}$ be a bifunction, and $\mathscr{D} : A \rightrightarrows B$. Assume that*

(i) *for $N_B = \{b^1, b^2, \ldots, b^n\} \in \mathscr{F}(B)$, there exists $N_A = \{a^1, a^2, \ldots, a^n\} \in \mathscr{F}(A)$ with $\mathrm{conv} N_A \subset A$ such that for each index set $I \subset \{1, 2, \ldots, n\}$ one has*

$$\mathrm{conv}\{a^i, i \in I\} \cap \bigcap_{i \in I} [\mathscr{D}^{-1}(b^i) \cap \mathrm{lev}_< \Phi(\cdot, b^i)] = \emptyset;$$

(ii) *for any $b \in B$, $\mathscr{D}^{-1}(b) \cap \mathrm{lev}_< \Phi(\cdot, b)$ is open.*

Then, there exists $\bar{a} \in A$ such that $\Phi(\bar{a}, b) \geq 0$ for all $b \in \mathscr{D}(\bar{a})$.

Proof Assume that for each $a \in A$, there is $b \in B$ such that $a \in \mathscr{D}^{-1}(b)$ and $\Phi(a, b) < 0$. Then,

$$A = \bigcup_{b \in B} [\mathscr{D}^{-1}(b) \cap \mathrm{lev}_< \Phi(\cdot, b)]. \tag{10.4}$$

By the compactness of A, (ii) and (10.4) together imply the existence of $N_B = \{b^0, b^1, \ldots, b^n\}$ such that

$$A = \bigcup_{i=0}^{n} [\mathscr{D}^{-1}(b^i) \cap \mathrm{lev}_< \Phi(\cdot, b^i)]. \tag{10.5}$$

Let $N_A = \{a^0, a^1, \ldots, a^n\}$ be corresponding to N_B in (i). By (10.5), the family

$$\mathscr{P} := \{\mathscr{D}^{-1}(b) \cap \mathrm{lev}_< \Phi(\cdot, b) \cap \mathrm{conv} N_A, \quad b \in N_B\}$$

is a finite open covering of the compact set $\mathrm{conv} N_A$. Let $\{\psi_i\}_{i=0}^n$ be a continuous partition of unity associated with the covering $\mathscr{P}$ and $\psi : \mathrm{conv} N_A \to \Delta_n$ be defined by $\psi(a) := \sum_{i=0}^n \psi_i(a) e_i$ for all $a \in \mathrm{conv} N_A$ (e_i is the ith standard unit vector). Then, ψ is continuous and hence $\psi \circ \varphi : \Delta_n \to \Delta_n$ is continuous, where $\varphi : \Delta_n \to A$, is defined by

$$\varphi(e) := \sum_{i=0}^{n} \lambda_i a^i \text{ for all } e = \sum_{i=0}^{n} \lambda_i e_i \in \Delta_n.$$

Brouwer's fixed-point theorem gives us a point $\bar{e} \in \Delta_n$ such that $\bar{e} = \psi(\varphi(\bar{e})) = \psi(\bar{a})$ (where $\bar{a} := \varphi(\bar{e})$). Let $J(\bar{a}) := \{j \in \{0, 1, \ldots, n\} | \; \psi_j(\bar{a}) \neq 0\}$ and $M(\bar{a}) := \{a^j \in N \mid j \in J(\bar{a})\}$. We have

$$\bar{e} = \psi(\bar{a}) = \sum_{j \in J(\bar{a})} \psi_j(\bar{a}) e_j,$$

and hence $\bar{a} = \varphi(\bar{e}) \in \text{conv}M(\bar{a})$. By (i), there exists $\bar{j} \in J(\bar{a})$ such that either $b^{\bar{j}} \notin \mathscr{D}(\bar{a})$ or $\Phi(\bar{a}, b^{\bar{j}}) \geq 0$. On the other hand, the definitions of the partition $\{\psi_i\}_{i=0}^n$ and of $J(\bar{a})$ give that

$$\bar{a} \in \{a \in \text{conv}N_A \mid \psi_{\bar{j}}(a) \neq 0\} \subseteq \mathscr{D}^{-1}(b^{\bar{j}}) \cap \text{lev}_{<}\Phi(\cdot, b^{\bar{j}}) \cap \text{conv}N_A,$$

i.e., $b^{\bar{j}} \in \mathscr{D}(\bar{a})$ and $\Phi(\bar{a}, b^{\bar{j}}) < 0$, a contradiction. □

Corollary 10.2 *Let A be a nonempty compact subset of a topological vector space, B be a nonempty set, R be a relation linking $a \in A$ and $b \in B$, and $\mathscr{D} : A \rightrightarrows B$. Assume that*

(i) *for $N_B = \{b^1, b^2, \ldots, b^n\} \in \mathscr{F}(B)$, there exists $N_A = \{a^1, a^2, \ldots, a^n\} \in \mathscr{F}(A)$ with $\text{conv}N_A \subset A$ such that for each index set $I \subset \{1, 2, \ldots, n\}$ one has*

$$\text{conv}\{x^i, i \in I\} \cap \bigcap_{i \in I} [\mathscr{D}^{-1}(b^i) \cap R_{b^i}^{-1}(0)] = \emptyset;$$

(ii) *for any $b \in B$, $\mathscr{D}^{-1}(b) \cap R_b^{-1}(0)$ is open.*

Then, there exists $\bar{a} \in A$ such that $R(\bar{a}, b)$ holds for all $b \in \mathscr{D}(\bar{a})$.

Proof For Φ_R defined by (10.3), we have $\text{lev}_{<}\Phi(\cdot, b) = R_b^{-1}(0)$ for all $b \in B$. Hence, by applying Lemma 10.1 for Φ_R, we obtain the conclusion. □

Definition 10.3 Let A be a convex subset of a vector space and B be a nonempty set.

(a) A binary relation R linking elements $a \in A$ and $b \in B$ is said to be GENERALIZED KKM if, for every $\{b^1, ..., b^n\} \subset B$, there exists a corresponding subset $\{a^1, \ldots, a^n\} \subset A$ such that for any $I \subset \{1, \ldots, n\}$ and $a \in \text{conv}\{a^j \mid j \in I\}$, there is $i \in I$ such that $R(a, b^i)$ holds. When $A \equiv B$, R is called KKM if, for every finite subset $\{a^1, a^2, \ldots, a^n\}$ of A and for every $a \in \text{conv}\{a^1, a^2, \ldots, a^n\}$, there is some index i such that $R(a, a^i)$ holds.
(b) A set-valued map $S : B \rightrightarrows A$ is said to be GENERALIZED KKM if, for every $\{b^1, ..., b^n\} \subset B$, there exists a corresponding subset $\{a^1, \ldots, a^n\} \subset A$ such that for any $I \subset \{1, \ldots, n\}$, $\text{conv}\{a^j \mid j \in I\} \subset \bigcup_{j \in I} S(b^j)$.

When $A \equiv B$, it is not hard to see that if R is KKM, then it is generalized KKM. Moreover, R is KKM if and only if Φ_R, defined by (10.3), is properly quasimonotone in the second component (i.e., $\Phi \in \mathscr{A}^2_{pr.qm}$). Similarly, a KKM map (see Exercise 2.1) is generalized KKM.

The following proposition provides some sufficient conditions for the assumption (i) in Corollary 10.2 to be satisfied.

Proposition 10.4 *Let A be a nonempty convex subset of a topological vector space, B be a nonempty set, R be a relation linking $a \in A$ and $b \in B$, and $\mathscr{D} : A \rightrightarrows B$. Assume that at least one of the following conditions is satisfied*

(i_1) *either R is generalized KKM, or the map $b \to A \backslash \mathscr{D}^{-1}(b)$ is generalized KKM;*
(i_2) *$A \equiv B$ and either $R(a, a) = 1$ and $(R^a)^{-1}(0)$ is convex for each $a \in A$, or the map $a \longmapsto \mathrm{conv}(\mathscr{D}(a) \cap (R^a)^{-1}(0))$ has no fixed points.*

Then, the assumption (i) in Corollary 10.2 is satisfied.

Proof *Case* (i_1). When R (resp., the map $b \to A \setminus \mathscr{D}^{-1}(b)$) is generalized KKM, for $N_B = \{b^1, b^2, \ldots, b^n\} \in \mathscr{F}(B)$, there exists $N_A = \{a^1, a^2, \ldots, a^n\} \in \mathscr{F}(A)$ with $\mathrm{conv}N_A \subset A$ such that for each index set $I \subset \{1, 2, \ldots, n\}$ one has

$$\mathrm{conv}\{a^i, i \in I\} \subset \bigcup_{i \in I} R_{b^i}^{-1}(1) \text{ (resp., } \mathrm{conv}\{a^i, i \in I\} \subset \bigcup_{i \in I}(A \setminus \mathscr{D}^{-1}(b^i))$$
$$= A \setminus \bigcap_{i \in I} \mathscr{D}^{-1}(b^i)),$$

or equivalently,

$$\mathrm{conv}\{a^i, i \in I\} \cap \bigcap_{i \in I} R_{b^i}^{-1}(0) = \emptyset \text{ (resp., } \mathrm{conv}\{a^i, i \in I\} \cap \bigcap_{i \in I} \mathscr{D}^{-1}(b^i) = \emptyset).$$

This implies (i) in Corollary 10.2.
Case (i_2). It is not hard to show that if $A \equiv B$, $R(a, a) = 1$, and $(R^a)^{-1}(0)$ is convex for each $a \in A$, then R is KKM (and hence, generalized KKM). If the map $a \longmapsto \mathrm{conv}(\mathscr{D}(a) \cap (R^a)^{-1}(0))$ has no fixed points, we take any $\{b^1, b^2, \ldots, b^n\} \subset A$, $\{a^1 = b^1, a^2 = b^2, \ldots, a^n = b^n\} \subset A$ and $I \subset \{1, 2, \ldots, n\}$. If $\mathrm{conv}\{a^i, i \in I\} \cap \bigcap_{i \in I}[\mathscr{D}^{-1}(b^i) \cap R_b^{-1}(0) = \mathrm{conv}\{a^i, i \in I\} \cap \bigcap_{i \in I}[\mathscr{D}^{-1}(a^i) \cap R_{a^i}^{-1}(0) \neq \emptyset$, then there exists $a \in \mathrm{conv}\{a^i, i \in I\}$ such that $a \in \mathscr{D}^{-1}(a^i) \cap R_{a^i}^{-1}(0)$ for all $i \in I$, i.e., $\{a^i, i \in I\} \subset \mathscr{D}(a) \cap (R^a)^{-1}(0)$. This implies that $a \in \mathrm{conv}\{a^i, i \in I\} \subset \mathrm{conv}(\mathscr{D}(a) \cap (R^a)^{-1}(0))$, which means that a is a fixed point of the map $a \longmapsto \mathrm{conv}(\mathscr{D}(a) \cap (R^a)^{-1}(0))$, a contradiction. Thus, we must have $\mathrm{conv}\{a^i, i \in I\} \cap \bigcap_{i \in I}[\mathscr{D}^{-1}(b^i) \cap R_b^{-1}(0) = \emptyset$, i.e., (i) of Corollary 10.2 is satisfied. □

Definition 10.5 For a subset K of a vector space $\mathbb{E}$ and $x \in \mathbb{E}$, the OUTWARD SET of K at x is denoted and defined as follows:

$$\mathbf{O}(K, x) := \bigcup_{\lambda > 1} (\lambda x + (1 - \lambda)K).$$

For the problem (VRP), let $P_R : B \rightrightarrows A$ be defined by

$$P_R(b) := [A \setminus \mathscr{D}_2^{-1}(b)] \cup [\mathrm{Fix}\mathscr{D}_1 \cap R_b^{-1}(1)] \text{ for all } b \in B.$$

For each $b \in B$, it is clear that $P_R(b)$ is closed if $\mathrm{Fix}\mathscr{D}_1$ is closed and $\mathscr{D}_2^{-1}(b)$ and $R_b^{-1}(0)$ are open.

Proposition 10.6 *Asssume that A is a convex subset of a vector space, B is a nonempty set and at least one of the following conditions is satisfied*

(i) *the map* $b \mapsto A \setminus \mathscr{D}_2^{-1}(b)$ *is generalized KKM;*
(ii) $\mathrm{Fix}\mathscr{D}_1 = A$ *and* R *is generalized KKM;*
(iii) $A \equiv B$, $\mathrm{conv}\mathscr{D}_2(a) \subset \mathscr{D}_1(a)$ *for all* $a \in A$, *and for every finite set* $\{a^1, a^2, \ldots, a^n\} \subset A$ *one has*

$$\mathrm{conv}\{a^i, i = 1, 2, \ldots, n\} \cap \bigcap_{i=1}^{n} [\mathscr{D}_2^{-1}(a^i) \cap R_{a^i}^{-1}(0)] = \emptyset$$

(in particular, either R *is KKM or the map* $a \longmapsto \mathrm{conv}(\mathscr{D}_2(a) \cap R^a)^{-1}(0))$ *has no fixed points);*
(iv) $A \equiv B$, $\mathrm{Fix}\mathscr{D}_1$ *is nonempty,* $\mathscr{D}_1$ *is convex-valued, and for each* $a \in \mathrm{Fix}\mathscr{D}_1$, $\mathscr{D}_2(a) \subset \mathscr{D}_1(a)$ *and* $R(a, b) = 1$ *for all* $b \in \mathscr{D}_2(a) \cap \boldsymbol{O}(\mathscr{D}_1(a), a)$.

Then, P_R *is generalized KKM.*

Proof Suppose that P_R is not generalized KKM. Then,

($*$) there exists $N_B = \{b^1, b^2, \ldots, b^n\} \in \mathscr{F}(B)$ such that for all $N_A = \{a^1, a^2, \ldots, a^n\} \in \mathscr{F}(A)$, there exists $I \subset \{1, 2, \ldots, n\}$ satisfying $\mathrm{conv}\{a^i, i \in I\} \cap \bigcap_{i \in I} [\mathscr{D}_2^{-1}(b^i) \cap ((A \setminus \mathrm{Fix}\mathscr{D}_1) \cup R_{b^i}^{-1}(0))] \neq \emptyset$.

Case (i). It follows from ($*$) that

$$\begin{aligned}\emptyset &\neq \mathrm{conv}\{a^i, i \in I\} \cap \bigcap_{i \in I} [\mathscr{D}_2^{-1}(b^i) \cap ((A \setminus \mathrm{Fix}\mathscr{D}_1) \cup R_{b^i}^{-1}(0))] \\ &\subset \mathrm{conv}\{a^i, i \in I\} \cap \bigcap_{i \in I} \mathscr{D}_2^{-1}(b^i).\end{aligned}$$

This implies that $\mathrm{conv}\{a^i, i \in I\} \not\subset A \setminus \bigcap_{i \in I} \mathscr{D}_2^{-1}(b^i) = \bigcup_{i \in I} (A \setminus \mathscr{D}_2^{-1}(b^i))$, i.e., the map $b \mapsto A \setminus \mathscr{D}_2^{-1}(b)$ is not generalized KKM. This contradicts (i).
Case (ii). It follows from ($*$) that, as $\mathrm{Fix}\mathscr{D}_1 = A$,

$$\emptyset \neq \mathrm{conv}\{a^i, i \in I\} \cap \bigcap_{i \in I} [\mathscr{D}_2^{-1}(b^i) \cap R_{b^i}^{-1}(0)] \subset \mathrm{conv}\{a^i, i \in I\} \cap \bigcap_{i \in I} R_{b^i}^{-1}(0),$$

which means that $\mathrm{conv}\{a^i, i \in I\} \not\subset \bigcup_{i \in I} (A \setminus R_{b^i}^{-1}(0)) = \bigcup_{i \in I} R_{b^i}^{-1}(1)$. This implies that R is not generalized KKM, a contradiction.

Case (iii). By taking $N_A = N_B$ in (∗), we have: for some $I \subset \{1, 2, \dots, n\}$,

$$\emptyset \neq \mathrm{conv}\{b^i, i \in I\} \cap \bigcap_{i \in I} [\mathscr{D}_2^{-1}(b^i) \cap ((A \setminus \mathrm{Fix}\mathscr{D}_1) \cup R_{b^i}^{-1}(0))]$$

$$= \mathrm{conv}\{b^i, i \in I\} \cap \Big(((A \setminus \mathrm{Fix}\mathscr{D}_1) \cap \bigcap_{i \in I} \mathscr{D}_2^{-1}(b^i)) \cup \bigcap_{i \in I} [\mathscr{D}_2^{-1}(b^i) \cap R_{b^i}^{-1}(0)] \Big).$$

This and (iii) imply that $\mathrm{conv}\{b^i, i \in I\} \cap \big((A \setminus \mathrm{Fix}\mathscr{D}_1) \cap \bigcap_{i \in I} \mathscr{D}_2^{-1}(b^i)\big) \neq \emptyset$. Then, there exists $b \in \mathrm{conv}\{b^i, i \in I\}$ such that $b \notin \mathrm{Fix}\mathscr{D}_1$ and $b \in \bigcap_{i \in I} \mathscr{D}_2^{-1}(b^i)$, i.e., $b \notin \mathscr{D}_1(b)$ and $\{b^i, i \in I\} \subset \mathscr{D}_2(b)$. It follows that $b \notin \mathscr{D}_1(b)$ and $b \in \mathrm{conv}\{b^i, i \in I\} \subset \mathrm{conv}\mathscr{D}_2(b) \subset \mathscr{D}_1(b)$, a contradiction.
Case (iv). We fix a point $a^0 \in \mathrm{Fix}\mathscr{D}_1$ and take $N_A = \{a^i = a^0, i = 1, 2, \dots, n\}$ in (∗). Then, for some $I \subset \{1, 2, \dots, n\}$, we have: $\emptyset \neq \mathrm{conv}\{a^i, i \in I\} \cap \bigcap_{i \in I} [\mathscr{D}_2^{-1}(b^i) \cap ((A \setminus \mathrm{Fix}\mathscr{D}_1) \cup R_{b^i}^{-1}(0))] = \{b^0\} \cap \bigcap_{i \in I} \big(\mathscr{D}_2^{-1}(b^i) \cap [(A \setminus \mathrm{Fix}\mathscr{D}_1) \cup R_{b^i}^{-1}(0)]\big)$. This and the fact that $a^0 \in \mathrm{Fix}\mathscr{D}_1$ imply that $a^0 \in \bigcap_{i \in I} \big(\mathscr{D}_2^{-1}(b^i) \cap R_{b^i}^{-1}(0)\big)$, i.e., $\{b^i, i \in I\} \subset \mathscr{D}_2(a^0) \cap (R^{a^0})^{-1}(0)$. Furthermore, $\{b^i, i \in I\} \subset \mathscr{D}_2(a^0) \subset \mathscr{D}_1(a^0)$. By the convexity of $\mathscr{D}_1(a^0)$, for any b^i $(i \in I)$ and $0 < \mu < 1$, $(1 - \mu)a^0 + \mu b^i := b \in \mathscr{D}_1(a^0)$. Then $b^i = (1 - \lambda)a^0 + \lambda b \in \mathbf{O}(\mathscr{D}_1(a^0), a^0)$, where $\lambda = \mu^{-1} > 1$. Thus, $\{b^i, i \in I\} \subset \mathscr{D}_2(a^0) \cap (R^{a^0})^{-1}(0) \cap \mathbf{O}(\mathscr{D}_1(a^0), a^0)$ contradicting (iv). □

Theorem 10.7 *Assume that A is a compact convex subset of a topological vector space, B is a nonempty set, P_R is generalized KKM and $P_R(b)$ is closed for each $b \in B$. Then, (VRP) has solutions.*

Proof Define a relation $\overline{R}$ linking $a \in A$ and $b \in B$ as follows:

$$\overline{R}(a, b) \text{ holds iff } a \in \mathscr{D}_1(a), \text{ and } R(a, b) \text{ holds.} \tag{10.6}$$

Direct calculations yield that, for all $a \in A$ and $b \in B$,

$$\mathscr{D}_2^{-1}(a) \cap \overline{R}_b^{-1}(0) = A \setminus P_R(b) = \mathscr{D}_2^{-1}(b) \cap [(A \setminus \mathrm{Fix}\mathscr{D}_1) \cup R_b^{-1}(0)], \tag{10.7}$$

For each $b \in B$, since $P_R(b)$ is closed, $\mathscr{D}_2^{-1}(a) \cap \overline{R}_b^{-1}(0) = A \setminus P_R(b)$ is open. Hence, the condition (ii) of Corollary 10.2 holds for $\mathscr{D}_2$ and $\overline{R}$. Moreover, by (10.7), we easily check that the generalized KKM property of P_R is equivalent to the following condition:

(i$_*$) for $N_B = \{b^1, b^2, \dots, b^n\} \in \mathscr{F}(B)$, there exists $N_A = \{a^1, a^2, \dots, a^n\} \in \mathscr{F}(A)$ such that for each index set $I \subset \{1, 2, \dots, n\}$ one has

$$\mathrm{conv}\{a^i, i \in I\} \cap \bigcap_{i \in I} [\mathscr{D}_2^{-1}(b^i) \cap \overline{R}_{b^i}^{-1}(0)] = \emptyset.$$

This (i_*) is (i) of Corollary 10.2 for $\mathscr{D}_2$ and $\overline{R}$. Thus, by Corollary 10.2, there exists $\bar{a} \in A$ such that $\overline{R}(\bar{a}, b)$ holds for all $b \in \mathscr{D}_2(\bar{a})$. This means that $\bar{a} \in \mathscr{D}_1(\bar{a})$ and $R(\bar{a}, b)$ hold for all $b \in \mathscr{D}_2(\bar{a})$, i.e., $\bar{a}$ is a solution of (VRP). □

Lemma 10.8 *Let A be a nonempty compact convex set and B be a nonempty convex set, each in a topological vector space, and $\Phi : A \times B \to \mathbb{R}$ be a bifunction. Assume that*

(i) *for each $b \in B$, $\mathrm{lev}_{\geq}\Phi(\cdot, b)$ is nonempty, closed and convex;*
(ii) *for each $a \in A$, the set $\mathrm{lev}_{<}\Phi(a, \cdot)$ is open and convex.*

Then, there exists $\bar{a} \in A$ such that $\Phi(\bar{a}, b) \geq 0$ for all $b \in B$.

When $A \equiv B$, Lemma 10.8 becomes Lemma 1.33.

Proof Suppose that for each $a \in A$, there is $b \in B$ such that $\Phi(a, b) < 0$. Then,

$$A = \bigcup_{b \in B} \{a \in A \mid \Phi(a, b) < 0\} = \bigcup_{b \in B} \mathrm{lev}_{<}\Phi(\cdot, b). \tag{10.8}$$

By the compactness of A, (ii) and (10.8) together imply the existence of $\bar{N} \in \mathscr{F}(B)$ such that

$$A = \bigcup_{b \in \bar{N}} \mathrm{lev}_{<}\Phi(\cdot, b),$$

or, equivalently,

$$\emptyset = \bigcap_{b \in \bar{N}} \mathrm{lev}_{\geq}\Phi(\cdot, b). \tag{10.9}$$

Since $\mathrm{lev}_{\geq}\Phi(\cdot, b)$ is nonempty for each $b \in B$, the cardinality $|\bar{N}| \geq 2$. Without loss of generality, we can assume that among the finite subsets of B satisfying (10.9), the cardinality of $\bar{N}$ is the smallest.

Now, let $b', b'' \in \bar{N}$ with $b' \neq b''$, and set

$$C = \begin{cases} \bigcap_{b \in \bar{N} \setminus \{b', b''\}} \mathrm{lev}_{\geq}\Phi(\cdot, b) & \text{if } |\bar{N}| > 2, \\ A & \text{if } |\bar{N}| = 2. \end{cases}$$

Define mappings $\gamma : [0, 1] \to B$ and $P, Q : [0, 1] \rightrightarrows A$ by, for $t \in [0, 1]$,

$$\gamma(t) = (1 - t)b' + tb'',$$

$$P(t) = \mathrm{lev}_{\geq}\Phi(\cdot, \gamma(t)) = \{a \in A \mid \Phi(a, \gamma(t)) \geq 0\},$$

$$Q(t) = P(t) \cap C.$$

By (ii), both P and Q have nonempty closed convex values. Moreover, assume that there exist $[t_1, t_2] \subset [0, 1]$ and $t \in [t_1, t_2]$ such that $P(t) \not\subset P(t_1) \cup P(t_2)$. Then, there exists $a \in P(t)$ such that $a \notin P(t_1) \cup P(t_2)$, i.e., $\Phi(a, \gamma(t_1)) < 0$ and $\Phi(a, \gamma(t_2)) < 0$. This and (iii) imply that $\Phi(a, \gamma(t)) < 0$, i.e., $a \notin P(t)$, a contradiction. Thus, for all $[t_1, t_2] \subset [0, 1]$,

$$P(t) \subset P(t_1) \cup P(t_2) \text{ for all} t \in [t_1, t_2].$$

In particular, $P(t) \subset P(0) \cup P(1)$ for all $t \in [0, 1]$. Hence, $Q(t) \subset Q(0) \cup Q(1)$ for all $t \in [0, 1]$, and thus $Q([0, 1]) \subset Q(0) \cup Q(1)$.

We show that P is upper semicontinuous. Because the values of P are compact (by (ii) and the compactness of A), we need to check that P is closed. Take any net $\{(t_\mu, a_\mu)\}_{\mu \in (I, \preceq)} \subset [0, 1] \times A$ such that $(t_\mu, a_\mu) \to (\bar{t}, \bar{a})$ and $a_\mu \in P(t_\mu)$. Suppose to the contrary that $\bar{a} \in A \setminus P(\bar{t}) = \{a \in A \mid \Phi(a, \gamma(\bar{t})) < 0\}$. Then,

$$\gamma(\bar{t}) \in \{b \in B \mid \Phi(\bar{a}, b) < 0\} =: U(\bar{a}),$$

which is an open set by (iii). Since γ is continuous, there exists a neighborhood $V_{\bar{t}}$ of $\bar{t}$ such that $\gamma(t) \in U(\bar{a})$ for all $t \in V_{\bar{t}}$. Then, there exist $t_1, t_2 \in V_{\bar{t}}$ such that $\bar{t} \in [t_1, t_2] \subset V_{\bar{t}}$. We have $\gamma(t_i) \in U(\bar{a})$, i.e., $\bar{a} \in A \setminus P(t_i)$ $(i = 1, 2)$. Hence, $\bar{a} \in A \setminus (P(t_1) \cup P(t_2))$ which is an open set. As $t_\mu \to \bar{t}$, there exists $\bar{\mu}$ such that $t_\mu \in (t_1, t_2)$ for all $\mu \succeq \bar{\mu}$. Then, $a_\mu \in P(t_\mu) \subset P(t_1) \cup P(t_2)$ for all $\mu \succeq \bar{\mu}$, contradicting the convergence $a_\mu \to \bar{a}$. Thus, P is upper semicontinuous. Since C is closed (by (ii)), we easily prove that Q is also upper semicontinuous. Being an image of a connected set via an upper semicontinuous map with connected values, $Q([0, 1])$ is connected. It follows that $Q(0) \cap Q(1) \neq \emptyset$. On the other hand, by (10.9),

$$Q(0) \cap Q(1) = \bigcap_{b \in \bar{N}} \{a \in A \mid \Phi(a, b) \geq 0\} = \emptyset,$$

and we obtain a contradiction. □

Using Lemma 10.8 and Theorem 2.14, we prove the following result for the problem (VRP).

Theorem 10.9 *For (VRP), assume that A is a compact convex subset of a locally convex topological vector space, B is a convex subset of a topological vector space and the following conditions are satisfied*

(i) *$\mathscr{D}_1$ is a closed map with convex values;*
(ii) *$\mathscr{D}_2$ is lower semicontinuous with convex values;*
(iii) *for each $a \in A$ and $b \in \mathscr{D}_2(a)$, $R(a', b) = 1$ for at least one $a' \in \mathscr{D}_1(a)$;*
(iv) *for each $b \in B$, $R_b^{-1}(1)$ is convex;*
(v) *for each $a \in A$, $(R^a)^{-1}(0)$ is convex;*
(vi) *the set $\{(a, b) \in A \times B \mid R(a, b) = 1\}$ is closed.*

Then, (VRP) has solutions.

Proof Let $S : A \rightrightarrows A$ be defined by

$$S(a) := \{a' \in \mathscr{D}_1(a) \mid R(a', b) = 1 \quad \text{for all } b \in \mathscr{D}_2(a)\} = \mathscr{D}_1(a) \cap \bigcap_{b \in \mathscr{D}_2(a)} R_b^{-1}(1).$$

We prove that S has a fixed point. Let $a \in A$. By (i) and the compactness of A, $\mathscr{D}_1(a)$ is nonempty, compact, and convex. By (ii), $\mathscr{D}_2(a)$ is nonempty and convex. Note that (vi) implies that for each $a \in A$ and $b \in B$, the set $R_b^{-1}(1)$ is closed and $(R^a)^{-1}(0)$ is open. Then, by (i), (iv), (vi), and the definition of S, $S(a)$ is closed and convex. For each $a \in A$, considering $\Phi_R^a := \Phi_R|_{\mathscr{D}_1(a) \times \mathscr{D}_2(a)}$, we have:

- for each $b \in \mathscr{D}_2(a)$, $\Phi_R(a', b) \geq 0$ for some $a' \in \mathscr{D}_1(a)$ by (iii), and hence $\text{lev}_{\geq} \Phi_R^a(\cdot, b)$ is nonempty;
- for each $b \in \mathscr{D}_2(a)$, $\text{lev}_{\geq} \Phi_R^a(\cdot, b) = \{a' \in \mathscr{D}_1(a) \mid \Phi_R(a', b) \geq 0\} = \mathscr{D}_1(a) \cap R_b^{-1}(1)$ is closed and convex in $\mathscr{D}_1(a)$ by (i), (iv) and (vi);
- for each $a' \in \mathscr{D}_1(a)$, $\text{lev}_{<} \Phi_R^a(a', \cdot) = \{b \in \mathscr{D}_2(a) \mid \Phi_R(a', b) < 0\} = \mathscr{D}_2(a) \cap (R^{a'})^{-1}(0)$ is open and convex in $\mathscr{D}_2(a)$ by (ii), (v) and (vi).

Then, by Lemma 10.8, there exists $\bar{a}' \in \mathscr{D}_1(a)$ such that $\Phi_R^a(\bar{a}', b) \geq 0$ for all $b \in \mathscr{D}_2(a)$, which means that $S(a) \neq \emptyset$. We prove the closedness of S. Let a net $\{(a^\alpha, u^\alpha)\}_{\alpha \in I} \subset \text{gph} S$ ($\subset \text{gph} \mathscr{D}_1$) converge to (a^0, u^0). Since $\mathscr{D}_1$ is closed, $u^0 \in \mathscr{D}_1(a^0)$. Take any $b \in \mathscr{D}_2(a^0)$. By the lower semicontinuity of $\mathscr{D}_2$, there exist a subnet $\{a^{\alpha_\beta}\}_{\beta \in J}$ of $\{(a^\alpha, u^\alpha)\}_{\alpha \in I}$ and a net $\{b^\beta\}_{\beta \in J}$ with $b^\beta \in \mathscr{D}_2(a^{\alpha_\beta})$ such that $\{b^\beta\}_{\beta \in J}$ converges to b. Since $u^{\alpha_\beta} \in S(a^{\alpha_\beta}) \subset \bigcap_{b \in \mathscr{D}(a^{\alpha_\beta})} R_b^{-1}(1)$, we have $(u^{\alpha_\beta}, b^\beta) \in \{(a, b) \in A \times B \mid R(a, b) = 1\} := M$. By (vi), $(u^0, b) \in M$, i.e., $u^0 \in R_b^{-1}(1)$. Thus, $u^0 \in \bigcap_{b \in \mathscr{D}_2(a^0)} R_b^{-1}(1)$. It follows that $u^0 \in \mathscr{D}_1(a^0) \cap \bigcap_{b \in \mathscr{D}_2(a^0)} R_b^{-1}(1) = S(a^0)$, i.e., $(a^0, u^0) \in \text{gph} S$. Thus, S is closed. By Theorem 2.14, S has a fixed point $\bar{a}$. It is clear that this $\bar{a}$ is also a solution of (VRP). □

In the remainder of this section, we will discuss some particular cases of (VRP). First we consider the problem (VRP$_\alpha$). For the problem (VRP$_\alpha$), we define $P_{\mathscr{R}}^\alpha : B \rightrightarrows A$ by

$$P_{\mathscr{R}}^\alpha(b) := [A \setminus \mathscr{D}_2^{-1}(b)] \cup [\text{Fix} \mathscr{D}_1 \cap \{a \in A \mid \alpha(y, \mathscr{P}(a, b)), \ \mathscr{R}(a, b, y) = 1\}]$$
$$\text{for all } b \in B.$$

By restating Theorem 10.7 for the variational relation R defined by (10.1), we obtain the following result.

Corollary 10.10 *Assume that A is a compact convex subset of a topological vector space, B is a nonempty set, $P_{\mathscr{R}}^\alpha$ is generalized KKM and $P_{\mathscr{R}}^\alpha(b)$ is closed for each $b \in B$. Then,* (VRP$_\alpha$) *has solutions.*

For each $b \in B$, by definition of $P^{\alpha}_{\mathscr{R}}$, $P^{\alpha}_{\mathscr{R}}(b)$ is closed if Fix $\mathscr{D}_1$ is closed, $\mathscr{D}_2^{-1}(b)$ is open and

$$M^{\alpha}_{\mathscr{R}}(b) = \{a \in A \mid \alpha(y, \mathscr{P}(a, b)),\ \mathscr{R}(a, b, y) = 1\}$$

is closed.

Proposition 10.11 *For each $b \in B$, $M^{\alpha}_{\mathscr{R}}(b)$ is closed if*

(i) $\alpha = \alpha_1$: *$\mathscr{P}(., b)$ is lower semicontinuous; $\alpha = \alpha_2$: $\mathscr{P}(., b)$ is upper semicontinuous with compact images;*
(ii) *the set $\{(a, y) \in A \times Y \mid \mathscr{R}(a, b, y) = 1\}$ is closed.*

Proof Let $a_\gamma \in M^{\alpha}_{\mathscr{R}}(b)$ be such that $a_\gamma \to a_0$. Then, $\forall y \in \mathscr{P}(a_0, b)$, by the lower semicontinuity of $\mathscr{P}(., b)$, $\exists y_\gamma \in \mathscr{P}(a_\gamma, b)$ such that $y_\gamma \to y$. Since $U := \{(a, y) \in A \times Y \mid \mathscr{R}(a, b, y) = 1\}$ is closed and $(a_\gamma, y_\gamma) \in U$ for all γ, we have $(a_0, y) \in U$. Hence, $a_0 \in M^{\alpha}_{\mathscr{R}}(b)$. □

Definition 10.12 Let A be a convex set in a vector space, B and Y be a nonempty sets, $\mathscr{P} : A \times B \rightrightarrows Y$ and $\alpha \in \{\alpha_1, \alpha_2\}$. A relation $\mathscr{R} \subset A \times B \times Y$ is said to be $(\alpha, \mathscr{P})$-GENERALIZED KKM if, for every $\{b^1, ..., b^n\} \subset B$, there exists a corresponding subset $\{a^1, \ldots, a^n\} \subset A$ such that for any $I \subset \{1, \ldots, n\}$ and $a \in \mathrm{conv}\{a^j \mid j \in I\}$, there is $i \in I$ such that $\alpha(y, \mathscr{P}(a, b^i)), \mathscr{R}(a, b^i, y)$ holds. When $A \equiv B$, $\mathscr{R} \subset A \times A \times Y$ is called $(\alpha, \mathscr{P}) - KKM$ if, for every finite subset $\{a^1, a^2, \ldots, a^n\}$ of A and for every $a \in \mathrm{conv}\{a^1, a^2, \ldots, a^n\}$, there is some index i such that $\alpha(y, \mathscr{P}(a, a^i)), \mathscr{R}(a, a^i, y)$ holds.

Clearly, relation $\mathscr{R}$ is $(\alpha, \mathscr{P})$-generalized KKM (resp., $(\alpha, \mathscr{P})$-KKM) if and only if, the binary relation R, given by (10.1), is generalized KKM (resp., KKM). It follows from Proposition 10.6 for the relation R defined by (10.1) that $P^{\alpha}_{\mathscr{R}}$ is generalized KKM if at least one of the following conditions is satisfied

(i) the map $b \mapsto A \setminus \mathscr{D}_2^{-1}(b)$ is generalized KKM;
(ii) Fix$\mathscr{D}_1 = A$ and $\mathscr{R}$ is $(\alpha, \mathscr{P})$-generalized KKM;
(iii) $A \equiv B$, $\mathrm{conv}\mathscr{D}_2(a) \subset \mathscr{D}_1(a)$ for all $a \in A$, and either $\mathscr{R}$ is $(\alpha, \mathscr{P})$-KKM or the map $a \longmapsto \mathrm{conv}(\mathscr{D}_2(a) \cap \{b \in A \mid \overline{\alpha}(y, \mathscr{P}(a, b)),\ \mathscr{R}(a, b, y) = 0\}$ has no fixed points.

The following corollary is a version of Theorem 10.9 for the variational relation R defined by (10.1).

Corollary 10.13 *For* (VRP$_\alpha$), *assume that A is a compact convex subset of a locally convex topological vector space, B is a convex subset of a topological vector space, and the following conditions are satisfied*

(i) *$\mathscr{D}_1$ is a closed map with convex values;*
(ii) *$\mathscr{D}_2$ is lower semicontinuous with convex values;*
(iii) *for each $a \in A$ and $b \in \mathscr{D}_2(a)$, there exists $a' \in \mathscr{D}_1(a)$ such that $\alpha(y, \mathscr{P}(a', b)), \mathscr{R}(a', b, y) = 1$;*

(iv) *for each* $b \in B$, *the set* $M_1^{\alpha}(b) := \{a \in A \mid \alpha(y, \mathscr{P}(a,b)),\ \mathscr{R}(a,b,y) = 1\}$ *is convex;*

(vi) *for each* $a \in A$, *the set* $M_0^{\alpha}(b) := \{b \in B \mid \bar{\alpha}(y, \mathscr{P}(a,b)),\ \mathscr{R}(a,b,y) = 0\}$ *is convex;*

(vii) *the set* $M_{\alpha} := \{(a,b) \in A \times B \mid \alpha(y, \mathscr{P}(a,b)),\ \mathscr{R}(a,b,y) = 1\}$ *is closed.*

Then, (VRP$_{\alpha}$) *has solutions.*

We discuss some sufficient conditions for assumptions of Corollary 10.13 to be satisfied.

Proposition 10.14 *Let* A, B, Y *be convex subsets of topological vector spaces. Then, the set*

$$M_{\alpha} := \{(a,b) \in A \times B \mid \alpha(y, \mathscr{P}(a,b)),\ \mathscr{R}(a,b,y) = 1\}$$

is closed if, the set

$$N := \{(a,b,y) \in A \times B \times Y \mid \mathscr{R}(a,b,y) = 1\}$$

is closed and

$\alpha = \alpha_1$: $\mathscr{P}$ *is lower semicontinuous;*
$\alpha = \alpha_2$: $\mathscr{P}$ *is upper semicontinuous with compact values.*

Proof Let $\{(a^{\nu}, b^{\nu})\}_{\nu} \subset M_{\alpha}$ be a net converging to (a,b). We must prove that $(a,b) \in M_{\alpha}$.

Case $\alpha = \alpha_1$. Choose any $y \in \mathscr{P}(a,b)$. We show that $\mathscr{R}(a,b,y) = 1$, and hence $(a,b) \in M_{\alpha_1}$. By the lower semicontinuity of the map $\mathscr{P}$, there exists a net $\{y^{\nu}\}_{\nu}$ with $y^{\nu} \in \mathscr{P}(a^{\nu}, b^{\nu})$ converging to y. Then, for all ν, $(a^{\nu}, b^{\nu}, y^{\nu}) \in N$ and $(a^{\nu}, b^{\nu}, y^{\nu}) \to (a,b,y)$. Since N is closed $(a,b,y) \in N$, and therefore, $\mathscr{R}(a,b,y) = 1$.

Case $\alpha = \alpha_2$. We show that there exists $y \in \mathscr{P}(a,b)$ such that $\mathscr{R}(a,b,y) = 1$, and hence $(a,b) \in M_{\alpha_2}$. For each ν, by the nonemptiness of $\mathscr{P}(a^{\nu}, b^{\nu})$, there exists $y^{\nu} \in \mathscr{P}(a^{\nu}, b^{\nu})$. Since $\mathscr{P}$ is upper semicontinuous with compact values, there is a subnet $\{y^{\nu_{\mu}}\}_{\mu}$ of the net $\{y^{\nu}\}_{\nu}$ such that $y^{\nu_{\mu}} \to y \in \mathscr{P}(a,b)$. Then, for all μ, $(a^{\nu_{\mu}}, b^{\nu_{\mu}}, y^{\nu_{\mu}}) \in N$ and $(a^{\nu_{\mu}}, b^{\nu_{\mu}}, y^{\nu_{\mu}}) \to (a,b,y)$. As N is closed, $(a,b,y) \in N$, i.e., $\mathscr{R}(a,b,y) = 1$. □

Definition 10.15 Let A and Y be convex subsets of vector spaces. A multivalued map $\mathscr{P} : A \rightrightarrows Y$ is said to be CONVEX if

$$\lambda \mathscr{P}(a^1) + (1-\lambda)\mathscr{P}(a^2) \subset \mathscr{P}(\lambda a^1 + (1-\lambda)a^2)$$

for every $a^1, a^2 \in A$ and all $\lambda \in [0,1]$. If the inclusion "$\subset$" is replaced by "$\supset$," then the multivalued map $\mathscr{P}$ is said to be CONCAVE.

Proposition 10.16 *Let A, B, Y be convex subsets of vector spaces. Then, for $b \in B$, the set*

$$M_1^{\alpha}(b) := \{a \in A \mid \alpha(y, \mathscr{P}(a,b)),\ \mathscr{R}(a,b,y) = 1\}$$

is convex if, the set

$$N_b := \{(a,y) \in A \times Y \mid \mathscr{R}(a,b,y) = 1\},$$

is convex and

$\alpha = \alpha_1$: *$\mathscr{P}(\cdot, b)$ is concave;*
$\alpha = \alpha_2$: *$\mathscr{P}(\cdot, b)$ is convex.*

Proof Let $a^1, a^2 \in M_1^{\alpha}(b)$ and $\lambda \in [0,1]$. We show that $a = \lambda a^1 + (1-\lambda)a^2 \in M_1^{\alpha}(b)$.

Case $\alpha = \alpha_1$. Taking any $y \in \mathscr{P}(a,b)$, we must show that $\mathscr{R}(a,b,y) = 1$. By the concavity of the map $\mathscr{P}(\cdot, b)$, $y \in \mathscr{P}(a,b) = \mathscr{P}(\lambda a^1 + (1-\lambda)a^2, b) \subset \lambda\mathscr{P}(a^1,b) + (1-\lambda)\mathscr{P}(a^2,b)$. Hence, there exist $y^1 \in \mathscr{P}(a^1,b)$ and $y^2 \in \mathscr{P}(a^2,b)$ such that $y = \lambda y^1 + (1-\lambda)y^2$. Moreover, for each index $i \in \{1,2\}$, $\mathscr{R}\left(a^i, b, y^i\right) = 1$ (since $a^i \in M_1^{\alpha_1}(b)$), i.e., $(a^i, y^i) \in N_b$. As N_b is convex, $(a,y) = \lambda(a^1,y^1) + (1-\lambda)(a^2,y^2) \in N_b$, and hence $\mathscr{R}(a,b,y) = 1$.

Case $\alpha = \alpha_2$. We show that there exists $y \in \mathscr{P}(a,b)$ such that $\mathscr{R}(a,b,y) = 1$. For each index $i \in \{1,2\}$, since $a^i \in M_1^{\alpha_2}(b)$, there exist $y^i \in \mathscr{P}(a^i,b)$ such that $\mathscr{R}(a^i,b,y^i) = 1$ (i.e., $(a^i,y^i) \in N_b$). Set $y := \lambda y^1 + (1-\lambda)y^2$. Then, by the convexity of the map $\mathscr{P}(\cdot,b)$, $y \in \lambda\mathscr{P}(a^1,b) + (1-\lambda)\mathscr{P}(a^2,b) \subset \mathscr{P}(\lambda a^1 + (1-\lambda)a^2, b) = \mathscr{P}(a,b)$. As N_b is convex, $(a,y) = \lambda(a^1,y^1) + (1-\lambda)(a^2,y^2) \in N_b$, and hence $\mathscr{R}(a,b,y) = 1$. □

Proposition 10.17 *Let A, B, Y be convex subsets of vector spaces. Then, for $a \in A$, the set*

$$M_0^{\alpha}(a) := \{b \in B \mid \overline{\alpha}(y, \mathscr{P}(a,b)),\ \mathscr{R}(a,b,y) = 0\}$$

is convex if the set

$$N^a := \{(b,y) \in B \times Y \mid \mathscr{R}(a,b,y) = 0\}$$

is convex and

$\alpha = \alpha_1$: *$\mathscr{P}(a, \cdot)$ is convex;*
$\alpha = \alpha_2$: *$\mathscr{P}(a, \cdot)$ is concave.*

Proof Let $b^1, b^2 \in M_0^{\alpha}(a)$ and $\lambda \in [0,1]$. We show that $b = \lambda b^1 + (1-\lambda)b^2 \in M_0^{\alpha}(a)$.

Case $\alpha = \alpha_1$. The proof is similar to that of the case "$\alpha = \alpha_2$" in Proposition 10.16. We show that there exists $y \in \mathscr{P}(a, b)$ such that $\mathscr{R}(a, b, y) = 0$. For each index $i \in \{1, 2\}$, since $b^i \in M_0^{\alpha_1}(a)$, there exist $y^i \in \mathscr{P}(a, b^i)$ such that $\mathscr{R}(a, b^i, y^i) = 0$ (i.e., $(b^i, y^i) \in N^a$). Set $y := \lambda y^1 + (1 - \lambda)y^2$. Then, by the convexity of the map $\mathscr{P}(a, \cdot)$, $y \in \lambda\mathscr{P}(a, b^1) + (1 - \lambda)\mathscr{P}(a, b^2) \subset \mathscr{P}(a, \lambda b^1 + (1-\lambda)b^2) = \mathscr{P}(a, b)$. As N^a is convex, $(b, y) = \lambda(b^1, w^1)+(1-\lambda)(b^2, y^2) \in N^a$, and hence $\mathscr{R}(a, b, y) = 0$.

Case $\alpha = \alpha_2$. The proof is similar to that of the case "$\alpha = \alpha_1$" in Proposition 10.16. Taking any $y \in \mathscr{P}(a, b)$, we must show that $\mathscr{R}(a, b, y) = 0$. By the concavity of the map $\mathscr{P}(a, \cdot)$, $y \in \mathscr{P}(a, b) = \mathscr{P}(a, \lambda b^1 + (1 - \lambda)b^2) \subset \lambda\mathscr{P}(a, b^1)+(1-\lambda)\mathscr{P}(a, b^2)$. Hence, there exist $y^1 \in \mathscr{P}(a, b^1)$ and $y^2 \in \mathscr{P}(a, b^2)$ such that $y = \lambda y^1 + (1 - \lambda)y^2$. Moreover, for each $i \in \{1, 2\}$, since $b^i \in M_0^{\alpha_2}(a)$, $\mathscr{R}\left(a, b^i, y^i\right) = 0$, i.e., $(b^i, y^i) \in N^a$. As N^a is convex, $(b, y) = \lambda(b^1, y^1) + (1 - \lambda)(b^2, y^2) \in N^a$, and hence $\mathscr{R}(a, b, y) = 0$. □

Next we state some results for the problem (IP$_{r\alpha}$). We define $P_{r\alpha} : B \rightrightarrows A$ by

$$P_{r\alpha}(b) := [A \setminus \mathscr{D}_2^{-1}(b)] \cup [\text{Fix}\mathscr{D}_1 \cap \{a \in A \mid \alpha(y, \mathscr{P}(a, b)),$$
$$r(F(y, b, x), G(y, a))\}] \text{ for all } b \in B.$$

Corollary 10.18 *Assume that A is a compact convex subset of a topological vector space, B is a nonempty set, $P_{r\alpha}$ is generalized KKM and $P_{r\alpha}(b)$ is closed for each $b \in B$. Then,* (IP$_{r\alpha}$) *has solutions.*

Proof Corollary 10.18 is a particular case of Corollary 10.10 with the relation $\mathscr{R}$ defined by (10.2). □

For each $b \in B$, $P_{r\alpha}(b)$ is closed if Fix$\mathscr{D}_1$ is closed, $\mathscr{D}_2^{-1}(b)$ is open and

$$M_{r\alpha}(b) = \{a \in A \mid \alpha(y, \mathscr{P}(a, b)),\ r(F(y, b, a), G(y, a))\}$$

is closed.

Proposition 10.19 *Let A, B, Y, Z be subsets of vector spaces. Let $F : Y \times B \times A \rightrightarrows Z$, $G : Y \times A \rightrightarrows Z$ and $\mathscr{P} : A \times B \rightrightarrows Y$ be set-valued maps. Then, for each $b \in B$, $M_{r\alpha}(b)$ is closed if*

- $r = r_1$, $\alpha = \alpha_1$: *$\mathscr{P}(., b)$ and $F(., b, .)$ are lower semicontinuous and G is closed;*
- $r = r_1$, $\alpha = \alpha_2$: *$\mathscr{P}(., b)$ is upper semicontinuous and compact-valued, $F(., b, .)$ is lower semicontinuous and G is closed;*
- $r = r_2$, $\alpha = \alpha_1$: *$\mathscr{P}(., b)$ is lower semicontinuous, $F(., b, .)$ is upper semicontinuous and compact-valued and G is closed;*
- $r = r_2$, $\alpha = \alpha_2$: *$\mathscr{P}(., b)$ and $F(., b, .)$ are upper semicontinuous and compact-valued and G is closed.*

Proof We prove for $r = r_1, \alpha = \alpha_1$; the other cases can be proved similarly. We verify that the following set is closed:

$$M_{r_1\alpha_1}(b) := \{a \in A \mid \forall y \in \mathscr{P}(a, b), F(y, b, a) \subseteq G(y, a)\}.$$

Let $a_\gamma \in M_{r_1\alpha_1}(b)$ be such that $a_\gamma \to a_0$. By the assumed lower semicontinuity, $\forall y \in \mathscr{P}(a_0, b), \forall z \in F(y, b, a_0), \exists y_\gamma \in \mathscr{P}(a_\gamma, b) : y_\gamma \to y, \exists z_\gamma \in F(y_\gamma, b, a_\gamma) \subseteq G(y_\gamma, a_\gamma)$ such that $z_\gamma \to z$. Since G is closed, $z \in G(y, x_0)$. Hence, for all $y \in \mathscr{P}(a_0, b)$, $F(y, b, a_0) \subseteq G(y, a_0)$, i.e., $a_0 \in M_{r_1\alpha_1}(b)$. □

Corollary 10.20 *For* ($\mathrm{IP}_{r\alpha}$), *assume that A is a compact convex subset of a locally convex topological vector space, B is a convex subset of a topological vector space, and the following conditions are satisfied*

(i) *$\mathscr{D}_1$ is a closed map with convex values;*
(ii) *$\mathscr{D}_2$ is lower semicontinuous with convex values;*
(iii) *for each $a \in A$ and $b \in \mathscr{D}_2(a)$, there exists $a' \in \mathscr{D}_1(a)$ such that*

$$\alpha(y, \mathscr{P}(a', b)),\ \ r(F(y, b, a'), G(y, a'));$$

(iv) *for each $b \in B$, $M_1^{r\alpha}(b) := \{a \in A \mid \alpha(y, \mathscr{P}(a, b)),\ r(F(y, b, a), G(y, a))\}$ is convex;*
(v) *for each $a \in A$, $M_0^{r\alpha}(b) := \{b \in B \mid \bar{\alpha}(y, \mathscr{P}(a, b)),\ \bar{r}(F(y, b, a), G(y, a))\}$ is convex;*
(vi) *$M_{r\alpha} := \{(a, b) \in A \times B \mid \alpha(y, \mathscr{P}(a, b)),\ r(F(y, b, a), G(y, a))\}$ is closed.*

Then, ($\mathrm{IP}_{r\alpha}$) *has solutions.*

Proposition 10.21 *Let A, B, Y, Z be subsets of vector spaces. Let $F : Y \times B \times A \rightrightarrows Z$, $G : Y \times A \rightrightarrows Z$ and $\mathscr{P} : A \times B \rightrightarrows Y$ be set-valued maps. Then, the set*

$$M_{r\alpha} := \{(a, b) \in A \times B \mid \alpha(y, \mathscr{P}(a, b)),\ r(F(y, b, a), G(y, a))\}$$

is closed if, G is closed and

$r = r_1, \alpha = \alpha_1$: $\mathscr{P}$ and F are lower semicontinuous;
$r = r_1, \alpha = \alpha_2$: $\mathscr{P}$ is upper semicontinuous with compact values, F is lower semicontinuous;
$r = r_2, \alpha = \alpha_1$: $\mathscr{P}$ is lower semicontinuous, $\mathscr{P}$ is upper semicontinuous with compact values;
$r = r_2, \alpha = \alpha_2$: $\mathscr{P}$ and F are upper semicontinuous with compact values.

Proof We prove for $r = r_1, \alpha = \alpha_1$; the other cases can be proved similarly. By Proposition 10.14, it is sufficient to prove that the set

$$N := \{(a, b, y) \in A \times B \times Y \mid F(y, b, a) \subset G(y, a)\}$$

is closed. Let $\{(a^\nu, b^\nu, y^\nu)\}_\nu \subset N$ be a net converging to (a, b, y). We show that $F(y, b, a) \subset G(y, a)$. Indeed, take any $z \in F(y, b, a)$. By the lower semicontinuity of F, there exists a net $\{z^\nu\}_\nu$ with $z^\nu \in F(y^\nu, b^\nu, a^\nu) \subset G(y^\nu, a^\nu,)$ converging to z. Since G is closed, $z \in G(y, a)$. □

Proposition 10.22 *Let A, B, Y, Z be convex subsets of vector spaces. Let $F : Y \times B \times A \rightrightarrows Z$, $G : Y \times A \rightrightarrows Z$ and $\mathscr{P} : A \times B \rightrightarrows Y$ be set-valued maps. Then, for $b \in B$, the set*

$$M_1^{r\alpha}(b) := \{a \in A \mid \alpha(y, \mathscr{P}(a, b)),\ r(F(y, b, a), G(y, b))\}$$

is convex if, G is convex and

$r = r_1, \alpha = \alpha_1$: $\mathscr{P}(\cdot, a)$ and $F\cdot, b, \cdot)$ are concave;
$r = r_1, \alpha = \alpha_2$: $\mathscr{P}(\cdot, b)$ is convex and $F\cdot, b, \cdot)$ is concave;
$r = r_2, \alpha = \alpha_1$: $\mathscr{P}(\cdot, b)$ is concave and $F\cdot, b, \cdot)$ is convex;
$r = r_2, \alpha = \alpha_2$: $\mathscr{P}(\cdot, b)$ and $F\cdot, b, \cdot)$ are convex.

Proof We prove for $r = r_1, \alpha = \alpha_1$; the other cases can be proved similarly. By Proposition 10.16, it is sufficient to prove that the set

$$N_b := \{(a, y) \in A \times Y \mid F(y, b, a) \subset G(y, a)\}$$

is convex. Let $(a^1, y^1), (a^2, y^2) \in N_b$ and $\lambda \in [0, 1]$. By the concavity $F\cdot, b, \cdot)$ and the convexity of G, we have: $F(\lambda y^1 + (1 - \lambda)y^2, b, \lambda a^1 + (1 - \lambda)a^2) \subset \lambda F(y^1, b, a^1) + (1 - \lambda)F(y^2, b, a^2) \subset \lambda G(y^1, a^1) + (1 - \lambda)G(y^2, a^2) \subset G(\lambda y^1 + (1 - \lambda)y^2, \lambda a^1 + (1 - \lambda)a^2)$. Thus, $\lambda(a^1, y^1) + (1 - \lambda)(a^2, y^2) \in N_b$. □

Proposition 10.23 *Let A, B, Y, Z be convex subsets of vector spaces. Let $F : Y \times B \times A \rightrightarrows Z$, $G : Y \times A \rightrightarrows Z$ and $\mathscr{P} : A \times B \rightrightarrows Y$ be set-valued maps. Then, for $a \in A$, the set*

$$M_0^{r\alpha}(a) := \{b \in B \mid \overline{\alpha}(y, \mathscr{P}(a, b)),\ \bar{r}(F(y, b, a), G(y, a))\}$$

is convex if, $G(\cdot, a)$ is concave and

$r = r_1, \alpha = \alpha_1$: $\mathscr{P}(a, \cdot)$ and $F(\cdot, \cdot, a)$ are convex;
$r = r_1, \alpha = \alpha_2$: $\mathscr{P}(a, \cdot)$ is concave and $F(\cdot, \cdot, a)$ is convex;
$r = r_2, \alpha = \alpha_1$: $\mathscr{P}(a, \cdot)$ is convex and $F(\cdot, \cdot, a)$ is concave;
$r = r_2, \alpha = \alpha_2$: $\mathscr{P}(a, \cdot)$ and $F(\cdot, \cdot, a)$ are concave.

Proof We prove for $r = r_1, \alpha = \alpha_1$; the other cases can be proved similarly. By Proposition 10.17, it is sufficient to prove that the set

$$N^a := \{(b, y) \in B \times Y \mid F(y, b, a) \not\subset G(y, a)\}$$

is convex. Let $(b^1, y^1), (b^2, y^2) \in N^a$ and $\lambda \in [0, 1]$. By the convexity $F\cdot, \cdot, a)$ and the concavity of $G(\cdot, a)$, we have: $F(\lambda y^1 + (1-\lambda)y^2, \lambda b^1 + (1-\lambda)b^2, a) \supset \lambda F(y^1, b^1, a) + (1-\lambda)F(y^2, b^2, a)$ and $\lambda G(y^1, a) + (1-\lambda)G(y^2, a) \supset G(\lambda y^1 + (1-\lambda)y^2, a)$. Since $\lambda F(y^1, b^1, a) + (1-\lambda)F(y^2, b^2, a) \not\subset \lambda G(y^1, a) + (1-\lambda)G(y^2, a)$, we have $F(\lambda y^1 + (1-\lambda)y^2, \lambda b^1 + (1-\lambda)b^2, a) \not\subset G(\lambda y^1 + (1-\lambda)y^2, a)$. Thus, $\lambda(b^1, y^1) + (1-\lambda)(b^2, y^2) \in N^a$. □

10.3 Stability Conditions

10.3.1 *About Semicontinuity of Set-Valued Maps*

Let X and Y be topological spaces. For $x \in X$, let $\mathcal{N}(x)$ stand for the set of neighborhoods of x. The basic semicontinuity concepts for $G : X \rightrightarrows Y$ are the following. G is called inner semicontinuous (isc in short) at $\bar{x}$ if $\text{liminf}_{x\to\bar{x}} G(x) \supset G(\bar{x})$, and outer semicontinuous (osc) at $\bar{x}$ if $\text{limsup}_{x\to\bar{x}} G(x) \subset G(\bar{x})$. Here liminf and limsup are the Painlevé-Kuratowski inferior and superior limits in terms of nets:

$$\text{liminf}_{x\to\bar{x}} G(x) := \{y \in Y : \forall x^\nu \to \bar{x}, \exists y^\nu \in G(x^\nu), y^\nu \to y\},$$

$$\text{limsup}_{x\to\bar{x}} G(x) := \{y \in Y : \exists x^\nu \to \bar{x}, \exists y^\nu \in G(x^\nu), y^\nu \to y\}.$$

Equivalently, G is isc at $\bar{x}$ if

$$\forall x^\nu \to \bar{x}, \forall \bar{y} \in G(\bar{x}), \exists y^\nu \in G(x^\nu), y^\nu \to \bar{y}.$$

If G is both outer and inner semicontinuous at $\bar{x}$, we say that G is continuous at this point. Close to the outer and the inner semicontinuity are the (Berge) upper and lower semicontinuity: The lower semicontinuity coincides with inner semicontinuity, while the upper semicontinuity differs from the outer semicontinuity, though the two are close to each other. G is called closed at $\bar{x}$ if for each net $(x^\nu, y^\nu) \in \text{gr}G := \{(x, y) : z \in G(x)\}$ with $(x^\nu, y^\nu) \to (\bar{x}, \bar{y})$, $\bar{y}$ must belong to $G(\bar{x})$. We say that G satisfies a certain property on $A \subset X$ if G satisfies it at every point of A. If $A = X$, we omit "in X." Observe that G is closed if and only if its graph is closed.

Next, we use the following notions to derive properties related to inner and outer openness of the set-valued map G.

$$\text{liminfo}_{x\to\bar{x}} G(x) := \{y \in Y : \exists U \in \mathcal{N}(\bar{x}), \exists V \in \mathcal{N}(y), \forall x \in U, V \subset G(x)\};$$

$$\text{limsupo}_{x\to\bar{x}} G(x) := \{y \in Y : \exists V \in \mathcal{N}(y), \exists x^\nu \to \bar{x}, \forall \nu, V \subset G(x^\nu)\}.$$

Notice further that the inferior and superior open limits were defined as follows (we add "st." and "w." in the notations to avoid confusion and write only st.limsup, by similarity):

$$\text{st.limsupo}_{x\to\bar{x}}G(x) := \{y \in Y : \exists V \in \mathcal{N}(y), \exists x^\nu \to \bar{x} : x^\nu \neq \bar{x}, \forall \nu, V \subset G(x^\nu)\}.$$

However, as more frequently met in the literature, we allow x to take the value $\bar{x}$ in this chapter.

By definition, the complement map G^c is given by

$$G^c(\lambda) = [G(\lambda)]^c = X\backslash G(\lambda),$$

the following result provides relationships between the above set limits.

Proposition 10.24 *The following relations hold.*

(a) $\text{liminfo}_{\lambda\to\lambda_0} G(\lambda) \subset \liminf_{\lambda\to\lambda_0} G(\lambda) \subset \limsup_{\lambda\to\lambda_0} G(\lambda)$*;*
(b) $\text{liminfo}_{\lambda\to\lambda_0} G(\lambda) \subset \text{limsupo}_{\lambda\to\lambda_0} G(\lambda) \subset \limsup_{\lambda\to\lambda_0} G(\lambda)$*;*
(c) $\text{liminfo}_{\lambda\to\lambda_0} G^c(\lambda) = \left[\limsup_{\lambda\to\lambda_0} G(\lambda)\right]^c$*;*
(d) $\liminf_{\lambda\to\lambda_0} G^c(\lambda) = \left[\text{limsupo}_{\lambda\to\lambda_0} G(\lambda)\right]^c$*.*

Proof The first and the second statements follow directly from the definition.

To prove statement (c), let x be an arbitrary element of the set limit on the left hand side, and U, V be two neighborhoods of λ_0 and x, respectively, defined as in the definition of the inferior open limit. Then, the intersection set $V \cap G(\lambda)$ is nonempty for all $\lambda \in U$, and hence we conclude that the vector x does not belong to the set $\limsup_{\lambda\to\lambda_0} G(\lambda)$. Conversely, let x be an element outside of the open inferior limit of $G^c(\lambda)$. Then, by the definition, for every neighborhoods U of λ_0 and V of x, there exists $\lambda_{U,V} \in U$ such that $V \not\subset G^c\left(\lambda_{U,V}\right)$. Therefore, there exists some vector $x_{U,V}$ belonging to the intersection $V \cap G\left(\lambda_{U,V}\right)$. Next we choose U and V from a basis of neighborhoods of λ_0 and x, respectively, so that the nets $\lambda_{U,V}$ and $x_{U,V}$ converge, respectively, to λ_0 and x. Then, the vector x belongs to the set $\limsup_{\lambda\to\lambda_0} G(\lambda)$. So, statement (c) is verified.

Finally we prove (d). Let x be an arbitrary element of the limit in the left hand side. Then, for any net λ^ν converging to λ_0, there exists some element $x^\nu \notin G\left(\lambda^\nu\right)$ such that the net of these elements x^ν converges to the vector x. Consequently, for every neighborhood V of x, we conclude that $V \not\subset G\left(\lambda^\nu\right)$ for all λ^ν sufficiently close to the element λ_0. Therefore, x does not belong to $\text{limsupo}_{\lambda\to\lambda_0}G(\lambda)$. For the converse, if x is outside of the set on the left-hand side, then there exists a net λ^ν converging to λ_0 and a neighborhood V of x such that $V \cap G^c\left(\lambda_\upsilon\right) = \emptyset$. Thus, $V \subset G\left(\lambda^\nu\right)$ for all ν, which implies that the vector x does not belong to the set on the right-hand side. The proof is complete. □

Remark 10.25 Observe that the following relations hold:

$$\text{limsupo}_{x\to\bar{x}}G(x) = \text{st.limsupo}_{x\to\bar{x}}G(x) \cup \text{int}G(\bar{x}),$$

$$\text{liminfo}_{x\to\bar{x}}G(x) = \text{w.liminfo}_{x\to\bar{x}}G(x) \cap \text{int}G(\bar{x}).$$

However, in the sequel, we will not use the semi-limits on the right-hand side of these relations. Here and later int$A$, cl$A$ and bd$A$ stand for the interior, closure and boundary of A, respectively.

A set-valued map G is called inner open (outer open) at $\bar{x} \in X$ if

$$\text{liminfo}_{x\to\bar{x}}\, G(x) \supset G(\bar{x})$$

$$(\text{limsupo}_{x\to\bar{x}}\, G(x) \subset G(\bar{x}),\ \text{resp}).$$

These concepts help one link semicontinuity of G and its complement G^c ($G^c(x) := Y \setminus G(x)$) and to characterize a map by its graph as follows.

Proposition 10.26 *The following assertions hold.*

(a) *G is outer open at λ_0 if and only if G^c is inner semicontinuous at λ_0.*
(b) *G is outer semicontinuous at λ_0 if and only if G^c is inner open at λ_0.*
(c) *G is outer semicontinuous and closed-valued (respectively, inner open and open-valued) on Λ if and only if its graph is closed (respectively, open).*
(d) *If G is outer semicontinuous at λ_0, then it is outer open there.*
(e) *G is inner open at λ_0, then it is inner semicontinuous there.*

Proof The first and the second assertions follow from the statements (d) and (c) of Proposition 10.24, assertion (d) follows from inclusion (b) of Proposition 10.24, and assertion (e) is quite easy to obtain from the definition.

Finally, we verify assertion (c). If G is closed, then it is easy to see that G is outer-continuous. Furthermore, for any λ, the set $G(\lambda)$ is closed, because it is the intersection of the closed set $\{\lambda\} \times X$ with the graph of G.

For the converse, assume that G is outer-continuous and closed-valued. Let $\{(\lambda^\nu, x^\nu)\}$ be an arbitrary net of the graph of G converging to $\left(\bar{\lambda}, \bar{x}\right)$. Then, one can find a subnet, that we also denote by the same symbol, such that either all λ^ν coincide with $\bar{\lambda}$, which yield $x^\nu \in G\left(\bar{\lambda}\right)$ for all ν, or $\lambda^\nu \neq \bar{\lambda}$. In the first case, we have $\bar{x} \in G\left(\bar{\lambda}\right)$ as the latter set is closed. In the second case, $\bar{x}$ belongs to the superior limit of $G\left(\lambda^\nu\right)$, which is included in $G\left(\lambda_0\right)$ by the outer-continuity of G. Hence, the element $\left(\bar{\lambda}, \bar{x}\right)$ belongs to the graph of G, and so G is closed.

For the second statement of the assertion (c), the case of open maps, it suffices to notice here that a map is open if and only if its complement is closed. □

We propose the following new definitions of semi-limits of set-valued maps

$$\text{liminf}^*_{x\to\bar{x}} G(x) := \{y \in Y : \exists U \in \mathcal{N}(\bar{x}), \forall x \in U, y \in G(x)\},$$

$$\text{limsup}^*_{x\to\bar{x}} G(x) := \{y \in Y : \exists x^\nu \to \bar{x}, \forall \nu, y \in G(x^\nu)\}.$$

It is known that the (Painlevé-Kuratowski) liminf and limsup of a map are always closed sets and that liminfo and limsupo of a map are always open. However, many examples in the remaining part of this section show that the above two new semi-limits may be neither open nor closed. The following relations show that the introduction of the two new semi-limits can be helpful.

Proposition 10.27 *For $G : X \rightrightarrows Y$, the following assertions hold:*

(a) $\text{limsupo}_{x\to\bar{x}} G(x) \subset \text{limsup}^*_{x\to\bar{x}} G(x) \subset \text{limsup}_{x\to\bar{x}} G(x)$;
(b) $\text{liminfo}_{x\to\bar{x}} G(x) \subset \text{liminf}^*_{x\to\bar{x}} G(x) \subset \text{liminf}_{x\to\bar{x}} G(x) \subset \text{cl}G(\bar{x})$;
(c) $\text{liminf}^*_{x\to\bar{x}} G(x) = [\text{limsup}^*_{x\to\bar{x}} G^c(x)]^c$;
(d) $G(\bar{x}) \subset \text{limsup}^*_{x\to\bar{x}} G(x)$ *and* $\text{cl}G(\bar{x}) \subset \text{limsup}_{x\to\bar{x}} G(x)$;
(e) $\text{liminf}^*_{x\to\bar{x}} G(x) \subset G(\bar{x})$.

Proof Statements (a) and (b) are derived from Proposition 10.24(a). Statements (c) and (d) are consequences of Proposition 10.24(b). Statement (d) is obtained directly from Proposition 10.24(c).

For (c), let $y \in \text{liminf}^*_{x\to\bar{x}} G(x)$. Suppose $y \in \text{limsup}^*_{x\to\bar{x}} G^c(x)$. There is a net $\{x^\nu\} \subset X$ converging to $\bar{x}$ such that $y \in G^c(x^\nu)$ for all ν. Since $y \in \text{liminf}^*_{x\to\bar{x}} G(x)$,

$$\exists U \in \mathcal{N}(\bar{x}), \forall x \in U, y \in G(x).$$

As $x^\nu \to \bar{x}$, there exists ν_0 such that $x^{\nu_0} \in U$, which implies that $y \in G(x^{\nu_0})$, contradicting the fact that $y \in G^c(x^\nu)$ for all ν. Hence,

$$\text{liminf}^*_{x\to\bar{x}} G(x) \subset [\text{limsup}^*_{x\to\bar{x}} G^c(x)]^c.$$

Conversely, suppose $y \in [\text{limsup}^*_{x\to\bar{x}} G^c(x)]^c$ but $y \notin \text{liminf}^*_{x\to\bar{x}} G(x)$. Then,

$$\forall U^\nu \in \mathcal{N}(\bar{x}), \exists x^\nu \in U^\nu, y \notin G(x^\nu).$$

Therefore, there is a net $\{x^\nu\} \subset X$ converging to $\bar{x}$ such that $y \in G^c(x^\nu)$ for all ν, which implies that

$$y \in \text{limsup}^*_{x\to\bar{x}} G^c(x).$$

This contradiction yields (c), since $\text{liminf}^*_{x\to\bar{x}} G(x) \supset [\text{limsup}^*_{x\to\bar{x}} G^c(x)]^c$. □

Correspondingly, we propose the following new kinds of semicontinuity.

Definition 10.28

(a) G is said to be STAR-OUTER SEMICONTINUOUS (star-osc) at $\bar{x} \in X$ if

$$\text{limsup}^*_{x\to\bar{x}} G(x) \subset G(\bar{x});$$

(b) G is said to be STAR-INNER SEMICONTINUOUS (star-isc) at $\bar{x} \in X$ if

$$\text{liminf}^*_{x\to\bar{x}} G(x) \supset G(\bar{x}).$$

It is known that G is osc at $\bar{x} \in X$ if and only if $\text{limsup}_{x\to\bar{x}} G(x) = G(\bar{x})$, and isc at $\bar{x} \in X$ if and only if $\text{liminf}_{x\to\bar{x}} G(x) = \text{cl}G(\bar{x})$. By Proposition 10.27(d) and (e), we have the first similar but different thing for the above new semicontinuity notions:

(a) G is star-osc at $\bar{x} \in X$ if and only if $\text{limsup}^*_{x\to\bar{x}} G(x) = G(\bar{x})$;
(b) G is star-isc at $\bar{x} \in X$ if and only if $\text{liminf}^*_{x\to\bar{x}} G(x) = G(\bar{x})$.

Now we prove relations between the mentioned kinds of semicontinuity.

Proposition 10.29 *The following assertions hold.*

(a) *If G is outer semicontinuous at $\bar{x}$, then G is star-outer semicontinuous at $\bar{x}$.*
(b) *If G is star-outer semicontinuity at $\bar{x}$, then G is outer open at $\bar{x}$.*
(c) *If G is star-inner semicontinuous at $\bar{x}$, then G is inner semicontinuous at $\bar{x}$.*
(d) *If G is inner open at $\bar{x}$, then G is star-inner semicontinuous at $\bar{x}$.*
(e) *If G is usc at $\bar{x}$, then G is star-outer semicontinuous at $\bar{x}$.*
(f) *G is star-inner semicontinuous if and only if G^c is star-outer semicontinuous.*

Proof Assertions (a) and (b) are derived from Proposition 10.27(a). Assertions (c) and (d) are consequences of Proposition 10.27(b). Assertion (f) is obtained directly from Proposition 10.27(c). For (e), suppose to the contrary the existence of $y \in \text{limsup}^*_{x\to\bar{x}} G(x)$ and $\{x^\nu\} \subset X$ converging to $\bar{x}$ such that $y \in G(x^\nu)$ for all ν, but $y \notin G(\bar{x})$. If U is a neighborhood of $G(\bar{x})$, then so is $U \setminus \{y\}$, as $y \notin G(\bar{x})$. Since G is usc at $\bar{x}$, there exists $V \in \mathcal{N}(\bar{x})$ such that $G(V) \subset U \setminus \{y\}$. There exists ν_0 such that $x^{\nu_0} \in V$. This implies that $G(x^{\nu_0}) \subset U \setminus \{y\}$, contradicting the fact that $y \in G(x^\nu)$ for all ν. □

Remark 10.30 We discuss the considered definitions of semicontinuity for the special case of $g(\cdot)$ being single-valued. All lower semicontinuity, upper semicontinuity, and continuity (in the sense of Berge) are equivalent to the usual continuity of a single-valued map. But, continuity in the sense of Rockafellar-Wets is weaker. Simply think of the real function $y = x^{-1}$ if $x \neq 0$ and $y(0) = 0$, which is both inner and outer semicontinuous at zero, but is discontinuous at zero. All these four definitions of semicontinuity have proved to be fundamental for set-valued maps.

However, in some cases, they are still not convenient in use. We explain this in simple examples.

Example 10.31 (With non-closed images, a "good" map may be neither usc nor osc) Let $G : \mathbb{R} \rightrightarrows \mathbb{R}$ be defined by $G(x) := (0, 2^x)$ for $x \in \mathbb{R}$. Then, at any point, G is neither usc nor osc, though it may even look smooth. In this case, G is outer open at each point.

Example 10.32 (with unbounded non-closed images, a "good" map may be neither usc nor osc) Let $G : \mathbb{R} \rightrightarrows \mathbb{R} \times \mathbb{R}$ be defined by $G(x) := \{(y, xy) \in \mathbb{R}^2 : y \in (0, +\infty)\}$ for $x \in \mathbb{R}$. Then, G is neither usc nor osc at any point. But, G is both outer open and star-osc at each point. Observe that if G is osc at $\bar{x}$, then $G(\bar{x})$ must be closed, which may be violated even when G has a constant open value (see also Example 10.35).

Unlike in the above examples, outer openness does not seem to describe well the behavior in the following.

Example 10.33 (With images having empty interior, a "bad" map may be outer open) Let $G : \mathbb{R} \rightrightarrows \mathbb{R} \times \mathbb{R}$ be defined by $G(x) := \{(y, 1) \in \mathbb{R}^2 : y \in \mathbb{R}\}$ for $x \neq 0$ and $G(0) = \{(y, 0) \in \mathbb{R}^2 : y \in \mathbb{R}\}$. Then, G is outer open at 0, but its behavior is "discontinuous" for our usual feeling. Observe that G is not star-osc at zero though this property is weaker than being osc.

To complete our discussion of the content of Remark 10.30, observe from the definition and Proposition 10.29 that any single-valued map is outer open and never inner open. The star-outer semicontinuity and star-inner semicontinuity notions are also not significant in this case, since the former is relatively too weak (weaker than the usual continuity) and the latter is too strong. Namely, a (single-valued) map, which is star-inner semicontinuous at a point, must be locally constant around it. Hence, these four notions are designed specially to insight local behavior of set-valued maps. Observe further that a complete "symmetry" of liminf* and limsup* given in Proposition 10.27(c) does not have counterparts neither for liminf and limsup nor for liminfo and limsupo.

Now we show that the reverse implications in assertions (a)-(e) of Proposition 10.29 do not hold in general indeed.

Example 10.34 (For (a) and (d), star-outer semicontinuity but not outer semicontinuity, and star-inner semicontinuity but not inner openness) Let $G(x) \equiv (-1, 0]$ for $x \in \mathbb{R}$. Then, G is star-outer semicontinuous at 0, since $\operatorname{limsupo}^*_{x\to 0} G(x) = (-1, 0] = G(0)$. But, G is not outer semicontinuous at 0, as $\operatorname{limsup}_{x\to 0} G(x) = [-1, 0]$. Furthermore, G is star-inner semicontinuous, since $\operatorname{liminf}^*_{x\to 0} G(x) = (-1, 0]$, but G is not inner open, because $\operatorname{liminfo}_{x\to 0} G(x) = (-1, 0)$.

Example 10.35 (For (b), outer openness but not star-outer semicontinuity) Let $G(x) = (-1, |x|)$ for $x \in \mathbb{R}$. Then, $\operatorname{limsupo}_{x\to 0} G(x) = (-1, 0) = G(0)$

and $\limsup^*_{x\to 0} G(x) = (-1, 0]$. Hence, at 0, G is outer open but not star-outer semicontinuous.

Example 10.36 (For (c) and (e), inner semicontinuity but not star-inner semicontinuity, and star-outer semicontinuity but not upper semicontinuity) Let $G(x) = \{(y, xy) \in \mathbb{R}^2 : y \in \mathbb{R}\}$ for all $x \in \mathbb{R}$. Then, G is inner semicontinuous at 0 as $\liminf_{x\to 0} G(x) = \{(y, 0) : y \in \mathbb{R}\} = G(0)$. But, G is not star-inner semicontinuous at 0, since $\liminf^*_{x\to 0} G(x) = \{(0, 0)\}$ does not contain $G(0)$. Furthermore, G is star-outer semicontinuous as $\limsup^*_{x\to 0} G(x) = G(0)$. G is not usc, because for an arbitrary neighborhood U of $G(0)$, one cannot find a neighborhood N of zero such that $G(N) \subset U$.

Next, we propose notions which are closely related to star-inner semicontinuity and star-outer semicontinuity. These notions will be used in the subsequent sections for studying semicontinuity properties of solution maps of our variational problems.

Definition 10.37 Let $G : X \rightrightarrows Y$ and $\theta \in Y$.

(a) G is said to have the θ-INCLUSION PROPERTY at $\bar{x}$ if, for any $x^\nu \to \bar{x}$,

$$[\theta \in G(x^\nu),\ \forall \nu] \Longrightarrow [\theta \in G(\bar{x})].$$

(b) G is said to have the θ-INCLUSION COMPLEMENT PROPERTY at x_0 if, for any $x^\nu \to \bar{x}$,

$$[\theta \in G(\bar{x})] \Longrightarrow [\exists \bar{\nu}, \theta \in G(x^{\bar{\nu}})].$$

To clarify above concepts, we consider some special cases. Let Y be a topological vector space, $C, U \subset Y$ with nonempty interior, C being closed, and $H : X \rightrightarrows Y$. Then, one can verify the following relations.

(a) For $G = H - (Y \setminus -\text{int}C)$, G^c has the 0-inclusion property (or G has the 0-inclusion complement property) at $\bar{x}$ if and only if H has the C-inclusion property at $\bar{x}$. At the same time, setting $G := H + \text{int}C$, G has the 0-inclusion property (or G^c has the 0-inclusion complement property) at $\bar{x}$ if and only if H has the strict C-inclusion property at $\bar{x}$ (by the mentioned definition).
(b) With $G = H - \text{int}U$, G^c has the 0-inclusion property (or G has the 0-inclusion complement property) at $\bar{x}$ if and only if H is U-lsc at $\bar{x}$. At the same time, setting $G := H - (Y \setminus \text{int}U)$, G has the 0-inclusion property (or G^c has the 0-inclusion complement property) at $\bar{x}$ if and only if H is U-usc at $\bar{x}$.

Concerning these inclusion properties, we have the following statement.

Proposition 10.38

(a) *G has the θ-inclusion property at $\bar{x}$ if and only if G^c has the θ-inclusion complement property at $\bar{x}$.*

(b) *The set* $\{x \in X : \theta \in G(x)\}$ *is closed if and only if* G *has the* θ*-inclusion property.*
(c) *The set* $\{x \in X : \theta \notin G(x)\}$ *is closed if and only if* G *has the* θ*-inclusion complement property.*
(d) G *is star-outer semicontinuous at* $\bar{x}$ *if and only if* G *has the* θ*-inclusion property at* $\bar{x}$ *for every* θ*.*
(e) G *is star-inner semicontinuous at* $\bar{x}$ *if and only if* G *has the* θ*-inclusion complement property at* $\bar{x}$ *for every* θ*.*

Proof Assertions (a)–(c) are obvious from definition. For (d), let $\{x^\nu\} \subset X$ converge to $\bar{x}$ so that $\theta \in G(x^\nu)$ for all ν. Then, $\theta \in \text{limsup}^*_{x\to x_0} G(x)$. The star-outer semicontinuity at $\bar{x}$ implies that $\text{limsup}^*_{x\to\bar{x}} G(x) \subset G(\bar{x})$. Hence, $\theta \in G(\bar{x})$. Conversely, if $\theta \in \limsup^*_{x\to\bar{x}} G(x)$, there exists $\{x^\nu\}$ converging to $\bar{x}$ such that $\theta \in G(x^\nu)$ for all ν. Since G has the θ-inclusion property at $\bar{x}$, $\theta \in G(\bar{x})$. Hence, $\text{limsup}^*_{x\to\bar{x}} G(x) \subset G(\bar{x})$, i.e., G is star-outer semicontinuous at $\bar{x}$. (e) is obvious from (f) of Proposition 10.29, and (a),(d), since one has the equivalence relations: G is star-inner semicontinuous at $\bar{x}$ $\Longleftrightarrow$ G^c is star-outer semicontinuous at $\bar{x}$ $\Longleftrightarrow$ G^c has the θ-inclusion property at $\bar{x}$ for every θ $\Longleftrightarrow$ G has the θ-inclusion complement property at $\bar{x}$ for every θ. □

The rest of this section is devoted to calculus rules of semi-limits and semicontinuity for intersections and unions of maps.

Proposition 10.39 *For* $F, G : X \rightrightarrows Y$*, the following containments and inclusions hold for* $\sharp$ *being any of "sup," "sup*," "supo," "inf," "inf*," "info."*

(a) $\lim\sharp_{x\to\bar{x}}(F \cap G)(x) \subset \lim\sharp_{x\to\bar{x}} F(x) \cap \lim\sharp_{x\to\bar{x}} G(x)$. *Moreover,*

$$\text{liminfo}_{x\to\bar{x}}(F \cap G)(x) = \text{liminfo}_{x\to\bar{x}} F(x) \cap \text{liminfo}_{x\to\bar{x}} G(x),$$

$$\text{liminf}^*_{x\to\bar{x}}(F \cap G)(x) = \text{liminf}^*_{x\to\bar{x}} F(x) \cap \text{liminf}^*_{x\to\bar{x}} G(x),$$

$$\text{liminf}_{x\to\bar{x}} F(x) \cap \text{liminfo}_{x\to\bar{x}} G(x) \subset \text{liminf}_{x\to\bar{x}}(F \cap G)(x).$$

(b) $\lim\sharp_{x\to\bar{x}}(F \cup G)(x) \supset \lim\sharp_{x\to\bar{x}} F(x) \cup \lim\sharp_{x\to\bar{x}} G(x)$. *Moreover,*

$$\text{limsup}_{x\to\bar{x}}(F \cup G)(x) = \text{limsup}_{x\to\bar{x}} F(x) \cup \text{limsup}_{x\to\bar{x}} G(x),$$

$$\text{limsup}^*_{x\to\bar{x}}(F \cup G)(x) = \text{limsup}^*_{x\to\bar{x}} F(x) \cup \text{limsup}^*_{x\to\bar{x}} G(x),$$

$$\text{limsupo}_{x\to\bar{x}}(F \cup G)(x) \subset \text{limsupo}_{x\to\bar{x}} F(x) \cup \text{limsup}_{x\to\bar{x}} G(x).$$

Proof Assertions (a) and (b) for the inferior limit and the superior limit are obvious. On the other hand, the assertions for the superior open limit and for the inferior open limit can be deduced directly from the definitions.

(a) Now let us verify the equality for the inferior open limit of the intersection. Obviously, the inclusion

$$\operatorname{liminfo}_{\lambda\to\bar\lambda} F(\lambda)\cap G(\lambda) \subset \operatorname{liminfo}_{\lambda\to\bar\lambda} F(\lambda)\cap \operatorname{liminfo}_{\lambda\to\bar\lambda} G(\lambda)$$

holds true. Let x be an arbitrary element of the set in the right-hand side. Then, there exist neighborhoods V_1 and V_2 of x and neighborhoods U_1 and U_2 of $\bar\lambda$ such that $F(\lambda)$ contains V_1 for all $\lambda \in U_1$ and $G(\lambda)$ contains V_2 for all $\lambda \in U_2$. By setting $V = V_1 \cap V_2$ and $U_1 \cap U_2$, we have $V \subset F(\lambda)\cap G(\lambda)$ for all $\lambda \in U$. Therefore, x belongs to the set in the left-hand side.
The proof of the inclusion

$$\operatorname{limsup}^*_{x\to\bar x}(F\cap G)(x) \subset \operatorname{limsup}^*_{x\to\bar x} F(x)\cap \operatorname{limsup}^*_{x\to\bar x} G(x)$$

is direct by checking the definition. For showing the equality

$$\operatorname{liminf}^*_{x\to\bar x}(F\cap G)(x) = \operatorname{liminf}^*_{x\to\bar x} F(x)\cap \operatorname{liminf}^*_{x\to\bar x} G(x),$$

first let y belong to the left-hand side, i.e., there exists a neighborhood U of $\bar x$ such that

$$y \in (F\cap G)(x) = F(x)\cap G(x) \text{ for all } x \in U.$$

Thus, y belongs to the right-hand side. Let y now be in the right-hand side. There are two neighborhoods U_1 and U_2 of $\bar x$ such that $y \in F(x)$ for all $x \in U_1$ and $y \in G(x)$ for all $x \in U_2$. Then, $y \in F(x)\cap G(x)$ for all $x \in U := U_1 \cap U_2$. Thus, y belongs to the left-hand side.
Passing to the inclusion

$$\operatorname{liminf}_{x\to\bar x} F(x)\cap \operatorname{liminfo}_{x\to\bar x} G(x) \subset \operatorname{liminf}_{x\to\bar x}(F\cap G)(x),$$

let y be in the left-hand side. For any net $x^\nu \to \bar x$, because $y \in \operatorname{liminf}_{x\to\bar x} F(x)$, there is $y^\nu \in F(x^\nu)$ such that $y^\nu \to y$. Since $y \in \operatorname{liminfo}_{x\to\bar x} G(x)$, there are $U \in \mathcal{N}(\bar x)$ and $V \in \mathcal{N}(y)$ such that $V \subset G(x)$ for all $x \in U$. Without loss of generality, we may assume that $(x^\nu, y^\nu) \in U \times V$ for all ν. This implies that $y^\nu \in F(x^\nu)\cap G(x^\nu)$ and converging to y. Thus y belongs to the right-hand side.

(b) We let x be an arbitrary element of the superior open limit of $F \cup G$ at $\bar\lambda$. It follows from the definition that there exist a neighborhood V of x and a net $\{\lambda^\nu\}$ converging to $\bar\lambda$ such that

$$V \subset F\left(\lambda^\nu\right) \cup G\left(\lambda^\nu\right) \quad \text{for all } \nu. \tag{10.10}$$

If x belongs to $\limsup_{\lambda\to\bar\lambda} G(\lambda)$, then we are done. If not, by Proposition 10.24(c), the vector x belongs to the set $\operatorname{liminfo}_{\lambda\to\bar\lambda} G^c(\lambda)$, and consequently there exist neighborhoods W of x and U of $\bar\lambda$ such that $W \subset G^c(\lambda)$

for all $\lambda \in U, \lambda \neq \bar{\lambda}$. This together with (10.10) implies that $V \cap W \subset F(\lambda^\nu)$ for all ν with $\lambda^\nu \neq \bar{\lambda}$. Equivalently, the vector x belongs to the set $\text{limsupo}_{\lambda \to \bar{\lambda}} F(\lambda)$. Finally, the last inclusion follows from statements (c) and (d) of Proposition 10.24.

Let us prove the equality

$$\text{limsup}^*_{x \to \bar{x}}(F \cup G)(x) = \text{limsup}^*_{x \to \bar{x}} F(x) \cup \text{limsup}^*_{x \to \bar{x}} G(x).$$

Let $y \in \text{limsup}^*_{x \to \bar{x}} F(x)$, i.e., there exists a net $\{x^\nu\}$ converging to $\bar{x}$ such that $y \in F(x^\nu)$ for all ν. Hence, $y \in (F \cup G)(x^\nu)$ for all ν. Thus, y belongs to the left-hand side. The case $y \in \text{limsup}^*_{x \to \bar{x}} G(x)$ is similar. Let now $y \in \text{limsup}^*_{x \to \bar{x}}(F \cup G)(x)$, i.e., there exists $\{x^\nu\}$ converging to $\bar{x}$ such that $y \in F(x^\nu) \cup G(\nu)$ for all ν. Therefore, there exists a subnet $\{x^{\nu_\beta}\}$ such that $y \in F(x^{\nu_\beta})$ for all β or $y \in G(x^{\nu_\beta})$ for all β. Then, $y \in \text{limsup}^*_{x \to \bar{x}} F(x)$ or $y \in \text{limsup}^*_{x \to \bar{x}} G(x)$. Thus, y belongs to the right-hand side. The inclusion

$$\text{liminf}^*_{x \to \bar{x}}(F \cup G)(x) \supset \text{liminf}^*_{x \to \bar{x}} F(x) \cup \text{liminf}^*_{x \to \bar{x}} G(x)$$

can also be verified by the definition.

Finally, we check the inclusion

$$\text{limsupo}_{x \to \bar{x}}(F \cup G)(x) \subset \text{limsupo}_{x \to \bar{x}} F(x) \cup \text{limsup}_{x \to \bar{x}} G(x).$$

If y belongs to the left-hand side, i.e, there exist $V \in \mathcal{N}(y)$ and a net $\{x^\nu\}$ converging to $\bar{x}$ such that $V \subset F(x^\nu) \cup G(x^\nu)$ for all ν. If y belongs to $\text{limsup}_{x \to \bar{x}} G(x)$, then we are done. If not, we can get that y belongs to $\text{liminfo}_{x \to \bar{x}} G^c(x)$, which means that there are neighborhoods W of y and U of $\bar{x}$ such that $W \subset G^c(x)$ for all $x \in U$. Since $V \subset F(x^\nu) \cup G(x^\nu)$ and $W \subset G^c(x^\nu)$ for all ν, then $V \cap W \subset F(x^\nu)$. Thus, $y \in \text{limsupo}_{x \to \bar{x}} F(x)$. □

The following three examples explain the limitations of several inclusions/equalities in Proposition 10.39.

Example 10.40 (The equality in Proposition 10.39(a) fails for $\sharp$ being "sup*") Let $F, G : \mathbb{R} \rightrightarrows \mathbb{R}$ be defined by

$$F(x) := \begin{cases} (-1, x) & \text{if} \quad x \geq 0, \\ (0, 1) & \text{if} \quad x < 0, \end{cases} \qquad G(x) := \begin{cases} (-1, 0) & \text{if} \quad x \geq 0, \\ (x, 1) & \text{if} \quad x < 0. \end{cases}$$

Then,

$$\text{limsup}^*_{x \to 0} F(x) = \text{limsup}^*_{x \to 0} G(x) = (-1, 1)$$

and

$$\text{limsup}^*_{x\to 0}(F\cap G)(x) = (-1,0)\cup(0,1).$$

Hence,

$$\text{limsup}^*_{x\to 0}(F\cap G)(x) \not\subset \text{limsup}^*_{x\to 0}F(x)\cap \text{limsup}^*_{x\to 0}G(x).$$

Example 10.41 (The equality in Proposition 10.39(b) fails for $\sharp$ being "inf*") Let $F, G : \mathbb{R} \rightrightarrows \mathbb{R}$ be defined by

$$F(x) := \begin{cases} [0,2] & \text{if} \quad x \geq 0, \\ [1,2] & \text{if} \quad x < 0, \end{cases} \qquad G(x) := \begin{cases} [1,2] & \text{if} \quad x \geq 0, \\ [0,2] & \text{if} \quad x < 0. \end{cases}$$

Then, $\text{liminf}^*_{x\to 0}F(x) = \text{liminf}^*_{x\to 0}G(x) = [1,2]$ and $\text{liminf}^*_{x\to 0}(F\cup G)(x) = [0,2]$. Hence,

$$\text{liminf}^*_{x\to 0}(F\cup G)(x) \not\supset \text{liminf}^*_{x\to 0}F(x)\cup \text{liminf}^*_{x\to 0}G(x).$$

Example 10.42 Related to Proposition 10.39(a), we show a case where

$$\text{liminf}_{x\to\bar{x}}F(x)\cap \text{lim}\sharp_{x\to\bar{x}}G(x) \not\subset \text{liminf}_{x\to\bar{x}}(F\cap G)(x)$$

for $\sharp$ being "inf*" or "inf." Let $F, G : \mathbb{R} \rightrightarrows \mathbb{R}$ be defined by $F(x) := (-\infty,-1]\cup[1-3^{-|x|},+\infty)$ for $x\in\mathbb{R}$ and

$$G(x) := \begin{cases} (-\infty, 1-2^{-|x|}]\cup[1,+\infty) & \text{if} \quad x \neq 0, \\ (-\infty, 0.5]\cup[1,+\infty) & \text{if} \quad x = 0. \end{cases}$$

We have

$$(F\cap G)(x) = \begin{cases} (-\infty,-1]\cup[1,+\infty) & \text{if} \quad x \neq 0, \\ (-\infty,-1]\cup[0,0.5]\cup[1,+\infty) & \text{if} \quad x = 0. \end{cases}$$

Then, $\text{liminf}_{x\to 0}F(x) = (-\infty,-1]\cup[0,+\infty)$ and $\text{liminf}_{x\to 0}G(x) = \text{liminf}^*_{x\to 0}G(x) = (-\infty,0]\cup[1,+\infty)$. Hence,

$$\text{liminf}_{x\to 0}F(x)\cap \text{liminf}_{x\to 0}G(x) = (-\infty,-1]\cup\{0\}\cup[1,+\infty),$$

$$\text{liminf}_{x\to 0}F(x)\cap \text{liminf}^*_{x\to 0}G(x) = (-\infty,-1]\cup\{0\}\cup[1,+\infty).$$

Since $\liminf_{x\to 0}(F\cap G)(x) = (-\infty, -1]\cup[1, +\infty)$, the mentioned inclusion does not hold for $\sharp$ being "inf*" or "inf" in this case.

Example 10.43 Related to Proposition 10.39(b), we show a case where

$$\text{limsupo}_{x\to\bar{x}}(F\cup G)(x) \not\subset \text{limsupo}_{x\to\bar{x}}F(x)\cup \text{lim}\sharp_{x\to\bar{x}}G(x)$$

for $\sharp$ being "limsupo" or "limsup*." Let $F, G : \mathbb{R} \to 2^{\mathbb{R}}$ be defined by $F(x) \equiv (-1, 0]$ and $G(x) \equiv (0, 1)$ for $x \in \mathbb{R}$. We have $(F \cup G)(x) = (-1, 1)$ for all $x \in \mathbb{R}$. Then, $\text{limsupo}_{x\to 0}F(x) = (-1, 0)$, $\text{limsupo}_{x\to 0}(F\cup G)(x) = (-1, 1)$, and $\text{limsupo}_{x\to 0}G(x) = \text{limsup}^*_{x\to 0}G(x) = (0, 1)$. Hence,

$$\text{limsupo}_{x\to\bar{x}}(F\cup G)(x) \not\subset \text{limsupo}_{x\to\bar{x}}F(x)\cup \text{lim}\sharp_{x\to\bar{x}}G(x)$$

for $\sharp$ being "limsupo" or "limsup*."

By applying Proposition 10.39(a), we obtain the following statements.

Proposition 10.44 *The following assertions hold.*

(a) *If F and G are outer semicontinuous, star-outer semicontinuous, outer open, inner open, or star-inner semicontinuous at $\bar{x}$, then so is their intersection.*
(b) *If F is inner semicontinuous and G is inner open at $\bar{x}$, then their intersection is inner semicontinuous at $\bar{x}$.*

Example 10.45 (Proposition 10.44(b) is no longer true if the inner openness of G is replaced by star-inner semicontinuity or inner semicontinuity) Let $F, G : \mathbb{R} \rightrightarrows \mathbb{R}$ be defined by

$$F(x) := (-\infty, -1] \cup [1 - 2^{-|x|}, +\infty),$$

and

$$G(x) := (-\infty, 0] \cup [1, +\infty),$$

for all $x \in \mathbb{R}$. We have

$$(F\cap G)(x) = \begin{cases} (-\infty, -1] \cup [1, +\infty), & \text{if} \quad x \neq 0, \\ (-\infty, -1] \cup \{0\} \cup [1, +\infty), & \text{if} \quad x = 0. \end{cases}$$

F is inner semicontinuous at 0 but $F \cap G$ is not, since

$$\liminf_{x\to 0}(F\cap G)(x) = (-\infty, -1] \cup [1, +\infty) \not\supset (F\cap G)(0).$$

The reason is that G is not inner open at 0 due to

$$\text{liminfo}_{x\to 0}G(x) = (-\infty, 0) \cup (1, +\infty) \not\supset G(0).$$

Observe that G is both star-inner semicontinuous and inner continuous at 0 by

$$\text{liminf}^*_{x\to 0}G(x) = \text{liminf}_{x\to 0}G(x) = G(0) = (-\infty, 0] \cup [1, +\infty).$$

From Proposition 10.39(b), we easily obtain the following statement.

Proposition 10.46 *The following assertions hold.*

(a) If F and G are outer semicontinuous, star-outer semicontinuous, inner open, inner semicontinuous, or star-inner semicontinuous at $\bar{x}$, then so is their union.
(b) If F is outer open and G is outer semicontinuous at $\bar{x}$, then their union is outer open at $\bar{x}$.

Example 10.47 (The outer openness in Proposition 10.46(b) does not hold if the outer semicontinuity of G is replaced by star-outer semicontinuity or outer openness) Let $F, G : \mathbb{R} \rightrightarrows \mathbb{R}$ be defined by $G(x) := (0, 1)$ for $x \in \mathbb{R}$ and

$$F(x) := \begin{cases} (-1, 0] & \text{if} \quad x \neq 0, \\ (-1, 0) & \text{if} \quad x = 0. \end{cases}$$

Then,

$$(F \cup G)(x) = \begin{cases} (-1, 1) & \text{if} \quad x \neq 0, \\ (-1, 0) \cup (0, 1) & \text{if} \quad x = 0. \end{cases}$$

Clearly F is outer open at 0 but $F \cup G$ is not, since

$$\text{limsupo}_{x\to 0}(F \cup G)(x) = \text{limsupo}^*_{x\to 0}(F \cup G)(x) = (-1, 1) \not\subset (F \cup G)(0).$$

The cause is that G is not outer semicontinuous at 0 (as $\text{limsup}_{x\to 0}G(x) = [0, 1] \not\subset G(0)$). However, in this case, G is both star-outer semicontinuous and outer open at 0 by

$$\text{limsupo}_{x\to 0}G(x) = \text{limsup}^*_{x\to 0}G(x) = G(0) = (0, 1).$$

Proposition 10.48 *The following assertions hold.*

(a) *If F is outer semicontinuous (resp., star-outer semicontinuous, outer open) at $\bar{x}$ and if*

$$\mathrm{limsup}_{x\to\bar{x}} G(x) \cap F(\bar{x}) \subset G(\bar{x})$$

$$(\text{resp, } \mathrm{limsup}^*_{x\to\bar{x}} G(x) \cap F(\bar{x}) \subset G(\bar{x}), \ \mathrm{limsupo}_{x\to\bar{x}} G(x) \cap F(\bar{x}) \subset G(\bar{x})),$$

then $F \cap G$ is outer semicontinuous (resp., star-outer semicontinuous, outer open) at $\bar{x}$.

(b) *If F is star-inner semicontinuous (resp., inner open) at $\bar{x}$ and if*

$$\mathrm{liminf}^*_{x\to\bar{x}} G(x) \supset G(\bar{x}) \cap F(\bar{x})$$

$$(\text{resp., } \mathrm{liminfo}_{x\to\bar{x}} G(x) \supset G(\bar{x}) \cap F(\bar{x})),$$

then $F \cap G$ is star-inner semicontinuous (resp., inner open) at $\bar{x}$.

(c) *If F is inner semicontinuous at $\bar{x}$ and if*

$$\mathrm{liminfo}_{x\to\bar{x}} G(x) \supset G(\bar{x}) \cap F(\bar{x}),$$

then $F \cap G$ is inner semicontinuous at $\bar{x}$.

Proof

(a) By Proposition 10.39(a), we have

$$\mathrm{limsup}_{x\to\bar{x}} (F \cap G)(\bar{x}) \subset \mathrm{limsup}_{x\to\bar{x}} F(\bar{x}) \cap \mathrm{limsup}_{x\to\bar{x}} G(\bar{x})$$

$$\subset F(\bar{x}) \cap \mathrm{limsup}_{x\to\bar{x}} G(\bar{x}) \subset F(\bar{x}) \cap G(\bar{x}),$$

where the second inclusion is due to the outer semicontinuity of F and the last one follows from the hypothesis on G. The proof for the star-outer semicontinuity and outer openness is similar.

(b) Also from Proposition 10.39(a), we have

$$\mathrm{liminf}^*_{x\to\bar{x}} (F \cap G)(\bar{x}) = \mathrm{liminf}^*_{x\to\bar{x}} F(\bar{x}) \cap \mathrm{liminf}^*_{x\to\bar{x}} G(\bar{x})$$

$$\supset F(\bar{x}) \cap \mathrm{liminf}^*_{x\to\bar{x}} G(\bar{x}) \supset F(\bar{x}) \cap G(\bar{x}),$$

where the second inclusion is obtained from the star-inner semicontinuity of F and the last one follows from the hypothesis on G. The proof for the inner openess is similar.

(c) Proposition 10.39(a) implies also that

$$\begin{aligned}\operatorname{liminf}_{x\to\bar{x}}(F\cap G)(\bar{x}) &\supset \operatorname{liminf}_{x\to\bar{x}}F(\bar{x})\cap \operatorname{liminfo}_{x\to\bar{x}}G(\bar{x})\\ &\supset F(\bar{x})\cap \operatorname{liminfo}_{x\to\bar{x}}G(\bar{x})\supset F(\bar{x})\cap G(\bar{x}),\end{aligned}$$

where the second inclusion is obtained from the inner semicontinuity of F and the last one from the hypothesis on G.

□

Example 10.49 Proposition 10.48(c) is no longer true if the inclusion

$$\operatorname{liminfo}_{x\to\bar{x}}G(x)\supset G(\bar{x})\cap F(\bar{x})$$

is replaced by

$$\lim\sharp_{x\to\bar{x}}G(x)\supset G(\bar{x})\cap F(\bar{x})$$

for $\sharp$ being "inf*" or "inf." Indeed, let $F, G : \mathbb{R} \rightrightarrows \mathbb{R}$ be defined by

$$F(x) := (-\infty, -1] \cup [1 - 2^{-|x|}, +\infty),$$

and

$$G(x) := (-\infty, 0] \cup [1, +\infty),$$

for $x \in \mathbb{R}$. We have

$$(F\cap G)(x)=\begin{cases}(-\infty,-1]\cup[1,+\infty) & \text{if } \quad x\neq 0,\\ (-\infty,-1]\cup\{0\}\cup[1,+\infty) & \text{if } \quad x=0.\end{cases}$$

Then, it is easy to see that F is inner semicontinuous at 0 but $F \cap G$ is not, since

$$\operatorname{liminf}_{x\to 0}(F\cap G)(x)=(-\infty,-1]\cup[1,+\infty)\not\supset(F\cap G)(0).$$

The cause is that

$$\operatorname{liminfo}_{x\to 0}G(x)=(-\infty,0)\cup(1,+\infty)\not\supset G(0)\cap F(0).$$

Although

$$\lim\sharp_{x\to 0}G(x)=(-\infty,0]\cup[1,+\infty)\supset G(0)\cap F(0),$$

for $\sharp$ being "inf*" or "inf."

10.3.2 *Stability Conditions for Variational Relation Problems*

Continuity Conditions for Variational Relation Problems

Throughout this section A, B and Y are nonempty sets, $\mathscr{D}_1 : A \rightrightarrows A$, $\mathscr{D}_2 : A \rightrightarrows B$ and $\mathscr{P} : A \times B \rightrightarrows Y$ are set-valued maps with nonempty values. Let $\mathscr{R}(a, b, y)$ be a relation linking elements $a \in A, b \in B$ and $y \in Y$. Then, $\mathscr{R}$ identifies a subset of the product space $A \times B \times Y$ in the sense that relation $\mathscr{R}(a, b, y)$ is satisfied if and only if the triple (a, b, y) belongs to that subset. In this way, a relation can be considered as equality/inequality of real functions, or as inclusion of set-valued maps of variables a, b and y. In this section, we consider a simple variational relation problem mentioned in the previous section:
(VRP) Find $\bar{a} \in A$ such that

(a) $\bar{a}$ is a fixed point of $\mathscr{D}_1$, that is, $\bar{a} \in \mathscr{D}_1(\bar{a})$;
(b) $\mathscr{R}(\bar{a}, b, y)$ holds for every $b \in \mathscr{D}_2(\bar{a})$ and $y \in \mathscr{P}(\bar{a}, b)$.

As it has already been mentioned in the previous section, weaker models (WVRPs) can be defined by requiring $\mathscr{R}(a, b, y)$ to be satisfied for all $b \in \mathscr{D}_2(a)$ and some $y \in \mathscr{P}(a, b)$, or for some $b \in \mathscr{D}_2(a)$ and $y \in \mathscr{P}(a, b)$. Because the methods and techiques of study of these models are similar to (VRP), our overall concern is going to be focused on (VRP) only. In the previous section, we have given a number of sufficient conditions for existence of solutions of (GVRP) and then the corresponding results for the problem (VRP) have also been provided, and hence the purpose of this section is only to investigate topological properties of the solution set of the problem (VRP) and its stability.

Throughout this section, the following notations are used.

- Sol(VRP) is the solution set of the problem (VRP).
- K is the set of all fixed points of the map $\mathscr{D}_1$.
- $\Gamma = \{a \in A \mid \mathscr{R}(a, b, y) \text{ holds for all } b \in \mathscr{D}_2(a) \text{ and } y \in \mathscr{P}(a, b)\}$.
- $\mathbb{P}_{\mathscr{R}}(b) = \{a \in A \mid \mathscr{R}(a, b, y) \text{ holds for all } y \in \mathscr{P}(a, b)\}$ for each $b \in B$.
- $\mathbb{P}(b) = A \setminus \mathscr{D}_2^{-1}(b) \bigcup \{K \cap \mathbb{P}_{\mathscr{R}}(b)\}$ for each $b \in B$.

For a subset D of A, its complement in A is denoted by D^c. Hence, for instance, K^c is the collection of all non-fixed points of $\mathscr{D}_1$, while Γ^c consists of those a of A for which $\mathscr{R}(a, b, y)$ does not hold for some $b \in \mathscr{D}_2(a)$ and $y \in \mathscr{P}(a, b)$. In our analysis, we also employ the following well-known formulas.

$$\text{Sol(VRP)} = K \cap \Gamma. \tag{10.11}$$

$$\text{Sol(VRP)} = \cap_{b \in B} \mathbb{P}(b). \tag{10.12}$$

Topological Properties of the Solution Set

In this section, we aim to find out under what kind of data the solution set of (VRP) has nice properties such as uniqueness, convexity, boundedness, and closedness. These properties are desired, because they are tightly related to convergence and efficiency of methods for computing solutions of the problem. As mentioned before, in this section, we target topological properties of the solution set and its stability only, and thus we are going to assume that the solution set of the problem (VRP) is nonempty in a neighborhood of a reference point. Let us begin with the uniqueness of the solution of the problem. We first recall here that a set-valued map F on a metric space (M, d) is said to be contractive if there is a positive number $\alpha < 1$ such that for every $u, v \in M$ one has

$$\max\left\{\sup_{x\in F(u)} d(x, F(v));\ \sup_{y\in S(v)} d(y, S(u))\right\} \leq \alpha d(u, v).$$

Proposition 10.50 *Problem* (VRP) *has at most one solution if either A is a metric space and $\mathscr{D}_1$ is a contraction or the following conditions hold*

(a) *the fixed point set K of $\mathscr{D}_1$ is contained in the image $\mathscr{D}_2(x)$ for all x being fixed points of $\mathscr{D}_1$;*
(b) *$\mathscr{R}$ is antisymmetric on K, that is, if $a, b \in K$ with $a \neq b$ and $\mathscr{R}(a, b, y)$ holds for all $y \in \mathscr{P}(a, b)$, then $\mathscr{R}(b, a, y)$ does not hold for some $y \in \mathscr{P}(b, a)$.*

Proof It is evident that if $\mathscr{D}_1$ is a contraction, then it has at most one fixed point, and therefore (VRP) cannot have more than one solution. Assume (a) and (b). Let a_1 and a_2 be two solutions of (VRP). In particular, both of them are fixed points of $\mathscr{D}_1$ and in view of (a), $a_1 \in \mathscr{D}_2(a_2)$ and $a_2 \in \mathscr{D}_2(a_1)$. This implies that $\mathscr{R}(a_1, a_2, y)$ holds for all $y \in \mathscr{P}(a_1, a_2)$, and $\mathscr{R}(a_2, a_1, y)$ holds for all $y \in \mathscr{P}(a_2, a_1)$. In view of (b), the solutions a_1 and a_2 must coincide. □

The antisymmetry of the relation $\mathscr{R}$ can be guaranteed, for instance, when $\mathscr{R}$ is determined by a strict inequality of a real-valued function or by a strict inclusion of a set-valued map. Therefore, let f be a real-valued function on $A \times B$, and let F be a set-valued map from $A \times B$ to some space Z. Assume that $\mathscr{R}(a, b, y)$ holds for $a \neq b$ if and only if $f(a) < f(b)$, or it holds for $a \neq b$ if and only if $F(a) \subset F(b)$ and $F(a) \neq F(b)$. Then, $\mathscr{R}$ is antisymmetric.

The following example of classical variational inequality problem gives a link between the antisymmetry of a relation and the strict monotonicity of an operator.

Example 10.51 Consider the Stampacchia variational inequality problem:

Find $\bar{a} \in A$ such that $\langle f(\bar{a}), x - \bar{a}\rangle \geqslant 0$ for all $x \in A$,

where A is a convex subset of a real Hilbert space H and f is an operator from H to itself. To express this problem in the form of (VRP), it suffices to set $B = A, Y = H, \mathscr{D}_1(x) = A, \mathscr{D}_2(x) = A$ and $\mathscr{P}(x, b) = \{b\}$ for all $x, b \in A$. The relation $\mathscr{R}$ is defined as follows:

$\mathscr{R}(x, b, y)$ holds if and only if $\langle f(x), y - x\rangle \geqslant 0$.

It is easy to see that the relation $\mathscr{R}$ is antisymmetric if the operator is strictly monotone in the sense that $\langle f(x) - f(y), x - y\rangle > 0$ when $x \neq y$.

Now we assume that A, B and Y are topological spaces. We say that $\mathscr{R}$ is closed if the set determining it is closed in the product space $A \times B \times Y$. This is often the case when $\mathscr{R}$ is defined by non-strict inequalities of real-valued continuous functions, or non-strict inclusions of set-valued maps. We say also that $\mathscr{R}$ is closed in the first variable if $\mathscr{R}(a, b, y)$ holds whenever $\mathscr{R}(a^\nu, b, y)$ holds for all ν with a^ν converging to a and with b and y being fixed. The next two propositions give conditions for the closedness of the solution set.

Proposition 10.52 *The solution set* Sol(VRP) *is closed provided that the following conditions hold:*

(a) *$\mathscr{D}_1$ is closed;*
(b) *$\mathscr{D}_2$ is inner semicontinuous;*
(c) *$\mathscr{P}$ is inner semicontinuous;*
(d) *$\mathscr{R}$ is closed.*

Proof Let $\{a^\nu\}$ be a net of solutions of (VRP) converging to some $a \in A$. We show that this limit is also a solution. Indeed, by (a), it belongs to the set K. For every $b \in \mathscr{D}_2(a)$ and $y \in \mathscr{P}(a, b)$, one can find $b^\nu \in \mathscr{D}_2(a^\nu)$ converging to b and $y^\nu \in \mathscr{P}(a^\nu, b^\nu)$ converging to y due to (b) and (c). We know that $\mathscr{R}(a^\nu, b^\nu, y^\nu)$ holds by assumptions. In view of (d), $\mathscr{R}(a, b, y)$ is satisfied. Hence, a belongs to Γ. By the relation (a) it is a solution of the problem. □

The closedness of the solution set can also be obtained by relaxing conditions (c) and (d), but strengthening condition (b).

Proposition 10.53 *The solution set* Sol(VRP) *is closed provided that the following conditions hold:*

(a) *$\mathscr{D}_1$ is closed;*
(b) *$\mathscr{D}_2$ has open inverse values;*
(c) *$\mathscr{P}$ is inner semicontinuous in the first variable;*
(d) *$\mathscr{R}$ is closed in the first and third variables.*

Proof As in the previous proof, let $\{a^\nu\}$ be a net of solutions of (VRP) converging to some $a \in X$. We show that this limit is a solution by using the second relation (10.12). Let b be any element of B. By (b), the set $A\backslash\mathscr{D}_2^{-1}(b)$ is closed, and by (a) the set K is closed. We prove now that $\mathbb{P}_{\mathscr{R}}(b)$ is closed. Indeed, let $\{a^\nu\}$ be a net of elements of $\mathbb{P}_{\mathscr{R}}(b)$ converging to to some $a \in A$. Let y be any element of $\mathscr{P}(a, b)$. By (c) there is $y^\nu \in \mathscr{P}(a^\nu, b)$ converging to y. Since $\mathscr{R}(a^\nu, b, y^\nu)$ are satisfied, in view of (d), $\mathscr{R}(a, b, y)$ is satisfied too. By this, a belongs to $\mathbb{P}_{\mathscr{R}}(b)$, and so $\mathbb{P}_{\mathscr{R}}(b)$ is closed. We conclude that $\mathbb{P}(b)$ is closed for every $b \in B$ and hence Sol(VRP) is closed. □

In order to study the convexity of the solution set and to make the presentation more readable, we recall some convexity concepts of set-valued maps.

Assume that A, B and Y are convex subsets of linear spaces.

- A set-valued map $\mathscr{D} : A \rightrightarrows Y$ is said to be convex (by inclusion) if for any $x_1, x_2 \in A$ and $t \in [0, 1]$ one has

$$tS(x_1) + (1 - t)S(x_2) \subset S(tx_1 + (1 - tx_2)).$$

  When inclusion " $\subset$" is replaced by inclusion " $\supset$," the map $\mathscr{D}$ is said to be concave.
- The relation $\mathscr{R}$ is said to be convex if whenever $\mathscr{R}(a_i, b_i, y_i)$ holds for $a_i \in A$, $b_i \in B$ and $y_i \in Y$, $i = 1, 2$, the relation

$$\mathscr{R}(ta_1 + (1 - t)a_2, tb_1 + (1 - t)b_2, ty_1 + (1 - t)y_2)$$

  is satisfied for all $t \in [0, 1]$. In other words, $\mathscr{R}$ is convex if the set determining it is convex in the product space $A \times B \times Y$.

In many cases, the maps $\mathscr{D}_1$ and $\mathscr{D}_2$ are constant with convex values, so that they are concave and convex at the same time. Moreover, if $\mathscr{R}$ is defined by the inequality $f(a, b, y) \leq 0$ where f is a real-valued function on $A \times B \times Y$ and A, B and Y are convex sets, then $\mathscr{R}$ is convex if and only if the function f is convex.

Proposition 10.54 *Assume that A, B and Y are convex subsets of linear spaces. Then the solution set* Sol(VRP) *is convex provided that the constraint map $\mathscr{D}_1$ is convex, the maps $\mathscr{D}_2$ and $\mathscr{P}$ are concave and the relation $\mathscr{R}$ is convex.*

Proof It follows from the convexity of $\mathscr{D}_1$ that K is convex. We show that Γ is convex. Indeed, let a_1 and a_2 be two elements of Γ. For each $t \in [0, 1]$ and $b \in \mathscr{D}_2(ta_1 + (1 - t)a_2)$, there exist $b_1 \in \mathscr{D}_2(a_1)$ and $b_2 \in \mathscr{D}_2(a_2)$ such that $b = tb_1 + (1 - t)b_2$. Let y be any element of $\mathscr{P}(ta_1 + (1 - t)a_2, tb_1 + (1 - t)b_2)$. Since $\mathscr{P}$ is concave, there exist $y_1 \in \mathscr{P}(a_1, b_1)$ and $y_2 \in \mathscr{P}(a_2, b_2)$ such that $y = ty_1 + (1 - t)y_2$. Then, $\mathscr{R}(a_i, b_i, y_i)$, $i = 1, 2$, hold. The convexity of $\mathscr{R}$ implies that

$$\mathscr{R}(ta_1 + (1 - t)a_2, tb_1 + (1 - t)b_2, ty_1 + (1 - t)y_2)$$

is satisfied. Hence, $ta_1 + (1 - t)a_2$ belongs to Γ. According to the relation (10.11), we conclude that Sol(VRP) is convex. □

When the image of every element $a \in A$ under $\mathscr{D}_2$ is the whole set B, the convexity requirement on $\mathscr{P}$ and $\mathscr{R}$ can be weakened.

Proposition 10.55 *Assume that A, B and Y are convex subsets of linear spaces. Then each of the following conditions is sufficient for the solution set* Sol(VRP) *to be convex:*

(a) *$\mathscr{D}_1$ is convex, $\mathscr{D}_2$ is constant with value B, $\mathscr{P}$ is convex in the first variable and $\mathscr{R}$ is convex in the first and the third variables;*
(b) *$\mathscr{D}_1$ is convex, $\mathscr{D}_2$ is constant with value B, $\mathscr{P}$ is constant with value Y and $\mathscr{R}$ is convex in the first variable.*

Proof Apply the same argument as in the proof of the preceding proposition with the use of (10.12) instead of (10.11). □

Example 10.56 Let A and f be as in Example 10.51. Consider the Minty variational inequality problem:

Find $\bar{a} \in A$ such that $\langle f(x), x - \bar{a}\rangle \geq 0$ for all $x \in A$.

To express this problem in the form of (VRP), it suffices to set

$B = A, Y = H, \mathscr{D}_1(x) = A, \mathscr{D}_2(x) = A$ and $\mathscr{P}(x, b) = A$ for all $x, b \in A$.

The relation $\mathscr{R}$ is defined as follows:

$\mathscr{R}(x, b, y)$ holds if and only if $\langle f(y), y - x\rangle \geq 0$.

It is clear that $\mathscr{R}$ is convex in the first variable, so the solution set of the Minty variational inequality problem is always convex if it is nonempty in view of Proposition 10.55. When f is continuous and pseudo-monotone in the sense that $\langle f(y), x - y\rangle \geq 0$ implies

$$\langle f(x), x - y\rangle \geq 0,$$

the solution sets of the Stampacchia and Minty problems coincide, hence the solution set of the Stampacchia problem is convex too.

We now turn to the boundedness of solution sets. To this end, a concept of coerciveness of the relation $\mathscr{R}$ is needed.

Definition 10.57 Assume that A is a subset of a normed space. We say that $\mathscr{R}$ is COERCIVE in the first variable (with respect to the constraints $(\mathscr{D}_1, \mathscr{D}_2, \mathscr{P})$) if for every sequence $\{x^k\}$ in K whose norm converges to ∞, one can find an integer k_0 such that whenever $k \geq k_0$ the relation $\mathscr{R}\left(x^k, b^k, y^k\right)$ does not hold for some $b^k \in \mathscr{D}_2\left(x^k\right)$ and $y^k \in \mathscr{P}\left(x^k, b^k\right)$.

It is clear that when $\mathscr{D}_1(A)$ is bounded, then K is bounded, and hence $\mathscr{R}$ is coercive. When $\mathscr{R}$ is defined by the inequality $f(x, b, y) \leq 0$, where f is a real-valued function on $A \times B \times Y$, it is coercive in the first variable provided that f is coercive in x in the sense that

$$\lim_{\|x\| \to \infty} f(x, b, y) = \infty$$

for all b and y.

Proposition 10.58 *Assume that A is an unbounded subset of a normed space. Then, the solution set* Sol(VRP) *is bounded if $\mathscr{R}$ is coercive in the first variable.*

Proof The conclusion follows directly from the hypothesis and from the formula (10.11). □

In the Stampacchia problem if the operator f is coercive on A in the sense that

$$\lim_{x\in A, \|x\|\to\infty} \langle f(x), x-a\rangle = \infty$$

for some fixed a, then it is clear that $\mathscr{R}$ is coercive in the first variable. Therefore, the solution set is bounded.

On the other hand, for the Minty problem, if all vectors $f(y)$, $y \in A$ are strictly positive on the recession cone of A without the origin, then $\mathscr{R}$ is coercive, and so the solution set is bounded too.

Convergence Conditions for (VRP)

Throughout this subsection $\Lambda, \mathbb{A}, \mathbb{B}$ and Y are topological spaces. For each $\lambda \in \Lambda$, we assume that $A^\lambda \subset \mathbb{A}$, $B^\lambda \subset \mathbb{B}$, $Y^\lambda \subset Y$ are nonempty sets; $\mathscr{D}_1^\lambda : A^\lambda \rightrightarrows A^\lambda$, $\mathscr{D}_2^\lambda : A^\lambda \rightrightarrows B^\lambda$, $\mathscr{P}^\lambda : A^\lambda \times B^\lambda \rightrightarrows Y^\lambda$ are set-valued maps with nonempty values; and $\mathscr{R}^\lambda(a, b, y)$ is a relation linking elements $a \in A^\lambda, b \in B^\lambda$ and $y \in Y^\lambda$. The variational relation problem with the data A^λ, B^λ, Y^λ, $\mathscr{D}_1^\lambda$, $\mathscr{D}_2^\lambda$, $\mathscr{P}^\lambda$ and $\mathscr{R}^\lambda$ is denoted by (VRP)$^\lambda$. The notations Sol(VRP)(λ), K^λ, Γ^λ, etc. are defined accordingly. In this section, we study the question how the solution set Sol(VRP)(λ) changes when the data of the problem vary. More precisely, we shall establish inner- and outer-continuity as well as inner and outer openness of Sol(VRP)(λ) as a set-valued map of the variable λ. We shall fix a value $\lambda_0 \in \Lambda$ and assume that the problem (VRP) corresponds to the case of (VRP)$^{\lambda_0}$. The complement of the map Γ^λ is understood in the space $\mathbb{A}$.

Theorem 10.59 *The set-valued map* Sol(VRP)(λ) *is outer open at* λ_0, *that is,*

$$\text{limsupo}_{\lambda\to\lambda_0} \text{Sol(VRP)}(\lambda) \subset \Sigma,$$

if the maps K^λ *are outer open at* λ_0 *and* $\text{limsupo}_{\lambda\to\lambda_0} \Gamma^\lambda \cap K \subset \Gamma$. *In particular, this is true under the following conditions*

(i) $\limsup_{\lambda\to\lambda_0} A^\lambda \subset A$;
(ii) *for every* $x \in A$, $x \in \mathscr{D}_1(x)$ *whenever* $x \in \mathscr{D}_1^{\lambda^\nu}(x)$ *for some net* λ^ν *converging to* λ_0;
(iii) *for every* $x \in K$,

(iii$_1$) $\liminf_{\lambda\to\lambda_0} \mathscr{D}_2^\lambda(x) \supset \mathscr{D}_2(x)$;
(iii$_2$) $\liminf_{\substack{\lambda\to\lambda_0 \\ b'\in\mathscr{D}_2^\lambda(x)\to b}} \mathscr{P}^\lambda(x, b') \supset \mathscr{P}(x, b)$;
(iii$_3$) $\mathscr{R}(x, b, y)$ *holds whenever there are some nets* λ^ν *converging to* λ_0, $b^\nu \in \mathscr{D}_2^{\lambda^\nu}(x)$ *converging to b and* $y^\nu \in \mathscr{P}^{\lambda^\nu}(x, b^\nu)$ *converging to y such that* $\mathscr{R}^{\lambda^\nu}(x, b^\nu, y^\nu)$ *holds for all* ν.

Proof According to (10.11), we have $\mathrm{Sol(VRP)}(\lambda) = K^\lambda \cap \Gamma^\lambda$. Hence, the first part of the theorem follows from Proposition 10.48(a). For the second part, we show that the set-valued map $\lambda \mapsto K^\lambda$ is outer open at λ_0. Let $x \in \mathrm{limsupo}_{\lambda\to\lambda_0} K^\lambda$. There is a net λ^ν converging to λ_0 and a neighborhood V of x with $V \subset A^{\lambda^\nu}$ and $z \in \mathscr{D}_1^{\lambda^\nu}(z)$ for all $z \in V$. It follows from (i) that $x \in A$, and from (ii) that $x \in \mathscr{D}_1(x)$. Therefore, $x \in K$.

Now, if $x \in K \cap \mathrm{limsupo}_{\lambda\to\lambda_0} \Gamma^\lambda$, then for some net λ^ν converging to λ_0 and some neighborhood V of x, one has $V \subset A^{\lambda^\nu}$ and the relation $\mathscr{R}^{\lambda^\nu}(z, b, y)$ holds for all $z \in V$, $b \in \mathscr{D}_2^{\lambda^\nu}(z)$ and $y \in \mathscr{P}^{\lambda^\nu}(z, b)$. Given $b \in \mathscr{D}_2(x)$ and $y \in \mathscr{P}(x, b)$, in view of (iii$_1$) and (iii$_2$), one can find $b^\nu \in \mathscr{D}_2^{\lambda^\nu}(x)$ and $y^\nu \in \mathscr{P}^{\lambda^\nu}(x, b^\nu)$ such that b^ν and y^ν converge, respectively, to b and y. Then $\mathscr{R}^{\lambda^\nu}(x, b^\nu, y^\nu)$ holds, and by (iii$_3$), $\mathscr{R}(x, b, y)$ holds too. In this way, $K \cap \mathrm{limsupo}_{\lambda\to\lambda_0} \Gamma^\lambda \subset \Gamma^\lambda$ and by Proposition 10.48(a), the proof is complete. □

Theorem 10.60 *The map* $\mathrm{Sol(VRP)}(\lambda)$ *is outer semicontinuous at* λ_0*, that is,*

$$\limsup_{\lambda\to\lambda_0} \mathrm{Sol(VRP)}(\lambda) \subset \mathrm{Sol(VRP)},$$

if K^λ *is outer semicontinuous at* λ_0 *and* $\left[\, limsup_{\ \lambda\to\lambda_0} \Gamma^\lambda\right] \cap K \subset \Gamma$*. In particular, this is true under the following conditions*

(i) $\limsup_{\lambda\to\lambda_0} A^\lambda \subset A$;
(ii) *for every* $x \in A$, $x \in \mathscr{D}_1(x)$ *whenever there are a net* λ^ν *converging to* λ_0 *and* $x^\nu \in \mathscr{D}_1^{\lambda^\nu}(x^\nu)$ *converging to* x;
(iii) *for every* $x \in K$,

(iii$_1$) $\liminf\limits_{\substack{\lambda\to\lambda_0 \\ x^\lambda\in A^\lambda\to x}} S_2^\lambda\left(x^\lambda\right) \supset S_2(x)$;

(iii$_2$) $\liminf\limits_{\substack{\lambda\to\lambda_0, x^\lambda\in A^\lambda\to x \\ b^\lambda\in\mathscr{D}_2^\lambda(x^\lambda)\to b}} \mathscr{P}^\lambda\left(x^\lambda, b^\lambda\right) \supset \mathscr{P}(x, b)$;

(iii$_3$) $\mathscr{R}(x, b, y)$ *holds whenever* $\mathscr{R}^{\lambda_\nu}(x^\nu, b^\nu, y^\nu)$ *holds for some* $x^\nu \in A^{\lambda^\nu}$, $b^\nu \in \mathscr{D}_2^{\lambda^\nu}(x^\nu)$ *and* $y^\nu \in \mathscr{P}^{\lambda^\nu}(x^\nu, b^\nu)$ *converging to* x, b *and* y*, respectively, and* λ_ν *converges to* λ_0.

Proof As in the proof of Theorem 10.59, the first part of the theorem follows from Proposition 10.48(a). For the second part, we note that due to (i) and (ii), the set-valued map $\lambda \mapsto K^\lambda$ is outer-continuous at λ_0. Now let $x \in \left[\limsup_{\lambda\to\lambda_0} \Gamma^\lambda\right] \cap K$. There are $x^\nu \in \Gamma^{\lambda^\nu}$ converging to x as λ^ν converging to λ_0. We show that x belongs to Γ. Let b be any element of $\mathscr{D}_2(x)$ and let y be any element of $\mathscr{P}(x, b)$. By (iii$_1$) and (iii$_2$), there are some $b^\nu \in \mathscr{D}_2^{\lambda^\nu}$ and $y^\nu \in \mathscr{P}^{\lambda^\nu}(x^\nu, b^\nu)$ which converge, respectively, to b and y. Because $\mathscr{R}^{\lambda^\nu}(x^\nu, b^\nu, y^\nu)$ holds for all ν, by (iii$_3$) one deduces that $\mathscr{R}(x, b, y)$ is satisfied, and so x belongs to Γ, completing the proof. □

Theorem 10.61 *The set-valued map* Sol(VRP)(λ) *is inner open at* λ_0*, that is,*

$$\operatorname{liminfo}_{\lambda\to\lambda_0} \operatorname{Sol(VRP)}(\lambda) \supset \operatorname{Sol(VRP)},$$

if the maps K^λ *are inner open at* λ_0 *and* $\operatorname{liminfo}_{\lambda\to\lambda_0} \Gamma^\lambda \supset \Gamma \cap K$*. In particular, this is true under the following conditions*

(i) $\operatorname{liminfo}_{\lambda\to\lambda_0} A^\lambda \supset A$*;*
(ii) *for every* $x \in A$, $\operatorname{liminfo}_{\substack{\lambda\to\lambda_0 \\ x'\in A^\lambda\to x}} \mathscr{D}_1^\lambda\left(x'\right) \supset \mathscr{D}_1(x)$*;*
(iii) *for every* $x \in K$,

(iii$_1$) $\mathscr{D}_2^\lambda\left(x'\right)$ *is upper semicontinuous and compact-valued in the variables* $\left(\lambda, x'\right)$ *at* (λ_0, x)*;*
(iii$_2$) $\mathscr{P}^\lambda\left(x', b'\right)$ *is upper semicontinuous and compact-valued in the variables* $\left(\lambda, x', b'\right)$ *at* (λ_0, x, b) *with* $b \in \mathscr{D}_2(x)$*;*
(iii$_3$) $\mathscr{R}(x, b, y)$ *does not hold whenever* $\mathscr{R}^{\lambda^\nu}(x^\nu, b^\nu, y^\nu)$ *does not hold for some* $x^\nu \in A^{\lambda^\nu}$ *converging to* x, $b^\nu \in \mathscr{D}_2^{\lambda^\nu}(x^\nu)$ *converging to* b, $y^\nu \in \mathscr{P}^{\lambda^\nu}(x^\nu)$ *converging to* y*, and* λ^ν *converging to* λ_0.

Proof The first part of the theorem follows from Proposition 10.48(b). For the second part, first we prove that the map $\lambda \mapsto K^\lambda$ is inner open at λ_0. Let x be an element of K, that is $x \in A$ and $x \in \mathscr{D}_1(x)$. By (i), there are neighborhoods U_1 of λ_0 and V_1 of x such that $V_1 \subset A^\lambda$ for all $\lambda \in U_1, \lambda \neq \lambda_0$. It follows from (ii) that there exist neighborhoods U_2 of λ_0, W_2 and V_2 of x such that $V_2 \subset \mathscr{D}_1^\lambda\left(x'\right)$ for all $\lambda \in U_2, \lambda \neq \lambda_0$, and $x' \in W_2 \cap A^\lambda$. By taking $U = U_1 \cap U_2$ and $V = V_1 \cap V_2 \cap W_2$, we deduce that for every $\lambda \in U, \lambda \neq \lambda_0$, and $x' \in V$, one has $x' \in A^\lambda \cap \mathscr{D}_1^\lambda\left(x'\right)$. Therefore, $V \subset K^\lambda$ for all $\lambda \in U, \lambda \neq \lambda_0$.

Now, to complete the proof, let $x \in \Gamma \cap K$. We have to show that $x \in \operatorname{liminfo}_{\lambda\to\lambda_0} \Gamma^\lambda$. Suppose on the contrary that x does not belong to the inferior open limit of Γ^λ at λ_0. According to Proposition 10.24(c), x belongs to the superior limit of $\left(\Gamma^\lambda\right)^c$. Let $x^\nu \in \left[\Gamma^{\lambda^\nu}\right]^c$ converging to some x as λ^ν converges to λ_0. For each ν, either x^ν does not belong to A^{λ^ν}, or it does belong to A^{λ^ν} and $\mathscr{R}^{\lambda^\nu}(x^\nu, b^\nu, y^\nu)$ does not hold for some $b^\nu \in \mathscr{D}_2^{\lambda^\nu}(x^\nu)$ and $y^\nu \in \mathscr{P}^{\lambda^\nu}(x^\nu, b^\nu)$. If the first case occurs for a subnet of λ^ν, then in view of (i) and Proposition 10.26(b), x belongs to neither A nor Γ, then in view of (i) and Proposition 10.26(b), x belongs to neither A nor Γ. In the other case, in view of (iii$_1$) and (iii$_2$), we may assume that b^ν and y^ν converge, respectively, to some $b \in \mathscr{D}_2(x)$ and $y \in \mathscr{P}(x, b)$. By (iii$_3$), $\mathscr{R}(x, b, y)$ does not hold, which yields $x \notin \Gamma$, a contradiction. The proof is complete. □

It is clear that the roles of the maps K^λ and Γ^λ in the first parts of the above theorems are symmetric, so that they can be swapped without altering the conclusion. We now turn to the inner continuity of the solution set.

Theorem 10.62 *The set-valued map* Sol(VRP)(λ) *is inner-semicontinuous at* λ_0, *that is,*

$$\operatorname{liminf}_{\lambda\to\lambda_0}\mathrm{Sol(VRP)}(\lambda) \supset \mathrm{Sol(VRP)},$$

if either K^λ *is inner open at* λ_0 *and* $\operatorname{liminf}_{\lambda\to\lambda_0}\Gamma^\lambda \supset \Gamma\cap K$, *or* K^λ *is inner semicontinuous at* λ_0 *and* $\operatorname{liminfo}_{\lambda\to\lambda_0}\Gamma^\lambda \supset \Gamma\cap K$. *In particular, this is true under the following conditions*

(i) $\operatorname{liminfo}_{\lambda\to\lambda_0} A^\lambda \supset A$;
(ii) *for every* $x\in A$, $\operatorname{liminfo}_{\substack{\lambda\to\lambda_0\\ x'\in A^\lambda\to x}}\mathscr{D}_1^\lambda(x') \supset \mathscr{D}_1(x)$;
(iii) *for every* $x\in K$,

(iii$_1$) $\mathscr{D}_2^\lambda(x)$ *is upper semicontinuous and compact-valued at* λ_0;
(iii$_2$) $\mathscr{P}^\lambda(x,b')$ *is upper semicontinuous and compact-valued at* (λ_0,b) *with* $b\in\mathscr{D}_2(x)$;
(iii$_3$) $\mathscr{R}(x,b,y)$ *does not hold whenever there are* λ^ν *converging to* λ_0, $b^{\lambda^\nu}\in\mathscr{D}_2^{\lambda^\nu}(x)$ *and* $y^{\lambda^\nu}\in\mathscr{P}^{\lambda^\nu}(x,b^{\lambda^\nu})$ *converging to* b *and* y, *respectively, such that* $x\in A^{\lambda^\nu}$ *and* $\mathscr{R}^{\lambda^\nu}(x,b^{\lambda^\nu},y^{\lambda^\nu})$ *does not hold for all* ν.

Proof Again, the first part of the theorem follows from Proposition 10.48(c). For the second part, observe that the map $\lambda\mapsto K^\lambda$ is inner open at λ_0 as proven in the preceding theorem. Let $x\in\Gamma\cap K$. We have to show that for every net λ^ν converging to λ_0, one can find some $x^\nu\in\Gamma^{\lambda^\nu}$ converging to x. Suppose to the contrary that this is not true; that is, there is a neighborhood V of x such that $V\cap\Gamma^{\lambda^\nu}=\emptyset$ for some net λ^ν converging to λ_0. In view of (1), we may assume that $V\subset A^{\lambda^\nu}$ for all ν. Therefore, there are some elements b^ν of $\mathscr{D}_2^{\lambda^\nu}(x)$ and y^ν of $\mathscr{P}^{\lambda^\nu}(x,b^\nu)$ such that $\mathscr{R}^{\lambda^\nu}(x,b^\nu,y^\nu)$ does not hold. By (iii$_1$) and (iii$_2$), one may assume that b^ν and y^ν converge, respectively, to $b\in\mathscr{D}_2(x)$ and $y\in\mathscr{P}(x,b)$. The last hypothesis (iii$_3$) yields that $\mathscr{R}(x,b,y)$ does not hold. This is in contradiction with $x\in\Gamma$, and the proof is complete. □

We finish this subsection with another kind of sufficient conditions for continuity of the map Γ^λ. These involve the complement of the relation $\mathscr{R}$. We say that a set $\Lambda_0\subset\Lambda$ is open (respectively, closed) at λ_0 if there is an open (respectively, closed) neighborhood U_0 of λ_0 such that $U_0\cap\Lambda_0$ is open (respectively, closed). We need the following lemma.

Lemma 10.63 *The following assertions are true.*

(a) Γ^λ *is outer open (resp., inner semicontinuous) at* λ_0 *if, for every* $x\notin\Gamma$ *(respectively,* $x\in\Gamma$*), the set*
$\mathscr{U}_x=\{\lambda\in\Lambda\mid R^\lambda(x,b,y)$ *does not hold for some* $b\in S_2^\lambda(x)$ *and* $y\in T^\lambda(x,b)\}$
is open (respectively, closed) at λ_0.

(b) *Γ^λ is outer semicontinuous (resp., inner open) at λ_0 if the set*

$\mathscr{U} = \Big\{(\lambda, x) \in \Lambda \times \mathscr{A} \mid$ *either* $x \notin A^\lambda$ *or* $R^\lambda(x, b, y)$ *does not hold for some* $b \in S_2^\lambda(x)$ *and* $y \in T^\lambda(x, b)\Big\}$

is open at (λ_0, x) *with* $x \notin \Gamma$ *(resp., closed at* (λ_0, x) *with* $x \in \Gamma$*).*

Proof To prove (a), let $x \notin \Gamma$. Then, $\lambda_0 \in \mathscr{U}_x$. If $\mathscr{U}_x$ is open at λ_0, then there is a neighborhood U of λ_0 which is contained in $\mathscr{U}_x$. Hence, x belongs to $\left[\Gamma^\lambda\right]^c$ for all $\lambda \in U$. Therefore, x belongs to $\liminf_{\lambda\to\lambda_0} \Gamma^\lambda$, which, by Proposition 10.26(a), implies that Γ^λ is outer open.

Let $x \in \Gamma$. Then, $\lambda_0 \notin \mathscr{U}_x$. If $\mathscr{U}_x$ is closed at λ_0, then there is a closed neighborhood U_0 of λ_0 such that $U_0 \cap U_x$ is a closed set. Consequently, there is some open neighborhood $U \subset U_0$ of λ_0 such that $U \cap \mathscr{U}_x$ is an empty set. So, x belongs to Γ^λ for all $\lambda \in U$. Hence, x belongs to liminf $_{\lambda\to\lambda_0}\Gamma^\lambda$, and Γ^λ is inner semicontinuous.

For the outer semicontinuity in assertion (b), let $x_0 \notin \Gamma$. Then, $(\lambda_0, x_0) \in \mathscr{U}$. Since the set $\mathscr{U}$ is open at this point, there are open neighborhoods U of λ_0 and V of x_0 such that $U \times V \subset \mathscr{U}$. This implies that $x \notin \Gamma^\lambda$ for all $x \in V$ and $\lambda \in U$. By definition, x_0 belongs to liminfo $_{\lambda\to\lambda_0}\left[\Gamma^\lambda\right]^c$ and by Proposition 10.26(b), Γ^λ is outer semicontinuous.

For the inner openness, let $x_0 \in \Gamma$. Then $(\lambda_0, x_0) \notin \mathscr{U}$. Since the set $\mathscr{U}$ is closed, there are closed neighborhoods U_0 of λ_0 and V_0 of x_0 such that $U_0 \times V_0 \cap \mathscr{U}$ is closed. We can find some small open neighborhoods $U \subset U_0$ of λ_0 and $V \subset V_0$ of x_0 such that $U \times V \cap \mathscr{U} = \emptyset$. This implies that $V \subset \Gamma^\lambda$ for all $\lambda \in U$. Therefore, Γ^λ is inner open at λ_0, and the proof is complete. □

Remark 10.64 In many models, it is assumed that the sets A^λ, B^λ and Y^λ are constant and that the set

$$\mathscr{W} = \left\{(\lambda, x, b, y) \in \Lambda \times A \times B \times Y \mid \mathscr{R}^\lambda(x, b, y) \text{ does not hold }\right\}$$

is open at (λ_0, x, b, y) for $y \in T(x, b)$, $b \in \mathscr{D}_2(x)$ and $x \notin \Gamma$. If in addition the map $\mathscr{D}_2^\lambda(x)$ is inner semicontinuous at (λ_0, x) and $\mathscr{P}^\lambda(x, b)$ is inner semicontinuous at (λ_0, x, b) for all $x \notin \Gamma$, $b \in \mathscr{D}_2(x)$, then the set $\mathscr{U}$ is open at (λ_0, x) for all $x \notin \Gamma$.

Indeed, let $x_0 \notin \Gamma$. Then, there are some $b_0 \in \mathscr{D}_2(x_0)$ and $y_0 \in \mathscr{P}(x_0, b_0)$ such that $\mathscr{R}(x_0, b_0, y_0)$ does not hold. By the hypothesis, there are open neighborhoods U_1 of λ_0, V_1 of x_0, W_1 of b_0 and Z_1 of y_0 such that $U_1 \times V_1 \times W_1 \times Z_1 \subset \mathscr{W}$. By the inner semicontinuity hypothesis of $\mathscr{P}^\lambda$, there are open neighborhoods $U_2 \subset U_1$ of λ_0, $V_2 \subset V_1$ of x_0 and $W_2 \subset W_1$ of b_0 such that $\mathscr{P}^\lambda(x, b) \cap Z_1 \neq \emptyset$ for all $(\lambda, x, b) \in U_1 \times V_1 \times W_1$. Similarly, by the inner semicontinuity of $\mathscr{D}_2^\lambda$, one can find open neighborhoods $U \subset U_2$ of λ_0 and $V \subset V_2$ of x_0 such that $\mathscr{D}_2^\lambda(x) \cap W_2 \neq \emptyset$ for all $(\lambda, x) \in U \times V$. It is clear now that for each $(\lambda, x) \in U \times V$, there are some $b \in \mathscr{D}_2^\lambda(x)$ and $y \in \mathscr{P}^\lambda(x, b)$ such that (λ, x, b, y) belongs to $\mathscr{W}$, that is, $\mathscr{R}^\lambda(x, b, y)$ does not hold. In this way, $U \times V \subset \mathscr{U}$.

Along the same lines, assume that the set $\mathcal{W}$ is closed at (λ_0, x, b, y), the maps $\mathcal{D}_2^\lambda$ and $\mathcal{P}^\lambda$ are outer semicontinuous and compact-valued at (λ_0, x) and (λ_0, x, b), respectively, for every $x \in \Gamma, b \in \mathcal{D}_2(x)$ and $y \in \mathcal{T}(x, b)$, and that the set Y is compact. Then $\mathcal{U}$ is closed at (λ_0, x) for all $x \in \Gamma$. In fact, let $x_0 \in \Gamma$. Then for every $b \in \mathcal{D}_2(x_0)$ and $y \in \mathcal{T}(x_0, b)$, $\mathcal{R}(x_0, b, y)$ does hold. By hypothesis, there are closed neighborhoods U_{by} of λ_0, V_{by} of x_0, W_{by} of b and Z_{by} of y such that $U_{by} \times V_{by} \times W_{by} \times Z_{by} \cap \mathcal{W} = \emptyset$. By the hypothesis, the set $\mathcal{P}(x_0, \mathcal{D}_2(x_0))$ is compact, and hence we can choose a finite number of points, say $b_1, \ldots, b_k \in \mathcal{D}_2(x_0)$ and $y_1, \ldots, y_k \in \mathcal{P}(x_0, \mathcal{D}_2(x_0))$ such that the sets $W_1 = \cup_{i=1}^k W_{b_i y_i}$ and $Z_1 = \cup_{i=1}^k Z_{b_i y_i}$ are closed neighborhoods of $\mathcal{D}_2(x_0)$ and $\mathcal{P}(x_0, \mathcal{D}_2(x_0))$, respectively. Set $U_1 = \cap_{i=1}^k U_{b_i y_i}$ and $V_1 = \cap_{i=1}^k V_{b_i y_i}$. They are closed neighborhoods of λ_0 and x_0, respectively, and the closed neighborhood $U_1 \times V_1 \times W_1 \times Z_1$ does not meet the set $\mathcal{W}$. Now, by the outer semicontinuity of $\mathcal{P}^\lambda$, there are closed neighborhoods $U_2 \subset U_1$ of λ_0, $V_2 \subset V_1$ of x_0 and $W_2 \subset W_1$ of $\mathcal{D}_2(x_0)$ such that $\mathcal{T}^\lambda(x, \mathcal{D}_2(x)) \subset Z_1$. Similarly, by the outer semicontinuity of $\mathcal{D}_2^\lambda$, there are closed neighborhoods $U \subset U_2$ of λ_0 and $V \subset V_2$ of x_0 such that $\mathcal{D}_2^\lambda(x) \subset W_2$ for all $\lambda \in U$ and $x \in V$. Consequently, the closed neighborhood $U \times V$ of (λ_0, x_0) does not meet $\mathcal{W}$ which means that it does not intersect $\mathcal{U}$. Equivalently, $\mathcal{U}$ is closed at (λ_0, x_0).

By applying Theorems 10.59–10.62, Lemma 10.63 and the above remark, we obtain sufficient conditions for continuity of the solution map Sol(VRP)(λ) in terms of the sets $\mathcal{U}_x$, $\mathcal{U}$ and $\mathcal{W}$ above.

Corollary 10.65 *The following assertions hold.*

(a) Sol(VRP)(λ) *is outer open at* λ_0 *provided that the map* K^λ *is outer open at* λ_0 *and the set* $\mathcal{U}_x$ *is open at* λ_0 *for all* $x \notin \Gamma$.
(b) Sol(VRP)(λ) *is outer semicontinuous at* λ_0 *provided that the map* K^λ *is outer semicontinuous at* λ_0 *and that the set* $\mathcal{U}$ *is open at* (λ_0, x) *for all* $x \notin \Gamma$, *which is true in particular when the maps* A^λ, B^λ *and* Y^λ *are constant, the maps* $\mathcal{D}_2^\lambda$ *and* $\mathcal{P}^\lambda$ *are inner semicontinuous at* (λ_0, x) *and* (λ_0, x, b), *respectively, and the set* $\mathcal{W}$ *is open at* (λ_0, x, b, y) *for* $x \notin \Gamma$, $b \in \mathcal{D}_2(x)$, *and* $y \in \mathcal{P}(x, b)$.
(c) Sol(VRP)(λ) *is inner open at* λ_0 *provided that the map* K^λ *is inner open at* λ_0 *and the set* $\mathcal{U}$ *is closed at* (λ_0, x) *for all* $x \in \Gamma$, *which is true in particular when* Y *is compact,* $\mathcal{D}_2^\lambda(x)$ *and* $\mathcal{P}^\lambda(x)$ *are outer semicontinuous, compact-valued, respectively, at* (λ_0, x) *and* (λ_0, x, b), *and* $\mathcal{W}$ *is closed at* (λ_0, x, b, y) *for every* $x \in \Gamma, b \in \mathcal{D}_2(x)$, $y \in \mathcal{P}(x, b)$.
(d) Sol(VRP)(λ) *is inner semicontinuous at* λ_0 *provided that the map* K^λ *is inner open (respectively, inner semicontinuous) at* λ_0 *and the set* $\mathcal{U}_x$ *(respectively,* $\mathcal{U}$*) is closed at* λ_0 *(respectively,* (λ_0, x)*) for all* $x \in \Gamma$.

Note that although the assumptions of Corollary 10.65 are difficult to check, they are properly weaker than the corresponding conditions of Theorems 10.59–10.62. Indeed, from the proof of this corollary, we see that the particular assumptions of Theorems 10.59–10.62 imply the assumptions of Corollary 10.65. The following example shows that the converse is not true and Corollary 10.65 may be applicable in cases where the second parts of the above mentioned theorems are not.

Example 10.66 Let $A^\lambda = B^\lambda = Y^\lambda = \mathbb{R}$, $\Lambda = [0, 1]$, $\lambda_0 = 0$, $\mathscr{D}_1^\lambda(x) = [0, \lambda]$,

$$\mathscr{D}_2^\lambda(x) = \begin{cases} (0, 1) & \text{if } \lambda = 0, \\ \{-1\} & \text{otherwise}, \end{cases}$$

$$\mathscr{P}^\lambda(x, b) = \begin{cases} (-1, 0) & \text{if } \lambda = 0, \\ \{1\} & \text{otherwise}, \end{cases}$$

$$F^\lambda(b, y) = [0, 1].$$

Let the relation $\mathscr{R}$ be defined by: $\mathscr{R}(x, b, y)$ holds if and only if $F^\lambda(b, y) \subset \mathbb{R}_+$.

Then, $K^\lambda = [0, \lambda]$ satisfies all the assumptions in (a)–(e) of Corollary 10.65. All the sets $\mathscr{U}_x$, $\mathscr{U}$ and $\mathscr{W}$ are empty and hence also satisfy the assumptions in (a)-(e). Direct calculations give Sol(VRP)$(\lambda) = [0, \lambda]$, which have all continuity and openness properties. But $\mathscr{D}_2^\lambda(x)$ and $\mathscr{P}^\lambda(x, b)$ do not fulfil any assumption of Theorems 10.59–10.62.

Hölder Calmness of the Solution Map of (VRP)

In this section, we consider a simple variational relation problem with two parameters and discuss the Hölder calmness of the solution map of the reference problem. The problem under consideration as well as some concepts employed in this section are presented below. Let X, M and Λ be metric spaces. Unless there is a risk of misinterpretation, the metrics on all of these spaces are denoted by d.

Let $\mathscr{D} : \Lambda \rightrightarrows X$ be a multivalued map with nonempty closed values. Let $\mathscr{R} \subset X \times X \times M$ be a relation linking elements $x, y \in X$ and depending on a parameter $\mu \in M$. For each pair of parameters (λ, μ), we consider the following variational relation problem

(VRP)(λ, μ) Find $x(\lambda, \mu) \in \mathscr{D}(\lambda)$ such that $\mathscr{R}(x(\lambda, \mu), y, \mu)$ holds for every $y \in \mathscr{D}(\lambda)$.

This is a particular case of the problem formulated and studied in the previous subsection. For each $(\lambda, \mu) \in \Lambda \times M$, we denote by Sol(VRP)$(\lambda, \mu)$ the solution set of the problem (VRP)(λ, μ), and we suppose that the solution set is nonempty in the neighborhood of the reference point. The aim of this section is to consider the calmness of the mapping $(\lambda, \mu) \mapsto$ Sol(VRP)(λ, μ), extending existing conclusions from equilibrium problems to variational relation problems.

For the convenience of the reader, we recall here some notions used in this subsection. For two nonempty subsets A, B of a metric space (X, d) and some point $a \in \mathbb{X}$, we denote by $d(a, B) := \inf_{b \in B} d(a, b)$ the distance between the point a and the set B, and by $e(A, B) = \sup_{a \in A} d(a, B)$, the excess of A beyond B and by

$H(A, B) = \max\{e(A, B), e(B, A)\}$ the Hausdoff-Pompeiu distance between A and B. By convention, $e(\emptyset, B) = 0$ for $B \neq \emptyset$.

When a is a real number, we will denote by

$$a_- = d(a, [0, +\infty[) = \max\{-a, 0\}, \quad \text{and} \quad a_+ = d(a,]-\infty, 0]) = \max\{a, 0\}.$$

Let X and Y be metric spaces. A function $g : X \to Y$ is said to be Hölder continuous if there exist two real numbers $k, \gamma > 0$ such that

$$d\left(g(x), g\left(x'\right)\right) \leq kd^{\gamma}\left(x, x'\right) \text{ for every } x, x' \in X.$$

The function g is said to be Hölder calm at a point $\bar{x} \in X$ if there exist two real numbers $k, \gamma > 0$ and a neighborhood $N(\bar{x})$ of $\bar{x}$ such that

$$d(g(x), g(\bar{x})) \leq kd^{\gamma}(x, \bar{x}) \text{ for every } x \in N(\bar{x}).$$

A set-valued map $T : X \rightrightarrows Y$ is said to be Hölder continuous if there exist two real numbers $k, \gamma > 0$ such that

$$e\left(T(x), T\left(x'\right)\right) \leq kd^{\gamma}\left(x, x'\right) \text{ for every } x, x' \in X.$$

The map T is said to be Hölder calm at a point $\bar{x} \in X$, $\bar{y} \in T(\bar{x})$, if there exist two real numbers $k, \gamma > 0$ and neighborhood $N(\bar{x})$ of $\bar{x}$, such that

$$e(T(x) \cap N(\bar{y}), T(\bar{x})) \leq kd^{\gamma}(x, \bar{x}) \text{ for every } x \in N(\bar{x}).$$

When $\gamma = 1$, we get the notions of Lipschitz continuity and calmness.

For given initial points $\bar{\lambda}$, $\bar{\mu}$ of parameters, and the solution set $\text{Sol(VRP)}(\bar{\lambda}, \bar{\mu})$, we are interested in the behavior of the map $(\lambda, \mu) \mapsto \text{Sol(VRP)}(\lambda, \mu)$ when the parameter (λ, μ) varies in some neighborhood of $(\bar{\lambda}, \bar{\mu})$. The following result provides sufficient conditions for the Hölder calmness of the solution map of the variational relation problem (VRP) (λ, μ).

Theorem 10.67 *Let $\bar{\mu} \in M$, and $\bar{\lambda} \in \Lambda$ be given. Assume that the following conditions are satisfied:*

(i) *$\mathscr{D}$ is Höder calm at $\bar{\lambda}$, that is, there exist two real numbers $l, \alpha > 0$ and a neighborhood $U(\bar{\lambda})$ of $\bar{\lambda}$ such that, for any $\lambda \in U(\bar{\lambda})$, we have*

$$H(\mathscr{D}(\lambda), \mathscr{D}(\bar{\lambda})) \leq ld^{\alpha}(\lambda, \bar{\lambda}); \tag{10.13}$$

(ii) *$\mathscr{R}$ is Hölder calm at $\bar{\mu}$ in the sense that there exist two real numbers $m, \beta > 0$ and a neighborhood $V(\bar{\mu})$ of $\bar{\mu}$ such that, if $\mathscr{R}(x, y; \bar{\mu})$ and $\mathscr{R}(y, x; \mu)$ are satisfied for $x, y \in X$, and $\mu \in V(\bar{\mu})$, then we have*

$$d(x, y) \leq md^{\beta}(\mu, \bar{\mu}); \tag{10.14}$$

(iii) *$\mathscr{R}$ is Höder continuous in the first and the second variables in the sense that there exist three real numbers $n, \delta, \eta > 0$ such that, if $\mathscr{R}(x_1, y_1; \bar{\mu})$ and $\mathscr{R}(x_2, y_2; \bar{\mu})$ are satisfied for $x_1, x_2, y_1, y_2 \in X$, then we have*

$$d^{\eta}(x_1, x_2) \leq n\left(d^{\delta}(x_1, y_2) + d^{\delta}(x_2, y_1)\right). \tag{10.15}$$

Then, the following statements hold true

(a) *the solution set* Sol(VRP) $(\bar{\lambda}, \bar{\mu})$ *is a singleton;*
(b) *for any $\lambda \in U(\bar{\lambda})$ and $\mu \in V(\bar{\mu})$,*

$$e(\mathrm{Sol(VRP)}(\lambda, \mu), \mathrm{Sol(VRP)}(\bar{\lambda}, \bar{\mu})) \leq k_1 d^{\gamma_1}(\lambda, \bar{\lambda}) + k_2 d^{\gamma_2}(\mu, \bar{\mu}),$$

where k_1, k_2, γ_1 and γ_2 depend on $m, n, l, \alpha, \beta, \delta, \eta$.

Proof We prove first that if the problem (VRP)$(\bar{\lambda}, \bar{\mu})$ is solvable, then it has the uninique solution in this neighborhood. Let $x, x' \in$ Sol(VRP)$(\bar{\lambda}, \bar{\mu})$. Then, the vectors x, x' belong to $\mathscr{D}(\bar{\lambda}) \cap \bar{W}$ and the relations $\mathscr{R}\left(x, x'; \bar{\mu}\right)$ and $\mathscr{R}\left(x', x; \bar{\mu}\right)$ hold true. Due to condition (ii), we have $d\left(x, x'\right) \leq m d^{\beta}(\bar{\mu}, \bar{\mu}) = 0$, and hence $x \equiv x'$.

Step 1. Let $\lambda \in U(\bar{\lambda})$, $x(\bar{\lambda}, \bar{\mu}) \in$ Sol(VRP)$(\bar{\lambda}, \bar{\mu})$ and $x(\lambda, \bar{\mu}) \in$ Sol(VRP)$(\lambda, \bar{\mu})$ be given. We will estimate the distance between two vetors $x(\bar{\lambda}, \bar{\mu})$ and $x(\lambda, \bar{\mu})$. Because $x(\bar{\lambda}, \bar{\mu}) \in \mathscr{D}(\bar{\lambda})$ and $x(\lambda, \bar{\mu}) \in \mathscr{D}(\lambda)$, it follows from condition (i) that there exist two vectors $x_1 \in \mathscr{D}(\lambda)$ and $x_2 \in \mathscr{D}(\bar{\lambda})$ such that

$$d\left(x(\bar{\lambda}, \bar{\mu}), x_1\right) \leq l d^{\alpha}(\lambda, \bar{\lambda}) \text{ and } d\left(x(\lambda, \bar{\mu}), x_2\right) \leq l d^{\alpha}(\lambda, \bar{\lambda}). \tag{10.16}$$

On the other hand, because $x(\lambda, \bar{\mu})$ is a solution of the variational problem with respect to the parameters $\lambda, \bar{\mu}$ and $x_1 \in \mathscr{D}(\lambda)$, we conclude that the relation $\mathscr{R}(x(\lambda, \bar{\mu}), x_1; \bar{\mu})$ is satisfied. Similarly, we can also derive that the relation $\mathscr{R}\left(x(\bar{\lambda}, \bar{\mu}), x_2; \bar{\mu}\right)$ is satisfied too. Combining (iii) and (10.16), we get the following estimations.

$$d^{\eta}(x(\bar{\lambda}, \bar{\mu}), x(\lambda, \bar{\mu})) \leq n\left(d^{\delta}(x(\lambda, \bar{\mu}), x_2) + d^{\delta}\left(x(\bar{\lambda}, \bar{\mu}), x_1\right)\right) \leq 2nl^{\delta} d^{\alpha\delta}(\lambda, \bar{\lambda}),$$

or equivalently

$$d(x(\bar{\lambda}, \bar{\mu}), x(\lambda, \bar{\mu})) \leq \left(2nl^{\delta}\right)^{\frac{1}{\eta}} d^{\frac{\alpha\delta}{\eta}}(\lambda, \bar{\lambda}). \tag{10.17}$$

Step 2. Let $\mu \in V(\bar{\mu})$ and $x(\lambda, \mu) \in$ Sol(VRP)$(\lambda, \mu) \cap W$ be fixed, we aim to estimate the distance between two vectors $x(\lambda, \bar{\mu})$ and $x(\lambda, \mu)$. Since these vectors are solutions of the variational relation problem, the relations $\mathscr{R}(x(\lambda, \bar{\mu}), x(\lambda, \mu); \bar{\mu})$ and $\mathscr{R}(x(\lambda, \mu), x(\lambda, \bar{\mu}); \mu)$ are satisfied. Then, by condition (ii), we have

$$d(x(\lambda, \bar{\mu}), x(\lambda, \mu)) \leq m d^{\beta}(\mu, \bar{\mu}). \tag{10.18}$$

Step 3. Combining (10.17) and (10.18), for any $x(\lambda, \mu)$ belonging to Sol(VRP) (λ, μ), we have the following estimations

$$\begin{aligned} d(x(\lambda, \mu), x(\bar{\lambda}, \bar{\mu})) &\leq d(x(\lambda, \mu), x(\lambda, \bar{\mu})) + d(x(\lambda, \bar{\mu}), x(\bar{\lambda}, \bar{\mu})) \\ &\leq \left(2nl^{\delta}\right)^{\frac{1}{\eta}} d^{\frac{ax}{\eta}}(\lambda, \bar{\lambda}) + md^{\beta}(\mu, \bar{\mu}), \end{aligned}$$

and consequently

$$e(\mathrm{Sol(VRP)}(\lambda, \mu), \mathrm{Sol(VRP)}(\bar{\lambda}, \bar{\mu})) \leq \left(2nl^{\delta}\right)^{\frac{1}{\eta}} d^{\frac{a\delta}{\eta}}(\lambda, \bar{\lambda}) + md^{\beta}(\mu, \bar{\mu}).$$

The proof is complete. □

By using the same techniques as in the proof of above theorem, we can also establish sufficient conditions for the local Hölder calmness of the solution map of the variational relation problem (VRP) (λ, μ).

Theorem 10.68 *Let* $\bar{\mu} \in M, \bar{\lambda} \in \Lambda$ *and* $x(\bar{\lambda}, \bar{\mu}) \in \mathrm{Sol(VRP)}(\bar{\lambda}, \bar{\mu})$ *be fixed. Assume that there exist neighborhoods* $U(\bar{\lambda})$, $V(\bar{\mu})$, *and* W *of* $\bar{\lambda}, \bar{\mu}$ *and* $x(\bar{\lambda}, \bar{\mu})$, *respectively, for which the following conditions are satisfied:*

(i) $\mathscr{D}$ *is Höder calm at* $\bar{\lambda}$, *that is, there exist two real numbers* $l, \alpha > 0$ *such that, for any* $\lambda \in U(\bar{\lambda})$,

$$H(\mathscr{D}(\lambda), \mathscr{D}(\bar{\lambda})) \leq ld^{\alpha}(\lambda, \bar{\lambda}); \tag{10.19}$$

(ii) $\mathscr{R}$ *is locally Hölder calm at* $\bar{\mu}$ *in the sense that there exist two real numbers* $m, \beta > 0$ *such that, if* $\mathscr{R}(x, y; \bar{\mu})$ *and* $\mathscr{R}(y, x; \mu)$ *are satisfied for* $x \in X$, $y \in W$ *and* $\mu \in V(\bar{\mu})$, *then we have*

$$d(x, y) \leq md^{\beta}(\mu, \bar{\mu}); \tag{10.20}$$

(iii) $\mathscr{R}$ *is Höder continuous in the first and the second variables in the sense that there exist three real numbers* $n, \delta, \eta > 0$ *such that, if* $\mathscr{R}(x_1, y_1; \bar{\mu})$ *and* $\mathscr{R}(x_2, y_2; \bar{\mu})$ *are satisfied for* $x_1, x_2, y_1, y_2 \in X$, *then we have*

$$d^{\eta}(x_1, x_2) \leq n\left(d^{\delta}(x_1, y_2) + d^{\delta}(x_2, y_1)\right). \tag{10.21}$$

Then, the following statements hold true

(a) *the set* Sol(VRP) $(\bar{\lambda}, \bar{\mu}) \cap W$ *is a singleton;*
(b) *for any* $\lambda \in U(\bar{\lambda})$ *and* $\mu \in V(\bar{\mu})$,

$$e(\mathrm{Sol(VRP)}(\lambda, \mu) \cap W, \mathrm{Sol(VRP)}(\bar{\lambda}, \bar{\mu})) \leq k_1 d^{\gamma_1}(\lambda, \bar{\lambda}) + k_2 d^{\gamma_2}(\mu, \bar{\mu}),$$

where k_1, k_2, γ_1 *and* γ_2 *depend on* $m, n, l, \alpha, \beta, \delta, \eta$.

Remark 10.69 Condition (ii) of Theorem 10.68 is a local calmness property of the relation; it is weaker than the corresponding one of Theorem 10.67; and hence when we follow this condition, we will obtain the local calmness property of the solution map, a weakened version of the conclusion of Theorem 10.67. In the special case, $\mu = \bar{\mu}$, condition (ii) becomes the antisymmetry property of the relation $\mathscr{R}$ at $\bar{\mu}$.

Example 10.70 Let $X = \mathbb{R}$ and $M = [-1, 1]$ be endowed with the Euclidean distance and $m, \beta > 0$ be some fixed constants. We consider a relation $\mathscr{R}$ defined by

$$\mathscr{R}(x, y; \mu) \text{ is satisfied if and only if } y - m|\mu|^\beta \le x \le 1 - |\mu|.$$

Then, the relation $\mathscr{R}$ satisfies conditions (ii) and (iii) of Theorems 10.67 and 10.68 for the parameter $\bar{\mu} = 0$.

Indeed, let x, y and μ be such that $\mathscr{R}(x, y; 0)$ and $\mathscr{R}(y, x; \mu)$ hold, that is, $y \le x \le 1$ and $x - m|\mu|^\beta \le y \le 1 - |\mu|$. Then

$$|x - y| = x - y \le m|\mu|^\beta = md^\beta(\mu, \bar{\mu}),$$

so (10.14) and (10.20) have been checked.

Let $x_1, x_2, y_1, y_2 \in \mathbb{R}$ be such that $\mathscr{R}(x_1, y_1; 0)$ and $\mathscr{R}(x_2, y_2; 0)$ hold, that is $y_1 \le x_1 \le 1$ and $y_2 \le x_2 \le 1$. If $x_1 \le x_2$, then $0 \le x_2 - x_1 \le x_2 - y_1$. If $x_2 \le x_1$, then $0 \le x_1 - x_2 \le x_1 - y_2$. These inequalities imply that $|x_1 - x_2| \le |x_1 - y_2| + |x_2 - y_1|$, and hence (10.15) and (10.21) are also satisfied.

Let $\mathscr{D}(\lambda) = [0, 1 - \lambda]$, with the parameter $\lambda \in \left[0, \frac{1}{2}\right]$. For $\bar{\lambda} = 0$, the condition (i) of Theorems 10.67 and 10.68 will be verified. For each (λ, μ), we consider the following variational relation problem.

(VRP)(λ, μ) Find $x \in [0, 1 - \lambda]$ such that $y - m|\mu|^\beta \le x \le 1 - |\mu|$ holds for any $y \in [0, 1 - \lambda]$.

When $\bar{\lambda} = \bar{\mu} = 0$, the solution of (VRP) is unique, in fact Sol(VRP)$(0, 0) = \{1\}$. For each $(\lambda, \mu) \in \Lambda \times M$, the solution set is given by

$$\text{Sol(VRP)}(\lambda, \mu) = \left[\{1 - \lambda - m|\mu|^\beta\}_+, \min\{1 - \lambda, 1 - |\mu|\}\right].$$

This set is nonempty if, and only if $|\mu| - m|\mu|^\beta \le \lambda$. If Sol(VRP)$(\lambda, \mu) \ne \emptyset$, we have the estimation

$$e(\text{Sol(VRP)}(\lambda, \mu), \text{Sol(VRP)}(0, 0)) = \begin{cases} 1, & \text{if } 1 - \lambda - m|\mu|^\beta \le 0, \\ \lambda + m|\mu|^\beta, & \text{otherwise.} \end{cases}$$

In both situations, we always have $e(\text{Sol(VRP)}(\lambda, \mu), \text{Sol(VRP)}(0, 0)) \le \lambda + m|\mu|^\beta$, which confirms the conclusion of Theorem 10.67. Of course, by choosing $W = \mathbb{R}$, the conclusion of Theorem 10.68 is also obtained.

Theorem 10.71 *The conclusions of Theorems 10.67 and 10.68 are still valid if the conditions (ii) and (iii) are replaced by the following condition:*

There exist real numbers $m, n, \beta, \delta > 0$ such that, if $\mathscr{R}(x_1, y_1; \mu_1)$ and $\mathscr{R}(x_2, y_2; \mu_2)$ hold for some vectors $x_1, x_2, y_1, y_2 \in X$, and $\mu_1, \mu_2 \in M$, then we have

$$d(x_1, x_2) \leq n\left(d^{\delta}(x_1, y_2) + d^{\delta}(x_2, y_1)\right) + md^{\beta}(\mu_1, \mu_2).$$

Proof Let $\lambda \in U(\bar{\lambda}), \mu \in V(\bar{\mu}), x(\lambda, \mu) \in \mathrm{Sol(VRP)}(\lambda, \mu)$. Because $x(\bar{\lambda}, \bar{\mu}) \in \mathscr{D}(\bar{\lambda})$ and $x(\lambda, \mu) \in \mathscr{D}(\lambda)$, condition (i) implies that there exist vectors $x \in \mathscr{D}(\lambda)$ and $x' \in \mathscr{D}(\bar{\lambda})$ such that

$$d\left(x(\lambda, \mu), x'\right) \leq ld^{\alpha}(\bar{\lambda}, \lambda) \text{ and } d(x(\bar{\lambda}, \bar{\mu}), x) \leq ld^{\alpha}(\bar{\lambda}, \lambda).$$

On the other hand, since $\mathscr{R}\left(x(\bar{\lambda}; \bar{\mu}), x'.\bar{\mu}\right)$ and $\mathscr{R}(x(\lambda, \mu), x; \mu)$ are satisfied, we have

$$d(x(\bar{\lambda}, \bar{\mu}), x(\lambda, \mu)) \leq n\left(d^{\delta}(x(\bar{\lambda}, \bar{\mu})), x) + d^{\delta}(x(\lambda, \mu)), x')\right) + md^{\beta}(\bar{\mu}, \mu).$$

Consequently,

$$d(x(\bar{\lambda}, \bar{\mu}), x(\lambda, \mu)) \leq 2nl^{\delta}d^{\alpha\delta}(\bar{\lambda}, \lambda) + md^{\beta}(\bar{\mu}, \mu).$$

The conclusion follows. □

We finish this subsection with an application of the above results to a variational inclusion problem.

Let $F : X \times X \times M \rightrightarrows Y$ and $G : X \times X \times M \rightrightarrows Y$ be two maps with nonempty values. Consider the following variational inclusion problem:
(VIP)(λ, μ) Find $\bar{x} \in \mathscr{D}(\lambda)$ such that

$$F(\bar{x}, y, \mu) \subset G(\bar{x}, y, \mu) \text{ for every } y \in \mathscr{D}(\lambda).$$

For $\lambda \in \Lambda$ and $\mu \in M$, we denote by $\mathrm{Sol(VIP)}(\lambda, \mu)$ the solution set of this problem, and we suppose that it is nonempty in some neighborhood of the reference point.

Theorem 10.72 *Let $\bar{\lambda} \in \Lambda$ and $\bar{\mu} \in M$ be given. Assume that there exist some neighborhoods $U(\bar{\lambda})$ and $V(\bar{\mu})$ of $\bar{\lambda}$ and $\bar{\mu}$, respectively, such that the following conditions are satisfied:*

(i') *there exist two real numbers $l, \alpha > 0$ such that, for all $\lambda \in U(\bar{\lambda})$,*

$$H(\mathscr{D}(\lambda), \mathscr{D}(\bar{\lambda})) \leq ld^{\alpha}(\lambda, \bar{\lambda});$$

(ii') *there exist two real number* $h, \eta > 0$ *such that, for all* $x, y \in X$,

$$hd^{\eta}(x, y) \leq e(F(x, y, \bar{\mu}), G(x, y, \bar{\mu})) + e(F(y, x, \bar{\mu}), G(y, x, \bar{\mu})); \tag{10.22}$$

(iii') *there exist three real numbers* $a, \zeta > 0$ *and* $\theta \geq 0$ *such that, for all* $\mu \in V(\bar{\mu})$, *and for all* $x, y \in X$, *with* $x \neq y$, *we have*

$$\begin{aligned} &\max\{e(F(x, y, \bar{\mu}), F(x, y, \mu)), e(G(x, y, \mu), G(x, y, \bar{\mu}))\} \\ &\leq ad^{\zeta}(\mu, \bar{\mu})d^{\theta}(x, y); \end{aligned} \tag{10.23}$$

(iv') *there exist two real numbers* $n, \delta > 0$ *such that, for any* $x, y_1, y_2 \in X$,

$$\begin{aligned} &\max\left\{e\left(F\left(x, y_1, \bar{\mu}\right), F\left(x, y_2, \bar{\mu}\right)\right), e\left(G\left(x, y_1, \bar{\mu}\right), G\left(x, y_2, \bar{\mu}\right)\right)\right\} \\ &\leq \frac{n}{2} d^{\delta}\left(y_1, y_2\right); \end{aligned} \tag{10.24}$$

(v') $\theta < \eta$.

Then, the following statements hold true:

(a) *the solution set* Sol(VIP) $(\bar{\lambda}, \bar{\mu})$ *is a singleton;*
(b) *for every* $\lambda \in U(\bar{\lambda})$ *and* $\mu \in V(\bar{\mu})$,

$$e(\text{Sol(VIP)}(\lambda, \mu), \text{Sol(VIP)}(\bar{\lambda}, \bar{\mu})) \leq k_1 d^{\gamma_1}(\lambda, \bar{\lambda}) + k_2 d^{\gamma_2}(\mu, \bar{\mu}),$$

where k_1, k_2, γ_1 *and* γ_2 *depend on numbers* $n, l, a, h, \delta, \alpha, \eta, \zeta, \ldots$.

Proof In order to apply Theorem 10.67, we are going to show that conditions (ii) and (iii) of the mentioned theorem are satisfied, for the relation $\mathscr{R} \subset X \times X \times M$ defined by:

$\mathscr{R}(x, y, \mu)$ holds if and only if $F(x, y, \mu) \subset G(x, y, \mu)$.

Let $\mu \in V(\bar{\mu})$, and $x, y \in X$ such that $\mathscr{R}(x, y, \bar{\mu})$ and $\mathscr{R}(y, x, \mu)$ hold true. Equivalently,

$$F(x, y, \bar{\mu}) \subset G(x, y, \bar{\mu}) \text{ and } F(y, x, \mu) \subset G(y, x, \mu). \tag{10.25}$$

By the definition of the function e, (10.25) implies that

$$e(F(x, y, \bar{\mu}), G(x, y, \bar{\mu})) = e(F(y, x, \mu), G(y, x, \mu)) = 0. \tag{10.26}$$

Due to conditions (ii'), (iii') and (10.26), in the case $x \neq y$, we have

$$\begin{aligned} hd^{\eta}(x, y) &\leq e(F(x, y, \bar{\mu}), G(x, y, \bar{\mu})) + e(F(y, x, \bar{\mu}), G(y, x, \bar{\mu})) \\ &\leq e(F(y, x, \bar{\mu}), G(y, x, \bar{\mu})) \\ &\leq e(F(y, x, \bar{\mu}), F(y, x, \mu)) + e(F(y, x, \mu), G(y, x, \mu)) \\ &\quad + e(G(y, x, \mu), G(y, x, \bar{\mu})) \\ &\leq e(F(y, x, \bar{\mu}), F(y, x, \mu)) + e(G(y, x, \mu), G(y, x, \bar{\mu})) \\ &\leq 2ad^{\zeta}(\mu, \bar{\mu})d^{\theta}(x, y). \end{aligned}$$

Therefore,

$$d(x, y) \leq \left(\frac{2a}{h}\right)^{\frac{1}{\eta-\theta}} d^{\frac{\varepsilon}{\eta-\theta}}(\mu, \bar{\mu}),$$

that is condition (10.14) holds. If $x = y$, then this condition is obvious.

We now check inequality (10.15). Let $x_1, x_2, y_1, y_2 \in X$ such that the relations $\mathscr{R}(x_1, y_1, \bar{\mu})$ and $\mathscr{R}(x_2, y_2, \bar{\mu})$ are satisfied, and hence the inclusions $F(x_1, y_1, \bar{\mu}) \subset G(x_1, y_1, \bar{\mu})$ and $F(x_2, y_2, \bar{\mu}) \subset G(x_2, y_2, \bar{\mu})$ hold.

By using (10.23), (10.24) and the triangle inequality, we obtain

$$\begin{aligned} &hd^{\eta}(x_1, x_2) \\ &\leq e(F(x_1, x_2, \bar{\mu}), G(x_1, x_2, \bar{\mu})) + e(F(x_2, x_1, \bar{\mu}), G(x_2, x_1, \bar{\mu})) \\ &\leq e(F(x_1, x_2, \bar{\mu}), F(x_1, y_1, \bar{\mu})) + e(F(x_1, y_1, \bar{\mu}), G(x_1, y_1, \bar{\mu})) + \\ &\quad + e(G(x_1, y_1, \bar{\mu}), G(x_1, x_2, \bar{\mu})) + e(F(x_2, x_1, \bar{\mu}), F(x_2, y_2, \bar{\mu})) + \\ &\quad + e(F(x_2, y_2, \bar{\mu}), G(x_2, y_2, \bar{\mu})) + e(G(x_2, y_2, \bar{\mu}), G(x_2, x_1, \bar{\mu})) \\ &\leq n\left(d^{\delta}(x_2, y_1) + d^{\delta}(x_1, y_2)\right), \end{aligned}$$

this brings the proof to the end. □

Similar to the above arguments, we also provide a weakened version of Theorem 10.72 as follows.

Theorem 10.73 *Let $\bar{\lambda} \in \Lambda, \bar{\mu} \in M$, and $x(\bar{\lambda}, \bar{\mu}) \in$ Sol(VIP)$(\bar{\lambda}, \bar{\mu})$ be fixed. Assume that the following conditions hold.*

(i') *$\mathscr{D}$ is Hölder calm at $\bar{\lambda}$, that is, there exist two real numbers $l, \alpha > 0$ and a neighborhood $U(\bar{\lambda})$ of $\bar{\lambda}$ such that, for all $\lambda \in U(\bar{\lambda})$,*

$$H(\mathscr{D}(\lambda), \mathscr{D}(\bar{\lambda})) \leq ld^{\alpha}(\lambda, \bar{\lambda});$$

(ii') *there exist two real numbers* $h, \eta > 0$ *such that, for all* $x, y \in X$,

$$hd^{\eta}(x, y) \leq e(F(x, y, \bar{\mu}), G(x, y, \bar{\mu})) + e(F(y, x, \bar{\mu}), G(y, x, \bar{\mu})); \tag{10.27}$$

(iii') *there exist real numbers* $a, \zeta > 0$, $\theta \geq 0$, *a neighborhood* $V(\bar{\mu})$ *of* $\bar{\mu}$, *and a neighborhood* W *of* $x(\bar{\lambda}, \bar{\mu})$ *such that, for all* $\mu \in V(\bar{\mu})$, $x \in \bar{W}$ *and* $y \in X$, *with* $x \neq y$,

$$\begin{aligned}&\max\{e(F(x, y, \bar{\mu}), F(x, y, \mu)), e(G(x, y, \mu), G(x, y, \bar{\mu}))\}\\&\quad\leq ad^{\zeta}(\mu, \bar{\mu})d^{\theta}(x, y);\end{aligned} \tag{10.28}$$

(iv') *there exist* $n, \delta > 0$ *such that for any* $x, y_1, y_2 \in X$,

$$\begin{aligned}&\max\left\{e\left(F\left(x, y_1, \bar{\mu}\right), F\left(x, y_2, \bar{\mu}\right)\right), e\left(G\left(x, y_1, \bar{\mu}\right), G\left(x, y_2, \bar{\mu}\right)\right)\right\}\\&\quad\leq \frac{n}{2}d^{\delta}\left(y_1, y_2\right);\end{aligned} \tag{10.29}$$

(v') $\theta < \eta$.

Then, the following statements hold true

(a) *the set* Sol(VIP) $(\bar{\lambda}, \bar{\mu}) \cap \bar{W}$ *is a singleton;*
(b) *for every* $\lambda \in U(\bar{\lambda})$ *and* $\mu \in V(\bar{\mu})$,

$$e(\mathrm{Sol(VIP)}(\lambda, \mu) \cap \bar{W}, \mathrm{Sol(VIP)}(\bar{\lambda}, \bar{\mu})) \leq k_1 d^{\gamma_1}(\lambda, \bar{\lambda}) + k_2 d^{\gamma_2}(\mu, \bar{\mu})$$

where k_1, k_2, γ_1 *and* γ_2 *depend on numbers* $n, l, a, h, \delta, \alpha, \eta, \zeta, \ldots$

Finally, we consider an example to illustrate the applicability of the above theorems.

Example 10.74 Let $X = \mathbb{R}^2_+$ and $M = [0, \infty[$. We consider the following inclusion in the Euclidean space $\mathbb{R}^2$,

$$B(\mathbf{0}, \mu) \subset B\left(x, \frac{1}{4}d(x, y) + \mu\right),$$

where $\mathbf{0} = (0, 0)$, x, y are points in X, $d(x, y)$ is the Euclidean distance between x and y and $B(x, r)$ denotes the closed ball centered at x of radius r. For $\Lambda \equiv M$, we prove that the following maps

$$\mathcal{D}(\lambda) \equiv X,$$

$$F(x, y, \mu) = B(\mathbf{0}, \mu),$$

and

$$G(x, y, \mu) = B\left(x, \frac{1}{4}d(x, y) + \mu\right)$$

verify conditions (i')–(iv') of Theorems 10.72 and 10.73, for $\bar{\lambda} = \bar{\mu} = 0$. Clearly, condition (i') of the above theorems holds true.

Let $x = (x_1, x_2)$, $y = (y_1, y_2) \in X$. We have

$$e\left(\{\mathbf{0}\}, B\left(x, \frac{1}{4}d(x, y)\right)\right) = \begin{cases} 0, & \text{if } d(\mathbf{0}, x) \leq \frac{1}{4}d(x, y), \\ d(\mathbf{0}, x) - \frac{1}{4}d(x, y), & \text{otherwise.} \end{cases} \tag{10.30}$$

Due to the triangle inequality, it is impossible that both

$$e\left(\{\mathbf{0}\}, B\left(x, \frac{1}{4}d(x, y)\right)\right) \quad \text{and} \quad e\left(\{\mathbf{0}\}, B\left(y, \frac{1}{4}d(x, y)\right)\right)$$

are zero. If the first excess is nonzero and the second one is zero, then by (10.30), we have

$$\begin{aligned} e\left(\{\mathbf{0}\}, B\left(x, \frac{1}{4}d(x, y)\right)\right) + e\left(\{\mathbf{0}\}, B\left(y, \frac{1}{4}d(x, y)\right)\right) &= d(\mathbf{0}, x) - \frac{1}{4}d(x, y) \geq \\ \geq d(\mathbf{0}, x) - \frac{1}{4}d(x, y) + d(\mathbf{0}, y) - \frac{1}{4}d(x, y) & \\ = \sqrt{x_1^2 + x_2^2} + \sqrt{y_1^2 + y_2^2} - \frac{1}{2}d(x, y) & \\ \geq \sqrt{(x_1 - y_1)^2 + (x_2 - y_2)^2} - \frac{1}{2}d(x, y) = \frac{1}{2}d(x, y), & \end{aligned}$$

and hence condition (ii') in the mentioned theorems is satisfied. The same can be obtained if the first excess is zero and the second one is nonzero or both terms of the sum are nonzero.

On the other hand, we have $e(\{\mathbf{0}\}, B(\mathbf{0}, \mu))) = 0$ and

$$e\left(B\left(x, \frac{1}{4}d(x, y) + \mu\right), B\left(x, \frac{1}{4}d(x, y)\right)\right) = \mu,$$

and so (10.23) holds with $a = \zeta = 1$ and $\theta = 0$.

Finally, to confirm condition (iv'), we notice here that

$$e\left(B\left(x, \frac{1}{4}d\,(x, y)\right), B\left(x, \frac{1}{4}d\,(x, z)\right)\right) = \begin{cases} 0, & \text{if } d\,(x, y) \leq d\,(x, z), \\ \frac{1}{4}d\,(y, z), & \text{otherwise.} \end{cases}$$

Therefore, by these theorems, the solution map of this problem satisfies the Hölder calmness conditions.

10.3.3 Stability Conditions for Variational Inclusion Problems

Let X and Λ be Hausdorff topological spaces and $\mathbb{Z}$ be a topological vector space. Let $\mathcal{D}_1, \mathcal{D}_2 : X \times \Lambda \rightrightarrows X$ be surjective, and $F : \mathbb{X} \times X \times \Lambda \rightrightarrows \mathbb{Z}$. For each $\lambda \in \Lambda$, we investigate the following variational inclusion problem,

$$(\text{QVIP}_\lambda) : \text{find} \bar{x} \in \mathcal{D}_1(\bar{x}, \lambda) \text{ such that, for each } y \in \mathcal{D}_2(\bar{x}, \lambda),\ 0 \in F(\bar{x}, y, \lambda).$$

Upper Semicontinuity Property of Solution Maps

For $\lambda \in \Lambda$, we denote the set of solutions of (QVIP_λ) by Sol(VIP)(λ). Let

$$\mathcal{E}(\lambda) := \{x \in X : x \in \mathcal{D}_1(x, \lambda)\}.$$

We assume that Sol(VIP)$(\lambda) \neq \emptyset$ and $\mathcal{E}(\lambda) \neq \emptyset$ for all mentioned λ in a neighborhood of $\bar{\lambda} \in \Lambda$. In this section, we investigate sufficient conditions for Sol(VIP)$(\cdot)$ to satisfy various upper semicontinuity properties.

Theorem 10.75 *Suppose for* $(\text{QVIP})_{\lambda \in \Lambda}$ *that*

(a) $\mathcal{D}_2(x, .)$ *is lsc at* $\bar{\lambda}$ *for all* $x \in \mathcal{E}(\bar{\lambda})$;
(b) $F(x, ., .)$ *has the* 0*-inclusion property in* $\mathcal{D}_2(\mathcal{E}(\bar{\lambda}), \bar{\lambda}) \times \{\bar{\lambda}\}$ *for all* $x \in \mathcal{E}(\bar{\lambda})$.

If $\mathcal{E}$ *is outer open or star-outer semicontinuous at* $\bar{\lambda}$*, then so is* Sol(VIP).

Proof By the similarity, we consider only the case of star-outer semicontinuity. Let $x \in \limsup^*_{\lambda \to \bar{\lambda}}$ Sol(VIP)(λ) be arbitrary, then there is $\{\lambda^\nu\} \subset \Lambda$ converging to $\bar{\lambda}$ such that $x \in$ Sol(VIP)(λ^ν) for all ν. As $x \in \mathcal{E}(\lambda^\nu)$, the star-outer semicontinuity of $\mathcal{E}$ implies that $x \in \mathcal{E}(\bar{\lambda})$. We claim that $x \in$ Sol(VIP)$(\bar{\lambda})$. Indeed, for $y \in \mathcal{D}_2(x, \bar{\lambda})$, the lower semicontinuity of $\mathcal{D}_2(x, .)$ at $\bar{\lambda}$ yields $y^\nu \in \mathcal{D}_2(x, \lambda^\nu)$ such that $y^\nu \to y$. Since $0 \in F(x, y^\nu, \lambda^\nu)$, it follows from (b) that $0 \in F(x, y, \bar{\lambda})$. □

Theorem 10.76 *If* $\mathcal{E}$ *is outer semicontinuous at* $\bar{\lambda}$*, so is* Sol(VIP)*, provided that*

(a) $\mathcal{D}_2$ *is lsc in* $\mathcal{E}(\bar{\lambda}) \times \{\bar{\lambda}\}$;
(b) F *has the* 0*-inclusion property in* $\mathcal{E}(\bar{\lambda}) \times \mathcal{D}_2(\mathcal{E}(\bar{\lambda}), \bar{\lambda}) \times \{\bar{\lambda}\}$.

Proof Let $x \in \limsup_{\lambda \to \bar{\lambda}}$ Sol(VIP)(λ). There are nets $\{\lambda^\nu\}$ converging to $\bar{\lambda}$ and $\{x^\nu\}$ converging to x with $x^\nu \in$ Sol(VIP)(λ^ν). By the outer semicontinuity of $\mathcal{E}$, $x \in \mathcal{E}(\bar{\lambda})$. To show that $x \in$ Sol(VIP)$(\bar{\lambda})$, let $y \in \mathcal{D}_2(x, \bar{\lambda})$. The lower semicontinuity of $\mathcal{D}_2$ in $\mathcal{E}(\bar{\lambda}) \times \{\bar{\lambda}\}$ implies the existence of $y^\nu \in \mathcal{D}_2(x^\nu, \lambda^\nu)$ with $y^\nu \to y$. Because $0 \in F(x^\nu, y^\nu, \lambda^\nu)$, assumption (b) implies that $x \in$ Sol(VIP)$(\bar{\lambda})$. □

Theorem 10.77 *The solution map* Sol(VIP) *of* $(\mathrm{QVIP})_{\lambda\in\Lambda}$ *is both usc and closed at* $\bar{\lambda}$, *if*

(a) $\mathscr{D}_2$ *is lsc in* $\mathscr{E}(\bar{\lambda}) \times \{\bar{\lambda}\}$;
(b) F *has the 0-inclusion property in* $\mathscr{E}(\bar{\lambda}) \times \mathscr{D}_2(\mathscr{E}(\bar{\lambda}), \bar{\lambda}) \times \{\bar{\lambda}\}$;
(c) $\mathscr{E}$ *is usc at* $\bar{\lambda}$ *and* $\mathscr{E}(\bar{\lambda})$ *is compact.*

Proof Suppose there is an open set $U \supset \mathrm{Sol(VIP)}(\bar{\lambda})$ such that

$$\forall \lambda^\nu \to \bar{\lambda}, \exists x^\nu \in \mathrm{Sol(VIP)}(\lambda^\nu), \forall \nu, x^\nu \notin U.$$

By the upper semicontinuity of $\mathscr{E}$ and the compactness of $\mathscr{E}(\bar{\lambda})$, one can assume that $x^\nu \to x \in \mathscr{E}(\bar{\lambda})$. We claim that $x \in \mathrm{Sol(VIP)}(\bar{\lambda})$. Indeed, for $y \in \mathscr{D}_2(x, \bar{\lambda})$, the lower semicontinuity of $\mathscr{D}_2$ in $\mathscr{E}(\bar{\lambda}) \times \{\bar{\lambda}\}$ yields $y^\nu \in \mathscr{D}_2(x^\nu, \lambda^\nu)$ with $y^\nu \to y$. Since $0 \in F(x^\nu, y^\nu, \lambda^\nu)$, (b) gives that

$$x \in \mathrm{Sol(VIP)}(\bar{\lambda}) \subset U,$$

which is a contradiction, since $x^\nu \notin U$, for all ν. Now let $(\lambda^\nu, x^\nu) \to (\bar{\lambda}, x)$ with $x^\nu \in \mathrm{Sol(VIP)}(\lambda^\nu)$. Arguing similarly as above, we see that $x \in \mathrm{Sol(VIP)}(\bar{\lambda})$. □

Remark 10.78 Assumption (c) in Theorem 10.77 can be replaced by the condition that X is compact, $\mathscr{D}_1$ is usc and closed-valued in $X \times \{\bar{\lambda}\}$. Indeed, let $x^\nu \in \mathscr{E}(\lambda^\nu)$ and $\lambda^\nu \to \bar{\lambda}$. We need in the proof of Theorem 10.77 that $x^\nu \to x$ for some $x \in \mathscr{E}(\bar{\lambda})$. Suppose $x^\nu \to x \notin \mathscr{D}_1(x, \bar{\lambda})$. Because $\mathscr{D}_1(x, \bar{\lambda})$ is closed, there are neighborhoods N of x and V of $\mathscr{D}_1(x, \bar{\lambda})$ such that $N \cap V = \emptyset$. Since $\mathscr{D}_1$ is usc at $(x, \bar{\lambda})$, without loss of generality, we may assume that $\mathscr{D}_1(x^\nu, \lambda^\nu) \subset V$ for each ν. Then, we have $x^\nu \in \mathscr{D}_1(x^\nu, \lambda^\nu) \subset V$ and hence $x^\nu \notin N$ for each ν, contradicting the convergence $x^\nu \to x$.

The following example shows that assumption (b) in Theorems 10.75–10.77 may be satisfied even when neither outer continuity nor other upper semicontinuity of F is fulfilled.

Example 10.79 Let $X = \mathbb{W} = \mathbb{R}$, $\Lambda = [0, 1]$, $\mathscr{D}_1(x, \lambda) \equiv \mathscr{D}_2(x, \lambda) \equiv [0, 1]$, $\bar{\lambda} = 0$, and

$$F(x, y, \lambda) = \begin{cases} \{0\}, & \text{if } \lambda = 0, \\ [-1, 1], & \text{otherwise.} \end{cases}$$

Then, it is not hard to see that all the assumptions of Theorems 10.75–10.77 are satisfied and, accordingly, Sol(VIP) is outer open, star-outer semicontinuous, outer semicontinuous, usc and closed at 0 (in fact Sol(VIP)(λ)=[0,1] for all $\lambda \in [0, 1]$). One easily checks that F is not outer open, not star-outer open, not star-outer semicontinuous, not outer semicontinuous, not usc at (0, 0, 0).

The following three examples illustrate Theorems 10.75 and 10.76.

Example 10.80 Let $X = \mathbb{W} = \mathbb{R}$, $\Lambda = [0, 1]$, $\mathcal{D}_1(x, \lambda) = (-1, \lambda)$, $\mathcal{D}_2(x, \lambda) \equiv [0, 1]$, $\bar{\lambda} = 0$, and

$$F(x, y, \lambda) = \begin{cases} \{0\}, & \text{if } \lambda = 0, \\ [-1, 1], & \text{otherwise.} \end{cases}$$

We have $\mathcal{E}(\lambda) = (-1, \lambda)$ for $\lambda \in [0, 1]$. Hence, $\mathcal{E}$ is outer open (but not star-outer semicontinuous, not outer semicontinuous, not usc) at 0. It is not hard to see that all the assumptions in Theorem 10.75 are satisfied, and, according to it, Sol(VIP) is outer open at 0 (in fact Sol(VIP)$(\lambda) = (-1, \lambda)$ for all $\lambda \in [0, 1]$). Evidently in this case, Sol(VIP) is not star-outer semicontinuous, not outer semicontinuous, not usc at 0.

Example 10.81 Let $X = \mathbb{R}^2$, $\mathbb{W} = \mathbb{R}$, $\Lambda = [0, 1]$, $\bar{\lambda} = 0$,

$$\mathcal{D}_1(x, \lambda) = \{(t, \lambda t) : t > 0\},$$

$$\mathcal{D}_2(x, \lambda) \equiv [0, 1] \times [0, 1],$$

and

$$F(x, y, \lambda) = \begin{cases} \{0\}, & \text{if } \lambda = 0, \\ [-1, 1], & \text{otherwise.} \end{cases}$$

Since $\mathcal{E}(\lambda) = \{(t, \lambda t) : t > 0\}$ for $\lambda \in [0, 1]$, $\mathcal{E}$ is star-outer semicontinuous (but neither outer semicontinuous nor usc) at 0. It is not hard to see that all the assumptions in Theorem 10.75 are satisfied, and, according to this statement, Sol(VIP) is star-outer semicontinuous at 0 (in fact Sol(VIP)$(\lambda) = \{(t, \lambda t) : t > 0\}$ for $\lambda \in [0, 1]$). Evidently in this case, Sol(VIP) is neither outer semicontinuous nor usc at 0.

Example 10.82 Let $X = \mathbb{R}^2$, $\mathbb{W} = \mathbb{R}$, $\Lambda = [0, 1]$, $\bar{\lambda} = 0$,

$$\mathcal{D}_1(x, \lambda) = \{(t, \lambda t) : t \in \mathbb{R}\},$$

$$\mathcal{D}_2(x, \lambda) \equiv [0, 1] \times [0, 1],$$

and

$$F(x, y, \lambda) = \begin{cases} 0 & \text{if } \lambda = 0, \\ [-1, 1] & \text{otherwise.} \end{cases}$$

Then, $\mathcal{E}$ is outer semicontinuous (but not usc) at 0, since $\mathcal{E}(\lambda) = \{(t, \lambda t) : t \in \mathbb{R}\}$ for $\lambda \in [0, 1]$. It is not hard to see that all the assumptions in Theorem 10.76

are satisfied and, accordingly, Sol(VIP) is outer semicontinuous at 0 (in fact $\mathrm{Sol(VIP)}(\lambda) = \{(t, \lambda t) : t \in \mathbb{R}\}$ for $\lambda \in [0, 1]$). Evidently in this case, Sol(VIP) is not usc at 0.

Lower Semicontinuity Property of Solution Maps

Theorem 10.83 *For* $(\mathrm{QVIP})_{\lambda\in\Lambda}$, *if* $\mathcal{E}$ *is inner open or lsc at* $\bar{\lambda}$, *then so is* Sol(VIP), *provided that*

(a) $\mathcal{D}_2$ *is usc and has compact values in* $\mathcal{E}(\bar{\lambda}) \times \{\bar{\lambda}\}$;
(b) F *has the 0-inclusion complement property in* $\mathcal{E}(\bar{\lambda}) \times \mathcal{D}_2(\mathcal{E}(\bar{\lambda}), \bar{\lambda}) \times \{\bar{\lambda}\}$.

Proof Because of the similarity, we check only the inner openness. Suppose to the contrary that there exists $x \in \mathrm{Sol(VIP)}(\bar{\lambda})$ such that $x \notin \mathrm{liminfo}_{\lambda\to\bar{\lambda}}\mathrm{Sol(VIP)}(\lambda)$. As $x \in \mathcal{E}(\bar{\lambda})$, by the inner openness of $\mathcal{E}$, one has $x \in \mathrm{liminfo}_{\lambda\to\bar{\lambda}}\mathcal{E}(\lambda)$. Then,

$$\exists U \in \mathcal{N}(\bar{\lambda}), \exists V \in \mathcal{N}(x), \forall \lambda \in U, V \subset \mathcal{E}(\lambda).$$

Since

$$\mathrm{liminfo}_{\lambda\to\bar{\lambda}}\mathrm{Sol(VIP)}(\lambda) = [\limsup_{\lambda\to\bar{\lambda}} S^c(\lambda)]^c,$$

we have

$$x \in \limsup_{\lambda\to\bar{\lambda}} S^c(\lambda).$$

Therefore, there exist a net λ^ν converging to $\bar{\lambda}$ and a net $x^\nu \in S^c(\lambda^\nu)$ converging x. We can assume that $(\lambda^\nu, x^\nu) \in U \times V$ for all ν, and hence $x^\nu \in \mathcal{E}(\lambda^\nu)$. Then, there is $y^\nu \in \mathcal{D}_2(x^\nu, \lambda^\nu)$ such that

$$0 \notin F(x^\nu, y^\nu, \lambda^\nu).$$

As $\mathcal{D}_2$ is usc at $(x, \bar{\lambda})$ and $\mathcal{D}_2(x, \bar{\lambda})$ is compact, one finds $y \in \mathcal{D}_2(x, \bar{\lambda})$ such that $y^\nu \to y$ (taking a subnet). As $x \in \mathrm{Sol(VIP)}(\bar{\lambda})$, we have $0 \in F(x, y, \bar{\lambda})$. Assumption (b) implies the existence of $\bar{\nu}$ such that $0 \in F(x^{\bar{\nu}}, y^{\bar{\nu}}, \lambda^{\bar{\nu}})$, a contradiction. □

Theorem 10.84 *The star-inner semicontinuity of* $\mathcal{E}$ *at* $\bar{\lambda}$ *implies the same property for* Sol(VIP), *if*

(a) $\mathcal{D}_2(x, .)$ *is usc at* $\bar{\lambda}$ *and* $\mathcal{D}_2(x, \bar{\lambda})$ *is compact for all* $x \in \mathcal{E}(\bar{\lambda})$;
(b) $F(x, ., .)$ *has the* 0-*inclusion complement property in* $\mathcal{D}_2(\mathcal{E}(\bar{\lambda}), \bar{\lambda}) \times \{\bar{\lambda}\}$ *for all* $x \in \mathcal{E}(\bar{\lambda})$.

Proof Suppose to the contrary the existence of $x \in \mathrm{Sol(VIP)}(\bar{\lambda})$ such that

$$x \notin \operatorname{liminf}^*_{\lambda \to \bar{\lambda}} \mathrm{Sol(VIP)}(\lambda).$$

Then, there exists λ^ν converging $\bar{\lambda}$ such that

$$x \notin \mathrm{Sol(VIP)}(\lambda^\nu),$$

for all ν. The star-inner semicontinuity of $\mathscr{E}$ implies that

$$x \in \operatorname{liminf}^*_{\lambda \to \bar{\lambda}} \mathscr{E}(\lambda).$$

Hence, there exists a neighborhood U of $\bar{\lambda}$ such that $x \in \mathscr{D}_1(x, \lambda)$ for all $\lambda \in U$. Assuming that $\lambda^\nu \in U$ for all ν, one has $x \in \mathscr{D}_1(x, \lambda^\nu)$ and $x \notin \mathrm{Sol(VIP)}(\lambda^\nu)$ for all ν. Therefore, there exists $y^\nu \in \mathscr{D}_2(x, \lambda^\nu)$ with

$$0 \notin F(x, y^\nu, \lambda^\nu).$$

Since $\mathscr{D}_2(x, .)$ is usc at $\bar{\lambda}$ and $\mathscr{D}_2(x, \bar{\lambda})$ is compact, one has $y \in \mathscr{D}_2(x, \bar{\lambda})$ such that $y^\nu \to y$ (taking a subnet if necessary). As $x \in \mathrm{Sol(VIP)}(\bar{\lambda})$, we have

$$0 \in F(x, y, \bar{\lambda}).$$

Assumption (b) yields some $\bar{\nu}$ such that $0 \in F(x, y^{\bar{\nu}}, \lambda^{\bar{\nu}})$, a contradiction. □

The following example indicates that assumption (b) in Theorems 10.83 and 10.84 may be satisfied even when many types of inner semicontinuity of F are not fulfilled.

Example 10.85 Let $X = \mathbb{W} = \mathbb{R}$, $\Lambda = [0, 1]$, $\mathscr{D}_1(x, \lambda) \equiv (0, 1)$, $\mathscr{D}_2(x, \lambda) \equiv [0, 1]$, $\bar{\lambda} = 0$, and

$$F(x, y, \lambda) = \begin{cases} [-1, 1] & \text{if } \lambda = 0, \\ \{0\} & \text{otherwise.} \end{cases}$$

Then, the assumptions of Theorems 10.83 and 10.84 are satisfied and, according to them, Sol(VIP) is inner open, star-inner semicontinuous and lsc at 0 (in fact $\mathrm{Sol(VIP)}(\lambda)=(0,1)$ for all $\lambda \in [0, 1]$). Checking directly, we see that F is not inner open, not star-inner continuous, not lsc at $(0, 0, 0)$.

To develop other conditions for lower semicontinuity of Sol(VIP), which are more suitable than the above results in some cases, we need the following definition. $G : X \times X \rightrightarrows \mathbb{W}$ is called generalized 0-convex on a convex set $A \subset X$ if, for all $x, y_1, y_2 \in A$, from $0 \in G(x, y_1)$ and $0 \in \operatorname{int}G(x, y_2)$, it follows that $0 \in \operatorname{int}G(x, (1-t)y_1 + ty_2)$ for all $t \in (0, 1)$.

Note that this is a modification of the generalized Δ-concavity. Indeed, let $g : X \times X \to \mathbb{W}$ be a single-valued map, $\Delta : \mathbb{X} \rightrightarrows \mathbb{W}$, and $A \subset X$. If we set

$$H(x, y) := g(x, y) - \Delta(x),$$

then H is generalized 0-convex on A if and only if g is generalized Δ-concave on A. We use the term "convex" instead of "concave" to suit the following known definition. $G : X \rightrightarrows \mathbb{W}$ is said to be convex (concave) on $A \subset X$ if, for each $x, y \in A$ and $t \in [0, 1]$,

$$(1-t)G(x) + tG(y) \subset G((1-t)x + ty)$$

$$(G((1-t)x + ty) \subset (1-t)G(x) + tG(y), \text{ resp.}).$$

We consider also the following problem $(\widetilde{\text{QVIP}_\lambda})$ as auxiliary to (QVIP_λ)

$$(\widetilde{\text{QVIP}_\lambda}) : \text{ find } \bar{x} \in \mathscr{D}_1(\bar{x}, \lambda) \text{ such that, for each } y \in \mathscr{D}_2(\bar{x}, \lambda),\ 0 \in \text{int}F(\bar{x}, y, \lambda).$$

Let $\widetilde{\text{Sol}}(\text{VIP})(\lambda)$ be the solution set of $(\widetilde{\text{QVIP}_\lambda})$. Clearly $\widetilde{\text{Sol}}(\text{VIP})(\lambda) \subset \text{Sol}(\text{VIP})(\lambda)$.

Theorem 10.86 *Assume for problem* $(\widetilde{QVIP_\lambda})$ *that* $\widetilde{\text{Sol}}(\text{VIP})(\lambda) \neq \emptyset$ *in a neighborhood of* $\bar{\lambda} \in \Lambda$ *and*

(a) $\mathscr{D}_2$ *is usc and has compact values in* $\mathscr{E}(\bar{\lambda}) \times \{\bar{\lambda}\}$; $\mathscr{D}_2(., \bar{\lambda})$ *is concave on* $\mathscr{E}(\bar{\lambda})$;
(b) $\text{int}F$ *has the 0-inclusion complement property on* $\mathscr{E}(\bar{\lambda}) \times \mathscr{D}_2(\mathscr{E}(\bar{\lambda}), \bar{\lambda}) \times \{\bar{\lambda}\}$;
(c) $\mathscr{E}$ *is lsc at* $\bar{\lambda}$ *and* $\mathscr{E}(\bar{\lambda})$ *is convex;*
(d) $F(., ., \bar{\lambda})$ *is generalized* 0*-convex on* $\mathscr{E}(\bar{\lambda}) \times \mathscr{D}_2(\mathscr{E}(\bar{\lambda}), \bar{\lambda})$.

Then, Sol(VIP) *is lsc at* $\bar{\lambda}$.

Proof First, we prove that $\widetilde{S}$ is lsc at $\bar{\lambda}$. Suppose to the contrary that

$$\exists x \in \widetilde{\text{Sol}}(\text{VIP})(\bar{\lambda}), \exists \lambda^\nu \to \bar{\lambda}, \forall x^\nu \in \widetilde{\text{Sol}}(\text{VIP})(\lambda^\nu), x^\nu \not\to x.$$

Since $\mathscr{E}$ is lsc at $\bar{\lambda}$, there is a net $\bar{x}^\nu \in \mathscr{D}_1(\bar{x}^\nu, \lambda^\nu)$, $\bar{x}^\nu \to x$. By the above assumption, there is a subnet $\bar{x}^\beta$ such that

$$\bar{x}^\beta \notin \widetilde{\text{Sol}}(\text{VIP})(\lambda^\beta),$$

for all β, i.e., for some $y^\beta \in \mathscr{D}_2(\bar{x}^\beta, \lambda^\beta)$,

$$0 \notin \text{int}F(\bar{x}^\beta, y^\beta, \lambda^\beta).$$

As $\mathscr{D}_2$ is usc at $(x, \bar{\lambda})$ and $\mathscr{D}_2(x, \bar{\lambda})$ is compact, one has $y \in \mathscr{D}_2(x, \bar{\lambda})$ such that $y^\beta \to y$ (taking a subnet if necessary). As $x \in \widetilde{\text{Sol}}(\text{VIP})(\bar{\lambda})$, we have $0 \in$

$\text{int}F(x, y, \bar{\lambda})$. Since $(\bar{x}^\beta, y^\beta, \lambda^\beta) \to (x, y, \bar{\lambda})$, assumption (b) implies the existence of an index $\bar{\beta}$ such that

$$0 \in \text{int}F(\bar{x}_{\bar{\beta}}, y_{\bar{\beta}}, \lambda_{\bar{\beta}}),$$

which is a contradiction.

Let $\bar{x} \in \text{Sol(VIP)}(\bar{\lambda}), \bar{x}^1 \in \widetilde{\text{Sol}}\text{(VIP)}(\bar{\lambda})$ and $x_t = (1-t)\bar{x} + t\bar{x}^1$ with $t \in (0, 1)$. By the convexity of $\mathscr{E}(\bar{\lambda})$, $x_t \in \mathscr{E}(\bar{\lambda})$. Since $\mathscr{D}_2(., \bar{\lambda})$ is concave, for all $y_t \in \mathscr{D}_2(x_t, \bar{\lambda})$, there exist $\bar{y} \in \mathscr{D}_2(\bar{x}, \bar{\lambda})$, $\bar{y}_1 \in \mathscr{D}_2(\bar{x}^1, \bar{\lambda})$ such that

$$y_t = (1-t)\bar{y} + t\bar{y}_1.$$

Since $F(\cdot, \cdot, \bar{\lambda})$ is generalized 0-convex,

$$0 \in \text{int}F(x_t, y_t, \bar{\lambda}),$$

i.e., $x_t \in \widetilde{\text{Sol}}\text{(VIP)}(\bar{\lambda})$. Hence,

$$\text{Sol(VIP)}(\bar{\lambda}) \subset \text{cl}\widetilde{\text{Sol}}\text{(VIP)}(\bar{\lambda}).$$

By the lower semicontinuity of $\widetilde{S}$ at $\bar{\lambda}$, Sol(VIP) has the same property, since

$$\begin{aligned}\text{Sol(VIP)}(\bar{\lambda}) &\subset \text{cl}\widetilde{\text{Sol}}\text{(VIP)}(\bar{\lambda}) \subset \text{liminf}\widetilde{S}(\lambda^\nu) \subset \text{Sol(VIP)}(\lambda^\nu) \\ &\subset \text{liminfSol(VIP)}(\lambda^\nu).\end{aligned}$$

□

The following example shows that the new assumption (d) is essential.

Example 10.87 Let $X = \mathbb{W} = \mathbb{R}$, $\Lambda = [0, 1]$, $\mathscr{D}_1(x, \lambda) = \mathscr{D}_2(x, \lambda) = [\lambda, \lambda + 3]$, $\bar{\lambda} = 0$, and

$$F(x, y, \lambda) = (-\infty, x - \lambda - 1] \cup [x, +\infty).$$

Then, it is easy to verify that (a), (b), and (c) of Theorem 10.86 are satisfied. But, $\text{Sol(VIP)}(0) = \{0\} \cup [1, 3]$ and $\text{Sol(VIP)}(\lambda) = [\lambda + 1, \lambda + 3]$ for all $\lambda \in (0, 1]$, and thus Sol(VIP) is not lsc at 0. The cause is that (d) is violated. Indeed, let $x_1 = 0$ and $x_2 = 2$. Then, for all $y \in \mathscr{D}_2(X, 0) = [0, 3]$, we have

$$F(0, y, 0) = (-\infty, -1] \cup [0, +\infty),$$

$$F(2, y, 0) = (-\infty, 1] \cup [2, +\infty),$$

and

$$F\left(\frac{1}{2}x_1 + \frac{1}{2}x_2, y, 0\right) = (-\infty, 0] \cup [1, +\infty).$$

Hence, $0 \in F(x_1, y, 0)$ and $0 \in \text{int}F(x_2, y, 0)$, but

$$0 \notin \text{int}F\left(\frac{1}{2}x_1 + \frac{1}{2}x_2, y, 0\right).$$

Theorem 10.86 is useful, while Theorem 10.83 is inapplicable in the following.

Example 10.88 Let $X = \mathbb{W} = \mathbb{R}$, $\Lambda = [0, 1]$, $\mathscr{D}_1(x, \lambda) \equiv \mathscr{D}_2(x, \lambda) \equiv [-2, 2]$, $\bar{\lambda} = 0$, and

$$F(x, y, \lambda) = (-\infty, x - \lambda].$$

Then, $\mathscr{D}_2(x, \lambda)$ satisfies assumption (a) of Theorem 10.86. The set

$$\{(x, y, \lambda) : 0 \notin \text{int}F(x, y, \lambda)\} = \{(x, y, \lambda) : x - \lambda \leq 0\}$$

is closed. Therefore, int$F$ has the 0-inclusion complement property on $\mathscr{E}(\bar{\lambda}) \times \mathscr{D}_2(\mathscr{E}(\bar{\lambda}), \bar{\lambda}) \times \{\bar{\lambda}\}$. Furthermore, $\mathscr{E}(\lambda) \equiv [-2, 2]$ fulfils (c). To check the generalized 0-convexity on $\mathscr{E}(0) \times \mathscr{D}_2(\mathscr{E}(0), 0)$ of $F(., ., 0)$ in (d), let $0 \in F(x_1, y_1, 0)$ and $0 \in \text{int}F(x_2, y_2, 0)$, i.e., $0 \in (-\infty, x_1]$ and $0 \in (-\infty, x_2)$. If $x_1 \geq 0, x_2 > 0$, for all $t \in (0, 1)$, we have $0 \in (-\infty, (1 - t)x_1 + tx_2),$, i.e.,

$$0 \in \text{int}F[(1 - t)(x_1, y_1, 0) + t(x_2, y_2, 0)].$$

According to Theorem 10.86, Sol(VIP) is lsc at 0 (in fact Sol(VIP)$(\lambda) = [\lambda, 2]$ for all $\lambda \in [0, 1]$). However, F does not have the 0-inclusion complement property on $\mathscr{E}(0) \times \mathscr{D}_2(\mathscr{E}(0), 0) \times \{0\}$. Indeed, let $(-\frac{1}{n}, 0, 0) \to (0, 0, 0)$. Due to

$$F(0, 0, 0) = (-\infty, 0], \text{ and } F(-\frac{1}{n}, 0, 0) = (-\infty, -\frac{1}{n}],$$

we have $0 \in F(0, 0, 0)$ but

$$0 \notin F\left(-\frac{1}{n}, 0, 0\right).$$

Therefore, we cannot apply Theorem 10.83.

Well-Posedness for Variational Inclusion Problems

We now discuss sufficient conditions and metric characterizations of well-posedness for variational inclusion problems, and so for the convenience of presentation, we assume that X, Λ are normed spaces. For each $x \in \mathbb{X}$ and each $r > 0$, we denote by $B(x, r)$ the closed ball centered at x with radius r. When $X = \mathbb{R}$, we denote by $B^+(0, r)$ the closed interval $[0, r]$. Now, we define concepts of approximate solution sequences, well-posedness, and well-posedness in the generalized sense for (QVIP).

Definition 10.89 Let $\lambda \in \Lambda$, and let $\{\lambda_n\} \subset \Lambda$ be any sequence such that $\lambda_n \to \lambda$. Let $e : X \to \mathbb{W}$ be a continuous mapping. A sequence $\{x_n\} \subset X$ is called an approximating solution sequence for $(\text{QVIP})_\lambda$ corresponding to $\{\lambda_n\}$ if there exists a sequence $\{\varepsilon_n\}$ of positive real numbers with $\varepsilon_n \to 0$ such that, for each $n \in \mathbb{N}$,

(a) $x_n \in \mathscr{D}_1(x_n, \lambda_n)$;
(b) $0 \in F(x_n, y, \lambda_n) + B^+(0, \varepsilon_n)e(x_n), \quad \forall y \in \mathscr{D}_2(x_n, \lambda_n)$.

Definition 10.90

(a) (QVIP) is said to be well-posed if for every $\lambda \in \Lambda$, $(\text{QVIP})_\lambda$ has a unique solution x_λ, and for every sequence $\{\lambda_n\} \subset \Lambda$ with $\lambda_n \to \lambda$, every approximating solution sequence for $(\text{QVIP})_\lambda$ corresponding to $\{\lambda_n\}$ converges to x_λ.
(b) (QVIP) is said to be well-posed in the generalized sense if for every $\lambda \in \Lambda$, $(\text{QVIP})_\lambda$ has a nonempty solution set $\text{Sol(VIP)}(\lambda)$, and for every sequence $\{\lambda_n\} \subset \Lambda$ with $\lambda_n \to \lambda$, every approximating solution sequence for $(\text{QVIP})_\lambda$ corresponding to $\{\lambda_n\}$ has a subsequence which converges to a point of $\text{Sol(VIP)}(\lambda)$.

For each $\lambda \in \Lambda$,, the approximating solution set for $(\text{QVIP})_\lambda$ is defined for all $\delta, \varepsilon > 0$ by

$$\Omega_\lambda(\delta, \varepsilon) = \bigcup_{\lambda' \in B(\lambda, \delta)} \widetilde{\text{Sol}}(\text{VIP})(\lambda', \varepsilon),$$

where $\widetilde{\text{Sol}}(\text{VIP}) : \Lambda \times \mathbb{R}^+$ is defined by

$$\begin{aligned} &\widetilde{\text{Sol}}(\text{VIP})(\lambda', \varepsilon) \\ &\quad := \{x \in \mathscr{D}_1(x, \lambda') : 0 \in F(x, y, \lambda') + B^+(0, \varepsilon)e(x), \forall y \in \mathscr{D}_2(x, \lambda')\}, \end{aligned} \tag{10.31}$$

for all $\lambda' \in \Lambda$, $\varepsilon \in \mathbb{R}^+$. Clearly, we have, for every $\lambda \in \Lambda$,

(a) $\text{Sol(VIP)}(\lambda) \equiv \widetilde{\text{Sol}}(\text{VIP})(\lambda, 0) \subset \Omega_\lambda(\delta, \varepsilon) \;\; \forall \delta, \varepsilon > 0$;
(b) if $0 < \delta_1 \leq \delta_2$ and $0 < \varepsilon_1 \leq \varepsilon_2$, then $\Omega_\lambda(\delta_1, \varepsilon_1) \subset \Omega_\lambda(\delta_2, \varepsilon_2)$.

Proposition 10.91 *For (QVIP), assume that*

(a) $\mathscr{D}_1$ *is closed and* $\mathscr{D}_2$ *is lower semicontinuous;*
(b) *for each* $x \in X$*, the multivalued map* $(y, \lambda) \mapsto F(x, y, \lambda)$ *is closed.*

Then

$$\mathrm{Sol(VIP)}(\lambda) = \bigcap_{\delta>0,\varepsilon>0} \Omega_\lambda(\delta, \varepsilon)$$

for each $\lambda \in \Lambda$.

Proof For any given $\lambda \in \Lambda$,, it is clear that $\mathrm{Sol(VIP)}(\lambda) \subset \cap_{\delta>0,\varepsilon>0}\Omega_\lambda(\delta, \varepsilon)$. Thus, we only need to show that $\bigcap_{\delta>0,\varepsilon>0} \Omega_\lambda(\delta, \varepsilon) \subset \mathrm{Sol(VIP)}(\lambda)$. Suppose on the contrary that there exists $x^* \in \bigcap_{\delta>0,\varepsilon>0} \Omega_\lambda(\delta, \varepsilon)$ such that $x^* \notin \mathrm{Sol(VIP)}(\lambda)$. Then for each $\delta > 0$ and each $\varepsilon > 0$,

$$x^* \in \Omega_\lambda(\delta, \varepsilon)\backslash \mathrm{Sol(VIP)}(\lambda).$$

In particular, for each $n \in \mathbb{N}$,, we have

$$x^* \in \Omega_\lambda\left(\frac{1}{n}, \frac{1}{n}\right)\backslash \mathrm{Sol(VIP)}(\lambda),$$

and so there exists $\lambda_n \in B(\lambda, \frac{1}{n})$ such that $x^* \in \mathscr{D}_1(x^*, \lambda_n)$ and

$$0 \in F(x^*, y, \lambda_n) + B^+\left(0, \frac{1}{n}\right)e(x^*), \quad \forall y \in \mathscr{D}_2(x^*, \lambda_n). \tag{10.32}$$

Obviously, $\lambda_n \to \lambda$. Since $\mathscr{D}_1$ is closed, $(x^*, \lambda_n) \to (x^*, \lambda)$ and $x^* \in \mathscr{D}_1(x^*, \lambda_n)$,, we have

$$x^* \in \mathscr{D}_1(x^*, \lambda).$$

We observe that for each $y \in \mathscr{D}_2(x^*, \lambda)$, since $\mathscr{D}_2$ is l.s.c. at (x^*, λ) and $(x^*, \lambda_n) \to (x^*, \lambda)$ there exists $y_n \in \mathscr{D}_2(x^*, \lambda_n)$ such that $y_n \to y$ as $n \to \infty$. Combining $y_n \in \mathscr{D}_2(x^*, \lambda_n)$ and (10.32), we have

$$0 \in F(x^*, y_n, \lambda_n) + B^+\left(0, \frac{1}{n}\right)e(x^*), \quad \forall n \in \mathbb{N}.$$

Thus, for each $n \in \mathbb{N}$, there exists a sequence $\{\gamma_n\} \subset B^+(0, \frac{1}{n})$ such that

$$0 \in F(x^*, y_n, \lambda_n) + \gamma_n e(x^*),$$

which gives that $0 - \gamma_n e(x^*) \in F(x^*, y_n, \lambda_n)$, that is,

$$((x^*, y_n, \lambda_n), -\gamma_n e(x^*)) \in \mathrm{gph} F.$$

It is clear that

$$\{((x^*, y_n, \lambda_n), -\gamma_n e(x^*))\} \to ((x^*, y, \lambda), 0).$$

The closedness of the mapping $(y, \lambda) \mapsto F(x, y, \lambda)$ implies that

$$((x^*, y, \lambda), 0) \in \mathrm{gph} F.$$

That is $0 \in F(x^*, y, \lambda)$ and so $x^* \in \mathrm{Sol(VIP)}(\lambda)$, which is a contradiction. Hence,

$$\bigcap_{\delta>0,\varepsilon>0} \Omega_\lambda(\delta, \varepsilon) \subset \mathrm{Sol(VIP)}(\lambda).$$

The proof is complete. □

The following example illustrates the case when Proposition 10.91 is applicable.

Example 10.92 Let $E = \mathbb{W} = \mathbb{R}$, $X = [0, +\infty)$ and $\Lambda = [-1, 1]$. For every $(x, y, \lambda) \in X \times X \times \Lambda$,, let $e(x) = x^2$,

$$\mathscr{D}_1(x, \lambda) = [\lambda^2, +\infty),$$

and

$$\mathscr{D}_2(x, \lambda) = [x^2 + \lambda^2, x^2 + 1].$$

Define a set-valued mapping $F : X \times X \times \Lambda \rightrightarrows \mathbb{W}$ by

$$F(x, y, \lambda) = [-(\lambda^2 + 1)y, 2x - y + 1].$$

Obviously, it is easy to see that all the assumptions of Proposition 10.91 are satisfied. Hence,

$$\mathrm{Sol(VIP)}(\lambda) = \bigcap_{\delta>0,\varepsilon>0} \Omega_\lambda(\delta, \varepsilon),$$

for each $\lambda \in \Lambda$. Indeed, for every $\lambda \in \Lambda$ and every $\delta > 0$, $\varepsilon > 0$,

$$\begin{aligned}
&\mathrm{Sol(VIP)}(\lambda) \\
&\quad = \{x \in \mathscr{D}_1(x, \lambda) : 0 \in F(x, y, \lambda), \;\; \forall y \in \mathscr{D}_2(x, \lambda)\} \\
&\quad = \{x \in [\lambda^2, +\infty) : 0 \in [-(\lambda^2 + 1)y, 2x - y + 1], \; \forall y \in [x^2 + \lambda^2, x^2 + 1]\} \\
&\quad = [\lambda^2, 2],
\end{aligned}$$

and

$$\begin{aligned}\mathrm{Sol(VIP)}(\lambda)(\varepsilon) &= \{x \in \mathscr{D}_1(x,\lambda) : 0 \in F(x,y,\lambda) + B(0,\varepsilon)e(x),\ \ \forall y \in \mathscr{D}_2(x,\lambda)\}\\ &= \{x \in [\lambda^2,+\infty) : 0 \in [-(\lambda^2+1)y, 2x-y+1] + B(0,\varepsilon)x^2,\\ &\qquad \forall y \in [x^2+\lambda^2, x^2+1]\}\\ &= \left[\lambda^2, \frac{2}{1-\varepsilon}\right].\end{aligned}$$

It follows that, for each $\delta > 0$ and $\varepsilon > 0$

$$\begin{aligned}\bigcap_{\delta>0,\varepsilon>0} \Omega_\lambda(\delta,\varepsilon) &= \bigcap_{\varepsilon>0}\left[\bigcap_{\delta>0}\Big(\bigcup_{\lambda'\in B(\lambda,\delta)} S_{\lambda'}(\varepsilon)\Big)\right]\\ &= \bigcap_{\varepsilon>0}\left[\bigcap_{\delta>0}\Big(\bigcup_{\lambda'\in B(\lambda,\delta)} \left[\lambda'^2, \frac{2}{1-\varepsilon}\right]\Big)\right] = [\lambda^2, 2] = \mathrm{Sol(VIP)}(\lambda).\end{aligned}$$

Lemma 10.93 *For (QVIP), assume that*

(a) *$\mathscr{D}_1$ is closed and $\mathscr{D}_2$ is lower semicontinuous;*
(b) *for each $\lambda \in \Lambda$, the map $(x, y) \mapsto F(x, y, \lambda)$ is closed.*

Then, for each $(\lambda, \varepsilon) \in \Lambda \times \mathbb{R}^+$, $\widetilde{\mathrm{Sol}}(\mathrm{VIP})(\lambda, \varepsilon)$ is a closed subset of X, where $\widetilde{\mathrm{Sol}}(\mathrm{VIP})$ is defined by (10.31).

Proof Let $(\lambda, \varepsilon) \in \Lambda \times \mathbb{R}^+$ be fixed. If $x \in \mathrm{cl}\widetilde{\mathrm{Sol}}(\mathrm{VIP})(\lambda, \varepsilon)$,, then there exists a sequence $\{x_n\} \subset \widetilde{\mathrm{Sol}}(\mathrm{VIP})(\lambda, \varepsilon)$ such that $x_n \to x$ as $n \to \infty$. It follows that, for each $n \in \mathbb{N}$, $x_n \in \mathscr{D}_1(x_n, \lambda)$ such that for each $y \in \mathscr{D}_2(x_n, \lambda)$,

$$0 \in F(x_n, y, \lambda) + B^+(0, \varepsilon)e(x_n). \tag{10.33}$$

Since $\mathscr{D}_1$ is closed, we have $x \in \mathscr{D}_1(x, \lambda)$. For each $y \in \mathscr{D}_2(x, \lambda)$, since $\mathscr{D}_2$ is l.s.c., there exists a sequence $\{y_n\}$ with $y_n \in \mathscr{D}_2(x_n, \lambda)$ such that $y_n \to y$ as $n \to \infty$. Combining $y_n \in \mathscr{D}_2(x_n, \lambda)$ and (10.33), we have

$$0 \in F(x_n, y_n, \lambda) + B^+(0, \varepsilon)e(x_n),\ \ \forall n \in \mathbb{N}.$$

Thus there exists a sequence $\{\nu_n\} \subset B^+(0, \varepsilon)$ such that

$$0 \in F(x_n, y_n, \lambda) + \nu_n e(x_n),\ \ \forall n \in \mathbb{N}.$$

Observe that $B^+(0, \varepsilon) := [0, \varepsilon] \subset \mathbb{R}$ is compact, one can assume that $\nu_n \to \nu$, $\nu \in B^+(0, \varepsilon)$ as $n \to \infty$. Since $(x, y) \mapsto F(x, y, \lambda)$ is closed, one has

$$0 \in F(x, y, \lambda) + \nu e(x) \subset F(x, y, \lambda) + B^+(0, \varepsilon)e(x).$$

Therefore $x \in \widetilde{\mathrm{Sol}}(\mathrm{VIP})(\lambda, \varepsilon)$,, and this implies that $\widetilde{\mathrm{Sol}}(\mathrm{VIP})(\lambda, \varepsilon)$ is a closed subset of X. The proof is complcted. □

Lemma 10.94 *For (QVIP), assume that*

(a) *Λ is finite-dimensional;*
(b) *$\mathscr{D}_1$ is closed;*
(c) *$\mathscr{D}_2$ is lower semicontinuous;*
(d) *F is closed.*

Then, for each $\lambda \in \Lambda$ and $\delta, \varepsilon > 0$, $\Omega_\lambda(\delta, \varepsilon)$ is a closed subset of X.

Proof For each $\lambda \in \Lambda$ and any $\delta, \varepsilon > 0$, let $\{x_n\} \subset \Omega_\lambda(\delta, \varepsilon)$ and $x_n \to x$ as $n \to \infty$.
It follows that, for each $n \in \mathbb{N}$, there exists a sequence $\lambda_n \in B(\lambda, \delta)$, such that $x_n \in \mathscr{D}_1(x_n, \lambda_n)$ and

$$0 \in F(x_n, y, \lambda_n) + B^+(0, \varepsilon)e(x_n), \quad \forall y \in \mathscr{D}_2(x_n, \lambda_n). \tag{10.34}$$

Since Λ is finite dimensional, we can suppose that $\lambda_n \to \bar{\lambda} \in \overline{B(\lambda, \delta)}$. By similar arguments as in the proof of Lemma 10.93, we can show that $x \in \mathscr{D}_1(x, \bar{\lambda})$ such that

$$0 \in F(x, y, \bar{\lambda}) + B^+(0, \varepsilon)e(x), \quad \forall y \in \mathscr{D}_2(x, \bar{\lambda}).$$

Thus, $x \in \Omega_\lambda(\delta, \varepsilon)$, and so $\Omega_\lambda(\delta, \varepsilon)$ is closed. This completes the proof. □

Now, we are in a position to state and prove our main results.

Theorem 10.95 *For (QVIP), assume that*

(a) *$\mathscr{E}$ is complete;*
(b) *$\mathscr{D}_1$ is closed;*
(c) *$\mathscr{D}_2$ is lower semicontinuous;*
(d) *F is closed.*

Then, (QVIP) is well-posed if and only if for every $\lambda \in \Lambda$,

$$\Omega_\lambda(\delta, \varepsilon) \neq \emptyset, \forall \delta, \varepsilon > 0, \text{ and } diam(\Omega_\lambda(\delta, \varepsilon)) \to 0 \text{ as } (\delta, \varepsilon) \to (0, 0). \tag{10.35}$$

Proof Supposed that (QVIP) is well-posed. Then, for every $\lambda \in \Lambda$, $(\mathrm{QVIP})_\lambda$ has the unique solution x_λ, and so $\Omega_\lambda(\delta, \varepsilon) \neq \emptyset$, for all $\delta, \varepsilon > 0$. Now we shall show that

$$\mathrm{diam}\,(\Omega_\lambda(\delta, \varepsilon)) \to 0 \text{ as } (\delta, \varepsilon) \to (0, 0). \tag{10.36}$$

Suppose to the contrary that there exist $r > 0$, sequences of positive real numbers $\{\delta_n\}$ and $\{\varepsilon_n\}$ with $(\delta_n, \varepsilon_n) \to (0, 0)$ as $n \to \infty$ and sequences $\{x_n\}$ and $\{x_n'\}$ with $x_n, x_n' \in \Omega_\lambda(\delta_n, \varepsilon_n)$ for each $n \in \mathbb{N}$ such that

$$d(x_n, x_n') > r, \quad \forall n \in \mathbb{N}. \tag{10.37}$$

For each $n \in \mathbb{N}$, since $x_n, x_n' \in \Omega_\lambda(\delta_n, \varepsilon_n)$,, there exist $\lambda_n, \lambda_n' \in B(\lambda, \delta_n)$ such that

$$x_n \in \mathscr{D}_1(x_n, \lambda_n) \text{ and } 0 \in F(x_n, y, \lambda_n) + B^+(0, \varepsilon_n)e(x_n), \quad \forall y \in \mathscr{D}_2(x_n, \lambda_n),$$

and

$$x_n' \in \mathscr{D}_1(x_n', \lambda_n') \text{ and } 0 \in F(x_n', y, \lambda_n') + B^+(0, \varepsilon_n)e(x_n'), \quad \forall y \in \mathscr{D}_2(x_n', \lambda_n').$$

Clearly, $\lambda_n \to \lambda$ and $\lambda_n' \to \lambda$ as $n \to \infty$. Hence $\{x_n\}$ and $\{x_n'\}$ are approximating solution sequences for $(\text{QVIP})_\lambda$ corresponding to λ_n and λ_n', respectively. By the well-posedness of $(\text{QVIP})_\lambda$, $\{x_n\}$ and $\{x_n'\}$ converge to the unique solution x_λ of $(\text{QVIP})_\lambda$, which gives a contradiction to (10.37). This implies that (10.36) holds.

Conversely, suppose that condition (10.35) holds. Let $\lambda \in \Lambda$ be fixed. Let $\{\lambda_n\}$ be any sequence in Λ with $\lambda_n \to \lambda$ as $n \to \infty$. Suppose that $\{x_n\}$ is an approximating solution sequence for $(\text{QVIP})_\lambda$ corresponding to $\{\lambda_n\}$, then there exists a nonnegative sequence $\{\varepsilon_n\} \downarrow 0$ such that for each $n \in \mathbb{N}$,

$$x_n \in \mathscr{D}_1(x_n, \lambda_n) \text{ and } 0 \in F(x_n, y, \lambda_n) + B^+(0, \varepsilon_n)e(x_n), \quad \forall y \in \mathscr{D}_2(x_n, \lambda_n). \tag{10.38}$$

For each $n \in \mathbb{N}$, let $\delta_n = d(\lambda_n, \lambda)$. Then, $\lambda_n \in B(\lambda, \delta_n)$ and $x_n \in \Omega_\lambda(\delta_n, \varepsilon_n)$ for each $n \in \mathbb{N}$, and $\delta_n \to 0$ as $n \to \infty$. It follows from (10.35) that $\{x_n\}$ is a Cauchy sequence, and so it converges to a point $x \in X$. It then follows from the closedness of $\mathscr{D}_1$ that $x \in \mathscr{D}_1(x, \lambda)$. We observe that for each $y \in \mathscr{D}_2(x, \lambda)$, since $\mathscr{D}_2$ is l.s.c., there exists a sequence $\{y_n\}$ with $y_n \in \mathscr{D}_2(x_n, \lambda_n)$ such that $y_n \to y$ as $n \to \infty$. Applying (10.38), one has

$$0 \in F(x_n, y_n, \lambda_n) + B^+(0, \varepsilon_n)e(x_n), \quad \forall n \in \mathbb{N}.$$

Thus, for each $n \in \mathbb{N}$,, there exists a sequence $\{\gamma_n\} \subset B^+(0, \varepsilon_n)$ such that

$$0 \in F(x_n, y_n, \lambda_n) + \gamma_n e(x_n),$$

which gives that

$$((x_n, y_n, \lambda_n), -\gamma_n e(x_n)) \in \text{gph} F.$$

It is clear that

$$((x_n, y_n, \lambda_n), -\gamma_n e(x_n)) \to ((x, y, \lambda), 0) \text{ as } n \to \infty.$$

Therefore, the closedness of the mapping F implies that

$$((x, y, \lambda), 0) \in \text{gph}F,$$

that is $0 \in F(x, y, \lambda)$. Hence x is a solution of $(\text{QVIP})_\lambda$. Next, we will show that $(\text{QVIP})_\lambda$ has a unique solution. Suppose to the contrary that $(\text{QVIP})_\lambda$ has two distinct solutions x_1 and x_2. It is easy to see that $x_1, x_2 \in \Omega_\lambda(\delta, \varepsilon)$ for all $\delta, \varepsilon > 0$. It follows that

$$0 < d(x_1, x_2) \leq \text{diam}(\Omega_\lambda(\delta, \varepsilon)), \forall \delta, \varepsilon > 0,$$

which gives a contradiction to (10.35). This implies that $(\text{QVIP})_\lambda$ has a unique solution. This completes the proof. □

Example 10.96 Let $X = \Lambda = \mathbb{W} = \mathbb{R}, X = [0, 1]$. For every $(x, y, \lambda) \in X \times X \times \Lambda$,, let $e(x) = 1$,

$$\mathscr{D}_1(x, \lambda) := \begin{cases} [0, \frac{1}{2}], & \text{if } \lambda \neq \frac{1}{2}, \\ [0, 1], & \text{if } \lambda = \frac{1}{2}, \end{cases}$$

and

$$\mathscr{D}_2(x, \lambda) := \begin{cases} [0, 1], & \text{if } \lambda \neq \frac{1}{2}, \\ [0, \frac{1}{2}], & \text{if } \lambda = \frac{1}{2}. \end{cases}$$

Define a set-valued mapping $F : X \times X \times \Lambda \rightrightarrows \mathbb{W}$ by

$$F(x, y, \lambda) := (-\infty, (\lambda^2 + 1)(y - x)].$$

It is easy to see that conditions (a)–(c) of Theorem 10.95 are satisfied. However, F is closed. Indeed, let $((x_n, y_n, \lambda_n), z_n)$ be any sequence in gph$F$ such that $((x_n, y_n, \lambda_n), z_n) \to (x, y, \lambda, z)$ as $n \to \infty$. Thus,

$$z_n \in F_n(x_n, y_n, \lambda_n) = (-\infty, (\lambda_n^2 + 1)(y_n - x_n)] \text{ for all } n \in \mathbb{N}. \tag{10.39}$$

We shall prove that $(x, y, \lambda, z) \in \text{gph}F$. By (10.39), we have

$$z_n \leq (\lambda_n^2 + 1)(y_n - x_n), \text{ for all } n \in \mathbb{N}.$$

Since $((x_n, y_n, \lambda_n), z_n) \to (x, y, \lambda, z)$ as $n \to \infty$, we conclude that $z \leq (\lambda^2 + 1)(y - x)$. Thus,

$$z \in (-\infty, (\lambda^2 + 1)(y - x)].$$

That is $z \in F(x, y, \lambda)$ and so $(x, y, \lambda, z) \in \text{gph}F$. This implies that F is closed. It follows that, if $\lambda = \frac{1}{2}$,

$$\begin{aligned}\text{Sol(VIP)}(\lambda) &= \{x \in \mathscr{D}_1(x, \lambda) : 0 \in F(x, y, \lambda), \ \ \forall y \in \mathscr{D}_2(x, \lambda)\} \\ &= \left\{x \in [0, 1] : 0 \in (-\infty, (\lambda^2 + 1)(y - x)], \ \ \forall y \in \left[0, \frac{1}{2}\right]\right\} \\ &= \{0\},\end{aligned}$$

if $\lambda \neq \dfrac{1}{2}$

$$\begin{aligned}\text{Sol(VIP)}(\lambda) &= \{x \in \mathscr{D}_1(x, \lambda) : 0 \in F(x, y, \lambda), \forall y \in \mathscr{D}_2(x, \lambda)\} \\ &= \left\{x \in \left[0, \frac{1}{2}\right] : 0 \in (-\infty, (\lambda^2 + 1)(y - x)], \ \ \forall y \in [0, 1]\right\} \\ &= \{0\}.\end{aligned}$$

For each $\delta > 0$ and $\varepsilon > 0$, if $\lambda = \frac{1}{2}$,

$$\begin{aligned}&\text{Sol(VIP)}(\lambda)(\varepsilon) \\ &= \left\{x \in \mathscr{D}_1(x, \lambda) : 0 \in F(x, y, \lambda) + B^+(0, \varepsilon)e(x), \ \ \forall y \in \mathscr{D}_2(x, \lambda)\right\} \\ &= \left\{x \in [0, 1] : 0 \in (-\infty, (\lambda^2 + 1)(y - x)] + B^+(0, \varepsilon), \ \ \forall y \in \left[0, \frac{1}{2}\right]\right\} \\ &= \left[0, \frac{\varepsilon}{\lambda^2 + 1}\right],\end{aligned}$$

if $\lambda \neq \dfrac{1}{2}$

$$\begin{aligned}&\text{Sol(VIP)}(\lambda)(\varepsilon) \\ &= \{x \in \mathscr{D}_1(x, \lambda) : 0 \in F(x, y, \lambda) + B^+(0, \varepsilon)e(x), \ \ \forall y \in \mathscr{D}_2(x, \lambda)\} \\ &= \left\{x \in \left[0, \frac{1}{2}\right] : 0 \in (-\infty, (\lambda^2 + 1)(y - x)] + B^+(0, \varepsilon), \ \ \forall y \in [0, 1]\right\} \\ &= \left[0, \frac{\varepsilon}{\lambda^2 + 1}\right].\end{aligned}$$

It follows that,

$$\Omega_\lambda(\delta,\varepsilon) = \bigcup_{\lambda' \in B(\lambda,\delta)} S_{\lambda'}(\varepsilon) = \begin{cases} \left[0, \dfrac{\varepsilon}{(\lambda-\delta)^2+1}\right], & \text{if } \lambda > 0, \\ [0,\varepsilon], & \text{if } \lambda = 0, \\ \left[0, \dfrac{\varepsilon}{(\lambda+\delta)^2+1}\right], & \text{if } \lambda < 0, \end{cases}$$

for sufficiently small $\delta > 0$. Therefore, for every $\lambda \in \Lambda$, $\mathrm{diam}(\Omega_\lambda(\delta,\varepsilon)) \to 0$ as $(\delta,\varepsilon) \to (0,0)$. By Theorem 10.95, $(\mathrm{QVIP})_\lambda$ is well-posed.

Remark 10.97 We cannot replace the supposed well-posedness in Theorem 10.95 by the generalized well-posedness. Therefore, we have to employ the Kuratowski measure of noncompactness to study characterizations of the well-posedness in the generalized sense for (QVIP).

Lemma 10.98 *Let (X,d) be a complete metric space. If (F_n) is a decreasing sequence of nonempty, closed, and bounded subsets of X such that $\lim_{n\to\infty} \mu(F_n) = 0$, then the intersection $F_\infty = \bigcap_{n=1}^{\infty} F_n$ is a nonempty and compact subset of X.*

Theorem 10.99 *Let $\mathscr{E}$ be complete and Λ be finite-dimensional. Assume that*

(a) *$\mathscr{D}_1$ is closed;*
(b) *$\mathscr{D}_2$ is lower semicontinuous;*
(c) *F is closed.*

Then, (QVIP) is well-posed in the generalized sense if and only if for every $\lambda \in \Lambda$,

$$\Omega_\lambda(\delta,\varepsilon) \neq \emptyset, \quad \forall \delta,\varepsilon > 0, \quad \textit{and} \quad \mu(\Omega_\lambda(\delta,\varepsilon)) \to 0 \quad \textit{as} \quad (\delta,\varepsilon) \to (0,0). \tag{10.40}$$

Proof Suppose that (QVIP) is well-posed in the generalized sense. Let $\lambda \in \Lambda$ be fixed. Then $\mathrm{Sol(VIP)}(\lambda)$ is nonempty. Now we show that $\mathrm{Sol(VIP)}(\lambda)$ is compact. Indeed, let $\{x_n\}$ be any sequence in $\mathrm{Sol(VIP)}(\lambda)$. Then, $\{x_n\}$ is an approximating solution sequence for $(\mathrm{QVIP})_\lambda$. By the well-posedness in the generalized sense of (QVIP), $\{x_n\}$ has a subsequence which converges to a point of $\mathrm{Sol(VIP)}(\lambda)$. Thus, $\mathrm{Sol(VIP)}(\lambda)$ is compact. Clearly, for each $\delta,\varepsilon > 0$,

$$\mathrm{Sol(VIP)}(\lambda) \subset \Omega_\lambda(\delta,\varepsilon),$$

and so $\Omega_\lambda(\delta,\varepsilon) \neq \emptyset$. Now we will show that

$$\mu(\Omega_\lambda(\delta,\varepsilon)) \to 0 \quad \text{as} \quad (\delta,\varepsilon) \to (0,0). \tag{10.41}$$

Observe that for every $\delta, \varepsilon > 0$,

$$\begin{aligned}
&\mathscr{H}(\Omega_\lambda(\delta, \varepsilon), \mathrm{Sol(VIP)}(\lambda)) \\
&\quad = \max\{H^*(\Omega_\lambda(\delta, \varepsilon), \mathrm{Sol(VIP)}(\lambda)), H^*(\mathrm{Sol(VIP)}(\lambda), \Omega_\lambda(\delta, \varepsilon))\} \\
&\quad = H^*(\Omega_\lambda(\delta, \varepsilon), \mathrm{Sol(VIP)}(\lambda)),
\end{aligned}$$

and $\mathrm{Sol(VIP)}(\lambda)$ is compact. Indeed, let $\{x_n\} := \{(x_n, \lambda_n)\}$ be an arbitrary sequence in $\mathrm{Sol(VIP)}(\lambda)$. Then, it is clear that $\{x_n\}$ is an approximating sequence of $(QVIP)$. Thus, it has a subsequence converging to a point in $\mathrm{Sol(VIP)}(\lambda)$. Therefore, $\mu(\mathrm{Sol(VIP)}(\lambda)) = 0$. Now for any $\nu > 0$, there are finite sets $A_1^\nu, A_2^\nu, \dots, A_{n^\nu}^\nu$ for some $n^\nu \in \mathbb{N}$ such that

$$\mathrm{Sol(VIP)}(\lambda) \subset \bigcup_{k=1}^{n_\nu} A_k^\nu \quad \text{and} \quad \operatorname{diam} A_k^\nu \le \nu, \quad \text{for all } k = 1, 2, \dots, n^\nu.$$

Next, for each $k \in \{1, 2, \dots, n_\nu\}$, we define the set

$$M_k^\nu = \{z \in X : d(z, A_k^\nu) \le \mathscr{H}(\Omega_\lambda(\delta, \varepsilon), \mathrm{Sol(VIP)}(\lambda))\}.$$

We show that $\Omega_\lambda(\delta, \varepsilon) \subset \bigcup_{k=1}^{n_\nu} M_k^\nu$. To this end, let $x \in \Omega_\lambda(\delta, \varepsilon)$ be given. Thus, we have

$$d(x, \mathrm{Sol(VIP)}(\lambda)) \le \mathscr{H}(\Omega_\lambda(\delta, \varepsilon), \mathrm{Sol(VIP)}(\lambda)).$$

Because $\mathrm{Sol(VIP)}(\lambda) \subset \bigcup_{k=1}^{n_\nu} A_k^\nu$,, we also get

$$d\left(x, \bigcup_{k=1}^{n_\nu} A_k^\nu\right) \le d(x, \mathrm{Sol(VIP)}(\lambda)) \le \mathscr{H}(\Omega_\lambda(\delta, \varepsilon), \mathrm{Sol(VIP)}(\lambda)).$$

Therefore, there exists $k_0 \in \{1, 2, \dots, n_\nu\}$ such that

$$d(x, A_{k_0}^\nu) \le \mathscr{H}(\Omega_\lambda(\delta, \varepsilon), \mathrm{Sol(VIP)}(\lambda)),$$

thereby yielding $x \in M_{k_0}^\nu$. Therefore, we get the desired inclusion. Futhermore, we see that, for any $k \in \{1, 2, \dots, n_\nu\}$,

$$\operatorname{diam} M_k^\nu \le \nu + 2\mathscr{H}(\Omega_\lambda(\delta, \varepsilon), \mathrm{Sol(VIP)}(\lambda)). \tag{10.42}$$

Indeed, for any $y, y' \in M_k^\nu$ and $m, m' \in A_k^\nu$,

$$d(y, y') \le d(y, m) + d(m, m') + d(m', y'),$$

which gives that

$$\begin{aligned} d(y, y') &\le \inf_{z\in A_k^\nu} d(y, z) + \inf_{z,z'\in A_k^\nu} d(z, z') + \inf_{z'\in A_k^\nu} d(z', y') \\ &= d(y, A_k^\nu) + \inf_{z,z'\in A_k^\nu} d(z, z') + d(y', A_k^\nu) \\ &\le \nu + 2\mathscr{H}(\Omega_\lambda(\delta, \varepsilon), \mathrm{Sol(VIP)}(\lambda)), \end{aligned}$$

which leads to the desired result (10.42). It follows from the definition of μ that

$$\mu(\Omega_\lambda(\delta, \varepsilon)) \le 2\mathscr{H}(\Omega_\lambda(\delta, \varepsilon), \mathrm{Sol(VIP)}(\lambda)) + \nu, \text{ for all } \nu > 0.$$

Therefore, we can conclude that

$$\mu(\Omega_\lambda(\delta, \varepsilon)) \le 2\mathscr{H}(\Omega_\lambda(\delta, \varepsilon), \mathrm{Sol(VIP)}(\lambda)) = 2H^*(\Omega_\lambda(\delta, \varepsilon), \mathrm{Sol(VIP)}(\lambda)).$$

To prove (10.41), it is sufficient to show that

$$H^*(\Omega_\lambda(\delta, \varepsilon), \mathrm{Sol(VIP)}(\lambda)) \to 0 \text{ as } (\delta, \varepsilon) \to (0, 0). \tag{10.43}$$

If (10.43) does not hold, then there exist $r > 0$, sequences of positive real numbers $\{\delta_n\}$ and $\{\varepsilon_n\}$ with $(\delta_n, \varepsilon_n) \to (0, 0)$ as $n \to \infty$ and a sequence $\{x_n\}$ with $x_n \in \Omega_\lambda(\delta_n, \varepsilon_n)$ for every $n \in \mathbb{N}$ such that

$$d(x_n, S_\lambda) > r, \ \forall n \in \mathbb{N}. \tag{10.44}$$

For each $n \in \mathbb{N}$, since $x_n \in \Omega_\lambda(\delta_n, \varepsilon_n)$,, there exists $\lambda_n \in B(\lambda, \delta_n)$ such that

$$x_n \in \mathscr{D}_1(x_n, \lambda_n) \quad \text{and} \quad 0 \in F(x_n, y, \lambda_n) + B^+(0, \varepsilon_n)e(x_n), \quad \forall y \in \mathscr{D}_2(x_n, \lambda_n).$$

Clearly $\lambda_n \to \lambda$ as $n \to \infty$. Hence $\{x_n\}$ is an approximating solution sequence for $(\mathrm{QVIP})_\lambda$ corresponding to $\{\lambda_n\}$. Then, by the well-posedness in the generalized sense of (QVIP), $\{x_n\}$ has a subsequence $\{x_{n_k}\}$ which converges to some point of Sol(VIP)(λ). This contradicts (10.44), and so (10.43) holds. Therefore, (10.41) is proved.

Conversely, suppose that condition (10.40) holds. We will show that (QVIP) is well-posed in the generalized sense. Let $\lambda \in \Lambda$ be fixed. Thus, by Proposition 10.91 and Lemma 10.93, we see that Ω_λ is closed. Further, $\mathrm{Sol(VIP)}(\lambda) = \bigcap_{\delta,\varepsilon>0} \Omega_\lambda(\delta, \varepsilon)$. Since $\mu(\Omega_\lambda(\delta, \varepsilon)) \to 0$ as $(\delta, \varepsilon) \to (0, 0)$, by Lemma 10.98, Sol(VIP)(λ) is a nonempty compact subset and

$$H^*(\Omega_\lambda(\delta, \varepsilon), \mathrm{Sol(VIP)}(\lambda)) \to 0 \text{ as } (\delta, \varepsilon) \to (0, 0). \tag{10.45}$$

Let $\{\lambda_n\}$ be any sequence in Λ with $\lambda_n \to \lambda$ as $n \to \infty$. Suppose that $\{x_n\}$ is an approximating solution sequence for $(\text{QVIP})_\lambda$ corresponding to $\{\lambda_n\}$,, then there exists a sequence of positive real numbers $\{\varepsilon_n\}$ with $\varepsilon_n \to 0$ such that, for each $n \in \mathbb{N}$,

$$x_n \in \mathscr{D}_1(x_n, \lambda_n) \quad \text{and} \quad 0 \in F(x_n, y, \lambda_n) + B^+(0, \varepsilon_n)e(x_n), \quad \forall y \in \mathscr{D}_2(x_n, \lambda_n).$$

For each $n \in \mathbb{N}$,, let $\delta_n = d(\lambda_n, \lambda)$. Then, $\lambda_n \in B(\lambda, \delta_n)$ and $x_n \in \Omega_\lambda(\delta_n, \varepsilon_n)$ for every $n \in \mathbb{N}$, and $\delta_n \to 0$ as $n \to \infty$. It follows from (10.45) that

$$d(x_n, \text{Sol(VIP)}(\lambda)) \leq H^*(\Omega_\lambda(\delta_n, \varepsilon_n), \text{Sol(VIP)}(\lambda)) \to 0, \text{ as } n \to \infty.$$

Since $\text{Sol(VIP)}(\lambda)$ is compact, for each $n \in \mathbb{N}$, there exists $\bar{x}_n \in \text{Sol(VIP)}(\lambda)$ such that

$$d(x_n, \bar{x}_n) = d(x_n, \text{Sol(VIP)}(\lambda)) \to 0, \text{ as } n \to \infty.$$

By the compactness of $\text{Sol(VIP)}(\lambda)$, $\{\bar{x}_n\}$ has a subsequence $\{\bar{x}_{n_k}\}$ which converges to a point $\bar{x} \in \text{Sol(VIP)}(\lambda)$. Hence, the corresponding subsequence $\{x_{n_k}\}$ of $\{x_n\}$ converges to $\bar{x}$. This implies that (QVIP) is well-posed in the genelized sense. This completes the proof. □

Using similar reasoning as in the proof of Theorem 10.99, we can obtain the following characterization by considering the Hausdorff distance between the solution set and the approximating solution set.

Theorem 10.100 *Let $\mathscr{E}$ be complete and Λ be compact (or a finite-dimensional normed space). Assume that*

(a) *$\mathscr{D}_1$ is closed;*
(b) *$\mathscr{D}_2$ is lower semicontinuous;*
(c) *F is closed.*

Then, (QVIP) is well-posed in the generalized sense if and only if for every $\lambda \in \Lambda$, $\text{Sol(VIP)}(\lambda)$ is nonempty and compact and $H^*(\Omega_\lambda(\delta, \varepsilon), \text{Sol(VIP)}(\lambda)) \to 0$ as $(\delta, \varepsilon) \to (0, 0)$.

The following theorem shows that under suitable conditions, the well-posedness of (QVIP) is equivalent to the existence of the unique solution.

Theorem 10.101 *Let $\mathscr{E}$ be finite-dimensional. Assume that*

(a) *$\mathscr{D}_1$ is closed;*
(b) *$\mathscr{D}_2$ is lower semicontinuous;*
(c) *F is closed;*
(d) *for every $\lambda \in \Lambda$, there exists $\varepsilon > 0$ such that $\Omega_\lambda(\varepsilon, \varepsilon)$ is nonempty and bounded.*

Then, (QVIP) is well-posed if and ony if for every $\lambda \in \Lambda$, $(QVIP)_\lambda$ has the unique solution.

Proof The necessity is obvious. For the sufficiency, assume that $(\text{QVIP})_\lambda$ has the unique solution x_λ for every $\lambda \in \Lambda$. Let $\lambda \in \Lambda$ be fixed. Let $\{\lambda_n\}$ be any sequence in Λ with $\lambda_n \to \lambda$ as $n \to \infty$. Suppose that $\{x_n\}$ is an approximating solution sequence for $(\text{QVIP})_\lambda$ corresponding to $\{\lambda_n\}$, then there exists a sequence of positive real numbers $\{\varepsilon_n\}$ with $\varepsilon_n \to 0$ such that, for each $n \in \mathbb{N}$,

$$x_n \in \mathscr{D}_1(x_n, \lambda_n) \quad \text{and} \quad 0 \in F(x_n, y, \lambda_n) + B^+(0, \varepsilon_n)e(x_n), \quad \forall y \in \mathscr{D}_2(x_n, \lambda_n). \tag{10.46}$$

For each $n \in \mathbb{N}$,, let $\delta_n = d(\lambda_n, \lambda)$. Then, $\lambda_n \in B(\lambda, \delta_n)$ and $x_n \in \Omega_\lambda(\delta_n, \varepsilon_n)$ for each $n \in \mathbb{N}$ and $\delta_n \to 0$ as $n \to \infty$. Let $\varepsilon > 0$ be such that $\Omega_\lambda(\varepsilon, \varepsilon)$ is nonempty and bounded. Then there exists some $n_0 \in \mathbb{N}$ such that

$$x_n \in \Omega_\lambda(\delta_n, \varepsilon_n) \subset \Omega_\lambda(\varepsilon, \varepsilon), \quad \text{for all } n \geq n_0.$$

Thus $\{x_n\}$ is bounded. Let $\{x_{n_k}\}$ be any subsequence of $\{x_n\}$ such that $x_{n_k} \to x$ as $k \to \infty$. Since $\mathscr{D}_1$ is closed, we have $x \in \mathscr{D}_1(x, \lambda)$. For each $y \in \mathscr{D}_2(x, \lambda)$, since $\mathscr{D}_2$ is l.s.c., there exists a sequence $\{y_{n_k}\}$ with $y_{n_k} \in \mathscr{D}_2(x_{n_k}, \lambda)$ such that $y_{n_k} \to y$ as $k \to \infty$. By (10.46), one has

$$0 \in F(x_{n_k}, y_{n_k}, \lambda_{n_k}) + B^+(0, \varepsilon_{n_k})e(x_{n_k}), \quad \forall k \in \mathbb{N}.$$

Thus, for each $k \in \mathbb{N}$,, there exists a sequence $\{\gamma_{n_k}\} \subset B^+(0, \varepsilon_{n_k})$ such that

$$0 \in F(x_{n_k}, y_{n_k}, \lambda_{n_k}) + \gamma_{n_k} e(x_{n_k}),$$

which gives that

$$((x_{n_k}, y_{n_k}, \lambda_{n_k}), -\gamma_{n_k} e(x_{n_k})) \in \text{gph} F.$$

It is clear that $\{((x_{n_k}, y_{n_k}, \lambda_{n_k}), -\gamma_{n_k} e(x_{n_k}))\} \to ((x, y, \lambda), 0)$. The closedness of the mapping F implies that

$$((x, y, \lambda), 0) \in \text{gph} F,$$

that is $0 \in F(x, y, \lambda)$. Thus, x is a solution of $(\text{QVIP})_\lambda$. By the uniqueness of the solution set for $(\text{QVIP})_\lambda$,, we have $x = x_\lambda$. Therefore, the sequence $\{x_n\}$ converges to x_λ, and hence (QVIP) is well-posed. □

The following theorem shows that under suitable conditions, the well-posedness in the generalized sense for (QVIP) is equivalent to the existence of a solution.

Theorem 10.102 *Let $\mathscr{E}$ be finite-dimensional. Assume that*

(a) *$\mathscr{D}_1$ is closed;*
(b) *$\mathscr{D}_2$ is lower semicontinuous;*
(c) *F is closed;*
(d) *for every $\lambda \in \Lambda$,, there exists $\varepsilon > 0$ such that $\Omega_\lambda(\varepsilon, \varepsilon)$ is nonempty and bounded.*

Then, $(QVIP)$ is well-posed in the generalized sense if and only if for every $\lambda \in \Lambda$, $(\text{QVIP})_\lambda$ has a nonempty solution set $\text{Sol}(\text{VIP})(\lambda)$.

Proof The necessity is obvious. For the sufficiency, assume that $(\text{QVIP})_\lambda$ has a nonempty solution set $\text{Sol}(\text{VIP})(\lambda)$ for every $\lambda \in \Lambda$. Let $\lambda \in \Lambda$ be fixed. Let $\{\lambda_n\}$ be any sequence in Λ with $\lambda_n \to \lambda$ as $n \to \infty$. Suppose that $\{x_n\}$ is an approximating solution sequence for $(QVIP)_\lambda$ corresponding to $\{\lambda_n\}$, then there exists a sequence of positive real numbers $\{\varepsilon_n\}$ with $\varepsilon_n \to 0$ such that, for each $n \in \mathbb{N}$,

$$x_n \in \mathscr{D}_1(x_n, \lambda_n) \quad \text{and} \quad 0 \in F(x_n, y, \lambda_n) + B^+(0, \varepsilon_n)e(x_n), \quad \forall y \in \mathscr{D}_2(x_n, \lambda_n). \tag{10.47}$$

For each $n \in \mathbb{N}$,, let $\delta_n = d'(\lambda_n, \lambda)$. Then, $\lambda_n \in B(\lambda, \delta_n)$ and $x_n \in \Omega_\lambda(\delta_n, \varepsilon_n)$ for each $n \in \mathbb{N}$ and $\delta_n \to 0$ as $n \to \infty$. Let $\varepsilon > 0$ be such that $\Omega_\lambda(\varepsilon, \varepsilon)$ is nonempty and bounded. Then there exists some $n_0 \in \mathbb{N}$ such that

$$x_n \in \Omega_\lambda(\delta_n, \varepsilon_n) \subset \Omega_\lambda(\varepsilon, \varepsilon), \quad \text{for all } n \geq n_0.$$

Thus $\{x_n\}$ is bounded. Let $\{x_{n_k}\}$ be any subsequence of $\{x_n\}$ such that $x_{n_k} \to x$ as $k \to \infty$. It follows from the closedness of $\mathscr{D}_1$ that $x \in \mathscr{D}_1(x, \lambda)$. For each $y \in \mathscr{D}_2(x, \lambda)$, since $\mathscr{D}_2$ is l.s.c., there exists a sequence $\{y_{n_k}\}$ with $y_{n_k} \in \mathscr{D}_2(x_{n_k}, \lambda)$ such that $y_{n_k} \to y$ as $k \to \infty$. Using (10.47), we obtain

$$0 \in F(x_{n_k}, y_{n_k}, \lambda_{n_k}) + B^+(0, \varepsilon_{n_k})e(x_{n_k}), \quad \forall k \in \mathbb{N}.$$

Thus, for each $k \in \mathbb{N}$,, there exists a sequence $\{\beta_{n_k}\} \subset B^+(0, \varepsilon_{n_k})$ such that

$$0 \in F(x_{n_k}, y_{n_k}, \lambda_{n_k}) + \beta_{n_k}e(x_{n_k}),$$

which gives that

$$((x_{n_k}, y_{n_k}, \lambda_{n_k}), -\beta_{n_k}e(x_{n_k})) \in \text{gph}F.$$

It is clear that $\{((x_{n_k}, y_{n_k}, \lambda_{n_k}), -\beta_{n_k}e(x_{n_k}))\} \to ((x, y, \lambda), 0)$. By virtue of the closedness of F, we have

$$((x, y, \lambda), 0) \in \text{gph}F,$$

that is, $0 \in F(x, y, \lambda)$. Therefore, x is a solution of $(\text{QVIP})_\lambda$ and so (QVIP) is well-posed in the generalized sense. The proof is completed. □

10.4 Exercises

Exercise 10.1 Let A be a nonempty, compact, and convex subset of a topological vector space, B a compact topological space, $\rho(a, b)$ a relation linking $a \in A$ and $b \in B$, and $S : A \rightrightarrows B$. Assume that the following conditions are fulfilled:

(i) ρ is closed in both variables (i.e., for any net $(a_\gamma, b_\gamma) \to (a, b)$, if $\rho(a_\gamma, b_\gamma) = 1$ for all γ, then $\rho(a, b) = 1$) and S has the closed values;
(ii) for each $a \in A$, there exists $x \in A$ such that $\rho(a, b) = 0$ for all $b \in S(x)$;
(iii) for each $a \in A$, $\{x \in A \mid \rho(a, b) = 0 \text{ for all } b \in S(x)\}$ is convex.

Prove that there exists $\bar{a} \in A$ such that $\rho(\bar{a}, b)$ does not hold for all $b \in S(\bar{a})$.

Exercise 10.2 Let A be a nonempty compact convex subset of a topological space, B convex subset of a topological vector space, $R(a, b)$ a relation linking $a \in A$ and $b \in B$, and $\mathcal{D} : A \rightrightarrows B$. Assume that the following conditions are fulfilled:

(i) for each $a \in A$, $\mathcal{D}(a) \cap (R^a)^{-1}(0)$ is open and convex;
(ii) for each $b \in B$, $(A \setminus \mathcal{D}^{-1}(b)) \cup R_b^{-1}(1)$ is closed and convex.

Prove that there exists $\bar{a} \in A$ such that $R(\bar{a}, b)$ holds for all $b \in \mathcal{D}(\bar{a})$.

Exercise 10.3 Let A be a nonempty and convex subset of a topological vector space and $T : A \rightrightarrows A$. We say that T is a relaxed-KKM-map (rKKM-map) if for all $N \in \langle A \rangle$, $\text{conv}N \subseteq \bigcup_{x \in N} (T(x) \cup N)$. For $\Omega : A \rightrightarrows A$, we define $\Omega^* : A \rightrightarrows A$ by $\Omega^*(x) = A \setminus \Omega^{-1}(x)$ for all $x \in A$. Prove the following assertion: *Let A be a nonempty and convex subset of a Hausdorff topological vector space and $\Omega : A \rightrightarrows A$ nonempty-valued. If Ω^* is rKKM and closed-valued, and $\Omega^*(x)$ is compact for some x, then there exists $\bar{x} \in A$ such that $\Omega(\bar{x}) = \{\bar{x}\}$.*

Exercise 10.4 Prove the following assertion: *Let A be a compact convex subset of a Hausdorff topological vector space and $\Omega : A \rightrightarrows A$ nonempty-valued. If $\Omega^*(x) \cup \{x\}$ and $\Omega(x) \setminus \{x\}$ are convex for each $x \in A$, $\Omega(x)$ and $\Omega^{-1}(x)$ are open for each $x \in A$, then there exists $\bar{x} \in A$ such that $\Omega(\bar{x}) = \{\bar{x}\}$.*

Exercise 10.5 Let A, B be nonempty sets and $F : A \times B \rightrightarrows \overline{\mathbb{R}}$. We say that the minimax equality holds for F if

$$F^* := \inf \bigcup_{b \in B} \sup \bigcup_{a \in A} F(a, b) = \sup \bigcup_{a \in A} \inf \bigcup_{b \in B} F(a, b) =: F_*. \tag{10.48}$$

If A, B are compact topological spaces and F is continuous and compact-valued, then the sets

$$F(A, b) := \bigcup_{a \in A} F(a, b) \text{ and } F(a, B) := \bigcup_{b \in B} F(a, b)$$

are compact for each $a \in A$ and $b \in B$. Hence, $\max \bigcup_{a \in A} F(a, b)$ and $\min \bigcup_{b \in B} F(a, b)$ exist. Furthermore, we can show that the (single-valued) functions $b \mapsto \max \bigcup_{a \in A} F(a, b)$ and $a \mapsto \min \bigcup_{b \in B} F(a, b)$ are continuous. Then, the sets $\bigcup_{b \in B} \max \bigcup_{a \in A} F(a, b)$ and $\bigcup_{a \in A} \min \bigcup_{b \in B} F(a, b)$ are compact and hence have a minimum and a maximum. Therefore, in this case, if the minimax equality holds for F, then it can be rewritten as

$$\min \bigcup_{b \in B} \max \bigcup_{a \in A} F(a, b) = \max \bigcup_{a \in A} \min \bigcup_{b \in B} F(a, b). \tag{10.49}$$

A point $(\bar{a}, \bar{b}) \in A \times B$ is called a saddle point of F if,

$$\max \bigcup_{a \in A} F(a, \bar{b}) = F(\bar{a}, \bar{b}) = \min \bigcup_{b \in B} F(\bar{a}, b).$$

Note that we always have $F^* \geq F_*$. Hence, if F has a saddle point, then

$$\inf \bigcup_{b \in B} \max \bigcup_{a \in A} F(a, b) = \sup \bigcup_{a \in A} \min \bigcup_{b \in B} F(a, b). \tag{10.50}$$

(a) Assume that $A \equiv B$ is a nonempty compact and convex subset of a topological vector space, $\delta \in \mathbb{R}$, and the following conditions are satisfied

  (i) for any $b \in A$, $\sup F(b, b) \leq \delta$ and the set $\{a \in A \mid \sup F(a, b) > \delta\}$ is convex;
  (ii) for any $a \in A$, $F(a, \cdot)$ is lower semicontinuous.

  Prove that there exists $\bar{b} \in A$ such that $\sup \bigcup_{a \in A} F(a, \bar{b}) \leq \delta$. In particular, (10.48) holds if for any $\delta > F_*$, (i) and (ii) are satisfied.

(b) Assume that A, B are nonempty convex subsets of topological vector spaces, B is compact, $\delta \in \mathbb{R}$, and the following conditions are satisfied

  (i) for all $b \in B$, the set $\{a \in A \mid \sup F(a, b) > \delta\}$ is convex and $F(\cdot, b)$ is lower semicontinuous;
  (ii) for all $a \in A$, the set $\{b \in B \mid \sup F(a, b) \leq \delta\}$ is convex and $F(a, \cdot)$ is lower semicontinuous.

  Prove that there exists $\bar{b} \in A$ such that $\sup \bigcup_{a \in A} F(a, \bar{b}) \leq \delta$. In particular, (10.48) holds if for any $\delta > F_*$, (i) and (ii) are satisfied.

(c) Assume that A, B are nonempty compact convex subsets of topological vector spaces and the following conditions are satisfied

(i) for all $M \in \mathscr{F}(A)$, $L \in \mathscr{F}(B)$ and $(x, y) \in \text{conv}M \times \text{conv}L$, there exists $(a, b) \in M \times L$ with $\sup F(a, y) \leq \inf F(x, b)$;
(ii) for all $(a, b) \in A \times B$, $F(a, \cdot)$ and $F(\cdot, b)$ are lower semicontinuous.

Prove that F has a saddle point.

(d) Assume that A, B are nonempty compact convex subsets of topological vector spaces and the following conditions are satisfied

(i) F is lower semicontinuous;
(ii) for all $(a, b) \in A \times B$, the sets $\{(x, y) \in A \times B \mid \sup F(a, y) > \inf F(x, b)\}$ and $\{(x, y) \in A \times B \mid \sup F(a, y) \leq \inf F(x, b)\}$ are convex.

Prove that F has a saddle point.

Exercise 10.6 Consider the game $\mathscr{G} = (A_1, A_2, \ldots, A_m; c_1, c_2, \ldots, c_m)$, where A_i is the strategy set, and $c_i : A := \prod_{i=1}^m A_i \to \mathbb{R}$ is the payoff function of the ith player. For a point $a \in A$, we denote by $a_{\hat{i}}$ its projection onto $A_{\hat{i}} := \prod_{j \neq i} A_j$. Recall that a point $\bar{a} := (\bar{a}_1, \bar{a}_2, \cdots, \bar{a}_m) \in A$ is called a Nash equilibrium point of $\mathscr{G}$ if for all $b_i \in A_i$ and $i = 1, 2, \cdots, m$, $c_i(\bar{a}) \geq c_i(\bar{a}_{\hat{i}}, b_i)$.

(a) Prove that $\mathscr{G}$ has a Nash equilibrium point if for each $i \in \{1, 2, \ldots, m\}$, A_i is a nonempty compact convex subset of a topological vector space and the following conditions are satisfied

(i) for all $M \in \mathscr{F}(A)$ and $a \in \text{conv}M$, there exists $b \in M$ with $c_i(a) \geq c_i(a_{\hat{i}}, b_i)$ for all $i = 1, 2, \cdots, m$;
(ii) for all $b \in A$, the set $\{a \in A \mid \min_{1 \leq i \leq m}(c_i(a) - c_i(a_{\hat{i}}, b_i)) \geq 0\}$ is closed.

(b) Check (i) and (ii) for the case where $A_1 = A_2 = \ldots = A_m = [0, 1]$, $c_i(a) = -a_i + \sum_{j \neq i} a_j$ for all $a = (a_1, a_2, \ldots, a_m) \in A$.

Exercise 10.7 Let X be a complete metric space, and let $\Gamma : X \rightrightarrows X$ be a set valued map. Prove that if the following conditions are satisfied

(a) for each $x \in X$, $x \in \Gamma(x)$ and $\Gamma(x)$ is closed,
(b) if $y \in \Gamma(x)$, then $\Gamma(y) \subset \Gamma(x)$,
(c) for any sequence $\{x_n\}$ in X for which $x_{n+1} \in \Gamma(x_n)$ for all $n \in \mathbb{N}$, one has

$$\lim_{n \to \infty} d(x_n, x_{n+1}) = 0,$$

then there exists some $\bar{x} \in X$ such that $\Gamma(\bar{x}) = \{\bar{x}\}$.

Exercise 10.8 Let X be a complete metric space, $T : X \rightrightarrows X$ be a set-valued map, and $\mathscr{R}(x, y)$ be a relation linking elements $x, y \in X$. Prove that if the following conditions are satisfied

(a) $\mathscr{R}$ is a quasi-ordered relation on $X \times X$ (i.e., $\mathscr{R}(x, x)$ holds for all $x \in X$, and if $\mathscr{R}(x, y)$ and $\mathscr{R}(y, z)$ hold, then $\mathscr{R}(x, z)$ holds),
(b) for each $x \in X$, there exists $y \in T(x)$ such that $\mathscr{R}(x, y)$ holds,
(c) for each $x \in X$, $\mathscr{R}(x, \cdot)$ is closed in the second variable,
(d) for any sequence $\{x_n\}$ in X for which $\mathscr{R}(x_n, x_{n+1})$ holds for all $n \in \mathbb{N}$, one has

$$\lim_{n\to\infty} d(x_n, x_{n+1}) = 0,$$

then for each $u \in X$, there exists some $\bar{x} \in X$ such that $\bar{x} \in T(\bar{x})$ and

◇ $\mathscr{R}(u, \bar{x})$ holds,
◇ $\mathscr{R}(\bar{x}, y)$ does not hold for all $y \in X \setminus \{\bar{x}\}$.

Exercise 10.9 Let A be a compact convex subset of a locally convex topological vector space, B be a convex subset of a topological vector space and $\mathbb{W}$ be a topological space, $Q : A \rightrightarrows \mathbb{W}$ be a set-valued map with nonempty values, and $R(x, y, z)$ be a relation linking elements $x \in A$, $y \in B$ and $z \in \mathbb{W}$. Prove that if the set-valued map Q is lower semicontinuous and the relation R is closed, then the set

$$\{(x, y, u) \in A \times B \times A \mid R(u, y, z) \text{ holds for all } z \in Q(x)\}$$

is closed in $A \times B \times A$.

Exercise 10.10 Let X be a complete metric space, $G : X \rightrightarrows \mathbb{R}$ be a set-valued map with nonempty values. Prove that if G is closed and $G(X)$ is bounded below, then for each $u \in X$ and $v \in G(u)$, there exist $\bar{x} \in X$, $\bar{y} \in G(\bar{x})$ such that

◇ $\bar{y} - v + \varepsilon d(u, \bar{x}) \leq 0$,
◇ $[(G(x) - \bar{y}) + \varepsilon d(\bar{x}, x)] \cap (-\mathbb{R}_+) = \emptyset$ for all $y \in X \setminus \{\bar{x}\}$.

Exercise 10.11 Let A be a nonempty compact convex subset of a locally convex Hausdorff topological vector space, and let $T : A \rightrightarrows A$, $Q : A \rightrightarrows A$ be set valued maps and $\mathscr{R}$ be a variational relation linking points of A. Prove that if

(a) T is upper semicontinuous with nonempty compact convex values, and Q is nonempty convex valued,
(b) for each $x \in A$, the set $\{y \in A \mid \mathscr{R}(x, y) \text{ does not hold}\}$ is convex,
(c) for each $y \in A$, the set $\{x \in A \mid y \in Q(x)\} \cap \{x \in A \mid \mathscr{R}(x, y) \text{ does not hold}\}$ is open in A,
(d) for each $x \in A$ and $y \in [\bigcup_{\lambda\geq 1}(\lambda x + (1 - \lambda)T(x))] \cap Q(x)$, $\mathscr{R}(x, y)$ holds,

then there exists $\bar{x} \in A$ such that $\bar{x} \in T(\bar{x})$ and $\mathscr{R}(\bar{x}, y)$ holds for all $y \in Q(\bar{x})$.

Exercise 10.12 Let A be a nonempty compact convex subset of a locally convex Hausdorff topological vector space, Y be a vector space, and let $T : A \rightrightarrows A$, $Q : A \rightrightarrows A$, and $F : A \times A \rightrightarrows Y$ be set valued maps. Prove that if

(a) T is upper semicontinuous with nonempty compact convex values, and Q is nonempty convex valued,
(b) for each $x \in A$, the set $\{y \in A \mid 0 \notin F(x, y)\}$ is convex,
(c) for each $y \in A$, the set $\{x \in A \mid y \in Q(x)\} \cap \{x \in A \mid 0 \notin F(x, y)\}$ is open in X,
(d) for each $x \in A$ and $y \in [\bigcup_{\lambda \geq 1}(\lambda x + (1 - \lambda)T(x))] \cap Q(x), 0 \in F(x, y)$,

then there exists $\bar{x} \in A$ such that $\bar{x} \in T(\bar{x})$ and $0 \in F(\bar{x}, y)$ for all $y \in Q(\bar{x})$.

Exercise 10.13 Let A be a nonempty compact convex subset of a real normed space, $T : A \rightrightarrows A$ be upper semicontinuous with nonempty compact convex values, and $Q : A \rightrightarrows A$ be a set valued map with nonempty convex values and open (in A) fibers. Prove that if $[\bigcup_{\lambda \geq 1}(\lambda x + (1 - \lambda)T(x))] \cap (Q(x) \setminus \{x\}) = \emptyset$ for all $x \in A$, then there exists $\bar{x} \in A$ such that $\bar{x} \in T(\bar{x}) \cap Q(\bar{x})$.

Exercise 10.14 Let X be a reflexive Banach space with its the dual space X^*, $K \subset X$ be nonempty closed and convex with the barrier cone

$$\mathrm{barr}(K) = \{x^* \in X^* \mid \sup_{x \in K}\langle x^*, x\rangle < \infty\}$$

is nonempty, and $T : K \rightrightarrows X^*$ be a maximal monotone set-valued map with nonempty values. Prove that the set $\{x \in K \mid 0 \in T(x)\}$ is nonempty and bounded if, and only if $K_\infty \cap T(K)^- = \emptyset$, where K_∞ is the recession cone of K and

$$T(K)^- = \{x^* \in X^* \mid \langle x^*, x\rangle \leq 0, \forall x \in T(K)\}.$$

Exercise 10.15 Let $X = \ell^2$ and so $X^* = \ell^2$. The inner product in ℓ^2 is defined by

$$\langle x, y\rangle := \sum_{n=1}^{\infty} x_n y_n, \quad \forall x, y \in \ell^2.$$

Let $C := \{x = (x_1, x_2, \ldots, x_n, \ldots) \in \ell^2 : x_n \geq 0, \forall n \in \mathbb{N}\}$. Define $T : C \to \ell^2$ by

$$T(x) := x, \quad \forall x \in C.$$

Prove that T is maximal monotone, and find all $x \in X$ such that $0 \in T(x)$.

10.5 Commentary

The variational relation problem was introduced in 2008, by Dinh The Luc [353], as a unified model for a large class of problems in nonlinear analysis and applied mathematics including optimization problems, variational inequalities, variational inclusions, equilibrium problems, etc. As this approach has proved to be a powerful tool for studying a wide class of problems in nonlinear analysis and applied mathematics, this problem has been studied in a number of papers; see, e.g., [85–87, 87, 290, 333, 340, 352]. However, as demonstrated in the first part of Sect. 10.2, a variational relation problem is practically equivalent to a generalized equilibrium problem. In Sect. 10.2, we provide extensions of Lemmas 1.31 and 1.33 and use them to establish some existence results for variational relation problems and variational inclusions. All results in this subsection are established under some conditions of convexity.

Section 10.3

This section provides an overview of stability results for the variational relation problem [353] and variational inclusion problem [236, 237, 288, 413]. We begin by presenting the concepts and properties of generalized continuity of maps given in [36, 82, 183, 283, 396]. Subsequently, we employ these results to establish convergence conditions and Hölder calmness conditions for variational relation problems [257, 283]. Additionally, we utilize these findings to investigate parametric variational inclusion problems and examine sufficient conditions for the semicontinuity properties of the solution maps [28, 29, 36, 52].

Furthermore, we provide the concept of well-posedness for general parametric quasi-variational inclusion problems (QVIP). We also delve into the corresponding notions of well-posedness in a generalized sense for (QVIP). To complement our exploration, we provide metric characterizations of well-posedness for (QVIP). We also demonstrate that under suitable conditions, well-posedness is equivalent to the existence of unique solutions [446–448]. It's worth noting that stability theory of such models is still evolving and remains an area that demands more attention of researchers. For more details on the relationships between the techniques and approaches in this section and the existing ones, we refer the reader to some typical works [142, 182, 277, 336, 370, 415, 440, 445].

Appendix A
Solutions

Chapter 1

Exercise 1.1 Assume that $\langle T(\bar{x}), y - \bar{x}\rangle \geq 0$ for all $y \in A$. Then,

$$\begin{aligned}\|(\bar{x} - T(\bar{x})) - y\|^2 &= \langle(\bar{x} - T(\bar{x})) - y, (\bar{x} - T(\bar{x})) - y\rangle \\ &= \langle T(\bar{x}), T(\bar{x})\rangle + 2\langle T(\bar{x}), y - \bar{x}\rangle + \langle y - \bar{x}, y - \bar{x}\rangle\end{aligned}$$

$$= \|T(\bar{x})\|^2 + 2\langle T(\bar{x}), y - \bar{x}\rangle + \|y - \bar{x}\|^2 \geq \|T(\bar{x})\|^2 \ \text{ for all } y \in A.$$

Conversely, assume that $\|T(\bar{x})\| \leq \|(\bar{x} - T(\bar{x})) - y\|$ for all $y \in A$. Then, taking any $y \in A$ and $t \in (0, 1)$, we have:

$$\|T(\bar{x})\|^2 \leq \|(\bar{x}-T(\bar{x}))-(\bar{x}+t(y-\bar{x}))\| = \|T(\bar{x})\|^2+t^2\|y-\bar{x}\|^2-2t\langle T(\bar{x}), y-\bar{x}\rangle.$$

It follows that

$$\langle T(\bar{x}), y - \bar{x}\rangle \geq -\frac{t}{2}\|y - \bar{x}\|^2 \ \text{ for all } t \in (0, 1).$$

Letting $t \to 0^+$, we obtain: $\langle T(\bar{x}), y - \bar{x}\rangle \geq 0$.

Exercise 1.2 Taking $(x, y) \in C \times D$, $(x', y'), (x'', y'') \in A_\phi(x, y)$, and $t \in (0, 1)$, when $\phi(x, \cdot)$ is concave and $\phi(\cdot, y)$ is convex, we have: $\phi(tx' + (1 - t)x'', y) - \phi(x, ty'+(1-t)y'') \leq t\phi(x', y)+(1-t)\phi(x'', y)-(t\phi(x, y')+(1-t)\phi(x, y'')) = t(\phi(x', y) - \phi(x, y')) + (1 - t)(\phi(x'', y) - \phi(x, y'')) < 0$. Hence, $(tx' + (1 - t)x'', ty' + (1 - t)y'') \in A_\phi(x, y)$. Thus, $A_\phi(x, y)$ is convex.

L. Q. Anh et al., *Equilibrium Problems*, Forum for Interdisciplinary Mathematics, https://doi.org/10.1007/978-3-031-89003-1

Exercise 1.3 Given $x \in A$, take any $y = (y_i)_{i \in I}, y' = (y'_i)_{i \in I} \in A$ and $t \in (0, 1)$. Since $c_i(x_{-i}, \cdot)$ is concave for each $i \in I$, we have

$$c_i(x) - c_i(x_{-i}, ty_i + (1-t)y'_i)) \leq c_i(x) - (tc_i(x_{-i}, y_i) + (1-t)c_i(x_{-i}, y'_i))$$

$$= t(c_i(x) - c_i(x_{-i}, y_i)) + (1-t)(c_i(x) - c_i(x_{-i}, y'_i)) \text{ for all } i \in I.$$

It follows that

$$\begin{aligned} N(x, ty + (1-t)y') &= \sum_{i=1}^{n} [c_i(x) - c_i(x_{-i}, ty_i + (1-t)y'_i))] \\ &\leq \sum_{i=1}^{n} [t(c_i(x) - c_i(x_{-i}, y_i)) + (1-t)(c_i(x) - c_i(x_{-i}, y'_i))] \\ &= t \sum_{i=1}^{n} [c_i(x) - c_i(x_{-i}, y_i)] + (1-t) \sum_{i=1}^{n} [c_i(x) - c_i(x_{-i}, y'_i)] \\ &= tN(x, y) + (1-t)N(x, y'). \end{aligned}$$

Thus, $N(x, \cdot)$ is convex.

Exercise 1.4 Hint.

(a) Apply Theorem 1.34 for Φ' with $\Phi'(x, y) = \Phi(x, y) - \alpha$.
(b) Apply part (a) with $\alpha = \inf_{x \in A} \Phi(x, x)$.

Exercise 1.5 We prove this for i = 1; the case i = 2 can be proved similarly. Assume that $\Phi \in \mathscr{A}^1_{qc}$. Let any $N = \{x^1, x^2, \ldots, x^n\} \in \mathscr{F}(A)$ and $x \in \text{conv}N$. Then, $x = \sum_{i=1}^{n} t_i x^i$, where $t_1, t_2, \ldots, t_n \in [0, 1]$ and $\sum_{i=1}^{n} t_i = 1$. Since $\Phi(\cdot, x)$ is quasiconvex,

$$\begin{aligned} \Phi(x, x) = \Phi\left(\sum_{i=1}^{n} t_i x^i, x\right) &\leq \max\{\Phi(x^1, x), \Phi(x^2, x), \ldots, \Phi(x^n, x)\} \\ &= \max_{x^i \in N} \Phi(x^i, x). \end{aligned}$$

Thus, $\Phi \in \mathscr{A}^i_{dqc}$.

Exercise 1.6 Hint. Check that $\Phi_q(x, y) \leq \Phi(x, y)$ for each $(x, y) \in A \times A$ and $\Phi_q(x, \cdot)$ is quasiconvex for each $x \in A$. Then, apply Theorem 1.34 and Proposition 1.29(a).

Exercise 1.7 **Hint.** Check that $\Phi_{\bar{q}}(x, y) \leq \Phi(x, y)$ for each $(x, y) \in A \times A$ and $\Phi_{\bar{q}}(x, \cdot)$ is quasiconvex and semicontinuous for each $x \in A$. Then, apply Theorem 1.50 and Proposition 1.29(a).

Exercise 1.8 **Hint.** For $\Phi(x, y) = \langle F(x), y - x\rangle$. From inclusions in Fig. 1.10(a), Proposition 1.29(d) and Example 1.39(a) we see that when conditions of Exercise 1.8 hold, $\Phi \in (-\mathscr{A}^1_{pr.qm}) \cap \mathscr{A}^2_{st.qcl} \cap \mathscr{A}^=_A$ and Φ has the property $\mathbb{P}^2$. Moreover, when T is hemicontinuous, it is easy to check that Φ is upper-sign continuous in the first argument. It is clear that $\text{lev}_{\leq}\Phi(x, \cdot)$ is closed for all $x \in A$. Thus, applying Theorem 1.50 (in the compact case), there exists $\bar{x} \in A$ such that $\Phi(\bar{x}, y) = \langle F(\bar{x}), y - \bar{x}\rangle \geq 0$ for all $y \in A$, i.e., $\text{VI}(A, T)$ has a solution.

Exercise 1.9 **Hint.**

(a) For $x \in A$ and $y \in A$, take $z = y$ and $U(x) = \text{lev}_{<\alpha}\Phi(\cdot, y)$.
(b) The inclusion $\subset$ is clear. For the inclusion $\supset$, take any $x \in \bigcap_{y\in A} \text{cl}(\text{lev}_{\geq\alpha}\Phi(\cdot, y))$. By using the α-transfer upper semicontinuity, show that if $x \notin \text{cl}(\bigcap_{y\in A} \text{lev}_{\geq\alpha}\Phi(\cdot, y))$, then there exists a point $z \in A$ such that $x \notin \text{cl}(\text{lev}_{\geq\alpha}\Phi(\cdot, z)) \supset \bigcap_{y\in A} \text{cl}(\text{lev}_{\geq\alpha}\Phi(\cdot, y))$.

Exercise 1.10 **Hint.**

(a) Let $\Phi' : A \times A \to \mathbb{R}$ be defined by, for all $x, y \in A$,

$$\Phi'(x, y) := \begin{cases} \;\;0 \text{ if } x \in \text{cl}(\text{lev}_{\geq}\Phi(\cdot, y)), \\ -1 \text{ otherwise.} \end{cases}$$

By Theorem 1.34 for Φ', there exists $\bar{x} \in A$ such that $\Phi'(\bar{x}, y) \geq 0$ for all $y \in A$, or equivalently, $\bar{x} \in \bigcap_{y\in A} \text{cl}(\text{lev}_{\geq}\Phi(\cdot, y)) = \text{cl}(\bigcap_{y\in A} \text{lev}_{\geq\alpha}\Phi(\cdot, y))$ (by Exercise 1.9). Hence, $\bigcap_{y\in A} \text{lev}_{\geq\alpha}\Phi(\cdot, y) \neq \emptyset$, i.e., there exists $x^* \in A$ such that $\Phi(x^*, y) \geq 0$ for all $y \in A$.
(b) Apply part (a) for Φ' with $\Phi'(x, y) = \Phi(x, y) - \alpha$.
(c) Apply part (b) with $\alpha = \inf_{x\in A} \Phi(x, x)$.

Exercise 1.11 **Hint.** Apply Exercise 1.10 and Proposition 1.4 (a).

Exercise 1.12 **Hint.**

(a) Argue similarly to part (a) of Exercise 1.10 but using Theorem 1.82 instead of Theorem 1.34.
(b) Apply part (a) for Φ' with $\Phi'(x, y) = \Phi(x, y) - \alpha$.

Exercise 1.13 **Hint.** A consequence of Proposition 1.29(b) and the fact that if $\Phi \in \mathscr{A}_{caqm}$, then $\Phi \in -\mathscr{A}_{qm}$.

Exercise 1.14 **Hint.** Apply Theorem 1.82 and Proposition 1.3.

Exercise 1.15 **Hint.** Use Proposition 1.4 (a) and part (b) of Exercise 1.12.

Excrcise 1.16 Hint.

(a) Use Corollary 1.93 with $\phi(x', y) - \phi(x, y') = \xi(y)\xi(y')(\frac{\varphi(x)}{\xi(y)} - \frac{\varphi(x')}{\xi(y')})$.

(b) Use Corollary 1.93 with $\phi(x', y) - \phi(x, y') = \frac{\varphi(x)\xi(y) - \varphi(x')\xi(y')}{(\varphi(x)+\xi(y'))(\varphi(x')+\xi(y))}$.

(c) Use Corollary 1.93. Here observe that

$$\phi(x', y) - \phi(x, y') = \sum_{i=0}^{n}(\xi(y)\xi(y'))^{2i+1}\left(\left(\frac{\varphi(x')}{\xi(y')}\right)^{2i+1} - \left(\frac{\varphi(x)}{\xi(y)}\right)^{2i+1}\right)$$

$$=\left(\frac{\varphi(x')}{\xi(y')} - \frac{\varphi(x)}{\xi(y)}\right)\sum_{i=0}^{n}(\xi(y)\xi(y'))^{2i+1}$$

$$\sum_{j=0}^{2i}\left(\frac{\varphi(x)}{\xi(y)}\right)^{2i-j}\left(\frac{\varphi(x')}{\xi(y')}\right)^{j}.$$

By the fact that $\sum_{j=0}^{2i} a^{2i-j}b^j \geq 0$ for all $a, b \in \mathbb{R}$, we have:

$$\mathbf{a}(x, y, x', y') := \sum_{i=0}^{n}(\xi(y)\xi(y'))^{2i+1}\sum_{j=0}^{2i}\left(\frac{\varphi(x)}{\xi(y)}\right)^{2i-j}\left(\frac{\varphi(x')}{\xi(y')}\right)^{j} \geq 0.$$

(d) Apply Exercise 1.15.

Chapter 2

Exercise 2.1 Hint.

(a) Let $\Phi : A \times A \to \mathbb{R}$ be defined by,

$$\Phi(x, y) := \begin{cases} 0 & \text{if } x \in S(y), \\ -1 & \text{otherwise.} \end{cases}$$

Then, $\bar{x} \in \bigcap_{y\in A} S(y)$ if and only if $\Phi(\bar{x}, y) \geq 0$ for all $y \in A$. Moreover, $\text{lev}_{\geq}\Phi(\cdot, y) = S(y)$ for all $y \in A$, and $\Phi \in \mathscr{A}^2_{pr.qm}$ if and only if S is KKM.

(b) Let $S : A \rightrightarrows A$ be defined by $S(y) := \text{lev}_{\geq}\Phi(\cdot, y)$. Then, S is KKM if and only if $\Phi \in \mathscr{A}^2_{pr.qm}$.

Exercise 2.2 Hint.

(a) • *Theorem 1.34 implies Theorem 2.93.* Let $\Phi : A \times A \to \mathbb{R}$ be defined by,

$$\Phi(x, y) := \begin{cases} 0 & \text{if } y \notin S(x), \\ -1 & \text{otherwise.} \end{cases}$$

Then, $S(\bar{x}) = \emptyset$ if and only if $\Phi(\bar{x}, y) \geq 0$ for all $y \in A$. Moreover, $\text{lev}_{\geq}\Phi(\cdot, y) = A \setminus S^{-1}(y)$ for all $y \in A$, and $\Phi \in \mathscr{A}^2_{pr.qm}$ if and only if $\text{conv}N \cap \bigcap_{y^i \in N} S^{-1}(y^i) = \emptyset$ for all $N \in \mathscr{F}(A)$.

• *The KKM-Fan lemma implies Theorem 2.93.* Let $T : A \rightrightarrows A$ be defined by $T(y) := A \setminus S^{-1}(y)$. Then, $S(\bar{x}) = \emptyset$ if and only if $\bar{x} \in \bigcap_{y \in A} T(y)$. Moreover, T is KKM if and only if $\text{conv}N \cap \bigcap_{y^i \in N} S^{-1}(y^i) = \emptyset$ for all $N \in \mathscr{F}(A)$.

(b) • *Theorem 2.93 implies Theorem 1.34* . Let $S : A \rightrightarrows A$ be defined by $S(x) := \{y \in A \mid \Phi(x, y) < 0\}$. Then, $\bar{x} \in \text{Sol(EP)}$ if and only if $S(\bar{x}) = \emptyset$. Moreover, $\text{lev}_{\geq}\Phi(\cdot, y) = A \setminus S^{-1}(y)$ for all $y \in A$, and $\Phi \in \mathscr{A}^2_{pr.qm}$ if and only if $\text{conv}N \cap \bigcap_{y^i \in N} S^{-1}(y^i) = \emptyset$ for all $N \in \mathscr{F}(A)$.

• *Theorem 2.93 implies Theorem 2.21.* Apply Theorem 2.93 for $S : A \rightrightarrows A$ defined by

$$S(x) := \begin{cases} \mathscr{D}(x) \cap \text{lev}_{<}\Phi(x, \cdot) & \text{if } x \in \text{cl(Fix}\mathscr{D}), \\ \mathscr{D}(x) & \text{if } x \notin \text{cl(Fix}\mathscr{D}) \end{cases} \qquad \text{for all } x \in A.$$

• *Theorem 2.93 implies Theorem 2.92* . Apply Theorem 2.93 for $T : A \rightrightarrows A$ defined by $T(x) := A \setminus S^{-1}(x)$ for all $x \in A$.

Exercise 2.3 Hint. Show that a required (QEP) is defined by $\Phi(x, y) := \varphi(y) - \varphi(x)$ and

$$\mathscr{D}(x) := \{y \in A \mid g(y, \omega) \leq 0 \ \text{ for all } \omega \in \Omega \text{ and } h(y, \lambda) = 0 \text{ for all } \lambda \in \Lambda\}.$$

Exercise 2.4

(a) This is clear.

(b) Let S be as in the part (a). We prove that S has a fixed point. Let $x \in A$. Since A is compact and $\mathscr{D}(x)$ is nonempty, closed (by the closedness of $\mathscr{D}$), and convex, $\mathscr{D}(x)$ is nonempty, compact, and convex. Moreover, as $\text{lev}_{\geq}\Phi$ is closed and $\Phi \in -\mathscr{A}^1_{qcl}$, $\text{lev}_{\geq}\Phi(\cdot, y)$ is closed and convex for each $y \in A$. Hence, by the definition of S, $S(x)$ is closed and convex. Since $\Phi \in \mathscr{A}^2_{pr.qm}$, $\Phi|_{\mathscr{D}(x) \times \mathscr{D}(x)}$ is properly quasimonotone in the second argument on $\mathscr{D}(x)$. Moreover, $\text{lev}_{\geq}\Phi|_{\mathscr{D}(x) \times \mathscr{D}(x)}(\cdot, y) = \mathscr{D}(x) \cap \text{lev}_{\geq}\Phi(\cdot, y)$ is closed for each $y \in \mathscr{D}(x)$. Hence, by Theorem 1.34, there exists $x' \in \mathscr{D}(x)$ such that $\Phi|_{\mathscr{D}(x) \times \mathscr{D}(x)}(x', y) \geq 0$ for all $y \in \mathscr{D}(x)$, which means that $S(x) \neq \emptyset$. We prove the closedness of S. Let a net $\{(x^\alpha, u^\alpha)\}_{\alpha \in I} \subset \text{gph}S$ ($\subset \text{gph}\mathscr{D}$) converge to (x^0, u^0). Since $\mathscr{D}$ is closed, $u^0 \in \mathscr{D}(x^0)$. Take any $y \in \mathscr{D}(x^0)$. By the lower semicontinuity of $\mathscr{D}$, there exist a subnet $\{x^{\alpha_\beta}\}_{\beta\in J}$ of $\{(x^\alpha, u^\alpha)\}_{\alpha\in I}$ and a net $\{y^\beta\}_{\beta\in J}$ with $y^\beta \in \mathscr{D}(x^{\alpha_\beta})$ such that $\{y^\beta\}_{\beta\in J}$ converges to y. Since $u^{\alpha_\beta} \in S(x^{\alpha_\beta}) \subset \bigcap_{y\in\mathscr{D}(x^{\alpha_\beta})} \mathrm{lev}_{\geq}\Phi(\cdot, y)$, we have $(u^{\alpha_\beta}, y^\beta) \in \mathrm{lev}_{\geq}\Phi$. As $\mathrm{lev}_{\geq}\Phi$ is closed, $(u^0, y) \in \mathrm{lev}_{\geq}\Phi$, i.e., $u^0 \in \mathrm{lev}_{\geq}\Phi(\cdot, y)$. Thus, $u^0 \in \bigcap_{y\in\mathscr{D}(x^0)} \mathrm{lev}_{\geq}\Phi(\cdot, y)$. It follows that $u^0 \in \mathscr{D}(x^0) \cap \bigcap_{y\in\mathscr{D}(x^0)} \mathrm{lev}_{\geq}\Phi(\cdot, y) = S(x^0)$, i.e., $(x^0, u^0) \in \mathrm{gph}S$. Thus, S is closed. By Theorem 2.14, S has a fixed point $\bar{x}$. It is clear that this $\bar{x}$ is also a solution of (QEP).

Exercise 2.5 **Hint.** Show that the set-valued mapping $x \mapsto M(x) = \{y \in \mathscr{D}(x) \mid \Phi(x, y) = m(x)\}$ has a fixed point $\bar{x}$ (by using Kakutani's theorem). Then, $\bar{x} \in \mathscr{D}(\bar{x})$ and $0 \leq \Phi(\bar{x}, \bar{x}) = m(\bar{x}) = \min_{y\in\mathscr{D}(\bar{x})} \Phi(\bar{x}, y)$. Hence, $\bar{x}$ is a solution of (QEP).

Exercise 2.6 It is sufficient to show that the mapping $H : A \rightrightarrows A$ in the proof of Theorem 2.28, which is defined for all $x \in A$ by

$$H(x) = \begin{cases} \mathrm{conv}(\mathscr{D}(x) \cap \mathrm{lev}_{<}\Phi(x, \cdot)) & \text{if } x \in \mathrm{Fix}\mathscr{D}, \\ \mathscr{D}(x) & \text{if } x \notin \mathrm{Fix}\mathscr{D}. \end{cases}$$

is lower semicontinuous. It is clear that $H(x) = \mathrm{conv}T(x)$, where $T : A \rightrightarrows A$ is defined for all $x \in A$ by

$$T(x) = \begin{cases} \mathscr{D}(x) \cap \mathrm{lev}_{<}\Phi(x, \cdot) & \text{if } x \in \mathrm{Fix}\mathscr{D}, \\ \mathscr{D}(x) & \text{if } x \notin \mathrm{Fix}\mathscr{D}. \end{cases}$$

By Proposition 2.8, we only need to verify that T is lower semicontinuous. Take any $x \in A$ and any open set V in A such that $T(x) \cap V \neq \emptyset$. Since $\mathscr{D}$ and $F = S \cap H$ are lower semicontinuous ($F = S \cap H$ is lower semicontinuous by Proposition 2.94), there exist open sets U_1 and U_2 in A such that

$$\mathscr{D}(x) \cap V \neq \emptyset \text{ for all } x \in U_1 \quad \text{and} \quad (\mathscr{D}(x) \cap \mathrm{lev}_{<}\Phi(x, \cdot)) \cap V \neq \emptyset \text{ for all } x \in U_2.$$

By choosing

$$U = \begin{cases} U_1 \cap U_2 & \text{if } x \in \mathrm{Fix}\mathscr{D}, \\ U_1 \cap (A \backslash \mathrm{Fix}\mathscr{D}) & \text{if } x \notin \mathrm{Fix}\mathscr{D}, \end{cases}$$

we have: $T(x) \cap V \neq \emptyset$ for all $x \in U$. Thus, T is lower semicontinuous.

Exercise 2.7 **Hint.** By Theorem 2.21, there exists $\bar{x} \in \mathrm{cl}(\mathrm{Fix}\mathscr{D})$ such that $\Phi(\bar{x}, y) \geq 0$ for all $y \in \mathscr{D}(\bar{x})$. Since $\mathrm{Fix}\mathscr{D} \subset \mathrm{Fix}\overline{\mathscr{D}}$ and $\mathrm{Fix}\overline{\mathscr{D}}$ is closed (because $\overline{\mathscr{D}}$ is closed), we have $\mathrm{cl}(\mathrm{Fix}\mathscr{D}) \subset \mathrm{Fix}\overline{\mathscr{D}}$.

Exercise 2.12 Hint.

(a) For $y \in A$, take

$$H(y) := [A \setminus \mathscr{D}_2^{-1}(y)] \cup [\mathrm{Fix}\mathscr{D}_1 \cap \mathrm{lev}_{\geq}\Phi(\cdot, y)]$$

and $\Phi_H : A \times A \to \mathbb{R}$ defined as

$$\Phi_H(x, y) := \begin{cases} 0 \text{ if } x \in H(y), \\ -1 \text{ otherwise.} \end{cases}$$

Then, show that the equilibrium problem defined by Φ_H is a required problem.

(b) Consider (QEP) with

$$\mathscr{D}(x) := \begin{cases} \mathscr{D}_2(x) \text{ if } x \in \mathrm{Fix}\mathscr{D}_1, \\ \mathscr{D}_1(x) \text{ if } x \notin \mathrm{Fix}\mathscr{D}_1. \end{cases}$$

(c) Apply Theorem 1.34 for Φ_H in (a) or Theorem 2.21 for $\mathscr{D}$ in (b).

Chapter 3

Exercise 3.1 Hint. Let $E \in \mathscr{E}(Y)$. Assume that there exists $e \in E \cap (-\mathrm{core}E)$. Then, for all $v \in Y$, there is $n_v \in \mathbb{N}$ such that $-e + \frac{1}{n_v}v \in E$ which means that $v \in n_v e + n_v E \subset E + E \subset E$. Hence, $E = Y$, a contradiction.

We have: $E \cap (-E) = [(E \backslash \mathrm{core}E) \cup \mathrm{core}E] \cap [((-E) \backslash (-\mathrm{core}E)) \cup (-\mathrm{core}E)] = [(E \backslash \mathrm{core}E) \cap ((-E) \backslash (-\mathrm{core}E))] \cup [\mathrm{core}E \cap ((-E) \backslash (-\mathrm{core}E))] \cup [E \cap (-\mathrm{core}E)] = (E \backslash \mathrm{core}E) \cap ((-E) \backslash (-\mathrm{core}E))$.

Exercise 3.2 Hint. Assume that $v - v' \in -\mathrm{core}E' \Leftrightarrow v' - v \in \mathrm{core}E'$. Then, there exists $\lambda > 0$ such that $(v' - v) - \lambda q \in E' \Leftrightarrow (v + \lambda q) - v' \in -E'$. This and the part (f) of Proposition 3.10 give that $r_E^q(v + \lambda q) \leq r_E^q(v')$, i.e, $r_E^q(v) + \lambda \leq r_E^q(v')$. Hence, $r_E^q(v) < r_E^q(v')$.

Exercise 3.3 Hint. As $q \in Q_E$, $\mathbb{R}_+ q + E = E \cup (\mathbb{R}_{++} q + E) = E \cup \mathrm{core}E = E$.

Exercise 3.4

(a) Take any $v \in Y$. When $q \in \mathscr{B}_E$, there exists $E' \in E_{\mathscr{E}(Y)}$ such that $q \in \mathrm{core}E'$. Then, there exists $n_v \in \mathbb{N}$ such that $\frac{1}{n_v}v + q \in E'$ and $-\frac{1}{n_v}v + q \in E'$, i.e., $v + n_v q \in n_v E' \subset E'$ and $v \in n_v q - n_v E' \subset n_v q - E' \subset \mathbb{R}q - E$. Hence, by the definition of r_E^q, $r_E^q(v) < +\infty$. If $r_E^q(v) = -\infty$, then there exists $\lambda > n_v$ such that $v \in -\lambda q - E$. Then, $v + n_v q \in -(\lambda - n_v)q - E \subset -\mathbb{R}_{++} q - E = -\mathrm{core}E$. Thus, $v + n_v q \in E' \cap (-\mathrm{core}E) = \emptyset$, we obtain a contradiction.

(b) **Hint.** This is a direct consequence of part (a) and Exercise 3.2.
(c) **Hint.** This is a direct consequence of Proposition 3.10 (d).

Exercise 3.5

(a) For all $q \in Y$, $E + E \subset E \subset \mathrm{vcl}_q E \subset \mathbb{R}_+ q + \mathrm{vcl}_q E$. Hence, $\mathrm{H}_E = Y$.
(b) As $E \in \mathscr{E}(Y)$, $E \in \mathscr{B}_E$. The equality $\mathscr{B}_E = \mathrm{core} E$ follows easily.

Exercise 3.6 If $q \in \mathrm{Q}_E$, then $q = q + 0 \in q + E \subset \mathbb{R}_{++} q + E = \mathrm{core} E$. Thus, $\mathrm{Q}_E \subset \mathrm{core} E$. By definition of $\mathrm{core} E$, it is easily to show that $\mathrm{core} E \subset \mathbb{R}_{++} q + E$ for all $q \in Y$. Moreover, for all $q \in \mathrm{core} E$, $\mathbb{R}_{++} q + E \subset \mathrm{core} E + E \subset \mathrm{core} E$. Thus, $\mathrm{core} E \subset \mathrm{Q}_E$.

Exercise 3.7 *(a) implies (b)*: Assume that h is E-lower semicontinuous. Take any $v \in Y$ and $x \in h^{-1}(v + \mathrm{int} E)$. Then, $h(x) \in v + \mathrm{int} E$, and hence, there exists a neighborhood U of x in A such that $h(y) \in v + \mathrm{int} E + E) = v + \mathrm{int} E$. It follows that $x \in U \subset h^{-1}(v + \mathrm{int} E)$, and therefore, $h^{-1}(v + \mathrm{int} E)$ is open.

(b) implies (c): Suppose that, for all $v \in \mathrm{int} E$, $h^{-1}(v + \mathrm{int} E)$ is an open subset of A. For each $x \in A$ and any $v \in \mathrm{int} E$, we have $x \in h^{-1}(h(x) - v + \mathrm{int} E)$, which is open. Hence, there exists an open neighborhood $U \subset A$ of of x such that $h(x') \in h(x) - v + \mathrm{int} E$ for all $x' \in U$.

(c) implies (a) as $0 \in \delta(\mathrm{int} E)$: Let $x \in A$ and let V be an open neighborhood of $h(x)$. Since $0 \in -h(x) + V$ and $0 \in \delta(\mathrm{int} E)$, $(-h(x) + V) \cap \mathrm{int} E \neq \emptyset$. Hence, there is $v \in \mathrm{int} E$ such that $h(x) - v \in V$. Then there exists an open neighborhood U of x such that $h(x') \in h(x) - v + \mathrm{int} E \subset V + \mathrm{int} E \subset V + E$ for all $x' \in U$. Thus, h is E-lower semicontinuous.

Exercise 3.8 **Hint.** Argue similarly as for Exercise 3.7.

Exercise 3.10 **Hint.** Use Exercise 3.7. For each $x \in A$ and any $v \in \mathrm{int} E$, by (c) of Exercise 3.7, there exist two open neighborhoods U_h and U_s of x such that $h(x') \in h(x) - \frac{1}{2} v + \mathrm{int} E$ for all $x' \in U_h$ and $s(x') \in s(x) - \frac{1}{2} v + \mathrm{int} E$ for all $x' \in U_s$. Setting $U = U_h \cap U_s$, we have: $(h+s)(x') = h(x') + s(x') \in h(x) + s(x) - v + \mathrm{int} E$ for all $x' \in U$, which shows that $s + h$ satisfies (c) of Exercise 3.7.

Exercise 3.11 **Hint.** Use Exercise 3.7. For $(x, z) \in A \times B$ and any $v \in \mathrm{int} E$, by (c) of Exercise 3.7, there exist two open neighborhoods U_h and U_s of x and z respectively such that $h(x') \in h(x) - \frac{1}{2} v + \mathrm{int} E$ for all $x' \in U_h$ and $s(z') \in s(z) - \frac{1}{2} v + \mathrm{int} E$ for all $z' \in U_s$. Setting $U = U_h \times U_s$, we have: $f((x', z')) = h(x') + s(z') \in h(x) + s(z) - v + \mathrm{int} E$ for all $(x', z') \in U$, which shows that f satisfies (c) of Exercise 3.7.

Exercise 3.12 **Hint.** Argue similarly as in the proof of Proposition 1.29.

Exercise 3.13

(a) Assume that f is not properly E-quasimonotone of type 1 in the first component (in the second component, resp.). Then, there exist $N = \{x^1, \ldots, x^n\} \in \mathscr{F}(A)$

and $x = \sum_{i=1}^{n} \lambda_i x^i \in \text{conv}N$ ($\lambda_i \geq 0$ for all $i = 1, \ldots, n$ and $\sum_{i=1}^{n} \lambda_i = 1$) such that $f(x^i, x) \in -E$ ($f(x, x^i) \in -E$, resp.) for all $i = 1, \ldots, n$. This implies that $\sum_{i=1}^{n} \lambda_i f(x^i, x) \in -E$ ($\sum_{i=1}^{n} \lambda_i f(x, x^i) \in -E$, resp.) because E is convex. Since $E \in \mathscr{E}(Y)$, it follows from the E-diagonally quasiconvexity of type 2 in the first component (in the second component, resp.) of f that $f(x, x) \in \sum_{i=1}^{n} \lambda_i f(x^i, x) - E \subset -E$ ($f(x, x) \in \sum_{i=1}^{n} \lambda_i f(x, x^i) - E \subset -E$, resp.), which contradicts to the assumption that $f(x, x) \notin -E$ for all $x \in A$.

(b) **Hint.** Arguing similarly as for (a).

Exercise 3.14 Let $f : A \times A \to Y$ be defined by $f(x, y) = l(x, y) + g(x, y) - l(x, x) - g(x, x)$ for all $x, y \in A$. Then, f is properly core$C$-quasimonotone of type 1 in the second component. Indeed, if not, then there exist $N = \{x^1, \ldots, x^n\} \in \mathscr{F}(A)$ and $x = \sum_{i=1}^{n} \lambda_i x^i \in \text{conv}N$ ($\lambda_i \geq 0$ for all $i = 1, \ldots, n$ and $\sum_{i=1}^{n} \lambda_i = 1$) such that $f(x, x^i) = l(x, x^i) + g(x, x^i) - l(x, x) - g(x, x) \in -\text{core}C$ for all $i = 1, \ldots, n$. It follows that $\sum_{i=1}^{n} \lambda_i f(x, x^i) = [\sum_{i=1}^{n} \lambda_i l(x, x^i) - l(x, x)] + [\sum_{i=1}^{n} \lambda_i g(x, x^i) - g(x, x)] \in -\text{core}C$. Hence, $\sum_{i=1}^{n} \lambda_i l(x, x^i) - l(x, x)] \in -\sum_{i=1}^{n} \lambda_i g(x, x^i) + g(x, x)] - \text{core}C$. By (ii), $-\sum_{i=1}^{n} \lambda_i g(x, x^i) + g(x, x) \in -C$. Hence, $\sum_{i=1}^{n} \lambda_i l(x, x^i) - l(x, x) \in -C - \text{core}C \subset -\text{core}C$. This contradicts (i).

Thus, using Theorem 3.46 (with $E = \text{core}C$) and $i = 1$) and the coercivity condition (C_{11}^{E}) (with $E = \text{core}C$), we have the conclusion.

Exercise 3.15 Let $x, y \in \text{lev}^u_{-\text{core}E} h := \{x \in A \mid h(x) \in u - \text{core}E\}$ and $t \in (0, 1)$. Then, for $z_t = tx + (1-t)y \in [x, y]$, by the properly E-quasiconvexity of h, we have either $h(z_t) \in h(x) - E \subset u - \text{core}E - E \subset u - \text{core}E$ or $h(z_t) \in h(y) - E \subset u - \text{core}E - E \subset u - \text{core}E$. Thus, $z_t \in \text{lev}^u_{-\text{core}E} h$, and hence $\text{lev}^u_{-\text{core}E} h$ is convex, i.e., h is core$E$-quasiconvex.

Exercise 3.16 Assume that $f + g$ is not properly core$C$-quasimonotone of type 1 in the second component. Then, there exist $N = \{x^1, \ldots, x^n\} \in \mathscr{F}(A)$ and $x = \sum_{i=1}^{n} \lambda_i x^i \in \text{conv}N$ ($\lambda_i \geq 0$ for all $i = 1, \ldots, n$ and $\sum_{i=1}^{n} \lambda_i = 1$) such that $f(x, x^i) + g(x, x^i) \in -\text{core}C$ for all $i = 1, \ldots, n$. This and (ii) imply that $f(x^i, x) \in \text{core}C$ for all $i = 1, \ldots, n$. It is not hard to prove that *if a vector function $h : A \to Y$ is properly C-quasiconcave, then $\{x \in A \mid h(x) \in \text{core}C\}$ is convex* (also see Exercise 3.16. Then, by (iii) we have $f(x, x) = f(\sum_{i=1}^{n} \lambda_i x^i, x) \in \text{core}C$, which contradicts (i).

Exercise 3.17 **Hint.** A consequence of Exercise 3.16, Theorem 3.46 ((with $f + g$ replacing f, $E = \text{core}C$) and $i = 1$) and the coercivity condition (C_1^{E}) (with $f + g$ replacing f and $E = \text{core}C$).

Exercise 3.18 **Hint.** Fix $e^1 \notin -E$ and $e^2 \in -E$. Let $f' : A \times A \to \mathbb{R}$ be defined for all $x, y \in A$ by

$$f'(x, y) := \begin{cases} e^1 \text{ if } x \in \text{cl}(\text{lev}^i_E f(\cdot, y)), \\ e^2 \text{ otherwise.} \end{cases}$$

Apply Theorem 3.46 (with $i = 1$) for f', by using the same arguings as in Exercises 1.9(b) and 1.10(a).

Exercise 3.19 **Hint.** It is clear that f is E_λ-cyclically anti-quasimonotone of type 1 if and only if g_λ is cyclically anti-quasimonotone. Moreover, $\text{lev}^1_E f(\cdot, y) = \{x \in A \mid g_\lambda(x, y) \geq 0\} := \text{lev}_{\geq} g_\lambda(\cdot, y)$, and $\overline{\text{lev}}^1_E f(\cdot, y) = \{x \in A \mid g_\lambda(x, y) < 0\} := \text{lev}_{<} g_\lambda(\cdot, y)$. Thus, by applying Theorem 3.71 (with $i = 1$) for g_λ and E_λ, we obtain the conclusion.

Exercise 3.20 **Hint.** For $f(x, y) := (T(x), y - x) = (\langle T_1(x), y - x\rangle, \ldots, \langle T_k(x), y - x\rangle)$ and $\lambda = (\lambda_1, \ldots, \lambda_k) \in \mathbb{R}^k$, $g_\lambda(x, y) = \langle \lambda, f(x, y)\rangle$.

(a) Apply Exercise 3.19 for $g_\lambda(x, y) = \langle \lambda, f(x, y)\rangle = (\varphi(y) - \varphi(x))\gamma_\lambda(x)$, where $\varphi(x) = \sum_{i=1}^m c^i_\lambda x_i$.
(b) Apply part (a) with $c^i_\lambda = \sum_{j=1}^k \lambda_j a_{ji}$, $i = 1, \ldots, m$, and $\gamma_\lambda = \gamma$.
(c) Apply Exercise 3.19 for $g_\lambda(x, y) = \langle \lambda, f(x, y)\rangle = (\varphi(y) - \varphi(x))\gamma_\lambda(x)$, where $\varphi(x) = \sum_{i=1}^m c^i x_i$ and $\gamma_\lambda(x) = \sum_{j=1}^k \lambda_j \gamma_j(x)$.

Exercise 3.21 **Hint.** For the traffic network $G = (\mathbf{N}, \mathbf{A}, c, \mathbf{W}, d)$ with capacity constraints on arcs, $g_\lambda(x, y)$ in Exercise 3.20 is calculated as folows.

$$g_\lambda(x, y) = \sum_{w \in W} \sum_{p \in P_w} (y_p - x_p) \sum_{j=1}^k \lambda_j \mathbb{T}_{jp}(x).$$

If the vector cost of vector flow x on arc $a \in \mathscr{A}$ is $t_a(x) = (t_{1a}(x), \ldots, t_{ka}(x))$, then $\mathbb{T}_{jp}(x) = \sum_{a \in \mathscr{A}} \delta_{ap} t_{ja}(x)$ and

$$g_\lambda(x, y) = \sum_{w \in W} \sum_{p \in P_w} (y_p - x_p) \sum_{a \in \mathscr{A}} \delta_{ap} \sum_{j=1}^k \lambda_j t_{ja}(x). \tag{A.1}$$

(a) From (A.1) we calculate: $g_\lambda(x, y) = (\varphi(y) - \varphi(x))\gamma_\lambda(x)$, where $\varphi(x) = \sum_{w \in W} \sum_{p \in P_w} (\sum_{a \in \mathscr{A}} \delta_{ap} r_a) x_p$ and $\gamma_\lambda(x) = \sum_{j=1}^k \lambda_j \gamma_j(x)$. Apply Exercise 3.19 for this g_λ.
(b) By using (A.1), $g_\lambda(x, y) = (\varphi(y) - \varphi(x))\gamma(x)$, where $\varphi(x) = \sum_{w \in W} \sum_{p \in P_w} (\sum_{j=1}^k \sum_{a \in \mathscr{A}} \delta_{ap} \lambda_j \alpha^j_a) x_p$. Apply Exercise 3.19 for this g_λ.

Exercise 3.22 **Hint.** Apply part (b) of Exercise 3.21 for $\lambda = (0, 1)$. Simple calculations yield, for $\lambda = (0, 1)$ and $(x, y) \in A \times A$,

$$g_\lambda(x, y) = (\varphi(y) - \varphi(x))\gamma(x),$$

where $\varphi(x) = 3x_1 + 2x_2 + x_3 + 2x_4 + x_5$.

Exercise 3.23 **Hint.** For $x, y \in A$ and $\lambda = (\lambda_1, \ldots, \lambda_k) \in \mathbb{R}^k$, we calculate

$$f(x, y) := N(x, y) = \sum_{i=1}^{n} [\mathbf{c}_i(x) - \mathbf{c}_i(x_{\hat{i}}, y_i)] = \big(\beta_1(x, y), \ldots, \beta_k(x, y)\big),$$

where $\beta_j(x, y) = \sum_{i=1}^{n} (c_{ij}(x) - c_{ij}(x_{-i}, y_i))$, and

$$g_\lambda(x, y) = \langle \lambda, f(x, y) \rangle = \sum_{i=1}^{n} (\sum_{j=1}^{k} \lambda_j c_{ij}(x) - \sum_{j=1}^{k} \lambda_j c_{ij}(x_{-i}, y_i))$$
$$= \sum_{i=1}^{n} h_i(x_i) - \sum_{i=1}^{n} h_i(y_i).$$

Apply Exercise 3.19 for this g_λ.

Problems of Chapter 4

Exercise 4.1 (See [326])
Hint. By using properties of a continuous function and the definition of the approximate solution set, the conclusion follows.

Exercise 4.2 (See [326])
Hint. For any real number δ satisfying $0 < \delta < \varepsilon_0$, consider $N(\varphi_0) := [\varepsilon_0 - \delta, \varepsilon_0 + \delta]$ as a given neighborhood of ε_0. Let $\varepsilon_1, \varepsilon_2 \in N(\varepsilon_0)$ with $\varepsilon_1 < \varepsilon_2$. For any number η satisfying $0 < \eta < \varepsilon_2 - \varepsilon_1$, by the continuity of f in the third variable, there exists a neighborhood $N(\mu_0)$ such that for each $x, y \in E$,

$$|f(x, y, \mu_1) - f(x, y, \mu_2)| \leq \eta, \forall \mu_1, \mu_2 \in N(\mu_0).$$

This together with the definition of approximate solutions implies the needed conclusion.

Exercise 4.3
Hint. By using the concavity assumption, for any $\lambda \in [0, 1]$, we have

$$f(\lambda x_1 + (1 - \lambda)x_2, y, \mu_0) + \varepsilon \geq \lambda(f(x_1, y, \mu_0) + \varepsilon) + (1 - \lambda)(f(x_2, y, \mu_0) + \varepsilon).$$

Combining this with the convexity of E, we get the conclusion.

Exercise 4.4
Hint. By Exercise 4.2, we have

$$S(\varepsilon_0 - \delta, \mu_0) \subset S(\varepsilon, \mu) \subset S(\varepsilon_0 + \delta, \mu_0), \forall \mu \in N(\mu_0), \forall \varepsilon \in [\varepsilon_0 - \delta, \varepsilon_0 + \delta].$$

Letting $\gamma := \varepsilon_2 + \theta(\varepsilon_1 - \varepsilon_2)$, where $1 < \theta < \varepsilon_2/(\varepsilon_2 - \varepsilon_1)$ and $\varepsilon_1, \varepsilon_2$ are defined as in Exercise 4.2, we have

$$\frac{1}{\theta}S(\gamma, \mu_0) + (1 - \frac{1}{\theta})S(\varepsilon_2, \mu_0) \subset S(\varepsilon_1, \mu_0).$$

The above conclusions imply that for a convex neighborhood B of the origin,

$$S(\varepsilon, \mu) \subset S(\varepsilon_0, \mu_0), \forall \mu \in N(\mu_0), \forall \varepsilon \in [\varepsilon_0 - \delta, \varepsilon_0 + \delta].$$

The conclusion follows.

Exercise 4.5 (See [99])
Hint. We first prove that via the proper quasiconvexity condition, the set

$$\bigcap_{x \in K(\lambda)} \{y \in \mathscr{K}(\lambda) \mid f(x, y) \leq \sup_{\Delta} f\}$$

is nonempty for any $\lambda \in \Lambda$, where $\Delta := \{(t, t) \mid t \in K(\lambda)\}$. Then, by the given conditions, we show that the aforementioned intersection set is also the solution set of the equilibrium problem.

Exercise 4.6
Hint. The conclusion follows from the fact that: If $\mathscr{D}$ is open at $\bar{\lambda}$ and h is lower semicontinuous on $\{\bar{\lambda}\} \times \mathscr{D}(\bar{\lambda})$, the map

$$\lambda \mapsto \sup_{y \in \mathscr{D}(\lambda)} h(y, \lambda)$$

is lower semicontinuous at $\bar{\lambda}$.

Exercise 4.7
Hint. The conclusion follows from the closedness of the solution map. Let $\{(x_n, \mu_n)\}$ converge to $(\bar{x}, \bar{\mu})$ with $x_n \in S(\mu_n)$. For $y \in K$ and $t \in [0, 1]$, we will have

$$0 = f(y_t, y_t; \mu) \leq \max\{f(y_t, y, \mu), f(y_t; \bar{x}, \mu)\},$$

where $y_t = ty + (1 - t)\bar{x}$. This implies that

$$0 \leq \limsup_{t \to 0^+} f(y_t, y, \mu) \leq f(\bar{x}, y, \mu),$$

and the needed conclusion follows.

Exercise 4.8
Hint. The conclusion can be obtained from the following Robinson–Ursescu theorem:

Let X and Y be Banach spaces, F be a proper closed convex map (i.e., map with closed convex graph) from X into Y the range of which has a nonempty interior Let $y_0 \in \operatorname{int}(R(F))$ and $x_0 \in F^{-1}(y_0)$ be chosen. There exists some $\gamma > 0$ such that

$$\forall x \in \operatorname{dom} F, \forall y \in y_0 + \gamma B(0,1), d(x, F^{-1}(y)) \leq \frac{1}{\gamma} d(y, F(x))(1 + \|x - x_0\|).$$

In particular, if F^{-1} is lower semicontinuous on $\operatorname{int}(R(F))$, then by taking $x = x_0$ we get

$$\forall y \in y_0 + \gamma B(0,1), d(x_0, F^{-1}(y)) \leq \frac{1}{\gamma} d(y, F(x_0)).$$

Exercise 4.9
Hint. It is easy to check that the map $f(\cdot, \cdot, \mu)$ is strictly pseudomonotone, and so the solution set $S(\mu)$ is a singleton, say $x(\mu)$. By direct computations, we have

$$d(x(\mu_1), x(\mu_2)) \leq k d(\mu_1, \mu_2)^{\rho},$$

where k and ρ are constants. Hence, we have the conclusion.

Exercise 4.10
Hint. First, we can prove that G is locally surjective. Let V_0 be a bounded subset of Y. Then, we can show that $|G^{-1}| < +\infty$ and

$$H(G^{-1}(v), G^{-1}(0) \leq |G^{-1}| \|v\|, \forall v \in V_0,$$

where $|G^{-1}|$ is the norm of the convex process G^{-1}. Furthermore, we also get a closed subset B of X such that $G^{-1}(V_0) \subset B$. Next, because X is a reflexive space, closed bounded sets are compact in X_σ. Finally, we prove that G^{-1} is closed, and then due to $G^{-1}(V_0)$ being a subset of a compact set, we will obtain the needed conclusion.

Exercise 4.11
Hint. We need only prove the upper semicontinuity of S on $Y \times X_\sigma$. Moreover, because $S(V_0)$ is contained in a weakly compact set B of X, it is sufficient to show that S is closed in $Y \times X_\sigma$.

Exercise 4.12
Hint. We first prove that the map

$$\zeta(e, x) := \max\{-f(x, x') \mid x' \in G^{-1}(e)\}$$

is usc on $Y \times X$. Let $\varepsilon > 0$ and W be open such that $W \cap S_\varepsilon(0) \neq \emptyset$. Let $\bar{x} \in W \cap S_\varepsilon(0)$. Then, $\bar{x} \in G^{-1}(0)$ and $\zeta(0, \bar{x}) \leq 0$. Moreover, there exist neighborhoods $V_1 \subset V_0$ of $0 \in Y$ and $U \subset W$ of $\bar{x}$ such that

$$\zeta(e, x) \leq \zeta(0, \bar{x}) + \varepsilon, \forall e \in V_1, \forall x \in U.$$

On the other hand, since G is surjective at $\bar{x}$, there exists a neighborhood $V_2 \subset V_1$ of $0 \in Y$ such that

$$g^{-1}(e) \cap U \neq \emptyset, \forall e \in V_2.$$

It follows from the earlier discussions that $W \cap S_\varepsilon(e) \neq \emptyset$ for all $e \in V_2$.

Problems of Chapter 5

Exercise 5.1 (See [198])
Hint. By direct computations, we have

$$\mathrm{gph}(T) = \left\{(x, y) \in \mathbb{R}^2 \mid (x-1)^2 + (y-1)^2 \leq 1\right\}.$$

This implies the needed conclusion.

Exercise 5.2 (See [198])
Hint. Applying properties of the Hausdorff distance H, we get that $(\mathscr{P}(E), d)$ is a metric space. Moreover, by the completeness property of E, we can prove that $(\mathscr{P}(E), d)$ is also complete.

Exercise 5.3 (See [198])
Hint. Let $\{x_k\}_k$ be an arbitrary sequence in E with $x_k \to x$. Then, we get

$$K(x_k) \to K(x), K'(x_k) \to K'(x).$$

By the continuity of the Hausdorff distance, we derive that $g(x_k) \to g(x)$.

Exercise 5.4
Hint. In order to prove that the map G is usc, it is sufficient to show that it is closed as E is compact. If G is not lsc, there exist $(x, K) \in E \times \mathscr{D}(E, \mathscr{P}(E))$, a sequence $(x^k, K_k) \to (x, K)$, and an open subset U of E with $U \cap G(x, K) \neq \emptyset$ such that

$$U \cap G(x^k, K_k) = \emptyset, \forall k.$$

Then, due to the definition of G and the properties of Hausdorff distance H, we get a contradiction.

Exercise 5.5
Hint. Applying the result of Exercise 5.2 and the definition of Ω, we get the conclusion.

Exercise 5.6
Hint. Since E is compact, it is sufficient to show that $\mathscr{E}$ is closed, which follows from Exercise 5.2.

Exercise 5.7
Hint. If $\mathscr{E}$ is not lsc at K, then there exists a sequence $K_k \to K$, $x_0 \in \mathscr{E}(K)$, and a real number $\varepsilon > 0$ such that

$$B(x_0, \varepsilon) \cap \mathscr{E}(K_k) = \emptyset,$$

for all n. This implies a contradiction.

Exercise 5.8
Hint. Since E is compact, it is sufficient to show that S is closed. For any $(f_0, K_0) \in P$, let $\{(f^k, K_k)\} \subset P$ and $x^k \in S(f^k, K_k)$ be sequences such that $(f^k, K_k) \to (f_0, K_0)$ and $x^k \to x_0$ as $k \to \infty$. Note that $x^k \in \mathscr{E}(K_k)$, and hence, we also have $x_0 \in \mathscr{E}(K_0)$. Finally, we obtain $x_0 \in S(f_0, K_0)$.

Exercise 5.9
Hint. The conclusion follows from Exercises 5.2, 5.3, and 5.8.

Exercise 5.10
Hint. Suppose that S is not lsc at (f, K). Then, there exist $x_0 \in S(f, K)$, a positive real number ε, and a sequence (f_k, K_k) converging to (f, K) such that

$$S(f_k, K_k) \cap B(x_0, \varepsilon) = \emptyset,$$

for all k. This together with the result of Exercise 5.7 yields a contradiction.

Problems of Chapter 6

Exercise 6.1 (See [101])
Hint. If S is not uhc at λ', then there exists a neighborhood $\mathscr{V}'\left(S\left(\lambda'\right)\right)$ of $S\left(\lambda'\right)$ such that for every neighborhood $\mathscr{U}'\left(\lambda'\right)$ of λ', $S\left(\mathscr{U}'\left(\lambda'\right)\right)$ is not included in $\mathscr{V}'\left(S\left(\lambda'\right)\right)$. Because X is first countable, one can find sequences $\left\{\lambda^k\right\} \subset \Lambda$ and $\left\{x^k\right\} \subset K$ such that $\lambda^k \to \lambda'$ and $x^k \in S\left(\lambda^k\right) \backslash \mathscr{V}'\left(S\left(\lambda'\right)\right)$. Due to the compactness of K, we can assume that $x^k \to \bar{x} \in K \backslash S\left(\lambda'\right)$, and then there exists $\bar{y} \in K$ such that

$$\alpha = f\left(\bar{x}, \bar{y}, \lambda'\right) \in -\operatorname{int} C.$$

In particularly, $\bar{x} \in \text{lev}_{<\alpha/2}\, f\left(\cdot, \bar{y}, \lambda'\right)$. Since the set $\text{lev}_{<\alpha/2} f\left(\cdot, \bar{y}, \lambda'\right)$ is open, there exists k_0 such that

$$f\left(x^k, \bar{y}, \lambda'\right) < \alpha/2, \forall k \geq k_0.$$

Choosing a neighborhood $\mathscr{V}(0)$ of the origin such that

$$\mathscr{V}(0) + \alpha/2 \subset -\operatorname{int} C,$$

and letting

$$\mathscr{V}\left(f\left(x^k, \bar{y}, \lambda'\right)\right) = \mathscr{V}(0) + f\left(x^k, \bar{y}, \lambda'\right),$$

one has $\mathscr{U}_{\bar{y}}\left(\lambda'\right)$ such that

$$f\left(x^k, \bar{y}, \lambda\right) \in \mathscr{V}(0) + f\left(x^k, \bar{y}, \lambda'\right) - C \subset -\operatorname{int} C, \forall \lambda \in \mathscr{U}_{\bar{y}}\left(\lambda'\right).$$

The aforementioned conclusions lead to a contradiction.

Exercise 6.2
Hint. Consider $\lambda^k \to \lambda'$ and $x^k \to \bar{x}$ such that $x^k \in S\left(\lambda^k\right)$, then for every $y \in K$, we have

$$f\left(x^k, y, \lambda^k\right) \not< 0, \quad f\left(y, x^k, \lambda^k\right) \not> 0, \quad f\left(y, \bar{x}, \lambda'\right) \not> 0.$$

It is not hard to see that $f\left(\bar{x}, y, \lambda'\right) \not< 0$ for every $y \in K$, i.e., $\bar{x} \in S\left(\lambda'\right)$. Finally, because the graph of S is closed and K is compact, S is uhc.

Exercise 6.3
Hint. It is easy to show that the solution is unique. Let us denote by $x(\lambda)$ the solution corresponding to the parameter λ. Then, for $\lambda, \lambda' \in \Lambda$,

$$\begin{aligned} f\left(x(\lambda), x\left(\lambda'\right), \lambda\right) \not< 0 \Rightarrow d_X^a\left(x(\lambda), x\left(\lambda'\right)\right) u \in \left(-C + \left\|\lambda - \lambda'\right\|^b v\right) \\ + (Y \setminus \operatorname{int}(C)) - C. \end{aligned}$$

Hence,

$$d_X^a\left(x(\lambda), x\left(\lambda'\right)\right) u \in Y \setminus \operatorname{int}(C) + \left\|\lambda - \lambda'\right\|^b v,$$

as $(Y \setminus \operatorname{int}(C)) - C = Y \setminus \operatorname{int}(C)$. Due to $u \in \operatorname{int}(C)$, we imply that

$$d_X^a\left(x(\lambda), x\left(\lambda'\right)\right) u - \left\|\lambda - \lambda'\right\|^b v \not> 0.$$

Let Y^* denote the topological dual space of Y, and let $C^* = \{w \in Y^* \mid \langle w, z\rangle \geq 0, \forall z \in C\}$. Because C is solid and different from Y, C^* has a weak* compact base, i.e., there exists $B \subset C^*$, B convex, weak* compact, such that $0 \notin B$ and $C^* = \bigcup_{t\geq 0} tB$. Then, we may gain, for instance, $B = \{w \in C^* \mid \langle w, c\rangle = 1\}$, where $c \in \operatorname{int} C$ is arbitrary. Fix such a base B and define

$$\psi(z) = \max_{w\in B}\langle w, z\rangle, \quad \varphi(z) = \min_{w\in B}\langle w, z\rangle, \forall z \in Y.$$

Then, we can prove that

$$d^a\left(x(\lambda), x\left(\lambda'\right)\right) \leq \frac{\psi(v)}{\varphi(u)}\left\|\lambda - \lambda'\right\|^b.$$

Using these results, we get the needed conclusion.

Exercise 6.4
Hint. Take (λ, μ), $\left(\lambda', \mu'\right)$, and consider

$$d_X\left(x(\lambda, \mu), x\left(\lambda', \mu'\right)\right) \leq d_X\left(x(\lambda, \mu), x\left(\lambda', \mu\right)\right) + d_X\left(x\left(\lambda', \mu\right), x\left(\lambda', \mu'\right)\right).$$

It follows from Exercise 6.3 that

$$d_X^a\left(x(\lambda, \mu), \bar{x}\left(\lambda', \mu\right)\right) \leq k\left\|\lambda - \lambda'\right\|^b.$$

By considering two cases that $f\left(x\left(\lambda', \mu\right), x\left(\lambda', \mu'\right), \lambda'\right) \not< 0$ and $f\left(x\left(\lambda', \mu\right), x\left(\lambda', \mu'\right), \lambda'\right) < 0$, we get some k_1 such that

$$d_X^a\left(x\left(\lambda', \mu\right), x\left(\lambda', \mu'\right)\right) \leq k_1\left\|\mu - \mu'\right\|^{g\xi}.$$

The needed conclusion follows.

Exercise 6.5 (See [197])
Hint. Let $\left\{f^k\right\}$ be an arbitrary Cauchy sequence in Φ. Because the function space $\mathscr{C}(K \times K, Y)$ is a complete metric space, there is $f^0 \in \mathscr{C}(K \times K, Y)$ such that $f^k \to f^0$. We need to prove $f^0 \in \Phi$, or equivalently $S\left(f^0\right) \neq \emptyset$.

Due to $S\left(f^k\right) \neq \emptyset$, there exists $x^k \in S\left(f^k\right)$ for each positive integer k, and then by the compactness of K we can assume that $x^k \to \bar{x} \in K$. Because of $x^k \in S\left(f^k\right)$, $f^k \to f^0$, $x^k \to \bar{x}$ and the equicontinuity of $\left\{f^k\right\}_{k=0}^{\infty}$, we bring proof to the end.

Exercise 6.6
Hint. Since K is compact, it is sufficient to show that S is closed on Φ. Let $\left\{f^k\right\}_{k=1}^{\infty} \subset \Phi$ and $x^k \in S\left(f^k\right)$ be sequences such that $f^k \to f^0 \in \Phi$ and

$x^k \to x \in K$ as $k \to \infty$. We now prove $x \in S\left(f^0\right)$. If $x \notin S\left(f^0\right)$, there exists some $y \in K$ such that

$$f^0(x, y) \in -\mathrm{int}C.$$

Therefore, there exists a positive real number $\epsilon > 0$ such that

$$N_\epsilon\left(f^0(x, y)\right) \subset -\mathrm{int}C.$$

Due to $f^k \to f^0$, $x^k \to x$ and the equicontinuity of $\left\{f^k\right\}_{k=0}^{\infty}$, there exists $k^0 \in \mathbb{N}$ such that

$$f^k\left(x^k, y\right) \in N_\epsilon\left(f^0(x, y)\right) \subset -\mathrm{int}C, \quad \forall k \geq k^0,$$

which is impossible.

Exercise 6.7
Hint. By applying Exercise 6.5, one sees that Φ is a Baire space. Then, due to Exercise 6.5, one gets the conclusion.

Exercise 6.8
Hint. The discussions are divided into two cases.

Case 1: $S(f) = \{x_f\}$ is a singleton. If S is not lsc at f, then there exist a sequence $f^k \to f$ and an open neighborhood U of x_f such that $S\left(f^k\right) \cap U = \emptyset$ for all k. Due to $S\left(f^k\right) \neq \emptyset$, for each k we can take $x^k \in S\left(f^k\right)$. Applying the compactness of K, we can assume that $x^k \to \bar{x}$, and then we have $\bar{x} = x_f$. This is impossible.

Case 2: $S(f)$ is not a singleton. Consider a open subset V of X such that $V \cap S(f) \neq \emptyset$. Then, we can take $x_1 \in V \cap S(f)$ and $x_2 \in S(f)$ with $x_2 \neq x_1$. For each $\lambda \in (0, 1)$, we put

$$x_\lambda := \lambda x_1 + (1 - \lambda)x_2.$$

Then, since $x \mapsto f(x, y)$ is strictly C-weakly quasiconcave, we imply that

$$f\left(x_\lambda, y\right) \notin -C, \quad \forall y \in K,$$

or equivalently $x_\lambda \in S(f)$ for all $\lambda \in (0, 1)$. Therefore, there exists $\bar{\lambda} \in (0, 1)$ such that $x_{\bar{\lambda}} \in V \cap \mathrm{conv}\{x_1, x_2\}$ and

$$f\left(x_{\bar{\lambda}}, K\right) \cap (-C) = \emptyset.$$

By the compactness of $f\left(x_{\bar{\lambda}}, K\right)$ and the closedness of $-C$, there exists a positive real number $\delta > 0$ such that

$$B\left(f\left(x_{\bar{\lambda}}, K\right), \delta\right) \cap (-C) = \emptyset,$$

where $B(A, \delta) := \{x \in Y \mid d(x, A) < \delta\}$.

Exercise 6.9
Hint. By using the strong pseudomonotonicity condition, we derive that $S(f)$ is a singleton, say x_f. For any $f' \in \Phi$ and $x' \in S\left(f'\right)$, by applying the strong pseudomonotonicity of f again, we can get some $u \in \text{int}C$ and a positive real number λ such that

$$d^{\lambda}\left(x_f, x'\right) u \in -f\left(x', x_f\right) - C = -f\left(x', x_f\right) + f'\left(x', x_f\right) - f'\left(x', x_f\right) - C.$$

Next, we take some $v \in -\text{int}C$ and some positive real number ε such that $B(v, \varepsilon) \subset -\text{int}C$. Then, we imply that

$$-f\left(x', x_f\right) + f'\left(x', x_f\right) \in -C - \left\|f - f'\right\| \frac{v}{\varepsilon}.$$

This together with earlier discussions would imply that

$$d^{\lambda}\left(x_f, x'\right) u + \left\|f - f'\right\| \frac{v}{\epsilon} \notin \text{int}C.$$

Let Y^* be the topological dual space of Y, and let $C^* = \{w \in Y^* \mid \langle w, z\rangle \geq 0, \forall z \in C\}$. Due to $\text{int}C \neq \emptyset$ and $C \neq Y$, there exists a convex weak* compact subset $B \subset C^*$ such that $0 \notin B$ and $C^* = \bigcup_{t \geq 0} tB$. Now we choose $B = \{w \in C^* \mid \langle w, e\rangle = 1\}$, where $e \in \text{int}C$. For such a base B, we define

$$\psi_B(z) = \max_{w \in B} \langle w, z\rangle, \quad \varphi_B(z) = \min_{w \in B} \langle w, z\rangle, \quad \forall z \in Y,$$

then we have

$$d^{\lambda}\left(x_f, x'\right) \leq \frac{\psi_B(-v)}{\epsilon \varphi_B(u)} \left\|f - f'\right\|.$$

Finally, we show that the set

$$\left\{\frac{\psi_B(-v)}{\epsilon \varphi_B(u)}\right\}_B$$

has a positive lower bound α, and then the needed conclusion follows.

Exercise 6.10 (See [471])
Hint. If there exists $\mu_0 \in M$ such that $S(\cdot)$ is not lsc at μ_0. Then, there exist a sequence $\{\mu^k\}$ with $\mu^k \to \mu_0$ and $x_0 \in S(\mu_0)$ such that for any $x^k \in S(\mu^k), x^k \nrightarrow x_0$. Since $x_0 \in S(\mu_0)$, we have

$$f(x_0, y, \mu_0) \notin -C\backslash\{0\}, \quad \forall y \in \mathscr{D}(\mu_0).$$

Because $\mathscr{D}(\cdot)$ is lsc at μ_0, there exists a sequence $\{\bar{x}^k\}, x^k \in \mathscr{D}(\mu^k)$ such that $\bar{x}^k \to x_0$. We can assume that $\bar{x}^k \in A(\mu^k)\backslash S(\mu^k)$, and hence, there exists $y^k \in S(\mu^k)$ such that

$$d\left(\bar{x}^k, y^k\right) \leq d\left(F\left(\bar{x}^k, y^k, \mu^k\right), Y\backslash - \text{int}C\right).$$

Due to the upper semicontinuity and compactness of $\mathscr{D}(\cdot)$ at μ_0 that there exist $y_0 \in \mathscr{D}(\mu_0)$ and a subsequence $\{y^{k_j}\}$ of $\{y^k\}$ such that $y^{k_j} \to y_0$. This implies a contradiction. Thus, the needed conclusion is obtained.

Exercise 6.11 (See [254])
Hint. The nonemptyness of the solution set $S(\lambda, \mu)$ with $(\lambda, \mu) \in \Lambda \times M$ is not hard to obtain from the results of the previous chapters, and furthermore, it is compact. We now prove that S is upper semicontinuous. Indeed, let a sequence $\{(\lambda^k, \mu^k)\} \subset \Lambda \times M$ be such that $(\lambda^k, \mu^k) \to (\lambda, \mu)$, and take a sequence $\{x^k\}$ with $x^k \in S(\lambda^k, \mu^k)$. Since the set $S(\lambda, \mu)$ is compact, we only need to show that there exist $x \in S(\lambda, \mu)$ and a subsequence $\{x^{k_j}\}$ of $\{x^k\}$ such that $x^{k_j} \to x$. Since $\mathscr{D}$ is usc and compact-valued, there are $x \in \mathscr{D}(\lambda)$ and a subsequence $\{x^{k_j}\}$ of $\{x^k\}$ such that $x^{k_j} \to x$. Then, we have $x \in S(\lambda, \mu)$, which completes the proof.

Exercise 6.12 (See [254])
Hint. It is easy to obtain the nonemptyness of the solution set $S(\lambda, \mu)$ for all $(\lambda, \mu) \in \Lambda \times M$ from the results of previous chapters. Let a sequence $\{(\lambda^k, \mu^k)\} \subset \Lambda \times M$ converge to (λ_0, μ_0) and $x_0 \in S(\lambda_0, \mu_0)$. We will show that there exists a sequence $\{x^k\}$ with $x^k \in S(\lambda^k, \mu^k)$ such that $x^k \to x_0$.

By the assumption on $\mathscr{D}$, for any closed convex neighborhood $V(x_0)$ of x_0, there is a neighborhood N of λ_0 with $N \subset U(\lambda_0)$ such that for any $\lambda \in N$, $\mathscr{D}(\lambda) \cap V(x_0) \neq \emptyset$. Due to $\lambda^k \to \lambda_0$ and $\mu^k \to \mu_0$, there exist a neighborhood B of μ_0 with $B \subseteq I(\mu_0)$ and a number $k_0 \in \mathbb{N}$ such that for all $k \geq k_0$, $\lambda^k \in N$ and $\mu^k \in B$. So, $\mathscr{D}(\lambda^k) \cap V(x_0) \neq \emptyset$. Then, for any $k \geq k_0$, we get some $x^k \in \mathscr{D}(\lambda^k) \cap V(x_0)$ satisfying

$$f(g(x^k), y, \mu^k) \notin -\text{int}\, C(x^k), \quad \forall y \in \mathscr{D}(\lambda^k),$$

or equivalently $x^k \in S(\lambda^k, \mu^k)$.

Exercise 6.13 (See [295])
Hint. Let $x \in \mathfrak{L}$ and $d \in (-\operatorname{int} C(x)) \cap (f(x) + \operatorname{int} C(x))$. Then, $d-\operatorname{int} C(x)$ is a neighborhood of $f(x)$. Hence, due to the $(-C(x))$-continuity of f at x, there exists a neighborhood U_1 of x such that

$$f(u) \in d - \operatorname{int} C(x), \text{ for all } u \in U_1.$$

Since $d - \operatorname{int} C(x) \subset d - \operatorname{cl} C(x)$ and $(c - \operatorname{cl} C(x))^c$ is a neighborhood of $W(x)$, the upper semicontinuity of W implies the existence of a neighborhood U_2 of x such that $d \in -\operatorname{cl} C(x) \subset -\operatorname{int} C(u)$ for all $u \in U_2$. By checking $U_1 \cap U_2 \subset \mathfrak{L}$, we derive the openness of $\mathfrak{L}$.

Exercise 6.14 (See [295])
Hint. Let $x \in \mathfrak{L}$ and let $d \in (-\operatorname{int} C(x)) \cap (f(x)+ \operatorname{int} D(x))$ be arbitrary. Then, $d-\operatorname{int} D(x)$ is a neighborhood of $f(x)$, and $d-\operatorname{int} D(x)-\operatorname{int} D(x) = -\operatorname{int} D(x)$. Because of $(-D(x))$-continuity of f at x, there exists a neighborhood U_1 of x such that

$$f(u) \in d - \operatorname{int} D(x), \text{ for all } u \in U_1.$$

Because K is closed, there exists a neighborhood U_2 of x such that $d \in -\operatorname{int} C(u)$ for all $u \in U_2$. By the definition of D, there exists a neighborhood U_3 of x such that $\operatorname{int} D(x) \subset \operatorname{int} C(u)$ for all $u \in U_3$. By proving $\bigcap_{i=1}^{3} U_i \subset \mathfrak{L}$, we get the needed conclusion.

Exercise 6.15 (See [295])
Hint. Since X is compact, it suffices to show that S is closed. For any $\lambda \in \Lambda$, consider sequences $\{\lambda^k\} \subset \Lambda$ converging to λ and $\{x^k\} \subset X$ with $x^k \in S(\lambda^k)$ for all k. Then, by the given conditions, we can assume that $x^k \to x \in \mathscr{D}(x, \lambda)$. If $x \notin S(\lambda)$, then there exists $y \in \mathscr{D}(x, \lambda)$ such that

$$f(x, y, \lambda) \in -\operatorname{int} C(x, \lambda).$$

Hence, there exists some neighborhood N of (x, y, λ) such that

$$f(x', y', \lambda') \in -\operatorname{int} C(x', \lambda') \text{ for all } (\lambda', x', y') \in N.$$

This implies a contradiction, and hence, S is usc on Λ.

Exercise 6.16 (See [295])
Hint. Let $(x, y, \lambda) \in \mathfrak{U}$, i.e., $f(x, y, \lambda) \notin -\operatorname{cl} C(x, \lambda)$ and let

$$d \in \left((-\operatorname{cl} C(x, \lambda))^c\right) \cap (f(x, y, \lambda) - \operatorname{int} C(x, \lambda)).$$

Then, the set $d+$ int $C(x, \lambda)$ is a neighborhood of $f(x, y, \lambda)$. Due to

$$d + \operatorname{int} C(x, \lambda) + \operatorname{int} C(x, \lambda) = d + \operatorname{int} C(x, \lambda),$$

the $C(x, \lambda)$-continuity of f at (x, y, λ) implies that there exists a neighborhood U of (x, y, λ) such that

$$f(u, v, \zeta) \in d + \operatorname{int} C(x, \lambda), \text{ for all } (u, v, \zeta) \in U.$$

Since $d+$ int $C(x, \lambda) \subset d + \mathrm{cl}C(x, \lambda)$ and $(d + \mathrm{cl}C(x, \lambda))^{\mathrm{c}}$ is a neighborhood of $C(x, \lambda)$, there exists a neighborhood V of (x, λ) such that

$$(d + \mathrm{cl}C(x, \lambda)) \cap (-\mathrm{cl}C(u, \zeta)) = \emptyset \text{ for all } (u, \zeta) \in V$$

as C is usc on $X \times \Lambda$. Finally, by proving $U \cap (V \times X) \subset \mathfrak{U}$, we bring the proof to its end.

Exercise 6.17
Hint. Let $(x, y, \lambda) \in \mathfrak{U}$ be arbitrary. Then, we have

$$f(x, y, \lambda) \notin -\mathrm{cl}C(x, \lambda).$$

Let $d \in (-clC(x, \lambda))^{\mathrm{c}} \cap (f(x, y, \lambda)-$ int $D(x, \lambda))$. Because f is $D(x, \lambda)$-continuous at (x, y, λ), there exists a neighborhood U of (x, y, λ) satisfying

$$f(u, v, \zeta) \in d + \operatorname{int} D(x, \lambda), \forall (u, v, \zeta) \in U.$$

Due to the closedness of C and the definition of D, there exists a neighborhood V of (x, λ) such that

$$(d + \mathrm{cl}D(x, \lambda)) \cap (-\mathrm{cl}C(u, \zeta)) = \emptyset, \forall (u, \zeta) \in V.$$

The needed conclusion is obtained from checking the inclusion $U \cap (V \times X) \subset \mathfrak{U}$.

Exercise 6.18
Hint. Let $\lambda \in \Lambda$ and V be an open subset of X with $V \cap S(p) \neq \emptyset$ and $x_1 \in V \cap S(p)$. If $x_1 \notin S'(\lambda) := \{x \in \mathscr{D}(x, \lambda) \mid f(x, y, \lambda) \notin -\mathrm{cl}C(x, \lambda)$ for all $y \in \mathscr{D}(x, \lambda)\}$. we can choose $x_2 \in S'(\lambda)$. Let $x_t = tx_1 + (1 - t)x_2$ for each $t \in (0, 1)$, then we obtain

$$f(x_t, y, \lambda) \notin -\mathrm{cl}C(x_t, \lambda), \forall y \in [t\mathscr{D}(x_1, \lambda) + (1 - t)\mathscr{D}(x_2, \lambda)].$$

Because $\mathscr{E}(p)$ is convex for each $\lambda \in \Lambda$, $x_t \in \mathscr{D}(x_t, \lambda)$, and so

$$\mathscr{D}(x_t, \lambda) \subset t\mathscr{D}(x_1, \lambda) + (1 - t)\mathscr{D}(x_2, \lambda),$$

as $\mathscr{D}(\cdot, \lambda)$ is concave for each $\lambda \in \Lambda$. Therefore, we have $x_t \in S'(\lambda)$ for all $t \in (0, 1)$, and hence there exists $x \in V \cap S'(\lambda) \cap \operatorname{conv}\{x_1, x_2\}$. If $x_1 \in S'(\lambda)$, then let $x = x_1$. For each $y \in \mathscr{D}(x, \lambda)$, due to the weak quasiconcavity condition, there exist corresponding neighborhoods N_y of λ, U_y of u, and V_y of y such that

$$f(u, v, \zeta) \notin -\mathrm{cl}C(u, \zeta), \forall (u, v, \zeta) \in U_y \times V_y \times \in N_y.$$

Since $\mathscr{D}(x, \lambda)$ is compact, there exists a subset $\{y_1, \ldots, y_k\} \subset \mathscr{D}(x, \lambda)$ such that

$$\mathscr{D}(x, \lambda) \subset \bigcup_{i=1}^{n} V_{y_i}.$$

Thus, we obtain that

$$f(u, v, \zeta) \notin -\mathrm{cl}C(u, \zeta), \text{ for all } u \in \bigcap_{i=1}^{n} U_{y_i}, v \in \bigcup_{i=1}^{n} V_{y_i} \text{ and } \zeta \in \bigcap_{i=1}^{n} P_{y_i}.$$

By the upper semicontinuity of $\mathscr{D}$, there exist neighborhoods U of x and N of λ such that

$$\mathscr{D}(u, \zeta) \subset \bigcup_{i=1}^{n} V_{y_i}, \text{ for all } u \in U \text{ and } \zeta \in N.$$

By the lower semicontinuity of the map $\mathscr{E}$, we get a neighborhood N' of λ such that for each $\zeta \in N'$, there is some $u' \in \mathscr{D}\left(u', \zeta\right)$ belonging to $\bigcap_{i=1}^{n} U_{y_i} \cap V$. Letting $\mathfrak{N} = \bigcap_{i=1}^{n} N_{y_i} \cap N \cap N'$, we have $S\left(p'\right) \cap V \neq \emptyset$ for each $\lambda' \in \mathfrak{N}$.

Exercise 6.19

Hint. We first prove that

$$Y = \cup\{tx - \mathrm{int}C \mid t \in \mathbb{R}^+ \setminus \{0\}\}.$$

Then, for any $y \in Y$, there exists $t \in \mathbb{R}^+ \setminus \{0\}$ such that

$$y \in tx - \mathrm{int}C.$$

Hence, there exists $c \in \mathrm{int}C$ satisfying $y = tx - c$. By choosing $n \geq [t] + 1$, where $[t]$ is the ceiling function giving the smallest integer greater than or equal to t, we get

$$y = nx - (n - t)x - c \in nx - \mathrm{int}C.$$

As a result, $y - nx \in -\mathrm{int}C$.

Exercise 6.20
Hint. If there are $a \in C$ and $b \notin -\text{int}C$ such that

$$a + b \in -\text{int}C.$$

Then, there exists $c \in -\text{int}C$ such that $a + b = c$. By the convexity of C, we have

$$b = -a + c \in -C - \text{int}C = -\text{int}C,$$

which is a contradiction. Therefore, $a + b \notin \text{int}C$.

Problems of Chapter 7

Exercise 7.11 $\text{sdl}\Phi = \text{sdl}\Phi^k = \{(0, 1) \mid 0 \le x \le 1\}$ for all $k \in \mathbb{N}$. So, we have a full presentation of e/h-convergence for the saddle points.

Exercise 7.12 The mai-lop limits and weak mai-lop limits coincide and are the unique bifunction $\Phi(x, y) = \ln(y/x)$ on $(0, e] \times (0, 1]$.

Exercise 7.18 The (unique) e/h-limit is

$$\Phi(x, y) = \begin{cases} 1 & \text{if}(x, y) \in [0, 1] \text{ and } x \neq y, \\ 0 & \text{if}(x, y) \in [0, 1] \text{ and } x = y. \end{cases}$$

To see the mis-lop fails, for $x = 0$ and any $x^k \in A^k \to x$, we take $y = 0$ and $y^k \in B^k \to 0$ such that $y^k \neq x^k$ for all k. Then, $\text{ls}_k \Phi^k(x^k, y^k) = 1 > 0 = \Phi(x, y)$.

Exercise 7.19

(a) All the variational limits coincide and equal $\Phi(x, y) = \ln x \ln y$ defined on $[1/2, 3/2]^2$.
(b) These Φ^k do not have saddle points. Choose

$$\begin{aligned} \varepsilon^k =& \max\{\ln(1 - k^{-1}) \ln 1/2 - \ln(1 - k^{-1}) \ln 3/2; \\ & \ln(1 + k^{-1}) \ln 3/2 - \ln(1 + k^{-1}) \ln 1/2\}. \end{aligned}$$

Then, for any k, ε^k-$\text{sdl}\Phi^k$ is the set consisting of the vertical intervals $\{1 - k^{-1}\} \times ([1/2, 1 - k^{-1}] \cup [1 + k^{-1}, 3/2])$ and $\{1 + k^{-1}\} \times ([1/2, 1 - k^{-1}] \cup [1 + k^{-1}, 3/2])$ and the horizontal intervals $([1/2, 1 - k^{-1}] \cup [1 + k^{-1}, 3/2]) \times \{1 + k^{-1}\}$ and $([1/2, 1 - k^{-1}] \cup [1 + k^{-1}, 3/2]) \times \{1 - k^{-1}\}$. The sets ε^k-$\text{sdl}\Phi^k$ converge to

$$\text{sdl}\Phi = (\{1\} \times [1/2, 3/2]) \cup ([1/2, 3/2] \times \{1\}.$$

(c)

$$\text{mami}\Phi^k = ([1/2, 1 - k^{-1}] \cup [1 + k^{-1}, 3/2]) \times \{1\}$$

tending to the whole set $\text{mami}\Phi = [1/2, 3/2] \times \{1\}$;

$$\text{mima}\Phi^k = \{1 - k^{-1}\} \times \{1/2\}$$

tending to the singleton $\{(1, 1/2)\} = \text{mima}\Phi$.

The aforementioned two sets of Φ are different; this fact also asserts the emptiness of $\text{sdl}\Phi$ by definition.

Problems of Chapter 9

Exercise 9.1 Sol(VwQEP) = $\emptyset$ and ε-Sol(VwQEP) = $[-1, 1]$.

Exercise 9.3 $f_1 = f_1^k$ and $f_2 = -1$ for $(x, y) \in [-1, 1] \times [0, 1]$.

Exercise 9.4 Sol(DVwQEP) = ε-Sol(DVwQEP) = $[0, 1]$.

Exercise 9.6 All the sides of the relations in assertion (c) of Theorem 9.14 are equal to $[-1, 1] \times \{0\}$.

Exercise 9.7 t^k-Sol(VsQEP^k) = Sol(VsQEP) = $[1, e]$.

Problems of Chapter 10

Exercise 10.1 We use Corollary 10.2 to deduce the conclusion. Let $\mathscr{D}(a) = A$ and for $(a, x) \in A \times A$, $R(a, x) = 1$ iff $\rho(a, b) = 1$ for some $b \in S(x)$. For any $x \in A$, the set $Q(x) := \{a \in A \mid \rho(a, b) = 1 \text{ for some } b \in S(x)\}$ is closed. Indeed, for any net $\{a_\alpha\}$ in $Q(x)$ such that $a_\alpha \to a_0$, there exists a net $\{b_\alpha\}$ with $b_\alpha \in S(x)$ and $\rho(a_\alpha, b_\alpha) = 1$ for all α. Since $S(x)$ is closed (by i)) and A is compact, $S(x)$ is compact. Hence, we can assume that $b_\alpha \to b_0 \in S(x)$. By (i), $\rho(a_0, b_0) = 1$, i.e., $a_0 \in Q(x)$. Thus, for all $x \in A$, $\mathscr{D}^{-1}(x) \cap R_x^{-1}(0) = \{a \in A \mid \rho(a, b) = 0 \text{ for all } b \in S(x)\} = A \setminus Q(x)$ is open. For all $a \in A$, by (iii), $(R^a)^{-1}(0) = \{x \in A \mid \rho(a, b) = 0 \text{ for all } b \in S(x)\}$ is convex. (ii) implies that (VRP) in this case has no solution. Thus, from Corollary 10.2, we see that (i) of Corollary 10.2 must be false, which implies that (i_2) of Proposition 10.4 must be false. Then, there exists $\bar{a} \in A$ such that $R(\bar{a}, \bar{a}) = 0$, which means that $\rho(\bar{a}, b)$ does not hold for all $b \in S(\bar{a})$.

Exercise 10.2
Hint. Define a relation R' linking $a \in A$ and $x \in A$ as follows

$$R(a, x) = 1 \text{ iff } R(a, b) = 1 \quad \text{for all } b \in \mathscr{D}(x).$$

Let $\Phi_{R'} : A \times A \to \mathbb{R}$ be defined by

$$\Phi_{R'}(a, x) = \begin{cases} 1 \text{ if } R'(a, x) = 1, \\ -1 \text{ if } R'(a, x) = 0. \end{cases} \quad \text{for all } a \in A \text{ and } x \in A.$$

Then, Apply Lemma 10.8 for this $\Phi_{R'}$.

Exercise 10.3 **Hint.** Apply Corollary 10.2 for $A = B$, $\mathscr{D} = \Omega$, and $a \in A$ linking $b \in A$ in relation R iff $b = a$.

Exercise 10.4 **Hint.** Apply Exercise 10.2 for $A = B$, $\mathscr{D} = \Omega$, and $a \in A$ linking $b \in A$ in relation R iff $b = a$.

Exercise 10.5 **Hint.** • For $A = B$, let $\mathscr{D}(b) \equiv B$ and

$$R(b, a) \text{ holding iff } \sup F(a, b) \leq \delta.$$

Then, $\sup \bigcup_{a \in A} F(a, \bar{b}) \leq \delta$ iff $R(\bar{b}, a)$ iff $R(\bar{b}, a)$ holds for all $a \in \mathscr{D}(\bar{b})$
• Let $\mathscr{D}' : A \times B \rightrightarrows A \times B$ by $\mathscr{D}'((a, b)) \equiv A \times B$, and R' linking $(a, b) \in A \times B$ and $(x, y) \in A \times B$ by

$$R'((a, b), (x, y)) \text{ holding iff } \sup F(x, b) \leq \inf F(a, y).$$

Then, $(\bar{a}, \bar{b})$ is a saddle point of F if and only if $R'(\bar{a}, \bar{b}), (x, y))$ holds for all $((x, y) \in \mathscr{D}((\bar{a}, \bar{b}))$.

(a) Use Corollary 10.2 for $\mathscr{D}$ and R. Prove that (i_2) of Proposition 10.4 holds (hence, (i) of Corollary 10.2 holds) when (i) holds, and the set $\{b \in B | \ \sup F(a, b) \leq \delta\}$ is closed for each $a \in A$ (hence, (ii) of Corollary 10.2 holds)when (ii) holds. Thus, by applying Corollary 10.2 we obtain the conclusion.
(b) Use Exercise 10.2 for $\mathscr{D}$ and R. Prove that the set $\{a \in A \mid \sup F(a, b) > \delta\}$ is open when $F(\cdot, b)$ is lower semicontinuous, and the set $\{b \in B | \ \sup F(a, b) \leq \delta\}$ is closed when $F(a, \cdot)$ is lower semicontinuous. Thus, by applying Exercise 10.2 we obtain the conclusion.
(c) Use Corollary 10.2 for $\mathscr{D}'$ and R'.
(d) Use Exercise 10.2 for $\mathscr{D}'$ and R'.

Exercise 10.6 **Hint.** Apply Corollary 10.2 for $A = B$, $\mathscr{D} = A$, and $R(a, b)$ holding iff $c_i(a) \geq c_i(a_{\hat{i}}, b_i)$ for all $i = 1, 2, \ldots, m$.

Exercise 10.7
Hint. If the distance d satisfies condition (c) then the equivalent distance

$$d' = \frac{d}{l + d}$$

also does, so we can suppose that d is bounded on X. Due to condition (a), we have $P(x) \neq \emptyset$ for all $x \in X$, and so we can construct a generalized Picard-iteration such that $x_1 = \hat{x}$, $x \in \Gamma(x_n - 1)$ and

$$d(x_n, x_{n-1}) \geq \frac{\mathrm{diam}\Gamma(x_{n-1})}{2} - \frac{1}{2^{n-1}}.$$

Therefore, by conditions (b) and (c), we have

$$\Gamma(x_{n-1}) \supset \Gamma(x_n) \text{ and } \mathrm{diam}\Gamma(x_n) \to 0 \text{ as } n \to \infty.$$

Due to the completeness of the reference space, the nonempty, closed sets $\Gamma(x_n)$ have a unique common point $\bar{x}$, i.e.,

$$\cap_{n=1}^{\infty} \Gamma(x_n) = \{\bar{x}\}.$$

The point $\bar{x}$ satisfies the conclusion. On the one hand, due to $\bar{x} \in \Gamma(x_n)$ and (b), we have

$$\Gamma(\bar{x}) \subset \cap_{n=1}^{\infty}\Gamma(x_n) = \{\bar{x}\},$$

and so it follows from condition (a) that $\bar{x} \subset \Gamma(\bar{x})$.

Exercise 10.8 (See [334])
Hint. Let $u \in X$ be fixed, we set

$$U := \{x \in X \mid R(u, x) \text{ holds }\},$$

and for each $x \in X$, we set

$$\Gamma(x) := \{y \in U \mid R(x, y) \text{ holds }\}.$$

Then, we show that there exists $\bar{x} \in U$ such that $\Gamma(\bar{x}) = \{\bar{x}\}$, and then by conditions (i) and (iv), we can conclude that this point satisfies the two conclusions.

Exercise 10.9 (See [86])
Hint. Let $\{(x^\nu, y^\nu, u^\nu)\}$ be a net in the set $M := \{(x, y, u) \in D \times X \mid R(u, y, z) \text{ holds for all } z \in Q(x)\}$, and $\{(x^\nu, y^\nu, u^\nu)\}$ converges to some point $(x, y, u) \in D \times X$. For each $z \in Q(x)$, since Q is lower semicontinuous, there exists

a net $\{z^\nu\}$ converging to z with $z^\nu \in Q(x^\nu)$ for all indices ν. Since $(x^\nu, y^\nu, u^\nu) \in M$, $R(u^\nu, y^\nu, z^\nu)$ holds. Because the relation R is closed, conclude that $(x, y, u) \in M$.

Exercise 10.10
Hint. Consider the graph of G endowed with the metric δ defined by

$$\delta((x_1, y_1), (x_2, y_2)) = d(x_1, x_2) + |y_1 - y_2|, \forall (x_1, y_1), (x_2, y_2) \in \text{grph}(G).$$

Because graph(G) is a closed subset of $X \times \mathbb{R}$, (graph$(G), \delta$) is a complete metric space. Let $\mathscr{R}$ be the relation on graph$(G) \times$ graph(G) defined by

$$\mathscr{R}((x_1, y_1), (x_2, y_2)) \text{ holds if and only if} y_2 - y_1 + \varepsilon d(x_1, x_2) \leq 0.$$

Then, $\mathscr{R}$ is a quasi-ordered relation. Further arguments are similar to those in Exercise 10.8

Exercise 10.11 (See [87])
Hint. For $y \in X$, let

$$G_y = \{x \in X : y \in Q(x) \text{ and } R(x, y) \text{ does not hold }\}.$$

Then $G_y = Q^-(y) \cap \{x \in X : R(x, y) \text{ does not hold }\}$, and the sets G_y are open. Let $G_0 = \{x \in X : x \notin T(x)\}$, then G_0 is also open.

If the conclusion is false, Then, for each $x \in X$, either $x \in G_0$ or $x \in G_y$, for some $y \in Q(X)$. Thus, $X = G_0 \cup \bigcup_{y \in Q(X)} G_y$. Since X is compact, there exists a finite set $\{y_1, \ldots, y_n\} \subseteq Q(X)$ such that $X = G_0 \cup \bigcup_{i=1}^n G_{y_i}$. For the sake of simplicity, we will put G_i instead of G_{y_i}. Let $\{\alpha_0, \alpha_1, \ldots, \alpha_n\}$ be a partition of unity on X subordinated to the open cover $\{G_0, G_1, \ldots, G_n\}$. For each $x \in X$, set

$$S(x) = \alpha_0(x)T(x) + \alpha_1(x)y_1 + \cdots + \alpha_n(x)y_n.$$

Then, $S(a)$ is nonempty compact and convex, and so we conclude that there exists $x_0 \in X$ satisfying $x_0 \in S(x_0)$. It is not hard to check that each of the cases $\alpha_0(x_0) = 0$, $\alpha_0(x_0) = 1$ and $\alpha_0(x_0) \in (0, 1)$ leads to a contradiction. Let $I = \{i \in \{1, \ldots, n\} : \alpha_i(x_0) > 0\}$. For each $i \in I$, $x_0 \in G_i$, and hence, $x_0 \in Q^-(y_i)$ and $R(x_0, y_i)$ does not hold. Since $\{y_i : i \in I\} \subseteq Q(x_0)$ and $Q(x_0)$ is convex, $\text{co}\{y_i : i \in I\} \subseteq Q(x_0)$.

If $\alpha_0(x_0) = 0$, then

$$x_0 = \sum_{i \in I} \alpha_i(x_0) y_i \in \text{co}\{y_i : i \in I\}.$$

Thus, $R(x_0, x_0)$ holds. On the other hand, since $R(x_0, y_i)$ does not hold for all $i \in I$, and so $R(x_0, x_0)$ does not hold; a contradiction.

If $\alpha_0(x_0) = 1$, it follows that $x_0 \in S(x_0) = T(x_0)$. On the other hand, since $\alpha_0(x_0) > 0$, $x_0 \in G_0$, that is, $x_0 \notin T(x_0)$; a contradiction again.

If $\alpha_0(x_0) \in (0,1)$, then there exists $y_0 \in T(x_0)$ such that

$$x_0 = \alpha_0(x_0)\, y_0 + \sum_{i \in I} \alpha_i(x_0)\, y_i .$$

Dividing the previous relation by $1 - \alpha_0(x_0)$ and denoting by $\lambda = \frac{1}{1-\alpha_0(x_0)}$, we get

$$\lambda x_0 + (1-\lambda) y_0 = \sum_{i \in I} \frac{\alpha_i(x_0)}{1-\alpha_0(x_0)} y_i \in \operatorname{co}\{y_i : i \in I\} .$$

We imply that $R(x_0, \lambda x_0 + (1-\lambda) y_0)$ holds. On the other hand, since $\lambda x_0 + (1-\lambda) y_0 \in \operatorname{co}\{y_i : i \in I\}$, we have $R(x_0, \lambda x_0 + (1-\lambda) y_0)$ does not hold. The contradiction is obtained.

Exercise 10.12 (See [87])
Hint. Apply Exercise 10.11 when $R(x, y)$ holds if and only if $0 \in F(x, y)$.

Exercise 10.13 (See [87])
Hint. Because Q has nonempty convex values and open fibers, we imply that Q has a continuous selection, that is, there exists a continuous map $f : X \to X$ such that $f(x) \in Q(x)$ for all $x \in X$. Let the relation R be defined as follows

$$R(x, y) \text{ holds } \Leftrightarrow \quad \|x - f(x)\| \leq \|y - f(x)\|.$$

It is easy to see that for each x, $\{y \in X : R(x, y)$ does not hold $\}$ is a convex set. Applying Exercise 10.11, there exists $\bar{x} \in X$ such that $\bar{x} \in T(\bar{x})$, and $\|\bar{x} - f(\bar{x})\| \leq \|y - f(\bar{x})\|$ for each $y \in Q(\bar{x})$. Taking $y = f(\bar{x}) (\in Q(\bar{x}))$ we get $\|\bar{x} - f(\bar{x})\| \leq 0$, that is $\bar{x} = f(\bar{x})$. It follows that $\bar{x} \in T(\bar{x}) \cap Q(\bar{x})$, and thus, we get the conclusion.

Exercise 10.14 (See [357])
Hint. The "only if" part: Firstly, we prove that there exists $x_0 \in C$ such that $0 \in T(x_0)$. Suppose that $C_\infty \cap T(C)^- = \{0\}$. Since T is maximal monotone, for $n \in \mathbb{N}$, let $x_n := (J + nT)^{-1}(0)$, i. e., $0 \in J(x_n) + nT(x_n)$. This means that for $u_n \in J(x_n)$ such that $-\frac{u_n}{n} \in T(x_n)$. Then, we can conclude that $\{x_n\}$ is bounded.

Without loss of generality, we can assume that $x_n \to x_0$. For any $z \in C$ and for any $z^* \in T(z)$, we obtain

$$\left\langle -z^*, x_0 - z \right\rangle \geq 0.$$

Hence, the maximal monotonicity of T implies that $0 \in T(x_0)$. Next, we show that $\{x \in C \mid 0 \in T(x)\}$ is bounded. If not, then for every $m > 0$, there exists $x_m \in C$ with $\|x_m\| \geq m$ such that $0 \in T(x_m)$. Without loss of generality, $\frac{x_m}{\|x_m\|} \to \tilde{d} \in C_\infty$. It follows from $\operatorname{int}(\operatorname{barr} C) \neq \emptyset$, we have $\tilde{d} \neq 0$. Let $\tilde{y} \in C$ and $\tilde{y}^* \in T(\tilde{y})$. Then, for all $\|\tilde{y}\| < m$, by the monotonicity of T, we have $\langle -\tilde{y}^*, x_m - \tilde{y}\rangle \geq 0$. It turns out that

$$\left\langle -\tilde{y}^*, \frac{x_m}{\|x_m\|} - \frac{\tilde{y}}{\|x_m\|} \right\rangle \geq 0.$$

Letting $m \to \infty$ in (4), we get $\left\langle \tilde{y}^*, \tilde{d} \right\rangle \leq 0$ for all $\tilde{y}^* \in T(C)$, and hence, $0 \neq \tilde{d} \in C_\infty \cap T(C)^-$. It is impossible, and so the set $\{x \in C \mid 0 \in T(x)\}$ is bounded.

The "if" part: Let $x_0 \in \{x \in C \mid 0 \in T(x)\}$. For any $\rho > 0$ with $\|x_0\| < \rho$. If the conclusion does not hold, then $C_\infty \cap T(C)^- \neq \{0\}$. Because $C_\infty \cap T(C)^-$ is a closed and convex cone, we can pick up $d_0 \in C_\infty \cap T(C)^-$ such that $\|d_0\| = 1$. Since $d_0 \in C_\infty$, we obtain $x_0 + 2\rho d_0 \in C$. For any $y \in C$ and for any $y^* \in T(y)$, by the monotonicity of T, we have

$$\left\langle -y^*, x_0 - y \right\rangle \geq 0.$$

Because $d_0 \in T(C)^-$, $\langle y^*, d_0\rangle \leq 0$. Hence, $\langle y^*, 2\rho d_0\rangle \leq 0$. Moreover, from the earlier discussions, we have $\langle -y^*, x_0 + 2\rho d_0 - y\rangle \geq 0$. Then, the maximal monotonicity of T implies that

$$0 \in T\left(x_0 + 2\rho d_0\right).$$

But,

$$\|x_0 + 2\rho d_0\| \geq 2\rho \|d_0\| - \|x_0\| > 2\rho - \rho = \rho.$$

Because $\|x_0 + 2\rho d_0\| > \rho$ and ρ is arbitrary, we obtain a contradiction as the set $\{x \in C \mid 0 \in T(x)\}$ is bounded.

Exercise 10.15 (See [357])
Hint. It is not hard to see that for all $x, y \in C$,

$$\langle T(x) - T(y), x - y\rangle = \langle x - y, x - y\rangle = \sum_{i=1}^{\infty} (x_i - y_i)^2 = \|x - y\|_2^2 \geq 0,$$

and

$$(I + \lambda T)\left(\ell^2\right) = \ell^2,$$

which means that T is maximal monotone. Applying Exercise 10.14, we conclude that

$$x \in C \mid 0 \in T(x) = \{0\}.$$

Glossary of Notation

Operation symbol

$\mathscr{F}(A)$	the class of the finite subsets of A
$\mathscr{A}_A^{=}$	the classes of bifunctions Φ defined on $A \times A$ satisfying $\Phi(x, x) = 0$ for all $x \in A$
$\mathscr{A}_A^{\geq}$	the classes of bifunctions Φ defined on $A \times A$ satisfying $\Phi(x, x) \geq 0$ for all $x \in A$
$\mathscr{A}_A$	the classes of bifunctions Φ defined on $A \times A$ satisfying: for each $x \in A$, $\Phi(x, y) \geq 0$ for some $y \in A$
$\text{dom}\varphi$, $\text{gph}\varphi$	effective domain of φ, graph of φ
$\text{epi}\varphi$, $\text{hyp}\varphi$	epigraph of φ, hypograph of φ
$\text{lev}_{\leq\beta}\varphi$, $\text{lev}_{<\beta}\varphi$	lower β-level sets of φ
$\text{lev}_{\geq\beta}\varphi$, $\text{lev}_{>\beta}\varphi$	upper β-level sets of φ
$\text{lev}_{\leq}\varphi$, $\text{lev}_{<}\varphi$	lower 0-level sets of φ
$\text{lev}_{\geq}\varphi$, $\text{lev}_{>}\varphi$	upper 0-level sets of φ
Sol(EP)	the solution set of the equilibrium problem (EP)
Δ_n	the standard n-simplex in $\mathbb{R}^{n+1}$
$\mathscr{A}_c^i$	the class of bifunctions which are convex in the ith component
$\mathscr{A}_{s.c}^i$	the class of bifunctions which are strictly convex in the ith component
$\mathscr{A}_{eqc}^i$	the class of bifunctions which are explicitly quasiconvex in the ith component
$\mathscr{A}_{qc}^i$	the class of bifunctions which are quasiconvex in the ith component
$\mathscr{A}_{s.qc}^i$	the class of bifunctions which are strictly quasiconvex in the ith component
$\mathscr{A}_{qcl}^i$	the class of bifunctions which are quasiconvex-like in the ith component

L. Q. Anh et al., *Equilibrium Problems*, Forum for Interdisciplinary Mathematics, https://doi.org/10.1007/978-3-031-89003-1

$\mathscr{A}^i_{s.qcl}$	the class of bifunctions which are strictly quasiconvex-like in the ith component
$\mathscr{A}^i_{ss.qc}$	the class of bifunctions which are semistrictly quasiconvex in the ith component
$\mathscr{A}^i_{st.qcl}$	the class of bifunctions which are strongly quasiconvex-like in the ith component
$\mathscr{A}_m$	the class of monotone bifunctions
$\mathscr{A}_{s.m}$	the class of strictly monotone bifunctions
$\mathscr{A}_{pm}$	the class of pseudomonotone bifunctions
$\mathscr{A}_{s.pm}$	the class of strictly pseudomonotone bifunctions
$\mathscr{A}_{qm}$	the class of quasimonotone bifunctions
$\mathscr{A}^i_{pr.qm}$	the class of bifunctions which are properly quasimonotone in ith component
$\mathscr{A}_\Delta$	the class of bifunctions which satisfy the triangle inequality
$\mathscr{A}_{cam}$	the class of cyclically anti-monotone bifunctions
$\mathscr{A}_{caqm}$	the class of cyclically anti-quasimonotone bifunctions
$\mathscr{A}_{tr}$	the class of the bifunctions which have the transitive property: for $x, y, z \in A$, if $\Phi(x, y) < 0$ and $\Phi(y, z) < 0$, then $\Phi(x, z) < 0$
(Σ, τ)	KKM-structure
$\text{conv}_\Sigma B$	Σ- convex hull of B
$\Sigma\mathscr{A}^i_{qc}$	the class of bifunctions which are Σ-quasiconvex in the ith component
$\Sigma\mathscr{A}^i_{s.qc}$	the class of bifunctions which are strictly Σ-quasiconvex in the ith component
$\Sigma\mathscr{A}^i_{qcl}$	the class of bifunctions which are Σ-quasiconvex-like in the ith component
$\Sigma\mathscr{A}^i_{s.qcl}$	the class of bifunctions which are strictly Σ-quasiconvex-like in the ith component
$\Sigma\mathscr{A}^i_{st.qcl}$	the class of bifunctions which are strongly Σ-quasiconvex-like) in the ith component
$\Sigma\mathscr{A}^i_{pr.qm}$	the class of bifunctions which are properly Σ-quasimonotone in the ith component
$\mathscr{A}_{KKM}$	the class of KKM-bifunctions
(Γ, τ_Γ) or Γ	connectedness structure
$\mathscr{A}_\Gamma$	the class of connectedness-bifunctions
$S : X \rightrightarrows Y$	a set-valued mapping S from X to Y
$\text{dom}S$	$\{x \in X \mid S(x) \neq \emptyset\}$- domain of a set-valued mapping S
$\text{rang}S$	$\bigcup_{x\in\text{dom}S} S(x)$-range of a set-valued mapping S
$\text{gph}S$	$\{(x, y) \in X \times Y \mid y \in S(x)\}$-graph of a set-valued mapping S
$S^{-1} : Y \rightrightarrows X$	inverse mapping of S
$S^+(B)$	$\{x \in X \mid S(x) \subset B\}$-strong inverse image of B under S
$S^-(B)$	$\{x \in X \mid S(x) \cap B \neq \emptyset\}$-weak inverse image of B under S

Sol(QEP)	the solution set of quasiequilibrium problem (QEP)
Fix$\mathscr{D}$	the set of fixed points of set-valued mapping $\mathscr{D}$
Sol(SSEP$_F$)	the solution set of strong set-valued equilibrium problem (SSEP$_F$)
Sol(WSEP$_F$)	the solution set of weak set-valued equilibrium problem (WSEP$_F$)
Sol(SSQEP$_{(F,\mathscr{D})}$)	the solution set of strong set-valued quasiequilibrium problem (SSQEP$_{F,\mathscr{D}}$)
Sol(WSQEP$_{(F,\mathscr{D})}$)	the solution set of weak set-valued quasiequilibrium problem (WSQEP$_{F,\mathscr{D}}$)
$\mathrm{Lev}_{\subset}^{\geqq} F(\cdot, y)$	$\{x \in A \mid F(x, y) \subset \mathbb{R}_+\}$ for $F : A \times A \rightrightarrows \mathbb{R}$
$\mathrm{Lev}_{\subset}^{<} F(\cdot, y)$	$\{x \in A \mid F(x, y) \cap (-\mathbb{R}_{++}) \neq \emptyset\}$ for $F : A \times A \rightrightarrows \mathbb{R}$
$\mathrm{Lev}_{\subset}^{\leqq} F(x, \cdot)$	$\{y \in A \mid F(x, y) \cap (-\mathbb{R}_+) \neq \emptyset\}$ for $F : A \times A \rightrightarrows \mathbb{R}$
$\mathrm{Lev}_{\subset}^{>} F(\cdot, x)$	$\{y \in A \mid F(y, x) \cap (-\mathbb{R}_+) = \emptyset\}$ for $F : A \times A \rightrightarrows \mathbb{R}$
$\mathrm{Lev}_{\subset}^{<} F(x, \cdot)$	$\{y \in A \mid F(x, y) \cap (-\mathbb{R}_{++}) \neq \emptyset\}$ for $F : A \times A \rightrightarrows \mathbb{R}$
$\mathrm{Lev}_{\cap}^{\geqq} F(\cdot, y)$	$\{x \in A \mid F(x, y) \cap \mathbb{R}_+ \neq \emptyset\}$ for $F : A \times A \rightrightarrows \mathbb{R}$
$\mathrm{Lev}_{\cap}^{<} F(\cdot, y)$	$\{x \in A \mid F(x, y) \subset -\mathbb{R}_{++}\}$ for $F : A \times A \rightrightarrows \mathbb{R}$
$\mathrm{Lev}_{\cap}^{<} F(x, \cdot)$	$\{y \in A \mid F(x, y) \cap \mathbb{R}_+ = \emptyset\}$ for $F : A \times A \rightrightarrows \mathbb{R}$
$\mathrm{Lev}_{\cap}^{\leqq} F(x, \cdot)$	$\{y \in A \mid F(x, y) \subset -\mathbb{R}_+\}$ for $F : A \times A \rightrightarrows \mathbb{R}$
$\mathrm{Lev}_{\cap}^{>} F(\cdot, x)$	$\{y \in A \mid F(y, x) \cap \mathbb{R}_{++} \neq \emptyset\}$ for $F : A \times A \rightrightarrows \mathbb{R}$
$\mathscr{S}^{\ell}\mathscr{A}_c^i$	the class of set-valued bifunctions which are $\mathbb{R}_+$-convex of type ℓ in ith component
$\mathscr{S}^{\ell}\mathscr{A}_{qc}^i$	the class of set-valued bifunctions which are $\mathbb{R}_+$-quasiconvex of type ℓ in ith component
$\mathscr{S}^{\ell}\mathscr{A}_{st.qcl}^i$	the class of set-valued bifunctions which are strongly $\mathbb{R}_+$-quasiconvex-like of type ℓ in ith component
$\mathscr{S}^{\ell}\mathscr{A}_{pr.qm}^i$	the class of set-valued bifunctions which are properly $\mathbb{R}_+$-quasimonotone of type ℓ in ith component
$\mathscr{S}^1\mathscr{A}_A^{\geqq}$	the classes of set-valued bifunctions $F : A \times A \rightrightarrows \mathbb{R}$ satisfying $F(x, x) \subset \mathbb{R}_+$ for all $x \in A$
$\mathscr{S}^2\mathscr{A}_A^{\geqq}$	the classes of set-valued bifunctions $F : A \times A \rightrightarrows \mathbb{R}$ satisfying $F(x, x) \cap \mathbb{R}_+ \neq \emptyset$ for all $x \in A$
$\mathscr{S}^2\mathscr{A}_A^{=}$	the classes of set-valued bifunctions $F : A \times A \rightrightarrows \mathbb{R}$ satisfying $F(x, x) \cap (-\mathbb{R}_+) = \{0\}$ for all $x \in A$
$\mathscr{S}^1\mathscr{A}_A^{=}$	the classes of set-valued bifunctions $F : A \times A \rightrightarrows \mathbb{R}$ satisfying $F(x, x) \cap \mathbb{R}_+ = \{0\}$ for all $x \in A$
$\mathrm{lev}_E^1 f(x, \cdot)$	the set $\{y \in A \mid f(x, y) \notin -E\}$ for $f : A \times A \to Y$, $E \subset Y$ and $x \in A$
$\mathrm{lev}_E^2 f(x, \cdot)$	the set $\{y \in A \mid f(x, y) \in E\}$ for $f : A \times A \to Y$, $E \subset Y$ and $x \in A$
$\overline{\mathrm{lev}}_E^1 f(x, \cdot)$	the set $\{y \in A \mid f(x, y) \in -E\}$ for $f : A \times A \to Y$, $E \subset Y$ and $x \in A$
$\overline{\mathrm{lev}}_E^2 f(x, \cdot)$	the set $\{y \in A \mid f(x, y) \notin E\}$ for $f : A \times A \to Y$, $E \subset Y$ and $x \in A$

$\mathrm{lev}^1_E f(\cdot, y)$	the set $\{x \in A \mid f(x, y) \notin -E\}$ for $f : A \times A \to Y$, $E \subset Y$ and $y \in A$
$\mathrm{lev}^2_E f(\cdot, y)$	the set $\{x \in A \mid f(x, y) \in E\}$ for $f : A \times A \to Y$, $E \subset Y$ and $y \in A$
$\overline{\mathrm{lev}}^1_E f(\cdot, y)$	the set $\{x \in A \mid f(x, y) \in -E\}$ for $f : A \times A \to Y$, $E \subset Y$ and $y \in A$
$\overline{\mathrm{lev}}^2_E f(\cdot, y)$	the set $\{x \in A \mid f(x, y) \notin E\}$ for $f : A \times A \to Y$, $E \subset Y$ and $y \in A$
E^c	the complement of E
$\mathrm{core} E$	algebraic interior of E
$\mathrm{vcl} E$	vectorial closure of E
$\mathrm{vcl}_q E$	vectorial closure of E in direction q
$\mathrm{cone} E$	cone generated by E
$\mathscr{K}$	the set of proper cones
$\mathscr{K}_c$	the set $\{D \in \mathscr{K} \mid D$ solid and convex$\}$
$\mathscr{K}^{core}_C$	the set $\{D \in \mathscr{K}_c \mid \mathrm{core} D \supset C \setminus \{0\}\}$
$\mathscr{E}(Y)$	the set $\{E \subset Y \mid E$ solid, $E \neq Y$, and $E + E \subset E\}$
Q_E	the set $\{q \in Y \mid \mathbb{R}_{++}q + E = \mathrm{core} E\}$
$\mathscr{E}_\bullet(Y)$	the set $\{E \in \mathscr{E}(Y) \mid E \cap (-E) \neq \emptyset\}$
$\mathscr{Q}_E$	the set $\mathrm{Q}_E \cap \mathrm{core} E$
$\mathscr{F}_D$	the set $\{E \subset Y \mid E + D \subset E\}$
$\mathrm{Sol}^{He}(\mathrm{VEP}_{(f,C)})$	the set of Henig-proper C-solutions of vector equilibrium problem defined by f and C
$\mathrm{Sol}^{Be}(\mathrm{VEP}_{(f,C)})$	the set of Benson-proper C-solutions of vector equilibrium problem defined by f and C
$\mathrm{Sol}^e(\mathrm{VEP}_{(f,C)})$	the set of C-solutions of vector equilibrium problem defined by f and C
$\mathrm{Sol}^w(\mathrm{VEP}_{(f,C)})$	the set of weak C-solutions of vector equilibrium problem defined by f and C
$\mathscr{E}^1_{He}(C)$	$\{E \subset Y \mid$ there exist $D \in \mathscr{K}^{core}_C$ with $C + \mathrm{core} D \subset E\}$ for $C \subset Y$
$\mathscr{E}^1_{Be}(C)$	$\{E \subset Y \mid \mathrm{vclcone}(C - E^c) \cap (-C_0) = \emptyset\}$ for $C \subset Y$
$\mathscr{E}^1_e(C)$	$\{E \subset Y \mid C_0 \subset E\}$ for $C \subset Y$
$\mathscr{E}^1_w(C)$	$\{E \subset Y \mid \mathrm{core} C \subset E\}$ for $C \subset Y$
$\mathscr{E}^2_{He}(C)$	$\{E \subset Y \mid$ there exist $D \in \mathscr{K}^{core}_C$ with $0 \notin C + \mathrm{core} D + E\}$ for $C \subset Y$
$\mathscr{E}^2_{Be}(C)$	$\{E \subset Y \mid \mathrm{vclcone}(C + E) \cap (-C_0) = \emptyset\}$ for $C \subset Y$
$\mathscr{E}^2_e(C)$	$\{E \subset Y \mid 0 \notin C_0 + E\}$ for $C \subset Y$
$\mathscr{E}^2_w(C)$	$\{E \subset Y \mid 0 \notin \mathrm{core} C + E\}$ for $C \subset Y$
$\mathscr{V}_E \mathscr{A}^=_A$	$\{f : A \times A \to Y \mid f(x, x) \in E \cap (-E)$ for all $x \in A\}$
$\mathscr{V}^1_E \mathscr{A}^\geq_A$	$\{f : A \times A \to Y \mid f(x, x) \notin -E$ for all $x \in A\}$
$\mathscr{V}^2_E \mathscr{A}^\geq_A$	$\{f : A \times A \to Y \mid f(x, x) \in E$ for all $x \in A\}$

$\mathscr{V}^\ell \mathscr{A}^i_{st.qcl}$	the class of vector bifunctions which are E-quasiconvex-like of type ℓ in the ith component
$\mathscr{V}\mathscr{A}_m$	the class of vector bifunctions which are E-monotone
$\mathscr{V}\mathscr{A}_{pm}$	the class of vector bifunctions which are E-pseudomonotone
$\mathscr{V}\mathscr{A}_{qm}$	the class of vector bifunctions which are E-quasimonotone
$\mathscr{V}^\ell \mathscr{A}^i_{pr.qm}$	the class of vector bifunctions which are properly E-quasimonotone of type ℓ in ith component
$\mathscr{V}^\ell_E \mathscr{A}_{caqm}$	the class of vector bifunctions which are E-cyclically anti-quasimonotoneof type ℓ.
$\mathscr{V}_E \mathscr{A}_\Delta$	the class of vector bifunctions, which satisfy the E-triangle inequality
$\mathscr{V}_E \mathscr{A}_{cm}$	the class of vector bifunctions which are E-cyclically monotone
$\mathrm{Sol}^{He}(\mathrm{VQEP}_{(f,C,\mathscr{D})})$	the set of Henig-proper C-solutions of vector quasiequilibrium problem defined by f, $\mathscr{D}$ and C
$\mathrm{Sol}^{Be}(\mathrm{VQEP}_{(f,C,\mathscr{D})})$	the set of Benson-proper C-solutions of vector quasiequilibrium problem defined by f, $\mathscr{D}$ and C
$\mathrm{Sol}^{e}(\mathrm{VQEP}_{(f,C,\mathscr{D})})$	the set of C-solutions of vector quasiequilibrium problem defined by f, $\mathscr{D}$ and C
$\mathrm{Sol}^{w}(\mathrm{VQEP}_{(f,C,\mathscr{D})})$	the set of weak C-solutions of vector equilibrium problem defined by f, $\mathscr{D}$ and C

Operation symbols of Chaps. 4–6 and 10

$\{(EP)_{\Lambda\times M}\}$	a family of equilibrium problems
Sol(DEP)	the solution map of dual equilibrium problem of (EP)
$\mathrm{Sol}(\mathrm{EP}^+)$	the solution map of the auxiliary problem of (EP)
$\widetilde{\mathrm{Sol}}(\mathrm{EP})$	the approximate solution map of problem (EP)
$d(a, P)$	the distance from a point a to a set P
$\mathbb{D}(P, Q)$	$\mathbb{D}(P, Q) = \sup_{p\in P,q\in Q} d(p, q)$
$\mathrm{ex}(P, Q)$	the excess of P from Q
$H(P, Q)$	the Hausdorff distance between P and Q
diam$A$	the diameter of a set A
$B(x, r)$	the closed ball of radius $r \geq 0$ centered at x
$\eta(M)$	the Kuratowski measure of a set M
$\Pi(\delta, \varepsilon, \lambda, \mu)$	the approximating solution set of (EP)(λ, μ)
Sol(OP)	the solution map of optimization problem
$\widetilde{\mathrm{Sol}}(\mathrm{OP})$	the approximate solution set of optimization problem

Sol(SVI), Sol(MVI)	the solution maps of the Stampacchia and Minty variational inequality problems, resp
$\widetilde{\mathrm{Sol}}$(SVI), $\widetilde{\mathrm{Sol}}$(MVI)	the approximate solution maps of the Stampacchia and Minty variational inequality problems, resp
$\widetilde{\mathrm{Sol}}$(FP)	the approximate solution maps of fixed-point problems
$\{(\mathrm{QEP})_{\Lambda\times M}\}$	a family of quasi-equilibrium problems
$\mathscr{E}(\lambda)$	the map of fixed points of (QEP)
Sol(QEP^1)	the solution map of quasiequilibrium problem (QEP^1)
Sol(QEP^{1+})	the solution map of auxiliary quasiequilibrium problem
Sol(QSVI), Sol(QMVI)	the solution maps of Stampacchia and Minty quasivariational inequalities, resp
$G = [N, L]$	the graph of a network
P_w	the set of paths connecting the origin-destination pair w
F_ℓ	the traffic flow on link ℓ
V_p	the nonnegative traffic flow on path p
$t_\ell(F)$	the user travel cost associated with the link ℓ
$\tau_p(V)$	the user travel cost on path p
Sol(TNP)	the solution map of parametric traffic network problems
Sol(SEP), Sol(WEP)	the solution map of equilibrium problems (SEP) and (WEP), resp
$\widetilde{\mathrm{Sol}}$(WEP)	the approximate solution map of equilibrium problems (WEP)
$\mathrm{L}_{\not\geq b}h$, $\mathrm{L}_{\geq b}h$, $\mathrm{L}_{\not\leq b}h$, $\mathrm{L}_{>b}h$	the level sets of the map h
r^e_C	the Gerstewitz function, $r^e_C(y) = \inf\{t \in \mathbb{R} \mid y \in te - C\}$
Sol(VQEP)	the solution map of vector quasi-equilibrium problems (VQEP)
$\delta_{\mathscr{D}}$	the oriented distance function, $\delta_{\mathscr{D}}(y) := d(y, \mathscr{D}) - d(y, Y \setminus \mathscr{D})$
$\mathscr{L}(X, Y)$	the set of continuous mappings from X into Y
ι_D	the indicator function of D, $\iota_D(x) = \begin{cases} 0, & \text{if } x \in D, \\ +\infty, & \text{if } x \in X \setminus D. \end{cases}$
Sol(RWEP)	the solution map of the regularized problem RWEP(f_ε, K)

Sol(RSEP)	the solution map of the regularized problem RSEP(f_ε, K)
Eff(p, K)	the efficient solution map of vector optimization problem VOP(p, K)
IEff(p, K)	the ideal solution map of vector optimization problem VOP(p, K)
$\mathscr{P}_0(Y)$	the family of nonempty subsets of Y
$A \preceq^l_C B$, $A \prec^l_C B$, $A \preceq^u_C B$, $A \prec^u_C B$	the l-type and u-type set less order relations wrt C
$\mathscr{P}_{0C}(Y)$	the family of all nonempty, C-proper and $-C$-proper subsets of Y
$g^u_{e',a}$, $G^u_{e',A}$, $\mathscr{G}^u_{e'}$	the u-type generalized Gerstewitz functions
$g^l_{e,a}$, $G^l_{e,A}$, $\mathscr{G}^l_e$	the l-type generalized Gerstewitz functions
$\mathscr{T}^l$, $\mathscr{T}^u$	the generalized Gerstewitz functions
$\widetilde{\mathrm{Sol}}(\mathrm{SOP})^l$, $\widetilde{\mathrm{Sol}}(\mathrm{SOP})^u$	the approximate solution maps of set optimization problem (SOP)
$\widetilde{\mathrm{Sol}}(\mathrm{EP})^\alpha_1$	the solution approximate map of auxiliary equilibrium problems (EP^1_α)
$\widetilde{\mathrm{Sol}}(\mathrm{EP})^\alpha$	the solution approximate map of equilibrium problems (EP_α)
$\mathscr{P}_{\mathrm{pro}\,C}(Y)$, $\mathscr{P}_{\mathrm{cl}\,C}(Y)$, $\mathscr{P}_{\mathrm{bd}\,C}(Y)$	the families of C-proper, C-closed, and C-bounded subsets of Y, resp
$A \sim^\alpha B$	the equivalence relation on $\mathscr{P}(Y)$
$[x]^\alpha_F$	the class of all decisions which are equivalent to x
α-Eff	the set of all Pareto efficient solutions of problem (α-SOP)
$\gamma^{(\alpha)}_A$	the α- marginal nonlinear scalarization function
$\widetilde{\mathrm{Sol}}(\mathrm{BVI})$	the approximate solution map of the Browder variational inclusion
s-Eff, Eff, w-Eff	the set of all strongly efficient solutions, efficient solutions, and weakly efficient solutions of the vector optimization problem (VOP), resp
$\widetilde{\mathrm{Sol}}(\mathrm{SVVI})$, $\widetilde{\mathrm{Sol}}(\mathrm{MVVI})$	the approximate solution maps of the Stampacchia and Minty vector variational inequality problem, resp
$\mathscr{L}^p([t_0, t_1], \mathbb{R}^n)$ $(1 \le p \le \infty)$	the Lebesgue space of the integrable functions defined on $[t_0, t_1]$
S_0	the solution map of the optimal control problem (OCP)

S^{QO}	the solution map of the quadratic programming
$\text{liminfo}_{x\to\bar{x}}G(x)$	the inferior open limit
$\text{limsupo}_{x\to\bar{x}}G(x)$	the superior open limit
$\text{st.limsupo}_{x\to\bar{x}}G(x)$	the strong superior open limit
$\text{w.liminfo}_{x\to\bar{x}}G(x)$	the weak inferior open limit
$\mathscr{R}(a, b, y)$	a relation linking elements $a \in A, b \in B$ and $y \in Y$
Sol(VRP)	the solution set of the variational relation problem (VRP)
Sol(VIP)	the solution map of variational inclution problem (VIP)
$\widetilde{\text{Sol}}$(VIP)	the solution map of $(\widetilde{\text{QVIP}}_\lambda)$

Operation symbols of Chaps. 7–9

$:=$ and $=:$	equal by definition
$\equiv$	identically equal
$\text{cl}A$, $\text{conv}A$, $\text{bd}A$	closure, convex hull, and boundary of a set A
$\text{fix}\varphi$	set of fixed points of function φ
$*$	indication of some dual/adjoint operation
$\xrightarrow{e}$, $\xrightarrow{i-e}$, $\xrightarrow{w-e}$, $\xrightarrow{w-i-e}$	epi-, inside epi-, weak epi-convergence, weak inside epi-convergence
$\text{e-lim}_k\varphi^k$, $\text{i-e-lim}_k\varphi^k$, $\text{w-e-lim}_k\varphi^k$, $\text{w-i-e-lim}_k\varphi^k$	epi-limmit (inside epi-limmit, weak epi-limmit, weak inside epi-limmit) of sequence $\{\varphi^k\}_k$
$\xrightarrow{h}$, $\xrightarrow{i-h}$, $\xrightarrow{w-h}$, $\xrightarrow{w-i-h}$	hypo-, inside hypo-, weak hypo-convergence, weak inside hypo-convergence
$\text{h-lim}_k\varphi^k$, $\text{i-h-lim}_k\varphi^k$, $\text{w-h-lim}_k\varphi^k$, $\text{w-i-h-lim}_k\varphi^k$	hypo-limmit (inside hypo-limmit, weak hypo-limmit, weak inside hypo-limmit) of sequence $\{\varphi^k\}_k$ of functions
ls_k, li_k	$\text{limsup}_{k\to+\infty}$, $\text{liminf}_{k\to+\infty}$
Ls_k, Li_k	$\text{Limsup}_{k\to+\infty}$, $\text{Liminf}_{k\to+\infty}$ (set-limits)
$\xrightarrow{P-K}$	Painlevé-Kuratoski tending to
$\text{fv-fcn}(X)$	set of finite-valued functions on space X
$\inf_A\varphi$, $\sup_A\varphi$	infimum-, supremum-value of φ over A
$\varepsilon\text{-argmin}_A\varphi$	$\{x \in A \mid \varphi(x) \leq \inf_A\varphi + \varepsilon\}$
$\varepsilon\text{-argmax}_A\varphi$	$\{x \in A \mid \varphi(x) \geq \sup_A\varphi - \varepsilon\}$
$\text{epi}\varphi$, $\text{hypo}\varphi$, $\text{gph}\varphi$	epigraph, hypograph, graph of φ
$\text{eli}\varphi^k(x)$, $\text{els}\varphi^k(x)$	lower and upper epi-limits of sequence $\{\varphi^k\}_k$
$\text{hli}\varphi^k(x)$, $\text{hls}\varphi^k(x)$	lower and upper hypo-limits of sequence $\{\varphi^k\}_k$
$\text{w-eli}\varphi^k(x)$, $\text{w-els}\varphi^k(x)$, $\text{w-hli}\varphi^k(x)$, $\text{w-hls}\varphi^k(x)$	the weak types of the preceding limits
$\mathscr{N}_\infty$	the filter of the neighborhoods of $+\infty$
$\mathscr{N}_\infty^\sharp$	the grill of the neighborhoods of $+\infty$

rec($X \times Y$), nrec($X \times Y$) the set of finite-valued bifunctions on rectangles (non-rectangles, resp.) in space $X \times Y$

$\xrightarrow{p}$ pointwise convergence

$\xrightarrow{g}$ graphical convergence

$\xrightarrow{c}$ continuous convergence

$\xrightarrow{e/h}$, $\xrightarrow{i-e/h}$, $\xrightarrow{w-e/h}$, $\xrightarrow{w-i-e/h}$ epi/hypo-convergence, inside epi/hypo-convergence , weak epi/hypo-convergence, weak inside epi/hypo-convergence

e/h-limit$_k \Phi^k$, i-e/h-limit$_k \Phi^k$, w-e/h-limit$_k \Phi^k$, w-i-e/h-limit$_k \Phi^k$ epi/hypo-limit (inside epi/hypo-limit, weak epi/hypo-limit, weak inside epi/hypo-limit) of sequence $\{\Phi^k\}_k$ of bifunctions

ehls$\Phi^k(x, y)$ upper e/h-limit of sequence $\{\Phi^k\}_k$ at $(x, y) \in A \times B$

ehls$\Phi^k(x)$ upper e/h-limit of sequence $\{\Phi^k\}_k$ at $x \notin A$

ehli$\Phi^k(x, y)$ lower e/h-limit of sequence $\{\Phi^k\}_k$ at $(x, y) \in A \times B$

ehli$\Phi^k(y)$ lower e/h-limit of sequence $\{\Phi^k\}_k$ at $y \notin B$

w-ehls$\Phi^k(x, y)$, w-ehls$\Phi^k(x)$, w-ehli$\Phi^k(x, y)$, w-ehli$\Phi^k(y)$ weak types of the preceding four limits

misΦ, maiΦ set of minsup-points of Φ, set of maxinf-points of Φ

minsupΦ, maxinfΦ minsup-value and maxinf-value of Φ, resp.

sdlΦ set of saddle points of Φ

wsdlΦ set of weak saddle points of Φ

ε-wsdlΦ, str-ε-sdlΦ set of ε-weak saddle points and strong ε-saddle points of Φ

ε^k-sdlΦ^k set of ε^k-saddle points of Φ

$\xrightarrow{mis-lop}$, $\xrightarrow{i-mis-lop}$, $\xrightarrow{w-mis-lop}$, $\xrightarrow{w-i-mis-lop}$ minsup-lop convergence, inside minsup-lop convergence, weak minsup-lop convergence, weak inside minsup-lop convergence

mis-lop-lim$_k \Phi^k$, i-mis-lop-lim$_k \Phi^k$, w-mis-lop-lim$_k \Phi^k$, w-i-mis-lop-lim$_k \Phi^k$ minsup-lop limit (inside minsup-lop limit, weak minsup-lop limit, weak inside minsup-lop limit) of sequence $\{\Phi^k\}_k$

w-lli$\Phi^k(x, y)$ weak lower lop-limit of Φ^k for $y \in B$ and $x \in X$

w-lls$\Phi^k(x, y)$ weak upper lop-limit of Φ^k for $y \in Y$ and $x \in A$

$\sqsubset$ strongly included in

Sol(EP), Sol(sEP) solution set of equilibrium problem (EP), solution set of strong equilibrium problem Sol(sEP)

ε-Sol(EP), ε-Sol(sEP) set of ε-solutions of equilibrium problem (EP), set of ε-solutions of strong equilibrium problem (sEP)

$\varepsilon\theta$-Sol(EP) set of $\varepsilon\theta$-solutions (second-kind of approximate solutions) of (EP)

ε-Sol(DEP), ε-Sol(DsEP) set of ε-solutions of dual equilibrium problem (DEP), resp, dual strong equilibrium problem (DsEP)

$\varepsilon\theta$-sdlΦ set of $\varepsilon\theta$-saddle points of Φ (second kind approximate saddle points)

ε-Sol(OP), ε-Sol(DOP) set of ε-solutions of optimization problem (OP), of dual optimization problem (DOP)
$\varepsilon\theta$-Sol(OP) ($\varepsilon\theta$-Sol(DOP)) set of $\varepsilon\theta$-solutions (second kind approximate solutions) of (OP) and (DOP), resp.
$d\varphi(x; h)$ directional derivative of φ at x in direction h
ε-Ne($\mathscr{G}$), ε-Ne($D\mathscr{G}$) set of ε-Nash equilibria of Nash game $\mathscr{G}$ and of dual Nash game $D\mathscr{G}$, resp.
ε-Ne($s\mathscr{G}$), ε-Ne($Ds\mathscr{G}$) set of strong ε-Nash equilibria of Nash game $\mathscr{G}$ and of dual Nash game $D\mathscr{G}$
$\varepsilon\theta$-Ne($\mathscr{G}$), $\varepsilon\theta$-Ne($D\mathscr{G}$) set of $\varepsilon\theta$-Nash equilibria of Nash game $\mathscr{G}$ and of dual Nash game $D\mathscr{G}$, resp.
Ep($\mathscr{E}$) Walras equilibrium price system of Walras pure exchange model $\mathscr{E}$
Sol(QEP), Sol(DQEP), Sol(sQEP), Sol(DsQEP) solution sets of four types of quasiequilibrium problems and dual quasiequilibrium problems
lsc-fcn(X) set of lower semicontinuous functions
$\mathsf{d}^e_\rho(\varphi, \psi)$ ρ-epi-distance between φ and ψ
$\mathsf{d}^e(\varphi, \psi)$ epi-distance between φ and ψ
usc-fcn(Y) set of upper semicontinuous functions
$\mathsf{d}^h_\rho(\varphi, \psi)$ ρ-hypo-distance between φ and ψ
$\mathsf{d}^h(\varphi, \psi)$ hypo-distance between φ and ψ
lsc-nrec(X, Y) set of bifunctions with lsc sup-projections
usc-nrec(X, Y) set of bifunctions with usc inf-projections
$\mathsf{d}^{ebi}(\Phi, \Phi')$ ebi-distance between two bifunctions Φ, Φ'
$\mathsf{d}^{hbi}(\Phi, \Phi')$ hbi-distance between two bifunctions Φ, Φ'
sdl-val$E$ saddle value of bifunction E
Sol(sQVI), Sol(wQVI) solution set of strong set-valued and weak set-valued quasivariational inequalities
ε-Sol(wQVI), ε-Sol(sQVI) ε-solution set of weak set-valued and strong set-valued quasivariational inequalities
$\varepsilon\theta$-Sol(wQVI) $\varepsilon\theta$-solution set of (wQVI)
$\partial h(x)$, $\partial_\varepsilon h(x)$ sudifferential and ε-subdifferential of convex function h at x
$N_\varepsilon(M, x)$ ε-normal cone of convex set M at x
$i(M, x)$ value of the indicator function of the set M at x
ε-$\mathscr{T}(x)$ ε-enlargement of $\mathscr{T}(x)$
Sol(DsQVI), Sol(DwQVI) solution set of the duals of strong set-valued and weak set-valued quasivariational inequalities, (sQVI) and (wQVI), resp
mix-ε-Sol(sQVI) mixed ε-solution set of (sQVI)
ε, γ-$K(H)$ set of ε, γ-(approximate) feasible vector flows
ε, γ-weq(TP) set of ε, γ-weak equilibrium flows of traffic problem (TP)
and ε, γ-seq(TP) set of ε, γ-strong equilibrium flows of traffic problem (TP)
cd-fcns($\mathbb{R}^m$) set of (cumulative) distribution functions on $\mathbb{R}^m$

usc-fcns($\mathbb{R}^m$; [0, 1])	set of upper semicontinuous (usc) functions from $\mathbb{R}^m$ to [0, 1]
Ne(VwQ$\mathscr{G}$)	set of weak Nash quasiequilibrium vectors
Ne(VsQ$\mathscr{G}$	set of strong Nash quasiequilibrium vectors
Ne(DVwQ$\mathscr{G}$)	set of dual weak Nash quasiequilibrium vectors
Ne(DVsQ$\mathscr{G}$)	set of dual strong Nash quasiequilibrium vectors
ε-Sol(VwQEP)	set of ε-approximate weak quasiequilibrium vectors
ε-Sol(VsQEP)	set of ε-approximate strong quasiequilibrium vectors
ε-Sol(DVwQEP)	set of ε-approximate dual weak quasiequilibrium vectors
ε-Sol(DVsQEP)	set of ε-approximate dual strong quasiequilibrium vectors
t-Sol(VsQEP)	set of third-kind t-approximate strong quasiequilibrium vectors
t-Sol(DVsQEP)	set of third-kind t-approximate strong quasiequilibrium vectors
ε-Ne(VwQ$\mathscr{G}$)	set of first-kind ε-approximate weak Nash quasiequilibrium vectors
ε-Ne(VsQ$\mathscr{G}$)	set of first-kind ε-approximate strong Nash quasiequilibrium vectors
ε-Ne(DVwQ$\mathscr{G}$)	set of first-kind ε-approximate dual weak Nash quasiequilibrium vectors
ε-Ne(DVsQ$\mathscr{G}$)	set of first-kind ε-approximate dual strong Nash quasiequilibrium vectors
t-Ne(VsQ$\mathscr{G}$)	set of third-kind t-approximate strong Nash quasiequilibrium vectors
t-Ne(DVsQ$\mathscr{G}$)	set of third-kind t-approximate dual strong Nash quasiequilibrium vectors
$r_C^e(\cdot)$ and $r_{-C}^{-e}(\cdot)$	two Tammer scalarization functions
$\varepsilon\theta$-Sol(VwQEP)	set of second-kind $\varepsilon\theta$-approximate weak quasiequilibrium vectors
$\varepsilon\theta$-Sol(DVwQEP)	set of second-kind $\varepsilon\theta$-approximate dual weak quasiequilibrium vectors
ε-dwsdlΦ	set of ε-dual weak saddle points
$\delta_H(z)$	Hiriart-Urruty scalarization function (signed distance or oriented distance) from z to set H

List of Statements

Chapter 1: Solution Existence for Scalar Equilibrium Problems

1.1 Proposition:	relationship between saddle points and minimax equality
1.2 Proposition:	relationship between saddle points and solutions of equilibrium problem
1.3 Proposition:	relationship between solutions of equilibrium problem and minimax equality
1.4 Proposition:	characterization of existence of saddle points in terms of solution existence of equilibrium problems; relationship between solutions of equilibrium problem and minimax problem
1.5 Proposition:	relationship between Nash equilibria and solutions of equilibrium problem
1.6 Definition:	equilibrium flows
1.7 Proposition:	relationship between equilibrium flows of a network and a solutions of equilibrium problem
1.8 Proposition:	relationship between solutions of a intersection problem and equilibrium problem
1.9 Definition:	finite solvability
1.10 Definition:	lower and upper semincontinuity of a function
1.11 Proposition:	relationship between the semicontinuity of a function and the closedness of its level sets
1.12 Definition:	simplices
1.13 Definition:	simplicial subdivisions
1.14 Definition:	proper and complete labels of simplices
1.15 Theorem:	Sperner's lemma
1.16 Theorem:	Brouwer's fixed-point theorem
1.17 Definition:	hemicontinuity of bifunctions

L. Q. Anh et al., *Equilibrium Problems*, Forum for Interdisciplinary Mathematics, https://doi.org/10.1007/978-3-031-89003-1

1.18 Definition:	sign-continuity of bifunctions
1.19 Definition:	generalized convexity
1.20 Example:	generalized convexity
1.21 Definition:	quasiconvexlikeness
1.22 Example:	quasiconvexlikeness
1.23 Proposition:	characterizations of generalized convexity
1.24 Proposition:	relations between types of generalized convexity under lower hemicontinuity
1.25 Definition:	monotonicity of bifunctions
1.26 Example:	monotonicity of bifunctions
1.27 Definition:	proper quasimonotonicity of bifunctions
1.28 Example:	proper quasimonotonicity of bifunctions
1.29 Proposition:	relations between some classes of bifunctions
1.30 Proposition:	the conservation of properties of bifunctions on a convex subset
1.31 Lemma:	a sufficient condition for the finite solvability of equilibrium problem under convexity and closedness of level sets
1.32 Corollary:	existence of solutions for continuous variational inequality; existence of fixed points for continuous mapping
1.33 Lemma:	a sufficient condition for the finite solvability of equilibrium problem under convexity and closedness of level sets
1.34 Theorem:	a sufficient condition for the solvability of equilibrium problem under convexity, closedness of level sets, and coercivity
1.35 Corollary	a sufficient condition for the solvability of equilibrium problem in reflexive Banach space under convexity, closedness of level sets, and coercivity
1.36 Corollary	a sufficient condition for the solvability of equilibrium problem in reflexive Banach space under convexity, closedness of level sets, and coercivity
1.37 Example	An illustration of Corollary 1.36
1.38 Definition	properties $\mathbb{P}$ and $\mathbb{P}^i$
1.39 Example	properties $\mathbb{P}$ and $\mathbb{P}^i$
1.40 Definition	dual property of a bifunction
1.41 Example	dual property of a bifunction
1.42 Proposition	a sufficient condition for dual property to be satisfied
1.43 Lemma	relationship between solutions of equilibrium problem and dual equilibrium problem
1.44 Definition	core of a subset relative to its supset
1.45 Lemma	relationship between solutions of equilibrium problem on a subset and on supset

1.46 Theorem	a sufficient condition for the solvability of equilibrium problem under convexity, closedness of level sets, coercivity, and the property $\mathbb{P}^2$
1.47 Corollary	a sufficient condition for the solvability of equilibrium problem in reflexive Banach space under convexity, closedness of level sets, coercivity, and property $\mathbb{P}^2$ i
1.48 Corollary	a sufficient condition for the solvability of equilibrium problem in reflexive Banach space under convexity, closedness of level sets, coercivity, and property $\mathbb{P}^2$
1.49 Theorem	a sufficient condition for the solvability of equilibrium problem under convexity, closedness of level sets, coercivity, the dual property
1.50 Theorem	a sufficient condition for the solvability of equilibrium problem under convexity, closedness of level sets, coercivity, the property $\mathbb{P}^2$ and upper-sign continuity
1.51 Corollary	a sufficient condition for the solvability of equilibrium problem in reflexive Banach space under convexity, closedness of level sets, coercivity, property $\mathbb{P}^2$ and upper-sign continuity
1.52 Corollary	a sufficient condition for the solvability of equilibrium problem in reflexive Banach space under convexity, closedness of level sets, coercivity, property $\mathbb{P}^2$ and upper-sign continuity
1.53 Example	existence of saddle points based on Corollary 1.52
1.54 Corollary	a sufficient condition for minimax equality to be hold
1.55 Corollary	Sion's minimax theorem
1.56 Corollary	a sufficient condition for existence of saddle points
1.57 Corollary	von Neumann's minimax theorem
1.58 Example	existence of Nash equilibria based on Corollary 1.57
1.59 Corollary	a sufficient condition for existence of Nash equilibria
1.60 Example	existence of Nash equilibria
1.61 Definition	triangle inequality
1.62 Proposition	properties of class of bifunctions satisfying triangle inequality
1.63 Theorem	Weierstrass' theorem for equilibrium problem
1.64 Corollary	Weierstrass's minimization theorem
1.65 Theorem	a sufficient condition for the solvability of equilibrium problem under triangle inequality property, lower semicontinuity, and coercivity
1.66 Theorem:	a sufficient condition for the solvability of equilibrium problem defined on a space in class $\mathscr{M}$ under triangle inequality property, lower semicontinuity, and coercivity
1.67 Corollary	a sufficient condition for the solvability of equilibrium problem defined on Banach space under triangle inequality property, weakly lower semicontinuity, and coercivity
1.68 Example	existence of solutions of an equilibrium problem with a proximinal domain
1.69 Theorem	Cantor's principle for equilibrium problem

Chapter 2: Existence of Solutions to Scalar Quasiequilibrium Problems

defined on a separable Banach space under convexity, lower semicontinuity, and the closedness of the set of all fixed points of constraint mapping

2.29 Corollary: consequence of Theorem 2.28

2.30 Lemma: an extension of Lemma 1.43

2.31 Theorem: sufficient conditions for the solvability of quasiequilibrium problem under convexity, openness of level sets, and dual property

2.32 Remark: consequences of Theorem 2.31

2.33 Corollary: a sufficient condition for the solvability of quasiequilibrium problem defined on a Hausdorff locally convex space under convexity, closedness of level sets, and dual property

2.34 Corollary: a sufficient condition for the solvability of quasiequilibrium problem defined on a Hausdorff locally convex space under convexity, semicontinuity, closedness of level sets, and dual property

2.35 Theorem: sufficient conditions for the solvability of quasiequilibrium problem under convexity, openness of level sets, upper-sign continuity, and property $\mathbb{P}^2$

2.36 Corollary: a sufficient condition for the solvability of quasiequilibrium problem defined on a Hausdorff locally convex space under convexity, closedness of level sets, upper-sign continuity, and property $\mathbb{P}^2$

2.37 Corollary: a sufficient condition for the solvability of quasiequilibrium problem defined on a Hausdorff locally convex space under convexity, semicontinuity, closedness of level sets, upper-sign continuity, and property $\mathbb{P}^2$

2.38 Example: applications of Theorem 2.35 and Corollary 2.37

2.39 Theorem: a sufficient condition for the solvability of quasiequilibrium problem defined on a separable Banach space under convexity, lower semicontinuity, the closedness of the set of all fixed points of constraint mapping, and dual property

2.40 Corollary: a sufficient condition for the solvability of quasiequilibrium problem defined on a separable Banach space under convexity, lower semicontinuity, the closedness of the set of all fixed points of constraint mapping, the closedness of graph of leval mapping, and dual property

2.41 Theorem: a sufficient condition for the solvability of quasiequilibrium problem defined on a separable Banach space under convexity, lower semicontinuity, the closedness of the set of all fixed points of constraint mapping, upper-sign continuity, and property $\mathbb{P}^2$

2.42 Corollary: a sufficient condition for the solvability of quasiequilibrium problem defined on a separable Banach space under convexity, lower semicontinuity, the closedness of the set of all fixed

2.85 Theorem:	the solution existence of Browder variational inclusion under properly quasimonotonicity and coercivity
2.86 Theorem:	the solution existence of Browder variational inclusion under upper semicontinuity and coercivity
2.87 Theorem:	the solution existence of Browder variational inclusion under upper hemicontinuity, properly quasimonotonicity and coercivity
2.88 Theorem:	the solution existence of Browder variational inclusion lower upper hemicontinuity, properly quasimonotonicity, and coercivity
2.89 Theorem:	the solution existence of Browder variational inclusion under upper semicontinuity and coercivity
2.90 Corollary:	the solution existence of Browder variational inclusion Hausdorff locally convex spac under upper semicontinuity and coercivity
2.91 Corollary:	the solution existence of Browder variational inclusion Hausdorff locally convex spac under upper hemicontinuity, properly quasimonotonicity and coercivity

Chapter 3: Solution Existence for Vector Equilibrium Problems

3.1 Definition:	algebraic interior, vectorial closure, algebraically solid set, algebraically open set, vectorially closed set, vectorial closure of a set in direction q, q-vectorially closed set
3.2 Example:	algebraic interior, vectorial closure
3.3 Proposition:	relationship between topological interior, algebraic interior, topological closure, and vectorial closure
3.4 Proposition:	some properties of algebraic interior and vectorial closure
3.5 Definition:	cone
3.6 Definition:	cone generated by a set
3.7 Proposition:	some properties of cones
3.8 Proposition:	a property of vectorially closed and solid convex cone
3.9 Definition:	nonconvex separation functional
3.10 Proposition:	properties of nonconvex separation functional
3.11 Definition:	(q, E)-upper/lower quasicontinuity, E-upper/lower quasicontinuity
3.12 Example:	(q, E)-upper/lower quasicontinuity of h does not imply the (q, E)-upper/lower quasicontinuity
3.13 Definition:	E-upper/lower semicontinuity
3.14 Proposition:	relationship between E-upper/lower quasicontinuity and E-upper/lower semicontinuity
3.15 Definition:	upper/lower E-hemicontinuity
3.16 Definition:	solution concepts of vector equilibrium problem
3.17 Example:	kinds of solutions of vector equilibrium problem
3.18 Example:	kinds of solutions of vector equilibrium problem
3.19 Example:	kinds of solutions of vector equilibrium problem

3.96 Definition: solution concepts of vector quasiequilibrium problem

3.97 Proposition: relationship between existence of kinds of solutions of vector quasiequilibrium problem and existence of solutions of scalar quasiequilibrium problem

3.98 Theorem: sufficient conditions for the solvability of vector quasiequilibrium problem under convexity, openness of level sets

3.99 Corollary: a sufficient condition for the solvability of vector quasiequilibrium problem defined on a Hausdorff locally convex space under convexity, closedness of level sets

3.100 Theorem: a sufficient condition for the solvability of vector quasiequilibrium problem defined on a separable Banach space under convexity, lower semicontinuity, and the closedness of the set of all fixed points of constraint mapping

3.101 Lemma: an extension of Lemma 3.53

3.102 Theorem: sufficient conditions for the solvability of vector quasiequilibrium problem under convexity, openness of level sets, and E-dual property of type ℓ

3.103 Corollary: a sufficient condition for the solvability of vector quasiequilibrium problem defined on a Hausdorff locally convex space under convexity, openness of level sets, and E-dual property of type ℓ

3.104 Theorem: a sufficient condition for the solvability of vector quasiequilibrium problem defined on a separable Banach space under convexity, lower semicontinuity, and E-dual property of type ℓ

Chapter 4: Stability of Scalar Equilibrium Problems

4.1 Definition: concepts of semicontinuity for a set-valued (multivalued) map

4.2 Lemma: equivalent forms of semicontinuity properties of a set-valued map

4.3 Definition: concepts of Hausdorff semicontinuity for a set-valued map

4.4 Lemma: relations between semicontinuity, closedness, and Hausdorff semicontinuity of a setvalued map

4.5 Theorem: upper semicontinuity and closedness of solution maps

4.6 Definition: sign-preserving function

4.7 Proposition: characterizations of the sign-preserving property

4.8 Theorem: upper semicontinuity and closedness of dual solution maps

4.9 Proposition: relationship between original and auxiliary solution sets

4.10 Lemma: lower semicontinuity of auxiliary solution maps

4.11 Theorem: lower semicontinuity of solution maps

4.12 Remark: upper and lower pseudocontinuous functions

4.13 Definition: Hausdorff upper and lower semicontinuity of multifunctions

4.14 Theorem: Hausdorff lower semicontinuity of solution maps

4.15 Theorem: Hausdorff continuity of approximate solution maps

5.27 Example: the generalized level closedness of the objective function cannot be dropped
5.28 Theorem: characterizations of unique well-posedness
5.29 Theorem: characterizations of well-posedness in terms of measures of noncompactness
5.30 Corollary: upper semicontinuity and closedness of the solution maps of Stampacchia and VMinty quasivariational inequalities
5.31 Remark: the generalized level closedness concept in dual spaces
5.32 Corollary: Hölder continuity of the solution maps of Stampacchia and Minty quasivariational inequalities
5.33 Example: Hölder continuity of the solution maps of Stampacchia and Minty quasivariational inequalities without the global boundedness assumption
5.34 Corollary: Hölder continuity of the solution maps of Stampacchia and Minty quasivariational inequalities via pseudostrong monotonicity
5.35 Example: incomparability of strong monotonicity and strong pseudomonotonicity together with quasimonotonicity
5.36 Definition: a user traffic equilibrium flow
5.37 Lemma: relationships between traffic network problems and quasivariational inequalities
5.38 Remark: the positive values condition of the travel demand
5.39 Lemma: relationships between parametric traffic network problems and parametric quasivariational inequalities
5.40 Lemma: a version of Hoffman's lemma
5.41 Lemma: properties of the feasible maps of parametric traffic network problems
5.42 Lemma: Hölder continuity of feasible maps of parametric traffic network problems
5.43 Corollary: upper semicontinuity and closedness of the solution maps of parametric traffic network problems
5.44 Corollary: continuity of the solution maps of parametric traffic network problems via quasimonotonicity conditions
5.45 Corollary: continuity of the solution maps of parametric traffic network problems via pseudomonotonicity conditions
5.46 Corollary: Hölder continuity of solution maps of parametric traffic network problems via strong monotonicity
5.47 Corollary: Hölder continuity of solution maps of parametric traffic network problems via pseudostrong monotonicity

Chapter 6: Stability of Vector Equilibrium Problems

6.1 Proposition: relationships between semicontinuity and level closedness of maps
6.2 Definition: generalized cone-semicontinuity

6.29 Theorem:	Hausdorff lower semicontinuity of solution maps of strong equilibrium problems
6.30 Definition:	cone-quasimonotonicity and cone-pseudomonotonicity
6.31 Remark:	relationships between two types of cone-quasimonotonicity and cone-pseudo-monotonicity
6.32 Example:	a counterexample for relationships between cone-quasimonotonicity and conepseudomonotonicity
6.33 Theorem:	continuity of solution maps of Sol(WEP)
6.34 Theorem:	continuity of solution map of strong equilibrium problems
6.35 Theorem:	continuity of solution map of weak equilibrium problems
6.36 Theorem:	continuity of solution maps of strong equilibrium problems
6.37 Remark:	a special case of conditions (W) and (S)
6.38 Theorem:	Hölder continuity of solution maps of weak and strong equilibrium problems
6.39 Definition:	Hölder calm maps
6.40 Remark:	a motivation of the assumption that the Hölder degree is less than 1
6.41 Definition:	Hölder calm vector maps
6.42 Theorem:	Hölder calmness of solution maps of weak and strong equilibrium problems
6.43 Example:	the Hölder calmness condition of the constraint map is essential
6.44 Example:	the Hölder calmness of the objective map cannot be dropped
6.45 Example:	the conditions related to monotonicity are crucial
6.46 Definition:	cone-Hölder continuous maps
6.47 Lemma:	Hölder continuity of approximate solution maps of weak equilibrium problems
6.48 Lemma:	Hölder continuity of approximate solution maps of strong equilibrium problems
6.49 Theorem:	Hölder continuity of approximate solution maps of parametric weak and strong equilibrium problems
6.50 Example:	cruciality of generalized concavity conditions
6.51 Definition:	approximating sequences for weak and strong equilibrium problems
6.52 Definition:	well-posedness under perturbations of weak and strong equilibrium problems
6.53 Definition:	generalized Hadamard well-posedness for weak and strong equilibrium problems
6.54 Lemma:	convergence characterizations of upper semicontinuity
6.55 Theorem:	closedness of solution sets of weak and strong equilibrium problems
6.56 Theorem:	upper semicontinuity of approximate solution maps of parametric weak and strong equilibrium problems
6.57 Theorem:	relationships between well-posedness and stability

6.58 Corollary:	well-posedness under perturbations and generalized Hadamard well-posedness of weak and strong equilibrium problems
6.59 Theorem:	relationships between stability and unique well-posedness
6.60 Corollary:	unique well-posedness under perturbations and Hadamard well-posedness of weak and strong equilibrium problems
6.61 Lemma:	the union of a family of lower semicontinuous set-valued maps
6.62 Lemma:	linear scalarization for interior of a solid convex cone
6.63 Lemma:	a version of Eidelheit's separation theorem
6.64 Theorem:	linear scalarization for weak equilibrium problems
6.65 Definition:	cone-strong monotonicity, cone-strong pseudomonotonicity
6.66 Definition:	cone-Hölder continuity of a vector-valued map
6.67 Lemma:	topological preserving properties via linear scalarization functions
6.68 Lemma:	cone-Hölder continuity via linear scalarization functions
6.69 Theorem:	Hölder continuity of solution maps of linear scalarized equilibrium problems
6.70 Theorem:	Hölder continuity of solution maps of weak equilibrium problems via linear scalarizations for a strongly pseudomonotone map
6.71 Theorem:	Hölder continuity of solution maps of weak equilibrium problems via linear scalarizations for a strongly convex map
6.72 Example:	the strong convexity is essential
6.73 Example:	the monotonicity is crucial
6.74 Theorem:	Hölder continuity of approximate solution maps of linear scalarized weak equilibrium problems
6.75 Theorem:	Hölder continuity of approximate solution maps of weak equilibrium problems via linear scalarization functions
6.76 Example:	essentialness of the concavity property of the objective map
6.77 Example:	the convexity of the constrained map cannot be dispensed
6.78 Example:	cruciality of the boundedness of the constraint map
6.79 Theorem:	Hausdorff continuity of approximate solution maps of linear scalarized weak equilibrium problems
6.80 Theorem:	Hausdorff continuity of approximate solution maps of weak equilibrium problems via linear scalarization functions
6.81 Example:	the cone-concavity of the objective map is crucial
6.82 Theorem:	Hausdorff continuity of approximate solution maps of weak equilibrium problems with fixed constrained set via linear scalarization functions
6.83 Remark:	convexlike maps in totally ordered spaces
6.84 Example:	Hausdorff continuity of solution maps of equilibrium problems without assuming continuity conditions
6.85 Definition:	the Gerstewitz function
6.86 Proposition:	equivalence between vector relations and their Gerstewitz function values relations

6.141 Example: an example of a convergence analysis for the strong form of a monotone (VEP)
6.142 Definition: Definition of explicitly cone pseudo convexlikeness
6.143 Proposition: a relationship between weak and efficient solutions of a vector optimization problem
6.144 Theorem: a result of the solvability of strong vector equilibriumproblems by the virtue of techniques appeared in vector optimization problems
6.145 Remark: some remarks on a coercivity condition for vector optimization problems
6.146 Theorem: a result of the solvability of noncompact strong vector equilibriumproblems by the virtue of techniques appeared in vector optimization problems
6.147 Theorem: a convergence result of the regularization method for the strong form of nonmonotone vector equilibrium problems

6.148 Example: a example of the regularization method for the strong form of nonmonotone vector equilibrium problems
6.149 Definition: cone-proper, cone-closed, cone-bounded, and cone-compact sets
6.150 Definition: level quasiconcave maps
6.151 Lemma: equivalent statements of 0-level quasiconcave conditions
6.152 Remark: relationships between level quasiconcavity and quasiconcavity properties
6.153 Example: a case of level quasiconcavity but not quasiconcavity
6.154 Remark: level-quasiconvexity and weak quasiconvexity
6.155 Example: level-quasiconvexity and weak convexity are incomparable
6.156 Definition: generalized Gerstewitz functions
6.157 Definition: dual generalized Gerstewitz functions
6.158 Proposition: estimations of types of Gerstewitz functions
6.159 Remark: characterizations of nonpositive values of the generalized Gerstewitz function
6.160 Lemma: finite values conditions for generalized Gerstewitz functions
6.161 Lemma: calculating Gerstewitz functions
6.162 Proposition: convexity and quasiconvexity of Gerstewitz functions
6.163 Proposition: convexity and quasiconvexity of generalized Gerstewitz functions
6.164 Proposition: continuity of generalized Gerstewitz functions
6.165 Proposition: Hölder continuity of generalized Gerstewitz functions
6.166 Definition: KKM maps
6.167 Lemma: the KKM-Fan lemma
6.168 Lemma: an extended version of KKM-Fan lemma
6.169 Lemma: KKM maps in terms of set less order relations
6.170 Lemma: characterizations of cone-quasiconvexity and cone-quasiconcavity via set less order relations

6.202 Lemma:	closedness conditions of solution sets of set optimization problems
6.203 Theorem:	well-posedness conditions for set optimization problems
6.204 Theorem:	global well-posedness conditions for set optimization problems
6.205 Remark:	cruciality of the closedness of the solution set of a set optimization problem
6.206 Example:	a special case of globally well-posed set optimization problems
6.207 Remark:	relations of cone convexity properties of a map
6.208 Corollary:	Hölder conditions for the Browder variational inclusion
6.209 Corollary:	Hausdorff conditions for the Browder variational inclusion
6.210 Definition:	strong, efficient, weak solutions of vector optimization problems
6.211 Corollary:	upper semicontinuity and closedness of strong and weak solution maps of vector optimization problems
6.212 Corollary:	Hölder continuity of strong and weak solution maps of vector optimization problems
6.213 Corollary:	Hausdorff continuity of approximate solution maps of Stampacchia and Minty vector variational inequalities
6.214 Corollary:	Hölder continuity of the solution maps of the Stampacchia vector variational inequality
6.215 Corollary:	Hölder continuity of the solution map of the Minty vector variational inequality
6.216 Definition:	saturated arc
6.217 Definition:	weak vector equilibrium flow
6.218 Lemma:	relationships between vector traffic equilibrium problems and vector variational inequalities
6.219 Corollary:	Hausdorff conditions for vector traffic equilibrium problems
6.220 Definition:	equicontinuous function
6.221 Lemma:	integrand inequality
6.222 Lemma:	Lipschitz conditions of state equations
6.223 Lemma:	compactness and convexity conditions of state equations
6.224 Corollary:	Holder conditions for fuel optimal frictionless horizontal motion of a mass point problem
6.225 Lemma:	sum of strong convexity and convexity functions
6.226 Lemma:	strong convexity of a quadratic function given by a positive definite matrix
6.227 Corollary:	Hölder conditions for a parametric quadratic programming

Chapter 7: Variational Convergence of Bifunctions and Approximations of Equilibrium Problems

7.1 Definition:	epi-convergence, inside epi-convergence
7.2 Definition:	weak epi-convergence, weak inside epi-convergence

7.140 Theorem: consistency of approximate statistical estimators

7.141 Theorem: complete consistency of statistical estimators

7.142 Theorem: epi/hypo-convergence of expectation bifunctions

7.143 Theorem: consistency for stochastic minimax problems

7.144 Theorem: any cluster point of approximate saddle points of Lagrangians is a solution-pair of primal-dual stochastic programs

Chapter 8: Approximations of Quasiequilibrium Problem

8.1 Definition: variational objects of bifunctions on general sets

8.2 Proposition: characterizations and properties of saddle points of general bifunctions

8.3 Example: examples of variational objects of bifunctions, illustrations of Proposition 8.2

8.4 Proposition: relationship between solutions of primal and dual problems and types of saddle points

8.5 Example: illustration of Proposition 8.4

8.6 Proposition: relationaships between approximate primal and dual solutions of quasiequilibrium problems and types of approximate saddle points

8.7 Definition: epi/hypo-convergence, inside epi/hypo convergence

8.8 Definition: weak epi/hypo-convergence, weak inside epi/hypo-convergence

8.9 Example: i-e/h-convergence holds, e/h-convergence fails

8.10 Example: e/h-convergence holds

8.11 Proposition: weak e/h-convergence versus weak epi- and hypo-convergence of slices

8.12 Definition: lower and upper e/h-limits, weak lower and upper e/h-limits

8.13 Theorem: formulas of the types of upper and lower e/h-limits

8.14 Theorem: characterization of the types of e/h-convergence

8.15 Example: a case that the e/h-limit is unique

8.16 Theorem: geometry of the types of lower and upper e/h-limits ending the equivalence class

8.17 Theorem: convergence of approximate weak saddle points under e/h-convergence

8.18 Theorem: condition (a) of a type of e/h-convergence implies (a) of the corresponding type of epi-convergence of sup-projections

8.19 Definition: types of tightness conditions

8.20 Theorem: epi-convergence of sup-projections

8.21 Proposition: epi-convergence of inf-projections

8.22 Theorem: convergence of minsup-points

8.23 Example:	the tightness condition in Theorem 8.22 is not necessary
8.24 Theorem:	convergence of approximate weak saddle points, a lower bound of approximate saddle points
8.25 Example:	illustrations of Theorem 8.24
8.26 Definition:	minsup-lopsided convergence, inside minsup-lopsided convergence
8.27 Definition:	maxinf-lopsided convergence, inside maxinf-lopsided convergence
8.28 Example:	mis-lop convergence holds; weak mai-lop and e/h-convergence fail
8.29 Example:	all e/h-, mis-lop-, and mai-lop-limits are equal to a singleton
8.30 Proposition:	characterizations of weak mis-lop convergence in terms of slices
8.31 Definition:	lower and upper weak lop-limits
8.32 Remark:	relationships between "semi" lop-limits and "semi" e/h-limits
8.33 Theorem:	formulas of lower and upper weak lop-limits
8.34 Theorem:	characterizations of weak lop-convergence
8.35 Remark:	comparisons between e/h- and lop-convergence
8.36 Theorem:	sufficient condition for mis-lop convergence
8.37 Example:	failure of weak mis-lop convergence under graphical convergence
8.38 Definition:	tightness of the types of mis-lop convergence
8.39 Proposition:	a type of epi-convergence of sup-projections in terms of the corresponding type of x-ancillary tight mis-lop convergence
8.40 Theorem:	convergence of minsup-points in terms of x-ancillary tight lop-convergence
8.41 Example:	x-ancillary tightness in Proposition 8.39 and Theorem 8.40 is essential; these statements are not necessary conditions
8.42 Theorem:	complete convergence of minsup-points under x-tight lop-convergence
8.43 Theorem:	complete convergence of approximate saddle points under tight lop-convergence
8.44 Lemma:	a case that e/h-,mis-lop, and epi-convergence are equivalent
8.45 Theorem:	quantification of variational convergence of bifunctions
8.46 Proposition:	convergence of approximate solutions of quasiequilibrium problems
8.7 Proposition:	convergence of approximate solutions of strong quasiequilibrium problems

Chapter 9: Approximations of Vector Quasiequilibrium Problems

Chapter 10: Variational Inclusion Problems and Variational Relation Problems

References

1. M. Adán, V. Novo, Weak efficiency in vector optimization using a closure of algebraic type under cone-convexlikeness. Eur. J. Oper. Res. **149**(3), 641–653 (2003)
2. B. Alleche, V.D. Rădulescu, Further on set-valued equilibrium problems in the pseudo-monotone case and applications to Browder variational inclusions. Optim. Lett. **12**(8), 1789–1810 (2018)
3. B. Alleche, V.D. Rădulescu, Set-valued equilibrium problems with applications to Browder variational inclusions and to fixed point theory. Nonlinear Anal. Real World Appl. **28**, 251–268 (2016)
4. B. Alleche, V.D. Rădulescu, M. Sebaoui, The Tikhonov regularization for equilibrium problems and applications to quasi-hemivariational inequalities. Optim. Lett. **9**(3), 483–503 (2015)
5. E. Allevi et al., Noncooperative games with vector payoffs under relative pseudomonotonicity. J. Optim. Theory Appl. **118**, 245–254 (2003)
6. E. Allevi et al., Infinite player noncooperative games with vector payoffs under relative pseudomonotonicity. J. Glob. Optim. **34**, 79–96 (2006)
7. L.Q. Anh, Semicontinuity of the solution maps to vector equilibrium problems with equilibrium constraints. Optimization **71**(3), 737–751 (2022)
8. L.Q. Anh, Various types of well-posedness for vector equilibrium problems with respect to the lexicographic order. Vietnam J. Math. **51**(2), 397–414 (2023)
9. L.Q. Anh, N.H. Danh, T.N. Tam, Continuity of solution maps to parametric set optimization problems via parametric equilibrium problems. Acta Math. Vietnam. **45**(2), 383–395 (2020)
10. L.Q. Anh, P.T. Duoc, T.T.T. Duong, Connectedness properties of the efficient sets and the nondominated sets to vector optimization problems. Optim. Lett. **16**(8), 2457–2468 (2022)
11. L.Q. Anh, P.T. Duoc, H.M. Linh, Scalar representations and Hausdorff continuity of solution mappings to parametric set optimization problems via set less order relations. Oper. Res. Lett. **53**, 107071 (2024)
12. L.Q. Anh, P.T. Duoc, T.N. Tam, Continuity of approximate solution maps to vector equilibrium problems. J. Ind. Manage. Optim. **13**(4), 1685–1699 (2017)
13. L.Q. Anh, P.T. Duoc, N.M. Tung, On Lipschitz continuity of solutions to equilibrium problems via the Hiriart-Urruty oriented distance function. Comput. Appl. Math. **41**(1), 57 (2022)
14. L.Q. Anh, T.Q. Duy, On penalty method for equilibrium problems in lexicographic order. Positivity **22**, 39–57 (2018)
15. L.Q. Anh, T.Q. Duy, Regularization of vector equilibrium problems. Optim. Lett. **17**(3), 699–720 (2023)

L. Q. Anh et al., *Equilibrium Problems*, Forum for Interdisciplinary Mathematics, https://doi.org/10.1007/978-3-031-89003-1

16. L.Q. Anh, T.Q. Duy, Tykhonov well-posedness for lexicographic equilibrium problems. Optimization **65**(11), 1929–1948 (2016)
17. L.Q. Anh, T.Q. Duy, D.V. Hien, Stability of efficient solutions to set optimization problems. J. Glob. Optim. **78**(3), 563–580 (2020)
18. L.Q. Anh, T.Q. Duy, D.V. Hien, Well-posedness for the optimistic counterpart of uncertain vector optimization problems. Ann. Oper. Res. **295**, 517–533 (2020)
19. L.Q. Anh, T.Q. Duy, P.Q. Khanh, Continuity properties of solution maps of parametric lexicographic equilibrium problems. Positivity **20**, 61–80 (2016)
20. L.Q. Anh, D.V. Hien, On well-posedness for parametric vector quasiequilibrium problems with moving cones. Appl. Math. **61**(6), 651–668 (2016)
21. L.Q. Anh, N.V. Hung, Gap functions and Hausdorff continuity of solution mappings to parametric strong vector quasiequilibrium problems. J. Ind. Manage. Optim. **14**(1), 65–79 (2018)
22. L.Q. Anh, N.V. Hung, Levitin–Polyak well-posedness for strong bilevel vector equilibrium problems and applications to traffic network problems with equilibrium constraints. Positivity **22**, 1223–1239 (2018)
23. L.Q. Anh, N.H. Danh, P.T. Duoc, Hausdorff continuity of solution maps to equilibrium problems via the oriented distance function. Optim. Lett. **18**, 1907–1924 (2024)
24. L.Q. Anh, P.Q. Khanh, Continuity of solution maps of parametric quasiequilibrium problems. J. Glob. Optim. **46**(2), 247–259 (2010)
25. L.Q. Anh, P.Q. Khanh, Hölder continuity of the unique solution to quasiequilibrium problems in metric spaces. J. Optim. Theory Appl. **141**, 37–54 (2009)
26. L.Q. Anh, P.Q. Khanh, On the Hölder continuity of solutions to parametric multivalued vector equilibrium problems. J. Math. Anal. Appl. **321**(1), 308–315 (2006)
27. L.Q. Anh, P.Q. Khanh, On the stability of the solution sets of general multivalued vector quasiequilibrium problems. J. Optim. Theory Appl. **135**(2), 271–284 (2007)
28. L.Q. Anh, P.Q. Khanh, Semicontinuity of solution sets to parametric quasivariational inclusions with applications to traffic networks I: upper semicontinuities. Set-Valued Anal. **16**, 267–279 (2008)
29. L.Q. Anh, P.Q. Khanh, Semicontinuity of solution sets to parametric quasivariational inclusions with applications to traffic networks II: lower semicontinuities applications. Set-Valued Anal. **16**(7–8) 943–960 (2008)
30. L.Q. Anh, P.Q. Khanh, Semicontinuity of the solution set of parametric multivalued vector quasiequilibrium problems. J. Math. Anal. Appl. **294**(2), 699–711 (2004)
31. L.Q. Anh, P.Q. Khanh, Sensitivity analysis for multivalued quasiequilibrium problems in metric spaces: Hölder continuity of solutions. J. Glob. Optim. **42**, 515–531 (2008)
32. L.Q. Anh, P.Q. Khanh, Sensitivity analysis for weak and strong vector quasiequilibrium problems. Vietnam J. Math. **37**, 237–253 (2009)
33. L.Q. Anh, P.Q. Khanh, Uniqueness and Hölder continuity of the solution to multivalued equilibrium problems in metric spaces. J. Glob. Optim. **37**(3), 449–465 (2007)
34. L.Q. Anh, P.Q. Khanh, Various kinds of semicontinuity and the solution sets of parametric multivalued symmetric vector quasiequilibrium problems. J. Glob. Optim. **41**, 539–558 (2008)
35. L.Q. Anh, P.Q. Khanh, D.T.M. Van, On well-posedness for perturbed quasi-equilibrium and quasi-optimization problems. Numer. Funct. Anal. Optim. **42**(5), 583–607 (2021)
36. L.Q. Anh, P.Q. Khanh, D.N. Quy, About semicontinuity of set-valued maps and stability of quasivariational inclusions. Set-Valued Var. Anal. **22**, 533–555 (2014)
37. L.Q. Anh, P.Q. Khanh, T.N. Tam, Continuity of approximate solution maps of primal and dual vector equilibrium problems. Optim. Lett. **13**, 201–211 (2019)
38. L.Q. Anh, P.Q. Khanh, T.N. Tam, On Hölder continuity of approximate solutions to parametric equilibrium problems. Nonlinear Anal. Theory Methods Appl. **75**(4), 2293–2303 (2012)
39. L.Q. Anh, P.Q. Khanh, T.N. Tam, On Hölder continuity of solution maps of parametric primal and dual Ky Fan inequalities. TOP **23**(1), 151–167 (2015)

40. L.Q. Anh, P.Q. Khanh, D.T.M. Van, Well-posedness under relaxed semicontinuity for bilevel equilibrium and optimization problems with equilibrium constraints. J. Optim. Theory Appl. **153**(1), 42–59 (2012)
41. L.Q. Anh, P.Q. Khanh, D.T.M. Van, Well-posedness without semicontinuity for parametric quasiequilibria and quasioptimization. Comput. Math. Appl. **62**(4), 2045–2057 (2011)
42. L.Q. Anh, A.Y. Kruger, N.H. Thao, On Hölder calmness of solution mappings in parametric equilibrium problems. TOP **22**(1), 331–342 (2014)
43. L.Q. Anh, K.T. Nguyen, T.N. Tam, On Hölder continuity of approximate solution maps to vector equilibrium problems. Turk. J. Math. **41**(6), 1591–1607 (2017)
44. L.Q. Anh, T.Q. Duy, D.V. Hien, Stability for parametric vector quasiequilibrium problems with variable cones. Numer. Funct. Anal. Optim. **40**(4), 461–483 (2019)
45. L.Q. Anh, T.N. Tam, Hausdorff continuity of approximate solution maps to parametric primal and dual equilibrium problems. Top **24**, 242–258 (2016)
46. L.Q. Anh, T.N. Tam, Sensitivity analysis for parametric vector equilibrium problems. J. Nonlinear Convex Anal. **18**(9), 1707–1716 (2017)
47. L.Q. Anh, T.N. Tam, The use of a family of Gerstewitz scalarization functions in the context of vector optimization with variable domination structures to derive scalarization results. Optim. Methods Software **39**(2), 368–383 (2024)
48. L.Q. Anh, T.N. Tam, N.H. Danh, On Lipschitz continuity of approximate solutions to set-valued equilibrium problems via nonlinear scalarization. Optimization **72**(2), 439–461 (2023)
49. L.Q. Anh, P. Thanh Duoc, T.N. Tam, On Hölder continuity of solution maps to parametric vector primal and dual equilibrium problems. Optimization **67**(8), 1169–1182 (2018)
50. L.Q. Anh, P.T. Duoc, T.T.T. Duong, A new scalarization approach and applications in set optimization. Optimization **72**(4), 1093–1117 (2023)
51. L.Q. Anh, N.V. Hung, Stability of solution mappings for parametric bilevel vector equilibrium problems. Comput. Appl. Math. **37**, 1537–1549 (2018)
52. L.Q. Anh, N.V. Hung, The existence and stability of solutions for symmetric generalized quasi-variational inclusion problems. Filomat **29**(9), 2147–2165 (2015)
53. L.Q. Anh, N.V. Hung, V.M. Tam, Regularized gap functions and error bounds for generalized mixed strong vector quasiequilibrium problems. Comput. Appl. Math. **37**, 5935–5950 (2018)
54. L.Q. Anh et al., Convergence of solutions to set optimization problems with the set less order relation. J. Optim. Theory Appl. **185**(2), 416–432 (2020)
55. L.Q. Anh et al., On Hölder calmness and Hölder well-posedness of vector quasiequilibrium problems. Vietnam J. Math. **41**, 507–517 (2013)
56. L.Q. Anh et al., Qualitative properties of solutions to set optimization problems. Comput. Appl. Math. **40**(2), 1–18 (2021)
57. L.Q. Anh et al., Stability analysis for set-valued equilibrium problems with applications to Browder variational inclusions. Optim. Lett. **15**, 613–626 (2021)
58. L.Q. Anh et al., The Tikhonov regularization for vector equilibrium problems. Comput. Optim. Appl. **78**, 769–792 (2021)
59. L.Q. Anh et al., Well-posedness for lexicographic vector equilibrium problems, in *Constructive Nonsmooth Analysis and Related Topics* (Springer, 2013), pp. 159–174
60. L.Q. Anh et al., Well-posedness for vector quasiequilibria. Taiwan. J. Math. **13**(2B), 713–737 (2009)
61. L.Q. Anh, T.Q. Duy, P.Q. Khanh, Levitin–Polyak well-posedness for equilibrium problems with the lexicographic order. Positivity **25**(4), 1323–1349 (2021)
62. P.K. Anh, N.H. Trinh, Splitting extragradient-like algorithms for strongly pseudomonotone equilibrium problems. Numerical Algorithms **76**(1), 67–91 (2017)
63. Q.H. Ansari, Vectorial form of Ekeland-type variational principle with applications to vector equilibrium problems and fixed point theory. J. Math. Anal. Appl. **334**(1), 561–575 (2007)
64. Q.H. Ansari, A. Farajzadeh, S. Schaible, Existence of solutions of strong vector equilibrium problems. Taiwan. J. Math. **16**(1), 165–178 (2012)

65. Q.H. Ansari, F. Flores-Bazán, Generalized vector quasi-equilibrium problems with applications. J. Math. Anal. Appl. **277**(1), 246–256 (2003)
66. Q.H. Ansari, E. Köbis, J.-C. Yao, *Vector Variational Inequalities and Vector Optimization* (Springer, 2018)
67. Q.H. Ansari, I.V. Konnov, J.-C. Yao, Existence of a solution and variational principles for vector equilibrium problems. J. Optim. Theory Appl. **110**(3), 481–492 (2001)
68. Q.H. Ansari, W. Oettli, D. Schlager, A generalization of vectorial equilibria. Math. Meth. Oper. Res.-ZOR **46**(2), 147–152 (1997)
69. Q.H. Ansari, N.-C. Wong, J.-C. Yao, The existence of nonlinear inequalities. Appl. Math. Lett. **12**(5), 89–92 (1999)
70. Q.H. Ansari, J.-C. Yao, On vector quasi—equilibrium problems, in *Equilibrium Problems and Variational Models. Nonconvex Optimization and Its Applications*, vol 68. (Springer, 2003), pp. 1–18
71. Q.H. Ansari, I. Konnov, J.-C. Yao, On generalized vector equilibrium problems. Nonlinear Anal. Theory Methods Appl. **47**(1), 543–554 (2001)
72. K.J. Arrow, General economic equilibrium: purpose, analytic techniques, collective choice. Am. Econ. Rev. **64**(3), 253–272 (1974)
73. K.J. Arrow, G. Debreu, Existence of an equilibrium for a competitive economy. Econometrica J. Econ. Soc. **22**(3), 265–290 (1954)
74. H. Attouch, *Variational Convergence for Functions and Operators*, vol. 1 (Pitman Advanced Publishing Program, 1984)
75. H. Attouch, D. Aze, R.J.-B. Wets, Convergence of convex-concave saddle functions: continuity properties of the Legendre-Fenchel transform with applications to convex programming and mechanics. Ann. Inst. Henri Poincaré **5**, 537–572 (1988)
76. H. Attouch, R.J.-B. Wets, Approximations of vector quasiequilibria and vector Nash quasiequilibria. C.R. Acad. Sci. Paris Sér. I Math. **296**, 657–660 (1983)
77. H. Attouch, R.J.-B. Wets, Epigraphical analysis, in *Annales de l'Institut Henri Poincaré C, Analyse non Linéaire*, vol. 6 (Elsevier, 1989), pp. 73–100
78. H. Attouch, R.J.-B. Wets, A convergence theory for saddle functions. Trans. Am. Math. Soc. **280**(1), 1–41 (1983)
79. H. Attouch, R.J.-B. Wets, Approximation and convergence in nonlinear optimization, in *Nonlinear Programming*, vol. 4 (Elsevier, 1981), pp. 367–394
80. H. Attouch, R.J.-B. Wets, Quantitative stability of variational systems: I. The epigraphical distance. Trans. Am. Math. Soc. **328**(2), 695–729 (1991)
81. J.-P. Aubin, I. Ekeland, *Applied Nonlinear Analysis: Pure and Applied Mathematics* (Courier Corporation, 1984)
82. J.-P. Aubin, H. Frankowska, *Set-Valued Analysis* (Springer, 2009)
83. D. Aussel, N. Hadjisavvas, On quasimonotone variational inequalities. J. Optim. Theory Appl. **121**(2), 445–450 (2004)
84. A. Bagh, Epi/hypo–convergence: The slice topology and saddle points approximation. J. Appl. Anal. **2**(1), 13–39 (1996)
85. M. Balaj, Existence theorems for a variational relation problem. Numer. Funct. Anal. Optim. **37**(4), 459–479 (2016)
86. M. Balaj, New existence criteria of the solutions for a variational relation problem. Optimization **66**(5), 675–689 (2017)
87. M. Balaj, L.-J. Lin, Generalized variational relation problems with applications. J. Optim. Theory Appl. **148**, 1–13 (2011)
88. T.Q. Bao, B.S. Mordukhovich, Extended Pareto optimality in multiobjective problems, in *Recent Developments in Vector Optimization* (Springer, 2012), pp. 467–515
89. E. Bednarczuk, Stability analysis for parametric vector optimization problems. Diss. Math. **442**, 1–126 (2007)
90. G. Beer, *Topologies on Closed and Closed Convex Sets*, vol. 268 (Springer Science & Business Media, 1993)

91. G. Beer, R. Rockafellar, R.J.-B. Wets, A characterization of epi-convergence in terms of convergence of level sets. Proc. Am. Math. Soc. **116**(3), 753–761 (1992)
92. M. Bianchi, N. Hadjisavvas, S. Schaible, Minimal coercivity conditions for variational inequalities. Application to exceptional families of elements. J. Optim. Theory Appl. **122**(1), 17 (2004)
93. M. Bianchi, N. Hadjisavvas, S. Schaible, Vector equilibrium problems with generalized monotone bifunctions. J. Optim. Theory Appl. **92**(3), 527–542 (1997)
94. M. Bianchi, S. Schaible, Generalized monotone bifunctions and equilibrium problems. J. Optim. Theory Appl. **90**(1), 31–43 (1996)
95. M. Bianchi, G. Kassay, R. Pini, Existence of equilibria via Ekeland's principle. J. Math. Anal. Appl. **305**(2), 502–512 (2005)
96. M. Bianchi, G. Kassay, R. Pini, Well-posed equilibrium problems. Nonlinear Anal. Theory Methods Appl. **72**(1), 460–468 (2010)
97. M. Bianchi, G. Kassay, R. Pini, Ekeland's principle for vector equilibrium problems. Nonlinear Anal. Theory Methods Appl. **66**(7), 1454–1464 (2007)
98. M. Bianchi, R. Pini, A note on equilibrium problems with properly quasimonotone bifunctions. J. Glob. Optim. **20**(1), 67–76 (2001)
99. M. Bianchi, R. Pini, A note on stability for parametric equilibrium problems. Oper. Res. Lett. **31**(6), 445–450 (2003)
100. M. Bianchi, R. Pini, Coercivity conditions for equilibrium problems. J. Optim. Theory Appl. **124**(1), 79–92 (2005)
101. M. Bianchi, R. Pini, Sensitivity for parametric vector equilibria. Optimization **55**(3), 221–230 (2006)
102. M. Bianchi, S. Schaible, Equilibrium problems under generalized convexity and generalized monotonicity. J. Glob. Optim. **30**(2–3), 121–134 (2004)
103. G. Bigi, A. Capaătaă, G. Kassay, Existence results for strong vector equilibrium problems and their applications. Optimization **61**(5), 567–583 (2012)
104. G. Bigi, M. Castellani, G. Kassay, A dual view of equilibrium problems. J. Math. Anal. Appl. **342**(1), 17–26 (2008)
105. G. Bigi et al., Existence and solution methods for equilibria. Eur. J. Oper. Res. **227**(1), 1–11 (2013)
106. G. Bigi et al., *Nonlinear Programming Techniques for Equilibria* (Springer, 2019)
107. P. Billingsley, *Probability and Measure* (John Wiley & Sons, 2008)
108. E. Blum, W. Oettli, From optimization and variational inequalities to equilibrium problems. Mathematics Student-India **63**(1), 123–145 (1994)
109. J.F. Bonnans, A. Shapiro, *Perturbation Analysis of Optimization Problems* (Springer, 2013)
110. A. Braides, *Local Minimization, Variational Evolution and Γ-Convergence*, vol. 2094 (Springer, 2014)
111. R. Brânzei et al., Approximate fixed point theorems in Banach spaces with applications in game theory. J. Math. Anal. Appl. **285**(2), 619–628 (2003)
112. H. Brézis, L. Nirenberg, G. Stampacchia, A remark on Ky Fan's minimax principle. Bollettino dell Unione Matematica Italiana **1**(2), 257–264 (2008)
113. F.E. Browder, Existence and approximation of solutions of nonlinear variational inequalities. Proc. Natl. Acad. Sci. **56**(4), 1080–1086 (1966)
114. F.E. Browder, The fixed point theory of multi-valued mappings in topological vector spaces. Mathematische Annalen **177**(4), 283–301 (1968)
115. M.J. Cánovas et al., Isolated calmness of solution mappings in convex semiinfinite optimization. J. Math. Anal. Appl. **350**(2), 829–837 (2009)
116. A. Capaătaă, Existence results for proper efficient solutions of vector equilibrium problems and applications. J. Glob. Optim. **51**(4), 657–675 (2011)
117. J. Caristi, Fixed point theorems for mappings satisfying inwardness conditions. Trans. Am. Math. Soc. **215**, 241–251 (1976)
118. J. Caristi, W.A. Kirk, Geometric fixed point theory and inwardness conditions. in *The Geometry of Metric and Linear Spaces* (Springer, 1975), pp. 74–83

119. M. Castellani, M. Giuli, On equivalent equilibrium problems. J. Optim. Theory Appl. **147**(1), 157–168 (2010)
120. M. Castellani, M. Giuli, Refinements of existence results for relaxed quasimonotone equilibrium problems. J. Glob. Optim. **57**(4), 1213–1227 (2013)
121. M. Castellani, M. Giuli, An existence result for quasiequilibrium problems in separable Banach spaces. J. Math. Anal. Appl. **425**(1), 85–95 (2015)
122. M. Castellani, M. Giuli, Approximate solutions of quasiequilibrium problems in Banach spaces. J. Glob. Optim. **64**(3), 615–620 (2016)
123. M. Castellani, M. Giuli, Ekeland's principle for cyclically antimonotone equilibrium problems. Nonlinear Anal. Real World Appl. **32**, 213–228 (2016)
124. M. Castellani, M. Giuli, M. Pappalardo, A Ky Fan minimax inequality for quasiequilibria on finite-dimensional spaces. J. Optim. Theory Appl. **179**(1), 53–64 (2018)
125. M. Castellani, M. Pappalardo, M. Passacantando, Existence results for nonconvex equilibrium problems. Optim. Methods Softw. **25**(1), 49–58 (2010)
126. L.-C. Ceng et al., Well-posedness for mixed quasivariational-like inequalities. J. Optim. Theory Appl. **139**, 109–125 (2008)
127. O. Chadli, Z. Chbani, H. Riahi, Equilibrium problems and noncoercive variational inequalities. Optimization **50**(1–2), 17–27 (2001)
128. O. Chadli, S. Schaible, J.-C. Yao, Regularized equilibrium problems with application to noncoercive hemivariational inequalities. J. Optim. Theory Appl. **121**(3), 571–596 (2004)
129. O. Chadli, Q.H. Ansari, S. Al-Homidan, Existence of solutions and algorithms for bilevel vector equilibrium problems: An auxiliary principle technique. J. Optim. Theory Appl. **172**(3), 726–758 (2017)
130. O. Chadli, N.-C. Wong, J.-C. Yao, Equilibrium problems with applications to eigenvalue problems. J. Optim. Theory Appl. **117**(2), 245–266 (2003)
131. O. Chaldi, Z. Chbani, H. Riahi, Equilibrium problems with generalized monotone bifunctions and applications to variational inequalities. J. Optim. Theory Appl. **105**(2), 299–323 (2000)
132. S.-S. Chang, Y. Zhang, Generalized KKM theorem and variational inequalities. J. Math. Anal. Appl. **159**(1), 208–223 (1991)
133. T.-H. Chang, C.-L. Yen, KKM property and fixed point theorems. J. Math. Anal. Appl. **203**(1), 224–235 (1996)
134. T.-H. Chang et al., On S-KKM property and-related topics. J. Math. Anal. Appl. **229**(1), 212–227 (1999)
135. D. Charles, Some results on multivalued mappings and inequalities without convexity, in *Nonlinear and Convex Analysis: Proceedings in Honor of Ky Fan*, vol. 107 (CRC Press, 1987), p. 99–106
136. B. Chen, N.-J. Huang, Continuity of the solution mapping to parametric generalized vector equilibrium problems. J. Glob. Optim. **56**, 1515–1528 (2013)
137. C.-R. Chen, L.-L. Li, M.-H. Li, Hölder continuity results for nonconvex parametric generalized vector quasiequilibrium problems via nonlinear scalarizing functions. Optimization **65**(1), 35–51 (2016)
138. C. Chen, S. Li, K.L. Teo, Solution semicontinuity of parametric generalized vector equilibrium problems. J. Glob. Optim. **45**(2), 309–318 (2009)
139. J. Chen, Q.H. Ansari, J.-C. Yao, Characterizations of set order relations and constrained set optimization problems via oriented distance function. Optimization **66**(11), 1741–1754 (2017)
140. J. Chen et al., Tikhonov type regularization methods for inverse mixed variational inequalities. Optimization **69**(2), 401–413 (2020)
141. Y. Cheng, D. Zhu, Global stability results for the weak vector variational inequality. J. Glob. Optim. **32**(4), 543–550 (2005)
142. C. Chidume, H. Zegeye, K.R. Kazmi, Existence and convergence theorems for a class of multi-valued variational inclusions in Banach spaces. Nonlinear Anal. Theory Methods Appl. **59**(5), 649–656 (2004)

143. D. Chiriaev, G.W. Walster, Interval arithmetic specification, in *Available from World Wide Web:* http://www.mscs.mu.edu/~globsol/walster-papers.html.[69] (1998)
144. T.D. Chuong, Lipschitz-like property of an implicit multifunction and its applications. Nonlinear Anal. Theory Methods Appl. **74**(17), 6256–6264 (2011)
145. T.D. Chuong, D.S. Kim, Hölder-like property and metric regularity of a positive-order for implicit multifunctions. Math. Oper. Res. **41**(2), 596–611 (2016)
146. T.D. Chuong, A.Y. Kruger, J.-C. Yao, Calmness of efficient solution maps in parametric vector optimization. J. Glob. Optim. **51**(4), 677–688 (2011)
147. G. Colombo, V.V. Goncharov, B.S. Mordukhovich, Well-posedness of minimal time problems with constant dynamics in Banach spaces. Set-Valued Var. Anal. **18**(3–4), 349–372 (2010)
148. H.W. Corley, Games with vector payoffs. J. Optim. Theory Appl. **47**, 491–498 (1985)
149. J. Cotrina, A. Hantoute, A. Svensson, Existence of quasi-equilibria on unbounded constraint sets. Optimization **71**(2), 337–354 (2022)
150. J. Cotrina, A. Svensson, The finite intersection property for equilibrium problems. J. Glob. Optim. **79**, 941–957 (2021)
151. J. Cotrina, M. Théra, J. Zúñiga, An existence result for quasi-equilibrium problems via Ekeland's variational principle. J. Optim. Theory Appl. **187**(2), 336–355 (2020)
152. J. Cotrina, J. Zúñiga, A note on quasi-equilibrium problems. Oper. Res. Lett. **46**(1), 138–140 (2018)
153. J. Cotrina, J. Zúñiga, Quasi-equilibrium problems with non-self constraint map. J. Glob. Optim. **75**(1), 177–197 (2019)
154. G.P. Crespi, M. Dhingra, C.S. Lalitha, Pointwise and global well-posedness in set optimization: a direct approach. Ann. Oper. Res. **269**(1), 149–166 (2018)
155. G.P. Crespi, M. Papalia, M. Rocca, Extended well-posedness of vector optimization problems: the convex case. Taiwan. J. Math. **15**(4), 1545–1559 (2011)
156. G.P. Crespi, M. Papalia, M. Rocca, Extended well-posedness of quasiconvex vector optimization problems. J. Optim. Theory Appl. **141**(2), 285–297 (2009)
157. G. Crespi, A. Guerraggio, M. Rocca, Well posedness in vector optimization problems and vector variational inequalities. J. Optim. Theory Appl. **132**, 213–226 (2007)
158. P. Cubiotti, Existence of solutions for lower semicontinuous quasi-equilibrium problems. Comput. Math. Appl. **30**(12), 11–22 (1995)
159. P. Cubiotti, Existence of Nash equilibria for generalized games without upper semicontinuity. Int. J. Game Theory **26**(2), 267–273 (1997)
160. P. Cubiotti, J.-C. Yao, Nash equilibria of generalized games in normed spaces without upper semicontinuity. J. Glob. Optim. **46**(4), 509–519 (2010)
161. G. Dal Maso, *An Introduction to Γ-Convergence*, vol. 8 (Springer, 2012)
162. P. Daniele, F. Giannessi, A. Maugeri, *Equilibrium Problems and Variational Models*, vol. 68 (Springer, 2003)
163. N. von Daniels, Tikhonov regularization of control-constrained optimal control problems. Comput. Optim. Appl. **70**(1), 295–320 (2018)
164. A. Daniilidis, N. Hadjisavvas, Coercivity conditions and variational inequalities. Math. Program. **86**(2), 433–438 (1999)
165. A. Danilidis, N. Hadjisavvas, On generalized cyclically monotone operators and proper quasimonotonictty. Optimization **47**(1–2), 123–135 (2000)
166. M. Darabi, J. Zafarani, Tykhonov well-posedness for quasi-equilibrium problems. J. Optim. Theory Appl. **165**, 458–479 (2015)
167. M. De Luca, Generalized quasi-variational inequalities and traffic equilibrium problem, in *Variational Inequalities and Network Equilibrium Problems* (Springer, 1995), pp. 45–54
168. G. Debreu, A social equilibrium existence theorem. Proc. Natl. Acad. Sci. **38**(10), 886–893 (1952)
169. G. Debreu, *Theory of Value: An Axiomatic Analysis of Economic Equilibrium*, vol. 27 (Yale University Press, 1959)
170. C. Dellacherie, P.-A. Meyer, *Probabilities and Potential, C: Potential Theory for Discrete and Continuous Semigroups* (Elsevier, 2011)

171. D. Dentcheva, R. Henrion, A. Ruszczynński, Stability and sensitivity of optimization problems with first order stochastic dominance constraints. SIAM J. Optim. **18**(1), 322–337 (2007)
172. M. Dhingra, C.S. Lalitha, Well-setness and scalarization in set optimization. Optim. Lett. **10**(8), 1657–1667 (2016)
173. H.T.H. Diem, Consistency of statistical estimating of solutions to stochastic optimization problems. J. Glob. Optim. **83**, 825–842 (2022)
174. H.T.H. Diem, Approximations stability and optimality conditions in nonsmooth optimization. Ph.D. thesis, Vietnam National University-Ho Chi Minh City (2015)
175. H.T.H. Diem, P.Q. Khanh, Approximations of optimization-related problems in terms of variational convergence. Vietnam J. Math. **44**(2), 399–417 (2016)
176. H.T.H. Diem, P.Q. Khanh, Approximations of quasi-equilibria and Nash quasi-equilibria in terms of variational convergence. Set-Valued Var. Anal. **32**(1), 1 (2024)
177. H.T.H. Diem, P.Q. Khanh, Approximations of quasi-variational inequalities and traffic networks. A view from variational convergence. Under review
178. H.T.H. Diem, P.Q. Khanh, Criteria for epi/hypo convergence of finite-valued bifunctions. Vietnam J. Math. **43**(2), 439–458 (2015)
179. H.T.H. Diem, P.Q. Khanh, Epi/Hypo-convergence of bifunctions on general domains and approximations of quasi-variational problems. Set-Valued Var. Anal. **28**(3), 519–536 (2020)
180. H.T.H. Diem, P.Q. Khanh, Global approximations of vector optimization problems in terms of variational convergence. J. Optim. Theory Appl. **203**(1), 83–110 (2024)
181. H.T.H. Diem, P.Q. Khanh, On types of variational convergence of bifunctions for approximation in optimization. Under review
182. X.P. Ding, Sensitivity analysis for generalized nonlinear implicit quasi-variational inclusions. Appl. Math. Lett. **17**(2), 225–235 (2004)
183. A.L. Dontchev, R.T. Rockafellar, R.T. Rockafellar, *Implicit Functions and Solution Mappings: A View from Variational Analysis*, vol. 616 (Springer, 2009)
184. A.L. Dontchev, T. Zolezzi, *Well-Posed Optimization Problems* (Springer, 2006)
185. R.M. Dudley. *Real Analysis and Probability* (CRC Press, 2018)
186. J. Dupacová, R.J.-B. Wets, Asymptotic behavior of statistical estimators and of optimal solutions of stochastic optimization problems. Ann. Stat. **16**(4), 1517–1549 (1988)
187. T.Q. Duy, Robust efficiency and well-posedness in uncertain vector optimization problems. Optimization **72**(4), 937–955 (2023)
188. T.Q. Duy, Hadamard well-posedness for a set optimization problem with an infinite number of constraints. Optimization **73**(4), 1–19 (2023)
189. T.Q. Duy, Levitin–Polyak well-posedness in set optimization concerning Pareto efficiency. Positivity **25**(5), 1923–1942 (2021)
190. F. Facchinei, C. Kanzow, Beyond monotonicity in regularization methods for nonlinear complementarity problems. SIAM J. Control. Optim. **37**(4), 1150–1161 (1999)
191. F. Facchinei, J.-S. Pang, *Finite-Dimensional Variational Inequalities and Complementarity Problems*, vol. I (Springer, 2007)
192. F. Facchinei, J.-S. Pang. *Finite-Dimensional Variational Inequalities and Complementarity Problems*, vol. II (Springer, 2007)
193. M. Fakhar, J. Zafarani, Equilibrium problems in the quasimonotone case. J. Optim. Theory Appl. **126**(1), 125–136 (2005)
194. M. Fakhar, J. Zafarani, Generalized equilibrium problems for quasimonotone and pseudomonotone bifunctions. J. Optim. Theory Appl. **123**(2), 349–364 (2004)
195. K. Fan, A generalization of Tychonoff's fixed point theorem. Mathematische Annalen **142**(3), 305–310 (1961)
196. K. Fan, A minimax inequality and applications. Inequalities **3**, 103–113 (1972)
197. X. Fan, C. Cheng, H. Wang, Sensitivity analysis for vector equilibrium problems under functional perturbations. Numer. Funct. Anal. Optim. **35**(5), 564–575 (2014)
198. X. Fan, C. Cheng, H. Wang, Stability analysis for vector quasiequilibrium problems. Positivity **17**, 365–379 (2013)

199. Y.-P. Fang, N.-J. Huang, J.-C. Yao, Well-posedness of mixed variational inequalities, inclusion problems and fixed point problems. J. Glob. Optim. **41**(1), 117–133 (2008)
200. A. Farajzadeh, J. Zafarani, Equilibrium problems and variational inequalities in topological vector spaces. Optimization **59**(4), 485–499 (2010)
201. E.A. Feinberg, P.O. Kasyanov, J.O. Royset, Epi-convergence of expectation functions under varying measures and integrands. arXiv preprint arXiv:2208.03805 (2022)
202. F. Flores-Bazán, Existence theorems for generalized noncoercive equilibrium problems: the quasi-convex case. SIAM J. Optim. **11**(3), 675–690 (2001)
203. F. Flores-Bazán, Existence theory for finite-dimensional pseudomonotone equilibrium problems. Acta Applicandae Mathematica **77**(3), 249–297 (2003)
204. F. Flores-Bazán, E. Hernández, A unified vector optimization problem: complete scalarizations and applications. Optimization **60**(12), 1399–1419 (2011)
205. F. Flores-Bazán, E. Hernández, Optimality conditions for a unified vector optimization problem with not necessarily preordering relations. J. Glob. Optim. **56**(2), 299–315 (2013)
206. J. Foraker, J.O. Royset, I. Kaminer, Search-trajectory optimization: Part I, formulation and theory. J. Optim. Theory Appl. **169**(2), 530–549 (2016)
207. M.P. Friedlander, P. Tseng, Exact regularization of convex programs. SIAM J. Optim. **18**(4), 1326–1350 (2007)
208. J.-Y. Fu, Vector equilibrium problems. Existence theorems and convexity of solution set. J. Glob. Optim. **31**(1), 109–119 (2005)
209. F. Giannessi, Theorems of alternative, quadratic programs and complementarity problems. Var. Inequal. Complement. Probl. **1**, 151–86 (1980)
210. F. Giannessi, On Minty variational principle, in *New Trends in Mathematical Programming* (Springer, 1998), pp. 93–99
211. F. Giannessi, A. Maugeri, *Variational Inequalities and Network Equilibrium Problems* (Springer, 1995)
212. F. Giannessi, A. Maugeri, P.M. Pardalos, *Equilibrium Problems: Nonsmooth Optimization and Variational Inequality Models*, vol. 58 (Springer, 2006)
213. I.L. Glicksberg, A further generalization of the Kakutani fixed theorem, with application to Nash equilibrium points. Proc. Am. Math. Soc. **3**(1), 170–174 (1952)
214. X. Gong, Continuity of the solution set to parametric weak vector equilibrium problems. J. Optim. Theory Appl. **139**(1), 35–46 (2008)
215. X. Gong, Efficiency and Henig efficiency for vector equilibrium problems. J. Optim. Theory Appl. **108**(1), 139–154(2001)
216. X.-H. Gong, Optimality conditions for Henig and globally proper efficient solutions with ordering cone has empty interior. J. Math. Anal. Appl. **307**(1), 12–31 (2005)
217. X.-H. Gong, Scalarization and optimality conditions for vector equilibrium problems. Nonlinear Anal. Theory Methods Appl. **73**(11), 3598–3612 (2010)
218. A. Göpfert et al., *Variational Methods in Partially Ordered Spaces*, vol. 17 (Springer, 2003)
219. G. Gürkan, J.-S. Pang, Approximations of Nash equilibria. Math Program **117**(1), 223–253 (2009)
220. C. Gutiérrez et al., Ekeland variational principles in vector equilibrium problems. SIAM J. Optim. **27**(4), 2405–2425 (2017)
221. C. Gutiérrez et al., Pointwise well-posedness in set optimization with cone proper sets. Nonlinear Anal. Theory Methods Appl. **75**(4), 1822–1833 (2012)
222. C. Gutiérrez, B. Jiménez, V. Novo, Improvement sets and vector optimization. Eur. J. Oper. Res. **223**(2), 304–311 (2012)
223. C. Gutiérrez, V. Novo, J.L. Ródenas-Pedregosa, A note on existence of weak efficient solutions for vector equilibrium problems. Optim. Lett. **12**(3), 615–623 (2018)
224. C. Gutiérrez et al., Approximate solutions of vector optimization problems via improvement sets in real linear spaces. J. Glob. Optim. **70**(4), 875–901 (2018)
225. C. Gutiérrez et al., Convergence of solutions of a set optimization problem in the image space. J. Optim. Theory Appl. **170**(2), 358–371 (2016)

226. C. Gutiérrez et al., Nonconvex separation functional in linear spaces with applications to vector equilibria. SIAM J. Optim. **26**(4), 2677–2695 (2016)
227. C. Gutiérrez et al., Scalarization in set optimization with solid and nonsolid ordering cones. J. Glob. Optim. **61**(3), 525–552 (2015)
228. J. Gwinner, On the regularization of monotone variational inequalities. Ops Res. Verf. **28**, 374–386 (1978)
229. N.H. Ha, D.V. Hieu, L.D. Muu, An accelerated regularization method for variational inclusions. J. Appl. Numer. Optim. **6**(1), 97–113 (2024)
230. T.X.D. Ha, A Hausdorff-type distance, a directional derivative of a set-valued map and applications in set optimization. Optimization **67**(7), 1031–1050 (2018)
231. T.X.D. Ha, A unified scheme for scalarization in set optimization. Optimization **73**(5), 1645–1679 (2024)
232. J. Hadamard, Sur les problemes aux derive espartielles et leur signification physique. Bull. Princeton University **13**, 1–20 (1902)
233. N. Hadjisavvas, S. Schaible, From scalar to vector equilibrium problems in the quasimonotone case. J. Optim. Theory Appl. **96**(2), 297–309 (1998)
234. N. Hadjisavvas, J.E. Martinez-Legaz, J.-P. Penot, Generalized convexity and generalized monotonicity, in *Proceedings of the 6th International Symposium*, Samos, Greece, Sept. 1999
235. N. Hadjisavvas, S. Schaible, Quasimonotonicity and pseudomonotonicity in variational inequalities and equilibrium problems, in *Generalized Convexity, Generalized Monotonicity: Recent Results* (Springer, 1998), pp. 257–275
236. N.X. Hai, P.Q. Khanh, The solution existence of general variational inclusion problems. J. Math. Anal. Appl. **328**(2), 1268–1277 (2007)
237. N.X. Hai, P.Q. Khanh, N.H. Quan, Some existence theorems in nonlinear analysis for mappings on GFC-spaces and applications. Nonlinear Anal. Theory Methods Appl. **71**(12), 6170–6181 (2009)
238. N.X. Hai, N.H. Quan, V.V. Tri, Some saddle-point theorems for vectorvalued functions. J. Glob. Optim. **86**(1), 141–161 (2023)
239. A.H. Hamel, Equivalents to Ekeland's variational principle in uniform spaces. Nonlinear Anal. Theory Methods Appl. **62**(5), 913–924 (2005)
240. Y. Han, Lipschitz continuity of approximate solution mappings to parametric generalized vector equilibrium problems. J. Optim. Theory Appl. **178**(3), 763–793 (2018)
241. Y. Han, N.J. Huang, Continuity and convexity of a nonlinear scalarizing function in set optimization problems with applications. J. Optim. Theory Appl. **177**(3), 679–695 (2018)
242. Y. Han, N.J. Huang, Well-posedness and stability of solutions for set optimization problems. Optimization **66**(1), 17–33 (2017)
243. E. Hernández, R. López, On epi-convergence for set-valued maps. Optimization **71**(2), 403–417 (2022)
244. E. Hernández, L. Rodríguez-Marín, Nonconvex scalarization in set optimization with set-valued maps. J. Math. Anal. Appl. **325**(1), 1–18 (2007)
245. E. Hernández, L. Rodríguez-Marín, M. Sama, On solutions of set-valued optimization problems. Comput. Math. Appl. **60**(5), 1401–1408 (2010)
246. D.V. Hieu, L.D. Muu, P.K. Quy, New iterative regularization methods for solving split variational inclusion problems. J. Ind. Manage. Optim. **19**(1), (2023)
247. D.V. Hieu et al., Regularization extragradient methods for equilibrium programming in Hilbert spaces. Optimization **71**(9), 2643–2673 (2022)
248. J.-B. Hiriart-Urruty, Lipschitz r-continuity of the approximate subdifferential of a convex function. Mathematica Scandinavica **47**(1), 123–134 (1980)
249. J.-B. Hiriart-Urruty, Tangent cones, generalized gradients and mathematical programming in Banach spaces. Math. Oper. Res. **4**(1), 79–97 (1979)
250. R.B. Holmes. *Geometric Functional Analysis and Its Applications*, vol. 24 (Springer, 2012)

251. S. Al-Homidan, Q.H. Ansari, J.-C. Yao, Some generalizations of Ekelandtype variational principle with applications to equilibrium problems and fixed point theory. Nonlinear Anal. Theory Methods Appl. **69**(1), 126–139 (2008)
252. S. Al-Homidan, Q.H. Ansari, G. Kassay, Vectorial form of Ekeland variational principle with applications to vector equilibrium problems. Optimization **69**(3), 415–436 (2020)
253. S. Hu, N. Papageorgiou, *Handbook of Multivalued Analysis, Volume I: Theory* (Kluwer, 1997)
254. N.J. Huang, J. Li, H. Thompson, Stability for parametric implicit vector equilibrium problems. Math. Comput. Model. **43**(11–12), 1267–1274 (2006)
255. P.G. Hung, L.D. Muu, The Tikhonov regularization extended to equilibrium problems involving pseudomonotone bifunctions. Nonlinear Anal. Theory Methods Appl. **74**(17), 6121–6129 (2011)
256. D. Inoan, J. Kolumbán, Existence theorems via duality for equilibrium problems with trifunctions. Optimization **67**(5), 537–547 (2018)
257. D. Inoan, Calmness of the solution mapping of parametric variational relation problems. Filomat **35**(10), 3541–3548 (2021)
258. A. Ioffe, R. Lucchetti, Typical convex program is very well posed. Math. Program. **104**(2), 483–499 (2005)
259. A.D. Ioffe, R. Lucchetti, J.P. Revalski, Almost every convex or quadratic programming problem is well posed. Math. Oper. Res. **29**(2), 369–382 (2004)
260. A.D. Ioffe, J.V. Outrata, On metric and calmness qualification conditions in subdifferential calculus. Set-Valued Anal. **16**(2), 199–227 (2008)
261. A.N. Iusem, G. Kassay, W. Sosa, On certain conditions for the existence of solutions of equilibrium problems. Math. Program. **116**(1–2), 259–273 (2009)
262. A.N. Iusem, W. Sosa, New existence results for equilibrium problems. Nonlinear Anal. Theory Methods Appl. **52**(2), 621–635 (2003)
263. S. Jafari et al., Existence results for-quasimonotone equilibrium problems in convex metric spaces. Optimization **66**(3), 293–310 (2017)
264. J. Jahn, Karush–Kuhn–Tucker conditions in set optimization. J. Optim. Theory Appl. **172**(3), 707–725 (2017)
265. J. Jahn. *Vector Optimization* (Springer, 2009)
266. J. Jahn, T.X.D. Ha, New order relations in set optimization. J. Optim. Theory Appl. **148**(2), 209–236 (2011)
267. B. Jiménez, V. Novo, A. Vílchez, Six set scalarizations based on the oriented distance: properties and application to set optimization. Optimization **69**(3), 437–470 (2020)
268. A. Jofré, R.J.-B. Wets, Continuity properties of Walras equilibrium points. Ann. Oper. Res. **114**(1), 229–243 (2002)
269. A. Jofré, R.J.-B. Wets, Variational convergence of bifunctions: motivating applications. SIAM J. Optim. **24**(4), 1952–1979 (2014)
270. A. Jofré, R.J.-B. Wets, Variational convergence of bivariate functions: lopsided convergence. Math. Program. **116**(1), 275–295 (2009)
271. O. Kada, T. Suzuki, W. Takahashi, Nonconvex minimization theorems and fixed point theorems in complete metric spaces. Mathematica Japonicae **44**(2), 381–391 (1996)
272. J.B. Kadane, C. Srinivasan, Bayes decision problems and stability. Sankhya¯: Indian J. Stat. Ser. A **60**(3), 383–404 (1998)
273. S. Kakutani et al., A generalization of Brouwer's fixed point theorem. Duke Math. J. **8**(3), 457–459 (1941)
274. P. Kall, Approximation to optimization problems: an elementary review. Math. Oper. Res. **11**(1), 9–18 (1986)
275. G. Kassay, M. Miholca, Existence results for vector equilibrium problems given by a sum of two functions. J. Glob. Optim. **63**(1), 195–211 (2015)
276. G. Kassay, V. Radulescu. *Equilibrium Problems and Applications* (Academic Press, 2018)

277. K.R. Kazmi, M. Bhat, Convergence and stability of iterative algorithms of generalized set-valued variational-like inclusions in Banach spaces. Appl. Math. Comput. **166**(1), 164–180 (2005)
278. A. Khaliq, S. Krishan, Vector quasi-equilibrium problems. Bull. Aust. Math. Soc. **68**(2), 295–302 (2003)
279. A.A. Khan, C. Tammer, C. Zălinescu, Regularization of quasi-variational inequalities. Optimization **64**(8), 1703–1724 (2015)
280. A.A. Khan, C. Tammer, C. Zălinescu. *Set-Valued Optimization* (Springer, 2015)
281. P.Q. Khanh, N.H. Quan, Existence results for nonconvex vector equilibrium problems with general preferences. Under review
282. P.Q. Khanh, Optimality conditions via norm scalarization in vector optimization. SIAM J. Control. Optim. **31**(3), 646–658 (1993)
283. P.Q. Khanh, D.T. Luc, Stability of solutions in parametric variational relation problems. Set-Valued Anal. **16**, 1015–1035 (2008)
284. P.Q. Khanh, L.M. Luu, On the existence of solutions to vector quasivariational inequalities and quasicomplementarity problems with applications break to traffic network equilibria. J. Optim. Theory Appl. **123**(3), 533–548 (2004)
285. P.Q. Khanh, L.M. Luu, Some existence results for vector quasivariational inequalities involving multifunctions and applications to traffic equilibrium problems. J. Glob. Optim. **32**(4), 551–568 (2005)
286. P.Q. Khanh, L.M. Luu, T.T.M. Son, Well-posedness of a parametric traffic network problem. Nonlinear Anal. Real World Appl. **14**(3), 1643–1654 (2013)
287. P.Q. Khanh, N.H. Quan, A topological characterization of the existence of fixed points and applications. Mathematische Nachrichten **287**(2–3), 281–289 (2014)
288. P.Q. Khanh, N.H. Quan, Existence results for general inclusions using generalized KKM theorems with applications to minimax problems. J. Optim. Theory Appl. **146**, 640–653 (2010)
289. P.Q. Khanh, N.H. Quan, Intersection theorems, coincidence theorems and maximal-element theorems in GFC-spaces. Optimization **59**(1), 115–124 (2010)
290. P.Q. Khanh, N.H. Quan, Topologically-based characterizations of the existence of solutions of optimization-related problems. Mathematische Nachrichten **288**(17–18), 2057–2078 (2015)
291. P.Q. Khanh, N.H. Quan, Versions of the Weierstrass theorem for bifunctions and solution existence in optimization. SIAM J. Optim. **29**(2), 1502–1523 (2019)
292. P.Q. Khanh, N.H. Quan, J.-C. Yao, Generalized KKM-type theorems in GFC-spaces and applications. Nonlinear Anal. Theory Methods Appl. **71**(3–4), 1227–1234 (2009)
293. S. Khoshkhabar-Amiranloo, Stability of minimal solutions to parametric set optimization problems. Appl. Anal. **97**(14), 2510–2522 (2018)
294. S. Khoshkhabar-Amiranloo, E. Khorram, Pointwise well-posedness and scalarization in set optimization. Math. Meth. Oper. Res. **82**(2), 195–210 (2015)
295. K. Kimura, J.-C. Yao, Semicontinuity of solution mappings of parametric generalized vector equilibrium problems. J. Optim. Theory Appl. **138**, 429–443 (2008)
296. K. Kimura, J.-C. Yao, Sensitivity analysis of solution mappings of parametric vector quasi-equilibrium problems. J. Glob. Optim. **41**(2), 187–202 (2008)
297. K. Kimura et al., Well-posedness for parametric vector equilibrium problems with applications. J. Ind. Manage. Optim. **4**(2), 313 (2008)
298. D. Kinderlehrer, G. Stampacchia, *An Introduction to Variational Inequalities and Their Applications* (SIAM, 2000)
299. J. Kindler, Topological intersection theorems. Proc. Am. Math. Soc. **117**(4), 1003–1011 (1993)
300. A.J. King, R.J.-B. Wets, Epi-consistency of convex stochastic programs. Stochastics Stochastics Rep. **34**(1–2), 83–92 (1991)
301. B. Knaster, C. Kuratowski, S. Mazurkiewicz, Ein Beweis des Fixpunktsatzes für n-dimensionale Simplexe. Fundamenta Mathematicae **14**(1), 132–137 (1929)

302. H. König, A general minimax theorem based on connectedness. Archiv der Mathematik **59**(1), 55–64 (1992)
303. I.V. Konnov, *Combined Relaxation Methods for Variational Inequalities*, vol. 495 (Springer, 2001)
304. I.V. Konnov, *Equilibrium Models and Variational Inequalities* (Elsevier, 2007)
305. I.V. Konnov, Equilibrium formulations of relative optimization problems. Math. Meth. Oper. Res. **90**(1), 137–152 (2019)
306. I.V. Konnov, On penalty methods for non monotone equilibrium problems. J. Glob. Optim. **59**(1), 131–138 (2014)
307. I.V. Konnov, Regularization method for nonmonotone equilibrium problems. J. Nonlinear Convex Anal. **10**, 93–101 (2009)
308. I.V. Konnov, Regularized penalty method for general equilibrium problems in Banach spaces. J. Optim. Theory Appl. **164**(2), 500–513 (2015)
309. I.V. Konnov, D.A. Dyabilkin, Nonmonotone equilibrium problems: coercivity conditions and weak regularization. J. Glob. Optim. **49**(4), 575–587 (2011)
310. I.V. Konnov, S. Schaible, Duality for equilibrium problems under generalized monotonicity. J. Optim. Theory Appl. **104**(2), 395–408 (2000)
311. A.S. Konsulova, J.P. Revalski, Constrained convex optimization problems-well-posedness and stability. Numer. Funct. Anal. Optim. **15**(7–8), 889–907 (1994)
312. A. Kristály, C. Varga, Set-valued versions of Ky Fan's inequality with application to variational inclusion theory. J. Math. Anal. Appl. **282**(1), 8–20 (2003)
313. K. Kuratowski, *Topology*, vol. 2 (Academic Press, 1968), 511 pp.
314. D. Kuroiwa, The natural criteria in set-valued optimization (Nonlinear analysis and convex analysis). RIMS Kokyuroku Kyoto Univ. **1031**, 85–90 (1998)
315. C.S. Lalitha, M. Mehta, Vector variational inequalities with cone-pseudomonotone bifunctions. Optimization **54**(3), 327–338 (2005)
316. M. Lassonde, On the use of KKM multifunctions in fixed point theory and related topics. J. Math. Anal. Appl. **97**(1), 151–201 (1983)
317. S. László, Vector equilibrium problems on dense sets. J. Optim. Theory Appl. **170**(2), 437–457 (2016)
318. B. Lemaire, C.O.A. Salem, J.P. Revalski, Well-posedness of variational problems with applications to staircase methods. Comptes Rendus de l'Académie des Sciences-Series I-Mathematics **332**(10), 943–948 (2001)
319. E.S. Levitin, B.T. Polyak, Convergence of minimizing sequences in conditional extremum problems, in *Doklady Akademii Nauk*, vol. 168 (Russian Academy of Sciences, 1966), pp. 997–1000
320. A.B. Levy, Calm minima in parameterized finite-dimensional optimization. SIAM J. Optim. **11**(1), 160–178 (2000)
321. G.H. Li, S.J. Li, M.X. You, Relationships between the oriented distance functional and a nonlinear separation functional. J. Math. Anal. Appl. **466**(1), 1109–1117 (2018)
322. S.J. Li, G. Chen, K. Teo, On the stability of generalized vector quasivariational inequality problems. J. Optim. Theory Appl. **113**(2), 283–295 (2002)
323. S.J. Li et al., Hölder continuity and upper estimates of solutions to vector quasiequilibrium problems. Eur. J. Oper. Res. **210**(2), 148–157 (2011)
324. S.J. Li et al., The Hölder continuity of solutions to generalized vector equilibrium problems. Eur. J. Oper. Res. **199**(2), 334–338 (2009)
325. S. Li, Z. Fang, Lower semicontinuity of the solution mappings to a parametric generalized Ky Fan inequality. J. Optim. Theory Appl. **147**(3), 507–515 (2010)
326. X.-B. Li, S. Li, Continuity of approximate solution mappings for parametric equilibrium problems. J. Glob. Optim. **51**(3), 541–548 (2011)
327. M.B. Lignola, J. Morgan, α-Well-posedness for Nash equilibria and for optimization problems with Nash equilibrium constraints. J. Glob. Optim. **36**, 439–459 (2006)
328. M.B. Lignola, J. Morgan, Well-posedness for optimization problems with constraints defined by variational inequalities having a unique solution. J. Glob. Optim. **16**, 57–67 (2000)

329. M.B. Lignola, Well-posedness and L-well-posedness for quasivariational inequalities. J. Optim. Theory Appl. **128**, 119–138 (2006)
330. M.B. Lignola, J. Morgan, Epiconvergence and continuous convergence. Atti Sem. Mat. Fis. Univ. Modena **42**, 467–475 (1994)
331. M.B. Lignola, J. Morgan, Generalized variational inequalities with pseudomonotone operators under perturbations. J. Optim. Theory Appl. **101**, 213–220 (1999)
332. L.-J. Lin, A KKM type theorem and its applications. Bull. Aust. Math. Soc. **59**(3), 481–494 (1999)
333. L.-J. Lin, Q.H. Ansari, Systems of quasi-variational relations with applications. Nonlinear Anal. Theory Methods Appl. **72**(3–4), 1210–1220 (2010)
334. L.-J. Lin, M. Balaj, Y.-C. Ye, Quasi-variational relation problems and generalized Ekeland's variational principle with applications. Optimization **63**(9), 1353–1365 (2014)
335. L.-J. Lin, H.-L. Chen, The study of KKM theorems with applications to vector equilibrium problems and implicit vector variational inequalities problems. J. Glob. Optim. **32**(1), 135–157 (2005)
336. L.-J. Lin, C.-S. Chuang, Well-posedness in the generalized sense for variational inclusion and disclusion problems and well-posedness for optimization problems with constraint. Nonlinear Anal. Theory Methods Appl. **70**(10), 3609–3617 (2009)
337. L.-J. Lin, W.-S. Du, Ekeland's variational principle, minimax theorems and existence of nonconvex equilibria in complete metric spaces. J. Math. Anal. Appl. **323**(1), 360–370 (2006)
338. L.-J. Lin, W.-S. Du, On maximal element theorems, variants of Ekeland's variational principle and their applications. Nonlinear Anal. Theory Methods Appl. **68**(5), 1246–1262 (2008)
339. L.-J. Lin, N.X. Tan, On quasivariational inclusion problems of type I and related problems. J. Glob. Optim. **39**(3), 393–407 (2007)
340. L.-J. Lin, S.-Y. Wang, Simultaneous variational relation problems and related applications. Comput. Math. Appl. **58**(9), 1711–1721 (2009)
341. L.-J. Lin, Z.-T. Yu, Fixed-point theorems and equilibrium problems. Nonlinear Anal. Theory Methods Appl. **43**(8), 987–999 (2001)
342. Z. Lin, The study of traffic equilibrium problems with capacity constraints of arcs. Nonlinear Anal. Real World Appl. **11**(4), 2280–2284 (2010)
343. A. Löhne, *Vector Optimization with Infimum and Supremum* (Springer, 2011)
344. X.-J. Long, N.-J, Huang, Metric characterizations of α-well-posedness for symmetric quasi-equilibrium problems. J. Glob. Optim. **45**, 459–471 (2009)
345. H. Longguang, Existence of solutions on weak vector equilibrium problems. Nonlinear Anal. Theory Methods Appl. **65**(4), 795–801 (2006)
346. R. López, Approximations of equilibrium problems. SIAM J. Control. Optim. **50**(2), 1038–1070 (2012)
347. R. López, Variational convergence for vector-valued functions and its applications to convex multiobjective optimization. Math. Meth. Oper. Res. **78**(1), 1–34 (2013)
348. R. López, C. Vera, On the set of weakly efficient minimizers for convex multiobjective programming. Oper. Res. Lett. **36**(5), 651–655 (2008)
349. P. Loridan, ε-solutions in vector minimization problems. J. Optim. Theory Appl. **43**, 265–276 (1984)
350. D.T. Luc, Existence results for densely pseudomonotone variational inequalities. J. Math. Anal. Appl. **254**(1), 291–308 (2001)
351. D.T. Luc, *Theory of Vector Optimization* (Springer, 1989)
352. D.T. Luc, E. Sarabi, A. Soubeyran, Existence of solutions in variational relation problems without convexity. J. Math. Anal. Appl. **364**(2), 544–555 (2010)
353. D. Luc, An abstract problem in variational analysis. J. Optim. Theory Appl. **138**, 65–76 (2008)
354. R. Lucchetti, F. Patrone, A characterization of Tyhonov well-posedness for minimum problems, with applications to variational inequalities. Numer. Funct. Anal. Optim. **3**(4), 461–476 (1981)
355. R. Lucchetti, F. Patrone, Hadamard and Tyhonov well-posedness of a certain class of convex functions. J. Math. Anal. Appl. **88**(1), 204–215 (1982)

356. R. Lucchetti, R.J. Wets, Convergence of minima of integral functionals, with applications to optimal control and stochastic optimization. Stat. Risk Model. **11**(1), 69–84 (1993)
357. X.-P. Luo, Y.-B. Xiao, W. Li, Strict feasibility of variational inclusion problems in reflexive Banach spaces. J. Ind. Manag. Optim. **16**(5), 2495–2502 (2020)
358. V.L. Makarov et al., *Mathematical Economic Theory: Pure and Mixed Types of Economic Mechanisms*, vol. 33 (Elsevier, 1995)
359. O.L. Mangasarian, The ill-posed linear complementarity problem, in *Complementarity and Variational Problems: State of the Art*, ed. by M.C. Ferris, J.S. Pang (SIAM Publications, 1997), pp. 226–233
360. M.A. Mansour, H. Riahi, Sensitivity analysis for abstract equilibrium problems. J. Math. Anal. Appl. **306**(2), 684–691 (2005)
361. M.A. Mansour, L. Scrimali, Hölder continuity of solutions to elastic traffic network models. J. Glob. Optim. **40**, 175–184 (2008)
362. M. Margiocco, F. Patrone, L. Pusillo Chicco, A new approach to Tikhonov well-posedness for nash equilibria: n-person games. Optimization **40**(4), 385–400 (1997)
363. G. Mastroeni, On auxiliary principle for equilibrium problems, in *Equilibrium Problems and Variational Models* (Springer, 2003), pp. 289–298
364. A. Maugeri, Variational and quasi-variational inequalities in network flow models. Recent developments in theory and algorithms, in *Variational Inequalities and Network Equilibrium Problems* (Springer, 1995), pp. 195–211
365. E. Michael, Continuous selections. I. Ann. Math. **63**(2), 361–382 (1956)
366. G.J. Minty, On the generalization of a direct method of the calculus of variations. Bull. Ame. Math. Soc. **73**(3), 315–321 (1967)
367. B.S. Mordukhovich, Multiobjective optimization problems with equilibrium con-straints. Math. Program. **117**(1), 331–354 (2009)
368. B.S. Mordukhovich, *Variational Analysis and Generalized Differentiation I: Basic Theory*, vol. 330 (Springer, 2006)
369. B.S. Mordukhovich, *Variational Analysis and Generalized Differentiation II: Applications*, vol. 331 (Springer, 2006)
370. B.S. Mordukhovich, J.V. Outrata, Coderivative analysis of quasi-variational inequalities with applications to stability and optimization. SIAM J. Optim. **18**(2), 389–412 (2007)
371. B.S. Mordukhovich, J.V. Outrata, M. Červinka, Equilibrium problems with complementarity constraints: case study with applications to oligopolistic markets. Optimization **56**(4), 479–494 (2007)
372. J. Morgan, V. Scalzo, Discontinuous but well-posed optimization problems. SIAM J. Optim. **17**(3), 861–870 (2006)
373. J. Morgan, V. Scalzo, Pseudocontinuity in optimization and nonzero-sum games. J. Optim. Theory Appl. **120**, 181–197 (2004)
374. J. Morgan, V. Scalzo, Variational stability of social Nash equilibria. Int. Game Theory Rev. **10**(01), 17–24 (2008)
375. U. Mosco, Convergence of convex sets and of solutions of variational inequalities. Adv. Math. **3**(4), 510–585 (1969)
376. U. Mosco, Implicit variational problems and quasi variational inequalities, in *Nonlinear Operators and the Calculus of Variations* (Springer, 1976), pp. 83–156
377. L.D. Muu, V. Nguyen, N. Quy, On Nash–Cournot oligopolistic market equilibrium models with concave cost functions. J. Glob. Optim. **41**(3), 351–364 (2008)
378. L.D. Muu, W. Oettli, Convergence of an adaptive penalty scheme for finding constrained equilibria. Nonlinear Anal. Theory Methods Appl. **18**(12), 1159–1166 (1992)
379. L.D. Muu, Stability property of a class of variational inequalities. Optimization **15**(3), 347–351 (1984)
380. M. Nasri, W. Sosa, Equilibrium problems and generalized Nash games. Optimization **60**(8–9), 1161–1170 (2011)
381. H. Nikaido, K. Isoda, Note on non-cooperative convex game. Pac. J. Math. Ind. **5**(5), 807–815 (1955)

382. Z. Nishnianidze, Fixed points of monotonic multiple-valued operators. Bull. Georgian Acad. Sci **114**, 489–491 (1984)
383. M.A. Noor, Auxiliary principle technique for equilibrium problems. J. Optim. Theory Appl. **122**(2), 371–386 (2004)
384. P. Oliveira, P. Santos, A. Silva, A Tikhonov-type regularization for equilibrium problems in Hilbert spaces. J. Math. Anal. Appl. **401**(1), 336–342 (2013)
385. S. Park, Fixed points and quasi-equilibrium problems. Math. Comput. Model. **32**(11–13), 1297–1304 (2000)
386. S. Park, H. Kim, Foundations of the KKM theory on generalized convex spaces. J. Math. Anal. Appl. **209**(2), 551–571 (1997)
387. T. Pennanen, Epi-convergent discretizations of multistage stochastic programs via integration quadratures. Math. Program. **116**(1), 461–479 (2009)
388. C. Phelps, J.O. Royset, Q. Gong, Optimal control of uncertain systems using sample average approximations. SIAM J. Control Optim. **54**(1), 1–29 (2016)
389. E. Polak, *Optimization: Algorithms and Consistent Approximations*, vol. 124 (Springer, 2012)
390. P. Preechasilp, R. Wangkeeree, A note on semicontinuity of the solution mapping for parametric set optimization problems. Optim. Lett. **13**(5), 1085–1094 (2019)
391. J. Revalski, N. Zhivkov, Well-posed constrained optimization problems in metric spaces. J. Optim. Theory Appl. **76**(1), 145–163 (1993)
392. J.P. Revalski, Hadamard and strong well-posedness for convex programs. SIAM J. Optim. **7**(2), 519–526 (1997)
393. R.T. Rockafellar, First-and second-order epi-differentiability in nonlinear programming. Trans. Am. Math. Soc. **307**(1), 75–108 (1988)
394. R.T. Rockafellar, Integral functionals, normal integrands and measurable selections, in *Nonlinear Operators and the Calculus of Variations* (Springer, 1976), pp. 157–207
395. R.T. Rockafellar, J.O. Royset, Random variables, monotone relations, and convex analysis. Math. Program. **148**(1), 297–331 (2014)
396. R.T. Rockafellar, R.J.-B. Wets, *Variational Analysis*, vol. 317 (Springer, 2009)
397. R.T. Rockafellar, R.J.-B. Wets, Variational systems, an introduction, in *Multifunctions and Integrands* (Springer, 1984), pp. 1–54
398. W. Rödder, A generalized saddlepoint theory: Its application to duality theory for linear vector optimum problems. Eur. J. Oper. Res. **1**(1), 55–59 (1977)
399. J.O. Royset, Approximations and solution estimates in optimization. Math. Program. **170**, 479–506 (2018)
400. J.O. Royset, Optimality functions in stochastic programming. Math. Program. **135**(1), 293–321 (2012)
401. J.O. Royset, Set-convergence and its application: a tutorial. Set-Valued Var. Anal. **28**(4), 707–732 (2020)
402. J.O. Royset, E. Pee, Rate of convergence analysis of discretization and smoothing algorithms for semiinfinite minimax problems. J. Optim. Theory Appl. **155**(3), 855–882 (2012)
403. J.O. Royset, R.J.-B. Wets, *An Optimization Primer* (Springer, 2021)
404. J.O. Royset, R.J.-B. Wets, Lopsided convergence: an extension and its quantification. Math. Program. **177**(1), 395–423 (2019)
405. J.O. Royset, R.J.-B. Wets, Multivariate epi-splines and evolving function identification problems. Set-Valued Var. Anal. **24**(4), 517–545 (2016)
406. J.O. Royset, R.J.-B. Wets, Optimality functions and lopsided convergence. J. Optim. Theory Appl. **169**(3), 965–983 (2016)
407. J.O. Royset, R.J.-B. Wets, Variational theory for optimization under stochas-tic ambiguity. SIAM J. Optim. **27**(2), 1118–1149 (2017)
408. J.O. Royset, R.J. Wets, Variational analysis of constrained M-estimators. Ann. Stat. **48**(5), 2759–2790 (2020)

409. A.M. Rubinov, R.N. Gasimov, Scalarization and nonlinear scalar duality for vector optimization with preferences that are not necessarily a pre-order relation. J. Glob. Optim. **29**(4), 455–477 (2004)
410. W. Rudin, *Functional Analysis* (McGrawHill, 1991)
411. P.H. Sach et al., Sensitivity in mixed generalized vector quasiequilibrium problems with moving cones. Nonlinear Anal. Theory Methods Appl. **73**(3), 713–724 (2010)
412. P.H. Sach, Solution existence in bifunction-set optimization. J. Optim. Theory Appl. **176**(1), 1–16 (2018)
413. P.H. Sach, L.J. Lin, L.A. Tuan, Generalized vector quasivariational inclusion problems with moving cones. J. Optim. Theory Appl. **147**, 607–620 (2010)
414. P.H. Sach, L.A. Tuan, New scalarizing approach to the stability analysis in parametric generalized Ky Fan inequality problems. J. Optim. Theory Appl. **157**(2), 347–364 (2013)
415. Salahuddin, Parametric generalized set-valued variational inclusions and resolvent equations. J. Math. Anal. Appl. **298**(1), 146–156 (2004)
416. G. Salinetti, Stability of Bayesian decisions. J. Stat. Plann. Inference **40**(2–3), 313–329 (1994)
417. O. Scherzer et al., *Variational Methods in Imaging* (Springer, 2009)
418. A. Shapiro, Consistency of sample estimates of risk averse stochastic programs. J. Appl. Probab. **50**(2), 533–541 (2013)
419. L.S. Shapley, F.D. Rigby, Equilibrium points in games with vector payoffs. Nav. Res. Logist. Q. **6**(1), 57–61 (1959)
420. M.-H. Shih, K.-K. Tan, Browder-Hartman-Stampacchia variational inequalities for multivalued monotone operators. J. Math. Anal. Appl. **134**(2), 431–440 (1988)
421. M. Sion, On general minimax theorems. Pac. J. Math. **8**(1), 171–176 (1958)
422. M.J. Smith, The existence, uniqueness and stability of traffic equilibria. Transp. Res. B Methodol. **13**(4), 295–304 (1979)
423. W. Takahashi, Existence theorems generalizing fixed point theorems for multival-ued mappings, in *Fixed Point Theory and Applications*, Vol. **252** (Wiley, 1991), pp. 397–406
424. C. Tammer, P. Weidner, *Scalarization and Separation by Translation Invariant Functions* (Springer, 2020)
425. C. Tammer, C. Zălinescu, Lipschitz properties of the scalarization function and applications. Optimization **59**(2), 305–319 (2010)
426. N.X. Tan, Random quasi-variational inequality. Mathematische Nachrichten **125**(1), 319–328 (1986)
427. G. Tian, On the existence of price equilibrium in economies with excess demand functions. Econ. Theory Bull. **4**, 5–16 (2016)
428. A.N. Tikhonov, On the solution of ill-posed problems and regularization method. Dokl. Akad. Nauk SSSR **152**, 501–504 (1963)
429. A.N. Tikhonov, On the stability of the functional optimization problem. USSR Comput. Math. Math. Phys. **6**(4), 28–33 (1966)
430. P. Tseng, Error bounds for regularized complementarity problems, in *Ill-posed Variational Problems and Regularization Techniques*, ed. by R. Théra, M. Tichatschke (Springer, 1999), pp. 247–274
431. H. Tuy, Pivotal methods for computing equilibrium points: unified approach and new restart algorithm. Math. Program. **16**(1), 210–227 (1979)
432. H. Tuy, Concave programming with linear constraints, in *Doklady Akademii Nauk*, vol. 159 (Russian Academy of Sciences, 1964), pp. 32–35
433. H. Tuy, Sur quelques propriétés des réseaux de transport et leurs applications. Bulletin de L'Académie Polonaise Des Sciences: Série des Sciences Mathématiques, Astronomiques, et Physiques **12**(7), 412–418 (1964)
434. B.V. Dinh, L.D. Muu, On penalty and gap function methods for bilevel equilibrium problems. J. Appl. Math **2011**(1), 646452 (2011)
435. D.V. Hieu, P.K. Anh, L.D. Muu, Strong convergence of subgradient extra-gradient method with regularization for solving variational inequalities. Optim. Eng. **22**, 2575–2602 (2021)

436. D.V. Hieu, J.J. Strodiot, et al., Strongly convergent algorithms by using new adaptive regularization parameter for equilibrium problems. J. Comput. Appl. Math. **376**, 112844 (2020)
437. D.V. Hieu et al., Explicit extragradient-like method with regularization for variational inequalities. Results Math. **74**(4), 137 (2019)
438. D.V. Hieu et al., Iterative regularization methods with new stepsize rules for solving variational inclusions. J. Appl. Math. Comput. **68**, 571–599 (2022)
439. D.V. Hieu et al., Strong convergence of new iterative projection methods with regularization for solving monotone variational inequalities in Hilbert spaces. Math. Methods Appl. Sci. **43**(17), 9745–9765 (2020)
440. R.U. Verma, Sensitivity analysis for generalized strongly monotone variational inclusions based on the (A, η)-resolvent operator technique. Appl. Math. Lett. **19**(12), 1409–1413 (2006)
441. M. Volle, Contributions à la dualité en optimisation et à l'épi-convergence. Ph.D. thesis, Pau (1986)
442. P.T. Vui, L.Q. Anh, R. Wangkeeree, Levitin–Polyak well-posedness for set optimization problems involving set order relations. Positivity **23**(3), 599–616 (2019)
443. P.T. Vui, L.Q. Anh, R. Wangkeeree, Well-posedness for set optimization problems involving set order relations. Acta Math. Vietnam. **45**(2), 329–344 (2020)
444. D.W. Walkup, R.J.-B. Wets, Continuity of some convex-cone-valued mappings. Proceedings of the American Mathematical Society **18**(2), 229–235 (1967)
445. S.-H. Wang, N.-J. Huang, Levitin-Polyak well-posedness for generalized quasi-variational inclusion and disclusion problems and optimization problems with constraints. Taiwan. J. Math. **16**(1), 237–257 (2012)
446. S.-H. Wang, N.-J. Huang, D. O'Regan, Well-posedness for generalized quasi-variational inclusion problems and for optimization problems with constraints. J. Glob. Optim. **55**(1), 189–208 (2013)
447. S.-H. Wang, N.-J. Huang, M.-M. Wong, Storng Levitin-Polyak well-posedness for generalized quasi-variational inclusion problems with applications. Taiwan. J. Math. **16**(2), 665–690 (2012)
448. R. Wangkeeree, L.Q. Anh, P. Boonman, Well-posedness for general parametric quasi-variational inclusion problems. Optimization **66**(1), 93–111 (2017)
449. J.G. Wardrop, Road paper some theoretical aspects of road traffic research. Proc. Inst. Civ. Eng. **1**(3), 325–362 (1952)
450. R. Wets, Convergence of convex functions, variational inequalities and convex optimization problems. Var. Inequal. Complement. Probl. **375**, 403 (1980)
451. R.J.-B. Wets, A formula for the level sets of epi-limits and some applications, in *Mathematical Theories of Optimization* (Springer, 1983), pp. 256–268
452. R. Wijsman, Convergence of sequences of convex sets, cones and functions. Bull. Am. Math. Soc. **70**(1), 186–188 (1964)
453. R.A. Wijsman, Convergence of sequences of convex sets, cones and functions. II. Trans. Am. Math. Soc. **123**(1), 32–45 (1966)
454. K.-C. Wong, Excess demand functions, equilibrium prices, and existence of equilibrium. Econ. Theory **10**, 39–54 (1997)
455. S.E. Wright, Consistency of primal-dual approximations for convex optimal control problems. SIAM J. Control Optim. **33**(5), 1489–1509 (1995)
456. W.-T. Wu, A remark on the fundamental theorem in the theory of games. Sci. Record (NS) **3**, 229–233 (1959)
457. Y. Xu, S. Li, A new nonlinear scalarization function and applications. Optimization **65**(1), 207–231 (2016)
458. Y. Xu, S. Li, On the solution continuity of parametric set optimization problems. Math. Meth. Oper. Res. **84**(1), 223–237 (2016)
459. N.D. Yen, Hölder continuity of solutions to a parametric variational inequality. Appl. Math. Optim. **31**(3), 245–255 (1995)

460. N.D. Yen, On an existence theorem for generalized quasi-variational inequalities. Set-Valued Anal. **3**(1), 1–10 (1995)
461. N.D. Yen, G.M. Lee, Solution sensitivity of a class of variational inequalities. J. Math. Anal. Appl. **215**(1), 48–55 (1997)
462. M. You, G. Li, Approximate properly solutions of constrained vector optimization with variable coradiant sets. Optim. Lett. **17**(3), 721–738 (2023)
463. R.C. Young, The algebra of many-valued quantities. Mathematische Annalen **104**(1), 260–290 (1931)
464. F. Yousefian, A. Nedić, U.V. Shanbhag, On smoothing, regularization, and averaging in stochastic approximation methods for stochastic variational inequality problems. Math. Program. **165**(1), 391–431 (2017)
465. A. Zaffaroni, Degrees of efficiency and degrees of minimality. SIAM J. Control. Optim. **42**(3), 1071–1086 (2003)
466. E. Zeidler, *Nonlinear Functional Analysis and Its Applications: Fixed-Point Theorems* (Springer, 1986)
467. J. Zeng et al., Stability results for convex vector-valued optimization problems. Positivity **15**(3), 441–453 (2011)
468. M. Zervos, On the epiconvergence of stochastic optimization problems. Math. Oper. Res. **24**(2), 495–508 (1999)
469. C.-L. Zhang, N.-J. Huang, Vectorial Ekeland variational principle for cyclically antimonotone vector equilibrium problems. Optimization **69**(6), 1255–1280 (2020)
470. C. Zhang, A class of equilibrium problems with lower and upper bounds. Nonlinear Anal. Theory Methods Appl. **63**(5–7), e2377–e2385 (2005)
471. W. Zhang, Z. Fang, Y. Zhang, A note on the lower semicontinuity of efficient solutions for parametric vector equilibrium problems. Appl. Math. Lett. **26**(4), 469–472 (2013)
472. X.Y. Zheng, K.F. Ng, Calmness for L-subsmooth multifunctions in Banach spaces. SIAM J. Optim. **19**(4), 1648–1673 (2009)
473. J.X. Zhou, G. Chen, Diagonal convexity conditions for problems in convex analysis and quasi-variational inequalities. J. Math. Anal. Appl. **132**(1), 213–225 (1988)
474. T. Zolezzi, Extended well-posedness of optimization problems. J. Optim. Theory Appl. **91**(1), 257–266 (1996)
475. T. Zolezzi, Well-posedness criteria in optimization with application to the calculus of variations. Nonlinear Anal. Theory Methods Appl. **25**(5), 437–453 (1995)
476. T. Zolezzi, Condition number theorems in optimization. SIAM J. Optim. **14**(2), 507–516 (2003)
477. T. Zolezzi, On well-posedness and conditioning in optimization. ZAMM **84**(7), 435–443 (2004)
478. T. Zolezzi, Well-posedness and optimization under perturbations. Ann. Oper. Res. **101**, 351–361 (2001)

Subject Index

L. Q. Anh et al., *Equilibrium Problems*, Forum for Interdisciplinary Mathematics, https://doi.org/10.1007/978-3-031-89003-1

T

U

V